THE RAY SOCIETY

INSTITUTED 1844

This Volume is No. 152 of the series
LONDON 1979

PARASITIC COPEPODA OF BRITISH FISHES

Z. KABATA, PH.D., D.SC. (ABERDEEN)

Department of the Environment
Fisheries and Marine Service
Pacific Biological Station
Nanaimo, B.C., Canada

Published 1979
ISBN 0 903874 05 9

Publications (Sales)
British Museum (Natural History)
Cromwell Road,
London SW7 5BD

For the Ray Society
c/o The British Museum (Natural History)
Cromwell Road, London, SW7 5BD England

The text has been set in Clowes Times
by Benham & Company Limited
and printed in Great Britain on C. Townsend Hook's
Fineblade coated cartridge paper
by Lowe & Brydone Printers Limited,
Thetford, Norfolk

The author dedicates this work
to the memory of his predecessors:
William Baird,
George Stewardson Brady,
Thomas and Andrew Scott
and Robert Gurney,
all of whom helped to chart
the paths he has trod.

Contents

Introduction . . . ix
Notes added in proof . . . xii
The impact of parasitism on copepod morphology . . . 1
Shape and structure . . . 6
Cephalization . . . 6
Trunk formation . . . 10
Appendages . . . 14
The first antenna . . . 17
The second antenna . . . 20
The mouth . . . 23
The mandible . . . 28
The first maxilla . . . 31
The second maxilla . . . 32
The maxilliped . . . 35
The swimming legs . . . 37
Other structures . . . 40
Classification . . . 43
A key to families of copepods parasitic on British fishes . . . 53
Diagnoses of families, descriptions of genera and species . . . 55
Suborder Poecilostomatoida . . . 55
Fam. Bomolochidae . . . 55
Fam. Taeniacanthidae . . . 68
Fam. Ergasilidae . . . 81
Fam. Chondracanthidae . . . 96
Fam. Philichthyidae . . . 135
Suborder Cyclopoida . . . 142
Fam. Lernaeidae . . . 142
Suborder Siphonostomatoida . . . 156
Fam. Caligidae . . . 156
Fam. Euryphoridae . . . 200
Fam. Trebiidae . . . 205
Fam. Pandaridae . . . 208
Fam. Cecropidae . . . 229
Fam. Dichelesthiidae . . . 238
Fam. Eudactylinidae . . . 250
Fam. Kroyeriidae . . . 264
Fam. Pseudocycnidae . . . 268
Fam. Hatschekiidae . . . 271
Fam. Lernanthropidae . . . 282
Fam. Pennellidae . . . 287
Fam. Sphyriidae . . . 315
Fam. Lernaeopodidae . . . 324
Host-parasite check list . . . 417
Bibliography . . . 421
Index of Latin names of copepods . . . 439
Index of Latin names of fishes . . . 460
Plates . . . 469

Introduction

Some years ago, during an autumn symposium held in London by the British Society for Parasitology, one of its eminent members, the late Dr. E. L. Taylor, introduced from the chair this author's paper on the systematics of parasitic Copepoda by referring to these parasites as "romantic animals". Dr. Taylor's well-known wit expressed, in capsule form, an average biologist's attitude to parasitic Copepoda. Most biologists have heard about them during their undergraduate days, when copepods were presented to them as examples of various types of adaptation to parasitism. They read some scanty descriptions which were accompanied by illustrations as picturesque as they were inaccurate. The memory of their undergraduate days is for many surrounded by a "good old days" halo, embedded in which are the half-forgotten strange shapes and complicated biological patterns of Copepoda, gleaned from long-obsolete books. Quaint, but hardly belonging to real life.

Having worked with this group of parasites for over 20 years, the author frequently has had to challenge this attitude by talking and writing about their real significance as enemies of fishes capable of causing serious economic damage. He has tried to draw attention to the very real need for a better knowledge of copepods parasitic on fishes, knowledge without which our attempts at controlling these dangerous pests cannot be fully successful.

In addition to the utilitarian value of such knowledge, it can lead us to a much better understanding of the nature of ectoparasitism, that form of coexistence of living organisms which has been overshadowed by our preoccupation with endohelminths and their biology. From the point of view of acquisition of pure biological knowledge, the study of parasitic copepods can be very rewarding.

The author is convinced that the best beginning to a study of any group of animals is a knowledge of their morphology and systematics. At present the morphology of parasitic copepods is only poorly known. One of the reasons for this state of affairs is the small size of these animals. Morphological details of small animals are often ignored by observers just because they are small. No effort has been spared in describing minute details of mammalian dentition and a lot of space has been devoted to the discussion of the biological significance of these details. In contrast, hardly any attention has been paid, for example, to the apical armature of the copepod first antenna. The author ventures to suggest that, had the copepod been the size of a cow, the tip of its first antenna would have become a topic for exhaustive studies. One tends to forget that the dimensional scale does not influence the biological importance. The world of the copepod is very different from ours and one cannot apply to it standards and criteria derived from other habitats and other size ranges. In his approach to the morphology of parasitic copepods, the author has tried to discern the functional significance of structures. His descriptions and illustrations are intended to present these parasites as living organisms rather than a row of pallid cadavers embalmed in their vials. The measure of his success will have to be assessed by the readers.

The study of the morphological minutiae of small arthropods bristles with difficulties too many to recount here. Much of it involves the need for interpretation rather than straight observation. Experience and background knowledge must be employed. This is why we sometimes find the same structure seen and understood differently by different authorities. This is why mistakes creep in despite the most stringent safeguards. The author is acutely aware of the fact that his own work is not free of errors and can only hope them to be few and relatively unimportant.

For most non-specialists systematics in general, and that of parasitic copepods in particular, is something that is used for labelling animals. The realization that the so-called authorities might be at loggerheads as to the correct way to label comes as an uncomfortable feeling. This author does not believe that the classification is worth much unless it attempts to reflect relationships within the group that is its object. He has long been dissatisfied with the existing classification of Copepoda. This book expresses his dissatisfaction by proposing a new scheme of relationships within this order. His views, as those of the proponents of other systems of classification, are of necessity speculative. There is nothing dogmatic about them. The reader is asked to weigh his arguments but should feel free to form his own conclusions.

In the course of the last two decades the author has often been approached by people in search of a book that might be used as an introduction to parasitic copepods. No such book has been written. Although this monograph is intended primarily as a guide to the parasitic copepods of British fishes, it was written with a wider circle of readers in mind. The British fauna of copepods is presented in it as the tip of an iceberg and the relationship between that tip and the larger submerged part, i.e. the copepod fauna of the world, is discussed for each family individually. Much information extraneous to the British region has been included to illustrate this relationship. This information and the general discussion of the impact of parasitism on copepod morphology might perhaps go some way towards making the monograph useful as an introduction to parasitic copepods in general.

In the course of its existence of over a century, the Ray Society has devoted a fair share of its efforts and resources to monographs dealing with copepods. Several of them include more or less extensive accounts of the parasitic members of that order. One of them (Scott and Scott, 1913) is restricted to parasitic copepods of fishes. The Scotts' monograph has become a widely accepted source of reference not only in Britain but in many other parts of the world. Its success is confirmed by the fact that it has recently been reprinted in the United States. After more than half a century of progress, however, the Scotts' classic is now largely outdated, and a new account is needed. The author is grateful to the Ray Society for giving him an opportunity of spotlighting in this way the group of parasites in which he has been for so long and so vitally interested.

The project of writing this monograph was first conceived in 1963, exactly half a century after the publication of the Scotts' book. It took many years to collect material and assemble the necessary literature. The progress of preparing it was delayed by the author's move from Scotland to Canada. The work on illustrations began in 1969 and took about $3\frac{1}{2}$ years to complete, while the writing of the text occupied a similar length of time.

A work of this scope dispels any illusions one might have had about the adequacy of one's individual resources. Without willing and unstinted help from many friends and colleagues, no author can hope to carry it out satisfactorily. This author was fortunate in having such help. He is keenly aware of his indebtedness to all those who in various ways assisted him in his efforts.

The moral support and encouragement he received during his work in Aberdeen from Mr. B. B. Parrish and from Dr. L. Margolis in Nanaimo, B.C., Canada, were more than essential to him, creating an atmosphere without which this somewhat extracurricular task could not have been contemplated. The truly devoted help of his Scottish colleagues in collecting, over many years, specimens of British parasitic copepods was equally indispensable. Dr. K. McKenzie and Mr. A. D. Campbell must be mentioned as outstanding among them. The stalwart assistance of the Aberdeen librarian, Mr. H. McCall, has to be acknowledged and special gratitude be expressed to Mr. J. D. Milne, who taught the author the art of scientific illustration.

Very important was the help that the author received from several musea, always ready to lend specimens, irreplaceable type material included. Foremost among them was the British Museum (Natural History), London. Many of its staff members were very helpful, to mention Dr. J. P. Harding, Dr. K. G. McKenzie, Mrs. Sayers, Dr. R. Lincoln and Mr. G. M. Bennell. The helpfulness and kindness of Dr. Macfarlane must also be acknowledged.

The Smithsonian Institution, Washington, D.C., in the person of the author's friend, Dr. R. Cressey, helped more than can be said in brief, generously making available the inestimable heritage of C. B. Wilson. Dr. Torben Wolff of the Universitetets Zoologiske Museum, København, lent specimens and translated old Danish references. Similar services were rendered by: Dr. R. Oleröd, Naturhistoriska Riksmuseet, Stockholm; Dr. W. Vervoort, Museum van Natuurlijke Historie, Leiden; Dr. C. H. Brandes, Ubersee-Museum, Bremen; Dr. H. E. Gruner, Zoologisches Museum, Humboldt-Universität, Berlin; and Prof. A. Capart, Institut Royal des Sciences Naturelles, Bruxelles.

Many other colleagues supported the author by lending or giving him specimens, assisting in the search for obscure references or commenting on parts of his manuscript. Conscious that he might have omitted some of them, the author wishes to record his gratitude to the following (in alphabetical order): Dr. F. Baudin Laurencin (Ivory Coast); Dr. O. N. Bauer (USSR); Dr. B. Berland (Norway); Dr. J. Chubb (UK); Dr. G. Fryer (UK); Dr. R. U. Gooding (Barbados); Dr. J. Grabda (Poland); Dr. J. Green (UK); Dr. A. V. Gusev (USSR); Dr. R. Hamond (Australia); Dr. G. C. Hewitt (New Zealand); Dr. A. G. Humes (USA); Dr. P. L. Illg (USA); Dr. V. N. Kazachenko (USSR); Dr. C. R. Kennedy (UK); Dr. W. L. Klawe (USA); Dr. R. R. Parker (Australia); Dr. J. D. Slinn (UK); Dr. J. H. Stock (The Netherlands); Dr. W. Templeman (Canada); Dr. R. Vik (Norway); and Dr. M. Walkey (UK).

Last but by no means least, the author wishes to acknowledge the irredeemable debt he owes his wife Mary. Over the years when most of his so-called spare time was claimed by the copepods, she carried, without complaint, a double burden of home affairs and was always cheerfully ready to help, editing the text, typing the manuscript or proof-reading. Nobody could have asked for a better helpmate.

Nanaimo, B.C., Canada
August, 1975

Notes added in proof

A time lag inherent in the production of a book of this size allows events to overtake it. Hence it is not uncommon for such a book to reach its readers in a somewhat out-of-date condition. These notes are intended to mend some holes punctured in the fabric of this book by recent developments.

The author proposed (p. 369) that *Pseudocharopinus pastinacae* be relegated to synonymy with *P. malleus*, basing his proposal on some distinctive morphological features these taxa have in common. The manuscript was in the publishers' hands before the author received a paper of Raibaut and Maamouri (1975), in which the same proposal had been made, quite independently. The French-Tunisian team, supporting the author's proposal, inadvertently stole his thunder.

In dealing with Sphyriidae, the author (p. 318) referred to *Paeonocanthus antarticensis* (Hewitt, 1965). Some explanation of the origin of this binomen is required. Hewitt (1965) described a new species of sphyriid, which he named *Periplexis antarcticensis*. His description was published on May 19th. In June of the same year, Kabata (1965) described *Paeonocanthus tricornutus*, a new genus of Sphyriidae. Detailed comparison shows that these taxa are conspecific. The nature of the respiratory cylinders does not permit inclusion of Hewitt's copepod in *Periplexis*, while the combination of processes and other features forces the establishment of a new genus to accommodate it. Kabata's *Paeonocanthus* must, therefore, stand. Hewitt's specific name *antarcticensis*, being chronologically prior by a few weeks, remains valid. Hence the use of *Paeonocanthus antarcticensis*.

The account of Eudactylinidae (p. 250) contains two closely related genera, *Protodactylina* Laubier, 1966, and *Bariaka* Cressey, 1966. The author shows some differences between their appendages, particularly maxillipeds. Cressey (personal communication), having carefully reexamined specimens of both genera, found them congeneric. The differences are definitely at the specific level. The reasons for his decision are outlined in a paper now in press. They are quite acceptable to this author, who derives some consolation from the fact that even without a detailed examination of the material he considered these genera as being close to each other.

The discussion of *Congericola* (p. 279) contains reference to specimens from New Zealand waters placed by Hewitt (1969) in *C. pallidus*. This author believed them to be a new species of *Congericola*. It is gratifying to note that since that discussion was written, Hewitt (1975) came to share the author's view and renamed the New Zealand copepod *Congericola kabatai*.

July 1977.

Hewitt, G. C. 1965. A new species of *Periplexis* (Sphyriidae, Copepoda) from the Southern Ocean. Trans. roy. Soc., N.Z., Zool., 6: 103–106.

Hewitt, G. C. 1969. Some New Zealand parasitic Copepoda of the family Eudactylinidae. Zool. Publ. Victoria Univ. Wellington, No. 49: 1–31.

Hewitt, G. C. 1975. New name for some *Congericola* (parasitic Copepoda) from New Zealand conger eels. N.Z. Jl mar. freshw. Res., 9: 563–565.

Kabata, Z. 1965. Parasitic Copepoda of fishes. Rep. BANZ antarct. Res. Exped., 8(6): 1–16.

Raibaut, A. and Maamouri, K. 1975. Remarques sur deux especes de copepodes parasites de selaciens de Tunisie. Bull. Must. nat. Hist. nat., Paris, ser. 3, Zool. 227 (No. 320): 1037–1047.

The Impact of Parasitism on Copepod Morphology

The polyphyletic origin of parasitic Copepoda is now a general tenet. It would be strange, indeed, to suppose that the astounding plethora of parasitic copepods derives from a single event of parasitic adaptation by a single ancestral species. It is also presumed that most parasitic copepods, especially those parasitic on fishes, probably have their origins among the representatives of cyclopoids. This presumption is couched in terms of the classical Sarsian classification. Later in this book, the author will propose that this classification be modified so as substantially to alter the present content of the term "cyclopoid".

There are two ways in which these hypothetical ancestors of the present-day parasites could have become preadapted to a life of more or less advanced intimacy with other living organisms. A fairly large number of free-living copepods (including both marine and freshwater species) are known to prey on, or harass, fish larvae. (Some instances of this behaviour have been reviewed by Kabata (1970a).) Attacking larvae larger than themselves, these copepods tear off strips of finfolds often exceeding their own size. Such micropredatory habits, calling for brief and intermittent periods of contact between the smaller attacker and larger prey, can lead, by temporal extension of those contacts, to the development of incipient host-parasite systems. A similar starting point for a would-be parasite of invertebrate hosts was described in an interesting paper by Fryer (1957).

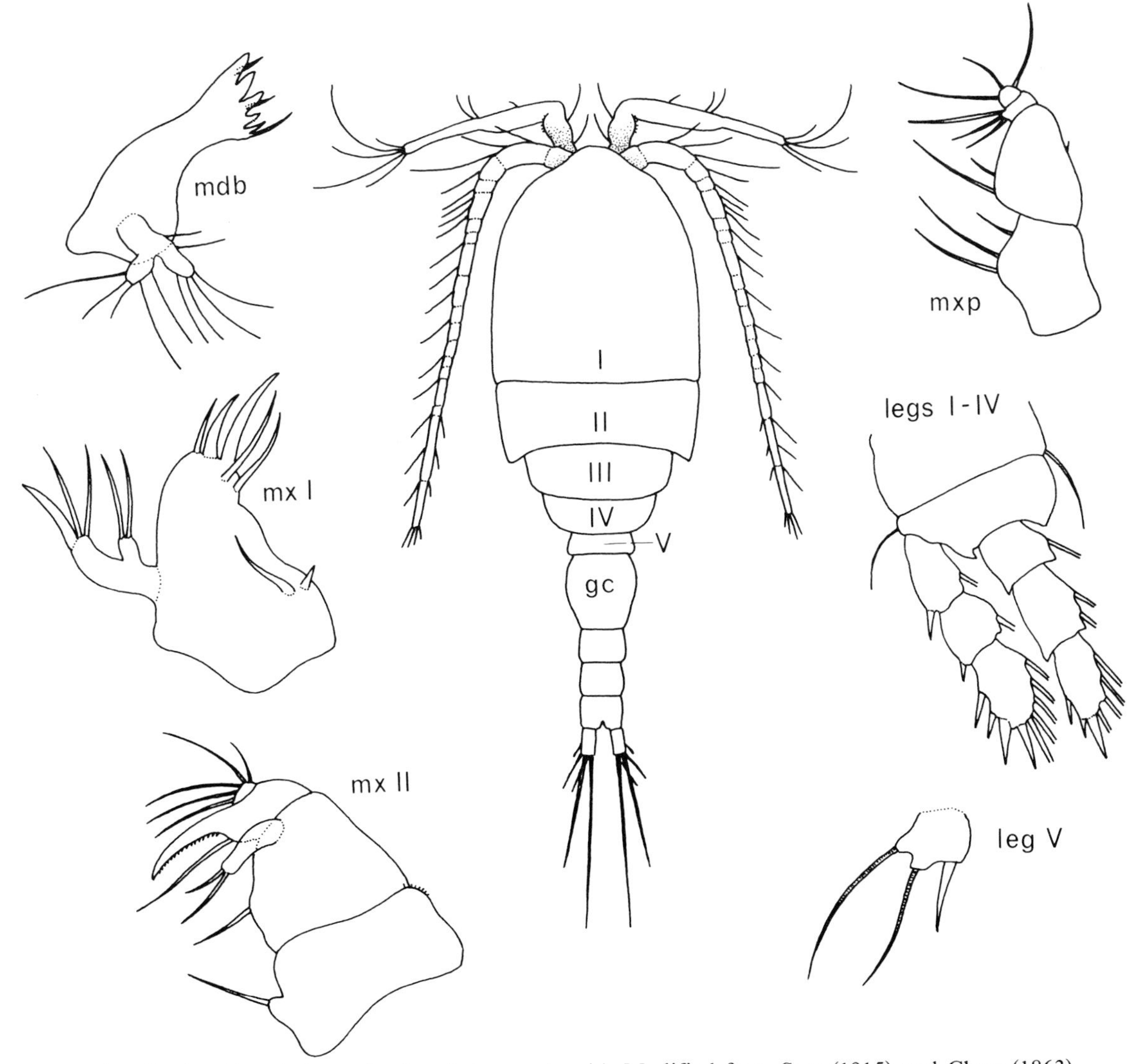

Text fig. 1. Morphology of a free-living cyclopoid. Modified from Sars (1915) and Claus (1863).

The other way might lead through the adaptation to a more leisurely and sedentary mode of life. Some copepods took to browsing or scavenging on the bottom and acquired buccal apparatus suitable for this purpose, as well as appendages better suited for maintaining hold of the nourishing substrate. Such copepods, brought by chance into contact with suitable bottom-dwellers, would be well placed to exploit those encounters and might end by forming more or less permanent associations culminating in the emergence of parasitism.

Among the mechanisms determining the evolution of parasitism, an important part is undoubtedly played by a combination of physiological and biochemical factors, some of which were speculated upon by Bocquet and Stock (1963). In this book we are concerned rather with the morphological expressions of various adaptive (and perhaps some non-adaptive) modifications that make a once free-living copepod a fitting partner in a host-parasite system.

To appreciate the extent of changes wrought in the morphology of copepods by their adaptation to a life of dependance on other animals, one should become acquainted with a free-living copepod. An example of such a copepod is probably best looked for among the cyclopoids, copepods that could be morphologically not too dissimilar from the presumptive ancestors of some present-day parasitic forms. Before engaging in this scrutiny, however, one must clearly realize that the modern free-swimming copepods are highly specialized for their mode of life. They must be seen as end-points of a line that in the distant past diverged from a common stem, from which probably are derived also to-day's parasites. A look beyond their respective specializations is necessary if one is to visualize the morphological matrix common to them both.

As can be expected, the morphology of a free-living copepod is moulded by the overriding demands of locomotory efficiency (Text fig. 1). Hence the bullet-shaped anterior tagma, the *cephalothorax*. This tagma, resulting from the process of cephalization, widespread among non-sessile animals, consists of the cephalon and more than one thoracic segment. (It is proposed here to use the term *cephalosome* for a tagma incorporating only one thoracic segment.) Among the free-living copepods one finds species in which the process of cephalization has not progressed beyond its very first stages. Claus (1863, p. 12) illustrates it in a table, where he shows the pontellid copepods as having the antennary region set apart from the rest of the body. In these copepods, therefore, even the *cephalon* (a tagma containing no thoracic segments) has not been fully formed. Some species of this group have passed on to the next stage. In the latter species the cephalon is complete, but the maxilliped-bearing first thoracic segment has retained its identity. There exist some poecilostome copepods, including a few parasitic species, in which the cephalosome condition has been reached, the anterior tagma extending to the posterior limit of the maxilliped-bearing segment. Much more common, however, is the cephalothorax, a tagma including two thoracic segments and, in consequence, bearing the first two thoracic appendages: the maxillipeds and the first pair of swimming legs.

Our generalized cyclopoid in Text fig. 1 shows this morphological condition. The cephalothorax is followed by four *free thoracic segments*, each bearing a pair of swimming legs. These segments vary in proportions from species to species but as a rule diminish in size from the second to the fifth leg-bearing segment, the last of them being usually much reduced. This segment is followed by the *genital complex*. The oft-used term *genital segment* is rejected here because this rather small tagma consists of more than one, commonly of two, segments. There is some disagreement as to the somatic origin of these segments, some authorities considering them as thoracic, while others believe they are derived from the abdomen. This author favours the former interpretation. If one accepts, as many do, that among the copepods no appendages are carried on abdominal segments, one must also accept the presence of a vestigial pair of legs on the genital complex as an indication of its at least partially thoracic nature.

Posterior to the genital tagma, the *abdomen* is usually cylindrical and consists of different numbers of segments. The body ends in *uropods*, two flat or subcylindrical sclerites of various lengths, provided with an array of pinnate setae. (The term uropod has been adopted by this author in acceptance of Bowman's (1971) arguments favouring it over other terms such as caudal furca, caudal rami, etc.)

It is clear that the animal shown in Text fig. 1 is well adapted to make its way through the water. Propelled by the flickering beat of its swimming legs and the sweeps of its first antennae, its fusiform anterior end cleaves the liquid medium, trailing behind it its slender posterior part whose feathery tip acts as a steering oar or vane. There is very little morphological difference between sexes, the male being distinguished from the female mainly by the structure and proportions of its genital complex, often by differences in the segmentation of its abdomen and by some modification of the antennae, which the male uses as its prehensile aids to mating.

A copepod can become dependent on another living organism without making many morphological concessions in acknowledgment of the change in its way of life. Text fig. 2 shows an asterocherid copepod associated with an echinoderm. Except for the proportions of the body and the loss of one abdominal segment, the free-swimming copepod in Text fig. 1 and the associated one in Text fig. 2 are very similar. (The differences in their cephalothoracic appendages will be discussed later.)

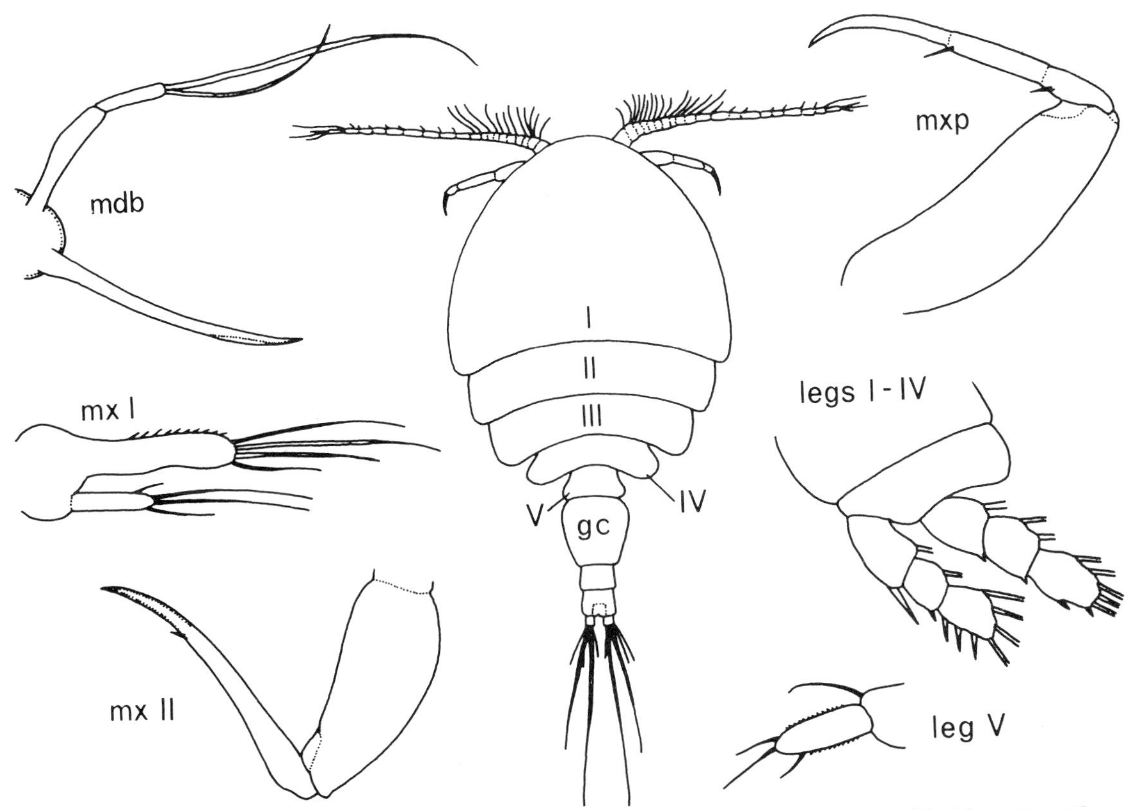

Text fig. 2. Morphology of an asterocherid copepod associated with an invertebrate. Modified from Sars (1915).

When one looks at the free-living young stages of parasitic copepods (Figs. 124–128, 383–390, 1338, 1341), one is struck not only by the extent of the differences between these stages and the adults, especially females of their species; even more striking is the similarity between the young stages of parasitic copepods and the adults of free-swimming copepods. This similarity becomes closer still, when comparison is made between the early ontogenetic stages of parasitic and free-living copepods. It appears as if the parasites, having made a switch from free to dependent existence, broke through the mould of the "normal" copepod structure in search of the morphological expression of that change. Their ontogenetic progression has been extended by one more step, a step which the adults of the free-living species did not have to take.

In becoming a parasite, a copepod, like any other animal undergoing the same process, exchanges one set of biological problems for another. For a free-living animal the most pressing problems are those relating to its survival as an individual. It must be able to obtain food in sufficient quantities and with sufficient frequency. The other side of the coin is the imperative of avoiding being eaten. It must be able either to escape or to defend itself from predators. Its biological success would not be assured, however, unless it could also solve the problems relating to the survival of its species. It must find a mate and reproduce. By and large, the search for food and its attendant difficulties loom as a greater problem than that caused by the reproductive imperative.

A parasite, in contrast, is assured of an abundant food supply, once its perilous quest for the host is successfully completed. The need of escaping predators becomes less acute, since it has been relegated to the host, usually a much larger animal with a narrower range of potential predators. On the other hand, because of the reduced mobility of the parasite, finding a mate becomes less easy in many instances. Hence the need for special biological devices aimed at increasing the chances of contact between the sexes.

All these factors become operative after the copepod has found its host. Until its last free-swimming stage, which might be referred to as the infective stage, there is little morphological difference between the non-parasites and most of the parasitic copepods. The influence of parasitism during those pre-parasitic stages is evident mainly in their reduction, both in duration and in the number of stages when compared with free-living species. The net effect of these changes is a progressively shorter period which the parasite spends outside its host. They occur, however, only in the more highly adapted parasitic families. In most others, no changes are evident.

The assumption of the parasitic habit significantly alters the entire life of the copepod. Firstly, as mentioned above, it acquires an easily accessible, secure and plentiful supply of food. Secondly, its locomotory activity is either reduced, permanently or for periods of varying duration, or even completely abandoned. The female parasite remains stationary, no longer spends its energy in search of food, and is able to divert its entire biological effort into reproduction. This is just as well, since the key to the survival of any parasitic species is in the production of the largest possible number of offspring.

Successful reproduction requires that the female be fertilized. In copepods relatively little affected by parasitism, such as Ergasilidae, mating takes place before the female has settled on the host. The male remains free-swimming throughout its brief existence. Where the female becomes permanently attached to the host at the beginning of its adult life, the onus of establishing contact between the sexes falls on the male. Hence, it must remain mobile at least until it has located the female. The morphological consequences of this necessity are obvious. The male is often less modified than the female, particularly in its locomotory ability. The foundations of sexual dimorphism which eventually reaches extravagant proportions are thus laid by the adoption of parasitism.

An immobile copepod has no need for a streamlined, well articulated body or for locomotory appendages. As can be expected, many parasitic species become compact or variously inflated and lose their segmentation, either partially or completely. Their locomotory appendages remain at the stage of development reached at the end of the free-swimming period and are frequently dwarfed by the growth of the copepod during its parasitic phase, while their musculature suffers disuse atrophy. The copepod is now in need of maintaining its hold on the host. From the point of view of morphology it means the need for attachment organs. These organs are duly developed, either based on the adaptation of "non-parasitic" appendages to a new role or arising as totally new structures of the anchor or holdfast type.

Plentiful food leads in many instances to increase in size, particularly among those species that are permanently fixed to their hosts. Although many parasites remain quite small, there are also many that greatly exceed the size of all known free-living copepods. An extreme example of gigantism is the length of a species of *Pennella* living on whales. Attaining two feet, this size can claim being a record among copepods and having few equals among the extant Crustacea. The size of a parasitic copepod often appears to be dependent on that of its host. In spite of many exceptions to this rule, it can be maintained that a large host is more likely than a small host to carry a large copepod. The size also depends to some extent on the location selected by the parasite on the host. Copepods living in habitats in which space is at a premium tend to remain small, or to become so folded as to crowd the greatest possible volume of tissue into the smallest possible space. No such restraints are imposed on the copepods attached to the external surface of the host. Size and shape are related to each other in a complex manner, their reciprocal influences being impossible to determine. A large animal is exposed to environmental pressures and forces of different magnitude and even of a different nature than those acting upon a small animal living in an identical place. (A sheep feeding in a meadow lives in a world vastly different from that of a grasshopper inhabiting the grass of the same meadow.) For example, the larger animal usually requires much more powerful and complicated attachment organs than those that suffice the smaller animal. Consequently, it has to undergo much more extensive modifications.

The greatly increased production of eggs leads to a comparative gigantism of the part of the body containing the reproductive organs, usually set apart as the genital complex. Looking at *Lernaeocera branchialis*, for example (Figs. 1330–1337), few observers are struck by the fact that they have before them an animal 90% of which consists of the genital complex. Some copepods appear to be disrupted into two parts, separated by a grossly inflated genital storage region, but remaining relatively unmodified in other respects. In others the genital trunk is less strikingly developed, while still dominating the other parts of the body.

It must be noted that at least some of the changes in the morphology of parasitic copepods might be non-adaptive. Freed of the rigours imposed by the need for efficient movement, these copepods might be less subject to selection against morphological innovations that would have spelled the doom of a free-living species. Looking, for example, at a series of equally successful species of *Chondracanthus*, some with a single pair of trunk processes (Figs. 196–198) and others studded with them until they look like veritable hedgehogs (Figs. 129–133), one cannot but wonder what adaptive significance can be attributed to these processes. Morphological changes can be also secondary in nature, arising in the wake of other, more fundamental modifications.

As shown by the above discussion, one of the paradoxes inherent in the demands placed upon an animal living parasitically is the fact that it has to reconcile the degree of immobility characteristic of most successful parasites with the necessity for each successful generation to find, and become attached to, a new host individual. In other words, the animal must change in the course of its life from an efficient swimmer into a sessile or semi-sessile organism, a change that calls for morphological transformation. The extent of that transformation depends on the degree of adaptation to parasitism. The pre-adult stages of the life cycle undergo morphological changes as the result of a series of moults, thus resembling their free-living relatives (though many of them go through several moults while attached to the fish). Having reached sexual maturity, these copepods cease to moult, though the entire process of morphological transformation lies still ahead. The mechanism by which this process is accomplished is *allometric growth*. It produces changes in the proportions of different parts of the body at different times. When these changes are small, the fully grown copepod does not look very unlike the young adult; when they are extensive, all resemblance between the two might be obliterated. In the latter case we speak of *metamorphosis*, though, in view of a complete gradation of changes, it is difficult to draw a line separating metamorphosis from a less far-reaching modification.

One of the simplest and most common ways in which copepods change after the attachment to the host and attainment of maturity is due to the process referred to by this author as "diphasic growth" (Kabata, 1960a). The most vigorously growing region of the copepod directly after it settles on the fish is its anterior part, i.e. the region most vital to prehension and feeding, functions enabling it to

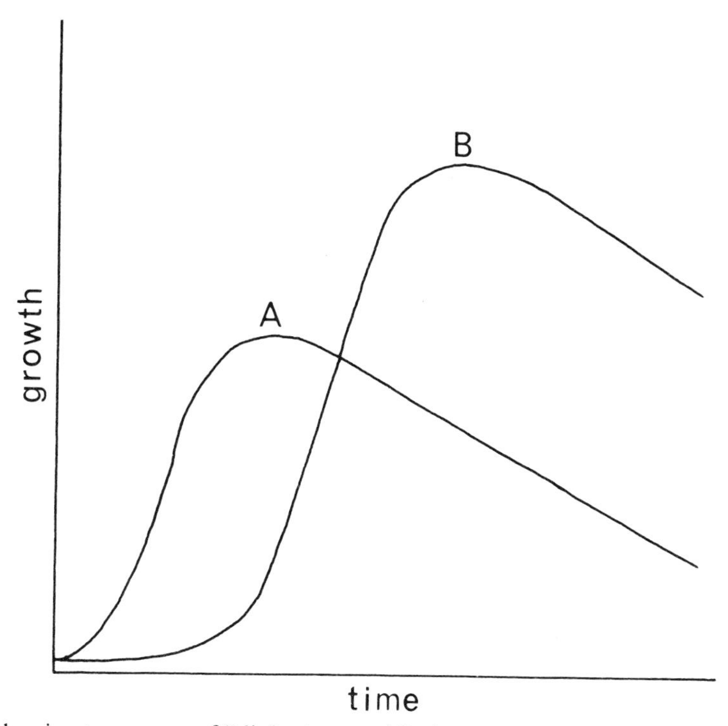

Text fig. 3. Graph showing two curves of "diphasic growth" of some parasitic copepods. Explanations in text.

retain contact with the source of food which is to provide energy for further development, as well as to ensure the food intake itself. This, the "first phase" of "diphasic growth", is shown in Text fig. 3 as curve A. As soon as the attachment has been made secure and the food supply becomes sufficient to support development, the "second phase" of "diphasic growth" begins. During that phase, shown in curve B in Text fig. 3, the posterior part of the copepod, which until then grew very slowly or not at all, enters upon a period of fast growth which eventually causes it to outstrip in size the previously larger anterior part. The posterior part serves reproduction, the second vital function.

The morphological changes due to diphasic growth in poecilostome copepods can be seen by comparing Fig. 179 with Fig. 175. The former shows a young adult *Chondracanthus lophii* Johnston, 1836. Compared with the fully grown specimen in Fig. 175, it appears distinctly top-heavy, its anterior part (up to the level of the second pair of legs) constituting about $\frac{2}{5}$ of the total body length. In the mature specimen this part accounts for only about $\frac{1}{10}$ of that length. An even more dramatic example is shown by changes occurring in the siphonostome *Clavella adunca* (Strøm, 1762). Figure 1868 shows it soon after it has completed its last moult. Only a small sac at the bottom of the picture represents its posterior part. In Fig. 1866 the first phase of growth (curve A) has just passed its peak and the upswing of curve B has begun, as evident in the proportionately larger posterior part of the body. By the time the second phase has run its course, the animal assumes its definitive shape, shown in Figs. 1861–1865. The posterior part becomes dominant.

Many copepods have more than two regions developing at different rates and different times. The concept of diphasic growth must, therefore, be extended to one of "multiphasic growth", related to the existence of several development centres. They will be discussed in greater detail below.

Having dealt with the principles and processes operative in effecting morphological changes produced in copepods by parasitism, we shall now review the changes themselves, dealing first with the broad ranges of shape and structure and ending with the appendages.

Shape and structure

To illustrate the impact of parasitism on the structure of copepods, we shall look at the outcome of several tendencies in modifications resulting from the progressive adaptation to that mode of life. We will recognize three categories of changes: (i) cephalization, i.e. the incorporation of successive segments into the cephalothorax, often associated with some loss of segmentation; (ii) the development of the trunk required to accommodate the greatly enhanced reproductive activity of the parasite, also commonly associated with extensive loss of external segmentation; and (iii) the development of anchoring devices for permanent attachment to the host. (In many instances, particularly frequent among those species that have retained the ability of moving over the surface of the host, no special holdfast structures are produced, the attachment to the host being entrusted to specialized prehensile appendages.) It is not uncommon to see in the same copepod more than one tendency, sometimes all three, in the morphological adaptations to parasitism.

Generally speaking, one would expect a direct relationship between the extent of parasitic adaptation and the progress of morphological changes expressed in these three tendencies. It must be remembered, however, that the parasites which have not progressed far along one or more of these paths are not necessarily primitive. It might mean no more than the fact that their particular type of host–parasite relationship did not call for any morphological concessions of this nature.

Cephalization

The process of cephalization can be traced in a group of siphonostome families related to Caligidae (Text fig. 4). The primitive among them is Dissonidae, exemplified by *Dissonus*. In this family only the first leg-bearing segment is incorporated into the cephalothorax, those bearing the second to the fourth being clearly delimited between that tagma and the genital complex. The only exception to this is the sometimes indistinct boundary between the second leg-bearing segment and the cephalothorax. The genital complex consists of two segments and bears vestiges of the fifth legs in the female, fifth and sixth in the male. This type of segmentation is undoubtedly primitive and occurs in the majority of free-living copepods (cf. Text fig. 1). The main difference between Dissonidae and the free-living copepods is the dorsal shield of the parasitic species. An obvious adaptation to life on the surface of the fish, the shield is roughly crescentic, its anterior margin

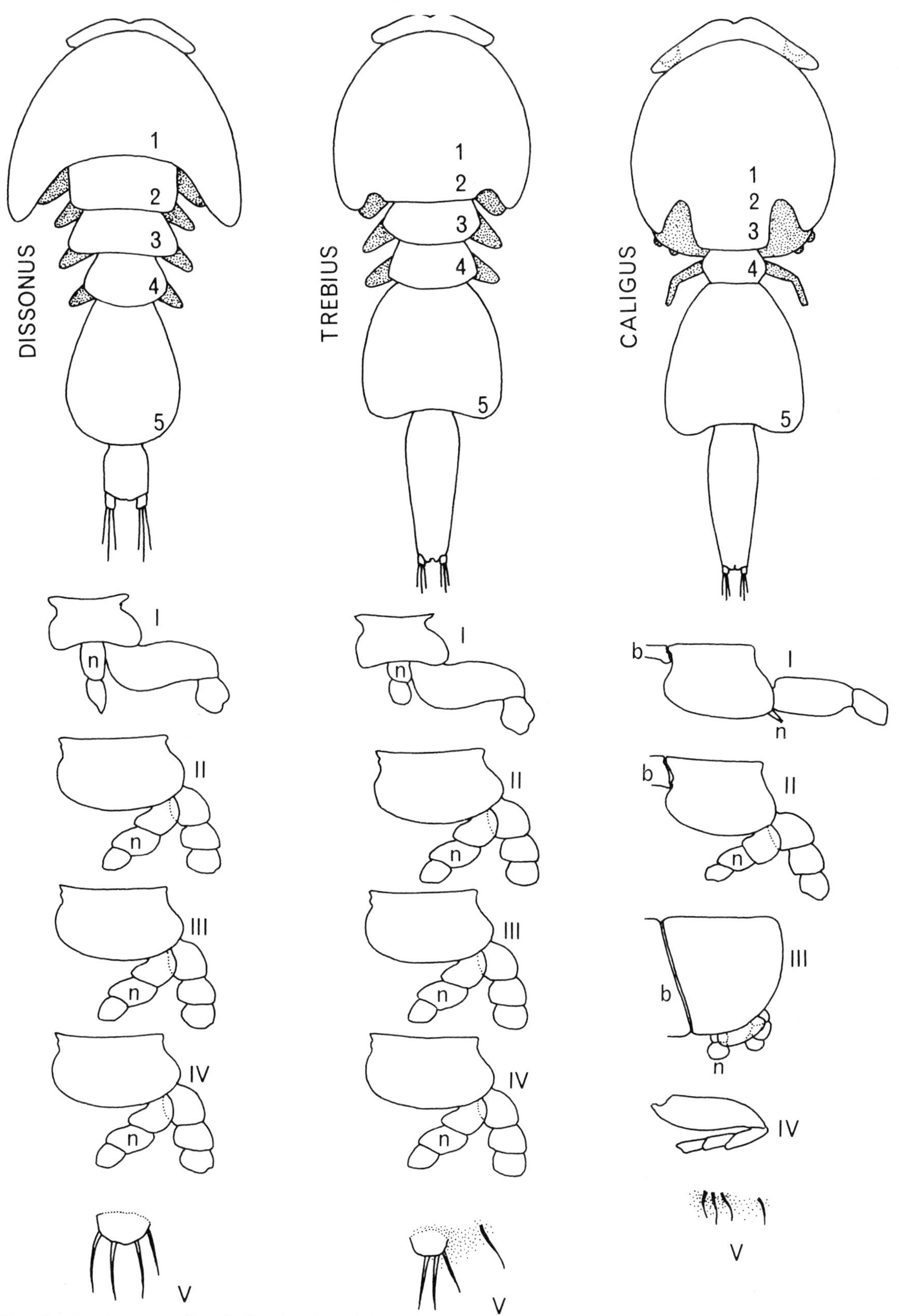

Text fig. 4. Progress of cephalization in caligiform copepods, with associated development of dorsal shield and changes in morphology of swimming appendages.

convex and its posterolateral corners drawn out in the posterior direction. The structure of the shield allows it to adhere closely to the substrate and permits the continuous stream of water, rushing past the flanks of the fish, to slip smoothly over its top. In some dissonids the second leg-bearing segment is equipped with aliform lateral expansions of the tergum which also add to the prehensile power of the shield.

The small family Trebiidae, containing only two genera (*Trebius* Krøyer, 1838 and *Kabataia* Kazachenko et al., 1972) is characterized by the incorporation of the second leg-bearing segment into the cephalothorax. Between the latter and the genital complex there remain only two free segments, as shown by the diagram of *Trebius* in Text fig. 4. The shield has been enlarged and perhaps has become more efficient. In most other respects the species of this family resemble those of Dissonidae.

One more step has been taken by Caligidae, exemplified by *Caligus* Müller, 1785 in Text fig. 4. In species of this family the third leg-bearing segment forms a part of the cephalothorax, only the small fourth leg-bearing segment remaining free in front of the genital complex. Even further expansion of the dorsal shield has taken place.

Among siphonostome copepods with caligiform facies but distinguished by the possession of dorsal and/or dorsolateral plates one finds a series of similar steps. Pandaridae (*Pandarus* in Text fig. 5) have the same segmentation as Dissonidae. Cecropidae (*Cecrops* in Text fig. 5) also have only the first leg-bearing segment incorporated in the cephalothorax, but their next two segments have fused together (except for *Orthagoriscicola*) forming a small tagma in front of the fourth leg-bearing segment. The segmentation of Euryphoridae (*Euryphorus* in Text fig. 5) is very similar to that of Caligidae.

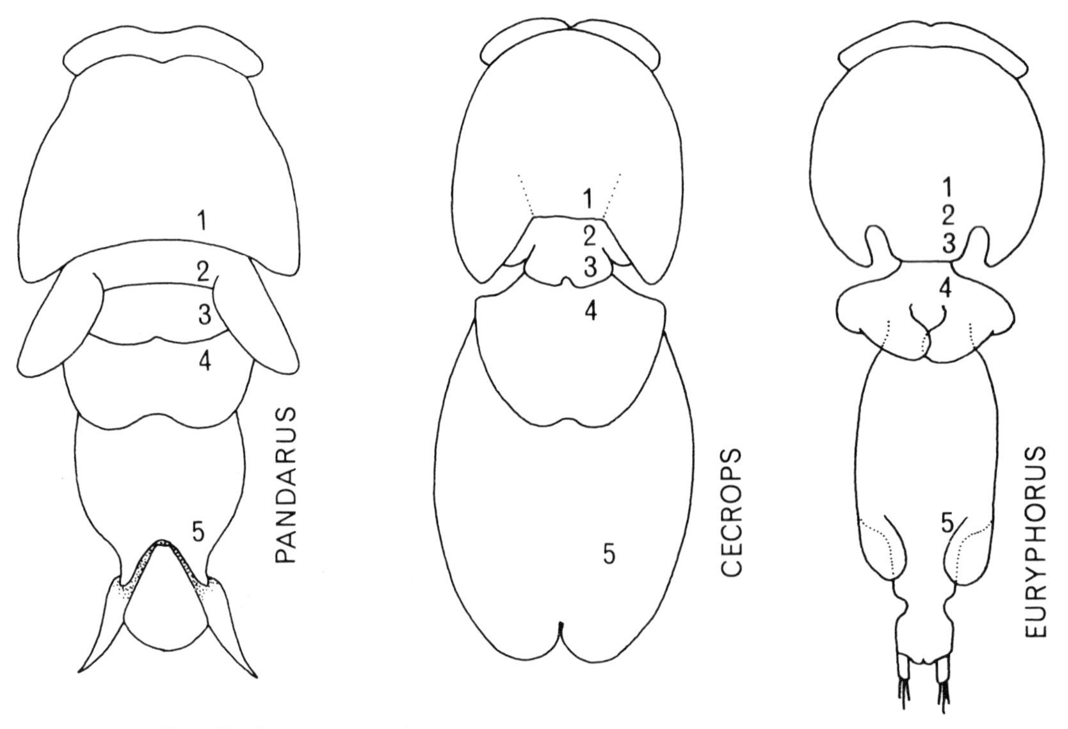

Text fig. 5. Progress of cephalization in Pandaridae and their relatives.

Cephalization made no progress among siphonostome copepods that bear no resemblance to the caligids (Text fig. 6). Neither is it in evidence among the poecilostomes (Text fig. 7). It appears to be the mechanism responsible for the development of the dorsal shield, a structure which has evolved as the means of adhering to the surface of the fish, without losing the ability to move freely over it. The parasites that live permanently attached to gill filaments or other parts of the host have no need for it, hence no cephalization.

On the strength of the above argument one might come to the conclusion that Dissonidae in Text fig. 4 are more primitive than Trebiidae, while the latter are less advanced than Caligidae. Similarly,

Pandaridae in Text fig. 5 might be considered more primitive than either Cecropidae or Euryphoridae. The argument finds further support also in the fact that among those families the "more primitive" are generally parasitic on more primitive hosts. For example, Dissonidae live on both elasmobranchs and teleost fishes. The same is true of Trebiidae (only a single species of *Kabataia* has a teleost host). Caligidae, on the other hand, live predominantly on the teleosts. A similar situation obtains in the families shown in Text fig. 5.

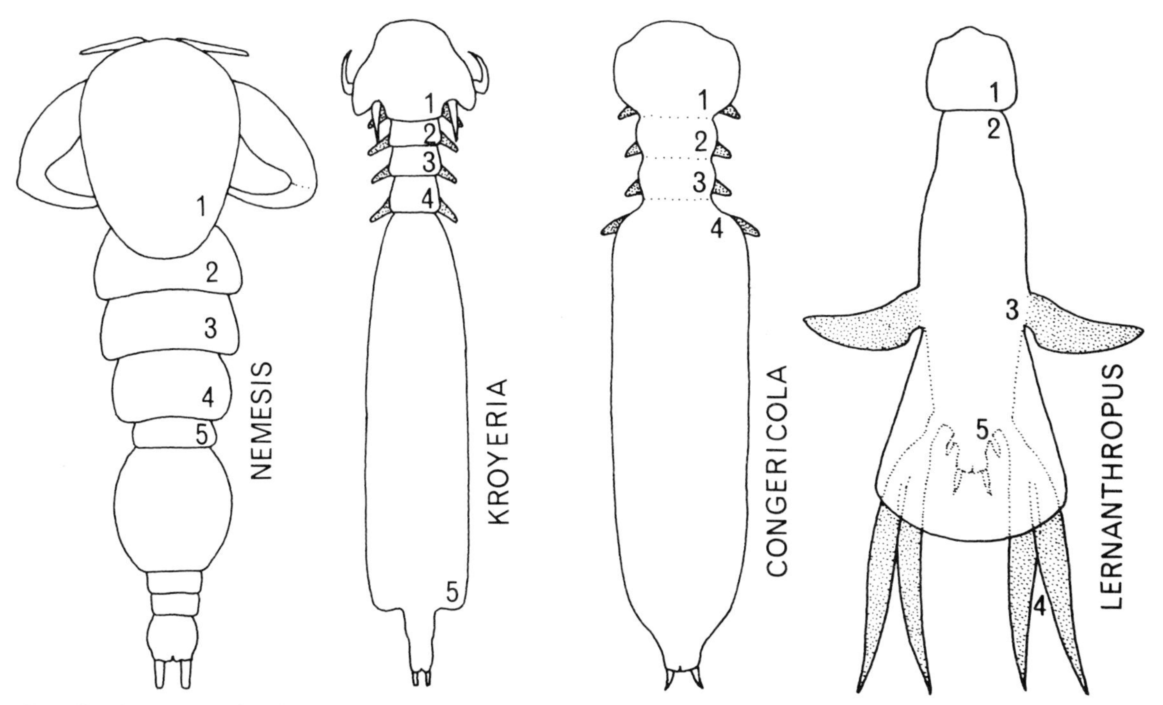

Text fig. 6. Progressive loss of external segmentation among siphonostome copepods other than those with caligoid facies.

The evolution of the dorsal shield through cephalization has not been without influence on the other structures of the copepods involved. In particular, its effects can be seen on the swimming appendages, presenting a clear example of correlated changes in morphology (see p. 7).

It is interesting to note that copepods like *Dissonus* and *Pandarus*, with only the first leg-bearing segment incorporated in the cephalothorax and with the posterior margin of the shield wide open in the absence of the apron of the modified third legs, have developed other aids to attachment. Most of them live on elasmobranchs. Perhaps the shagreen texture of the elasmobranch skin makes adhesion by suction difficult and does not stimulate selection for the process of cephalization. Be it as it may, *Dissonus* gets by with the aid of posteriorly directed spines, with which the ventral surface of its genital complex is equipped (except for *D. nudiventris* Kabata, 1965). *Pandarus*, on the other hand, has developed a set of so-called adhesion pads (p. 42) with coarse surfaces facilitating retention of contact with the host. Here we might caution against the conclusion that the level of phylogenetic development can be measured by the progress of cephalization. The ostensibly primitive *Dissonus* has recently been found to have a nauplius with modified appendages and reduced swimming ability (Anderson and Rossiter, 1969). Resembling in some ways the quiescent nauplius of Lernaeopodidae, this stage appears to be on the way out of the life cycle, a feature most unusual for this group of copepods. Primitive morphology, apparently, can go hand in hand with extensive biological adaptations.

The fact that biological similarities lead to parallelism in morphological characteristics is also demonstrable in some instances among parasitic Copepoda. We have already described how the process of cephalization results in evolution of a cup-like dorsal shield in Caligidae. The third pair of legs, borne upon a segment incorporated in the cephalothorax, becomes a broad apron effectively sealing off the posterior margin of the cup and improving its suction power (Text fig. 8B). Among the poecilostomes, Bomolochidae and their relatives also have a concave ventral surface of the

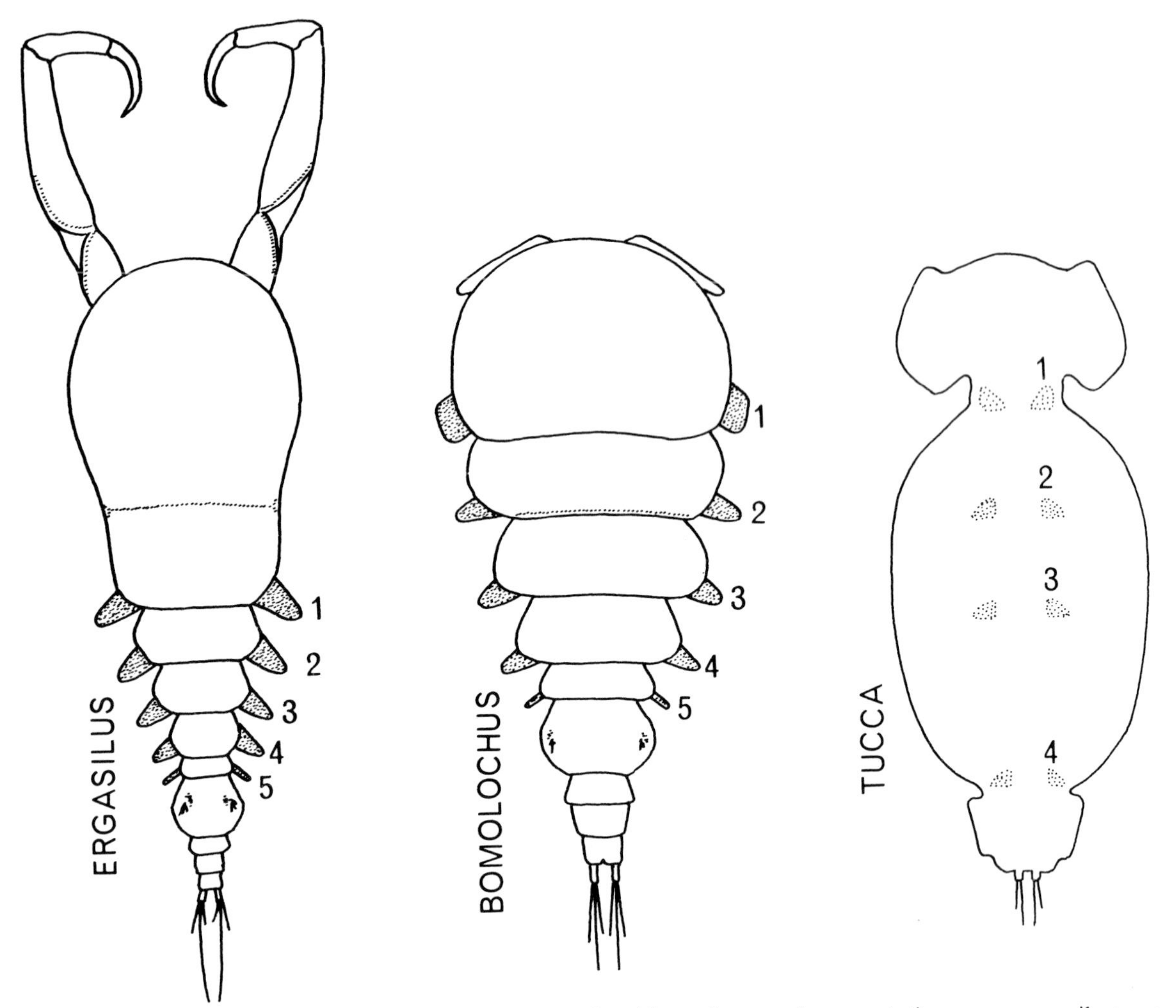

Text fig. 7. Some stages in trunk formation and associated loss of external segmentation among poecilostome copepods.

cephalothorax adapted to act as a suction cup (Text fig. 8A). In these copepods the process of cephalization stopped at the first leg-bearing segment. Their third legs are far behind the posterior margin of the cephalothorax. It is, therefore, the first legs that must act as a watertight barrier. To do this, they have become modified, though much less so than the caligid third pair (see also p. 62). Although the bomolochid cephalothorax appears less efficient as a sucker than its caligid counterpart, both are perfectly functional. The same aims were achieved by different morphological means.

Trunk formation

Another common way in which external segmentation is lost among the less modified parasitic copepods is the process of trunk formation, during which the genital complex becomes greatly enlarged and fuses with the segment or segments anterior to it to form a large tagma, often dominating the entire appearance of the copepod.

The siphonostome copepods with caligoid facies (i.e. both Caligidae with their relatives and Pandaridae with theirs) have genital complexes consisting of two segments only, those bearing the fifth and, in the males, the sixth pair of legs. This tagma might become very large in relation to the rest of the body (Text fig. 48A–G), but it never incorporates additional segments anterior to it. In *Caligus* and related genera with similar segmentation there is, in fact, little room for the anterior progress of fusion, since only one free segment is present between the cephalothorax and the genital complex. (The position is less clear with regard to the abdominal segments. *Caligus*, for example,

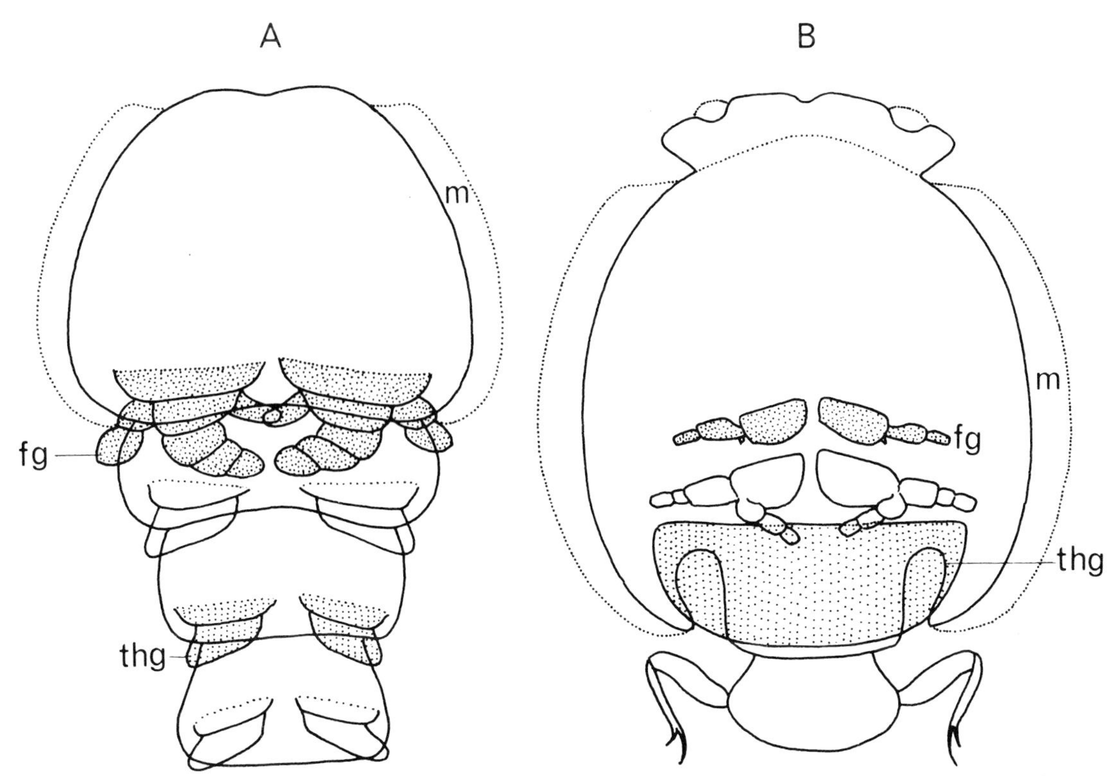

Text fig. 8. Parallelism in parasitic adaptation. Cephalothorax modified into suction cup in poecilostome Bomolochidae (A) and siphonostome Caligidae (B).

fg — first leg; m — marginal membrane; thg — third leg.

contains species with one-segmented and five-segmented abdomina. The diminution in the size of the abdomen might come about either by incorporation into the genital segment or by suppression of one or more segments. The latter is much more likely, since there is no change in the position of genital orifices or vestigial legs in species with reduced abdomina, a change that might be expected to occur with incorporation of the posterior segments.)

The sedentary siphonostomes formerly placed in the family Dichelesthiidae *sensu lato* can be arranged in a series illustrating several stages of trunk formation. The genus *Nemesis* Risso, 1826 (Text fig. 6), for example, is characterized by intact segmentation between the cephalothorax and the genital complex. Leg-bearing segments two to five are clearly identifiable, though the fifth is usually reduced in size and bears reduced legs. In *Kroyeria* van Beneden, 1853 (Text fig. 6) the fifth leg-bearing segment has been included in the genital complex, now forming a tagma fully deserving the name of trunk. Most of the trunk volume is due to the great expansion in the length of that segment. In this feature *Kroyeria* bears some resemblance to the most primitive of the caligiform siphonostomes, *Dissonus* (Text fig. 4), which it also resembles by incipient formation of a cephalothoracic shield. Only three segments, those bearing the second to the fourth legs, remain distinct between the cephalothorax and the trunk. *Congericola* van Beneden, 1851 (Text fig. 6) is left with only two such segments between these tagmata, the segmental boundaries being recognizable only as shallow constrictions. The evidence for the tendency towards further loss of segmentation in a group of genera related to *Congericola* can be seen in the example of *Hatschekia* Poche, 1902 (Figs. 1101, 1116, 1128, 1136, 1141). In some species of that genus the segmental boundaries in the region between the two tagmata are very difficult, or even impossible, to find. In others there is only one segment (the second leg-bearing segment) distinguishable in front of the trunk, or even that segment might become partly absorbed, leaving only a narrow constriction. No free segments are left in *Lernanthropus* Blainville, 1822 (Text fig. 6), a copepod genus in which the trunk directly follows the cephalothorax, and sometimes accounts for as much as 80% of the entire body length.

The development of a trunk is also common in the poecilostome copepods parasitic on fishes. While genera such as *Ergasilus* and *Bomolochus* von Nordmann, 1832, remain fully segmented, *Tucca*

Krøyer, 1837 (Text fig. 7) has developed a trunk comprising all leg-bearing segments, with the possible exception of the first one, situated in the region of a narrow neck-like constriction marking the transition between the cephalothorax and the trunk. A similar trunk is prevalent also in the family Chondracanthidae.

There is, however, one important difference between the trunk formation of the siphonostomes and the poecilostomes. In the former its development starts posteriorly at the level of the genital segment, which is always included in the trunk.* In the siphonostone copepods one might refer, therefore, to this tagma as the *genital trunk*. In the poecilostomes, on the other hand, the posterior end of the trunk is the boundary between the genital complex and the segment anterior to it (e.g. *Acanthochondria*, Text fig. 33). Hence, the poecilostomes have a *pregenital trunk*. These distinctions are obliterated in copepods which have become so completely modified by their adaptation to parasitism as to have lost even the borders between the individual tagmata (e.g. the cyclopoid Lernaeidae).

It is worthwhile at this point to mention the conflict of opinion between Wilson (1910) and Calman (1926). The former authority postulated that modification proceeds from behind forwards. In other words, the posterior part of the body and its appendages are likely to be less primitive than the anterior part. The latter authority, on the other hand, suggested that it is a "general rule among Arthropods that specialization begins anteriorly and works backwards; we should expect the posterior limbs to be the more primitive". As we have seen above, both processes occur, in fact, among the parasitic copepods. Calman's broader view took in all the arthropods and his statement has, perhaps, a more general validity, especially with regard to the ancestral free-living forms in which the cephalon was in the process of evolution. Wilson concentrated on the parasitic copepods he knew best and saw mainly the development of the trunk.

The loss of external segmentation makes it difficult to determine the fate of individual segments, and in particular to determine how much each of them contributed to the tagmata that engulfed them. In some instances (e.g. in Lernaeopodidae) the trunk musculature provides a clue, but the external landmarks of former segmentation can be found only in the position of segmental appendages, when these appendages have been retained. Sometimes the general appearance of two copepods might be similar, although the somatic origin of their individual body regions is different. For example, *Congericola* (Fig. 1159) does not differ greatly from *Hatschekia* (Fig. 1136). The position of the fourth pair of legs in *Congericola* shows, however, that its trunk is formed, in addition to the genital complex, mainly of the fifth, and perhaps partly of the fourth, leg-bearing segment. In *Hatschekia* the position of the vestigial third legs (minute setae halfway down the lateral margins) indicates that the third and perhaps part of the second leg-bearing segment are also involved. Similarity in the shapes of the adult females of *Lernaea* L. (Fig. 391) and *Lernaeocera* Blainville, 1822 (Figs. 1330–1337) led to erroneous systematic conclusions that might have been avoided had the segmental structure of their trunks been taken into account. The position of the four pairs of biramous swimming legs of *Lernaea* (visible as pairs of small dots in Fig. 391) shows that all the leg-bearing segments except the first have contributed about equally to the development of the trunk. Originally close together (Fig. 390), the legs move apart during the process of growth. The legs of the young female *Lernaeocera* are also close to one another (Fig. 1342), but in this copepod they remain close throughout the life cycle (Fig. 1376). The trunk of *Lernaeocera* receives no contribution from the four leg-bearing segments. Similar morphological results are obtained in different ways.

The last example introduces us to one of the most extensive morphological changes which the copepod can undergo during its ontogeny. So far, our discussion of changes resulting from the processes of allometric growth has dealt with modifications which, in the final result, produce a copepod still recognizable as a form derived from its free-swimming juvenile stage. With *Lernaea* and *Lernaeocera* we meet copepods, the adult females of which cease to bear any resemblance to the young stages of their species. Changes of that nature and extent can no longer be expressed by the curves of the diphasic growth process. The allometric growth that produces them has all the characteristic features of metamorphosis.

The process of metamorphosis differs from one copepod species to another, since it is always directed towards the attainment of morphological characteristics demanded by the conditions peculiar to the host–parasite relationship in which the copepod is involved. This being the case, the discussion of the metamorphosis will be tackled individually for each family. A single example will suffice at this point.

* With the exception of *Lernanthropus* and its relatives.

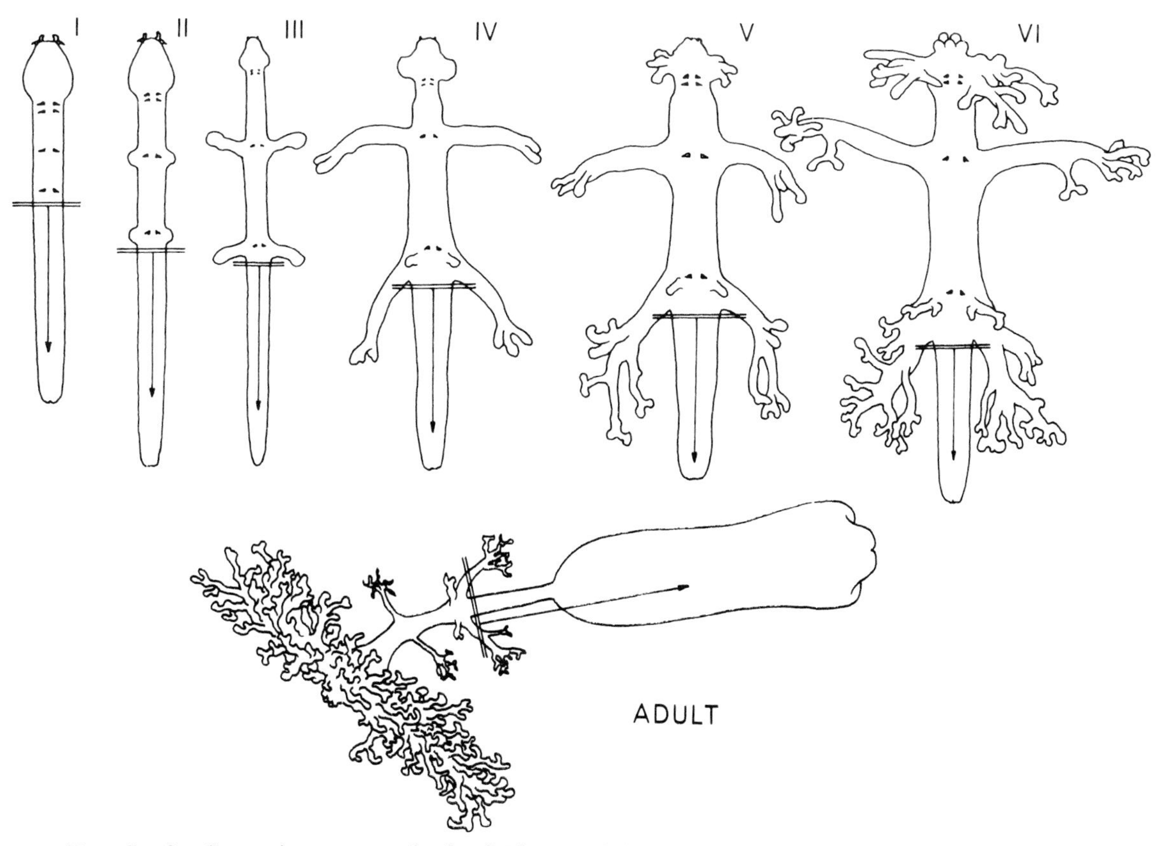

Text fig. 9. Stages in metamorphosis of *Phrixocephalus cincinnatus*. Modified from Kabata (1969a).

Text fig. 9 shows the progress of metamorphosis in a pennellid species, *Phrixocephalus cincinnatus* Wilson, 1908. Its free-swimming stages are unknown, but they are in all probability similar to those of *Lernaeocera* (Figs. 1338–1344). Its life history from the moment of attachment to the definitive host was described by Kabata (1969a), who divided it into seven stages, the first six of which are passed within the eye of that host, the flatfish *Atheresthes stomias*. The youngest stage found in the eye (stage I) has three morphological regions: a somewhat bulbous cephalothorax, a thoracic neck formed mainly by the third and fourth leg-bearing segments (the first and the second do not participate in its development, as can be seen by the position of the legs they carry), and a genital-abdominal region, marked off by two transverse lines and an arrow. The last part comprises about half the total length and bears no appendages other than the uropods. During the early growth period (stages II and III) the length proportion of the genital-abdominal region and the anterior part of the body remains about the same. Not much new development occurs in the posterior half, while the anterior half begins to develop the rudiments of the attachment organ, the holdfast. At stage II, three areas show slight lateral swellings: the posterior half of the cephalothorax (at the level of the maxilliped-bearing segment, which in the pennelid females carries no appendages) and the third and fourth leg-bearing segments. At stage III these swellings are enlarged, those of the leg-bearing segments becoming digitiform processes of some length. Stage IV is marked by increased growth rate of the anterior region, now comprising more than half the total length. Vigorous growth continues in the region of the posterior leg-bearing segments, the processes becoming quite long and beginning to branch. The lateral swellings of the cephalothorax now form about half its total volume. All these trends become even further accentuated at stage V. The genital-abdominal region is now reduced to about $\frac{1}{3}$ of the total length. The cephalothoracic swellings begin to sprout antler-like processes and the anterior margin pushes out subspherical swellings associated with the antennary and buccal regions. By the time stage VI has been reached, the anterior part has almost completed its extension in length. The lateral processes of the posterior leg-bearing segments (the secondary holdfast) (HS) have nearly attained their definitive sizes but the cephalothoracic expansion continues at an unabated pace. The genital-abdominal region has been left far behind in development. Sexual maturity is yet to be attained and attachment more firmly secured. For the remainder of the ontogenetic development the genital-abdominal region becomes the most energetically growing part of the body, expanding not only in length but also in girth, bursting out of the eye of the fish and becoming the familiar cylindrical trunk of the adult female. At the same time the cephalothoracic processes become a bush-like canopy of dendritically branching and intertwined processes, its diameter much larger than that of the trunk.

The development of *P. cincinnatus* can be expressed in the form of growth curves shown in Text fig. 10. Two of these curves are similar to the two curves of diphasic growth (see Text fig. 3), curve A tracing the rate of development of the anterior half of the body, while curve B describes that of the genital-abdominal region. There are, however, two more curves, one marking the progress of the primary holdfast (HP) formed by the cephalothorax and the other referring to the growth of the secondary holdfast (HS) produced by the outgrowths of the two leg-bearing segments. Each of these four regions develops at its own pace and at different times. Although their development must be synchronized with the sequence of biological needs, an impression is created of four independent development centres, the activity of each being prompted by different stimuli and subject to different conditions.

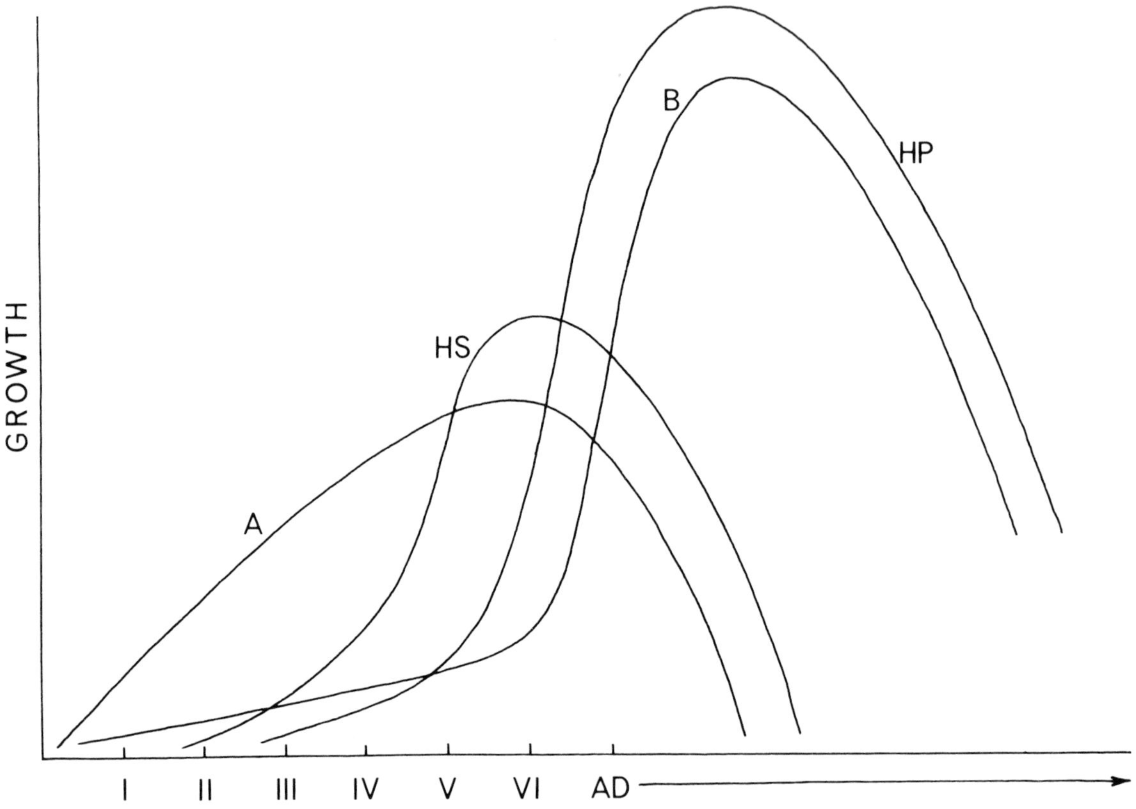

Text fig. 10. Growth curves describing metamorphosis of *Phrixocephalus cincinnatus*. Explanations in text.

A similar set of curves can be produced for all parasitic copepods. The development of some might call for more than four curves, but all are reducible to the activity of several development centres. Expressed in these terms, the entire process of metamorphosis becomes comparable to the simpler diphasic growth, from which it differs only in the degree of complexity imposed by the greater number of development centres.

Appendages

The change from a free-swimming existence to one of more or less permanent association with another living organism exerts a profound effect on the appendages of the copepods. Some functions previously served by these appendages are no longer being performed, other and new functions assume paramount importance. An appendage might have to change because its function has changed, though under the new conditions it has lost none of its importance. On the other hand, with the disappearance of its original function the appendage might find itself far less needed and nearly functionless. Some appendages might undergo little or no change, having been preadapted to the new role by the similarity between the old and the new function. (For example, the appendage that served the free-living copepod to grab its prey might be used by the parasite to hold on to its host.) The extent of the structural modification depends on the degree of the functional change. The impact of these changes on the individual appendages of copepods parasitic on fishes will be discussed below. Before proceeding with this review, however, it is necessary to identify these appendages and to discuss briefly some difficulties which exist in our understanding of their homologies.

As regards their origin, the appendages can be divided into two categories: primary and secondary. The former are the original appendages of every appendage-bearing segment, though in the main they have been much altered in the course of evolution. The disappearance of segmental boundaries during the process of tagmosis sometimes makes the connection of these appendages with their segments difficult to establish and is largely responsible for much of our confusion as to their true nature. The secondary appendages are phylogenetically late developments, evolved through aristogenesis in direct response to specific biological needs. It is with the primary appendages that we will be concerned below.

It is generally accepted that the first segment of the copepod head bears no appendages. The second and the third carry the first and second antennae respectively. (It might be mentioned here that Ferris and Henry (1949) challenged this view and postulated that in all Crustacea the two pairs of antennae arise from a single segment. Their challenge has been ignored.) There has been no agreement as to the homology of appendages posterior to the antennae, different workers seeing them in a different light. This lack of agreement on the homology of the cephalic appendages of Copepoda was summed up by Lang (1946) in the table reproduced below, with some modifications.

Table 1. Designation of copepod appendages by various authors (modified from Lang, 1946).

Author	Copepods	1	2	3	4	5
Claus (1863)	All copepods	Mdb	Mx	Mxp1	Mxp2	
Hansen (1893)	All copepods	Mdb	Mxe	Mx	Mxp	
Giesbrecht (1893)	All copepods	Mdb	Mx1	Mx2	Mxp	
Sars (1913–1921)	All copepods, except poecilostomes	Mdb	Mx	Mxp1	Mxp2	
T. Scott and A. Scott (several papers)	Free-living copepods	Mdb	Mx	Mxp1	Mxp2	
Heegaard (1947a)	Free-living copepods	Mdb	—	Mx2	Mxp1	Mxp2
A. Scott (1901a)	*Lepeophtheirus, Lernaeocera*	Mdb	Mx1	Mx2	Mxp	Mxp2
Wilson (1905)	Caligidae	Mdb	Mxe	Mx	Mxp1	Mxp2
Sars (1913–1921)	Poecilostomes	—	Mx	Mxp1	Mxp2	
Heegaard (1947a)	Caligidae, Ergasilidae	Mdb	Mx1	Mx2	Mxp1	Mxp2

Mdb = Mandible; Mx = Maxilla; Mxe = Maxillule; Mx1 = First maxilla; Mx2 = Second maxilla; Mxp = Maxilliped; Mxp1 = First maxilliped; Mxp2 = Second maxilliped

Lang (1946) himself believed that the poecilostome copepod has no mandible. He opposed Heegaard (1945, 1947a) and Scott and Scott (1913) who, like Claus (1863), postulated the existence of two pairs of maxillipeds.

The presence of mandibles in all copepods, poecilostomes included, has been by now convincingly demonstrated. Even grossly modified species, in which no mandible can be found during the adult stage, possess it during the earlier stages of ontogeny. Bocquet and Stock (1963) reviewed the evidence which appears to this author to settle the question definitely in favour of the presence of mandibles in Poecilostomatoida.

The other point at issue centres around the first maxilla (maxillule). Wilson (1905) and Heegaard (1945, 1947a) believed that the postantennary process of Caligidae, Bomolochidae and Taeniacanthidae (the last two included by them in Ergasilidae) represents the first maxilla. This meant that the appendage which other authors interpreted as the first maxilla had to be displaced backwards in the series and labelled second maxilla. The remaining two pairs of appendages had, therefore, to become the first and second maxillipeds respectively. The series of limbs was increased by one pair. (Claus (1863), who also believed that there are two pairs of maxillipeds, based his view on the mistaken belief that the last two pairs of the series arise from a single rudiment. His series included only one pair of maxillae. Although later (Claus, 1895) he acknowledged his mistake, some authors followed his earlier views. A. Scott (1901) and Wilson (1905), in particular, accepted them with some modifications and reservations.)

The protagonists of the view that the postantennary process is the first maxilla base their belief on two arguments. The first, and a very obvious one, is the general similarity in appearance between the process and the "other first maxilla", situated close to the mouth. Heegaard (1947a, p. 205) was clearly influenced by this similarity. The second argument is based on the fact that the process is supplied by a nerve which arises from the suboesophageal ganglion immediately in front of the one that runs to the "other first maxilla". This argument was put forward by A. Scott (1901) and Heegaard (1945, 1947a). (Though Scott was not entirely convinced that this innervation proves the spine to be a segmental appendage.) Wilson (1905) designated the postantennary process as the first maxilla without going into reasons for his decision. Later (Wilson, 1911) he changed his mind and suggested that the postantennary process is the exopod of the first maxilla which migrated outwards in Caligidae. In the bomolochid and taeniacanthid copepods, on the other hand, it was the endopod that migrated inwards, leaving the spine-like exopod in its original external position. According to Wilson, his view is justified by the fact that in *Lepeophtheirus edwardsii* Wilson, 1905, the nerves supplying the process and the first maxilla are united for some distance from their origin, "while in some *Caligus* species they are distinctly branches from a common trunk" (Wilson, 1911, p. 281).

The homology of the postantennary process with the first maxilla is opposed by many workers, who find the arguments in its favour unacceptable. Similarity of appearance is dismissed on the grounds that other spiniform structures occurring in the caligid copepods cannot be interpreted in similar terms. One notable example of such a structure is the sternal furca. (Though Ferris and Henry (1949), having found it supplied by a nerve from the suboesophageal ganglion, were consistent enough to claim that it, too, is a segmental appendage.)

The innervation argument was rejected by Lang (1946, 1948), who also put forward two counter-arguments. The first one was based on a borderline established between the cephalic and thoracic appendages. The latter can be identified by the possession of a narrow sclerite connecting their bases. In some copepods, Lang stated, such a connecting lamella has been found between the maxillipeds (Heegaard's second maxillipeds) but never between the pair of appendages that precedes them (Heegaard's first maxillipeds, Lang's second maxillae). The cephalic nature of that pair was held to be thus demonstrated. The second counter-argument is that of ontogenetic sequence. If the postantennary spine is a first maxilla, then it should appear during the course of postembryonic development as the appendage next after the mandible and before the second maxilla. According to Lang, this does not happen. The postantennary process makes its appearance much later, only after appendages posterior to it in the series have been in existence for some time.

Neither side convincingly demonstrated the validity of their concepts. Similarities in appearance are inacceptable as evidence. The innervation argument suffers from the fact that no dissection of the nervous system of a parasitic copepod has been detailed enough. Moreover, the ganglia of the central nervous system supply, by nerves arising from them, parts of the body other than the appendages. For example, in *Caligus* the nerve running to the frontal plate arises just anteriorly to the nerve supplying the postantennary process, while the first maxilla is innervated by a nerve arising in front of the one running to the lateral musculature. Wilson's (1911) statement that the postantennary process and the first maxilla are supplied by two branches of a single nerve serves only to increase uncertainty. Before any conclusions can be made on the basis of such information, more abundant and accurate data are clearly required.

The ontogenetic sequence counter-argument was opposed by Heegaard (1947) in a statement that a structure which is well on the way to complete disappearance (most copepods have no postantennary process) might be late in emerging during the course of ontogeny, so that the orderly sequence of appearance can be broken. He quoted as an example the mandible of Lernaeopodidae, supposedly appearing only during the pupal stage in *Clavella*. The example is spurious, since the mandible was found in the copepodid stage of that genus by Gurney (1934). Among other members of the family, *Salmincola californiensis* (Dana, 1852) was found by Kabata and Cousens (1973) to have the mandible at the copepodid stage. In *Pseudocharopinus dentatus* (Wilson, 1912) the author found this appendage in the nauplius. Heegaard's position could be much better defended by the presence of the postantennary process in the copepodid of *Caligus clemensi* Parker and Margolis, 1964, studied by Kabata (1972a). The process was very small, disappeared when the copepodid moulted into the first chalimus stage and did not reappear until development reached the stage of chalimus III.

In the argument about the ontogenetic sequence, sight has been lost of the fact that the moult that follows the last nauplius stage marks a transition between two vastly different morphological forms. From a small, subtriangular creature with only three pairs of appendages, the copepod turns into an elongate, fusiform, segmented animal equipped with a full set of appendages up to and including the second swimming legs (and the rudiments of the third). All the mouth parts (and sometimes also the postantennary process) appear simultaneously. There is, in fact, no chronological succession of oral appendages posterior to the mandible. There is also nothing to indicate that the postantennary process has undergone any change in position. At the earliest stage at which it has ever been seen (i.e. the copepodid of *Caligus clemensi*), it appears near the lateral margin of the cephalothorax, in the place where it will remain for the duration of the life cycle. The extensive morphogenetic restructuring which must intervene between the nauplius and the copepodid stages passes unnoticed within the naupliar cuticle. Only when its secrets are probed will we be able to give the final answer to the vexing problem of the postantennary process and its implications for the homology of the appendages. In the meantime, it is best, in the view of this author, to adhere to the concept of the appendages held by the majority. (This attitude represents a change of mind, since in his early work the author followed Heegaardian terminology of the appendages.) In the account which follows, we shall accept the existence of two pairs of antennae, mandible, two pairs of maxillae and one pair of maxillipeds in all the copepod species that are going to be reviewed and described.

The first antenna

In the free-living copepods the first antennae are almost always much longer than the second antennae. In those included by Pesta (1908) in the "floating type" (Schweber) they frequently exceed the length of the entire body, while in the more active swimmers (Schwimmern), though relatively shorter, they still retain their superiority in length over the second antennae. Structurally, they form a single row of more or less cylindrical segments, gradually diminishing in size towards the distal end. Only very rarely a diminutive second branch might be present (e.g. Lilljeborg's *appendix membranacea*). The number of segments increases during the process of development and in some species exceeds 20 during the adult stage. The first antennae are involved in locomotion and, in the males, assist in mating by developing, in one or both members of the pair, a geniculation used as a locking hinge for holding the female during copulation.

The first antenna of copepods living parasitically on fishes is much more insignificant in size. The developmental cycles are known for only a few species, but, in some of them at least, the number of segments in the first antenna increases with successive moults, as it does in their free-swimming relatives. For example, Grabda (1963) found three segments in the first antenna of the copepodid stage of *Lernaea cyprinacea* L. One segment is added at each moult until, at the stage immediately preceding the pre-metamorphosis female and the adult male, the appendage has six segments. This author examined the copepodid stage of *Dichelesthium oblongum* (Abildgaard, 1794), in which the first antenna was only two-segmented, while in the adult of that species it has six segments. It is not unreasonable to presume that the addition of segments accompanies the development of the first antennae in many parasitic copepods, if not in the majority.

There exists, however, a large group of parasitic copepods in which this does not take place. The group consists of six families, the species of which have two-segmented first antennae at the first post-nauplius stage (the copepodid) and retain that number throughout their entire life cycle. The families are: Dissonidae, Trebiidae, Caligidae (Fig. 439), Euryphoridae (Fig. 743), Pandaridae (Figs. 773, 774) and Cecropidae (Fig. 907) (although in the last of these families there are some species with three segments at the adult stage). In this group the first antennae change but little during ontogeny. The most important change is the proportional increase in size of the basal segment. In many genera, particularly among Caligidae, it tends to become flat and its base fits neatly under the frontal plate. The anterior margin acquires equipment of both armed and naked setae, constant in number within genera and even within some families. The short cylindrical distal segment is equipped with apical armature, also surprisingly uniform in composition throughout the entire group. The sensory function of the first antennae in these six families was demonstrated by A. Scott (1901a), who found these appendages very richly innervated, each seta receiving a separate branch of the antennular nerve. This author saw the tips of the first antennae in *Caligus* repeatedly raised and lowered while the copepod was in motion. It appeared to palpate the substrate. A marked change in function of the first antenna has taken place as a result of parasitic adaptation.

The first antenna of the copepodid of *Dichelesthium oblongum* is strikingly similar to those found in the corresponding stages of Caligidae. From this, as well as from the fact that in the adult *D. oblongum* the first antenna is quite different in structure, one might tentatively conclude that in Caligidae and their allies the development of this appendage has been arrested at the copepodid stage. The two-segmented condition is best to meet the necessity, common to members of these families, of adhering closely to the surface of the fish. Crawling movements, during which the frontal region must glide smoothly over the substrate, further accentuate the need for a short, rather flat first antenna that would not form an obstacle and not be exposed to undue friction during the forward motion of the copepod. Although during the adult life of the female many members of this group (e.g. Pandaridae or Cecropidae) remain largely stationary, the need for locomotion during the early adult stage remains an important biological factor. The structure of the first antenna cannot escape its influence.

A large group of poecilostome copepods, generally considered as relatively little modified by their adaptation to parasitism, is associated with the surfaces of their hosts. They are capable of changing the site of attachment but exercise that ability only seldom, if ever, and are considerably less mobile than the copepods of the group just described. Their sedentary existence might find at least partial explanation in the fact that most of them live on the gills, where feeding conditions are uniformly good over a large area. A change of place confers on them scanty benefits, as well as being fraught with the danger of losing the all-important grip on their hosts and being swept away. The first antenna of the members of this group is usually a simple series of unmodified cylindrical segments, well armed with setae. The number of segments varies from four (among Telsidae) to seven (in some Ergasilidae and Taeniacanthidae). The segments are usually fairly well delimited and richly armed with setae. The appendage is fairly large in relation to the size of the body (at least by comparison with the preceding group). Typical examples are the first antenna of *Anchistrotos onosi* (T. Scott, 1902) (Fig. 56) and of *Ergasilus sieboldi* von Nordmann, 1832 (Fig. 90). In addition to the three families already mentioned, the author also includes in this group the Bomolochidae and Tuccidae. The former are not entirely typical, inasmuch as they frequently live on surfaces other than gills and are more mobile than other members of the group. Significantly, their first antennae show signs of differentiation that, if carried to its typical conclusion, would make them similar to those of Caligidae. The proximal part of the appendage expands, becomes flattened and either partially or completely loses its segmentation (Fig. 3). The three distal segments remain cylindrical and well differentiated (flagellum of Vervoort, 1962). In some genera one or more setae on the proximal part become modified (Text fig. 26) but their structure offers no clues as to the functional reasons for the modification.

The siphonostome copepods include several families that form a group similar to the one described above. Although in the past included in a single family Dichelesthiidae *sensu lato*, these copepods are morphologically much more heterogenous than corresponding poecilostomes. Almost all of them live on the gills, but *Anthosoma crassum* (Abildgaard, 1794) burrows in the flesh of the host, leaving exposed only the posterior part of the thorax and abdomen. Among the members of the group the locomotory activity during the adult phase of life is either severely restricted or ceases altogether. Their first antennae range in structure much more broadly than among the poecilostomes of the preceding group. In Dichelesthiidae *sensu stricto* they are a simple series of six well distinguishable segments, poorly equipped with setae except for the apical armature (Figs. 1007, 1036). In other families the segmentation is both less distinct and more variable, though the antenna remains a simple cylindrical appendage (Kroyeriidae, Fig. 1085; Pseudocycnidae, Fig. 1313; and Hatschekiidae, Fig. 1131). Some structural modifications suggest that in several members of the group the first antennae have been pressed into service as auxiliary organs of attachment. In *Eudactylina* van Beneden, 1853, (Eudactylinidae) for example, this appendage has developed a geniculate flexion near the base and carries one or more powerful claw-like spines (Fig. 1180). Spines of similar type occur also on the first antennae of some species of *Hatschekia* Poche, 1902 (Hatschekiidae) (Fig. 1119). In some members of this group the first antenna is associated with curious structures of unknown homology. In *Lernanthropus* Blainville, 1822, a so-called parabasal flagellum arises close to the base of the appendage (Fig. 1058), whereas in *Anthosoma crassum* (Dichelesthiidae) a trough-like outgrowth offers support to the basal part of the first antenna (Fig. 1034). The general impression made by the first antennae of this group of copepods is that they have been deprived of their original function and, being only marginally useful under new conditions, have been neglected by evolution.

The next group consists of siphonostomes the adult females of which are permanently attached to the surface of the fish by modified second maxillae. These copepods are surface browsers and are able to cover a more or less extensive surface area during their feeding activities, while remaining safely anchored to the host. They are greatly modified in their adaptation to parasitism, the extent of modification being clearly perceived in differences between the adult females and free-swimming copepodids, where these are known. Such data on their life history as are available indicate that in the copepods of this group the first antennae either remain more or less unchanged in the course of development, undergoing only reduction in their setal armature, or lose their segmentation and become small digitiform pegs. The group comprises the families Lernaeopodidae, Naobranchiidae and Tanypleuridae, the first-named being the largest and most diversified of the three. The lernaeopodid first antenna is always very short and usually slightly inflated at the base. Its segmentation is usually indistinct, but never exceeds four identifiable segments. Armature other than apical is almost completely reduced (Fig. 1506). The apical armature may also become vestigial, especially in the genus *Salmincola* Wilson, 1915, where it is often absent due to wear *in vivo*. When present, it is usually quite uniform within genera or groups of genera. In several species of *Pseudocharopinus* Kabata, 1964, the base of the appendage carries a prominent papilliform process of unknown function (Fig. 1624). (For more details on the lernaeopodid first antenna see p. 342.) While generally similar to the first antenna of Lernaeopodidae, that of Naobranchiidae is distinguished by the absence of apical armature. Its segmentation is obscure and its distal half carries a fairly long, robust, digitiform process (Kabata, 1968a, Fig. 5D). In the only known member of Tanypleuridae, *Tanypleurus alcicornis* Steenstrup and Lütken, 1861, the first antenna is unsegmented, short and reminiscent of the lernaeopodid appendage in its armature (Kabata, 1969b, Fig. 6). The adult females belonging to this group appear to have but little use for their first antennae.

The poecilostome copepods have produced no group equivalent to the one described above. Therodamasidae, Pharodidae and Chondracanthidae contain those poecilostome species that have become most extensively modified in their adaptation to parasitism. Therodamasidae have burrowing habits and are sessile as adults. Some ability to move over the surface of the host is known to exist among Pharodidae; Chondracanthidae remain stationary throughout their adult lives. Among Therodamasidae and Pharodidae the first antennae are in about the same condition as among the less modified poecilostomes. They are simple series of more or less distinct segments, variously armed. In Chondracanthidae they have become unsegmented, with inflated basal and cylindrical terminal parts. All armature is reduced, very small and often difficult to see. Apical armature is frequently lost due to damage. The relative size of the first antennae varies from very large and prominent (Fig. 130) to diminutive (Fig. 338). The claim that some of the chondracanthid species have lost their first antennae needs verification. Our knowledge of the biology of these copepods being very scanty, the role of the first antennae in this group is not clear.

The family Pennellidae contains many species that can be classified as mesoparasitic (see p. 290). All known members of Sphyriidae (Siphonostomatoida) and most of Lernaeidae (Cyclopoida) also belong to this category. The entire head of the mesoparasite is buried in the tissues of the host. The pre-metamorphosis females of Pennellidae have cylindrical, well armed, though obscurely segmented, first antennae (Figs. 1350, 1351) that are practically unchanged from those of the copepodids. In Lernaeidae (Fig. 393) they differ from those of the earlier stages only by the larger number of segments, when discernible. The process of metamorphosis leaves these appendages mainly unaffected, though they, like the remaining appendages, become dwarfed by the gigantic growth of the copepod. The extent of modification cannot be judged in the case of Sphyriidae, since the early developmental stages of these copepods are all but unknown, but the highly aberrant form of the first antenna in *Sphyrion lumpi* (Krøyer, 1845) (Fig. 1446) suggests that it must be considerable. The suggestion is supported by the fact that in the males of the genus *Lophoura* Kölliker, 1853 (Fig. 1450) and *Tripaphylus* Richiardi, 1880 (Fig. 1461) this appendage has the more conventional cylindrical structure. It appears that in Pennellidae and Lernaeidae the first antenna, having lost such function as it might have had, is not being diverted to other uses and persists only as a functionless link with the free-living pre-metamorphosis period.

There are also no special modifications of the first antenna among the endoparasitic copepods. It might become obsolete, as in the female *Sarcotaces* Olsson, 1872, but whenever retained it is relatively unmodified, well armed and clearly segmented (Fig. 418). Nothing is known about its functional significance in these copepods.

The second antenna

During their nauplius stage, all copepods possess a biramous second antenna which, together with the first antenna and the mandible, serves as a locomotory appendage. Although in the nauplii of some parasitic copepods its segmentation might be partially indistinct, the exopod, about equal in length to the endopod, invariably contains a greater number of segments than the latter ramus. This primitive type of second antenna has been retained throughout adult life by some free-living copepods, exemplified by *Calanus finmarchicus*. The second antenna of that copepod has a short, two-segmented sympod, two-segmented endopod and seven-segmented exopod, the latter ramus consisting of four very short, and three slightly longer, segments. Primitive in structure, the second antenna of *C. finmarchicus* appears to have retained also its function as a locomotory appendage, though it has been overshadowed by inordinate development of the first antenna.

Even among the copepods related to *Calanus* there are, however, some species in which the structure of the adult second antenna differs from that of the nauplius stage. The change consists of reduction in the size of the exopod. Among the harpacticoid copepods this trend has become even more accentuated, the exopod shrinking to quite insignificant size, compared with the endopod. Another development is the alignment of the sympod and the endopod along the same longitudinal axis, so that the diminutive exopod has become little more than a palp-like process arising from the lateral side of the second sympodial segment.

The end-point of this trend might be represented by the free-living cyclopid copepods. Their second antenna is a four-segmented appendage (two segments of sympod and two of endopod), completely devoid of the other ramus. It has become much smaller than the first antenna, a development suggesting a change and perhaps some loss of functional importance. (It should be mentioned that in some free-swimming copepods the opposite tendency is manifest, the exopod becoming dominant.)

Gradually losing its function as an organ of locomotion, the second antenna emerges with increasing frequency as an appendage adapted for prehension. In the genus *Sapphirina*, for example, while still relatively small, it is definitely prehensile. Its basal segment is short and a geniculate joint is interposed between it and the much longer second segment; the third and fourth segments have become equivalent with the subchela and claw, being reflected and opposable against the second segment. Through several intermediate stages, this type of second antenna develops into a prominent grasping limb, its strong terminal claw eminently adapted for prehension. In *Corycaeus* it has reached a condition resembling that existing in *Ergasilus*, a poecilostome parasitic copepod. The differences between sexes in the structure of the second antenna suggest that the male *Corycaeus* uses this appendage during mating. The second antenna of this copepod and of its relatives appears, therefore, to have taken over the function that in other free-living copepods (e.g. Cyclopidae) is performed by the first antenna.

The development of an unciform claw in *Corycaeus* has been parallelled in numerous instances among Copepoda parasitic on fishes. The overwhelming majority of these parasites possess claw-tipped second antennae. More often than not they fulfil the part of the main, or even the only, organ of attachment. Even those mesoparasitic copepods that, for the maintenance of contact with their hosts, depend on special anchoring structures developed during metamorphosis use their second antennae for attachment at the infective copepodid stage. The great variety in host-parasite relationships is naturally reflected in the variety of structure of the second antenna. Some types of this appendage are shown diagrammatically in Text fig. 11.

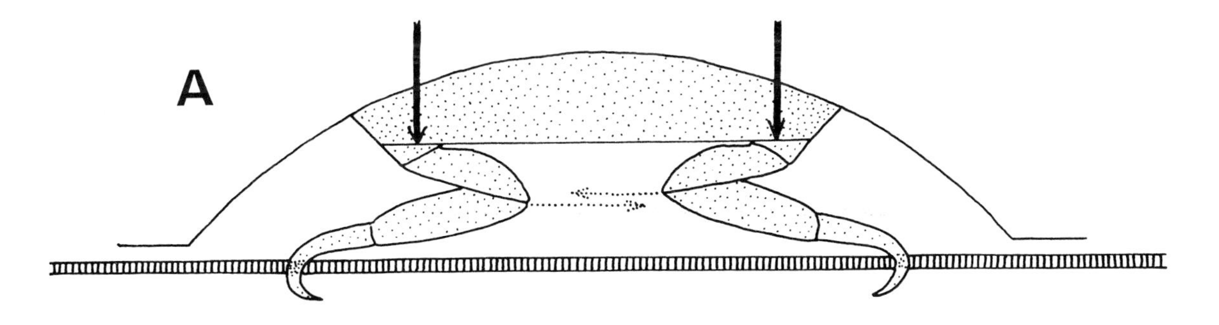

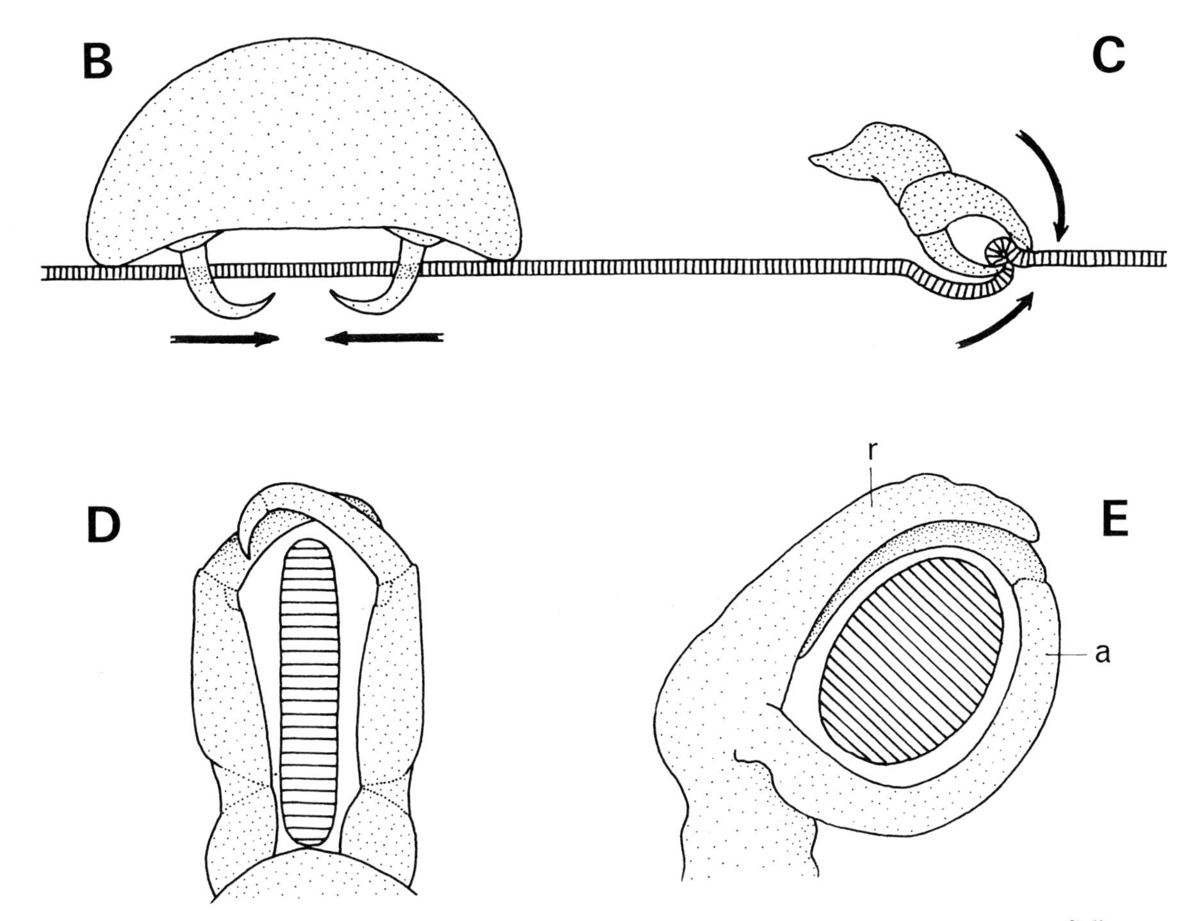

Text fig. 11. Types of second antennae (diagrammatic). A. Piercing second antennae of Caligidae (Siphonostomatoida) and of some Poecilostomatoida; B. Piercing second antennae of Chondracanthidae; C. Clasping second antennae of Pennellidae; D. Clasping second antennae of some *Ergasilus* species; E. Clasping second antennae of *Shiinoa*.

a — second antenna; r — rostrum.

Both siphonostome and poecilostome copepods that retain their ability to move over the surface of the fish, become modified so as to use their cephalothoracic tagmata like suckers. The suction is, however, supplemented by the anchoring power of the second antennae, at least during those times when the copepod is stationary. A group of siphonostome families (Dissonidae, Trebiidae, Caligidae, Euryphoridae, Pandaridae and, in some measure, Cecropidae) differs from all other siphonostomes parasitic on fishes in having second antennae in which the proximal halves, extending mediad from their bases, are overlaid by reflected distal halves. An example of this type is the second antenna of *Lepeophtheirus pectoralis* (Müller, 1777) (Fig. 631). Similarly folded second antennae exist in the poecilostome families Bomolochidae, Taeniacanthidae, Tuccidae and Telsidae, exemplified by that of *Bomolochus solae* Claus, 1864 (Fig. 4). The siphonostome appendage is clearly derived from a structure different from the one that gave rise to the poecilostome appendage. The former ends in a single massive claw, the latter is equipped with several much smaller ones. They differ also in other essential features. We are faced here with a clear instance of parallel evolution resulting from similar host-parasite relationships. The manner in which the suction of the cephalothorax is synchronized with the holding power of the second antenna is illustrated in Text fig. 11A. To produce suction, the saucer-like cephalothorax must be pressed down, as indicated by the arrows in the figure. Downward pressure on the bases of the second antennae translates itself into mediad displacement of the elbow-like joint of the appendage and locks in the claws that are inserted into the tissues of the fish. To loosen this grip the suction must be released and the joint unflexed. The copepod is then ready to move. Characteristically enough, in those copepods in which suction is less efficient and other means of attachment have been developed, the second antennae are also differently affected. For example, in the genus *Pandarus* Leach, 1816, parasitic on elasmobranchs, the attachment is aided by the so-called adhesion pads (Fig. 770). The claws of the second antennae are poorly developed and blunt (Fig. 775). On the other hand, in *Phyllothyreus cornutus* (Edwards, 1840) the adhesion pads are rudimentary and the cephalothorax is not as efficient in suction as, for example, those of Caligidae. These deficiencies are offset by the development of very powerful second antennae (Fig. 880).

The unciform claws of the second antennae can be used in one of two ways. They can anchor the parasite to the host by piercing and hooking into the integument of the host. Alternatively, they can assist in transforming the appendage into a structure which, together with its opposite member, acts as a vice enclosing a slender outgrowth of the host's body (e.g. a gill filament). They can be referred to as piercing or clasping antennae respectively. The great majority of copepods parasitic on fishes have second antennae of the piercing type. One variant of the piercing second antennae has already been described, but there are several others.

The simplest and most primitive second antennae are those that have retained their slender and simple shape and sometimes even traces of the original four-segmented structure. Their prehensile function is evident only in the presence of a terminal claw, in some copepods accompanied by other apparently prehensile processes. Nothing is known of the way in which these appendages perform their function, or how efficient they are, but they give an impression of being rather ineffectual. Antennae of this type can be present in both primitive and highly modified copepods. Among the former is the genus *Nemesis* Risso, 1826 (Fig. 1273), while a good example of the latter is provided by *Lernaea* L. (Fig. 394). In the latter instance the second antenna has been arrested in development at the pre-metamorphosis stage of the copepod.

An interesting variety of this primitive second antenna occurs in the male of *Philichthys xiphiae* Steenstrup, 1862. By acquiring a strongly hooked claw on the penultimate segment, in addition to the more usual terminal claw, the antenna became bifid in appearance (Fig. 420) and has been, indeed, referred to as a special type of second antenna by Gerstaecker (1866–1879).

Almost all other types of the second antenna have been derived from a fairly primitive appendage similar to that of *Nemesis* or *Lernaea*. A very common trend in evolution was the one that resulted in enlargement of the terminal claw, accompanied by shortening of other parts of the appendage (fusion or loss of segments). The outcome of this process is a more compact, robust, subchelate antenna with greatly enhanced prehensile efficiency. Antennae of this type exist in Pseudocycnidae (Fig. 1314), Hatschekiidae (Fig. 1120), Lernanthropidae (Fig. 1060) and, among Poecilostomatoida, in Therodamasidae. This trend reaches its end-point in the second antenna of Chondracanthidae (Text fig. 11B), which consists of nothing but an uncinate claw articulating with small basal elements. Inserted deeply into the integument, the two antennae together tightly staple the head of the copepod to the surface of the fish, with the mouth resting flat on the nutritive substrate.

The second antenna of Dichelesthiidae *sensu stricto* can be derived from the primitive appendage by elongation and loss of segmentation. Only its terminal parts remain rigidly sclerotized. The claw is short and sharply hooked, subchelate. The long, unsegmented stem of the appendage is clad in rather thin cuticle and is highly contractile. When contracted, the antenna appears to consist of only its distal part (Fig. 1004), the long and flexible stem being visible only in fully extended appendages

(Fig. 1029). In *Anthosoma* this contractility is an obvious adaptation to burrowing into the flesh of the host but in *Dichelesthium* the need for it is not presently understood.

Lernaeopodidae and Naobranchiidae attach themselves to the fish with the aid of their second maxillae. In many of them the mouth, situated at the end of a long cephalothorax, must be held in the feeding position by prehensile second antennae. Naturally, less power is required to maintain this hold than to anchor the entire parasite. The antennae, therefore, are less well developed, but they are able to perform a great variety of movements and quite capable of hooking into the integument of the fish and keeping the mouth in position for a sufficient period (Fig. 1573).

The second antennae of the clasping type exist in Ergasilidae. Species of the genus *Ergasilus* can be arranged in a series as regards the size of their antennae. In some they are too short to enclose a gill filament between them. In these species the antennae function as another variant of the piercing type, the tips of the terminal claws engaging the sides of the filament. The reach of the antennae increases until in many species they can clasp the filament. The increase in size is accompanied by relative elongation of the second segment, subchela and claw (Text fig. 11D). The clasp formed by the antennae can become permanent.

While in *Ergasilus* and its relatives the second antennae must close against each other to enclose a piece of the host's tissue, in the genus *Shiinoa* Kabata, 1968, they act in parallel with each other. A prominent, roof-like rostrum has developed from the pre-antennary part of the cephalothorax. The claws of both second antennae fit under it, side by side, and become fused with it. The appendage loses its segmentation. A complete circle closes like a padlock around the gill filament of the host and the parasite swings securely around it (Text fig. 11E).

To achieve pincer-like action, both second antennae are required in Ergasilidae. The same effect can be achieved by one second antenna, if it becomes chelate (Text fig. 11C). The advantage of this arrangement is obvious. Chelate second antennae exist in Pennellidae, in which they are used only during the early developmental stage (copepodid), to be replaced as attachment organs by the secondarily produced cephalothoracic antlers in the adult. The tenacity of their grip was vividly described by Sproston (1941), who stressed in particular the flexibility of their articulation with the base, enabling them to rotate within an arc of 180°. No longer used by the adult, the antennae are retained by them unchanged (Figs. 1349, 1353). *Kroyeria* is equipped with similar second antennae (Fig. 1087) but in that genus they act, together with the maxillipeds, as main organs of attachment also at the adult stage. *Pseudohatschekia* Yamaguti, 1939, a genus *incerte sedis*, also possesses chelate second antennae.

Finally, there are few copepods parasitic on fishes with second antennae that are not modified for prehension. In the mesoparasitic female *Sphyrion* they have become vestigial and obviously functionless (Fig. 1446). In *Tanypleurus* they are unsegmented, peg-like and perhaps sensory, if one were to judge from their armature (Kabata, 1969b). In *Catlaphila* Tripathi, 1960, they are two-segmented and rather resemble the first antennae. Unfortunately, we know almost nothing about the development of those copepods and the possible course of ontogenetic changes in the structure of their second antennae.

The mouth

Copepods parasitic on fishes can be divided into two groups by the type of their mouths. The structure of the mouth and oral appendages was the main reason for splitting off Siphonostomatoida and Poecilostomatoida from the Sarsian order Cyclopoida (see classification). Although both types have many varieties, these are based only on differences in details of their component parts. This author knows of no copepods with mouths that can be considered as an intermediate stage between the two basic types.

The poecilostome mouth is, in effect, a transverse slit. The anterior side of the slit is overhung by the labrum, a plate-like structure varying in shape from a transverse rectangle in the chondracanthids (Text fig. 38; Fig. 137) to semicircular or subtriangular projections in the bomolochids and their allies. The posterior margin of the rectangular labrum is often concave. In many chondracanthid species it is armed with denticles along that margin and its distolateral corners might be extended by the presence of rounded processes. In some species of that family the labrum shows sexual dimorphism, that of the male having a median process on its anterior surface (Ho, 1970). Fine denticulation or setation is also common on the labra of the bomolochids, in which it often constitutes a diagnostic feature at the specific level.

The lateral sides of the mouth are open to admit the oral appendages. The posterior side is delimited by the labium, often an ill-defined structure, appearing as a rather shapeless, sometimes lobate protuberance. Some authors are doubtful about the homology of this swelling and refer to it as the labium in quotes. The labial swelling is usually lower than the labrum. It has been suggested that, in poecilostomes, the primitive mode of feeding involves the passing of food particles forwards into the gaping buccal orifice from the posterior position. This would provide a functional reason for the low rise of the labium above the ventral surface of the cephalothorax.

Humes (1967) described a poecilostome parasite of antipatharian coelenterates in Madagascar, *Vahinius petax*, a copepod distinguished by the prominence of its mouth. Both labrum and labium are highly elevated and together form a cone superficially resembling a siphonostome mouth. The cone, however, is split by a transverse gap and the lips show no signs of fusion, or even of contact.

The mouth of the Cyclopoida is similar to that of Poecilostomatoida in all essential features.

In contrast, the siphonostome mouth is formed by partial or complete fusion of labrum and labium. As suggested by its name, this type of mouth is a tube or syphon, built around and above the oesophageal opening and separating it from the buccal rim by a distance equal to the length of the lips. (Four families parasitic on invertebrates (Stellicomitidae, Nanaspidae, Cancerillidae and Micropontiidae), although siphonostome in the structure of their buccal appendages, have their mouth syphons in a barely rudimentary condition.) The lateral margins of the lips remain separate from each other for a short distance at the base, creating a small triangular opening through which the mandibles enter the buccal cone. In some species the fusion is incomplete also near the tips, which very often are armed with marginal strips of membrane, rows of setae, or similar structures. In Pennellidae the cone formed by the lips bears distally a complicated buccal tube (Text fig. 59, Figs. 1355, 1357) consisting of rings of hard cuticle capable of being partly telescoped into one another, and armed distally with transparent marginal membrane. The pennellid buccal cone can turn into an inordinately long proboscis, as in the genus *Ophiolernaea* Shiino, 1958. In some siphonostome families (Pennellidae, Lernaeopodidae, Caligidae, possibly others) the roof of the labrum carries a pair of buccal stylets, sometimes with two distal spines or setae (Fig. 1356).

The buccal cone of the caligids and their allies, as well as that of pandarids and their allies, is capable of movement in the anteroposterior plane and when not in use is folded against the ventral surface of the body. No copepod parasitic on fishes can equal the relative length of the mouth cone occurring in siphonostome copepods parasitic on, or associated with, invertebrates (except for the peculiar pennellid genus *Ophiolernaea*). Nevertheless, in some species of dorsoventrally flattened pandarids it appears to be too long to be placed at right angles to the surface of the host. In such copepods the tip of the mouth drags behind the base, as shown diagrammatically in Text fig. 12. For the buccal opening to remain flat on that surface in this position, it must be set obliquely to the long axis of the mouth cone. This is achieved by differences in length between labrum and labium, the latter being appropriately longer (Figs. 777, 778).

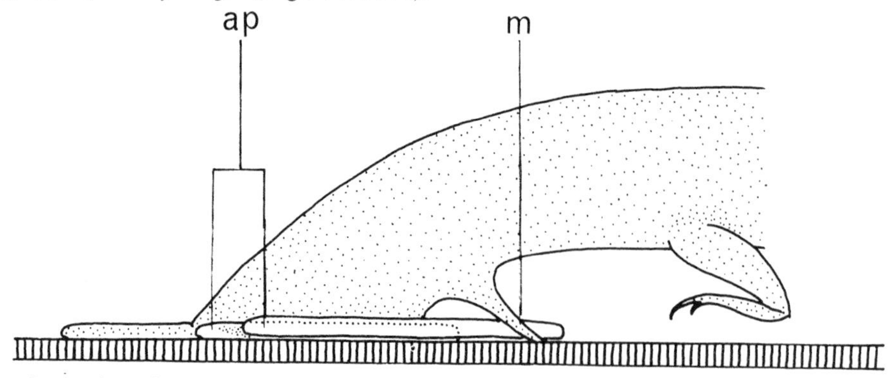

Text fig. 12. Sagittal section through anterior end of a pandarid copepod (diagrammatic). ap — adhesion pad; m — mouth.

In the highly advanced copepods of the caligid type the mouth cone is relatively shorter than that of the pandarids and both lips are of the equal length (Text fig. 13). In *Caligus* and *Lepeophtheirus*, and possibly in other caligid genera, the tip of the labium carries a structure described by Kabata (1974) and shown diagrammatically in Text figs. 15 and 16. The structural differences between the caligid and pandarid mouth cones appear to be relatively small. The shorter length of the caligid

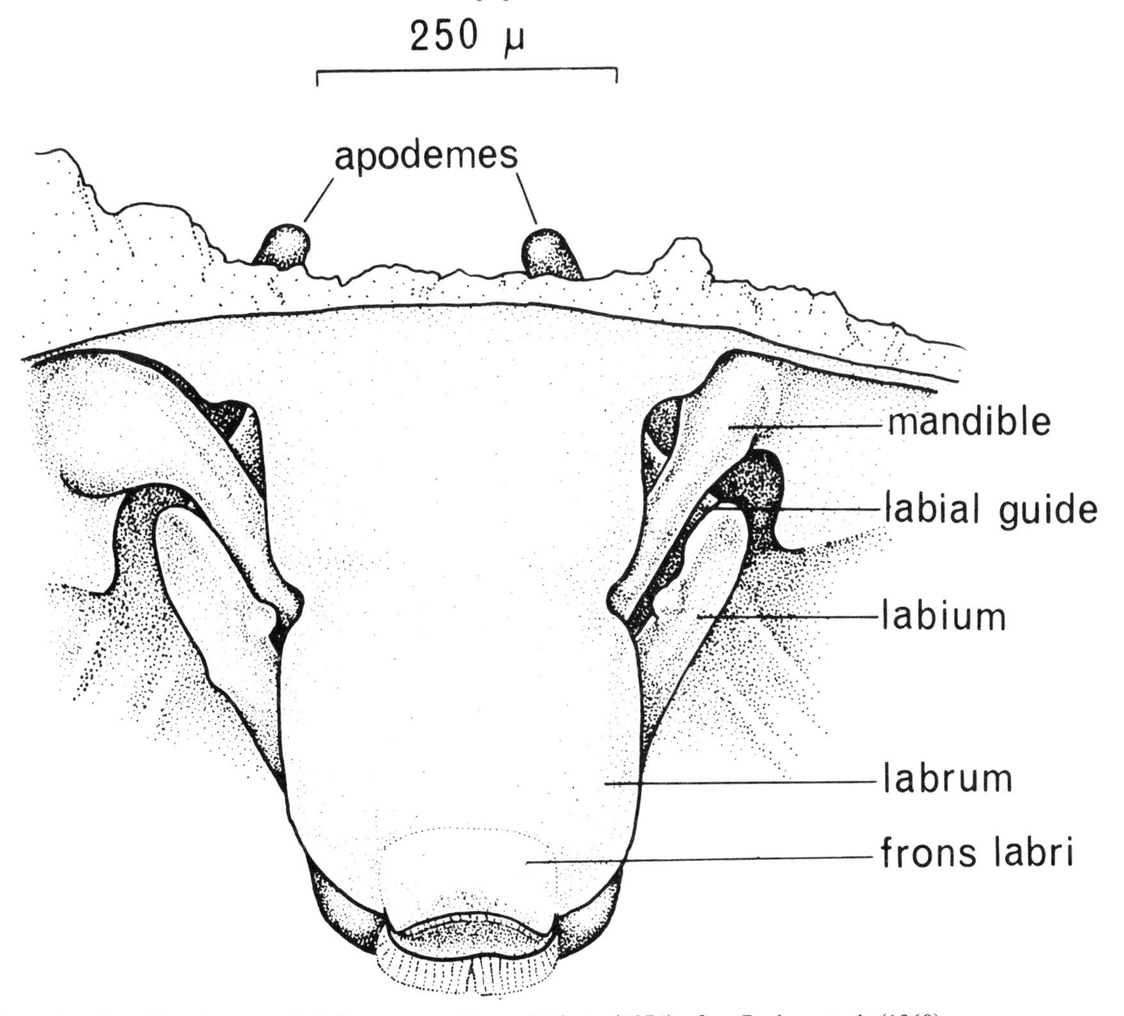

Text fig. 13. Mouth cone of *Caligus curtus*. From Kabata (1974) after Parker et al. (1968).

cone allows it, however, to swing into a position perpendicular to the surface of the fish, a fact of considerable functional significance. Based on "morphological speculation", the mode of action of the caligid mouth cone was described by Kabata (1974). With slight modifications, the description is quoted below.

"To feed, the copepod must place the cone in a position perpendicular to its body. This could be effected by contractions of the *levatores oris* (e, Text fig. 14), the force of which is transmitted to the cone through the apodemes (ad, Text fig. 14) acting as levers. In that position the mouth can be applied to the surface of the fish. The marginal membrane spreads around it and seals it from outside. The membrane, striated and greatly frayed, is equipped in some species with a row of short setae along the base, further increasing its insulating properties. A slight drop in intrabuccal pressure is needed to maintain firm contact between the mouth and the skin of the fish. It could be produced by the contraction of all or some of the muscles in the *compressores labri* groups (a, Text fig. 14). (The tip of the mouth cone could also be pushed into the skin by some mechanisms extrinsic to the mouth itself, e.g. by depression of the central part of the dorsal shield.) Whatever the mechanism, pressing of the mouth into the skin would push away the labial fold (Text fig. 15) and expose to the skin a divided bar of *strigil*, armed with about 100 fine and sharp teeth (Text figs. 15, 16). A rhythmic fluctuation in the pressure pushing the tip of the labium into the skin would cause the two halves of the *strigil* to move away from each other, pivoting on a common base, each divergent move being followed by a convergent one. As can be supposed from their structure, the teeth of the *strigil* would execute sawing movements, particularly effective during their inward stroke. The resulting accumulation of debris can be picked up from the surface by the movement of the mandibles, the mandibular teeth acting as conveyors moving the fish tissue into the buccal cavity.

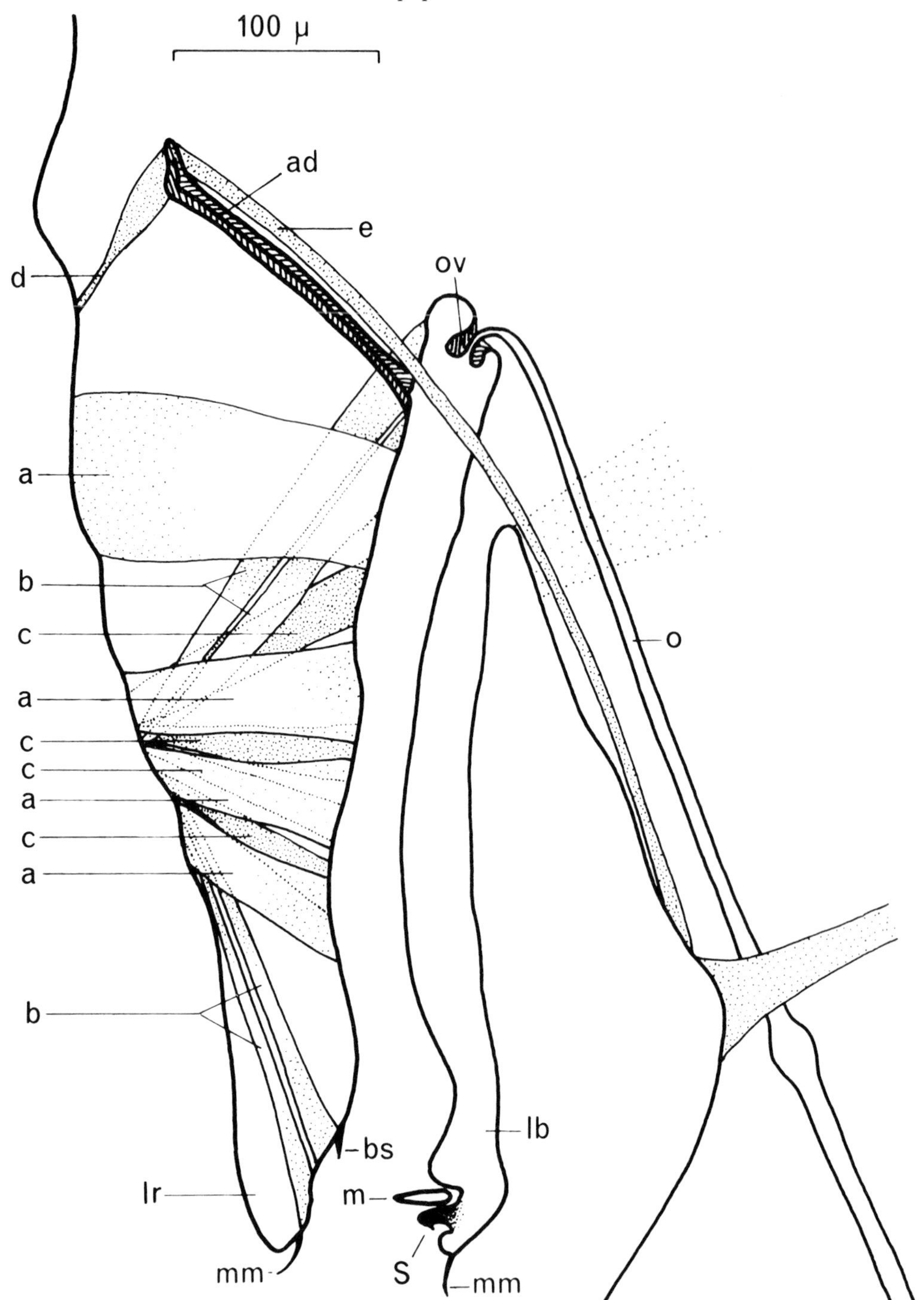

Text fig. 14. Sagittal section through mouth cone of *Caligus curtus*. From Kabata (1974).

a — compressores labri laterales; ad — apodeme; b — contractores labri; bs — buccal stylets; c — compressores labri mediales; d — depressor oris; e — levator oris; lb — labium; lr — labrum; m — mandible; mm — marginal membrane; o — oesophagus; ov — oesophageal valve; S — strigil.

Some change in the position of the mouth might be necessary for the mandibles to come in contact with the debris produced by the action of the *strigil*. Such change could be produced by a combined and complicated action of the musculature. *Contractores labri* (b, Text fig. 14), acting in unison, can lift slightly the *frons labri* (Text fig. 13) and break the seal, allowing the mouth to move. *Depressor oris* (d, Text fig. 13) and *levator oris* (e, Text fig. 14) would act jointly to control the backward slide of the mouth cone, while the *contractores labri* maintain sufficient pressure within the buccal cavity to prevent a complete loss of contact between it and the surface of the host. A sequential contraction of *contractores labri* would be capable of producing a peristaltic pressure wave further propelling the particles of the fish tissue into the buccal cavity. Some mechanism of this kind is required, because the movements of the mandibles themselves can convey these particles only near the tip of the mouth. The movements must be limited by the length of the mandibles and by lack of sufficient flexibility of their bases, not capable of withdrawing for distances that would allow the mandibular teeth to traverse the entire length of the mouth cone."

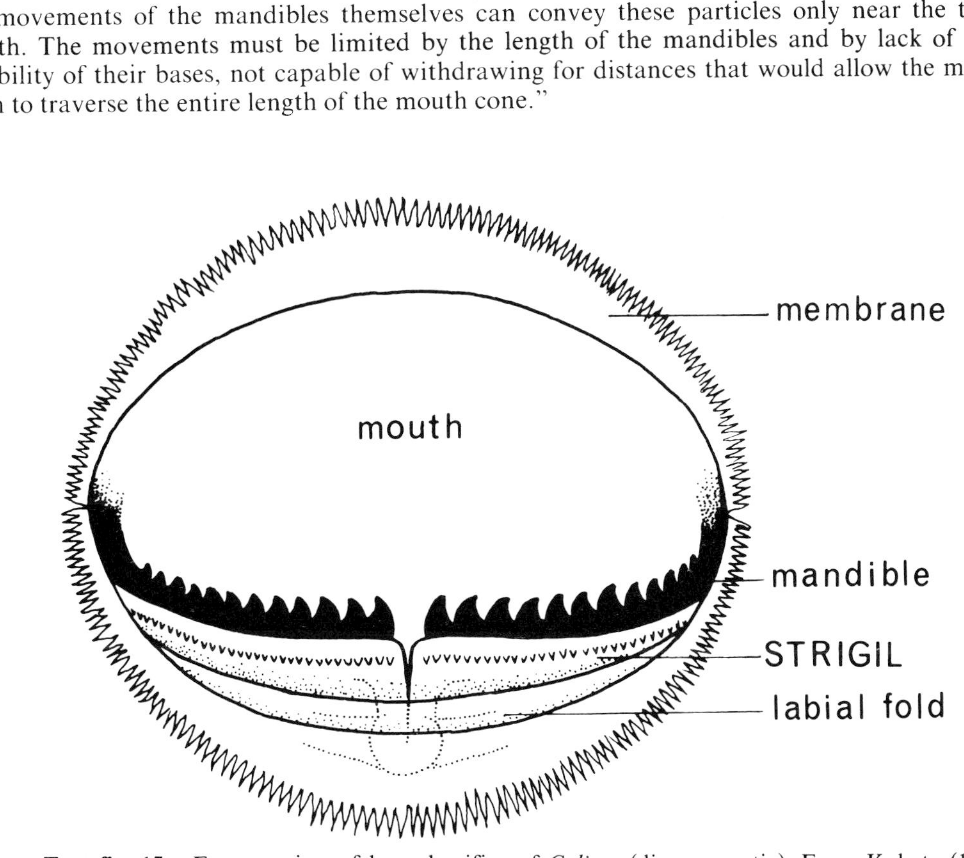

Text fig. 15. Face-on view of buccal orifice of *Caligus* (diagrammatic). From Kabata (1974).

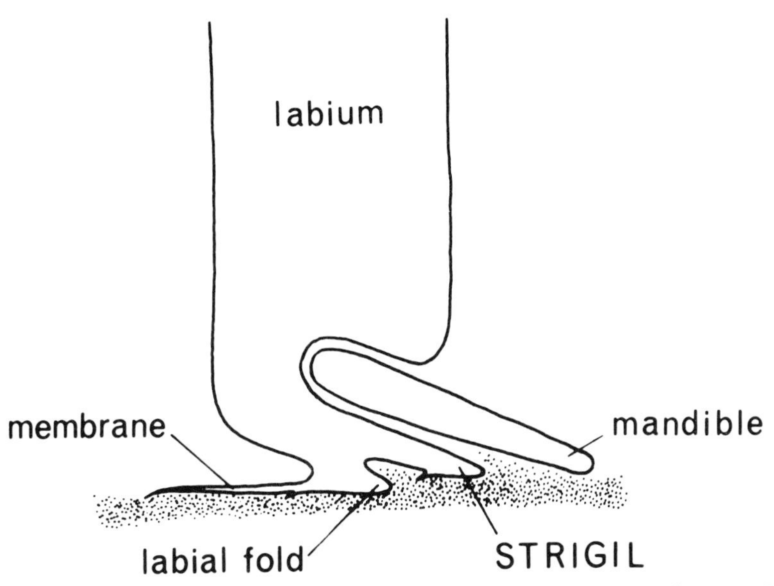

Text fig. 16. Sagittal section through tip of caligid labium (diagrammatic). From Kabata (1974).

No such speculation has been attempted for the pandarid mode of feeding. In all likelihood these copepods feed in a manner similar to that of the caligids, though in the absence of the strigil they must be able to pry off superficial cells of their hosts' integument in some other way. Unless, of course, they live only on mucus easily scooped up by their mandibles.

These comments show how little is known about the biology of parasitic copepods. Hopefully, they underscore also the fact stated at the beginning of this section, that is to say the presence of two radically different types of mouth, both with numerous variations. As regards the influence of parasitism on the buccal structures, it is most evident in the existence of the siphonostome mouth. So far as is known, there are hardly any free-living siphonostomes. It can be speculated that this type of mouth evolved from the primitive open type by elongation and fusion of lips, but in almost complete absence of intermediate stages there is no evidence that this really was its origin. In contrast, there are numerous free-living poecilostomes with mouths which are quite similar to those of their associated or parasitic relatives. In fact, the poecilostomes are a minority among copepods parasitic on fishes, though they predominate among those parasitic on invertebrates. The impact of parasitism on the structure of the copepod mouth appears, therefore, to have been slight.

The mandible

The nauplius stage of parasitic copepods, like that of their free-living relatives, usually carries three pairs of appendages, the most posterior of which is the mandible. The exception to this rule is some species of Lernaeopodidae. In that family the nauplius stage is passed through within the egg membrane and the mandible is either absent (as in *Clavella*), or greatly reduced (as in *Pseudocharopinus*).

By the structure of their mandible, the nauplii of the parasitic copepods can be divided into three groups, as shown in Text fig. 17. (It must be remembered that nauplial stages are known for only a few species and that other types might yet be discovered.) The most commonly occurring nauplii

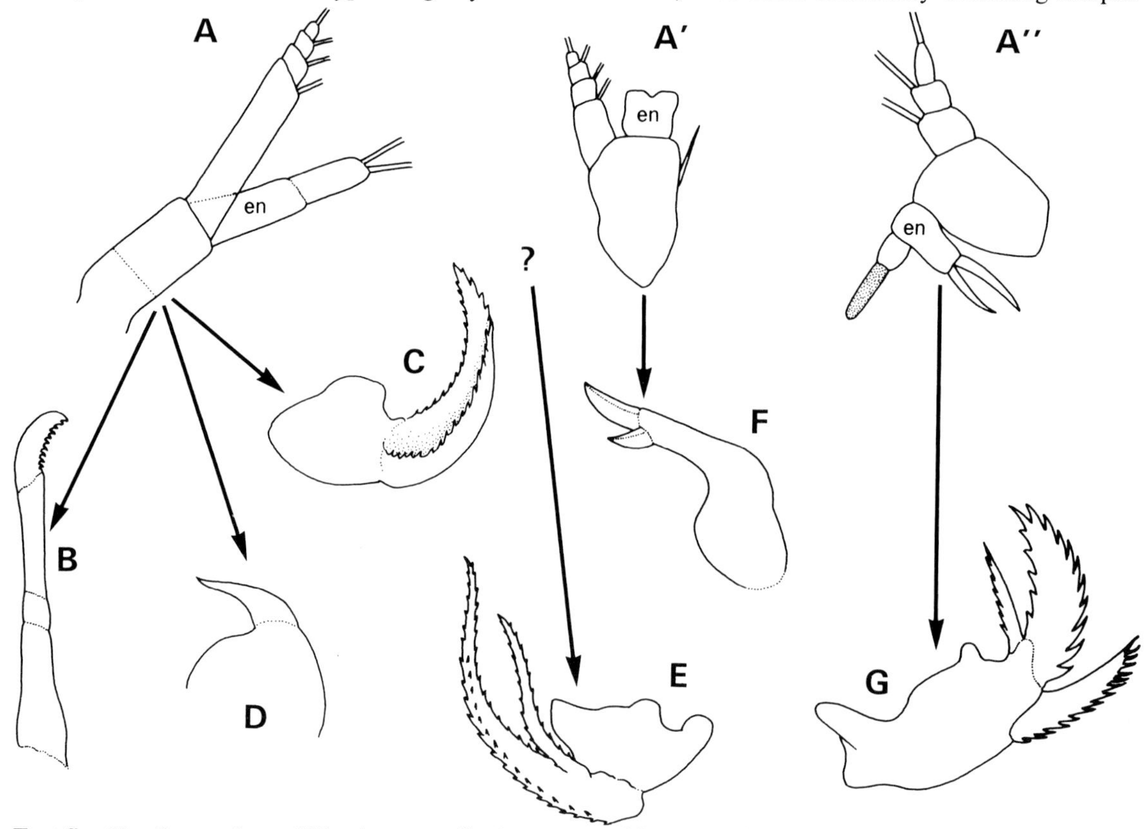

Text fig. 17. Types of mandible. A — nauplii of many parasitic copepods; A′ — nauplius of *Bomolochus*; A″ — nauplius of *Ergasilus*; B — adult Siphonostomatoida; C — some adult Poecilostomatoida; D — adult *Lernaea*; E — adult *Shiinoa*; F — adult Bomolochidae and allies; G — adult Ergasilidae.

have mandibles shown in Text fig. 17A. The literature contains contradictory data as to its segmentation. Its sympod is said to consist of two segments and its exopod of four, the first one providing most of the length of the ramus. Each exopod segment is armed with one sturdy seta. Heegaard (1947a) believed that the mandibles of this type were divisible into two categories: one, in which the endopod consisted of three segments, was typical of his Pectinata, the other had a two-segmented endopod and characterized his Fistulata. (These two groups were roughly equivalent to the more generally accepted Sarsian Caligoida and Cyclopoida respectively.) This author's own studies have led him to believe that the segmentation of the endopod is obscure. Segmental divisions are often absent from it, as is the boundary between the sympod and the rami. It seems probable that some authors took artifacts for boundaries. The nauplial cuticle tends to be rather loose and wrinkles easily in the fixative. Unlike the exopod, the endopod is not armed with setae that can be taken as definite markers of segmental boundaries. It seems probable that in most instances the endopod has two segments, in some species perhaps only one. Its length, however, is usually only slightly exceeded by that of the exopod.

Another type of nauplial appendage is that occurring in the genus *Bomolochus* (A′, Text fig. 17). In this type the endopod is much shorter than the exopod, almost rectangular, with a shallow indentation in its distal margin, and most of its setous armature modified into short, claw-like spines (omitted in the drawing). The exopod remains largely unchanged, while the sympod is short and apparently unsegmented.

The third type (A″, Text fig. 17) is characterized by modification of the first endopod segment into a strong gnathobase. With the appendage in its normal position, the gnathobase points in a medial direction. The second segment is small and carries a structure that might be a sensory cylinder. The exopod consists of either three or four segments.

The three types might be seen as a series, with the mandible of *Bomolochus* being its middle member and that of the A″ type, typical of Ergasilidae, showing the greatest departure from the primitive.

The A type mandibles occur in the nauplii of Siphonostomatoida, Poecilostomatoida and Cyclopoida. With the metamorphosis of the last nauplius stage into the copepodid and during the subsequent stages the mandible changes radically, eventually assuming in the caligiform copepods the form shown diagrammatically in Text fig. 17B. The adults of the caligiform siphonostomes are fairly uniform in the structure of this appendage. All have rather long, some very long, uniramous and subcylindrical mandibles with a flat distal end. A series of either uniform or varied teeth is usually borne on that blade-like tip, though in some Hatschekiidae and Pennellidae the teeth might be absent. In Caligidae the mandible consists of four parts which might or might not correspond to segments. In other families mandibular segmentation is often obscure or absent.

The change from a biramous and setiferous nauplial mandible into a very specialized, uniramous, tooth-bearing adult appendage is very striking. Among the siphonostomes parasitic on fishes there are no species with adults possessing mandibles of intermediate type. Heegaard (1947a, Fig. 86) described the copepodid of *Haemobaphes cyclopterina* (Fabricius, 1780) with mandibles consisting of a short and stocky sympod, spatula-like endopod and three-segmented exopod. This structure is, indeed, intermediate between those of the nauplial and adult mandibles of the caligiform copepods. Heegaard's description, however, calls for confirmation. This author examined copepodids of *Haemobaphes diceraus* Wilson, 1917, and found its mandible to be a scaled-down replica of the adult appendage. No trace of exopod was present. On the other hand, Izawa (1969) showed that the mandible of *Caligus spinosus* Yamaguti, 1939, has an outgrowth at the base, possibly an exopod. Shown in his Fig. 9D, it was not mentioned in the text.

(Among siphonostome copepods associated, or presumably associated, with invertebrates there are species that can be arranged into a series showing a gradual change from the nauplial type of mandible to the adult type. The mandibles of their adults are shown in Text fig. 18. *Ascomyzon asterocheres* Boeck (Text fig. 18A) has mandibles with a very short sympod, a blade-carrying, unsegmented endopod and a two-segmented exopod armed with two very long terminal setae. In *Ascomyzon simulans* Scott (Text fig. 18B) the endopod appears to have fused with the sympod and the exopod is proportionately shorter than that of the preceding species. The mandible of *Rhynchomyzon purpurocinctus* (Scott) (Text fig. 18C) has an even shorter exopod and a row of fine denticles on the blade of the endopod. A similar mandible occurs in *Collocheres gracilicauda* (Brady) (Text fig. 18D). In a myzopontiid species *Neopontius angularis* Scott (Text fig. 18E) no trace remains of the exopod and the mandible shows many features of similarity with a corresponding appendage of the caligids.

This similarity is even more accentuated in *Bradypontius magniceps* (Brady) (Text fig. 18F, F′). All those species belong to, or are related to, the copepods from which some authors derive Caligidae and related fish parasites.)

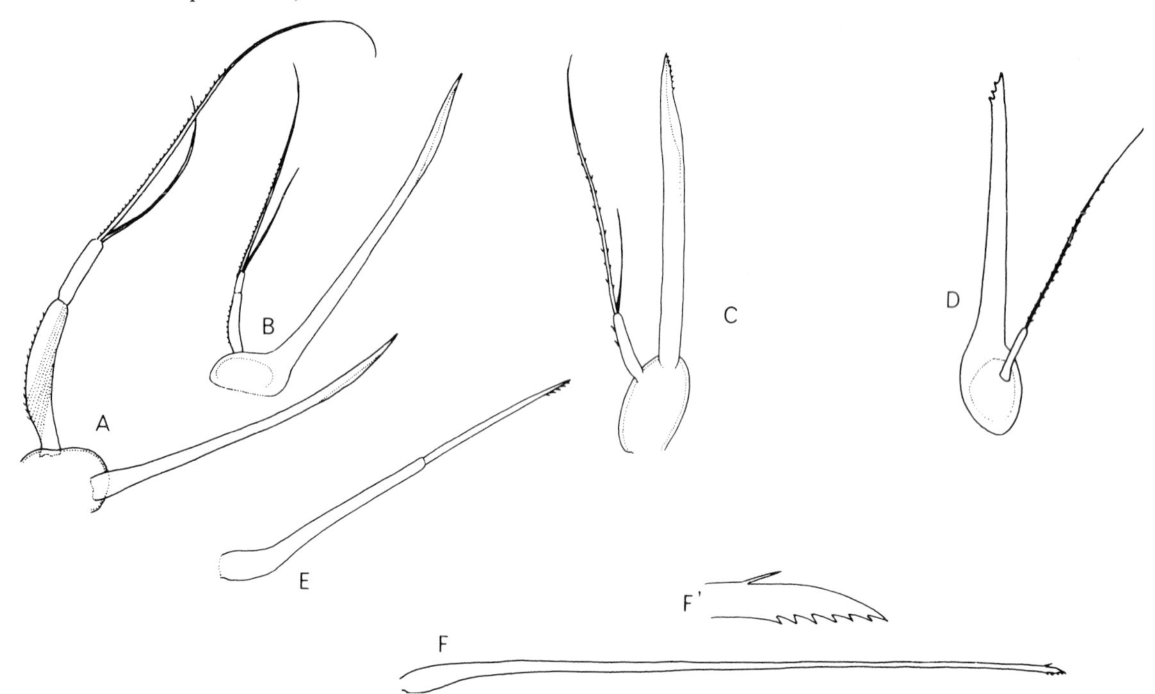

Text fig. 18. Mandibles of some siphonostome copepods associated with invertebrates. A — *Ascomyzon asterocheres*; B — *Ascomyzon simulans*; C — *Rhynchomyzon purpurocinctus*; D — *Collocheres gracilicauda*; E — *Neopontius angularis*; F, F′—*Bradypontius magniceps*. Redrawn from Sars (1915).

The type A mandible of poecilostome nauplii metamorphoses into an adult appendage quite dissimilar from it and from that of the adult siphonostomes. Consisting of a short, stocky basal part and a falciform blade bearing two marginal rows of denticles (Text fig. 17C), it occurs in all species of a large family of fish parasites, Chondracanthidae, and their relatives. Another, simplified version of this type is found in the cyclopoid genus *Lernaea* (Text fig. 17D), also derived from nauplial mandible type A. The unique genus *Shiinoa* has similar mandibles, differing from type C by possession of a second, smaller falciform blade (Text fig. 17E). The developmental cycle of this genus is unknown, hence we cannot establish the type of nauplial mandible from which it originates. Neither can we find another like it among parasites of fishes, though some copepods parasitic on invertebrates have similar mandibles (e.g. Pseudomyicolidae).

As far as is known, the mandible type A′ gives rise to only one type of adult appendage. This appendage, obscurely segmented, is distinguished by a robust base and more or less cylindrical shaft (usually at an angle with the base) bearing two blade-like processes (Text fig. 17F). This type of mandible occurs in four poecilostome families with species parasitic on fishes: Bomolochidae, Taeniacanthidae, Tuccidae and Telsidae.

The nauplial mandible type A″ has been found in the family Ergasilidae (*Ergasilus* and *Thersitina*). Its adult counterpart is essentially like the chondracanthid mandible with added secondary falciform blade and palp (Text fig. 17C). In addition to Ergasilidae, similar mandibles occur in Therodamasidae and in many copepods associated with invertebrates (e.g. Clausidiidae).

In summing up this brief review one must point out two important facts: firstly, there are really no intermediate forms between the siphonostome and poecilostome appendage; secondly, while the structure of the mandible is rather uniform among the siphonostomes, several different types occur among the poecilostomes. It is also important to note that, in the majority of instances, the parasitic copepods have relatives with mandibles of similar type among free-living or associated copepods. This suggests that very little change was required to make these appendages functionable under the conditions of parasitism. All this is important to remember when searching for phylogenetic relationships of various copepod groups. The matter will be brought up again later.

The first maxilla

This is the first segmental appendage which appears in the post-nauplial stages of the copepods. Between its emergence during the copepodid stage and its definitive form in the adult it undergoes only relatively modest changes, mainly changes in the proportions of its individual components. In some copepods (e.g. *Dichelesthium oblongum*) it appears to acquire additional armature with the progress of development; this is a rare occurrence.

The controversial homology of this appendage has already been discussed (p. 16). The best recent review and summary of arguments was made by Lewis (1969), who concluded that in "Caligoidea" at least the structure consisting of a spine and a setiform node or papilla lateral to the mouth cone is a true first maxilla.

In contrast with the mandible, the first maxilla of the poecilostome copepods is fairly uniform in structure. On the other hand, the siphonostomes are characterized by a wide range of structural types of this appendage. Studying this range among the "Caligoidea", Lewis (1969) arrived at the conclusion that it might be regarded as a series showing progressive absorption and loss of the base of the sympod, with gradual separation of the two rami of an originally biramous appendage. (He was cautious enough to deny any implications that his series represented phylogenetic relationships. Rather did it point to a general tendency in the evolution of the first maxilla among the parasitic siphonostomes, a tendency expressed independently and to a varying degree.) The primitive siphonostomes of the family Artrotrogidae (associated with invertebrates) show this biramous structure quite unmistakably. Among the caligiform siphonostomes, those of the family Dissonidae have the sympod and endopod of the first maxilla aligned along the same long axis, fused and sometimes spiniform, with the reduced exopod protruding laterally or dorsolaterally and bearing one to three setae (Text fig. 19A). A somewhat similar type of first maxilla is found among Dichelesthiidae (Text fig. 19B), in which it is laterally compressed, with dorsal or dorsolateral exopod and either one or both rami carrying some armature. To this type, too, belong the first maxillae of Pandaridae (Text fig. 19C), Eudactylinidae (Fig. 1188), Kroyeriidae (Fig. 1090), some Pseudocycnidae and Lernanthropidae (Fig. 1075). A modified version of this type occurs also in Cecropidae.

A completely different first maxilla is the appendage of Lernaeopodidae (Text fig. 19D), in which the main corpus is a fused sympod and endopod, ending in two or three setiferous papillae and bearing either a lateral or ventral exopod at various stages of reduction. In Naobranchiidae, a similar first maxilla has completely lost its exopod. The pennellid appendage (Text fig. 19E) shows a more advanced stage in absorption of the sympod, while in Tanypleuridae the exopod is completely reduced. The family Hatschekiidae, and particularly the species of the genus *Hatschekia*, show a further stage in the disappearance of the base of the first maxilla, until in some of them the two rami have become separated and arise independently from the surface of the cephalothorax (Text fig. 19F). The complete separation of the rami is also characteristic of Caligidae, Trebiidae and Euryphoridae, in which the endopod has become a spiniform, heavily sclerotized, either simple or two- or three-pronged spine (often dissimilar in the two sexes), while the exopod is a small papilla surmounted by one to three setae (Text fig. 19G).

In Sphyriidae, particularly in the females of the family, the reduction of the first maxilla is at its most advanced, there remaining only a tiny papilliform outgrowth, either unarmed or bearing some very short setae.

The poecilostomes of the family Chondracanthidae have very small, rather flat first maxillae, previously believed to be mandibular palps due to their position close to the base of the mandible. The appendage has concave medial and convex lateral margins; its distolateral tip bears two processes, one much smaller than the other (Text fig. 19H). In many species additional armature is present on the medial margin and sometimes on the flat surface of the appendage. Essentially similar first maxillae occur in Pharodidae and Ergasilidae. Those of the latter family are distinguished by the length and slenderness of their armature. In another group of poecilostomes (Bomolochidae, Taeniacanthidae, Tuccidae) the first maxilla has become a small, short outgrowth tipped with three or four setae, some of which might be comparatively very long (Text fig. 19K). By comparison, in related Telsidae the armature is very short, the general structure remaining similar (Text fig. 19J). The lernaeid first maxilla has become reduced to a simple short spine, or a small spine-bearing papilla.

The impact of parasitism on the copepod first maxilla can be indirectly gauged by comparing this

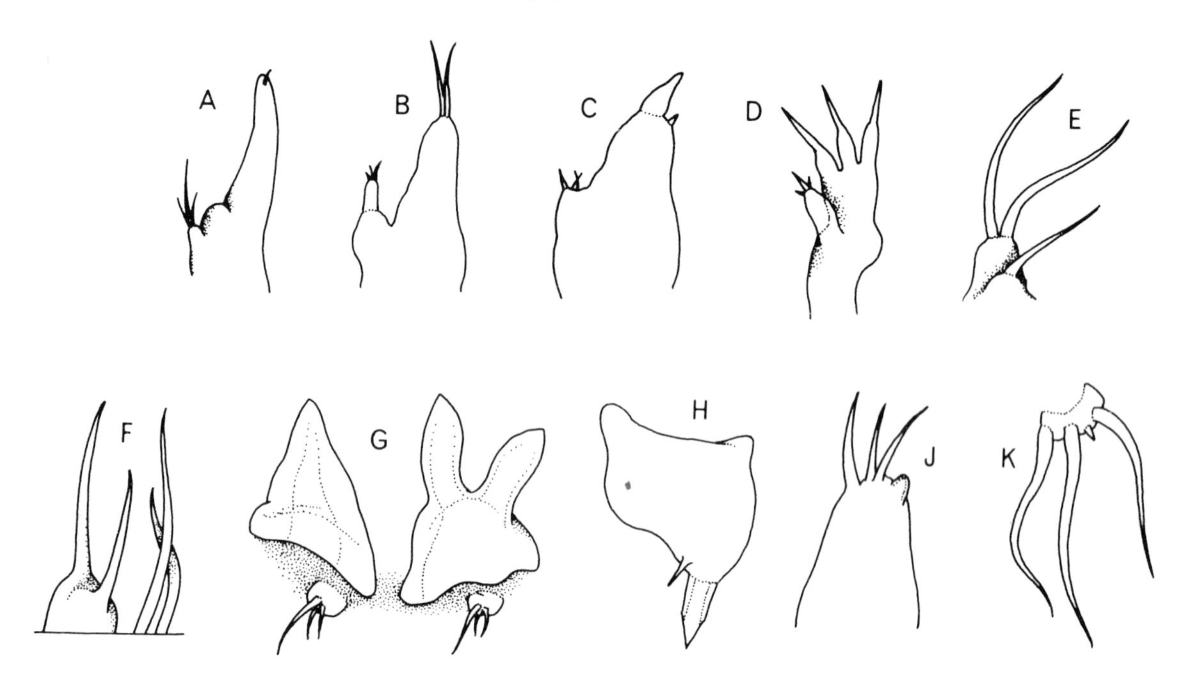

Text fig. 19. Types of first maxilla. A — Dissonidae; B — Dichelesthiidae; C — Pandaridae; D — Lernaeopodidae; E — Pennellidae; F — *Hatschekia*; G — caligiform Siphonostomatoida; H — Chondracanthidae; J — Telsidae; K — Bomolochidae and allies.

appendage, as it occurs in the parasitic species, with that of free-living copepods. The first general observation one makes on such comparison is that in the parasites the first maxilla tends to become reduced in size. In the free-living species it is a robust appendage, adapted to take part in raptatory feeding (Fryer, 1957). In some cyclopids at least, the two members of the pair are asymmetrical and tightly opposable, to provide a better prehensile apparatus. Judging from the size and structure of the first maxilla in the parasites, this appendage is no longer directly involved in feeding, certainly not in the manipulation of food. It is not opposable with the other member of the pair and (in siphonostomes, at any rate) has limited opportunity of contact with the particles of food being ingested. On the other hand, its setous armature might have a sensory function. It might, therefore, be construed as a kind of "taster". In the caligiform siphonostomes the spiniform part of the appendage is more likely to be used as an auxiliary grapnel holding the copepod on the surface of the host. These suggestions are, of course, purely speculative. The author, however, feels justified in making them, in view of the fact that no direct observations have been made on the function of the first maxilla in parasitic copepods. The structure is being used as a possible clue to that function.

The second maxilla

We know rather more about the function of this appendage, at least among some siphonostome parasites, than we do about that of the first maxilla. The author considers it a "parasitic appendage", inasmuch as it appears to be more directly involved in typically parasitic aspects of copepod biology. In Lernaeopodidae it has become greatly influenced by the parasitic mode of life and, in turn, exerts a profound influence on the entire morphology of the copepod.

The parasites of the families Hatschekiidae and Lernanthropidae probably use their second maxillae as prehensile appendages which help to maintain them on the gills of their hosts. These appendages are of subchelate type, i.e. their distal part is reflected over and is opposable against the proximal, the two together acting as a vice imprisoning in it a portion of the host's tissue. Examples of this type are the second maxillae of *Lernanthropus* (Fig. 1065) and of *Hatschekia* (Fig. 1110). The subchelate structural plan is fairly widespread among various copepod appendages and, indeed, is known in other Crustacea. It is interesting to note that among parasitic siphonostomes it occurs in the second maxillae of only two families with members living exclusively on the gills of teleosts. For

Hatschekiidae, which are devoid of maxillipeds, the grasping second maxillae are particularly important as organs of prehension.

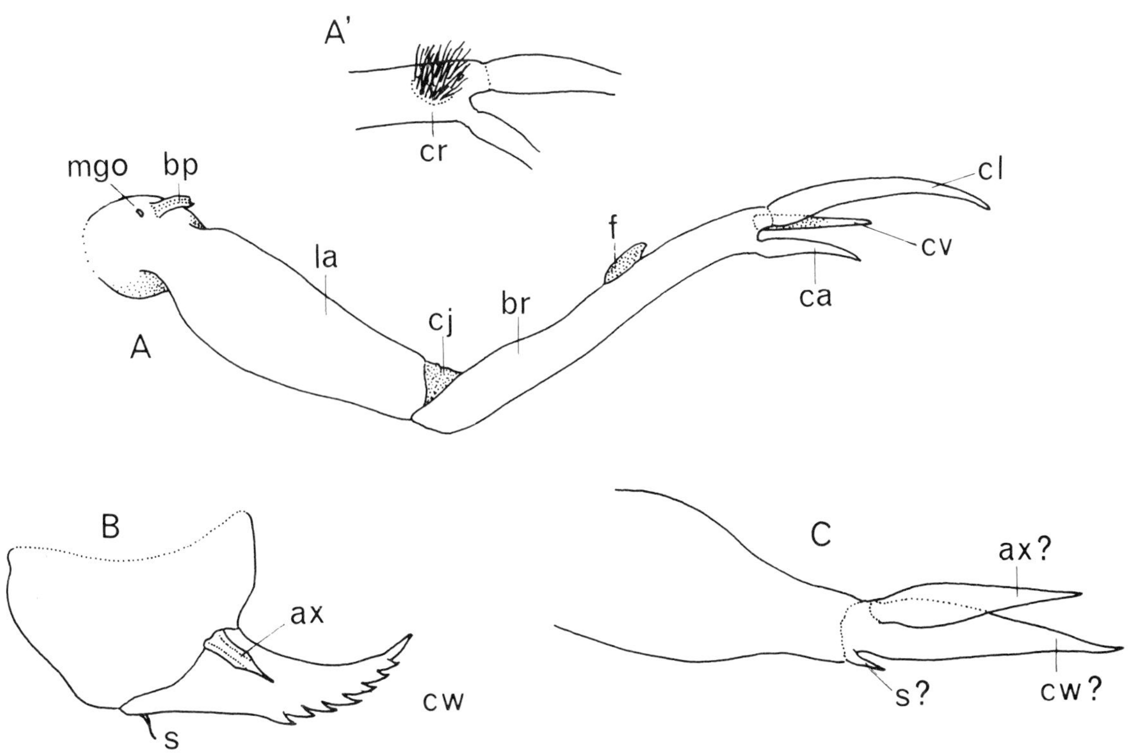

Text fig. 20. Types of second maxilla. A — brachiform second maxilla of many siphonostome copepods; A′ — same, with slightly different apical armature; B — second maxilla of many poecilostomes, particularly of Chondracanthidae; C — second maxilla of Bomolochidae and allies.

ax — auxiliary process; bp — basal process; br — brachium; ca — canna; cj — cubital joint; cl — calamus; cv — clavus; cw — claw; f — flabellum; la — lacertus; mgo — orifice of maxillary gland; s — seta.

Much more prevalent among the siphonostome copepods parasitic on fishes is the type of second maxilla referred to by this author as *brachiform*, in recognition of the superficial resemblance its structure bears to that of a human arm (Text fig. 20A). Because of the lack of appropriate terminology, the descriptions of this appendage tend to be rather wordy and cumbersome. It is proposed here, therefore, to adopt for various parts descriptive terms partly drawn from vertebrate morphology. The chances of introducing confusion thereby are insignificant and the resulting convenience is appreciable. The proximal, more robust part of the appendage is described as *lacertus*, or upper arm. It consists of a small basal section, not always identifiable and not distinctly divided from the main section. The basal section carries the orifice of the maxillary gland (see Lewis, 1969). It might also carry a small basal process either setiform or appearing as a flat, distally truncated flap of cuticle. The basal process is probably always present in Caligidae. In the genus *Pseudanuretes* Yamaguti, 1936, its length rivals that of the entire appendage, but in all other genera it is small enough to escape notice of all but a careful observer. Pandaridae and their allies appear not to have basal processes on their first maxillae. By means of the *cubital joint* the lacertus articulates with the *brachium*, or lower arm. This part is usually longer and more slender than the lacertus, more or less cylindrical or slightly flattened. In Caligidae and their allies it carries, on the inner margin, at about mid-length, a flap of striated membrane, the *flabellum*. The distal end of the brachium is connected with a third part, which might or might not represent a separate segment, flexible, tapering and variously armed *calamus*, or rod. A similar, though shorter and subterminally positioned rod, *canna*, arises from the brachium on the outer side of the calamus. In Caligidae, Trebiidae and Dissonidae these two processes are the only ones present at the tip of the second maxilla, but in some species of *Dissonus* there is a tuft of setae at the base of the calamus. In Euryphoridae, Pandaridae and

Cecropidae, on the side of the tip between the bases of the calamus and canna, a third process, *clavus*, might arise, either armed or unarmed and extending at an angle of about 30° from the long axis of the brachium. Alternatively, the same place on the brachium is occupied by a low swelling covered by closely crowded setae or denticles. This structure is referred to here as the *crista*, or plume (Text fig. 20A′). In the family Dichelesthiidae, the tip of the brachiform second maxilla has been curiously modified (Figs. 1016–1019, 1043, 1044), its structure suggesting at least some capacity for prehension. The second maxilla of Pennellidae, Eudactylinidae, Kroyeriidae and Pseudocycnidae, otherwise similar to other brachiform second maxillae, ends in a single claw-like structure, often bearing complicated armature. The pennellid appendage is also characterized by the possession of two very strong, unciform spines on the surface of the lacertus.

Morphologically, the subchelate and brachiform second maxillae are superficially similar. The main difference between them is the relationship between the brachium and the part distal to it. Functionally, however, they are utterly dissimilar. The effective action of the subchelate appendage depends on the flexing movement and the joint between the subchela and the part proximal to it might be designated as its "functional pivot". To pinch and to hold a fold of the host's tissue is the purpose of this type of appendage. In contrast the brachiform second maxilla is most frequently used to manipulate the frontal filament during the developmental stages of most siphonostomes in which they are known. This function calls for the shifting of the "functional pivot" from the cubital joint to the tip of the appendage. Thus, extension, rather than flexion becomes the working movement. In adult Caligidae, the second maxilla also assists in movements over the surface of the fish and in copepods with the normal locomotory apparatus out of action it is capable of taking over as the only locomotory appendage (Kabata and Hewitt, 1971).

The second maxilla of Lernaeopodidae has become permanently fused with the organ of attachment. Chelate or subchelate during the early stages of development, this appendage, in the course of ontogeny, is associated in turn with two organs of attachment: the frontal filament of the larva and the bulla of the adult. In some genera it has, itself, become an organ of attachment. (For detailed discussion see p.329.) By sprouting profusely branching dendrites from its tip, *Tanypleurus* has similarly modified its second maxilla as the organ of attachment of the adult stage. The development of this copepod is unknown. A unique, ribbon-like structure of the second maxilla of Naobranchiidae also adapts this appendage to serve as an attachment organ of the clasping type.

The second maxilla of the sphyriid copepod has become reduced, particularly in the female, in which it might be nothing more than a small tubercle with a seta or two on its apex (Fig. 1448). Among the cyclopoids, the reduction of this appendage is evident in *Lernaea*. In poecilostome Telsidae it has assumed the structure of a rather misshapen subchela.

By the morphology of their second maxillae, most of the remaining poecilostomes fall into one of two groups. Chondracanthidae, Pharodidae, Ergasilidae and Philichthyidae have second maxillae of the type shown in Text fig. 20B. This appendage consists of a robust but short basal part articulating with a strong, inward-pointing claw, either armed with denticles or naked. At the base of the claw there is an auxiliary process of varying length, while on its posterior margin near the proximal end there is a short seta, often difficult to notice. In Bomolochidae, Taeniacanthidae and Tuccidae this appendage (Text fig. 20C) has a longer and more slender base. Distally it carries two processes of subequal length and variously armed (no armature shown in diagram), as well as a third process, very often much shorter than the other two but sometimes rivalling them in length. The long axes of the processes are more or less in line with that of the base, while in the chondracanthid appendage the long axis of the claw is almost at right angles with that of the basal part. It can be speculated that the three processes of the appendages shown in Text fig. 20C are homologous with the claw, auxiliary process and seta of the appendage shown in B, as indicated in the lettering of the figure. This question, however, would have to be examined thoroughly in depth before a final conclusion could be reached.

The assumption of the parasitic mode of life does not appear to have exerted a great influence on most second maxillae. In the course of evolution this appendage was probably most extensively influenced by the development of the siphonostome type of mouth, a process that in all likelihood antedated parasitism and perhaps can even be considered as preadaptive to this mode of life. With the concentration of the food gathering activity on the surface enclosed by the rim of the buccal orifice, the second maxilla became functionally divorced from feeding and in the caligiform copepods was pressed into service as an appendage auxiliary to attachment. The position is quite different in the poecilostomes. The wide-open buccal cavity and, as has been suggested (Kabata,

1970), the resort to extrabuccal digestion make the assistance of the second maxilla in the process of feeding indispensable. (Halisch (1940) suggested that in *Ergasilus* even the first pair of swimming legs, more distant from the mouth than the second maxillae, aids in pushing the food towards the mouth.) In neither group, however, does the second maxilla show a great variety of structure, regardless of the type of host-parasite relationship.

The maxilliped

This is the most anterior appendage of thoracic origin. It is borne upon a segment usually fused with the cephalon. In most instances known it functions as a prehensile limb, though in some copepods it is probably associated with feeding.

In the siphonostome copepods the maxilliped is a subchelate appendage. A generalized siphonostome maxilliped is shown in Text fig. 21A.

Among this group of copepods the maxilliped of this type occurs in many varieties, differing from one another in the proportions of their component parts and details of armature. In descriptions of the siphonostomes much space is devoted to this often complicated structure. To facilitate descriptions and to save both space and time, the author proposes to use a set of new names for the parts of the maxilliped.

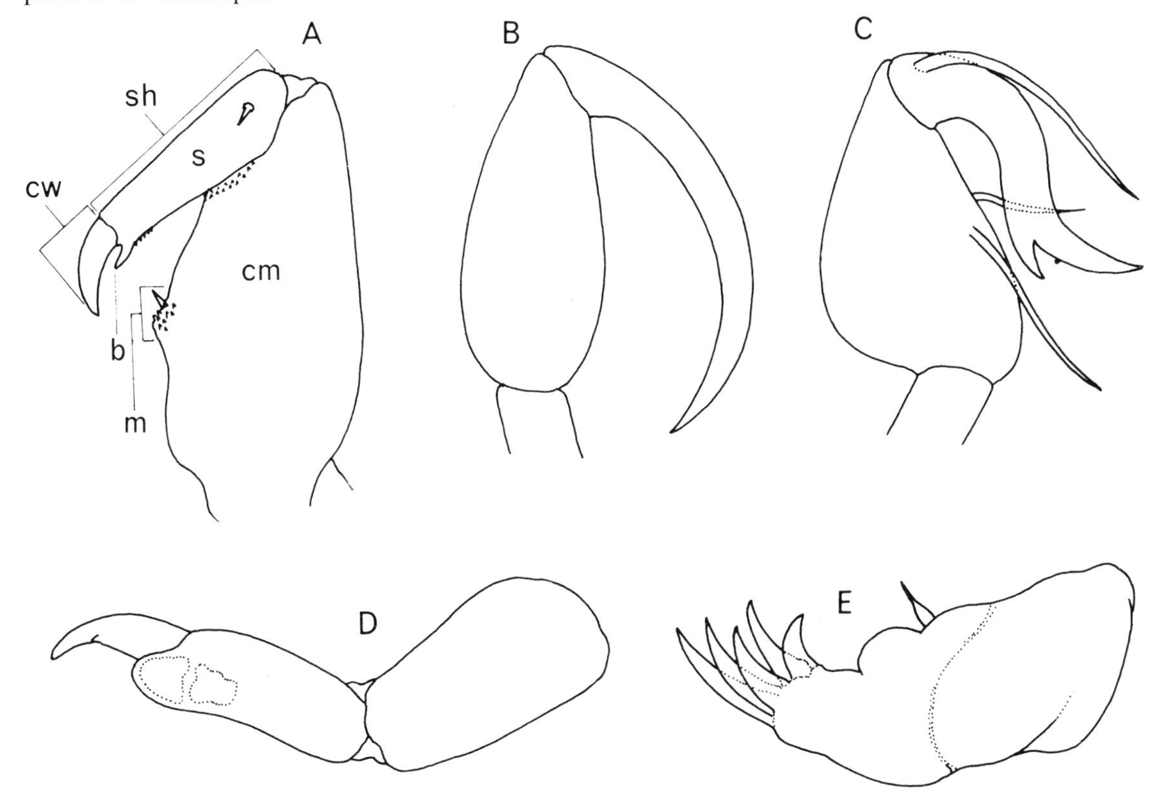

Text fig. 21. Types of maxilliped. A — generalized siphonostome maxilliped; B — Bomolochidae and allies (males); C — Bomolochidae and allies (females); D — Chondracanthidae; E — Lernaeidae.

b — barb; cm — corpus; cw — claw; m — myxa; s — subchela; sh — shaft.

The main body, or *corpus maxillipedis* (cm), is a robust structure, usually unsegmented, though sometimes bearing traces of primitive segmentation suggesting that it originally consisted of more than one (probably two) segments. In many siphonostomes the medial margin of the corpus has, near its mid-length, the *myxa* (m), a low swelling often armed with a patch of denticles and/or a short spiniform seta or spine. In some species the entire myxal area protrudes in a medial direction, and forms a platform against which the tip of the claw rests when the subchela is in the closed position (e.g. in *Caligus centrodonti* Baird, 1850). The protrusion might become prominent enough to give the entire appendage a chelate appearance. It is this feature that prompted the name *Caligus chelifer* Wilson, 1905. The distal end of the medial margin is also sometimes beset with fine denticles, particularly in some genera of Lernaeopodidae. The corpus articulates with the *subchela* (s), a cylindrical, often tapering or uneven structure of varying length. The subchela consists of the *shaft* (sh) and *claw* (cw), the latter usually well delimited from the former. The shaft often bears a spiniform seta near the base. Traces of subdivisions can be found on it in some species and additional armature is occasionally present. At the base of the claw the inner margin of the shaft produces a short digitiform process, the barb (b), either armed with denticles or tubercles, or naked. Alternatively, a simple seta might be present in this position. Just proximal to the barb, the inner margin of the shaft is covered by denticulation, either in the form of a patch or an oblong palisade-like enclosure. Some denticulation might also be present on other surfaces of the shaft near the base of the claw. The latter is always somewhat curved and sometimes carries one or more secondary denticles on its inner margin. In some species its surface might be sculptured by fine longitudinal ridges. In rare instances (in the subgenus *Brevibrachia*, genus *Salmincola*, fam. Lernaeopodidae) the claw is substantially reduced, being present as little more than a stub.

(The maxilliped of *Kroyeria* (Fig. 1092) is something of an exception, in that its corpus consists of two fairly clearly delimited segments and its subchela is not divided into shaft and claw.)

Many species of siphonostomes are distinctly dimorphic as regards the shape of their maxillipeds (e.g. Figs. 514, 515), probably as the reflection of functional differences of this appendage in the two sexes. The male uses it during mating to grasp the female and might use it also to manipulate and transfer the spermatophores. The female maxilliped, on the other hand, is never used in any activity associated with reproduction. In Pennellidae the dimorphism has reached the point where only the male possesses the maxilliped. There is also no maxilliped in the female of *Tanypleurus*, but the male of this copepod has not been found as yet and the degree of dimorphism cannot be assessed. In Hatschekiidae the maxilliped appears to have been lost in both sexes.

Many poecilostomes are also dimorphic in the structure of their maxillipeds. In the relatively little modified families Bomolochidae, Taeniacanthidae and Telsidae the males, as a rule, have subchelate maxillipeds. These appendages, however, resemble those of the siphonostomes only superficially. All of them have a corpus consisting of two segments (Text fig. 21B) and a very long subchela not differentiated into shaft and claw. The concave margin of the subchela is often armed with denticles or serrations (Fig. 63) and might bear prominent secondary outgrowths and/or one or more setae. The maxillipeds of the female bomolochids and telsids differ from those of the males in the structure of their claws (Text fig. 21C). Nothing is known of their function or the way in which it is performed. The peculiar, sigmoid shape of the claw offers no clues as to the way in which it might be employed. The females of Taeniacanthidae show a wide range of variation in the morphology of their maxillipeds (Text fig. 29), indicating progressive loss of functional ability. The loss culminates in the disappearance of the maxilliped in the genus *Echinirus*. There is also no maxilliped in the females of Ergasilidae, the family in which the males have pseudosubchelate maxillipeds with unusually long subchelae (Fig. 110). The maxilliped of the females of *Tucca* (Tuccidae) has lost much of its segmentation. It appears to be partly reduced and to have diminished functional efficiency. No maxillipeds are known for the males of these copepods, nor for both sexes of *Therodamas*. (The female of the genus *Paeonodes*, co-familial with *Therodamas*, has unusual maxillipeds, not yet described in sufficient detail.)

There is one more type of maxilliped in this group of Copepoda. The chondracanthid appendage (Text fig. 21D) is typically brachiform. The robust lacertus articulates through a cubital joint with the brachium which usually has an inflated pad near the tip. The pad is subdivided by a transverse groove and carries two patches of denticles. It overlaps the claw, a simple structure with one or more secondary teeth on its concave margin. The position and structure of this appendage suggests that its function is to assist in feeding. There is no dimorphism. The copepods belonging to the family Philichthyidae appear to have lost the maxilliped completely.

The cyclopoid Lernaeidae have maxillipeds of a completely different type (Text fig. 21E). They are indistinctly segmented and armed with multiple claws on their apices. They, too, appear to serve the function of feeding and show no dimorphism.

As was the case with many second maxillae, the adaptation to parasitism appears not to have affected the maxilliped to any great extent. The associated or free-living species related to those parasitic on fishes or invertebrates are often easily recognizable as relatives by the morphology of their maxillipeds. The only clear-cut change is the reduction or loss of this appendage among the parasites. Since it occurs in some copepods and not in others to which they are obviously related and which they resemble in their host-parasite relationship, it would be difficult to point to the influence of parasitism as the cause of this phenomenon. The reasons for it might well be sought elsewhere.

The swimming legs

As their name implies, the thoracic appendages carried on the six segments posterior to the maxilliped-bearing segment either are, or formerly were, functioning as locomotory appendages. Closely similar in function, they might be expected to be equally similar morphologically. A typical primitive copepod leg (Text figs. 1, 2) consists of a flat, two-segmented sympod bearing two rami, the endopod and the exopod, both consisting of three segments. The medial margins of both rami are armed with long, pinnate setae. The lateral margin of the endopod seldom carries more than outgrowths in the distal corners of the segments, while that of the exopod is equipped with spiniform armature, particularly well developed on its distal segment.

However, far from all of the so-called swimming legs serve locomotion, even in the free-swimming copepods. Neither are they all morphologically similar. To begin with, the majority of non-parasitic species appear to have only five pairs of legs. In very many the fifth pair is either reduced or otherwise modified. In some, the change of function, with the consequent structural modification of the fifth leg, occurs only in the male, thus introducing distinct dimorphism in the appendages. This occurs mainly when the male fifth leg is adapted to assist in reproductive activities (grasping of the female, manipulation of the spermatophores, etc.). The changes are often accompanied by loss of symmetry. Many examples of this modification can be quoted among the calanoid and cyclopoid copepods. At the anterior end of the series of legs, the first pair is also frequently modified. The difference might be only a slightly smaller size and reduction of the endopod to two segments, as in many calanoids. On the other hand, those legs might undergo structural changes that make them resemble the maxillipeds. Like many fifth legs, the first legs of some harpacticoid copepods are no longer symmetrical, the left member of the pair being modified. Similar changes occur, though less frequently, in the legs of the second pair (e.g. in *Canthocamptus*). These modifications occur most commonly in those genera the members of which already possess prehensile appendages (maxillipeds, second antennae) and presumably serve to increase the copepod's ability to grasp or become attached.

It seems clear that most free-swimming copepods are able to achieve and maintain a required level of locomotory efficiency without the help of one or two pairs of their usual set of five pairs of legs. The parasitic copepods, many of which have need of efficient locomotion only during the early dispersal stages, have even less reason for the retention of their swimming appendages in the primitive condition. A detailed description of changes that the legs of various parasitic copepods undergo will be given when dealing with those copepods individually. It would be useful at this point, however, to give an overview of these changes. As mentioned earlier (p. 14), we can divide them into three categories: (i) changes associated with the loss of function; (ii) changes associated with a change of function; and (iii) changes which result by correlation with those occurring in other limbs.

(i) The changes associated with the loss of function conform in most cases to one of three change patterns and are clearly related to the type of host–parasite relationship of the copepod and to other morphological changes resulting from that relationship. Curiously enough, the legs are usually least affected by the extreme type of morphological adaptation, metamorphosis. In copepods that undergo this most drastic structural change, the legs develop fairly normally during the pre-metamorphosis period. Their rami, one-segmented at the first post-nauplial stage, often acquire their full complement of three segments by the time the infective stage has been reached. In pre-metamorphosis females and adult males of *Lernaea*, for example, the first four pairs of legs could be taken for those of a free-living copepod (Figs. 401, 402). (There are also two pairs of leg vestiges.) What might be termed a "retroactive influence of parasitism" is perhaps manifest in the failure of some copepods belonging to this group to develop full segmentation, as in some Lernaeidae (e.g. *Lernaeogiraffa* and *Dysphorus*). Other copepods might also fail to develop one ramus, invariably the endopod, on their third and fourth legs (e.g. *Lernaeocera*, Fig. 1364). Since, however, these departures from the primitive structural plan are finalized during the free-swimming period of life and since they somewhat resemble certain changes that take place in the legs of various free-living copepods, the role of parasitism as the reason behind them can be open to question.

With the beginning of the morphologically explosive process of metamorphosis, the legs are relegated to the status of biological discards. It is as if, no longer needed to propel the copepod through the water in its non-sentient search for the host fish, they are left behind to be dwarfed by vigorous growth of the body. It must be stressed, though, that they are not subject to any further changes (barring muscular atrophy) and usually remain attached to the disproportionately large body of the parasite as reminders of its free-living past and copepod affinities. Their definitive positions on the body depend on the location and levels of activity of various development centres instrumental in metamorphosis. In Pennellidae, for example, the leg-bearing segments are not deeply involved in this process and the legs remain in more or less the same positions as they were during the free-swimming stage. In Lernaeidae, on the other hand, each leg-bearing segment contributes substantially to the formation of the trunk and the pairs of legs become separated from one another by wide spaces.

The copepods belonging to Sphyriidae are exceptional among the metamorphosing parasites, in that during that process they lose all traces of their swimming legs. Our knowledge of the ontogeny of this family is practically non-existent, so that we do not know what condition their legs attain before they disappear.

Both the siphonostome and the poecilostome copepods include genera or families in which the change to the parasitic way of life resulted in few or no alterations in their primitive segmentation. Among the former Eudactylinidae, and among the latter Bomolochidae, can serve as examples of this category. Generally speaking, the legs of the parasitic copepods with primitive segmentation differ from those of the free-swimming copepods in modified armature (more numerous, larger and more complicated sets of spines) and sometimes in their reduced complement of segments. The first pair of legs, borne upon a segment that has been incorporated in the cephalothorax, tends to become smaller than the following three pairs (unless they are adapted to a new function). The fifth legs are not only smaller but uniramous, while the sixth, if present at all, are represented by mere vestigial setae. A difference in leg armature might exist between the sexes in some genera. In *Nemesis*, for example, the female appendages have lost all pinnate setae (Fig. 1300), while the males have retained some of them in a near-primitive condition (Figs. 1304–1306). The extent of armature modification depends also on the type of host–parasite relationship. The legs of the permanently sessile Eudactylinidae (Figs. 1202, 1247, 1300) have their armature reduced and modified much more extensively than the legs of the relatively mobile Bomolochidae (Figs. 20–23).

The third type of change in this category is that associated with changes in segmentation. As we have seen earlier (p. 10), among parasitic copepods the most common change in segmentation is associated with the progressive evolution of the genital trunk, in the course of which one segment after another is incorporated in the genital complex, in a postero-anterior sequence. The inclusion of a segment in that complex means that it has become subservient to the function of reproduction. It must be refitted for that purpose (changes in musculature, invasion of coelomic spaces by genital storage organs, etc.), and in the process of refitting the legs are either seriously reduced or completely eliminated. In Siphonostomatoida the locomotory appendages remain on the trunk in the form of a small seta, a set of setae or, at best, a small setiform sclerite. In the poecilostomes (e.g. Tuccidae), on the other hand, the legs on the trunk remain unchanged, though their growth is arrested at the pre-adult size.

It is generally true that the number of pairs of legs is equal to the number of leg-bearing segments which remain anterior to the trunk (disregarding those appendages that have changed their function). The only exception that comes to mind is the fourth leg of *Lernanthropus* (Figs. 1056, 1073), an appendage that develops into a very large, bilobate, foliaceus structure. Its evolution might, however, have as yet unsuspected functional significance. The first two pairs in this genus are also somewhat better developed, but their association with segments included in the trunk is reflected in their diminutive size.

The legs belonging to segments that have lost their identity due to the process of cephalization (see p. 6) are usually exempt from the devastating effect that is exerted on the appendages of segments swallowed by the trunk. Unless subject to functional changes, they tend to remain more or less unaltered.

(ii) Changes caused by the adaptation of swimming legs to functions other than locomotion are not easily classified. Their nature depends on the function which the leg is to serve, as well as on the structure on which they are superimposed. Our knowledge of the biology of parasitic copepods is usually insufficient to give us many clues as to the parts played by their individual appendages. The functions here ascribed to variously modified legs are, in the main, only the author's interpretation of the functional potential of their structure. Keeping in mind the somewhat speculative character of this account, we can identify four possible functions for which the swimming legs of parasitic copepods have been adapted: (a) prehension associated with attachment and feeding; (b) prehension associated with mating (in males only); (c) protection; and (d) storage of food reserves.

Both siphonostomes and poecilostomes include sizeable groups of species which maintain their contact with the host by suction. The entire cephalothorax of these copepods becomes a suction cup. In Caligidae the efficiency of this suction device is enhanced by the pair of modified third legs, which together form the posterior margin of the cup and ensure complete sealing off from the external space. To fulfil this function, the legs undergo several extensive changes: the interpodal bar becomes greatly expanded so as to fill in completely the space between the sympods of the two legs; the sympods themselves become enlarged and acquire special outgrowths, or *vela*, which project into the interval between the rami; the latter are reduced in size (Fig. 458). A similar effect is achieved in the poecilostome Bomolochidae and Taeniacanthidae by rather different means (Text fig. 6). In that family the first leg plays the part of the posterior margin of the suction cup. Unlike the caligid third leg, it does not undergo marked expansion of the sympod, but its rami are splayed out and flattened to the point of grossness (Figs. 9, 65). Probably less efficient than the watertight leg-barrier of the caligids, the bomolochid first leg serves a much smaller copepod. One can presume that the suction force required to keep a *Bomolochus* or a *Taeniacanthus* on the surface of the fish is also much less than that required by a *Caligus* or a large *Lepeophtheirus*.

A unique adaptation of a biramous leg for the purpose of prehension occurs in the genus *Nemesis* (Eudactylinidae). The first leg of both sexes in this genus has curiously modified rami (Fig. 1281). In the process of modification the exopod becomes a hook capable of enclosing space by fitting into a fork produced by the endopod (Text fig. 22). The sympod is slightly broadened but otherwise unaffected by this adaptation. Far more thorough is the change suffered by the third legs of *Lernanthropus* (Lernanthropidae). The legs of the female lose all semblance of their original structure, becoming hard, shoehorn-like processes protruding from the trunk at angles which vary from species to species, and assisting attachment by inserting themselves between the gill filaments of the host fishes (Figs. 1056, 1071).

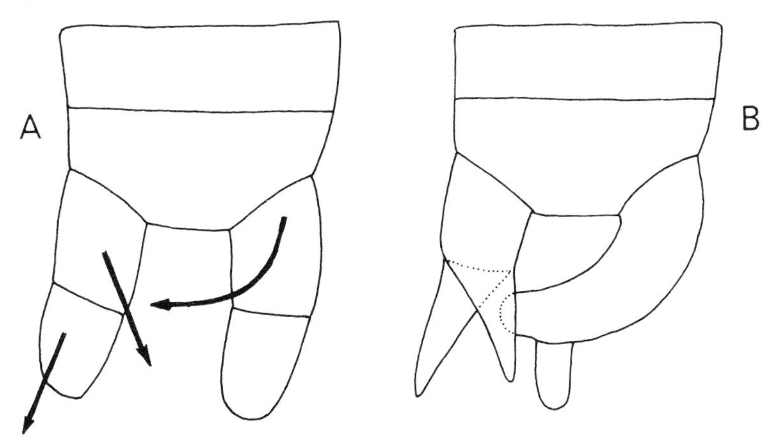

Text fig. 22. Modification of first swimming leg in *Nemesis*. A — Unmodified biramous leg (arrows showing directions in which segments must develop to produce change characteristics of *Nemesis*); B — Leg of *Nemesis*.

When the structural change occurs only in the appendages of the males, one is justified in presuming that it is associated with reproductive activities. We encounter such dimorphism in the second antennae of the caligid copepods (see p. 162). It occurs also in the third leg of some copepods. For example, the third leg of a female *Nemesis* resembles closely its second one (Fig. 1282). In contrast, that of the male has its endopod modified into a long hook (Fig. 1290). The two hook-like endopods meet and overlap each other, thus forming an efficient grasping mechanism, presumably used to hold the female during mating. The third endopod of several pandarid and cecropid copepod males is also modified (Figs. 846, 875, 955, 982), but in their case the mode of employment is not so obvious from their structure as it is in *Nemesis*. The same is true of the second endopod of the male *Dichelesthium* (Fig. 1023).

In one instance known to the author, the legs of a parasitic copepod have assumed a protective function. *Anthosoma crassum* (Abildgaard, 1794) burrows in the flesh of its elasmobranch host, burying itself in it almost completely. The site of attachment varies, but in most cases the copepod is tightly surrounded by the musculature of the fish. The sympods of its legs expand enormously and cover the posterior part of the body by a skirt of elastic, overlapping plates, supplemented in the females by elytra produced by several segments (Fig. 1027). The female of the species has completely lost the rami of its legs (Figs. 1047, 1049), but the first two pairs of legs of the male retain them in a vestigial condition (Figs. 1050–1053).

The fourth leg of *Cecrops latreilli* Leach, 1816, is the only copepod leg known to serve as a storage depot. In some specimens of this copepod the leg is only a rounded sclerite with reduced, unsegmented rami and a longitudinal depression in the centre (Fig. 922); its dorsal side carries a wrinkled, empty bag of cuticle (Fig. 923). In others the bag is distended (Fig. 924) by a homogeneous, caseous substance. The nature of the substance is unknown but it might well be a food reserve.

There are some modifications of legs that are not readily explicable in terms of function. For example, the first two pairs of legs in *Chondracanthus*, borne upon a segment forming a neck-like constriction between the cephalothorax and trunk, do not become miniature, completely segmented remnants of the preadult morphology. Instead, they keep pace with the growth of the copepod, though at the expense of complete morphological transformation. They become large, bilobed (less often trifid or undivided) structures (Figs. 186, 187), liberally covered with minute denticles. The very fact of their growth and metamorphosis suggests that they do not outlive their usefulness for the adult *Chondracanthus*, but it is difficult to see what their function could be. The same is true for one or more pairs of legs of other chondracanthids.

Another example of modification with no apparent purpose is the second leg of *Eudactylina*. While in some species of that genus the second leg has three-segmented rami of normal proportions (Fig. 1237), in others the exopod becomes modified by elongation and/or swelling of its basal segment (Fig. 1199), the other two segments being crowded at its tip. We have no clue why closely related congeners, living in similar habitats (gills of elasmobranch fish), should so differ in this morphological feature.

(iii) The changes of the third category, i.e. those that do not result from a direct response to a functional need but arise indirectly in the wake of modifications occurring in other parts of the body, are usually difficult to detect. In view of our poor knowledge of copepod biology one might, indeed, contend that we cannot detect them at all. However, given the axiom that the animal evolves as a harmonious whole, at least some of its component parts must be affected by changes in other parts. As we learn more about live copepods, we will undoubtedly discover many correlations between structural changes, those of the swimming legs included. In the meantime, one example of this category is recounted below.

In Text fig. 4 we saw three caligiform siphonostomes showing three stages in the process of cephalization. The most primitive of them was represented by a member of the genus *Dissonus*. The swimming legs of *Dissonus* are modified from the primitive conditions only at both ends of the series. The first leg has two-segmented rami, the endopod (n) being much smaller than the exopod, the basal segment of which is abnormally long. The fifth leg is a semicircular sclerite bearing four short setae. In the genus *Trebius*, in spite of the fusion of the second leg-bearing segment with the cephalothorax, there is no change in the condition of legs, except for further reduction in the size of the first endopod and the fifth leg. Second to fourth legs in both genera are identically developed and appear to form a single functional, probably locomotory, unit.

Fusion of the third leg-bearing segment with the cephalothorax, as in *Caligus*, brings with it extensive modification of the series of swimming legs. The endopod (n) of the first leg is retained as a mere vestige, but the leg is otherwise similar to those of *Dissonus* and *Trebius*. The second leg shows no changes, but the third has been profoundly modified. Expansion of the sympods of this pair, accompanied by a great increase in the width of the interpodal bar (b) results in the development of a broad apron, which extends right across the posterior margin of the dorsal shield. Both rami are reduced and the space between them filled by an oval cuticular flap, the *velum*. Made watertight along the anterior and lateral margins by a hyaline strip of marginal membrane, the shield is now an inverted saucer-like structure perfectly suited to act as a sucker. The first two pairs work together as locomotory appendages, propelling the copepod along the surface of the fish or through the water (Kabata and Hewitt, 1971). The fourth leg remains outside the cephalothoracic shield and is functionally divorced from the preceding legs. It loses its endopod and its exopod is greatly modified.

In some caligid genera (*Pseudocaligus* Scott, 1901 and *Pseudolepeophtheirus* Markevich, 1940) there is only a small stump left in place of the fourth leg and in *Caligopsis* Markevich, 1940, it has disappeared altogether. (A stage intermediate between *Trebius* and *Caligus* is occupied by *Euryphorus* Edwards, 1840 (Text fig. 5). Although its segmentation is comparable to that of *Caligus*, the modification of its third pair of legs is less advanced (Fig. 757) and the fourth leg (Fig. 760) is at a half-way stage between those of *Trebius* and *Caligus*. It appears as if the modification of the fourth leg was directly related to that of the third.)

Other structures

In addition to segmental appendages, copepod parasites have some other structures that appear to have arisen as adaptations to the parasitic mode of life. Even more than in the case of those segmental appendages, we have to depend on indirect inference to interpret their functional purpose. In most instances they are cuticular outgrowths of various kinds and are probably largely useful as aids in attachment to the host. These structures can be divided into several categories, which will be enumerated below. Each will be illustrated by one or more examples but no attempt will be made to give exhaustive lists.

Cuticular spines are the most common structures present in parasitic copepods that are attributable to the influence of parasitism. This is not to say that all cuticular spines are parasitic adaptations. For example, rostral spines occur also among free-living copepods. Those spines, however, that to the author's knowledge have their antecedents among non-parasites, will be left out of this account.

Probably the best known and certainly the most controversial cuticular spine is the one known as the postantennary process of Caligidae and its equivalent in Taeniacanthidae. It was mentioned above (p. 16) and will be described in detail for each British species that possesses it. Here it will suffice to say that in Caligidae the postantennary spine is a simple, somewhat hooked sclerite with an inflated base (Fig. 638), occasionally with a second tine arising from that base (Fig. 507) and always associated with some sensory setae. In some Taeniacanthidae it resembles the simple caligid spine very closely (Fig. 56) but it is never associated with sensory setae.

The other well known cuticular structure is the caligid *sternal furca*. Situated on the ventral surface in the interval between the maxillipeds and the first pair of legs, the sternal furca is a more or less rectangular box of cuticle with an open dorsal wall, two dorsal outgrowths protruding into the body at the anterior end, and two processes attached more or less parallel to each other at the posterior end (Figs. 452, 453). By the appropriate action of two pairs of antagonistic muscles attached to the dorsal outgrowths the box can be tilted and the two posterior processes placed at different angles in relation to the surface of the body. Ferris and Henry (1949) suggested that the furca is a modified pair of segmental appendages, but their view has not been generally accepted. The function of the furca is not known, but it has been said to act as a brake, when the copepod is in danger of slipping backwards over the host or to assist in raising the arch of the cephalothoracic shield to reduce pressure underneath and augment the force of its suction. The shape of the furcal tines varies from digitiform (Fig. 543) to spatulate (Fig. 577). Some tines are bifid (Fig. 656). The shape tends to change with age and might show sexual dimorphism. The tips of the tines not infrequently bear signs of wear.

Not so common are simple cuticular outgrowths present in a few caligoid species, one at each side of the furca. They can be sharp and subtriangular as in *Lepeophtheirus nordmannii* Edwards, 1840 (Fig. 669) or almost semicircular as in *Caligus coryphaenae* Stp. & Ltk., 1861 (Fig. 524). Their function is unknown but their position suggests that they might serve as auxiliaries to the braking activities of the furca.

Many members of the siphonostome family Pandaridae have fairly prominent dentiform or other outgrowths, sometimes with grooved surfaces, situated posteromedially to the bases of the maxillipeds and often with bands of thin cuticle running between them and those appendages. The thin cuticle covers powerful muscles, which probably constitute part of the extrinsic musculature of the maxillipeds. Its relationship with the outgrowths, if any, is not known. Neither is the function of the outgrowths, though the grooved surfaces of some of them have earned them the name of *adhesion pads*. Examples of these structures are shown in Figs. 812, 813, 831, 862 and 883.

Some siphonostome copepods have dagger-like, posteriorly directed processes of hard cuticle associated with their interpodal bars. In the genus *Kroyeria* van Beneden, 1853, there are two such processes on each bar, arising near their points of articulation with the bases of the sympods (Fig. 1095). In other genera (e.g. *Eudactylina* van Beneden, 1853) there might be a single median process.

The genus *Kroyeria* is also equipped with a pair of stylets, situated in the posterior sinuses of their dorsal shields (Figs. 1083, 1084). These stylets, unlike many other cuticular spines, articulate with the body of the copepod via a rather complicated joint. It appears that they help their possessors to wedge in more firmly in their position on the gills of the fish.

Some cuticular spines are little more than prolongations of the corners of individual segments. They do occur in some free-living copepods, in which their function cannot be discerned. The outgrowths of the postero-lateral corners of the third leg-bearing segment of the male *Philichthys xiphiae* Steenstrup, 1862, however, appear capable of helping it maintain its position in the lateral line canal of the host (Figs. 417, 422).

Buccal stylets resemble other cuticular spines but are situated within the buccal tube, on the inner surface near the tip of the labrum. They carry two setiform processes at their distal ends. A pair of buccal stylets has been described in several species of Pennellidae. This author observed them also in some Caligidae and Lernaeopodidae. They are probably much more prevalent among the siphonostomes than is now believed. At least two primitive siphonostomes with no known host affiliations were found to have them also (gen. *Megapontius* Hulsemann, 1965). Neither their anatomy nor function is known. Some authors have suggested that they are modified segmental appendages (mandibles of Sproston (1941); first maxillae of Kabata (1962a)). These suggestions cannot be accepted.

Cuticular outgrowths can also be plate- or wing-like, in which case they are referred to as *dorsal*, or *lateral plates*, *elytra*, or *alae*. A segment might produce a single dorsal plate, arising near its anterior margin, covering it and often extending over the dorsal surface of the following margin. Alternatively, there can be a pair of dorsolateral plates. The margin of a single plate might be deeply cleft in the centre, suggesting that a fusion of two plates has taken place there. The plates are usually absent during early life and develop only from the pre-adult stage onwards. In the males of the plate-bearing species the plates are much smaller or do not develop at all. The plates are most common on the thoracic segments between the cephalothorax and the genital complex, though all segments seem able to produce them. The dorsal shield of the caligiform siphonostomes is probably the result of the fusion of several such plates. The plates are particularly prevalent in copepods belonging to the family Pandaridae and their relatives (Figs. 741, 769, 801, 822, 850 and 878). They are probably protective in function, if, indeed, they are functional at all. Almost certainly protective are the dorsal plates of the female *Anthosoma crassum* (Fig. 1027), another and more primitive siphonostome.

Another type of structure with presumably adhesive function is the so-called *adhesion pad*, already mentioned in another connection (p. 9). The adhesion pads develop either in association with appendages, or on the ventral surface of the body, usually along the margins of segmental plates or similar outgrowths. Both types occur in *Pandarus* (Fig. 770). The pads can have either a grooved or stippled surface, the former being more common. Their number and distribution are specific for all species that possess them. When present in association with the bases of the second antennae, those par excellence prehensile appendages, they seem to take over the function of prehension from them. Such antennae are often weak and their claws are much reduced (e.g. that of *Echthrogaleus coleoptratus* (Guerin-Meneville, 1837), Fig. 857). The adhesion pads, too, are characteristic of Pandaridae and their allies.

Yet another aid to attachment are small, cup-like suckers, the *lunules*, present in a few caligid genera. The lunules develop from the anterior margin of the cephalothoracic frontal plates, or, more exactly, from part of their marginal membranes. The cups of the lunules are of different depths; in some species they are very shallow, with incomplete margins. It is difficult to assess their importance to the copepod. Of the two caligid genera, both otherwise very similar and equally successful, *Caligus* has lunules and *Lepeophtheirus* has not. Clearly, much more detailed study is required in this area.

The siphonostome family Pennellidae is characterized by the propensity of their members to form outgrowths on various parts of their bodies. The occurrence and shape of these processes in various genera of the family will be described below (p. 237). At this point, however, a brief discussion of their function might be useful and they do fall within the category of secondary structural modifications dealt with in this section. The most obvious function for branching processes embedded in the tissues of the host is attachment. They form anchoring holdfasts. Some other functions, however, might also be served by these appendages. Monterosso (1925) found that the cephalothoracic processes ("appendici rizoidi cefaliche") of *Peroderma cylindricum* Heller, 1865, contained two vessels, running in parallel the entire length of the process, and reacting differently to stains. This prompted him to postulate for them some absorptive function. The present author (Kabata, 1970a) found similar vessels in the processes of *Phrixocephalus cincinnatus* Wilson, 1908. They were also present, according to Monterosso, in the abdominal brush processes of *Pennella* (Figs. 1423, 1432) which are not embedded in the host tissue and can hardly be absorptive. On the other hand, cephalic papillae of *Cardiodectes*, enclosed in the lumen of bulbus arteriosus of the fish heart and bathed by a continuous stream of blood, have no vessels within. It is difficult to imagine that they are positioned in that site without being capable of absorbing something from the percolating blood. It might well be that several types of anatomically different processes exist among Pennellidae and each might be able to function in more than one way, depending on its location and association with the environment of the copepod. The same might be true, though to a lesser degree, of the posterior processes of Sphyriidae (Figs. 1443–1445).

Two curious structures, referred to as *parabasal* structures, occur among the members of two genera belonging to the former family Dichelesthiidae *sensu lato*. The first antenna of *Lernanthropus* has near its base a slender *parabasal flagellum* (Fig. 1058), and a *parabasal papilla* of differing shapes occurs next to the base of the second antenna in various species of the genus *Hatschekia* (Fig. 1106, 1120, 1139, 1146 and 1154). The functions of these structures are unknown.

Finally, one could also place among these special structures the *bulla*, the attachment organ of the females in the family Lernaeopodidae. Unlike all others previously mentioned, it is a structure of internal origin, produced in the frontal region which, in most parasitic siphonostomes, is responsible also for the production of the frontal filament, the larval attachment organ. A detailed description of this structure is given on p. 329.

Classification

Sorting, or classification, of objects or phenomena is among the most basic human intellectual processes. As far back as our records can take us, we uncover attempts at the understanding of mutual relationships between various components of man's abiotic, as well as biotic, environment. Not surprisingly, the appearance, the shape, was the most common criterion on which such arrangements were based. For the zoologist, morphology was, and is, the corner-stone of animal classification. Often derived from superficial observations, his knowledge of morphology, however, failed him only too frequently. Even worse, before the idea of evolution became accepted, he had been deprived of the conceptual framework within which various degrees of relationship, and the processes of which they are the outcome, could be conveniently accommodated. Morphological features used as systematic discriminants were seen in the abstract, as it were, as structures without a past and often with only the vaguest idea of purpose.

It is at first rather surprising to find that the present state of copepod systematics is but little removed from the condition they attained under those circumstances. We have a reasonably good grasp of many interspecific and intergeneric relationships. Even some interfamilial links are within reach of our understanding. When it comes to relationships between larger taxa, however, we tend to retire in some confusion to the positions taken, and views bequeathed upon us, by the systematists of more than half a century ago. The system of classification proposed early in the twentieth century by G. O. Sars and broadly accepted by the students of copepods is quite obviously based on a pigeon-hole principle of putting like with like. The post-Sarsian contributions, with the possible exception of Lang's (1948), were little more than attempts to improve on Sars here and there. No all-embracing new schemes came forth.

The reasons for it are not far to seek. The most important among them was the newly-awakened realization that our knowledge of copepod morphology was deplorable. As the consequence of this realization, the workers in the field threw themselves with mounting zeal into the study of morphology down to its finest minutiae. From the briefest and the most sketchy, the descriptions of the new taxa became lengthy and detailed. This praiseworthy diligence, however, had one unwelcome effect. It consumed time and energy to the point at which little of either was left for broader views and speculations. The era of the specialist has its less desirable sides.

The systematics of copepods from the family level downwards are discussed individually for each family later in this book. This part is an attempt to take a fresh look at systematics at the subordinal level. To present the background against which his views grew and took shape, the author proposes first to outline the criteria he accepted and then to recount briefly those of the earlier classification systems on which he drew.

It is almost superfluous to repeat that systematics must be based primarily on morphology. It must be very clearly understood, however, that morphological differences are useful for systematics only inasmuch as they provide a measure of the phylogenetic distance between the species compared. It might be argued that small differences denote close relationship, while great ones bespeak the absence or a low degree of kinship. A modern systematist, and particularly one that concerns himself with the classification of parasitic animals, must, however, be on guard against the acceptance of superficial similarities. He must see the morphological structures as end points of long evolutionary processes. These processes might have run their courses in parallel with those taking place in other animals, not necessarily related. Phylogenetically distant species, faced with similar biological problems, are known to have found similar morphological solutions. (The similarity of the eye of a vertebrate to that of a cephalopod mollusc presents a well-known example of convergence.) Conversely, close relatives pitched into different modes of life might with time become morphologically dissimilar. With morphology as a yardstick, the phylogeny of a group of animals might be, therefore, rather difficult to unravel, particularly when that group, like the copepods, comprises members scattered through diverse ecological niches and living very different lives. And yet, there is no more reliable yardstick available.

From the standpoint of systematics, morphological features can be said to fall into two categories: primary and secondary, or primitive and advanced. An ectoparasitic copepod and a planktonic, free-swimming one might be very different from each other. Their differences are mainly due to the accumulation of specialized features acquired in the process of adaptation to their respective modes of life. This divergence does not mean, however, that they cannot share one or several morphological characteristics which each of them inherited from a common ancestor. The advanced, secondary characteristics overlay, as it were, the basic copepod design of primary or primitive features. In searching for phylogenetic relationships, one should seek out and use these primitive characteristics as indicators more important and reliable than the structures that are the outcome of later responses to evolutionary pressures. The question that must be answered here, is: how does one distinguish between these two categories of characteristics? Not easy to find, this answer is crucial to the entire problem of systematics. It might well be different for each group of animals.

There are at least two clues that point to the primitive nature of a morphological structure. If it occurs in more than one group of copepods that are otherwise different, both morphologically and biologically, it is more likely to be primitive than the one with the narrower occurrence, particularly occurrence in a single specialized group. If a structure clearly betrays its specialized function, it is more likely to be secondary or advanced than the one which is apparently subservient to a basic, generalized function.

One might be accused of indulging in speculation, formulating judgements of this kind but, speculative though they might be, they need not be unfounded, when careful morphological studies are their background.

It has been frequently suggested that, although the same organ or appendage in two groups of copepods is morphologically quite different, it cannot be used as a criterion that separates them, if species also exist with that organ or appendage morphologically intermediate between the two groups. The author is prepared to give this proposition only a qualified approval. Indeed, should two groups of copepods, each several hundred species strong, be distinguishable by a particular character and should we find a third, equally abundant one, with that character in a condition intermediate between them, then, indeed, there is little ground for applying different labels to the two extremes of what becomes a continuous series. On the other hand, the connecting link between two large taxa represented by one or very few species is no reason for repudiating their distinctness. There are few sharp boundaries in nature. A parent-and-offspring relationship between these taxa could have conceivably resulted in some animals that "fell between two stools". A lateral, or sibling, relationship between them might have had similar consequences. In his pursuit of phylogenetic connections, the systematist should treat these tenuous links as very useful clues.

A student of parasitic copepods faces a particularly hard task. The animals which form the object of his studies became parasitic at various times and at different points in the evolution of copepods. They are final expressions of two different processes: one that might be termed the general process of evolutionary change and another, constituting express adaptation to parasitism. The relative importance of these processes, as manifest in the morphology of different groups of copepods, varies from one group to the next, depending on the point along the evolutionary path at which these copepods parted company with their free-living relatives. It is important that some assessment of these two processes and their effects be made, when one approaches the problems of classification of parasitic copepods. Hence, it is imperative that the free-living copepods be taken into account as an indispensable component of any classification scheme including parasitic copepods. The concept of "Copepoda parasitica" as a taxon has about the same scientific merit as that of phlogiston.

The gross morphological modification of many parasitic copepods has often been seen as an obstacle to their incorporation in a general scheme of classification. This is far from being universally true. The preceding part of this book should have made it clear that the adaptation to parasitism is only an extra step that was taken by many copepods beyond the morphological condition attained by their non-parasitic brethren. What is needed here to make our systematics acceptable is a step back to the pre-parasitic morphology of these copepods. The studies of life histories, from the systematic point of view, are an attempt to extend our morphological comparisons to more comparable stages in the life cycles. The general value of these comparisons is not invalidated by the fact that only too often the life histories are unknown, or by the fact that in some copepods the effects of parasitism are imprinted very deeply on all the stages of the developmental cycle. Here and there, at points strategically important for classification, one might discern relationships implied in the morphological similarities between the pre-adult forms. Every systematist would be well advised to keep an eye open for these clues.

The author's thinking, as outlined above, compelled him to disagree with the philosophy on which "Sarsian classification" was based and to reject a substantial part of the classification itself. It placed him at variance with the views of some eminent copepodologists. For example, the evaluation by Wilson (1910) of the criteria used in the classification of copepods is directly opposite to that made by the author. Wilson sees as the most valuable aids to the recognition of systematic subdivisions those secondary or advanced characteristics that appear in the history of the group with the progress of their adaptation to parasitism. The "testimony of degeneration" (Wilson, 1910, p. 619) consists of precisely those features which the present author would not hesitate to classify as secondary. Wilson's assessment of the systematic significance of the secondary morphological features was not, in his eyes, incompatible with the acceptance of the principles on which "Sarsian systematics" were based.

It should be mentioned at this point that the classification proposed by Sars (1903) and enjoying wide currency at present is based on seven genera, representing "seven distinct types among the Copepoda, which may indicate as many great divisions or sub-orders". The genera are: *Calanus, Harpacticus, Cyclops, Notodelphys, Monstrilla, Caligus* and *Lernaea* (= *Lernaeocera*) and the seven "great divisions" bear names derived from these generic appellations. As was mentioned earlier, these divisions have been formed by adding to the type genus forms which resemble it more than they do any other of the remaining six type genera. Wilson's belief in the "testimony of degeneration" made it necessary for him to postulate the existence in the Sarsian scheme of yet another large taxon, Lernaeopodoida, erected to accommodate the copepods showing this "degeneration" in the extreme degree. Cheek by jowl in this taxon were found the siphonostome Lernaeopodidae and the poecilostome Chondracanthidae, two unrelated families which represent the end-points of two separate lines of evolution and have in common only their ephemeral dwarf males and the extent to which their females are modified. The systematic enormity of Lernaeopodoida is a glaring example of the worthlessness of the secondary morphological characteristics as aids to classification. (Curiously enough, Wilson (1910) originally accepted the affinities of Chondracanthidae as being much closer to Cyclopoida, which was a position more correct from the phylogenetic point of view.)

Sars proposed his scheme of classification as the result of rejection of the earlier schemes. Two of these schemes merit scrutiny and reconsideration, as regards the principles on which they were based.

Thorell (1859) proposed subdivision of Copepoda into three "series" by the structure of their mouths. The series Gnathostoma contained copepods with biting mandibles, fairly short, broad, apically dentiferous and provided with palps. They had large, widely open buccal cavities. Poecilostoma had similar mouths but were allegedly devoid of mandibles. Siphonostoma had stylet-like mandibles, either dentiferous or not, enclosed in more or less elongated, conical syphon-like mouths formed jointly by labrum and labium. One of the novel features of this classification was its disregard of the mode of life and grouping of the free-living and parasitic species together on the basis of their similarity of mouth-parts.

Thorell's classification received widespread support initially but was subsequently rejected by most carcinologists. Confidence in its validity was undermined when Thorell's allegation of the absence of the mandible in Poecilostoma was proved incorrect by Claus (1862), who redistributed the species of that series among the remaining two series. The final dismissal of this classification came with the statement of Sars (1903) that: "quite gradual transitions in the structure of the oral parts have been found to exist, and moreover, by accepting these groups, otherwise evidently nearly allied forms would be separated from each other in an unnatural manner . . ."

Sars' comments raise two objections to the use of the mouth-parts as aids to classification. The first one is the alleged existence of quite gradual transitions between the three types of mouths proposed by Thorell, and the other, which is the result of the acceptance of the first, is that the system calls for separation of "evidently nearly allied forms".

If we look for facts that would support the first objection raised by the eminent Norwegian authority, we soon discover, as mentioned earlier (p. 23), that there are at best very few copepods we would consider as intermediate in oral morphology between the three major types (numbering among these types the poecilostomes which are now known to have falcate mandibles). There exist many, sometimes quite appreciable differences within each type, but seldom can one make a mistake as to which type any copepod belongs. Some features intermediate between the gnathostome and poecilostome type might be claimed for the genus *Hemicyclops*. A few species of, for example, Nanaspidae, Stellicomitidae or Cancerillidae have mouths with poorly accentuated siphonostome characteristics and might be construed as intermediate species between that type and the gnathostomes. The existence of these very infirm connecting links should be seen, however, rather as an exception which reaffirms the distinctness of the three main types of mouth parts. Sars' assertion must, therefore, be found wanting. One certainly cannot acclaim it as a basis for rejection of the systematic value of oral morphology.

The question next arises, on what evidence can we assert that the division of copepods by the mouth parts separates "evidently nearly allied forms"? Can forms with different mouth-parts be nearly allied? Sars does not amplify his statement but it can be assumed that his evidence of the affinity between these copepods is based on their general morphology and, perhaps, the structure of appendages other than oral. There is nothing to suggest that these features are not secondary morphological characteristics which impart superficial similarity to unrelated forms, particularly when these forms clearly show their lack of affinity by possessing mouths of different types. On the contrary, if two forms are morphologically similar to the extent that they can be suspected of close relationship to each other, then it is very likely that they are also similar biologically. If, in spite of these similarities, they have mouths of completely different structure, they might well have converged morphologically in the course of evolution. Sars' second objection, therefore, can hardly be used as a foundation for firm taxonomic decisions.

Giesbrecht (1892) divided Copepoda into two divisions on the basis of segmentation. This classification was based exclusively on the study of planktonic copepods, a fact that in itself made it unattractive to all those who were keenly aware of the biological and ecological diversity of Copepoda. One of the divisions, Gymnoplea, comprised copepods with bodies divided into two parts by a major articulation passing behind the fifth leg-bearing segment. The fifth leg, therefore, was situated on the anterior part of the body. It was always fully segmented and unmodified, though in some species reduced in size by comparison with the preceding legs. The other division, Podoplea, was distinguished from the first by the position of the major articulation, which passed in front of the fifth leg-bearing segment. Hence, in copepods of this group the fifth leg was situated on the posterior part of the body. It was always reduced, usually to a uniramous, two-segmented appendage, clearly unfit for its original locomotory function.

The assessment of Giesbrecht's classification given by Sars (1903) can be seen as typical for most copepodologists, with the possible exception of those few whose interests were restricted to the planktonic forms (e.g., Rose, 1933). It can be summed up in this quotation: "The establishment by Dr. Giesbrecht of the two great divisions, or sub-orders, Gymnoplea and Podoplea, is certainly more natural; but these divisions do not nearly suffice for the comprising of all the Copepoda, as they apparently relate only to the pelagic forms examined by that author."

This somewhat summary review of antecedents should suffice to set the scene for the author's own views on the way in which the copepods are divided into major groups, and how they came to be so divided. As most classification systems, the one proposed below must necessarily rest on some degree of conjecture. It cannot lay claims to completeness. Nevertheless, it should provide a reasonable framework for most copepods now known to science, leaving outside comparatively few extravagants that will have to be satisfied, for the time being, with the label of *incerte sedis*.

The most prevalent mode of life among Crustacea, as well as among all the other aquatic arthropods, is that of a benthic or nearly benthic animal. It can be presumed without being unreasonable that the demersal habits are primitive among Crustacea, their swimming representatives having only secondarily acquired their ability to break the ties binding them to the bottom. One can equally reasonably presume that the primitive ancestral "archicopepod" also lived on or near the bottom. More likely than not, it was an animal with a more elongated body than those of extant primitive members of the group. The process of cephalization had already provided it with a well-developed head, either a cephalon or a cephalosome (see p. 2). It had rather short, poorly armed first and non-prehensile second antennae. Its mouth and oral appendages were probably not unlike those of the modern gnathostomes (Text fig. 1). (Even to-day the gnathostome mandible is the most common among various subdivisions of Copepoda, occurring in species with widely diverse types of biology and much more likely to be primitive than either of the other two types known.) Its body, poorly differentiated and not subdivided into tagmata, consisted of a series of similar segments, as is usual in primitive metamerism. Many, perhaps most, of these segments carried typical biramous legs, fully segmented but probably poorly armed with spines or setae. With undulating movements made possible by inter-segmental articulations, these ancestors of modern copepods slithered over the bottom like some aquatic centipedes, feeding on detritus, small prey, or both.

The hypothetical fate of the progeny of these copepods and, incidentally, an outline of classification based on it, are shown diagrammatically in Text fig. 23. One obvious direction in which evolutionary forces were likely to push these primitive animals was the mid-water life of a swimmer.

To remake the undifferentiated bottom dweller into a planktonic animal, evolution had to provide it with a more rigid body. The directional forces pushing towards it were assisted by the general tendency among meristically simple animals to develop a greater degree of inter-segmental

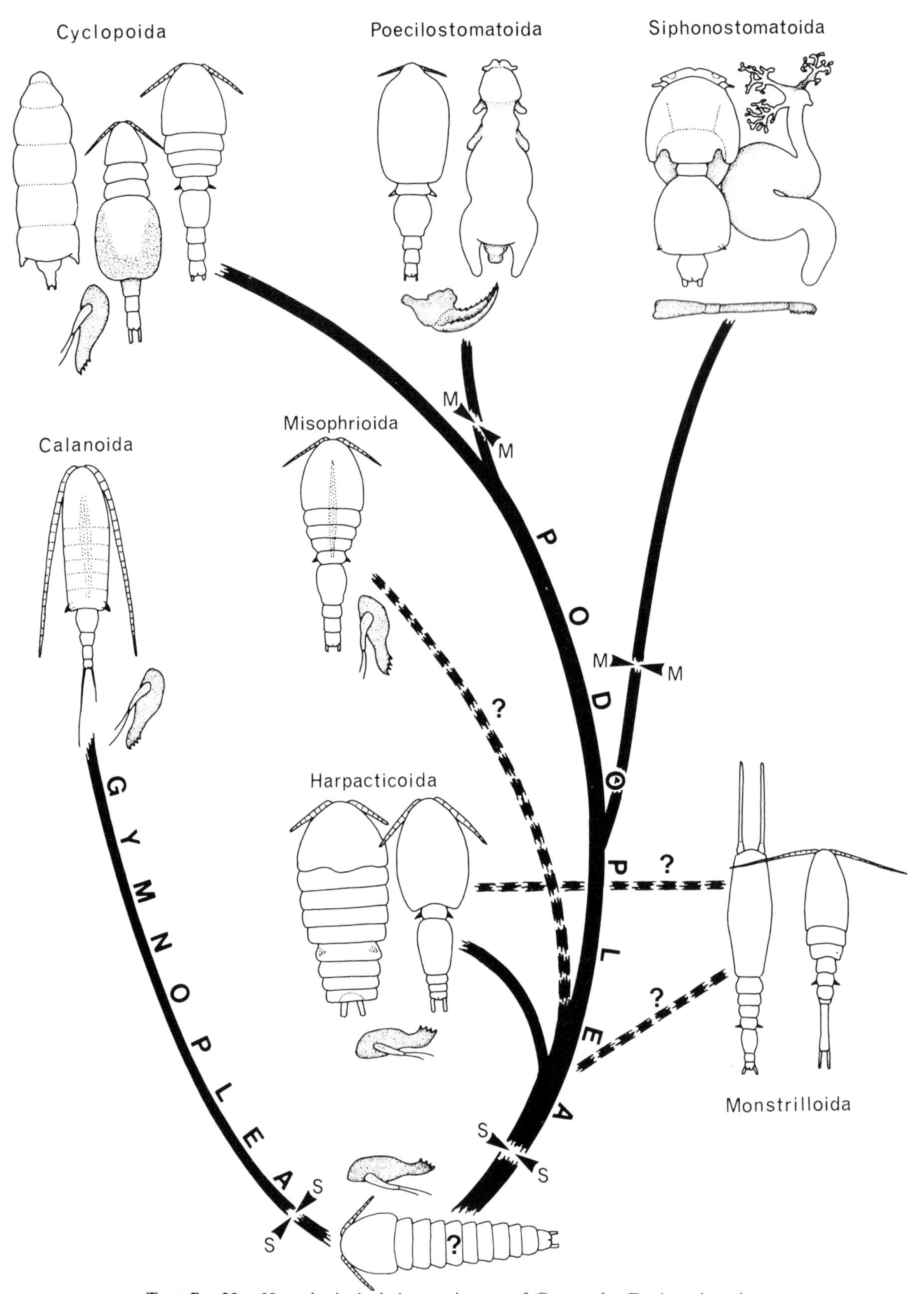

Text fig. 23. Hypothetical phylogenetic tree of Copepoda. Explanations in text.

integration. The most common outcome of this tendency among free-swimming copepods is the subdivision of the body into two tagmata, with a joint or major articulation between them. The development of that major articulation might be termed the first "major event" in the evolution of Copepoda, an event that occurred probably on more than one separate occasion.

In one group of copepods the major articulation developed in the "gymnoplean position" behind the fifth leg-bearing segment, giving rise to Giesbrecht's (1892) Gymnoplea, coextensive with Sars' (1903) Calanoida. The copepods belonging to this order represent an epitomy of adaptation to planktonic life, more floating than swimming in mode of progression. Their slender, rigid bodies, with comparatively short (due to the position of the major articulation) posterior part capable of acting as a depth rudder, hover in mid-water, head up, suspended with the aid of very long, multi-segmented first antennae and often luxuriantly developed armature of the uropods. This is not to say that there are no active and vigorous swimmers among Calanoida. The floating habit prevails, however, and so successful proved their adaptation to this habit that Calanoida came to occupy a dominant position in marine copepod plankton, both in number of species and in abundance of individuals. Although diverse in their feeding habits, they could specialize without any drastic changes in their oral apparatus. The mouth parts of the calanoids are still the standard gnathostome appendages, perhaps not far removed structurally from those of their hypothetical demersal ancestors.

There can be little doubt that the gymnoplean, gnathostome Calanoida form a compact, well-defined group. Some copepods from other groups, particularly some Cyclopoida, do show certain morphological features that mimic those of the calanoids, but these superficial similarities can well be attributable to convergent evolution on the background of the pelagic existence. Gurney (1931) deemed the most serious objection to Giesbrecht's system to be the fact that it separates Calanoida in a fundamental manner from all the rest. This author is of the opinion that Calanoida, an eminently valid taxon, became early separated from the main stem of copepod evolution. By exploiting their potential in the planktonic environment, they came to occupy an important position in it, but at the price of evolutionary sterility. They spawned no "affiliated" groups and, so far as is known to this author, produced no true parasites or associates.

Having evolved into the veritable epitomy of plankton-dwellers, the calanoids have retained many primitive features. Their fifth leg, however, although less modified than in other copepods, has lost its primitive natatory function.

When other copepods took off from the bottom it was to become active swimmers, sometimes indifferent ones. As opposed to passively floating planktonic species, most of these copepods, when not in motion, tend to assume a horizontal position in mid-water. One reason for this posture is the relative shortness of their first antennae, which fail to supply sufficient buoyancy to the anterior part of the body. The other is the proportion between the respective lengths of the anterior and posterior parts of the body, which are more evenly balanced than in Calanoida. The extension in the length of the posterior part is achieved by forward displacement of the major articulation. This results in the "podoplean" type of segmentation, in which the fifth leg finds itself associated with the posterior part of the body. It is usually more or less vestigial.

The contact with the bottom, commonly retained by active swimmers, and the podoplean segmentation, provided flexibility which proved to be fertile from the evolutionary point of view. This author believes that all extant groups of Copepoda other than Calanoida are descended from the primitive podopleans, even if some of them have lost their major articulation or have it secondarily displaced.

Before taking up the story of the main line of copepod evolution, one should take a brief look at an aberrant offshoot of the podopleans, represented by the sub-order Monstrilloida. It is impossible to say at which point in evolution they became separated from their relatives, though some clues suggest that it must have been rather early in copepod history. Firstly, they are all parasitic on invertebrates (usually polychaetes, exceptionally echinoderms) and have succeeded in developing a very intimate host–parasite relationship. Both these facts are indicative of the ancient nature of their parasitism. They are unusual among the parasitic copepods in becoming infective at the nauplius stage, i.e. earlier in the life history than any other copepods. We shall never know how or why the monstrillid nauplius came to be infective, but it is clear that this event determined the course of evolution for Monstrilloida. Relieved of the need for active feeding during their very brief pre-parasitic stage, they jettisoned all the morphological appurtenances of that need. They have no mouth or functional gut, no oral appendages and no second antennae (mandatory for prehension in infective copepodids). Their free-swimming adults have their major articulation in the podoplean position (except for *Thespesiopsyllus* in which it runs one segment further forward). The primitive structure of their locomotory appendages can be taken as another clue of their early breach with other podoplean copepods. All pairs of legs situated in front of the major articulation are fully developed and have clearly three-segmented rami.

The most primitive branch in the mainstream of copepod evolution is undoubtedly represented by species encompassed in the sub-order Harpacticoida. These gnathostome copepods have moved but little from their ancestral habitats. To-day they are dominant on or near the sea bottom, as well as in its interstices. They also predominate in the littoral zone, which is another way of exploiting demersal habits. Some harpacticoids became associated with aquatic vegetation, relatively few entered into relationship with other animals, even fewer became parasitic on vertebrates. The primitive nature of Harpacticoida is suggested by the lack of progress in the process of tagmosis, a fact which, in free-living and otherwise "normal" copepods is not likely to be attributable to secondary modification. Good examples of harpacticoids in which the major articulation is barely noticeable can be found, e.g., in the genera *Canuella* Sars, 1903, *Ectinosoma* Boeck, 1864, and *Microsetella* Brady and Robertson, 1873. (It is interesting to speculate whether the nauplii of Harpacticoida, with their "appendages modified for creeping, clasping and browsing" (Dudley, 1966) have the potential to become parasitic and could have, by appropriate biological adaptation, given rise to Monstrilloida.)

The suborder Misophrioida, an enigmatic little group of planktonic species, is also likely to have separated quite early from the main stem of Podoplea. Lang (1948) suggested even that it evolved separately from the podopleans, presumably because, unlike them, its members possess a "heart". The copepods of this group, however, have distinctly podoplean segmentation and many morphological features reminiscent of the cyclopoids (though in others they do bear certain resemblance to the calanoids). Until more is known about them, they are tentatively considered here as a possible early offshoot of the podoplean stock.

The most advanced group of main podoplean stem comprises the suborder Cyclopoida, a taxon the delineaments of which are seen by this author rather differently than they have been drawn previously. Its members have podoplean segmentation either throughout their life cycles or during the pre-adult ontogenetic stages. Their buccal apparatus is gnathostome. This means that of the Sarsian suborder Cyclopoida only Cyclopoida Gnathostoma can be retained in it. The author places in it also the parasitic family Lernaeidae and the entire suborder Notodelphyoida.

The free-living cyclopoids were able to radiate into various types of habitat. They achieved dominance in freshwater plankton but abound also in other freshwater niches and are conspicuously present in marine plankton. They display all the classical features of podoplean segmentation which prompted Giesbrecht to set up his system of classification. These features are less clearly pronounced in cyclopoids associated with invertebrates and all but obliterated in adult parasites. These different types of Cyclopoida evolved divergently, each specializing for a different mode of life. The pelagic species acquired a fused, rigid anterior tagma, while in the benthic, crawling species the region between the genital segment and the posterior margin of the cephalothorax retained a greater measure of intersegmental articulation, to meet the demands of their locomotory needs. Their podoplean nature is evident, however, from the fact that their fifth leg is reduced in the manner typical of the free-living cyclopoids and forms a component of the posterior part of the body. The dorsoventral compression of the body can also be seen as the outcome of their benthic adaptations.

The lernaeid parasites, clearly podoplean during the pre-metamorphosis stages of the female and in adult males, break through the bounds of "normal" copepod morphology after settling on their hosts and completely lose their external segmentation. The lernaean mouth parts during their entire ontogenetic cycle are modified to the point at which their resemblance to others of gnathostome type is rather tenuous. Their affinity with the gnathostomes is still, however, evident in the structure of the maxilliped (Text fig. 21E, Fig. 398). Another reason for associating them with Cyclopoida is their almost exclusively freshwater distribution. The cyclopoids are freshwater par excellence and are more likely than any other copepod group to be ancestors of Lernaeidae.

A very distinct position within Cyclopoida is occupied by copepods placed by Sars in the suborder Notodelphyoida. This group of greatly modified copepods, mainly associates of invertebrates, shows morphological diversity that baffled many systematists. Their segmentation is often obscured by extreme modifications resulting from adaptations to their dependent, sedentary life. The pre-metamorphosis stages, however, are unmistakably podoplean. The situation of the mouth parts appears even more confused than that of segmentation. Some species, while adult, resemble in this respect other cyclopoids. Others have oral appendages partly, or almost completely, reduced (e.g. Enterocolidae) and in some characteristics resemble somewhat similarly modified poecilostomes.

The validity of Notodelphyoida as a distinct suborder has been questioned in the past. Lang (1948) placed most of them with the cyclopoids among Cyclopoida Gnathostoma Notodelphyidiformes, reserving a place among Cyclopoida Poecilostoma for some of their more modified members. The excellent work of Dudley (1966) established, however, that the allegedly poecilostome members of "Notodelphyoida" have no direct affinity with those copepods and that the reduction of their mouth parts is of a different nature than that occurring among Poecilostomatoida. Bocquet and Stock (1963) also suggested that it might become necessary to dismantle Notodelphyoida. By now the view of the spurious nature of that suborder has become current among the workers most familiar with it. The author subscribes to this view and firmly places "Notodelphyoida" in the suborder Cyclopoida. Their position within the suborder is left for the experts to determine.

With the exception of mouthless Monstrilloida and some much-modified Cyclopoida, all the copepods reviewed so far were unmistakably gnathostome. Interestingly enough, with the exception of the family Lernaeidae, about 70 species strong, they evolved no species parasitic on fishes. It would almost seem as if the gnathostome biting mouth parts were particularly suited for a raptatory mode of feeding and tended to serve best as buccal apparatus for actively feeding animals. It would also seem that, to succeed as fish parasites, the copepods had to await the event which this author likes to refer to as the second major event in copepod evolution: the development of the non-gnathostome type of mouth. (The first event, it will be recalled, was the development of the major articulation.)

This second event appears to have occurred on two separate occasions, at two different times in the history of Copepoda. The probably more recent occasion was the development of the poecilostome mouth, characteristic of the copepods grouped by this author in the suborder Poecilostomatoida. The poecilostomes probably branched off from a stem common with cyclopoids not very far back in the evolutionary past and retain in some instances a close resemblance to the cyclopoids. Indeed, one might perceive a connecting link between Poecilostomatoida and Cyclopoida in the genus *Hemicyclops* Boeck, 1873 (Clausidiidae: Poecilostomatoida).

Like their cyclopoid relatives, the poecilostome copepods are podoplean, though they, too, often lose their major articulation in the adult stages of the more advanced forms. The position and shape of the fifth leg in the adults of non-metamorphosed species and in pre-adults of those that undergo metamorphosis suffice to disclose their podoplean affinity. The structure of the poecilostome mouth was described earlier (p. 23) and is illustrated in Text fig. 38 and Fig. 137. Like the gnathostome mouth, it has a wide-open buccal cavity with a moderately developed labrum covering only part of it at the anterior margin. It differs from the gnathostome mouth, however, in the type of the oral appendages, particularly in its falcate mandible (Text fig. 17C).

Poecilostomatoida include numerous copepods associated with invertebrates and exemplified by Lichomolgidae and their diverse relatives. Also included are the parasites of fishes, variously advanced in the intimacy of their host–parasite relationship. The little modified Ergasilidae, Bomolochidae and their allies, as well as the greatly changed Chondracanthidae, are the members of this suborder. (Incidentally, the close relationship between Poecilostomatoida and the cyclopoids is impressively suggested by the similarity between the naupliar morphology of both groups, particularly that of the mandible of *Ergasilus* and *Cyclops*.) Unlike the cyclopoids, however, the poecilostomes are predominantly marine. Of their number, only the family Ergasilidae contains some species that became successfully established in freshwater habitats.

The other occasion on which a new mouth developed from the primitive gnathostome type occurred probably much earlier in copepod history. The new mouth was of siphonostome structure (Text fig. 13). There are two reasons for supposing that its origin is more ancient than that of the poecilostome mouth. The first is the structure of the mouth itself. The development of a siphon-like extension in which both lips are intimately involved is a complicated process calling for more evolutionary effort than the simple modification of oral appendages required to convert a primitive gnathostome mouth into a more modern poecilostome one. Consequently, more time might have been needed to evolve it. The second reason suggests itself when one studies the mode of action of a siphonostome mouth. This type of mouth can only be used when the animal possessing it is stationary and in contact either with the bottom or with another organism or floating object (for the mode of feeding of a caligid siphonostome see p. 25). Contact with the bottom, the benthic habit, has already been suggested as being more primitive for copepods than mid-water existence. The development of the siphonostome mouth therefore might be supposed to predate the development of the swimming habits. It is not unlikely also that this big evolutionary event provided pre-adaptation to associated and parasitic modes of life. There are hardly any free-swimming siphonostomes.

At this point it might be worth considering the possibility that the siphonostome mouth evolved independently more than once. Should this question be answered in the affirmative, the effect on the proposed system of classification would be obviously great and worthy of the closest scrutiny. This author presumes that the siphonostomes have evolved together from a single stock. If we disregard the rudimentary siphons of Nanaspidae and similar copepods, we find that the siphonostome mouths can be placed in one of two groups. To the first group belong all the siphonostomes associated with the invertebrates, as well as the lernaeopodids, caligids and their relatives. In this type the two lips remain often fairly loosely connected near the tip of the siphon, as shown in Text fig. 13. The second group comprises such morphologically dissimilar copepods as the parasitic Pennellidae and presumably free-living Megapontiidae (known only from deepwater plankton of both the Atlantic and Pacific). The mouth of this group (Text fig. 59) consists of the buccal cone surmounted by an elaborate buccal tube developed from the distal part of the labium. In spite of these differences, copepods of both groups possess buccal stylets (see p. 41), structures that have caused some controversy in the past. Our knowledge of the siphonostome mouth is far from satisfactory, but the buccal stylets are known in Megapontiidae and Pennellidae of the latter group, as well as in Caligidae and Lernaeopodidae of the former. One might argue more than one origin for the mouth. It becomes more difficult to argue for the polyphyletic origin of both the mouth and the buccal stylets. It is also unnecessary to regard the possession of the frontal filament by some, but not all, siphonostomes as evidence of their dual origin. The frontal filament is a form of larval attachment used by many siphonostome parasites of fishes previously placed in the Sarsian suborder Caligoida. The very fact that some species with either type of siphonostome mouth are equipped with it suggests that it is more likely to have evolved on a single occasion. On the balance, there appears to be no need to repudiate the possibility of the common origin of the siphonostomes.

The suborder Siphonostomatoida, as envisaged by this author, comprises all copepods with siphonostome mouths, i.e., those previously placed in Cyclopoida Siphonostoma, as well as the former suborders Caligoida and Lernaeopodoida (the last-named without its poecilostome chondracanthid component). The members of this suborder associated with invertebrates superficially resemble the cyclopoids with which they share their mode of life. Both groups have primitive podoplean segmentation with four rather freely articulating segments between the cephalothorax and genital complex and with typically podoplean reduction of the fifth leg. Both are depressed dorsoventrally and tend to be short and broad, with suborbicular outline of the cephalothorax (Text fig. 2). It is from copepods like these that the ectoparasites of fishes are probably derived, similarly segmented at adult stage and without the complicated buccal tube at the tip of the siphon-like mouth. The predominantly mesoparasitic pennellids are probably descended from copepods similar to Megapontiidae (cf. Geptner, 1968).

The siphonostome mouth gave its possessors a key to success in the exploitation of the parasitic mode of life, and particularly of fishes as hosts. The abundance of species and tremendous variety of morphological and biological adaptations provide a good index of that success. Using Yamaguti's (1963) compendium as a rough guide (1,500 species listed), it is easy to ascertain that approximately 75% of copepods parasitic on fishes belong to Siphonostomatoida (20% to Poecilostomatoida, 5% to Cyclopoida). It is not surprising, therefore, that this book will deal mainly with the representatives of this suborder.

To sum up this review of copepod classification, a list is given below of all the families of copepods parasitic on British fishes, placed in their respective suborders:

Suborder	*Family*	*Suborder*	*Family*
Poecilostomatoida	Bomolochidae	Siphonostomatoida	Caligidae
	Taeniacanthidae		Euryphoridae
	Ergasilidae		Trebiidae
	Chondracanthidae		Pandaridae
	Philichthyidae		Cecropidae
Cyclopoida	Lernaeidae		Dichelesthiidae comb. nov.
			Eudactylinidae comb. nov.
			Kroyeriidae fam. nov.
			Pseudocycnidae comb. nov.
			Hatschekiidae fam. nov.
			Lernanthropidae fam. nov.
			Pennellidae
			Sphyriidae
			Lernaeopodidae

A Key to Families of Copepoda Parasitic on British Fishes

The diagnosis of each family of Copepoda parasitic on British fishes, given in this book, is based on the combined characteristics of all their members, regardless of their geographic distribution. This key, however, is intended to be a guide to the copepods occurring in British waters. Hence the range of characters used to define individual families is narrower than it would be if non-British members of these families were included.

Some families appear in the key more than once. This is necessary in those instances where they contain members with widely divergent characteristics, or when their members change their characteristics from pre-metamorphosis to mature adult state. In sexually dimorphic families, where adult males are not more or less permanently associated with females and might be found on their own, they are treated separately. Stages younger than pre-metamorphosis adults are not included in the key.

1 Body with some sharp and definite intersegmental divisions in relaxed specimens (e.g. Fig. 1082) . . . 2
Body without sharp and definite intersegmental divisions (e.g. Figs. 406, 1334) . . . 13

2 Body dorsoventrally flattened, anterior part covered by dorsal shield (e.g. Fig. 505) . . . 3
Body not dorsoventrally flattened, dorsal shield absent (e.g. Fig. 1213) or present (e.g. Fig. 1082) . . . 7

3 Dorsal or dorsolateral plates present on one or more segments (e.g. Fig. 741) . . . 4
Dorsal or dorsolateral plates absent . . . 6

4 Cephalothorax incorporating only first leg-bearing segment (e.g. Fig. 769) . . . 5
Cephalothorax incorporating third leg-bearing segment (Fig. 741) . . . **Euryphoridae**

5 Female maxilliped with squat, robust corpus (Figs. 784, 813, 830, 862, 883) . . . **Pandaridae**
Female maxilliped with long, slender corpus (Figs. 912, 945, 968, 979, 993) . . . **Cecropidae**

6 Cephalothorax incorporating second leg-bearing segment (e.g. Fig. 722) . . . **Trebiidae**
Cephalothorax incorporating third leg-bearing segment (e.g. Fig. 592) . . . **Caligidae**

7 Six pairs of legs present, sixth pair vestigial (e.g. Fig. 14) . . . 8
Five pairs of legs present . . . 10
Four pairs of legs present . . . 12
Three pairs of legs present, either foliaceous and covering posterior part of body (Fig. 1028), or with rami reduced or absent (Figs. 1022, 1024) . . . **Dichelesthiidae**
Two pairs of biramous legs present (Fig. 427); body long and slender, consisting of 11 clearly delimited parts, with well developed uropods (Fig. 417) . . . **Philichthyidae** (males)
Legs absent, body with numerous lateral and other processes (Figs. 406–411) . . . **Philichthyidae** (females)

8 Second antenna three-segmented, not folded, armed apically with one short hook and several setae (Fig. 394) . . . **Lernaeidae** (pre-metamorphosis)
Second antenna folded (Fig. 57) . . . 9

9 Maxilliped posterior to mouth (Fig. 73) . . . **Taeniacanthidae**
Maxilliped lateral to mouth (Text fig. 25) . . . **Bomolochidae**

10 First leg modified (Fig. 1321); long cylindrical trunk forming greatest part of body (Fig. 1309) . . . **Pseudocycnidae** (females)
Fourth leg flagelliform, at tip of digitiform lateral process (Fig. 1325) . . . **Pseudocycnidae** (males)
First leg not modified, with well First leg not modified, with well-developed rami . . . 11

11 Cephalothorax constituting about $\frac{1}{2}$ of body length (Fig. 88), sometimes inflated (Fig. 108); second antenna subchelate (Figs. 102, 113); maxilliped absent in female; parasites of teleosts . . . **Ergasilidae**
Cephalothorax constituting less than $\frac{1}{4}$ of body length (Fig. 1175); second antenna not subchelate, with short apical claw (Fig. 1182); maxilliped present in female, either chelate (Fig. 1193) or subchelate (Fig. 1279); parasites of elasmobranchs . . . **Eudactylinidae**

12 All four pairs of legs biramous (Fig. 1093); cephalothorax covered by flat dorsal shield (Fig. 1083) . . . **Kroyeriidae**
First two pairs of legs biramous (Fig. 1363), second two pairs uniramous (Fig. 1364); cephalothorax without flat dorsal shield (Figs. 1341, 1342) . . . **Pennellidae** (pre-metamorphosis)

13 Anterior part deeply buried in tissues of host . . . 16
Anterior part not buried in tissues of host, though sometimes covered by proliferating epithelium . . . 16

14 Part buried in host tissue comprising antennary region of cephalothorax and neck-like pre-mandibular region; mouth above surface of host (Figs. 361, 362) . . . . . . **Chrondracanthidae** (Lernentominae)
Entire cephalothorax, oral region included, buried in host tissue 15

15 Holdfast consisting of two pairs of processes, usually one branched and one simple; body subcylindrical, broader posteriorly; four pairs of legs separated from one another by long intervals (Fig. 391); eggs multiseriate . . . . **Lernaeidae** (metamorphosed)
Holdfast consisting of one pair of simple processes (Figs. 1398, 1426), three simple (Fig. 1421) or branching processes (Fig. 1334), or paired lobes with secondary holdfast processes (Fig. 1389); body straight, with (Fig. 1423) or without abdominal brush (Fig. 1398), or with sigmoid flexure (Figs. 1372, 1388); four pairs of legs close together (Fig. 1376); egg-sacs string-like, eggs uniseriate . . **Pennellidae** (metamorphosed)
Holdfast of variously shaped lobes (Figs. 1441, 1452); anterior region slender, posterior broad and flat, with clusters of posterior processes (Figs. 1445, 1456); legs absent, egg-sacs straight, eggs multiseriate . . . . **Sphyriidae**

16 Body consisting of only two distinguishable parts (e.g. Fig. 317); small cephalothorax and large trunk with or without intervening narrow neck 17
Body consisting of more than two distinguishable parts 18

17 Cephalothorax with (Fig. 159) or without processes (Fig. 129); trunk with (Fig. 130) or without processes (Fig. 253); second antennae simple hooks (Fig. 136); legs unsegmented, trifid (Fig. 146), bifid (Fig. 186) or undivided (Fig. 345); eggs multiseriate . . . . **Chondracanthidae** (Chondracanthinae)
Cephalothorax without processes (Fig. 1137) or with simple median outgrowth (Fig. 1127); trunk subcylindrical, without processes (Fig. 1128) or with pair of small posterolateral outgrowths (Fig. 1144); second antenna subchelate (Fig. 1132); two pairs of biramous legs (Fig. 1133) eggs uniseriate . . . . **Hatchekiidae**

18 Posterior part of body covered by dorsal plate in female (Figs. 1055, 1072); exposed in male (Fig. 1078); fourth pair of legs bilobed, foliaceus . . . . **Lernanthropidae**
Body divisible into three regions: cephalothorax, trunk and modified arm-like second maxillae fused at tips with bulla (Fig. 1462); second maxillae sometimes fused together and reduced (Fig. 1861) . . . . **Lernaeopodidae**

Diagnoses of Families, Descriptions of Genera and Species

Suborder Poecilostomatoida

Family Bomolochidae

Morphology

The members of this family have made relatively few concessions to the parasitic mode of life, their structure retaining an unmistakable likeness to that of their free-living poecilostome relatives. A great deal of general similarity exists also between the bomolochids and the taeniacanthids, introducing a measure of taxonomic doubt still not entirely resolved. The hallmark of the bomolochid (and taeniacanthid) structure is its clearly preserved segmentation, only a few original segmental boundaries having been dispensed with during the process of adaptation to parasitism.

The general outline of the gross morphology of the now-recognized 11 bomolochid genera is shown diagrammatically in Text fig. 24. In dorsal aspect, three types of facies can be distinguished, linked by intermediate forms. Probably the most typical is the so-called "cyclopoid" facies (Vervoort, 1962), in which the most anterior tagma, the cephalothorax, is much broader than the rest of the body, imparting to it a top-heavy, tadpole-like appearance. Examples of this type of facies are found in the genus *Bomolochus* von Nordmann, 1832 and in at least six other genera (*Nothobomolochus* Vervoort, 1962; *Holobomolochus* Vervoort, 1969; *Pumiliopes* Shen, 1957; *Pumiliopsis* Pillai, 1967; *Pseudorbitocolax* Pillai, 1971 and *Acanthocolax* Vervoort, 1969). Less common is the "elongate" facies, characterized by gradual narrowing of the body in the posterior direction and exemplified by species of *Dicrobomolochus* Vervoort, 1969 and *Pseudoeucanthus* Brian, 1902. The third type of facies is

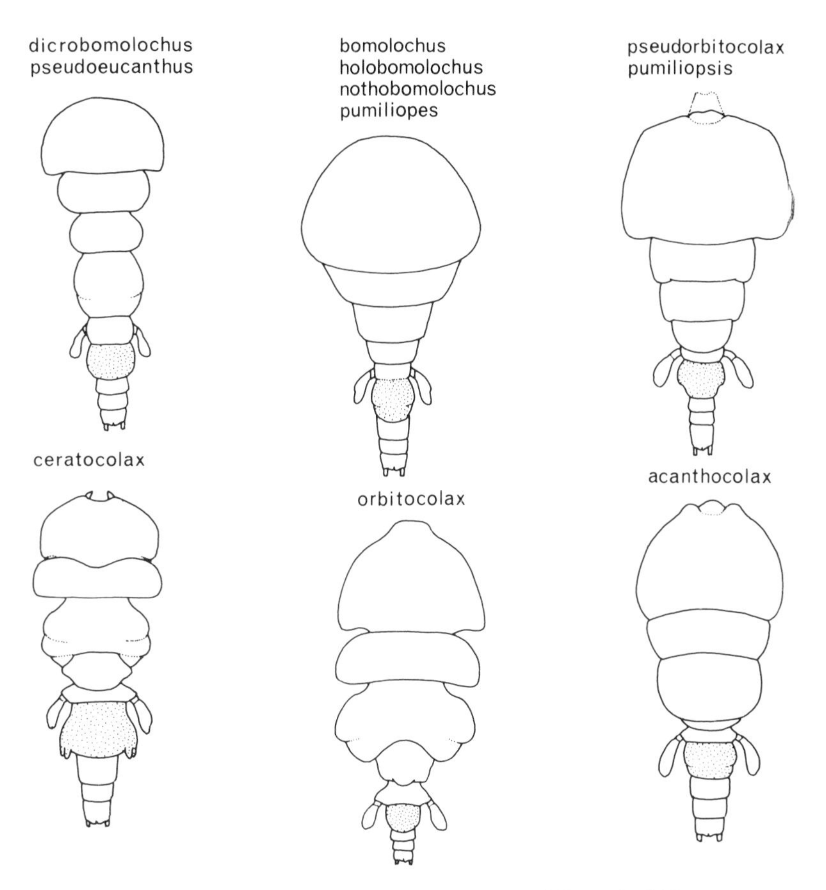

Text fig. 24. Morphological types of Bomolochidae.

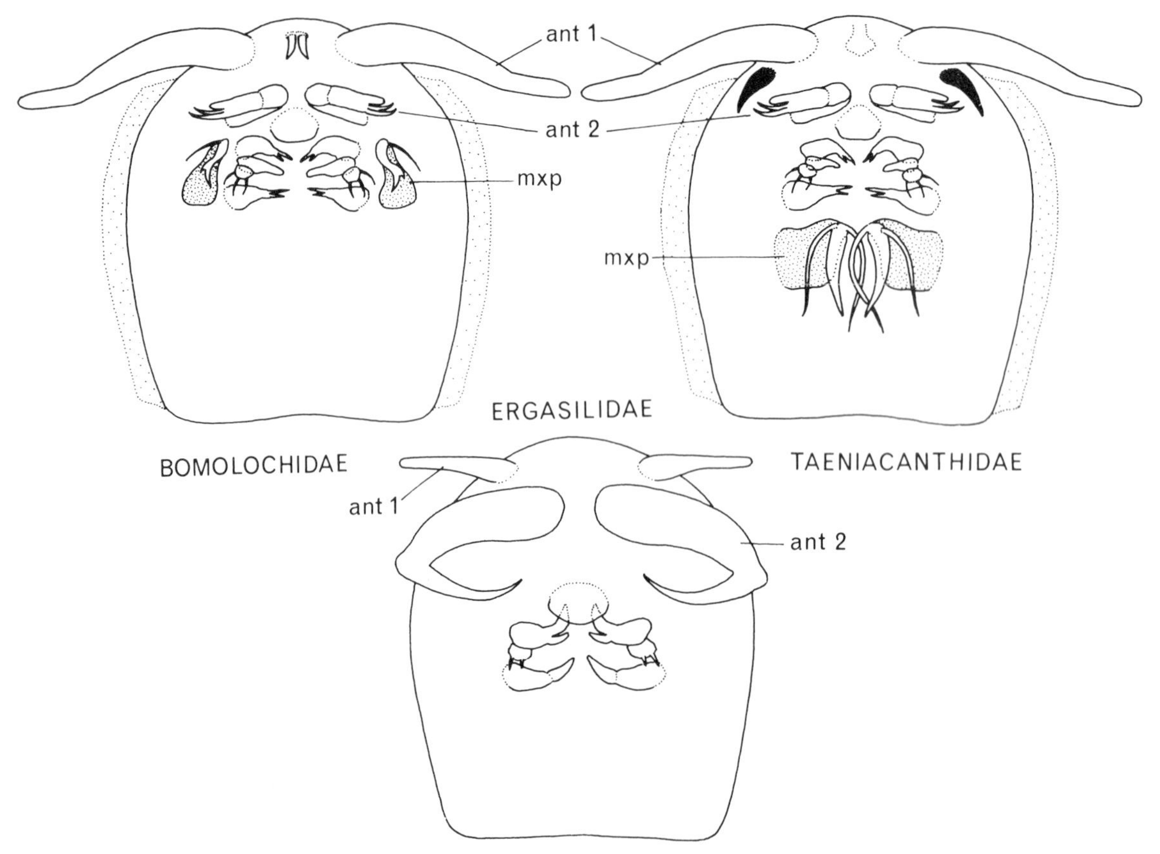

Text fig. 25. Cephalothoraces of three families of poecilostome copepods. Diagrams of ventral surfaces. ant 1 — first antenna; ant 2 — second antenna; mxp — maxilliped.

represented by *Orbitocolax* Shen, 1957. The typical feature of this facies is the width of the first two free thoracic segments. About as wide as the cephalothorax, these segments are followed by a third that is less than $\frac{1}{3}$ of their width. As the rest of the body is also very narrow, the animal with this type of facies appears to consist of two parts: a much broader and longer anterior and a short, very narrow posterior part.

The cephalothorax incorporates the first leg-bearing segment and constitutes from less than $\frac{1}{4}$ (in *Dicrobomolochus, Pseudoeucanthus* and *Ceratocolax* Vervoort, 1965) to about $\frac{1}{3}$ of the total length (e.g. in some species of *Bomolochus* and *Acanthocolax*). Varying in size and shape, the cephalothorax has a convex dorsal surface, sometimes bisected longitudinally by a more or less well marked groove. Its anterior margin might be smoothly rounded or it might project forward in its central, rostral area (e.g. in *Pumiliopes* and *Pseudorbitocolax*). The projecting part may, or may not, be divided from the rest by two posterolateral grooves. The posterior margin, usually transversely truncated, sometimes overlaps part of the following segment. The lateral margins are provided with strips of transparent membrane. The ventral surface is slightly concave. This is an adaptive feature of paramount importance, making the cephalothorax an inverted cup, a sucker providing the main means of attachment to the fish. The strips of membrane laterally and the expanded first legs (Fig. 9) posteriorly help to seal off the sucker, enhancing its efficiency. Within the cup, on the ventral surface of the cephalothorax, are the antennae, mouth and all the buccal appendages (see Bomolochidae, Text fig. 25).

The cephalothorax is followed by a series of four leg-bearing segments, clearly delimited from one another and generally decreasing in size from the first to the fourth. It is clear from Text fig. 24, however, that there are many exceptions to this general rule. The second leg-bearing segment might be smaller than the succeeding ones, so that a neck-like constriction is introduced into the appearance of the copepod. In *Orbitocolax* and *Acanthocolax*, for example, the third leg-bearing segment might be larger than the second. In some species the second and the third might be partially fused. The morphological variability of this region of the bomolochid body is increased by the development of posterodorsal swellings in many species. Extending backwards, these swellings partly or completely obscure the next segment in the series and might be responsible for an apparent sigmoid flexure of the copepod, seen in its lateral aspect. Such flexure distinguishes bomolochids with the "body of the 'Artacolax' type" (Vervoort, 1962, p. 18), the term derived from the now-rejected genus of Wilson (1908). The posterolateral swelling is most common in the third leg-bearing segment, occurring in the genera *Bomolochus*, *Holobomolochus*, *Acanthocolax*, *Nothobomolochus*, *Ceratocolax* and *Pumiliopes*. In *Pseudorbitocolax*, however, it is the fourth leg-bearing segment that covers much of the fifth from its dorsal side, while a slight overhang of this segment can be found in *Orbitocolax*. In most genera the segments of this region, when seen in their dorsal aspect, form smoothly confluent lateral margins. In some, however, there are pronounced constrictions between them, either because of the narrowing of the anterior and posterior ends of individual segments (e.g. in *Dicrobomolochus*), or because of a deep constriction at the anterior end of the third leg-bearing segment (in *Ceratocolax* and *Orbitocolax*). The fifth leg-bearing segment is usually much smaller than other segments in the series and is invariably broader than long.

The genital complex is formed by the fusion of the seventh thoracic and first abdominal segments. (In some species of *Bomolochus* it may also include the sixth thoracic segment.) It is always small, more or less subspherical and, with one exception, devoid of outgrowths or processes. The exception is the genital complex of *Ceratocolax* (Text fig. 24), not only larger than those of other genera, but also provided with lateral plates. These structures serve as covers for the genital orifices and by protruding from the posterolateral corners give this somite a somewhat curious appearance.

The abdomen is always three-segmented, its constituent segments differing in length from one another. It is a short part of the body, never exceeding $\frac{1}{4}$ of the total length; in some genera (*Orbitocolax*) it is quite diminutive. The last abdominal segment carries a pair of uropods (one on either side of the anal slit), each provided with five marginal and one dorsal setae. In most genera the fourth marginal seta (counting from the lateral side) is thickened and lengthened, though in *Pseudoeucanthus* it remains fairly short. In *Bomolochus*, *Holobomolochus* and *Acanthocolax* the third marginal seta is modified like the fourth.

A rostral plate extends forwards between the bases of the first antennae, or it might curve ventrad between them. The forward prolongation is quite long in *Pumiliopsis*, with a bifid and serrated apex. The ventral margin might be smooth and thickened in the form of a ridge, or it may be equipped with a pair of rostral tines (Fig. 41). These tines overhang a subrostral plate (Fig. 40). The shape of the rostral plate, the presence or absence of tines and the shape of the subrostral plate are specific features and can differ significantly within generic boundaries.

The first antenna consists of an indistinctly segmented base and usually three-segmented cylindrical distal part, referred to by Vervoort (1962, 1969) as the "flagellum". In addition to various long and slender setae, the base bears 15 sturdy setae, usually plumose. In most genera these setae differ but little from one another as exemplified by *Holobomolochus* (Text fig. 26). Some of these setae might become reinforced by strips of heavy cuticle. In *Bomolochus* (Text fig. 26) the fourth seta is often longer than its neighbours, hard and distinctly hooked. In *Dicrobomolochus* (Text fig. 26), the second and third setae are modified to form a short, bifid structure on a raised base, while in *Nothobomolochus* (Text fig. 26) very long and hard cuticular outgrowths are present in place of setae 3–5. In *Acanthocolax* (Text fig. 26) a rounded cuticular protuberance arises near the anterior margin in the centre of the base. *Ceratocolax euthynni* Vervoort, 1965, has its first antenna associated with a prominent, tooth-like dorsal plate which, however, is more closely fused with the wall of the cephalothorax than with the appendage itself. The armature of the "flagellum" appears to differ from genus to genus, and perhaps also from species to species (not all descriptions are accurate enough to determine this fact).

The second antenna (Fig. 4) is generally believed to be uniramous due to the complete reduction of its exopod. The sympod consists of two segments, a long basal one (coxa) and a much shorter terminal one (basis). The endopod of two fused segments folds back over the sympod, its tip pointing in a lateral direction. A complicated set of armature, including a lamelliform process, originates near the border of the two segments.

The mouth is of a poecilostome type, with prominent, often ornamented labrum. The buccal appendages comprise mandibles, paragnaths and the first and second maxillae, arranged in a more or less anteroposterior series. The mandibles have strong bases, cylindrical shafts and two tapering

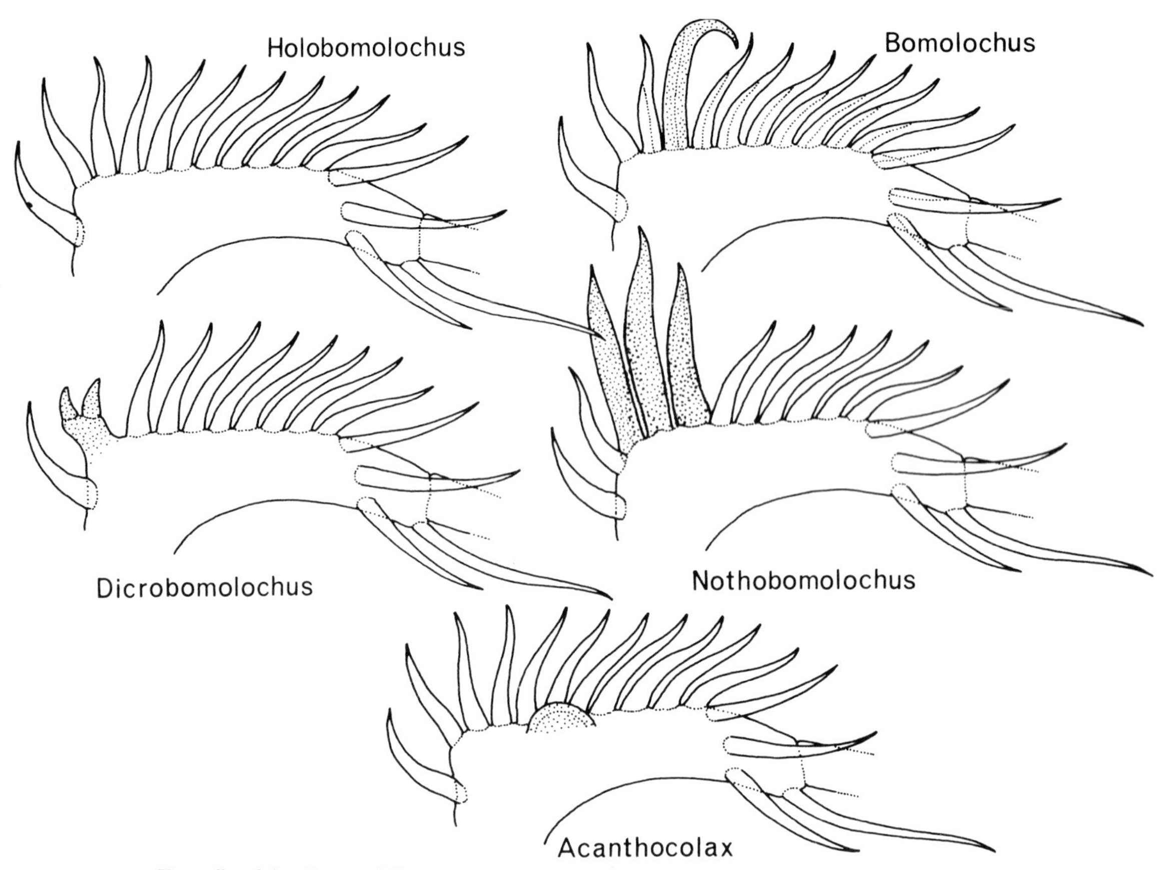

Text fig. 26. Base of first antenna in different bomolochid genera (diagrammatic).

apical processes, varying in shape and armature from species to species. The paragnaths are elongate, with robust bases. The first maxillae are small and rounded; they carry several long and short setae (their number characteristic for the species). The second maxillae in all instances are provided with two terminal processes (claw and auxiliary process?), varying somewhat in shape and armature from one species to the next.

The maxillipeds are strongly developed and lateral to the buccal region (Bomolochidae, Text fig. 25). They are two-segmented, with a powerfully developed second segment articulating with a strong sigmoid claw (endopod). The claws differ between species in details of structure and number of setae.

The members of the family have six pairs of legs. The first four pairs are biramous. The rami of the first leg are flattened and expanded, their segments, partly or completely fused with one another, together providing a watertight screen, sealing off the posterior margin of the sucker-like concavity of the cephalothorax. The other three pairs are relatively unmodified. In most genera the rami of these legs are distinctly three-segmented. *Pumiliopes* and *Pumiliopsis* have legs with two-segmented rami (in the latter genus the fourth endopod is three-segmented). *Acanthocolax*, with otherwise three-segmented rami, has only two segments in the first exopod. The same is true of *Pseudorbitocolax*. *Pseudoeucanthus* has the first leg with a one-segmented exopod; exopods 2–4 are three-segmented, while all four endopods are either two- or three-segmented. The fifth leg is always small and uniramous, usually with a small proximal segment and a much larger, spatulate distal segment. In *Orbitocolax*, on the other hand, this leg is one-segmented. The sixth leg is a mere vestige, being represented by small setae in the vicinity of the genital orifices. The number of setae varies from one (in *Orbitocolax*) to three (in most genera), four setae occurring very occasionally.

No males have been described for *Acanthocolax*, *Dicrobomolochus*, *Pseudoeucanthus* and *Pumiliopes*. The extent of sexual dimorphism is not great and the males, when known, have distinctly bomolochid morphology. The differences between them and the females might involve the shape and relative size of the cephalothorax and of the leg-bearing thoracic segments. In some species (e.g.

Orbitocolax oniscoides Vervoort, 1969) these differences lead to a different general appearance of the two sexes.

The male abdomen consists of two segments only. The first antenna does not show the modifications of the plumose setae exhibited by the corresponding female appendages. The number of the plumose setae on the base might become somewhat reduced. The maxillipeds, lateral in position as in the females, are clearly subchelate, the endopodal claw being slightly hooked and often serrated but never sigmoid.

The first leg is not modified and resembles, more clearly than does its female counterpart, the legs of the following three pairs. The four pairs of biramous legs often differ from those of the females in armature and, though less frequently, in segmentation. In most genera the fifth leg is one-segmented; the sixth leg is absent.

History and systematics

The earliest history of Bomolochidae passed within the boundaries of the family Ergasilidae and will be recounted when that family is being discussed (p. 83). They were first recognized as an independent family by Claus (1875). A gradual change in his views on the relationship of these copepods to Ergasilidae was influenced by morphological differences between these two groups, differences diagrammatically represented in Text fig. 25. Claus' views (1875, p. 340–341), couched in somewhat archaic morphological terms, are worth quoting in full.

"The *Bomolochus* group, now known to us from a series of clearly identifiable genera, can hardly remain linked with Ergasilidae. Although I believed it possible until recently, the differences are too significant to permit their retention within Ergasilidae as a subfamily. While in *Ergasilus* the posterior antennae are the sole prehensile appendages and the second maxillipeds are missing altogether, in the bomolochids the posterior antennae are much shorter and of completely different structure; the second maxilliped is present, very lateral in position and in the males assuming abnormally large size. Added to this are striking modification of the first pair of legs arising from the cephalothorax and presence of accessory hooks on the ventral surface of the cephalothorax, quite similar to those that are found in Caligidae. The anterior antennae are much more elongate and carry long setae. On their bases, these setae are crowded together in a comb-like fashion. The cuticular armour is as a rule much thicker, approaching that of Caligidae, though the body usually remains cyclopiform. At the same time, there occur in Bomolochidae swellings and deformities, not only of cephalothorax but of all thoracic segments. With relative reduction of the abdomen, their body might come to resemble that of *Ergasilina* and *Eudactylina* (*Taeniacanthus*). To Bomolochidae belong the genera *Bomolochus* Burm., *Eucanthus* Cls., *Taeniacanthus* Sph., but not, as I believed for a long time, *Sabelliphilus* Sars, for which Claparede presents an insufficiently detailed description. That genus undoubtedly belongs to Lichomolgidae."

This eminently correct reasoning of the famous Austrian carcinologist had to wait a long time for general approval. Well into the 20th century Wilson (1911) still treated Bomolochidae as a subfamily of Ergasilidae. By the third decade of this century, however, they were almost universally accepted as an independent family group. (Incidentally, Yamaguti (1963) was incorrect in crediting the establishment of this family to Sumpf (1871).)

There was, however, far less agreement as to the membership of Bomolochidae. The main point at issue was for a long time the inclusion into the family of *Taeniacanthus* Sumpf, 1871 and other genera related to it. Wilson (1911) included Taeniacanthinae as one of the three subfamilies of his Ergasilidae. Having decided later (Wilson, 1932) to remove the bomolochids from Ergasilidae, he pulled out also the taeniacanthids and gave them, as Taeniacanthidae, a full familial status, without giving at the same time any diagnosis or discussion of reasons. His action was accepted by Yamaguti (1939a), who, however, reverted later (Yamaguti, 1963) to the concept of Taeniacanthinae as a subfamily of Bomolochidae. Current authorities (Vervoort, 1962; Ho, 1969) accept Taeniacanthidae as an independent family.

On the other hand, Yamaguti (1963) excluded from Bomolochidae the genus *Tucca* Krøyer, 1837 (which Vervoort (1962) made the basis of a bomolochid subfamily Tuccinae), and made it the type genus of a new family Tuccidae. With the abstraction by Ho (1967) of *Telson* Pearse, 1952, from Bomolochidae to the newly-founded Telsidae, the present-day contents of Bomolochidae were established.

At the generic level, the development of the systematics of Bomolochidae was also less than smooth. The small size of the bomolochid copepods made their study difficult and resulted in many inadequate descriptions. These, in their turn, gave rise to misunderstandings and incorrect diagnoses. In most instances the root cause was insufficient appreciation of the bomolochid morphology and the resulting inability to establish valid systematic criteria. Not unnaturally, undue prominence was given to the proportions of the body, while more constant features of morphology, because of their minuteness, were undervalued or completely disregarded. The best example of the use of false criteria was the establishment of the genus *Artacolax* by Wilson (1908), on the basis of various degrees of swelling exhibited by individual genera, a feature influenced by the state of maturity and altogether unstable. The lack of systematic foundations led to the description of other genera (Pseudobomolochus Wilson, 1913; *Bomolochoides* Vervoort, 1962; *Parabomolochus* Vervoort, 1962), now discarded in the light of the more detailed knowledge available. While the relative sizes of the individual parts of the body cannot be disregarded, the morphological criteria now mainly used in the systematics of Bomolochidae are of a different nature. The structure of the female first antenna, the shape and armature of the mouth parts, the segmentation of the legs and their armature formula, are among the features calling for the attention of the student of this family.

The diagnosis of Bomolochidae, as currently constituted, reads as follows:

Female: Cephalothorax largest part of body, varying in shape, commonly broader than long, with or without rostral tines, always with concave ventral surface and lateral strips of marginal membrane; posterior margin transversely truncated, rarely overlapping following segment or overlapped by it. Four leg-bearing segments (legs two to four) distinctly delineated from one another; in some species second partially fused with third; first segment as broad as cephalothorax in *Ceratocolax* and *Orbitocolax*, otherwise mainly narrower; remaining segments usually gradually smaller. Genital complex of two segments (including first abdominal segment), small, subcircular. Abdomen three-segmented, segments well distinguishable. Well developed uropods present. Egg sacs oval or elongated. First antenna uniramous, variously armed, in some genera with modified setae; second antenna uniramous, with endopod two-segmented, reflected over sympod. Mouth poecilostome. Mandible small, with two terminal processes; paragnaths present. First maxilla small, setiferous; second maxilla elongated, with two terminal processes. Maxillipeds lateral to oral region, with endopod in shape of sigmoid claw. First four pairs of legs biramous; first leg flattened and enlarged, others not modified; rami three- or two-segmented. Fifth leg uniramous, two-segmented; sixth leg vestigial.

Male: General structure resembling that of female, cephalothorax usually comparatively smaller. Genital complex incorporating another abdominal segment. Abdomen two-segmented. First antenna not modified. Maxilliped subchelate, claw not sigmoid. First leg not modified. Fifth leg one-segmented; sixth absent.

Type genus: *Bomolochus* von Nordmann, 1832.

In the British fauna Bomolochidae are represented by two genera: *Bomolochus* and *Holobomolochus*, distinguishable from each other with the aid of the following key.

First antenna of female with hooked fourth seta (Fig. 26) . . . . ***Bomolochus***
First antenna of female without modified setae (Fig. 43) . . . . ***Holobomolochus***

Genus Bomolochus von Nordmann, 1832

Female: Cephalothorax usually broader than long, broadest part of body, with or without longitudinal dorsal groove, in some species posterior margin overlapping next segment or overlapped by it. Rostrum with or without tines. Four free leg-bearing segments gradually narrowing, none as broad as cephalothorax. Genital complex small, suborbicular. Abdomen three-segmented. Uropods well armed. Basal part of first antenna with 15 stout setae, fourth seta hooked, often larger than others. Second antenna uniramous, with two-segmented endopod armed with strong, hooked setae. Mouth parts and maxilliped as in family diagnosis. First four pairs of legs biramous, rami three-segmented; first leg and second endopod flattened. Sixth leg of three, rarely four setae.

Male: Cephalothorax relatively smaller than in female. Genital complex of three segments. Abdomen two-segmented. First antenna with 15 unmodified plumose setae on basal part.

Maxillipeds subchelate. Armature and segmentation of legs often different from that of female. Fifth leg uniramous, sixth absent.

Type species: *Bomolochus soleae* Claus, 1864.

Comments: This genus was established by von Nordmann (1832) for *B. parvulus*, a species he described at length but imprecisely. Since he published no figures of that species and since it cannot be identified with any certainty, it is no better than a species inquirenda. As such, it is ill-suited to serve as a type species of its genus. Wilson (1911) suggested that it be replaced in that capacity by *B. bellones* Burmeister, 1835, as the next *Bomolochus* species in chronological order. The most recent reviewer of Bomolochidae (Vervoort, 1962, 1969), however, struggling with the taxonomic chaos existing in this group, was worried by the fact that *B. bellones* had a modified fourth basal seta on its first antenna, while many species then counted among the members of the genus had not. He proposed therefore that *B. soleae* Claus, 1864, be designated as the type species of *Bomolochus*. Vervoort's suggestion was accepted by the International Commission for Zoological Nomenclature (Anonymous, 1965). *B. soleae* thus became the type species of this genus, superseding all other suggestions and possible designations, although it is now clear that *B. bellones* could have filled this part equally suitably.

At the time of writing, the genus *Bomolochus* comprises 19 species whose generic affiliation can be considered beyond doubt. Vervoort (1969) included in his list of *Bomolochus* 13 of these species, unaccountably leaving out *B. bellones*. That species, other omissions and subsequent additions make up the number. More than 10 others are known in the literature but are either clearly not congeneric (e.g. *B. jonesi* Bennet, 1967) or cannot be identified due to the incomplete description. The list of species accepted by this author as belonging to *Bomolochus* and their geographic distribution is shown below.

Table 2. Geographic distribution of *Bomolochus*.

Species	North Atlantic	South Atlantic	North Pacific	South Pacific	Indian Ocean
B. anonymous (Vervoort, 1965)		x			
B. bellones Burmeister, 1835	x	x	x	x	x
B. concinnus Wilson, 1911	x				
B. constrictus Cressey, 1970			x	x	
B. cuneatus Fraser, 1920			x		
B. decapteri Yamaguti, 1936			x		
B. ensiculus Cressey, 1970			x	x	
B. exilipes Wilson, 1911	x				
B. globiceps (Vervoort & Ramirez, 1968)		x			
B. hemirhamphi Pillai, 1965					x
B. hyporhamphi Yamaguti & Yamasu, 1959			x		
B. megaceros Heller, 1865			x		
B. mycterobius (Vervoort, 1965)		x			
B. psettobius (Vervoort, 1965)		x			
B. selaroides Pillai, 1965					x
B. sinensis Cressey, 1970			x	x	
B. soleae Claus, 1864 (type species)	x				
B. tumidus Shiino, 1957			x		
B. unicirrus Brian, 1902	x				

It should be recognized that the above list is far from closed. Many more species are likely to be discovered in all regions in the future.

The members of the genus *Bomolochus* are parasitic on the gills, in the branchial chambers and occasionally also in the nasal capsules of their marine teleost hosts. As the list shows, most of them are restricted in their distribution (though this can be stated only tentatively, in view of the incompleteness of the existing records).

In British waters two species of *Bomolochus* are presently known. This author believes that a systematic search is likely to increase this number. He himself on one occasion found a single and damaged specimen of *Bomolochus* in the gill chamber of *Regalecus glesne* taken in Scottish waters. The specimen clearly did not belong to either of the two species. Unfortunately, its condition made a description impossible.

The two British species can be distinguished from each other as follows.

Second maxilla with terminal process of unequal thickness (Fig. 7); maxilliped with very small seta on basal segment (Fig. 8); lateral margin of first exopod with three spines (Fig. 9) . . . . ***B. soleae***

Second maxilla with terminal processes of almost equal thickness, finely denticulated on one side, with spiniform setae on other (Fig. 32); maxilliped with long seta on basal segment; lateral margin of first exopod segment with six spines (Fig. 33) . . . . ***B. bellones***

Bomolochus soleae Claus, 1864

(Figs. 1–24)

Syn.: nec *Bomolochus solae*; of T. Scott (1893), et auct.

Female (Figs. 1, 2): Cephalothorax broader than long, reniform, with rounded anterior margin; lateral margins also rounded, with strips of transparent membrane (not shown in Figs. 1, 2); rostral plate prominent anteriorly between bases of first antennae; dorsal surface gently convex. Second to fifth leg-bearing segments broader than long, progressively smaller, fifth only $\frac{1}{3}$ width of second, all with noticeable constrictions between them; lateral margins of 2–4 leg-bearing segments more or less covering bases of legs. Second leg-bearing segment dorsally overhanging posterior margin of cephalothorax. Third leg-bearing segment inflated, its lateral parts sometimes set off from central part (depending on condition of ovaries). Genital complex larger than preceding segment, with lateral margins rounded or almost triangular. Abdomen three-segmented, about equal in length to genital complex and fifth leg-bearing segment combined; its segments progressively smaller. Uropods slightly shorter than third abdominal segment, longer than wide, slightly tapering, with thickened and lengthened fourth and fifth setae, latter longer than abdomen. Egg sacs with few moderately large eggs, about as long as genital complex and abdomen together. Total length without uropodal setae about 1.5 mm (after Vervoort, 1969). Longest uropodal seta about 0.4 mm.

First antenna (Fig. 3) with indistinctly segmented basal part and three-segmented, cylindrical distal part (flagellum of Vervoort (1969)); basal part with eight slender, naked setae (one very long) in addition to 15 robust plumose setae; fourth seta from base of appendage modified into strong hook; setae 3–14 with one margin reinforced by thick cuticular strip; first and second segments of distal part with three setae each, terminal apparently with six (three apical and three subapical). Second antenna (Fig. 4) with two-segmented sympod; first segment long, robust, subcylindrical, with long slender seta near distal end; second segment short; long first segment of endopod with six rows of small denticles and one row of long, slender spiniform setae continuing along margin of lamelliform process; distal end with four spiniform, curving setae and two straight and slender setae; second segment robust, as large as lamelliform process, with three rows of denticles continuing over its surface from first segment. Mandible (Fig. 6) with flat, rounded base and cylindrical shaft; large subtriangular process with sharp ventral edge and smaller but similar process carried at apex. Paragnath with inflated base and bluntly pointed apical process. First maxilla (Fig. 5) small, rounded, with two long and two shorter setae, all plumose (plumules omitted in drawing). Second maxilla (Fig. 7) with long, tapering base; distal end with two strong spiniform processes, one thicker than other and with serrated margin; spinule or denticle near base of thicker process. Maxilliped (Fig. 8) with two-segmented sympod, first segment short, subcylindrical, with setule near distal end; second segment powerfully developed, muscular, with two long plumose setae; claw sigmoid, with sharp-pointed tip, hooked auxiliary outgrowth and long plumose seta near base (all plumules omitted in drawings).

First to fourth legs with two-segmented sympods, biramous. Sympod of first leg (Fig. 9) strongly flattened, its medial margin with long lamelliform process (modified seta) fringed by setules; lateral margin similarly fringed; second segment unarmed. Sympods 2–4 with seta on medial margin of first segment and setules on lateral margin of second segment. Rami with armature as follows:

	Endopod			Exopod		
	1	2	3	1	2	3
Leg 1	1-0	1-0	5	0-I	6-II	
Leg 2	1-0	2-0	3,II	0-I	1-I	5,IV
Leg 3	1-0	1-0	2,II	0-I	1-I	5,III
Leg 4	1-0	1-0	1,I	0-I	1-I	5,III

First exopod apparently two-segmented (Fig. 9), its distal segment showing signs of fusion of two components; endopod three-segmented, with finely fringed lateral margins; all setae pinnate. Second endopod (Fig. 10) enlarged and flattened, both rami of second leg with setules on lateral margin; lateral spines of legs 2–4 with fringe on lateral and serrations on medial margins, distally with spur and flagelliform seta. Rami of third leg (Fig. 11) unmodified. Endopod of fourth leg (Fig. 12) somewhat narrower and longer than that of preceding limb. Fifth leg (Fig. 13) two-segmented; first segment short, with spinulated lateral margin and single distal seta; second segment with terminal seta flanked by two spiniform setae; fourth seta subterminal on lateral margin; both margins fringed with setules. Sixth leg (Fig. 14) reduced to three slender setules. Uropod (Fig. 15) shorter than last abdominal segment, slightly tapering, with fringe of setules on lateral margin and with five marginal setae; third and fourth setae from lateral side enlarged, former $\frac{2}{3}$ length of latter; one seta on dorsal surface.

Male (Figs. 16, 17) with relatively narrow cephalothorax. Free thoracic segments with sides barely covering basal parts of legs, gradually narrowing posteriorly. Genital complex slightly longer than wide, subrectangular, somewhat inflated. Abdomen two-segmented. Total body length, without uropodal setae, about 0.85 mm (after Vervoort, 1969).

Bases of first antennae covered by semicircular sclerites. First antenna (Fig. 18) five-segmented; anterior margin of segments 1–3 with 15 uniform plumose setae (three on first segment, eleven on second and one on third); three slender unarmed additional setae on second and one on third segment; one seta on fourth segment and four on terminal. Maxilliped subchelate (Fig. 19); second segment less inflated than in female, with two rows of small denticles some distance from medial margin and two setae at its mid-length; subchela long, curving, with serrated inner margin and two setae near base.

Second segment of first sympod (Fig. 20) with spinulated medial margin, one spine enlarged, with flagelliform seta at tip. Armature of four biramous pairs of legs as follows:

	Endopod			Exopod		
	1	2	3	1	2	3
Leg 1	1-0	1-0	5,I	0-I	1-I	4,III
Leg 2	1-0	2-0	3,II	0-I	1-I	5,III
Leg 3	1-0	1-0	2,II	0-I	1-0	5,III
Leg 4	1-0	1,I,1		0-I	1-0	5,II

Rami of first leg (Fig. 20) not flattened or enlarged, those of second and third legs (Figs. 21, 22) similar to those of first. Endopod of fourth leg with two segments only (Fig. 23). Fifth leg (Fig. 24) one-segmented, spatulate, with two terminal setae and short fringe of setules on distomedial margin. Sixth leg absent.

Comments: The above description is given *fide* Vervoort (1969). The author was unable to secure specimens of *B. soleae* for examination.

About 30 years following its discovery, *B. soleae* became confused with the species now known as *Holobomolochus confusus* (Stock, 1953). T. Scott (1893) identified that species as *B. soleae* and his identification was accepted by most subsequent authors. As the result of this, it is now virtually impossible to disentangle the records of these two species. How widespread this confusion became can be exemplified by the fact that a large collection of supposed *B. soleae*, accumulated over a long time at the Smithsonian Institution, Washington, D.C., was found by this author to consist exclusively of *Holobomolochus confusus*. A similar situation prevailed in other collections, including that made by the author himself.

We are indebted to Stock (1953) for recognizing this confusion of identities. Under the circumstances, however, the best course of action is to wipe the slate clean and to begin compiling our data on *B. soleae* from scratch. For doubtful records of this species, the interested reader is referred to Vervoort (1962, pp. 24–25).

The original description by Claus (1864) was from *Solea solea*. Although it is likely that other fishes also harbour this copepod, no reliable records of such hosts exist as yet. The parasite lives on *Solea solea* along the Atlantic seaboard of Europe and in the Mediterranean. Vervoort's (1969) records came from the Dutch coast. There is no reason to suppose that *Solea solea* in British waters is free from this parasite. (Although Slinn (1970) found none on this fish in the Port Erin area of the Irish Sea.)

Bomolochus bellones Burmeister, 1835

(Figs. 25–37)

Syn.: *Parabomolochus bellones* (Burmeister, 1835)

Female (Fig. 25): Cephalothorax width about twice its length, general outline in dorsal aspect slightly crescentic; anterior margin convex, with deep indentation centrally, posterior margin concave, not overhanging following segment; dorsal surface somewhat inflated (degree of inflation depending on sexual maturity), with well developed longitudinal groove. Rostrum (Fig. 28) with foliolate, sharp-tipped tines connected by strongly sclerotized bar. Second leg-bearing segment short, slightly crescentic, narrower than cephalothorax, dorsally inflated; third leg-bearing segment longer but narrower than second, convex dorsally, sometimes with posterodorsal protrusion; fourth smaller than third, anteriorly narrower (in partly contracted specimens anterior part often not visible), with gently rounded posterior margin, occasionally with posterodorsal protrusion; fifth similar to, but smaller than, fourth. Genital complex about as long as fourth leg-bearing segment, suborbicular. Abdomen three-segmented, segments gradually diminishing in size and tapering. Uropods and egg sacs similar to those of type species. Total length without uropodal setae 1.74–1.87 mm; length of longest uropodal seta 0.46–0.53 mm.

First antenna similar to that of type species but longer and more slender; base covered by sclerite with digitiform outline (Fig. 26); armature similar to that of type species except for that of terminal segment (Fig. 27), consisting of three long and one short apical setae, three subapical setae and one seta at mid-length of posterior margin. Second antenna similar to that of type species. Labrum covered with minute setules. Mandible as in *B. soleae* but with subterminal process relatively smaller; terminal process with finely serrated posterior margin (Fig. 29). Paragnath (Fig. 30) with strong, oval base and digitiform process; narrow, striated ridge running along two margins of process. First maxilla (Fig. 31) small, rounded, with two robust and two slender pinnate setae (pinnules partly omitted in drawing). Second maxilla with broad proximal part and subcylindrical distal part; latter (Fig. 32) with two spiniform setae, one much shorter than other, near distal end and carrying apically two pointed processes with minute denticles on one side and densely crowded spiniform setules on other (only partly visible in drawing). Maxilliped similar to those of type species, but with much longer seta on basal segment.

Armature formula of first four pairs of legs as follows:

	Endopod			Exopod		
	1	2	3	1	2	3
Leg 1	1-0	1-0	5	0-I	6,V	
Leg 2	1-0	2-0	3,II	0-I	1-I	5,IV
Leg 3	1-0	1-0	2,II	0-I	1-I	5,III
Leg 4	1-0	1-0	1,I,1	0-I	1-I	4,III

First leg strongly flattened and enlarged; lateral spine of first exopod segment fairly long (Fig. 33), with finely drawn tip; second exopod segment with five lateral spines, first and last simple, second to fourth varying in shape, with thin flange along one margin (Fig. 34); lateral margin of endopod with fringe of fine setae. (Cressey and Collette (1970) described this leg as having three-segmented exopod, in contrast to Vervoort (1962), who considered it two-segmented. This author found segmentation of exopod indistinct and variable.) Second leg with greatly expanded endopod, third and fourth unmodified. Lateral margins of all endopod segments fringed with setae. Lateral spines of all exopod segments (Fig. 35) similar to one another, slender, with comb-like spiniform setules along lateral margins, blunt distolateral projection and subterminal flagelliform setae (Fig. 36). Fifth leg (Fig. 37) two-segmented; basal segment short, with slender seta on dorsal surface and several coarse denticles on lateral margin; terminal segment long and narrow, in some specimens slightly spatulate, with short spine at about midlength of lateral margin and irregularly spaced coarse denticles proximal to it, distal margin with two spiniform setae flanking longer and slender seta, three patches of spinules on distolateral surface near tip. (Fifth leg shown by Cressey and Collette (1970) differs slightly in details of armature from this description.) Sixth leg three naked setules of about same length and one shorter setule anterior to them.

Male: Not seen. Only available description appears to be that of Hartmann (1870).

Comments: This species, well distinguishable by its general appearance and by the details of its appendages, was discovered in the Helgoland area by Burmeister (1835). It has never been seen in areas other than the Atlantic coast of Europe and the Mediterranean, until Cressey and Collette (1970) searched through many collections of Belonidae, while working on the copepods of these fishes. The results of their researches radically altered our view of *B. bellones* as a species with *Belone bellone* as the only host and with fairly limited distribution range. They found this copepod in samples from numerous North and South Atlantic localities, as well as in those from the Pacific and from Australian waters. Far from being specific to a single host species, it was found on the following belonid fishes: *Ablennes hians, Belone bellone, Hyporhamphus regularis, H. melanochir, B. svetovidovi, Platybelone argalus, Strongylura anastomella, S. leiura, S. incisa, S. marina, S. notata, S. senegalensis, S. strongylura, S. timucu, S. urvillii, Tylosurus acus, T. crocodilus* and *Lhotskia gavialoides*.

In Britain, *B. bellones* was found in the Plymouth area and might be expected to extend wherever its British host, *Belone bellone*, is to be found.

Genus Holobomolochus Vervoort, 1969

Female: Cephalothorax broadest part of body, broader than long, with inflated dorsal surface, sometimes overlapping succeeding segment; with or without rostral tines and dorsal longitudinal groove. Four free leg-bearing segments gradually narrowing, often with dorsal inflations and posterodorsal projections. Genital complex of last thoracic and first abdominal segments fused together. Abdomen three-segmented, sometimes barely longer than genital complex (in *H. scutigerulus*). Uropods well developed and armed. Egg sacs cylindrical, with rounded ends. First antenna with indistinctly segmented basal and three-segmented terminal parts; no modification of 15 plumose setae on basal part. Other cephalic appendages and maxillipeds as in *Bomolochus*. First four pairs of legs biramous; first endopod two-segmented (but three-segmented in *H. confusus*); first exopod one- to three-segmented; all other rami three-segmented (but fourth exopod two-segmented in *H. attenuatus*); species differing from one another in armature of these legs. Fifth and sixth leg as in *Bomolochus*.

Male: Similar to those of *Bomolochus*. Maxilliped subchelate, with claw-like terminal segment. Fourth endopod two-segmented. Leg structure poorly known.

Type species: *Holobomolochus nothrus* (Wilson, 1913).

Table 3. Geographic distribution of *Holobomolochus*.

Species	North Atlantic	South Atlantic	North Pacific	South Pacific	Indian Ocean
H. acutus (Gnanamuthu, 1948)					x
H. albidus (Wilson, 1932)	x				
H. ardeolae (Krøyer, 1863)	x		x		
H. attenuatus (Wilson, 1913)*					
H. confusus (Stock, 1953)	x				
H. longicaudus (Cressey, 1969)			x		
H. nothrus (Wilson, 1913)	x				
H. occultus Kabata, 1971			x		
H. palleucus (Wilson, 1913)	x				
H. prolixus (Cressey, 1969)			x		
H. scutigerulus (Wilson, 1936)	x				
H. spinulus (Cressey, 1969)			x		
H. venustus Kabata, 1971			x		

*The distribution of this species extends along both Pacific and Atlantic coasts of Central America, along the former as far south as Equador.

Comments: This genus was established by Vervoort (1969) during the final stages of his painstaking groping towards an understanding of the intrafamilial relationships of Bomolochidae. It gathers within its boundaries all bomolochid species which closely resemble *Bomolochus* in their morphology but have females without modified setae on the basal part of their first antennae.

In addition to the type, 12 other species have been hitherto included in this genus. It seems certain that many more remain to be discovered. The geographic distribution of currently known species is shown in Table 3.

In British waters only one species of this genus has been found so far. This sole representative, *H. confusus*, might have as yet unknown relatives in the region.

Holobomolochus confusus (Stock, 1953)

(Figs. 38–52)

Syn.: *Bomolochus solae*; of T. Scott (1893) and auct.
Bomolochus confusus Stock, 1953

Female (Figs. 38, 39): Cephalothorax broader than long, anteriorly rounded and with protruding rostral lobe; posteriorly concave; transverse posterolateral welt (Fig. 39) slightly overlapping succeeding segment; no mid-dorsal groove. Rostral tines diverging (Fig. 40), sharp-tipped, abruptly curving, ventrally with three denticles connected with one another by thin flanges (Fig. 41). Second leg-bearing segment only slightly narrower than cephalothorax, with lateral ends rounded and often inflated; posterior margin concave, overlapping third leg-bearing segment. Latter $\frac{5}{8}$ width of second segment, about as long, with convexly rounded posterior margin dorsally covering all, or almost all, of fourth segment and imposing sigmoid curve on body in lateral view (Fig. 39). Fourth leg-bearing segment little more than $\frac{1}{2}$ width of third, short. Fifth segment as wide as fourth. Genital complex suborbicular, length equal to that of first two abdominal segments combined. Abdomen three-segmented, second segment shortest; large part of ventral surface of third segment covered by closely crowded denticles. Uropod (Fig. 52) longer than broad, slightly tapering, with more than $\frac{1}{2}$ of its ventral surface covered with sharp denticles; five marginal setae and one dorsal; third and fourth marginal setae enlarged; fourth double length of third. Total length about 1.5 mm.

First antenna (Fig. 43) with somewhat flattened basal part longer than cylindrical distal part; armature of base 15 unmodified plumose setae, six ventral and two long dorsal simple setae (four ventral and four dorsal according to Pillai (1966)); first segment of distal part with one plumose and two simple setae (three simple setae according to Pillai (1966)); second segment with three setae; terminal segment with four apical, three slender subapical setae and one short seta at midlength of posterior margin. Second antenna with long robust basal segment bearing one slender seta distally; second segment very short, with one seta; denticulation of third and fourth segment arranged in longitudinal rows; lamelliform process with comb-like row of spiniform setae extending to third segment; terminal armature of five hooked spiniform setae and two shorter and slender ones. Labrum, except for midventral line, covered with hair-like setules. Mandible similar to that of *B. bellones* (Fig. 29) but with finer and longer serrations on margins of apical processes. Paragnath (Fig. 42) stout, apically rounded, with flange at base covered with hair-like setules; margin of distal half with single row of minute denticles, its ventral surface with broad strip of fine, fairly long setules; patch of denticles on dorsal surface. First maxilla (Fig. 45) with three long pinnate setae and one short spiniform seta. Second maxilla large, with robust base; apical armature (Fig. 46) of one short spine and two processes: smaller process with rows of slender, fairly long denticles along two margins, larger with posterior margin denticulated (some denticles possibly also present on dorsal surface). Maxilliped with basal segment as stout as second, with naked seta distally; two plumose setae on second segment; sigmoid claw (Fig. 44) with long, slender tip, auxiliary outgrowth at about mid-length and sharp triangular process on proximal half. Armature formula of biramous legs as follows:

	Endopod			Exopod		
	1	2	3	1	2	3
Leg 1	1-0	1-0	5	0-1	1-I	5,III
Leg 2	1-0	2-0	3,II	0-I	1-I	5,IV
Leg 3	1-0	2-0	2,II	0-I	1-I	5,III
Leg 4	1-0	1-0	I,1,I	0-I	1-I	5,III

First leg flattened and expanded, with fairly well marked segmental boundaries on rami; lateral margin of endopod with fringe of setae; lateral spines of exopod of three types (Fig. 47): those of first and second segment with transversely crenellated surfaces and long fine tips; first two of third segment curving, with serrated flanges along convex margins, third straight and simple. Endopod of second leg not enlarged, with fringe along lateral margin; lateral spines of second exopod (Fig. 48) with prominent striated flanges along two margins, conical tip and robust subterminal flagellum; fringe of setae on lateral margin of first exopod segment; lateral seta of third exopod segment half-pinnate. Third leg similar to second, but with different lateral spines of exopod (Fig. 49). Fourth leg with narrow and long third endopod segment and with characteric lateral spines on exopod (Fig. 50). Fifth leg (Fig. 51) two-segmented; short basal segment with slender pinnate seta on dorsal surface and patch of coarse spinules in distolateral corner; spatulate terminal segment with setiferous medial margin; strong spine at about mid-length of lateral margin, proximal to it large patch of coarse setules, distal to it smaller patch of fine setules; tip with two spines (both with serrated margins not shown in drawing) flanking longer pinnate seta (pinnules omitted in drawing). Sixth leg represented by three setae.

Male: Cephalothorax suborbicular in dorsal aspect, with prominent rostral lobe; no pronounced dorsal inflation or overlapping of following segment. Four free leg-bearing segments gradually narrowing, not inflated. Genital complex suboval. Abdomen two-segmented, second segment with proximal transverse row and two distal triangular patches of setae on ventral surface. Total length without uropodal setae 0.7–1.3 mm.

Appendages as in female but maxilliped subchelate; its penultimate segment with strip of blunt denticles along medial margin and two slender setae, one near midlength, other on proximal half; subchela with denticulated inner margin and slender seta near base. Legs differing somewhat in armature and segmentation from those of female; fourth endopod two-segmented, second segment elongated. (Details of leg structure imperfectly known.)

Comments: This description differs in some details from that of Pillai (1966), though based on the same material. The differences, however, are those between the observers rather than between the copepods themselves and cannot be assigned any taxonomic significance.

The author has never seen the male of *H. confusus*. The description above is given *fide* Bainbridge (1909) and Wilson (1911).

H. confusus is a common parasite of the nasal cavities of the cod, *Gadus morhua*. This author found it also in *Melanogrammus aeglefinus*, as many as six specimens being present in a single nasal capsule. (A. Scott (1929) claimed that up to 30 can be collected from a single large cod.) Stock (1959) found it in *Conger conger*. It was, however, the work of Boxshall (1974a) that gave us an idea of the abundance and broad specificity of this parasite. The list of hosts in the North Sea given by that author included the following species: *Gadus morhua* (92% infected); *Merlangius merlangus* (36%); *Cyclopterus lumpus* (36%); *Molva molva* (14%); *Pollachius virens* (4%); *Melanogrammus aeglefinus* (3%); *Trisopterus luscus* (1.5%); *Pollachius pollachius* and *Labrus bergylta* (isolated specimens).

This species has been found so far only in the European part of the Atlantic and in the North Sea and adjacent areas. It is very common in British waters.

Family Taeniacanthidae

Morphology

Numbering currently less than 60 species, the members of this family are very closely related to Bomolochidae. Their morphology, none the less, exhibits some features not encountered among the bomolochids and, on the whole, tends to be rather more variable. This variability might be at least partially due to the fact that the taeniacanthids are more catholic in their choice of hosts. Taeniacanthidae are among those relatively few families that contain species parasitic both on fishes (elasmobranchs and teleosts) and on invertebrate hosts (echinoids).

Like Bomolochidae, all taeniacanthid females possess a cephalothorax, up to four free leg-bearing segments, a genital complex and an abdomen, varying in the number of segments. The most generalized and, at the same time, most prevalent type of morphology is the one shown in Text fig. 27A and B. In this type there are four distinct, leg-bearing, thoracic segments between the cephalothorax and the genital complex, all of fairly uniform structure and all clearly observable in dorsal aspect. Within this relatively simple plan, considerable differences of habitus can result from mere changes in proportions of the free leg-bearing segments. These changes are illustrated in Text fig. 27A and B, the former being a diagram of *Taeniacanthus carchariae* Sumpf, 1871, the latter of several species of *Taeniacanthus* Sumpf, 1871 and of *Parataeniacanthus* Yamaguti, 1939. At one extreme (A) the first three free leg-bearing segments are very large, rivalling the cephalothorax in size, while the fourth and the part of the body posterior to it are greatly reduced. At the other extreme (B) there is a regular reduction in size between the first and the fourth free leg-bearing segments and the posterior part constitutes about half of the total body length. In both, a small genital segment, formed by the coalescence of the last thoracic and the first abdominal segment, is interposed between the thorax and the abdomen. The abdomen is four-segmented, cylindrical or slightly tapering, often carrying denticular ornamentation on its ventral surface and bearing on its posterior extremity a pair of well developed and armed uropods. There is little or no dorsal swelling or lateral expansion in any segment.

There are some noticeable differences from this type in the general appearance of *Taeniastrotos* Cressey, 1969 and some species of *Anchistrotos* Brian, 1906 (Text fig. 27C). Here the cephalothorax often tends to be rather large, with a tendency also to have a shield-like outline, resulting from the protrusion of its posterolateral corners. It may or may not have an anteriorly prominent rostral lobe. The free leg-bearing segments have lobiform lateral expansions in *Taeniastrotos*, but retain their simple structure in *Anchistrotos*. Moreover, in the former genus, the fourth leg-bearing segment expands dorsally and covers most of the fifth segment, while in the latter the fifth segment is fully exposed on its dorsal side. The genital complex and the four-segmented abdomen show no special modifications. Somewhat similar, though differing in proportions, is the general appearance of *Pseudotaeniacanthus* Yamaguti and Yamasu, 1959 (Text fig. 27D).

An interesting modification of the taeniacanthid structural plan is found in the genus *Metataeniacanthus* Pillai, 1963 (Text fig. 27E). The appearance of its sole member, *M. synodi* Pillai, 1963, is dominated by the size of its abdomen, constituting more than half of the total body length. Its first two segments are larger than the cephalothorax and together with their two more modestly-sized companions impart to the female of *M. synodi* a curiously grub-like look.

The type species of the genus *Taeniacanthodes* Wilson, 1936, *T. gracilis* Wilson, 1936, is marked by the coalescence of segments, greatly affecting its general appearance (Text fig. 27F). The third and fourth leg-bearing segments of this species have fused and form a moderately inflated incipient trunk. Fusion of the third and fourth abdominal segments has also occurred, resulting in a three-segmented abdomen. Ho (1972a) placed in *Taeniacanthodes* another species, *T. haakeri*, which differs in its general morphology from the type species and resembles rather the structural plan shown in Text fig. 27D, except for its three-segmented abdomen. Ho's action was prompted by the close similarity between the maxillipeds of the two species and by the presence in both of characteristic posterolateral cephalothoracic processes, unique to these species. Whether these two characters should be allowed to override the essential systematic importance of differences in segmentation is a moot point. At this stage of our knowledge no firm decision can be made on the validity of Ho's treatment of *T. haakeri* and the genus *Taeniacanthodes* must remain a taxon consisting of two rather ill-matched species.

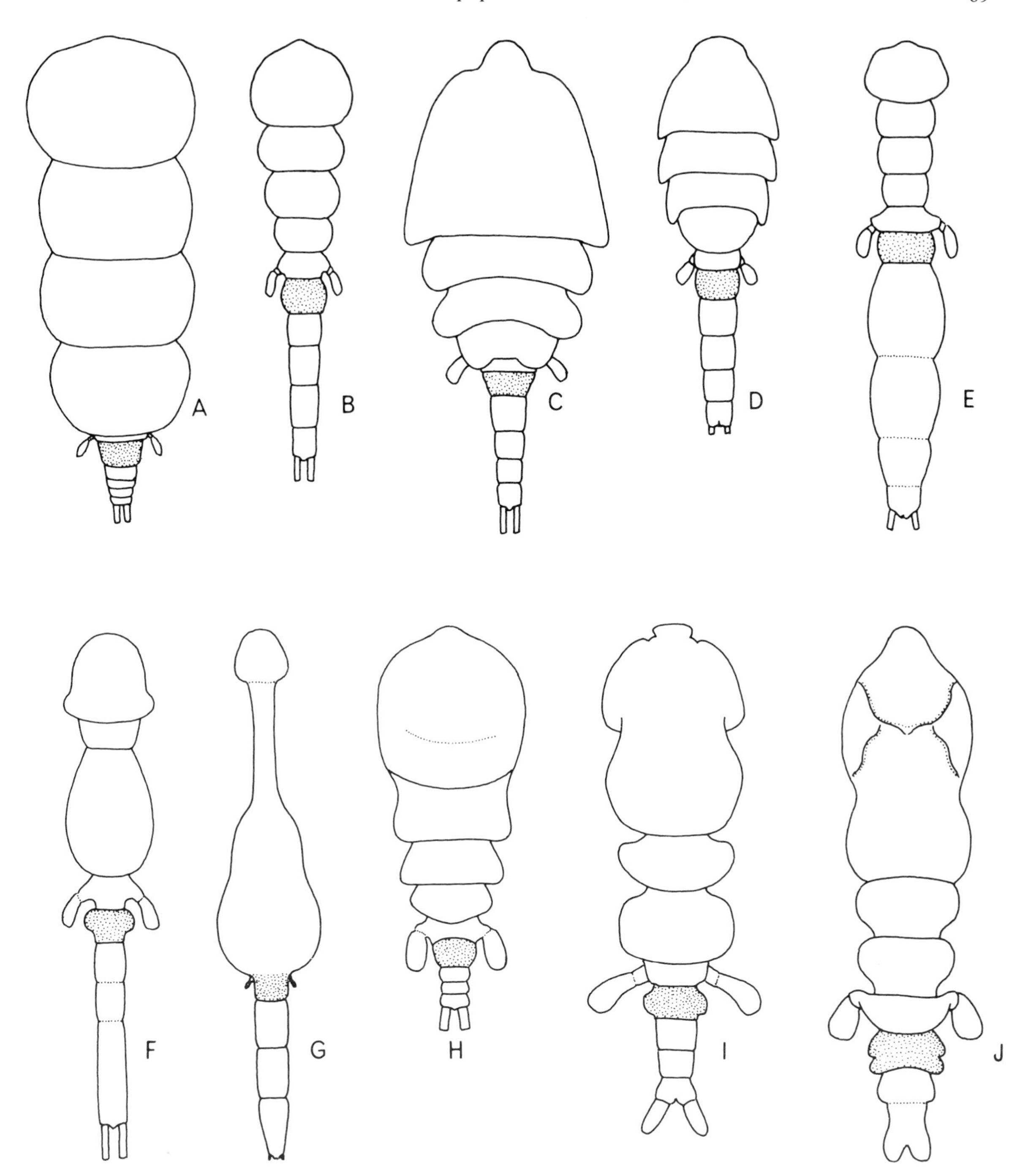

Text fig. 27. Morphological types of Taeniacanthidae. A. *Taeniacanthus carchariae*; B. Several species of *Taeniacanthus* and *Parataeniacanthus*; C. Some species of *Anchistrotos*; D. *Pseudotaeniacanthus*; E. *Metataeniacanthus*; F. *Taeniacanthodes gracilis*; G. *Scolecicara*; H. *Echinosocius*; I. *Echinirus*; J. *Clavisodalis*.

The most extensive modification, indicative of adaptation to parasitism, is shown by the genus *Scolecicara* Ho, 1969 (Text fig. 27G). The anterior part of its only species, *S. humesi* Ho, 1969, up to and including the genital complex, has become completely unsegmented. Its small cephalothorax is followed by a thoracic neck formed mainly from the second leg-bearing segment (as shown by the position of the second pair of legs at its base). The neck expands gradually into a pyriform trunk the narrower anterior part of which is formed by the third and the broader posterior part by the fourth leg-bearing segments. The fifth is telescoped into the posterior extremity of the trunk and is discernible only by the position of the fifth legs at the point where the trunk narrows abruptly into the genital complex. That small tagma, not clearly separate from the trunk, has apparently incorporated also the second abdominal segment. The abdomen, thus, has retained only three segments with clearly drawn borders. It is cylindrical, gently tapering at the tip and constitutes about $\frac{1}{3}$ of the total body length.

The three remaining taeniacanthid genera, all parasitic on echinoids, show a somewhat different type of modification, with a tendency towards enlargement of the cephalothorax by incorporation of the second leg-bearing segment. This tendency is at its weakest in *Echinosocius* Humes and Cressey, 1959 (Text fig. 27H), in which the border between those two parts of the body is still distinguishable, particularly on the dorsal surface. The abdomen is very small and has retained only three segments. The general outcome of these changes is a rather top-heavy appearance.

In the genus *Echinirus* Humes and Cressey, 1959 (Text fig. 27I), the fusion of the cephalothorax with the very large second leg-bearing segment is more complete, the border between them being evident mostly in deep lateral constrictions. As in *Echinosocius*, the abdomen is three-segmented, but it is proportionately larger. The resulting general appearance is more evenly balanced.

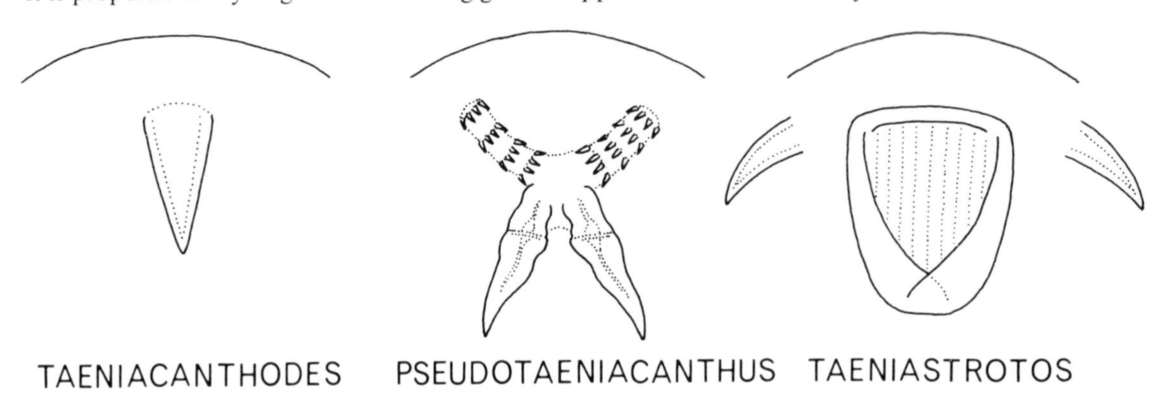

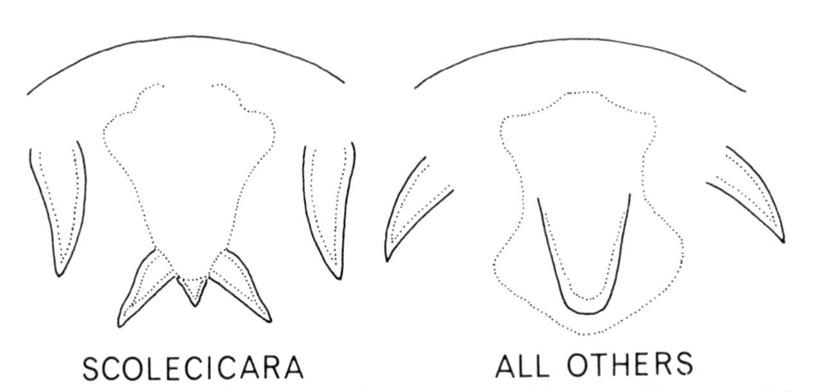

Text fig. 28. Rostral areas in various taeniacanthid genera (diagrammatic).

In *Clavisodalis* Humes, 1970 (Text fig. 27J), the anterior tagma of the body, long and with graceful lateral indentations, demonstrates its composite nature only by the presence of dorsal suture lines, rigidly fused and no longer articulating. There are only three free leg-bearing segments distinguishable. Changes unique among the taeniacanthids have taken place in the abdomen. It consists of only two imperfectly delimited segments. The second segment has become completely fused with broadly rounded uropods, which still retain their complete armature.

One of the significant morphological differences between Taeniacanthidae and their close relatives, Bomolochidae, is the structure of their rostral regions. The bomolochids, as a rule, have rostra equipped with prominent paired tines, while they are equally commonly devoid of postantennary processes. The opposite is true of the taeniacanthids, which also display a greater diversity of structure in this region. Five types of rostrum encountered in this family are shown in

Text fig. 28. In seven genera the rostral plate is devoid of tines, though it usually has a low cuticular ridge, differing in shape and prominence from species to species. All of them are equipped with postantennary processes. In *Taeniastrotos* a shield-like plaque of heavy cuticle occupies the rostral region. In *Scolecicara* there are three rostral tines, a small medial one intervening between two longer ones. Two genera have no postantennary processes. One of them is *Taeniacanthodes*, equipped with a spike-like medial tine in the rostral area. The other is *Pseudotaeniacanthus*. In that genus the two rostral tines are supplemented by two diverging ridges, each with four rows of transversely arranged hooklets.

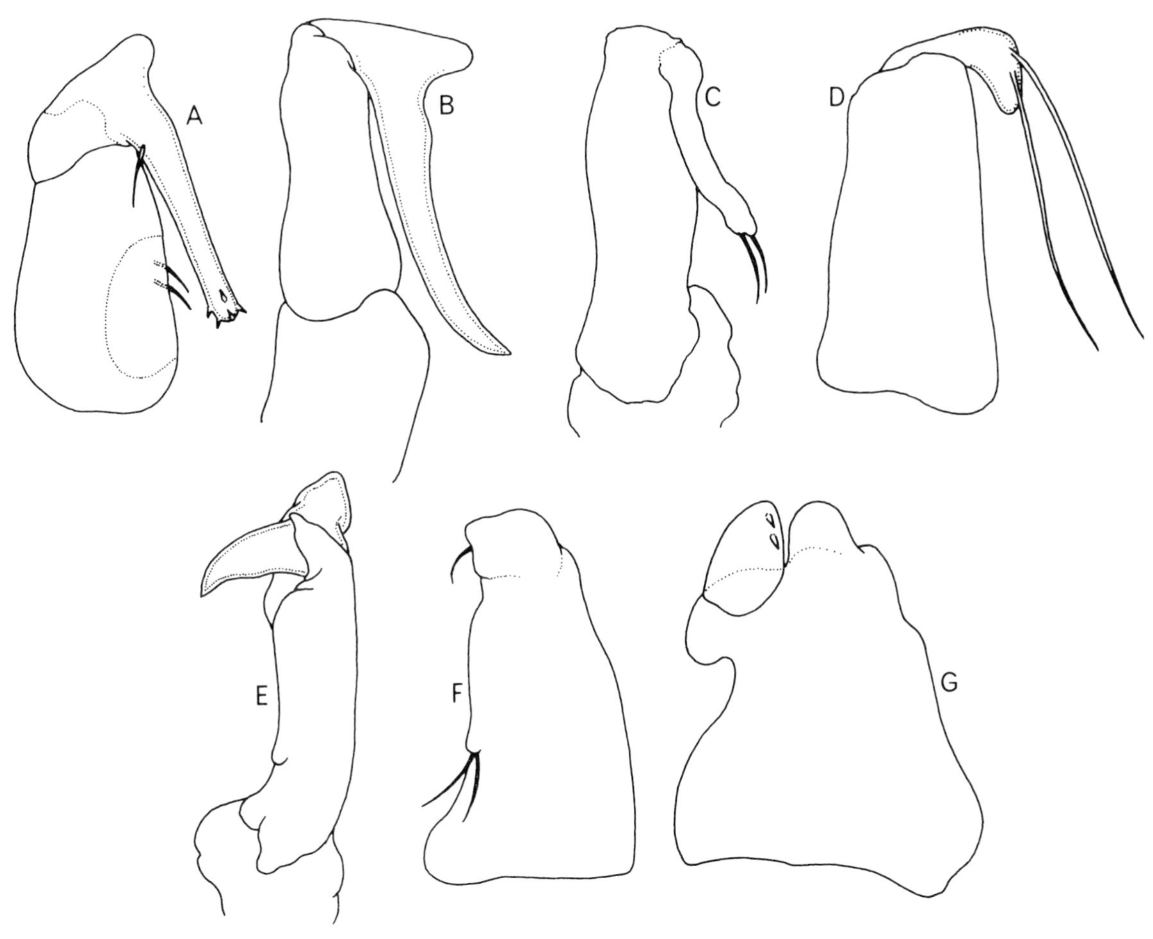

Text fig. 29. Maxillipeds of taeniacanthid copepods. A. *Clavisodalis*; B. Many species of *Taeniacanthus* and *Parataeniacanthus*; C. Some species of *Taeniacanthus*; D. Some species of *Taeniacanthus*; E. *Taeniacanthodes*; F. *Metataeniacanthus, Pseudotaeniacanthus, Echinosocius* and some species of *Taeniacanthus* and *Anchistrotos*; G. *Scolecicara*.

The structure of the taeniacanthid first antenna is also more variable than that of the corresponding bomolochid appendage. The taeniacanthid antenna often, though not always, lacks the long, fused basal part. It tends to be more distinctly segmented, its first two or three segments being usually broader than the cylindrical more distal segments. The number of segments varies from five to seven. (The descriptions are often contradictory and even more often lacking in sufficient detail to allow any certainty in formulating these generalizations.) The second antenna and the mouth parts closely resemble those of Bomolochidae. The maxillipeds, however, are different and distinctly taeniacanthid. To begin with, they are posterior to the mouth rather than lateral as in Bomolochidae (Text fig. 25). They also exhibit a much wider range of structural variations, as shown in Text Fig. 29.

According to Humes (1970), the maxilliped of *Clavisodalis* (Text fig. 29A) is four-segmented, or at least bears traces of its original four-segmented structure. The boundaries between the proximal and distal pairs are much less distinct than the border between the second and third segments. The appendage is subchelate, with a straight subchela, claw-like and bearing denticles on its otherwise blunt tip. Text fig. 29B shows the most prevalent type of taeniacanthid maxilliped, exemplified by many species of *Taeniacanthus* and *Anchistrotos*. This appendage is three-segmented, its first short, unarmed and sturdy segment articulating with the more slender second. The terminal segment is a claw, varying in shape from straight to sickle-like, often ornamented with denticles, serrations or striations. It may have a small secondary process near its tip. The base of the claw has a prominent outgrowth which gives it a likeness to the letter T; it also gives rise to two very long setae (omitted in drawing). In some species, maxillipeds of this type have been described as two-segmented. It is difficult to decide whether this difference in segmentation is real or whether it means nothing more than inaccurate observation. An interesting variety of this type is the maxilliped found in some species of *Taeniacanthus* (Text fig. 29C). The terminal segment of this appendage is not claw-like and appears to be less heavily sclerotized. Its blunt tip bears two setae, which further detracts from its similarity to a claw. The terminal claw of maxillipeds might become reduced, as shown in Text fig. 29D. Another type of maxilliped with a reduced claw occurs in the genus *Taeniacanthodes* (Text fig. 29E). The second segment of this maxilliped is slender and long, comprising most of the length of the appendage. If one accepts this series of types as suggesting a tendency towards reduction of the claw and of the prehensile function of the maxilliped, one can consider as its endpoint the type of appendage shown in Text fig. 29F and occurring in *Metataeniacanthus*, *Pseudotaeniacanthus*, *Echinosocius* and in some species of *Taeniacanthus* and *Anchistrotos*. Described variously as one- or two-segmented, it is clearly no longer prehensile. A rather oddly shaped maxilliped of *Scolecicara* (Text fig. 29G) does not fit easily into this series. It has a broad and squat basal segment bearing distally a small conical protuberance and next to it a segment resembling that protuberance in size and shape and equipped with two apparently sharp denticles. The available description of this appendage does not allow one to judge whether or not that segment is opposable and whether or not, in consequence, the appendage has retained some of its prehensile function. The tendency towards simplification and possibly reduction of the maxilliped among Taeniacanthidae is further implied by the fact that *Echinirus* has lost it altogether.

The number of swimming legs is the same as, and their structure similar to, those of the legs of Bomolochidae. The first legs are flattened and expanded; they form the posterior sealing margin of the subcephalothoracic hollow. The segmentation of their rami is not always clear, but when discernible it shows those rami to be either two- or three-segmented. The remaining legs are not modified and have three-segmented rami, with the exception of *Taeniacanthodes* which has legs with two-segmented endopods and three-segmented exopods. (Humes (1970) reported two-segmented exopods in some abnormal legs of *Clavisodalis*.) The fifth leg is uniramous and two-segmented. The sixth is represented by three setae, in *Metataeniacanthus* borne upon the tip of a small outgrowth, in other genera arising directly from the surface of the genital complex. The structure and armature of the uropods are similar to those of Bomolochidae.

No males have been described for *Metataeniacanthus*, *Taeniacanthodes*, *Taeniastrotos* and *Scolecicara*. Where males are known, their morphology bears a similar relationship to that of the females as has been described for the bomolochids (p. 58). They are usually smaller than the females and their free leg-bearing segments tend to diminish in size more regularly than in the females. For this reason, the male taeniacanthids, more frequently than the females, tend to belong to the "elongate" rather than the "cyclopoid" type of habitus. Their segmentation is usually the same as in the females, though in some genera (e.g. *Anchistrotos*) the male abdomen has one segment less than the corresponding part of the female. Except for the maxilliped, the appendages of the males are also very similar to those of the females. The male maxilliped is typically subchelate, its structure varying from genus to genus mainly in the shape and size of the penultimate segment, which can be short and squat or elongate and slender. In some species the medial margin of this segment carries an elaborately armed and often long process. The swimming legs of the male often have armature differing from that of the female of the species, but their segmentation appears to be the same in both sexes.

History and systematics

While outlining the development of the present-day Bomolochidae (p. 59), the author included a short history of the separation of Taeniacanthidae from that family. When first grouped together as a subfamily of Ergasilidae by Wilson (1911), it contained, in addition to its present type genus,

Taeniacanthus, three other genera: *Irodes* Wilson, 1911, *Phagus* Wilson, 1911 and *Anchistrotos*. Wilson distinguished them with the aid of the following key.

a. Each of the first three free segments as large as the cephalothorax, the four together fully four-fifths of the entire length . . . . **Taeniacanthus**

a. Each of the first three free segments much smaller than the cephalothorax, the four together about half the entire length . . . . b

b. Maxillipeds little larger than the second maxillae, their terminal joint seta-like, pointing inwards and forwards, and covered with hairs, with one or two accessory plumose setae . . . . **Irodes**

b. Maxillipeds much larger than the second maxillae, and armed with a curved claw, or with smooth spines folded back against the basal joint c

c. Mandibles and second maxillae bipartite; maxillary hooks wanting . . . . **Phagus**

c. Mandibles and second maxillae simple; maxillary hooks large and often two-jointed . . . . . **Anchistrotos**

Two of these genera are no longer considered valid. *Phagus* was proposed for *Bomolochus muraenae* Richiardi, incompletely described by Brian (1906) and, at best, acceptable as a species inquirenda. *Irodes* was rejected by Ho (1969) on the grounds that it was never properly defined. He stated: "So far as I am aware, there is no known species of taeniacanthid that can fit perfectly Wilson's definition of *Irodes*, not even the type-species which Wilson assigned to it." The species placed later in *Irodes* made it a quite heterogenous assembly. The author entirely concurs with Ho's decision.

The validity of *Parataeniacanthus* was questioned by Pillai (1963a), who pointed out that its only distinguishing feature is incomplete fusion between the cephalothorax and the first leg-bearing segment. The matter calls for further study. Another genus that would have to be counted among Taeniacanthidae, if it were valid, is *Assecula* Gurney, 1927. A. Scott (1929) correctly pointed out that grounds for its erection were not acceptable.

Wilson's key shown above stressed unduly the taxonomic significance of bodily proportions and relative sizes of appendages. While still acceptable at the specific level, these characters cannot be accepted as discriminants at the generic level. Should one attempt construction of a key to the 11 taeniacanthid genera now known, one would have to base it on the segmentation combined with nature of the postantennary hooks (Wilson's "maxillary hooks"), when present, and with the structure of the appendages. Similar characters were employed by Yamaguti (1963) in his key to seven taeniacanthid genera (including the rejected *Phagus* and *Irodes*).

Since Wilson (1911) first defined his subfamily Taeniacanthinae, no definition of this taxon, now raised to the familial level, has been published. The author proposes this definition to run as follows.

Female: Cephalothorax followed by four leg-bearing segments, in most genera free and well defined, but reduced to three in *Taeniacanthodes* (through fusion of second with third), *Echinirus* and *Clavisodalis* (through incorporation of second leg-bearing segment), or fused into unsegmented neck and trunk (in *Scolecicara*). Genital complex of two to four segments, followed by two- to four-segmented abdomen. Uropods present, well developed and armed, in *Clavisodalis* fused with last abdominal segment. Rostral region with or without tines, postantennary process usually present (absent in *Taeniacanthodes* and *Pseudotaeniacanthus*). First antenna uniramous, with varied, sometimes indistinct segmentation. Second antenna and buccal appendages as in Bomolochidae. Maxilliped posterior to oral region, one- to four-segmented, terminal segment in appendages of more than one segment claw-like (except in *Scolecicara*); no maxilliped in *Echinirus*. Four pairs of biramous swimming legs, first flattened and expanded; fifth leg uniramous, sixth reduced to three setae.

Male: Unknown in *Metataeniacanthus*, *Taeniacanthodes*, *Taeniastrotos* and *Scolecicara*. Segmentation and general structure of known males similar to those of females (abdomen with one segment less in some genera). Appendages similar to those of female but maxilliped typically subchelate. Sixth leg absent.

Type genus: *Taeniacanthus* Sumpf, 1871.

The family is represented in the British fauna by members of two genera: *Anchistrotos* and *Taeniacanthus*. The present state of our knowledge of Taeniacanthidae leaves, for this author at any rate, a question mark over criteria used to distinguish between them. As regards the British members of the family, one can recognize *Anchistrotos* by the presence of a distinct rostral lobe in the centre of the anterior margin of the cephalothorax, while *Taeniacanthus* has no such lobe.

Genus Anchistrotos Brian, 1906

Female: Habitus elongate. Cephalothorax with small but distinguishable rostral lobe. Four leg-bearing segments gradually diminishing in size. Genital complex fused from two segments; abdomen four-segmented. First antenna six- or seven-segmented. Postantennary process well developed, claw-like. Second antenna and mouth parts typical for family. Maxilliped prehensile or not. Swimming legs 1–4 biramous, rami of first two- or three-segmented, others three-segmented.

Male: Similar to female but with three-segmented abdomen and subchelate maxilliped. Sixth leg missing.

Type species: *Anchistrotos gobii* Brian, 1906.

Comments: Brian (1906) described *A. gobii* as a type species of a subgenus constituting part of the genus *Bomolochus*. Wilson (1911) raised it to the generic status, including in it also *Bomolochus ostracionis* Richiardi, 1870, *Eucanthus balistae* Claus, 1864 and *E. marchesetti* Della Valle, 1884. The last of these three was never described.

It has been mentioned above that some difficulties might be experienced in distinguishing *Anchistrotos* from *Taeniacanthus*. Yamaguti (1963) used the presence of the rostral lobe as the main characteristic of *Anchistrotos*. Lewis (1967) pointed out, however, that *A. occidentalis* Wilson, 1924, has no such lobe and that it is not apparent in descriptions of several other species. He also suggested that the uropods of *Anchistrotos* tend to be shorter than those of *Taeniacanthus* but, aware of the variability of this characteristic, doubted its value as a generic discriminant. Some differences in the segmentation of the first antennae have also been suggested as possible future discriminants (Lewis, 1967). More accurate descriptions are, however, necessary for this to be of any value.

Held together by rather tenuous links, the genus *Anchistrotos* now contains 13 accepted species, with one exception parasitic on the gills or in the gill chamber of marine teleosts. *A. gracilis* (Heller, 1865) lives on the gills of an elasmobranch, *Zygaena malleus*. The geographic distribution of the species of *Anchistrotos* is shown below.

Table 4. Geographic distribution of *Anchistrotos*.

Species	North Atlantic	North Pacific	Indian Ocean
A. balistae (Claus, 1864)	x		
A. callionymi Yamaguti, 1939		x	
A. gobii Brian, 1906	x		
A. gracilis (Heller, 1865)		x	
A. hamatus Rounds, 1960	x		
A. laqueus Leigh-Sharpe, 1935	x		
A. moa Lewis, 1967		x	
A. occidentalis Wilson, 1924	x		
A. onosi (T. Scott, 1902)	x		
A. ostracionis (Richiardi, 1870)			x
A. pleuronichthydis Yamaguti, 1939		x	
A. sauridi Pillai, 1969			x
A. zeugopteri (T. Scott, 1902)	x		

(The North Atlantic is understood here as including adjacent seas, such as the Mediterranean.)

While doubts might be entertained as to the generic differences between *Anchistrotos* and *Taeniacanthus*, it is usually quite easy to distinguish between species of *Anchistrotos*. Probably the best characteristic of each species is its maxilliped. This is certainly true of the three British species of the genus, distinguishable with the aid of the following key (to females only).

Key to the British species of *Anchistrotos*

Claw of maxilliped with hooked tip (Fig. 64) . . . . **A. onosi**

Claw of maxilliped bent at about mid-length at obtuse angle, with slender tip and denticles on distal half (Fig. 75) . . . . **A. laqueus**

Claw of maxilliped almost straight, with rather blunt tip, possibly with several denticles near tip (Fig. 78) . . . . **A. zeugopteri**

Anchistrotos onosi (T. Scott, 1902)

(Figs. 53–72)

Syn.: *Bomolochus onosi* T. Scott, 1902

Female (Figs. 53, 54): Cephalothorax with noticeable mid-anterior rostral lobe, rounded lateral and straight posterior margin; mid-dorsal groove extending almost entire length of tagma; dorsal surface almost flat (Fig. 54), ventral concave; strips of transparent membrane along lateral margins; rostral region (Fig. 62) with V-shaped low cuticular ridge and three small cuticular papilliform swellings; first antennae visible in dorsal aspect, extending beyond lateral limits of tagma. Free leg-bearing segments gradually narrowing, imparting regularly elongate general appearance. Second segment with gently rounded lateral margin, in some specimens with incipient posterolateral lobiform projections, posterior margin straight or slightly indented; third segment not more than half length of second; fourth shorter than third, about half width of second; fifth narrower than fourth, with small lateral projections under bases of fifth legs. Genital complex suborbicular, rather shorter than combined length of two preceding segments. Abdomen four-segmented, segments of about equal length, only slightly tapering. Uropods narrow, about as long as, or slightly longer than, last abdominal segment. Total length about 1.3 mm.

First antenna (Fig. 56) of six segments, first two larger than others, second longer than first, remaining four cylindrical and short; first segment with five plumose setae on anterior and proximal margins (plumules omitted); second with nine setae on anterior margin, three on dorsal and two on ventral surface; third with seven on anterior margin and one on dorsal surface; setae all naked on subsequent segments; four on fourth, two on fifth and seven in apical armature of sixth (three long and four short). Postantennary process (Fig. 56) with sturdy base and slender curving tine. Second antenna (Fig. 57) uniramous, with two-segmented sympod; first segment constituting about half of total length of appendage, bearing single seta distally; second short with similar seta; endopod of two segments completely fused, reflected over sympod, with slender, needle-like denticles along outer margin of entire ramus; first endopod segment carrying distally lamelliform process with marginal row of denticles, three hooked spiniform setae with denticle near tip and four slender setae of different lengths. Mouth poecilostome. Mandible (Fig. 58) indistinctly two-segmented, first segment stronger than cylindrical second, distally bearing two laminiform processes, subapical more slender than apical. Paragnath (Fig. 59) digitiform, unsegmented and apparently unarmed. First maxilla (Fig. 61) papilliform, with rounded apex bearing two short naked setae and four long pinnate ones (pinnules omitted). Second maxilla (Fig. 60) two-segmented; first segment robust, second tapering, bearing distally one short spiniform seta and two long processes, one with two rows of denticles, other apparently unarmed. Maxilliped (Fig. 64) three-segmented; first segment short, with one seta; second long, powerful, with two setae on medial margin; third segment claw-like, unciform, with short row of denticles along hook's curvature and two very long setae arising near base, proximal one surmounting small swelling.

First four pairs of swimming legs with two-segmented sympods, first segment with long pinnate seta in distomedial corner, second with lappet-like outgrowth fringed with setules (first leg, Fig. 65) or short seta in same place (legs 2–4, Fig. 66); pinnate seta on medial margin of first leg, no armature in others. First endopod two-segmented, all other rami three-segmented. Armature formula below.

	Endopod			Exopod		
	1	2	3	1	2	3
Leg 1	1-0	7		0-1	1-1	6
Leg 2	1-0	2-I	3,III	0-I	1-I	6,III
Leg 3	1-0	2-I	2,III	0-I	1-I	6,II
Leg 4	1-0	1-I	I,1,II	0-I	1-I	6,II

First leg with rami enlarged and flattened, other legs normal. Setal fringes on lateral margins of endopod segments (Figs. 66, 68) and tufts of setules above lateral spines of exopods (Fig. 67). Spines (Fig. 69) with apical spur and subapical short but sturdy seta, medial margin with several setae and lateral serrated. Interpodal bars of third and fourth legs with fringes of setules. Fifth leg (Fig. 71) with rather indistinct division between its two segments; short basal segment with one slender seta on dorsal surface; slightly peduncular terminal segment with slender pinnate apical seta (pinnules omitted) flanked by spiniform setae with flagelliform tips; similar seta on lateral margin. Sixth leg (Fig. 70) of three slender setae. Uropod (Fig. 72) with four marginal pinnate setae (third and fourth lengthened and thickened) and one seta on dorsal surface.

Male (Fig. 55): Cephalothorax more orbicular than that of female, with more prominent rostral lobe but without posterolateral lobes. Genital complex rather more prominent than in female. Abdomen three-segmented. Total length about 1 mm.

Appendages other than maxilliped similar to those of female. Maxilliped (Fig. 63) three-segmented; basal segment short, squat, with single seta; second segment forming corpus maxillipedis, long, slightly tapering, with two subequal setae near mid-length of medial margin, latter serrated for most of its distal half; terminal segment forming very long, strong subchela, with slightly inflated base bearing two long setae (one each on dorsal and ventral surfaces), most of its inner margin with irregular and coarse denticulation. Border between two segments of fifth leg less distinct than in female; sixth leg absent.

Comments: Described by T. Scott (1902) as *Bomolochus*, this copepod was recognized as a member of the genus *Anchistrotos* by Stock (1953). Its most characteristic features are the female maxilliped and cuticular sculpturing of the rostral region. Taken together they serve to distinguish this species from other *Anchistrotos* in British waters.

The description given above differs in some details from the original description of T. Scott (1902), particularly in the segmentation and armature of the first antenna. Having looked at Scott's material, however, the author is quite convinced that his specimens are conspecific with those of the discoverer.

The second to fourth swimming legs have no armature on medial margins of their second sympodal segments. They all have one long pinnate seta on medial margins of their first endopodal segments. The first leg, on the other hand, has one pinnate seta on the medial margin of the second sympodal segment and only two-segmented endopod. This position of setae suggests that the first endopod segment of the first leg became incorporated in the sympod, helping in the expansion required by the function of the leg.

The specific name of this copepod recognizes the contemporary generic name of its two rockling hosts, *Onos*. In present-day nomenclature, its only known hosts are *Ciliata mustela* and *Rhinonemus cimbrius*. The distribution of *A. onosi*, as shown by finds recorded so far, is restricted to the northern British waters (Firth of Forth, Moray Firth and Shetlands).

Anchistrotos laqueus Leigh-Sharpe, 1935

(Figs. 73–75)

Female (Fig. 73): Cephalothorax bluntly rounded anteriorly, with very small rostral lobe; lateral margin converging posteriorly, posterior margin straight. Rostral region (Fig. 74) with suborbicular ridge of cuticle interrupted at mid-anterior point. Second leg-bearing segment slightly narrower than cephalothorax, short. Remaining three leg-bearing segments tapering gradually, fifth less than half width of second. Genital complex as long as fourth and fifth leg-bearing segments together. Abdomen four-segmented, segments broader than long, tapering, fourth slightly more than half width of first. Uropods as in *A. onosi*. Total length about 0.8 mm.

Appendages, except for maxilliped, not adequately described, given *fide* Leigh-Sharpe (1935a). First antenna six-segmented. Postantennary process similar to that of *A. onosi*. Second antenna and buccal appendages not well known. Maxilliped (Fig. 75) three-segmented (only two segments shown in drawing); second segment long, tapering, with two setae on distal half of medial margin; third segment claw-like, bent at obtuse angle near mid-length, with margin of distal half denticulated, tip slender, attenuated, long conical process near base surmounted by robust seta as long as entire appendage, another similar seta distal to first.

First swimming leg apparently with two-segmented rami, second to fourth with three-segmented. Armature not described. No fringe of setae on third and fourth interpodal bars.

Male: Unknown.

Comments: In spite of its scanty description, *A. laqueus* is quite readily identifiable by the sculpturing of its rostral region and the claw of its maxilliped. The author examined only one wholemount specimen, which did not allow a study of its morphological details. Habitus drawing by

Leigh-Sharpe, redrawn here as Fig. 73, does not show clearly the division between the first leg-bearing segment and the cephalothorax but that segment is obviously separate from the anterior tagma.

A. laqueus is known from two finds. Leigh-Sharpe (1935a) found its type specimens on the gills of *Serranus cabrilla*, taken off Plymouth. Oakley found it again in 1937 on the same host and in the same locality (Leigh-Sharpe, 1939).

Anchistrotos zeugopteri (T. Scott, 1902)

(Figs. 76–78)

Syn.: *Bomolochus zeugopteri* T. Scott, 1902

Female (Fig. 76): Cephalothorax with fairly prominent rostral lobe; lateral margins rounded, with strips of transparent membrane (not shown in drawing); incipient posterolateral lobes in some specimens; dorsal surface convex, with longitudinal medial groove, ventral surface concave. Rostral region (Fig. 77) with shield-like cuticular plaque. Second leg-bearing segment about $\frac{2}{3}$ width of cephalothorax, broader than long, sometimes with small posterolateral lobes. Third leg-bearing segment slightly narrower than second, about $\frac{1}{2}$ its length; fourth segment diminished in similar proportions. Fifth leg-bearing segment narrower anteriorly, with lateral expansions under bases of fifth legs. Genital complex suborbicular, in relaxed specimens with narrow anterior neck. Abdomen four-segmented, segments broader than long, tapering. Uropods well developed, armed as in *A. onosi*. Total length about 0.8 mm.

Except for maxilliped, description of appendages given *fide* Leigh-Sharpe (1939). First antenna six-segmented, first three segments widened and flattened. Postantennary process smooth, elongated and curved. Second antenna four-segmented, with three claws in armature of penultimate segment. Mandible with two subequal terminal blades. Paragnath curving, digitiform. First maxilla with three setae. Second maxilla with one denticulated and one smooth terminal process. Maxilliped (Fig. 78) with short, apparently unarmed first segment; second segment robust, with two setae on distal half of medial margin; third segment heavily sclerotized claw, tapering and almost straight, bearing two setae.

First leg with two-segmented, legs 2–4 with three-segmented rami, armature not described. Fifth leg uniramous, two-segmented; sixth of three setae.

Male: Generally resembling female but with three-segmented abdomen. Maxilliped similar to that of male *A. onosi*.

Comments: Leigh-Sharpe (1939) saw a row of denticles along the distal half of the maxilliped's claw. This author was unable to find them in the scanty material at his disposal. The male maxillipeds, according to Leigh-Sharpe, are asymmetrical, one of the pair being longer than the other and its subchela with different curvature. This author was unable to see, on the dorsal surface of the cephalothorax, the pink eyes described by Leigh-Sharpe.

A. zeugopteri lives on the upper surface of the flatfish, *Zeugopterus punctatus*, especially on its dorsal and ventral fins. It has been recorded from the Clyde estuary (T. Scott, 1902) and from Roscoff in France (Leigh-Sharpe, 1939). The latter find consisted of specimens slightly larger than those constituting the type material, but Leigh-Sharpe found no differences between the two groups of specimens which would warrant their being placed in two different species. A parasite of a rather uncommon flatfish, *A. zeugopteri* must be included in the category of rare species.

Genus Taeniacanthus Sumpf, 1871

Female: Habitus elongate. Cephalothorax without rostral lobe. Four free leg-bearing segments diminishing in size, mode of decrease differing from species to species. Genital complex of two fused segments. Abdomen four-segmented. First antenna five- to seven-segmented. Postantennary process unciform, well developed. Second antenna and mouth parts typical for family. Maxillipeds prehensile or not. Four pairs of biramous swimming legs, rami of first leg two- or three-segmented, others three-segmented.

Male: Similar to female, but with three-segmented abdomen and subchelate maxilliped. Sixth leg missing.

Type species: *Taeniacanthus carchariae* Sumpf, 1871.

Comments: The group of copepods now contained in the genus *Taeniacanthus* consists of 24 species, parasitic mainly on external surfaces and in branchial chambers of various marine teleosts. Several members of the genus parasitize elasmobranchs (*T. albidus* Wilson, 1911; *T. carchariae* Sumpf, 1871; *T. flagellans* Wilson, 1913; *T. narcini* Pillai, 1963 and *T. wilsoni* A. Scott, 1929). They can be said to form two morphological groups (not coinciding with the two types of hosts), depending on the size of their four free leg-bearing segments. At one extreme are species exemplified by the type of the genus, *T. carchariae* (Text fig. 27A), in which the first three free segments are subequal in size and not much smaller than the cephalothorax, while the fourth is diminutive by comparison. At the other extreme are species in which the free segments decrease in size by about equal increments from the first to the last (Text fig. 27B). The groups are not clearly drawn, with intermediate stages present. One such stage is illustrated by *T. sauridae* Yamaguti and Yamasu, 1959, in which only the first two free segments are long, the abrupt change in size coming between the second and the third.

The systematics of this genus rest on rather poor foundations, such important discriminants as the structure of the maxilliped and the armature of the swimming legs having been often inadequately described or neglected in comparisons. Many descriptions are unsuitable for comparative purposes and create conditions favourable for the multiplication of spurious species. A good example of such inadequate description is the type species, *T. carchariae*, described well enough to determine its generic characteristics but in terms general enough to make specific identification difficult. The best proof of this is in the fact that in over 100 years since its description only one worker (Capart, 1959) thought he recognized it (and tentatively at that) in his material. The validity of his claim has been challenged by Ho (1969), mainly on the grounds that Capart's specimens were parasitic on a teleost, while the original description was based on a parasite of a shark. The challenge itself was suggestive of our poor knowledge of the genus, having been couched in terms that had largely nothing to do with morphology. A careful revision of *Taeniacanthus* is obviously needed.

The geographic distribution of the now known species of the genus is shown below.

Table 5. Geographic distribution of *Taeniacanthus*.

Species	North Atlantic	South Atlantic	North Pacific	Indian Ocean
T. acanthocepolae Yamaguti, 1939			x	
T. albidus Wilson, 1911	x			
T. canthigasteri Izawa, 1967			x	
T. carchariae Sumpf, 1871*				
T. coelus Wilson, 1922			x	
T. dentatus Sebastian, 1964				x
T. flagellans Wilson, 1913	x			
T. fugu Yamaguti and Yamasu, 1959			x	
T. indicus Pillai, 1963				x
T. kitamakura Yamaguti and Yamasu, 1959			x	
T. lagocephali Pearse, 1952	x	x?	x	x
T. longicaudus Pillai, 1963				x
T. longichela Yamaguti and Yamasu, 1959			x	
T. monacanthi Yamaguti, 1939			x	
T. narcini Pillai, 1963				x
T. neopercis Yamaguti, 1939			x	
T. pectinatus Yamaguti and Yamasu, 1959			x	
T. pteroisi Shen, 1957			x	
T. sauridae Yamaguti and Yamasu, 1959			x	
T. sebastichthydis Yamaguti, 1939			x	
T. sebastisci Yamaguti, 1939			x	
T. tetradonis (Bassett-Smith, 1898)			x	x
T. upenei Yamaguti, 1954			x	
T. wilsoni A. Scott, 1929	x			

*No locality given in the original description, no other reliable records present.

There is no doubt that more species of the genus *Taeniacanthus* are still to be discovered and that the last word has not been said on the host ranges and distribution of those already known.

Only one species, *T. wilsoni*, is known to occur in British waters.

Taeniacanthus wilsoni A. Scott, 1929

(Figs. 79–87)

Female (Fig. 79): Cephalothorax almost regularly semicircular, about $\frac{1}{3}$ wider than long, with incipient posterolateral lobes and strips of membrane along lateral margins; first antenna visible in dorsal aspect, reaching well beyond lateral limits of cephalothorax. Rostral region (Fig. 80) with low circular ridge of cuticle. Four free leg-bearing segments gradually diminishing in posterior direction; first and second with deep lateral notches between each other and adjacent segments; first slightly narrower than cephalothorax, second only $\frac{2}{3}$ width of first; third and fourth segments with only shallow lateral notches at boundaries, fourth only about $\frac{1}{2}$ width of cephalothorax. Genital complex slightly larger than preceding segment. Abdomen four-segmented, tapering, its third segment being shortest. Uropods about as long as last abdominal segment. Total length 1.22–1.44 mm.

First antenna six-segmented, its third segment longer than others; armature not known in detail. Postantennary process (Fig. 82) long and slender, gently hooked; its moderately inflated base bearing narrow longitudinal flange. Second antenna with well-spaced, slender spines along margin of both endopod segments (Fig. 81) (second segment obscured by lamelliform process in drawing); in addition to process, endopod bearing three hooked, robust spiniform setae and four slender ones. Mandible with robust base; its longer terminal process with single row of denticles along one margin. Paragnath not described. First maxilla with two long and two shorter setae. Second maxilla with three terminal processes (one shorter than other two), all with rows of spinules along two margins. Maxilliped (Fig. 83) with robust corpus bearing two setae on distal half of medial margin; terminal claw slender, slightly curving in distal half, denticulated near tip and bearing two setae (one twice as long as other) near base.

Rami of first four pairs of swimming legs armed as shown below.

	Endopod			Exopod		
	1	2	3	1	2	3
Leg 1	1-0	5,II		0-I	1-I	5,II
Leg 2	1-0	2-0	3,III	0-I	1-I	6,III
Leg 3	1-0	2-0	2,III	0-I	1-I	6,II
Leg 4	1-0	1-0	I,1,II	0-I	1-I	6,II

All setae pinnate except lateralmost half-pinnate setae on distal exopod segments; all spines apparently simple. Second spine on distal segment of first endopod (Fig. 84) easily overlooked. Distal segment of fourth endopod (Fig. 86) elongated, with modified armature (pinnules omitted from long terminal seta). Fifth leg (Fig. 85) two-segmented; first segment short, with one seta on dorsal surface and group of denticles in posterolateral corner of ventral surface; second segment broadly spatulate with indentation in distal margin; group of denticles in distomedial corner, one pinnate seta (pinnules omitted) and two spines on distal margin, one spine on distal half of ventral surface; all spines with serrated margins. Uropods (Fig. 87) slender, tapering, with six marginal setae: first on lateral margin, second to fifth on distal and sixth on medial; third and fourth setae lengthened and thickened.

Male: Unknown.

Comments: The above description is based largely on the original one (A. Scott, 1929). There has been no subsequent description. The author had at his disposal some badly preserved wholemount specimens of the type material. His examination showed some differences between Scott's description and details observed. For example, the spine of the maxilliped, according to the discoverer, is armed with setae along its entire margin (in contrast, see Fig. 83). Other differences also exist and were illustrated, where possible. More material is needed to redescribe *T. wilsoni* accurately.

This copepod was originally described from the gills of *Raja fullonica* taken in the Irish Sea. There were no subsequent records. Perhaps the small size of the parasite and relative inaccessibility of the gills of its host cause it to be overlooked. The author himself was unable to find it on any species of

Raja examined in Scottish waters. Boxshall (1974a) found *Taeniacanthus* sp. on the gills of *Raja naevus* in the North Sea. In his view, it differed sufficiently from the original description of *T. wilsoni* to warrant doubts as to its specific identity. Considering, however, that the description was not very accurate, Boxshall's find might prove to have been *T. wilsoni* after all and so extend both the distribution and the host ranges of this parasite.

Family Ergasilidae

Morphology

The family Ergasilidae groups now over 100 species, parasitic mainly on marine and freshwater teleost fishes. It includes also seven species parasitic on bivalve molluscs. The ergasilids belong to fish parasites morphologically least affected by their dependence on other living organisms. The absence of gross modifications is due to their type of parasitism, phylogenetically probably rather recent and requiring only fairly loose association with the host. The male ergasilids remain free-swimming throughout their lives, while the females, having mated during their free-swimming phase, settle on the outer surfaces and, particularly, on the gills of the fish. The species of the genus *Ostrincola* Wilson, 1933, inhabit the mantle cavity of their bivalve hosts and *Teredophilus inflatus* Harding, 1964, has become ensconced within the pericardium of an African mussel.

The morphology of the ergasilid female does not greatly depart from that of a free-living cyclopoid. At its most primitive (Text fig. 30A), the female has the cephalosome incorporating the maxillipedal segment (almost always devoid of that appendage), but separated more or less distinctly from the first leg-bearing segment. That segment is followed by four free thoracic leg-bearing segments, a genital complex which is a tagma formed by the fusion of the last two thoracic segments and by a three-segmented abdomen. The general appearance of the copepod depends on the proportions of the constituent segments. The type shown in Text fig. 30A might be equated with the "cyclopoid" facies of Bomolochidae (see p. 55), while an equivalent of the "elongated" facies is shown in Text fig. 30B, otherwise being very much like the type of structure shown in Text fig. 30A. (The second leg-bearing segment in this figure is indicated by a dotted line and the genital segment by stippling.) Forms intermediate between A and B, and other similar forms also occur in the family. They are found among the species of *Ergasilus* von Nordmann, 1832, as well as in *Paraergasilus* Markevich, 1937, *Neoergasilus* Yin, 1956, *Ostrincola* and *Sinergasilus* Yin, 1949. (The segmentation of the last-named genus is sometimes obscured by the presence of so-called "false rings".)

In many species of *Ergasilus* the border between the cephalosome and the first leg-bearing segment disappears, partially or completely, and a true cephalothorax is formed (Text fig. 30C). In some species of that genus this cephalothorax expands into a trunk-like tagma accommodating large internal genitalia (Text fig. 30D). Such expansion is sometimes accompanied by crowding of the remaining leg-bearing segments. The fifth leg-bearing segment, usually the smallest of the free thoracic segments, occasionally loses its border with the preceding segment. In other species the third and fourth leg-bearing segments become nearly superimposed upon each other in the dorsoventral plane, so that the fourth leg becomes more dorsal than posterior to the third.

The development of the trunk reaches a very high degree in the genus *Thersitina* (Norman, 1905). The only species of this genus *T. gasterostei* (Pagenstecher, 1861), when seen in dorsal aspect, appears nearly spherical, but its lateral view (Text fig. 30E) shows that its trunk is longest along its dorsoventral axis and is not as uniformly shaped as its dorsal view suggests. The trunk of *T. gasterostei* incorporates also the second leg-bearing segment (as suggested by Gurney (1933)). In some individuals of that species the fifth leg-bearing segment becomes indistinctly separated from its predecessor.

Incorporation of the free thoracic segments into the cephalothoracic tagma is at its greatest in *Teredophilus* (Text fig. 30F). In this genus only the fifth leg-bearing segment remains discreet, all the others being recognizable only by their terga, placed on the dorsal surface of the smooth, lemon-shaped tagma.

All the ergasilid genera reviewed above could be fitted into a more or less logical series, showing progressive incorporation of the leg-bearing segments into the cephalothorax in the process commonly referred to as cephalization. The next two genera cannot be fitted into this series. One of them, the monotypic genus *Nipergasilus* Yin, 1949 (Text fig. 30H) is characterized by an "out of sequence" fusion of the fourth and fifth leg-bearing segments and their enlargement into a modestly-sized trunk. On the other hand, the first leg-bearing segment remains free of the cephalosome. The other genus, *Pseudergasilus* Yamaguti, 1936 (Text fig. 30G), is distinguished by an almost complete confluence of all its segments. One could see in it a logical next step to follow the morphological condition of *Teredophilus*, were it not for the fact that no trunk-like expansion has taken place in any of its segments. On the other hand, it could have arisen as the result of a simultaneous weakening of all segmental boundaries, the effect of a single genetic process.

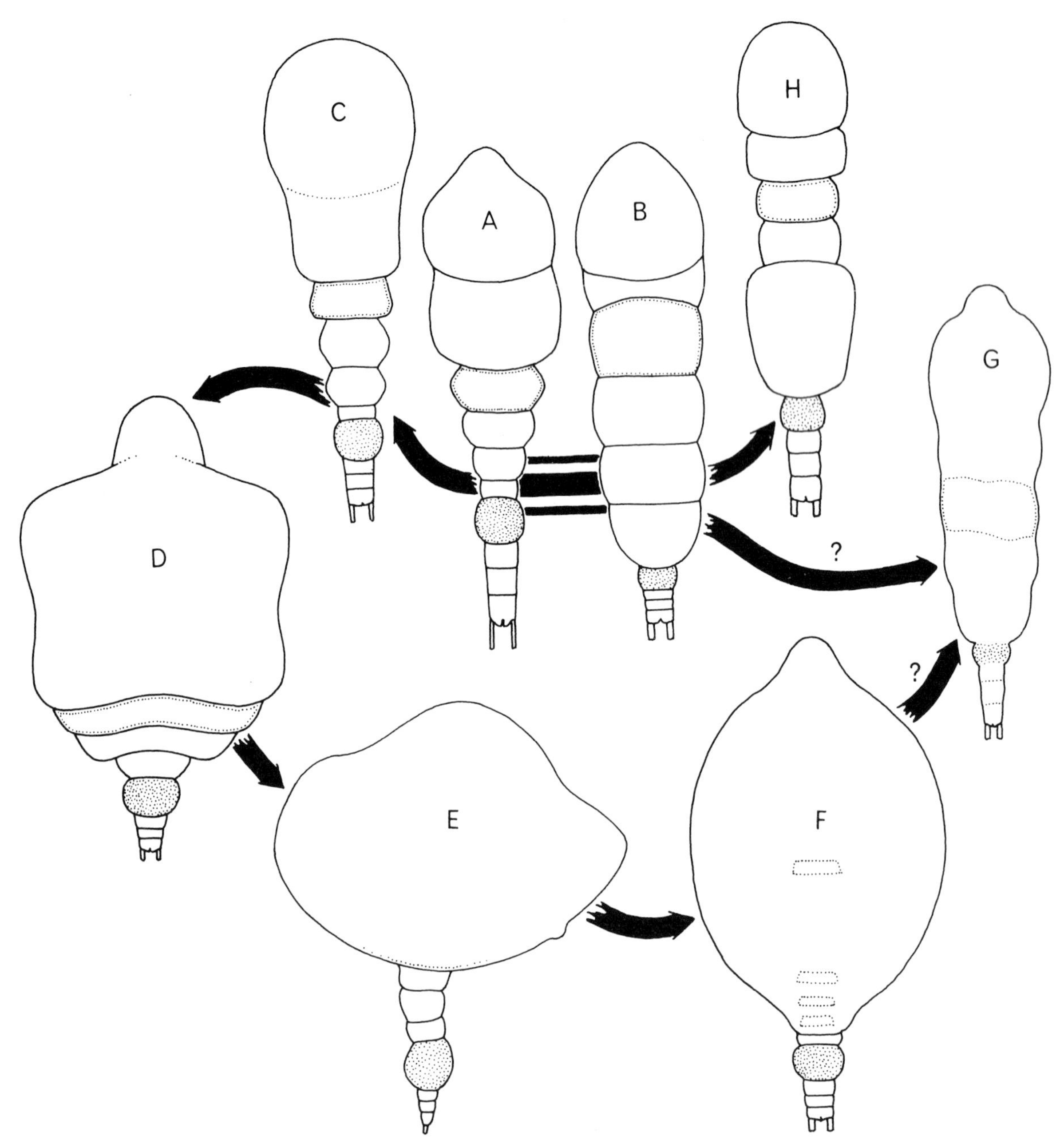

Text fig. 30. Morphological types of Ergasilidae. A and B. Forms occurring in *Ergasilus* and some other genera; C and D. Some species of *Ergasilus*; E. *Thersitina*; F. *Teredophilus*; G. *Pseudergasilus*; H. *Nipergasilus*.

It must be remembered that the above discussion applies only to fully mature females, often very unlike the young adults, particularly in those genera that develop large trunks.

In the structure of their appendages, the females of this family differ but little from one another. As an unfortunate consequence of their diminutive size, they have often been inadequately or incorrectly described and this renders their comparisons difficult. A broad picture can, however, be drawn.

The first antenna is a simple uniramous appendage that can be exemplified by that of *Ergasilus sieboldi* von Nordmann, 1832 (Fig. 90). It is always well armed with setae and is most commonly six-segmented, though five-segmented first antennae occur among species of *Ergasilus*, as well as in *Thersitina* and *Teredophilus*. *Ostrincola* has a seven-segmented first antenna. The second antenna is also uniramous (Fig. 91) and subchelate, the subchela being the third segment, bearing a claw and opposable against the two-segmented stem, with free articulation between the first and the second segments. The differences between the second antennae of various genera and species are mainly in the shapes, sizes and proportions of their component segments. (Some authors regard the claw as the fourth segment. This opinion is not acceptable, since the second antenna of *Paraergasilus* has on its subchela three thin claws in place of the more usual single one. The claw, then, is only the armature carried by the subchela segment and not an independent segment.) The mouth parts are of poecilostome nature. They are minute, usually located near the centre of the ventral surface or somewhere in its anterior half, at the apex of a small prominence. The mandible is two-segmented, with a falcate terminal segment, and may bear a palp on the basal segment. The first maxilla is short, suborbicular and usually has two distal setae and a pointed distomedial projection. The second maxilla is also falcate, armed with dentiform setae and fairly uniform in structure throughout the family. The maxilliped is missing in all females. The only remnants of this appendage are present in the form of small cuticular swellings in *Sinergasilus*, while in *Teredophilus* only faint marks on the cuticle are said to point out the place where this appendage should have been situated.

The structure of the four biramous pairs of swimming legs is also much the same in all ergasilid genera. Their structural pattern is that of a two-segmented sympod and three-segmented rami, except for the two-segmented exopod of the fourth legs. (For examples of typical leg structure see Figs. 95 and 97.) There are exceptions to this prevailing number of segments. In *Ergasilus* several species have a two-segmented first endopod. The fourth leg of *Neoergasilus* has a one- to two-segmented exopod and a one- to three-segmented endopod; the same limb of *Nipergasilus* has both rami two-segmented. Uniformly three-segmented rami occur on all the biramous legs of *Ostrincola*. Some legs display unusual structural features. For example, the first leg of *Neoergasilus longispinosus* Yin, 1956, has a prominent extension on the second sympodial segment between the two rami. (This structure is not unique to *N. longispinosus*.) The second exopod segment of that leg has a flap-like process extending laterally and as large as the segment itself. The fifth leg varies from a short seta, sometimes with a papilliform base (e.g. in *Thersitina*), to a two-segmented appendage with spatulate second segment (in *Ostrincola*). There is no sixth leg. All genera have well developed uropods.

History and systematics

The history of this family begins with the description of the first two species of *Ergasilus*, *E. sieboldi* and *E. gibbus*, by von Nordmann (1832). In his view this genus of copepods "appears at first sight to have much in common with *Cyclops* and to form, to some extent, a form intermediate between the orders Lophyropoda and Poecilopoda. Since its first pair of legs has been modified into true attachment organs, while the remaining appendages serve the function of swimming, it must be numbered among the members of the latter order." Nordmann did not create a family group taxon for *Ergasilus* and expressed no other views than the above quoted one as to its affinities. In the same publication there appeared a description (not accompanied by illustrations) of *Bomolochus parvulus*, thus establishing the genus *Bomolochus*, destined to be associated with *Ergasilus* in the same family for the next few decades. Wilson (1911) interpreted Nordmann as recognizing the close relationship between the two genera. It is rather doubtful if one can place this interpretation on the following sentence of Nordmann's (1832, p. 135), the only one referring to the affiliations of *Bomolochus*: "A parasitic copepod which forms the best intermediate stage between the two above-named genera [*Ergasilus* and *Caligus*] and which in more than one way resembles also the genus *Nemesis*."

A higher taxon for *Ergasilus* was, however, created within the next three years by Burmeister (1835), whose scheme of copepod classification contained a family Ergasilina. His definition of the family, being the first in the field, is worth quoting in full.

"The genera of this family, of which only the females are known, have large cephalothoraces, their bodies consisting of, at the most, eight segments; as the position of the egg sacs shows, five of these segments belong to the thorax and three to the tail or hindbody. On the cephalothorax there are present one or two pairs of feelers consisting of several segments, the posterior ones commonly having the appearance of hooked attachment organs, while the anterior consist of four to twelve segments. Eyes are usually missing, rarely one or two being present on the forehead. The mouth, situated between the most anterior legs, is beak-like; either short, barely elevated, or long, thin and equipped with tentacles. One observes four to six pairs of feet. The four segmented pairs of legs, present on the first four segments behind the cephalothorax, are hardly ever missing. On the other hand, the hooked feet on the cephalothorax itself are not always present. The egg sacs are either saccular or thin, filiform, decreasing in thickness posteriorly. They are suspended from the fifth segment behind the cephalothorax. The filiform egg sacs are divided into compartments; the saccular have separate shells for each egg. The young are born with three pairs of swimming legs; their further development is not yet known."

This definition was broad enough to accommodate in the family, in addition to *Ergasilus*, also *Nicothoe*, *Bomolochus*, *Lamproglena*, *Anthosoma*, *Dichelesthium* and *Nemesis*. The last three, later to become members of the family Dichelesthiidae *sensu lato*, were grouped in a separate division B of Ergasilina, the remaining genera constituting division A. The members of the two divisions were distinguished by the structure of the mouth, those of division A having short, slightly protruding mouths, not proboscides, while those belonging to B had mouths of definite proboscis type. These features, later recognized as having discriminant value at the subordinal level, were used by Burmeister as a mark of intrafamilial rank.

The well known Danish researcher, Krøyer (1837–1838) appears to have adopted Burmeister's concept of the family, adding several species and contributing to our knowledge mainly by his discovery of the males of *Ergasilus*. His later work (Krøyer, 1863–1864) did not add anything new.

Edwards (1840) created for Ergasilina a new family name: Pachycephala. He divided the family into two tribes, corresponding almost exactly to Burmeister's divisions A and B. The infrafamiliar name "Ergasiliens" was restricted to *Ergasilus*, *Bomolochus* and *Nicothoe*. *Lamproglena* was transferred to "Dichelesthiens" and remained there, with *Anthosoma*, *Dichelesthium* and *Nemesis*, until well into the 20th century.

The name Ergasilidae was used in its present form by Baird (1850) and Dana (1852, 1853). The former author, who dealt with British fauna only, included in it, as its sole British representative, the genus *Nicothoe* and gave scanty diagnosis that did not add to the development of the family concept. Dana used the term "tribe" as a suprafamilial designation. He created a tribe Ergasiloidea, consisting of three families: Monstrillidae, Ergasilidae and Nicothoidae. In his view, Ergasilidae were close to the free-swimming Corycaeidae. Dana was the first to express doubts about the affinities of *Bomolochus*. The differences in the structure of the second antennae of *Bomolochus* (Fig. 4) and *Ergasilus* (Fig. 91) provided ample grounds for those doubts.

In Thorell's (1859) well-known work, Ergasilidae found themselves in Poecilostoma. Though he did not deal with this family at any length, he did include in it a newly discovered genus *Lichomolgus*, associated with the ascidians.

At about the same time, Steenstrup and Lütken (1861), who grouped copepods according to the type of their egg sacs, placed their "Ergasilerne" together with other copepods carrying saccular egg sacs with multiseriate eggs. Besides *Ergasilus*, the family contained *Bomolochus*, *Doridicola* and *Nicothoe*, clearly a heterogenous assemblage and a retrograde step in comparison with Dana's (1852, 1853) concept of the ergasilids.

Worthy of note also is the fact that in the same year, Pagenstecher (1861) described *Thersites gasterostei*, the type of the present-day *Thersitina*, still a valid genus of Ergasilidae.

Claus (1864), who looked at the morphology of the appendages in greater detail than most of his predecessors or contemporaries, saw clearly the unique position of *Ergasilus*. He was the first to reject *Bomolochus* and its relatives as members of Ergasilidae. He also denied this relationship to *Nicothoe* and did not accept Thorell's decision to place *Lichomolgus* in this family.

The big monograph of Heller (1865), on the other hand, retained Ergasilidae (as Ergasilina) in the form of a collecting unit for many poecilostome genera then known. Without defining the family, he included in his key to it *Ergasilus*, *Bomolochus*, *Doridicola*, *Thersites* (= *Thersitina*), *Lichomolgus*, *Artrotrogus*, *Asterocheres* and *Nicothoe*. Claus' knowledge was overlooked also by Gerstaecker (1866–1879) in his big compendium, where Ergasilidae, in addition to the above-mentioned genera, included also *Sepicola*, *Eolicola*, *Terebellicola* and *Sabelliphilus*. In the key to this heterogeneous family, however, he picked upon one feature which remains a good discriminant even to-day. By dividing its genera according to the position of their mouths, he separated *Ergasilus* and *Thersites* (mouth near the centre of ventral surface of cephalothorax) from all the others (mouth in anterior position). Apparently, he attached little importance to this difference and left the family very broadly defined.

Hesse (1871) described two genera: *Megabrachinus* and *Macrobrachinus*, which might be said to show some resemblance to the ergasilid structure pattern. Neither his description nor his highly stylized drawings make them identifiable to-day and no finds were reported subsequent to Hesse's publication. (Pearse's (1947) *Macrobrachinus felichthys* was later reduced to synonymy with

Ergasilus.) Under the circumstances, they are best treated as mere historical curios. Yamaguti's (1963) retention of these genera in Ergasilidae, without comments or reservations, is rather puzzling.

No novel approach to Ergasilidae appeared in the literature for the remainder of the century, most authors retaining earlier family structure without, or with only slight, modifications.

The first worker of the 20th century to revise Ergasilidae was Wilson (1911), who maintained the broad concept of the extended family, subdividing it into three subfamilies with the aid of the following key.

a. Second antennae much elongated, forming stout clasping organs, ending in a single strong claw; maxilliped entirely wanting in the female . . . . **Ergasilinae**
a. Second antennae normal size, terminal joint roughened and ending in several claws; maxillipeds present in both sexes . . . . b
b. Maxillipeds turned forward outside, and partly in front, of the other mouth-parts; maxillary hooks not present . . . . **Bomolochinae**
b. Maxillipeds in their normal position behind the other mouth-parts; maxillary hooks present . . . . **Taeniacanthinae**

The later family group taxa were already drawn here, though at a taxonomically lower level, the differences between them being the same as those diagrammmatically shown in Text fig. 25. Wilson's Ergasilinae, the only taxon of interest at this point, contained *Ergasilus*, *Thersitina* and, unaccountably, *Macrobrachinus* (but not *Megabrachinus*).

Further development of our concept of Ergasilidae amounted to gradual vindication of Claus' forgotten views, though without reference to that author. G. O. Sars (1918), having defined the family, added the following comment. "This family also is here taken in a more restricted sense than done by most other authors. Thus I find it necessary to remove the genus *Bomolochus* of Nordmann, which is generally included in the present family, but which seems to me to differ in many points so materially from the other genera, that it scarcely can be associated with them. In the restriction here adopted the family as yet comprises 3 genera, viz., *Ergasilus* Nordmann, *Ergasiloides* G. O. Sars and *Thersitina* Norman (*Thersites* Pagenstecher)."

Sars was almost right. No other genera now contained in the family were known at that time. He could not foresee that his own *Ergasiloides* would be reduced to synonymy with *Ergasilus* by Fryer (1968a).

By the third decade of the 20th century, Wilson (1932) came to the conclusion that his three subfamilies of Ergasilidae should be elevated to the full familial status. His decision found general acceptance with the students of the semiparasitic Cyclopoida, with the notable exception of Yamaguti (1963) who still treats them as subfamilies of Ergasilidae *sensu lato*.

As one would expect, the intrafamilial systematics of Ergasilidae developed with our increased knowledge of those copepods, coupled with discoveries which revealed new members of the family and resulted in amendments to the family diagnosis. Morphological features such as the structure and armature of the swimming appendages, structure of the second antennae, and often the shape of the cephalothorax, constitute the most common discriminants at the specific level. Differences between genera are more often based on the differences in segmentation of the body, though views differ as to the taxonomic value of that feature. Fryer (1965a), for example, attaches little importance to the one-segment difference in the abdomen, originally taken by G. O. Sars (1918) as the main difference between *Ergasilus* and *Ergasiloides*. While this author concurs with Fryer's action of relegating the latter to synonymy with the former, he would hesitate to place in the same genus a species which has only the first leg-bearing segment incorporated in the cephalothorax and another with two such segments incorporated. The reason for this hesitation is the author's belief that the segmental changes in these two examples had different origins. The gradual process of cephalization by absorption of consecutive segments into the anterior tagma is a process of evolution common to parasites and free-living copepods, its every step signifying changes of magnitude that should be almost always recognized as being in the generic category. Other segmental changes are of decidedly minor import and occur more or less haphazardly as unrelated events. An example of such changes are differences in the segmentation of the posterior part of the body among copepods unmistakably belonging to the same genus (e.g. *Ergasilus*).

The matter is not quite straightforward. There is an instance of a single difference in the structure of the second antenna constituting a generic difference (between *Ergasilus* and *Paraergasilus*). We know of some congeners with the first leg-bearing segment either incorporated into the cephalothorax, or only partially fused with it, or even entirely free. These differences can be accommodated within the generic limits, if only because they form a series of gradual changes that is not easily bisected by a definite generic boundary. Even when the first leg-bearing segment appears to be clearly delimited from the anterior tagma, the border between them tends to be obscured on the ventral surface.

A further complication is introduced by frequent occurrence of trunk development, accompanied by cephalization. The process of trunk formation, distending the anterior part of the body, tends also to obliterate boundaries between the component parts of the trunk. A word of caution should be added here as regards the taxonomic value of that structure. Its presence in completely mature females, if not accompanied by segmental differences, can be taken only as a token of specific distinctness, never generic. An example of exaggerated importance attached to the trunk as a taxonomic feature is the creation of *Markewitschia* by Yamaguti (1963). *Ergasilus auritus* Markevich, 1940, an undoubted member of its genus, was removed to this new taxon mainly on the grounds of possession of a large trunk. The error of this decision was pointed out by Roberts (1970).

As stated earlier, the trunk of ergasilid females develops solely in the region of the cephalothorax. In relation to the total body size, it is at its largest in *Teredophilus* which also has the greatest number of segments incorporated in the cephalothorax. Indeed, as reference to Text fig. 30 will show, there are no ergasilids with more than one leg-bearing segment fused with the head that have not developed trunks.

The segmentation, or lack of it, in *Pseudergasilus* is a rather special case. The main feature of that genus is its almost complete loss of all segmental borders *simultaneously*, without cephalization or trunk formation. It seems to this author reasonable to suppose that this change came about through the operation of a mechanism quite distinct from those responsible for all other segmental modifications in the family. Hence he favours retention of this genus, the validity of which was questioned by Fryer (1965a).

As a general principle, the systematics of Ergasilidae should be based on morphological differences, such as segmentation, structure of the appendages and trunk formation, in that order, with attention to interspecific and intergeneric similarities. In particular, the prevalence and constancy of these differences should be kept in mind, when taxonomic decisions are being taken.

Based on its present composition and the foregoing review, the definition of the family can be presented as follows.

Female: Cephalosome separate from first leg-bearing segment, or fused into cephalothorax including that segment, with or without following segments up to fourth leg-bearing segment. Cephalothorax slender, pyriform or fusiform, or inflated into large trunk-like tagma of varying shapes. Leg-bearing thoracic segments, when not incorporated into cephalothorax, always distinctly delimited from one another, except for fifth segment sometimes fused with genital complex (or with fourth leg-bearing segment in *Nipergasilus*). Genital complex of two or more segments, usually fairly small and subspherical. Abdomen of one to three segments. Uropods present. First antenna uniramous, five- to seven-segmented. Second antenna uniramous, three-segmented, subchelate, ending in powerful claw (three claws in *Paraergasilus*); claws sometimes fused more or less permanently in mature females. Oral region in centre of ventral surface of cephalothorax. Mandibles two-segmented, falcate, with palp. First maxilla small, suborbicular, with two terminal setae and distomedial process. Second maxilla two-segmented, falcate. Maxillipeds absent. Four pairs of biramous swimming legs, rami one- to three-segmented. Fifth leg uniramous, one- to two-segmented, commonly reduced to one or more setae. Sixth leg absent.

Male: Generally resembling female but never forming trunk and with abdomen one segment more than in female. Appendages similar to those of female but strong, subchelate maxilliped present.

The family is represented in the British fauna by two genera, distinguishable with the aid of the key below.

Cephalothorax slender, its longest axis in anteroposterior plane; second leg-bearing segment free (Fig. 89) . . . . **_Ergasilus_**

Cephalothorax inflated, subspherical, its longest axis in dorsoventral plane; second leg-bearing segment fused with cephalothorax (Fig. 108) . . . . **_Thersitina_**

Genus Ergasilus von Nordmann, 1832

Female: Cephalosome separate from first leg-bearing segment or fused with it to form cephalothorax; latter oval, pyriform, or with indented lateral margins, usually broader anteriorly; in some species inflated and trunk-like, sometimes broader than long. Four leg-bearing segments separate from one another but fifth sometimes indistinctly divided from, or incorporated in, genital complex; segments gradually decreasing in width posteriorly, or subequal, with abrupt change in width between them and genital complex. Latter two- or three-segmented, usually small and suborbicular (much longer than broad in *E. coleus*). Abdomen two- or three-segmented, segments usually clearly delimited, subequal and diminishing posteriorly. Uropods subcylindrical, with four distal setae (only three setae and long tapering process in *E. coleus* and *E. semicoleus*). First antenna five- or six-segmented. Second antenna with single terminal claw. Mouth parts as in family diagnosis. Biramous swimming legs with two- or three-segmented endopods and one- to three-segmented exopods. Fifth leg uniramous, one- or two-segmented, or reduced to one to three setae.

Male: As in family diagnosis.

Type species: *Ergasilus sieboldi* von Nordmann, 1832.

Comments: The great majority of Ergasilidae belong to this genus, now numbering about 80 species. Within the general limits defined by the generic diagnosis, they display a fairly broad range of morphological variability, beginning with the presence or absence of segmental boundary between the cephalosome and the first leg-bearing segment and extending through the proportions of various parts of the body and the number of abdominal segments to differences in the structure of the appendages.

The most striking feature distinguishing some *Ergasilus* species is a comparatively massive trunk, shown in Text fig. 30D for *E. auritus*. It occurs also, in different sizes and shapes, in *E. cotti* Kellicott, 1892, *E. nerkae* Roberts, 1963, *E. tumidus* Markevich, 1940, *E. wilsoni* Markevich, 1933, and several others. In some species the inflation of the anterior tagma is demonstrable without attaining the size that would justify its classification as trunk (e.g. in *E. coleus* Cressey, 1970).

The free leg-bearing segments are almost always of simple structure, but in some species the first one of them (the second leg-bearing segment) is sometimes adorned with a pair of dorsolateral processes, small and more or less rounded (e.g. in *E. nerkae*).

It has been mentioned earlier that three is the most common number of abdominal segments for the female *Ergasilus*. There are, however, three types of genital-abdominal region. It is at its most abbreviated (Text fig. 31C) in those species that were originally described as *Ergasiloides*. Those species have compact suboval genital complexes, including an additional abdominal segment, and two-segmented abdomina. (The original description (Sars, 1909) showed only a one-segmented abdomen, but later work (Fryer, 1965a) established the existence of two segments.) The genital complex is most extensive in *E. coleus* (Text fig. 31B), in which it incorporates the fifth leg-bearing segment and one additional abdominal segment to form a substantial tagma distinguished by transverse wrinkling. The majority of species, however, have compact genital complexes and three-segmented abdomina (Text fig. 31A).

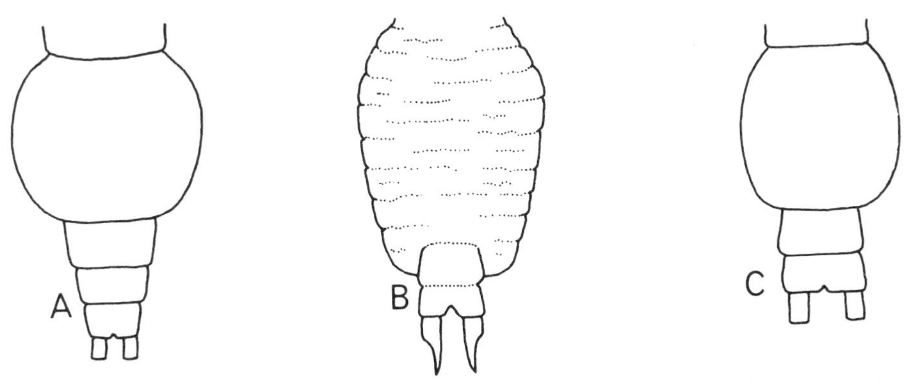

Text fig. 31. Three types of abdomen in genus *Ergasilus*. A. Most species; B. *E. coleus*; C. Species formerly placed in *Ergasiloides*.

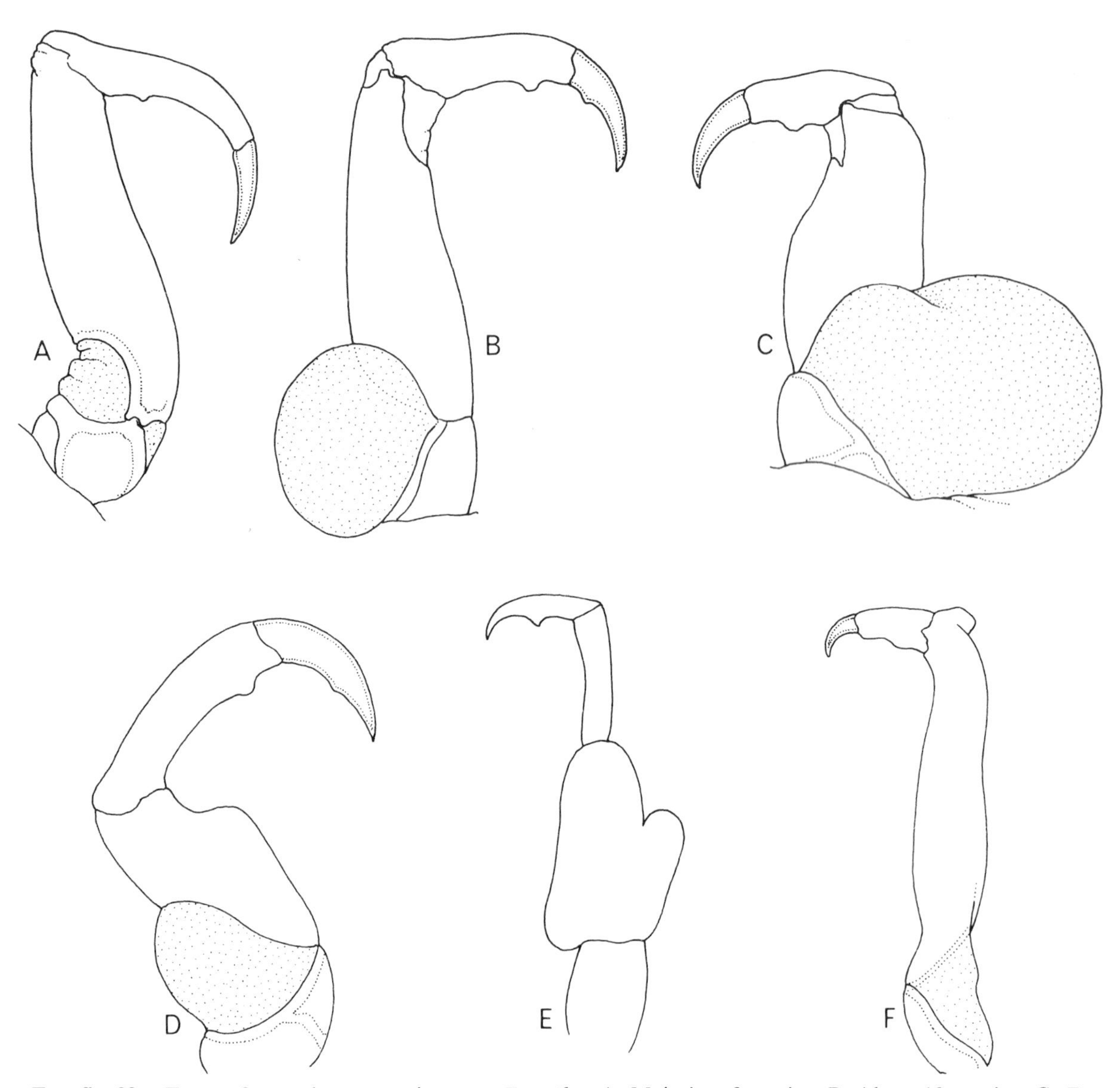

Text fig. 32. Types of second antennae in genus *Ergasilus*. A. Majority of species; B. About 10 species; C. *E. auritus* and *E. cyprinaceus*; D. *E. caeruleus*; E. *E. funduli*; F. *E. amplectens*, *E. cunula* and *E. tenax*.

The morphology of the second antenna is quite variable and constitutes one of the best characteristic features for many species. Five, or perhaps six different types of this appendage can be recognized (Text fig. 32), with some intermediates between them. About half of all known species of *Ergasilus* have second antennae of the type shown in Text fig. 32A. Type A antennae vary in length, thickness and proportions of their component segments, particularly in the degree of curvature of the subchela and its claw. (In *E. coleus* and *E. semicoleus* Cressey, 1970, the antennae, otherwise belonging to type A, are enclosed in wrinkled cuticular sheaths. In the former only the claws emerge from the sheaths, piercing the sheaths of the opposite antennae and providing a firmly locking grip. In the latter species the entire subchela protrudes from the sheath.) About ten species have antennae shown diagrammatically in Text fig. 32B. These differ from the type A appendages in having enlarged posterior parts of the first segment (to use Roberts' (1970) terminology). The enlargement is subspherical and varies in size from species to species. Its highest development is achieved in species such as *E. auritus* and *E. cyprinaceus* Rogers, 1969 (Text fig. 32C). *E. caeruleus* Wilson, 1911, has second antennae with characteristic curvature of the medial margin of the second segment (Text fig. 32D). A puzzling second antenna was described by Krøyer (1863) for his *E. funduli*. In the absence of more recent descriptions it is difficult to interpret its structure (Text fig. 32E). Unless one accepts the two distal divisions as the subchela and claw distorted by handling, this antenna has one segment more than all the others in the genus. On the other hand, if these divisions are subchela and claw, then the second antenna of *E. funduli* might prove to be not very different from the antenna type D. Finally, three species (*E. amplectens* Dogiel and Akhmerov, 1952, *E. cunula* (Cressey, 1970) and *E. tenax* Roberts, 1965) have long, slender, second antennae with very short subchelae and claws (Text fig. 32F). Cressey's species has a shield-like process arising at the base of the subchela and surrounding its entire inner surface and a part of the claw (not shown in diagram). Roberts' *E. tenax* has subchela and claw designed to interlock with those of the opposite member of the pair, so as to form a firmly sealed ring of antennae around the gill filament of the host. In *E. amplectens* these appendages are permanently fused with each other in mature females. It should be mentioned also that the antennae are armed with setae and protuberances (on all but the first segment) that vary from species to species and further help in identification.

In contrast with the second antennae, the oral appendages of *Ergasilus* are fairly uniform throughout the genus and are of minor interest from the taxonomic point of view. A good example of their structure will be found in the description of the type species. The main differences occur in the armature of the mandible (e.g. three-pronged spiniform process in the mandible of *E. hypomesi* Yamaguti, 1936) and of the second maxilla.

The rami of the four biramous swimming legs are most commonly three-segmented, except for the two-segmented fourth exopod. A group of more than 10 species, however, differs in having a two-segmented first endopod. There are several species with unusual leg segmentation. In *E. ceylonensis* Fernando and Hanek, 1973, *E. felichthys* (Pearse, 1947) and *E. mendisi* Fernando and Hanek, 1973, the fourth exopod is three-segmented and in the South American *E. iheringi* Tidd, 1942, both rami of the fourth leg are reduced, the endopod being two-segmented and the exopod one-segmented.

The male *Ergasilus* differs from the female mainly in different segmentation of the abdomen and the possession of a subchelate maxilliped. Other minor differences are also present. For many species the males are still unknown, since they are free-swimming and less likely to be captured by the methods used in parasite studies.

The systematics of the genus can be seen as a segment of the familial history and taxonomy discussed earlier. Little can be added to this topic, except to mention that some specialists might take issue with three taxonomic decisions made by this author. The first is the acceptance of Fryer's (1965a) relegation of *Ergasiloides* to synonymy with *Ergasilus*; the second is his retention of *Pseudergasilus* as a valid genus, though some suggestions have been made that its species should be accommodated in *Ergasilus*; the third is the inclusion of *Acusicola* Cressey, 1970, in *Ergasilus*, following the suggestion of Johnson and Rogers (1973).

The members of the genus are widespread both in freshwater habitats and in the oceans of the world. At the time of writing, their numbers in various regions can be summarized as follows.

Freshwater		Marine	
Europe	8	Atlantic	14
Asia	22	Pacific	7
Africa	11	Indian Ocean	3
North America	20	Mediterranean	5
South America	3	Black Sea	4
Australia	1		

Only three species of *Ergasilus* have been recorded from Great Britain so far. Two are freshwater (*E. sieboldi* and *E. gibbus* von Nordmann, 1832) and one mainly marine (*E. nanus* van Beneden, 1870). They can be identified with the aid of the following key.

1. Second antenna with prominent swelling on basal segment (Fig. 2020) . . . . **E. gibbus**
   Second antenna with basal segment not inflated (Fig. 91, 102) 2

2. Fifth leg two-segmented, with two setae on distal segment (Fig. 99) . . . . **E. sieboldi**
   Fifth leg unsegmented, with two apical and one subapical setae (Fig. 107) . . . . **E. nanus**

Ergasilus sieboldi von Nordmann, 1832

(Figs. 88–100)

Syn.: *Ergasilus trisetaceus* von Nordmann, 1832
?*Ergasilus esocis* Sumpf, 1871
Ergasilus surbecki Baumann, 1912
Ergasilus hoferi Borodin, 1915
Ergasilus baicalensis Messjatzeff, 1926

Female: (Figs. 88, 89): Cephalothorax varying in shape from pyriform to violin-shaped, with lateral constriction in posterior half (sometimes elongate oval), more or less depressed dorsoventrally, with small oral cone in centre of anterior half of ventral surface, distant from antennary area; anterior margin rounded, posterior more or less truncated. Second to fifth leg-bearing segments narrower than cephalothorax, gradually diminishing posteriorly; fourth segment shorter than preceding two, fifth very short and narrow. Genital complex from suborbicular to broader than long, broader than fifth leg-bearing segment, with dorsolateral oviduct orifices and four transverse, complete or incomplete rows of denticles on ventral surface (one of them along posterior margin). Abdomen three-segmented, about as long as genital complex, slightly tapering; second segment often shortest; third segment deeply incised in centre of posterior margin, its border with uropods denticulated; rows of denticles along posterior margins of first and second segments also. Uropods (Fig. 100) about half length of abdomen, much longer than broad, somewhat spatulate; apical armature of four setae: medial corner of apex with thickened seta more than twice length of uropod, lateral with thinner seta more than uropod's length, ventral to them two much thinner and shorter setae. (Gurney (1933) described five setae in armature of uropod; his description was not corroborated by this author's observations.) Total length 1.0–2.0 mm.

First antenna (Fig. 90) six-segmented, segments not very clearly delimited, tapering, about equally long except for smaller apical segment; first segment bearing three setae, second eleven setae, third five setae, fourth four, fifth three and sixth six setae (longest seta of fifth segment was described by Markevich (1931, 1956) as belonging to sixth segment); terminal segment bearing three about equally long setae near centre of apex, two shorter ones at anterior side of apex and single seta near midlength of posterior margin. Second antenna (Fig. 91) with short, robust first segment devoid of armature, uninflated, with geniculate bend between it and second segment; latter long, slender, tapering, with short digitiform process on distal half of medial margin; subchela about as long as second segment, slightly curving, somewhat expanded at base and tip, bearing two small papilliform outgrowths on inner margin and powerful, sharp, hooked claw distally. Mouth poecilostome. Mandible (Fig. 92) unsegmented, long, with two distal falciform blades: longer denticulated along both margins, shorter along one margin; on posterior margin of appendage long narrow palp with one row of denticles; on anterior margin sclerotized protuberance differing in shape and size from specimen to specimen (often much larger than in drawing). First maxilla (Figs. 92, 94) very small, suborbicular, with narrow basal part and tapering, fine-pointed distomedial process; apex with two setiform processes of about equal length. Second maxilla (Figs. 92, 93) obscurely segmented, with robust base and tapering, falciform distal process covered with spiniform setae on dorsal surface; setae seen ventrally as two marginal rows. (The author was unable to see division between base and process, described and figured by earlier workers, though such division probably exists.)

Four pairs of biramous swimming legs with two-segmented sympods. First sympod (Fig. 95) with two rows of denticles on lateral margin of first segment and slender seta lateral to base of exopod on second segment. Sympods of other three pairs also with denticulation on posterior margin and ventral surface in median corner of second segments (Fig. 97). Rami three-segmented, except for two-segmented fourth exopod (Fig. 98). Armature of rami as in formula below.

	Endopod			Exopod		
	1	2	3	1	2	3
Leg 1	1-0	1-0	4,II	0-I	1-I	5,II
Leg 2	1-0	2-0	4,I	0-I	1-0	6,I
Leg 3	1-0	2-0	4,I	0-I	1-0	6,I
Leg 4	1-0	2-0	3,I	0-I	5	—

Lateral margins of both rami partially or completely covered by rows of denticles, outer margins of lateral spines of exopod segments finely serrated (Fig. 96). First exopod with additional row of denticles some distance from lateral margins of all three segments. (Markevich (1931, 1956) showed also fringe of setules on medial margin of basal segment of first exopod. Not observed by this author.) All setae pinnate, except for halfpinnate lateralmost setae on terminal segments of all exopods. Fifth leg (Fig. 99) two-segmented; basal segment very small, with lateral pinnate seta reaching distal half of second segment; latter long, suboval, with two long pinnate setae near apex.

Male: Generally resembling female, though smaller and more slender, with narrower cephalothorax; division between cephalothorax and first leg-bearing segment observable in young specimens. Free leg-bearing segments as in female. Genital complex ovate, relatively larger than in female, often with small posterolateral lobes. Abdomen four-segmented. Largest seta of uropod exceeding combined length of uropod, abdomen and genital complex. Total length of body about 1.0 mm.

Second antenna relatively smaller and more slender than that of female, with more prominent processes on subchela and second segment. Other appendages as in female. Maxilliped apparently five-segmented, with short first, long second segment and three-segmented, slender subchela, its first two segments very short; medial margin of second segment denticulated near base of subchela; second segment of subchela bearing single short seta.

Comments: The variability in shape and pigmentation of the cephalothorax, encountered in the course of development and maturation, has plagued the taxonomic history of this species with many doubts and resulted in descriptions of several spurious species, now reduced to synonymy with *E. sieboldi*. The very widespread and common occurrence of this copepod exposed it to the attention of many casual observers and was partly responsible for scanty and inaccurate descriptions which further aggravated the difficulties. Following the recent work of Roberts (1970), which established sound criteria for specific identification of *Ergasilus*, few doubts need be entertained as to the identity of *E. sieboldi*.

This copepod is a parasite of freshwater fishes in Eurasia. Its host species come from many families, the greatest number of them belonging to Cyprinidae. Among other families are Salmonidae, Thymallidae, Coregonidae, Percidae, Esocidae, Siluridae, Gobiidae and Gasterosteidae. There are also records (from the Baltic and Black Seas) of *E. sieboldi* parasitic on migratory fishes taken in brackish waters. Markevich (1931) recorded it from a sturgeon, Petrushevski (1957) from *Caspialosa*, Giesbrecht (1882) from herring (Clupeidae), and Reshetnikova (1955) from a mullet (Mullidae). Möller (1974) listed it as parasitic on *Platichthys flesus* in the eastern part of the Baltic.

Parasitizing many species of economically important fishes, *E. sieboldi* has become a serious pest, especially in the pond fisheries of Europe and Asia, on occasions inflicting serious losses by causing large-scale mortalities. Its harmful effects are augmented by the fact that it is capable of building up large populations on the gill filaments of its hosts (though it is found on other sites on its hosts as well). As many as 5,431 specimens on a single fish have been recorded (Heinemann, 1934). In view of its economic importance, *E. sieboldi* has attracted the attention of many workers, who have studied its biology, especially its host-parasite relationships. It is all the more surprising that its complete life history has not been described until very recently (Zmerzlaya, 1972).

Although the Eurasian character of *E. sieboldi* seems evident, some views have been expressed of the possible occurrence of this species in other continents. Johnson and Rogers (1973) suggested that *E. arthrosis* Roberts, 1969, a North American species, "will eventually be shown to be a part of a polytypic species of *E. sieboldi* or perhaps a member of a closely related complex." Szidat (1956) described *E. sieboldi* var. *patagonicus* from South America. In the absence of sufficiently convincing evidence, no firm opinion can be formed on these suggestions at present.

Common as it is on the European continent, *E. sieboldi* was not recorded from the British Isles until very recently. In his monograph on British freshwater copepods, Gurney (1933) described it as unknown in the British Isles. He mentioned, however, that "it is most probable that it actually does occur or that it may in future be introduced." Fryer (1969a) was the first to make a record of *E. sieboldi* in Yorkshire (Don drainage area), where the parasite was found on the gills of a dead *Salmo trutta*. Since the fish had been recently introduced into the reservoir where it was found dead and since it was uninfected at the time of introduction, *E. sieboldi* must have been present there in some strength. Kennedy (1974) listed *Salmo trutta, Tinca tinca* and *Abramis brama* as hosts of this copepod in Britain and Ireland.

Ergasilus nanus van Beneden, 1871

(Figs. 101–107)

Female (Fig. 101): Cephalothorax pyriform, with constriction marking former border between cephalothorax and first leg-bearing segment (in younger specimens apparently dorsal furrow present at same level); anterior margin rounded, posterior straight or concave. Four free leg-bearing segments gradually diminishing, last of them very small, in some specimens indistinctly separated from genital complex. Latter suborbicular, bearing four incomplete rows of spinules on ventral side. Abdomen three-segmented, tapering, slightly longer than genital complex, all segments with single row of spinules on posterior margin ventrally. Uropods subquadrangular, length about that of last abdominal segment, longer than broad; armature similar to that of type species but largest seta comparatively much longer; crescentic rows of spinules on ventral surface near apex. Total body length 0.8–1.2 mm.

First antenna similar to that of type species, though with less distinct segmentation (Scott and Scott (1913) described it as "apparently only five-jointed", but the setation of their first segment consists of 15 setae, or one more than the number of setae on first two segments in *E. sieboldi.*) Second antenna (Fig. 102) less robust and more slender than that of *E. sieboldi*, with short, slender seta between tip of subchela and base of claw. Mandible (Fig. 104) with larger falciform blade distinguished by pronounced size discrepancy between two rows of denticles; smaller falciform blade with two parallel rows of denticles (only one shown in drawing); teeth of palp blunt and coarse. First maxilla (Figs. 103, 105) with slender distomedial process and apical setiform processes of unequal length. Second maxilla (Fig. 105) with dorsal surface of distal process covered by slender, spiniform setae; border between distal process and base more distinct than in *E. sieboldi.*

Biramous legs as in type species. Spines in armature of rami with broad and serrated cuticular flanges (Fig. 106). Fifth leg (Fig. 107) with no distinct division between its two segments, much longer than broad, distinctly truncated and bearing two apical pinnate setae, one subapical and one basal.

Male: Unknown.

Comments: This copepod has not been studied in sufficient detail. Its type material perished in Louvain during World War II. The author had at his disposal only two damaged females.

As presented in Figs. 103–105, the mouth parts of *E. nanus* differ substantially from those described by Markevich (1956). In particular, the first maxilla of Markevich's description has three, not two, apical setiform processes, and is thus different from all other species of the genus.

E. nanus was first found by van Beneden (1871) on *Mugil chelo* off the coast of Belgium. Two other species of *Ergasilus* are known to parasitize the mugilid fishes: *E. lizae* Krøyer, 1863 and *E. mugilis* Vogt, 1877. Like *E. nanus*, these species are imperfectly known. *E. mugilis* is hardly more than a species inquirenda and *E. lizae* received its first competent treatment only very recently (Roberts, 1970). Faced with three ill-defined species of *Ergasilus* living on the mugilids, Roberts concluded "that *E. lizae* is probably a cosmopolitan parasite of *Mugil* spp." Leaving *E. mugilis* out of consideration, and accepting Roberts' supposition as true, one must, nevertheless, differentiate between *E. nanus* and the chronologically prior *E. lizae*. If nothing else, the presence of a conical process on the second sympodial segment of the second leg of *E. lizae* (observed by Roberts) and its absence in *E. nanus* (observed by this author) indicate that they are specifically distinct from each other.

E. nanus is a parasite with a fairly wide range of hosts. Its best known host species are the mugilids and it has been recorded from five species of Mugilidae. The other large group of its hosts are the gobiids. It has been found on the gills of five species of *Neogobius*, one of *Gobius* and one of *Pomatoschistus* (Sea of Azov). *E. nanus* is one of the few parasitic copepods able to tolerate transition from the marine to a freshwater environment. Its hosts include some clupeid fishes (*Alosa, Caspialosa*) that migrate up rivers, as well as their exclusively marine relatives (*Engraulis*). This tour de force is completed with parasitization by *E. nanus* of completely freshwater species of fish (*Clarias anguillaris, Gobio gobio, Phoxinus phoxinus*).

Most records of *E. nanus* come from the European Atlantic, the Mediterranean, the Black and Azov Seas. In the heartland of the European continent it has been found in Hungary (Pónyi and Molnár, 1969) and Czechoslovakia (Kašták, 1956). Wilson (1923a) recorded it from the Nile (Cairo). It is difficult to decide whether Wilson's (1935a) record from the American side of the Atlantic,

Dry Tortugas, is correct. Causey (1953a) was the only other author who claimed to have found it (under an incorrect name *E. nana*) in that part of the world, off Louisiana. There is a good chance that these two records refer, in fact, to *E. lizae*. More careful taxonomic work is needed if this and similar confusions are to be cleared up.

E. nanus is a rare visitor to British waters, where it occurs on the local species of Mugilidae. T. Scott (1901) found it on *Crenimugil labrosus* off Aberdeen. This author found it in the same area on an undetermined species of *Mugil*.

Ergasilus gibbus von Nordmann, 1832

(Figs. 2017–2031)

Female (Figs. 2017, 2018): Cephalothorax constituting more than half total body length, elongate and narrow, with rounded anterior margin; lateral margins more or less parallel to each other, in many specimens with slight constriction in posterior half; dorsal surface somewhat uneven ("saddle-shaped"), not inflated, ventral with buccal eminence in centre of anterior half. No trace of division between cephalothorax and first leg-bearing segment. Second and third leg-bearing segments much larger than third and fourth. Genital complex small, subspherical, with several incomplete transverse rows of minute denticles on ventral surface. Abdomen (Fig. 2031) three-segmented, total length less than that of third leg-bearing segment, middle segment shortest, third with deep notch in centre of posterior margin; single rows of minute denticles in posterior half of ventral surface (incomplete in third segment). Uropods (Fig. 2031) subquadrangular, with transverse rows of denticles on ventral surface near posterior margin; armature of four setae, one lengthened and thickened, one about half length of first, two short and slender. Total length (without uropod setae) 1.2–2.0 mm.

First antenna (Fig. 2019) six-segmented, segments progressively smaller; first segment with three short setae, second with four short and three very long, third with three long and one short, slender on ventral surface, fourth with two small on anterior margin and two very long on posterior; fifth with two small on anterior and one long on posterior margin; sixth with one short subapical seta and apical armature of four long and three short setae. Second antenna (Fig. 2020) with semispherical inflation of basal segment; second segment long, tapering, with papilliform outgrowth on medial margin; subchela sturdy, with papilliform process near tip on inner margin; claw hooked, smooth (no setule observed between claw and subchela). Mandible (Fig. 2021) long, sturdy, distally with two blades, larger spatulate, smaller tapering, both with setiferous or serrated margins; palp falciform, with one row of denticles; near centre prominent protuberance surmounted by somewhat clavate process. (Impossible to determine whether process hollow or with longitudinal groove on surface.) First maxilla (Fig. 2022) very small, narrow, with two apical processes of subequal length. Second maxilla (Figs. 2023, 2024) robust, subtriangular, with falciform distal blade; lateral surface of blade (Fig. 2023) covered with spiniform setae.

Sympods of four swimming legs two-segmented. Basal segment of first leg (Fig. 2025) with crescentic row of denticles in lateral corner, that of second and third (Fig. 2027) with two similar rows, that of fourth (Fig. 2029) probably with only one (not possible to determine with certainty). Second segment of first sympod with single row of denticles medial to base of endopod and one long seta lateral to base of exopod; those of remaining legs with two rows of denticles. With exception of two-segmented fourth exopod, all rami three-segmented. Armature formula below.

	Endopod			Exopod		
	1	2	3	1	2	3
Leg 1	1-0	1-0	4,II	0-I	1-0	5,II
Leg 2	1-0	1-0	4,I	0-I	1-0	6,I
Leg 3	1-0	1-0	4,I	0-I	1-0	6,I
Leg 4	1-0	1-0	3,I	0-I	5,I	—

Lateral margins of all segments with strips of serrated membrane, also present along one margin of endopod spines (Fig. 2026); rows of minute denticles at base of membranes (not shown in drawings); distal segment of exopod of second-fourth legs with very small spine, often obscured by membrane (Fig. 2028). Fifth leg (Fig. 2030) reduced to small papilla bearing long, slender seta.

Male: Unknown.

Comments: Previous descriptions of this species were too generalized to provide morphological details useful for comparative purposes. The only illustration of the habitus known to the author is that given by the discoverer (von Nordmann, 1832). Markevich (1956) reproduced his earlier figures showing the second antenna and the posterior extremity. The former did not reveal clearly the nature of the basal swelling. The latter exaggerated the lateral ends of the rows of denticles on the last abdominal segment, showing them as quite prominent spines. This is an illusion caused by superimposition of several denticles (not unknown in other copepods). The "spines" disappear on slight rotation of the abdomen. Markevich also differed from this author in his armature formula of the swimming legs.

E. gibbus is a parasite living on the gills of the common European eel, *Anguilla anguilla*. Most of its records come from the coasts of the Baltic and the North Sea, from brackish coastal regions. In or near the Gulf of Neva about 50% of all the eels examined were found to be infected (Markevich, 1956). The claim of Gadd (1904) that he found *E. gibbus* on *Leuciscus rutilus* (= *Rutilus rutilus*) must be considered suspect, since it was not confirmed by extensive subsequent searches in the same area. In Britain it has been recorded very recently (Canning et al., 1973) on an eel in Slapton Ley Nature Reserve.

Genus Thersitina Norman, 1905

Closely related to *Ergasilus*, this genus is monotypic. Its diagnosis, therefore, is the same as that of its only species and will be found below. The name *Thersitina* was first used by Norman (1905) as a replacement for the preoccupied name *Thersites*, originally given to it by Pagenstecher (1861). The general similarity between *Thersitina* and *Ergasilus* prompted some authors to consider them as congeneric. However, morphological differences between them (e.g. structure of the second antenna) are large enough to warrant their retention as independent taxa.

Thersitina gasterostei (Pagenstecher, 1861)

(Figs. 108–128)

Syn.: *Thersites gasterostei* Pagenstecher, 1861
Ergasilus gasterostei Krøyer, 1863
Ergasilus biuncinatus Gadd, 1901
Thersites gasterosteus T. Scott, 1901
Thersitina biuncinata (Gadd, 1901); of Yamaguti (1963)

Female (Fig. 108): Cephalothorax incorporating first and second leg-bearing segments, much inflated, sublenticular, its long axis at right angles to that of remainder of body; anterior surface often bearing prominent hump; antennary region ventral, mouth posteroventral, distant from antennary region; in dorsal aspect shape suborbicular. Three free leg-bearing segments diminishing in size, sometimes not clearly delineated, especially fifth leg-bearing segment. Genital complex (Fig. 111) subquadrangular or narrower posteriorly, with dorsal genital slits and ventral surface armed with about eight transverse rows of denticles. Abdomen (Fig. 111) three-segmented, second segment shortest, third with deep midposterior cleft. Uropods (Fig. 111) about equally long and wide, length about that of last abdominal segment, with truncated distolateral corners; armature of four setae, one thickened and longer than abdomen. Total length of adult female (measured along long axis of posterior part of body and excluding uropod setae) 0.6–1.0 mm.

First antenna (Fig. 112) uniramous, five-segmented, with segments decreasing in size distally; anterior and dorsal surface bearing numerous slender setae of various lengths; first segment with 16 setae, second with four, third with five, fourth with three and fifth with eight; apical armature consisting of four long and three shorter setae, one seta at midlength of segment. Second antenna (Fig. 113) three-segmented, subchelate; basal segment short, unarmed, with geniculate joint between it and second segment; latter long, robust, tapering, also unarmed; subchela very short, distally bearing powerful claw more than twice its own length, curving and sharp-pointed; secondary short claw arising dorsally on distal end of subchela. Mouth poecilostome, at apex of small eminence. Mandible (Fig. 114) strong, curving, distally with large falciform blade bearing few (about 10) coarse teeth along one margin; second blade short, spiniform, toothless; prominent swelling on anterior margin bearing frayed cuticular flange; palp on posterior margin about as large as larger blade, with single row of peg-like denticles. First maxilla similar to that of *Ergasilus*. Second maxilla (Fig. 115) with robust base and narrow but sturdy distal process, turning up at tip and armed apically with five to nine denticles. Maxilliped absent.

Four biramous swimming legs with three-segmented rami, except for two-segmented fourth exopod (Fig. 116). Sympods two-segmented; first segment unarmed, second with slender seta lateral to base of exopod; in first leg also single row of denticles on posterior margin medial to endopod (Fig. 117), in remaining pairs two such rows (Fig. 118). Armature formula of rami as follows.

	Endopod			Exopod		
	1	2	3	1	2	3
Leg 1	1-0	1-0	4,II	0-I	1-0	5,II
Leg 2	1-0	2-0	4,I	0-I	1-0	6
Leg 3	1-0	2-0	4,I	0-I	1-0	6
Leg 4	1-0	2-0	4,I	0-I	6	—

According to Gurney (1933) the second segment of the fourth exopod carries only five setae. Spines and lateral margins of some segments with strips of serrated membrane. Fifth leg (Fig. 119) papilliform, with one seta at apex and one at base.

Male (Figs. 109, 110) (*fide* Gurney, 1933): Cephalothorax fusiform, not inflated, with rounded anterior and truncated posterior margin; only first leg-bearing segment incorporated. Borders between free leg-bearing segments rather indistinct. Genital complex longer than broad, widening posteriorly, with incipient posterolateral lobes. Abdomen five-segmented (Gurney, 1933) or four-segmented (Markevich, 1956). Uropod as in female, but long seta comparatively much longer. Total length without uropod setae about 1.0 mm.

Appendages similar to those of female. Maxilliped (Fig. 110) subchelate, with very long and slender subchela. Single seta on posterolateral corners of genital complex (possibly representing vestigial sixth legs).

Comments: *T. gasterostei* is a common species and is probably better known than most parasitic poecilostomes. Its biology has been studied in detail mainly because its common hosts, the sticklebacks, make very good laboratory animals. Its life history comprises four nauplius stages (Figs. 120–123), followed by five copepodid stages of increasing complexity (Figs. 124–128). The ontogeny of *T. gasterostei* was investigated by Gurney (1913, 1933), the physiology of its host-parasite relationships by Walkey et al. (1970).

Krøyer (1863) and Gadd (1901) placed this species in *Ergasilus*, from which it must be excluded for reasons stated above. Interestingly, having described the species independently from Pagenstecher (1861), Krøyer gave it an identical specific name. Yamaguti (1963) considered Gadd's (1901) species *Ergasilus biuncinatus* as the second species of *Thersitina* but differences between Gadd's specimens and *T. gasterostei* are due more to differences in descriptive conventions than to real morphological dissimilarities.

Until recently, *T. gasterostei* was known only from the Old World, where it occurs mainly in the branchial chambers of *Gasterosteus aculeatus* and *Pungitius pungitius*, in localities as far apart from each other as the British Isles, the Caspian Sea and the Pacific coast of the U.S.S.R. Beside sticklebacks, other fishes are known to harbour this copepod. Markevich (1933) mentioned its occurrence in the Caspian Sea on *Chalcaburnus chalcoides*, *Atherina pontica*, *Perca fluviatilis*, *Gobius fluviatilis*, *Masogobius* sp., *Syngnathus nigrolineatus* and *Pygasteus platygaster*.

Krøyer (1863) recorded *T. gasterostei* in Greenland. Recently, Threlfall (1968) and Hanek and Threlfall (1969, 1970a, 1970b) found it on the Atlantic coast of Canada on *Gasterosteus aculeatus*, *G. wheatlandi* and *Pungitius pungitius*. This author saw it on the three-spined stickleback from a small lake on Vancouver Island, British Columbia. Lester (1974) recorded it on the western Canadian mainland, near Vancouver. With Valdez's (1974) discovery of this copepod on Amchitka (Aleutian Is.) the circumpolar ring has been closed. It is now clear that *T. gasterostei* is cosmopolitan in the higher latitudes of the northern hemisphere.

In Britain, this parasite has been frequently found on *Gasterosteus aculeatus* and *Pungitius pungitius* in many localities, from Scotland to the south of England.

Family Chondracanthidae

Morphology

The often bizarre appearance of an adult female chondracanthid provides little evidence that it has been derived from the primitive podoplean structure by few and relatively simple processes: loss of external segmentation, simplification and/or reduction of natatory appendages and the development of a massive trunk, necessary to accommodate the output of eggs on a scale compatible with the demands of parasitic life. The final outcome of these processes is shown in Text fig. 33. A chondracanthid female consists of three tagmata. The anterior tagma is a cephalosome, or a true cephalothorax. The second and by far the largest tagma is the trunk which consists of the second to fifth leg-bearing segments, or first to fifth in species with the cephalosome as the anterior tagma. The third, small tagma is the genito-abdomen, derived from the genital segment and abdomen of the ancestral free-living poecilostome and usually displaying its dual origin quite clearly. There are many minor and some major departures from this structural plan. They are discussed below in turn for each tagma.

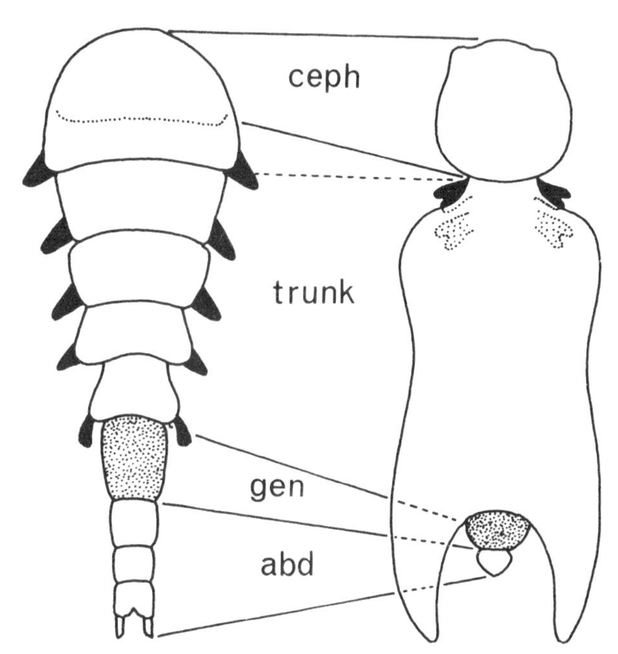

Text fig. 33. Structural alterations required to change a primitive podoplean into a chondracanthid copepod.

Cephalosome or cephalothorax. As many as twenty types of structure of this tagma can be distinguished among the 33 chondracanthid genera now known. In Text fig. 34 they have been arranged, in order, from the most simple to the most complicated. Text fig. 34A shows the most simple type, which is always a cephalosome, usually flat dorsally and covered by a fairly thick dorsal shield with a more or less marked groove in the mid-dorsal line, at least in its anterior half. (Some species have shields without grooves and there is at least one with three such grooves.) It bears no processes, though in some instances it might have small rounded swellings on its ventral surface. Its shape varies quite broadly, but usually there is no great difference between the length and the width of the tagma. As well as being simple, this type is one of the most common in the family, occurring in the genera: *Acanthochondria* Oakley, 1927, *Acanthochondrites* Oakley, 1930, *Andreina* Brian, 1939, *Ceratochondria* Yu, 1939, some species of *Chondracanthodes* Wilson, 1932 and *Chondracanthus* Delaroche, 1811, *Medesicaste* Krøyer, 1863, *Protochondria* Ho, 1970, and *Pseudoblias* Heegaard, 1962. By a pronounced elongation and reduction in width, this type might change into that shown in Text fig. 34B. Examples of such a "long head" are *Heterochondria longicephala* (Yu and Wu, 1932) and *Acanthochondria argutula* (Markevich, 1940). All intermediate stages between them can be found, mainly among the species of *Acanthochondria*.

Similar to the above, but differing from them by partial separation of the anterior part of this tagma from the rest, is the cephalothorax of *Blias* Krøyer, 1863 (Text fig. 34C). In *Chondracanthus ornatus* T. Scott, 1900 (Text fig. 34D) the antennary region is set apart fairly distinctly from the rest of the tagma and cleaved by a rather deep mid-dorsal groove. In addition, this species has a prominent rounded swelling on the dorsal surface. The distinct structure of this type prompted Oakley (1930) to establish a separate genus, *Disphaerocephalus*, for this species. Current views, however, deny the validity of this separation.

In the genus *Scheherezade* Leigh-Sharpe, 1934 (Text fig. 34E) the head is not delineated from the trunk and there is no separate cephalosome. Only by the position of the first (and only) pair of legs can the division between it and the trunk be drawn. The anterior margin is rounded, with some elevation of the antennary region, but the lateral margins continue in a smooth line with those of the anterior, cylindrical part of the trunk.

In the next three genera the anterior tagma is modified by possession of swellings or expansions in the anterolateral corners. In *Berea* Yamaguti, 1963 (Text fig. 34F), these corners are inflated into subspherical protuberances, so that the anterior margin is much broader than the posterior margin of the tagma. In *Pseudochondracanthus diceraus* Wilson, 1908 (Text fig. 34G) these corners are produced into longer and narrower swellings and in the genus *Pseudacanthocanthopsis* Yamaguti and Yamasu, 1959, they form rounded, anteroventrally pointing lobes (Text fig. 34H). A special type, not shown in Text fig. 34, is the cephalothorax of *Pseudodiocus* Ho, 1972. It might be considered as a cross of the types shown in Text fig. 34H and 34J, in that its anterolateral corners expand into prominent trilobed outgrowths, while the posterolateral corners carry simple, short processes.

In some species of *Chondracanthus* (*Ch. lophii* Johnston, 1836 and *Ch. horridus* Heller, 1865) the lateral margins are pushed out into short, posteriorly directed processes (Text fig. 34I), but it is much more common to find in that genus processes of various lengths arising from the posterolateral corners, such as those in *Ch. pinguis* Wilson, 1912 (Text fig. 34J). Similar processes occur also in *Chondracanthodes, Cryptochondria* Izawa, 1971, *Immanthe* Leigh-Sharpe, 1934, *Praecidochondria* Kabata, 1968, *Prochondracanthopsis* Shiino, 1960, *Protochondracanthus* Kirtisinghe, 1950, and *Prochondracanthus* Yamaguti, 1939. In the last-named genus these processes are very small. On the other hand, in *Lateracanthus* Kabata and Gusev, 1966, they are long enough to reach half way down the trunk and broad enough at the base to occupy the entire lateral margins.

In some genera these processes are double, as in *Humphreysia* Leigh-Sharpe, 1934 (Text fig. 34K) and in *Juanettia* Wilson, 1921 (Text fig. 34L). In the latter, however, another pair of small processes flanks the oral region, arising from the ventral surface of the tagma. Unlike those in the above two examples, the posterolateral processes of *Chondracanthus distortus* Wilson, 1922 (Text fig. 34M) are quite separate from each other, the shorter pair being anterior to the longer.

An unusual pair of posterodorsal processes, in addition to the posterolateral ones, occurs in *Pterochondria* Ho, 1973 (Text fig. 34N). (The structure of the anterior tagma of this genus appears to differ considerably with the age of the female. Young modified specimens of its only species, *P. alatalongicollis* (Heegaard, 1940) have no processes on their anterior tagma, its shape being very similar to that shown in Text fig. 34A).

A partial subdivision of the cephalosome into two parts, one comprising the antennary region and the other the oral region, occurs in two chondracanthid genera. In *Brachiochondria* Shiino, 1957, the anterior part carries two prominent processes, often pointing inwards (Text fig. 34O); in similar *Neobrachiochondria* Kabata, 1969 (Text fig. 34P), the oral part of the cephalosome has another pair of processes, one on either side of the mouth and both much larger than the anterior pair.

Rhynchochondria Ho, 1967 (Text fig. 34Q) has a prominent swelling protruding from the centre of the anterior margin in front of the antennae. Two smaller processes are present in the posteroventral corners and a prominent, finger-like process on the ventral surface behind the mouth. This process, extending well beyond the boundary of the tagma has been instrumental in selecting the name for the genus.

In *Diocus* Krøyer, 1863 (Text fig. 34R) the cephalosome, with its small anterolateral tubercles, is separated by a groove from a much larger subspherical swelling formed by the first leg-bearing segment; the latter is itself sharply delimited from the following segments. Perhaps the most peculiar structure of the anterior end of the body is found in the genus *Strabax* von Nordmann, 1864 (Text fig. 34S). A diminutive cephalon is perched atop a complicated structure consisting of three pairs of swellings, two subspherical and one with deeply cleft tips, all separated from the trunk proper by a long cylindrical neck.

The structure of the anterior end is the chief distinguishing feature of the subfamily

Text fig. 34. Types of cephalothorax in Chondracanthidae. A. Simplest type, occurring in many genera; B. "Long head" type of *Heterochondria* and some species of *Acanthochondria*; C. *Blias*; D. *Chondracanthus ornatus*; E. *Scheherezade*; F. *Berea*; G. *Pseudochondracanthus diceraus*; H. *Pseudocanthocanthopsis*; I. Some species of *Chondracanthus*; J. *Chondracanthus pinguis* and some species of several other genera; K. *Humphreysia*; L. *Juanettia*; M. *Chondracanthus distortus*; N. *Pterochondria*; O. *Brachiochondria*; P. *Neobrachiochondria*; Q. *Rhynhochondria*; R. *Diocus*; S. *Strabax*; T. Subfamily Lernentominae.

Lernentominae. All premandibular segments are greatly elongated to form a "cephalic neck" with a bulbous anterior expansion in the antennary region. The mouth and all the buccal appendages are separated from that region by the length of this neck, all the postmandibular segments of the cephalon being absorbed into the trunk so that they cannot be delimited. Two presently recognized genera of this subfamily are: *Lernentoma* de Blainville, 1822 (Text fig. 34T) and *Brachiochondrites* Markevich, 1940. A clear view of the position of the mouth in the former can be seen in Figs. 361, 362.

Trunk. In all genera with the cephalosome as their anterior tagma the trunk consists of five segments (leg-bearing segments one to five); where a cephalothorax has been formed, the first leg-bearing segment remains outside the limits of this tagma. The trunk might or might not be separated from the cephalosome by a narrow "neck" (Ho, 1970, recognizes this as an independent structure but it is in reality only the anterior end of the trunk); most genera have very short necks but in some they have become elongated.

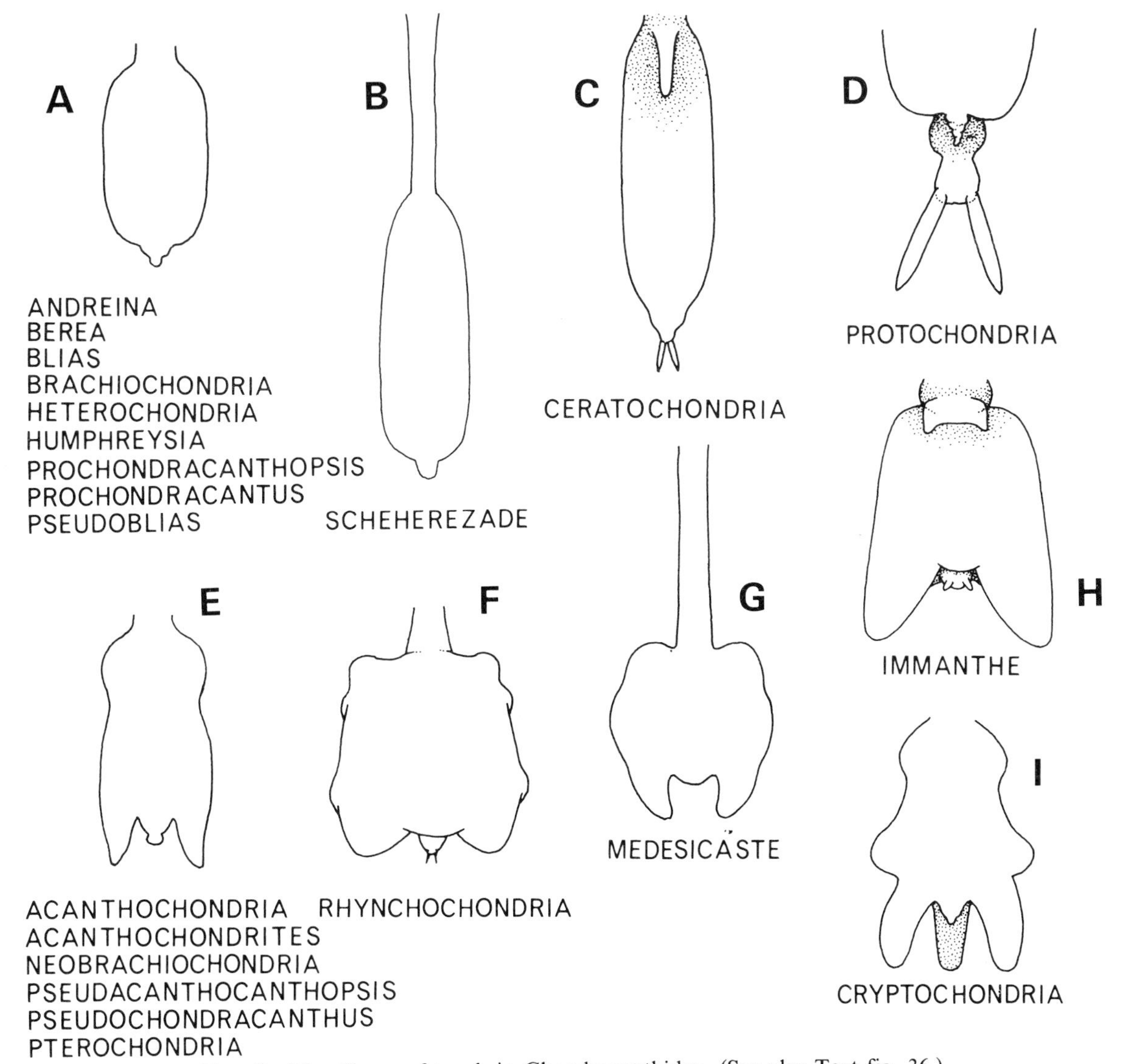

Text fig. 35. Types of trunk in Chondracanthidae. (See also Text fig. 36.)

The genital segment invariably remains outside the trunk, which is, therefore, a pre-genital formation, although its size and main features of its structure are a direct response to the reproductive needs of the parasite.

According to their morphology, the chondracanthid trunks can be divided into three groups. The first two of them are shown in Text fig. 35. One series (Text fig. 35A-D) comprises trunks usually completely devoid of processes. The most representative and most common type is shown in Text fig. 35A. Varying in length and width, the trunks of this type are simple in outline and have no processes of any kind. Their anterior necks are usually short. A variety of this type (Text fig. 35B), represented by *Scheherezade*, is very elongate and possesses a very long neck. While paired processes are never present in this type of trunk, median processes are sometimes found, either on the dorsal side near the neck, as in *Ceratochondria* (Text fig. 35C), or on the ventral side at the border with the genital segment, as in *Protochondria* (Text fig. 35D). The latter process is very small and is used by the male for attachment.

The second group (Text fig. 35E–I) can be distinguished from the first mainly by the possession of a pair of posterior processes, or, more accurately, posterolateral lobes, usually pointed and extending backwards beyond the limits of the abdomen. The most typical example of this group is shown in Text fig. 35E. There are many different shapes of this type, from short and squat to very slender and long, the former occurring in some species of *Acanthochondria*, the latter being exemplified particularly well by *Pterochondria*. In some members of this group the neck might be well delimited and moderately long, as in *Rhynchochondria* (Text fig. 35F), which is also characterized by incipient lobes on its lateral margins. A very long, well demarcated neck is also present in *Medesicaste* (Text fig. 35G). Usually there are no other processes on the trunks of members of this group, but a single median posterior process occurs in *Cryptochondria* (Text fig. 35I). In *Immanthe* (Text fig. 35H) the ventral surface of the neck region carries a peculiar plate-like process of unknown function and homology.

The third group (Text fig. 36) is a motley assemblage of trunks with only two features in common: all have posterior processes and all are equipped with at least one pair of lateral processes. The posterior processes might be little more than indistinct lobes, barely classifiable as processes, as in *Praecidochondria* (Text fig. 36B), or they can become inflated so as to constitute a major part of the trunk, as in *Diocus* (Text fig. 36E). A simple example of this group is *Protochondracanthus* (Text fig. 36A), with one pair of lateral processes and short, rounded posterior processes. The number of lateral processes is most commonly two pairs (Text fig. 36C–F). In *Lateracanthus* (Text fig. 36D) they are associated with the thoracic legs which are situated on them, close to their median dorsal processes, while in *Juanettia* (Text fig. 36F) and some species of *Diocus* a pair of dorsolateral processes is present.

The genus *Pseudodiocus* (Text fig. 36G) is distinguished by possession of unique "processes upon processes", even its posterolateral processes bearing secondary outgrowths. It is also peculiar in having a pair of posterior processes dorsal to the genito-abdominal tagma.

The genus *Strabax* (Text fig. 36H) has been placed here merely for the sake of convenience, though it is completely atypical. With one posterodorsal and two pairs of deeply bifid lateral processes, this genus is characterized most of all by its "neck", a dominant feature of its morphology.

(No account was taken in this review of the genus *Chondracanthus*, some species of which would fit into two of these three groups. It will be discussed in detail under a separate heading later.)

The genito-abdominal region. The smallest tagma of the chondracanthid body, this region is usually more or less clearly divided into its two components: genital segment and abdomen. The former is always compact and covered by hard, scale-like cuticle, usually with a distinct mid-dorsal groove (Fig. 271), slashed on each side by the slits of the oviduct orifices (Fig. 170). Lacunae of thinner cuticle are often found on the ventral side of these orifices. The abdomen might be separated from the genital segment by a constriction (Fig. 188) or might be almost completely fused with it (Fig. 346), though still distinguishable by its thinner and paler cuticle. It varies in shape from simple subspherical to elongate. In some species it might be equipped with club-shaped (Figs. 147, 148) or rounded processes. Uropods, more or less modified, are always present. Although Chondracanthidae, as a rule, have genito-abdominal tagma in the posterior position, in *Praecidochondria* it is displaced to a subterminal, ventral position. In this feature it resembles Pharodidae, a closely related family.

The appendages of Chondracanthidae clearly betray their poecilostome nature. In the females the first antennae are usually present (except in *Strabax* and perhaps in *Acanthochondrites japonicus* Gusev, 1951). In most genera they are devoid of segmentation, though they consist of recognizable basal and terminal parts, the former fleshy and sometimes fairly complicated in structure, the latter invariably cylindrical, with rounded tip and varied in length. In some genera, however, these appendages are distinctly segmented (e.g. in *Diocus* and *Rhynchochondria*). The segmented first

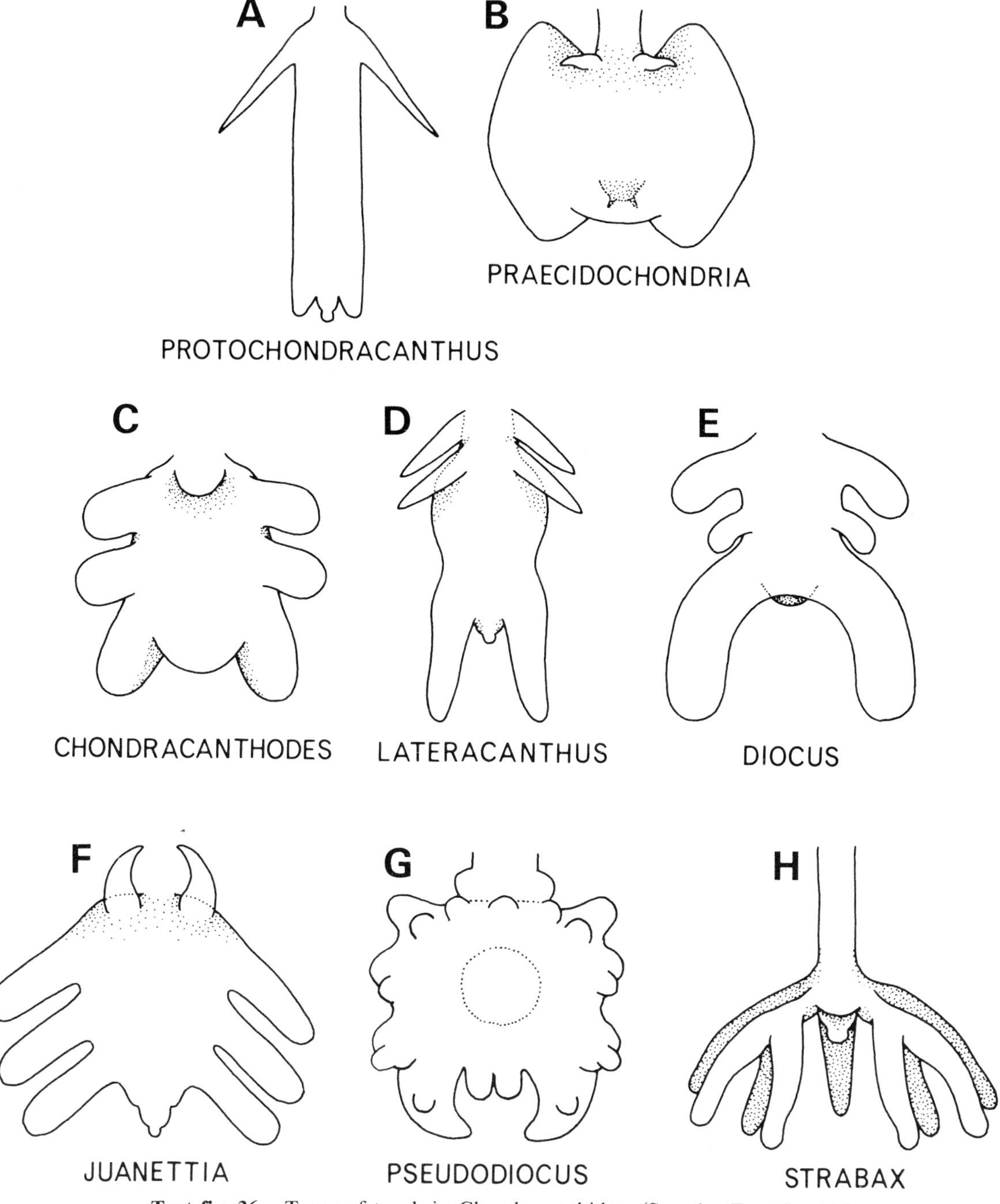

Text fig. 36. Types of trunk in Chondracanthidae. (See also Text fig. 35.)

antennae are setiferous, but apical armature is usually retained also in unsegmented ones, their basal parts carrying also several modified setae of presumably sensory character. The incompletely known first antenna of *Juanettia* appears to consist of a very long, unsegmented basal part from which the terminal part arises subapically. The latter part is presumably segmented, but the only specimens known so far had that part broken off, with only one setiferous segment remaining.

The second antenna (Text fig. 37) is the only attachment organ of the adult female and it is usually prominent and modified into an uncinate, heavily sclerotized structure, indistinctly divided into four components: basal frame, base, shaft and hook (see *Chondracanthus*, Text fig. 37). The homology of these parts is not quite clear. The only information we have on the chondracanthid ontogeny is that provided by the work of Heegaard (1947a). That author derives all but the basal frame of the second antenna from the terminal claw of the three-segmented endopod, the other two segments of which are also equipped with similar claws, though smaller. In Heegaard's view, the small two-segmented exopod of the copepodid antenna is lost, the first endopod segment fused with the one-segmented sympod and eventually reduced with it in length, until all that remains is the outermost ring of the basal frame. The second and third endopod segments are equally reduced and in the adult detectable only as the other two rings of the basal frame. The claw of the third endopod segment, on the other hand, grows and becomes the appendage of the adult. This interesting interpretation does not take account of several facts. In some species a small, digitiform and unsegmented outgrowth arises from the base of the antenna and is usually tipped with several setae. It has been referred to by some authors as the exopod. Ho (1970) prefers to use for it the name "auxiliary antennule", not accepting its homology as proven. Many antennae also have fine single setules arising near the border of base and shaft. It does not seem reasonable to postulate for these structures an origin later than during the copepodid stage. Since, as far as is known, no such structures are present at that stage on the

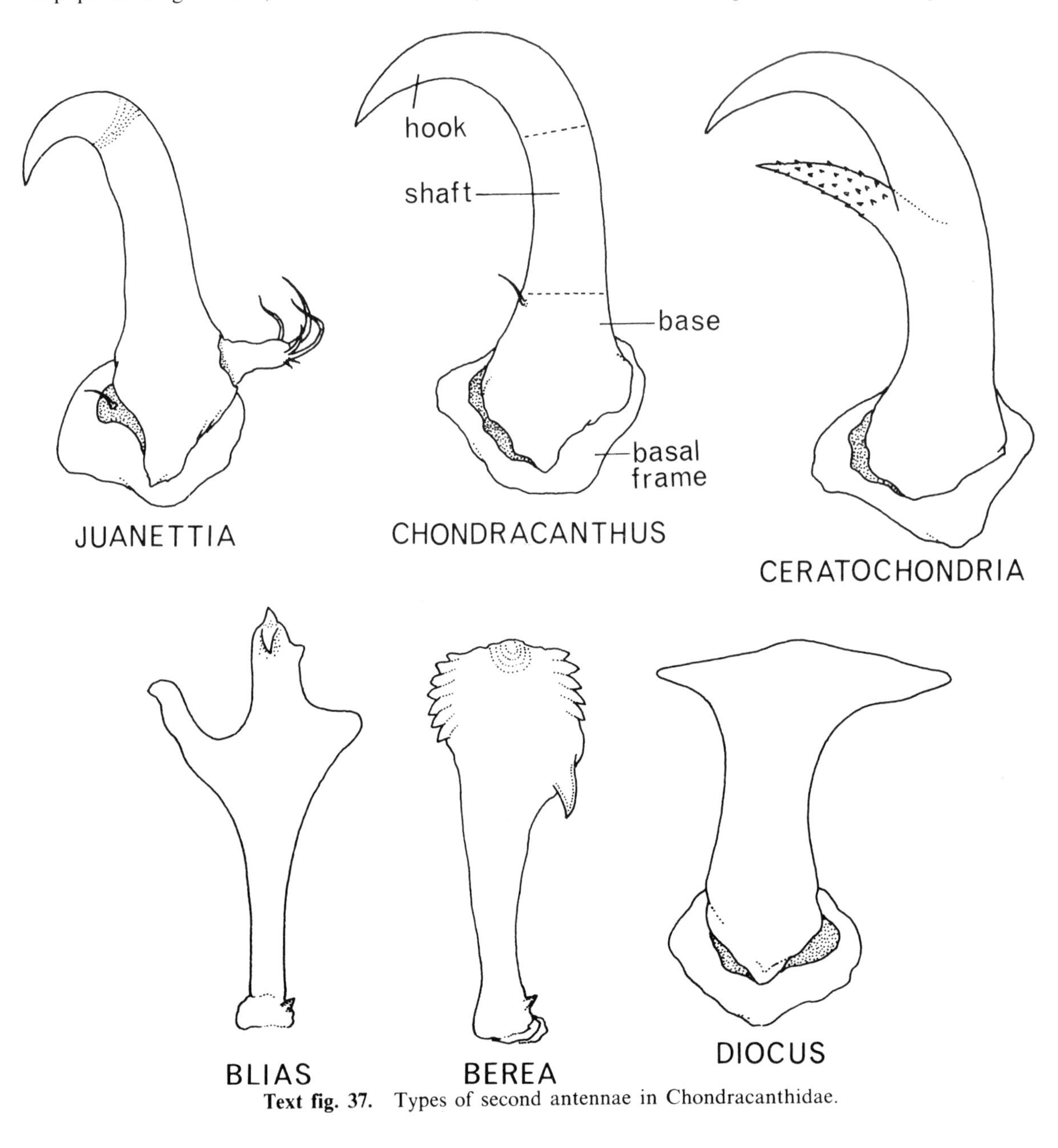

Text fig. 37. Types of second antennae in Chondracanthidae.

terminal claw of the endopod, it appears more than likely that the adult appendage is formed by a fusion of that claw with the remaining endopod segments and the sympod. In some species the traces of fusion can still be seen and in others a separate basal segment remains (Fig. 136). It is this author's view that one incurs less danger of error by accepting the armed outgrowth as a vestige of the exopod than by assuming it to be a completely new structure developed from the terminal claw. Ho's (1970) assertion that it occurs only in the most primitive genera is not unchallengable. It has been described in *Juanettia* (Text fig. 37), *Neobrachiochondria*, *Immanthe*, *Humphreysia*, *Prochondracanthus*, *Protochondria* and *Scheherezade*. The majority of the chondracanthid females, however, have second antennae without exopods. The length of the shaft varies from species to species, as does that of the hook. The latter often has a band of transverse striation and might be twisted at an angle to the plane of the long axis of the shaft (Fig. 202). The antenna of *Ceratochondria* (Text fig. 37) has a secondary prong, finely denticulated, arising from the shaft about where it passes into the hook. The antennae of several genera are not uncinate. That of *Blias* (Text fig. 37) has a shaft expanding distally and subdividing into three unequal branches, with some secondary denticulation. In *Berea* (Text fig. 37) the shaft is clavate with one major tooth and two series of smaller ones near its mace-like tip. Both these antennae have small sharp denticles arising from the base. Finally, the antenna of *Diocus* has developed a distal transverse bar and its appearance (Text fig. 37) resembles that of the letter T. The basal frame consists of three concentric rings, the base of the antenna being planted in the central ring, surrounded by the second and the third, which is rather pyriform due to its medial extension that links with the one of the opposite antenna.

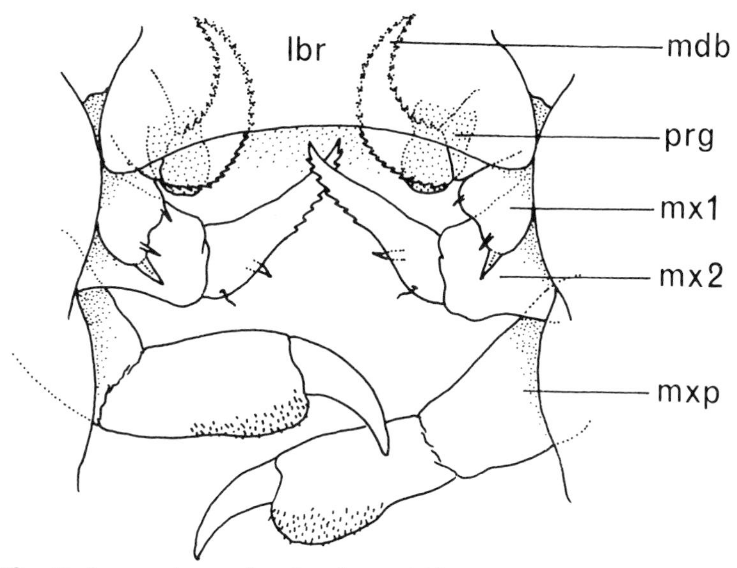

Text fig. 38. Oral appendages of a chondracanthid copepod in situ (diagrammatic). lbr — labrum; mdb — mandible; mxl — first maxilla; mx2 — second maxilla; mxp — maxilliped; prg — paragnath.

The oral region and its appendages are surprisingly uniform throughout the family (Text fig. 38, Fig. 137). A small, distinct labrum and a less distinct labium are present. The labrum is a cuticular transverse flap carrying various types of ornamentation (spinules, papillae, etc.) that could prove of great help in taxonomy. In view of their minute size, however, new methods of observation (electron microscopy) would have to be used for this purpose. So far, they have not been adequately studied and are disregarded in this account of the family. The flap of the labrum leaves most of the oral gape exposed and the appendages at least partly visible. Those that are most completely secreted are the mandibles (Text fig. 38, mdb). They consist of two parts, a compact basal and falcate terminal, the latter serrated along both margins. The length and slenderness (or lack of it) of the falcate part has been used as a specific feature in the taxonomy of Chondracanthidae, the number and size of its marginal denticles also varying from species to species, as well as within the species. In most instances, the denticulation differs also between the ventral and dorsal margin of the same mandible (Fig. 199). Obscured from view by the mandible is the paragnath (Text fig. 38, prg), a small, suboval structure, often armed with denticulation or setation. There are currently very few descriptions of this appendage, only Ho (1970) having paid any attention to its structure. In this work they have been ignored, though it must be admitted that they might be of help for identification purposes.

Next in the sequence of appendages are the first maxillae (Text fig. 38, mxl). They are very small and somewhat heart-shaped in most instances, though they might be described as reniform or suboval in some species. They are invariably armed with two or three spinules along their distomedial margins and very occasionally bear some delicate denticulation or setation on their surfaces. The second maxillae (Text fig. 38, mx2) are bipartite. Their basal part is a powerfully muscular, squat segment articulating with the terminal part, claw-like, armed with one row of marginal denticles and carrying also an accessory spinule (with or without ornamentation) and a slender, short seta near the base. The maxillipeds (Text fig. 38, mxp) are tripartite and brachiform, resembling the second maxillae of the caligid copepods in their structural plan. (The only exception here is the genus *Pseudodiocus* in which the maxilliped is vestigial.) Otherwise, the basal part is short, muscular and subcylindrical, always unarmed; the middle part is longer, usually somewhat inflated distally and armed with one or two subapical groups of denticles. The terminal part is a claw, more or less sharp, with stout base and in some species with one or two subapical denticles. When *in situ*, all these appendages (Fig. 137) are usually turned inwards and their true shape is not readily apparent. They have been presented in Text fig. 38 so as to display their true shapes.

The members of this family vary quite extensively in the number and structure of their thoracic legs. These appendages show various stages of adaptation to the parasitic mode of life, from the relatively unchanged, primitive podoplean limb to the vestigial condition. The author has attempted to show several series of structural modifications of chondracanthid legs by arranging them in Text fig. 39. It must be stressed that no phylogenetic significance is attached to these series.

In the following discussion, the capital letters in brackets refer to different figures in Text fig. 39. The simplest and least modified legs occur in the genera *Cryptochondria*, *Juanettia* and *Rhynchochondria* (A), in which the sympod is still recognizably two-segmented, and both rami have also retained their two-segmented structure and well developed armature. This primitive condition is somewhat less well retained in *Humphreysia* and *Prochondracanthus* (B), where the segmentation is less distinct and the armature of the rami reduced. From this condition the modification can proceed in two directions. One of them is the complete reduction of the sympod, all that remains of the leg being small vestiges of its rami, as in *Immanthe* (C). The other is the gradual disappearance of segmentation, usually more pronounced in the endopod than exopod, as in *Andreina*, *Diocus*, *Pseudodiocus* and *Neobrachiochondria* (D). In these genera the endopod has become one-segmented, while at least partial division of the exopod into two segments persists. Modification along this line leads to a leg structure such as that found in some species of *Prochondracanthus*, *Protochondracanthus* and *Pseudacanthocanthopsis* (E). Both rami are one-segmented but the exopod is larger than endopod and has retained some of its armature. An unusual modification of this structure is that of the legs of *Lateracanthus* (R), where both rami are unsegmented and unarmed, while the sympod has become incorporated into the prominent lateral trunk processes (cf. Text fig. 36D). A further trend in this direction occurs in the legs of some *Protochondracanthus* species (F), where the endopod is no more than a pointed outgrowth of the sympod, while the exopod is short, one-segmented and poorly armed. The last step in this series is that exemplified by some species of *Pseudacanthocanthopsis* (G) that have legs without the endopod, though a fairly large and still armed exopod is present.

From the structural condition illustrated in (D), the modification can take another direction, that of gradual loss of segmentation without disappearance of the rami. In one species of *Pseudacanthocanthopsis* (H) the legs are in that condition, the only unusual feature being the presence of a peculiar lobe arising from the outer division of the leg and perhaps representing the original distal segment of the exopod. Another modification might be that shown by the legs of *Protochondria* (I), in which both rami are one-segmented, though still armed. The unusual feature of these legs is the elongation of their sympods, which are very prominent and impart the characteristic appearance to the trunk of the parasite. The progressive fusion of the component parts leads to the development of a bifid structure, representing probably the most characteristic and best known type of chondracanthid limb. It occurs in eight genera, including *Chondracanthus* and *Acanthochondria* (J). The bifurcation becomes less pronounced in *Scheherezade* (K), where the presence of vestigial armature suggests that the limb has been derived from the primitive ancestral condition independently of (J). In several species the legs become tripartite (L). This structure has been found in some species of *Chondracanthus*. In the absence of ontogenetic evidence, it is not possible to decide the homology of

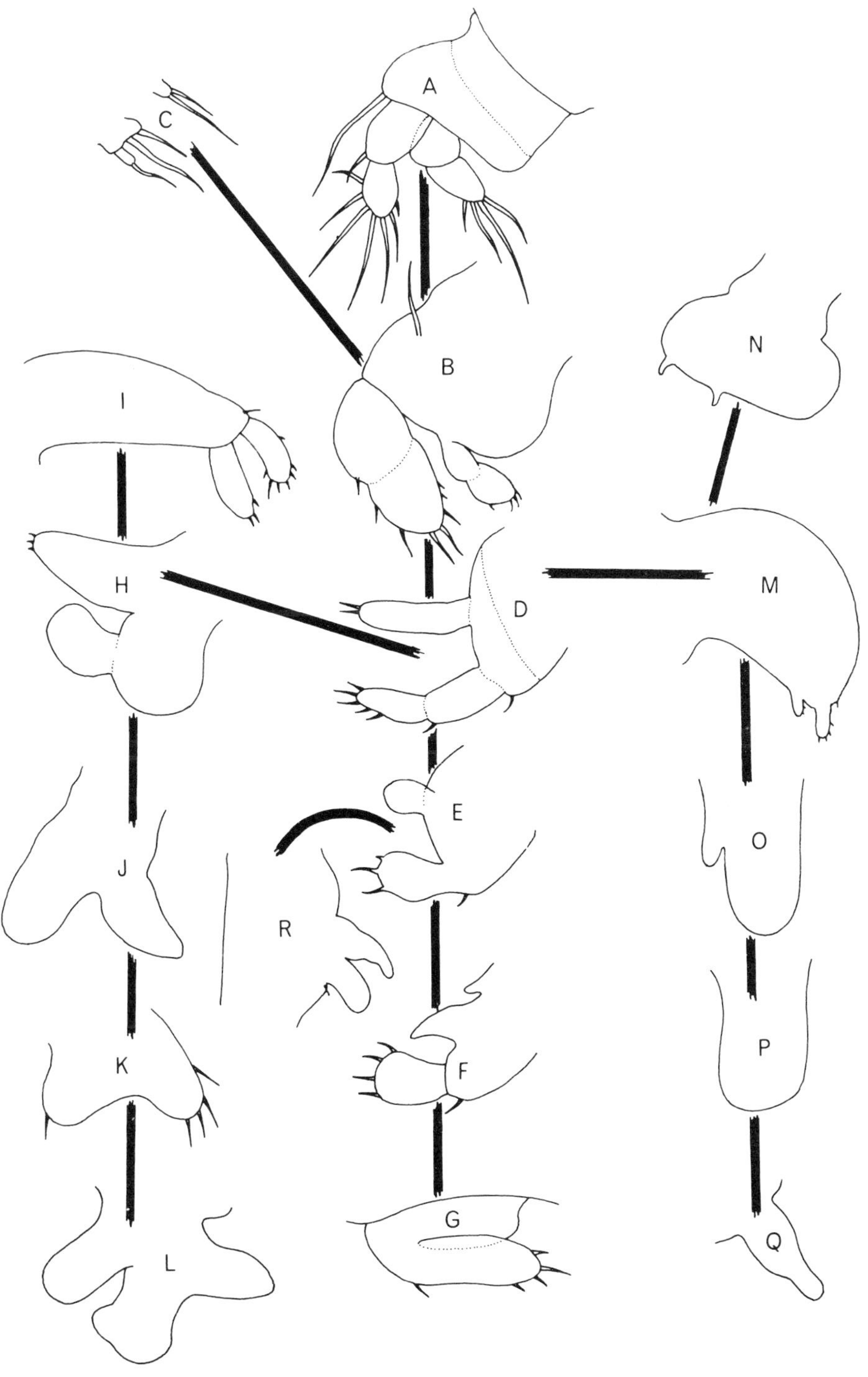

Text fig. 39. Types of legs in Chondracanthidae. A. *Cryptochondria*, *Juanettia* and *Rhynhochondria*; B. *Humphreysia* and *Protochondracanthus*; C. *Immanthe*; D. *Andreina*, *Diocus*, *Pseudodiocus* and *Neobrachiochondria*; E. *Prochondracanthus*, *Protochondracanthus* and *Pseudacanthocanthopsis*; F. Some species of *Protochondracanthus*; G. Some species of *Pseudacanthocanthopsis*; H. One species of *Pseudacanthocanthopsis*; I. *Protochondria*; J. Eight genera, including *Chondracanthus* and *Acanthochondria*; K. *Scheherezade*; L. Some species of *Chondracanthus*; M. *Blias*, *Chondracanthodes* and some species of *Pseudacanthocanthopsis*; N. *Brachiochondrites*; O. *Pseudoblias*; P. *Acanthochondrites*, *Heterochondria* and some species of *Protochondracanthus*; Q. *Praecidochondria*; R. *Lateracanthus*.

various parts of this limb. The central boot-shaped part might be an adventitious third swelling arising between the lobes of the original rami; on the other hand, that part might be the original leg with rami partially fused, whereas the other two lobes might be simply produced by swelling of the original sympod.

The third line of modification might be that leading to the disappearance of the rami with retention, or even enlargement (at least initially) of the sympod. The first step in this series is the limb of *Blias*, *Chondracanthodes* and some species of *Pseudacanthocanthopsis* (M), in which the rami are one-segmented and much reduced in relation to the sympod. One variant of this condition is that shown by *Brachiochondrites* (N), with the rami in a vestigial state and the sympod showing substantial terminal swelling. Progressive fusion of the different parts of the leg, accompanied by the reduction of the rami, brings about leg structure such as that shown in (O) and exemplified by *Pseudoblias*. Eventually, the legs are reduced to the state of a simple stump, as in *Acanthochondrites*, *Heterochondria* and some species of *Protochondracanthus* (P). The stump itself might become much reduced, as in *Praecidochondria* (Q).

It should be kept in mind that, parallel to these structural changes, one might also observe changes

Table 6. Number and structure of legs in the female Chondracanthidae.

Genus	Leg I	Leg II	Leg III	Leg IV	Leg V	Leg VI
Prochondracanthus	B-2	B-2	B-1	1s	3s	—
Rhynchochondria	B-2	B-2	B-2	—	—	—
Cryptochondria	B-2	B-2	—	—	2s	—
Berea	BL	BL	—	—	UL?	1s?
Blias	B-1	B-1	—	—	—	2s?
Immanthe	3s	2s	—	—	—	2s?
Diocus	B-1/2	B-1/2	—	—	—	—
Andreina	B-1	B-1	—	—	—	—
Protochondria	B-1	B-1	—	—	—	—
Chondracanthus	BL(TL)	BL(TL)	—	—	—	—
Acanthochondria	BL	BL	—	—	—	—
Chondracanthodes	BL	BL	—	—	—	—
Lateracanthus	BL	BL	—	—	—	—
Lernentoma	BL	BL	—	—	—	—
Pseudoblias	BL	BL	—	—	—	—
Pterochondria	BL	BL	—	—	—	—
Ceratochondria	BL	BL(v)	—	—	—	—
Pseudacanthocanthopsis	BL	UL(v)	—	—	—	—
Acanthochondrites	UL	UL	—	—	—	—
Heterochondria	UL	UL	—	—	—	—
Brachiochondrites	UL*	UL*	—	—	—	—
Prochondracanthopsis	UL(v)	UL(v)	—	—	—	—
Humphreysia	B-2	—	—	—	—	—
Juanettia	B-2	—	—	—	—	—
Neobrachiochondria	B-1**	—	—	—	—	—
Pseudodiocus	B-1(v)***	—	—	—	—	—
Protochondracanthus	B-1	—	—	—	—	—
Pseudochondracanthus	BL	—	—	—	—	—
Scheherezade	BL	—	—	—	—	—
Praecidochondria	UL(v)	—	—	—	—	—
Medesicaste	UL(v)	—	—	—	—	—
Brachiochondria	—	—	—	—	—	—
Strabax	—	—	—	—	—	—

B — biramous (figures refer to number of segments in rami); BL — bilobed; TL — trilobed; UL — unilobed; s — seta (accompanied by number indicating number of setae); v — vestigial.

Notes: The two pointed processes occurring on the ventral surface of *Berea*, near the border between trunk and genital segment, might or might not be vestiges of the fifth pair of legs.

*In *Brachiochondrites* the unilobate leg also possesses minute peg-like outgrowths that can be interpreted as vestiges of rami.

**In *Neobrachiochondria* the rami still retain some traces of segmental division, but not clear enough to classify them as two-segmented.

***In *Pseudodiocus* the endopod is only a lobe with a filiform apex.

in the size of the limbs, the legs of the same structure varying in their relative size very substantially. The number of legs is also affected. Most genera have less than the original full complement of these appendages. The first and second pairs are commonly retained. The gradual process of disappearance of these originally natatory appendages is shown in table 6.

The table makes it clear that the swimming legs were being reduced and lost from the posterior end of the trunk forwards, indicating the sequence in which the thoracic segments were incorporated into the increasingly large trunk. In some genera the vestigial sixth legs have been tentatively identified as the small setae in the vicinity of the oviduct orifices on the genital segment. It is fair to assume that these setae are vestigial legs, if only by analogy with the less modified podoplean copepods. One might also assume that, on careful examination, many more genera will be found to possess them.

The most striking feature of the chondracanthid male is its size. The males of all genera in this family are dwarfs, many times smaller than the females of their species. Unlike the diminutive males of Lernaeopodidae, however, they are not ephemeral and are usually found attached to the females near the genital region. In some genera they grip with their second antennae special processes arising from the surface of the female trunk in that region (e.g. *Protochondria*, Text fig. 35D). As far as is known, the male never leaves the female and has no direct contact with the host after the initial process of settlement and attachment. It has been generally accepted that the absence of prolonged contact with the host is responsible for the lesser degree of morphological modification of the male, when compared with the female. Among the chondracanthid males, however, can also be found a gradation in the degree of departure from the original structural plan of the free-swimming podoplean. The most primitive male type is exemplified by *Juanettia* (Text fig. 40), distinguishable from its free-swimming relatives only by its possession of the prehensile appendages. The males of *Juanettia* are completely and distinctly segmented, while their appendages bear hardly any traces of the influence of parasitism (other than the prehensile nature of the second antennae). The process of adaptation involves a gradual loss of external segmentation and general shortening of the body. In *Juanettia* the first leg-bearing segment is partly fused with the cephalosome. In *Andreina, Immanthe, Protochondria* and *Rhynchochondria*, the general loss of external segmentation did not extend to the fusion of that segment and left their anterior end in the cephalosome condition. In most other genera the fusion is complete. A good example of the initial steps in morphological response to parasitism is shown by the male *Protochondria* (Text fig. 40), in which some traces of segmentation are still discernible. The fusion of thoracic and abdominal segments leads to abbreviation of the body. The male *Blias* (Text fig. 40), the abdomen of which has become vestigial, demonstrates an intermediate condition. The end-point is represented by males such as that of *Medesicaste* (Text fig. 40), with virtually no abdomen left.

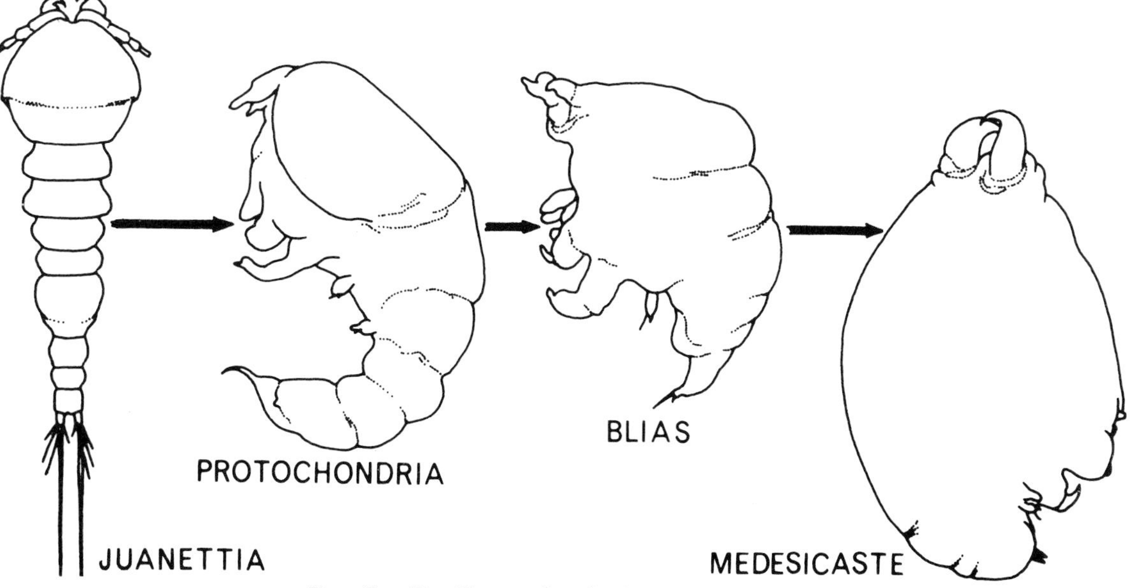

Text fig. 40. Types of males in Chondracanthidae.

The first antennae are similar in type to those of the female, though the males retain the antennular segmentation much more readily; their armature is also much better preserved. In some genera, however (*Brachiochondria, Medesicaste* and *Pseudochondracanthus*), the first antennae are absent. The second antennae are usually quite similar to those of the females, except in those genera in which the female second antenna diverges from the usual uncinate structure. They tend to be shorter and heavier than those of the females and are much more commonly equipped with an exopod.

The oral appendages are very similar to those of the females, though they tend to be less abundantly denticulated.

With the process of fusion and abbreviation of the thoracic segments, the swimming legs also undergo changes, simplification, reduction is size and eventual disappearance. The most common type of chondracanthid male leg is shown in Fig. 212, but many small differences from that type do exist. The variation in the number and structure of legs are shown in table 7.

Table 7. Number and structure of legs in the male Chondracanthidae.

Genus	Leg I	Leg II	Leg III	Leg IV	Leg V	Leg VI
Juanettia	B-2	B-2	B-2	U-2	2s	1sp, 1s
Rhynchochondria	B-2	B-2	BL	—	1s	2s
Diocus	B-1/2*	B-1/2*	2s	—	2s	2s
Strabax	UL	UL	2s	—	2s	2s
Pseudacanthocanthopsis	B-2	B-2	—	—	1s	1s
Cryptochondria	B-1	B-1	2s	—	1s	—
Berea	BL	BL	—	—	—	1s
Blias	BL**	BL**	—	—	—	2s
Chondracanthodes	B-1	B-1	—	—	—	—
Lateracanthus	B-1	B-1	—	—	—	—
Pseudodiocus	B-?	B-?	—	—	—	—
Andreina	U(v)	U(v)	—	—	—	—
Acanthochondria	BL	BL	—	—	—	—
Chondracanthus	BL	BL	—	—	—	—
Medesicaste	BL	BL	—	—	—	—
Protochondria	BL	BL	—	—	—	—
Pterochondria	BL	BL	—	—	—	—
Brachiochondrites	BL	BL	—	—	—	—
Lernentoma	BL	BL	—	—	—	—
Prochondracanthopsis	BL/UL(v)	BL	—	—	—	—
Acanthochondrites	UL	UL	—	—	—	—
Immanthe	(v)	(v)	—	—	—	—
Ceratochondria	U-?	—	—	—	—	—
Protochondracanthus	BL	—	—	—	—	—
Brachiochondria	—	—	—	—	—	—
Heterochondria	—	—	—	—	—	—
Pseudochondracanthus	—	—	—	—	—	—

U — uniramous; other symbols as in table for female legs. In *Ceratochondria* the segmentation of rami is not clear.

*The segmentation of rami cannot be ascertained.

**A short protuberance on medial margin, described by Ho (1970) as a tooth, is interpreted here as vestigial lobe of endopod.

The swimming legs are reduced in parallel with the loss of segmentation, which, as in the females, proceeds from the posterior end of the thorax forwards. The sixth leg, situated on the genital segment, and not involved in trunk formation, tends to persist in a vestigial form in genera with fairly well developed trunks (see *Berea* and *Blias* in the table). The fifth leg also persists much better than the fourth, the latter being the first to disappear (see *Rhynchochondria*). The most commonly retained legs are those of the first and second pairs. At the end of the series are completely legless genera.

The male described by Kabata (1969c) for *Neobrachiochondria* was not fully adult and its morphology cannot be used for comparison with that of adult males. The structure of the legs of male *Pseudoblias* is also not known with accuracy and has been left out of the table. No males are known for *Humphreysia*, *Praecidochondria*, *Prochondracanthus* and *Scheherezade*.

History and systematics

Most copepod species extensively modified as the result of parasitism were placed by early taxonomists in the genus *Lernaea*. Even when subsequent investigators removed some of those species and placed them in other genera, they left them, as a rule, in a group loosely defined as the "lernaeid copepods". No exception was made for the chondracanthid copepods known at that time. Various groups of parasitic copepods became gradually recognized as independent taxa, their degree of relationship with the lernaeids more or less tenuous, but the chondracanthids retained their spurious association with the lernaeids long enough badly to confuse their taxonomy.

A good example of the criteria used and the views held by the early 19th century taxonomists is the classification used by Burmeister (1835), who recognized five families of parasitic copepods. One of them was the family Lernaeoda, diagnosed as follows: "With feelers and jointed feet, two pairs of prehensile feet behind the proboscis." Lernaeoda were subdivided into three groups, according to the type of their attachment organs.

A. With sucker-like attachment organ at the junction of neck and body (here belongs *Clavella*).
B. With elongated, arm-like attachment organ fused at the tip (here belong other lernaeopodids).
C. Without arm-like attachment organ. Genus *Chondracanthus* (*Ch. triglae* v.N.; *Ch. cornutus* v.N.; *Ch. tuberculatus* v.N.; *Ch. zei* Cuv.). Genus *Lernanthropus* (*L. musca* Blainv.; *L. paradoxus* Burm.).

Interestingly enough, no real lernaeids were incorporated in this family. They were all placed in Pennelina.

The first writer to accord a familial status to the chondracanthids was Edwards (1840), who established for them the "famille des Chondracanthiens". Even then, however, he placed them in the order "Lernéides", together with "Lernéopodiens" and "Lernéoceriens". The diagnosis of "Chondracanthiens" was as follows: "Females fixed to their prey with the aid of maxillipeds which are strong and form hooks." (A mistaken reference to the second antennae.) "Antennae generally distinct, several pairs of thoracic appendages present, more or less rudimentary but free."

With the thoracic appendages as his main taxonomic criteria, Edwards recognized as members of this family the following genera: *Selius*, *Aethon*, *Cycnus*, *Clavella* (= *Hatschekia*), *Tucca*, *Peniculus*, *Lernanthropus* and *Chondracanthus*. The family was, therefore, clearly a composite taxon, only the genus *Chondracanthus* now remaining in it from the above list.

No changes were introduced into this taxonomic structure by Krøyer (1837). More species were added, however, to the genus *Chondracanthus*, which was destined to grow and to become a catch-all taxon within the family until the third decade of this century.

Baird (1850) divided his order Lerneadae into three tribes, according to the mode of attachment, thus clearly following the taxonomies of Burmeister and Edwards. The tribe containing the family Chondracanthidae was named Anchorastomacea and its diagnosis of the females was almost exactly that of Edwards' Chondracanthiens. A brief diagnosis of the males was added: "Free and unattached; very small and differing totally in appearance from the females."

Restricting his studies to British fauna, Baird made no mention of other families in the tribe. He was, however, the first author to use the name of this family in its present form and to limit it (at least for the British species) to *Chondracanthus* and *Lernentoma*, either by accident or design establishing it as an all-chondracanthid taxon. The family diagnosis ran as follows: "Organs representing thoracic feet, in form of considerable-sized, cartilaginous-looking, not articulated appendages; generally three pairs in number. Three pairs of foot-jaws."

Dana (1853), who made a great contribution to the taxonomy of parasitic Copepoda, on his own admission had no first-hand knowledge of Chondracanthidae and followed Edwards in his arrangement of this group. He kept the chondracanthids with Ancorellidae (the lernaeopodids) and Penellidae (the pennellids) in his order Lernaeoidea. His diagnostic key to the family included three subfamilies.

"Four or more cephalothoracic appendages, attachment by larger or smaller hooks.

Subfamily Selinae: Anterior antennae and posterior thoracic feet slender, more or less prehensile. Genus *Selius* Krøyer.

Subfamily Chondracanthinae: Anterior antennae slender or very short. Posterior thoracic legs short and roughly prehensile. *Chondracanthus* de la Roche; *Lernanthropus* Blainv.; *Lernentoma* Blainv.; *Cycnus*.

Subfamily Clavellinae: Anterior antennae obsolete. Posterior thoracic legs stout and short. *Clavella, Tucca, Peniculus, Aethon.*"

Clearly, Dana's acceptance of Edwards' systematics extended to his approval of Chondracanthidae as a grossly composite family.

The noted Danish workers Steenstrup and Lütken (1861) and Krøyer (1863–64) made no great changes in our concept of this family. Their "Gruppe Chondracanthierne" was still a composite group. Some new valid genera were added (e.g. *Diocus*) but *Medesicaste* was placed in Lernaeocerina. Von Nordmann (1864) followed the Danes, adding some more genera to the roster. Some of them (*Lamippe, Sphaerosoma*) are unrelated to the chondracanthids, others (*Strabax*) are still ranked among the members of the family, others still (*Pseudulus*) were based on damaged specimens, never again recorded and not identifiable.

The amassed literature of continental Europe bearing on the topic of Chondracanthidae was summarized and systematized in a comprehensive work of Heller (1865). Since it reflects contemporary views on the taxonomic criteria, Heller's key to the genera of his Chondracanthina is given below *in extenso*.

"I. Body not distinctly articulated.
  1. Second antennae uncinate.
    A. Body elongate.
      a. Two pairs of uniramous, articulated legs. *Blias* Krøyer.
      b. Two or more pairs of legs, simple or lobed, not articulated.
        i. Anterior part of body attenuated, forming long neck.
          *Genital part equipped with two racemose posterior processes. *Lesteira* Kr.[1]
          **Genital part without posterior processes. *Medesicaste* Kr.
        ii. Anterior part of body not forming long neck.
          *First antennae long, three- or four-segmented (in males seven-segmented); monstrous shape of body. *Diocus* Kr.
          **First antenna short, two- or rarely three-segmented.
          Body form slender. *Chondracanthus* Delar.
    B. Body short, ovate; antennae setiferous, six-segmented. Three pairs of uniramous, segmented legs. *Selius* Kr.[2]
  2. Second antennae forming horns with clavate, trifid tips. *Trichthacerus* Kr.

II. Body distinctly articulated.
  1. Three pairs of legs, two first pairs bifurcate. *Anteacheres* Sars.
  2. Two pairs of legs, not forked at tips. *Staurosoma* Will.[3]

[1]*Sphyrion* Cuv. is probably identical with this genus, as is perhaps *Lophoura* Köll.
[2]Here one might perhaps include *Lamippe*.
[3]If *Philichthys xiphiae* is, in fact, a crustacean, it should be placed somewhere near here."

It is interesting to note that a simple criterion of the absence of external segmentation sufficed in this instant to separate the chondracanthids fairly well from other, non-related genera grouped under point II in the key. With the exception of *Selius* and a genus inquirendus *Trichthacerus*, all the others are valid and still within the family. The key is for the females and, since there are no definitely segmented females in the family, those grouped under II cannot qualify for membership in the family.

Gerstaecker (1866–1879) again reviewed and summed up the existing taxonomic knowledge of Chondracanthidae. Impressed by the influence of parasitism on copepod morphology, he arranged his taxonomy according to the extent of parasitic adaptations. His "Fünfte Reihe" consisted, therefore, of the two most modified families, Lernaeopodidae and Chondracanthidae (the latter also referred to as Condracanthina (sic)). Gerstaecker's well-known compendium perpetuated in this way the old concept of Lernaeidea, later to be known as Lernaeopodoida, a major taxon combining the end-points of two evolutionary series, families of two different suborders.

Gerstaecker's diagnosis of Chondracanthidae was as follows:

"Male and female strikingly different in shape and size, the former a pygmy with inflated cephalothorax and barrel-like, clearly segmented abdomen, the latter relatively large, stout, imperfectly and mainly indistinctly segmented, of monstrous appearance due to irregular swellings and protuberances of individual parts. Anterior feelers short, two- or three-segmented, posterior mainly as hooks. Jaws on the side of the mouth, free; maxilliped small, with terminal claw. Legs reduced to two pairs, vestigial or lobe-like, sometimes absent. Reproductive organs paired, in adult female voluminous and often extending into protuberances of the body. Egg sacs paired, saccular, multiserial. Sessile parasites; dwarf males attached only to the bodies of females."

The genera included in the family were: *Lesteira*, *Medesicaste*, *Strabax*, *Trichthacerus*, *Blias*, *Chondracanthus* (considered a senior synonym of *Lernentoma*), *Splanchnotrophus* and *Diocus*. (Other genera, placed here more tentatively, were *Ismaila*, *Tanypleurus*, *Lamippe*, *Nereicola*, *Selius*, *Chelodiniformis*, *Anteacheres* and *Staurosoma*. None of them is a chondracanthid.)

At about the same time, two authors published papers which showed a completely different concept of chondracanthid affinities. Thorell (1859) placed Chondracanthidae in a group with free-living, semi-parasitic and other parasitic copepod families on the grounds of their mouth structure. This group, characterized by the absence of a suctorial mouth and (erroneously) mandible, included also Ergasilidae. Vogt (1877b) independently discovered an affinity between Chondracanthidae and Ergasilidae, contrasting the former with Lernaeopodidae and doubting the existence of a relationship between them. The differences included the morphology of their nauplii and of the males. The work of Thorell and Vogt, however, anticipating the final placing of Chondracanthidae in the order Cyclopoida, was largely disregarded by the copepod specialists of their time.

The next major work on parasitic Copepoda was that of Scott and Scott (1913). Their taxonomy was derived from Baird and, in the final source, from Burmeister. The family Chondracanthidae, together with Lernaeidae and Lernaeopodidae, was placed in the tribe Lernaeoida. The familial diagnosis included correct, if incomplete, descriptions of the appendages. The difference between the chondracanthid mandible and that of Lernaeopodidae (and Caligidae) was pointed out (Scott and Scott, 1913, p. 167), but no conclusions were drawn from that difference.

One of the most renowned copepod experts, Wilson (1932), also adhered to this traditional view of chondracanthid taxonomy. Following his earlier compatriot, Dana (1853), he saw Chondracanthidae as a family that shared the suborder Lernaeopodoida with Lernaeopodidae, Sphyriidae and Antheacheridae. His familial diagnosis was as follows: "Body of both sexes usually rigidly fused and showing no movable articulation and often no trace of segmentation; with or without dorsal, lateral or posterior processes. Sexual dimorphism universally present, resulting in great disparity of size between sexes and a corresponding dissimilarity in structure. Carapace often present, but never any paired plates as in the Caligoida."

Yamaguti (1963) upgraded the old composite taxon to the ordinal level. His Lerneopodidea contained Chondracanthidae, Lernaeopodidae, Naobranchiidae and Sphyridae. The family Chondracanthidae was divided into 3 subfamilies according to the following key:

1. Abdomen and caudal rami greatly reduced, shifted forward to ventral side of central portion of thorax . . . . . **Pharodinae**
   Abdomen and caudal rami more or less reduced, never shifted forward to ventral side of central portion of thorax, usually ventroterminal 2
2. Cephalic neck present, at the base of which the mouth opens . . . . . **Lernentominae**
   Cephalic neck absent; mouth opening on ventral surface of head . . . . . **Chondracanthinae**

Yamaguti added little new to the familial diagnosis. His view of the chondracanthids contained, however, some surprising details. For example, the genus *Andreina* was removed from the family and placed in a new order Andreinidea, which thus contained only one species, *Andreina synapturae*, a chondracanthid of unusually slender shape, but in all other respects clearly a member of this family.

Wilson and Yamaguti persisted in misplacing Chondracanthidae in a spurious order even though the family's proper place was determined by Oakley (1927, 1930), who transferred it to the Sarsian Cyclopoida. The structure of the mouth parts convinced Oakley, who was also aware of the work of Vogt (1877b), that it is necessary to dissociate them from the lernaeopodids and other families of clearly siphonostome type. Oakley's work has become the foundation of the modern taxonomy of Chondracanthidae. He established the subfamily Lernentominae and was the first to recognize the need for a revision of the genus *Chondracanthus*, from which he carved out four new genera, not all now accepted (*Acanthochondria*, *Acanthochondrites*, *Chondracanthopsis* and *Disphaerocephalus*).

Full account of Oakley's work was taken by most subsequent workers, including major reviewers such as Markevich (1956).

The subfamily Pharodinae Illg, 1948, was later raised to the familial level by Ho (1971a), leaving the chondracanthids with two subfamilies.

The latest and most thorough review of Chondracanthidae is that published by Ho (1970), to whom this writer is greatly indebted for his views on the family.

The intrafamilial taxonomy of Chondracanthidae has been made difficult by our inability to decide on the dependable diagnostic criteria. Our understanding of the relationship between morphology and phylogenetic links among various genera and species of this family still remains hazy. The present taxonomy of Chondracanthidae is based on the adult females. The structure of the mouth parts is uniform throughout the family, making them unsuitable for this purpose. The diagnoses are, therefore, based mainly on the following two points: (i) shape of the body, number, shape and distribution of cephalothoracic and trunk processes, and (ii) number and structure of swimming legs.

Prior to Oakley's (1930) work it was customary to lump all species with the "typical" trunk structure (Text figs. 35E, 36D) in the genus *Chondracanthus*. Probably partly under the influence of Leigh-Sharpe, who was the most celebrated "splitter" in the field of parasitic Copepoda, Oakley parcelled out that genus into five genera, mainly in accordance with the number and arrangement of processes. This fragmentation continues until the present day. Ho (1970) pointed out that out of 36 chondracanthid genera listed by Yamaguti (1963), 20 are monotypic. It made him wonder whether we do not attach undue importance to those processes. His reservations are even more justified when one takes into account the fact that the size and even the number of those processes are subject to differences with age (compare a juvenile female of *Chondracanthus lophii* in Fig. 179 with the adult one in Fig. 175). Moreover, a fair degree of intraspecific variability in those morphological characters is known to exist, reflecting either true genetic differences or influences of conditions at the site of attachment. Individual processes might become suppressed or distorted. On infrequent occasions, spurious processes might appear at various places, particularly on the trunk. The difficulties experienced by the taxonomists because of this lack of firm foundation can be exemplified by the following two cases. One of the common features of the chondracanthid trunk is the existence of posterolateral processes. In the genus *Chondracanthus* they are almost invariably present. *Ch. narium* Kabata, 1969, however, has none. The diagnosis of the genus *Acanthochondria* demands the absence of any cephalothoracic or trunk processes. One valid species of this genus is *A. lepidionis* Barnard, 1955, distinguished by a long cephalothorax and characteristically dark dorsal shield. Kabata (1970b) described a chondracanthid which corresponded with *A. lepidionis* in all aspects, but possessed a pair of very small, anterolateral processes. Their presence made it impossible to accommodate the parasite in the genus *Acanthochondria*, so that it was described as *Chondracanthus lepidionis*. Both copepods came from hosts belonging to the genus *Lepidion*. The differences between them were small and yet diagnostically significant. Is then the diagnosis at fault? We have no answer to this question.

Not all is clear, either, as regards the diagnostic value of the modified thoracic legs. A good example of it is the separation of *Acanthochondria* from *Acanthochondrites* mainly on the grounds of leg structure, bilobed in the former (Fig. 269) and unilobed in the latter (Fig. 345). On the other hand, *Chondracanthus zei*, the type species of its genus, has peculiar trilobed legs (Fig. 145). Kabata (1968b) considered this discrepancy in our evaluation of the thoracic legs as an "awkward fact" of chondracanthid systematics. Ho (1970) tried to minimize it by suggesting that the differences between bilobed and trilobed legs are much smaller than those between the unilobed and bilobed. However, it is difficult to evaluate objectively how much more significant taxonomically is the fact of reduction and fusion of modified rami than the development of completely new structures, one or two lobes, from the original sympod of the leg. Until this has been agreed upon, we must put up with the present chondracanthid taxonomy, always remembering that it is often artificial and tentative.

Chondracanthidae are a family restricted to, and broadly distributed on, marine fishes: elasmobranchs, holocephalans and teleosts. Almost all species parasitize the gill cavities of their hosts, where they are permanently hooked by their second antennae. In some instances they occur in protected parts of the hosts' cloaca, but have only rarely been found on exposed surfaces. At least 2 species (*Chondracanthus narium* Kabata, 1969 and *Ch. triventricosus* Sekerak, 1970) are known only from the nasal cavities of their hosts. (*Acanthochondrites annulatus*, occurring in nasal fossae, has been found also in the branchial chamber). The irritation caused by the process of attachment and by the feeding of the copepod often produces proliferation of the host's tissues around the site of attachment, sometimes leading to growth around the cephalothorax of the parasite that might cover it partially or completely. The species of Lernentominae burrow in the host's tissue and insert their premandibular segments, the "cephalic neck", quite deeply, though their mouths remain above the skin of the fish. In his earlier publication, Kabata (1970a) commented: "It is difficult to see what advantages are conferred on the parasite by the insertion of the antennary region so deeply into the host. There is no doubt that a very secure anchorage is achieved, but the majority of the related species living in very similar habitats do not appear to be in need of this excessive security." The buried part is usually surrounded by a thick capsule of the host's connective tissue.

Chondracanthidae are well represented in the British fauna. A key to their British members follows.

1. Cephalothorax fairly short, cephalic neck absent, oral region short distance behind antennae. Chondracanthinae . . . 2
   Cephalothorax with long cephalic neck between antennae and oral region. Lernentominae . ***Lernentoma***
2. Trunk with lateral, usually posterolateral (posterior), possibly dorsal and ventral processes, cephalothorax with or without processes . . . ***Chondracanthus***
   Trunk with posterior processes only, no processes on cephalothorax . . . 3
3. Thoracic legs bilobed . . . ***Acanthochondria***
   Thoracic legs unilobed . . . ***Acanthochondrites***

Genus *Chondracanthus* Delaroche, 1811

Female: Cephalosome or cephalothorax with or without processes. Anterior part of trunk (its first one or two segments) usually as wide as, or somewhat narrower than, cephalosome (or cephalothorax), forming neck of varying lengths. Trunk proper longer than broad, sometimes with indistinct transverse constriction at about midlength, i.e. corresponding with former border between last and penultimate segments of trunk; varying numbers of lateral, dorsal and ventral processes or swellings, both paired and unpaired, of different shapes, sizes and arrangement. Genito-abdominal tagma small, more or less distinctly divided into genital and abdominal components, situated near centre of posterior margin of trunk, usually (but not always) flanked by posterolateral processes of trunk. Uropods pointed, bearing three or four setae on basal part. Egg sacs cylindrical, straight or twisted, occasionally laterally compressed; eggs subspherical, multiseriate. First antenna usually with broad basal and cylindrical terminal part, sometimes both parts cylindrical, indistinctly divided; second antenna uncinate. Mandible falcate, with two rows of marginal denticles. First maxilla reniform or suboval, with two or three terminal and subterminal (rarely more) spinules and setules, occasionally with batch of minute setules on outer surface. Second maxilla with short, squat basal part and claw-like terminal part with single row of marginal teeth. Maxilliped brachiform; lacertus robust, cylindrical, unarmed; brachium subcylindrical, often distally inflated, with two subterminal batches of denticles; terminal claw gently curved, sometimes with one or two secondary teeth. Two pairs of legs, bilobed or trilobed, with or without terminal setules.

Male: Cephalothorax inflated, imperceptibly passing into narrower posterior part. External segmentation indistinct; genito-abdominal region indistinguishably fused with preceding segments, rounded posteriorly, with uropods similar to those of female. First antenna slender, cylindrical; second antenna uncinate, with or without vestigial exopod. Oral appendages similar to those of female, but with sparser denticulation. Two pairs of flat, lappet-like, bifid legs.

Type species: *Chondracanthus zei* Delaroche, 1811.

Comments: As already stated, the genus *Chondracanthus* developed in the course of its history into a dumping ground for sundry chondracanthid species and received much of its present form in the work of Oakley (1930), which established the foundations of modern taxonomy for this family. Oakley's generic diagnosis was very brief: "Female. Chondracanthinae with cephalic barbs, dorsal processes, and ventral processes all present. Thoracic legs bipartite or tripartite. Male. Very small. Female : Male = 20 (average)." This diagnosis is hardly adequate now, when one takes into account the great variety of structure in this genus. The type species, for example, has no cephalothoracic "barbs", while at least two species (*Ch. distortus* and *Ch. neali*) have two pairs of such processes.

When considering the morphology of the adult female *Chondracanthus*, one should remember that the trunk, however fused and modified, consists of five segments (in species with cephalosome) or four segments (in those with cephalothorax) and that one or two anterior ones tend to form the narrow neck of that trunk. The distribution of the trunk processes is segmental in character. The dorsal processes, almost always single and arranged in the mid-dorsal line, seldom exceed five. A greater number can be produced only by multiplying processes on each or some segments (as in *Ch. zei* and *Ch. lophii*). Many species have no dorsal processes. The paired lateral processes are most common of all. The posterior segment is rarely without them and on it they are referred to as posterolateral or posterior processes. Perhaps the most frequent number of lateral processes (posterolateral included) is three pairs, occurring in about a quarter of all hitherto known species. The greatest number is ten pairs (in *Ch. neali*) but in that species two are linked with their neighbours at the bases and could be seen as double, reducing their number to eight. Multiple divisions of processes occur only in the type species, *Ch. zei*. The ventral processes are the least common, a single one occurring in four species, two median ones in two species. (There is logic behind this phenomenon, since the ventral side is the one in permanent contact with the surface of the host.)

It is also worth pointing out that five species of *Chondracanthus* have trilobate legs (*Ch. horridus*, *Ch. nodosus*, *Ch. ornatus*, *Ch. wilsoni* and *Ch. zei*), a condition not occurring outside the genus.

The genus *Chondracanthus*, as defined here, includes also *Disphaerocephalus* Oakley, 1930, *Chondracanthopsis* Wilson, 1932, *Acanthocanthopsis* Heegaard, 1945, *Barnardia* Yamaguti, 1963 and *Protochondracanthoides* Yamaguti, 1963. The reasons for the rejection of these genera and for their inclusion in *Chondracanthus* were discussed by Ho (1970). The genus contains about 40 species (with some species inquirendae and some possibly invalid), all parasitic on marine teleosts. They are about equally divided between the Atlantic (with adjacent seas) and Pacific Oceans, seldom occurring in both. Only one species, *Ch. trilobatus* Pillai, 1964, is known from the Indian Ocean.

Six species of *Chondracanthus* are known from British waters. They can be identified with the aid of the key below.

1. Thoracic legs bilobed (Fig. 169) 2
   Thoracic legs trilobed (Fig. 146) 4

2. Cephalothorax with two pairs of lateral processes (Fig. 159); in addition to posterior processes of trunk, eight lateral pairs, four unpaired dorsal (Fig. 160) and one unpaired ventral (Fig. 159) process present . . . . ***Ch. neali***
   Cephalothorax with one pair of short posterolateral processes (Figs. 177, 197) 3

3. In addition to posterior processes of trunk, four pairs of lateral, seven unpaired dorsal (Fig. 175) and two unpaired ventral processes (Fig. 177) present . . . . ***Ch. lophii***
   In addition to posterior processes of trunk, one pair of short anterolateral and one of long lateral processes present (Fig. 198) . . . . ***Ch. merluccii***

4. Both simple and subdivided trunk processes present (Fig. 132) . . . . ***Ch. zei***
   Only simple trunk processes present 5

5. In addition to posterior processes of trunk, six pairs of lateral, one of ventrolateral processes, four unpaired dorsal tubercles and one ventral tubercle present (Figs. 213–215) . . . . ***Ch. nodosus***
   Dorsal conical swelling on cephalothorax (Fig. 234); in addition to posterior processes of trunk, three pairs of lateral and three unpaired dorsal processes present (Figs. 232–234) . . . . ***Ch. ornatus***

Chondracanthus zei Delaroche, 1811

(Figs. 129–158)

Syn.: *Lernacantha delarochiana* de Blainville, 1822
Lernacanthus delarochiana Blainville; of Valle (1880)
Chondracanthus zeus in Bassett-Smith (1899)
Chondracanthus delarochiana Blainv.; in Scott and Scott (1913)

Female (Figs. 129–133): Cephalosome dorsoventrally flattened, roughly quadrangular, equally long and broad, with distinctly notched lateral margins (Fig. 129) and well delimited dorsal shield bisected by mid-dorsal groove. Trunk longer than broad, with numerous processes. Dorsal surface

(Fig. 129) with four groups of processes (Text fig. 41K, *a–d*); group *a* of four, one of them bifid, in some specimens only three; group *b* of five processes, four in semicircle around fifth; group *c* of seven similarly arranged processes, two lateral on each side sometimes fused at bases; group *d* of six, two lateral pairs and two median single processes, posterior median extending backwards well beyond tip of abdomen. Two dorsolateral groups (Text fig. 41K, *e*, *f*) consisting of two and three processes respectively (Fig. 129). Three prominent lateral outgrowths (Text fig. 41K, *g–j*), flat and with several secondary processes each; outgrowth *g* with nine processes, two fused at bases (Figs. 131, 132; Text fig. 41G); outgrowth *h* with seven processes, one bifid at tip (Fig. 133, Text fig. 41H); outgrowth *j* homologous with posterior processes, divided into two parts (Text fig. 41J), ventral with seven processes, three single and two pairs fused for some distance, dorsal with eight processes exhibiting various degrees of fusion. Genito-abdominal tagma small (Figs. 147, 148), with well delimited components. Genital segment broader than long, dorsally with hard cuticle marked by mid-dorsal groove, ventrally with lacunae of thin cuticle; ventrolateral crescentic slits of oviduct orifices. Abdomen unsegmented, subcylindrical, about equally long and broad, posterolaterally with three pairs of lobes; ventral pair short, plump, with rounded tips, dorsal pair shorter than ventral and more pointed (Fig. 147); between them lateral pair, much longer, heavy and clavate. Uropods between ventral pair (Fig. 144), uneven in diameter; rounded apex surmounted by tapering process with frayed or subdivided tip; two setae on ventral and one on dorsal surface (also small tubercle on dorsal surface in some specimens). Egg sacs circular, laterally compressed (Fig. 133), screened by posterior process (outgrowth *j*) from lateral side. Length of cephalosome about 2 mm, trunk about 10 mm.

First antenna (Fig. 134) high in ventral aspect, flat, with pedunculate base; narrowing distally, with deep notch on lower margin between proximal and distal parts; tip cylindrical, with rounded apex; apical armature (Fig. 135) of nine setae, one of them somewhat subapical, on lower margin. Second antenna (Fig. 136) with short, stocky basal segment; terminal segment with short, stout shaft and abruptly tapering hook. Mouth parts typical for chondracanthids (Fig. 137). Mandible (Fig. 138) with short proximal part and falcate distal part; denticles on outer margin larger than those on inner margin. First maxilla (Fig. 139) with rounded apex and slightly convex medial margin; apical process with inflated base and slender tip, subapical short and fine; medial margin with small soft papilla, differing in shape from one specimen to another. Second maxilla (Fig. 143) with very strong, broad basal part; terminal claw with upturned tip and series of 10–12 denticles on outer margin; auxiliary process near base of claw on outer surface long, slender, unarmed; small seta on lower margin fine, moderately long. Maxilliped (Figs. 140–142) with robust unarmed lacertus; brachium apparently capable of rotating around its long axis, distally inflated, with two patches of subterminal denticles. Claw uneven, with slender tip and one or two secondary denticles on concave margin.

First and second legs (Figs. 145, 146) trilobate, with middle lobe "boot-shaped" and two lateral ones simple and rounded.

Male (Figs. 149, 150): Cephalosome subcircular in dorsal aspect, with convex dorsal and flat ventral surfaces. Trunk narrower than cephalosome, indistinctly segmented, posteriorly rounded, incorporating genito-abdomen. Body usually flexed, with ventral surface forming concave margin. Uropod (Fig. 157) tapering, with apparently frayed tips, two ventral setae and one dorsal on proximal half, as well as a minute setule on ventral side of base. Total length 0.5–0.75 mm.

First antenna (Fig. 151) cylindrical, with somewhat inflated base and rounded tip. (The material at this author's disposal did not allow him to ascertain the exact number of setae.) Second antenna (Fig. 152) with relatively broad base, short shaft and rather blunt hook; fine setule on concave margin; exopod short, cylindrical, apparently with two minute apical prickles and two slightly longer subapical setae. Mandible (Fig. 153) with less than 10 denticles on outer margin (one or more of them missing in some specimens); denticles on inner margin much smaller, often extending over only half of segment's length. First maxilla (Fig. 154) with rounded tip and processes relatively longer than those of female. Second maxilla (Fig. 155) with claw relatively shorter than that of female, devoid of denticles. Maxilliped with smaller batches of denticles on brachium and usually only one secondary tooth on claw (Fig. 156).

First and second legs of similar structure, though first leg larger than second. Legs bilobed (Fig. 158), lobes representing rami. Lateral lobe (exopod) with four apical setules, medial lobe (endopod) unarmed; long seta on lateral margin at base of lateral lobe.

Comments: *Ch. zei*, known for more than a century, has a relatively simple taxonomic history, its

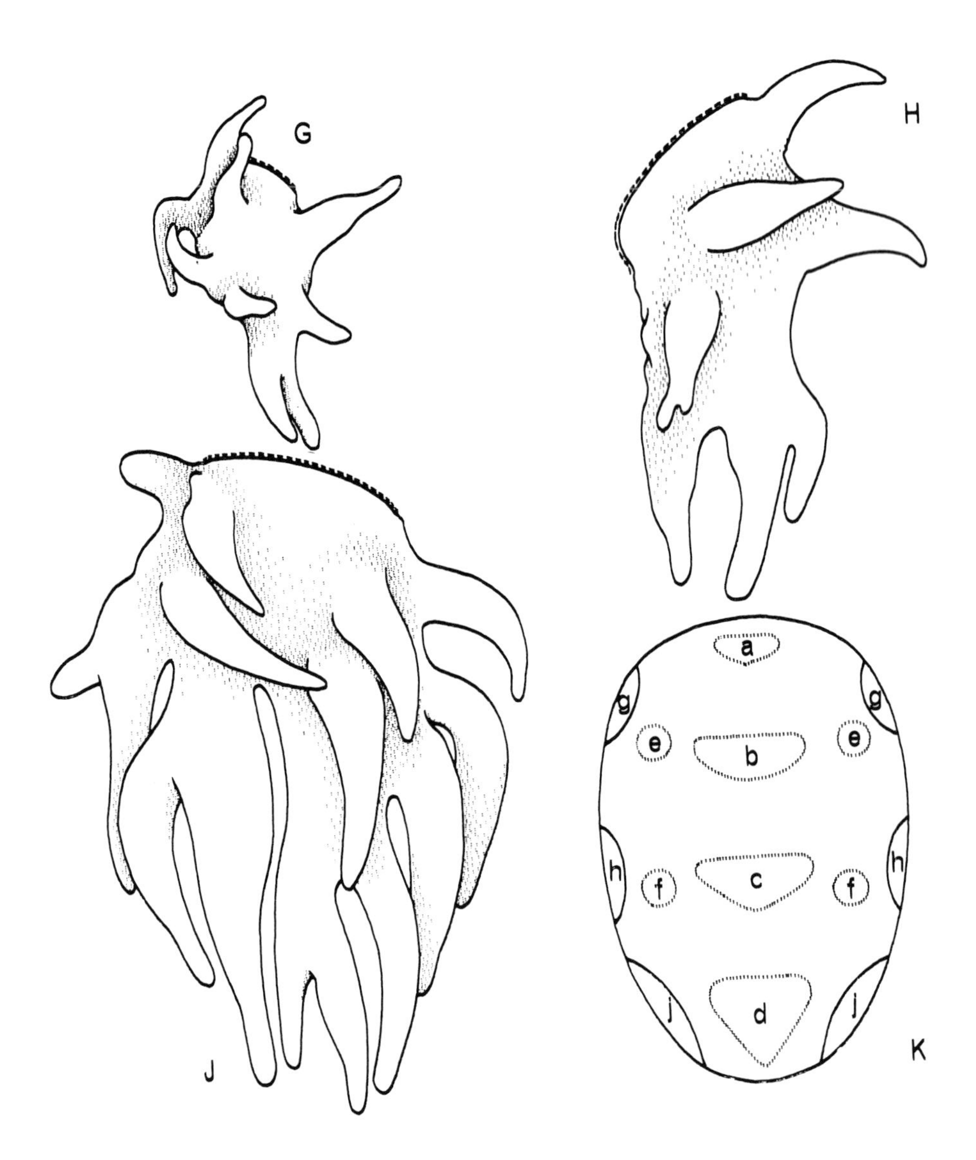

Text fig. 41. Distribution of processes on the trunk of *Chondracanthus zei* (diagrammatic). Explanations in text.

synonymy consisting of differently spelled versions of two names, one given to it by the discoverer and the other 11 years later by de Blainville (1822). The characteristic morphology of its trunk makes it readily identifiable and relatively immune to confusion. The only case of mistaken identity in its history is Bassett-Smith's (1899) suggestion that *Chondracanthus tuberculatus* von Nordmann, 1832, is a junior synonym of *Ch. zei*. (Bassett-Smith's mistake might have been due to the absence of illustrations from von Nordmann's description; it was rectified by Ho (1970, 1972c)).

Ch. zeï is a parasite of *Zeus faber*, with only one record from other species of fish. Capart (1959) found it on *Zenopsis conchifer*, another zeiid fish, in the South Atlantic. (De Blainville's (1822) statement that it occurs on "les branchies du thon" and on *Squalus* is patently unfounded.) The distribution covers the Atlantic sea-board of south-western Europe and as far South as the west coast of Africa. It has been found repeatedly in the Mediterranean and at least once in the Adriatic (Valle, 1880). Markevich (1951) mentioned its presence in the Black Sea, but did not confirm it in his later work (Markevich, 1956). *Ch. zei* is well known in British waters, though it cannot be considered common here, if only because of the comparative rarity of its host. Leigh-Sharpe (1934) commented on its scarcity in September samples off Plymouth, but added that reports had indicated it might have been more abundant earlier in the year. The distribution does not extend to the western side of the Atlantic, in view of the absence of *Zeus faber* from that area. Yamaguti (1939c) found *Ch. zei* on the same host species in Japanese waters.

The attachment site, always in the gill cavity of the fish, is marked by a prominent swelling, the so-called tumour of attachment, enveloping the entire cephalosome of the parasite. This tagma, held in rigid position, is completely incapable of motion. In many specimens it is at right angles to the long axis of the trunk (Figs. 132, 133).

Chondracanthus neali Leigh-Sharpe, 1930

(Figs. 159–174)

Female (Figs. 159–161): Cephalosome broader than long, dorsoventrally flattened; hard dorsal shield on anterior half of dorsal surface with mid-dorsal groove; posterolateral corners with long, tapering processes; anterolateral to them pair of similar though shorter processes (Fig. 159). Trunk longer than broad, with narrow anterior end forming division between it and cephalosome. Four processes in mid-dorsal line, increasingly larger in posterior direction (Fig. 160); ten lateral processes, posterior four arranged in two pairs, at least partially fused at bases. One small ventral process slightly posterior to second pair of legs. Genito-abdominal tagma (Figs. 162, 170) with short, compact genital segment bearing two small, blunt processes on midventral surface; abdomen separated by shallow constriction from genital segment, its proximal part inflated, posterior long, subcylindrical, with rounded tip. Uropod tapering, with well delimited terminal part covered by fine setules; two setae on proximal half. Egg sacs loosely coiled, much longer than body. Dimensions of cephalosome (without processes) 1.2 × 1.3 mm, body length 7.4 mm (after Ho (1972c)).

First antenna (Fig. 163) relatively large, with inflated basal and cylindrical terminal part; apical armature of at least six setae. Second antenna (Fig. 164) with short basal segment; base of terminal segment not greatly inflated, shaft tapering, hook (broken in drawing) sharp and slender. Labrum (Fig. 165) with smooth posterior margin. Mandible with 103 teeth in single row along outer margin of falcate distal part; two rows of teeth along inner margin, with 67 and 19 teeth respectively. First maxilla (Fig. 166) with rounded tip and slightly concave medial margin; apical process about $\frac{1}{3}$ longer than subapical; papilla on medial margin surmounted by two to five small digitiform protuberances. Second maxilla with 14 teeth on outer margin of claw (Fig. 167) and small subterminal tooth on opposite margin; auxiliary process robust, fairly short, with finely drawn tip, unarmed. Maxilliped resembling that of type species.

First leg (Fig. 169) smaller than second (Fig. 168), both bilobed, with short setae at base of lateral lobe.

Male (Fig. 172), of chondracanthid type, about 0.66 mm long and with cephalosome 0.3 mm wide. Minute setules on lateral margins of second leg-bearing and following segments. Uropod tapering, with three setae on basal half.

First antenna (Fig. 173) cylindrical, with six setae along anterior margin and eight in apical armature. Second antenna (Fig. 173) with very small exopod and setule on concave margin of terminal segment. Mouth parts similar to those of female (Fig. 171). Labrum with medial projection on anterior margin. Mandible with two rows of teeth on inner margin (two to six and seven to thirteen teeth respectively) and 13–21 teeth in single row on outer margin. First maxilla with long and slender armature (Fig. 171). Second maxilla (Fig. 174) with short, toothless claw; auxiliary process short, apparently with minute prickles.

Legs bilobed, with bifid outer and simple inner lobes.

Comments: The most characteristic features in the morphology of *Ch. neali* are the two pairs of processes on the cephalosome, which distinguish it from all its congeners except *Ch. distortus* Wilson, 1922, and a double row of teeth along the inner margin of the mandible. The latter feature appears to be unique in the genus.

A fairly recently discovered species, *Ch. neali* was taken on only three reported occasions, once "in deep water from S.W. Ireland" (Leigh-Sharpe, 1930) and twice in South Africa. Only one female was collected from Ireland but the South African finds contained over 60 females with about 40 males in all. On all three occasions the host was *Malacocephalus laevis*, a macrurid fish. The North and South Atlantic specimens are very similar, both in size and form. Ho (1972c) gave the number of lateral processes of South African females as eight, presumably counting the two posterior pairs as single processes. His Fig. 4B, however, shows the same number of lateral processes as Fig. 161 of this account.

Having at his disposal only the single British specimen, the author was unable to examine its appendages in detail. Their descriptions are given here *fide* Ho (1972c).

Chondracanthus lophii Johnston, 1836

(Figs. 175–195)

Syn.: *Lernentoma Dufresnii* de Blainville, 1822
Chondracanthus Lophius Risso, 1826
Chondracanthus Delarochiana Cuvier, 1829
Chondracanthus gibbosus Krøyer, 1837
Chondracanthus Delarochiana (pars); in Edwards (1840)
Lernentoma Lophii; in Baird (1850)
Lernentoma lophii, Johnston; in Thompson (1893)
Chondracanthus gibbus; in Schimkewitsch (1896)
Chondracanthus Lophii, Johns.; in Hansen (1923)
Chondracanthus abdominalis Heegaard, 1943
nec *Chondracanthus lophii* Johnston; of Barnard (1955a)

Female (Figs. 175–180): Cephalosome with anterior margin narrower than posterior, latter usually forming transverse welt (Fig. 178); posterolateral corners with short digitiform processes; well developed dorsal shield with mid-dorsal groove. Trunk five to six times longer than cephalosome, length much greater than width; seven processes in mid-dorsal line, four longer ones alternating with three shorter ones (Fig. 176); four pairs of lateral processes arranged in two more or less dorsoventral groups; posterolateral processes massive, broad, with bluntly pointed tips; two simple digitiform processes in midventral line (Fig. 177). Genito-abdominal tagma (Fig. 188) small, its components very clearly delimited; genital segment subspherical; abdomen with very narrow base, pyriform. Uropod (Fig. 189) with inflated basal part bearing two setae on ventral and one on dorsal surface; distal part slender, tapering, with small prickles on surface. Egg sacs several times longer than body, irregularly twisted (Fig. 176). Total length about 12 mm.

First antenna (Fig. 181) large, its larger basal part somewhat flattened, projecting in medial direction beyond pedunculate base; shorter distal part (Fig. 182) cylindrical, with rounded tip. Second antenna similar to that of type species. Mandible with very small and numerous denticles on both margins of distal falcate part; denticles of outer margin about as large as those of inner margin. First maxilla (Fig. 183) broad, with round, unarmed prominence on medial margin; subterminal process long and robust. Second maxilla with long and slender terminal claw (Fig. 185); usually more than 40 very small denticles on outer margin, none on inner margin; auxiliary process long, with fine tip, unarmed. Maxilliped similar to that of type species; terminal claw (Fig. 184) characteristically slender, with inflated base and usually with two secondary denticles on concave margin (some specimens with one denticle only).

First leg (Fig. 186) smaller than second (Fig. 187), both bilobed and "boot-shaped", with lateral lobe usually shorter and rounded at tip and medial longer and tapering; small seta at base of lateral lobe on lateral margin (omitted in drawings).

Male (Fig. 190) typical of its genus, though apparently more slender than that of type species and with external segmentation almost completely absent. Uropod (Fig. 194) prominent, slender, with fine, sharp tip; in addition to three setae near base, short cylindrical papilla present in some specimens. Total length about 0.5 mm.

First antenna similar to that of type species. Second antenna (Fig. 191): terminal segment with very broad base and short, sturdy shaft; small seta on concave margin; exopod short, subcylindrical, with two digitiform protuberances. Mandible as in female, but with fewer denticles. First maxilla (Fig. 193) with relatively very long apical armature and rounded prominence on medial margin. Second maxilla with terminal claw armed with relatively large five to ten denticles (Fig. 195) on outer margin, none on inner; auxiliary process very long and slender, unarmed. Maxilliped as in female but with fewer denticles in two pads of brachium and usually with only one secondary tooth on concave margin of claw.

First leg larger than second, bifid (Fig. 192); lateral lobe tapering, with one strong seta and one (sometimes two) slender setules; medial lobe shorter, surmounted by apparently soft papilla. In some specimens papilliform outgrowth present on medial margin near base; long lateral seta present. Second leg similar to first but never with medial papilliform outgrowth.

Comments: *Ch. lophii* is a common parasite of *Lophius piscatorius*, the bottom living angler fish. Its distribution extends over the North Sea, British waters and along the south-western European seaboard. Reports of its occurrence have come also from the Faroes to the North of this area and from the Mediterranean and Adriatic from the South and South-East. (Beside *Lophius piscatorius*, *L. budegasse* was recorded as a host of *Ch. lophii* in the Mediterranean.) The records often declare this copepod to be parasitic on the gill of its host, but its site of predilection is rather the gill pouch of the angler fish, where it is found in abundance. This author never saw *Ch. lophii* attached in any other place on the fish.

The relative abundance of this copepod has made it one of the better known species. Early stages of the female, fairly recently attached to the host and not yet fully developed, have been found frequently. One such specimen is illustrated in Figs. 179 and 180. Its specific morphological features are still in a rudimentary condition. The cephalosome is far advanced in its development but the posterolateral processes have not yet made their appearance. The first antennae are more slender, especially in their basal parts, their armature much better preserved than in the adult. The two pairs of the lateral processes of the trunk are beginning to bud, the two pairs of thoracic legs are already in their definitive form. Beginning with this stage, only further increase in the size of the trunk and the development of the trunk processes will mark the progress of female maturity.

Silas and Ummerkutty (1967), on the authority of Edwards (1840) and Pesta (1934), list *Thynnus* sp. tentatively among the hosts of *Ch. lophii*. This erroneous listing stems from the work of de Blainville (1822), who confused *Ch. zei* with *Ch. lophii* and gave host records inappropriate to either of these species. Thompson's (1893) record of "cod, ling . . ." as hosts of *Ch. lophii* in Liverpool Bay is also clearly incorrect.

Chondracanthus merluccii (Holten, 1802)

(Figs. 196–212)

Syn.: *Lernaea merluccii* Holten, 1802)
Chondracanthus xyphiae Guerin-Meneville, 1829
Chondracanthus merlangi Holt; of Krøyer (1863)
Chondracanthus stramineus Wilson, 1923
Chondracanthus sp.; of Wilson (1923a)

Female (Figs. 196–198): Cephalosome longer than broad, roughly conical, its anterior margin less than half width of posterior; dorsal shield with well marked posterior limit and mid-dorsal groove in its anterior half; posterior margin gently rounded, small pointed processes in posterolateral corners (Fig. 197). Trunk (excluding posterolateral processes) about four times length of cephalosome, much longer than wide; anterior end (first leg-bearing segment) forming neck-like constriction separating cephalosome from trunk; at level of second legs shoulder-like lateral expansions of trunk with small semispherical protuberances (Figs. 197, 198); lateral margins roughly parallel to each other, with pair of long ventrolateral processes at about midlength, extending posteriorly beyond tip of abdomen (Fig. 196); posterolateral processes similar in shape to, and somewhat shorter than, lateral pair; no dorsal or ventral processes. Genito-abdominal tagma (Fig. 206) small, compact, its components poorly delimited. Genital segment semispherical, its hard cuticle with mid-dorsal groove. Abdomen much smaller than genital segment, without dividing constriction, conical. Uropod with two ventral and one dorsal seta on basal part and minute prickles on tapering terminal part. Total length about 12 mm.

First antenna (Fig. 200) small, its basal part flattened, with rounded thin and translucent flange along ventral margin and two small papillae distally; terminal part short, cylindrical, indistinctly divided into two and usually deflected dorsad; apical armature of eight setae (six only in antenna illustrated). Second antenna (Figs. 201, 202) partially covered by and fitting into shallow groove on posterior surface of flange of first antenna; basal segment short, inflated, at angle with shaft of terminal segment; shaft with uneven diameter and nipple-like protuberance on posterior surface; hook tapering, sharp, deflected posteriorly from plane of shaft's axis. Labrum with denticles on posterior surface. Mandible with about 30 teeth on outer and about 20 on inner margin of falcate part; outer denticles (a) much larger than inner (b) (Fig. 199). First maxilla (Fig. 203) with terminal process long, tapering; subterminal process much shorter; no armature detected on medial margin. Second maxilla with long, only slightly curving claw, its outer margin with seven to ten large teeth, its inner margin with slight indentation near tip (Fig. 204); auxiliary process with fine tip and subterminal denticulation; basal setule very short. Claw of maxilliped (Fig. 205) rather straight, with uneven diameter and single tooth on concave margin; usual two patches of denticles on distal, inflated end of brachium.

First leg much smaller than second, both bilobed (Fig. 196); lateral lobe somewhat longer than medial, both rounded at tip.

Male (Fig. 207) with inflated subcircular cephalosome and cylindrical, indistinctly segmented posterior half of body, rounded at tip and with prominent uropods. Total length about 1.0 mm.

First antenna subcylindrical, tapering, with six setae on anterior margin and eight in apical armature. Second antenna (Fig. 208) with basal segment short, inflated base of terminal segment, short shaft and robust hook; setule on concave margin present but exopod absent. Mandible with fewer teeth than that of female. First maxilla (Fig. 209) with long and sturdy apical process and subapical process of almost equal length; (in some specimens medial margin with conical protuberance). Second maxilla with short, upturned and toothless claw (Fig. 210); auxiliary process unarmed. Claw of maxilliped (Fig. 211) almost straight, with single secondary tooth.

First leg larger than second, both bilobed; lateral lobe with one strong and one slender terminal seta; medial lobe unarmed (in some specimens with small notch, but more commonly pointed); lateral seta long (fig. 212). Small outgrowth present near base of medial margin of first leg, as in *Ch. lophii*.

Comments: *Ch. merluccii* is a parasite of the species of *Merluccius* inhabiting the Atlantic Ocean. It occurs on *M. merluccius* in the eastern North Atlantic, *M. bilinearis* in the western North Atlantic and *M. capensis* in the eastern South Atlantic. It is also harboured by hake species dwelling along the coast of Africa. The position is less clear as regards *M. hubbsi* in the western South Atlantic. According to Szidat (1955), the place of *Ch. merluccii* in that fish is occupied by *Ch. palpifer* Wilson, 1912, a related species known also from the North Pacific, where it lives on *M. productus*. Carvalho (1951), however, claimed to have found *Ch. merluccii* on *M. hubbsi* off the coast of Brazil, and Evdokimova (1973) made the same claim about that fish taken on the Patagonian shelf.

The wide distribution range of *Ch. merluccii* is accompanied by some regional differences in its morphology. Ho's (1971b) description of this species from *M. bilinearis* shows that it is considerably smaller than its relatives from the European hake. It has a shorter cephalosome with less prominent posterolateral processes. The anterolateral processes of the trunk are either smaller or absent and a rounded protuberance is present on the ventral surface of the trunk anterior to the genital segment. The first maxilla of the female appears to have, as a constant feature, a papilla on its median margin and the claw of the maxilliped shows several minute denticles near the base. Some of these differences might be attributable to the observer's errors or omissions, but others are undeniable. Similar differences prompted Wilson (1923b) to erect *Ch. stramineus* as a separate species for the parasite of the South African hake, *M. capensis*.

It seems that *Ch. merluccii* is capable of parasitizing other fishes, at least occasionally. This is suggested by Leigh-Sharpe's (1934) record of this copepod from *Trisopterus luscus* taken off Plymouth.

Ch. merluccii is quite common on hake in British waters.

Chondracanthus nodosus (Müller, 1776)

(Figs. 213–231)

Syn.: *Lernaea nodosa* Müller, 1776
Lernentoma nodosa (Müller); of Blainville (1822)
Chondracanthus williamsoni T. Scott, 1909
Chondracanthopsis nodosus (Müller); of Oakley (1930)
nec *Chondracanthus nodosus* (Müller); of Wilson (1935a)

Female (Figs. 213–215): Cephalosome broader than long, dorsoventrally flattened; dorsal shield with firm though uneven margins and mid-dorsal groove. In relaxed specimens fairly long and neck-like first leg-bearing segment; trunk about four times longer than cephalosome, less than twice its width; dorsal surface with four low, rounded swellings (Fig. 213) in mid-dorsal line; lateral margins with six short, blunt processes, usually with gap between first and second; similar process ventrolateral to and in gap between third and fourth processes (Fig. 214); another, less distinct, dorsolateral to sixth process; rounded swelling in centre of ventral surface (Fig. 215); posterolateral processes somewhat longer than lateral, digitiform, only slightly projecting beyond tip of abdomen. Genito-abdominal tagma (Figs. 217, 218) with roughly semispherical genital segment encased in hard cuticle, with mid-dorsal groove and thin ventrolateral lacunae; pair of short, slender outgrowths in midventral position. Abdomen with constriction at base, elongated, with rounded apex and uropods at about midlength. Uropod with slender, tapering tip carrying minute prickles; inflated proximal half with one dorsal seta, two ventral setae and small, papilliform ventral outgrowth. Total length about 8 mm.

First antenna (Fig. 220) small, with long, subcylindrical basal part in most specimens with slight geniculate flexion near base; narrow, short, cylindrical terminal part flexed dorsally; eight setae in apical armature, about six along anterior margin. Second antenna with broad basal segment, fairly short shaft and sharp, slender hook. Labrum with smooth posterior margin. Mandible with teeth equally large on both margins of falcate part; outer margin with 47–54 and inner with 40–43 teeth. First maxilla (Fig. 216) long, narrow, with rounded apex and robust apical process; subterminal process short, slender; small protuberance on medial margin near base. Second maxilla with heavy claw (Fig. 221) not much curved; 9–12 teeth on outer margin, single small tooth near tip on inner margin; auxiliary process subcylindrical, unarmed, with short, fine tip. Claw of maxilliped (Fig. 222) uneven, blunt, with one or two secondary teeth (in some specimens apparently fused at base).

Two pairs of about equally large trilobed legs (Figs. 223, 224); central, boot-shaped lobe representing tip of leg and both rami, lateral part (exopod) usually larger than medial, with usual small seta on lateral side of base; on either side large rounded lobes, representing outgrowths of sympod.

Male (Fig. 225) with relatively large, inflated cephalosome and small, subcylindrical posterior part with rounded apex; segmentation indistinct. Uropods prominent, tapering, with two ventral and one dorsal setae and minute prickles on distal half. Total length about 0.7 mm, width of cephalosome 0.35 mm.

First antenna cylindrical, with two stout setae on proximal half of anterior margin and 12 setae on distal half (including apical armature). Second antenna (Fig. 226) with small seta on dorsal side of basal segment; terminal segment with broad base, strong, short hook; small seta on concave margin; apparent lacuna of thin cuticle on convex margin. Mandible similar to that of female but with fewer marginal teeth (about 25 larger teeth on outer and 10–15 smaller on inner margin). First maxilla (Fig. 228) short, with very long apical and subapical processes; papilla on medial margin long, cylindrical, with tapering tip. Second maxilla varying in denticulation of claw, from five teeth (Fig. 230) to none (Fig. 231). Claw of maxilliped (Fig. 227) with broad base abruptly tapering and blunt tip; no secondary teeth; denticulation of brachium arranged in narrow strip.

Legs (Fig. 229) with lateral lobe rounded and short, surmounted by spiniform process (bifid in some specimens); medial lobe smaller, sometimes with fine tip; lateral seta often as long as leg.

Comments: *Ch. nodosus* parasitizes fish belonging to the genus *Sebastes* (Scorpaenidae) in the North Atlantic. It has been recorded from the European and North American coasts, as well as from Iceland, Greenland, the Faroes, the North and Barents Seas. Edwards' (1840) statement, repeated by Bassett-Smith (1899), that it is found on the gills of various pleuronectids, has not been corroborated by subsequent investigations. Wilson's (1935) record from *Sparisoma viridis* was later identified by Ho (1971) as a distinct species, *Ch. wilsoni.* A quantitative study (Williams, 1963) established the ends of the gill arches (both ventral and dorsal) as the sites of predilection for this copepod, though some specimens were found attached to the arches themselves.

The characteristic and readily identifiable appearance of this species is capable of great changes, depending on the degree of contraction. A good example of such change is shown in Fig. 219, to be compared with Fig. 213, a completely relaxed specimen. In view of the extensive distribution range of *Ch. nodosus* one might also expect some regional differences in morphology. Ho's (1971) careful description, based presumably on specimens from the western North Atlantic (though that author examined also material from the European seaboard), might be taken as some indication that such differences exist, when compared with the present description (based on British material).

Some reports indicate that significant differences exist in the incidence of *Ch. nodosus* in different stocks of *Sebastes* (cf. Sindermann, 1961; Yanulov, 1963; Williams, 1963). The possibility exists, therefore, that *Ch. nodosus* might be useful as an indicator species in stock discrimination studies of its host fish.

Leigh-Sharpe and Perkins (1924) described the host-parasite relationships between *Ch. nodosus* and *Sebastes*, reporting a very peculiar structure, the conjunctive tube, produced by the tissues of the fish. This tube, connecting the female genital orifice with a male enclosed in a tumour and not in direct contact with the female, was interpreted by them as a device for sperm transfer (though they found only one instance where it actually was attached to the female vulva. Ho (1971) vigorously rejected this concept as being completely at variance with the normal reproductive procedures in Chondracanthidae. It is certain that further careful investigations of this structure are required before any credence can be given to the view of the conjunctive tube as a structure produced by the host and aiding fertilization of the parasite.

Chondracanthus ornatus T. Scott, 1900

(Figs. 232–252)

Syn.: *Disphaerocephalus ornatus* (T. Scott, 1900)

Female (Figs. 232–235): Cephalosome broader than long, anterior margin with protruding antennary region, lateral margins with rectangular lateral lobes, posteriorly broader than anteriorly; dorsal surface with prominent swelling, partly bisected by fairly deep mid-dorsal groove; dorsal shield not apparent; ventral surface flat. First leg-bearing segment about half width of cephalosome, fused with and posteriorly expanding into trunk. Latter about 2.5 times longer than cephalosome, dorsally with three very large processes, cylindrical or slightly clavate, with bluntly rounded tips (Fig. 234); lateral margins also with three processes (Fig. 233) increasing in size in posterior direction; posterolateral processes small, uneven, only slightly projecting beyond tip of abdomen; no ventral processes. Genito-abdominal tagma and uropods similar to those of *Ch. lophii*. Total length about 7.5 mm (after Scott and Scott, 1913).

First antenna (Fig. 236) very small and inconspicuous, cylindrical, tapering distally, not clearly divided into two parts; apical armature (Fig. 237) of eight setae, one of them much shorter than others, with inflated base and drawn-out tip; at least three more setae and three papillae present. Second antenna (Fig. 238): terminal segment with robust shaft and slender hook; ring of transverse ridges near base of hook, small orifice with swollen rim near base of shaft. Mandible resembling that of type species. First maxilla (Fig. 240) fairly long, narrow, with robust and rather short apical process and very short subapical process; no armature on medial margin. Claw of second maxilla (Fig. 239) short, with about 10 teeth on outer margin and single tooth near tip on inner margin; auxiliary process relatively long, with short tip, unarmed. Brachium of maxilliped distally with two strip-like patches of denticles, those in distal patch distinctly larger than those in proximal; claw with one tooth on concave margin.

First leg (Fig. 241) much smaller than second (Fig. 242), both trilobed; terminal lobes as in bilobed legs of *Chondracanthus*; subterminal lobe on lateral side of sympod.

Male (Fig. 243) with inflated cephalosome and rather heavy posterior half; segmentation indistinct; uropods similar to those of type species, in most specimens with frayed and worn tips. Total length less than 0.5 mm.

First antenna similar to that of type species, no details available. Second antenna (Fig. 244) with broad basal segment; terminal segment broad at base, with short shaft and hook; setule on concave margin; no exopod. Mandible (Fig. 245) distinguished by relatively large numbers of teeth on both margins of falcate part (19 larger on outer, 23 smaller on inner margin). First maxilla (Fig. 246) long, with apical and subapical processes of about equal length and thickness; prominent rounded swelling on medial margin. Claw of second maxilla (Fig. 247) short, tapering, toothless; auxiliary process long and sturdy, unarmed. Maxilliped (Fig. 248) miniature replica of female appendage.

Legs (Fig. 249) similar to those of *Ch. nodosus*, with small, unarmed medial lobe and long lateral seta.

Juvenile male (Fig. 250), found by author attached to one of females, distinguished by relatively slender posterior part, with four segments distinguishable. Second antenna (Fig. 251) with exopod still present. Legs (Fig. 252) with rami better developed than in adult, though already unsegmented and unarmed; lateral seta relatively short.

Comments: Originally described (T. Scott, 1900) under its present name, this copepod was intended by Wilson to be transferred to *Chondracanthopsis*. According to Oakley (1930), Wilson's intention, expressed in a manuscript note, was changed after consultation with him and the species was placed in Oakley's *Disphaerocephalus*. The main reason for this decision appears to have been the copepod's possession of one pair of bilobed and one of trilobed legs. In fact, both pairs are trilobed and no other features exist that would necessitate the removal of *Ch. ornatus* from its original genus.

This copepod is a parasite of *Callionymus maculatus* and *C. festivus*, along the Atlantic and western Mediterranean shores of Europe. It was originally discovered in Scottish waters (records from both east and west coasts) and can be expected to occur in other British locations occupied by its host. It has never been found on *C. lyra* or *C. reticulatus*, in spite of their close relationship to the host species and regardless of the fact that their distribution overlaps that of the host species.

Genus Acanthochondria Oakley, 1927

Female: Cephalosome dorsoventrally flattened, varying in length and width; dorsal shield present, processes absent. Anterior end of trunk (first leg-bearing segment) forming narrow neck between cephalosome and posterior end of trunk. Latter from two to six times longer than cephalosome, slightly flattened dorsoventrally, often with marked transverse constriction at midlength; postero-lateral processes present, other processes absent. Appendages as in *Chondracanthus*. Legs always bilobed.

Male: Resembling that of *Chondracanthus*.

Type species: *Acanthochondria cornuta* (Müller, 1776).

Comments: This genus was separated from *Chondracanthus* by Oakley (1927) to accommodate species devoid of cephalosomatic and trunk processes. When Oakley (1930) wrote about it soon afterwards, he counted 17 species as its members. This author's recent count included 55 nominal species of *Acanthochondria*, four of them clearly either invalid or wrongly placed in the genus. It is quite certain that several others would not stand up to a closer examination.

The diagnosis of the genus *Acanthochondria* is straightforward enough to allow no doubts as to what constitutes the characteristic features of its species. Why, then, should we find taxonomic inadequacies in the literature dealing with it? The main source of error lies undoubtedly in the scarcity of material on which many species were established and in the intraspecific variability of those abundant species that have been repeatedly investigated. The absence of cephalosomatic and trunk processes, depriving the taxonomist of useful recognition marks, coupled with the relative uniformity of the structure of the mouth parts in the genus, compounds the difficulty. He is usually left with the shape of the head and trunk and the general proportions of the body, on which to base his decisions. While these criteria serve well enough to distinguish from one another very dissimilar species, it becomes well nigh impossible to distinguish those closely resembling one another. Given the intraspecific variability and large-scale changes in appearance due to various degrees of contraction of preserved specimens, we are placed in danger of taking phenotypic characters and artifacts for taxonomically significant features. (One case of confusion resulting precisely from these causes will be discussed when dealing with the type species.)

Another difficulty is associated with morphological changes due to age. Although little is known of the ontogeny of Chondracanthidae, changes in the proportions of the body with age have been observed. As mentioned earlier (p. 119), in *Chondracanthus* they include the gradual appearance and growth of processes that continue to enlarge for some time after maturity has been reached. Taxonomic doubts, perhaps associated with this phenomenon, are found within *Acanthochondria*. One example is afforded by *A. lepidionis* Barnard, 1955, a species with a characteristic shape of head and, particularly, dorsal shield (p. 112).

An example of a similar difficulty is *A. deltoidea* (Fraser, 1920), which has on its trunk a pair of very small ventrolateral swellings, too small to be regarded as processes by previous investigators. This author found that on properly relaxed specimens these swellings become unmistakable, though admittedly short, processes. The species, then, should be excluded from *Acanthochondria*. Shiino's (1955) *A. bicornis*, with a cephalosome very similar to that shown in Text fig. 34G, is also a doubtful member of this genus.

Some species have been placed in *Acanthochondria* on the basis of old illustrations and inadequate texts impossible to verify on specimens. *A. chilomycteri* (Thomson, 1890) was listed in this genus by Yamaguti (1963), though it is stated to have a segmented, unmodified first pair of legs. *A. psetti* (Krøyer, 1838) has a pair of prominent lateral processes on its trunk and should be returned to *Chondracanthus*.

Ideally, all nominal species of *Acanthochondria* should be redescribed on new material, given the same treatment, i.e. on completely relaxed specimens. Only fully mature females should be used and the intraspecific variability range established for all species. Clearly, it will be a long time before these conditions are fulfilled. In the meantime, great care must be exercised in describing new and identifying old species.

The structural variety in this relatively featureless genus is provided by the proportions of the body (with all the above reservations in mind). From species with short trunks (head : trunk ratio = 1 : 2) such as *A. brevicorpa* Yü, 1935, we move along the range to long-bodied species such as *A. platycephala* Heegaard, 1940 (head : trunk = 1 : 6). Some species have particularly long cephalosomes, in others they are broader than long. In a few, uncommonly long necks are present as the anterior part of the trunk, e.g. in *A. tchangi* Yü, 1935. Note should also be taken of such features as constantly present swellings of the cephalosome (Fig. 284) or trunk (Fig. 298). A good example of the latter is *A. laemonemae* Capart, 1959, with a cushion-like inflated area on dorsal side of the trunk immediately behind the cephalosome. Prominent enough to be easily noticed, such swellings must remain small enough not to be classifiable as processes. Distinguishing marks can be provided by hitherto disregarded features, such as the shape of the first antennae and of the genito-abdominal tagma, both probably less subject to artificial distortions than the trunk.

The members of the genus *Acanthochondria* are parasites of the gill cavities of various teleost fishes (with one or two species exceptionally parasitic on the gills or claspers of holocephalan hosts). Of the 50 species recognized by this author as either valid or currently unchallenged, a substantial majority of 36 have been found in the Pacific Ocean. Only 14 are known in the Atlantic. There is only one record of *Acanthochondria* from the Indian Ocean (Kirtisinghe, 1964).

The genus is represented in the British fauna by four species, identifiable with the aid of the key below.

Key to British species of *Acanthochondria*

1. Two pairs of small conical swellings on ventral surface of cephalosome (Fig. 284) . . . . ***A. clavata***
   No swellings on ventral surface of cephalosome 2
2. Pair of prominent rounded swellings on anterodorsal part of trunk (Fig. 298) . . . . ***A. limandae***
   Anterodorsal part of trunk without swellings 3
3. Length of second legs about equal to that of half of trunk (without posterolateral processes) (Fig. 317) . . . . ***A. soleae***
   Second thoracic legs much shorter than half length of trunk (Fig. 253) . . . . ***A. cornuta***

Acanthochondria cornuta (Müller, 1776)

(Figs. 253–281)

Syn.: *Lernaea cornuta* Müller, 1776
Anops cornuta; of Oken (1816)
Entomoda cornuta; of Lamarck (1816)
Lernentoma cornuta, of de Blainville (1822)
Chondracanthus cornutus; of Latreille (1829)
Chondracanthus flurae Krøyer, 1863
Chondracanthus pallidus v. Ben.; of Brian (1899a)
Chondracanthus depressus T. Scott, 1905
Chondracanthus depressus var. *oblongus* T. Scott, 1905
Acanthochondria flurae; of Oakley (1930)
Acanthochondria depressa; of Oakley (1930)
Acanthochondria soleae; of Schuurmans Stekhoven (1935) et *auct.*
Acanthochondria depressa var. *oblongata*; of Yamaguti (1963)

Female (Figs. 253–260): Cephalosome wider than long (Figs. 259, 260), its divergent lateral margins curving ventrally, with protruding posterior expansions (Fig. 258); posterior margin convex; dorsal surface with medial bisecting groove and shorter reinforcing ribs running from bases of first antennae (Fig. 259); oral region at posterior end of ventral surface. Trunk with neck-like anterior end formed by both leg-bearing segments, dorsal surface of second with or without shallow transverse groove; length of trunk 2.5 to 5 times greater than that of cephalosome, with lateral margins roughly parallel to each other, usually (though not always) with transverse constriction slightly behind midlength; length : width ratio variable; posterolateral processes robust, blunt-tipped, usually straight or convexly flexed, tips converging, more or less projecting beyond apex of abdomen. Genito-abdominal tagma (Figs. 271, 272) small, its components well differentiated. Genital segment with mid-dorsal groove and ventrolateral lacunae in its hard cuticle; two minute setiform processes in midventral area. Abdomen somewhat clavate, with a blunt rounded apex extending well beyond tips of uropods. Uropod (Fig. 273) with inflated base bearing two ventral and one dorsal setae; minute prickles on tapering distal part. Total length 5–9 mm.

First antenna (Fig. 262) prominent, subcircular in cross-section; proximal part reniform, with pedunculate base; distal part small, conical, deflected dorsally from long axis of appendage (Fig. 263); apical armature (Fig. 261) of eight setae, two of them papilliform, smaller than others; four other setae on distal part. (In many specimens distal part relatively longer than illustrated.) Second antenna (Fig. 264) with short, broad basal segment; distal segment with fairly narrow base, moderately long shaft and long, slender hook; ring of transverse ridges near base of hook. Labrum with very small conical outgrowths on lateral margins; posterior margin smooth. Falcate part of mandible with 45 teeth on outer and 40 on inner margin. First maxilla (Fig. 265) short, with robust apical and shorter subapical process with inflated base; medial margin concave, with irregularly shaped, rugose papilla at about midlength. Claw of second maxilla (Figs. 266, 267) straight, with upturned tip; 9–15 teeth on outer margin; single tooth on inner margin near tip (absent in some specimens); auxiliary process long, robust, with finely drawn tip bearing minute prickles. Maxilliped similar to that of *Chondracanthus*; two patches of denticles on brachium often coalescing; claw (Fig. 268) with broad base, sharp, curved tip and one tooth (sometimes two) on concave margin.

Second leg (Fig. 269) larger than first (Fig. 270), both bilobed; outer lobe (exopod) shorter, rounded; inner lobe (endopod) longer, pointed; cuticle of both lobes bearing minute spinules.

Male (Fig. 274): Cephalosome inflated, subcircular. Posterior part fused, roughly conical, indistinctly segmented. Main flexure between first and second legs. Uropod (Fig. 281) prominent, slender; otherwise as in female. Total length about 1 mm.

First antenna long, cylindrical, indistinctly segmented, with eight setae in apical armature and six more along anterior margin. Second antenna (Fig. 275) with short basal segment and small seta between it and terminal segment; latter with robust base, short shaft and blunt hook; fine seta on concave and apparent lacuna of thin cuticle on convex margin. (Ho (1970) described another short seta at that place.) Falcate part of mandible (Fig. 276) similar to that of female but proportionately shorter; 15 larger teeth on outer and 10–14 smaller on inner margin. First maxilla (Fig. 277) similar to that of female, with firm, subspherical medial papilla. Claw of second maxilla (Fig. 278) short, sharp, toothless; auxiliary process relatively large, apparently unarmed. Maxilliped (Fig. 279) as in female.

First leg (Fig. 280) with lateral lobe bearing 2–4 robust setae and medial lobe unarmed; lateral seta long. Second leg larger than first, lateral lobe apparently with two apical setae (only one in some specimens).

Comments: The reason for the long list of synonyms for this species is twofold. At the generic level, it suffered taxonomic vicissitudes common to many chondracanthid species discovered long before the development of modern concepts in this field had reached its present stage. At the specific level, it was a subject of confusion resulting from its lack of easily recognizable morphological features and a wide range of individual variations. The fact that it occurs on many host species created further difficulties and favoured a tendency to taxonomic fragmentation. Since particular morphological types tended to be associated with particular hosts, the temptation existed to give them separate taxonomic status, subsequently recognized by many workers (until recently including also this author). A typical *A. cornuta* (Fig. 253) is long and relatively slender; that described as *Chondracanthus flurae* by Krøyer (1863) is more stocky, with broader trunk but proportionately more slender posterolateral processes (Fig. 255). The type originally known as *Chondracanthus depressus* T. Scott, 1905 (Fig. 256) is intermediate between the other two. Specimens placed in *Ch. depressus* were so varied that T. Scott (1905) had to add to his new species a variety *oblongus*. The name *depressus* was meant to highlight the fact that this copepod showed pronounced dorsoventral flattening. In fact, specimens from *Platichthys flesus* are not more flattened than those from other fish and cannot be told apart from other forms by this character, as can be seen in Fig. 254 (a typical *A. cornuta*), when compared with *A. depressa* (Fig. 257).

Bassett-Smith (1899) was the first author to consider *A. flurae* a junior synonym of *A. cornuta*, adding also *A. soleae* to the synonymy of that species. The same view was held by Brian (1906). Hansen (1923) agreed with the relegation of *A. flurae* to synonymy, as did Heegaard (1947) and Ho (1970). This author now agrees with the decisions of his predecessors, as summed up by Ho. He suggests, however, that there is merit in retaining the name *flurae* in trinomial nomenclature used also for *Clavella adunca* (q.v.). One would recognize, then, *A. cornuta f. cornuta* and *A. cornuta f. flurae*. *A. depressa* simply cannot be distinguished.

A. cornuta is a parasite of flatfishes in the North Atlantic, known from both the European and the American side of the ocean. The taxonomic difficulties discussed above throw doubts on the validity of many of its records. Thompson and Scott (1903) claimed to have found it in Ceylon on *Cynoglossus oligolepis*, a record never again repeated in spite of subsequent investigations in that area. Wilson (1935b) recorded it from *Hippoglossus hippoglossus* (presumably *H. stenolepis*) in Alaska. Ho (1970) confirmed this puzzling record by re-examining Wilson's material. (One must take into account the possibility of mislabelling samples in this case. This author examined many specimens of *H. hippoglossus* in the North Atlantic without finding a single chondracanthid specimen.) Bere (1930a) reported finding *A. cornuta* (labelled *Ch. depressus*) on the west coast of North America, in the Vancouver Island region, on *Hexagrammos stelleri*, a very doubtful record from an unusual host and locality.

The list of other hosts is as follows: *Hippoglossoides platessoides*, *Eucitharus linguatula*, *Glyptocephalus cynoglossus*, *Glyptocephalus* sp., *Lepidorhombus whiffiagonis*, *Limanda limanda*, *Microstomus kitt*, "*Platessa halm*", *Pleuronectes pallasi*, *P. platessa*, *Pleuronectes* sp., *Pseudopleuronectes americanus*, *Scophthalmus maximus*, *Solea solea*, *Terhops oligolepis*, *Xiphias gladius*. Bresciani and Lützen (1962) listed "gadids and flounders" as hosts of this copepod.

There is no doubt that many of these species have been recorded in error. For example, Schuurmans Stekhoven (1935) found *Acanthochondria*, which he identified as *A. soleae*, on *Pleuronectes platessa*, *Platichthys flesus* and, unaccountably, on *Gairdrops mediterraneus*. It is clear from his figures, however, that the copepod was *A. cornuta*. This author (Kabata, 1959) found that in the northern North Sea, *A. cornuta f. cornuta* occurs only on *Glyptocephalus cynoglossus* and, rather rarely, on *Pleuronectes platessa*. *A. cornuta f. flurae*, on the other hand, can be found mainly on *Hippoglossoides platessoides* and *Platichthys flesus*. Out of 254 specimens of *Lepidorhombus whiffiagonis* examined, only one carried a single specimen of this copepod. *A. cornuta* was never found by the author on many examined *Solea solea*, *Microstomus kitt* or *Limanda limanda*, all of them being parasitized by different species of *Acanthochondria*. *Scophthalmus maximus* was completely free of chondracanthids. So were all the species of gadoid fishes.

In the same report, Kabata (1959) described differences in the host–parasite relationships between *A. cornuta* and its different hosts. For example, on *Pleuronectes platessa* it was found only in the narrow space between the pseudobranch and the gills, while on *Glyptocephalus cynoglossus* it was distributed along the branchiostegal pocket. The author concluded that *A. cornuta*, in common with other species of *Acanthochondria* found in that area, selected confined space as its preferential habitat.

The morphological plasticity of *A. cornuta* has led Ho (1970) to believe that several other species of *Acanthochondria* are, in fact, *A. cornuta* and should be placed in synonymy with it. He included among them, albeit tentatively, *A. barnardi* Capart, 1959, *A. chilomycteri* (Thomson, 1890), *A. compacta* Markevich, 1956, *A. deltoidea* (Fraser, 1920) and *A. gemina* Heegaard, 1962. The acceptance of Ho's suggestion would greatly extend both the host and geographical range of this species. This author considers such action inadvisable under the present conditions of our knowledge about *Acanthochondria*.

Acanthochondria clavata (Bassett-Smith, 1896)

(Figs. 282–296)

Syn.: *Chondracanthus clavatus* Bassett-Smith, 1896

Female (Figs. 282–284): Cephalosome about equally wide and long; dorsal shield well developed, with bisecting mid-dorsal groove and two shorter ribs from bases of first antennae; anterior margin with rounded protrusions in lateral corners, similar protrusions on lateral margins near anterior end (Fig. 283); ventral surface with four rounded swellings, one pair next to tips of first antennae, second pair posterior to it; oral region near posterior margin (Fig. 284). Both leg-bearing segments forming narrow neck of trunk. Trunk in relaxed adult females about 4.5 times longer than cephalosome (middle specimen in Fig. 282); usually broader posteriorly, with or without shallow transverse constriction at midlength and with posterior processes usually curving inwardly. Genito-abdominal tagma small, resembling that of type species. Uropods as in type species. Total length 5–8 mm.

First antenna (Fig. 285) with pronounced distoventral swelling on proximal part; distal part short, bulbous; apical armature as in type species. Second antenna and mandible similar to those of type species. First maxilla (Fig. 286) relatively large, with prominent apical process; length of subapical process more than $\frac{1}{2}$ length of apical, base inflated; medial margin unarmed. Claw of second maxilla (Fig. 287) long and slender, straight except at tip; 7–13 teeth on outer margin; auxiliary process robust, unarmed. Claw of maxilliped (Fig. 288) resembling that of type species, usually with one tooth on inner margin (in some specimens two, arranged one behind other or side by side, as in Fig. 289).

First leg (Fig. 290) shorter than second (Fig. 291); both bilobed, lobes of first leg resembling each other, those of second less alike, medial lobe often more pointed.

Male (Fig. 292), with inflated cephalosome, suborbicular in dorsal aspect; posterior half narrower, indistinctly segmented, with rounded extremity. Uropods prominent, similar to those of type species. Total length about 0.6 mm.

First antenna as in type species. Second antenna (Fig. 293) with short, robust shaft and relatively slender, long hook; small seta on concave margin, lacunae of thin cuticle on convex margin. Mandibular teeth fewer than in female appendage. First maxilla (Fig. 294) fairly broad and short, with long, slender apical process; subapical process $\frac{2}{3}$ length of apical; rounded protuberance on medial margin. Claw of second maxilla (Fig. 295) short and straight, toothless; auxiliary process relatively long, unarmed, with fine point. Maxilliped as in type species.

Two pairs of legs (Fig. 296); median lobe small, unarmed; lateral lobe larger, rounded, with two apical setae; lateral seta long.

Comments: Unlike the type species of this genus, *A. clavata* has been mainly free of taxonomic confusion. Even though the range of its morphological variability is fairly broad, as indicated in Fig. 282, one can recognize it readily by the structure of its cephalothorax and by the characteristic inward curve of its posterolateral processes (though the latter feature is not invariable).

A. clavata lives on a single host species, the flatfish *Microstomus kitt*. Kabata (1959) found that it lives attached to the inner wall of the anterior half of the host's branchiostegal pocket. It has been found in the North Sea and adjacent waters, as well as in the Faroes. The distribution range of its host extends from the White Sea in the North to the Bay of Biscay in the South and Iceland in the West. Perhaps *A. clavata* will be found in all of these localities, but at present it appears that it does not occupy the entire range of *Microstomus kitt*. For example, a thorough survey in the White Sea by Shulman and Shulman (1953) failed to find it there.

It is all the more surprising that Wilson (1921a) claimed to have found it in the Pacific (Juan Fernandez) on *Gobiesox sanguineus*, a completely unrelated host. Wilson's (1935b) even more improbable record from the claspers of a holocephalan *Hydrolagus colliei* off the coast of California was shown to have been made in error (Kabata, 1968b).

A. clavata is common in British waters.

Acanthochondria limandae (Krøyer, 1863)

(Figs. 297–316)

Syn.: *Chondracanthus limandae* Krøyer, 1863

Female (Figs. 297–300): Cephalosome about equally long and wide, its dorsal shield with mid-dorsal groove and two shorter ribs running from bases of first antennae; posterior end broader, lateral margins forming ventrally turned flanges with notches near anterior end (Fig. 297). Short neck of trunk formed by first leg-bearing segment; second leg-bearing segment with two evenly rounded dorsolateral swellings (Fig. 298), particularly noticeable in contracted specimens (Fig. 300). Trunk about three to four times longer than cephalosome, often somewhat narrower posteriorly, with or without shallow transverse constriction at about midlength; posterolateral processes straight, sometimes converging, usually not projecting far beyond tip of abdomen. Genito-abdominal tagma and uropod as in type species. Total length about 6 mm.

First antenna (Figs. 301, 302) with proximal part almost reniform, not greatly inflated; distal arising subterminally from proximal, slightly clavate and deflected dorsally from long axis of appendage; apical armature as in type species. Second antenna (Figs. 303, 304) with shaft slightly narrower at base, sometimes curving in dorsoventral plane; hook often (though not always) with marked constriction near base. Mandible with teeth of about equal size on both margins of falcate part; tip fairly long, slender. First maxilla (Fig. 305) long and narrow, with long apical and short, slender subapical process; medial margin with conical papilla. Claw of second maxilla (Fig. 306) long, curving, with indentation near tip on inner margin and with series of 10–15 teeth on outer margin; auxiliary process with fine tip bearing several minute spinules. Claw of maxilliped (Fig. 307) with broad base, slender tip and one tooth on concave margin.

Legs short and squat; first leg (Fig. 308) smaller than second (Fig. 309); both bilobed, lobes small, with shallow division between them.

Male (Fig. 310): Cephalosome inflated, suborbicular in dorsal aspect. Posterior part unsegmented or indistinctly segmented, tapering. Uropod as in type species. Total length 0.6–0.7 mm.

First antenna (Fig. 312) cylindrical, indistinctly segmented, with several setae along anterior margin (exact number not known); apical armature (Fig. 311) of four long and three short setae (one possibly missing in Fig. 311). Second antenna (Fig. 312) with very robust base, short shaft and sharp hook; small seta on concave margin, small lacuna of thin cuticle on convex margin. Mandible with characteristically large number of teeth. First maxilla (Fig. 313) about equally long and wide, with robust, long apical and slender, shorter subapical process; conical protuberance on medial margin. Claw of second maxilla (Fig. 314) short, only slightly curving, toothless; auxiliary process with long, fine tip, unarmed. Claw of maxilliped (Fig. 315) similar to that of female.

Legs (Fig. 316): both pairs similar; lateral lobe longer, with two apical short setae; medial lobe small, conical; lateral seta long.

Comments: The presence of the two rounded swellings on the anterior part of the dorsum (Fig. 298) renders this species easily recognizable. Consequently, its taxonomy is free of the confusion found in many other species of the genus. *A. limandae* lives in the branchiostegal pocket of the gill cavity of *Limanda limanda*, the common dab. It has been found in the North Sea, the Irish Sea and along the coast of Europe as far North as Murmansk. There are also records from the Faroes. Markevich (1956) reports finding it in Iceland.

Hansen (1923) identified the host of *A. limandae* in the Faroes as *Platichthys flesus*. This record is doubtful. *P. flesus* and *Limanda limanda* can easily be mistaken one for the other and the former, if present in the Faroes at all, must be extremely rare in that area.

Kabata (1959) found that of the 704 specimens of *Limanda limanda* in the North Sea only 17 (2.4%) carried *A. limandae*. Out of 70 fish in the Faroes 25 (35.7%) were parasitized. Examination of 105 fish in the Icelandic waters, however, failed to find this copepod.

Acanthochondria soleae (Krøyer, 1838)

(Figs. 317–334)

Syn.: *Chondracanthus soleae* Krøyer, 1838
nec *Acanthochondria soleae*; of Schuurmans Stekhoven (1935)

Female (Figs. 317–319): Cephalosome of about equal length and width, in some specimens rather broader posteriorly, its dorsal shield well developed, not covering entire dorsal surface; shallow indentations in anterior margin on each side of mid-dorsal groove. Trunk (posterolateral processes included) 2.5–3.5 times longer than cephalosome, its narrow neck moderately long and formed by both leg-bearing segments; length greater than width, slight transverse constriction at midlength usually present; posterolateral processes about $\frac{1}{4}$ to $\frac{1}{3}$ length of trunk, slender, often divergent, with rounded tips. Genito-abdominal tagma and uropods as in type species. Total length 7–9 mm.

First antenna (Figs. 320, 321) indistinctly divided into proximal and distal parts; proximal part with rounded swelling protruding in medial direction from base and another swelling of variable size on ventral surface (Fig. 321, vs); band of hard cuticle on concave margin; distal part short, conical, somewhat deflected dorsally from long axis of appendage; apical armature as in type species.

Terminal segment of second antenna (Fig. 322) with moderately expanded base; shaft long, with uneven diameter; hook with marked constriction at point of flexure. Falcate part of mandible with teeth of about equal size on both margins (Fig. 323). First maxilla (Fig. 324) with robust apical and very short, slender subapical process; medial margin concave, with small conical papilla on proximal side of concavity. Claw of second maxilla (Fig. 325) long, markedly curved, tip with shallow indentation, outer margin with series of about 16 teeth; auxiliary process markedly long and slender, unarmed. Claw of maxilliped (Fig. 326) slender, its base less inflated than in other species; one tooth on concave margin (two in some specimens).

First leg (Fig. 327) smaller than second (Fig. 328), both bilobed, relatively large, with long lobes; lateral lobes longer than medial, those of first leg more rounded than those of second. Lobes of second leg longer than basal part of leg. Second leg long enough to reach level of transverse constriction of trunk.

Male (Fig. 329): Cephalosome inflated, suborbicular; posterior part unsegmented or indistinctly segmented, narrow, with rounded tip. Uropods as in type species. Total length 0.6–0.7 mm.

First antenna as in type species. Second antenna (Fig. 330) with robust base, short shaft and sharp, tapering hook; small seta on concave margin and area of thin cuticle on convex margin near base. Mandible as in type species. First maxilla (Fig. 331) subquadrangular, its apical process long and slender, subapical process almost as long as apical; relatively large rounded protuberance on medial margin. Second maxilla (Fig. 332) with concave medial margin; claw strongly curving, toothless, with long, unarmed auxiliary process. Claw of maxilliped (Fig. 333) resembling that of female.

Legs (Fig. 334) suboval, with two short apical setae and pointed, small medial lobe; lateral seta relatively short.

Comments: *A. soleae* is easy to distinguish from all other British representatives of its genus. It is larger than those species and its legs are conspicuously longer than those of most other species of *Acanthochondria*. This is particularly true of the second leg and is clearly illustrated by comparison of Fig. 317 with Fig. 253. The female is also characterized by the shape of its first antenna and the slenderness of the claws of the second maxilla and maxilliped. In some male specimens the author found that the labrum is deflected outwards so as to display the buccal cavity and appendages. Although this might be an artifact of preservation, the author failed to observe it in males of other chondracanthid species he studied.

As suggested by its specific name, *A. soleae* is parasitic mainly on the sole, *Solea solea*. Its distribution covers the Atlantic seaboard of Europe, except for its most northern part. It occurs in the North and Irish Seas but, in common with its host, it becomes rare in the northern reaches of its range. In British waters it has been taken as far North as the Firth of Clyde on the west coast. Boxshall (1974a) reported it from the neighbourhood of Whitby, Yorkshire. Most of its British records come from the Channel coast. Brian (1924) found it on the Mauritanian coast of Africa and Capart (1959) entered another African record on *Solea* sp.

The literature contains a number of erroneous or doubtful host listings copied from several original records. As mentioned earlier (p. 126), Schuurmans Stekhoven (1935, 1936a) misidentified *A. cornuta* as *A. soleae*, adding three wrong hosts to the list. His drawings (Schuurmans Stekhoven, 1935) show that he had *A. cornuta* and his text clinches the diagnosis with the statement "thoracopods as in *A. depressa*", a species synonymous with *A. cornuta*. There are, therefore, no authenticated records of *Platichthys flesus*, *Pleuronectes platessa* and *Gairdrops mediterraneus* harbouring *A. soleae*.

Ronald (1959) lists *Limanda limanda* as a host for this copepod, quoting Schuurmans Stekhoven (1936a) as his authority. The quoted paper, however, does not mention this fish. Scott and Scott (1913) refer to dab (presumably *Limanda limanda*) being listed as a host of *A. soleae* in the records of the Museum Normanianum. T. Scott (1900) reported finding it also on *Scophthalmus maximus*, but there is no mention of this find in the later monograph (Scott and Scott, 1913). Boxshall (1974a) found it on *Scophthalmus rhombus* in the North Sea. It appears that *A. soleae* might, on occasions, live on hosts other than *Solea solea* but that these occasions are, at best, very rare.

Genus Acanthochondrites Oakley, 1930

Female: Cephalosome small, with narrow anterior and broader posterior part, dorsoventrally flattened, with dorsal shield but without outgrowths or processes. Trunk subcylindrical, four to six times longer than cephalosome, with somewhat uneven lateral margins, with or without posterolateral processes; no other outgrowths or processes present; anterior end only slightly narrowed at level of first leg-bearing segment. Genito-abdominal tagma small, with components not sharply delimited. Uropods present. First antenna very small, subcylindrical, with basal part not greatly inflated. Second antenna unciform. Oral appendages as in *Chondracanthus*. Two pairs of modified, unilobed legs.

Male: Very similar to that of *Acanthochondria* but of somewhat heavier build. Appendages resembling those of female. Two pairs of vestigial unilobate legs.

Type species: *Acanthochondrites annulatus* (Olsson, 1869).

Comments: The genus was established by Oakley (1930) for chondracanthids, the females of which bear no outgrowths on either cephalosome or trunk and have two pairs of unilobed legs. The copepods belonging here are parasitic exclusively on elasmobranchs (skates and rays). While easily identifiable at the generic level, these copepods are morphologically fairly variable and present some taxonomic problems as yet not completely resolved. It is entirely possible that the genus is monotypic and represented only by its type species, *A. annulatus*. The species is present in the British fauna. The taxonomic problems it presents are discussed following its description.

Acanthochondrites annulatus (Olsson, 1869)

(Figs. 335–356)

Syn.: *Chondracanthus annulatus* Olsson, 1869
Chondracanthus inflatus Bainbridge, 1909
Acanthochondrites japonicus Gusev, 1951
Acanthochondrites inflatus (Bainbridge); of Boxshall (1974b)

Female (Figs. 335–338): Cephalosome with posterior part about twice as wide as anterior, dorsoventrally flattened; anterior margin straight, anterolateral sometimes slightly concave; shield with ill-defined posterior margin, covering about $\frac{2}{3}$ of dorsal surface; oral region in posterior half of ventral surface. Anterior part of trunk only slightly narrowed at level of first leg-bearing segment; trunk four to six times longer than cephalosome, subcylindrical, often with indistinct transverse constrictions in various positions, especially at either end of first leg-bearing segment; posterolateral processes rounded, barely projecting beyond tip of abdomen, in some specimens absent; two small processes (difficult to see) on ventral surface at border with genital segment. Genito-abdominal tagma (Fig. 346) small, without constriction between genital segment and abdomen; latter conical. Uropod short, with inflated base; proximal part with three setae, distal with short, blunt process. Total length 5–16 mm.

First antenna (Fig. 339) very small, easily overlooked, much shorter than second (Fig. 338); base somewhat inflated, distal end tapering, no distinct division into two parts; apical armature (Fig. 340) of seven terminal and three subterminal setae; several setae along anterior margin (exact number not determined). Second antenna: terminal segment with fairly short shaft and sharp hook. Labrum with posterolateral corners slightly protruding, rounded; one small papilla on each lateral margin; no denticulation. Mandible with falcate part short and broad; 19–25 teeth on outer and 15–20 on inner margin, outer teeth much larger than inner. Paragnath small, reniform, with setules on surface. First maxilla (Fig. 341) suboval; apical process short and squat, subapical with inflated base, only slightly shorter than apical (in some specimens processes fused, separate at tips only). Claw of second maxilla (Fig. 342) rather straight, with 8–12 teeth on outer margin; auxiliary process robust, with fine tip, apparently unarmed. (Ho (1970) shows fine prickles near tip.) Claw of maxilliped (Fig. 343) wide at base, with comparatively blunt tip; secondary tooth present or absent; appendage of heavy build, though otherwise similar to those of other chondracanthids.

Two pairs of unilobate legs (Figs. 344, 345), second larger than first; in some specimens tips slightly indented, in others rounded.

Male (Fig. 347), with heavily inflated anterior part and cylindrical posterior, strongly flexed in dorsal direction. Uropod (Fig. 356) prominent, tapering, with two setae on ventral side of basal half and two (one short) on dorsal side. Total length 0.6–1.5 mm.

First antenna (Fig. 348) thicker at base than at tip, with at least five setae along anterior margin; setae of apical armature prominent, exact number not established. Second antenna (Fig. 349):

terminal segment with unusually broad base; shaft virtually absent, hook small; setule on concave margin; small exopod present, with two minute apical processes observable in some specimens (delicate cuticle makes it difficult to dissect it undamaged). Labrum as in female. Falcate part of mandible (Fig. 350) with 12–17 teeth on outer and 8–14 on inner margin; inner teeth slightly smaller. First maxilla (Fig. 351) similar to that of female but with comparatively longer, more slender apical armature. Claw of second maxilla (Fig. 352) comparatively shorter than that of female and with fewer teeth. Claw of maxilliped (Fig. 353) slender, with one secondary tooth on concave margin; small conical protuberance on same margin near base.

Two pairs of unilobed legs (Figs. 354, 355) tipped with two short setae; lateral seta long.

Comments: This relatively uncommon species, clearly recognizable by its lack of cephalosomatic and trunk processes and by its unilobed legs, has had its share of taxonomic difficulties. To begin with, at least six authors (most recently Ho (1970)) believed that *Chondracanthus laevirajae* Valle, 1880, was synonymous with it, although Valle (1880) never described it. It is not correct to include a nomen nudum in the synonymy of a species only because the taxon it purports to denote is mentioned in association with a host related to that of the species in question. Another point of uncertainty arises from a comment by Norman and Brady (1909), who examined type material of *A. annulatus* and of *Chondracanthus pallidus* van Beneden, 1851 and concluded that these two taxa are identical. Should this ever be confirmed, van Beneden's name, being chronologically prior, would have to replace *annulatus*. One would also have to take another look at *Congericola pallidus* van Beneden, 1854.

Bainbridge (1909) described *Chondracanthus inflatus* as a species distinct from *A. annulatus*. The essential points of difference between them were the absence of first antennae and posterolateral processes in *Ch. inflatus*. The second antennae of that species were, on the other hand, supposed to be larger than those of *A. annulatus*. Bainbridge's observations were at fault as regards the first antennae. These appendages are present in all specimens of the genus *Acanthochondrites* known so far, though they are easy to overlook because their small size and fragility render them vulnerable to handling. The apparent discrepancy in sizes of the second antennae is no more than a semantic misunderstanding. There is no doubt, however, that Bainbridge's specimens were devoid of posterolateral processes. Her specimens were also much smaller than those of typical *A. annulatus*, measuring 5–6 mm, in contrast to about 16 mm of those specimens. (Describing specimens without posterolateral processes, both Ho (1970) and Boxshall (1974b) refer to a posteromedian lobe of the trunk. This author failed to observe any such lobe. It is possible that contraction of the trunk might produce a welt in this region and an artificial "lobe".)

Subsequent authors (except for Bere, 1930b) reduced *Ch. inflatus* to synonymy with Olsson's species, their reasons having been recently summed up by Ho (1970). Boxshall (1974b), however, re-opens the issue, supporting the validity of Bainbridge's species, under the name *Acanthochondrites inflatus*. That author argues his case mainly on the grounds of the allegedly "biramous" structure of the leg of *A. inflatus* (as opposed to the single cylindrical structure shown by Ho) and its lack of posterolateral processes. In the latter instance, Boxshall quotes Kabata (1968b) on the overriding importance of trunk processes as specific diagnostic features. Quite justifiably, he discounts the differences in size as being host-dependent.

It must be noted, however, that only the trunk processes of fully adult females can be employed as specific taxonomic characters. In some species at least they might well be size-dependent. It is entirely possible that the absence of posterolateral processes in a 5–6 mm long female *Acanthochondrites* is a consequence of its small size. As regards the "biramous" legs of Boxshall's specimens, their structure has been misunderstood by that author. The slight apical indentation cannot be construed as a division into two lobes. The lateral seta, which marks the primitive base of the lateral lobe (exopod) is a clear indication that the former rami are fused together. This is indirectly confirmed by the unilobed legs of the male. Still to be explained is the presence in the male second maxilla of an "accessory spur", a process on the outer side of the claw not observed by other investigators. It does not seem likely, however, that Bainbridge's species, defended by Boxshall, can stand up to a closer scrutiny.

Gusev (1951) described *Acanthochondrites japonicus* which was supposed to differ from *A. annulatus* by its lack of the first antennae. Here also, the lack is more than likely only apparent. There is no reason to suppose that Gusev's taxon is anything but a synonym of *A. annulatus*.

The copepods of this species are parasitic on various members of the genus *Raja*. The literature records their occurrence on *R. batis*, *R. laevis* and *R. radiata* in the North Atlantic and *R. kenojei* in the North Pacific. (*R. laevis* and *R. radiata* apparently harbour the smaller specimens without posterolateral processes.) This author found several specimens of *A. annulatus* in the nasal fossa of *R. rhina*, off the west coast of Vancouver Island, on the Pacific coast of Canada. (The Pacific specimens, like their Atlantic relatives, did possess second antennae.) An unusual host, a shark, was recorded by Leigh-Sharpe (1933b) under the name *Galeus vulgaris*. It might have been a case of aberrant host association for this copepod.

If Valle's (1880) record of *Ch. pallidus* is ever confirmed as referring to this species, its host list would be extended by an additional host, *R. oxyrinchus* and its distribution to the Adriatic (and, presumably, western Mediterranean).

Genus Lernentoma de Blainville, 1822

A member of the subfamily Lernentominae, this genus is monotypic. Its diagnosis is, therefore, identical with that of its sole representative, *L. asellina* (L.).

Lernentoma asellina (L.)

(Figs 357–379)

Syn.: *Lernaea asellina* L., 1758
Lernentoma asellina; of Blainville (1822)
Lernaeomyzon Triglae Blainville, (1822)
Chondracanthus triglae; of von Nordmann (1832)
Chondracanthus Triglae Cuv.; of Krøyer (1838–39)
Chondracanthus Triglae, de Blainv.; of van Beneden (1815a)
Chondracanthus gurnardi Kr. (sub *Ch. triglae*); of Olsson (1869)
Chondracanthus gurnardi Van Ben.; of van Beneden (1871)
?*Medesicaste triglarum* Krøyer; of Valle (1880)
Oralien asellinus; of Bassett-Smith (1899)
Oralien asellinus (Linne); of Norman and Scott (1906)
Medesicaste asellinum (Linn.); of Scott and Scott (1913)
nec *Oralien triglae* (Blainville); of Wilson (1932)
Oralien triglae (Blainville, 1822); of Leigh-Sharpe (1933a)
Lernentoma asellina (Linn., 1761); of Leigh-Sharpe (1933a)
Lernentoma trigla de Blainville, 1882; of Ho (1970)

Female (Figs. 357–362): Cephalosome with long cylindrical neck separating antennary from oral region. Antennary region projecting anteriorly from bulbous expansion; latter broader than long, with rounded lateral margins, fairly well developed dorsal shield and transverse posterodorsal welt; neck contractile, differing in length from specimen to specimen, capable of flexing in all directions, in relaxed specimens often equalling length of trunk, basal part of neck usually with bulb-like expansion, oral region on its ventral surface (Fig. 362). Trunk depressed dorsoventrally, its anterior end slightly deflected in ventral direction, long axis of trunk usually parallel and dorsal to that of cephalic neck (Figs. 357, 361); deep waist-like constriction at about midlength dividing trunk into two distinct parts; anterior part with shoulder-like anterolateral corners carrying dorsally fairly short pointed processes (Fig. 359); posterior part with pointed anterolateral processes, varying in number from one (Fig. 359) to two pairs (Fig. 360), second pair larger or much smaller than first; in specimens with two pairs of processes sometimes small conical protuberances present on lateral margins behind second pair; lateral margin of posterior half rounded and converging to fairly short posterolateral processes. Genito-abdominal tagma (Fig. 372) with its two components well demarcated; genital segment conical, broader than long, with mid-dorsal groove in hard cuticle and ventrolateral lacunae of thinner cuticle associated with slit-like orifices of oviducts. Abdomen with narrow base, slightly inflated basal and longer, cylindrical terminal part with rounded apex; one dorsal and two ventral setae on basal part. Egg sacs straight, cylindrical; eggs multiseriate. Total length 6–13 mm.

First antenna (Fig. 363) very small, digitiform, with rounded tip; eight apical and four subapical setae; no armature determined on anterior margin. Second antenna (Fig. 365) with very short shaft and strong hook; tip of hook deflected from plane of long axis of appendage (Fig. 364). Falcate part of mandible (Fig. 366) comparatively short and broad, with 24–27 larger teeth on outer convex margin and similar number of smaller ones on inner concave margin. Paragnath small, with rounded tip and several apical prickles. First maxilla (Fig. 367) with rounded tip; apical process short, with inflated base, subapical half as long, subtriangular; medial margin with low rounded protuberance on basal part. (Ho (1970) reported presence of small prickles on distal half of this appendage, but they were not observed by this author.) Claw of second maxilla (Fig. 368) fairly long, with slender upturned tip; 8–10 teeth on outer margin; auxiliary process relatively short and squat, unarmed; basal seta very short (sometimes no more than a small knob). Claw of maxilliped (Fig. 369) fairly long and slender, with only slightly inflated base and one tooth on concave margin.

Two pairs of bilobed legs, those of second pair lateral to those of first (Fig. 358). First leg (Fig. 370) slightly smaller than second (Fig. 371). Notch between lobes shallow, lateral lobe (exopod) more rounded than medial (endopod).

Male (Fig. 373), with cephalosome suborbicular in dorsal aspect and cylindrical, ventrally flexed, indistinctly segmented posterior part. Uropods prominent, resembling those of female. Total length 0.75–0.85 mm.

First antenna (Fig. 374) cylindrical, with rounded apex; armature difficult to determine (but Ho (1970) reported apical armature of eight apical setae and six setae along anterior margin). Second antenna short, with broad base of terminal segment passing almost directly into sharp, sturdy hook; setule on concave margin, another at base of terminal segment; lacuna of thin cuticle on convex margin. Mandible as in female but with fewer teeth on falcate part. First maxilla (Fig. 375) broad at base, rather short, with both terminal and subterminal armature long, slender and of similar length; rounded swelling on medial margin. Claw of second maxilla (Fig. 376) relatively shorter than that of female, with only three to four teeth on outer margin and slender auxiliary process. Claw of maxilliped (Fig. 377) similar to that of female.

First leg (Fig. 378) larger than second (Fig. 379), both with well developed, unarmed medial lobe and more rounded lateral lobe armed with two apical setae (in some appendages only one seta present).

Comments: The taxonomic history of this species is probably more confused than that of any other chondracanthid. The confusion exists at both the specific and the generic level. At the former, the literature contains records of two species, distinguishable from each other by the number of lateral processes on the posterior half of the trunk (see Figs. 359, 360). At the latter, a succession of generic names was used and the two alleged species were eventually assigned to two genera. The origin of this chain of errors is traceable to Blainville (1822), who transferred Linneus' (1758) *Lernaea asellina* to a new genus *Lernentoma* but described another form of this species as *Lernaeomyzon*. The latter was placed later by Bassett-Smith (1899) in *Oralien*. The generic confusion was cleared up by Ho (1970), who rejected *Oralien* and recognized *Medesicaste* Krøyer, 1838 (a valid genus mistaken for *Lernentoma*) as a member of the subfamily Chondracanthinae, unrelated to this genus.

This author does not accept the existence of two species of *Lernentoma*. Their separate identity was doubted by many authors, some of the early ones being Krøyer (1838–39) and Edwards (1840). The only difference detectable between specimens with one pair of lateral processes on the posterior half of the trunk (Fig. 359) and those with two pairs, often with an additional pair of small tubercles (Fig. 360), is just the number of processes. The author found no marked and consistent differences in the structure of the legs, which was mentioned by Ho (1970). Neither in host affiliation nor distribution range were these two forms significantly different. Moreover, the form with two pairs of processes appears to differ in the degree of development of the second pair. A good example of an only partially developed second pair is shown in Pl. XXXV, Fig. 4 of Baird (1850). The author's decision to regard both forms as belonging to one species was finally swayed by his find of both of them on a single fish. Further studies are desirable to determine the relative frequency of occurrence of these two forms, and the degree of variability in the development of the second pair of processes and the small tubercles.

L. asellina is a parasite of fishes belonging to the family Triglidae and lives in the gill cavity of these fishes. Its cephalosome is buried in the host tissues up to the base of the cephalic neck, so that the mouth is in position just above the surface of the skin. The hosts recorded so far are: *Aspitrigla cuculus*, *Eutrigla gurnardus*, *Trigla capensis*, *T. lucerna*, *T. lyra* and *Trigloporus lastoviza*. There are several odd host records. T. Scott (1900) stated that he obtained "*Oralien asellinus*" on "plaice, halibut and other fishes". Scott and Scott (1913), however, retained only *Pleuronectes platessa* of these unusual hosts. A record of *Gadus* sp., the source of which the author was unable to trace, is repeated several times in the literature, sometimes with an expression of doubt (Markevich, 1956). The author found a small sample of *L. asellina*, consisting of six very large specimens, in the collection of the Zoology Department, University of Aberdeen, labelled "From the gills of whiting". It is possible that nothing more than a labelling error was involved. On the other hand, a very occasional occurrence of *L. asellina* on hosts other than Triglidae must not be dismissed altogether.

The distribution of *L. asellina* is confined to the eastern Atlantic. Ho (1970) discovered that Wilson's (1932) record of this species from the American Atlantic seaboard was a case of mistaken identity. The range extends into the western Mediterranean, from which the author had some samples taken by Dr. L. Euzet off the coast of France. Valle's (1880) record from the Adriatic is uncertain, since he used the name *Medesicaste* and possibly had a true *Medesicaste*. The northern distribution limit appears to be the southern edge of the Norwegian Sea, the southern limit was drawn by Barnard's (1955b) record from South Africa.

L. asellina can be found on gurnards in British waters but, as elsewhere, it is far from abundant.

Family Philichthyidae

Morphology

The family Philichthyidae comprises currently about 30 species, the most salient common feature of which is their endoparasitic or nearly endoparasitic mode of life. Its members are unique among the copepods parasitic on fishes in having become adapted to living completely enclosed within the tissues of their hosts. Most of them inhabit subcutaneous spaces such as mucus ducts or the lateral line canal, where they cause local swellings, usually visible externally. (One type of such swelling, produced on the frontal surface of the heads of *Cossyphus diplotaenia* and *Centrolabrus exoletus* by the presence of *Leposphilus labrei* Hesse, 1886, was considered until recently as an expression of normal and taxonomically significant variability (Quignard, 1968)). Some species, however, produce pouch-like invaginations, either from the walls of the alimentary canal, often in the rectal area, or directly from the flank of the fish. In the latter instance, a burrow is excavated between the scales and extends for about the length of three scales, the area above it being markedly swollen. The philichthyids, especially the pouch-dwellers, differ from the true endoparasites in that they permanently retain their contact with the external environment through a pore that marks their point of entry.

As in all other parasitic copepods, the mode of life of the philichthyids finds its reflection in their morphology, particularly in that of their females. Their adaptation to a highly specialized type of host-parasite relationship has resulted in numerous modifications which obscure their original structural plan and make their morphological homologies difficult to interpret. It is most unfortunate that the ontogeny of Philichthyidae is all but unknown, so that we are denied the benefit of a backward glance which would help us to trace the origins of many bizarre adult characteristics. (An excellent attempt by Izawa (1973) showed that *Sarcotaces komai* Shiino, 1953, a member of this family, has five nauplius stages preceding the copepodid stage. The copepodid, however, is still generalized enough not to give us any of the clues we seek.) Under the circumstances, the best way to gain some understanding of the philichthyid morphology is to take a look first at their comparatively less modified males.

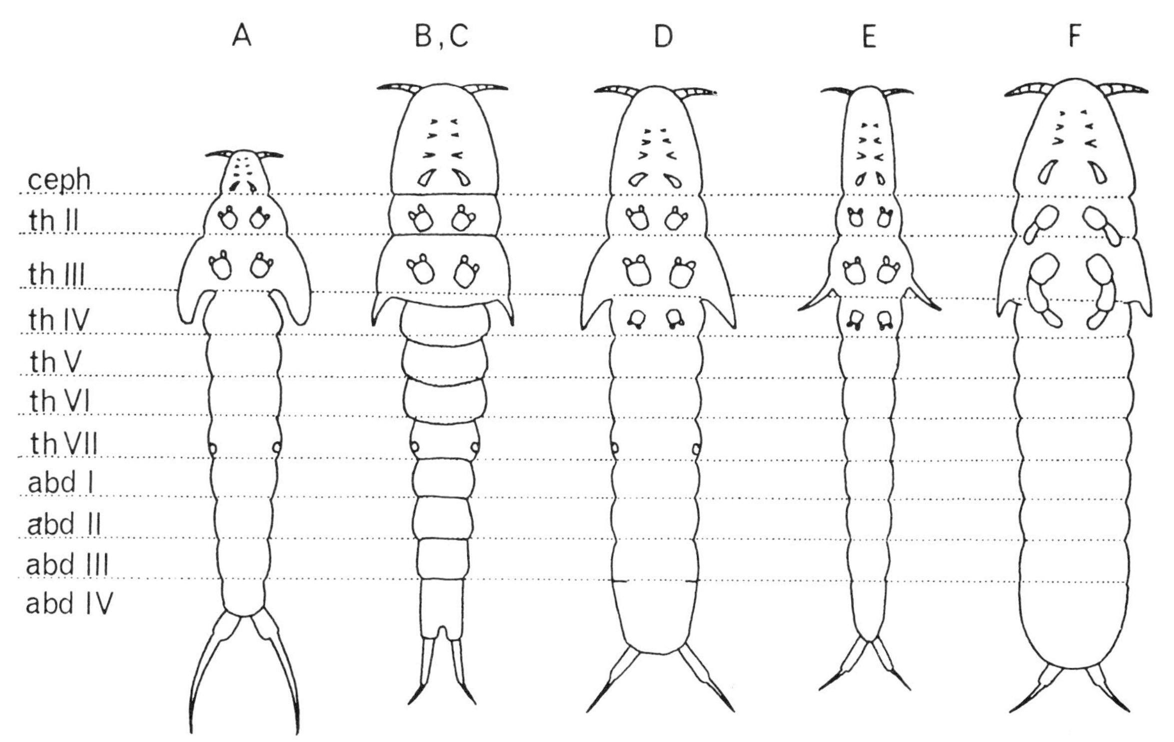

Text fig. 42. Segmentation of male Philichthyidae. A. *Sarcotaces*; B. *Colobomatus*; C. *Philichthys*; D. *Ichthyotaces*; E. *Leposphilus*; F. *Lernaeascus*. Modified from Delamare Deboutteville (1962).

The family consists of seven genera: *Philichthys* Steenstrup, 1862; *Leposphilus* Hesse, 1866; *Colobomatus* Hesse, 1873; *Sphaerifer* Richiardi, 1876; *Lernaeascus* Claus, 1886; *Sarcotaces* Olsson, 1872 and *Ichthyotaces* Shiino, 1932. The male of *Sphaerifer* has not yet been discovered. The morphology of the males of the other six genera was discussed by Delamare Deboutteville (1962) and is shown here in Text fig. 42. In all these males the first leg-bearing segment is fairly well delimited from the head, which can thus be described as the cephalosome. The only modification of the second thoracic (first leg-bearing) segment in some genera is a slight expansion in width (e.g. in *Sarcotaces*, Text fig. 42A). The third thoracic (second leg-bearing) segment, on the other hand, is recognizable in all instances by its comparatively large size and by possession of a pair of well developed posterolateral processes, most highly developed in the male *Sarcotaces* and smallest in *Lernaeascus* (Text fig. 42F). The next morphological landmark is the genital orifices. Delamare places them on the first abdominal segment. According to the interpretation accepted by this author, they are on the last thoracic segment, which would be the seventh thoracic in Delamare's segmental count.

In Text fig. 42 and in Delamare's account, the segments between the seventh thoracic and the "winged" third thoracic are clearly discernible and can be easily counted. This is not entirely correct. They are, indeed, well delimited in the male *Philichthys* (Fig. 417; Text fig. 42C), but in *Sarcotaces* the body posterior to the third thoracic segment is either without any signs of external segmentation or is indistinctly and variably subdivided by shallow transverse constrictions. Delamare took the uniformity of the male morphology as the most important and unifying familial characteristic. In many cases, however, one can only infer this uniformity by presuming homology with the structure of the male *Philichthys*. Text fig. 42 is rather idealized.

The abdomen of the male philichthyids is almost always four-segmented. Even Delamare, however, shows a partial loss of division between the penultimate and the last segments in the abdomen of *Ichthyotaces* (Text fig. 42D) and its complete loss in *Leposphilus* (Text fig. 42E).

In approaching our discussion of the morphology of the female, we are now armed with several at least tentative clues: (i) the first leg-bearing segment is probably separated from the head; (ii) an expansion might be present in the general area of the third thoracic segment; and (iii) the genital orifices would be most likely located on the seventh thoracic segment.

As mentioned previously, philichthyid copepods can be divided into two categories, according to their habitat: those inhabiting natural ducts or similar spaces in the body of the host and those that produce their own spaces to inhabit. The essential difference between these two types of habitat is that the latter form small, blind-ending sacs, whereas the former constitute parts of a larger system of spaces. It is not known how active is the movement of liquids within such spaces, but one can assume that they present more difficulties for a secure lodgement of the copepod than the close fitting sacs. Of the seven philichthyid genera, four are parasitic in mucus ducts. *Colobomatus* (Text fig. 43A) is distinguished by having several pairs of prominent processes; at least two pairs are present on the tagma fused from the fourth and fifth thoracic segments (according to Delamare (1962); in some species the third segment is also included), one pair on the genital segment and one pair, usually smaller, on the cephalosome, where also a single medial process might be present. In *Philichthys* (Text fig. 43E) there are more similar processes. In *Leposphilus* (Text fig. 43B) there are no processes but a prominent swelling extends from the third to the fifth thoracic segments. Finally, *Sphaerifer* (Text fig. 43C) has spherically inflated third to fifth thoracic segments and a pair of very large processes arising laterally from the swelling. It is not unlikely that the development of these processes, or swellings, or both, is a morphological response to the need for maintaining the position of the copepod within the duct. The pouch-dwellers have not developed such anchoring processes. *Lernaeascus* (Text fig. 43G) has an almost vermiform appearance and *Sarcotaces* (Text fig. 43D) is nearly ovate, with pointed posterior extremity and richly verrucose surface. *Ichthyotaces* (Text fig. 43F) is the only member of that group to possess several paired and single protuberances which impart to it a superficial resemblance to Chondracanthidae.

The segmentation of the females is much less clear than implied by Text fig. 43. In the externally unsegmented *Lernaeascus*, the retention of the swimming legs offers clear morphological landmarks that allow us to identify various segments. The second to the fourth thoracic segments form almost half of the entire body length, while the fifth to the seventh are compressed between the fourth and the abdomen, contributing a much smaller component of length. It is next to impossible to determine the segmentation of the female *Leposphilus*, where only the position of the seventh (?) thoracic segment can be identified by the presence of the genital orifices. The situation is not much better in *Colobomatus* and *Sphaerifer*. Delamare's (1962) interpretation of the segmentation in *Sarcotaces*

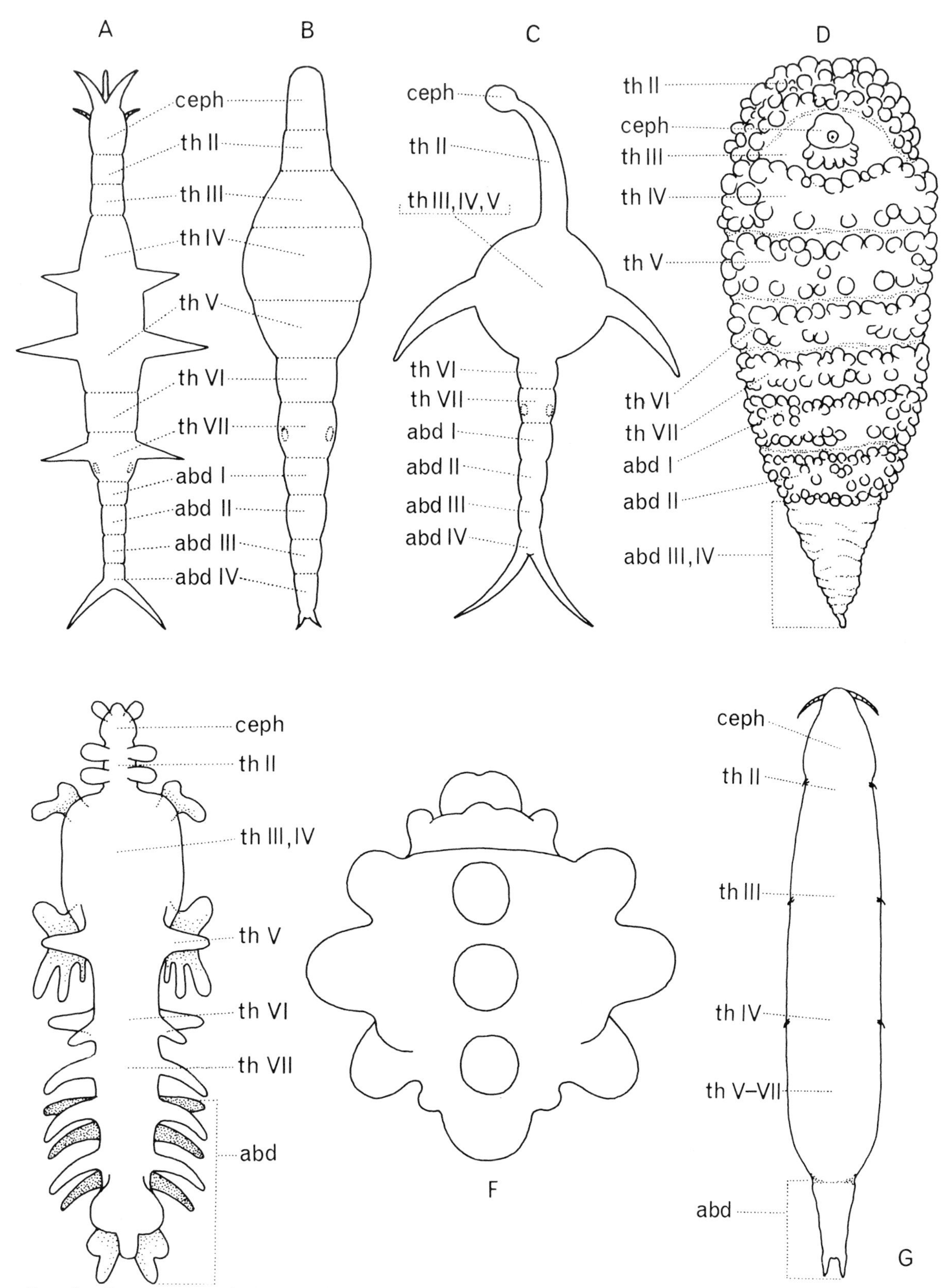

Text fig. 43. Morphological types of female Philichthyidae. A. *Colobomatus*; B. *Leposphilus*; C. *Sphaerifer*; D. *Sarcotaces*; E. *Philichthys*; F. *Ichthyotaces*; G. *Lernaeascus*. A–D modified from Delamare Deboutteville (1962).

(Text fig. 43D) is difficult to verify, especially as one finds it extremely difficult, often impossible, to locate the genital orifices on specimens. The transverse bands of verrucose protuberances are also frequently much more ill-defined than is shown in this figure. Similarly difficult to determine is the segmentation of *Ichthyotaces*. That of *Philichthys* will be discussed later.

The appendages of Philichthyidae are of the poecilostome general type. They are not sufficiently well known to be reviewed at the familial level.

History and systematics

Leydig (1851) was the first to describe a member of this family. *Sphaerosoma corvinae* Leydig, 1851, for over a decade remained an odd crustacean parasite with uncertain taxonomic position. A period of rapid discovery of the philichthyids was ushered in by the description of *Philichthys xiphiae* by Steenstrup (1862), a parasite of the swordfish, *Xiphias gladius*. Bergsoe (1864) described in detail the general morphology of this copepod, though he did not produce any descriptions of its appendages (except for the swimming legs of the male). He also stated that *Philichthys* does not fit into any of the known families of parasitic Copepoda. Two years later, Hesse (1866) described another philichthyid genus, *Leposphilus*, and proposed that it be placed in a separate family group taxon Lerneosiphonostomiens, a name never used subsequently.

Most of the currently known species of the family were described within 20 years of the discovery of *Ph.xiphiae*, mainly by Richiardi (1876a, 1876b, 1877a, 1877b, 1880, 1883). In addition to describing the majority of the species of *Colobomatus* (placed by him in the genus *Philichthys*), Richiardi also established the genus *Sphaerifer*. This name he applied to Leydig's *Sphaerosoma corvinae*, preoccupied by a coleopteran genus *Sphaerosoma* Samouelle, 1819.

The period covered by Richiardi's work produced a growing understanding that this highly unusual group of copepod parasites should be placed in a separate family. Hesse (1873), establishing his new genus *Colobomatus*, proposed for it also a new family, "Lerne-apodiant", referring to the absence of the swimming legs in the adult females. A paper of importance in establishing the foundations of the family was, however, that of Vogt (1877a), who proposed the name Philichthyidae and gave the first definition of the family, as follows.

"Males distinctly articulated, carrying two pairs of antennae, two pairs of maxillae (first one modified into powerful hook) and two pairs of swimming legs, both biramous. Dorsal cutaneous processes on second thoracic segment. Occasionally one pair of rudimentary swimming legs.

Females more or less articulated, devoid of articulated locomotory appendages but frequently equipped with lateral, non-articulated, poorly developed appendages. Antennae and buccal appendages more or less rudimentary."

As proposed by Vogt, the family comprised *Philichthys*, *Leposphilus*, *Colobomatus* and *Sphaerifer*. Vogt also suggested that Richiardi's (1876a) *Philichthys sciaenae* be removed to a different genus. Richiardi (1877b) took up the suggestion by transferring it to a new genus *Polyrhynchus*, later synonymized with *Colobomatus*, a genus which did not find favour with that author.

Claus (1886) added to the now well established family yet another monotypic genus *Lernaeascus*.

Brian (1906) accepted Richiardi's genus *Polyrhynchus* and reduced *Colobomatus* to synonymy with it, if only by ignoring it and referring all Richiardi's species of *Philichthys* to *Polyrhynchus*.

The present status of the family was rounded off by Delamare Deboutteville (1962), who added to it *Sarcotaces* and *Ichthyotaces*, mainly on the grounds of similarities of the male morphology between these two genera, on the one hand, and the remaining genera of Philichthyidae, on the other.

The most recent definition of the family was that given by Delamare Deboutteville (1962): "Female: body elongated, more or less segmented but without segmented locomotory appendages. Lateral lobes often present, either long and flexuous or spiniform. Antennae and buccal appendages more or less rudimentary. Five thoracic and five abdominal segments.

"Male: elongated, distinctly segmented, consisting of head, six thoracic segments and four abdominal segments. Posterior processes on third thoracic segment. Genital orifices on first abdominal segment. Three pairs of swimming legs, the posterior one very degenerate. Maxilliped with strong hooks.

"Biological characteristics: Living in mucus ducts on the heads of fishes or in lateral line canals, where they provoke formation of tumours (*Leposphilus*). Eggs multiserious. Egg sacs lightly attached, sometimes separating from the female and lying free along its flanks."

Apart from the fact that the segmentation of both sexes, as given in the diagnosis, does not conform to Delamare's own interpretation in the same paper, the diagnosis itself might conveniently be used for the purposes of identification at the familial level.

Most workers paid no attention to the place of the family in the systematics of parasitic copepods. It was generally and rather vaguely assumed that its place is with or near Cyclopoida. A completely different attitude was adopted by Yamaguti (1963) who carved out two new orders from the philichthyid genera. The order Philichthyidea contained in his scheme two families: Philichthyidae and Lerneascidae. The former contained *Philichthys*, *Leposphilus*, *Sphaerifer* and *Colobomatus* (the last-named subdivided into two subgenera: *Colobomatus* and *Polyrhynchus*). The order Sarcotacidea consisted of a single family Sarcotacidae, with two subfamilies: Sarcotacinae and Ichthyotacinae, both with nominate genera only.

While the poorly known mouthparts of Philichthyidae show their affinity with the poecilostome copepods, their segmentation (at least as shown by the male *Philichthys* and as attributed by Delamare Deboutteville to both sexes of other genera and species) differs from that of other poecilostomes parasitic on fishes. The philichthyids possess one more thoracic segment than, for example, Ergasilidae or Taeniacanthidae. These families have genital orifices on the segment that bears the vestigial sixth legs (i.e. on the fifth free thoracic segment). In Philichthyidae the first leg-bearing segment is free and the genital orifices are on the seventh. The need to keep them in a separate family is clearly demonstrable by this fact alone. At this stage, however, there is no reason to postulate for them a separate order. This author places them, therefore, in Poecilostomatoida.

The copepods of this family are particularly well protected from the eye of an observer. Their presence in their sheltered habitats is only sometimes betrayed by the formation of tumours visible on the surface of the fish. It is certain that they are much more common and abundant than their present records suggest. Were one to accept the literature as a good indicator of their distribution, one would probably come to the conclusion that they are predominantly Mediterranean. In fact they have been found in that sea in greater numbers only because Richiardi searched for them with particular diligence. This is especially true of *Colobomatus*, the largest genus of the family (25 out of 31 philichthyid species known). More recent records make this genus appear almost cosmopolitan, its presence having been established on both Asian and American shores of the Pacific, as well as in the Indian Ocean. This author is convinced that these records are but the tip of an iceberg.

In the British fauna the family is represented so far only by *Philichthys xiphiae*. Closer search is very likely to uncover more philichthyid species in British waters. The host of Hesse's (1873) *Colobomatus bergyltae*, *Labrus bergylta*, is not uncommon here. It might well carry that parasite. Search for the philichthyid copepods in home waters might prove to be very rewarding for future British workers.

Genus Philichthys Steenstrup, 1862

At one time incorporating most of the species presently belonging to the genus *Colobomatus*, *Philichthys* is now a monotypic genus. Brian (1905) mentioned finding an unnamed species without indicating either its host or its locality. While writing that paper, Brian was aware of the fact that many species had been placed in *Philichthys* incorrectly; he would not have identified any *Colobomatus* (in his nomenclature *Polyrhynchus*) as a member of this genus. Since, however, he gave no description or illustration, the validity of his identification cannot be verified and *Philichthys* must be considered as monotypic.

Philichthys xiphiae Steenstrup, 1862

(Figs. 406–432)

Female (Figs. 406–411): Cephalosome of two indistinctly separate parts: anterior very small and conical, dorsally bearing one pair of clavate outgrowths diverging obliquely forwards; posterior much larger and expanding in width posteriorly. Neck-like part narrower than and following cephalosome, equipped with two pairs of clavate outgrowths extending laterally from their bases. Posterior to neck broadly expanded part of thorax with flat lateral areas and welt-like central area indistinctly divided in four by transverse ventral grooves (Figs. 407, 409, 411); dorsal surface flat, with boot-shaped anterolateral processes pointed obliquely forwards and with tripartite postero-lateral processes (central part deeply cleft, its branches more slender and longer than outer parts). Long, narrow, thoraco-abdominal part. Dorsal surface of its thoracic section smooth (Fig. 408), ventral with transverse (segmental?) divisions and four pairs of lateral processes; first pair straight, sturdy, second short and blunt, directly ventral to third, longer and flexing dorsally; fourth equally long and flexing at level of genital orifices. Abdominal section with five pairs of similar processes, two ventral and curving dorsally, three arising dorsally and curving ventrally. Posterior extremity (Figs. 412–414) with T-shaped process ventral to the last pair of abdominal processes and tripartite processes (uropods?) on either side of longitudinal anal slit; branches of these processes pointing posteriorly, ventrally and laterally. Egg sacs multiserial, cylindrical, attached to genital orifices at about midlength (in contrast to terminal attachment of most other copepods). Total length 9–12 mm.

Cephalosomatic appendages poorly known. First antenna (*fide* Claus, 1887) uniramous, small, indistinctly four-segmented, with setae on anterior margin and apex. Second antenna apparently missing. Mandible and second maxilla unknown. First maxilla (Fig. 415) one-segmented, short, with group of about seven spiniform setae in distomedial corner and claw on distal margin; claw (Fig. 416) with sharp, upturned tip and row of about 10 teeth along one margin. Maxilliped vestigial, two-segmented. Mouth of poecilostome type; labrum centrally divided by narrow, rounded flap.

Swimming legs absent.

Male (Fig. 417): Cephalosome with evenly rounded anterior and nearly parallel lateral margins; posterior end transversely truncated. Antennary region separated from rest by heavy cuticular ridge on ventral surface (Fig. 424). Six free thoracic segments well delimited from one another; first leg-bearing segment as wide as cephalosome, anteriorly narrower, short; second leg-bearing segment about equally wide, longer, with parallel lateral margins and prominent posterolateral processes (Figs. 417, 423) protruding backwards to level of fourth free segment; third segment smaller than preceding two; fourth and fifth larger than third; genital (sixth free thoracic) segment longer than preceding one, with short setae (vestigial legs?) in posterolateral corners. Abdomen four-segmented, first three segments about equally long, gradually narrowing; fourth segment as long as preceding two combined, slender. Uropod (Fig. 432) long, slender; distal margin with two long and two short setae; one short seta on medial margin of distal half, another near lateral margin of proximal half. Total length (with uropods) about 2.5 mm.

First antenna (Fig. 418) six-segmented, first three segments partially fused but distinguishable by plaques of heavy cuticle; basal segment with single seta; second segment with five and third with three setae; three setae each on fourth and fifth segments; apical armature (Fig. 419) of three long, robust setae on anterior side, four very long and slender ones on posterior side of apex; single short seta on ventral side of three robust ones. Second antenna (Fig. 420) four-segmented; basal segment robust, with small blunt process near distal border; second segment longer and more slender, with similar process at midlength; third segment very short (Fig. 421), carrying strong hooked spine and two setae of unequal length; fourth with similar though longer spine, another very short hook and two setae of unequal length on opposite margins of segment. Mouth (Fig. 424) poecilostome, with fairly long labrum. Mandible (Fig. 426) short, one-segmented, tipped with two falciform blades of unequal length, both with two serrated margins (longer blade with only part of one of its margins serrated). Second maxilla (Fig. 425) large, one-segmented and bearing powerful terminal claw.

Two pairs of swimming legs, both biramous. Sympods two-segmented (Fig. 427), with slender short seta on second segment lateral to exopod. Rami two-segmented. Armature as shown below.

	Endopod		Exopod	
	1	2	1	2
Leg 1	1-0	4,II	0-I	4,III
Leg 2	1-0	2,III	0-I	4,II

Setae all pinnate (pinnules omitted in Figs. 427–429). Spines (Fig. 430) short, with needle-like serrations on both margins, except for longer and smooth medial spine of second segment of endopod. Lateral margins of first exopod segments denticulated. Third leg (Fig. 431) short, one-segmented, with four setae of about equal length.

Comments: The morphology of this species is still imperfectly known. In particular, the segmentation of the female and the mouth parts of both sexes require further study. In this account the interpretation of the female segmentation was based largely on Delamare Deboutteville (1962), with a change in designation of the genital segment from abdominal to thoracic. It should be remembered, however, that, in the absence of any knowledge of the developmental stages, this interpretation can at best be tentative. The segmentation of the anterior end of the female is especially obscure. The narrow neck between the cephalosome and the broad thoracic part might be a fusion of two segments rather than a single segment. The divisions of the cylindrical part of the thorax and of the abdomen are based on the existence of the lateral processes and transverse constrictions along the dorsal side. On the other hand, similar constrictions on the broad anterior part of the thorax (Figs. 409, 411), which would suggest that as many as four segments are included there, are disregarded. Only two segments are assumed to have formed that part (Text fig. 43E). The segmentation of the abdomen is equally unclear. The presence of the three pairs of dorsal processes on the abdomen might indicate three anterior segments of the abdomen, with the terminal segment carrying greatly modified uropods. The peculiar T-shaped ventral process must not be interpreted as a true segment. Its relationship to other structures of the posterior extremity of the female is clear from Figs. 412–414. (The difficulty of interpreting the structure of the female is best illustrated by the fact that Bergsoe (1864), who described it in considerable detail, reversed the dorsal and ventral surfaces. This mistake was at least in part caused by the fact the *Ph. xiphiae* is rather unusually flexed, with the dorsal side concave and the ventral convex.)

The segmentation of the male is much easier to understand. An interesting clue to the function of the posterolateral processes of the second leg-bearing segment is offered by Fig. 422. It is possible that these processes aid the male in maintaining its position in the mucus duct of the fish. (A similar position of these processes was observed by Vogt (1877b) in *Leposphilus labrei.*)

There is also some difficulty in finding homologies of the cephalosomatic appendages of *Ph. xiphiae*. Their designation in this description is based in part on Claus (1887). The absence of the second antenna in the female was explained in his work by their incorporation into the system of chitinous bars and struts constituting the framework of the mouth. The first maxilla of the female (Fig. 415) bears some resemblance to the corresponding appendage of other poecilostomes, but could also claim similarities with their second maxillae. The male mandible is reminiscent of that appendage in Shiinoidae (a poecilostome family recently established by Cressey (1975)). Its position, however (Fig. 424) could be interpreted as placing it posteriorly to the limb designated as the second maxilla. In our present state of knowledge no firm decisions on these homologies can be made.

Ph. xiphiae is a parasite of the ducts and sinuses in the frontal bones of the swordfish, *Xiphias gladius*. According to the present distribution records, it is largely confined to the northern Atlantic. It was found in the Kattegat (Bergsoe, 1864), in the western Mediterranean (Brian, 1905) and in the Adriatic (Valle, 1880). Its records along the coast of New England were summarized by Wilson (1932). Thomson (1890) recorded its occurrence in the same host in New Zealand waters and his record was accepted by Hewitt and Hine (1972). Thomson's drawing, however, differs in a number of essential points from the specimens examined by this author. The New Zealand copepod appears not to have lateral expansion of the anterior part of the thorax; it has strange two-segmented outgrowths behind the cephalosome. The number and shape of the lateral thoracic and abdominal processes differ from those of the Atlantic specimens. Thomson's material should be re-examined before the New Zealand record of this species is accepted.

In British waters *Ph. xiphiae* has been taken on a single occasion reported by Scott and Scott (1913). These authors based their report on a letter received from Dr. S. F. Harmer, who claimed to have found two specimens of *Ph. xiphiae* in a swordfish taken probably near Lowestoft. (Strangely enough the same letter is invoked by Hamond (1969), as having been written by S. F. Harmer to Patterson). At best, this parasite must be considered very rare here.

Suborder Cyclopoida

Family Lernaeidae

Morphology

Among the cyclopoid copepods parasitic on fishes, the highest degree of adaptation to parasitism has been reached by a small group of about 50 species belonging to the family Lernaeidae. Extensive morphological modifications in all genera except *Lamproglena* von Nordmann, 1832, and *Lamproglenoides* Fryer, 1964, make the morphology of fully adult females difficult to understand. They also tend to obscure the phylogenetic affinities of the family. The true nature of the lernaeid copepods can be revealed only through knowledge of their developmental stages, currently either unknown or known very imperfectly. A notable exception to this unfortunate rule is provided by the genus *Lernaea* L. which, because of its economic importance, has been studied probably more intensively than most parasitic copepods. The life cycle of its type and best known species, *L. cyprinacea* L. is known in detail. The most recent and most comprehensive work dealing with it was published by Grabda (1963). Since it is very probable that most lernaeids do not differ greatly from one another as far as ontogeny is concerned, a brief account of the morphology of the developmental stages of *L. cyprinacea* might be useful at this point as an introduction to lernaean morphology in general.

The first three stages in the life cycle of *L. cyprinacea* are nauplii (Figs. 380–382), differing from one another only slightly, except for the progressive increase in length. The third nauplius contains within its cuticle the first copepodid stage. There are five copepodid stages resembling those of other cyclopoid families. At the stage of copepodid V, the sex of the parasite is clearly distinguishable. This stage is succeeded by the young adult male and female stages, sometimes referred to as "cyclopoids". The author will apply to them the terms "pre-metamorphosis female" and "adult male".

The morphological development from one stage to the next is marked mainly by the addition of segments and appearance of additional swimming appendages. It is illustrated in Figs. 383–390 and shown in the table below.

Table 8. Sequence of segmental changes in development of *Lernaea cyprinacea*.

Stage	No. of segments		Fig. No.
	Free thoracic	Abdominal	
Copepodid I	3	1	383
Copepodid II	3	2	384
Copepodid III	4	2	385
Copepodid IV	4	3	386
Copepodid V, female	5	3	390
Copepodid V, male	5	3	387
Pre-metamorphosis female	5	4	388
Adult male	5	4	389

The pre-metamorphosis female and the adult male of *Lernaea* have segmentation identical with those of many poecilostomatoids such as *Anchistrotos* (Figs. 53, 76) and *Taeniacanthus* (Fig. 79). We do not know what this segmentation is in most other lernaeids. The species of the genus *Lamproglena*, however, which do not undergo metamorphosis, have similar segmentation when adult, although they do not have more than three abdominal segments.

One can, therefore, without too much risk of gross error, visualize a pre-metamorphosis female (and an adult male) lernaeid as having segmentation like that shown in Text fig. 44. The anterior end of the copepod is represented by its cephalothorax, its recent fusion with the first leg-bearing segment marked by a more or less noticeable groove on the dorsal surface. It is followed by five free thoracic segments, the most posterior of them being the genital segment. Next follow the four abdominal segments, gradually diminishing in size, the last of them bearing well developed uropods.

For the male, this stage marks the end of the life cycle. The female, on the other hand, is now confronted with its greatest challenge, the need for entering into close and intimate relationship with the host fish and the consequent morphological changes, extensive enough to merit the name of metamorphosis. (It must be remembered that no metamorphosis occurs in *Lamproglena* and

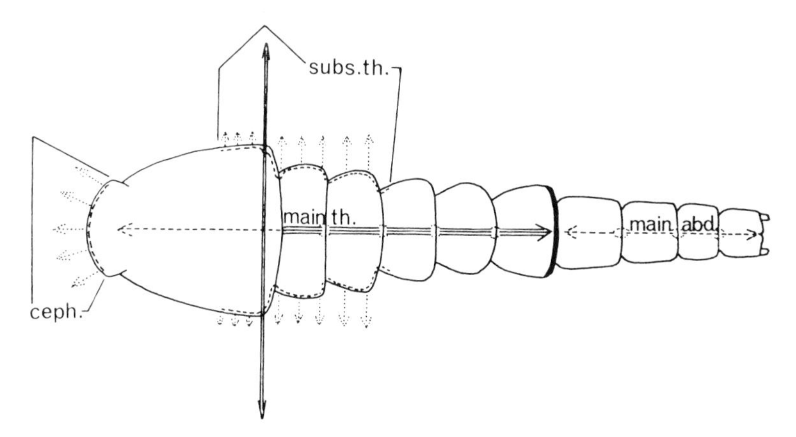

Text fig. 44. Development centres of female lernaeids.
ceph. — cephalic centre; main abd. — main abdominal centre; main th. — main thoracic centre; subs. th. — subsidiary thoracic centre.

Lamproglenoides.) These changes in size and appearance are the sum total of the processes of differential growth.

Although the definitive shapes attained by the females of different lernaeid genera can be distinctly different, they can all be seen as the outcome of the activity of several *development centres*. Varying in the levels of their activity from one genus to another, they appear to have a fairly constant location in them all. The location of the development centres is shown in Text fig. 44. A weak cephalic centre gives rise to small processes, a part of the holdfast structure in some lernaeids. The most active one is the main thoracic centre, its activity being responsible for the great extension in length of the body during metamorphosis. Since it extends from the cephalothorax to the posterior end of the genital segment, it might be termed a *development axis*. Usually its activity is at its most intensive between the third free segment and the part of the thorax posterior to it, though in many species its vigour is undiminished in the two anterior free thoracic segments. The growth along this axis is much less active in the cephalothoracic region, hence the anterior end of the lernaeid copepod does not increase in size to any extent during the course of metamorphosis. Growth of the body is invariably accompanied by the complete loss of external segmentation. Somewhat less vigorous, but not less important, is the activity of the subsidiary thoracic centre, working at right angles to the main axis of the body, and concentrated mainly in the posterior part of the cephalothorax (maxilliped- and first leg-bearing segments). It is this centre that is responsible for the development of the holdfast, the anchoring structure embedded in the host tissue. The activity of the centre diffuses, as it were, over the first two free thoracic segments, which in some genera produce a secondary holdfast or similar structure. Far less important is the main abdominal centre, producing some longitudinal extension, increase in girth and loss of external segmentation of the abdomen.

The genera of Lernaeidae fall into two distinct groups: those that undergo metamorphosis and those that do not. The latter are ectoparasitic living mainly on the gills of their hosts. The former become mesoparasites, their anterior ends buried in the host and their posteriors protruding above the site of penetration. The ectoparasitic group comprises *Lamproglena* and *Lamproglenoides* (Text fig. 45).

The genus *Lamproglena*, containing more than 20 species all parasitic on freshwater teleosts, is the most primitive member of Lernaeidae. Following the attainment of maturity, the females of *Lamproglena* increase in size only moderately and, though their external segmentation is at least partially lost, do not show any significant tagmosis. In *L. markewitschi* Sukhenko and Allamuratov, 1966, the segments between the cephalothorax and the genital segment fuse together and expand in girth to the point at which one can speak of incipient trunk formation, but this is exceptional. A more typical *Lamproglena*, *L. cleopatra* Humes, 1957, is shown in Text fig. 45. All segments present following the moult of the last copepodid stage are still more or less recognizable and their proportions have been largely retained. The difference between it and the segmentation shown in Text fig. 44 is in the three-segmented condition of the abdomen, contrasting with the four-segmented abdomen of pre-metamorphosis *Lernaea*. The genital segment is usually separated from the segments anterior to it by the small fifth leg-bearing segment that forms a waist-like constriction.

This condition resembles the exclusion of the genital segment from the poecilostome pregenital trunk, such as exists in Chondracanthidae (p. 100). It is at odds, however, with the aberrant developments in that part of the body exhibited by the metamorphosing genera of its own family.

The sizeable morphological gulf separating *Lamproglena* from other lernaeid genera was bridged by Fryer's (1964) discovery of *Lamproglenoides vermiformis*. This copepod, ectoparasitic on the gills of its freshwater teleost host, can be seen as a stage intermediate between metamorphosed and non-metamorphosed lernaeids (Text fig. 45). Though its external segmentation is partially retained, the extension in length, with distinct evidence of differential growth, disturbs the "cyclopoid" body proportions. The general appearance is sufficiently altered to allow one to talk about partial metamorphosis.

The seven mesoparasitic lernaeid genera have adult females profoundly modified as the result of metamorphosis. The essence of morphological changes imposed upon them by their mode of life consists in: (i) development of a holdfast intended to secure the position of the head within the host, (ii) elongation of the body necessary for the penetration of the host, on the one hand, and retention

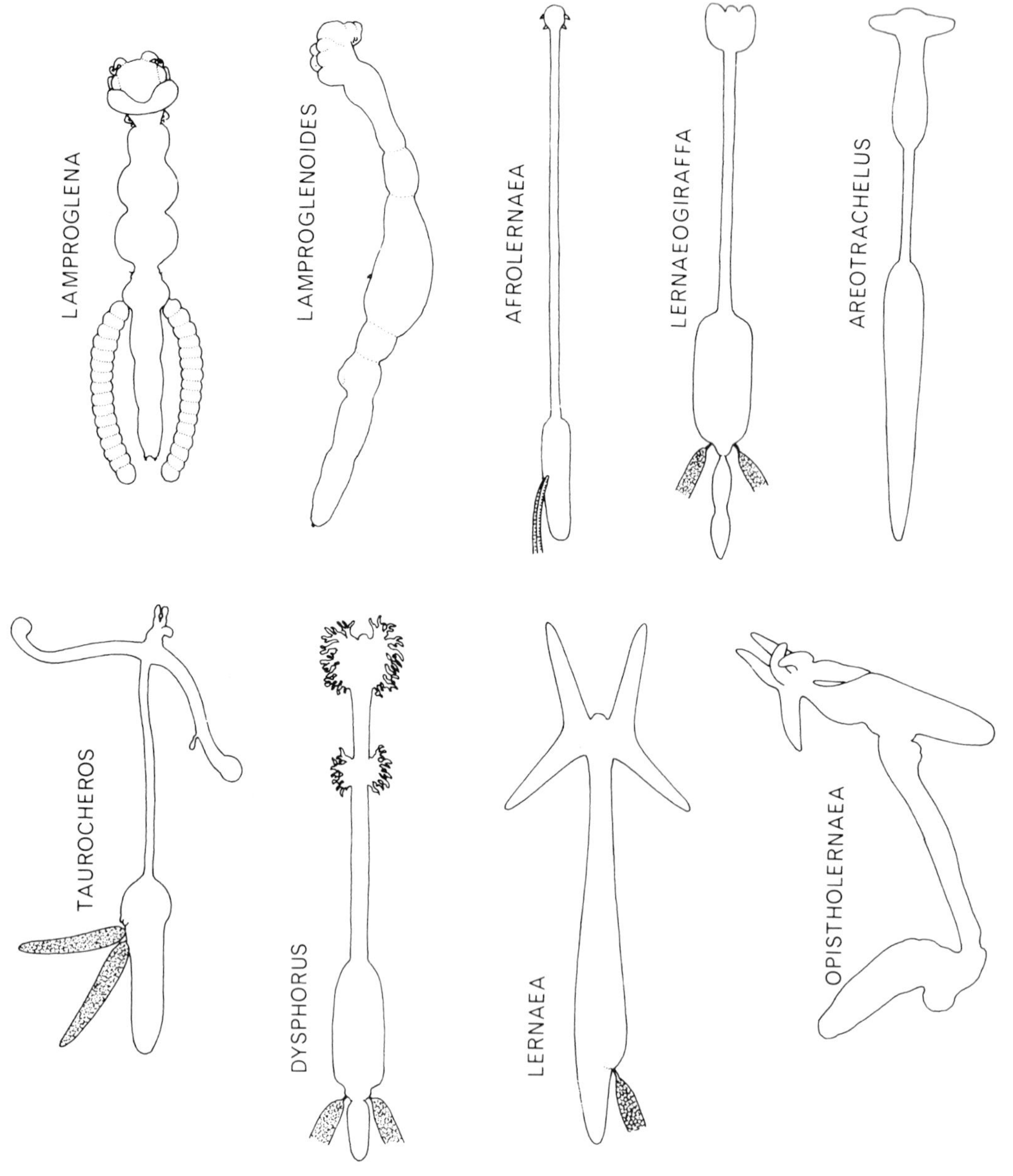

Text fig. 45. Types of morphology of adult female Lernaeidae.

of contact with the external environment, on the other, and (iii) expansion of the genital region to accommodate increased reproductive efforts. In consequence of these changes the original segmentation and appearance are lost and the body of the female becomes roughly divisible into three regions: the cephalothorax (with a more or less prominent holdfast), the more or less elongated neck, cylindrical or gradually expanding posteriorly, and the posterior part of variable structure (Text fig. 45). In discussing these regions below, attention has been paid to the position of the swimming legs, often retained though dwarfed by the prodigious growth of the female during the metamorphosis.

The cephalothorax of the genus *Afrolernaea* Fryer, 1956, is rather moderately large, subspherical, with a small anterior extension. Its holdfast is limited to two pairs of sharp-ending processes protruding laterally from the ventral surface of the cephalothorax. (The anterior of these two pairs is formed by modified second maxillae.) The posterior end of this region is marked by the posterior pair of sclerotized processes, representing the first pair of swimming legs. In *Lernaeogiraffa* Zimmermann, 1922, the holdfast consists of suboval lobes situated on each side of the head. The posterior limit of the lobes marks the limit of the cephalothorax, as indicated by the position of the first pair of swimming legs. The genus *Areotrachelus* Wilson, 1924, is based on a single find and its morphology is but poorly known. It resembles *Lernaeogiraffa* in having an only moderately developed holdfast but there the resemblance ends. The posterior end of the cephalothorax cannot be found (appendages of *A. truchae* (Brian, 1902) were never described). The transverse expansion of the holdfast is followed by a cylindrically inflated part that might be either the posterior half of the cephalothorax or the anterior part of the neck. The lateral stretch of the holdfast becomes much greater in the genus *Taurocheros* Brian, 1924, in which the thoracic subsidiary development centre produces a holdfast of four lobes, a small pair of anteriorly directed ones and a very long pair of cylindrical arms ending in bulbous swellings. The genus *Dysphorus* Kurtz, 1924, is unique among the lernaeids in possessing a holdfast characterized by numerous dendritically branching processes. In addition, a subsidiary holdfast is present on the neck. It is similarly dendritic and consists of four lobes arranged in a cruciform pattern. The holdfast of the genus *Lernaea* is well developed. It is bilaterally symmetrical as a rule (inasmuch as the conditions at the site of attachment allow normal development), usually consisting of a bi- or trifurcated branch on each side, though in some species it is more ramified (e.g. in *L. octocornua* Yin, 1960). The holdfast of *Opistholernaea* Yin, 1960, is generally similar to that of *Lernaea*.

The neck is usually a thin cylindrical structure. Judging from the position of the swimming legs, when present, it is composed of the elongated leg-bearing thoracic segments, the posterior of them tending to expand in length more vigorously than the anterior ones. In *Afrolernaea* it comprises almost $\frac{3}{4}$ of the total body length, in *Lernaeogiraffa*, *Taurocheros* and *Dysphorus* about half. In *Areotrachelus*, on the other hand, it is relatively short, constituting less than a quarter of the total body length, sandwiched between anterior and posterior inflated regions. As mentioned earlier, the neck of *Dysphorus* sports a much-branched, cruciform secondary holdfast. An expansion of the neck at about the level of the second swimming legs is also present in the species of *Opistholernaea*. The genus *Lernaea*, strictly speaking, has no clearly delimited neck, its middle region gradually expanding in girth and passing insensibly into the posterior region. In *L. tuberosa* Harding, 1950, the region behind the cephalothorax is beset with numerous peg-like outgrowths.

As regards the configuration of the posterior region, the mesoparasitic lernaeid genera can be divided into three groups with the following types of structure: (i) "posterior sac", (ii) well-defined trunk and abdomen and (iii) ill-defined trunk and abdomen.

The term "posterior sac" was coined by Fryer (1956) for the posterior part of the body of *Afrolernaea*. Morphologically, this part constitutes a fusion of the genital segment (as can be judged by the position of the oviduct orifices) and the abdomen. As the female *Afrolernaea* has not retained any swimming legs it is not possible to determine by external examination whether the anterior limit of the "sac" coincides with the anterior margin of the genital segment. It is possible that part of this segment remains incorporated in the neck, or that some of the pregenital segments form part of the "sac". Other members of this group are *Areotrachelus* and *Taurocheros*. The former has a posterior sac constituting about half of the total body length. The absence of all structural landmarks does not allow one to establish the morphological homologies of that region. In *Taurocheros* the posterior sac is inflated at the level of the genital segment and elegantly rounded at the posterior extremity. Here too, the morphological homology cannot be determined at the anterior limit, due to the absence of suitable structural landmarks.

The second group consists of *Lernaeogiraffa* and *Dysphorus*. Both genera have a cylindrical, well delimited trunk consisting of the genital segment fused with the segments anterior to it up to and including at least part of the fourth leg-bearing segment (the fourth leg is situated close behind the anterior limit of the trunk). The posterior end of the trunk, marked by the orifices of the oviducts, is sharply set off from the abdomen. The latter, much narrower than the trunk, is comparatively longer in *Lernaeogiraffa*, in which it also shows traces of segmental division in the form of a broad, shallow constriction.

The least coherent group is the one in which are included *Lernaea* and *Opistholernaea*. In *Lernaea* the anterior end of the trunk is not clearly marked off from the neck, which expands gradually and merges with it. The greatest width coincides with the level of the oviduct orifices (i.e. the posterior end of the trunk). At that level there is a more or less abrupt indentation on the ventral surface of the body, sometimes preceded by single or paired conical swellings (Fig. 391). Although no change can be found on the dorsal surface, the presence of the ventral constriction makes the abdomen much narrower than the region in front of it. It is also short, conical and usually has well developed uropods. A more definite anterior end of the trunk might be present in some species of *Opistholernaea*, which might also display a sigmoid or irregular curvature in this region. The overall effect creates a superficial resemblance to *Lernaeocera* (q.v.). The abdomen of *Opistholernaea* is greatly enlarged, as the consequence of a posterior shift in the position of the reproductive organ, now lodged, at least partly, in this part of the body.

As has been mentioned earlier, in all lernaeid genera that undergo metamorphosis the trunk, if developed, incorporates the genital segment. This situation is in contrast to that among Poecilostomatoida, where the pregenital trunk is more standard, and must be attributed to the morphological changes brought about by the process of metamorphosis.

The metamorphosed female lernaeids grow to relatively large sizes. Their appendages, however, do not become larger or undergo extensive structural changes once the pre-metamorphosis adult stage has been reached. Consequently, they become dwarfed and often difficult to find. In addition, the anterior ends of the mesoparasitic females, where most of the appendages are located, are ensheathed in strong connective tissue capsules deposited around them by the defensive reactions of the fish. The process of removal from the host and of stripping away the tissue debris only too often results in damage to, or complete destruction of, the parasite's appendages. The study of the appendages of the pre-metamorphosis stages obviates these difficulties but, unfortunately, these stages are known in only a few instances. All in all, our knowledge of the lernaeid appendages is far from complete.

The appendages of the rare and poorly known genus *Areotrachelus* have never been described. In some other genera they are only partially and imperfectly known.

The first antenna is a simple, uniramous appendage, variously and often indistinctly segmented. It is small and armed with setae along its anterior margin; its apical armature is not sufficiently well known. A typical lernaeid first antenna is shown in Fig. 393. Dollfus (1960) described *Delamarina* (= *Afrolernaea*) *nigeriensis* with first antenna short and carrying a powerful terminal claw. This appendage is structurally incompatible with the lernaeid first antenna. According to Fryer (1956) there are no antennae in *Afrolernaea*. His description of *A. longicollis* interprets the appendage corresponding to Dollfus' "first antenna" as "maxilla". The author favours Fryer's view on the homology of this appendage and accepts the absence of antennae in *Afrolernaea*.

The second antenna is also uniramous and often smaller than the first (e.g. in *Lamproglena*). It might be two-segmented (e.g. in *Lernaeogiraffa* and *Taurocheros*), three-segmented (e.g. in *Lernaea* and *Opistholernaea*) or indistinctly segmented. Various species of *Lamproglena* have been described as having second antennae of one to four segments. The setation along the anterior margin is more sparse than in the first antenna, sometimes absent. In some genera the apical armature includes a strong, claw-like hook (Fig. 395). The second antenna of *Lernaeogiraffa heterotidicola* Zimmermann, 1923, is a short, cylindrical, unsegmented and unarmed appendage.

The smallest and least well known lernaeid appendage is the mandible. It is often left out of descriptions, or passed over by a brief and imprecise statement, without accompanying illustrations. It has been referred to as "vestigial" (e.g. in *Afrolernaea nigeriensis*) or "hook-like". It consists most commonly of a squat, small segment, bearing an equally small claw.

The first maxilla has not been accurately described in *Lamproglena*. It is unknown in *Lernaeogiraffa*, *Taurocheros* and *Dysphorus*. Where known, it is a small claw, either arising directly from the surface of the cephalothorax, as in *Lernaea* or *Opistholernaea*, or mounted on a small but muscular segment, as in *Afrolernaea* (Dollfus' first antenna). A somewhat aberrant first maxilla is present in *Lamproglenoides*. It is an unsegmented, fleshy lobe with a geniculate flexion at about midlength. There are some short setae on the proximal half but the distal claw is absent.

Generally similar to the first maxilla though rather larger is the second maxilla. It is unknown in *Taurocheros*. Its most common structure is that of a claw surmounting a more or less well developed basal segment. In *Lamproglena* that segment is usually much larger than the claw, sometimes with a complicated shape. In *L. jordani* Paperna, 1964, however, the claw is quite prominent and rivals the basal segment in size. In some species of *Lernaea* the claw has a subterminal secondary tine (Fig. 397). The appendage of *Lernaeogiraffa* is two-segmented and carries two rather weak apical claws. There has been some doubt as to the second maxilla of *Afrolernaea*. Fryer (1956) believed it absent in *A. longicollis* (at least he recognized only one pair of maxillae). He described, however, a pair of strong processes, sharp-pointed and apparently unarmed. These he interpreted as the posterior pair of outgrowths, together with a similar anterior pair constituting the holdfast structure. *A. nigeriensis* (Dollfus, 1960) has similar processes, interpreted by the discoverer as the second antennae. They show vestigial segmental divisions and remnants of armature. It appears that they are greatly sclerotized and modified appendages that could be homologous with the second maxillae, judging from their general structure. As regards this appendage, *Lamproglenoides* is an odd member of its family. Its second maxilla is indistinctly segmented, with a weak apical claw and inflated base carrying a conical fleshy lobe.

The maxilliped, inasmuch as it is known, is probably the most uniform of the lernaeid appendages. It appears to be absent in *Afrolernaea*. In all other genera it is a fairly prominent limb, usually indistinctly segmented, often with a geniculate flexion at or near its midlength. In *Lamproglenoides* it carries two weak apical claws; in many species of *Lamproglena* there are three rather stronger claws (only one in *L. cornuta* Fryer, 1965). The most common number of claws, however, is five (Fig. 398), differing in length and shape from species to species.

Many lernaeid females retain some at least of their swimming legs at the post-metamorphosis stage. The number and condition of legs retained appears to bear no definite relationship to the general degree of parasitic adaptation, as can be seen in the table below. The symbols in the table are the same as those used for a similar table in Chondracanthidae (cf. p. 000).

Table 9. Number and structure of legs in female Lernaeidae.

Genus	Leg I	Leg II	Leg III	Leg IV	Leg V	Leg VI
Lernaea	B-3	B-3	B-3	B-3	v	v
Opistholernaea	B-3	B-3	B-3	B-3	v	v
Lamproglena	B-2	B-2/3	B-2/3	B-2/3	v	—
Lernaeogiraffa	B-2	B-2	B-2	B-2	v	—
Dysphorus	B-2	B-2	B-2	B-2	v	—
Lamproglenoides	B-?	B-?	B-?	B-?	—	—
Taurocheros	B-3	—	—	—	—	—
Afrolernaea	U	—	—	—	—	—

Lamproglena, the genus with the most primitive segmentation, appears to have lost the sixth pair of legs, while the greatly modified *Lernaea* has retained it. In the latter genus, however, due to the elongation of the leg-bearing segments, the legs are far apart on the body (Figs. 391, 405). The fifth and the sixth pair, when present, are reduced to small single segments, armed with several setae on the distal margin and often with a seta at the base. The first four pairs are usually similar to one another, though in *Lamproglena* the first pair often has different segmentation than that of the other pairs. In *L. jordani* and *L. intercedens* Fryer, 1965, the first two pairs are modified. In *Lamproglenoides* the rami are indistinctly segmented but some armature has been retained. *Taurocheros* has only one pair of legs, but that pair has three-segmented and fully armed rami. *Afrolernaea* retains only one pair of modified, sclerotized legs, although its pre-metamorphosis female has a full complement of these appendages. Even at that stage, however, their rami are only two-segmented (Fryer, 1956).

The morphology of the male lernaeids is similar to that of the pre-metamorphosis females, both as regards segmentation and the structure of the appendages. Differences between the sexes can be found only in the proportions of the genital segments and in the generally heavier build of the female (cf. Figs. 388, 389).

History and systematics

The recorded discovery of the first lernaeid is credited to Linnaeus, who in 1746 described and figured *Lernaea tentaculis quatuor*, later (Linnaeus, 1758) to be renamed *Lernaea cyprinacea*. From the very beginning the systematics of *Lernaea* were plagued by confusion at every taxonomic level, largely caused by a "nomenclatural switch" originated by Blainville (1822). As the result of this switch, the names *Lernaea* and *Lernaeocera* were exchanged and applied to inappropriate genera. Though belonging to two different suborders, these two highly modified genera were usually placed in the same higher taxon. Their joint systematic history is reviewed in the section on systematics of Pennellidae (see p. 293). The first authors to separate *Lernaea* from *Lernaeocera* (using these names in their incorrect sense) were Steenstrup and Lütken (1861), though that separation was based on spurious assumptions. Their compatriot and follower, Krøyer (1863, 1864) reserved for *Lernaea* (as *Lernaeocera*) a place in Lernaeocerini, a taxon embracing also the siphonostome *Tanypleurus* and the aberrant *Herpyllobius*.

Although Wilson (1917) restored to *Lernaea* its original name, he failed to recognize that it is unrelated to *Lernaeocera*. His family Lernaeidae was a composite taxon, in which the subfamily Lernaeinae formed a cyclopoid enclave among siphonostome taxa. Gurney (1933) established lack of affinity between *Lernaea* on the one hand and the siphonostome pennellids on the other and proposed that they be placed in different suborders. His proposal appears to have been insufficiently emphatic. Many subsequent authors disregarded it and continued to uphold the Wilsonian concept of Lernaeidae, lumping *Lernaea* in a taxon with a preponderance of siphonostome members. The realization that Lernaeidae are a cyclopoid family is clearly reflected in an important paper by Sproston et al. (1950). Studying the life cycle of *Lamproglena*, these authors established its cyclopoid nature and removed it from Dichelesthiidae *sensu lato*, a siphonostome taxon, to place it in the subfamily Lamprogleninae, a component of Lernaeidae. The other subfamilies of that family group were the nominate Lernaeinae and Ventriculininae. The latter subfamily, comprising cyclopoid copepods parasitic on invertebrates, is not really related to the lernaeids and was left out of the family by later workers.

Yin et al. (1963) saw Lernaeidae as consisting of three subfamilies: (i) Lernaeinae, with *Lernaea* and *Opistholernaea*; (ii) Lamprogleninae, with *Lamproglena* and (iii) Afrolernaeinae, with *Afrolernaea*.

A great contribution to our understanding of the lernaeids has been made by the work of Fryer (1956, 1958, 1959, 1961a, 1961b, 1964, 1965a, 1965b, 1968a), dealing with the freshwater Crustacea parasitic on African fishes. Rightly abstaining from dividing Lernaeidae into subfamilies, he sees them as having evolved from the ancestral lernaeid stock in three lines of descent. One of them is represented by *Lernaea* itself, a very successful genus with abundant specific radiation. The second line involves all but one of the metamorphosed genera: *Lernaeogiraffa*, *Dysphorus* and *Opistholernaea*, as well as the South American *Taurocheros*. (*Areotrachelus*, another South American lernaeid, is left out as insufficiently known.) The third line, based on the most primitive lernaeid genus, *Lamproglena*, gives rise to *Afrolernaea* (with *Lamproglena intercedens* as a connecting link) and to *Lamproglenoides* (possibly derived from the same stock as *Lamproglena cornuta* Fryer, 1965).

Fryer's analysis of the lernaeid phylogeny is, by its very nature, tentative. There can be little doubt, however, that it is largely, if not wholly, correct. Fryer took into account the fact that the lernaeid adaptations to parasitism can result in convergent, as well as in divergent, developments and tried to assess the relative primitivity of various morphological features on which his scheme was constructed. This approach, seldom resorted to in the systematics of parasitic copepods, is the only one which offers any hope of establishing a scheme of classification reflecting the true relationships within this group of parasites.

As understood here, then, the family Lernaeidae consists of nine genera, with quite varied morphology at the adult female stage. To accommodate them all in the familial fold, the following family diagnosis is proposed.

Pre-metamorphosis female (definitive stage for *Lamproglena* and *Lamproglenoides*): Cephalothorax small, well delimited from following segment, devoid of outgrowths or processes. Second to fifth thoracic leg-bearing segments gradually diminishing in width, well delimited, without outgrowths or processes (except in *Lamproglena cornuta*). Genital segment moderately enlarged, bearing sixth pair of legs. Abdomen varying in number of segments, but not exceeding four. Uropods present. Segmentation partially lost in *Lamproglenoides* and some species of *Lamproglena*. First antenna uniramous, variously or indistinctly segmented, usually bearing setae along anterior margin and with apical armature. Second antenna uniramous, two- or three-segmented, or with indistinct segmentation; setation sparse or absent along anterior margin; claw-like hook present or absent in apical armature. Mandible usually small, one-segmented, bearing apical claw. First maxilla usually one-segmented, with apical claw, sometimes subchelate (in *Lamproglenoides* unsegmented fleshy lobe with geniculate flexion). Second maxilla similar to first but usually larger. Maxilliped indistinctly segmented, with geniculate flexion; apical armature of one to five claws. First four pairs of legs biramous, rami two- or three-segmented (indistinctly segmented in *Lamproglenoides*, modified into sclerotized processes in some species of *Lamproglena*). Fifth and sixth pair, when present, reduced to single small segments with several apical setae.

Post-metamorphosis female: Consisting of cephalothorax, long thoracic neck and expanded posterior part. Cephalothorax with holdfast consisting of processes formed by modified appendages (in *Afrolernaea*), of branches or lobes, smooth, branched or with numerous dendritic processes. Neck cylindrical, smooth or bearing secondary holdfast (in *Dysphorus*), or with short, peg-like outgrowths in anterior half (in *Lernaea tuberosa*); in *Lernaea* passing imperceptibly into expanded posterior part. Latter forming posterior sac of abdomen and genital segment (in *Taurocheros*, *Areotrachelus* and *Afrolernaea*), or divided into distinct genital trunk and abdomen (in *Dysphorus* and *Lernaeogiraffa*), or consisting of indistinctly divided trunk and abdomen (in *Lernaea* and *Opistholernaea*). Antennae absent in *Afrolernaea*, otherwise cephalothoracic appendages as in pre-metamorphosis female. Full complement of legs retained only in *Lernaea* and *Opistholernaea*; sixth leg absent in *Lamproglena*, *Lernaeogiraffa* and *Dysphorus*; only four pairs present in *Lamproglenoides* and only one in *Taurocheros* and *Afrolernaea*.

Male: Resembling pre-metamorphosis female in segmentation and structure of appendages.

Lernaeidae are represented in the British fauna by the type species of the genus *Lernaea*.

Genus Lernaea L., 1758

Pre-metamorphosis female: Cephalosome fused with first leg-bearing segment, dorsoventrally flattened, very definitely separated from rest of body. Second to fifth leg-bearing thoracic segments clearly delimited, progressively smaller in posterior direction, bearing no outgrowths or processes. Genital segment narrower and longer than preceding segment. Abdomen clearly four-segmented, segments diminishing in size posteriorly; last segment deeply cleft by anal slit. Uropods well developed. First antenna uniramous, varying in number of segments, bearing setae along anterior margin and on apex. Second antenna uniramous, usually three-segmented, with only terminal segment armed, armature including one claw. Mandible one-segmented, with apical claw. First maxilla very small, wedge-shaped. Second maxilla indistinctly segmented, with simple or bifid claw. Maxilliped indistinctly segmented, with five apical claws and one spiniform seta on medial margin. Six pairs of legs present, first four biramous, rami three-segmented; fifth leg reduced to single setiferous segment, sixth to single seta.

Post-metamorphosis female: Small cephalothorax followed by well developed holdfast. Latter more or less symmetrical, of two simple or subdivided branches on each side; branches subcylindrical and long or short and lobiform. Neck usually smooth (with peg-like outgrowths on anterior part in *L. tuberosa*), expanding in posterior direction and imperceptibly passing into posterior part. More or less indistinct division between posterior end of genital segment and abdomen. Posterior extremity (abdomen) short, conical. Egg sacs multiseriate, saccular, eggs spherical. Appendages as in pre-metamorphosis female.

Male: Resembling pre-metamorphosis female in shape, segmentation and structure of appendages.

Type species: *L. cyprinacea* L., 1758.

Comments: The rich and confused history of this genus was described in the discussion of Lernaeidae (p. 148) and Pennellidae (p. 293). The members of the genus include some of the most dangerous parasites of economically important freshwater fishes. Some species of *Lernaea* have been responsible for serious losses to the fishing industry, both in the Old and New World and are a matter of interest and concern to many people.

The genus currently comprises about 40 nominal species, including some species inquirendae (*L.*

basteri Blainville, 1822; *L. multicornis* Cuvier, 1830; *L. lotae* Hermann, 1783 and *L. squamicola* Hermann, 1783). The presently recognized species of *Lernaea* and their distribution in various continents are shown in table 10.

The largest group of species (17) occurs in Africa, while eight have been recorded in North America and seven in far-eastern Asia. Fewer species occur in Europe and India. Until recently, the only *Lernaea* ever found in South America was a single record of *L. lagenula*, a doubtful species possibly belonging to *Opistholernaea* (cf. Fryer, 1968a). The first authenticated record (*L. argentinensis*) has been made within the last few years. The overwhelming majority of species are restricted in their distribution to a single continent. A notable exception is provided by the type species, *L. cyprinacea*. Its spread from one continent to another can be attributed, however, more to the intervention of man than to natural processes of dispersal. Fish husbandry activities are also responsible for the spread of *L. elegans* from its far-eastern range to the waters of the European

Table 10. Geographic distribution of *Lernaea*.

Species	EU	AF	IND	FEA	NAM	SAM
L. arcuata Soejanto, 1965			x			
L. argentinensis Paggi, 1975						x
L. bagri Harding, 1950		x				
L. barbicola Leigh-Sharpe, 1930		x				
L. barilii Harding, 1950		x				
L. barnimiana (Hartmann, 1865)		x				
L. bengalensis Gnanamuthu, 1956			x			
L. bistricornis Harding, 1950		x				
L. brachycera Yü, 1938				x		
L. catostomi (Krøyer, 1863)					x	
L. cruciata (Le Sueur, 1824)					x	
L. cyprinacea L., 1758	x	x	x	x	x	
L. diceracephala (Cunnington, 1914)		x				
L. dolabroides Wilson, 1918					x	
L. elegans Leigh-Sharpe, 1925				x		
L. haplocephala (Cunnington, 1914)		x				
L. hardingi Fryer, 1956		x				
L. hesarangattensis Srinivasachar & Sundarabai, 1974			x			
L. inflata Fryer, 1961		x				
L. insolens (Wilson, 1919)					x	
L. lagenula (Heller, 1865)						x
L. lophiara Harding, 1950		x				
L. minima Otte, 1965	x					
L. octocornua Yin, 1960				x		
L. orizophila Monod, 1932				x		
L. palati Harding, 1950		x				
L. parasiluri Yü, 1938		?x		x		
L. phoxinacea (Krøyer, 1863)	x					
L. polymorpha Yü, 1938				x		
L. pomatoides (Krøyer, 1863)					x	
L. rhodei Hu, 1948				x		
L. senegali Zimmermann, 1923		x				
L. tenuis (Wilson, 1916)					x	
L. tilapiae Harding, 1950		x				
L. tuberosa Harding, 1950		x				
L. variabilis (Wilson, 1916)					x	
L. werneri (Kurtz, 1922)		x				

EU — Europe; AF — Africa; IND — India and Southeast Asia; FEA — Far East Asia; NAM — North America; SAM — South America.

USSR. There is some doubt as to the distribution of *L. parasiluri*, a far-eastern species. Ho (1961) synonymized it with *L. piscinae* Harding, 1950, an African form, thus extending the range of the species to another continent. Heckman and Farley (1973) listed *L. piscinae* as occurring in California. No description or illustration was given. Should these authors' identification prove correct (which is rather doubtful), the distribution range of *L. parasiluri* would be extended even further.

It is clear that *Lernaea* is a very successful genus, capable of exploiting mesoparasitism to the full and of adapting to a large variety of habitats offered by different hosts and by different sites on these hosts. With successful adaptation comes vigorous speciation and a wide intraspecific morphological plasticity. The great diversity in the structure of the anterior part of the female, developing under the constraint imposed on it by the confining tissues of the host, is reflected particularly in the shape of the holdfast. Fryer (1961a) discussed some of the ways in which the site of attachment influences the development of the holdfast. Text fig. 46A shows the "angle of entry" of *L. barnimiana* becoming attached to the flank of its host. Sandwiched, as it were, between two scales, its holdfast develops in the anteroposterior plane (Text fig. 46B), to attain the definitive shape shown in Text fig. 46C. A female penetrating the roof of the mouth has the "angle of entry" more of less perpendicular to the surface of the host (Text fig. 46D). Its holdfast must develop within a fairly thin layer of tissue

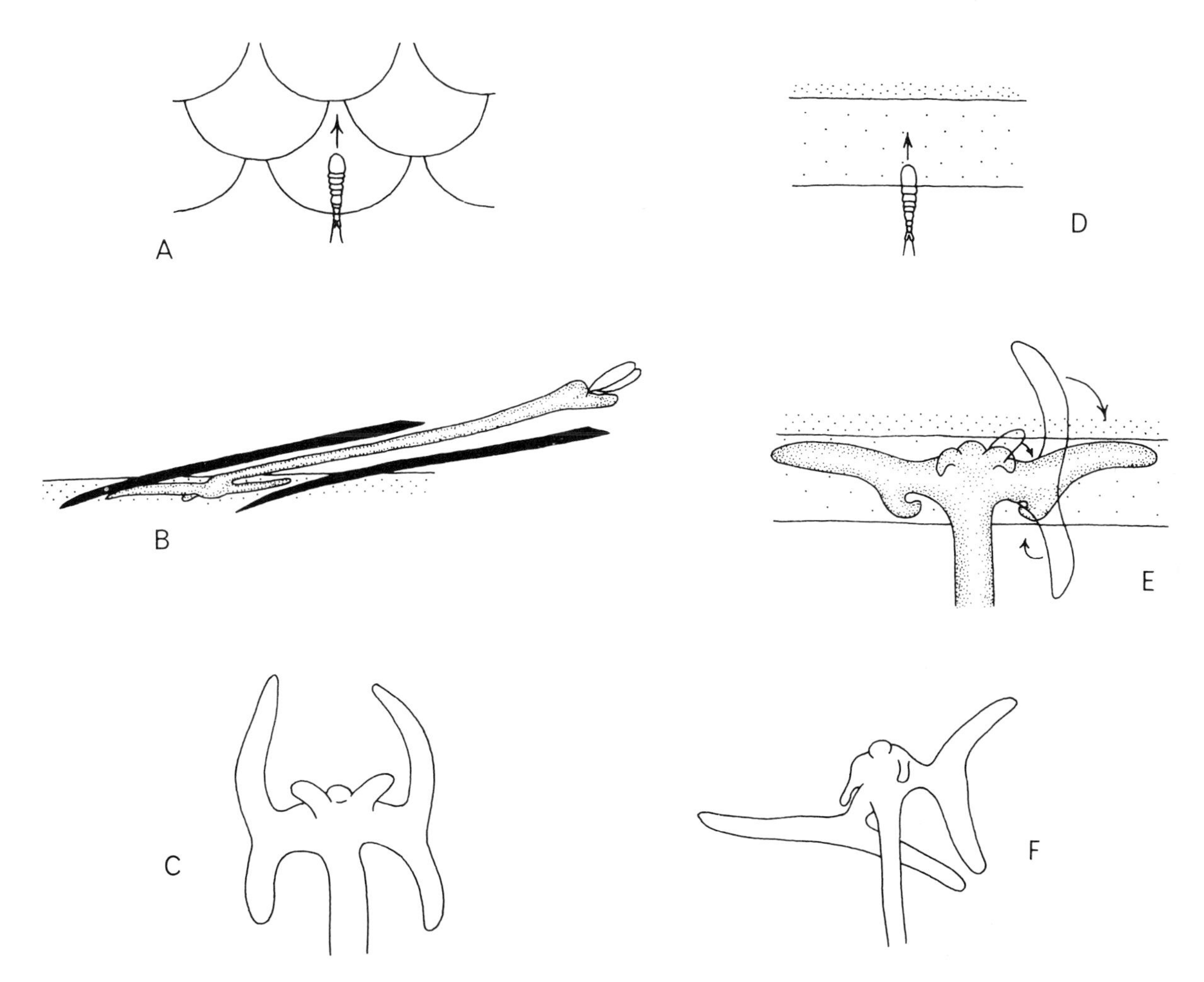

Text fig. 46. The effects of host-parasite relationship on morphology of the lernaean holdfast. A. Entry of *Lernaea barnimiana* between scales of its host; B. Growth of holdfast between scales; C. Definitive shape of holdfast developed under conditions as in B; D. Entry into roof of host's mouth; E. Growth of holdfast at site as in D; F. Definitive shape of holdfast grown under conditions present at E. Modified from Fryer (1961a).

limited distally by hard bone (Text fig. 46E), so that it grows in the plane at right angles to the anteroposterior plane and assumes a shape similar to that shown in Text fig. 46F.

Operating within the host-parasite system, mechanisms of similar nature might be regarded as having contributed to the specific radiation of the genus. Fryer (1968a, p. 90) came to the conclusion that "much of the adaptive radiation . . . within the genus *Lernaea* is concerned with elaboration and modification of anchor form". "All that can be said with confidence is that adaptive radiation has given rise to a great diversity in anchor form and that some of the differences are correlated with the nature of the tissues in which the anchor is embedded." Text fig. 47, illustrating various types of holdfast within the genus, was intended by Fryer (1968a, p. 89) to offer a tentative scheme of relationships within the group of the African species of *Lernaea*. It shows *L. lophiara*, a parasite characterized by the small size of its holdfast, as being closely related to *L. tilapiae*, whose holdfast has very long branches. The former species occurs on the fins, where the substrate for attachment is at a premium, the latter on the roof of the mouth, of their respective hosts. This example clearly demonstrates the necessity of bearing the host-parasite relationships constantly in mind, when any taxonomic decisions are being made involving parasitic Copepoda in general and *Lernaea* in particular. With morphological polymorphism at its broadest extent, factors other than morphology must be pressed into service. Not the least of them are the geographic distribution and the host affiliations, both double-edged weapons to be used with great caution. An additional serious difficulty encountered by taxonomists dealing with *Lernaea* is the fact that the structure of the

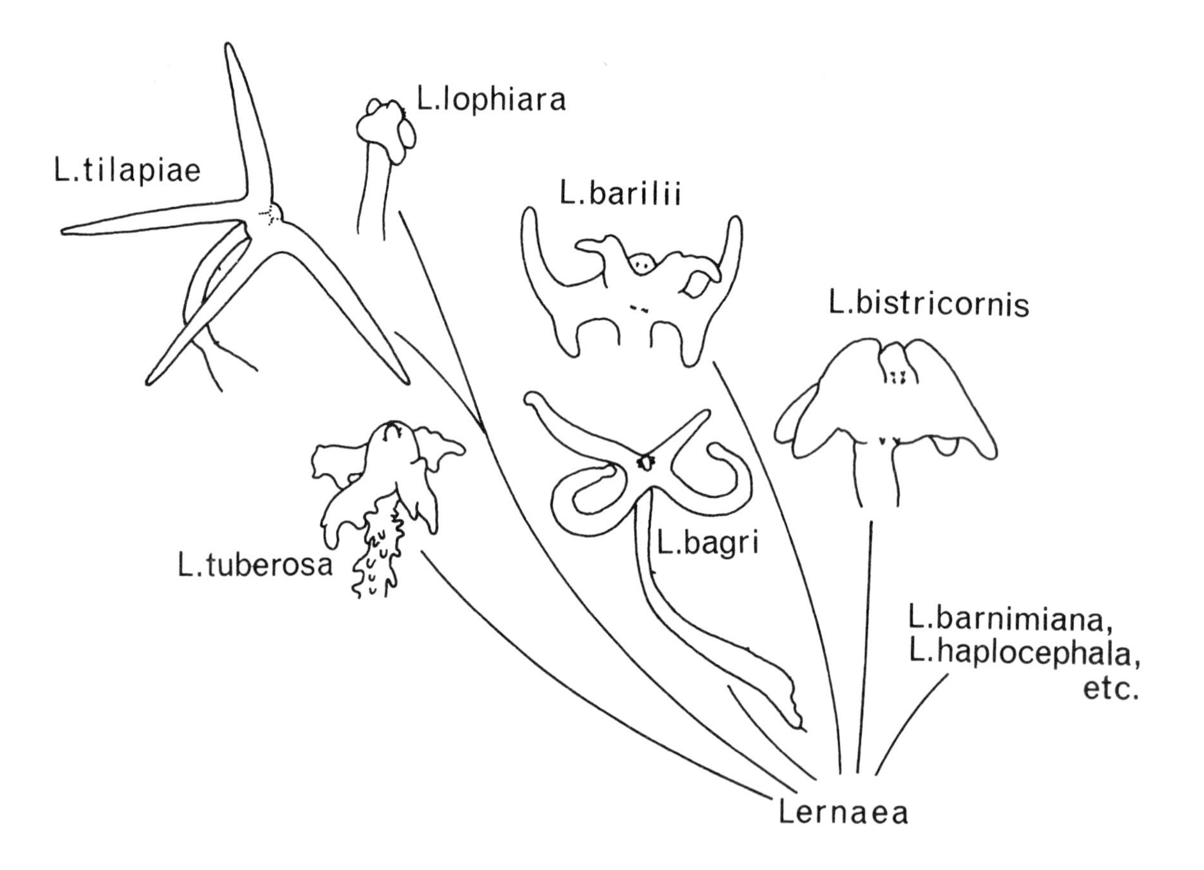

Text fig. 47. Phylogenetic tree of the genus *Lernaea* in Africa. Modified from Fryer (1968a).

appendages appears to be completely uniform throughout the genus and can be of no assistance in diagnosis.

The most interesting recent development in our approach to the systematics of the genus *Lernaea* came with the work of Poddubnaya (1973, 1974). Her experiments have revolutionized our concept of the lernaean species and might result in a complete revision of the genus. More about this work will be mentioned in the discussion of *L. cyprinacea*.

In the British fauna the genus is represented by its type species.

Lernaea cyprinacea L., 1758

(Figs. 380–405)

Syn.: *Lernaea tentaculis quatuor* L., 1746
Lernaeocera cyprinacea; of Blainville (1822)
Lernaeocera esocina Burmeister, 1835
nec *Lernaeocera cyprinacea*; of Nordmann (1832)
Lernaeocera gasterostei Bruhl, 1860
Lernaea esocina (Burmeister, 1835)
Lernaea ranae Stunkard and Cable, 1931
Lernaea carassii Tidd, 1933
?*Lernaea chackoensis* Gnanamuthu, 1951

Pre-metamorphosis female (Fig. 388): Cephalothorax dorsoventrally somewhat flattened, with rounded anterior, more or less parallel lateral and straight posterior margins; dorsal surface (especially in younger specimens) bearing more or less distinct, continuous or discontinuous, U-shaped groove, marking former border between cephalosome and first leg-bearing segment; eyes near centre of area anterior to suture. Between cephalothorax and genital segment four well delimited leg-bearing segments, gradually decreasing in width and length. Genital segment longer, but no wider, than preceding segment, its length : width proportions about 1 : 2, its anterolateral part bearing incipient suboval lobes. Abdomen four-segmented, segments gradually narrowing. Uropod (Fig. 403) subcylindrical, bearing small seta laterally, one medially and two in posterior corners; very long terminal seta in centre of posterior margin, with sturdy base separated by transverse suture and another some distance behind first. Total length 1.2–1.4 (without uropod 0.9–1.1) mm.

First antenna uniramous, six-segmented; basal segment unarmed, second to fifth segments carrying four setae each (one long seta on third and fifth segments, three very short ones on fourth); terminal segment with 11 setae: three on posterior margin, one near middle of anterior margin and seven at apex (three fine setae on posterior side, and four of different lengths and thicknesses on anterior side). Second antenna (Fig. 394) uniramous, three-segmented, first two segments together equalling length of third, both unarmed; third segment with three setae along posterior margin; apical armature (Fig. 395) of six variously long, slender setae and claw-like, powerful spine. Labrum (Fig. 396) with triangular point and smaller triangular outgrowths in distolateral corners. Mandible one-segmented, small, surmounted by slender, hooked spine. First maxilla very small, one-segmented, wedge-shaped. Second maxilla indistinctly segmented (two or three segments), ending in strong claw with secondary tine (only distal segment and claw in Fig. 397). Maxilliped (Fig. 398) of two indistinctly separated segments; basal segment robust (perhaps coalesced of two or three segments), prominent spine with papilliform base on medial margin near distal end; terminal segment much shorter and thinner, with five apical claws differing in length, slenderness and degree of flexure from one another.

First four pairs of legs biramous, rami three-segmented. Sympod two-segmented; proximal segment with seta in distomedial corner; distal segment with fine seta lateral to base of exopod, in first leg (Fig. 402) also with claw-like spine on medial margin. Armature formula of rami as follows.

	Endopod			Exopod		
	1	2	3	1	2	3
Leg 1	1-0	1-0	4,II	1-I	1-I	5,II
Leg 2	1-0	2-0	4,II	1-I	1-I	5,III
Leg 3	1-0	2-0	4,II	1-I	1-I	5,III
Leg 4	1-0	2-0	3,II	1-I	1-I	5,III

Fifth leg vestigial, one segmented; one slender seta at lateral side of base, four similar setae on distal margin of segment. Sixth leg reduced to one short, slender seta.

Post-metamorphosis female (Fig. 391): Cephalothorax very small, semispherical (Fig. 399), situated in centre of holdfast system developed from maxilliped- and first leg-bearing segments. Holdfast variable in size and shape, consisting of two pairs of arms; dorsal pair much larger than ventral, dividing into two branches some distance from their bases; ventral pair slender, usually simple (additional branches sometimes present on both dorsal and ventral arms, Fig. 392). Arms arranged either in anteroposterior plane (as in Figs. 391, 392) or at right-angles to it. Neck consisting of second to fourth leg-bearing segments, more or less circular in cross-section, expanding in girth posteriorly and imperceptibly passing into genital segment. No division between genital segment and abdomen observable on dorsal surface, ventrally marked by posterior end of semispherical or bilobed genital prominence ("pregenital prominence" of various authors). Abdomen conical, usually displaced dorsally from long axis of body, in some specimens with shallow transverse constriction. Uropod present, similar to that of pre-metamorphosis female. Total length 10–20 mm.

Appendages as in pre-metamorphosis female but segmentation of first antenna (Fig. 393) indistinct.

Male (Fig. 389): Resembling female in general appearance and segmentation. Genital segment relatively smaller than in female, suborbicular. Total length about 1.1 (without uropod 0.7) mm.

Appendages similar to those of female, with following exceptions. First antenna (Fig. 400) eight-segmented; short basal segment unarmed; 26 setae along anterior margin; terminal segment with 11 setae, two setae of apical armature (one long and flat, other slender and shorter) fused at base. First leg (Fig. 401) with bifid rather than simple claw on medial margin of sympod. Fifth leg with vestigial basal segment bearing one seta in posterolateral corner and terminal segment with six setae. Sixth leg vestigial, with three setae arising from tip of small swelling.

Comments: The complicated history of this species has been discussed earlier, when dealing with the family (p. 148) and genus (p. 149) to which it belongs. The morphological variability imposed by the host and site of attachment makes it difficult to delimit with precision the boundaries of the species. As illustrated by Harding (1950) and Fryer (1968a), this is particularly true of the shape of the holdfast, a structure important in diagnosis. Since the female continues to grow even after becoming ovigerous, further changes are imposed on its shape by the processes of growth.

The difficulties facing taxonomists became painfully obvious after publication of the work of Poddubnaya (1973, 1974). Rearing larvae from the same batch of eggs and infecting with them different host species, she was able to produce *L. elegans*, *L. ctenopharyngodonis* and *L. quadrinucifera* from the offspring of the same female. Poddubnaya's work also determined the validity of *L. elegans*, long thought to be synonymous with *L. cyprinacea*. These two species, undoubtedly closely related, are very polymorphous and their relationship with other species in the region is yet to be determined.

The relationship between *L. cyprinacea* and *L. esocina* (Figs. 404, 405) constitutes an unresolved taxonomic problem. The latter species was based by Burmeister (1835) on von Nordmann's description of *L. cyprinacea* parasitic on *Esox lucius*. The description was rather generalized and Burmeister's Latin diagnosis scanty. The differences between the newly proposed species and *L. cyprinacea* were small and no account was taken of the morphological variability of the lernaeids. The identification of *Lernaea* as either *cyprinacea* or *esocina* by subsequent authors in many instances appears to have been affected by the identity of the host. The difficulty of distinguishing between these two nominal species can be seen by comparing Harding's (1950) Fig. 10, showing the holdfast of *L. cyprinacea*, with Fig. 404 of this book, illustrating the same part of *L. esocina* from *Esox lucius* taken in Poland. (Indeed, the specimen depicted by Harding was originally labelled *L. esocina* in Canon Norman's collection.) The general appearance of *L. esocina* (Fig. 405) is particularly reminiscent of that of a young ovigerous specimen of *L. cyprinacea*, which it also resembles in size, being usually smaller than the fully adult *L. cyprinacea*. In the absence of a clear morphological distinction between them and in view of Poddubnaya's findings, one can assume that these two taxa are, in fact, conspecific. Experimental confirmation is, however, required to verify this assumption.

L. cyprinacea occurs in all six areas shown in Table 10. Its single African record was published by Fryer (1961b) and it does not appear to be common in that continent. Its otherwise widespread occurrence is coupled with, and undoubtedly fostered by, its ability to live on many unrelated fish hosts, as well as tadpoles, adult frogs and salamanders. Its most common host in the European area is *Carassius carassius*; its general preference for the cyprinid hosts is evident in records of not fewer than 46 species of Cyprinidae on which it was found. It has also been recorded on other Cypriniformes (Catostomidae, Cobitidae). Other hosts come from the following orders: Acipenseriformes (Acipenseridae), Anguilliformes (Anguillidae), Channiformes (Channidae), Cyprinodontiformes (Cyprinodontidae, Poecillidae), Gadiformes (Gadidae), Gasterosteiformes (Gasterosteidae), Perciformes (Anabantidae, Apogonidae, Centrarchidae, Cichlidae, Gobiidae, Mastacembalidae, Osphronemidae, Percidae), Salmoniformes (Esocidae, Plecoglossidae, Salmonidae, Umbridae) and Siluriformes (Bagridae, Ictaluridae, Siluridae). In all, over 100 species have been recorded as hosts of this copepod.

Although Baird (1850) mentioned *L. cyprinacea* in his monograph of British Entomostraca, a fully documented record of this parasite in the British fauna was published only recently (Fryer, 1968b). *L. cyprinacea* was found in several localities in south-eastern counties of England, where it parasitized *Carassius auratus*, *Cyprinus carpio*, *Leuciscus idus* and *Rutilus rutilus*, all in garden ponds. Some years ago this author was shown several specimens of *L. cyprinacea* collected in Scotland (Aberdeenshire). Unfortunately, neither the host fish nor the locality of this find was recorded. Kennedy (1974) added *Gasterosteus aculeatus* to the list of British hosts of this species.

Suborder Siphonostomatoida

Family Caligidae

Morphology

All members of the family Caligidae are parasites, clinging to their hosts' surfaces with the aid of prehensile appendages and capable of free movement over these surfaces. It is clear that maintenance of hold on a slippery surface swept by a current of water is best served by a low-profile shape, designed to disturb the water flow as little as possible. All Caligidae are of similar shape. The name of their second largest genus, *Lepeophtheirus* Nordmann, 1832, is derived from Greek words meaning "scale" and "louse". Vernacular names pointing out this similarity with lice are often used for the caligids by lay observers and show that flatness is the most striking feature of the caligid morphology.

In addition to its flatness, the body of the caligid copepod is characterized by being composed of four tagmata. The structure of the genus *Caligus* Müller, 1785 (Text fig. 48A) can be used as a typical example.

The anterior tagma, the cephalothorax (CPH), is a true hall-mark of all the caligoid copepods. At its most common this tagma, formed by the fusion of the cephalon with the thoracic segments up to and including the third leg-bearing segment, is a roundish or oval structure, covered by a slightly convex dorsel shield, its rims overhanging, as it were, most of the appendages and applied to the surface of the fish in the manner of an inverted saucer. Anteriorly, the margin of the shield is extended into two transversely arranged frontal plates, fused in mid-line and distinguishable from each other by a shallow median notch. In some genera the ventral surfaces of these plates are expanded near their lateral ends to form subcircular cups of different depths, acting as accessory suckers. These cups are known as *lunules* (Text fig. 50, lun). Along the anterior margins of the frontal plates and the lateral margins of the shield itself run strips of marginal membranes, transparent ribbons of cuticle, narrow anteriorly and fairly broad laterally. Spread on the surface of the host, these membranes help to seal off the concavity beneath the shield and to facilitate its adhesive function. The dorsal surface of the shield is scored by sutures, arranged in the shape of the letter H. Anterior to the transverse limb of the H lies the cephalic zone (1), with eyes visible through its cuticle. In the genus *Caligulina* Heegaard, 1972, near the anterior margin of the shield is a pair of prominent, refractile swellings, the *conspicilla*. Behind the transverse suture lies the thoracic zone (3). (It should be noted that the names "cephalic" and "thoracic", proposed by Wilson (1905), are not meant to imply that the border between these two zones is coincident with the boundary between the true cephalon and the thoracic segments fused with it.) To the right and left of the longitudinal sutures lie the lateral zones (2). In very few species (e.g. *Caligus coryphaenae* Steenstrup and Lütken, 1861, cf. Figs. 519, 520) another transverse suture, usually behind the eyes, links the anterior ends of the lateral sutures, dividing the cephalic zone into two unequal parts. Two more or less deep indentations in the posterolateral corners of the thoracic zone are known as the posterior sinuses. A flap of membrane, similar to the marginal membranes, is present in each sinus and acts as a one-way valve, permitting the flow of water from beneath the shield to the outside but preventing its movement in the opposite direction. The flat posterior end of the thoracic zone often protrudes above the next tagma, the fourth leg-bearing segment.

The ventral surface of the cephalothorax (Text fig. 50) has a central longitudinal ridge, most markedly developed in its central part, and two longitudinal lateral concavities, barely perceptible in the anterior part of the cephalothorax. With the exception of the anterolateral postantennary processes (pant), the bases of the appendages concealed under the dorsal shield are located in the region of the central ridge. The most anterior ones of these are the second antennae (ant2). Posteromedial to them is the mouth cone (mt), in most specimens its orifice pointing in the posterior direction. It is flanked by the first maxillae (mx1), with the bases of the second maxillae (mx2) close to them in the posterolateral direction. The prominent maxillipeds (mxp) are posterior to the second maxillae. The sternal furca (sf) has the same spatial relationship to the bases of the maxillipeds as the mouth cone has to the bases of the second antennae. Some distance behind the sternal furca the central ridge is traversed by the interpodal bar linking the legs of the first pair (l1). Close behind them is the second pair (l2), its interpodal bar broader than that of the first. The bar of the third legs (l3) is broad enough to fill the entire space between foliate sympods and to form with them an apron (ar), the function of which has been discussed earlier. The rami of the first two pairs of legs are located in the area of the greatest concavity and are ensured of a sufficient space in which to move. The posterior margins of the third legs protrude beyond that of the dorsal shield and lie under the fourth leg-bearing segment or even under the ventral surface of the genital complex. The other appendages which can be seen from the ventral aspect are the first antennae (ant1), located in the anterolateral corners of the shield and partly overlapped by the lateral ends of the frontal plates (fp).

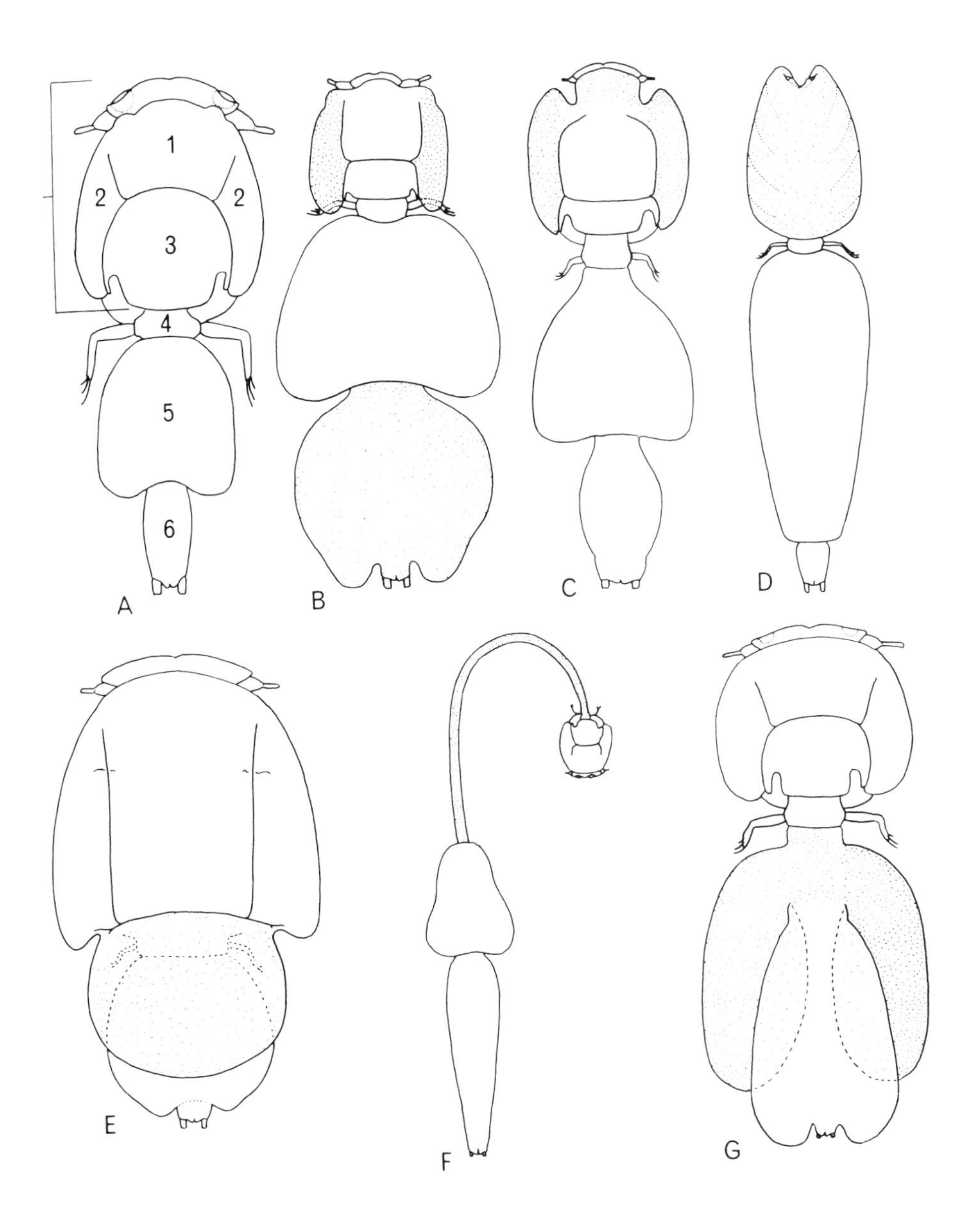

Text fig. 48. Morphological type of Caligidae. A. *Caligus*; B. *Alicaligus*; C. *Abasia*; D. *Hermilius*; E. *Mappates*; F. *Echetus*; G. *Pseudopetalus*. Explanations in text. (See also Text fig. 49.)

The second and smallest tagma (Text fig. 48A, 4) is the fourth leg-bearing segment of the thorax. It is the only segment of the caligid body situated between the cephalothorax and the genital complex and not fused with either of them. It is often referred to as the free thoracic segment.

Next follows the genital complex (Text fig. 48A, 5). Of greatly variable shape and size, it is always dorsoventrally depressed and much larger in the females of all species. Its shape in that sex is also greatly affected by age and stage of maturity, so that sometimes specific identification has been difficult. The genital complex consists of two completely fused segments and bears vestiges of the fifth (in females) and both fifth and sixth legs (in males). (For the possibility of the presence of sixth leg vestiges in the female see Parker et al., 1968.) On the ventral side, in the centre of the posterior margin of this tagma are paired vaginal orifices, sometimes found sealed by the spermatophores. On the same margin, some distance from the centre and on either side of the abdomen are the openings of the oviducts which also provide points of attachment for the egg strings. In many species they are situated at the apices of semispherical cavities accommodating the tips of these strings, filled with single rows of flattened eggs.

The posterior tagma is the abdomen (Text fig. 48A, 6). Joined with the genital complex in the central part of its posterior extremity, the abdomen varies greatly from species to species and sometimes has been used as a generic discriminant. It can be narrow and cylindrical, consisting of one or several segments arranged in an anteroposterior direction. The greatest number of segments recorded in the caligid abdomen is five, occurring in *Caligus quinqueabdominalis* Heegaard, 1962. It is sometimes broad and dorsoventrally depressed. Whatever its structure, however, its posterior margin is always bisected by a slit-like anus, on each side of which is a short uropod, always armed with a set of pinnate setae.

(It should be noted that the demarcations between individual tagmata are not always as clear as indicated in Text fig. 48 and in many drawings of various authors. The division between the cephalothorax and the fourth leg-bearing segment is always clearly discernible because of the definite and sharp end of the thoracic zone of the former. The real boundary between these two tagmata, however, is often concealed under that margin, particularly in partially contracted specimens. The divisions between the genital complex and the two adjacent tagmata are often without any external cuticular markings and can be deduced only from the general topography of the body. The same is often true of divisions between the individual segments of the abdomen, sometimes barely noticeable and sometimes present only in some specimens of a species known to have more than one abdominal segment.)

All species of the genus *Caligus* conform fairly closely to the structural outline described above. A very similar gross morphology is found in the genera: *Pseudocaligus* Scott, 1901; *Caligopsis* Markevich, 1940; *Caligulina* Heegaard, 1972; *Lepeophtheirus* Nordmann, 1832; *Pseudolepeophtheirus* Markevich, 1940; and *Caritus* Cressey, 1967.

In spite of the broad resemblance within the family, every tagma has a range of differences which sometimes greatly alter the appearance of the copepod. The cephalothorax, a structure admirably adapted to the function it fulfils, cannot be expected to display a broad range of morphological diversity in animals with fairly similar modes of life. Members of some caligid genera, however, have become specialized for sedentary life on gill filaments. Some of them show a tendency to modification of the cephalothorax, with the dorsal shield more or less folding its lateral zones ventralwards to embrace the filament of the host's gill. This tendency is evident in the genus *Alicaligus* Shiino, 1955 (Text fig. 48B). In *Abasia* Wilson, 1908 (Text fig. 48C), this enveloping movement is further facilitated by the presence of anterolateral clefts, partially separating the cephalic zone from the anterior ends of the lateral zones. (Some authors refer to the central part of *Abasia's* dorsal shield as a rostrum, a term not strictly applicable to a broad structure carrying on its anterior margin the frontal plates and first antennae.) The process of ventral folding has reached its ultimate degree in the genus *Hermilius* Heller, 1865 (Text fig. 48D). It differs from the preceding examples in having only one "hinge" in the mid-dorsal line. Folding along that line involves the entire shield, which closes almost completely by approximating the lateral margins so as to enclose all the appendages under it in a manner reminiscent of a bivalve shell. The only other noteworthy modification of the shield is the enlargement of the thoracic zone shown in Text fig. 48E. Extending backwards so as to overlap a part of the genital complex, this zone is expanded in a similar way in some species of the genus *Pseudanuretes* Yamaguti, 1936 and *Haeniochophilus* Yamaguti and Yamasu, 1959. Its greatest expansion occurs, however, in the genus *Mappates* Rangnekar, 1958, where it is associated with posterior extension of the cephalic zone to about the level of the posterior tips of the lateral zones. In *M. alter* Kabata, 1964, the thoracic zone is long enough to cover completely the entire posterior half of the body.

Although the fourth leg-bearing segment is a small and simple tagma with little potential for excessive modifications, it, too, has produced a characteristic feature for one caligid genus. In *Echetus* Krøyer, 1864, it has become elongated so as to form a thin neck extending between the cephalothorax and genital complex and accounting for about half of the total body length (Text fig. 48F). A similar, though less spectacular, extension might be present also in the genus *Parechetus* Pillai, 1961. Absence of a definite boundary between this segment and the genital complex does not allow one to determine with accuracy what contribution to this extension is made by either of these tagmata.

The next tagma, the genital complex, commonly referred to as the genital segment, shows a fair degree of variability in shape and relative size. As mentioned earlier, it consists of two segments fused with each other, at least a part of its variability being accounted for by age and maturity variations. Distension at the pre-parturition stage in gravid females is in contrast to the marked decrease in size and changes in contour resulting from extrusion of the egg masses. By comparison with the genus *Caligus* (Text fig. 48A), many caligid genera are much modified with regard to this tagma. Relative increase in size and lateral expansion occur in *Alicaligus* (Text fig. 48B) and considerable elongation is common in *Hermilius* (Text fig. 48D). Moreover, in *H. youngi* Kabata, 1964, flat aliform expansions are found in anterolateral regions, a feature rather unique in the family. In the genus *Pseudopetalus* Pillai, 1962 and *Dartevellia* Brian, 1939 (Text fig. 49C), the genital complex is much longer than the dorsal shield and attains a width greater than that of the latter, particularly in its posterior part. Lobes of modest size are present in its posterolateral corners. In the genus *Sciaenophilus* Beneden, 1852, the comparative length of the genital segment is equally great, though its width is less pronounced (Text fig. 49E). The posterolateral lobes mentioned above develop to a much greater degree in several genera. In *Parapetalus* Steenstrup and Lütken, 1861 (Text fig. 48G), these lobes constitute by far the major part of the tagma, extend its length by more than three times and overlap the abdomen for most of its length. A similar shape of the genital segment occurs in some species of *Caligodes* Heller, 1865 (Text fig. 49A) and in *Parechetus* (Text fig. 49F). In the latter, however, the lobes are relatively smaller and the anterior end of the tagma is slender and attenuated. The length of the posterolateral lobes probably achieves its extreme in the genera *Pupulina* Beneden, 1892 (Text fig. 49B) (not applicable to *P. minor* M. S. Wilson, 1952) and *Synestius* Steenstrup and Lütken, 1861 (Text fig. 49D). In both of them the lobes reach either to, or beyond, the level of the posterior extremity of the abdomen. Characteristic of the latter genus is the presence of a second pair of posterolateral lobes, ventral to and shorter than the first. A feature very unusual for Caligidae is the presence of a posterodorsal, medial lobe in *Caligodes megacephalus* Wilson, 1905. This lobe is much shorter than the more usual posterolateral pair and overhangs the basal part of the abdomen. In some species of the genus *Anuretes* Heller, 1865 (Text fig. 49G), the genital complex is the posterior tagma, the abdomen being diminutive and completely or almost completely incorporated in it. In species to which the latter applies the uropods appear to arise directly from the genital complex.

The fourth and posterior tagma, the abdomen, also displays a great diversity of structure. In *Alicaligus* (Text fig. 48B) it rivals in size, and somewhat resembles in shape, the genital complex. In *Echetus* (Text fig. 48F) it is longer than that tagma. In *Parapetalus* (Text fig. 48G) it assumes massive proportions, accounting for about half of the total length, expanding in width posteriorly and developing a pair of posterolateral lobes. A similar abdomen is found in *Caligodes* (Text fig. 49A), though it is rather less voluminous and devoid of lobes. In *Pseudopetalus* and *Dartevellia* (Text fig. 49C) it is not only very long (about half of total length) but also broad. Its width surpasses that of all other tagmata, exceeding that of the dorsal shield by more than three times. The abdomen is also markedly elongated in *Synestius* (Text fig. 49D) and *Sciaenophilus* (Text fig. 49E), in the latter accounting for more than half of the body length, though still remaining slender and cylindrical. Posterolateral lobes, already mentioned, are more prominent in *Alicaligus* and especially in *Pseudopetalus* and *Parechetus*. In the last-named genus (Text fig. 49F) they extend far beyond the level of the uropods. In many species of *Caligus* and *Lepeophtheirus* the abdomen is reduced, its length becoming much less than that of the thoracic zone of the dorsal shield. As already noted, this reduction becomes extreme in some species of *Anuretes* and *Pseudanuretes* Yamaguti, 1936 (Text fig. 49G).

All these morphological modifications occur among the females of Caligidae. The morphology of the males is much more uniform, which makes it often quite difficult to distinguish between caligid genera by their males. With the exception of *Abasia*, in which the male dorsal shield resembles that of the female, no modifications of that structure have been recorded. The genital complex is quite uniform in most genera, at least in its major characters. The male abdomen, sometimes consisting of a different number of segments than that of the female of the species, is seldom modified. (Though

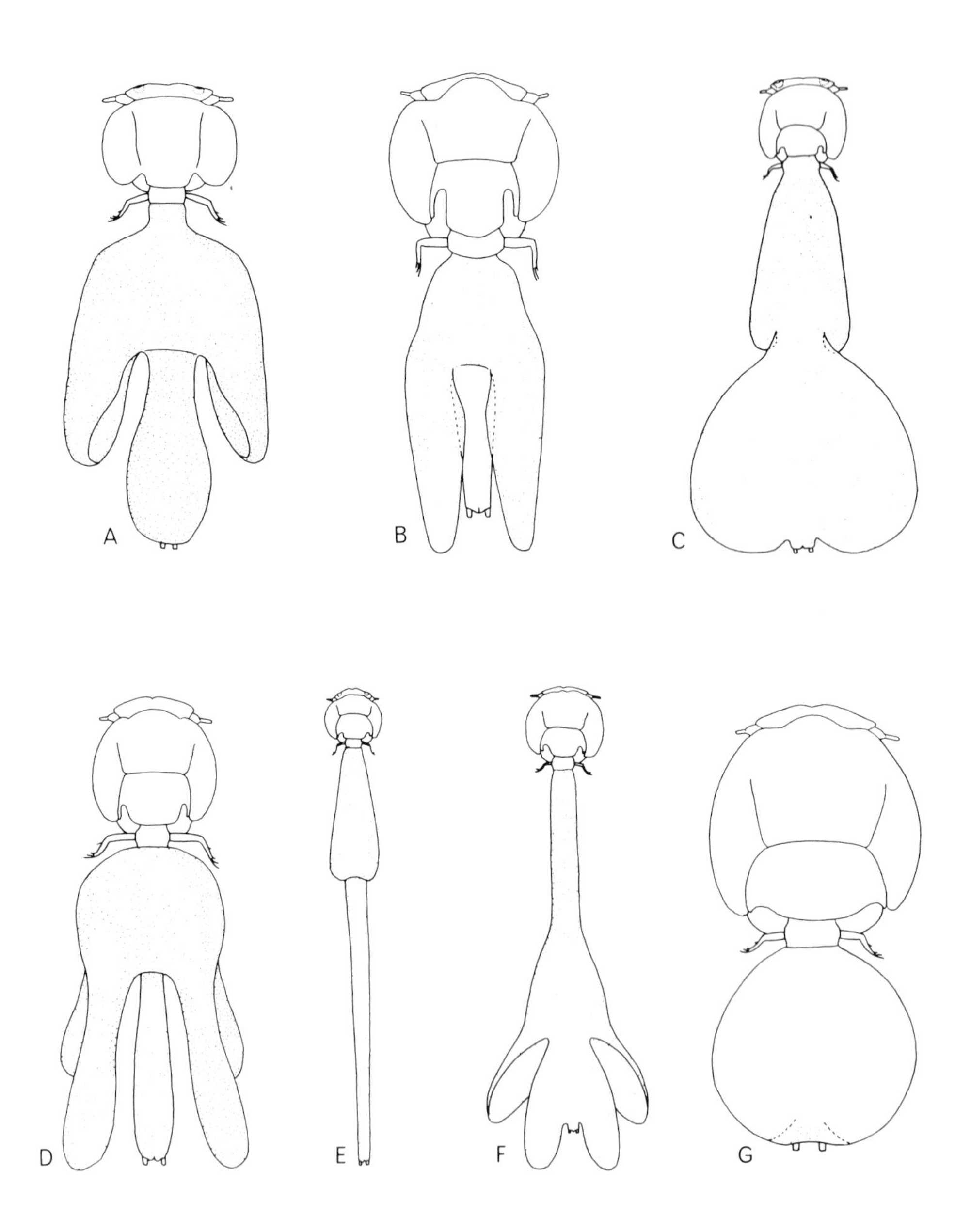

Text fig. 49. Morphological types of Caligidae. A. *Caligodes*; B. *Pupulina*; C. *Dartevellia*; D. *Synestius*; E. *Sciaenophilus*; F. *Parechetus*; G. *Anuretes*. (See also Text fig. 48.)

some degree of reduction occurs in certain species of *Lepeophtheirus* and *Anuretes*.) It should be kept in mind, however, that all these comments are based on information much more scanty than our knowledge of the females. The males are often scarce. No male has been described for *Dartevellia*, *Diphyllogaster*, *Hermilius*, *Haeniochophilus*, *Mappates*, *Parechetus* and *Pseudopetalus*. Judging from illustrations (Markevich, 1956), the male described for the genus *Caligopsis* might be a juvenile female.

Reviewing the structure of the appendages of the caligid copepods (Text fig. 50), one becomes aware of their uniformity. The differences from genus to genus, and even more from species to species, are usually only in the proportions of the individual components of various appendages. Less frequently, one of the "secondary appendages" (postantennary process or furca), or a part of the first maxilla are suppressed.

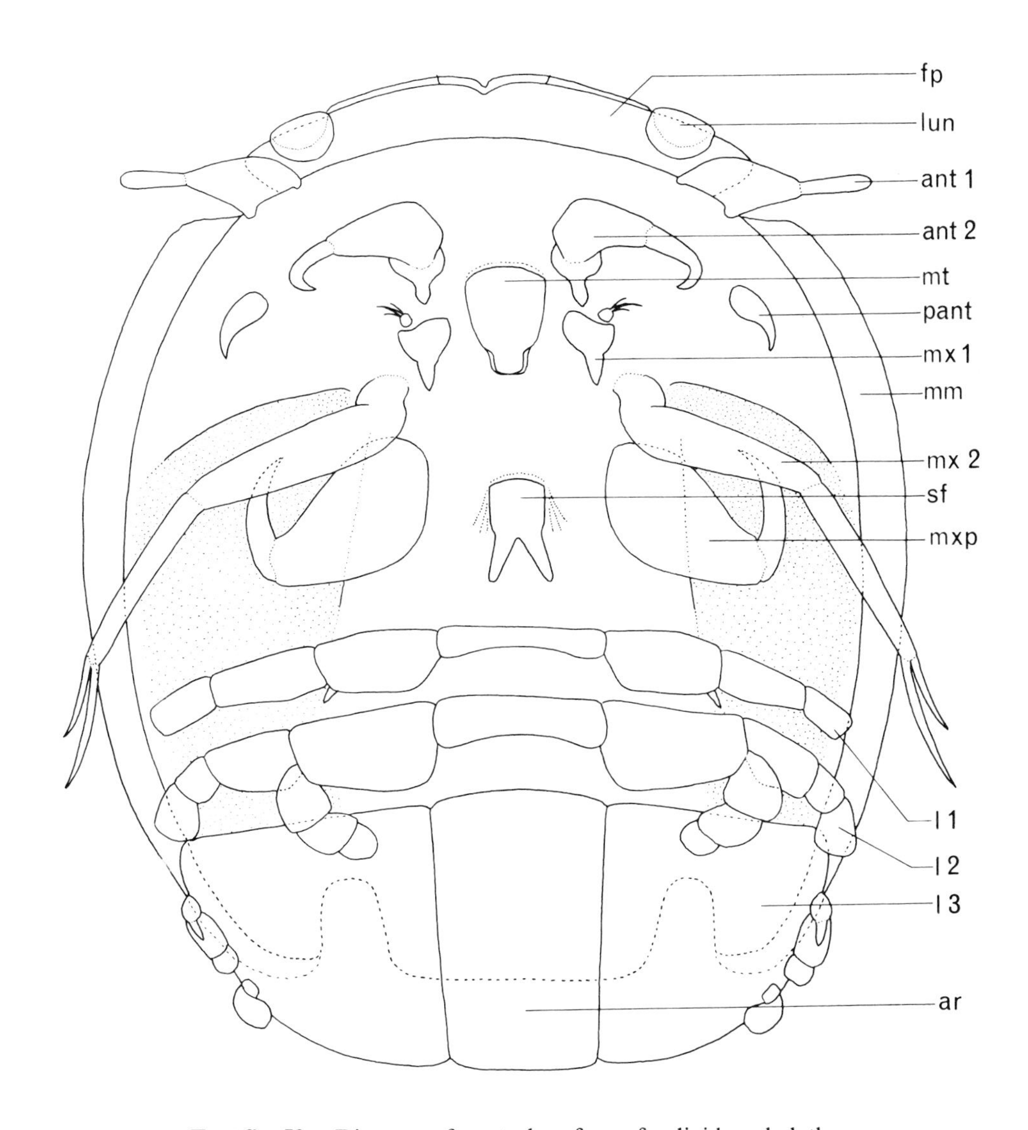

Text fig. 50. Diagram of ventral surface of caligid cephalothorax.

The first antenna is a two-segmented appendage, with flattened basal segment, differing in shape and armed with setae along its anterior margin. The number of setae tends to be constant within genera. In some caligids the posterodistal corner of the segment carries a spiniform outgrowth or two (e.g. in *Lepeophtheirus cuneifer* Kabata, 1974) or a long, stiff, plumose seta (e.g. in *Alicaligus tripartitus* Shiino, 1955). The distal segment is cylindrical or slightly clavate and its armature is the same throughout the family, consisting of one slender seta on the posterior margin and of the apical armature. The latter (Fig. 630) consists of two groups of setae. On the anterior end of the apex there are eight sturdy setae of unequal length, at least one of them with bluntly rounded end. The posterior side carries five much longer and more slender setae; two of them are fused at the base, two others are close to each other and one is single. This group of five setae varies in length. In some species (e.g. *Caligus coryphaenae* Stp. and Ltk., 1861; Fig. 522) they exceed the length of the setae of the anterior group by several times and might be plumose. This last feature might mean only the retention of juvenile characteristics. Most of the known early stages of the caligids have much longer setae in their apical armature, sometimes branching one or several times.

The second antenna of the female (Figs. 440, 441, 631) is four-segmented (if one considers the terminal claw, a structure consisting of a fusion of several segments, as one). Its structure is characterized by the reflection of the distal end over the proximal one. The functional significance of this arrangement was discussed earlier (p. 22). The small basal segment is unarmed and is often ignored in descriptions (omitted in Figs. 440, 441). In some species it appears partially fused with the second (Fig. 631). The second segment usually bears a posterior process of different shapes and sizes. The terminal claw is a simple hamulus, with one seta at its base and another some distance from it. The length, thickness and curvature of the claw vary from species to species. In some genera (*Hermilius* Heller, 1865 and *Pseudanuretes* Yamaguti, 1936) the claw has another tine some distance from the tip. A similar tine occurs in the second antenna of *Caligus coryphaenae* (Fig. 521). The male appendage differs from that of the female by inflation of the penultimate segment and by the complicated structure of the short terminal claw, at least in some genera. Another special male feature is the presence of large adhesion pads (Figs. 442, 443).

The mouth of caligid copepods was described in detail earlier (p. 25). The mandible (Figs. 445, 446) is a piston-like rod, consisting of four parts, possibly corresponding to four segments. The first three are circular in cross-section, the terminal fourth is flat and carries a series of 12 more or less uniform denticles along one margin. (Some authors reported on mandibles with 11 and 13 teeth, but within this family 12 is probably the most usual number.) The denticulated part of the mandible curves inwards to fit the guide groove along the margin of the labium, teeth towards the buccal lumen, and is flexible enough to bend and straighten up during its forward and backward movements respectively.

The first maxilla of Caligidae consists usually of two parts, separated from each other. In the genus *Caligus* Müller, 1785, the first maxilla of the female is represented by a hard, sclerotized process, wider at the base and tapering towards the tip. Close to, but separate from, the base is a papilliform outgrowth surmounted by three setae, one longer than the other two. The shape and the size of the process vary from species to species (Figs. 474, 496, 508, 523). The first maxilla shows sexual dimorphism, the maxillary process of the male often differing from that of the female (compare Fig. 496 with 497, 540 with 541, 584 with 585). In *Caligus minimus* Otto, 1821, the maxillary process of the male is vestigial and the papilla appears to carry only two setae (Fig. 585). In *Lepeophtheirus* von Nordmann, 1832, the maxillary process of the female is commonly bifid (Fig. 641), while that of the male has an additional third outgrowth on its median side. In many species of this genus a small adhesion pad is present near the maxillary process of the male (Fig. 640). In some genera both simple and bifid maxillary processes can be found. *Mappates plataxus* Rangnekar, 1955, has fleshy maxillary processes larger than the mouth and almost covering it between them. On the other hand, in *Caligulina* Heegaard, 1972, the only trace left of the process is a short and narrow flange. It is much reduced in *Haeniochophilus* Yamaguti, 1936, which also lacks the maxillary papilla. The first maxilla of *Pseudanuretes* Yamaguti, 1936, has no process, while the papilla is absent in *Alicaligus* Wilson, 1955. Both parts are present and of about equal size in *Synestius* Stp. & Ltk., 1861. In some caligids a small denticle is present on the surface of the cephalothorax near the tip of the maxillary process (e.g. in *Pupulina* van Beneden, 1892 or in *Lepeophtheirus sturionis* (Krøyer, 1838) (Fig. 532)).

The second maxilla of Caligidae is a typical brachiform appendage, described earlier (p. 33) and illustrated in Text fig. 20. Its tip is armed with two processes (calamus and canna), the latter of which is sometimes greatly reduced (e.g. in the genus *Hermilius*). The second maxilla of *Mappates* has two oblique strips of membrane near the tip of the brachium, in place of the more usual flabellum. In *Caligulina* the flabellum seems to have been replaced by an "accessory finger" (Heegaard, 1972) and in *Alicaligus* by a transverse row of setae. The genus *Pseudanuretes* is unique in having a whip-like process, as long as the second maxilla itself, on the surface of the cephalothorax near the base of this appendage. It is probably the "basal process" occurring in vestigial form in other caligids (e.g. in *Caligus*) on the base itself.

The maxilliped is always a subchelate appendage, often different in the two sexes of the same species. Its general structure was described on p. 35 and shown in Text fig. 21A. The maxilliped of caligids differs greatly in the size and shape of corpus maxillipedis, especially in that of its myxal area. In some species that area is inflated and displaced anteriorly so as to make the appendage nearly chelate (e.g. in *Caligus chelifer* Wilson, 1905). The maxilliped of *Parechetus* bears two claw-like processes on its corpus, while one very long and powerful process is found on it in *Pupulina*. The terminal claw of the subchela is usually simple, though it differs in length and curvature from one species to another. In *Hermilius* it has a secondary tooth near the tip.

The so-called "secondary appendages" (the lunules, postantennary processes and sternal furca) were described earlier (p. 42). Not all the caligid genera have all three of them. Some have none. Their occurrence is shown in Table 11.

Table 11. Occurrence of "secondary appendages" in caligid genera.

	Lunule	Postantennary process	Sternal furca
Abasia	−	−	−
Alicaligus	+	+(m) −(f)	−
Anuretes	−	+	+ −
Caligodes	+	+	+
Caligopsis	?	?	?
Caligulina	+	+	−
Caligus	+	+ −	+ −
Dartevellia	−	+(v)	−
Diphyllogaster	−	?	+ −
Echetus	+	−	−
Haeniochophilus	−	−	+
Hermilius	−	+ −	+ −
Lepeophtheirus	−	+	+
Mappates	−	−	−
Parapetalus	+	+	+
Parechetus	+	+	+
Pseudanuretes	−	−	−
Pseudocaligus	+	+ −	+ −
Pseudolepeophtheirus	−	+	+
Pseudopetalus	+	+	+
Pupulina	−	+	−
Sciaenophilus	+	+	+ −
Synestius	+	+	+

(m) = male; (f) = female; (v) = vestigial.

The thoracic legs, or swimming legs, are also uniform throughout the family, but for minor details in armature. The first leg (Fig. 455) has a two-segmented sympod, its segments overlapping in an oblique fashion. The endopod is vestigial, represented only by a very small, papilliform outgrowth. (In *Haeniochophilus* is appears to be absent altogether.) The exopod is robust, longer than the sympod, two-segmented. The basal segment, much larger than the terminal, is only poorly armed. The terminal segment carries four structures on its distal margin, their nature varying among species and genera, and usually, though not always, has three large pinnate setae on the posterior margin. Like other thoracopods, the first leg is linked with its opposite number by an interpodal bar (Text fig. 50). The sympod of the second leg (Fig. 456) is similarly two-segmented, but the extent of the overlap of the two segments is much greater. Both exopod and endopod are fully developed, three-segmented, their armature almost constantly uniform within the family. The most characteristic feature of the second leg is the presence of two large, transparent membranes attached to the upper (or anterior) margin of the sympod and the basal segment of the exopod and suspended from them along the dorsal (or posterior) surface of the leg (Fig. 457). The role played by these membranes in the combined propulsive action of the first two pairs of legs was described recently by Kabata and Hewitt (1971). The hall-mark of Caligidae, however, is their third pair of legs. Modified by expansion in width of the interpodal bar and of the sympods, these legs form an uninterrupted wall (Figs. 458, 459) sealing off the posterior end of the cavity formed by the cephalothorax. Both rami are greatly reduced in size. The exopod is provided with a prominent spine of the basal segment and consists of three segments, sometimes with rather indefinite boundaries. The endopod is reduced even more and is sometimes held to consist of only two segments, though Parker et al. (1968) argued the existence of an indistinct third segment in *Caligus curtus* Müller, 1785. In *Caligulina ocularis* Heegaard, 1972, a small, but distinct and well delimited segment is, indeed, present on the posterior margin of the second segment of the endopod, which is, then, truly three-segmented. The basal segment of this ramus is devoid of the usual claw. The fourth leg (Fig. 460) is uniramous, its sympod being one-segmented. The single ramus present (usually taken to be the exopod) varies somewhat in structure, being usually either two or three segmented. The boundaries between its segments are often obscure and sometimes can be discerned only by the presence of a spine marking the distal limit of segments. (For further discussion of this appendage see p. 40). The fifth leg (Figs. 462, 463), located on the ventral side of the posterolateral corner of the genital complex, is vestigial and attains some prominence only in those species of the genus *Lepeophtheirus* which are placed by some authors in *Dentigryps*. In those species it assumes the structure of a spiny process, armed with several setae and projecting well beyond the margin of the genital complex. The sixth leg is present normally only in the males of the family and is also vestigial, comprising only three or four short setae.

The structure of the uropods is uniform throughout the family, with minor differences in the contours but with complete constancy of armature (Fig. 464).

History and systematics

The concept of the familial unity of Caligidae was only slowly arrived at in the course of a long and complicated process, contributed to by the work of many authors. As is usual in the development of taxonomic understanding, the blurred outlines of a taxon, at times grossly inclusive, at others too restrictively exclusive, cleared only gradually. It is virtually impossible to grant credit for the establishment of Caligidae to any single author.

The roots of the family can be said to reach back into the XVIII century, to the date on which Müller (1785) established the genus *Caligus* for the parasite with a rather complicated and obscure earlier history. It is, however, in the early XIX century that one must seek the initial attempts at placing *Caligus* in a larger taxon. In an early work of Leach (1816) we find *Caligus* and a spurious genus *Binoculus* (survivor of the earliest attempts at classification of lower Crustacea), placed with all copepods equipped with a flat dorsal shield in the tribe of Entomostraca named Thecata, family Pseudonura. Later, Leach (1819) elaborated and modified his scheme. Division I of Entomostraca, identified only by the numeral, contained all species equipped with a flat, horizontal dorsal shield and sessile eyes. The nature of the shield was used as the next discriminant, the present-day caligids belonging to a subgroup with shield consisting of a single part. This subgroup was, in turn, divided into species with jaws (*Apus*) and those without jaws but with "rostrum". The latter fell into two groups again: those with four antennae (*Argulus*) and those with two (*Cecrops*, *Caligus*, *Pandarus*, *Anthosoma*). With the exception of the last-named, all these genera were to be placed subsequently in Caligidae and remained in that family for some time.

Contemporaneous with the work of Leach was that of Lamarck (1818) in which *Caligus*, *Cecrops*, *Argulus* and *Dichelesthium* were placed in "Branchiopodes parasites", a branch of the entomostracan division of "Branchiopodes". Although the name would suggest that the morphology of swimming legs was decisive in ascribing species to this division, it is worth noting that all genera in it possess a flat dorsal shield, smaller in *Dichelesthium*, larger in the remaining genera.

The first more precise definition of the genus *Caligus* was given by von Nordmann (1832), who framed it as follows: "Body consists of two main parts, the larger shield-like head and the narrower, either quadrate or heart-shaped hindbody, connected to each other by means of a short and narrow breast-piece, to which the seventh pair of feet is attached. Of the seven pairs of feet, the three anterior ones are jaw-like, the fourth and the seventh are simple, the fifth and the sixth are cleft and provided with plumose setae." Failing to recognize the difference between the thoracic legs and the appendages borne upon segments anterior to the first leg-bearing segment, this definition recognized that the genus *Caligus* has only one independent thoracic segment between the cephalothorax and the genital complex. This fact continued to be disregarded for some time after von Nordmann's work, in assessment of relationships among species with superficially similar morphology. Another fact pointed out by the same author remained unrecognized for some time subsequently: the existence of the genus *Lepeophtheirus*, based on its lack of lunules as a distinguishing feature. This was perhaps due to the fact that von Nordmann made a conspicuous error in describing them as the eyes (Randaugen, p. 31) and confused the issue. In spite of this error, his work was of high quality for the period. A keen insight into the difficulties of correctly systematizing parasitic copepods was shown by his refusal to indulge in the production of any schemes of classification. He was aware, he said, of the possibility that he might be blamed for this attitude, but with so many forms yet to be described (some in his possession) which did not fit into the schemes proposed by his contemporaries, he felt it prudent to await the accumulation of more illuminating data.

The name Caligina as a pseudofamilial, plural cognomen, applied to *Caligus* and its allies, appears in the paper by Burmeister (1835). The family is defined as being equipped with feelers and jointed legs. Its members have three pairs of jaw-feet and four of jointed legs posterior to the "proboscis" (mouth tube). Burmeister believes that his Caligina are related to Ergasilina, a group characterized by the possession of multisegmented "inner feelers" (first antennae), while the corresponding appendages of caligids are composed of only two or three segments. Burmeister's work is of particular interest because of his attempt at intrafamilial taxonomy, presented in the form of a key. Caligina are divided into two groups by the presence or absence of eyes. In the eyeless group, also distinguished by scale-like dorsal expansions on the hindbody, is the genus *Cecrops*. The group equipped with eyes is divided into two by the structure of the "last legs of hindbody" (fourth thoracic legs): uniramous and biramous. The former group includes the genera *Chalimus* (a juvenile stage of several caligid genera), *Caligus* and *Lepeophtheirus*; in the latter are *Pandarus* and *Dinematura*. It is interesting to note that all the copepods remaining presently in the family Caligidae (*Caligus* and *Lepeophtheirus*) do remain together in this system and the first tentative lines of division between them and other caligoid genera begin to be timidly drawn.

The familial name Caligina becomes current during the first half of the XIX century, the ending "ina" denoting a plural name derived from the name of the type genus of the family (though type genera were not formally designated at that time). Krøyer (1837, 1838–39) also used this name. Not trying to emulate Burmeister in the arrangement of intrafamilial groupings, even in the form of a key, he included in his Caligina, in addition to the genus *Caligus*, also *Chalimus*, *Lepeophtheirus*, *Pandarus*, *Cecrops* and *Dinemura* (= *Dinemoura*).

It should be remembered that during that period terms like "family" and "order" were often understood differently from their present day meaning. A taxon considered to have a familial rank by one investigator would be nominated differently by another, though the taxon itself might be constituted in the same or a similar way in both instances. For example, Caligina, ranked as a family by Burmeister (1835), became a tribe in the system of Edwards (1840). That renowned author divided his entomostracous order Siphonostoma into two families: Peltocephales and Pachycephales, once again supporting belief in the diagnostic importance of the dorsal shield. The Pachycephales were split into two tribes: Ergasiliens and Dichelesthiens. (This is the first instance of the latter group being separated from the caligids.) The Peltocephales consisted of the tribes Caligiens and Pandariens, another division foreshadowing later taxonomic boundaries. Edwards defined his Caligiens as follows: "Without lamelliform processes on dorsal surface of thorax. No suckers associated with the second maxillae." Further subdivision of the Caligiens was:

A. Fourth legs uniramous, ambulatory.
   a. No mediofrontal* process (*Caligus*).
   b. Mediofrontal* process present anteriorly (*Chalimus*).

B. Fourth legs biramous, natatory.
   a. Thorax with three distinct segments (*Trebius*).
   b. Thorax with four distinct segments (*Nogagus*).

*Frontal filament, by which the developmental stages, now known as chalimus stages, are attached to the host.

Edwards clearly did not recognize the taxonomic importance of lunules and disregarded von Nordmann's (1832) genus *Lepeophtheirus*. Although numbers of thoracic segments were used in his system, he did not use the single independent thoracic segment of the caligids as a noteworthy distinguishing mark. On the other hand he accepted the genus *Chalimus*. The tribe Caligiens remained a composite taxon, with species possessing both uniramous and biramous fourth legs.

At this junction Baird (1850) produced his major work, the first monograph on British Crustacea published by the Ray Society. Of interest here is Baird's order Siphonostomata (belonging to the legion Poecilopoda), tribe Peltocephala. In view of the fact that this work is the first one to use the now-accepted familial endings "dae" for his family group taxa, it should be well remembered by those who insist on assigning credits for the creation of families to individual authors. Baird's Peltocephala consisted of four families: Argulidae (*Argulus*), Caligidae (*Caligus*, *Lepeophtheirus*, *Chalimus* and *Trebius*), Pandaridae (*Pandarus*, *Dinemoura*) and Cecropidae (*Cecrops*, *Laemargus*). Caligidae were defined as having: "Head in form of a buckler (large), having anteriorly large frontal plates. Four pairs of feet, which are furnished with long plumose hairs. Antennae small, flat and two-jointed. Second pair of foot-jaws of two articulations, and not in form of a sucking disc. Thoracic segments uncovered." The reference to the absence of suckers on second maxillae is understandably necessary to distinguish Caligidae from the argulids, which were lumped with them in close taxonomic relationship. Significantly, Baird made no mention in his diagnosis of the single independent thoracic segment, noted earlier as the caligid characteristic feature by von Nordmann (1832). In the absence of this feature, inclusion of *Trebius* with Caligidae became possible. The flat shield (buckler) is still the salient feature of the family.

The next important work on copepod taxonomy in which Caligidae were dealt with, and in a somewhat novel fashion, was that of Dana (1852, 1853). Dana's Caligidae were a part of the tribe Caligacea which contained also the families Argulidae, Dichelesthiidae, Ergasilidae and Nicothoidae. (Edwards' (1840) decision to separate the ergasilids and the dichelesthiids from the caligids was thus reversed.) The diagnosis also contained some novel features. "Body anteriorly covered with a shield. External ovaries tubular, straight, with one series of eggs. Four first anterior pairs of legs subprehensile. Posterior antennae covered by carapace." It is in this work, too, that an attempt was made to introduce into the caligid taxonomy subfamilies, in their more or less modern sense. Three subfamilies were recognized in 1852 and distinction made between them in the following key:

1. Caliginae. Buccal tube subovate, stout. Maxillae at distance from tube, posteriorly tapering and elongate. External ovaries straight. Body broader anteriorly. (*Caligus*, *Lepeophtheirus*, *Chalimus*, *Caligeria*, *Calistes*.)
2. Pandarinae. Buccal tube slender, pointed. Maxillae close to tube, minute, lamelliform. External ovaries straight. Body sometimes broader posteriorly. (*Pandarus*, *Trebius*, *Nogagus*, *Specilligus*, *Dinemoura*, *Phyllophora*, *Euryphorus*, *Lepidopus*.)
3. Cecropinae. Buccal tube slender, pointed, Maxillae close to tube. External ovaries coiled under carapace. (*Cecrops*, *Laemargus*.)

In 1853, Dana added yet another subfamily to Caligidae, by removing from Pandarinae and elevating to subfamilial status the genus *Specilligus*. Insufficiently well described, this genus is now believed to be at least in part synonymous with *Nogagus*, another spurious genus.

It is worth noting that Dana was inconsistent in his diagnoses. Although in the diagnosis of Caligidae the possession of straight egg sacs was considered as typical for the family, the subfamily Cecropinae was defined as having coiled egg sacs. Nevertheless, Dana's work marked an important stage in the shaping of Caligidae. He was the first to divide them into subfamilies (using subfamilial names ending in modern "inae"), isolating in Caliginae the two genera (*Caligus* and *Lepeophtheirus*) that later provided the nucleus for the family in its present form. (*Caligeria* and *Calistes* are not caligid and *Chalimus*, as mentioned earlier, represents only juvenile stages of caligid species.)

Neither Baird's nor Dana's work exerted much influence on the development of the caligid taxonomy. Much more recognition was awarded to the two new systems of copepod classification, published some time later. Steenstrup and Lütken (1861) divided copepods, both free-living and parasitic, into three parallel rows of species, depending on the nature of their egg sacs. In this system *Caligus* found itself, with *Pandarus*, *Dichelesthium*, *Clavella* (= *Hatschekia*) and *Pennella*, in the series of species possessed of two uniseriate egg sacs. Thorell (1859) produced three parallel rows of families, dividing them by the nature of their buccal apparatus. In the row of those possessing "proboscis which usually incloses two mandibles" and with "maxillae from 3 to 0 pairs", in addition to Caligidae there were also Ascomyzontidae, Nicothoidae, Dichelesthiidae, Lernaeopodidae and Lernaeidae.

Claus (1862) recognized Caligidae as a family into which he included *Caligus, Synestius, Parapetalus, Calistes, Trebius, Dysgamus, Gloiopotes, Caligeria, Sciaenophilus, Elytrophora* and *Euryphorus*.

After more than three decades of silence, cautious von Nordmann (1864) returned to the scene with half-hearted acceptance of the scheme of Steenstrup and Lütken (1861), combining it with Edwards' (1840) system. Peltocephala of the latter writer were placed in the third row or series with two unisteriate egg sacs and divided into (a) Caligini, containing *Caligus* (with *Lepeophtheirus* as subgenus), as well as other genera which belonged to this family in Claus' (1862) list and (b) Pandarini. Another noted author at that time publishing a large work after long silence was Krøyer (1863, 1864), who was critical of Dana and of his compatriots Steenstrup and Lütken. Among other suggestions he made was the one that *Synestius* and *Parapetalus* should be relegated to synonymy with *Caligus*, and another that *Calistes* and *Dysgamus* are synonymous with *Trebius*, while *Elytrophora* is synonymous with *Euryphorus*. He now accepted the validity of the genus *Lepeophtheirus*, albeit with some apparent reluctance. It is worth noting that even the author of that genus, von Nordmann (1832), in his later work (1864) relegated it to the status of the subgenus of *Caligus*.

Heller (1865), in his monographic work placed Caligina in the order Siphonostoma and divided it into two groups: Caligina *sensu stricto* and Pandarina. The former were in two groups, distinguishable by the number of thoracic segments fused with the cephalothorax. With two segments so fused were *Trebius, Dysgamus* (male) and *Alebion* (male), the second now invalid, the other two left outside Caligidae. The group with three segments fused was divided into two by the presence or absence of alae on the "fourth abdominal segment" (fourth leg-bearing segment). Those with an alate segment are now outside the family, those without alae (*Caligus, Synestius, Parapetalus, Caligodes, Lepeophtheirus, Anuretes, Hermilius* and *Calistes*) are true caligids, except for *Calistes*, a genus inquirendum.

Gerstaecker (1866–1879), having attempted a rather ill-defined phylogenetic grouping of families, placed his Caligidae in association with Dichelesthiidae and Lernaeidae. The family was enlarged by inclusion of all the pandarid genera then accepted, as well as those placed previously in Cecropidae (or Cecropinae). Among the principal features he used in his key to the genera were the presence or absence of the alate outgrowths and the structure of the thoracic legs.

Between the date of Gerstaecker's publication and the end of the century two authors deserve mention, though their work was of clearly regional scope and their listings of species and genera in the larger taxa they proposed were restricted to those occurring locally. The first of them was Olsson (1869), who dealt with the parasitic copepods of Scandinavian fauna. Crediting the establishment of Caligidae to Burmeister, he gave the following definition of the family: "Body depressed, with cephalothorax expanded like a shield and fused with adjoining segment, or anterior abdomen. First segment of postabdomen (= genital segment), especially in female, large, tail poorly developed, bearing two caudal rami armed with setae. Anterior antennae with two free segments, on lateral ends of frontal plates. Posterior antennae hamate, four maxillipeds covered by shield. Proboscis enclosing two mandibles. Eight abdominal legs, four or more of which are biramous, natatory and respiratory, with many plumose setae. External egg strings with flat eggs, uniseriate. Male and female parasitic on skin or gills of fishes, sometimes free swimming. Biarticulate antennae and frontal laminae on their basal parts are among essential characters, but *Laemargus* and *Phyllophorus* have three-articulate antennae." The family was divided into two tribes, Caligini and Pandarini. The former were defined as having "proboscis ovate blunt; only last abdominal segment free (in *Trebius* also adjacent segment); animals sometimes equipped with dorsal plates (elytra). Swimming legs with plumose setae (*Caligus, Lepeophtheirus, Trebius*)". Clearly, it still appeared difficult to accept without reservations von Nordmann's principle of a single free segment being an important diagnostic feature, particularly when one was confronted with animals as similar in their gross morphology as *Caligus* (Fig. 538) and *Trebius* (Fig. 722).

Carus (1885) was concerned with the fauna of the Adriatic region. His familial definition had in it much taken from Olsson. "Body depressed, shield-like, of few segments, thoracic segment II and III, very often also I, fused with head; posterior to ample genital segment rudimentary abdomen; segments sometimes with aliform appendages (elytra); first antennae short, two- or three-segmented, second unciform; mouth conical, suctorial, mandibles styliform, maxillae with very large palp; maxillipeds unciform, prehensile; fifth leg vestigial or absent." The family was divided into tribes Pandarini and Caligini and the latter were defined very much as in Olsson and credited to him. To the three genera included in Caligini by that author, Carus added *Elytrophora*.

The next major work on Caligidae appeared at the beginning of the present century. In a series of detailed publications, Wilson (1905, 1907a, 1907b) revised the family at greater length than hitherto. He recognized it as consisting of five subfamilies (Caliginae, Pandarinae, Cecropinae, Euryphorinae and Trebinae), obviously drawing liberally on the work of his earlier compatriot, Dana (1852, 1853). To distinguish among them, Wilson used the following key.

1. Three anterior segments of thorax fused with the head; fourth and genital segments free 2
1. Two anterior thoracic segments fused with the head; third, fourth and genital segments free . . . **Trebinae**

2. Fourth segment without dorsal plates or any appendages except the fourth legs . . . . **Caliginae**
2. Fourth segment with a pair of dorsal plates which usually overlap the genital segment . . **Euryphorinae**

3. Frontal plates distinct; egg cases visible their entire length . . . . **Pandarinae**
3. Frontal plates fused with the carapace; egg cases convoluted, entirely hidden . . . . . **Cecropinae**

The subfamily Caliginae was defined as follows: "Carapace broad and always flattened dorso-ventrally; free-thorax segment without plates or appendages of any sort except the fourth pair of legs Genital segment enlarged, but usually smaller than the carapace; never much larger except in the genus *Echetus*. First and fourth thoracic legs uniramose, second and third biramose; fifth pair rudimentary but often visible as a pair of small papillae at the posterior corners of the genital segment. Adults active, most of the females as well as the males capable of swimming about freely."

The key to the genera, however, is sometimes at variance with the diagnosis of the subfamily. It includes *Calistes*; a genus not identifiable but alleged to have only the first legs uniramous, and *Dysgamus* with all four pairs biramous. Also included is the genus *Homoiotes* Wilson, 1905, defined as having dorsal plates covering most of the "abdomen" (= genital segment). (The latter genus was proved by Kabata (1973a), to be a composite taxon, consisting of three species of *Lepeophtheirus*, devoid of any kind of elytral structures.) At variance with the subfamilial diagnosis is also the genus *Calina* van Beneden, 1892, with only the fourth legs uniramous. In addition to the genus *Caligus*, Wilson included in Caliginae, beside the above-mentioned, also the genera *Parapetalus*, *Echetus*, *Anchicaligus* (the member of the family parasitic on an invertebrate), *Pseudocaligus*, *Sciaenophilus*, *Synestius*, *Caligodes*, *Diphyllogaster*, *Lepeophtheirus*, *Anuretes* and *Hermilius*. Generally speaking, however, Wilson's scheme provided foundations for the present concept of Caligidae. In his later work (Wilson, 1932), he upgraded the status of the subfamilies by giving familial rank to Cecropidae, Euryphoridae, Pandaridae and Trebiidae. This action was not discussed and its antecedents (e.g. Claus, 1862) were not mentioned. The process of final "pruning" of Caligidae of alien elements was completed.

Not all authors were prepared to accept this concept of the family. As late as 1956, Markevich reverted to the idea of "extended family", as shown in the following key.

1(4) Cephalon fused with three anterior thoracic segments; fourth and genital segments remain free.
2(3) Fourth thoracic segment smooth and, apart from the fourth pair of legs, bearing no other appendages . . . . **Caliginae**
3(2) Fourth thoracic segment bearing a pair of dorsal plates which usually overlap genital segment . . . . **Euryphorinae**
4(1) Cephalon fused with less than three thoracic segments.
5(8) Cephalon fused with first thoracic segment only, other segments free; one (sometimes more) free segments bearing dorsal plates.
6(7) Frontal plates clearly distinguishable; egg sacs visible throughout their length . . . . . . **Pandarinae**
7(6) Frontal plates fused with shield, egg sacs concealed . . . . **Cecropinae**
8(5) Cephalon fused with two thoracic segments, other segments free . . . . **Trebinae**

Yamaguti (1963), while accepting Wilson's Caligidae at the familial level, attempted to subdivide it into subfamilies with the aid of completely novel criteria, illustrated in the key below.

1. Lunules present . . . . **Caliginae**
   Lunules absent 2

2. Free thoracic segment produced into a long neck . . . . **Echetinae**
   Free thoracic segment short 3

3. Abdomen normal, usually small . . . . **Lepeophtheirinae**
   Abdomen reduced or lacking . . . . **Anuretinae**
   Abdomen fused ventrally with genital segment, both covered anteriorly by prominent posterior median lobe of carapace . . . . **Mappatinae**

Having reviewed the rather tortuous course travelled by the concept of Caligidae as a family unit, one becomes aware of a conflict between two tendencies. The first is to group together all species with flat, horizontal shields and similar mouth parts. Consequent upon it is the "extended family" idea upheld as recently as 1956 by Markevich. The second tendency is to bring into the definition more structural features and, with the increased precision of definition, to make the concept a more restrictive one. The introduction into the diagnosis of the number of segments fused with the cephalothorax and of the absence of even rudimentary elytra, suffices to delineate Caligidae *sensu stricto*. The latter tendency has now gained general currency and is shared by this author.

The definition of the family, therefore, should run as follows.

Body of four tagmata: cephalothorax, fourth leg-bearing segment, genital complex and abdomen. Cephalothorax incorporating thoracic segments up to third leg-bearing one, covered by horizontal, flat dorsal shield (longitudinally folded in *Abasia* and *Hermilius*). Fourth leg-bearing segment small (elongated in *Echetus*). Genital complex of two fused segments, dorsoventrally flattened, shape ranging from small, orbicular or oval to very large, sometimes with posterolateral lobes, rarely with posteromedial or anterolateral lobes. Abdomen dorsoventrally flattened or cylindrical, of one to five segments, usually short but sometimes large and/or laterally expanded, rarely (*Sciaenophilus*) accounting for largest proportion of body length. Frontal plates present, well delimited, with or without lunules. First antennae two-segmented, short, second antennae hamate. Postantennary processes present or absent. Mouth cone short, blunt, enclosing dentiferous mandibles. First maxillae sometimes much reduced, second long, slender. Maxilliped prehensile, subchelate or pseudochelate. First leg with or without vestigial endopod. Second leg biramous. Third leg fused with its opposite number, fourth leg uniramous. Fifth leg vestigial. Sixth leg usually absent in female, vestigial in male. Uropod flat or cylindrical, setiferous.

Yamaguti's (1963) proposal of new subfamilies for Caligidae has not yet been commented upon. It seems, however, that it will not gain general acceptance, based as it is in part on faulty information. Ho (1966a) pointed out, for example, that *Echetus* does possess lunules, contrary to Yamaguti's key and that it should, therefore, be included in Caliginae. A broader question is still to be answered, that of the value of subfamilial taxa as an aid to our understanding of intrafamilial relationships. In this author's view subfamilies of Caligidae, at least as formulated by Yamaguti, are artificial and superfluous. With the continuing discovery of new species and new morphological characters, their limits are bound to shift and our views of interrelationships of the caligids are bound to change more than once in coming years.

To the progress of future discoveries we must also leave the final mapping out of the "gray areas" in caligid taxonomy at the generic level. One of them is the question of the abdomen as the diagnostic criterion. Pillai (1967b) pointed out that no clear-cut border can be drawn between *Anuretes* and *Lepeophtheirus*, two genera distinguishable from each other mainly by the condition of the abdomen. In *Anuretes*, as the generic name implies, this tagma is fused with the genital complex so as to be completely indistinguishable. Nevertheless, some species of that genus have been described with distinct abdomina, though too small to be conveniently placed in *Lepeophtheirus*. A continuous series of species with progressively smaller abdomina thus exists and cannot clearly be broken by a generic boundary. Also devoid of abdomen is the only known representative of *Haeniochophilus*.

This problem is extended, as it were, in the opposite direction by the genus *Sciaenophilus*, whose morphology is dominated by the abdomen. The genus *Caligus* can be arranged into a series of species with increasingly larger abdomina, though none as long as in *Sciaenophilus*. The length of the abdomen being the sole difference between the females of *Caligus* and *Sciaenophilus* (their males are indistinguishable), the independence of these two genera from each other might be called into question. Some recent authors (e.g. Capart, 1959) have, indeed, refused to recognize the validity of *Sciaenophilus*.

One of the generally accepted generic discriminants is the fourth leg. The vestigial condition of this leg (Fig. 627) is a diagnositc feature of *Pseudocaligus* and *Pseudolepeophtheirus*, genera otherwise apparently indistinguishable from *Caligus* and *Lepeophtheirus* respectively. (The absence of the fourth leg marks off the genus *Caligopsis*). The vestigial fourth leg was also made a part of the generic diagnosis of *Pseudanuretes* Yamaguti, 1936. Another unique feature of *Pseudanuretes* is the presence of a long flagelliform process at the base of the second maxilla. In *P. fortipedis* Kabata, 1965, however, the fourth leg is normally developed. This species, almost identical with *P.*

chaetodonti Yamaguti, 1936 and, like it, equipped with a flagelliform process at the base of the second maxilla, is clearly its congener and yet differs in the condition of the fourth leg. If the caligid genera can encompass species with both normal and vestigial fourth legs, what is the basis for the retention of *Pseudocaligus* and *Pseudolepeophtheirus*? There appears to be none. This example spotlights another gray area in caligid taxonomy. The author is inclined to think that *Pseudolepeophtheirus* and *Pseudocaligus* should be synonymized with their parent genera. *Pseudocaligus* is retained in this account, however, pending more detailed morphological information.

The fifth leg presents similar difficulty. The genus *Dentigryps* has been accepted by many authors and contains several species, mainly described by Lewis (1964a, 1964b). The chief difference between it and *Lepeophtheirus* is the structure of the fifth leg, in *Dentigryps* always prominent and in dorsal aspect protruding well beyond the posterior margin of the genital complex. There are, however, some species of *Lepeophtheirus* in which the fifth leg of the female also protrudes, though not so much as in *Dentigryps*. A length gradation series exists here also and is difficult to divide by a generic boundary. The problem was discussed by Hewitt (1971), who concluded that the absence of such clear division deprives *Dentigryps* of a valid morphological basis and that all the species now in it should be synonymized with *Lepeophtheirus*.

Difficulties of like kind arise in connection with the structure of the sternal furca, presence or absence of the postantennary process and size of the spiniform process of the first maxilla. This author tentatively adopts the attitude of following the current majority opinion in each such instance, while pointing out the difficulty to the readers' attention. A special case is presented by *Midias lobodes* Wilson, 1911. Differing from typical *Caligus* only in having small lobes on the first abdominal segment, it is only a rather unusual species of that genus and treated as *Caligus lobodes* (Wilson, 1911). It is not allied to Euryphoridae, as sometimes suggested.

In British waters, Caligidae are represented by four genera, identifiable with the aid of the following key.

1.	Lunules absent	***Lepeophtheirus***
	Lunules present	2
2.	Abdomen in female equal to, or more than, ½ of body length	***Sciaenophilus***
	Abdomen in female less than ½ of body length	3
3.	Fourth leg well developed, of more than one segment	***Caligus***
	Fourth leg vestigial, of one segment only	***Pseudocaligus***

(This key does not distinguish between indistinguishable males of *Caligus* and *Sciaenophilus*).

Genus Caligus Müller, 1785

Caligidae. Cephalothorax suborbicular or oval, usually slightly longer than broad and anteriorly narrower (seldom posteriorly narrower). Lunules present. Dorsal sutures in H-shaped pattern (in *C. coryphaenae* second transverse suture behind eyes). Posterolateral ends of lateral zones each with one shallow depression. Female genital complex of varying size and shape, from suborbicular to elongate; male usually oval and relatively smaller. Abdomen usually cylindrical, one- to five-segmented, occasionally subquadrate or broader than long.

First antenna two-segmented, second four-segmented, hamate. Postantennary process hamate, sometimes vestigial or absent. Mouth cone short, blunt. Mandible of four parts, dentiferous. First maxilla with single, rarely bifurcate dentiform process. Second maxilla brachiform. Maxilliped subchelate. Sternal furca present or absent. First leg with vestigial endopod and two-segmented exopod, second leg with both rami well developed, three-segmented. Third leg fused to form apron, with rami reduced. Fourth leg three- or four-segmented. Fifth leg vestigial, consisting of four setules (in males vestigial sixth leg of three setules also present).

Type species: *Caligus curtus* Müller, 1785.

Comments: One of the most successful genera of parasitic copepods, *Caligus* consists of about 200 species, distributed throughout the oceans and seas of the world. A synopsis of the genus was published by Margolis et al. (1975). Only one of its species, *C. lacustris* Steenstrup and Lütken, 1861, occurs commonly and exclusively in fresh waters, though some have been occasionally recorded in brackish waters and rivers (e.g. *C. epidemicus* Hewitt, 1971). *Caligus* parasitizes innumerable species of fishes, mainly teleosts, though some are known only from elasmobranchs (*C. willungae* Kabata, 1965), or can be found on both types of hosts (*C. elongatus* Nordmann, 1832). Some species have very broad ranges of hosts, others are known from a single host species. On cursory examination, the genus appears to be better represented in tropical and subtropical waters than in the higher latitudes. In British waters, *Caligus* is represented by 13 species, including the type species.

1. Fourth leg three-segmented, with two apical setae (Fig. 516) 2
   Fourth leg three-segmented, with three apical setae (Fig. 488) 3
   Fourth leg four-segmented, with three apical setae (Fig. 528) 5
2. Sternal furca with tapering tines (Fig. 511) . . . . **C. centrodonti**
   Sternal furca with spatulate tines (Fig. 577) . . . . **C. labracis**
3. Second and third terminal setae of first exopod with secondary processes (Fig. 485) . . . **C. brevicaudatus**
   All four terminal setae of first exopod simple (Fig. 454) 4
4. Dentiform process of first maxilla short in both sexes, with broad base and sharp-pointed tip (Fig. 444) . . . . **C. curtus**
   Dentiform process of female first maxilla digitiform (Fig. 584), male vestigial (Fig. 585) . . . **C. minimus**
5. Postantennary process absent or vestigial, cuticular outgrowths on both sides of sternal furca (Fig. 524) . . . . **C. coryphaenae**
   Postantennary process well developed, no outgrowths on sides of sternal furca 6
6. Fourth seta of first exopod about as long as second and third (Fig. 500) 7
   Fourth seta of first exopod much longer than second and third (Fig. 477) 8
7. Postantennary process about as large as hamulus of female second antenna . . . . **C. diaphanus**
   Postantennary process very short (Fig. 503) . . . . **C. pelamydis**
8. Lateral margin of second endopod with fringe of fine setules (Fig. 554) 9
   Lateral margin of second endopod strongly spinulated (Fig. 479) 11
9. First seta of first exopod much shorter than second or third (Fig. 610) . . . . **C. zei**
   First seta of first exopod about as long as second and third (Fig. 553) 11
10. Male fourth legs reaching posterior end of body (Fig. 560), female longer than genital segment (Fig. 561), base of abdomen overlapping dorsal surface of genital segment (Fig. 561) . . . . **C. gurnardi**
    Male fourth leg not reaching posterior end of body (Fig. 550), female shorter than genital segment (Fig. 549), base of abdomen not overlapping dorsal surface of genital segment (Fig. 549) . . . . **C. elongatus**
11. Distal segment of first exopod with three long pinnate setae on posterior margin (as in *C. curtus*, Fig. 455) . . . . **C. bonito**
    Distal segment of first exopod with posterior margin unarmed (Fig. 599) . . . . **C. productus**

Caligus curtus Müller, 1785

(Figs. 433–468)

Syn.: *C. mülleri* Leach, 1816
C. bicuspidatus Nordmann, 1832
C. americanus Pickering and Dana, 1838
C. diaphanus; of Baird (1850)
C. elegans van Beneden, 1851
C. fallax Krøyer, 1863
C. aeglefini Krøyer, 1863

Female (Figs. 433–435): Cephalothorax longer than wide, broader posteriorly, with free margin of thoracic zone protruding some distance beyond posterior ends of lateral zones, dorsoventrally flattened. Lunules of moderate size, shallow. Fourth leg-bearing thoracic segment short and broad. Genital complex roughly rectangular, with sloping anterolateral and rounded or lobate posterolateral corners, in mature specimens as long as, or slightly longer than, thoracic zone of dorsal shield. (Differences in shape with maturity shown in Figs. 467, 468.) Abdomen of about same length and width, one-segmented, less than $\frac{1}{2}$ length of genital complex, slightly pedunculate at base. Total body length 5.1–10.1 mm.

First antenna (Fig. 439) with robust basal segment bearing 26 or 27 setae on anterior margin, terminal segment moderately inflated at apex, shorter than basal. Apical armature as described on p. 162. Second antenna (Figs. 440, 441) with sturdy hamulus, prominent posterior process of second segment and adhesion pad on dorsal surface of third segment. Postantennary process well developed. Mouth cone (Fig. 444) short, blunt, labrum with notched margins at about midlength. Four-segmented mandible (Figs. 445, 446) with 12 blunt, posteriorly curving teeth. First maxilla (Fig. 444) with dentiferous process (endopod) very broad at base, abruptly tapering to short, slender tip; basal papilla (exopod) with one long and two shorter setae. Second maxilla (Figs. 447, 448) with calamus much longer than canna; brachium longer and more slender than lacertus, with well developed flabellum. Maxilliped (Fig. 449) with long, slender corpus; medial margin with rounded, low swelling near midlength; subchela about $\frac{1}{2}$ length of corpus, with slightly curving robust shaft; claw less than $\frac{1}{2}$ length of shaft, curving, tapering, with short barbel at base. Sternal furca (Figs. 452, 453) with rectangular, long box and divergent tines; latter shorter than box, digitiform, blunt, devoid of flanges.

First leg (Figs. 454, 455) with vestigial endopod; second segment of exopod with terminal setae 1–3 of about equal length, with rows of fine denticles along parts of one margin and some across tips; seta 4 longer and more slender, unarmed; three long, pinnate setae on posterior margin of distal segment (pinnules omitted in Fig. 455). Second leg (Figs. 456, 457): lateral margin of endopod with fringe of fine setules on all three segments. Third leg (Figs. 458, 459) typical for family. Fourth leg (Figs. 460, 461) with robust sympod (accounting for about $\frac{1}{2}$ length of appendage) and two-segmented exopod; distal segment of latter with three setae, all with basal pectens; seta 1 much longer than other two, with single row of denticles, seta 2 slightly longer than seta 3, both with two rows of delicately serrated membranes; seta of proximal segment similar to short setae of distal; border between two segments of exopod often indistinct. Fifth leg (Fig. 462) vestigial, consisting of two small papillae surmounted by short pinnate setules, two and one respectively. Uropod (Fig. 464) subquadrangular, with oblique anterior margins and fringe of setules on medial margins; apical armature of three subequal pinnate setae and three setules, one medial and two lateral.

Male (Figs. 436–438): Cephalothorax and fourth leg-bearing segment similar to those of female. Genital complex relatively smaller (shorter than length of thoracic zone of dorsal shield), suborbicular, with two partly overlapping lobes in each posterolateral corner. As in female, genital complex differing in size and shape with maturity (Figs. 465, 466). Abdomen only slightly shorter than genital complex, otherwise as in female. Total length of body 5.4–12.3 mm.

Second antenna (Figs. 442, 443) with inflated penultimate segment equipped with one large, transverse, and one small adhesion pad; claw with two sharp, short tines diverging from base and with short, robust seta borne upon small subterminal tubercle. Maxilliped (Fig. 451): corpus much broader and more robust than in female, with two large pointed processes on medial margin; subchela and claw as in female. In young males, maxilliped (Fig. 450) much less robust, rather like female appendage.

Fifth leg (Fig. 463) on tip of outer posterolateral lobe, larger than, but resembling that of female. Sixth leg almost completely reduced, on tip of inner (slightly ventral to outer) posterolateral lobe.

Comments: In spite of its length, the history of *C. curtus* does not include a great deal of confusion. Having been removed by Müller (1785) from Linnaeus' composite taxon *Monoculus piscinus* and placed in a still rather ill-defined new species, *C. curtus* was widely accepted, though the ambiguities of the earliest descriptions left room for misinterpretations. It is to these early inaccuracies that one has to attribute descriptions of several taxa now placed in synonymy with *C. curtus*, all of them dating back to the nineteenth century. A full historical review of the species was given by Parker et al. (1968).

The distribution of *C. curtus* extends over the Arctic-Boreal region of the Atlantic and adjacent waters. Its records from the southern Atlantic (Tilesius, 1815), the Mediterranean (Brian, 1898, 1906), the Adriatic Sea (Heegaard, 1943), the Baltic (Markevich, 1956), the northern Pacific (Tilesius, 1815) and the Indian Ocean (Kirtisinghe, 1964) are either demonstrably erroneous or cannot be substantiated. Also inadmissible is Yamaguti's (1963) relegation of *C. bifurcus* Shen, 1958, to synonymy with *C. curtus*, an action that would extend the distribution range of this species to the Pacific region.

Although *C. curtus* is generally thought of as a parasite of the gadid fishes, its host range is very broad and includes species from diverse groups. Among the elasmobranchs, it parasitizes the dogfish, *Squalus acanthias*, and a number of species of *Raja* (*R. batis*, *R. fullonica*, *R. laevis*, *R. montagui*, *R. oxyrinchus*, *R. naevus* and *R. radiata*). It has been recorded from *Chimaera monstrosa*.

Among the gadid fishes it was found on *Gadus morhua*, *Trisopterus minutus*, *Brosme brosme*, *Melanogrammus aeglefinus*, *Merlangius merlangus*, *Microgadus tomcod*, *Molva molva*, *M. dypterygia*, *Pollachius pollachius*, *P. virens* and *Phycis blennoides*. It has been found on *Merluccius merluccius* on the European and on *M. bilinearis* on the American side of the northern Atlantic. The host range includes also percomorph fishes (*Sebastes marinus*, *Eutrigla gurnardus*, *Mugil cephalus* and *Crenimugil labrosus*) and flatfishes (*Scophthalmus maximus*, *S. rhombus*, *Hippoglossus hippoglossus*, *Microstomus kitt* and *Pleuronectes platessa*). *C. curtus* has also been found on the angler fish, *Lophius piscatorius*, and has been taken in plankton nets (Wilson, 1936). Its records from freshwater salmonids (Baird, 1850) is at least questionable. In British waters, *C. curtus* is common and widespread.

Caligus bonito Wilson, 1905

(Figs. 469–480)

Syn.: *C. kuroshio* Shiino, 1959

Female (Fig. 469): Cephalothorax suborbicular, with well developed frontal plates and moderately large lunules. First antenna with tip extending laterally as far as margins of shield. Free margin of thoracic zone projecting backwards beyond level of posterior ends of lateral zones. Fourth leg-bearing segment somewhat narrower than genital complex, with border between them indistinct or absent. Genital complex with length about twice greater than width, narrower anteriorly, with gentle posterolateral lobes, its length greater than that of thoracic zone of shield. Abdomen narrow, cylindrical, as long as, or slightly longer than, genital complex, very indistinctly divided into two segments (division often recognizable as slight indentations of lateral margins only). Total length about 8 mm.

Mandible (Fig. 472) similar to that of type species, with distal part bearing 12 teeth on one margin and smooth opposite margin. First maxilla (Fig. 474) with digitiform dentiform process, and exceptionally large basal papilla, its longest seta reaching near tip of dentiform process. Maxilliped with unarmed medial margin of corpus and long terminal claw of subchela. Sternal furca (Fig. 476) with strong subrectangular box (lateral margins slightly convex) and tines of about equal length, straight, with truncated tips and devoid of flanges.

First leg (Fig. 477): distal segment of exopod with seta 1 longer than 2 or 3, single, with strip of serrated membrane along one margin; setae 2 and 3 with very large secondary processes (Fig. 478) and row of denticles along convex margin. Seta 4 pinnate (pinnules omitted in Fig. 477), about $2\frac{1}{2}$ times as long as seta 1; three pinnate, long setae on posterior margin of same segment with inflated bases and between four and six short, dentiform, flat processes near base of lateral margins; those on distalmost seta longer and narrower than on other two. Second leg distinguished by armature of its endopod (Fig. 479); entire lateral margin of second segment with prominent, strong spines, flat and broad at bases, crowded and overlapping one another; small group of similar, though more slender, spines on lateral margin of basal segment, near junction with second segment. Fourth leg with exopod two-segmented (Fig. 480); distal segment comprising fused two segments; apical armature with three setae, first longer than other two, all with pectinate bases; on outer margin, at point marking former division, another and similar seta; basal segment well delimited, its distal corner with long seta, longer than other setae of leg, except first; all five setae with two narrow, marginal strips of serrated membrane (omitted in Fig. 480).

Male (Fig. 470): Dorsal shield and fourth leg-bearing segment resembling those of female. Genital complex oval or subquadrangular, longer than wide, lateral margins slightly convex, posterolateral corners bearing small protuberances. Length of complex shorter than that of thoracic zone of shield, its border with fourth leg-bearing segment often indistinct. Abdomen two-segmented, about as long as genital complex. First segment shorter than second, segmental border between them, as well as between first segment and genital complex, usually quite distinct. Total length about 5.5 mm.

Second antenna (Fig. 471) with complicated terminal claw, its tip consisting of three flat, subtriangular points, shorter point between two longer ones; closer to base of claw two sturdy setae, one on each side (only tip of one visible in Fig. 471). Postantennary process comparatively longer than in female. Dentiform process of first maxilla (Fig. 473) roughly similar to that of female, but with narrow, short flange on distal half and some transverse striations near that level; medial margin of maxilliped (Fig. 475) with three processes; basal process subtriangular, simple; distal near base of subchela, also simple; central of two spiniform outgrowths and rounded swelling.

Comments: A morphologically distinctive species, *C. bonito* has always been easily identifiable and the literature contains few doubtful entries relating to it. One source of confusion was the description of its mandible contained in Wilson's (1905) original account of the species. According to Wilson, the tip of the mandible opposite its dentiferous margin was crenated into about a dozen rounded projections, its apex being transversely truncated. No mandible of this type was ever reported from other caligids, or from any other parasitic copepod. This fact prompted Shiino (1959a) to establish a new species, *C. kuroshio*, for specimens with normal caligid mandibles identified as *C. bonito* by Yamaguti (1936). Other authors accepted Shiino's decision (Pillai, 1961, 1963b, 1967a; Kabata, 1965a; Lewis, 1964a), though Lewis (1967) later suspected its validity and Pillai (1971) came to the conclusion that *C. kuroshio* must be synonymous with *C. bonito*. To settle the question, this author examined Wilson's type material preserved in the Smithsonian Institution, Washington, D.C. Taking at random specimens from an abundant sample, he found that they all possessed normal mandibles, with tips comparable to that shown in Fig. 472. There can be no doubt that *C. kuroshio* must be placed in synonymy with *C. bonito*. The mandible described by Wilson must have been a monstrosity, or a result of damage due to handling or similar causes.

Recently, Thomas (1967) described a new species, *C. krishnai*, for specimens which Pillai (1963b) described as *C. kuroshio*. The validity of that species seems doubtful, since its claim to distinctness is based on criteria as unreliable as the shape and proportions of the genital complex, slight differences in the shape of the sternal furca and somewhat smaller size of processes on the medial margin of the maxilliped. In a species as widespread as *C. bonito* such morphological discrepancies might simply reflect differences between geographically distant populations.

The most common hosts of *C. bonito* are large scombrid fishes (*Sarda sarda*, *S. velox*, *S. chilensis*, *S. orientalis*, *Sarda* sp., *Euthynnus pelamis*, *E. lineatus*, *E. affinis*, *E. alleteratus*, *Katsuwonus pelamis*, *Thunnus thynnus*, *Scomberomorus maculatus*, *S. cavalla* and *Cybium* sp.). The host range includes also representatives of two other suborders of Perciformes, Mugiloidei and Percoidei. The former is represented by *Mugil cephalus* found parasitized with *C. bonito* in the Gulf of Mexico by Bere (1936). From the latter suborder come *Pomatomus saltatrix*, *Oligoplites saurus*, *Cratinus agassizi*, *Lutianus griseus*, *L. novemfasciatus* and *Lutianus* sp. This author found it on one occasion in the branchial cavity of *Sarda sarda* taken in 1966 in Scottish waters, off the mouth of South Esk. A single female, 7.5 mm long, constitutes a new record for British waters. The distribution range of *C. bonito* is as impressive as its host range. It has been found in the northern and southern Atlantic, the Mediterranean and even in the Black Sea (Reshetnikova, 1955). There are many records of its occurrence in the northern and southern Pacific and in the Indian Ocean. Barnard (1955a) found it in South African waters.

Caligus brevicaudatus A. Scott, 1901

(Figs. 481–488)

Female (Fig. 481): Cephalothorax suborbicular, with slightly narrower anterior end, narrow frontal plates and small lunules. Tip of first antenna not reaching as far as lateral limit of shield. Posterior sinuses shallow, free margin of thoracic zone projecting beyond tips of lateral zones. Genital complex always longer than thoracic zone of shield (in adult specimens), length greater than width, anterolateral corners rounded, posterolateral sometimes forming small, rounded lobes; lateral margins of genital complex with small, but clearly noticeable, setae, about 7–10 on each margin. Abdomen one-segmented, slightly broader than long, its length less that $\frac{1}{4}$ of genital complex. Anterior third of abdomen pushed into, but fairly well delimited from, genital complex. Uropod longer than $\frac{1}{2}$ length of abdomen. Total length about 5 mm.

Dentiform process of first maxilla (Fig. 482) with broad subtriangular base and sturdy, pointed, curving tip bearing small flange near apex. Corpus of maxilliped with smooth medial margin; terminal claw of subchela (Fig. 483) short, slender and with sharp point; barbel at base of claw more than half its length. Sternal furca (Fig. 484) with square basal box; tines not longer than box, slightly inturned, blunt, with narrow flanges on margins.

Apical armature of first exopod (Fig. 485) with all four setae of about equal length; seta 1 simple, with one strip of finely serrated and striated membrane along one margin; setae 2 and 3 with spiniform secondary processes equalling $\frac{1}{2}$ length of setae; anterior margins with serrated flanges; seta 4 robust, tapering, unarmed; posterior margin with three long pinnate setae. Endopod of second leg (Fig. 486) with fringe of fine setules on lateral margins of all three segments; basal segment with proximal short fringe and distal longer one. Fourth leg (Fig. 488) with distal tip of sympod bearing one fairly long and pinnate seta and one short multiple setule; exopod two-segmented; distal segment armed with three apical setae, gradually diminishing in length, all armed with strips of serrated, faintly striated membrane on two margins (Fig. 487); apical seta of basal segment similar to third seta of distal segment. Fifth leg apparently represented by single seta in posterolateral corner of genital complex.

Male: Unknown.

Comments: *Caligus brevicaudatus* is a rare species. Discovered in the Irish Sea more than 70 years ago, it has been recorded subsequently only a few times, always on the gurnards of the family Triglidae. Its host list comprises *Eutrigla gurnardus*, *T. lucerna* and *T. capensis*. (Reichenbach-Klinke's (1956) record from *Peristedion cataphractum* requires verification.) All British records come from the Irish Sea (A. Scott, 1901b; T. Scott and A. Scott, 1913). This author searched many years for *C. brevicaudatus* in Scottish waters, the North and Irish Seas without discovering one specimen. The rarity of the species is best evidenced by the fact that its male has not yet been found. (This description is based on examination of A. Scott's material.) Nevertheless, the range of the species is quite extensive. Markevich (1956) recorded it from the Barents Sea, Reichenbach-Klinke (1956) found it in the Gulf of Naples and Barnard (1955b) in South African waters.

Readers' attention should be drawn to the fact that *C. brevicaudus* Pillai, 1963, is not related to *C. brevicaudatus*.

Caligus centrodonti Baird, 1850

(Figs. 505–518)

Syn.: *C. abbreviatus* Krøyer, 1863

Female (Fig. 505): Cephalothorax suborbicular, with narrow frontal plates and rather small lunules. Posterior tips of lateral zones at about same level as free margin of thoracic zone. Tip of first antenna barely reaching lateral limit of shield. Fourth leg-bearing segment very short, about as wide as ½ free margin of thoracic zone, visibly narrower at base, sometimes with indistinct border between it and genital complex. Latter broader than long, shorter than thoracic zone of shield, with rounded anterolateral and slightly bulging posterolateral corners, with re-entrant posterior margin. In dorsal aspect lateral margin beset with short setules, about six each. Fifth leg protrudes beyond limit of genital complex. Abdomen one-segmented, diminutive, not more than ¼ length of genital complex, with indistinct border between them, barely extending beyond posterior limit of genital complex. Length about 4 mm.

Postantennary process (Fig. 507) with slender, sharply curving hamulus and straight process at base, parallel to tip of hamulus. Dentiform process of first maxilla (Fig. 508) short, with broad, subtriangular base and blunt, somewhat curving tip; narrow transparent flange between lateral corner of base and tip, some distance from apex. Maxilliped (Fig. 514) small, with very slender corpus; myxal area with small, pointed protrusion; subchela sturdy, with pronounced curvature in proximal half; claw short (less than ¼ length of subchela), with barbel less than ½ its own length. Sternal furca (Fig. 511) with box wider anteriorly, quadrangular posteriorly; tines diverging at base, with tips curving in, tapering and broad marginal flanges greatly adding to their width.

Distal segment of first exopod with four setae of about equal length (Fig. 512) ; all four setae simple; setae 1 and 4 apparently unarmed, setae 2 and 3 with narrow flanges running along ventral and dorsal margins near apex; posterior margin of distal segment with three long setae (base of one shown in Fig. 512), all with medial margins pinnate, lateral equipped with strips of broad, transparent membrane and short row of coarse setules near base; setules densely crowded (some omitted for clarity in Fig. 512). Endopod of second leg with lateral margin (Fig. 513) provided with fringe of fine setules on all three segments; on segment 2 fringe covers also broad strip of ventral surface near margin. Fourth leg (Fig. 516) with short, robust sympod, a long seta in its distal corner; exopod two-segmented ; distal segment with only two setae as apical armature, both with two strips of serrated membrane (not illustrated) and basal pectens; proximal segment much shorter than, and distinctly delimited from, distal one; its single apical seta longer than ½ length of distal segment and similar to other two setae, though apparently not pectinate. Fifth leg of three short setules surmounting very small swelling (Fig. 517). Uropod similar to that of male.

Male (Fig. 506): Cephalothorax and fourth leg-bearing segment similar to those of female. Genital complex much broader than long, shorter than ½ length of thoracic zone of shield; anterolateral corner rounded, posterolateral forming prominent lobes extending as far as posterior limit of abdomen (discounting uropods). Lateral margins setose as in female. Total length in order of 5 mm.

Second antenna with complicated terminal claw, its posterior side with two slender, laterally projecting, secondary tines (Fig. 509), as well as robust basal seta, anterior side covered by triangular membranous flap, exposing only tip of claw and second, subterminal seta. Maxilliped (Fig. 515) with robust corpus so much inflated on medial margin as to appear bifid, about equally long and wide. Two papilliform outgrowths near tip of myxa, claw of subchela closing in between them. Fifth leg, similar to that of female (Fig. 518), at tip of posterolateral lobes of genital complex. Uropod short, broad (Fig. 518), with oblique anterior and posterior margins. No trace of sixth leg discovered.

Comments: Soon after its discovery, C. *centrodonti* was independently described as *C. abbreviatus* by Krøyer (1863, 1864). Krøyer's illustrations, particularly of male and female habitus and of the fourth leg, leave no doubt that his specimens were conspecific with those of Baird. The synonymy was first suspected by T. Scott (1905), who later accepted it as definite (T. Scott and A. Scott, 1913). Of subsequent writers, only some Russian authors used Krøyer's name in preference to that of Baird (Osmanov, 1940; Reshetnikova, 1955).

The specific name of *C. centrodonti* commemorates its original and most common host, *Pagellus centrodontus*, a percoid fish now known as *P. bogaraveo*. Other known hosts are *Labrus bergylta*, *Labrus* sp., *Crenilabrus ocellatus, C. pavo* and *Sebastes marinus*, all but the last being members of the order Perciformes.

British records come from the Irish Sea, south coast of England and Cornwall, and from the Moray Firth in Scotland. Boxshall (1974a) found it in the North Sea. Krøyer's specimens were collected off Bergen (Norway) and Russian from the Black Sea. Surprisingly, the species has not been found in the Mediterranean.

Caligus coryphaenae Steenstrup and Lütken, 1861

(Figs. 519–530)

Syn.: ?*C. scutatus* Edwards, 1840
C. thymni Dana, 1852
?*C. bengoensis* Scott, 1895
C. aliuncus Wilson, 1905
nec *C. coryphaenae*; of Brian (1935)
nec *C. coryphaenae*; of Yamaguti (1936)
C. elongatus Heegaard, 1943
C. tesserifer Shiino, 1952
nec *C. coryphaenae*; of Barnard (1955a)

Female (Fig. 519): Cephalothorax oval, with well developed frontal plates and deep posterior sinuses. Free margin of thoracic zone about level with tips of lateral zones. Tip of first antenna reaching lateral limit of shield, or nearly so. In addition to usual sutures, second transverse suture present, running behind eyes, linking anterior tips of longitudinal sutures. Fourth leg-bearing segment about as wide as free margin of thoracic zone, short, with narrow anterior neck. Genital complex longer than thoracic zone, anteriorly as broad, or somewhat broader than, fourth leg-bearing segment, often not definitely delimited from latter; length greater than width, lateral margins diverging posteriorly, posterolateral corners forming lobes, sometimes rounded but often angular. Division between genital complex and abdomen indistinct. Latter about as long as genital complex, much longer than broad, usually slightly broader at base, with uneven diameter and indistinct segmentation (most often three segments recognizable, though some authors record as many as five); posterior margin V-shaped; uropod subquadrangular. Total length about 5.5–8.0 mm.

First antenna short, stocky, two plumose setae of basal segment extending beyond tip of terminal segment; latter (Fig. 522) club-shaped; eight anterior setae of apical armature as in all species of *Caligus*, posterior five longer than terminal segment, plumose. Postantennary process much reduced or absent. Dentiform process of first maxilla (Fig. 523) short, subtriangular or of similar shape. Corpus of maxilliped with smooth medial margin; subchela robust, with very long terminal claw (about ½ length of subchela) and short, sturdy barbel. Sternal furca (Fig. 524) with subquadrangular box and diverging tines about as long as box; tines straight, tapering, with marginal flanges (width and extent of latter very variable). At level of posterior end of box pair of semicircular cuticular flanges arising from surface of cephalothorax, one on each side of furca.

First leg: terminal armature of second segment of exopod (Fig. 526) consisting of four setae, fourth usually somewhat shorter than remaining three; seta 1 with smooth flange on anterior and serrated on posterior margin, broad and simple; setae 2 and 3 with both flanges serrated and with small, smooth secondary processes; seta 4 slender, with two rows of minute prickles and (in contrast with other three) devoid of basal pecten; posterior margin of segment with three long, pinnate setae. Endopod of second leg with fringe of flat, lanceolate setules on lateral margin (Fig. 527), extending over second, as well as parts of first and third segments. Sympod of fourth leg with particularly long and robust apical seta (in some specimens second, smaller seta present); exopod of three well delimited segments (Fig. 528), with three setae in apical armature, diminishing in length from seta 1 to 3; seta 1 (Fig. 529a) with one very narrow strip of serrated membrane, seta 2 (Fig. 529b) with two similar strips, posterior broader than anterior, seta 3 (Fig. 529c) with row of denticles on anterior and broad strip of serrated membrane on posterior margin; apical setae of first and second segments long, armed like terminal seta 3. Fifth leg (on ventral surface of posterolateral corner of genital complex) of four plumose setae, one much longer than other three.

Male (Fig. 520): Cephalothorax similar to that of female, but with comparatively larger frontal plates and lunules. Genital complex suborbicular, shorter than thoracic zone of shield, about as wide as fourth leg-bearing segment and indistinctly separated from it. In dorsal aspect, fifth and sixth legs protrude from below posterior margin of genital complex. Boundary between latter and abdomen often indefinite. Abdomen two-segmented, somewhat shorter than genital segment; second segment slightly longer than first, its posterior margin V-shaped. Uropod subquadrangular. Total length in order of 4.5–6.0 mm.

Second antenna (Fig. 521) with terminal claw like that of female, except for prominent secondary tine on concave margin near base of claw; adhesion pads small. Corpus of maxilliped with small corrugated adhesion pad in myxal area (Fig. 525). Fifth leg (Fig. 530) of four plumose setae, three short ones grouped together and one long one at some distance. Sixth leg (Fig. 530) medial to fifth, of three plumose setae, two short and one long.*

Comments: World-wide in distribution and catholic in its host association, *C. coryphaenae* is predictably distinguished by broad morphological variability. This, in turn, leads to errors of identification, as witnessed by several synonyms. On the other hand, similarity between *C. coryphaenae* and some other species of *Caligus* has resulted in errors of the opposite kind. Brian (1935) and Barnard (1955a) identified specimens of *C. euthynus* Kurian, 1961, as *C. coryphaenae*, while Yamaguti (1936) made a similar mistake in relation to *C. quadratus* Shiino, 1954. One of the most notably variable features is the segmentation of the abdomen, mentioned above in description. Another is the absence or great degree of reduction of the postantennary process. When one adds to it such details as the presence of a second transverse suture on the dorsal shield and of cuticular outgrowths flanking the sternal furca, one must consider *C. coryphaenae* a very atypical member of its genus. None of these details, however, is sufficient to warrant a removal of it from *Caligus*, especially in view of similar species, easily confused with it, being accommodated within this genus.

C. coryphaenae has been recorded from all warm-water regions of the world's oceans, as well as from the lower latitudes of temperate regions. In addition to its type host, *Coryphaena hippurus*, it has many hosts among far-ranging, fast-swimming scombrid fishes (*Sarda orientalis*, *Sarda* sp., *Katsuwonus pelamis, K. vagans, Katsuwonus* sp., *Euthynnus affinis, E. alleteratus, E. lineatus, Germo albacora, Thunnus thynnus, T. obesus, Acanthocybium solanderi, Gramatorcynus bilineatus, Neothunnus macropterus, Parathunnus sibi*). Of other perciform fishes it was found on some carangids (including *Caranx hippos*, *C. melampygus* and *Elegatus bipinnulata*), on species of remora (*Echeneis* sp.), on *Rachycentron canadum*, and on *Polydactylus opercularis*. Surprisingly enough, it occurs also on Tetraodontiformes (*Sphaeroides inermis* and *Balistes* sp.), as well as having been recorded on sharks (*Squalus acanthias* and *Isurus oxyrinchus*).

There is, as yet, no British record of *C. coryphaenae*. However, Tiews (1957) found it on the red tuna (*Thunnus thynnus*) in the North Sea. No locality was given, but some of Tiews' tunas were taken off the east coast of England and Scotland. Consequently, *C. coryphaenae* might be expected to visit British waters.

*These setae are easily lost or damaged, so that fewer numbers are found in many specimens.

Caligus diaphanus Nordmann, 1832

(Figs. 538–548)

Syn.: nec *C. diaphanus*; of Baird (1850)
?*C. torpedinis* Heller, 1865

Female (Fig. 538): Cephalothorax orbicular, with shallow posterior sinuses and small lunules. Tips of first antennae extending about as far laterally as limits of shield. Free margin of thoracic zone about level with posterior ends of lateral zones. Fourth leg-bearing segment narrower than free margin of thoracic zone, with narrow anterior neck. Genital complex about equally long and wide, with sloping anterolateral and rounded posterolateral corners, longer than thoracic zone of shield. Posterior margin usually straight, boundary between genital complex and abdomen usually indistinct. Latter cylindrical, two-segmented, with very indistinct segmental borders (often recognizable only as slight lateral indentations). Second segment less than $\frac{1}{2}$ length of tagma. Length of abdomen greater than that of genital complex. Total length of body in order of 4.5 mm.

First maxilla with digitiform dentiform process (Fig. 540), tip rounded, no secondary processes or surface markings. Sternal furca (Fig. 543) with square box and slightly divergent tines, longer than box, with rounded tips and devoid of flanges. Maxilliped with slender corpus and very long subchelar claw; medial margin smooth.

First leg: distal segment of exopod (Fig. 546) with four simple setae, densely crowded and overlapping, seta 1 set much closer to base of segment than other three and bearing single row of minute prickles; other setae of about equal length, with slender, pointed tips, apparently unarmed; posterior margin of segment with three long, pinnate setae. Endopod of second leg (Fig. 547) bearing on second and third segments patches of short, bristle-like setules, occupying lateral part of ventral surface and lateral margins. Exopod of fourth leg (Fig. 548) three-segmented, with distinct intersegmental boundaries; segments 1 and 2 carry one long seta at posterodistal corners, distal segment with three similar setae; all five setae of about same length and all with two fringes of smooth membrane; bases with pectens, that of second segment extending along entire posterior margin of segment; no pecten at base of proximal spine, but strip of membrane along margin of segment present. Uropod slightly narrower at base, longer than wide.

Male (Fig. 539): Cephalothorax similar to that of female. (In Fig. 539 anterior part of shield and frontal plates distorted by contraction.) Genital complex shorter than thoracic zone of shield, with straight anterior and posterior, convex lateral margins, longer than broad. Fifth leg visible in dorsal aspect on distal half of lateral margin, sixth in posterolateral corner. Abdomen cylindrical, two-segmented, shorter than genital complex. First segment shorter than second. Uropod similar to that of female. Total length about 2 mm.

First antenna with terminal claw (Fig. 544) in form of simple hamulus with secondary tine on concave margin near base and sturdy seta at base; adhesion pad small. (This appendage resembles second antenna of male *C. coryphaenae*.) First maxilla with dentiform process (Figs. 541, 542) slightly curving and bearing subterminal secondary outgrowth; latter with subtriangular flap of transparent, delicate membrane. Corpus of maxilliped (Fig. 545) broader than that of female, with prominent, shelf-like expansion in myxal area, slightly indented at tip; subchela similar to that of female, with very long claw and barbel nearly reaching tip of claw.

Comments: *C. diaphanus* is a distinctive and easily recognizable species. This became evident when Krøyer (1837) identified it from the original description (Nordmann, 1832) which contained no illustrations. (Nordmann's text contains reference to Fig. 15, but that figure was never published.) It appears that the identity of *C. diaphanus* was clear to most subsequent authors. Baird (1850) mistook *C. curtus* for it and, according to some authorities, Edwards (1840) misidentified it also. The text of the latter author, however, is not informative enough to make a definitive judgement on the identity of the animal described as *C. diaphanus*. Some controversy has developed more recently around the validity of *C. multispinosus* Shen, 1957, and its possible relationship with *C. diaphanus*. Pillai's (1961) description of *C. multispinosus* was placed in synonymy with *C. diaphanus* by Kirtisinghe (1964). Pillai (1967) rejected Kirtisinghe's views. Comparison of Shen's and Pillai's figures leaves no doubt that they deal with the same species. Comparison of *C. multispinosus* with *C. diaphanus* shows them to be very similar to each other. In particular, the general habitus, the structure of the fourth leg and (as far as one can make valid comparisons of drawings) of the male second antenna. There are, however, definite differences in smaller details. The shape of the dentiform process of the first maxilla differs somewhat in females of the two species (male not illustrated in *C. multispinosus*). The name of the latter species was prompted by dense spinulation present on the terminal setae of the exopod of the first leg and on all but the first of the setae of the fourth leg. This spinulation is definitely absent from *C. diaphanus*. Though these details are small, they suffice to prompt caution in deciding on the relationship between *C. diaphanus* and *C. multispinosus* until a more thorough examination can be made. The possible synonymy of *C. torpedinis* Heller, 1865, with *C. diaphanus* also requires corroboration.

The distribution range of the species is broad, extending over the European Atlantic seaboard from Spain to Norway (but not further North), the western Mediterranean and the Adriatic. Two areas outside that range have also been recorded: Ceylon (Thompson and Scott, 1903) and Panama Bay (Wilson, 1937). The records from those areas also require confirmation.

Recorded hosts of *C. diaphanus* are many and varied. The most common hosts are fishes of the family Triglidae (*Eutrigla gurnardus, Trigloporus lastoviza, Aspitrigla cuculus, Trigla milvus, T. lyra, T. lucerna, Trigla* sp. and *Lepidotrigla cavillone*). Several familes of Perciformes are also represented: Sparidae (*Pagellus bogaraveo, P. mormyrus, P. erythrinus* and *P. acarnae*), Carangidae (*Trachurus trachurus, Caranx hippos*), Scombridae (*Scomber scombrus*), Theraponidae (*Autisthes puta*) and Centropomidae (*Lates calcarifer*). Gadoid fishes are included (*Merluccius merluccius, Gadus morhua, Pollachius virens* and *Molva molva*), as are flatfishes (*Scophthalmus maximus, Lepidorhombus whiffiagonis, Solea solea* and "*Platessa passer*"). There is also a record from *Belone bellone*.

Caligus elongatus Nordmann, 1832

(Figs. 549–558)

Syn.: *C. kroeyeri* Edwards, 1840
C. rissoanus Edwards, 1840
nec *C. rapax* Edwards, 1840
C. rapax; of Baird (1850) (partim)
C. gurnardi Krøyer, 1863 (partim)
C. lumpi Krøyer, 1863
C. trachypteri Kollar, in Krøyer (1863)
C. latifrons Wilson, 1905
C. rabidus Leigh-Sharpe, 1936
nec *C. elongatus* Heegaard, 1943
C. arcticus Brandes, 1956

Female (Fig. 549): Cephalothorax oval, with well developed frontal plates, large lunules and deep posterior sinuses. Free margin of thoracic zone projecting well beyond level of tips of lateral zones. Tip of first antenna reaching lateral limit of shield. Fourth leg-bearing segment narrow, short, not distinctly delimited from genital complex. Latter ovate, as long as, or longer than, thoracic zone of shield. Posterolateral corners sometimes forming gently rounded lobes. Boundary between genital complex and abdomen indistinct. Latter one-segmented, about $\frac{1}{2}$ length of genital complex, cylindrical, longer than wide, often slightly narrower at base and apex. Uropod about $\frac{1}{4}$ length of abdomen, longer than wide. Total length of body about 5–6 mm.

Dentiform process of first maxilla (Fig. 556) with broad base and simple, tapering tip, without secondary processes or superficial markings. Maxilliped (Fig. 555) with robust, stocky corpus; medial margin smooth; subchela with very long terminal claw and short barbel. Sternal furca (Fig. 558) with short box and strong, diverging tines; tines fused at bases, tapering, with rounded tips, devoid of flanges.

Distal segment of first exopod (Fig. 553) with three terminal setae of equal length and fourth much longer; seta 1 with narrow strip of transparent membrane on distal half of anterior margin; setae 2 and 3 appearing chelate, due to presence of secondary processes arising at midlength, and slightly shorter than setae; seta 4 about twice as thick at base and twice as long as other setae; anterior margin with short, sparse setules in distal half, posterior with very narrow strip of membrane, difficult to see; posterior margin of segment with three long pinnate setae. Endopod of second leg (Fig. 554) with abundant fringe of fine setules on lateral margin of second segment and parts of first and third segments. Exopod of fourth leg (Fig. 552) two-segmented, distal segment fused of two and retaining original apical seta of second segment; first of three terminal setae about twice as long as other two, armed with two rows of densely crowded, minute prickles (only one usually visible);

remaining four setae of exopod about equal in length, all armed with two strips of narrow serrated membrane; division between first and second segment usually quite definite. Fifth leg of three setae, two together and third further apart, on posterolateral corner of genital complex.

Male (Fig. 550): Anterior half of body as in female. Genital complex oval, shorter than thoracic zone of shield. Abdomen two-segmented, cylindrical, about as long as genital complex. Anterior segment shorter than posterior, boundary usually well delimited. Total length of body about 4–5 mm.

Second antenna with bifid claw (Fig. 551), parts of its tips covered by overlapping cuticular flap; one strong seta on each side of claw, close to base; adhesion pads well developed. Dentiform process of first maxilla (Fig. 557) generally similar to that of female, but with short lateral outgrowth on second half and with broad band of transverse striations on middle third. Sternal furca with tines often relatively longer than those of female. Fifth leg of three and sixth of two setae.

Comments: The length of its synonymy, accrued during more than a century of research publications, reflects also the ubiquity of *C. elongatus* and the extraordinary range of its known hosts. Throughout most of its recorded history, it has been confused with *C. rapax* Edwards, 1840, as the result of misidentification by Baird (1850). The confusion was not unravelled until Parker (1969) published his excellent account of the species.

Under its incorrect name, *C. elongatus* has been recorded from most regions of the world's oceans. Its host list includes more than 80 species of fish, both teleosts and elasmobranchs, representing 17 orders and 43 families. In British waters its hosts comprise *Gadus morhua*, *Melanogrammus aeglefinus*, *Merlangius merlangus*, *Molva molva*, *Trisopterus minutus*, *T. luscus*, *Pollachius pollachius*, *P. virens*, *Brama brama*, *Dicentrarchus labrax*, *Trachurus trachurus*, *Scomber scombrus*, *Cyclopterus lumpus*, *Clupea harengus*, *Pomatoschistus minutus*, *Liza ramada*, *Crenimugil labrosus*, *Eutrigla gurnardus*, *Trigla lucerna*, *Aspitrigla cuculus*, *Zeus faber*, *Mola mola*, *Salmo trutta*, *S. salar*, *Lophius piscatorius*, *Limanda limanda*, *Scopthalmus maximus*, *Pleuronectes platessa*, *Microstomus kitt*, *Platichthys flesus*, *Raja batis*, *R. clavata*, *R. naevus*, *R. radiata*, *R. montagui*, *R. "blanda"*, and *Squalus acanthias*. It is probably the most common species of parasitic copepod in British waters.

Caligus gurnardi Krøyer, 1863

(Figs. 559–571)

Syn.: *C. gurnardi* Krøyer, 1863 (partim)
C. nanus Krøyer, 1863
nec *C. gurnardi*; of Brian (1898)
nec *C. gurnardi*; of Wilson (1908)
nec *C. gurnardi*; of Fraser (1920)

Female (Fig. 559): Generally resembling that of *C. elongatus*. Cephalothorax suborbicular to oval, with well developed frontal plates, large lunules, and moderately deep posterior sinuses. Tip of first antenna slender and long, reaching lateral limit of dorsal shield. Free margin of thoracic zone extending beyond level of tips of lateral zones. Fourth leg-bearing segment about equally long and wide, with narrower anterior neck, usually well delimited from the genital complex. Latter longer than thoracic zone of shield, suborbicular to oval, with sloping anterolateral corners and evenly rounded posterolateral ones; posterior margin transverse, straight. Abdomen about as long as, or slightly shorter than, $\frac{1}{2}$ length of genital complex, one-segmented, its base inserted somewhat to dorsal side of posterior extremity of preceding tagma, cylindrical, base somewhat wider than posterior end. Uropod longer than wide, less than $\frac{1}{2}$ length of abdomen. Total length of body about 6–7 mm.

Dentiform process of first maxilla (Fig. 563) tapering to rounded tip, often somewhat curved, but on occasions straight; no secondary processes or surface markings. (In some specimens distal half bears some transverse notches, perhaps artifacts.) Maxilliped similar to that of *C. elongatus*. Sternal furca with strong, quadrangular box (Fig. 565) and divergent tines, slightly curving, with rounded tips and sometimes narrow flanges at tips. No other flanges or membranes.

Distal segment of first endopod (Fig. 566) similar to that of *C. elongatus*; setae 1–3 of about same length, seta 4 about twice that length; seta 1 simple and apparently unarmed; setae 2 and 3 with chelate tips (Fig. 567); no armament discernible on long seta 4. Endopod of second leg (Fig. 568) with fringe of setules on lateral margins; setules of two lengths on basal segment, distally longer; those on third segment extending in tuft over part of ventral surface. Fourth leg (Fig. 561) of relatively great length, when stretched reaching posterior end of body; structure and armature of exopod (Fig. 569) like that of *C. elongatus*. Fifth leg (Fig. 570) of three setae, two in one group and third at some distance from them.

Male (Fig. 560): Anterior half similar to that of female. Genital complex oval, shorter than thoracic zone of shield, well delimited both anteriorly and posteriorly. Abdomen two-segmented, shorter than genital complex, cylindrical; first segment shorter than second, segmental boundary definite. Uropod about as long as first segment. Total length of body up to 6 mm.

First antenna with terminal claw (Fig. 562) bifid, tips covered by cuticular flap; two sturdy, long setae near base, one on either side of claw. Dentiform process of first maxilla (Fig. 564) similar to that of female, but with small lateral outgrowth on distal half and band of transverse grooves on middle third. Fourth leg relatively longer than that of female, when stretched might protrude beyond posterior limits of body. Fifth leg (Fig. 571) as in female; sixth leg medial to fifth, consisting of two setae surmounting low swelling.

Comments: Having examined Krøyer's type material of *C. gurnardi*, Parker (1965) found that it contained specimens of two species, some belonging to *C. elongatus*, others distinct from it. The latter became the foundation of *C. gurnardi*. Other specimens of that species were described in the same paper (Krøyer, 1863) as *C. nanus*. Accepting that species as a synonym of *C. gurnardi*, Parker (1965) acted as the first reviewer in the meaning of that term defined by the Code of Zoological Nomenclature, article 24(a), his action binding on subsequent workers. Some authors subsequent to Krøyer mistook other species for *C. gurnardi*. Brian (1898) recorded under its name specimens of *Caligus* which he later recognized as *C. minimus* (though that second identification is also suspect). Wilson (1908) and Fraser (1920) believed they discovered *C. gurnardi* off the Pacific coast of North America. The former record still calls for confirmation, the latter has now been identified as referring to *C. clemensi* Parker and Margolis, 1964.

The description given above points out the great similarity between this species and *C. elongatus*. The two species are distinguishable, though with some difficulty. The main diagnostic features of *C. gurnardi* are: the length of the fourth leg, exceeding that of *C. elongatus* (comp. Fig. 550 with Fig. 560) and the position of the base of the abdomen. This close similarity was at the source of confusion in Krøyer's original material. Some authors (Timm, 1894; Gadd, 1904) believed *C. gurnardi* to be a subspecies of *C. elongatus* (as *C. rapax*). Under a similar impression, Olsson (1869) included *C. gurnardi* in his account of *C. rapax* and combined the hosts of both in a common list, later quoted by Parker (1965) for the former species. Krøyer's specimens of *C. nanus* (= *C. gurnardi*) allegedly came from cod and haddock. However, the only ascertained hosts of this species are *Eutrigla gurnardus*, *Trigla lucerna* and *Aspitrigla cuculus*, no reliable records being made from any gadoid species.

C. gurnardi is rare. This author, having searched for years all species of Triglidae in Scottish waters was unable to find any specimens and had to base his description on material kindly made available by the British Museum.

The distribution range of *C. gurnardi* covers the European Atlantic seaboard.

Caligus labracis T. Scott, 1902

(Figs. 572–581)

Female (Fig. 572): Cephalothorax nearly orbicular, with dorsal shield somewhat narrower anteriorly, with moderately large lunules and tip of first antenna not reaching lateral limit of shield. Posterior sinuses shallow, free margin of thoracic zone slightly protruding beyond level of tips of lateral zones. Fourth leg-bearing segment about as long as broad, not distinctly delimited from genital complex. Latter usually longer than broad, more than length of thoracic zone, with sloping anterolateral corners (sometimes forming slight anterior protrusions) and rounded posterolateral corners. Posterior margin straight. No firm boundary between genital complex and abdomen. Latter apparently one-segmented (though in some specimens lateral indentations present, suggesting possible segmental boundary), shorter than $\frac{1}{2}$ length of genital complex, slightly longer than broad. Uropod more than $\frac{1}{2}$ length of abdomen. Total length 3.5 mm.

Postantennary process (Fig. 575) with sharply hooked hamulus and broad base; straight secondary tine present on base, on concave margin of process. Dentiform process of first maxilla (Fig. 578) slightly curving, tapering to fine tip, devoid of secondary outgrowths or surface sculpture. Sternal furca (Fig. 577) with subrectangular box and divergent, markedly spatulate tines (posterior process on second segment of second antenna similarly spatulate).

Terminal armature of first exopod (Fig. 579) crowded together, setae overlapping one another; seta 1 shortest, pointed and unarmed; setae 2 and 3 progressively longer, simple, bearing two ridges on either side of tips; seta 4 shorter than 3, unarmed; posterior margin of segment with three long, pinnate setae. Endopod of second leg (Fig. 580) with margin of fine setules on second segment; first segment with margin of short, stiff, bristle-like setules. Exopod of fourth leg (Fig. 581) two-segmented, terminal setae two, second shorter than first, both with strip of finely serrated membrane running their entire circumference; basal pectens prominent; apical seta of first segment similar in structure, but with smaller pecten; boundary between segments distinct. Fifth leg of three setae surmounting small tubercle on posterolateral corner of genital complex.

Male (Fig. 573): Cephalothorax as in female. Fourth leg-bearing segment of about equal length and width, lateral walls convex, boundary with genital complex indistinct. Latter narrow at base (about as wide as posterior margin of fourth leg-bearing segment), expanding in posterior direction, shorter than thoracic zone of shield. Each posterolateral corner with two small lobes, projecting posteriorly, one ventromedial to other (Fig. 576). Division between complex and abdomen indistinct. Latter subquadrate, about $\frac{1}{2}$ length of genital complex, one-segmented. Uropod only slightly shorter than abdomen. Total length 2.6 mm.

Terminal claw of second antenna (Fig. 574) trifid; when in situ, main claw pointing mediad, longest and most slender; two secondary tines shorter, blunt; short, pointed spinules present at base of these tines; a longer, slender seta near base of claw; two adhesion pads on penultimate segment. Fifth leg (Fig. 576) at tip of posterolateral lobe, corner of genital complex, consisting of three short, pinnate setae. No armature found on tip of second, ventromedial lobe which should be site of sixth leg (possibly broken off by handling).

Comments: *C. labracis* is a distinctive species, identifiable without difficulty. Its characteristic features (spatulate tines of sternal furca, basal tine of postantennary process, exopod of fourth leg) assure its immediate recognition even on cursory examination.

Discovered in British waters (Irish Sea), *C. labracis* is known almost exclusively around the British Isles, where it occurs apparently only on two host species: *Labrus bergylta* and *L. mixtus*. A single and surprising record from outside this region is that of Barnard (1955a) from South African waters, where it was found on *Clinus superciliosus*, a blennioid fish. Record includes no description or figure; it calls for confirmation.

Caligus minimus Otto, 1821

(Figs. 582–591)

Syn.: *C. minutus* auct.
?*C. gurnardi*; of Brian (1898)
C. curtus; of Brian (1906)
C. minimus var. *mugilis* Brian, 1935

Female (Fig. 582): Cephalothorax suborbicular to oval, with frontal plates indented at mid-dorsal line, lunules large, posterior sinuses shallow; posterior margin of thoracic zone projects beyond level of posterior tips of lateral zones; first antenna slender, reaching limit of lateral margin of shield. (In Fig. 582 dorsal shield somewhat contracted, its margins curling in ventral direction.) Fourth leg-bearing segment broader than long, expanding posteriorly, well delimited from genital complex. Latter similar to that of *C. labracis*, its anteriormost part often set off into neck-like portion; length of tagma greater than thoracic zone of shield, lateral margins convex, diverging posteriorly; posterior margin straight or slightly re-entrant at centre, posterolateral corners rounded, boundary with abdomen indistinct. Latter one-segmented, less than $\frac{1}{2}$ length of genital complex, subquadrate, with slightly peduncular base. Uropod prominent, almost as long as abdomen. Total length 4–5 mm.

Dentiform process of first maxilla (Fig. 584) long, tapering, with bluntly rounded tip, devoid of secondary outgrowths or surface sculpturing. In specimens examined by author, basal papilla of this appendage with only two subequal setae. Sternal furca (Fig. 587) with long rectangular box and short tines, curving and divergent, rimmed right round with fairly broad flanges.

Terminal armature of first exopod (Fig. 588) resembling that of *C. labracis*; setae 1–3 about equally long, seta 4 slightly longer; setae 1 and 4 simple, unarmed; setae 2 and 3 with two terminal flanges (Fig. 589) and minute denticles near attenuated tip. Endopod of second leg (Fig. 590) with fringe of fine setules on lateral margin of second segment and parts of first and third. Fourth leg (Fig. 591) with robust sympod; exopod two-segmented; terminal armature of three setae, first much longer than other two, all apparently unarmed, with pectinate bases; apical seta of first segment similar to short terminal setae; division between segments indistinct, marked by change in width of ramus. Fifth leg of three setules on small tubercle borne upon ventral surface, posterolateral corners of genital complex.

Male (Fig. 583): Anterior half resembling that of female, with fourth leg-bearing segment relatively longer. Genital complex obovate, shorter than thoracic zone of shield, barely wider than fourth leg-bearing segment. Abdomen more than $\frac{1}{2}$ length of genital complex, one-segmented, longer than wide, anteriorly narrower. Uropod slightly shorter than abdomen. Total length about same as that of female.

Terminal claw of second antenna (Fig. 586) trifid, one simple, slender tine and two arising from common base; sturdy seta at base of claw; adhesion pad on penultimate segment crescentic, another as transverse band across segment surface. Dentiform process of first maxilla (Fig. 585) greatly reduced, shorter than long seta of basal papilla. Sixth leg similar to that of *C. labracis*.

Comments: This species shares many features with *C. labracis* (general habitus, male second antenna, armature of first exopod). It differs from that species in structure of male first maxilla, sternal furca and terminal armature of the fourth leg.

Nordmann (1832) commented on the unsuitability of the specific name of *C. minimus*. He was certain that new species would be discovered to which the term "minimus" might be more properly applied. Nevertheless, the name must stand, he concluded. This did not prevent him from listing the species as *C. minutus*, in the same paper. The incorrect name was adopted by many subsequent authors (e.g. Burmeister, 1835; Krøyer, 1837; Heller, 1865; Stossich, 1880; Gouillart, 1937). Brian (1898) found specimens of *Caligus* which he identified as *C. gurnardi*, but amended his diagnosis later (Brian, 1906) to *C. minimus*. His second diagnosis is also doubtful, in view of the unusual host (*Alosa fallax*) on which that *Caligus* was found. *C. minimus* var. *mugilis* Brian, 1935, is not distinct enough from *C. minimus* to warrant taxonomic recognition.

C. minimus occurs in British waters and the North Sea, its distribution covering waters to the South as far as Spain and extending over the western Mediterranean and the Adriatic. Its recorded hosts are perciform fishes: *Dicentrarchus labrax*, *Pagellus bogaraveo*, *Umbrina cirrosa* and *Mugil cephalus*. Thompson (1847) claimed to have found it on *Hippoglossus hippoglossus* off Northern Ireland. No subsequent finds having been made on that host, Thompson's record must be seen as suspect.

Caligus pelamydis Krøyer, 1863

(Figs. 489–504)

Syn.: nec *C. scombri*; of J. V. Thompson, in W. Thompson (1856)
C. scomberi Bassett-Smith, 1896
C. scombri Bassett-Smith; of T. Scott (1901)

Female (Fig. 491): Cephalothorax orbicular, with narrow frontal plates and small lunules; posterior sinuses moderately deep, tip of first antenna reaching lateral limit of shield. Second transverse suture present, running behind eyes. Free margin of thoracic zone projecting beyond level of posterior tips of lateral zones. Fourth leg-bearing segment of about equal length and width, indistinctly divided from genital complex. Latter about as long as cephalothorax, ovate, with lateral margins diverging from narrow base towards posterior extremity. Length greater than width, posterolateral corners rounded, border with abdomen indistinct. Latter apparently one-segmented (short posterior part in many specimens indistinctly marked off by lateral constrictions), cylindrical, longer than genital complex. Uropod longer than wide. Total length 3–5.5 mm.

Dentiform process of first maxilla (Fig. 496) with narrow base, long and slender, without secondary processes or surface sculpturing. Maxilliped with medial margin of corpus unarmed (Fig. 498); subchela long, with very long, slender claw, poorly delimited from shaft. Sternal furca (Fig. 489) with long, quadrangular box; tines as long as box, curving, with narrow, indistinct flanges on lateral margins.

Endopod of first leg bulbous (Fig. 502), peduncular, with minute digitiform apical processes. Terminal armature of exopod (Fig. 500) crowded, setae overlapping one another; seta 1 (dorsalmost) longer than other three, with slender tip (Fig. 501) and one row of minute denticles, very difficult to observe; setae 2 and 3 about equally long, their distal parts with two rows of denticles along margins; seta 4 with only one row of denticles rather longer than those of other setae; posterior margin of segment with three long, pinnate setae. Endopod of second leg (Fig. 492) with patches of short, bristle-like setules along lateral side of ventral surfaces on second and third segment; no bristles on first segment. Fourth leg short, robust; exopod (Fig. 504) distinctly three-segmented, with three terminal setae; third segment with pointed distal end; apical setae of first and second segment in line with three terminal setae at almost even intervals, all setae of similar length; anterior margins of all setae bearing strip of serrated membrane, posterior bearing several rows of densely crowded setules. Fifth leg (Fig. 493) represented by four setae, three together and one further apart, borne upon posterolateral corner of genital complex.

Male (Fig. 490): Anterior half similar to that of female. Genital complex oval, shorter than thoracic zone of cephalothorax, often indistinctly divided from fourth leg-bearing segment or abdomen. Latter indistinctly two-segmented, about as long as genital complex, first segment about $\frac{1}{3}$ of abdomen length, second segment slightly tapering posteriorly. Uropod long, about $\frac{1}{3}$ of abdomen length. Total length of body under 3 mm.

Terminal claw of second antenna (Fig. 495) bifid, tines of claw uneven in length, marked by longitudinal grooves. Postantennary process (Fig. 503) in contrast with that of female, much reduced. Dentiform process of first maxilla (Fig. 497) relatively longer and more slender than in female, its middle third marked by bands of transverse striations not quite extending around entire circumference. Small swelling present on lateral side and short flange on ventrolateral side at about $\frac{2}{3}$ length from base. Maxilliped more robust than that of female, myxal area with two outgrowths on medial margin (Fig. 499), opposite tip of claw; latter not clearly delimited from shaft of subchela, with long barbel. Fifth leg (Fig. 494) similar to that of female, sixth on apex of small rounded outgrowth in posterolateral corner of genital complex, represented by two setae, longer than those of fifth.

Comments: Distinctive features of *C. pelamydis* have assured for it a fairly straightforward taxonomic history. With the exception of *C. scomberi* Bassett-Smith, 1896 (spelled *C. scombri* or even *C. scrombri* by various authors), no other name appears in its synonymy. W. Thompson (1856), writing up notes left by deceased J. V. Thompson, listed, among other species, *C. scombri* found by that worker off Southern Ireland on an unidentified host. No description or figures were given. Being prior to Bassett-Smith's (1896a) description, this name must be considered a *nomen nudum*.

Like other species which include among their hosts large scombrid fishes, *C. pelamydis* is very widespread, having been found in the northern and southern Atlantic, British waters and the North Sea, and also in the Pacific (New Zealand, Hawaii, off California). It is common in the Mediterranean and the Adriatic and there is even a record of its occurrence in the Black Sea (Reshetnikova, 1955).

The scombrid hosts of this species comprise *Euthynnus affinis*, *Katsuwonus pelamis*, *Scomberomorus maculatus*, *S. cavalla*, *Sarda sarda*, *Scomber scombrus* and *S. diego*. Other hosts are: *Trachurus symmetricus* (Carangidae), *Pogonias cromis* (Sciaenidae), *Pomatomus saltatrix* (Pomatomidae), *Brama brama* (Bramidae), *Trigla capensis* (Triglidae), *Thyrsites atun* (Gempylidae) and *Seriolella maculata* (Stromateidae). In British waters it is very common on mackerel (*Scomber scombrus*). Boxshall (1974a) found it also on *Trachurus trachurus*. In the experience of this writer, who sifted through samples containing hundreds of females, the sex ratio of *C. pelamydis* is such that one seldom finds a male.

Caligus productus Dana, 1852

(Figs. 592–603)

Syn.: nec *C. productus* Müller, 1785
C. monacanthi Krøyer, 1863
C. productus Dana; of Wilson (1905) (partim)
C. lobatus Wilson, 1935
C. katuwo Yamaguti, 1936
C. monacanthi Krøyer, 1863; of Wilson (1937) (partim)

Female (Fig. 593): Cephalothorax oval to ovate, narrower anteriorly, with large lunules and shallow posterior sinuses. Tip of first antenna not reaching lateral limit of shield. Free margin of thoracic zone about level with posterior tips of lateral zones. Fourth leg-bearing segment short, much broader than long, often indistinctly separated from genital complex. Latter longer than thoracic zone of shield, with diverging lateral margins and prominent, rounded lobes in posterolateral corners, overlapping anterior part of abdomen. Boundary between genital complex and abdomen usually distinct. Latter two-segmented, cylindrical, about as long as genital complex, segments of about equal length. (Lewis (1967) shows one-segmented abdomen for this species.) Uropod longer than broad, about ½ length of second abdominal segment. Total length 4.0–5.5 mm.

Dentiform process of first maxilla (Fig. 595) with moderately narrow base, tapering to acute apex; basal papilla with three setae, one much longer and stouter than the other two. Myxa of maxilliped with short conical process (Fig. 598); subchela with very short shaft and long claw, indistinctly delimited from it; barbel short. Sternal furca (Fig. 597) with long, rectangular box and slightly divergent, lanceolate tines longer than box; tines gently tapering, with narrow flanges on all margins.

Terminal armature of first exopod (Fig. 599): seta 1 with single row of diminutive setules (possibly remnants of narrow strip of membrane), longer than setae 2 and 3; those setae with secondary processes (arising in seta 2 at about ½ length, in seta 3 near base); distal halves with one margin serrated. Seta 4 twice as long as 1, with row of fine setules on anterior and strip of membrane on posterior margin; posterior margin of segment unarmed. Endopod of second leg (Fig. 600) with double row of strong spines (6 to 8 spines in each) on lateral margin of second segment; first segment with marginal tuft of stiff setules; third segment apparently without such armature. Fourth leg with two-segmented exopod (Fig. 601), segments distinctly delimited; second segment fused from original two, seta 1 of terminal armature longer than other two, with single strip of serrated membrane; setae 2 and 3 about equally long, with two such strips; fourth seta, marking apex of former second segment, similar to preceding two; apical seta of first segment slightly longer than preceding one; all setae with pectinate bases. Fifth leg (Fig. 602) apparently of three small setules, one at distance from other two, all near posterolateral corner of genital complex.

Male (Fig. 592): Anterior half similar to that of female. Genital complex oval, longer than wide, shorter than thoracic zone of shield, often poorly delimited from fourth leg-bearing segment but usually well from abdomen. Latter two-segmented, cylindrical, about as long as genital complex. First segment about ½ length of second. Uropod rectangular, longer than broad, about as long as first abdominal segment. Total length of body slightly less than that of female.

Second antenna with simple terminal claw (Fig. 594), sharply hooked, closing against adhesion pad of preceding segment; one long, sturdy seta on each side of claw, near its base; tip of claw with delicate membranous flanges on each side. Postantennary process similar to that of female but comparatively longer. Dentiform process of first maxilla (Fig. 596) with short lateral outgrowth on small swelling; several transverse grooves in same region. Fifth leg (Fig. 603) of four setae, three together and one separated. Sixth leg posterior to fifth, of two similar setae.

Comments: Although *C. productus* is a common and widespread species, many authors in the past had difficulties in identifying it correctly, due to inadequacies of the early descriptions. Hence the multiplicity of names contained in the synonymy of *C. productus*. A superficial resemblance between it and other species of *Caligus* occurring on the same or related hosts compounded the difficulties even further. Not all of them are reflected in the synonymy. For example, Wilson (1905) considered *C. scombri* a synonym of *C. productus*, as did Leigh-Sharpe (1926a), who used the correct spelling *C. scomberi*. Both versions are, in fact, synonyms of *C. pelamydis*. Capart (1959) considered *C. coryphaenae* of Yamaguti (1936) as synonymous with this species, but Shiino (1959b) and Pillai (1963c) accepted it as a synonym of *C. quadratus* Shiino, 1954. One source of confusion was the fact that *C. productus* was a homonym of Müller's (1785) species, later transferred to the pandarid genus *Dinemoura*. Some of this confusion remained in Yamaguti's (1963) host list of *C. productus*, which mentions the species parasitized by *Dinemoura producta*.

The most important group of fishes harbouring *C. productus* are Scombridae (*Scomberomorus cavalla*, *S. sierra*, *S. maculatus*, *Sarda sarda*, *S. orientalis*, *Katsuwonus pelamis*, *Katsuwonus* sp., *Thunnus thynnus*, *T. albacares*, *Euthynnus affinis* and *Auxis thazard*). Another common host is *Coryphaena hippurus* (and *Coryphaena* sp.) (Coryphaenidae). Suborder Percoidei is represented by *Pogonias cromis* (Sciaenidae), *Calamus brachysomas* and *Sparus aurata* (Sparidae), *Centropomus* sp. (Centropomidae), *Seriola dorsalis* (Carangidae), *Paralabrax clathratus* and *P. maculatofasciatus* (Serranidae) and *Lutjanus* sp. (Lutjanidae). Also parasitized are: *Polydactylus opercularis* (Polynemidae) and *Sphyraena argentea* (Sphyraenidae). Hosts of orders other than Perciformes are: *Elops saurus* (Elopiformes), and *Balistes polylepis* and *Balistes* sp. (Balistoidei).

C. productus has not been recorded so far from British waters, its only claim to inclusion in the British fauna being Leigh-Sharpe's (1926a) incorrect record. Its description is worth including here, however, if only as an aid to identification of other caligid species occurring on big scombrids. Since some of those scombrids do visit British waters, it is not unreasonable to suppose that *C. productus* will eventually be found here.

Caligus zei Norman and T. Scott, 1906

(Figs. 604–615)

Female (Fig. 604): Cephalothorax ovate, with anterior end narrower than posterior, lunules small, tip of long first antenna usually reaching lateral limit of shield. Posterior sinuses shallow, free margin of thoracic zone projecting beyond level of posterior tips of lateral zones. Fourth leg-bearing segment short, broad, often indistinctly delimited from genital complex. Latter suborbicular, with rounded lateral margins and posterior margin slightly re-entrant at centre; length greater than that of thoracic zone of shield. Abdomen one-segmented, subquadrangular, about $\frac{1}{3}$ length of genital complex. Uropod small, almost square, less than $\frac{1}{2}$ length of abdomen. Total length of body about 6 mm.

Dentiform process of first maxilla (Fig. 607) with broad base tapering to fine point, devoid of secondary outgrowths or surface sculpturing. Myxa of maxilliped smooth; claw of subchela and barbel long. Sternal furca (Fig. 609) with long, irregular box and tines longer than box, irregularly curving, often asymmetrical, devoid of flanges. (In Fig. 609 one tine showing traces of damage.)

Terminal armature of first exopod (Fig. 610) with seta 1 shortest, unarmed, inserted away from distal margin; seta 2 about $2\frac{1}{2}$ length of 1, with secondary process and serrated posterior margin; seta 3 similar, though shorter; seta 4 more than twice length of seta 2, slender, pinnate (sometimes only one row of pinnules observable); posterior margin of segment with three long pinnate setae. Endopod of second leg (Fig. 613) with fringe of fine setules on lateral margins of all three segments. Exopod of fourth leg (Fig. 611) long, slender, two-segmented; second segment fusion of original two, three setae of its terminal armature of about equal size, all with two strips of finely serrated membrane (not visible in Fig. 611); seta on posterior margin, marking former apex of middle segment, similar to terminal; apex of first segment also with similar seta; all setae with basal pectens mounted on prominent lappets. Fifth leg (Fig. 614) of three setae, on posterolateral corner of genital complex.

Male (Fig. 605): Cephalothorax similar to that of female. Genital complex shorter than thoracic zone of shield, oval, often notched on lateral margins near base. Abdomen two-segmented, slightly shorter than genital complex, first segment $\frac{1}{2}$ length of second. Uropod shorter than first abdominal segment. Total length of body under 5 mm.

Claw of second antenna (Fig. 606) bifid, both tips rounded, overlapping one another; one strong seta on each side of base; adhesion pad extensive. Dentiform process of first maxilla (Fig. 608) with broad base and fine, slightly curving tip; distal half with lateral swelling, covered by transverse striations; small, triangular process near distal margin of swelling. Maxilliped similar to that of female but with prominent short process in myxal area (Fig. 612). Fifth leg (Fig. 615) of three setae, on posterolateral corner of genital complex; sixth leg of two setae, medial to fifth leg (Fig. 615).

Comments: *C. zei* is not common. Its host, *Zeus faber*, is itself far from common and only rarely carries its namesake *Caligus*. Since its discovery (Norman and T. Scott, 1906), it was recorded in British waters less than 10 times. This writer, in spite of examining *Z. faber* at every opportunity for many years, was not fortunate enough to add it to his collection. There are no records from other areas, with the exception of a surprising one made by Barnard (1955b), who claimed to have identified *C. zei* on a snoek (*Thyrsites atun*) in South African waters. No description or figures were given. Barnard suggested that his identification was rather tentative and required corroboration.

Genus Pseudocaligus A. Scott, 1901

Caligidae. Diagnosis as for *Caligus*, but fourth leg vestigial, one-segmented (two-segmented in *P. similis*).

Type species: *P. brevipedis* (Bassett-Smith, 1896).

Comments: This author is not entirely convinced that *Pseudocaligus* should be retained as a separate taxon. (For discussion of validity, see p. 169). The genus is small, containing only six species, all but one from the Indo-Pacific region, mainly from India and Japan. Its members parasitize fishes of the family Tetraodontidae (*Pseudocaligus fugu* Yamaguti, 1936, *P. laminatus* Rangnekar, 1955 and *P. parvus* (Bassett-Smith, 1898)) and Fistularidae (*Pseudocaligus fistulariae* Pillai, 1961 and *P. similis* Lewis, 1968). There is a doubtful record (Wilson, 1924a) of *P. parvus* from the Galapagos Isles. The Mediterranean species *P. apodus* Brian, 1924, which lacks even vestigial 4th legs, belongs properly in the genus *Caligopsis*.

Only the type species of *Pseudocaligus* occurs in British waters.

Pseudocaligus brevipedis (Bassett-Smith, 1896)

(Figs. 616–627)

Syn.: *Caligus brevipedis* Bassett-Smith, 1896
Pseudocaligus brevipedes; of A. Scott (1901)
Pseudocaligus brevipes, of Norman and T. Scott (1906)

Female (Fig. 616): Cephalothorax orbicular, with broad frontal plates and large lunules. Tip of first antenna usually extending to lateral limit of shield. Posterior sinuses shallow. Free margin of thoracic zone about level with posterior tips of lateral zones. Fourth leg-bearing segment subquadrate, poorly delimited from genital complex. Latter as long as broad (sometimes broader), with rounded anterolateral and posterolateral corners; posterior margin more or less straight. Abdomen subquadrate, one-segmented, less than $\frac{1}{2}$ length of genital complex, with obliquely truncated posterolateral corners. Uropod large, more that $\frac{1}{2}$ length of abdomen. Total length of body about 3.5 mm.

Second antenna with spatulate posterior process on second segment. Digitiform process of first maxilla (Fig. 618) with broad base and tapering point, without secondary processes or surface sculpturing. Second maxilla with two strips of serrated membrane on canna (Fig. 619); calamus with four similar strips. Corpus of maxilliped broader at base, with prominent subtriangular myxa; subchela with long claw and barbel. Sternal furca (Fig. 624) with rounded box and divergent rami; divergence differs in different specimens, length of rami occasionally reaching level of first interpodal bar; no flanges. Short transverse cuticular ridge on each side of furca.

Terminal armature of first exopod (Fig. 625) of four setae, about equally long, simple and apparently unarmed; posterior margin with three long pinnate setae. Endopod of second leg (Fig. 626) with fringe of fine setules on second segment and parts of first and third. Fourth leg (Fig. 627) reduced, one-segmented, with three terminal setae, one longer than other two and pinnate, others unarmed. Fifth leg of three small setae in posterolateral corner of genital complex.

Male (Fig. 617): Cephalothorax as in female. Fourth leg-bearing segment equally long and wide, usually indistinctly separated from genital complex. Latter suborbicular, with two lobes in each posterolateral corner. Division between this tagma and abdomen indistinct. Length of genital complex less than that of thoracic zone of shield. Abdomen one-segmented, slightly longer than, or as long as, wide, only slightly shorter than genital complex. Uropod prominent, about $\frac{1}{2}$ length of abdomen. Total length of body less than that of female.

Second antenna (Fig. 623) with three-pronged claw, two longest tines being at about right angles to each other, shorter with straight short tine arising obliquely from base; fourth, small outgrowth present at base in some specimens, as well as one robust seta (two setae in some specimens). Ventral surface of cephalothorax slightly distomedial to tip of mouth cone with two transversely corrugated, longitudinally elongated adhesion pads (Fig. 621). Second maxilla with bifid canna (Fig. 620), serrated strips of membrane and denticulation on both branches; row of small denticles across their common base. Maxilliped more robust than that of female, with broader corpus; myxa with prominent, two-tipped outgrowth and small tubercle near base of claw. Fifth leg (Fig. 622) on tip of narrower, dorsal lobe in posterolateral corner of genital complex, represented by three small, pinnate setae. Number of setae of sixth leg was not determined. In Fig. 622 tip of broad ventral lobe bearing two small tubercles, possibly bases of broken setae of that appendage.

Comments: Both in its distribution and its host affiliations, *P. brevipedis* is atypical of its genus, of which it is a type species. Its most common host is a gadid fish, the three-bearded rockling *Gaidropsaurus mediterraneus*. The five-bearded rockling *Ciliata mustela* has also been found harbouring it, though less commonly. Leigh-Sharpe (1934) recorded it from *Gobius* (*Macrogobius*) *paganellus*.

P. brevipedis is not uncommon in British waters, occurring all around the British Isles. There are also some records of its occurrence off the coasts of France (Concarneau: Guiart, 1913; Gulf of Gascogne: Faure-Fremiet and Guilcher, 1948).

Genus Sciaenophilus van Beneden, 1852

Caligidae. Female resembling *Caligus* but with abdomen as long as, or longer than, rest of body, one- or two-segmented. Male indistinguishable from that of *Caligus* in general habitus. Sternal furca reduced, vestigial or absent. In *S. bennetti* second leg uniramous. Exopod of fourth leg two-segmented.

Type species: *Sciaenophilus tenuis* van Beneden, 1852.

Comments: Six nominal species are being currently included in this genus of questionable validity. Differing from one another in the presence or absence of sternal furca and including one unique species with uniramous second leg, the members of this genus have in common the extraordinary length of their abdomina, setting them apart from other caligids. Heegaard (1966) placed *Sciaenophilus* in a separate family, Sciaenophilidae, though he gave no definition of that family. Neither did he discuss the reasons for this taxonomic innovation.

Sciaenophilus bennetti (Causey, 1953) occurs in the Gulf of Mexico, *S. pharaonis* (Nordmann, 1832) in the Mediterranean and Indian waters, *S. macrurus* was recorded on only one occasion (Heller, 1865) in Java and *S. nibae* (Shen, 1957) off the coast of China. Only the type species occurs in British waters.

Sciaenophilus tenuis van Beneden, 1852

(Figs. 713–721)

Syn.: *Caligus* (*Sciaenophilus* van Ben.) *benedeni* Bassett-Smith, 1898
Sciaenophilus benedeni (Bassett-Smith, 1899)
Caligus benedeni (Capart, 1941)

Female (Fig. 713): Cephalothorax suborbicular, with well developed frontal plates and lunules. Tip of first antenna usually not extending to lateral limit of shield. Free margin of thoracic zone projecting well beyond level of posterior tips of lateral zones of shield. Fourth leg-bearing segment about as long as wide, not distinctly divided from genital complex. Latter about three times as long as cephalothorax (though length variable from one specimen to another, depending on age and maturity) lateral margins diverging posteriorly, posterior extremity more than twice as wide as anterior; posterolateral corners rounded and posterior margin straight. Abdomen narrow, cylindrical, longer than rest of body. In some specimens posterior end marked off by indistinct constriction, construed by some authors as evidence of two-segmented nature of this tagma. Total length of body 6–15 mm.

First antenna as in *Caligus*. Terminal hamulus of second antenna distinguishable by its length, slenderness and curvature. Mouth tube and mandibles as in *Caligus*. Dentiform process of first maxilla (Fig. 715) straight, tapering, devoid of secondary outgrowth or surface sculpturing. Second maxilla as in *Caligus*. Corpus of maxilliped (Fig. 716) robust, slightly curving, with smooth myxal area (some authors (Pillai, 1961) report small, pointed process in that area); subchela with indistinct division between shaft and claw, barbel small, tip robust, long, sharply curved. Sternal furca absent.

Terminal armature of first exopod (Fig. 717) on obliquely slanting distal margin; seta 1 short, not more than $\frac{1}{3}$ length of others, unarmed, with delicate basal pecten; setae 2 and 3 long, slender, with one serrated margin on distal quarter; seta 4 slightly longer than other two, more slender and apparently unarmed and devoid of pecten. First endopod (Fig. 718) pyriform, with two apical processes. Endopod of second leg (Fig. 719) with lateral margin of second segment and part of that of third segment equipped with fine fringe of setules; first segment with margin apparently unarmed. (Fringe on medial margin of second segment shown thinner, for sake of clarity.) Third leg similar to that of *Caligus*. Exopod of fourth leg (Fig. 720) two-segmented; distal segment fused from original two, with three setae in terminal armature; setae pectinate, progressively shorter from first to third, first with one strip of serrated membrane, others with two such strips; similar seta on posterior margin; apical seta of first segment also similar; all setae pectinate. Fifth leg (Fig. 721) of three setae surmounting conical papilla on posterolateral corners of genital complex.

Male (Fig. 714): Anterior part similar to that of female. Fourth leg-bearing segment shorter than in female, but otherwise similar. Genital complex longer than wide, about as long as thoracic zone of shield; lateral margins convex, with slight indentations in posterior halves. Abdomen two-segmented, cylindrical, longer than genital complex; first segment comprising about $\frac{1}{3}$ of total length of tagma. Uropod long, about $\frac{1}{4}$ length of abdomen. Total length less than that of female.

Appendages as in female. Maxilliped relatively shorter, its myxal area with prominent protuberance armed with fine denticle; barbel relatively longer.

Comments: This author had no opportunity of examining the male of *S. tenuis*. Fig. 714 and the description are based on work of Pillai (1961).

The taxonomic position of this species remains uncertain. No types survive, the original description is not useful for comparative purposes. Four decades after its description, Bassett-Smith (1898), describing a new species of *Sciaenophilus*, reduced it to a subgenus of *Caligus*. Later (Bassett-Smith, 1899) he returned *Sciaenophilus* to its generic status, but his decision has not been generally followed. *S. benedeni* Bassett-Smith, 1898, is a species which to this author does not appear different enough from *S. tenuis* to be retained as a separate taxon. Until a careful examination has been made it seems appropriate to refer to these two taxa as one, under the name of the chronologically older *S. tenuis*.

Taken *sensu lato*, then, *S. tenuis* is a widespread species, occurring in the northern and southern Atlantic (European seaboard, coasts of the southern USA, the North Sea and the Baltic, the Mediterranean and the coasts of Africa), as well as in the Indian Ocean (India, Ceylon). It has not been recorded from the Pacific. Although widespread, this species is neither common nor abundant, more than few specimens seldom occurring on one fish. Males are particularly scarce, none having been found so far in British waters.

S. tenuis is parasitic on fishes of the family Sciaenidae: *Argyrosomus regium*, *Sciaena diacanthus*, *S. angolensis*, *Umbrina valida*, *U. cirrosa*, *Johnius hololepidotus*, *Johnius* sp., *Larimus fasciatus* and *Pogonias cromis*. In British waters it has been found on a few occasions on *Argyrosomus regium* and *Umbrina cirrosa*.

Genus Lepeophtheirus von Nordmann, 1832

Caligidae. Generic diagnosis as for *Caligus*, but lunules absent. Dentiform process of first maxilla most commonly (though not always) bifid in female, bifid with soft medial third tine in male.

Type species: *Lepeophtheirus pectoralis* (Müller, 1777).

Comments: Parasitic exclusively on marine fishes, this common genus numbers some 90 species, scattered throughout the seas and oceans of the world, occurring predominantly on teleost hosts. Some species are narrowly specific, others include many fish species in their host range. *Lepeophtheirus* is particularly prevalent in temperate latitudes or cooler waters, in contrast with *Caligus*. The genus is represented in British waters by seven species.

Key to British species

Females

1. Abdomen about as long as, or longer than, genital complex (Fig. 676) 2
   Abdomen shorter than genital complex (Fig. 628) 5

2. Spiny outgrowth on cephalothorax surface near tip of dentiform process of first maxilla (Fig. 532), posterior margin of fourth leg with scaly denticulation (Fig. 536) . . . . ***L. sturionis***
   No outgrowth on cephalothorax surface near tip of dentiform process of first maxilla, posterior margin of fourth leg unarmed or with strip of membrane 3

3. Tines of sternal furca spatulate (Figs. 694, 695) . . . . ***L. salmonis***
   Tines of sternal furca not spatulate (Figs. 680, 705) 4

4. Spherical outgrowth of cephalothorax surface near base of postantennary process very large (Fig. 707) . . . . ***L. thompsoni***

Spherical outgrowth of cephalothorax surface near base of postantennary process relatively much smaller than in *L. thompsoni* . . . . **L. pollachius**

5. Sternal furca with bifid tines (Fig. 656) . . . . **L. hippoglossi**
   Sternal furca with simple tines 6

6. Pointed cuticular outgrowths on either side of sternal furca (Fig. 669) . . . . **L. nordmanni**
   No outgrowths on either side of sternal furca . . . . **L. pectoralis**

Males

(Male of *L. sturionis* not included in this key)

1. Abdomen indistinctly (Fig. 677) or distinctly (Fig. 663) two-segmented 2
   Abdomen short, one-segmented 3

2. Pointed cuticular outgrowths on either side of furca (Fig. 669) . . . . **L. nordmanni**
   No outgrowths on either side of furca . . . . **L. pollachius**

3. Sternal furca with bifid tines (Fig. 656) . . . . **L. hippoglossi**
   Sternal furca with simple tines 4

4. Tines of sternal furca spatulate (Fig. 694) . . . . **L. salmonis**
   Tines of sternal furca not spatulate 5

5. Spherical outgrowth of cephalothorax surface near base of postantennary process very large (Fig. 707) . . . . **L. thompsoni**
   Spherical outgrowth of cephalothorax near base of postantennary process smaller than base of process (Fig. 638) . . . . **L. pectoralis**

Lepeophtheirus pectoralis (Müller, 1777)

(Figs. 628–647)

Syn.: *Lernaea pectoralis* Müller, 1777
Caligus (*Lepeophtheirus*) *pectoralis*; of Krøyer (1838)
Lepeophtheirus pectoralis; of Baird (1850)

Female (Fig. 628): Cephalothorax suborbicular, with well developed frontal plates. Tip of first antenna reaching lateral limit of shield. Posterior sinuses shallow, free margin of thoracic zone at about level of posterior tips of lateral zones. Fourth leg-bearing segment about equally long and wide, not delimited definitely from genital complex. Latter as long, or almost as long, as dorsal shield, with rounded, sloping anterolateral corners, anteriorly narrower than posteriorly, posterolateral corners sometimes forming mild lobes, posterior margin re-entrant in centre, no definite boundary with abdomen. Abdomen one-segmented (some specimens with bulging anterior part suggesting segmental division), less than $\frac{1}{2}$ length of genital complex, longer than broad, slightly tapering in posterior direction. Uropod about $\frac{1}{4}$ length of abdomen, longer than broad. Total length of body about 5 mm.

First antenna as in *Caligus*; apical armature of distal segment consisting of two groups of setae (Fig. 630): anterior eight setae, shorter and sturdier and posterior five longer and more slender (two fused together at base). Second antenna (Fig. 631) indistinguishable from that of *Caligus*. Postantennary process (Fig. 638) with sturdy base and moderately short, blunt tip. Small spherical outgrowth on surface of cephalothorax near base of process (serving as support of tip of second antenna, when at rest). Mouth tube and mandible as in *Caligus* (see Figs. 444–446). First maxilla (Fig. 641) with bifid dentiform process; base of process semispherical, medial tine longer, with narrow marginal flanges and one obliquely across ventral surface, lateral tine with more prominent marginal flanges. Second maxilla as in *Caligus* (apparently both calamus and canna with two strips of membrane) (Fig. 639). Maxilliped (Fig. 646) with corpus broader at base, tapering distally; prominent tubercle in myxal area; shaft of subchela short, claw long and sharply curving, barbel short. Sternal furca (Fig. 636) with long, quadrangular box; tines shorter than box, slightly divergent, usually straight, tapering distally, with broad marginal flanges.

Terminal armature of first exopod (Fig. 642) with setae diminishing in length from 1 to 4; seta 1 with thick base, its diameter diminishing abruptly at about $\frac{1}{4}$ length, posterior margin with single row of very fine denticles; setae 2 and 3 with secondary processes in distal halves, their posterior margins proximal to bases of secondary processes with single rows of denticles; seta 4 short, slender, pinnate; setae 1 to 3 with pectinate bases; endopod (Fig. 643) small, slightly tapering towards tip, bifid, with two apical processes. Endopod of second leg (Fig. 644) with lateral margin of second segment, as well as of parts of first and third segments, equipped with fringe of fine setules. Exopod of fourth leg (Fig. 645) two-segmented; terminal setae of distal segment diminishing in length by about equal increments from seta 1 to 3, all with pectinate bases and two strips of marginal membrane; apical seta of first segment short, but otherwise similar to other three; segmental boundary distinct. Fifth leg (Fig. 634) subtriangular, with three pinnate setules along posterolateral side; fourth similar seta some distance to lateral side on surface of genital complex.

Male (Fig. 629): Cephalothorax similar to that of female. Fourth leg-bearing segment broadest at midlength, tapering both anteriorly and posteriorly, not clearly delimited from genital complex. Latter oval, shorter than thoracic zone of shield, indistinctly separated from abdomen. Latter less than $\frac{1}{2}$ length of genital complex, slightly narrower at base, about equally long and wide. Uropod more than $\frac{1}{2}$ length of abdomen, rectangular. Total length of body up to 3 mm.

Claw of second antenna (Figs. 632, 633) simple, sharply curved, with two transparent crescentic flanges on one side and one similar flange on opposite side of distal half; sturdy seta at base and long, slender seta present at about $\frac{1}{2}$ length of claw; tip of claw closing against small adhesion pad; another and larger adhesion pad present on penultimate segment. Dentiform process of first maxilla (Fig. 640) with medial tine shorter than lateral, both slender and with very small flanges on distal halves of medial tine; posteromedial to base of process surface of cephalothorax bearing oval, transversely corrugated and obliquely arranged adhesion pad. Maxilliped similar to that of female but often with slightly narrower corpus. Sternal furca (Fig. 637) often differing from that of female in having curving tines (intermediate forms between those shown in Fig. 636 and 637 often present). Fifth leg (Fig. 635) subtriangular, with three pinnate setules at apex and posterolateral side; fourth similar setule on lateral side of base. Sixth leg (Fig. 635) posterior to fifth, represented by three pinnate setules surmounting low swelling.

Comments: In common with other long-established species of *Lepeophtheirus*, this species has to its credit abundant literature, as voluminous as it is confusing. Due to the difficulties experienced by the early investigators in distinguishing between several similar species of the genus, no reliance can be placed on the old records. On the basis of his own experience and of records accepted by him as reliable, this author believes *L. pectoralis* to be a species parasitic predominantly on pleuronectid flatfishes (*Platichthys flesus*, *Pleuronectes platessa*, *Limanda limanda*, *Liopsetta glacialis*) along the Atlantic seaboard of Europe, as well as in the Faroes, Iceland and the White Sea. Many host records found in the literature are either instances of misidentification or result from transfer of *L. pectoralis* from its host to other fishes in mixed trawl catches. It is due to these reasons, probably, that the host list of *L. pectoralis* includes other flatfishes (*Scophthalmus maximus*, *Hippoglossus hippoglossus*) or fishes of other groups (*Anarhichas lupus*, *Melanogrammus aeglefinus*, *Scomber scombrus*, *Conger conger* and *Callionymus lyra*), elasmobranchs (*Raja radiata*) or even invertebrates (*Asterias rubens*). (However, Boxshall (1974a) did find this copepod on *Raja clavata* and *Gadus morhua* in the North Sea.) Records of this species from the Mediterranean region require verification.

L. pectoralis is often present on its host in large numbers, its site of predilection being the inner surface of pectoral fins where it aggregates in close ranks, the cephalothoraces of the copepods being arranged in longitudinal rows (Fig. 647), their posterior halves overlapping one another. Other parts of the hosts' surfaces are, however, also frequented.

Lepeophtheirus hippoglossi (Krøyer, 1837)

(Figs. 648–661)

Syn.: *Caligus curtus*; of Desmarest (1825)
Caligus hippoglossi Krøyer, 1837
Caligus hippoglossis; of Edwards (1840)
Lepeophtheirus hippoglossi; of Baird (1850)
Lepeophtheirus obscurus Baird, 1850 (male)
nec *Lepeophtheirus obscurus* (?) Baird; of Bassett-Smith (1899)
nec *Lepeophtheirus hippoglossi*; of Brian (1898, 1899b)

Female (Fig. 648): Cephalothorax orbicular, with well developed frontal plates. Tip of first antenna not reaching lateral limit of shield. Posterior sinuses shallow, free margin of thoracic zone at about level of posterior tips of lateral zones. Fourth leg-bearing segment with narrow anterior end, much broader than long, only slightly narrower than free margin of thoracic zone, very indistinctly delimited from genital complex. Latter longer than thoracic zone of shield, slightly broader than fourth leg-bearing segment, with convex lateral margins and lobes of posterolateral corners extending backwards to about $\frac{1}{2}$ length of abdomen. Latter one-segmented, less than $\frac{1}{4}$ length of genital complex, indistinctly divided from genital complex, longer than broad, somewhat tapering posteriorly. Uropod very small, less than $\frac{1}{4}$ length of abdomen, subquadrangular. (Genital complex and abdomen undergoing extensive changes in course of ontogeny, mainly consisting of relatively faster growth of genital complex and gradual dwarfing of abdomen. Fig. 658 shows that part of body in juvenile female.) Total length of body of adult female about 12.5 mm, but greatly variable.

Dentiform process of first maxilla (Fig. 651) with relatively prominent conical base and two tines shorter than base; medial tine longer than lateral, both with narrow marginal flanges (medial flange of medial tine in Fig. 651 showing damage sustained in vivo). Calamus of second maxilla with four strips of membrane, canna with two strips, smooth on anterior margin and deeply serrated on posterior margin. Maxilliped with smooth myxal area and long claw of subchela, with long barbel. Sternal furca (Fig. 656) with subquadrate box and tines about as long as box, bifid in distal halves; tines devoid of flanges, their tips often showing signs of wear and tending to be shorter in older specimens. (In juveniles bifid nature of tines less apparent, secondary tip developing with age.)

Terminal armature of first exopod (Fig. 654) of four about equally long setae. Seta 1 with two strips of striated (apparently not serrated) membrane; setae 2 and 3 (only seta 2 shown in Fig. 654) with similar strip of membrane along part of anterior margin, with secondary process arising at about midlength; process slender, nearly reaching tip of seta, unarmed, with narrow strip of membrane at base; seta 4 more slender, pinnate; setae 1–3 with pectinate bases. Endopod of second leg (Fig. 655) with fringes of fine setules along lateral margins of second, first and part of third segments. Exopod of fourth leg (Fig. 660) three-segmented, segment 2 being longest; terminal armature of three setae, seta 1 about twice as long as seta 3, carrying single row of strong, slanting denticles along posterior margin (Fig. 661a), seta 2 of length intermediate between that of setae 1 and 3; setae 2 and 3 with serrated strips of membrane along two margins (Fig. 661b); apical seta of second segment resembling preceding two, that of first greatly reduced and curved, claw-like; all five with pectinate bases. Fifth leg (Fig. 657) saddle-shaped, with three pinnate setules along posterolateral margin; fourth, similar seta on surface of genital complex, lateral to sclerite of leg.

Male (Fig. 649): Cephalothorax similar to that of female. Fourth leg-bearing segment short and broad, greatest width at midlength, tapering in both directions. Genital complex suborbicular, shorter than thoracic zone of shield, somewhat broader than fourth leg-bearing segment, with pair of prominent outgrowths in each posterolateral corner; legs five and six, borne upon tips of these outgrowths, well noticeable in dorsal aspect. Abdomen subquadrate, one-segmented, shorter than $\frac{1}{2}$ length of genital complex. Uropod relatively more prominent than in female, about $\frac{1}{2}$ length of abdomen. Total length of body about 6–7 mm.

Terminal claw of second antenna (Fig. 650) with single, strongly curving tip, armed with flanges on either side of distal half; short, robust seta at base and longer, flagelliform one some distance from base on opposite side. Dentiform process of first maxilla (Fig. 652) with prominent conical base and two tines, shorter than base, medial tine shorter than lateral; narrow, short flange on medial margin of medial tine; medial process on medial side of base of medial tine, about $\frac{1}{2}$ length of latter. Maxilliped similar to that of female, but with three small tubercles near base of subchela (Fig. 653). Fifth leg of specimens examined by author (Fig. 659) consisted of only two pinnate setules, probably as a result of damage (usual number should be four), sixth leg posterior to fifth, represented by three similar setules (one much shorter in Fig. 659, probably due to damage).

Comments: Discovered in Danish waters, this species stands out among all the European species of *Lepeophtheirus* by possessing a sternal furca with bifid tines. Consequently, it has only seldom been misidentified. Two instances of mistaken identity recorded in the literature are those of Baird (1850), who failed to appreciate the extent of sexual dimorphism of this species and described its male under a different name, and that of Brian (1898, 1899b), the reason for which is not obvious. Brian's identification of *Lepeophtheirus* found on *Mola mola* as *L. hippoglossi* was clearly erroneous and corrected later by himself (Brian, 1906).

The principal host of *L. hippoglossi* is the Atlantic halibut, *Hippoglossus hippoglossus*. Its distribution covers the range of that host, records having been made from Greenland, Iceland and the Barents Sea in the North, along the Atlantic coast of North America as far South as Massachussets, and along the European sea coasts. Markevich (1956) found some specimens in the Bering Sea in the western hemisphere.

In addition to *Hippoglossus hippoglossus*, other species have been found to carry *L. hippoglossi* on rare occasions, perhaps fortuitously. Among them is a relative of halibut, *Reinhardtius hippoglossoides* and non-related flatfishes, *Scophthalmus maximus* and *S. rhombus*. Other doubtful recorded hosts are *Somniosus microcephalus* and *Centrophorus squamosus*.

Lepeophtheirus nordmanni (Edwards, 1840)

(Figs. 662–675)

Syn.: *Caligus nordmanni* Edwards, 1840
Caligus ornatus Edwards, 1840
Lepeophtheirus nordmanni M. Edwards; of Baird (1850)
Lepeophtheirus ornatus; of Bassett-Smith (1899)
Lepeophtheirus hippoglossi; of Brian (1898, 1899b)
Lepeophtheirus ornatus Nordmann, 1832; of Wilson (1905)
Lepeophtheirus insignis Wilson, 1908
nec *Lepeophtheirus insignis*; of Barnard (1948)

Female (Fig. 662): Cephalothorax orbicular, with fairly narrow frontal plates. Tip of first antenna not reaching lateral limit of shield. Second transverse suture linking anterior ends of longitudinal sutures directly behind eyes. Posterior sinuses shallow, free margin of thoracic zone not protruding beyond tips of lateral zones. Fourth leg-bearing segment short and broad, narrower anteriorly, its greatest width slightly less than that of the free margin of thoracic zone. Border between this segment and genital complex usually distinct. Latter longer than thoracic zone of shield, suboval, with rounded lobes in posterolateral corners and convex lateral margins. Abdomen one-segmented, not clearly delimited from genital complex (at least in some specimens), less that $\frac{1}{2}$ length of genital complex, longer than wide. Uropod rectangular, longer than broad, about $\frac{1}{2}$ length of abdomen. Total length of body about 12 mm.

Second antenna (Fig. 664) with unusually long and slender terminal hamulus, its tip curving very sharply. Postantennary process (Fig. 666) also very long and slender, its hamulus narrowing abruptly from broad base and about twice as long as that base. First maxilla (Fig. 668) with conical base of dentiform process equipped with small lateral tubercle; tines of process long and slender, about equally long and thick, longer than base, devoid of flanges. Maxilliped with slender corpus, its myxal area unarmed; total length of subchela equalling that of corpus, its claw sharply curved, with tip reaching base of corpus, when closed. Sternal furca (Fig. 669) with longitudinally arranged oval box and very long tines (extending across first interpodal bar and reaching $\frac{1}{2}$ way to second); tines divergent, with bluntly rounded tips, their length variable, but most commonly with bands of transverse striations about midlength, either slightly curving in distal $\frac{1}{2}$ or straight. Subtriangular, sharp denticles present, one on each side of furca, arising from ventral surface of cephalothorax.

Terminal armature of first exopod resembling that of type species (cf. Fig. 642), but seta 1 with strip of membrane on posterior margin. Second endopod (Fig. 671) with fringe of fine setae on lateral margin of second and parts of first and third segments; second segment with pronounced lateral bulge. Exopod hook of third leg uncommonly long and slender, rami of leg also more than usually long. Fourth leg (Fig. 672) with three-segmented exopod; setae of distal segment diminishing in length from seta 1 to 3, all with pectinate bases; seta 1 armed with row of strong denticles along posterior margin (Fig. 673a), seta 2 with denticles on both margins (Fig. 673b), seta 3 with two strips of serrated membrane (Fig. 673c); both first and second segments with setae in their posterodistal corners similar to seta 3; posterior margins of all three segments with strips of membrane. Fifth leg (Fig. 675) subtriangular, with uneven sides; three small pinnate setae on distal $\frac{1}{2}$ of lateral side, fourth seta some distance towards lateral margin.

Male (Fig. 663): Cephalothorax similar to that of female. Fourth leg-bearing segment relatively narrower than that of female, distinctly delimited from genital complex. Latter shorter than thoracic zone of shield, longer than broad, with two prominent lobes on distal halves of both lateral margins. Abdomen distinctly two-segmented, shorter than genital complex and clearly demarcated from it. Segments cylindrical, of about equal length. Uropod rectangular, longer than posterior segment of abdomen. Total length of body about 6 mm.

Terminal claw of second antenna (Fig. 665) with strong secondary dentiform outgrowth on concave margin at about midlength; distal to its base narrow flanges on both sides of claw; long, flagelliform seta arising from claw near base of secondary process; shorter, sturdy seta on opposite side near base. Dentiform process of first maxilla (Fig. 667) simple and slender, with two short secondary processes, one at about midlength, other in distal half. Maxilliped (Fig. 670) similar to that of female, but with conical, short process in myxal area. Sternal furca relatively much smaller than that of female, with straight tines; triangular denticles on either side of furca relatively larger than those of female. Fifth leg (Fig. 674) resembling that of female, sixth posteromedial to fifth, represented by oval lobe equipped with three pinnate setae similar to those of leg 5.

Comments: *L. nordmanni* is a species with many characteristic morphological features that make it easy to identify. Even a cursory examination reveals characteristic slenderness of its appendages. The hamulus of the second antenna (Fig. 664), the subchela of the maxilliped (Fig. 670), the postantennary process (Fig. 666) and the dentiform process of the first maxilla (Fig. 668) cannot be mistaken for those of any other species from the British region. Even more distinctive are the slenderness of the tines of its sternal furca and the presence of denticles on either side of this appendage. The male genital complex affords another good distinguishing character.

Nevertheless, *L. nordmanni* was confused in the past with other species and described under different names, as evident from its synonymy. In common with some other species discovered early in the 19th century, it suffered from the inadequacies of the early descriptions. The identity of its host and its ubiquitous distribution were also contributing factors.

L. nordmanni is a parasite of *Mola mola* and its records come from areas as widely separated from one another as do those of its host. The northern Atlantic records come from Iceland and along both European and American coasts; the parasite was found in the Mediterranean and the Adriatic. South Atlantic records come from the Gulf of Guinea. In the North Pacific it is known in Japanese waters and off California. It has also been captured in New Zealand waters.

Lepeophtheirus pollachius Bassett-Smith, 1896

(Figs. 676–688)

Syn.: *Lepeophtheirus pollachii* Bassett-Smith, 1899
Lepeophtheirus innominatus Wilson, 1905

Female (Fig. 676): Cephalothorax oval or slightly ovate, with broad frontal plates. Tip of first antenna not extending to lateral limit of shield. Posterior sinuses shallow, free margin of thoracic zone not extending beyond level of posterior tips of lateral zones. Fourth leg-bearing segment not clearly delimited from genital complex, narrower than free margin of thoracic zone, expanding about midlength to form lateral protuberances accommodating bases of fourth legs. Genital complex about as long as dorsal shield, subquadrangular or oval, its posterior margin often broader than anterior. Abdomen cylindrical, as long as or slightly longer than genital complex, often with slight constriction about midlength suggesting obsolete segmental division, slightly tapering posteriorly. Uropod quadrangular, longer than broad, less than $\frac{1}{4}$ length of abdomen. Total length of body about 8–9 mm.

First and second antennae without distinguishing features. Dentiform process of first maxilla (Fig. 679) with strong conical base imperceptibly passing into tines apparently fused at base, slightly tapering, blunt and rounded at tips, devoid of flanges or surface markings. Second maxilla resembling that of type species but with particularly large flabellum. Corpus of maxilliped with unarmed myxal area; its subchela long ($\frac{3}{4}$ of its length represented by claw), barbel short. Sternal furca (Fig. 680) with strong, elongate box (anterior margin triangular, with apex pointing forwards); tines shorter than box, divergent, with bluntly rounded tips and narrow flanges along inner and often parts of outer margins.

Terminal armature of first exopod (Fig. 682), with setae decreasing in length from seta 1 to 3, seta 4 only slightly longer than seta 3. Setae 1–3 with pectinate bases; seta 1 simple with serrated posterior margin and a row of denticles near anterior margin, often extending only part of its length; setae 2 and 3 (Fig. 683) with secondary outgrowth in posterior half and with denticulations on both anterior and posterior margins (middle third on latter); seta 4 pinnate, not pectinate at base. Exopod of second leg (Fig. 684) with fringe of setules on lateral margin of second and parts of first and third segments; setules particularly long on first segment. Exopod of fourth leg (Fig. 685) three-segmented;

three terminal setae decreasing in length from seta 1 to 3, all with pectens at bases and with two strips of narrow, finely serrated membrane; segmental divisions clear; posterodistal setae of first and second segments similar to terminal seta 3; posterior margin of second and third segments with strips of striated membrane. (Setae of first and second segments often very short and claw-like.) Fifth leg (Fig. 687) quadrangular, arising from tip of small swelling, armed with the usual three setae. (Fourth seta, lateral to leg, was not observed in specimens examined.)

Male (Fig. 677): Cephalothorax and fourth leg-bearing segment similar to those of female. No distinct divisions between genital complex and other tagmata. Complex longer than thoracic zone of shield, narrow and long, with almost parallel lateral margins in many specimens. Bases of fifth and sixth legs inconspicuous, often not noticeable in dorsal aspect. Abdomen about as long as genital complex, cylindrical, indistinctly two-segmented. (Segmental boundary observable as lateral constrictions and slight furrow at about $\frac{1}{3}$ of length of tagma.) Uropod prominent, rectangular, about as long as first abdominal segment. Total length of body about 4.5 mm.

Claw of second antenna (Fig. 681) sharply hooked, simple, with narrow flanges on either side of distal half, with long, flagelliform seta at midlength and shorter one at base (on side opposite to that shown in Fig. 681); distal adhesion pad on penultimate segment prominent. Dentiform process of first maxilla (Fig. 678) with broad, conical base and single tine arising from its apex; tine with transverse furrow at about $\frac{1}{4}$ length, at which point slender secondary process present on medial side, smooth, not reaching tip of tine; latter blunt, rounded, devoid of flanges. Maxilliped (Fig. 688) similar to that of female, but with broad, bifid process near base of subchela. Sternal furca similar to that of female but often with less divergent tines. Fifth leg (Fig. 686) small, relatively longer and narrower than that of female, with three setae on apex and fourth near base. Sixth leg (Fig. 686) represented by three setae borne upon tip of small tubercle, posterior to leg 5.

Comments: The synonymy of *L. pollachius* is short, bearing testimony to its morphological distinctness. Bassett-Smith (1899) altered the ending of its original specific name to put it in the more common genitive case and the modified name was generally adopted. However, according to the International Code for Zoological Nomenclature, Article 11(g)(i)(2), the original name fulfils conditions laid down for the formation of names. Bassett-Smith's subsequent action cannot, therefore, be construed as a justified emendation and the original name should be restored.

L. pollachius belongs to the "long-tailed" species of its genus which are in the minority but which constitute more than half (4 out of 7) of the British species of *Lepeophtheirus*. It can be distinguished from the three other "long-tailed" British species by several morphological details. It differs from *L. salmonis* by the shape of its sternal furca and the dentiform process of its first maxilla; from *L. sturionis* by the same characters; from *L. thompsoni* mainly by the morphology of the male.

The three common hosts of this species, in order of their importance, are the gadoids *Pollachius pollachius*, *P. virens* and *Molva molva*. Wilson (1905) reported the occurrence of *Lepeophtheirus innominatus* (later relegated to synonymy with *L. pollachius*) on a "salmon", on the basis of a label made by Rev. A. M. Norman for specimens collected in Cornwall. Frequently repeated, this record produces an impression of host–parasite affinity between *L. pollachius* and the salmon, an impression which is quite misleading. No subsequent records of *L. pollachius* from salmon were made. Another unusual record is that of Dollfus (1956), who found one small male specimen of *L. pollachius* on the herring (*Clupea harengus*) in Scottish waters. Dollfus observed morphological differences between that specimen and those from the more usual hosts, which in his view were sufficient to designate it as *L. pollachii* forma *harengi*. In spite of extensive searching, this author failed to find any specimens of *L. pollachius* on Scottish herring. Since Dollfus gave no detailed account of the differences, no valid judgement can be made regarding this supposed infraspecific taxon.

L. pollachius appears to have fairly limited distribution, unlike its hosts, occurring only in British waters, both in the South (Plymouth, Cornwall, English Channel) and the North (Scottish waters, North Sea).

Lepeophtheirus salmonis (Krøyer, 1838)

(Figs. 689–700)

Syn.: *Caligus salmonis* Krøyer, 1838
Caligus vespa Edwards, 1840
Caligus strömii Baird, 1848
Lepeophtheirus strömii (Baird, 1850)
Caligus pacificus Gissler, 1883
Lepeophtheirus vesper? of Milne Edwards; in Bassett-Smith (1896b)
Caligus vesper? M-E; of Bassett-Smith (1899)
Lepeophtheirus pacificus (Gissler, 1883); of Wilson (1905)
Lepeophtheirus uenoi Yamaguti, 1939

Female (Fig. 689): Cephalothorax orbicular or somewhat ovate. Tip of first antenna not extending to lateral limit of dorsal shield. Posterior sinuses shallow; free margin of thoracic zone about level with posterior tips of lateral zones. Fourth leg-bearing segment short, narrower than free margin of thoracic zone, not clearly demarcated from genital complex, with lateral protuberances forming bases for attachment of fourth legs. Genital complex longer than thoracic zone of shield, with rounded anterolateral corners, parallel lateral margins and prominent, rounded lobes in posterolateral corners. Abdomen one-segmented, cylindrical, about as long as genital complex. Total length of body 7.4–18.2 mm.

Postantennary process with expanded base shorter than narrow, tapering tip. Spherical swelling for attachment of tip of second antenna fairly close to base of process. Dentiform process of first maxilla (Fig. 692) with medial tine slightly shorter than lateral, often with irregular inner margin of medial tine. Maxilliped (Fig. 698) with robust corpus, unarmed myxa; subchela with claw longer than shaft, robust, with short barbel. Sternal furca (Fig. 694) with longitudinally rectangular box and tines shorter than box, spatulate, blunt and divergent. In some older specimens tines unusually short (Fig. 695) due to wear in vivo.

Terminal armature of first exopod exactly as in *L. pollachius* (see Fig. 682), also second endopod (see Fig. 684). Exopod of fourth leg (Fig. 699) very indistinctly three-segmented; terminal setae decreasing in length from seta 1 to 3, all three with pectinate bases and with serrated anterior and posterior margins; seta marking obsolete boundary between second and third segment similar to terminal seta 3; seta marking barely discernible division between first and second segments (Fig. 700) greatly reduced, often partly covered by prominent pecten, sometimes difficult to observe, claw-like; posterior margins of second and third segments with narrow strips of membrane. Fifth leg (Fig. 696) roughly similar to that of *L. pollachius*. Distal setae of uropod characterized by inflated bases, sometimes quite prominent.

Male (Fig. 690): Cephalothorax similar to that of female. Fourth leg-bearing segment very short and broad (about $\frac{1}{2}$ width of free margin of thoracic zone), indistinctly delimited from genital complex. Latter shorter than thoracic zone of shield, with truncated anterior and posterior margins and slightly convex lateral margins. Abdomen one-segmented, shorter than genital complex, cylindrical. Total length of body about 5–7 mm.

Claw of second antenna (Fig. 691) simple, sharply flexing in distal half, with narrow flanges on either side and with two setae, flagelliform seta at about midlength and short, sturdy one near base. (For additional illustration see Kabata, 1973c.) Dentiform process of first maxilla (Fig. 693) with solid base and lateral tine longer and stronger than medial; medial process slender, longer than medial tine. Maxilliped with subchela relatively longer than in female, more slender and less curved; claw about as long as shaft, barbel long; corpus similar to that of female but with two groups of fine, blunt denticles on dorsal wall, next to base of subchela. Fifth leg (Fig. 697) resembling that of female, with fourth seta close to base. Sixth leg (Fig. 697) with three setae on tip of lobe, posterior to fifth. Uropod as in female, but apparently without basal inflations of distal setae.

Comments: This large and usually strikingly dark-coloured parasite belongs to the longest-known copepods recorded in scientific literature. It is very probable that it was included by writers such as Linnaeus and Fabricius in their composite species *Monoculus* and *Binoculus piscinus*. Hence, its early synonymy is not altogether clear. It has become customary to date *L. salmonis* to the work of Krøyer (1838), though it had been almost certainly known prior to that date. Its synonymy subsequent to Krøyer is as long as the recorded history of the species. Some of the confusion can be traced to the inaccuracies of descriptions. For example, it appears (as pointed out by Margolis, 1958) that

Yamaguti (1939b) established *L. uenoi* only because he found in the specimens so designated a small claw-like seta on the first segment of the exopod of the fourth leg, whereas Scott and Scott (1913) omitted that seta in their description.

Numerous records of *L. salmonis* show that the species is circumpolar in the northern hemisphere and that it occurs on most species of salmonids of the genera *Salmo* and *Oncorhynchus*. It is common on British salmonids. There are also some records from non-salmonid hosts. Gusev (1951) found it on *Leuciscus brandti* in the far-eastern waters of the USSR. Kazachenko et al. (1972) recorded it on *Sebastes rubrivinctus* off the Pacific coast of USA. The author's own collection includes a small male *L. salmonis* taken from *Ammodytes hexapterus* in British Columbia, Canada. He also collected some mature female specimens from the skin of *Acipenser transmontanus* in the Underwater Gardens in Victoria, B.C., Canada. The latter instance was a clear example of accidental transfer from salmon. In other instances, however, transfer might have taken place under natural conditions. Kazachenko et al. (1972) clearly state that in their case accidental transfer has been ruled out. At any rate, the occurrence of *L. salmonis* on non-salmonid hosts must be considered unusual and it is quite likely that hosts other than salmonids offer it no chance for development and survival. It flourishes on its salmonid hosts only during the marine phase of their lives and survives the host's entry into rivers by only a short time.

Lepeophtheirus sturionis (Krøyer, 1838)

(Figs. 531–537)

Syn.: *Caligus sturionis* Krøyer, 1838

Female (Fig. 531): Cephalothorax suborbicular. Tip of short first antenna far from reaching lateral limit of dorsal shield. Posterior sinuses shallow, free margin of thoracic zone about level with posterior tips of lateral zones. Fourth leg-bearing segment short, about half width of free margin of thoracic zone, sometimes indistinctly delimited from genital complex. Latter slightly longer than thoracic zone of shield, subrectangular, with sloping anterolateral corners, divergent lateral margins and rounded posterolateral corners. Greatest width about same as that of free margin of thoracic zone of shield. Abdomen not clearly delimited from genital complex, about as long as complex, indistinctly two-segmented (lateral constriction observable about $\frac{2}{3}$ distance from base). Uropod very small, subquadrate. Total length of body about 14 mm.

Dentiform process of first maxilla (Fig. 532) prominent, its inflated base tapering into long, single tine, dividing into two slender, tapering processes in its distal third. No flanges or membranes. Lateral side of base bearing small tubercle. Basal papilla very small. Sharp-pointed cuticular process present on ventral surface of cephalothorax about level with tips of tine. Claw of maxilliped very long. Sternal furca (Fig. 533) with long, regularly rectangular box and short tines (less than $\frac{1}{2}$ length of box), sharp and divergent, with very narrow flanges on both margins.

Terminal armature of first exopod (Fig. 534): seta 1 longer than 2 or 3, but slightly shorter than 4, apparently unarmed (some narrow strips of membrane might remain undetected); seta 2 with secondary process in distal half and slightly serrated distal end of anterior margin and some serrations in middle third of posterior margin; seta 3 similar to seta 2 but shorter; seta 4 slender, simple, pinnate. Three long pinnate setae on posterior margin of second segment of exopod; second endopod (Fig. 535) with fringe of setules on lateral margins of first, second and part of third segment. Fourth leg with three-segmented exopod (Fig. 537); segmental divisions distinct; terminal seta 2 about as long as 1, seta 3 much shorter, all with pectens at bases, setae 2 and 3 with both margins serrated, seta 1 with serrations only on posterior margin; second segment with posterodistal seta similar to terminal seta 3, that of first segment greatly reduced; posterior margins of all three segments with several rows of denticles (Fig. 536).

Male: Unknown.

Comments: A parasite of the sturgeon (*Acipenser sturio*), a fish which is itself fairly uncommon, *L. sturionis* is a rare and little known species. Its male has not yet been described and of its female only a few specimens are preserved in some museum collections. The author had an opportunity of examining one of these specimens, courtesy of the British Museum, London. All drawings of *L. sturionis* reproduced in this book are of that single specimen.

L. sturionis has been found in several localities along the European Atlantic seaboard (British waters, North Sea, Denmark, the Baltic, Baie de Seine). It occurs also along the Pacific and Atlantic coasts of North America. Thompson (1847) recorded its occurrence on two species of gurnards

(*Trigla lucerna* and *Aspitrigla cuculus*) in Belfast. This record is very probably based on a misidentification. Future finds are likely to extend the geographical range of *L. sturionis*.

Lepeophtheirus thompsoni Baird, 1850

(Figs. 701–712)

Syn.: *Caligus gracilis* van Beneden, 1851
nec *Caligus gracilis* Dana, 1852
Caligus branchialis Malm; of Steenstrup and Lütken (1861)
Lepeophtheirus rhombi Krøyer, 1863
Lepeophtheirus gibbus Krøyer, 1863
Lepeophtheirus gracilescens Krøyer, 1863
Lepeophtheirus gracilis Van Ben.; of Carus (1885)
?*Lepeophtheirus obscurus* Bassett-Smith, 1896
Lepeophtheirus branchialis; of Bassett-Smith (1899)
Lepeophtheirus (?) *obscurus* Baird; of T. Scott (1900)

Female (Fig. 701): Cephalothorax suborbicular. Tip of first antenna not reaching lateral limit of dorsal shield. Posterior margin truncated in appearance. Posterior sinuses very shallow, free margin of thoracic zone either level with or extending slightly beyond tips of lateral zones. Fourth leg-bearing segment with indistinct posterior border, apparently about equally long and broad, with lateral protuberances at bases of fourth legs, its width about half that of free margin of thoracic zone. Genital complex much longer than thoracic zone, longer than broad, with narrower anterior end and expanding in width posteriorly. Anterolateral corners sloping, posterolateral forming rounded lobes slightly overlapping abdomen. Latter cylindrical, about as long as, sometimes somewhat longer than, genital complex, in some specimens with indistinct dividing line across dorsal surface at about $\frac{3}{4}$ length from base. Uropod relatively very small. Total length of body about 8 mm.

Postantennary process (Fig. 707) with broad, massive base and short, sharply hooked shaft; spherical outgrowth near base of process relatively very large (compare with Fig. 638). Dentiform process of first maxilla (Fig. 709) with very prominent conical base and clearly delimited tines; medial tine longer and less robust than lateral; former often irregular in diameter, latter often with narrow flanges on both margins. Maxilliped without special distinguishing features. Sternal furca (Fig. 705) with long box, ovate in outline and tines shorter than box; tines blunt, divergent, sometimes slightly curving, with easily observable flanges on both margins.

Terminal armature of first exopod (Fig. 706) with seta 1 equipped with minute serrations along posterior margin and row of larger denticles on distal half near anterior margin; setae 2 and 3 progressively shorter, with secondary process arising out of distal $\frac{1}{4}$ and with denticulation on distal $\frac{1}{2}$ of anterior and middle $\frac{1}{3}$ of posterior margins; seta 4 about as long as 1, pinnate. Second endopod as in *L. pollachius* (cf. Fig. 684). Exopod of fourth leg (Fig. 710) two-segmented, distal segment formed by fusion of two; terminal setae three, seta 1 twice as long as seta 2 and about six times longer than seta 3, all three with pectinate bases and all with both margins equipped with strips of serrated membrane; at midlength of posterior margin of second segment another seta, similar to seta 3, marking position of distal limit of original middle segment; intersegmental divisions clear; posterodistal corner of first segment with short, often hooked seta, also with pecten and strips of membrane. In some specimens posterior margins, or parts of margins, of both segments with narrow strips of membrane. Fifth leg (Fig. 712) subquandrangular, fairly prominent, with three short pinnate setae on distal margin; fourth similar seta arising out of small tubercle on surface of posterolateral corner of genital complex some distance lateral to leg 5.

Male (Fig. 702): Cephalothorax similar to that of female, but first antennae appearing relatively longer. Fourth leg-bearing segment relatively shorter, not delimited clearly from genital complex. Latter longer than thoracic zone of shield, suborbicular, with small lobes for fifth and sixth legs of each lateral margin. Posterior margin usually transversely truncated, often poorly delimited from abdomen. Abdomen one-segmented, less than $\frac{1}{2}$ length of genital complex, about equally long and broad, slightly pedunculate. Uropod quadrangular, about $\frac{1}{2}$ length of abdomen. Total length of body about 4 mm.

Second antenna (Fig. 703) with simple claw, sharply curved in distal half and equipped with two flanges on each side (Fig. 704), as well as with flagelliform seta at midlength and short, sturdy one at base. Dentiform process of first maxilla (Fig. 708) with conical base constituting half its length; tines short, blunt, apparently devoid of flanges, about equally long; medial process longer than tines; posteromedial to tip of process small, oval adhesion pad present on ventral surface of cephalothorax. Fifth leg (Fig. 711) similar to that of female but relatively more prominent and with longer setae; fourth seta close to lateral side of base. Sixth leg (Fig. 711) posterior to fifth, armed with three similar setae.

Comments: The long and complicated synonymy of this species is due to two main reasons. *L. thompsoni* is a common parasite and often forms very large populations on its hosts, not infrequently composed of specimens of different ages and stages of development. In addition, it shows a considerable range of individual morphological variation, particularly in the shape of its genital complex. Krøyer's (1863) description of several species based on the same material bears ample testimony to the difficulties created by this morphological plasticity. The other reason lies in the fact that *L. thompsoni* had attracted the attention of several independently working observers at roughly the same time. Difficulties with exchanges and availability of pertinent literature compounded the problem. For example, Steenstrup and Lütken (1861) recognized their specimens as being identical with van Beneden's (1851a) *Caligus gracilis*, but changed their name in the belief that the name was preoccupied by Dana's species of *Caligus*, which in fact was two years later in chronology. Valle (1882) mentioned a *L. trygonis*, without describing it, as a new species occurring on an Adriatic stingray. The reasons that prompted Yamaguti (1963) to include this nomen nudum into the synonymy of *L. thompsoni* are not clear. It is also difficult to understand how Brian (1906) could construe *Caligus piscinus, sensu* Guerin-Meneville (1829–1843) and Edwards (1840), as synonyms of this species.

L. thompsoni parasitizes the gills of *Scophthalmus maximus* and *S. rhombus*. Its distinctive younger individuals can be found upon the skin of their hosts, often moving about very actively. On the other hand, most mature individuals are wedged in between the gill filaments, where they remain immobile, cephalothorax towards the gill arch. In heavy infections one finds three or four specimens in a row between two filaments, always attached to the same surface. The distribution of *L. thompsoni*, in addition to numerous records from British waters, extends over the North Sea, the western Baltic and the western Mediterranean. Some specimens have been found in the Faroes.

The literature contains a number of unusual records of *L. thompsoni*. Hansen (1923) found it on *Hippoglossus hippoglossus* in the Faroes (possibly an inadvertent transfer in a trawl catch). Wilson (1908) and Causey (1960) claimed to have found it on *Cynoscion nobilis* off the Pacific coast of the USA and Mexico. *Lophius americanus* and *Dasyatis centrura* were recorded as harbouring it off the Atlantic coast of the USA (Wilson, 1905; Sumner et al., 1913). Boxshall (1974a) found it on *Solea solea*. The most puzzling record, however, is that of *L. thompsoni* on *Arius venosus* in Ceylon (Thompson and Scott, 1903).

Family Euryphoridae

Morphology

The copepods belonging to this family bear a very close resemblance to the caligids. The anterior tagma, the cephalothorax, is similar enough in both families to be without any value in distinguishing between them. The main difference between Caligidae and Euryphoridae is the possession by the females and most males of the latter family of a pair of aliform dorsal plates, arising from the sides of the fourth leg-bearing segment.

The posterior half of the female body differs from one genus to another (Text fig. 51). The genital complex of *Euryphorus* Edwards, 1840, bears short, rounded plates in the posterolateral corners. *Gloiopotes* Steenstrup and Lütken, 1861, has those corners extended into digitiform lobes and armed with prominent styliform processes. In *Alebion* Krøyer, 1863 and *Paralebion* Wilson, 1911, the posterolateral lobes are usually long and pointed but no styliform processes are present. No lobes or processes occur on the genital complex of *Tuxophorus* Wilson, 1908, though the fairly prominent fifth legs can be seen, in dorsal aspect, protruding from underneath its posterolateral corners.

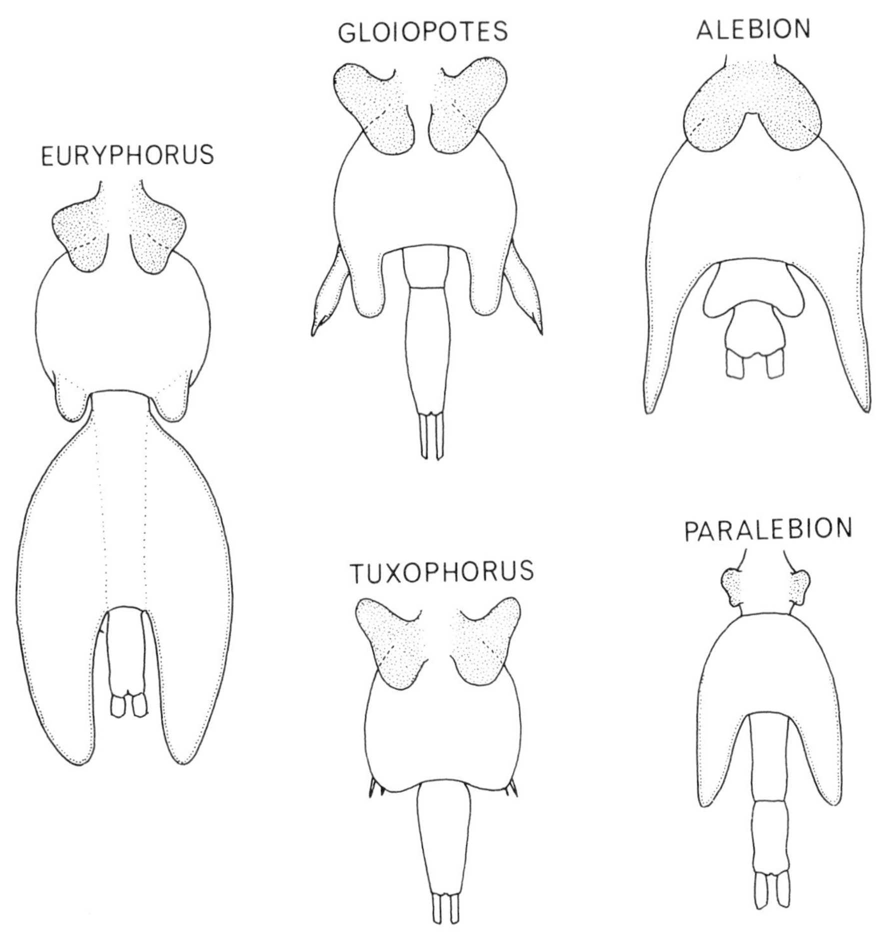

Text fig. 51. Morphological differences in posterior half of body among Euryphoridae.

The abdomen is usually two-, less often one-segmented. In *Euryphorus* it might equal the combined lengths of the rest of the body and its first segment has lateral alae, sometimes of comparatively very large size (Text fig. 51). In *Alebion* the first abdominal segment has lateral expansions, digitiform, often well developed and sometimes drawing level with the posterior limit of the abdomen.

In the males the familiar characteristics are less well developed, the genital complex in particular being devoid of prominent processes or lobes (though the styliform processes are present in male *Gloiopotes*).

The appendages are also very similar to those of Caligidae in structural pattern. In most species they do not provide any features that might be considered as familial discriminants. The exception is provided by the genus *Euryphorus*, which will be discussed in more detail below.

History and systematics

The history of Euryphoridae marks the course of a gradual recognition of their distinctness within the family Caligidae and the process of their separation, first as a subfamily Euryphorinae and then as an independent familial taxon. Wilson (1905), beginning his work on Caligidae, gave a brief definition of Euryphorinae as follows: "Three anterior segments of thorax fused with the head, fourth and genital segment free. Fourth segment with a pair of dorsal plates which usually overlap the genital segment." Soon afterwards (Wilson, 1907a), the subfamily received a thorough treatment in a paper devoted to it and Trebiinae as constituents of Caligidae. Wilson's concept of the subfamily is best illustrated by his key to the euryphorid genera.

1. Three thorax segments fused with the head; the fourth segment only free 2
1. Only the first thorax segment fused with the head, the others free; no dorsal plates; all the swimming legs biramose, the rami three-jointed . . . . **Dissonus**

2. One or more pairs of legs uniramose, the others biramose 3
2 All four pairs of swimming legs biramose 4

3. First legs only uniramose, their terminal claws curved and simple; setae on anal laminae short and non-plumose . . . . **Caligeria**
3. First and fourth legs uniramose; claws on first pair straight and three-parted; setae on anal laminae long and plumose . . . . **Gloiopotes**

4. Exopod of fourth legs three-jointed, endopod two-jointed 5
4. Both rami of fourth legs with the same number of joints 6

5. Setae on fourth legs plumose; first abdomen joint much larger than second and covered with a dorsal wing or with two lateral wings . . . . **Euryphorus**
5. Setae on fourth legs non-plumose; abdomen joints about the same size, without wings . . . . **Elytrophora**

6. Rami of fourth legs small, the two joints fused; rami of first three pairs two-jointed, without horny processes . . . . **Dysgamus**
6. Fourth legs rudimentary, hidden; exopods of first three pairs with horny processes . . . . **Alebion**

The above key shows that the original definition of the subfamily has been violated by the inclusion of *Dissonus* Wilson, 1906, a genus with segmentation of a different type and without the dorsal plates. This might have resulted from Wilson's idea of Euryphorinae as a group of species intermediate between Caliginae and Pandarinae, then also subfamilial units of the same family, some showing more resemblance to the former, others to the latter. The ground was prepared for the inclusion of other species, roughly answering this description, even if not compatible with the original definition. A good example of it was *Midias lobodes* Wilson, 1911, a caligoid in most features but equipped with moderate lateral expansions on the basal segment of the two-segmented abdomen. These and other minor morphological details, according to Wilson, meant that it "must be placed with other intermediate forms in the Euryphorinae". In the same publication the genus *Paralebion* was added to the subfamily.

It is interesting to add that at that stage the structure of the swimming legs was accepted as the best discriminant at the generic level within the subfamily. Sound in principle, this discriminant proved ineffective, mainly because some of the genera were either poorly known (*Euryphorus* and, particularly, *Dysgamus*), or interpreted only from sketchy descriptions and totally inadequate figures (*Caligeria*).

Without any further discussion, Wilson (1932) upgraded Euryphorinae to the familial status. He was followed in this by Yamaguti (1936). It is curious to note that the latter author (Yamaguti, 1963) later treated Euryphoridae as a new family, created by himself at that date. In his combination, the family consisted of two subfamilies: Euryphorinae (without lunules) and Tuxophorinae (with lunules). The latter comprised *Midias* and *Tuxophorus* Wilson, 1908.

One cannot help thinking that, confronted with the rather vague attitude of his predecessor,

Yamaguti (1963) was uncertain as to what Euryphoridae really were. With full justification, he excluded from them *Dissonus*, making it the type genus of a new family, Dissonidae. None the less, he made a proviso for it in the euryphorid family diagnosis, thus confusing the issue again. At the generic level his main criterion remained the structure of the legs, though other criteria for identification were added, as shown in his key to the genera of the subfamily Euryphorinae.

1. One or more pairs of legs uniramose, others biramose 2
   All anterior pairs of legs biramose 3

2. Fourth leg only uniramose.
   A Fourth leg rudimentary; exopods of first three pairs with horny claws; maxillary hook and furca absent . . . . ***Alebion***
   B Fourth leg 4-segmented; maxillary hook and furca present, latter double . . . . ***Paralebion***
   First and fourth legs uniramose; maxillary hook and furca present, both compound in female; genital segment with a pair of serrate styliform appendages . . . . ***Gloiopotes***
   First and fourth legs uniramose; maxillary hook absent, furca simple; genital segment without paired serrate appendages . . . . ***Caligeria***

3. Basal segment of abdomen with two lateral wings in female, with conical posterior lobes in male; caudal rami with plumose processes containing pulp; secondary furca present . . . . ***Euryphorus***
   Basal segment of abdomen with posterior lobes in female but no lateral wings; caudal rami with plumose setae, not containing pulp; no secondary furca . . . . ***Dysgamus***

The lack of certainty as to what are the diagnostic features of Euryphoridae resulted in some spurious additions to the family. *Midias carangis* Rangnekar, 1956, proved to be a synonym of a *Caligus* species. *Diphyllogaster aliuncus* Rangnekar, 1955, was transferred to *Paralebion* by Pillai (1967a), though it has no dorsal plates on its fourth leg-bearing segment; *Paralebion curticaudis* Wilson, 1913, was added, in spite of the absence of these plates. Neither is a euryphorid.

Apart from Yamaguti's revision, no attempt has been made recently to look at the family as a whole. Some of its genera, however, were thoroughly revised, with considerable benefits to our understanding of the intrafamilial relationships. Hewitt (1964a) and Cressey (1967b) revised *Gloiopotes*, cutting down the number of its species. *Alebion* was similarly reviewed by Cressey (1972). Finally, Heegaard (1972) took a close look at the genus *Euryphorus* and recognized it as a senior synonym of *Elytrophora*. *Dysgamus* was found by him to be an artificial genus, composed of juvenile stages of other genera, mainly *Euryphorus*.

There remain two unsolved problems. One is that of the identity of *Midias lobodes*. Devoid of dorsal plates, it cannot remain in Euryphoridae. Except for the shape of the abdomen and the rather prominent fifth legs, it could become a typical member of the genus *Caligus*. (*C. constrictus* Heller, 1865, as illustrated by Pillai (1961, p. 94, fig. 4) has a somewhat similar abdomen.) The other problem concerns the genus *Caligeria*, never again recorded after the initial description of *C. bella* Dana, 1852. Judging from Dana's (1855) pl. 94, fig. 8, it has many features in common with *Euryphorus*, *sensu* Heegaard (1972). Though a definitive decision is impossible, this author proposes to treat *C. bella* as a tentative synonym of *Euryphorus brachypterus*.

As recognized by this author, therefore, Euryphoridae at the time of writing contain five genera and 21 species. Their most abundant genus is *Alebion*, comprising eight species, widespread parasites of sharks. Next in abundance is *Gloiopotes* with five species, all on marine teleosts. There follows *Tuxophorus* with four species on teleostean hosts, *Paralebion* with two (one on teleosts and one on sharks) and *Euryphorus*, also with two species.

Only the last-named genus occurs in British waters.

Genus Euryphorus Edwards, 1840

Female: Euryphoridae. Cephalothorax caligiform. Fourth leg-bearing segment with pair of dorsolateral plates, partly overlapping genital complex. Latter with plates in posterolateral corners. Abdomen two-segmented, with pair of large or small, lateral aliform expansions. Appendages similar to those of Caligidae, but fourth swimming legs biramous, rami two- or three-segmented. Uropods present.

Male: Similar to female. First abdominal segment with or without lateral expansions.

Type species: *Euryphorus nordmanni* Edwards, 1840.

Comments: Ubiquitous in distribution, this genus has a very complicated taxonomic history. Descriptions of species and even genera from different geographical regions have been found to deal with one of the two species of *Euryphorus*. As mentioned earlier in the discussion of the family, *Dysgamus* and *Elytrophora* are now regarded as synonymous with *Euryphorus*.

At least part of the difficulty lies in the morphological plasticity of these two species (*E. nordmanni* and *E. brachypterus* (Gerstaecker, 1853)), difficulty perhaps concomittant with their distribution and their host range. Moreover, morphological changes in the course of their ontogeny are much more pronounced than e.g. in Caligidae, particularly as regards the aliform expansions. Some of these changes were discussed by Kabata and Gusev (1966) and Heegaard (1972).

Both species of the genus tend to parasitize large scombrid fishes as their main hosts, but many records have been made from unrelated species of marine teleosts. Only *E. brachypterus* occasionally occurs in British waters.

Euryphorus brachypterus (Gerstaecker, 1853)

(Figs. 741–766)

Syn.: ?*Caligeria bella* Dana, 1852
Elytrophora brachyptera Gerstaecker, 1853
Arnaeus thynni Krøyer, 1863
Elytrophora hemiptera Wilson, 1922
Dysgamus longifurcatus Wilson, 1923
Elytrophora atlantica Wilson, 1932
Elytrophora indica Shiino, 1958
Dysgamus sagamiensis Shiino, 1958
Elytrophora brachyptera brachyptera Hewitt, 1968

Female (Fig. 741): Cephalothorax with suborbicular dorsal shield; lateral margins slightly indented at level of transverse suture between cephalic and thoracic zones, provided with narrow strip of marginal membrane; lateral zones protruding beyond straight free margin of thoracic zone; second transverse suture deflected posteriorly behind eyes. Frontal plates relatively narrow, with circular lacunae divided by longitudinal cuticular bar (Fig. 745) in lateral ends; first antenna not reaching lateral limit of dorsal shield; lunule absent. Fourth leg-bearing segment with dorsal plates overlapping anterior end of genital complex and often each other in mid-dorsal line. Genital complex oval, its length about equal to distance between free margin and anterior transverse suture of dorsal shield; rounded posterolateral plates partly overlapping base of abdomen; posterior end of ventral surface (Fig. 742) with bilobed processes associated with vaginal openings; orifices of oviducts dorsolateral to posterior processes, dorsally covered by plates. Egg sacs long, uniseriate, eggs flat. Abdomen two-segmented, segments about equal in size, basal with small posterolateral expansions, second subquadrate; length of abdomen about half that of genital complex. Total length of body about 7–10 mm.

First antenna (Fig. 743) two-segmented; proximal segment broad at base, tapering distally, with 26 setae on distal half of anterior margin and ventral side of distal margin; setae sparsely plumose (in many specimens some missing); distal segment club-shaped, narrower at base, shorter than proximal, its armature similar to that of Caligidae, all setae unarmed. Second antenna (Fig. 744) similar to that of Caligidae; posterior process of second segment subtriangular, blunt; terminal claw long, slender, sharply curved, with seta near base and longer one some distance from it. Mouth tube (Fig. 746) short and broad; labrum indented at points of entry of mandibles, its distal margin truncated, with very short setules; labium longer, with longer fringe. Mandible (Fig. 747) of two parts; basal shorter and broader, terminal slender, with slightly expanded tip; dentiferous margin (Fig. 748) with 12 teeth gradually shorter towards base. Spiniform process of first maxilla (Fig. 749) with broad base, lateral side concave; basal papilla with one long and two shorter setae. Second maxilla (Fig. 750) brachiform; lacertus less than $\frac{1}{2}$ length of appendage, sturdy, unarmed; brachium longer, more slender, with small pectinate adhesion pad at about midlength on dorsal surface (Fig. 751) and distal to it relatively large flabellum, sigmoid and with serrated margin; tip similar to that of caligid copepods. Maxilliped (Fig. 752) with robust corpus; slight protuberance in myxal area and two rounded strips of cuticle on ventral surface in proximal half; subchela with short shaft and very long, slender, sharply curving claw, longitudinally grooved in distal half and with short barb at base. Sternal furca with long and slender box; tines straight, divergent, without membranes or flanges.

General structure of swimming legs similar to that of Caligidae, but first exopod less reduced (Fig. 753), third sympod less expanded and with narrower interpodal bar (Fig. 757), fourth leg biramous (Fig. 760). Small secondary furca on first interpodal bar. Armature formula as shown below.

	Endopod			Exopod		
	1	2	3	1	2	3
Leg 1	0-0	3	—	0-I	3-1,III	—
Leg 2	1-0	2-0	6	1-I	1-I	6,II
Leg 3	1-0	2-0	4	1-I	1-I	6,II
Leg 4	1-0	1-0	3	0-I	1-I	4,III

Distal margin of first exopod (Fig. 754) with one spine serrated along posterior margin and two along both margins; spines 2 and 3 arranged in dorsoventral plane, pecten at base of spine 3; seta 4 twice as long as spines, pinnate (pinnules omitted in Fig. 754). Lateral spines of second exopod (Fig. 756) with both margins serrated, deeply notched strips of cuticle along bases. Spines of second and third segments extending beyond distal limit of latter; second spine on third segment with only one margin serrated, lateralmost seta semipinnate, with serrated membrane along one margin. Lateral margins of second endopod with fine setae on first segment and spiniform on second. Third endopod (Fig. 758) similar to second; third exopod spines (Fig. 759) with fine tips and serrated marginal membranes; ventral surface of sympod with two patches of fine spinules (marked with dotted outlines in Fig. 757). Distomedial corner of fourth sympod (Fig. 760) with small patch of denticles; second and third segments of fourth endopod partly fused; exopod twice as long as endopod, setae on its medial margin short, spiniform. Uropod (Fig. 761) narrower at base; distal margin with four short, pinnate setae (only one shown pinnate in drawing), pinnate setule between two lateralmost setae, another in distomedial corner.

Male (Fig. 762) similar to female, its genital complex relatively smaller, devoid of posterolateral plates; abdomen narrower, without lateral expansions on basal segment. Total length 5–7 mm.

Claw of second antenna (Fig. 763) with short, blunt tine, setule on prominent base and long, slender seta, all on or near inner margin of proximal half; long flange on outer margin of distal half. Corpus of maxilliped (Fig. 765) with inward flexion, prominent protuberance in myxal area and one rounded flange near centre of ventral surface. Sternal furca (Fig. 764) narrower and longer than that of female, with less divergent rami. Medial setae on fourth exopod and distal on endopod relatively longer than those of female, pinnate. Fifth leg (Fig. 766) reduced to small protuberance on lateral margin of genital complex, surmounted by three small, pinnate setae; one similar seta anterior to base of leg.

Comments: An extensive range of morphological variability of *E. brachypterus* is suggested by the fact that many authors described it under several different names. This variability is mainly shown in the proportions of the body, structure of the appendages and the size and shape of the aliform plates. *Caligeria bella* has not been made a definite synonym by this author because not enough information is available on its structure. Should it ever become established that it is identical with *E. brachypterus*, its name would have priority over the one currently in use, a rather undesirable situation in view of the very common use of the present binomen.

For the discussion of the leg structure of this species, see p. 40.

All records of the adult *E. brachypterus* come from the big scombrids, the following hosts having been reported: *Thunnus alalunga*, *T. albacares*, *T. orientalis*, *T. thynnus*, *Thunnus* sp., *Parathunnus obesus*, *P. sibi* and *Makaira mitsukuuri*. Juvenile stages (under the name *Dysgamus sagamiensis*) were reported also from non-scombrid fishes: *Caranx uraspis* and *Spheroides inermis*. Records come from all the oceans of the world inhabited by the tunas, though they are most common from the Atlantic. In British waters the parasite appears occasionally with the migrating tuna (Bassett-Smith, 1896b; Tiews, 1957).

Family Trebiidae

The morphology of this family is sufficiently similar to that of Caligidae and Euryphoridae not to require any detailed discussion. Its main distinguishing feature is the presence of two free segments (third and fourth leg-bearing segments) betweeen the cephalothorax and the genital complex. Its members are also more primitive than those of the other two families in the condition of their third and fourth legs (Text fig. 4). Without fusion of the third legs there is no apron to cut off the posterior margin of the cup, formed by the dorsal shield of the cephalothorax, from the external environment. The suction exerted by the cephalothorax is, therefore, far less efficient than that in Caligidae and the attachment to the surface of the host less tenacious. The second, third and fourth legs form one uninterrupted series of natatory appendages. Although the locomotory mechanisms of Trebiidae have not been explored, one might hazard a guess that on closer examination they will prove to be substantially different from those of Caligidae (cf. Kabata and Hewitt, 1971), in which only the first two pairs of legs take part in locomotion. This might, indeed, provide an excellent example of relatively small morphological changes resulting in major functional consequences.

The history of Trebiidae, like that of the preceding family, is associated with the gradual dismemberment of Caligidae *sensu lato*. Trebiinae were made a subfamily within Caligidae by Wilson (1905), who used the presence of the free third leg-bearing segment as its main characteristic. Later (Wilson, 1932) he raised them to the familial rank. Until recently, Trebiidae contained only its type genus, *Trebius* Krøyer, 1838, and the family diagnosis was identical with that of the genus. Some changes in that diagnosis were necessitated by the recent discovery of a second trebiid genus, *Kabataia* Kazachenko et al., 1972. *K. ostorhinchi* is a parasite of *Ostorhinchus conwaii*, a teleost fish from the Great Australian Bight. It differs from *Trebius* in having lateral plates on the free, third leg-bearing segment and in lacking the sternal furca. Briefly, the family diagnosis must now read as follows:

Siphonostomatoida. Cephalothorax caligiform. Third leg-bearing segment free, with or without lateral plates. Sternal furca present or absent. First leg with two-segmented rami, endopod reduced; second to fourth legs similar to one another, with three-segmented rami of about equal size.

In British waters the family is represented by one species of *Trebius*.

Genus Trebius Krøyer, 1838

Female: Caligiform. Lunules absent. Third leg-bearing segment free, devoid of lateral plates. Genital complex with or without posterolateral processes. Abdomen one- to three-segmented (covered with spinules in *T. latifurcatus*). Cephalic appendages and maxillipeds similar to those of Caligidae. Sternal furca present. First leg with rami two-segmented, endopod reduced. Second to fourth legs similar to one another, rami three-segmented. Fifth leg vestigial; sixth absent. Uropods present.

Male similar to female, but with relatively smaller genital complex. Vestigial sixth leg present.

Type species: *Trebius caudatus* Krøyer, 1838

Comments: The genus *Trebius* contains currently 13 species, all parasitic on elasmobranchs, mainly on skates and rays (9 species), some on sharks (2 species) and one on *Squatina*. In addition one has been recorded from both sharks and rays. Wilson (1921b) reported the occurrence of *T. latifurcatus* on a flatfish, *Paralichthys*, in addition to its more normal elasmobranch hosts. It must be supposed that the occurrence of *Trebius* on that teleost was purely accidental.

The greatest number of species have been found in the Atlantic Ocean (*T. bilobatus* Brian, 1912, *T. caudatus*, *T. elongatus* Capart, 1953, *T. minutus* Capart, 1959, *T. nunesi* Capart, 1959 and *T. tenuifurcatus* Rathbun, 1887), followed by the Indian Ocean (*T. exilis* Wilson, 1906, *T. javanicus* Hameed and Pillai, 1973, *T. kirtii* Hameed and Pillai, 1973 and *T. sepheni* Hameed and Pillai, 1973) and the Pacific Ocean (*T. akajeii* Shiino, 1954, *T. latifurcatus* Wilson, 1921 and *T. longicaudatus* Shiino, 1954). More are likely to be discovered.

British fauna contains only the type species, *T. caudatus*.

Trebius caudatus Krøyer, 1838

(Figs. 722–740)

Syn.: ?*Trebius spinifrons* Edwards, 1840
nec *Trebius caudatus* Krøyer, 1838; of Lewis (1966)

Female (Fig. 722): Cephalothorax suborbicular, with shallow posterior sinuses; posterior tips of lateral lobes protruding beyond free margin of thoracic zone; latter less than $\frac{1}{2}$ length of cephalic zone; frontal plates well developed; first antennae not extending to outer limits of lateral margins. Third leg-bearing segment shorter than thoracic zone of dorsal shield, width slightly less than $\frac{1}{2}$ that of shield, with convex posterior margin, well delimited from following segment. Fourth leg-bearing segment slightly longer than preceding one, about $\frac{2}{3}$ width of that segment, with clearly drawn convex posterior margin. Genital complex $\frac{2}{3}$ length of dorsal shield (frontal plates included), narrower anteriorly, with sloping anterolateral and rounded posterolateral corners; latter (Fig. 733) with three spiniform processes dorsal to oviduct orifices and one similar process ventral to it (number of dorsal processes varying in some specimens, as many as five being sometimes present). Abdomen subcylindrical, indistinctly three-segmented, segments progressively smaller from first to third; in some specimens no border observable between second and third segments. Total length (frontal plates and uropods included) about 10 mm.

First antenna similar to that of *Caligus*; inner margin of proximal segment with two distal denticles; apical armature (Fig. 724) distinguished by bifurcation of four out of five slender setae on posterior side of apex (number of bifurcations not constant). Second antenna (Fig. 725) also similar to that of *Caligus*, with flat adhesion pad on second segment and prominent pad on third; unciform terminal claw with lateral flexion at midlength, with sturdy seta at base (Fig. 726), adhesion pad and slender seta at midlength and flanges on recurved distal half. Postantennary process (Fig. 728) bifid; stout base with short, triangular tine, two strips of narrow membrane on ventral surface, two on margins and two on dorsal surface; terminal tine about as long as base, curving ventrally (curvature foreshortened in drawing), tapering; branching sensory seta on base and lateral margin. Mouth tube (Fig. 727) caligid, with spiniform processes on each side of base of labrum. Mandible as in *Caligus*, with 12 teeth on dentiferous margin. First maxilla (Fig. 730); tine with inflated base and slender, distally bifid shaft; branches short, unarmed, medial shorter than lateral. Second maxilla as in Caligidae but without flabellum on brachium. Maxilliped (Fig. 731) with long, slender corpus; myxal area close to base, with small sclerotized ridge; subchela slender, cylindrical, shaft some $\frac{2}{3}$ of its length, with short seta and papilliform tubercle on distal half; claw slender, gently curved, with barb less than $\frac{1}{2}$ its length. Sternal furca with oblong box and straight, divergent tines with narrow flanges on tips.

First four pairs of legs biramous, with two-segmented sympods; first sympod with seta lateral to base of exopod and one medial to base of endopod (Fig. 735); second with marginal membrane on posterior and broad dorsal membrane on lateral margin; short seta at base of exopod (Fig. 736); third similar but with fringe of setae on posterior margin; fourth (Fig. 740) without lateral membrane. First leg with rami two-segmented; endopod reduced; legs 2–4 with three-segmented rami. Armature formula as follows.

	Endopod			Exopod		
	1	2	3	1	2	3
Leg 1	0-0	3	—	0-1	4,III	—
Leg 2	1-0	2-0	6	1-I	1-I	6,I
Leg 3	1-0	2-0	4,I	1-I	1-I	5,III
Leg 4	1-0	1-0	3,I	1-I	1-I	4,III

All lateral and most medial margins of endopods with fringes of fine setules (Fig. 739); some medial margins of exopod segments also fringed (Fig. 738). First and second exopod segments of second and third legs with strips of serrated membrane (Fig. 738); endopod spines (Fig. 739) unarmed. All setae pinnate, except lateral seta of third exopod segments (legs two and three) with lateral strip of membrane (Fig. 737). Fifth leg (Fig. 733) represented by three setules in lateral corner of genital complex. Uropod rectangular, with armature resembling that of Caligidae.

Male (Fig. 723): Anterior part similar to that of female. Genital complex oval, about as wide as fourth leg-bearing segment, less than $\frac{1}{2}$ length of dorsal shield of cephalothorax. Abdomen two-segmented, slightly longer than genital complex; first segment shorter than second, both broader posteriorly. Total length about 5 mm.

First maxilla (Fig. 729); tine with inflated base, shaft with rounded protuberance at about midlength and two minute denticles along distal half. Fifth leg (Fig. 732) vestigial, consisting of two pinnate and one naked setule arising from small swelling, with fourth naked setule anterior to base of leg. Sixth leg (Fig. 732) similar but with setae shorter and fourth seta lacking. Uropod (Fig. 723) relatively larger than that of female.

Comments: *T. caudatus* is not only the type, but apparently also the commonest, species of its genus. It has been found on many elasmobranch hosts, mainly rays (*Raja* sp., *R. alba*, *R. batis*, *R. blanda*, *R. clavata*, *R. fullonica*, *R. microcellata*, *R. montagui* and *Gymnura micrura*). There are some records from sharks and dogfishes (*Galeorhinus galeus*, *Squalus acanthias* and *S. fernandinus*). One tentative record includes *Raja macrorhynchus* (cf. Brian, 1902, 1906).

T. caudatus occurs throughout the North Atlantic, most of the localities reported being situated in its eastern part (the North Sea included). There are records from Iceland (Hansen, 1923; Stephensen, 1940) and the Gulf of Mexico (Bere, 1936). The species is known also in the western part of the Mediterranean and in the Adriatic. Nunes-Ruivo (1956) found it off the coast of Southwest Africa. In British waters it is fairly common, having been found in many localities around the British Isles.

Lewis (1966) described *T. caudatus* from Hawaii, where it was found on a species of ray, probably *Aetobates narinari*. His accurate description and careful illustrations reveal a number of differences between his specimens and *T. caudatus*. In addition to the general slenderness of shape, not characteristic of *T. caudatus*, they have, again unlike this species (Fig. 728), the postantennary process without a basal tine; their sternal furcae are more slender and longer. Judging from the illustrations, these specimens resemble most closely *T. longicaudatus* and should be assigned to that species.

Edwards (1840) very briefly described *T. spinifrons*, a species never subsequently encountered. His habitus drawing is very generalized and fits also *T. caudatus*. The host was *Squalus acanthias* (also listed for *T. caudatus*); no locality was given. No definite decision can be made without examining Edwards' material, but a strong possibility exists that it is identical with the present species.

Family Pandaridae

Morphology

The most recent authoritative review of this family (Cressey, 1967a) recognizes it as consisting of 12 genera. Definitely caligiform in their general structure, the female members of the family are characterized by having three distinct thoracic segments between the cephalothorax (including the first leg-bearing segment) and the genital complex, as well as paired dorsal or dorsolateral plates on at least one of these segments. Some of them also possess paired, fused or separate, or single plates on the abdomen. Another characteristic feature is the presence on their ventral surfaces of a system of adhesion pads, extensive in some, scanty in others (absent in *Phyllothyreus* and some species of *Dinemoura*). These structures do not serve as active attachment organs but their surface striations appear to create friction and so facilitate maintenance of the copepod's hold on the host fish. Typical also is a complete lack of body spines, though spinules of various kinds are present on some appendages.

In all members of the family the cephalothorax is readily recognizable as caligiform, covered by a characteristic dorsal shield. Unlike Caligidae, however, the pandarid copepods have shields that are divided usually into three zones by longitudinal sutures, the lateral zones sending rounded lobes well past the posterior margin of the central cephalic-thoracic zones. In some genera, notably *Pandarus* Leach, 1816, no sutures are evident and the shield is one heavy piece of armour. The three segments interposed between the large tagmata provide the best generic discriminants within this family. The genera fall into two groups: those with all three segments provided with dorsal or dorsolateral plates and those in which the second free segment carries no plates. The members of the first group have solid, one-piece appearances, while those of the second have silhouettes divided into anterior and posterior parts, due to the presence of a slender neck. There are six genera in each group. In a diagrammatic generalization, their structure is shown in Text fig. 52. The first, or *Pandarus*-group, contains two genera in which the plates of the second free segment either protrude beyond, or are even with, the posterior limit of the plates of the first. They are *Pandarus* (Text fig. 52A) and *Perissopus* Steenstrup and Lütken, 1861 (Text fig. 52B). The third free segment of *Pandarus* is covered by paired, fused dorsal plates, overlapping the base of the genital complex. The greatest part of that tagma is, however, visible in dorsal aspect. The abdomen is covered dorsally by a small, median plate. In *Perissopus* the genital complex is broad, with sharply pointed posterolateral corners. The base of its abdomen is displaced some distance along the ventral surface of the complex, so that the entire abdomen is ventral to it and cannot be seen in dorsal aspect. The other four genera of this group are distinguished by second free thoracic segments with paired, often partially fused dorsal plates, extending well past the posterior tips of the plates of the first free segment. In *Phyllothyreus* Norman, 1903 (Text fig. 52C) the plates of all three free segments are foliaceous, those of the first overlapping a substantial part of those of the second. The anterior part of the genital complex is covered by the plates of the third free segment but its posterior part is visible in dorsal aspect. A subcircular median dorsal plate covers the abdomen. In *Gangliopus* Gerstaecker, 1854 and *Pannosus* Cressey, 1967 (Text fig. 52D-E) the plates of the first free segment are small and mainly lateral to those of the second segment. The almost completely fused plates of the third segment cover much of the genital complex, while a median dorsal plate overlays the abdomen. (*Gangliopus* and *Pannosus* have nearly identical general appearance and can be distinguished by details in the structure of the appendages. Cressey (1967a) erected *Pannosus* for *Gangliopus japonicus* Shiino, 1960, because it differed from *G. pyriformis* Gerstaecker, 1854, in the adhesion pads of the antennary region, the dorsolateral plates of the first free segment and the subchela of the maxilliped. Both genera are monotypic.) The structure of *Pseudopandarus* Kirtisinghe, 1950 (Text fig. 52F) is similar to those of the preceding two genera, but more of the genital complex is exposed and the structure of the abdomen is different.

The genera of the second, or *Dinemoura*-group all have first free segments with fairly small dorsolateral plates, usually with tips overlapped by the lateral zones of the dorsal shield. (In *Pagina* Cressey, 1964, these plates are rudimentary, according to Cressey's illustrations, but in his text he considers them absent.) In *Dinemoura* Latreille, 1829 (Text fig. 52G) the dorsal plates of the last free (fourth leg-bearing) segment are moderately developed, partially fused and overlap the anterior part of the genital complex. Its distinctive generic feature is its posterior extremity, the morphology of which will be described in detail when the genus is dealt with. Similar to *Dinemoura* is *Demoleus*

Heller, 1865 (Text fig. 52H), though in this genus the posterior extremity is overlapped by the posterior lobes of the genital complex and the abdominal dorsal plate. In *Echthrogaleus* Steenstrup and Lütken, 1861 (Text fig. 52J) the dorsal plates of the last free segment cover a large part of the genital complex (more than half in *E. torpedinis* Wilson, 1907) and the abdomen is completely under

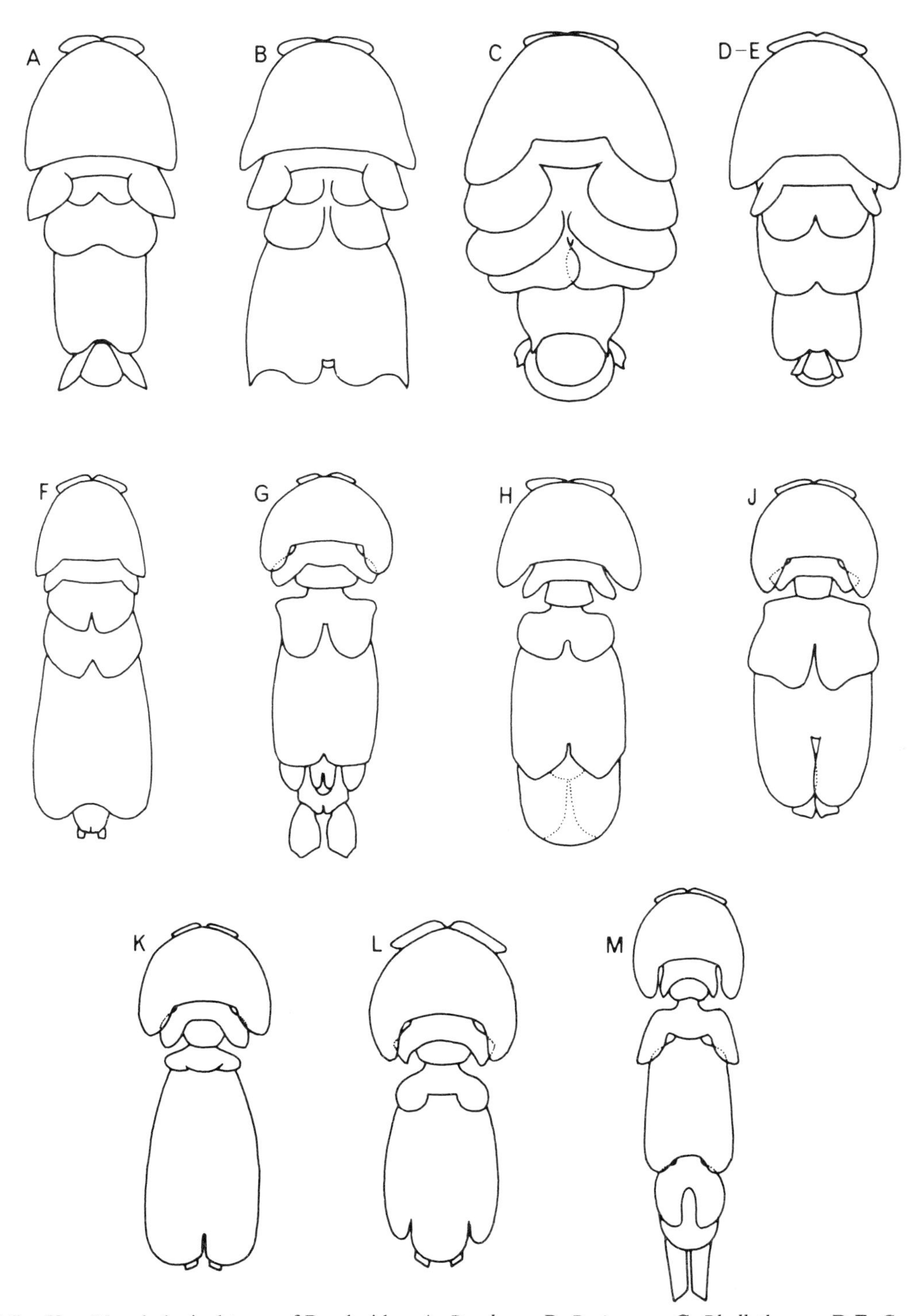

Text fig. 52. Morphological types of Pandaridae. A. *Pandarus*; B. *Perissopus*; C. *Phyllothyreus*; D,E. *Gangliopus* and *Pannosus*; F. *Pseudopandarus*; G. *Dinemoura*; H. *Demoleus*; J. *Echthrogaleus*; K. *Nessipus*; L. *Paranessipus*; M. *Pagina*.

Table 12. Segmentation of legs in Pandaridae.

	First leg		Second leg		Third leg		Fourth leg	
	En.	Ex.	En.	Ex.	En.	Ex.	En.	Ex.
Pandarus	2	2	2	2	2	2	1	1(2?)
Perissopus	2	2	2	2	1	1	1	1
Phyllothyreus	2	2	2	2	2	2	1	1
Gangliopus	2	2	2	2	2	2	1	1
Pannosus	2	2	2	2	2	2	1	1
Pseudopandarus	2	2	2	2	2	2	1	1
Dinemoura	2	2	(2)3	3	3	(2)3	1	1
Demoleus	2	2	2	2	2	2	2	2
Echthrogaleus	2	2	2(3)	3	2	3	1	1
Nessipus	2	2	2	2	2	2	1	1
Paranessipus	2	2	3	3	3	3	2	3
Pagina	2	2	3	3	3	3	3	3

the ventral surface of the genital complex, a tagma with well developed posterolateral lobes. *Nessipus* Heller, 1865 (Text fig. 52K) is unique among the pandarids in that the dorsal plates on the third free segment are either rudimentary or absent. In the monotypic *Paranessipus* Shiino, 1955 (Text fig. 52L) they are rather better developed. The posterior extremity is trilobed, the median lobe covering the abdomen from the dorsal side. *Pagina* (Text fig. 52M) has, on the last free segment, aliform dorsolateral plates which barely overlap the genital complex. The abdomen in this genus is terminal in position, two-segmented, both segments with dorsal plates. The plate of the first segment is deeply notched and overlaps the rounded median plate of the second.

The cephalothoracic appendages of Pandaridae are of a caligid type. It might be worth noting, however, that the two groups of genera within the family are distinguished from each other by the condition of the tip of the second maxilla, equipped with clavus in the *Pandarus*-group but with crista in the *Dinemoura*-group. In all genera the first four pairs of legs are biramous. The interpodal bars of the third legs are more or less broad, but the rami of those legs are not reduced. Reduction is shown, however, in the rami of the fourth legs in some genera. In contrast with Caligidae, the pandarids are not uniform in their leg structure. The numbers of segments in the rami of legs in various genera are shown in table 12.

As the table shows, the differences in leg segmentation are not only from genus to genus; in *Dinemoura* and *Echthrogaleus* they exist also within the generic limits.

Most pandarid females are equipped with adhesion pads, situated both on the ventral surface of the cephalothoracic region and on the swimming legs. Their distribution on the former is presented in table 13 (modified from Cressey, 1967a), in which presence is marked by a plus and absence by a minus sign.

Table 13. Distribution of adhesion pads on cephalothoraces of Pandaridae.

	A1	A2	Base of maxilliped	Between maxillipeds	Distal corner of dorsal shield
Pandarus	+	+	+	−	+
Pseudopandarus	+	+	+	−	+
Perissopus	+	+	+	−	−
Paranessipus	+	+	+	−	−
Pannosus	+	+	+	−	−
Nessipus	hooked	+	+	−	−
Dinemoura producta and *D. latifolia*	double	+	+	−	−
D. ferox and *D. discrepans*	−	−	−	−	−
Demoleus	+	+	+	+	−
Pagina	+	−	−	−	−
Echthrogaleus	+	+	hooked	−	−
Gangliopus	−	−	+	−	−
Phyllothyreus	−	−	−	−	−

All pandarids are equipped with uropods. The relative size and the structure of these appendages are, however, more diverse in this family than in any siphonostomes previously discussed here. This diversity can be appreciated by comparing the uropods of *Pandarus* (Figs. 771, 772) and *Phyllothyreus* (Fig. 894) on the one hand with those of *Dinemoura* (Fig. 803) and *Demoleus* (Fig. 851) on the other.

As in most caligiforms, the egg sacs of Pandaridae are long tubes with flat, uniseriate eggs. In most species the sacs are straight and long, but in *Demoleus heptapus* and *Dinemoura discrepans* they are folded into three strands each. In *Paranessipus incisus* they are folded several times and hidden beneath the genital segment. They are similarly concealed in *Demoleus latus*, though the folding there is more spiral.

The males of this family are less well known than the females. None has been found in *Pannosus* and *Paranessipus*. The pandarid males differ from the females in the absence of dorsal plates. All of them, however, have dorsolateral plates on the first free thoracic segment. Males of all genera are fairly similar to one another. They have small genital complexes, usually with no outgrowths or processes other than the vestigial fifth and sixth legs (there are some exceptions to this, e.g. the genital complex of the male *Dinemoura discrepans*). The biramous legs often have segmentation different from that of the female. Their more primitive condition is evident in better preserved setation.

It should be kept in mind that the immature females of Pandaridae tend to resemble the males much more closely than they do mature females, if only because of the imperfect development of their dorsal plates (compare Figs. 793 and 794).

History and systematics

The history of pandarid taxonomy, which started with a description of the first species of *Pandarus* by Leach (1816, 1819), continued for a good part of its course along that of Caligidae. The second genus, now belonging to Pandaridae, was described by Latreille (1829). This genus, *Dinemoura*, together with *Pandarus*, was placed by Burmeister (1835) in his family Caligina, both differing from other genera within it by the biramous nature of their fourth legs. (Burmeister also added that they had small eyes, unlike the species of other genera, believed to be blind.)

Edwards (1840) was the first to establish a definite taxon for the pandarids. His tribe "Pandariens" of the family "Peltocephales" was a sizeable unit, comprising *Pandarus*, *Dinemoura*, *Phyllophora* (= *Phyllothyreus*), *Cecrops* and *Laemargus* (= *Orthagoriscicola*). The two last-named genera, now members of Cecropidae, were separated from the others in Edwards' key by having egg sacs concealed under the dorsal plates of the posterior extremity. This distinction, valid at the time, could not be used now as a constant in a diagnosis, since we know at least three pandarid species with eggs also either partially or completely concealed, while there are two cecropid genera with long, straight egg sacs.

Dana (1852) treated pandarids as a subfamily of Caligidae and included in it *Nogagus*, *Specilligus*, *Pandarus*, *Dinemoura*, *Lepidopus* (= *Perissopus*), *Trebius*, *Phyllothyreus* and *Euryphora* (sic!). He diagnosed this subfamily as follows: "Mouth tube slender, pointed. Maxillae pressed close to mouth tube, small, lamellate. Ovisacs external, straight. Body sometimes broader posteriorly." *Cecrops* and *Orthagoriscicola* were split off for the first time into a rudimentary subfamily Cecropinae.

In a book published two years earlier, Baird (1850) established Pandaridae as a family group taxon. His family definition was very brief: "Head in form of a small buckler, provided with frontal plates. A series of one or more pairs of lamellar appendages, like the elytra of many insects, extended along the dorsal surface of the thorax. Oviferous tubes straight, external." Interested only in the British representatives of the family, Baird included in it only *Pandarus* and *Dinemoura*. It is interesting to note that in his synonymy of the family, Baird quoted Pandarinae Dana (the date was not given, but the reference, identical with that quoted for Cecropinae, dated back to 1849). It appears that Baird had available the reference which is now customarily quoted as Dana (1852). The volume in which that paper was published is stated on the title page to have covered a period from May 1848 to May 1852. Though Dana's diagnosis was not invoked, his taxon must have exerted some influence on Baird.

Baird's decisions were disregarded by many subsequent workers. For example, Steenstrup and Lütken (1861) reverted to the Edwardsian concept of systematics, placing the pandarids in a separate tribe in the family Peltocephala; they also added to the group new genera *Perissopus* and *Echthrogaleus*.

Heller (1865) included pandarids in his family Caligina, placing them in a subgroup Pandarina, distinguished by: "rostrum [mouth tube] long, slender, palps [first maxillae] articulate, foliaceous." Like Dana, he saw in the mouth parts the most characteristic features of the pandarids. His list of Pandarina consisted of: *Nessipus*, *Echthrogaleus*, *Dinemoura*, *Demoleus* (new), *Perissopus*, *Phyllothyreus*, *Gangliopus*, *Pandarus*, *Cecropsina* (= *Luetkenia*), *Cecrops* and *Orthagoriscicola*. More than a decade after Dana and Baird, Cecropidae and Pandaridae were thrown together again.

Gerstaecker (1866–1879) in his major treatise on Crustacea, retained the pandarids as a subdivision of Caligidae, referring to them as Nogagina, an appellation most unfortunate in view of the spurious nature of the genus *Nogagus* (see below). The group was a mixture of pandarid, cecropid and caligid genera, some pandarids having been left outside its limits.

A prolific and controversial worker of that time, Hesse (1883) added to the list yet another pandarid genus, *Lepimacrus*. His illustrations to the description of that genus show a highly stylized likeness to *Echthrogaleus*, of which it is probably a synonym. No positive decision can be reached on this point, however, because the type material is not extant and neither the description nor the drawings are useful for comparisons.

In his account of the Mediterranean fauna, Carus (1885) kept pandarids as a tribe Pandarini of Caligidae, including in it also the cecropid genera (*Orthagoriscicola*, *Cecrops* and *Luetkenia*). It is worth noting that he used the segmentation of the body as a taxonomic character. His definition of Pandarini was as follows: "Mouth tube slender, long; two or three posterior segments free, females with dorsal elytra; legs with rami armed, foliaceous; setae short, simple."

The present taxonomic status of the group was foreshadowed by the work of Wilson (1905, 1907b), clearly influenced by his American predecessor, Dana. The pandarids remained a subfamily of Caligidae, separate from the cecropids and defined as follows: "First thoracic segment only fused with the head, the others free; one or more of them with paired dorsal plates; all four pairs of legs biramose. Frontal plates distinct; egg cases visible their entire length." (The latter fact not strictly correct.) In his subsequent work, Wilson (1932) upgraded the pandarids to the familial status, without discussing his decision or giving a diagnosis of the family. Concerned only with the representatives of the fauna of Woods Hole, Massachusetts, he included in it *Perissopus*, *Echthrogaleus*, *Dinemoura* (which he designated as *Dinematura* Burmeister, 1833), *Pandarus*, *Nessipus* and *Parapandarus* (latter relegated by Cressey (1967a) to synonymy with *Phyllothyreus*).

It is rather strange to note that Scott and Scott (1913), who claimed that their work was based on Baird's earlier book in the same series of Ray Society monographs, left pandarids within Caligidae, omitting even to reserve for them any form of intrafamilial recognition. The same is true of the classification scheme proposed in a later monograph by Gurney (1933). Markevich (1956) left the pandarids in subfamilial status.

Bere (1936) described as a pandarid a new copepod which she named *Entephorus laminipes*. It was transferred by Yamaguti (1963) to Cecropidae. Yamaguti uses Pandaridae as a new name, emended from Wright's (1877) Pandarina. In fact, the term Pandarina had been used earlier by several workers (e.g. Heller, 1865) and Pandaridae had been a family group name applied by Baird (1850). Yamaguti's Pandaridae contain 14 genera, some of them now rejected (*Pholidopus*, *Lepimacrus* and *Achtheinus*) and one which, at best, can be considered a genus inquirendum (*Prosaetes*).

In his study of Pandaridae, this author largely accepts the views of Cressey (1967a). Cressey's Pandaridae consist of *Perissopus*, *Pandarus*, *Pannosus* (new), *Phyllothyreus*, *Gangliopus*, *Pseudopandarus*, *Dinemoura*, *Pagina*, *Echthrogaleus*, *Demoleus*, *Paranessipus* and *Nessipus*. The family diagnosis runs as follows:

"Female. Body caligiform, usually with dorsal plates. First thoracic segment fused with cephalon. Thoracic segments 2–4 free. Genital segment conspicuous. Abdomen of 1 or 2 segments with or without dorsal plates. Oral area with or without adhesion pads. First antenna 2-segmented. Mandible in form of stylet with 10–12 apical teeth. Mandible inserted within mouth tube. Maxilliped with terminal claw. Legs 1–4 biramose. Leg 5 reduced. Egg strings consisting of long strings of eggs, arranged in linear series.

"Male. Body caligiform, without dorsal plates. First thoracic segment fused with cephalon. Thoracic segments 3–4 free. Oral area generally as in female. Legs 5 and 6 present. Abdomen 1- or 2-segmented. Caudal rami large."

This diagnosis is broadly acceptable to the present author. One must examine it, however, as to its usefulness in separating Pandaridae from Cecropidae, with which the pandarid genera have been confused for so long. To do this, one must look at and compare the chief characteristics of both families. One of the most distinguishing features of Pandaridae is the presence of three free segments between the cephalothorax and the genital complex. Yamaguti (1963) contrasts it with "second and third segments more or less fused together " in Cecropidae. This, however, is not true for the cecropid genus *Orthagoriscicola* (Fig. 936), in which the second and third leg-bearing segments are clearly separate. On the other hand, among Pandaridae some *Echthrogaleus* have them fused together. As mentioned earlier, neither can the structure of the egg sacs be used for this purpose. Wilson (1905) distinguished between these two groups by the condition of their frontal plates, allegedly freely articulating with the dorsal shield in the pandarids, while fused completely in the cecropids. This characteristic is not an invariably good discriminant. In most cecropids the line of fusion is clearly distinguishable and in *Philorthagoriscus* the plates appear to be capable of articulation just as much as do those of their pandarid relatives. The presence of the adhesion pads cannot be taken as a constant pandarid characteristic, either. Some pandarids have no such structures.

The difficulty in drawing a boundary between these two families is perhaps best attested to by the fact that *Entepherus laminipes*, considered a pandarid by Bere (1936), was placed in Cecropidae by Yamaguti (1963), a decision with which this author concurs. In fact, most features are shared by both families to some extent. Casting around for a clear-cut distinction, one comes across differences in host affiliations, a characteristic which would rightly be shunned by most parasitologists. Pandaridae are parasites of elasmobranchs, whereas Cecropidae live on teleosts, predominantly on *Mola mola* (three out of five genera). Even here, however, the dividing line is somewhat blurred. Yamaguti and Yamasu (1960) reported a pandarid, *Gangliopus tetrapturi* (=*G. pyriformis* Gerstaecker, 1854) on a teleost, *Tetrapturus mitsukurii* and Dollfus (1946) found *Pandarus bicolor* on *Mola*. On the other hand, we have a record (Richiardi, 1880) of a cecropid, *Luetkenia integra* (species inquirenda) from two Mediterranean sharks.

One can add to the list the absence of marginal serrations on the cephalothoracic shield and dorsal plates of Pandaridae and their presence in Cecropidae. By this standard, *Entepherus* belongs to the latter, but *Luetkenia* would have to be excluded. Also, the posterior margin of the cephalothoracic shield in *Pandarus* (Fig. 769) often shows some serrations.

Perhaps the only feature which can be used without reservations to separate these two groups of copepods is the condition of the female corpus maxillipedis. In Cecropidae it is slender, though its myxal area is sometimes provided with well developed processes (Figs. 912, 945, 968, 979). In female pandarids, on the other hand, it is much modified and squat, its subchela, when closed, being at more or less right angles to the long axis of the corpus (Figs. 782, 813, 830, 862, 883). The fact remains, however, that the two families are distinguishable only with difficulty, a testimonial, perhaps, to a close relationship.

In spite of their distinctive habitus and marked characteristic features, Pandaridae have a long history of confusing descriptions, many of their species having accumulated long synonymies. As an example, we can quote *Perissopus oblongatus* (Wilson, 1908), originally described as *Achtheinus*. No fewer than nine other species of *Achtheinus*, described in the United States, South America, Japan, China and one by an Austrian worker, proved to be synonymous with this species (Cressey, 1967a). At least part of the responsibility for this state of affairs rests with a broad range of intraspecific variability, made even broader by pronounced morphological changes associated with the age differences. As happens only too often with the parasites of elasmobranchs, the material for a detailed study is woefully inadequate and conclusions based on it fail to survive a closer scrutiny. One of the most vexing examples of inadequate knowledge, compounded by confusing descriptions of scattered and scanty material, is that of the genus *Nogaus* Leach, 1819, based on male specimens. Several species have been subsequently described, all based on free-swimming males. *Nogaus*, without any doubt, is a composite genus. Some of its species have been eventually connected with females of various pandarid genera. Some, unfortunately, still remain unidentifiable and haunt the pandarid taxonomy with the vestiges of *Nogaus*. Even more unfortunately, two of them have been recorded in the British fauna. They will be dealt with below as Pandaridae incerte sedis.

Key to British genera of Pandaridae

1. Second free segment with paired dorsal plates (Figs. 769, 878) 2
   Second free segment without dorsal plates (Figs. 801, 822, 850) 3

2. Plates of second free segment equal to (Fig. 769) or shorter than (Text fig. 52A) those of the first **Pandarus**
Plates of second free segment longer than those of the first, which overlap their major part (Fig. 878) . . . . . **Phyllothyreus**

3. Abdomen two-segmented (Fig. 803) . . . . . **Dinemoura**
Abdomen one-segmented 4

4. Rami of fourth legs one-segmented, foliaceous (Fig. 839) . . . . . **Echthrogaleus**
Rami of fourth legs two-segmented (Figs. 864, 867) . . . . . **Demoleus**

The males of this family are not known well enough to make a similar key possible.

Genus Pandarus Leach, 1816

Female: Pandaridae. Cephalothorax caligiform, without sutures on dorsal shield; frontal plates well delimited. Dorsal plates of second leg-bearing segment either equal to or longer than those of third segment. Plates of fourth leg-bearing segment fused at base, overlapping anterior part of genital complex. Latter with concave posterior margin. Abdomen one-segmented, posteriorly rounded, with subcircular dorsal plate. First antenna as in Caligidae, with large adhesion pad at base; second antenna four-segmented, with terminal claw and large adhesion pad at base. Mouth tube long, with slender tip. Oral appendages similar to those of Caligidae. Claw of maxilliped spatulate. Adhesion pads at bases of maxillipeds, between first legs and in posterolateral corners of dorsal shield. First to fourth legs biramous, rami two-segmented in first to third and one-segmented in fourth legs; fifth leg vestigial. Uropods prominent. Egg sacs long, straight.

Male: No dorsal plates. Abdomen two-segmented. Rami of first four pairs of legs two-segmented. Third endopod unmodified. Fifth and sixth legs vestigial.

Type species: *Pandarus bicolor* Leach, 1816.

Comments: Copepods belonging to this genus live on sharks, occurring on the outer surfaces, predominantly fins, of their hosts, sometimes in great numbers. They are generally widespread, though some recently discovered species have been recorded so far only from one or a few localities. Of the 30 species of *Pandarus* described so far, only 11 are currently recognized as valid, others having been relegated to synonymy, mainly by Cressey (1967a). Of these, three have been recorded from the Atlantic, Pacific and Indian Oceans (*P. cranchii* Leach, 1819, *P. satyrus* Dana, 1852 and *P. smithii* Rathbun, 1886), two from the Atlantic and Pacific (*P. bicolor* Leach, 1816 and *P. zygaenae* Brady, 1883), three from the Pacific only (*P. carcharini* Ho, 1960, *P. katoi* Cressey, 1967 and *P. oblongus* Shiino and Izawa, 1966), two from the Atlantic only (*P. floridanus* Cressey, 1967 and *P. sinuatus* Say, 1817) and one from the Indian Ocean (*P. niger* Kirtisinghe, 1950). Four descriptions (all by Hesse, 1883) are in the category of species inquirendae (*P. carcharii-glauci*, *P. musteli-laevis*, *P. spinacii-acanthii* and *P. unicolor*). British fauna contains only the type species.

Pandarus bicolor Leach 1816

(Figs. 769–800)

Syn.: *Pandarus boscii* Leach, 1816
Caligus bicolor; of Lamarck (1818)
Pandarus fissifrons Edwards, 1840
Pandarus lividus Frey and Leuckart, 1847
Nogagus angustulus Gerstaecker, 1854 (male only)
Nogagus latus T. Scott, 1907 (male only)

Female (Fig. 769): Dorsal shield with rounded anterior and divergent lateral margins, latter with narrow marginal membranes; posterior margin with several irregular dentiform processes; except for marginal areas and part of mid-dorsal line, shield darkly pigmented, extent of pigmented area and density of its pigmentation greatly variable (in some specimens almost no pigmentation present). Frontal plates wide, well delimited from shield. Tips of first antennae not reaching lateral limits of dorsal shield. Second and third leg-bearing segments with pair of dorsal plates, about equally long; plates of second segment lateral to those of third; plates of third segment fused at base, both with irregular spots of dark pigmentation in centre (in some specimens connected by dark pigment stripes); both pairs overlap base of fourth leg-bearing segment. Latter covered by pair of dorsal plates fused for greater part of their lengths, extending over proximal part of genital complex. Lateral margins of complex converging posteriorly, deep indentation in posterior margin. Abdomen (Fig. 771) rounded posterolaterally, with small protuberances in anterolateral corners, dorsally covered by subcircular plate. Uropods projecting posterolaterally between genital complex and dorsal plate of abdomen. Transverse row of four adhesion pads in association with base of first and second antennae (Fig. 770); pair of smaller pads at bases of maxillipeds; transverse row of three pads at level of posterior margin of shield, one oval in each corner and one subcircular between bases of first legs, behind first interpodal bar. Total length 8–10 mm.

First antenna (Fig. 773) two-segmented; basal segment more than twice as long as terminal, its anterior margin with small swelling in proximal half and 26 plumose setae in distal half and on distal margin; terminal segment cylindrical, with rounded tip and one slender seta on posterior margin. Apical armature (Fig. 774) of 13 setae; five long, pointed setae on posterior side of apex, three long, blunt and five short digitiform setae on anterior side. Second antenna (Fig. 776) small, apparently four-segmented, partially obscured from view by adhesion pad; basal segment cylindrical, unarmed; second segment with bulbous, sclerotized protuberance (providing attachment for intrinsic muscles); third segment much narrower, subcylindrical (Fig. 775), with short, grooved process at midlength and longer, more slender one at distal end; fourth segment unciform, short subchela, blunt and with longitudinal grooves in distal half. Mouth tube (Fig. 777) long, distally slender; labrum with lateral indentations near midlength and lateral strips of membrane, tip (Fig. 778) with central subtriangular lobe flanked by shorter, narrower lappets; labium longer, broadly pointed, with circle of denticles on either side, extending over part of outer surface. Mandible (Figs. 778, 779) up to 0.5 mm long, slender, apparently two-segmented; basal part about $\frac{1}{4}$ length of total; dentiform margin with series of six blunt teeth, sixth usually much smaller than others, often followed by some irregular serrations. First maxilla (Fig. 781) pressed against base of mouth tube, sometimes tips of both maxillae meeting above labrum; basal half broader, abruptly narrowing at about midlength, with three short setae arising from anterior margin; terminal half tapering, apically with robust spiniform process and small, softer process on its posterior side. Second maxilla (Fig. 780) with muscular, unarmed lacertus and laterally compressed, slender brachium; canna almost as long as calamus, both with at least six longitudinal rows of setules; clavus short, plumose. Maxilliped (Figs. 782–784) with heavily sclerotized corpus at right angles to long axis of body; medial surface with two shelf-like areas; subchela flat, bifid (Fig. 785), one tip rounded, other longer and pointed; semispherical swelling on upper surface near base (Fig. 782). Powerful muscle running along bulge between base and adhesion pad.

First four legs with sympods obscurely segmented; legs 2–4 with medial margins of sympods closer together; adhesion pads in medial part of ventral surfaces (Figs. 786, 787, 790), those of legs 2–4 forming three pairs, progressively narrower posteriorly (Fig. 770). First to third leg also with patches of denticles. Armature formula as follows.

	Endopod		Exopod	
	1	2	1	2
Leg 1	0-0	3	0-1	3,I,3
Leg 2	0-0	5	0-1	VI,4
Leg 3	0-0	2	0-1	III,3,I
Leg 4	1	—	6	—

Proximal segment of first endopod with adhesion pad on ventral surface; pinnate setae of second segment subterminal (Fig. 786), tip and lateral margin with fringe of minute spinules. Second segment of first exopod with patch of spinules on ventral surface and part of lateral margin; all setae pinnate. Short basal segment of second endopod (Fig. 787) with two patches of spinules on ventral surface; long terminal segment with broad strip of spinules along lateral margin and around tip; setae finely denticulated. Basal segment of second exopod with large patch of spinules on ventral surface; similar spinules on lateral margin and near tip of terminal segment; short, sharp spines on medial margin, followed by papilliform seta and three longer setae, all with prickly surfaces. Third leg similar to second, details of apical armature in Figs. 788, 789. Fourth endopod with fringe of spinules on lateral margin and apex. Segmentation of fourth exopod obscure. (Cressey (1967a) described it as one-segmented. Author's own specimens showed partial subdivision (Fig. 790); even without it, however, abrupt difference in width and presence of seta at that point, corresponding to distolateral corner of basal segment in other legs, suggest that originally at least that ramus was two-segmented.) Fifth leg (Fig. 791) vestigial, conical, with short spine on apex and three pinnate setules, two subapical and one basal. Uropod (Fig. 771) subtriangular, blade-like, on anterior margin flat shelf at right angles to plane of blade (Fig. 772); two spinules on blade, two on dorsal margin of shelf and one apical.

Male (Figs. 792, 794): Cephalothorax differing from that of female by absence of pigmentation, presence of sutures separating lateral zones; farther backward protrusion of tips of those zones and broader marginal membranes; tips of first antennae almost reaching lateral limits of dorsal shield. Second leg-bearing segment with dorsolateral aliform plates; third leg-bearing segment unarmed, short, narrower than second; fourth also unarmed, longer and narrower than third. Genital complex longer than broad, about equalling combined length of three free segments; lateral margins convex, with slight protuberances at bases of fifth legs; posterior margin transversely truncated. Abdomen one-segmented, short, somewhat broader posteriorly. Total length about 6 mm.

Maxilliped similar to that of female but subchela with sharp tip (*fide* Scott and Scott, 1913). First four pairs of legs biramous, all rami two-segmented. Armature formula as below.

	Endopod		Exopod	
	1	2	1	2
Leg 1	0-0	3	0-I	3,IV
Leg 2	1-0	8	1-I	?5,IV
Leg 3	1-0	5	1-I	?5,IV
Leg 4	1-0	4	1-I	?5,IV

First leg (Fig. 795) resembling that of female, but with longer pinnate setae (pinnules omitted) and with fringes of fine setae on lateral margin of distal endopod and medial margin of proximal exopod segments. Other legs similar to one another (Fig. 796), differing only in number of pinnate setae on distal segments of endopods (Figs. 797, 798). Fifth leg (Fig. 799) of four setules on rounded protuberance; sixth leg of two similar setules. Uropod (Fig. 800) with four distal pinnate setae and fringe on medial margin.

Comments: The male of this species is not well known. Its descriptions by various authors contain many discrepancies, particularly in the armature formula of the swimming legs. For example, this author found that there are five pinnate setae on the distal segments of exopods two, three and four. Scott and Scott (1913) described six setae in these positions. Hewitt (1967) found four setae on the distal segment of the first endopod (three in the author's specimens) and only four on the distal segment of the second and third exopods (five in this author's material). Most surprising, however, is the fact that Hewitt's male had one-segmented rami on the fourth leg, with one seta fewer on the exopod. It is not possible to account for these discrepancies.

It has been mentioned earlier that *Pandarus* undergoes considerable morphological changes in the course of its ontogeny. This is best illustrated by the comparison of a juvenile female (Fig. 793) with an adult one (Fig. 769). The difference is particularly noticeable in the clearer segmentation of the younger specimen (sutures on the dorsal shield and between the first and second free segments), in the size of the genital complex and in the degree of development of the dorsal plates, particularly the medial abdominal plate. There exists also extensive variation among adult females as regards the characteristic pigmentation. In some specimens it might be almost absent.

B. bicolor is a widespread parasite of sharks, settling predominantly on the skin and fins of its hosts. The host list includes *Carcharinus* sp., *Carcharinus falciformis*, *Eulamia* sp., *Galeorhinus australis*, *G. galeus*, *Galeorhinus* sp., *Isurus mako*, *I. oxyrinchus*, *Mustelus canis*, *M. mustelus*, *Notorhynchus cepedianus*, *Prionace glauca*, *Scyliorhinus stellaris*, *Squalus acanthias* and *Squalus* sp. Records come from many localities in the Atlantic and Pacific Oceans, but apparently no finds have been reported from the Indian Ocean. Dollfus (1946) mentioned the occurrence of *P. bicolor* on *Mola mola*, a teleost fish. This single instance of association with a non-elasmobranch host must be seen as accidental.

In British waters, *P. bicolor* is quite common on the local sharks.

Genus Dinemoura Latreille, 1829

Female: Pandaridae. Cephalothorax caligiform, with sutures separating lateral zones; frontal plates distinct; posterior tips of lateral zones projecting to about second free segment. First free segment with dorsolateral aliform plates; second unarmed; third with paired dorsal plates overlapping base of genital complex. Latter large, with posterolateral lobes and with or without rudimentary dorsal plates. Abdomen two-segmented; first segment very short, with paired dorsal plates fused at base and with or without posteroventral plates; second segment larger, with single dorsal plate. Uropods prominent. Adhesion pads present or absent. First antenna two-segmented; second three-segmented, distal segment unciform claw. Oral appendages and maxilliped siphonostome. First to fourth legs biramous, rami of first leg two-segmented, those of second and third three-segmented, those of fourth one-segmented, laminiform. Fifth and sixth legs present, vestigial.

Male: Cephalothorax and first two free segments as in female; third free segment with dorsal plates rudimentary or absent. Genital complex small, oval, with or without posterolateral processes. Abdomen two-segmented, devoid of plates. Uropods prominent. Rami of first to fourth legs two-segmented; terminal segment of third endopod modified. Fifth and sixth legs vestigial.

Type species: *Dinemoura producta* (Müller, 1785).

Comments: There has been much confusion regarding the correct spelling of the generic name of *Dinemoura*, the form *Dinematura*, introduced by Burmeister (1835), being used with Latreille's original spelling in about the same frequency. The confusion was settled by Cressey (1967a) in favour of the original spelling. In his treatment of this genus, Cressey, however, did not discuss another confusing point, that of the structure of the posterior extremity of *Dinemoura*. Some authors (Wilson, 1932; Yamaguti, 1963) considered that in this genus the genital *segment* is followed by a "sixth segment" and a one-segmented abdomen. Cressey (1967a), on the other hand, describes the abdomen as two-segmented.

The view held by the earlier workers cannot be reconciled with the fact that in siphonostome, as well as in cyclopoid, parasitic copepods the genital complex consists of the fifth and sixth leg-bearing segments. When it remains a *segment*, not a complex, it is the sixth leg-bearing segment that carries the genital orifices and is, therefore, the genital segment. Consequently, it cannot be followed by a "sixth segment". Cressey identified the sixth leg near the border between the genital complex and the following, the first abdominal, segment. We have, then, a two-segmented abdomen, both its segments with an elaborate armature of plates.

The species of the genus *Dinemoura* are parasitic on the outer surfaces of their shark hosts. It is interesting to note that they appear to compete with members of the genus *Pandarus* and appear to be the dominant party in this antagonism. Cressey (1968) reported that *Pandarus*, invariably present on the skin and fins of its hosts, was found in the buccal cavity when *Dinemoura* was present in its customary habitats.

The genus *Dinemoura* contains four valid species, two of them with long synonymies resulting from many spurious descriptions. The only species not reduced to synonymy with one of the four is *D. hamiltoni* Thomson, 1890 (male), the description of which defies comparisons. Hewitt (1967) made no mention of it, while Cressey (1967a) suggested that it belonged to *Echthrogaleus*. It must remain a species inquirenda.

Dinemoura latifolia (Steenstrup and Lütken, 1861) and *D. ferox* (Krøyer, 1838) are both widespread parasites of sharks, but *D. discrepans* Cressey, 1967, is known so far only from its original record from the Indian Ocean, where it was found on two species of *Alopias*. The type species, *D. producta*, is the only species occurring in British waters.

Dinemoura producta (Müller, 1785)

(Figs. 801–821)

Syn.: *Caligus productus* Müller, 1785
Dinemoura producta; of Latreille (1829)
Binoculus productus von Nordmann, 1832
Dinematura producta; of Burmeister (1835)
Dinematura gracilis Burmeister, 1835
Pandarus lamnae Johnston, 1835

Nogagus gracilis Edwards, 1840
Dinemoura lamnae Baird, 1850
Nogagus productus Gerstaecker, 1853
?*Dinematura elongata* van Beneden, 1857
Dinematura lamnae; of Krøyer (1863)
Dinematura affinis Thomsen, 1949

Female (Fig. 801): Anterior and lateral margins of dorsal shield suborbicular in outline, posterior tips of lateral zones extending sometimes near boundary between third and fourth leg-bearing segments. Anterior central area often (though not always) darkly pigmented, with eye spots behind pigmented region. Lateral zones rimmed with marginal membranes. Frontal plates narrow, tips of first antennae not reaching lateral limits of shield. Two adhesion pads between bases of first and second antennae on both sides, one at bases of maxillipeds and one double pad between bases of first and second legs. First free segment with posterolateral aliform plates, their lateral margins with long, narrow adhesion pads on ventral surfaces; second free segment unarmed, in some specimens with central indentation in posterior margin. Third free (fourth leg-bearing) segment about as broad as central zone of cephalothorax, with narrow neck anteriorly and with pair of dorsal plates overlapping base of genital complex. Latter longer than cephalothorax and three free segments together, narrow, with parallel lateral margins and prominent posterolateral lobes; rudimentary dorsal plates medial to lobes (Fig. 802, rp). First abdominal segment with very prominent ventrolateral plates (Figs. 802, 803, vlp) and paired dorsal plates fused at base (Fig. 802, dp1); second abdominal segment often not clearly delimited from first, with single dorsal plate (Fig. 802, dp2) largely overlapped by plates of first segment. Total length between 13 and 18 mm, egg sac length up to 30 mm.

First antenna (Fig. 804) two-segmented; basal segment twice as long as terminal, broad, with pointed outgrowth on anterior margin near base; distal margin and distal half of anterior margin beset with 26 setae; large and plumose (marked by stippling in Fig. 804) and small, unarmed; longest plumose seta reaching tip of terminal segment; latter cylindrical, with one seta at midlength of posterior margin; apical armature (Fig. 805) of 13 setae, five in two groups on posterior side and eight on anterior side. Second antenna (Fig. 806) apparently three-segmented; basal segment short, cylindrical, with flat conical boss on ventral side; similar boss on second, largely thin-walled segment; terminal segment (Fig. 807) strongly hooked, with sharp tip, one short seta near base and one longer near midlength. Mouth tube very long and slender; labrum shorter than labium; tip of labrum (Fig. 808) with large, bifid flap, fringed by fine setules and with short apical outgrowth on both lobes of tip; row of transverse setules across base of flap, proximal part of flap with scattered fine setules; sharp-pointed process on either side of flap and strips of membrane along lateral margins of lip. Tip of labium foliaceous, broad, with serrated margins, central indentation anteriorly and two digitiform outgrowths in midventral line. Mandible long and slender, with 11 teeth in dentiferous margin, gradually diminishing towards base (Fig. 809). First maxilla (Fig. 810) with stout, unsegmented sympod; exopod short, bulbous, with three apical processes, two robust and with fine tips, one slender, tapering; endopod two-segmented, segments cylindrical, distal half as long as proximal, thinner, with minute apical papilla. Second maxilla (Fig. 811) brachiform, lacertus muscular, unarmed, brachium more slender; calamus somewhat peduncular, tapering, with short longitudinal or oblique rows of setules on entire surface. Crista small, with fairly long setae; canna pedunculate, about half as long as calamus, its anterior and dorsal (but not ventral) surfaces with minute, strong denticles. Maxilliped with apparently bipartite corpus; lateral wall heavily sclerotized, bulging (Fig. 812), anteromedially at base with area of thin cuticle (tc) covering proximal half of corpus and extending to unciform adhesion pad (ap); distal half of medial wall sclerotized more heavily than proximal; subchela unciform, short, with seta near base; tip closing against prominent flange of corpus (Fig. 813).

First four pairs of legs biramous; armature formula below.

	Endopod			Exopod		
	1	2	3	1	2	3
Leg 1	0-0	3	—	0-I	3,IV	—
Leg 2	1-0	2-0	6	1-I	1-I	5,III
Leg 3	0-0	2-0	4	1-I	1-I	5,III
Leg 4	modified			modified		

Sympods of first three pairs with adhesion pads on ventral surfaces, that of first leg (Fig. 814) with two pads on medial half, that of second (Fig. 815) and third (Fig. 816) with one pad. Second interpodal bar broader than first, third forming apron with expanded sympods of legs. Adhesion pad also on first segment of first endopod; first exopod with expanded basal segment; second endopod with oval flange in distomedial corner of first segment. Bases of rami in third leg (Fig. 816) overlapping, endopod apparently incapable of lateral movement. Three setae on lateral margin of third exopod's distal segment in Fig. 817. Sympod of fourth leg (Fig. 818) with oval medial lobe larger than itself and smaller lateral lobe (folded upon itself in drawing); rami not distinctly delimited from sympod, unsegmented; exopod slightly shorter than endopod, both with fringe of fine marginal setules and a few minute spinules. Fifth leg (Fig. 820) vestigial, small tubercle anterolateral to base of abdomen, with short apical spine and two subapical plumose setules. Sixth leg small lobe close to base of abdomen, usually obscured by its ventrolateral plates. Uropod (Fig. 819) prominent, with convex medial, transversely truncated posterior and straight lateral margins; four short pinnate or semipinnate setae in distolateral corner.

Male (Fig. 821): Cephalothorax, first and second free segments similar to those of female; third free segment larger than second, with small dorsolateral processes. Genital complex length about equal to that of three free segments combined; lateral margins straight or slightly convex, anterolateral and posterolateral corners rounded. Abdomen two-segmented, first segment smaller than second. Total length 10–14 mm.

Second antenna with small adhesion pad on second segment; claw short, blunt, with ridged area on inner margin proximal to base of prominent seta. First maxilla with only two setae on exopod.

Legs similar to those of female, but basal segment of third endopod with pinnate seta on inner margin; rami of fourth leg with armature better developed, endopod with six pinnate setae on medial margin, exopod with six pinnate setae on medial and four short spines on lateral margin.

Comments: The author was unable to secure any male specimens of this species for examination, the description above being given *fide* Hewitt (1967). Hewitt states explicitly that the third swimming leg of the male is similar to that of the female. If this is correct, then *D. producta* differs from some other species of the genus (e.g. *D. discrepans*) in the absence of a modified third endopod in the male. The matter is recommended for special attention to future investigators in British waters, who should collect and carefully study the males of *D. producta*.

This species has been described under many names in the course of its history. Its synonymy is largely agreed upon among the more recent authorities, with the exception of the position of van Beneden's (1857) taxon, *D. elongata*. Accepted by some (e.g. Scott and Scott, 1913) as a synonym of *D. producta*, this taxon has been placed by Cressey (1967a) in synonymy with *D. ferox*, no reasons being given for this transfer. As usually happens in the case of long and involved synonymy, the reasons are to be found in a broad range of intraspecific morphological variability. It is probably best illustrated by the fact that Hewitt (1967) was prompted to describe two forms of *D. producta*, found on two different sharks in New Zealand waters. Form A, from *Carcharodon carcharias*, had a much broader genital trunk than form B, from *Cetorhinus maximus*. They also differed in the proportions of various structures in their posterior extremities.

D. producta is widespread in the Atlantic (with Mediterranean and adjacent regions) and Pacific Oceans, but its only record from the Indian Ocean is that of Wilson (1923a) who reported it taken near Durban, South Africa. Its host list includes *Alopias vulpinus*, *Carcharodon carcharias*, *Cetorhinus maximus*, *Isurus oxyrinchus*, *Lamna ditropis*, *L. nasus*, *Prionace glauca*, *Squalus acanthias* and "tiburon" (?*Hypoprion signatus*). Whether *Somniosus microcephalus* is also added to the list depends on the acceptance or rejection of *D. elongata* as a synonym of *D. producta*, because the only record from that shark is the original description of *D. elongata*.

D. producta is common in British waters on the local shark species.

Genus Echthrogaleus Steenstrup and Lütken, 1861

Female: Pandaridae. Cephalothorax caligiform, with tips of lateral zones reaching as far back as second free (third leg-bearing) segment; frontal plates distinct; tips of first antennae nearly reaching lateral limits of shield. Adhesion pads at bases of first and second antennae and maxillipeds, as well as between bases of first legs. First free segment with dorsolateral plates, partly overlapped by tips of lateral zones of shield; second unarmed (in some species first and second segments fused together); third free segment with paired dorsal plates, usually covering large part of genital complex but rudimentary in *E. pectinatus*. Genital complex usually long, with very prominent posterolateral lobes covering most of abdomen and uropods (subquadrate, with abdomen and uropods exposed in *E. pectinatus*). Abdomen one-segmented. Uropods usually prominent. First antenna two-segmented; second three-segmented, terminal segment unciform claw. Oral appendages siphonostome. Maxilliped subchelate, tip of claw simple or bifid (in *E. pellucidus*). Legs 1–4 biramous, rami of first leg two-segmented; endopods of second and third legs two-segmented, exopods three-segmented; rami of fourth leg one-segmented (possibly two-segmented in *E. pectinatus*). Fifth and sixth legs vestigial.

Male: Cephalothorax and first two free segments similar to those of female, third segment with rudimentary dorsolateral plates. Genital complex oval or subrectangular, about half length of shield, with or without prominent protuberances at bases of fifth legs. Legs 1–4 biramous; rami of first leg two-segmented, remaining pairs with two-segmented endopod and three-segmented exopod. Uropods prominent. (Males unknown for *E. pectinatus* and *E. torpedinis*.)

Type species: *Echthrogaleus coleoptratus* (Guerin-Meneville, 1837).

Comments: The genus *Echthrogaleus*, in addition to its type species, contains currently four other species: *E. denticulatus* Smith, 1874, cosmopolitan on many shark species; *E. pellucidus* Shiino, 1963, recently discovered on *Alopias* sp. off the coast of Equador; *E. pectinatus* Kirtisinghe, 1964, taken on *Rhineodon typus* off Colombo, Ceylon and *E. torpedinis* Wilson, 1907. The last-named species, found only twice off the Atlantic coast of the USA, is the only member of this genus living on elasmobranchs other than sharks (*Torpedo occidentalis*). Kirtisinghe's (1964) *E. pectinatus* has many features unusual for the genus and should be investigated in greater detail before its final acceptance in *Echthrogaleus*.

In addition, Krøyer's (1863) *Dinematura indistincta* was transferred to this genus by Bassett-Smith (1899) as *E. indistinctus*. Krøyer's description is not very informative and is not accompanied by illustrations. The identity of the species cannot be established on the evidence available. For the time being, *E. indistinctus* should be treated as a species inquirenda.

British fauna contains only the type species of this genus.

Echthrogaleus coleoptratus (Guerin-Meneville, 1837)

(Figs. 822–849)

Syn.: *Pandarus alatus* "M. Edwards"; of Johnston (1835)
Dinemoura alata Guerin-Meneville, 1837
Dinemoura coleoptrata Guerin-Meneville, 1837
Dinemoura affinis Edwards, 1840
Dinematura braccata Dana, 1852
Echthrogaleus braccatus; of Heller (1865)
Nogagus lutkenii Norman, 1868
Echthrogaleus perspicax Olsson, 1869
Echthrogaleus affinis; of Brady (1883)
Echthrogaleus lutkenii; of Norman and Scott (1906)
Dinematura coleoptrata Guerin; of Scott and Scott (1913)

Female (Fig. 822): Dorsal shield of cephalothorax caligiform, longitudinal sutures marking off lateral zones; tips of latter extending nearly to anterior border of third free (fourth-leg bearing) segment; membranes along margins of lateral zones; frontal plates distinct, narrow; tips of first antennae nearly reaching lateral limits of dorsal shield. First free (second leg-bearing) segment with long dorsolateral plates, overlapped by tips of lateral zones of shield; posteriorly with marginal membranes. First and second segments fused completely or nearly completely in some older specimens; second segment unarmed; third free segment with very large paired dorsal plates covering nearly half of genital complex (plates commonly pale yellow, with scattered refractile dots on dorsal surface and pointed outgrowths, not visible in Fig. 822, in anterolateral corners). Genital complex longer and often broader than cephalothorax shield, with prominent posterolateral lobes protruding beyond level of uropods; lateral margins convex. Abdomen (Fig. 826) one-segmented, subquadrangular. Total length 10–14 mm.

Distribution of adhesion pads similar to that of preceding genus, but pads generally less well developed. First antenna (Fig. 823) with proximal segment constituting more than $\frac{2}{3}$ total length, its greatest width in distal half, distal margin and distal half of anterior margin with 26 plumose setae (plumules barely indicated in Fig. 823), longer on margin, shorter on adjacent ventral surface; distal segment cylindrical, with rounded tip, one naked seta on posterior margin and apical armature of 13 setae (as in *Dinemoura*, cf. Fig. 805). Second antenna (Fig. 825) indistinctly three-segmented, with small adhesion pad (ap) at base; second segment short, muscular and longitudinally flattened; third segment (Fig. 824) unciform claw, its shaft with sigmoid curve, tapering, sharp; one short seta near base, one longer on side of shaft. Mouth tube long, slender, labrum shorter than labium (Fig. 827), with apical bifid flap (bent over in figure) covered by setules, two anterolateral aliform processes; strips of membrane along lateral margins; labium with rounded tip and loops of fine spinules on either side near tip. Mandible similar to that of *Dinemoura producta*, its dentiferous margin with series of 11 teeth. First maxilla (Fig. 828) short, longitudinally flattened, with broad base and tapering tip; segmentation obscure; exopod small swelling surmounted by three setules; rounded apex of endopod with three minute tubercles. Second maxilla (Fig. 829) brachiform; lacertus long and stout, unarmed; brachium slender; calamus and canna tapering, with several longitudinal rows of denticles; crista large, subcircular, covered by strong denticles, with short digitiform process distally. Maxilliped (Figs. 830, 831) with complicated corpus structure; heavily sclerotized spike and strong cuticular bar on ventral surface, rounded pad with two cuticular flanges (one spiniform) where tip of subchela meets corpus; base connected by powerful muscle with dentiform, prominent adhesion pad; subchela not divided into shaft and claw, with robust base, one heavy spike on ventral side and slender seta near it, tip slender, sharply bent and pointed.

First four pairs of legs biramous, rami one- to three-segmented. Armature formula below.

	Endopod			Exopod		
	1	2	3	1	2	3
Leg 1	0-0	3	—	0-I	4,III	—
Leg 2	I-0	II,6	—	I-I	I-I	I,4,III
Leg 3	1-0	6	—	1-I	5,IV	—
Leg 4	modified			modified		

First leg (Fig. 833) with unsegmented sympod, pair of prominent interpodal pads (ipp) between bases; ventral surface of sympod with three adhesion pads and with one seta each near bases of rami; basal segment of endopod with one large pad; lateral margins of distal segments denticulated in both rami. Second leg (Fig. 834) with seven pads (four corrugated, three denticulated, one of latter bifid) on ventral surface of sympod, strong spiniform process (Fig. 835) on medial margin; both segments of endopod and basal of exopod with parts of ventral surfaces denticulated; second and third exopod segments not well delimited. Third leg (Fig. 836) with broad, indistinctly two-segmented sympod and broad interpodal bar, three adhesion pads on ventral surface of sympod (one denticulated along lateral margin); endopod (Fig. 838) with fringe of setae and denticulation on lateral margin of first segment, similar fringe on second segment; exopod (Fig. 837) two-segmented (but see comments), with fine denticulation on lateral margins of segments. Fourth leg (Fig. 839) with small denticulated pad on sympod and one-segmented foliaceous rami, their margins with varying numbers of denticles. Fifth leg (Fig. 840) subcylindrical, short, tapering, in some specimens with tip marked off by transverse constriction; three pinnate setae on one side. Sixth leg incorporated into area of spermatophore attachment. Uropod (Fig. 832) narrower at base, with convex lateral and medial margins and six small setae on uneven posterior margin.

Male (Fig. 842): Cephalothorax and first two free thoracic segments similar to those of female. No dorsal plates on third free segment. Genital complex (Fig. 843) longer than combined length of free segments, lateral margins slightly divergent, corners rounded. Abdomen two-segmented; first segment short, cylindrical, second broader posteriorly, with posterolateral corners obliquely truncated. Total length 6–8 mm.

First four pairs of legs biramous. First and second legs (Figs. 844, 845) similar to those of female; third leg with modified terminal endopod segment (Fig. 846), equipped with bulbous adhesion pad and two strong, claw-like spines; fourth leg (Fig. 847) with two-segmented endopod and three-segmented exopod; fifth leg (Fig. 849) digitiform protuberance with three plumose setae and one short spine; sixth leg similar but smaller (Fig. 848), with one seta and spine. Uropod (Fig. 843) narrower than in female, with longer setae and fringe on medial margins.

Comments: The author was unable to secure male specimens for examination, the above description is given *fide* Cressey (1967a).

As in other pandarid genera, *Echthrogaleus* undergoes considerable morphological changes during its ontogeny. The above descriptions apply only to the adults of both sexes. An immature female is shown in Fig. 841. It differs from the adult by underdeveloped dorsal plates of the third free segment and a relatively smaller genital complex, its posterolateral lobes still small, not covering the abdomen.

It appears that there might be some variability in the leg structure of *E. coleoptratus*. The third legs of the specimens examined by the author had two-segmented exopods (Fig. 837), though other recent authors (Shiino, 1954a; Hewitt, 1967) describe this ramus as three-segmented. Shiino's illustrations show the division between the second and third segment as being indistinct; Hewitt is quite definite in showing firm segmental boundaries. Fig. 834 in this work shows the division between the corresponding segments of the second leg as being also fairly indistinct. It is possible that this species tends towards reduction of the natatory appendages, manifest in the occasional disappearance of intersegmental boundaries. Variability is also evident in the proportions of the body, giving rise to pronounced differences in habitus and, in consequence, a large list of synonyms.

E. coleoptratus is a common parasite of sharks, living normally on the outer surfaces of its hosts (skin, fins). Its host list includes *Carcharinus floridanus*, *C. milberti*, *Carcharodon rondelletti*, *Centrophorus granulosus*, *Isurus glaucus*, *I. oxyrinchus*, *Lamna ditropis*, *L. nasus*, *Lamna* sp., *Prionace glauca*, *Squalus acanthias* and "tiburon" (?*Hypoprion signatus*). This list is almost certainly incomplete. Dollfus (1946) adds to it also *Mola mola*, without quoting the source of his record.

The distribution of *E. coleoptratus* covers the Atlantic and Pacific Oceans, where it has been found in many localities, both in the northern and southern hemispheres. It appears to be less abundant in the Indian Ocean. In British waters it is not uncommon, occurring in particular on the porbeagle shark, *Isurus oxyrinchus*, in Scottish waters and on the spiny dogfish, *Squalus acanthias*, there and in other regions.

Genus Demoleus Heller, 1865

Female: Pandaridae. Cephalothorax shield caligiform, with lateral zones partially marked off by longitudinal sutures; frontal plates distinct, with free lateral wings; tips of first antennae not reaching lateral limits of shield; lateral zones with marginal membrane. Adhesion pads at bases of first antennae, on second antennae, at and between bases of maxillipeds and between bases of first legs. First free (second leg-bearing) segment with dorsolateral plates; second unarmed; third with paired dorsal plates overlapping base of genital complex. Latter longer than cephalothorax, with more or less parallel lateral margins and prominent posterior lobes. Abdomen one-segmented, with large dorsal plate. First antenna two-segmented; second indistinctly segmented, with weak terminal claw. Oral appendages siphonostome. Maxilliped subchelate. First four pairs of legs biramous, rami two-segmented. Fifth and sixth legs vestigial. Uropods prominent. Egg strings either triply folded or in flat coils.

Male: Cephalothorax and first two free segments similar to those of female. Third free segment unarmed. Genital complex suboval, with or without small posterolateral lobes. Abdomen one- or two-segmented. Distal segment of third endopod modified.

Type species: *Demoleus heptapus* (Otto, 1821).

Comments: In addition to the type species, only one other is currently included in this genus. *D. latus* Shiino, 1954, had been discovered in Japanese waters and was subsequently recorded twice more (Hewitt, 1967; Cressey, 1967a) from the Pacific Ocean. It differs from the type species mainly in small details of the appendages and in the large size of its abdominal dorsal plate (Text fig. 52H). There appears to be some doubt as to the segmentation of the abdomen in the male of *D. latus*. Hewitt (1967) described it as having only one abdominal segment, a situation somewhat at odds with the more usual two-segmented abdomen structure, both in the male of *D. heptapus* and in males of other pandarid genera. Hewitt's specimens, by his own admission, greatly resembled those described by Shiino (1954a) as juvenile females. Further data are required.

Only the type species is present in British fauna.

Demoleus heptapus (Otto, 1821)

(Figs. 850–877)

Syn.: *Caligus heptapus* Otto, 1821
Caligus paradoxus Otto, 1828
Binoculus sexsetaceus Nordmann, 1832
Dinematura sexsetacea; of Burmeister (1835)
Demoleus paradoxus; of Heller (1865)

Female (Fig. 850): Cephalothorax with caligiform dorsal shield, lateral zones delimited by definite sutures, tips of zones extending backwards to level of second free (third leg-bearing) segment; frontal plates distinct, with prominent lateral wings overlapping basal segments of first antennae; tips of latter not reaching lateral limits of shield; margins of lateral zones and posterior margin of shield with narrow membrane strips. First free (second leg-bearing) segment with dorsolateral aliform plates, overlapped by tips of lateral zones, posterior end with marginal membrane. Second free segment narrower than first, unarmed. Third free segment with neck-like anterior end, paired dorsal plates covering anterior $\frac{1}{4}$ of genital complex. Latter longer than cephalothorax, dorsoventrally flattened, with parallel lateral margins and two prominent posterior lobes. Abdomen (Fig. 851) one-segmented, covered by semicircular dorsal plate. Egg strings triply folded. Total length (including uropods) 11–16 mm.

First antenna (Fig. 853): Basal segment robust, with small protuberance surmounted by seta in midlength of anterior margin, distal half of that margin and adjacent part of ventral surface armed with varying number of setae (about 30); setae of two types, shown in Figs. 854, 855; apical armature (Fig. 852) not fully ascertained. Second antenna (Fig. 857) of two indistinctly delimited parts; basal part robust, shapeless, with large striated adhesion pad on ventral surface; distal part slender, apparently flexible, apically with short, blunt claw (Fig. 856), long barb at its base and small papilliform protuberance some distance from it. Mouth tube long, slender, with labrum shorter than labium (Fig. 858), apically pointed and fringed with setae; two subterminal aliform processes, partially or completely covered by lip; labium with transversely truncated tip, small central protuberance and strips of narrow serrated membrane. Mandible as in other pandarid genera. First maxilla (Fig. 859) small, broad at base, tapering distally, laterally compressed; exopod small, displaced dorsad, with three apical processes (one seta?); distal half of dorsal margin with finely serrated surface; tip blunt, with large papilliform process, often sloping obliquely outwards, and small pointed outgrowth on its ventral side. Second maxilla brachiform; lacertus strong, unarmed; brachium slender; calamus (Fig. 860) pedunculate, tapering, with ring of setae round base and many longitudinal rows of setules, sometimes not running full length of calamus; crista represented by tuft of long setae; canna shorter than calamus, pedunculate, with denticulated surface. Maxilliped with pedunculate corpus; lateral wall (Fig. 861) heavily sclerotized, with flange in distal corner obscuring subchela; on medial side at base of flange (Fig. 862) papilliform process arresting movement of subchela; base with thinner cuticle extending towards transversely arranged adhesion pad (ap); subchela not divided into shaft and claw, with slender seta near base and gently bent, tapering tip.

First four pairs of legs biramous, rami two-segmented. Armature formula below.

	Endopod		Exopod	
	1	2	1	2
Leg 1	0-0	3	0-I	4,III
Leg 2	1-0	8	1-I	6,III
Leg 3	1-0	6	1-I	6,III
Leg 4	1-0	5	1-I	6,III

First leg (Fig. 863) much smaller than others; sympod with three corrugated areas, not clearly delimited; similar area on basal segment of endopod; lateral margin of second endopod and medial of basal exopod segment with fringes of setae; lateral margin of basal exopod segment broadly ridged; that of terminal with comb-like fringe of apparently fused heavy setules. Exopods of second to fourth legs similar to one another (Fig. 864). Second endopod (Fig. 865) with second segment much longer than first and with marginal fringes of setae on both margins; third endopod (Fig. 866) similar to second (fringe on medial margin of second segment omitted); fourth endopod (Fig. 867) with two segments of more equal sizes. Fifth leg (Fig. 868) cylindrical peg with rounded tip and two short apical setae; sixth leg (Fig. 869) flat, rounded outgrowth at margin of genital orifice. Uropod (Fig. 851) spatulate, with convex lateral and medial margins; posterior margin irregular, with six short setae.

Male (Fig. 870): Cephalothorax and first two free segments similar to those of female; third free (fourth leg-bearing) segment narrow, subcircular, devoid of dorsal plates. Genital complex about as long as two preceding segments together, with sloping anterolateral corners, parallel lateral and transversely truncated posterior margins. Abdomen two-segmented, less than $\frac{1}{2}$ length of genital complex; first segment cylindrical, second broader posteriorly. Total length 8–10 mm.

First antenna relatively larger than that of female. Second antenna (Fig. 871) three-segmented; first segment with cuticular ridge, second with striated area near distal end; third segment (Fig. 872) unciform claw with broader base and slender tip; papilliform process at base, slender seta at point of narrowing and small rounded protuberance on concave margin near seta. Maxilliped (Fig. 873) with long, pedunculate corpus, bearing three prominent rounded and ridged processes on mediolateral margin; at base of central process on medial side (Fig. 874) small papilla with minute apical setule; subchela not clearly divided into shaft and claw, with slender sharp tip and long seta near base.

First four pairs of legs similar to those of female, with exception of distal segment of third endopod (Fig. 875). Latter with medial flange, bearing on dorsal surface digitiform tapering process and having another process at base on ventral side (Fig. 876); narrow strip of membrane marking base of flange on dorsal surface; small protuberance on margin of segment between first and second pinnate setae. Fifth leg (Fig. 877) similar to that of female, but with longer setae; sixth similar to fifth, at boundary between genital complex and abdomen. Uropod (Fig. 870) less pedunculate and narrower than that of female.

Comments: The natatory appendages of *D. heptapus* have uniformly two-segmented rami, but the endopods of its second and third legs (Figs. 865, 866) show incipient division of the distal segments into two, tending to make the ramus three-segmented. This is rather interesting, since in Pandaridae it is the exopod, rather than endopod, that is more often three-segmented.

Scott and Scott (1913) placed *Nogagus grandis* Steenstrup and Lütken, 1861, in synonymy with this species, but Cressey (1967a) considered it synonymous with *Phyllothyreus cornutus*. Having compared Steenstrup and Lütken's (1861) Plate X, Fig. 19 with Figs. 870 and 896 of this book, the author agrees with Cressey's decision.

The literature contains many instances of misspelling of the specific name *heptapus*, the forms *heptatus* and *hepatus* having been used by various authors.

D. heptapus is primarily a parasite of the six-gilled sharks of the genus *Hexanchus*, having been recorded from *H. griseus*, *H. californicus* and *Hexanchus* sp. Leigh-Sharpe (1933a) reported its occurrence on *Squalus acanthias*. The distribution range covers mainly the North Atlantic and the Mediterranean, from whence come most of the records. Wilson (1935b) found it off the coast of California and Lewis (1966) in Hawaii. It is a rare visitor in British waters, where it is occasionally taken with *H. griseus*, itself rather uncommon. The author knows of no instance of its occurrence on the spiny dogfish and views Leigh-Sharpe's (1933a) record with considerable doubt.

Genus Phyllothyreus (Edwards, 1840)

This is a monotypic genus and its diagnosis is identical with that of its only species. *Parapandarus nodosus* Wilson, 1924, considered by Monod and Dollfus (1938) as a species of *Phyllothyreus* distinct from the type species, was synonymized with it by Cressey (1967a).

The original name given to this genus by Edwards (1840) was *Phyllophora*, preoccupied by three other genera. Norman (1903) suggested its replacement by *Phyllothyreus*. Wilson (1907b) stated that Poche (1902) had proposed for it *Laminifera*, a puzzling statement in view of the complete absence of any reference to this genus in Poche's paper. *Laminifera*, accepted by several later authors as the name for this genus, is, in any event, preoccupied by *Laminifera* Boettger, 1863, a molluscan genus. Brian (1946) was the first to observe and report on its invalidity for this copepod. Norman's name must stand. It has been used by some authors, with at least two variations of spelling.

Phyllothyreus cornutus (Edwards, 1840)

(Figs. 878–904)

Syn.: *Phyllophora cornuta* Edwards, 1840
Nogagus grandis Steenstrup and Lütken, 1861 (male only)
Phyllophorus cornutus M. Edw.; of Bassett-Smith (1899)
Phyllothyreus cornutus (H. Milne Edwards); of Norman (1903)
Laminifera cornuta Poche, 1902; of Wilson (1907b)
Phyllothreus cornutus (M. Edwards); of Scott and Scott (1913)
Parapandarus nodosus Wilson, 1924
Laminifera cornuta (Milne Edwards 1840); of Leigh-Sharpe (1933b)
Phyllothyreus nodosus (Ch. Br. Wilson, 1924); of Monod and Dollfus (1938)
Laminifera doello-juradoi Brian, 1944
Phyllothereus cornutus (Milne-Edwards, 1840); of Cressey (1967a)

Female (Fig. 878): Pandaridae. Dorsal shield of cephalothorax not divided into zones by sutures, with posterolateral tips extending beyond level of posterior margin and overlapping first free (second leg-bearing) segment; frontal plates narrow; no marginal membrane observed. First free segment with prominent, oval dorsolateral plates, largely covering those of following segment. Both second and third free segments with similar plates, those of third covering base of genital complex. Latter suborbicular, with two small, conical, posterolateral processes. Abdomen (Fig. 879) small, one-segmented, dorsally covered by small semicircular plate, ventrally by much larger one, with subtriangular anterolateral processes. Total length 12–16 mm.

First antenna (*fide* Cressey, 1967a) two-segmented, basal segment with 22 unarmed setae on anterior margin, terminal cylindrical with 12 unarmed setae. Second antenna (Fig. 880) very prominent, three-segmented; first segment short, with sclerotized plaque (adhesion pad?); second longer, unarmed; third powerful hook with short spine at base of inner margin and short seta some distance from it. Mouth tube long, slender, similar to that of *Pandarus bicolor* but with small triangular processes in anterolateral corners of labrum. Mandible with 10 or more fairly uniform teeth. First maxilla (Fig. 881) laterally compressed, squat, with exopod swelling displaced dorsally, bearing three short setae; endopod tip blunt, with prominent conical outgrowth and small denticle posterior to its base. Second maxilla brachiform; lacertus robust, unarmed; brachium slender, subcylindrical; calamus (Fig. 882) tapering, with six longitudinal rows of fine setules and some small denticles near base; clavus short, tapering, plumose; canna with only two rows of setules and small denticulated patch near tip. Maxilliped (Fig. 883) prominent, squat, with slightly pedunculate corpus, expanded into prominent flange in distal corner; lacuna of thin cuticle extending between base of corpus and unciform, prominent adhesion pad; subchela with robust base, fine tip and short seta near base.

First four pairs of legs biramous, rami of first three two-segmented, those of fourth one-segmented. Armature formula below.

	Endopod		Exopod	
	1	2	1	2
Leg 1	0-0	3	0-1	VII
Leg 2	0-0	VII	0-I	IX
Leg 3	0-0	IV	0-I	VII
Leg 4	III	—	VII	—

Sympods indistinctly two-segmented; first (Fig. 884) with three spinulated pads on ventral surface, second (Fig. 887) with four, third and fourth (Figs. 888, 890) with three. Third sympod with prominent medial lobe, overlapping broad flap of interpodal bar (Fig. 888); fourth with very large, suborbicular lateral expansion of basal segment (Fig. 890). Lateral margins of rami largely denticulated. Details of apical armature of first leg in Figs. 885, 886, those of third exopod in Fig. 889, those of fourth leg in Figs. 891, 892. Fifth leg (Fig. 893) subtriangular lobe with one setule on lateral side of base. Sixth apparently missing. Uropod (Fig. 894) broad, armed with short, blunt processes, setae and spines (detail in Fig. 895).

Male (Fig. 896): Cephalothorax shield caligiform, lateral zones partially delimited, their tips extending backwards as far as third free segment, margins with strips of membrane; frontal plates narrow; posterior margin slightly concave. First free segment (second leg-bearing) with dorsolateral plates overlapped by tips of dorsal shield; second free segment short, unarmed; third segment with small anterolateral processes and rounded posterior margin. Genital complex with sharp antero-lateral corners and small rounded posterolateral lobes. Abdomen two-segmented; first segment narrow, cylindrical; second broader, with obliquely truncated posterolateral corners. Total length 7–10 mm.

First antenna similar to that of female. Second antenna (Fig. 897) apparently four-segmented, with small adhesion pad on somewhat inflated third segment; terminal claw relatively smaller than that of female. Maxilliped (Fig. 898) with oval corpus, myxal area prominent, with three small adhesion pads; base of subchela more slender than that of female, basal seta longer. First four pairs of legs biramous, rami two-segmented. Armature formula below.

	Endopod		Exopod	
	1	2	1	2
Leg 1	0-0	3	0-1	3,IV
Leg 2	1-0	8	1-I	6,IV
Leg 3	1-0	6	1-I	5,IV
Leg 4	1-0	5	1-I	5,IV

Sympods indistinctly two-segmented; first sympod (Fig. 899) with three corrugated adhesion pads, second with two (Fig. 900), remaining sympods with one. Basal segment of first endopod (Fig. 899) with very prominent adhesion pad. Exopods of third and fourth legs similar to each other (Fig. 901), endopods (Fig. 902) differing only in number of terminal setae. Fifth leg (Fig. 903) vestigial, with one short spine and three plumose setae; sixth leg (Fig. 904) with one spine and one seta only. Uropod (Fig. 896) somewhat penduculate, with four pinnate setae and two naked setules on posterior margin.

Comments: The female pandarids are considered by Cressey (1967a) to possess a vestigial sixth leg, situated near the boundary between the genital complex and abdomen and often associated with the genital orifice. The fifth leg of this species (Fig. 893) appears to possess features characteristic of the sixth leg of other genera. If this structure is, indeed, a sixth leg, rather than a fifth, then the fifth has either disappeared, rather out of sequence, or has been missed by all investigators, present author included. More studies are required.

Unlike most pandarids, *P. cornutus* is parasitic on the gills and gill arches of sharks. Its host list includes *Carcharinus floridanus*, *C. milberti*, *Isurus nasus*, *I. oxyrinchus*, *Lamna nasus*, *Prionace glauca* and *Sphyrna zygaena*. According to Cressey (1967a), its most common host is *Prionace glauca*. Cressey and Lachner (1970) found *P. cornutus* in the stomach contents of the echeneid fishes associated with *Prionace glauca* and *Carcharinus floridanus*.

The distribution of *P. cornutus* is cosmopolitan, records having been made in the Atlantic and Pacific, as well as in the Indian Ocean. Specimens from different geographical areas show some discrepancy in sizes.

The species occurs occasionally in British waters on *Prionace glauca*.

Pandaridae incertae sedis

The genus *Nogaus* Leach, 1816, spelled *Nogagus* by many subsequent authors, was established for pandarid males, most of which were found free-swimming. The specimens were often not completely mature, which added to the difficulties of assessing their identities and affiliations. Wilson (1907b), in his discussion of the subfamily Pandarinae, which he treated as a part of Caligidae, stated that "much the hardest task of all is the discussion of the genus *Nogaus*, which is made entirely of males, 34 species, described by many authors in seven different languages". He transferred some of these males to the genus *Nessipus* Heller, 1865; several other genera received their share of *Nogaus*-type males at one time or another. Later reviewers, Cressey (1967a) included, steered clear of this "hardest task of all", leaving a disconcerting grey area in the taxonomy of Pandaridae.

The parasitic copepod fauna of British waters contains two examples of "*Nogaus*". Both of them have been shown to have parasitic connections with fishes and merit inclusion in this account. Neither of them, however, can be treated as anything more than a species inquirenda. Their descriptions below are necessarily incomplete.

Nogagus borealis Steenstrup and Lütken, 1861

(Fig. 767)

Syn.: *Nessipus borealis* (Steenstrup and Lütken, 1861); of Wilson (1907b)

Dorsal shield of cephalothorax caligiform, with lateral zones clearly delimited by longitudinal sutures and cephalic zone separated from thoracic by equally distinct transverse suture; frontal plates well delimited, with prominent lateral outgrowth overhanging bases of first antennae; marginal membrane broad; prominent refractile conspicilla at midanterior point of shield. First free (second leg-bearing) segment with posterolateral corners protruding posteriorly to level of third segment. Second segment smaller, with conical posterolateral protrusions; third segment smaller than second, with convex lateral margins and small lateral protuberances. Genital segment longer than two previous segments combined, slightly convergent posteriorly, with incipient posterolateral lobes in some specimens. Abdomen one-segmented, less than $\frac{1}{2}$ length of genital complex, with uropods lateral rather than posterior in position. Total length about 10 mm.

Cephalic appendages resembling those of other pandarid males. First four pairs of legs biramous, rami of first three two-segmented, those of fourth one-segmented. Fifth leg vestigial, sixth not observed.

Female: Unknowñ.

Comments: It appears that *N. borealis* is morphologically fairly variable, differences being observable particularly in the shape of the free thoracic segments and the distinctness of the transverse suture of the dorsal shield, apparently absent in some specimens (see Wilson, 1907b). The generic affinity of this species cannot be determined. Wilson (1907b) transferred it to the genus *Nessipus* Heller, 1865. Cressey (1967a), on the other hand, questioned the validity of this decision on the basis of comparison between *N. borealis* and the males of *Nessipus crypturus* Heller, 1865 and *N. tigris* Cressey, 1967. Not enough is known about either the male *Nessipus* or *Nogagus borealis* to decide whose views are more in accord with the truth.

Like most of the *Nogaus*-type males, this copepod was described from free-swimming specimens found in the North Atlantic. It is from that region that most records of *N. borealis* come. Wilson (1907b) found it in the Gulf of Alaska, where this author also found it in plankton samples. Hewitt (1967) recorded it from New Zealand waters and this author identified a few specimens in a plankton sample from New South Wales, Australia. It would appear, therefore, that the specific name *borealis* is something of a misnomer.

Markevich (1956), unaccountably, described the female of this species without giving any illustrations. His description is as follows. "Cephalothoracic shield of females transversely oval, clearly delimited from posterior part of body; its width much greater than length. Frontal plates well developed. Abdomen small, one-segmented, attached to ventral surface of genital segment and invisible in dorsal aspect. Caudal laminae broad, plate-like. First four pairs of legs biramous, rami of first three pairs two-segmented; rami of fourth one-segmented. Length of body about 10–11 mm. Egg sacs longer than body."

Hewitt (1967) was the only author ever to have found *N. borealis* on a host, on the body surface of *Isurus oxyrinchus*. Markevich (1956) claims that this copepod lives in "the buccal cavity of sharks" and adds the Bering Sea to its distribution range. The basis for his assertions has not been made clear.

Nogagus ambiguus T. Scott (1907) (provisional name)

(Fig. 768)

Dorsal shield of cephalothorax caligiform, with lateral zones partially delimited by longitudinal sutures; tips of lateral zones extending nearly to level of posterior end of first free (second leg-bearing) segment; broad marginal membranes; frontal plates well delimited; tips of first antennae reaching lateral limits of shield. First free segment with posterolateral processes reaching anterior end of third segment; second segment unarmed, about as broad as first; third segment much narrower, small. Genital complex not much larger than third leg-bearing segment, with triangular posterolateral processes. Abdomen one-segmented, about $\frac{1}{2}$ length of genital complex. Uropods posterolateral in position, broad. Total length 5.5 mm.

Oral and other cephalic appendages similar to those of other pandarid males. First four pairs of legs biramous, all rami two-segmented except for one-segmented fourth exopod.

Female: Unknown.

Comments: T. Scott (1907) described two new species of *Nogagus* collected from the skin of the spiny dogfish, *Squalus acanthias*, taken in the North Sea. One of them, *N. latus*, was subsequently (Scott and Scott, 1913) recognized as a juvenile form of *Pandarus bicolor*. The other, *N. ambiguus*, was very tentatively left as an independent species. There is little doubt that it is also a juvenile pandarid. With our scanty knowledge of pandarid development, however, we are not able to find any connection between it and any pandarid species.

Leigh-Sharpe (1934) listed another find of this enigmatic copepod, this time from the Plymouth area. Curiously, Brady (1910) described a very similar specimen, caught presumably in plankton, in Kerguelen, in the subantarctic waters. His specimen, however, differed from the British parasite in having one-segmented rami of the fourth leg. The identity of both Brady's and Scott's specimens still awaits clarification.

Family Cecropidae

Morphology

Cecropidae is a small family in some ways unique among its caligiform allies. Of its five constituent genera, three are obviously very closely related and parasitic on the same host species, to which they appear to be rigidly limited. The host in question is *Mola mola*, the sunfish, and the genera are *Cecrops* Leach, 1816, *Orthagoriscicola* (Krøyer, 1837) and *Philorthagoriscus* Horst, 1897. The names of the last two are derived from *orthagoriscus*, the name variously used (both specifically and generically) for the fish now known as *Mola mola*. The other two genera, *Luetkenia* Claus, 1864 and *Entepherus* Bere, 1936, do not occur on that fish and some of their structural characteristics make them, in a way, odd members of the family.

In view of their close relationship with one another, particularly in the case of the first three genera, it is not surprising that the members of the family are morphologically fairly uniform. They all have caligiform, cephalothoracic shields, with or without prominent posterolateral lobes. The frontal plates are fused with the anterior margin of the shield but are usually delineated by superficial suture lines. In *Cecrops* and *Orthagoriscicola* the shield is not subdivided, except for broad, shallow grooves on some specimens of the former genus. In other genera the lateral zones are at least partially delimited. The marginal membrane of the shield might be absent (*Cecrops*) or divided into numerous short sectors by dentiform marginal serrations (*Philorthagoriscus*).

The first and second free thoracic segments are fused together in *Cecrops*, *Philorthagoriscus* and *Luetkenia*, but are separate in *Orthagoriscicola* and *Entepherus*. The first segment always has prominent dorsolateral plates. In *Entepherus* it has also, apparently, a dorsal plate overlapping the second segment. In all genera, the third free segment has large paired dorsal plates, fused to a greater or lesser extent and, with the exception of *Luetkenia* and *Cecrops*, serrated along their margins. Plate-like expansions are also universal on the genital complex, usually notched at the midposterior point and completely covering the one-segmented abdomen. In *Cecrops*, these plates, together with the abdomen, form something of a brood pouch, within which the convoluted egg sacs are stored and hidden from view. Convoluted egg sacs occur also in *Orthagoriscicola*, but the remaining genera have the usual long, uniseriate and straight egg sacs common in caligiform copepods. The presence of spines or spiniform processes on the margins of the dorsal shield or segmental plates is characteristic of the family (except for *Luetkenia* and *Cecrops*). In *Orthagoriscicola* spines are also present on the dorsal surface of the cephalothoracic shield.

The cephalothoracic appendages are siphonostome and not greatly remarkable. Their one common feature is the slenderness of their maxillipeds. The first four pairs of legs are biramous. In *Orthagoriscicola* the first two pairs of legs have two-segmented rami, the other two one-segmented. In all other genera the rami of the first three pairs are two-segmented, those of the fourth are one-segmented.

The males resemble the females quite closely, but their genital complex tagmata are devoid of plate-like expansions. The abdomina, however, due to their anterior displacement, are at least partly concealed in dorsal aspect. This similarity extends also to the structure of the appendages, with the exception of the modified endopod of the third legs. In *Orthagoriscicola* this pair has two-segmented rami, unlike that in the female of the species.

History and systematics

The gradual development of the concept of Cecropidae as a family group taxon can be traced by following the history of Pandaridae (q.v.), with which their genera were originally included. Edwards (1840) placed *Cecrops* and *Orthagoriscicola* (under the name *Laemargus*), the only cecropids then known, in his tribe "Pandariens". Dana (1852) treated them as a subfamily of Caligidae, Baird (1850) as an independent family Cecropidae. Dana's influence on Wilson, the most prolific and authoritative writer on parasitic Copepoda, perpetuated their subfamilial status until Wilson (1932) himself upgraded them to a familial rank. (As recently as 1956, Markevich classified them as a subfamily.) No other changes took place, except for the gradual addition of three genera, the last of which, *Entepherus*, was transferred to the family from Pandaridae by Yamaguti (1963).

In the British fauna this family is represented by four of its five genera, the absent member being *Entepherus*.

Key to the British genera of Cecropidae

Females

1. First and second free thoracic segments fused together (Fig. 905) 2
   First and second free thoracic segments separate (Fig. 936), third with large, fused dorsal plates . . . . ***Orthagoriscicola***
2. First free thoracic segment with dorsolateral plates, second unarmed (Figs. 961, 986) 3
   First free thoracic segment with dorsolateral plates, second with small dorsal plate (Fig. 905) . . ***Cecrops***
3. Lateral margin of cephalothorax shield and margin of dorsal plates coarsely serrated (Fig. 961) . . . . ***Philorthagoriscus***
   Lateral margins of cephalothoracic shield and margins of dorsal plates smooth (Fig. 986) . . . ***Luetkenia***

Males

1. Margins of cephalothoracic shield and of dorsal plates smooth (Figs. 928, 1001) 2
   Margins of cephalothoracic shield and of dorsal plates serrated (Figs. 953, 977) 3
2. Dorsal plate of third free thoracic segment only slightly indented posteriorly, genital segment with transversely truncated anterior margin, semicircular (Fig. 928) . . . . ***Cecrops***
   Dorsal plates of third free thoracic segment distinctly paired, abdomen oval (Fig. 1001) . . . . ***Luetkenia***
3. Surface of dorsal shield with noticeable tubercles, without sutures, margin of genital complex serrated (Fig. 953) . . . . ***Orthagoriscicola***
   Surface of dorsal shield smooth, with longitudinal and oblique sutures, two subtriangular processes on both lateral margins of genital complex (Fig. 977) . . . . ***Philorthagoriscus***

Genus Cecrops Leach, 1816

Only one incontrovertibly valid species can be placed in this genus. The species is *C. latreillii* Leach, 1816. Consequently, the generic diagnosis is the same as that of *C. latreillii*. Hesse (1883) described *C. achantii-vulgaris* (sic!) from a single female specimen, apparently collected from *Squalus acanthias*. The description is inadequate, but sufficient to discard it as a possible member of the genus *Cecrops*. Commenting on the swimming legs of this species, as illustrated by Hesse, Wilson (1907b) remarked that "even a chirographic expert could not decipher them in the figures". Risso (1826) described *Cecrops desmaresti*, which is also rejected as a member of this genus by Wilson (1923c), on the basis of its leg structure. Both these species, at best, can be treated as species inquirendae. The only other nominal species, *C. exiguus* Wilson, 1923, is considered by this author as a tentative synonym of *C. latreillii*.

Cecrops latreillii Leach, 1816

(Figs. 905–935)

Syn.: ?*Cecrops exiguus* Wilson, 1923

Female (Fig. 905): Dorsal shield of cephalothorax narrower anteriorly, with prominent posterolateral lobes extending backwards as far as third free thoracic (fourth leg-bearing) segment, notched at about midlength of lateral margins; frontal plates rigidly fused with anterior margin but usually with superficial suture lines distinct; tips of first antennae almost reaching lateral limits of shield; no division into zones but often two shallow grooves extending from medial limits of posterolateral zones obliquely in anterolateral direction; no marginal membranes observed. First and second free thoracic segments fused together, first with dorsolateral plates largely overlapped by posterolateral lobes of shield, second with small dorsal plate, indented posteriorly and overlapping base of third free segment. Third free segment more than three times broader than second, with large dorsal plate covering some $\frac{1}{3}$ of genital complex, semicircular, with small midposterior notch. Genital complex larger than dorsal shield of cephalothorax, irregularly oval, dorsally covered by large plate (probably fused paired plates), with midposterior notch of variable depth; margins of plate soft, curved, ventrally round margins of genital complex, posteriorly enclosing fairly large space containing convoluted egg sacs, when present. Abdomen one-segmented, small, with narrow base; in young females with rounded corners, in mature females (Fig. 906) with anterolateral corners subtriangular, projecting anteriorly; posterolateral corners rounded, with narrow marginal flanges. Total length 20–30 mm.

First antenna (Fig. 907) two-segmented; basal segment longer and more robust than terminal, inflated somewhat on ventral side near junction with latter; setae on swelling and along distal half of anterior margin (exact number not ascertained); terminal segment cylindrical, with single naked seta on posterior margin and apical armature of undetermined composition. Second antenna (Fig. 908) indistinctly segmented (three-segmented; Wilson, 1907b); basal part robust, large, slender terminal claw sharply hooked, as long as remainder of appendage, with short basal process and longer seta some distance from base. Mouth cone (Fig. 909) long, with slender tip armed with fringe of fine setae. Mandible slender, with 11 teeth. First maxilla almost as long as mouth cone (Fig. 909), fleshy, laterally compressed, with base narrower than tip (Fig. 910); latter rounded, with patch of denticles and robust dentiform outgrowth on anterior side. Second maxilla (Fig. 911) brachiform; lacertus robust, unarmed; brachium slender, subcylindrical; crista small patch of very fine setules on small, rounded swelling; calamus with about six longitudinal rows of fine setules on distal half; canna less than $\frac{1}{4}$ length of calamus, its surface covered with densely crowded, small denticles. Maxilliped subchelate (Fig. 912); corpus slender, with area of thin cuticle on proximal part of medial margin and extending in welt-like swelling towards transversely situated, small adhesion pad (Fig. 913); rounded swelling of varying prominence in myxal area; subchela very long, slender, with sharply curving tip.

First four pairs of legs biramous. Sympods of first two pairs indistinctly two-segmented, apparently unarmed (Figs. 914, 916), except for small setae at bases of exopods; that of third pair much enlarged, rounded, with obscure segmentation and two prominent cuticular ridges on ventral surface (Fig. 919); lateral margin and adjacent part of ventral surface densely denticulated. Sympod of fourth leg of irregular shape (Fig. 922), apparently unsegmented, with irregular cuticular folds on dorsal surface (Fig. 923) and longitudinal depression on ventral surface (in some specimens dorsal folds inflated by accumulation of caseous substance, projecting dorsad as in Fig. 924). Rami of first three pairs two-segmented, those of fourth one-segmented. Armature formula below.

	Endopod		Exopod	
	1	2	1	2
Leg 1	0-0	3	0-I	3,IV
Leg 2	I-0	7	I-I	5,IV
Leg 3	I-0	4	0-I	4,III
Leg 4	3	—	3,IV	—

Basal segment of first endopod distally with denticulated swelling on ventral surface (Fig. 914); terminal with denticles on lateral margin and near bases of distal setae and spines; setae of distal segment with serrated margins, spine with or without subterminal secondary spikes (Fig. 915). Similar setae on distal segment of second endopod (Fig. 917); basal segment of exopod with very large distolateral spine (Fig. 918). Rami of third leg (Figs. 920, 921) with denticulated margins; exopod spines short. Fourth leg with rami unsegmented, not clearly delimited from sympod (Fig. 925). Fifth leg not located; sixth (Fig. 926) at margin of genital orifice, represented by subtriangular sclerotized fold. Uropod (Fig. 927) small, with narrow base, suborbicular, with small naked seta in posterolateral and posteromedial corners and four larger, finely denticulated setae on posterior margin.

Male (Fig. 928): Cephalothorax and first two free thoracic (second and third leg-bearing) segments as in female. Third free segment with large semicircular plate covering entire genital complex and most or all of abdomen. Genital complex transversely oval, with narrow anterior neck. Abdomen (Fig. 935) one-segmented, less than $\frac{1}{2}$ length of genital complex, oval, broader than long. Total length 14–17 mm.

Second antenna and maxilliped relatively more prominent than in female.

First leg as in female. Second exopod with relatively longer, curving posterolateral spine on basal segment and curiously modified posterior seta next to shortest spine of distal segment (Fig. 929). Sympod of third leg relatively less expanded than in female, with one cuticular ridge on ventral surface (Fig. 930); distal endopod segment (Fig. 931) sigmoid, with three distal setae and three lateral cylindrical processes set in dorsoventral plane. Sympod of fourth leg apparently unsegmented (Fig. 932), with two patches of fine setules and one strip of denticles; rami one-segmented, short; details of exopod armature in Fig. 933. Fifth leg (Fig. 934) represented by conical spine and simple naked seta in posterolateral corner of genital complex. Sixth leg not observed.

Comments: *C. latreillii*, also spelled occasionally *C. latreilli* and *C. latreillei*, is a gill parasite of *Mola* and has been recorded from many parts of the distribution range of this cosmopolitan fish. (A common difficulty in this regard is the fact that the taxonomy of *Mola* is rather uncertain, various specific names being used for one, or two, species of this genus by different authorities.) The literature contains also a rather large number of records from unusual hosts, some of them rather improbable. At least some are due to mislabelling of samples. Most of them show their spurious nature by the fact that they have never been repeated. Brian (1906) mentions a record from *Diodon* in California, a very improbable host affiliation for *C. latreillii*. Heller (1866) recorded it from *Thynnus thynnus* in the Adriatic; Heegaard (1943) mentioned it on the same host in the Atlantic, adding that he suspected an error in host designation. Lucas (1887) also mentioned it occurring on the tuna, *Pleuronectes rhombus* and *P. maximus* (turbot and halibut?), as well as on the usual *Mola mola*. His own find of *C. latreillii* came from the last-named species; the association between this parasite and the two flatfishes was only an anecdotal report. O'Riordan (1966) quoted a possible occurrence on *Pristiurus* sp. in Irish waters.

Wilson's (1923c) *C. exiguus* was described from an unidentified shark off Florida. The main reason for setting it up as a new species was the different proportions of the body and segmental plates. The differences Wilson claimed for them, as well as for the armature of the legs and cephalothoracic appendages, do not appear to be great enough to warrant a different taxon for his specimens. The matter merits further study, especially in view of the host difference between Wilson's and Leach's species. (Kensley and Grindley (1973), however, identified as *C. exiguus* specimens found by them on *Mola lanceolata* in South African waters.)

An interesting new contribution to our knowledge of this species is a paper by Grabda (1973), describing one of its late chalimus stages. Generally similar to the adult, it is attached to the gills of its host by a characteristic double frontal filament. Its fourth leg is less modified than that described above.

Genus Orthagoriscicola Poche, 1902

Orthagoriscicola muricatus (Krøyer, 1837)

(Figs. 936–960)

Syn.: *Laemargus muricatus* Krøyer, 1837
Orthagoriscicola muricata Krøyer, 1837); of Poche (1902)
Orthagoriscicola szidati Stekhoven and Stekhoven, 1956
Orthagoriscicola wilsoni Stekhoven and Stekhoven, 1956

Female (Fig. 936): Cephalothoracic shield trapezoidal, with somewhat rounded anterior margin and prominent swellings over bases of first antennae; lateral margins divergent in posterior direction, beset with spinules in posterior halves; posterior sinus shallow, posterolateral lobes only slightly protruding beyond level of posterior margin; frontal plates fused with shield; no suture lines on dorsal surface of shield but symmetrically arranged, blunt spinules in anterior half, along lateral margins and on posterolateral lobes (differences between specimens in number and arrangement of spinules possible). First free thoracic (second leg-bearing) segment about $\frac{2}{3}$ width of dorsal shield, with short lateral processes; second slightly longer and narrower. Third free thoracic segment with broad paired dorsal plates fused proximally and with at least part of margins serrated (extent of fusion, size and shape of plates greatly variable). Genital complex broad, with prominent rounded posterior plates, usually overlapping in mid-dorsal line and with posterolateral margin coarsely serrated. Abdomen (Fig. 937) one-segmented, dorsally covered by plates of genital complex; base narrow, width about equal to that of genital complex, anterior margin straight, posterolateral lobes extending well beyond tips of uropods; margins other than anterior with fine serrations. Total length about 20 mm.

First antenna (Fig. 938) three-segmented; basal segment about $\frac{1}{2}$ length of appendage, robust, broader at distal end, undetermined number of setae on distal half of anterior margin; second segment cylindrical, unarmed; third segment narrower than second, cylindrical, with two setae on posterior margin; apical armature (Fig. 939) with four long and six very short, papilliform setae (exact number not confirmed on sufficiently large material). Second antenna (Fig. 940) apparently three-segmented; basal segment short, unarmed; second muscular, robust, unarmed; third unciform, hooked 180°, robust at base, with slender tip; short seta at base, longer seta near midlength. Mouth tube long, slender; labrum shorter than labium, with pair of anterolateral aliform processes and marginal setae; margins of labium also setose. Mandible similar to that of *Cecrops* (number of teeth not determined). First maxilla (Fig. 941) laterally compressed, with broad base, narrowing distally, anterior margin sloping more than posterior and with one seta at midlength (vestigial exopod, anteriorly displaced); triangular apical part marked off by transverse suture or groove; small protuberance posterior to base of apical part. Second maxilla brachiform; lacertus robust, unarmed; brachium subcylindrical, slender; crista (Fig. 942) with tuft of flat, spiniform setae; calamus with slender denticles on anterior surface and denticulated, broad flanges along ventral and dorsal margins (Fig. 943); canna slightly shorter than calamus, with two marginal rows of denticles and one short row across base (Fig. 942). Maxilliped (Fig. 944) with cuticular wall longer on medial than on lateral surface; two spiniform processes in myxal area, one pointing anteriorly, one medially; one bulbous process near base of subchela (Fig. 945); diagonal cuticular ridge on medial surface beginning at myxal area; subchela indistinctly divided into robust shaft and strong, tapering claw, with short, stubby barbel at base.

First four pairs of legs biramous, greatly increasing in size from first to fourth. Sympods apparently unsegmented; that of first pair (Fig. 946) with three bulging adhesion pads, two corrugated (on medial margin and ventral surface) and one denticulated (on lateral margin); that of second pair (Fig. 948) with additional corrugated pad on ventral surface at base of exopod. Only one corrugated pad near medial margin of third sympod (Fig. 949), none on fourth (Fig. 950), but latter with curving flange on lateral margin in some specimens. Rami of first two pairs two-segmented; endopods with corrugated adhesion pads on basal segments; terminal segment of first endopod oval, with denticulated margins; distal segment of second with one terminal seta and crenated margins; basal segments of exopods with distolateral spines; terminal segments with strong distal spine; in first leg (Fig. 947) spiniform seta medial to spine and two blunt spines with short, flagelliform secondary setae near tips; in second three such spines. Rami of third and fourth legs one-segmented, foliaceous, very large, with uneven margins and varying armature. Fifth and sixth legs not observed. Genital orifices (Fig. 951) with rounded knobs on anterior margins. Uropod (Fig. 952) small, suborbicular, with four short spiniform setae posteriorly and one slender setule lateral and medial to them.

Male (Fig. 953): Cephalothoracic shield similar to that of female, more coarsely denticulated along lateral margins, differing in size and number of spines on dorsal surface. First two free thoracic segments without lateral flanges. Third free segment with dorsal plates more extensively fused, with only fairly shallow midposterior notch; margins of plates serrated. Genital complex with smaller dorsal plates, exposing part of abdomen in dorsal view; pair of genital plates on ventral surface at base of abdomen. Latter (Fig. 960) one-segmented, pyriform, with posterolateral lobes on either side of anus and uropods set dorsoventrally. Total length 10–15 mm.

First antenna relatively longer than that of female; second with comparatively larger and more slender claw. Maxilliped with one conical process and two subspherical corrugated pads in myxal area (Fig. 954) and one similar pad at base of subchela.

Third pair of legs (Fig. 956) with sympod broad, unsegmented; large adhesion pad on margin; rami two-segmented; exopod similar to those of preceding legs, but with simple terminal spines (Fig. 957); basal segment of endopod short, unarmed; terminal segment (Fig. 955) apically with long, digitiform process, slender curving spine about equally long and naked, somewhat shorter seta. Fifth and sixth legs not observed. Uropod (Figs. 958, 959) narrow, with straight lateral and medial margins and triangular posterior margin; posterior armature similar to that of female; small spiniform process with two setae at base on ventromedial surface near tip.

Comments: In spite of its long history, *O. muricatus* has not acquired a long list of synonyms. Its original generic name, *Laemargus*, having been found preoccupied, was replaced by Poche (1902) with its present name. Schuurmans Stekhoven and Schuurmans Stekhoven (1956) added two new species (*O. szidati* and *O. wilsoni*) to the genus, both clearly synonymous with *O. muricatus*. The differences they used as discriminants were mainly proportions of the various parts of the body and the nature and extent of spinulation, all characters with no taxonomic value. Their decision was influenced in no small measure by the fact that their *specimens* differed from Wilson's (1907b) *description* in several details. (Similar action was taken by Schuurmans Stekhoven (1936a, 1936b) with regard to Wilson's description of *Lernaeocera branchialis* (q.v.) from the American shores of the

Atlantic.) Difference of a similar nature (extent of marginal spinulation) between South African *specimens* of *O. muricatus* and Wilson's *description* were noted by Barnard (1955a).

With the exception of a single record (Wilson, 1907b) from *Selene vomer*, this species has always been found on *Mola mola*, clustering sometimes in large numbers on the skin of the host and producing extensive lesions, on occasions reaching into the subcutaneous layers and musculature. Most of the records come from the Atlantic Ocean and the Mediterranean, but the species has been recorded on a number of occasions in the Pacific (e.g. Shiino, 1954b; Kazachenko, 1972).

Genus Philorthagoriscus Horst, 1897

Philorthagoriscus serratus (Krøyer, 1863)

(Figs. 961–985)

Syn.: *Dinematura serrata* Krøyer, 1863

Female (Fig. 961): Dorsal shield of cephalothorax about equally long and broad, heavy, with lateral zones delimited and protruding backwards to level of third free thoracic (fourth leg-bearing) segment, posterior sinus deep, posterior margin straight to concave; lateral zones indistinctly divided into two parts by short transverse depressions; lateral margins serrated, with narrow marginal membrane; frontal plates fused with anterior margin; tips of first antennae almost reaching lateral limits of shield. First and second free thoracic segments fused together, first with prominent dorsolateral plates partly covered by tips of lateral zones. Third free thoracic segment about as broad as cephalothoracic shield, with narrow anterior end and large, paired dorsal plates, fused at base; plates with sloping anterior margins, dentiform outgrowths in anterolateral corners and serrated margins. Genital complex about as large as cephalothorax, half-covered by plates of third free segment; posterior end with two prominent, rounded plates covering abdomen; margins serrated. Abdomen (Fig. 962) one-segmented, broader than long, displaced forward along ventral surface of genital complex; base neck-like, subtriangular small lobes in posterolateral corners. Total length about 7–8 mm.

First antenna two-segmented; basal segment narrow, with undetermined number of denticles on anterior margin; terminal segment cylindrical, long, with single seta on posterior margin and undetermined apical armature. Second antenna (Fig. 963) three-segmented; basal segment subcylindrical, short, unarmed; second segment robust, somewhat inflated; third segment (Fig. 964) unciform, with short seta at base and longer one some distance from it. Mouth cone long, slender; labium longer than labrum, with fringe of spiniform setae; labrum (Fig. 965) with terminal flap bearing small bifid outgrowth and often deflected, anterolateral alae and long lateral strips of membrane. Mandible two-segmented, long and slender, with 12 teeth on dentiferous margin. First maxilla (Fig. 966) laterally compressed, narrow, distally tapering; exopod vestigial, represented by three setules on anterior margin; apex with digitiform process and small protuberance near its base. Second maxilla brachiform; lacertus robust, unarmed; brachium long, slender, cylindrical; clavus (Fig. 967) more than $\frac{1}{2}$ length of calamus, with broad base and blunt tip, surface beset with robust denticles; calamus with flange of serrated membrane on ventral surface and scattered denticles above it; canna about as long as calamus, with single row of spiniform setae along outer margin and two rows (one of very short denticles) on inner margin; dorsal surface (not seen in Fig. 967) spinulated. Maxilliped (Fig. 968) with corpus fairly long and slender, prominent conical outgrowth and sclerotized shelf in myxal area; subchela greatly tapering, curved, not divided into shaft and claw and apparently unarmed (Shiino, 1954b, observed division into claw and shaft and small barbel at base of claw).

First four pairs of legs biramous. Sympods indistinctly segmented; that of first leg (Fig. 969) with two corrugated adhesion pads (one on medial margin, one on ventral surface); slender seta on medial side of endopod base and another lateral to that of exopod. Sympod of second leg (Fig. 971) with pair of corrugated pads on medial margin, also with fringe of short setae and one prominent seta near base; paired pads on ventral surface near anteromedial corner, single pad near base of exopod. Sympod of third leg (Fig. 973) broadly expanded, with semicircular posterior flap and two poorly delimited pads on ventral surface; posterior margin with two strips of membrane meeting at centre near marginal papilla; spinulated pad near base of exopod. Sympod of fourth leg (Fig. 976) oval, unarmed, much smaller than preceding three. Rami of first three pairs two-segmented, those of fourth one-segmented. Armature formula below.

	Endopod		Exopod	
	1	2	1	2
Leg 1	0-0	3	0-I	3,IV
Leg 2	1-0	7	1-I	5,IV
Leg 3	1-0	4	0-I	4,III
Leg 4	3	—	4,IV	—

Basal segment of first endopod with adhesion pad on ventral surface, terminal with fringe of setae on lateral margin; three distal spines of first exopod with serrated margins, one smooth (Fig. 970). Lateral margin of second endopod fringed with setae, distal segment with short setae on medial margin also (Fig. 971); exopod with denticulated lateral margin of basal segment, terminal segment with rounded denticles (Fig. 972). Third endopod (Fig. 975) with fringes on all margins except medial of basal segment; exopod (Fig. 974) with medial margins fringed, lateral denticulated; small setiferous papilla on margin of basal segment. Armature of fourth exopod (Fig. 976) of small spines and setae interspersed in irregular series. Fifth and sixth legs not observed. Uropod (Fig. 962) longer than wide, with six setae (two very short, four longer, all pinnate) on posterior margin.

Male (Fig. 977): Cephalothorax similar to that of female, with more distinct transverse sutures on lateral zones; tips of first antennae reaching up to or beyond lateral limits of shield. First two free thoracic segments (second and third leg-bearing) also similar to those of female. Third free segment with paired dorsal plates, small and overlapping base of genital complex; posterior margins of plates serrated. Genital complex of about equal length and width, nearly $\frac{1}{2}$ length of cephalothorax; lateral margins either about parallel or converging posteriorly, with prominent triangular processes in anterior half and in posterolateral corners; small posterior lobes partly covering abdomen in dorsal view. Abdomen one-segmented, less than $\frac{1}{2}$ length of genital complex. Total length 4–5 mm.

First antenna (Fig. 978) relatively longer than that of female; first segment with more than 20 setae (number varying due to damage) on distal half of anterior margin and adjacent part of ventral surface; second segment cylindrical, with rounded apex, one seta on posterior margin and apical armature (details not determined). Second antenna and oral appendages as in female. Maxilliped (Fig. 979) with prominent conical outgrowth and two pads in myxal area and small pad near base of subchela; latter long, with prominent secondary outgrowth near base and very slender, sharp tip.

First two pairs of legs similar to those of female (cf. Fig. 980 with Fig. 969), but some exopod spines with secondary subterminal spines (Fig. 981). Third leg (Fig. 982) with relatively less expanded sympod; endopod modified, with segmental boundary obliterated (Fig. 983); powerful claw largely concealed in pocket-like cavity; sharp, prominent denticle and soft papilla on rim of cavity near tip of claw, with slender, blunt seta distal to it; distal margin with three long, pinnate setae, medial with short fringe. Fifth and sixth legs not observed. Uropod (Fig. 985) relatively longer and narrower than that of female, with setae comparatively longer.

Comments: In more than a century of its recorded history, *P. serratus* has been completely clear of taxonomic difficulties. Its incorrect generic affiliation with *Dinemoura* was rectified by Horst (1897). There were no misidentifications or invalid specific designations. The only literature record approaching nomenclatural confusion is mentioned by Scott and Scott (1913). According to them, *P. serratus* is recorded "under the Rev. A. M. Norman's MS. name *Momina fimbriata* in the appendix to Smiles' "Life of Thomas Edward" (p. 437, 1876), as one of the many creatures that keen-sighted naturalist added to the fauna of Scotland." Norman's manuscript name was not used by any subsequent writers, or, apparently, validly published.

P. serratus is a parasite of *Mola* and is often taken on the skin of that fish together with *Orthagoriscicola muricatus* and *Cecrops latreillii*. Its distribution records come predominantly from the Atlantic Ocean, but it occurs also in the Pacific (Shiino, 1954b). Brian (1906) notes its presence in the Mediterranean, pointing out that some confusion existed in his earlier papers between it and *Orthagoriscicola muricatus*. Wilson (1922b) reported a find on *Squalus acanthias* taken at Newfoundland. Although he mentioned that three lots of specimens were collected from that fish, the record appears doubtful, particularly as most of these specimens were immature and could have been incorrectly identified.

Genus Luetkenia Claus, 1864

Female: Cecropidae. Cephalothoracic shield caligiform, with lateral zones firmly delimited, divided into two parts by transverse sutures, extending posteriorly to, or nearly to, level of third free (fourth leg-bearing) segment; marginal membrane present or absent; frontal plates fused with anterior margin, suture discernible or not. First and second free thoracic segments fused together, first with dorsolateral plates. Third free segment with paired dorsal plates covering about half of genital complex, plates either fused only at base or almost completely. Genital complex about as long, or more than twice as long, as cephalothorax, with paired posterior lobes covering abdomen dorsally. Abdomen one-segmented, width about equal to, or double the length. Egg sacs straight. First antenna two-segmented, second three-segmented, with unciform claw. Oral appendages siphonostome. Maxilliped subchelate. Four pairs of biramous legs, rami of first three two-segmented, fourth one-segmented. Fifth and sixth legs absent. Uropods present.

Male: Cephalothorax as in female. Plates of third free segment small, covering only base of genital complex. Latter shorter than cephalothorax, with short posterolateral lobes. Abdomen visible in dorsal view, small, one-segmented. Appendages similar to those of female but third endopod modified. Fifth (or sixth) leg vestigial.

Type species: *Luetkenia asterodermi* Claus, 1864.

Comments: The genus *Luetkenia* contained only its type species until recently. Shiino (1963a) discovered *L. elongata* off the coast of California on the skin of *Luvarus imperialis*. Generally resembling *L. asterodermi*, Shiino's species can be distinguished from it by the pronounced elongation of its genital complex and by dorsal plates of the third free thoracic segment. Its maxilliped is more robust and squat. Richiardi (1880) added to the genus the name *L. integra*, applied to a parasite never described and taken from two species of sharks in the Mediterranean. Technically a nomen nudum, this species was considered by Brian (1906) to be a synonym of *Nessipus orientalis* Heller, 1865.

Luetkenia asterodermi Claus, 1864

(Figs. 986–1002)

Syn.: *Lütkenia asterodermi* Claus, 1864
Cecropsina glabra Heller, 1865
Lütkenia glabra Heller; of Valle (1880)
Luetkenia astrodermi; of Bassett-Smith (1899)

Female (Fig. 986): Cephalothoracic shield about equally wide and long, with notch in centre of anterior margin and one in both lateral margins; frontal plates fused with shield, suture indistinct or absent; tips of first antennae not reaching lateral limits of shield; lateral zones clearly delimited, divided by transverse suture into two parts, posteriorly reaching level of third free thoracic (fourth leg-bearing) segment; marginal membrane apparently absent; posterior sinus deep, posterior margin concave. First and second free thoracic segments fused; first with posterolateral plates. Third free segment with paired dorsal plates almost completely fused, covering about half of genital complex. Latter narrower than cephalothoracic shield, with prominent rounded posterior lobes, sometimes overlapping each other in mid-dorsal line and always covering abdomen dorsally; length of genital complex (posterior lobes included) somewhat more than that of cephalothorax. Abdomen (Fig. 987) one-segmented, about equally wide and long, with narrow base, rounded anterolateral corners and straight posterior margin. Total length about 10 mm.

First antenna (Fig. 988) two-segmented; basal segment robust, with setae on distal half of anterior margin and adjacent ventral surface (exact number not determined on sufficient material); terminal segment cylindrical, with rounded tip and single seta on posterior margin; details of apical armature not completely ascertained (in specimens examined nine apical setae). Second antenna (Fig. 989) three-segmented; first segment with sigmoid flexure, unarmed; second muscular, slightly inflated near distal end, unarmed; third unciform, with short seta at base and similar seta at about half length. Mouth cone long, slender; labrum shorter than labium, with anterolateral aliform processes; details of labium not ascertained. Mandible long, slender, with 11 teeth in dentiform margin (Fig. 990). First maxilla (Fig. 991) laterally compressed, tapering, with small slender process on anterior margin (vestigial exopod) and long, conical apical process, sometimes laterally inclined. Second maxilla (Fig. 992) brachiform; lacertus robust, unarmed; brachium long, cylindrical; calamus tapering, with several longitudinal rows of denticles; canna about half length of calamus, similarly armed; clavus about as long as canna, tapering, denticulated. Maxilliped (Fig. 993) with slender, indistinctly two-segmented corpus and small depression in myxal area; subchela long, slender, not clearly divided into shaft and claw, with one setule close to base.

Four pairs of biramous legs, first and fourth pair much smaller than others. Sympods of first three pairs indistinctly two-segmented; fourth unsegmented. First sympod (Fig. 994) with seta at base of both rami; second (Fig. 995) with fringe of short setae and one spine on medial margin; third (Fig. 996) expanded, with fringed medial margin and prominent lateral spinulated lobe; fourth (Fig. 999) narrow, with spiniform seta lateral to base of exopod and distolateral small denticulated swelling. First three pairs with rami two-segmented; fourth one-segmented. Armature formula below.

	Endopod		Exopod	
	1	2	1	2
Leg 1	0-0	3	0-I	7
Leg 2	1-0	7	1-I	9
Leg 3	0-0	4	0-I	7
Leg 4	4	—	7	—

All setae and spines very short and difficult to distinguish from one another, some with smooth, others with denticulated margins. First endopod much smaller than exopod (Fig. 994); basal segment of second exopod with small conical papilla and abundant spinulation on lateral margin (Fig. 995); both rami of third leg with fringed margins (Figs. 997, 998); endopod of fourth leg much shorter than exopod (Fig. 999). Fifth and sixth legs not observed. Uropod (Fig. 987) suboval, distally with four longer and two short pinnate setae (Fig. 1000).

Male (Fig. 1001): Cephalothorax and first two free segments similar to those of female. Third free segment with paired, rounded, dorsal plates covering base of genital complex. Latter oval, longer than wide, about half length of cephalothorax, with rounded posterolateral lobes. Abdomen one-segmented, not projecting beyond level of posterior end of genital complex. Total length about 6 mm.

Appendages similar to those of female. Legs (Fig. 1002) with pinnate setae much longer than corresponding spiniform setae of female. Third endopod modified, with strong lateral claw resembling that of *Philorthagoriscus* (cf. Fig. 983). Uropod narrow and longer than wide.

Comments: The author had no opportunity to examine any male specimens. The above description is based in part on Heller's (1865) and in part on comparison with the male of *L. elongata* Shiino, 1963.

L. asterodermi is a rare species. Originally described off Messina (Mediterranean), it has been most frequently reported from that sea and the Adriatic. The original host, on which the specific name of the parasite is based, was *Asterodermus coryphaenoides*, later identified as a juvenile form of *Luvarus imperialis*. It is that fish that is the main, indeed probably the only, host of *L. asterodermi*. Claus (1864) and Heller (1865) discovered it almost at the same time, hence Heller's lack of knowledge of Claus' work and his designation of the new copepod as *Cecropsina glabra*. Brian (1906) reported a find on *Asterodermus elegans*. Interestingly, Brian's record appears to have been the last from the Mediterranean. In the same year, Norman and Scott (1906) reported a find on *Luvarus imperialis*, the luvar, in Cornwall. To this author's knowledge, no later records appeared in the literature, though the old ones have been repeated several times. The rarity of *L. asterodermi* is directly related to that of its host, the luvar (see Wheeler, 1969). It can be assumed that the distribution of *L. asterodermi* is much broader than indicated by its reported finds. The author examined four females of this species collected in 1969 from the gills of a luvar washed ashore on Mexico Beach, Florida and kept in the collection of the Smithsonian Institution, Washington, D.C. (No. 285653).

With so little material to work on, the species is not sufficiently known. In particular, the range of intraspecific variability cannot be established. The author's Fig. 986, for example, compared with Heller's (1865) Pl. XIX, Fig. 1, shows that some variability exists at least in the extent of elongation of the genital complex. Heller's specimen is clearly more elongate than those examined by the author. Should this range prove to be broader than indicated by the material hitherto available, the validity of *L. elongata* would have to be reconsidered. Its appendages are virtually the same as those of *L. asterodermi* and it does live on the same host species. Intermediate stages between the two species may well yet be found.

Family Dichelesthiidae comb. nov.

In its new combination, presented below, the family Dichelesthiidae consists of only two extant genera: *Dichelesthium* and *Anthosoma*, both combining primitive as well as advanced morphological features. A recently reported fossil copepod, parasitic on a Lower Cretaceous fish found in Brazil (Cressey and Patterson, 1973), is also a member of this family. Both extant genera are monotypic, so that Dichelesthiidae in their present strict sense comprise only two species. Under the circumstances the discussion of the morphology of the members of the family can be conveniently left until the individual species are dealt with.

The beginnings of our knowledge of Dichelesthiidae go back to the immediate post-Linnaean period, when in a single paper Abildgaard (1794) described its two present members under the names *Caligus crassus* and *C. oblongus*. The latter became the type and the only species of the genus *Dichelesthium* Hermann, 1804, whereas the former came to occupy the same position in the genus *Anthosoma* Leach, 1816.

Leach (1816) was the first to attempt placing *Anthosoma* in a broader system of classification. His entomostracan division I ("body covered by a horizontal shield") contained two groups of species, both characterized by a shield composed of "but one part". One of these two groups was "with jaws", the other possessing no jaws but being equipped with a rostrum. The latter group was further subdivided into species possessing "antennae four" (*Argulus*) and those with "antennae two" (*Cecrops*, *Caligus*, *Pandarus* and *Anthosoma*). Although by that time the genus *Dichelesthium* had been established, Leach made no mention of it in his scheme, perhaps still believing that *Caligus oblongus* was in its rightful genus.

Von Nordmann (1832), following the classification suggested by Latreille (1829), recognized two families within Siphonostomata. His scheme was as follows.

I. Caligides.
   Tribe 1: Pinnodactyla (*Argulus*, *Caligus*, *Pandarus*, *Pterygopoda*, ?*Nogaus*).
   Tribe 2: Hymenopoda (*Dinemura*, *Anthosoma*, *Cecrops*).
II. Lernaeiformes (*Nicothoe*, *Dichelesthium*, *Nemesis*).

Curiously enough, the second family included neither *Lernaea* nor any true lernaeids, at that time sometimes left out of Crustacea altogether. The genus *Nemesis* Risso, 1826, later placed in Dichelesthiidae *sensu lato* by many authors is linked by von Nordmann with *Dichelesthium*. *Anthosoma*, however, is placed with two other genera, now known to be unrelated but sharing with it the eye-catching presence of foliaceous cuticular structures. This all-too-obvious but superficial distinction was to exert an influence on all future systematists of Dichelesthiidae.

Von Nordmann's work was followed almost immediately by that of Burmeister (1835). With a good deal of original thought, Burmeister regrouped the parasitic copepods known at that time. The dichelesthiid genera found themselves in the family Ergasilina, which was divided into two groups of genera, those with and those without a proboscis-like mouth. The only genera of the latter group were *Anthosoma*, *Dichelesthium* and *Nemesis*, the leaf-like expansions of the appendages of the first of them not being considered by Burmeister as significant enough to exclude it from this grouping. Burmeister's work exerted some influence on his successors, notably on Krøyer (1837) who retained these three genera as one of the two subdivisions in the family Ergasilina.

A novel concept of affinities between genera of this group was introduced by Edwards (1840), who was the first author to see them as a separate, if heterogeneous, group, distinguishable from other groups. His definition of the "tribu des Dichelestiens" was as follows: "Easily distinguishable from the Ergasiliens by the elongate form of their bodies, small size of their heads and vestigial condition of their abdomina. It should also be noted that their legs are less well developed. On the other hand, their organs of attachment are better developed and indicate that they are more advanced as parasites." The "tribe" contained *Anthosoma* (considered as showing affinities with the pandarids), *Nemesis* (bearing many features in common with the ergasilids) and *Dichelesthium and Lamproglena* (showing features transitional towards the lernaeids). With the work of Edwards, a taxon specifically intended to stress the underlying affinity to one another of the dichelesthiids *sensu lato* has made its appearance for the first time. Although its name did not receive the currently acceptable familial ending, it would be proper to view Edwards as the author of the family now known as Dichelesthiidae.

The later experts can be divided into two categories: those who were concerned with the overall classification of parasitic copepods and those who deliberately restricted themselves to their representatives within certain geographical regions. The work of the latter authors often contains gaps created by the absence, from the region of their interest, of species representing some taxonomic groups and leaves us in doubt as to the writers' views on the affinities within the parasitic copepods. One such author was Baird (1850), who dealt with the dichelesthiids only through their British representatives known to him. Like Edwards, he separated Ergasilidae from the dichelesthiids and, believing only *Anthosoma* to be present in British waters, established the family Anthosomadae, containing only its type genus and including in its synonymy Edwards' Dichelestidae (sic) (partim), Leach's Caligidae race I (partim) and Burmeister's Ergasilina (partim). No definition of the family was given, neither was it made clear what was Baird's view on the position of the other genera considered at various times as dichelesthiid.

At about the same time, the renowned American carcinologist, Dana (1853), published his major work which included the systematics of the family Dichelesthiidae. In view of its importance for the subsequent development of our concept of the family, Dana's Latin text is translated below.

Fam. III. Dichelestidae.

Subfam. 1. Dichelestinae. Body slender, segmented, not armed with elytra or similar structures.

Genus 1. *Dichelestium*, Hermann. Body slender, segmented, with non-alate segments. Second antennae long, stout, with two apical digits. Second legs (= maxillipeds, auth. comm.) prehensile, with tapering tips. Four anterior swimming legs very small, biramous; third pair in form of medium sized lobes; fourth pair obsolete.

Genus 2. *Nemesis*, Roux. Body slender, usually straight, segmented, with segments subequal, non-alate. Second legs (= maxillipeds, auth. comm.) with simple claws; first swimming legs simple, usually unarmed; second, third and fourth short, biramous, unarmed.

Subfam. 2. Anthosomatinae. Body slender, foliate and indistinctly segmented

Genus 1. *Anthosoma*, Leach. Cephalothorax two-segmented, first segment long, second with posterior elongate and bilobed alae. Second antennae long, with uncinate tips. Second legs (= maxillipeds, auth. comm.) with tapering tips. Swimming legs provided with lateral laminae, covering body from both sides in posterior part.

Though aware of Edwards' placing of *Lamproglena* von Nordmann, 1832, with Dichelesthiidae, Dana was doubtful of its affinity with this family and pointed out many differences between it and the other genera within Dichelesthiidae, beginning with segmentation and ending with the structure of the appendages and the type of the egg sac. Dana's classification was the first one in which an explicit demarcation line was drawn between *Dichelesthium* and *Anthosoma* at higher than generic level. In this it was in agreement with Baird's (1850) implicit division of similar kind.

Steenstrup and Lütken (1861) accepted "Dichelestiner" as a valid higher taxon. They included in it more genera than any previous authors (*Kroyeria*, *Pagodina*, *Eudactylina*, *Congericola*, *Ergasilina*, *Lernanthropus*, *Dichelesthium*, *Nemesis*, *Lamproglena* and *Anthosoma*), thus rendering the family unit rather similar to the later catch-all taxon which it remained until recently. Most of the newly-added genera were described by van Beneden (1851a, 1853a, 1853b, 1854), a prolific Belgian worker.

The later work of Krøyer (1863, 1864) did not add significantly to the development of the concept of Dichelesthiidae. That of von Nordmann (1864) derived from the older authors' ideas on the systematics and accepted the structure of the egg sacs as the basic division of major taxa. The group with filiform, uniseriate sacs (as distinct from saccular, multiseriate) included "Dichelestini" (*Dichelesthium*, *Kroyeria*, *Pagodina*, *Eudactylina*, *Congericola*, *Ergasilina*, *Lernanthropus*, *Nemesis*, *Lamproglena*, *Stalagmus*, *Donusa*, and *Anthosoma*). (Of these, *Lamproglena* is now known to belong to Cyclopoida, whereas *Ergasilina*, *Stalagmus* and *Donusa* are not valid genera.) No attempt was made by von Nordmann to search for internal relationships of the group.

Heller (1865), by providing a key to the family which he referred to as "Familia Dichelestina", offered us a scheme, at least a tentative one, of intrafamilial grouping. His key was as follows.

I. Body without foliaceous appendages.
  1. Body clearly segmented, with simple posterior legs.
    A. Second antennae cheliform.
      a. All abdominal legs biramous, rami three-segmented . . . . **Kroeyeria**
      b. Second and third pair of legs biramous, rami never segmented, third pair expanded, fourth pair absent . . . . **Dichelestium**
    B. Second antenna uncinate.
      a. Two pairs of legs . . . . **Clavella**
      b. Four pairs of legs.

i. Second maxillipeds cheliform . . . . ***Eudactylina***
ii. Second maxillipeds uncinate.
** Second maxillipeds slender . . . . ***Cycnus***
** Second maxillipeds massive . . . . ***Nemesis***
C. Second antenna simple, slender, setiferous . . . . ***Lamproglena***
2. Body poorly segmented, posterior legs foliaceous, laciniate . . . . ***Lernanthropus***
II. Body with two foliaceous elytra . . . . ***Anthosoma***

In addition to the above genera, Heller quoted van Beneden's view that *Enterocola* (a cyclopoid) should also be included in this family. He agreed that *Pagodina* was barely discernible from, and should be synonymized with, *Nemesis*, but he was not aware of the fact that *Ergasilina* was also a synonym of that genus. (The uniramous nature of the legs of *Ergasilina*, as suggested in this key, was postulated erroneously.) He also agreed that the name *Cycnus*, as preoccupied, should be replaced by *Congericola*. His use of the name *Clavella*, common at that time, was *sensu* Oken (1816) and applied to those species which later (Poche, 1902) were identified as *Hatschekia*.

The main interest of Heller's key is its use of several morphological criteria in conjunction, although the relative value of these criteria was not always correctly gauged. Not enough attention was paid to the segmentation of the body, undue importance was attached to the presence of elytra in *Anthosoma*, but at least the number of swimming legs and the structure of the maxillipeds (Heller's "second maxillipeds") and of the second antennae were taken into account. Heller also added a new genus *Pseudocycnus* to the family.

Similar systematization was adopted in Gerstaecker's (1866–1879) general account of Copepoda. His family "Dichelesthiina" was diagnosed as follows: "Body as a rule elongate, depressed, usually incompletely segmented, with abdominal segments free (only exceptionally armed with dorsal elytra). Anterior feelers (= first antennae, auth. comm.) free, thin, setiform, multisegmented (up to 15 segments), seldom quite short and only two- or three-segmented. Posterior feelers (= second antennae, auth. comm.) almost exclusively unciform or chelate, usually extending beyond anterior margin of cephalothorax. Mouthparts and maxillipeds as in Caligidae. Usually only four pairs of swimming legs, often small and stumpy, sometimes absent or partly deformed into lamelliform elytra. Eyes simple, median or absent. Males and females similar in shape and size or only slightly different; both, or at least the latter, sessile parasites."

Attempting to accommodate many strikingly different genera within its framework, the diagnosis is rather vague and overly inclusive. The genera included in the key to the family are: *Anthosoma*, *Tucca* (a poecilostome), *Norion*, *Epachthes* (syn. of *Lernanthropus*), *Lernanthropus*, *Stalagmus* (syn. of *Lernanthropus*), *Dichelesthium*, *Kroyeria*, *Pseudocycnus*, *Baculus* (a larval stage of the pennellid *Pennella*), *Philichthys* (type genus of *Phylichthyidae*), *Clavella* (= *Hatschekia*), *Nemesis*, *Congericola*, *Eudactylina*, *Lamproglena*, *Aethon*, *Ergasilina* (syn. of *Nemesis*) and *Donusa* (parasite of annelids, possibly a cyclopoid).

It is clear that with the continuing discovery and description of new species and genera which could not be fitted into other well-defined families, many of them found their way by default, as it were, into the family Dichelesthiidae, a taxon with rather vaguely drawn borders. Each new addition compounded the problem, bringing with it its own morphological features. This situation was carried over from the end of the 19th into the early 20th century. Brian (1906), who dealt only with the parasitic copepods of Italian fishes, included in Dichelesthiidae the genera *Anthosoma*, *Lernanthropus*, *Dichelesthium*, *Kroyeria*, *Hatschekia*, *Nemesis*, *Cycnus* (= *Congericola*), *Pseudocycnus*, *Eudactylina* and *Lamproglena*. Scott and Scott (1913) placed in this family *Dichelesthium*, *Anthosoma*, *Lernanthropus*, *Hatschekia*, *Kroyeria*, *Congericola* and *Eudactylina*, all representatives of the family in British waters.

The first author who attempted a review of intrafamilial organization of Dichelesthiidae was Wilson (1922a). As an American, he was close to the work of Dana (1853), so that it comes as no surprise to see his subfamilial divisions running more or less along the lines resembling Dana's, at least as regards the separation of *Dichelesthium* and *Anthosoma*.

Wilson's key to the subfamilies was as follows.

1. One or more body regions furnished with plates or wings, or both . . . . **Anthosominae**
No plates or wings on any body region 2

2. All four pairs of swimming legs present and equally developed . . . . **Eudactylinae**
All four pairs of swimming legs present, but one or more of them modified or rudimentary **Pseudocycninae**
One or more pairs of swimming legs lacking; those present usually modified or rudimentary **Dichelesthiinae**

Two criteria were used for the purpose of this subdivision: the presence of elytra or similar structures and the number and structure of swimming legs, greater value being attached to the former. In his subsequent account of the Copepoda of the Woods Hole Region, Wilson (1932) upgraded the status of his four subfamilies and established four families, distributing among them former dichelesthiid genera as follows.

Anthosomidae (*Anthosoma*, *Lernanthropus*);
Eudactylinidae (*Kroyeria*, *Kroeyerina*, *Nemesis*, *Eudactylina*, *Eudactylinodes*, *Eudactylinella*);
Pseudocycnidae (*Pseudocycnus*);
Dichelesthiidae (*Hatschekia, Pseudohatschekia, Pseudocongericola* and *Lamproglena*)

Several genera among those enumerated above were first established by Wilson himself. The notable absence of *Dichelesthium* from Dichelesthiidae was simply due to the fact that Wilson had no record of this genus from the Woods Hole region.

Wilson's work exerted great influence over subsequent students of parasitic copepods, so that his system of subfamilial (or familial) subdivision was followed by some important authors. Yamaguti (1939b) accepted the familial rank of the original four subfamilies, while Markevich (1956) retained them at the subfamilial level and transferred the genus *Lamproglena* to Eudactylinidae.

The most recent major revision of the group came with the work of Yamaguti (1963). It contains, however, many puzzling features, not the least of which is the fact that it claims the authorship of several new families (Anthosomatidae, Eudactylinidae and Pseudocycnidae), though these taxa were proposed by Wilson (1932) and later used by Yamaguti himself (1939). The main decision taken in this revision was the establishment of a new superfamily Dichelesthioidea, which, by gathering together all the splinter families of the former Dichelesthiidae *sensu lato*, implies the existence of a relationship between them. In addition to the four families, previously placed here, Yamaguti also adds to his superfamily a family Catlaphilidae Tripathi, 1960. This addition of a single and very badly described species is highly suspect and cannot be accepted without much more evidence of affinity between *Catlaphila elongata* Tripathi, 1960 and the rest of the group. The broad criteria for subdivisions into families are, in general, derived from Wilson (1922a, 1932), and can be seen as the direct outcome of a century-long confusion to which little original thought was contributed by the distinguished author. A good example of this confusion is *Metahatschekia congeri* Yamaguti and Yamasu, 1959, a species parasitic on the Japanese conger and placed in a new genus. Comparison with *Congericola pallidus* (van Beneden, 1854), parasitic on the European conger, shows clearly that the Japanese copepod is another species of *Congericola*, from which it differs mainly in the armature of the swimming legs. In Yamaguti's scheme, however, the genus *Metahatschekia* belongs to Dichelesthiidae, whereas *Congericola* is a member of Eudactylinidae.

The taxonomic situation of the group is clearly unsatisfactory, the only certain fact in it being the absence of affinity between various groups of genera and the necessity for breaking up the old Dichelesthiidae *sensu lato*. The present author compared all the genera of that family and made an assessment of their morphological differences and similarities with a view to reclassification and revision.

The first step to the setting up of any taxonomic system is a firm assessment of the diagnostic values of various morphological features. This is possible only with full knowledge of the comparative morphology of the revised group and with a reasonably good appreciation of the functional meaning of various morphological structures. Structures which are known to have a direct adaptive importance can vary greatly in two closely related species, depending as they do, to a large measure, on the conditions encountered by the copepod on its host and in its particular habitat. On the other hand, structural features of a more basic nature and, apparently at least, devoid of immediate functional significance are more likely to be an indication of relationships, since the reason for their differences must be sought in much more profound genetic divergence. For example, the type of segmentation, coupled with the general habitus of the body, is much more likely to be of taxonomic significance than the type of armature on appendages, particularly prehensile appendages. Among the caligiform copepods, the difference in the number of segments present between the cephalothorax and the genital complex determines the familial status of the species (p. 6). There is no reason why the same feature should be left out in determining the taxonomic status of the former dichelesthiid species. At present, however, the segmentation of species belonging to different splinter families of Dichelesthiidae *sensu lato* is either disregarded or misrepresented.

The most primitive type of segmentation known in this group is exemplified by *Nemesis* (Text fig.

6; Figs. 1295–1298). In this genus the cephalothorax contains the first leg-bearing segment, but thoracic segments bearing the second to the fifth legs are clearly demarcated. (The fifth leg-bearing segment is usually noticeably smaller than the preceding segment or the following genital segment.) In the female the abdomen has often one or more segments less than in the male. In addition to *Nemesis*, this type of segmentation occurs in *Eudactylina*, *Eudactylinella*, *Eudactylinodes*, *Eudactylinopsis*, *Protodactylina* and *Bariaka*.

The second group comprises genera with three (exceptionally four) segments distinguishable between the cephalothorax and the genital complex. As in all other groups, the first leg-bearing segment is fused with the cephalothorax, but the original fifth leg-bearing segment has become fused with the genital segment. Hence the resulting tagma is referred to as the genital complex (similar to that occurring in Caligidae and their allies). There are two distinct groups with this type of segmentation. One of them comprises *Dichelesthium* and *Anthosoma* (though in the former the female has four indistinct segments between the cephalothorax and genital segment, while in the latter the first leg-bearing segment shows traces of separation from the cephalothorax only on the ventral side and only in some, particularly younger, specimens). The segmentation is more distinct in the males than in the females and can be seen in Fig. 1006 (*Dichelesthium*) and Fig. 1030 (*Anthosoma*). In another group with this type of segmentation are the genera *Kroyeria* (Text fig. 6; Fig. 1082) and *Kroeyerina*. The most important difference between these two groups lies in the possession by *Dichelesthium* and *Anthosoma* of three pairs of partially modified swimming legs, while *Kroyeria* and *Kroeyerina* possess four pairs of well-developed biramous legs.

Further loss of segmentation is shown by the group in which only one or two segments are present (when distinguishable) between the cephalothorax and the genital complex. The segments intervening between the cephalothorax and the posterior tagma often form, in this group, a narrow neck region. This type of segmentation is shown by two sets of genera. The first set comprises *Pseudocycnus*, *Pseudocycnoides* and *Cybicola*, all three with poorly marked segmental boundaries. In *Pseudocycnoides* the boundaries are indistinguishable and, according to Yamaguti (1963), the first leg-bearing segment is not a part of the cephalothorax. (The latter fact needs confirmation.) In the second set of genera it becomes impossible to determine with certainty which segments take part in the formation of the neck region, sometimes only transverse wrinkles being taken by various authors as evidence of the existence of distinct segmentation. To this group belongs *Hatschekia* (apparently second and third leg-bearing segments in the neck region), *Prohatschekia* (first and second), *Congericola* (second and third) (Text fig. 6), *Pseudocongericola* (second and third) and *Bassettithia* (first to fourth).

Lastly, there are genera in which no free segments remain behind the cephalothorax, only a narrow constriction separating it from the next tagma. In some species of this group a deep constriction between the fourth and fifth leg-bearing segment divides the genital complex into a pregenital trunk and a genito-abdominal region. The group comprises *Lernanthropus* (Text fig. 6), *Lernanthropodes*, *Sagum*, *Aethon* and *Norion*. (In the last two, traces of segmentation are observable in the dorsal region of the trunk immediately behind the cephalothorax.)

Using the type of segmentation as the primary criterion for classification, one can, then, distinguish six groups of genera within the miscellaneous assemblage previously constituting Dichelesthiidae. These groups were set out in Text figs. 53, 54, which review the types of appendages (second antenna, second maxilla and maxillipeds). Comparison shows that the type of appendage within each group is fairly similar, though there are differences, sometimes major ones, from one group to the next.

The most common type of second antenna is a simple subchelate appendage, with compact corpus and simple uncinate subchela, ending in a claw, usually not clearly demarcated from the shaft. This type occurs in *Pseudocycnus* and its group (genera No. 12–14 in Text fig. 53). A special feature in this group is the presence of secondary tines on the subchela (detailed view of this antenna is shown in Fig. 1314). A similar antenna, though usually without secondary tines, is found in *Lernanthropus* and its allies (genera No. 15–19 in Text fig. 54), as well as in *Hatschekia* and its group (genera No. 20–24 in Text fig. 54). In the latter genus, however, the antenna is distinguished by the presence, at its base, of a papilliform process (indicated in outline in Text fig. 54, and shown in Figs. 1106, 1120, 1132, 1146 and 1154). An exclusive situation as regards the second antenna exists in *Anthosoma* and *Dichelesthium*, parasites with the second antenna retractable and apparently with the corpus more than one-segmented (genera No. 1, 2 in Text fig. 53). For details of these antennae see Figs. 1009, 1010, 1037 and 1038. The second antenna in the genus *Eudactylina* and its allies (No. 3–9 in Text fig.

53) is totally different, being five-segmented, with the third, and sometimes second, segment armed with hooked outgrowths. The terminal segment is a claw, usually with a secondary process. An even more striking difference occurs in the second antenna of *Kroyeria* and *Kroeyerina* (No. 10, 11 in Text fig. 53), which is chelate and surprisingly reminiscent of the maxilliped of *Eudactylina* (Fig. 1087).

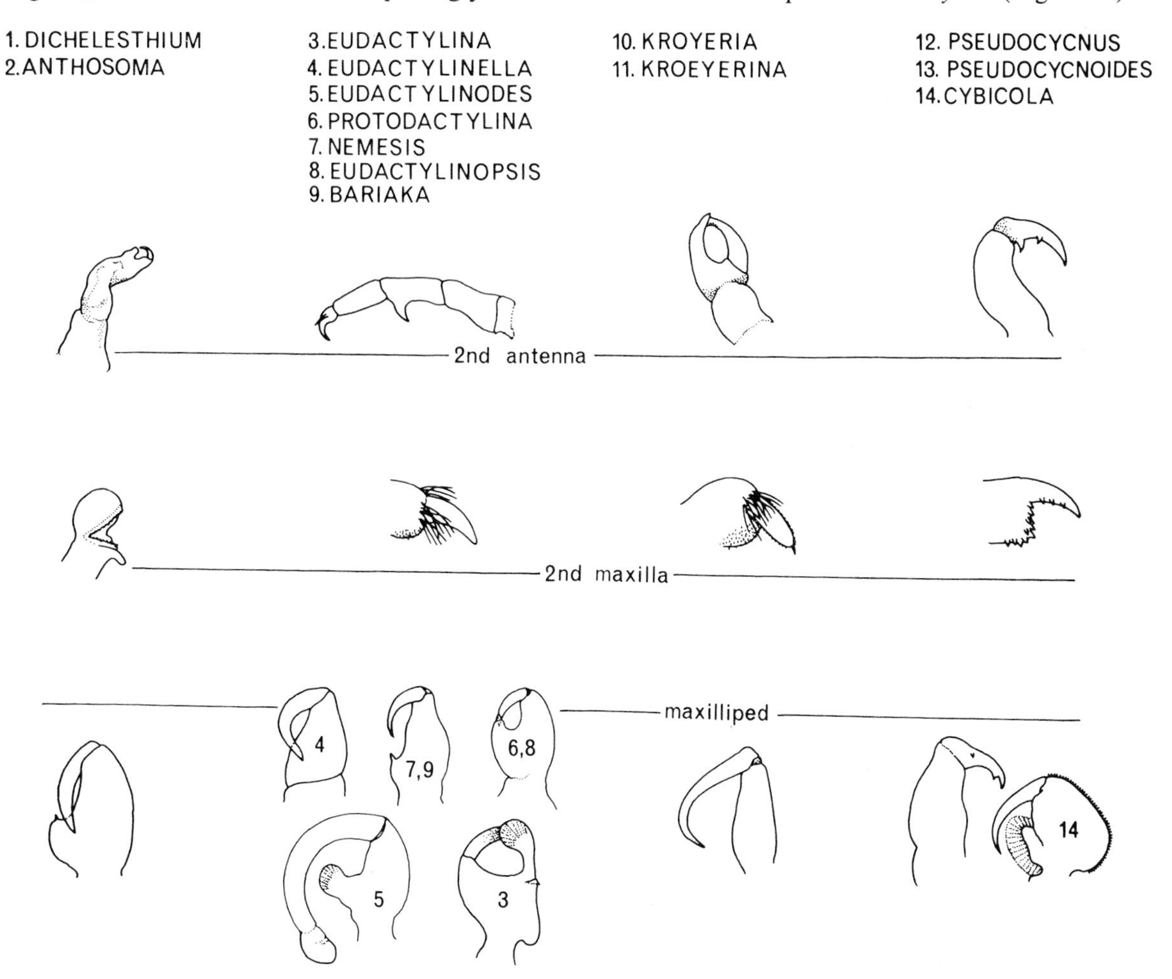

Text fig. 53. Diagrammatic comparison of salient morphological features of copepods formerly included in Dichelesthiidae *sensu lato* (part 1). (See also Text fig. 54.)

All genera have second maxilla of the usual brachiform structure (p. 33). Differences exist among groups as regards the shape of the terminal segment, though within each group a fair degree of resemblance is retained. The most unique type of structure is found in *Dichelesthium* and *Anthosoma*. Details are shown in Figs. 1016–1019 and 1043, 1044. With differences in details between them, both genera have terminal segments which suggest the ability of grasping, because of their apparent ability of apposing the distal segment (not shaped like a claw) to a platform-like outgrowth of the distal end of the brachium. In *Eudactylina* and its allies the distal segment is a claw with many cuticular folds (Fig. 1191) and bunches of setae at the base. In *Kroyeria* and *Kroeyerina* it is a claw with denticulation and a tenuous apex (Fig. 1091), in *Pseudocycnus* and its allies a denticulated claw, not clearly delimited from the brachium and apparently capable of a good deal of motion (Fig. 1318). *Lernanthropus* and its group usually have claws of differing length and with different armature, whereas in *Hatschekia* and allies the claw is bifid (Fig. 1122).

Similar differences between the groups of genera can be found in the structure of the maxilliped. In most groups it is a simple, subchelate structure, more or less the same within each group, except for *Eudactylina* and its allies. *Dichelesthium* and *Anthosoma* (Figs. 1020 and 1045) have maxillipeds with robust corpus and subchela closing against a well developed myxa. In *Kroyeria* (Fig. 1092) the

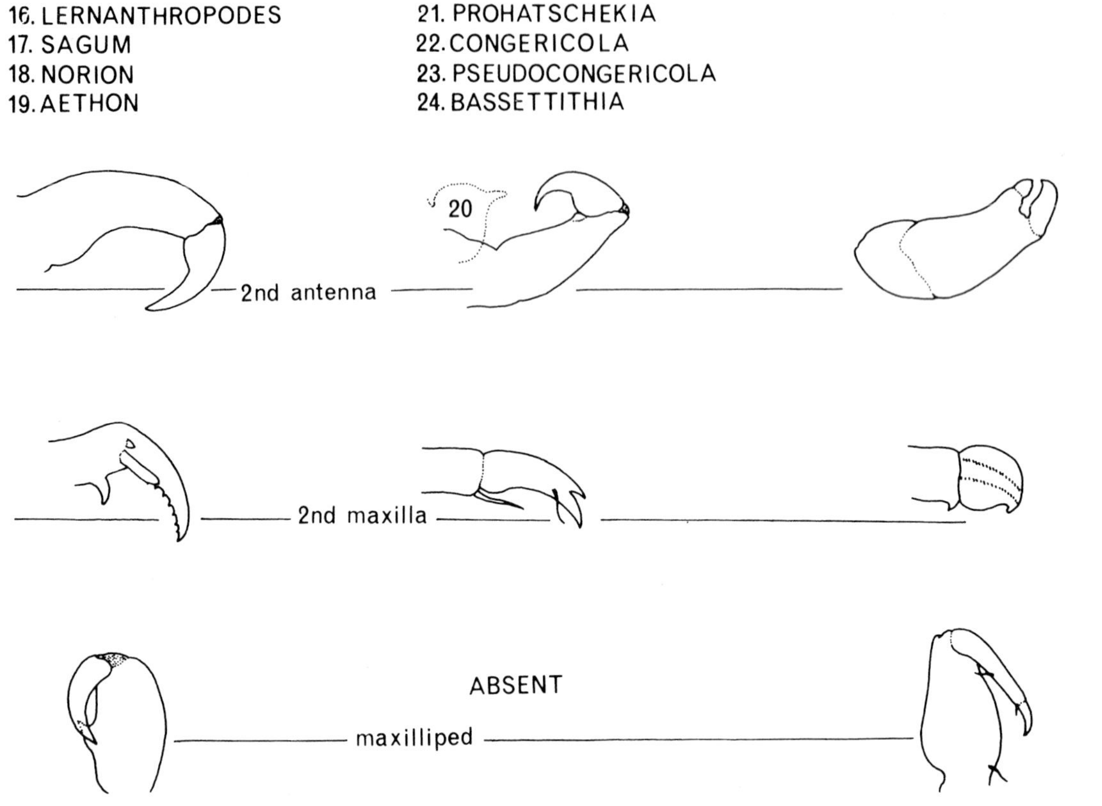

Text fig. 54. Diagrammatic comparison of salient morphological features of copepods formerly included in Dichelesthiidae *sensu lato* (part 2). (See also Text fig. 53.)

corpus is slender and unarmed, while the subchela is very long, slender and sharply curved. In the *Pseudocycnus* group the structure of the maxilliped shows a rather greater variability. Whereas *Pseudocycnus* and *Pseudocynoides* have a simple unarmed corpus and a relatively short subchela, in *Cybicola* (genus No. 14 in Text fig. 53) the corpus is modified by an outgrowth in the myxal area, with a broad membranous rim. *Lernanthropus* and its relatives have simple maxillipeds (Figs. 1067, 1077), with or without light armature in the myxal area, and with or without additional structure on the subchela itself. The poorly defined claws tend to be marked by fine longitudinal striations. In the *Eudactylina* group, however, each genus is distinguished by a different type of maxilliped, the genera of the group falling into a series in which transitional stages, from a subchelate to a chelate appendage, can be observed (genera No. 3–9 in Text fig. 53). *Eudactylinella* has a simple, subchelate appendage, with no armature on the corpus. In *Nemesis* and *Bariaka*, the myxal area is more prominent and armed with a spine, in *Protodactylina* that area is very prominent and juts out in an anterior direction, presenting a halfway stage to chelate structure. A similar structure occurs in *Eudactylinopsis*. *Eudactylinodes* has a remarkable maxilliped, with individual parts resembling those of *Eudactylina* but assembled together in completely different proportions. The myxal protuberance resembles the spoon-like structure of *Eudactylina* and the tip of an enormously long and greatly curved subchela bears a modified claw also reminiscent of that in *Eudactylina*. The latter genus, finally, has a fully chelate maxilliped (Fig. 1193). (These differences apply to females only, the males (when known) have simple subchelate maxillipeds.) In contrast to all the other groups, *Hatschekia* and its allies have no maxillipeds.

A fair degree of uniformity exists also in each group in regard to the number of swimming legs. Both *Dichelesthium* and *Anthosoma* have three pairs, whereas all the genera in the *Eudactylina* group have five, and *Kroyeria* and *Kroeyerina* four. There are four pairs in *Pseudocycnus* and *Cybicola*, though their team member, *Pseudocycnoides*, appears to have only three. The *Lernanthropus* group is characterized by four pairs (though some female *Lernanthropus* possess a fifth pair). Four pairs are also common in *Hatschekia* and its group, though the third and fourth pairs are commonly reduced to simple setae and are often lost in processing and handling the specimens. In *Pseudocongericola*, also a member of this group, only three pairs are present. The similarity in numbers is usually countered by differences in structure from one group to another. For example, though *Kroyeria* and *Hatschekia* have the same number of legs, in the former all the four pairs are biramous and well developed, while in the latter the third and fourth pairs are vestigial.

The author believes that Yamaguti (1963) was right in upholding the fragmentation of the family Dichelesthiidae *sensu lato*. His division, however, was based on unsound criteria and has to be revised. On the basis of the arguments set out above, the author proposes, therefore, that family units, into which the old family should be divided, are best separated in accordance with the following key (to females only).

1. Four distinct segments between cephalothorax and genital segment (Fig. 1213) . . . . . **Eudactylinidae**
   Three segments (exceptionally four) between cephalothorax and genital trunk 2
   Segmentation indistinct, neck present between cephalothorax and genital trunk 3
   No free segments or neck between cephalothorax and genital trunk, dorsal plate usually present on trunk (Figs. 1055–1057) . . . . **Lernanthropidae**
2. Free segments distinct, cephalothorax with caligiform dorsal shield, four pairs of biramous legs (Fig. 1082) . . . . **Kroyeriidae**
   Free segments rather indistinct, dorsal shield of cephalothorax not caligiform, three pairs of variously modified legs, elytra absent (Fig. 1005) or present (Fig. 1027) . . . . **Dichelesthiidae**
3. Second maxilla with bifid claw (Fig. 1122), maxilliped absent . . . . **Hatschekiidae**
   Second maxilla with simple, denticulated claw (Fig. 1318), maxilliped present, subchelate (Fig. 1328) . . . . **Pseudocycnidae**

This grouping into new family combinations leaves out one genus that cannot be accommodated with certainty in any of them. The genus is *Pseudohatschekia* Yamaguti, 1939. Both its segmentation and the structure of its appendages (genus No. 25 in Text fig. 54), as represented by its describer, prevent its inclusion into the larger taxa proposed above. More information is required before a valid decision can be taken in this matter. There remains as well the question of Yamaguti's (1963) superfamily Dichelesthioidea. It appears to this author that a taxon implying close relationship between copepods as different as, for example, *Lernanthropus* and *Kroyeria* is redundant. Each of the six newly-proposed families can take its independent place in the order Siphonostomatoida without any need for a closer linkage with its former associates.

In its new combination, Dichelesthiidae can be defined as follows.

Female: Cephalothorax fused more or less completely with the first leg-bearing segment, dorsal shield well developed. Three (or four) rather indistinctly delimited segments between cephalothorax and genital complex. Dorsolateral elytra on second leg-bearing segment present in *Anthosoma*, absent in *Dichelesthium*. Abdomen one- or two-segmented. First antenna six-segmented, second antenna subchelate and retractile, indistinctly segmented. Buccal region of siphonostome structure. Second maxilla with prehensile apex, maxilliped subchelate. Three pairs of modified swimming legs, with foliaceous sympods (*Anthosoma*) or without them (*Dichelesthium*), with or without reduced rami.

Male: As female, abdomen indistinctly two-segmented in *Dichelesthium*, dorsolateral elytra always absent from second leg-bearing segment.

Following the development of ideas on the affinity between *Dichelesthium* and *Anthosoma*, one notices that the close similarity between their appendages and general structure was overshadowed by the presence of foliaceous legs in the latter genus. Together with a pair of dorsolateral elytra on the second leg-bearing segment in the female, the scale-like sympods of all six legs alter the habitus of *Anthosoma* (Figs. 1026–1030) to render it superficially unlike that of *Dichelesthium* (Figs. 1003–1006). Remove those legs and one is struck by the similarity of the two genera to each other and to the lower Cretaceous fossil (Cressey and Patterson, 1973). A closer look at the legs of *Dichelesthium* shows that they, too, betray a tendency to expansion, incipient in the first two pairs (Figs. 1021, 1022) but quite obvious in the third (Fig. 1024). (It is interesting to note that elytra-like cuticular expansions were also present in the ancestral Cretaceous dichelesthiid.)

Those differences in leg size are most likely associated with the type of relationship entered into by the two genera with their hosts. *Dichelesthium* lives on the gills of the sturgeon, above the surface of the host. On the other hand, the burrowing habits of *Anthosoma* cause it to bury itself partially in the tissues of its host shark. Sometimes only the posterior extremity of the parasite protrudes from the cavity it scrapes out. It is clear to the observer that the scale-like legs and the elytra, overlapping one another around the copepod's body, protect it in some measure from the crushing pressure of the surrounding tough tissues of the shark. No great taxonomic significance could be attached to a difference of this type.

Key to the genera of Dichelesthiidae

First two pairs of legs biramous, rami one-segmented (Figs. 1021, 1022); third pair flipper-like, devoid of rami (Fig. 1024); elytra absent from both sexes . . . . ***Dichelesthium***
All three pairs of legs with foliaceously expanded sympods (Fig. 1028); females with two dorsolateral elytra on second leg-bearing segment (Fig. 1027) . . . . ***Anthosoma***

Genus Dichelesthium Hermann, 1804

The diagnosis of this genus is the same as that of its type and only species, *D. oblongum* (Abildgaard, 1794).

Dichelesthium oblongum (Abildgaard, 1794)

(Figs. 1003–1025)

Syn.: *Caligus oblongus* Abilgaard, 1794
Dichelesthium sturionis Hermann, 1804

Female (Figs. 1003–1005): Cephalothorax including first leg-bearing segment, pyriform, with well defined dorsal shield, expanding in width towards its posterior end. Posterolateral corners of the shield with or without distinct indentations. Second leg-bearing segment (first free thoracic) much broader than long, its lateral parts forming digitiform projections with apices pointing posterolaterally. Second free thoracic segment similar, though narrower and with lateral processes less well developed. Third free thoracic segment about as long as preceding two, somewhat bell-shaped in dorsal aspect (Fig. 1005). Borders between second and third, as well as between third and fourth segments, usually fairly indistinct. Latter about as long as third, subquadrangular, with lateral margins often slightly concave and sometimes (Fig. 1003) with slight bulges on anterior halves. Genital segment narrower than all preceding segments, as long as third and fourth together. In some specimens slight indentations in anterior half of lateral margins, or shallow groove across dorsal surface, giving impression of subdivision. Posterior margin straight. Genital orifices on ventral side near posterior margin. Abdomen (Fig. 1014) indistinctly two-segmented, with anterior segment smaller and almost completely obscured from dorsal aspect (Fig. 1005); anterior segment short, posterior subquadrangular, about twice length of anterior, with slightly convex lateral margins and truncated posterior. Total length 12.6–12.9 mm (in 10 specimens measured by the author).

First antenna (Figs. 1007, 1008) six-segmented, segments diminishing in girth (though not in length) from base to apex. Basal segment long, with surface markings sometimes mistaken for segmental boundary (Fig. 1007). One short, strong seta on anterior margin, slender, much longer seta on ventral surface near border with second. Latter cylindrical, much shorter than first, with two short setae on ventral surface, close together. Segments three to five unarmed. Terminal segment with rounded tip and with fine, slender seta on posterior margin near apex. Apical armature (Fig. 1008) with setae of three types: four very short setae in two pairs, one on anterior and another on posterior side of apex, sharp-tipped, thick-walled; two types of setae on anterior half of apex, in group of four; one pair long, slender, other shorter, thick, digitiform (aesthetes?), two more pairs of long, slender setae on posterior half of apex. Second antenna (Figs. 1009, 1010) retractile, long when fully extended (half retracted in Fig. 1004); corpus thin-walled, often transversely or obliquely wrinkled or crumpled due to partial contraction, strongly muscular; tip isolated in fairly well defined segment (Fig. 1010) of bifid appearance; medial margin with very prominent myxal outgrowth strongly indented at tip; subchela unciform, without division into claw and shaft or any secondary armature, short, tip sharply curved; long axis of distal segment oblique to that of main part of appendage. Mouth tube siphonostome (Fig. 1011), long, slender, with both lips of about equal length, distally with fringe of setae, laterally with slits for admission of mandibles. Mandible resembling in structure that of *Caligus* (Fig. 445), in length matching that of mouth tube, slender; apical part with dentiferous margin carrying 10 short, slightly curving teeth, increasing in size from apical to basal (Fig. 1012). First maxilla biramous (Fig. 1015), its long endopod oval in cross-section, tapering towards tip, surmounted by two slender setae (apparently fused at base and with two rows of fine plumules, not shown in Fig. 1015, running length of seta); exopod short, cylindrical, apparently two-segmented, with distal segment longer and much finer, ending in three setules. Second maxilla (Figs. 1016–1019) brachiform, with lacertus robust, constituting about half of appendage; brachium slender, with two subapical, peg-like processes (Fig. 1018); terminal part opposable against prominent digitiform process on apex of brachium and against sturdy, long setae in bunches on each side of process; two rows of denticles extending from subapical point, one to each side of base, and engaging setae, when closed (Figs. 1017, 1018); in addition, terminal part armed with a single, short process on one side, six rows of minute setules along outer surface (Fig. 1019) and varying number of similar rows on the opposite surface of the curving tip (Fig. 1018). Maxilliped (Fig. 1020) very prominent, subchelate; corpus suboval, broad, with two sharp-tipped points in myxal area; latter expanded into flange, screening tip of subchela from view, when closed; subchela with well delimited claw but no other armature.

First swimming leg (Fig. 1021) with short, squat, subquadrangular sympod, armed with slender seta lateral to base of exopod, short but robust pointed outgrowth on the medial side near base of endopod and four patches of minute denticles on ventral surface; endopod one-segmented, longer than broad, usually with slightly convex lateral margin and with armature consisting of marginal denticulation distally and distolaterally and of two short, stubby setae on distal margin, both with short rows of denticles; exopod longer than endopod, one-segmented, somewhat pedunculate, apparently capable of inward flexion, with five apical setae decreasing in length from medial to lateral, and sixth seta on lateral margin some distance from apical setae; two medial setae with one row of minute denticles, others with two such rows; fine denticulation present also on margins of ramus. Second leg (Fig. 1022) larger than first, with sympod broader and shorter, devoid of pointed process near base of endopod but with patches of denticles on ventral surface; endopod one-segmented, short, with rounded outline in ventral aspect (Fig. 1022), armed with two short setae and with marginal denticulation; exopod much longer, also one-segmented, with five thick, short and blunt setae (four terminal and one lateral) and with marginal denticulation, especially on lateral margin. Third leg (Fig. 1024) one-segmented, flipper-like structure, with three small papillae and two setules distally and with scattered minute papillae on ventral surface. Uropod (Fig. 1014) half length of second abdominal segment, suboval (armature, if any, not observed).

Male (Fig. 1006) generally similar to female but smaller and often with more distinct segmentation. Cephalothorax constituting about $\frac{1}{3}$ of total length ($\frac{1}{4}$ or less in female). Lateral processes of second and third leg-bearing segments relatively smaller and less well developed than in female. Third free segment smaller than preceding two, devoid of lateral processes. Fourth free segment absent. Genital segment obovate, about equal in length to that of both preceding segments. Abdomen (Fig. 1013) indistinctly two-segmented, with two genital plates overlapping its anterior border on ventral surface. Five specimens measured by author varied from 7.6 to 11.9 mm in length.

Endopod of second leg (Fig. 1023) short, with rounded outline and with subquadrangular projection in posterolateral quarter, ending with four rounded processes; two similar processes on ventral surface, near base of projection. Subapical swelling on dorsal surface armed with two setae, projecting well beyond margin of segment. Third leg (Fig. 1025) similar to that of female but much narrower and apparently devoid of ventral papillae.

Comments: In spite of its long history, the taxonomy of this species is almost free of confusion. Having been described originally as *Caligus oblongus* by Abildgaard (1794), it was rediscovered by Hermann (1804), who placed it in its present genus. Unaware of Abildgaard's paper, Hermann gave his species a specific name *sturionis*. Subsequent authors, even some of those who recognized that the two taxa are synonymous, continued to use Hermann's junior synonym. White (1850) restored to the species its rightful name, but the name *sturionis* still appears occasionally in the literature.

D. oblongum is a parasite of acipenserid fishes, a group of hosts suitably ancient for a parasite with so many primitive features and with demonstrated Cretaceous antecedents. Its host list comprises *Acipenser gueldenstadti*, *A. nacarii* (a Mediterranean species of questionable validity), *A. oxyrhynchus*, *A. ruthenus*, *A. stellatus* and *A. sturio* (the most commonly recorded host), as well as *Huso huso*.

The distribution range of *D. oblongum* covers the North Atlantic, the Mediterranean, Adriatic and Black Seas. Essentially a marine species, it is found occasionally on sturgeons migrating up the rivers, both in Europe (mainly in the rivers of the southern U.S.S.R.) and in North America

(Mississippi; Causey, 1955). In Britain, finds come from the South (Polperro), and the North (off Aberdeen), as well as from some other localities.

Genus Anthosoma Leach, 1816

A monotypic genus, its diagnosis is identical with that of its type species, *A. crassum* (Abildgaard, 1794).

Anthosoma crassum (Abildgaard, 1794)

(Figs. 1026–1054)
Syn.: *Caligus crassus* Abildgaard, 1794
Caligus imbricatus Risso, 1816
Anthosoma smithii Leach, 1816
Otrophesia imbricata (Risso, 1826)

Female (Figs. 1026–1028): Cephalothorax with dorsal shield constituting in dorsal aspect about $\frac{1}{2}$ total length, roughly ovoid, with deep notches on either side near anterior end. Lateral edges of shield curving in ventral direction. Posterior part and lateral margins of shield pale brown, central area dark purple to black, with longitudinal streaks. First leg-bearing segment fused with cephalothorax (traces of segmental boundary sometimes discernible). Three ill-defined segments between cephalothorax and genital segment, all somewhat broader than long, of about same size. Genital complex longer than broad, with convex margins, narrowing towards posterior extremity. Body from posterior margin of cephalothorax and up to second half of genital complex covered by overlapping, greatly expanded sympods of swimming legs and dorsally by equally overlapping and similarly large elytra of dorsolateral margins of second leg-bearing segment. (In some specimens entire posterior half covered.) Abdomen one-segmented (Figs. 1031, 1032), subquadrate, short, partly incorporated into genital complex. Genital orifices (Fig. 1031) ventral, near posterior end of genital complex, at margin of rectangular sclerite with crescentic extensions on either side; oviduct openings and attachments of egg sacs dorsal, overhung by fold of genital complex (Fig. 1032). Total length 8.5–15 mm.

First antenna (Figs. 1035, 1036) six-segmented, segments cylindrical, with gradually diminishing diameter; surface of basal segment with knob-like projection and heavy ridge near anterior margin, small papilla near border with second segment; second segment with minute setule on anterior margin, segments 3 and 4 unarmed, fifth segment with minute setule on anterior margin near tip; apical armature (Fig. 1035) apparently of nine setae; four short setae on anterior half (one aesthete?), four longer, more slender setae on posterior half. Below base of appendage shoehorn-like process, evident in dorsal aspect in notches on lateral margins of shield (Fig. 1027). Function unknown, but possibly supporting base of first antenna (Fig. 1034). Second antenna (Figs. 1037, 1038) retractile, its basal $\frac{2}{3}$ cylindrical, thin-walled, often irregularly wrinkled, small papilla at about half its length; distal $\frac{1}{3}$ well defined, sclerotized segment, narrowing distally, with sharp process on margin of second half, subchelate; subchela (Fig. 1038) not clearly divided into shaft and claw, its distal half sharply curved, with sharp tip; small barbel near indistinct base of claw, subtriangular flanges on each side near tip of shaft. Mouth tube (Fig. 1039) with lips fused distally; basal part of labrum with lateral flaps expanding laterally, covering exposed parts of mandible, distal end of this lip transversely truncated, with short, fine setules in lateral corners; tip of labium rounded, exposed in anterior aspect. Mandible (Fig. 1041) caligiform, of four parts, its terminal dentiferous part armed with series of 18 teeth, roughly uniform in size and shape (Fig. 1040). First maxilla (Fig. 1042) near base of mandible, biramous, extending almost to tip of mouth tube; sympod squat, roughly rectangular; endopod one-segmented, flat, tapering distally, surmounted by two slender setae fused at base; exopod two-segmented, cylindrical, with shorter basal segment unarmed, distal more slender, with three terminal setules. Second maxilla (Figs. 1043, 1044) brachiform, lacertus robust, unarmed, brachium subcylindrical, relatively slender, with two longitudinal flanges along posterior margin and with distal process, prominent and blunt, projecting posteriorly; tip bulbous with outer surface covered by longitudinal ridges and with two rows of strong denticles (Fig. 1044) marking off its inner, opposable surface and joining two similar rows on opposite surface of brachium. On medial side of that surface subspherical papilla covered with short, stubby setules. Maxilliped (Figs. 1045, 1046) with corpus roughly oval, except for prominent protrusion of myxal area, completely covering claw of subchela in posterior aspect and armed with one small, sharp outgrowth. Near base of subchela another flange-like expansion of medial margin. Subchela cylindrical, with apparently unarmed shaft and poorly delimited claw, tapering and curving. No barbel observed. (Flanges sometimes much broader than in Fig. 1046 and almost completely covering subchela.)

All three pairs of swimming legs modified into subcircular aliform plates, overlapping somewhat in midventral line (Fig. 1028), either one uppermost; rami absent; surfaces pitted with small refractile depressions. First and second legs (Fig. 1047) with single notches on medial margins; latter uneven (Fig. 1048a), with small protuberances, lateral margins (Fig. 1048b) with more pronounced serrations. Third leg (Fig. 1049) similar but without notch. Uropod (Figs. 1031, 1032) not distinctly delimited from abdomen, cylindrical, with rounded tips; armature not observed.

Male (Figs. 1029, 1030) similar to female, but with segmental boundaries somewhat better marked. No dorsolateral elytra on second leg-bearing segment. Genital complex about as long as three preceding segments together, tapering posteriorly; ventral surface near posterior end (Fig. 1033) with prominent genital plates. Abdomen similar to that of female. Length 7–10 mm.

Thoracic legs with some vestiges of rami. First leg (Fig. 1051) with vestigial rami in anteromedial corner; exopod (Fig. 1050) longer than endopod, one-segmented, with four terminal setae and two rows of small papillae surmounted with bristle-like setules (one row on lateral margin and one on ventral surface, also single papilla on medial margin); short, stiff seta lateral to base of ramus; endopod with two terminal setae and several marginal setules; near medial side of base short, stiff seta and papilliform outgrowth. Second leg (Fig. 1052) subcircular, with vestigial rami; endopod short, one-segmented (Fig. 1053), with three spiniform terminal setae; medial to ramus prominent papilliform swelling, larger than endopod and hollowed out on side nearer to it; exopod much larger than endopod, somewhat peduncular, suboval, with five distal setae and several papilliform outgrowths on margins; small, slender seta lateral to base of ramus. Third leg (Fig. 1054) devoid of rami, with oval thickened area in anteromedial quarter. Uropod (Fig. 1033) as in female.

Comments: Originally described (Abildgaard, 1794) as *Caligus crassus*, this copepod was later described independently by two authors in 1816. Leach (1816) established for it its present genus but also gave it a new specific name, *smithii*, a name that persisted in the literature for some time. Risso (1816) treated it as a new species of the genus *Caligus*, *C. imbricatus*, though he later transfered it to a new genus *Otrophesia* Risso, 1826. The striking appearance of this species prevented further confusion, at least at the generic level, but the family affiliation of *A. crassum* changed repeatedly, until the present author returned it to the family Dichelesthiidae.

A. crassum is a parasite of sharks belonging to the suborder Galeoidea, having been recorded from *Carcharias ferox*, *C. littoralis* (? = *C. taurus*), *Carcharias* sp., *Carcharodon carcharias*, *Cetorhinus maximus*, *Galeorhinus australis*, *G. galeus*, *Heptranchias perlo*, *Isurus bideni*, *I. glaucus*, *I. mako*, *I. oxyrinchus*, *Isurus* sp., *Lamna nasus*, *L. ditropis*, *Lamna* sp. and *Prionace glauca*. Unidentified sharks have also been recorded as hosts, but a really strange host for this species is a teleost, *Mola mola*, listed by Dollfus (1946), Delamare Deboutteville (1948), Birkett and Burd (1952) and Timon-David and Musso (1971).

The distribution range of *A. crassum* is as extensive as those of its many hosts, covering the North and South Atlantic (British waters, Belgian and French coasts, Danish Straits, the coast of the USA, Angola, South Africa and Argentina). Several records come from the Mediterranean and the Adriatic. It has been recorded in the North and South Pacific (Vancouver Island region, California, western equatorial part, central region, coasts of Chile and Peru, Hawaii and Japan). There is also a record from the western Indian Ocean. The occurrence of *A. crassum* on *Mola mola*, however, is known, so far, only in the Mediterranean and in British waters.

Family Eudactylinidae comb. nov.

Morphology

All seven genera included in Eudactylinidae in this combination resemble one another in their general appearance and can be described as having "eudactylinid facies", exemplified in Fig. 1216. All have the same segmentation in the pre-abdominal part of the body. (Pillai (1968a) described *Eudactylinopsis* as having its fifth leg-bearing segment fused with the genital segment. However, his Figs. 10A, B and 11D show that the genital segment is fairly well delimited.) The number of abdominal segments varies from one to three in the female and from three to four in the male (when known). The members of this family are usually characterized by the abundance of their cuticular adornments in the form of scaly or spiny flaps, or flat, boss-like encrustations, present particularly on the dorsal shield, terga of other segments and surfaces of appendages, especially the swimming legs. These cuticular structures, often of considerable complexity, are still very imperfectly known, their transparency and/or minute size rendering them very difficult to observe. Further studies with the aid of modern techniques might well reveal that they are of taxonomic importance. In *Eudactylinodes* and the male *Eudactylinopsis* the posterior margin of the dorsal shield is equipped with one or two pairs of styliform projections reminiscent of the female *Kroyeria*. Posterolateral or lateral expansions of the dorsal shield and of the terga occur in some species of *Eudactylina*.

As regards the structure of the first antenna, the eudactylinid genera fall into two groups, those with first antennae straight (*Eudactylinella*, *Eudactylinopsis*, *Protodactylina*, *Nemesis* and *Bariaka*), and those whose first antennae have lateral geniculate flexion between the first and the second or between the second and third segments (*Eudactylina*, *Eudactylinodes*). In the last two genera the first antennae are often armed with prominent and strong spines. The second antenna is usually five-segmented, the basal segment being very short. The differences between the genera and species in regard to this appendage are limited to minor structural details, as will become clear from the following descriptions of the British representatives of the family. This applies also to the structure of the buccal tube and appendages. The opposite holds true of the maxilliped, an appendage of different structure in each genus (cf. p. 244 and Text fig. 53). The scant sexual dimorphism in Eudactylinidae is probably best exemplified by the maxilliped, particularly in *Eudactylina*.

Sexual dimorphism is also evident in the structure of the swimming legs. In the males they are usually more primitive, as shown by their segmentation and by the retention of many elements of armature lost from the female appendages. The most primitive structure of the first four pairs of legs is found in *Protodactylina* (male unknown) in which these legs are regularly biramous, with two-segmented sympod and three-segmented rami, both well armed with long pinnate setae in addition to the lateral spines. In *Nemesis* the rami are two-segmented, while in *Eudactylina* they vary greatly in structure and armature. The fifth leg is a single lobe and the sixth is represented by two or three setae in the posterolateral corner of the male genital segment.

History and systematics

The reasons for revising Eudactylinidae and for giving it its present combination are set out in the discussion of the history of Dichelesthiidae. The first mention of Eudactylinidae as a family group taxon is found in Wilson (1932). In addition to the genera now included in it (*Eudactylina*, *Eudactylinella* Wilson, 1932, *Eudactylinodes* Wilson, 1932, *Nemesis* Risso, 1826), Wilson placed in it also *Kroyeria* and *Kroeyerina*. The only subsequent reviewer of the family was Yamaguti (1963) who added to the complement of Wilson's genera also *Congericola* (placed here in Hatschekiidae) and *Lamproglena* (a cyclopoid). Subsequent additions enlarged it by *Bariaka* Cressey, 1966, *Protodactylina* Laubier, 1966 and *Eudactylinopsis* Pillai, 1968.

In its new combination, the diagnosis of the family is as follows.

Female and male: Cephalothorax fused with first leg-bearing segment, followed by four free leg-bearing segments, genital segment (apparently partly fused with preceding segment in *Eudactylinopsis*) and abdomen of one or more segments. First antenna indistinctly segmented, uniramous; second antenna five-segmented, uniramous, with unciform terminal claw. Mouth tube and buccal appendages siphonostome. Second maxilla brachiform, maxilliped subchelate to chelate in female, subchelate in male. First to fourth leg biramous, fifth vestigial, sixth (when present) setiform. Parasitic on elasmobranchs.

Type genus: *Eudactylina* van Beneden, 1853.

In British waters Eudactylinidae are represented by the genera *Eudactylina* (five species) and *Nemesis* (two species). (It is also possible that *Protodactylina*, a parasite of *Hexanchus griseus*, a shark occasionally taken in British waters, will be found eventually. However, the present author found none in several specimens of this shark examined over the years.)

Key to British genera

First antenna with geniculate flexion between second and third segment (Fig. 1180); maxilliped chelate in female (Fig. 1193), subchelate in male (Fig. 1196); first swimming leg not modified (Fig. 1197) . . . . **Eudactylina**

First antenna straight (Fig. 1270); maxilliped subchelate in both sexes (Fig. 1279); first leg prehensile (Fig. 1281) . . . .**Nemesis**

Genus Eudactylina van Beneden, 1853

Eudactylinidae: Female. Cephalothorax covered by well delineated dorsal shield; four succeeding thoracic segments with well developed terga, usually with cuticular adornments. Abdomen two-segmented. First antenna with geniculate flexion between second and third segments. Second antenna five-segmented, with uncinate terminal claw. Mouth and buccal appendages siphonostome. Maxilliped chelate. First four pairs of legs biramous, rami from one- to three-segmented, often modified (especially exopod of second leg). Interpodal bars with sternal stylets. Fifth leg one-segmented, devoid of rami.

Male. Similar to female. Abdomen three- to four-segmented. Maxilliped subchelate. First four pairs of legs less modified, with better developed armature. Sixth pair of legs sometimes present, setiform.

Type species: *Eudactylina acuta* van Beneden, 1853.

Comments: in spite of the general uniformity of habitus, the species of this genus differ from one another in a multitude of structural details, from the type of cuticular adornment to the segmentation of the swimming legs. The latter is probably the best discriminant at the specific level and was used as a basis for the best existing key to the genus, published by Laubier (1968). Containing 21 species known at that time, this key is by now out of date but it remains a good example of a right approach to the taxonomy of *Eudactylina*. Some unusual features distinguish a few species. For example, *E. lancifera* Pillai, 1968 and *E. alata* Pillai, 1968, are unique in having a pair of anterolateral, styliform outgrowths on their dorsal shield. The latter of the two has also prominent posterolateral lobes on the tergum of its first free thoracic segment. Unusual features of a similar nature occur in other species.

As at present constituted, *Eudactylina* comprises 27 nominal species. Three of them (*E. carchariae-glauci* Hesse, 1884, *E. musteli-laevis* Hesse, 1884 and *E. squatinae-angeli* Hesse, 1884) have not been sufficiently well described to be recognized and must be considered species inquirendae. Several other species are also inadequately described, though their identity can be traced with some difficulty. Most species come from the Atlantic region (including the Mediterranean, Adriatic and North Seas). As many as 18 species were recorded here. Five species have been described from the Indian Ocean and four from the Pacific. Of the latter, only one (*E. papillosa* Kabata, 1970) is an exclusively Pacific species, the other three (*E. acanthii* Scott, 1901, *E. similis* Scott, 1902 and *E. aspera* Heller, 1865) being known also from the Atlantic. This list cannot be considered definitive. As facilities improve for the study of the elasmobranchs, more species are certain to be added to it, particularly in the Indian and Pacific Oceans.

In British waters, the genus is represented by five species: *E. acuta*, *E. acanthii*, *E. similis* Scott, 1902, *E. insolens* Scott and Scott, 1913 and *E. minuta* Scott, 1904.

Key to British species

1. First four pairs of legs with rami three-segmented; exopod of second leg modified (Fig. 1199) 2
   First legs with rami two segmented (Fig. 1247); exopod of second leg modified (Fig. 1248) 3

First four pairs of legs with rami three-segmented; exopod of second leg not modified (Fig. 1237) . . . . ***E. acanthii***

2. Large subterminal spine (Fig. 1181, s) on first antenna; uropod suboval (Fig. 1205) . . . . ***E. acuta***
Small subterminal spine (Fig. 1220, s) on first antenna; uropod subtriangular (Fig. 1227) . . . . ***E. similis***

3. Auxiliary spine on claw of second antenna less than ½ length of claw (Fig. 1244); third exopod segment of second leg with three about equally long processes (Fig. 1248) . . . . ***E. insolens***
Auxiliary spine on claw of second antenna more than ½ length of claw (Fig. 1257); third exopod segment of second leg with one long, one short process and one denticle (Fig. 1260) . . . . ***E. minuta***

This key can be used to identify only females of the above species. The males have been described so far only for two of them (*E. acuta* and *E. similis*).

Eudactylina acuta van Beneden, 1853

(Figs. 1175–1212)

Syn.: ?*Eudactylina complexa* Brian, 1924

Female (Figs. 1175, 1176): Cephalothorax with well delimited dorsal shield, expanding posteriorly, with lateral margins curving ventrally; surface covered with transversely arranged membranous flaps and occasional setae (Fig. 1179). Similar structures occurring on terga of following four segments, particularly in lateral quarters. Thoracic segments bearing legs 2–4 broader than long, tapering towards their anterior ends. (Figs. 1175, 1176 show relaxed specimens. In contracted specimens distances between terga of successive segments are much shorter and appearance of copepod is affected.) First three free thoracic segments subequal in size, slightly increasing in posterior direction. Fourth leg-bearing segment narrower than preceding one and subquadrate in dorsal aspect. Genital segment smaller than fourth leg-bearing segment, also subquadrate. Abdomen two-segmented, first segment larger than second. Total length about 2.8–3.2 mm.

First antenna (Figs. 1180, 1181) indistinctly segmented (probably of five segments), with geniculate flexion dividing two-segmented base from shaft; basal segment with one small seta on outer margin; second segment with two subequal setae on distal end of outer margin and with small seta near margin on dorsal side; strong spine and long, slender seta also present on dorsal surface; near geniculate flexion, outer margin armed with powerful claw-like spine with several serrations on convex margin; proximal part of shaft with two setae on ventral and two on dorsal side of claw (latter two omitted from Fig. 1180); other armature comprising one flagelliform and two short and stout setae, as well as two strong spines, one usually parallel to distal half of shaft, the other obliquely posterior, both with varying number of denticles along one margin; armature of distal part consisting of one seta on ventral side, one on posterior margin and four structures near anterior margin: two long setae (one aesthete?), one short and blunt and one claw-like spine (Fig. 1181, s); apical armature (Fig. 1181) of three long, slender setae, one shorter and stouter on anterior side of apex and group of five flagelliform setae on posterior side; on dorsal side another short, blunt seta present (aesthete?). On ventral side between bases of first antennae sharp rostral tine present; subrostral plate broad, with two lateral extensions (Fig. 1192). Second antenna (Figs. 1182, 1183) with short, unarmed basal segment; second segment with short, straight, styliform process on posterior margin, third segment with much longer, curving process and two slender setae at its base; medial side of second segment with semicircular membranous flaps (minute scaly denticles on anterior margin omitted in Fig. 1183); fourth segment unarmed, narrower than preceding one; terminal segment claw-like, with broad base and abruptly tapering, pointed, slender tip; on lateral side of tip auxiliary spine longer than ½ length of segment, on posterior margin of base two slender setae. Mouth cone short, with broad base; labrum with two peg-like setae one on each side of tip; labium (Fig. 1185) with two flaps of cuticular membrane at tip and row of stiff setae on each side in distal half. Mandible (Figs. 1186, 1187) caligiform, of two parts (about 90 μ long overall); terminal part with dentiferous margin armed with series of seven teeth, diminishing in size towards base. First maxilla (Fig. 1188) biramous, rami one-segmented, cylindrical, endopod shorter than exopod; latter with two setae, one half length of other, both with parts of or entire margins finely serrated; exopod more slender than endopod, surmounted by one longer and two very short setae. Second maxilla (Figs. 1189–1191) brachiform; lacertus longer than brachium, strong, with some membranous flaps near distal end (Fig. 1189), brachium with several rows of similar flaps along posterior margin and with bunch of prominent setae near base of claw; latter with sharply flexed point and many serrated flaps of membranes on medial side and posterior margin (Fig. 1191). Maxilliped (Figs. 1193–1195) chelate, indistinctly segmented, with basal part forming short, flexible peduncle, expanding into broad, sturdy segment; posterior margin of latter with short, transverse flange; posterolateral corner prolonged into spoon-like structure (Fig. 1194) for reception of opposable segment; latter cylindrical, curving, with two setae: one on outer margin near base, other on inner, near tip, ending in unciform claw (Fig. 1195) with narrow strip of membrane along convex margin; arising at base of claw flap of strong membrane entirely covering it from one side.

First four pairs of legs with three-segmented rami and two-segmented sympods. Armature of rami as follows.

	Endopod			Exopod		
	1	2	3	1	2	3
Leg 1	0-0	0-0	2	0-1	0-1	4
Leg 2	0-0	0-0	2	modified		
Leg 3	0-0	0-0	1	0-1	0-1	4
Leg 4	0-0	0-0	1	0-1	0-1	4

All four pairs of legs with two-segmented sympods. Basal segment of sympod of first leg with ventral surface partly covered by crescentic cuticular flaps (Fig. 1197); terminal with one small batch of similar flaps in medial corner and with two setae: one medial to base of endopod, other lateral to that of exopod; ventral surfaces of segments of both rami with similar cuticular flaps (Fig. 1198). Similar flaps abundant on ventral surfaces of sympod and endopod of second leg (Fig. 1199); small seta on sympod next to base of exopod; latter devoid of flaps, with one row of denticles along medial margin of greatly enlarged first segment; segments two and three crowded at tip of ramus, indistinctly delimited; second segment with one short, sturdy seta on ventral side (Fig. 1200); third with tapering tip and with group of four short, blunt setae on ventral side (Fig. 1225). Third and fourth legs very similar to each other (Fig. 1202); spines of terminal segment of exopod with irregularly branching tips (Fig. 1203), lateral margins of endopod segments with small, pointed spinules, obscured partly by cuticular flaps (Fig. 1201). Fifth leg (Fig. 1204) oval; lateral surface covered with cuticular flaps, distally three slender setae and one similar seta at base of leg (omitted in Fig. 1204); medial surface with several (usually five) transverse rows of fine setae. Uropod (Fig. 1205) suboval, with two distal denticles, one lateral and one dorsal seta; ventral surface with irregularly scattered, sharp denticles. Ventral surface of distal segment of abdomen with four groups of cuticular flaps (Fig. 1205).

Male (Figs. 1177, 1178) similar to female, but more slender and with three-segmented abdomen. Total length about 1.5 mm.

First antenna resembling that of female but more clearly segmented and with larger number of setae; distal half distinctly three-segmented; apex subtriangular, more pointed than in female. Second antenna with medial side of third segment armed with four elongate, slender processes; outgrowth on posterior margin relatively longer and more curved than in female (Fig. 1184). Claw of second maxilla relatively longer and more slender than in female. Maxilliped (Fig. 1196) subchelate; corpus two-segmented, basal segment short, robust, unarmed, second segment with one long, sturdy spine (with secondary spinules near apex) and two spinules near base on medial margin; more distal on that margin cuticular flaps; subchela with long shaft, curving near distal end and with one seta near base, other on concave margin nearer tip; claw short, blunt with very short barbel.

First four legs biramous. rami three-segmented with exception of two-segmented exopod of first leg. Armament formula as follows:

	Endopod			Exopod		
	1	2	3	1	2	3
Leg 1	0-0	0-0	2	0-1	5	—
Leg 2	1-0	0-0	2	0-1	0-1	3
Leg 3	0-0	0-0	2	0-1	0-1	3
Leg 4	0-0	0-0	2	0-1	0-1	3

Sympods of all four pairs two-segmented; that of first leg (Fig. 1206) with only few denticles on ventral surface near lateral margin; second segment with long seta medial to base of endopod and other near base of exopod. Second leg (Fig. 1207) more abundantly denticulated, denticles on rami particularly along or near to lateral margins, also armed with fringes of fine setae (Fig. 1208); distomedial corner of first segment of endopod with long, blunt process, extending as far as middle of third segment. Corresponding segment of third leg (Fig. 1209) without that structure. Distal segments with denticles along margins but without fringes of setae (Fig. 1210). Fifth leg (Fig. 1211) very small, one-segmented, distally with three setae longer than itself and with similar seta at base, on dorsal side. Distal segment of abdomen (Fig. 1212) with two rows of denticles on ventral surface. Uropod with two long spines, one apical, other subapical, lateral; small denticle at base of latter; one slender seta with denticle at base laterally, one dorsally, but extending in posteromedial direction.

Comments: *E. acuta* occurs mainly on the angel fish, *Squatina squatina*. Some records (Valle, 1880; Brian, 1906; Oorde-de Lint and Schuurmans Stekhoven, 1936) maintain that it parasitizes also the common dogfish, *Squalus acanthias*, but this author has never seen it on that fish. The distribution range of *Squatina squatina* covers the British waters, extending into the Mediterranean and Adriatic and as far South as the Canaries. *E. acuta* has been recorded from most parts of its range. Brian (1924) described another species of *Eudactylina*, *E. complexa*, from *S. squatina* taken off the coast of Mauretania. As far as one can judge from the rather meagre description, this species is synonymous with *E. acuta*. (Unable to secure Brian's specimens for comparison, the author quotes it in synonymy with a query.) This opinion finds support in the fact that no further records of *E. complexa* have been made since. It is worth noting that *Squatina dumeril*, inhabiting the Atlantic seaboard of North America, is parasitized by *Eudactylina spinula* Pearse, 1950, a species unmistakably different from *E. acuta*.

Eudactylina similis T. Scott, 1902

(Figs. 1213–1228)

Syn.: *Eudactylina rachelae* Green, 1958

Female (Figs. 1213–1216): Dorsal shield of cephalothorax and lateral parts of terga of four succeeding thoracic segments with cuticular flap similar to those of *E. acuta* (Fig. 1179). Shield longer than broad, slightly broader posteriorly. Next three thoracic segments broader than cephalothorax, two shorter than broad, third about equally long and broad, longer than preceding segments. Fourth free segment narrower than cephalothorax, about as long as broad. Genital segment smaller than preceding one. Abdomen two-segmented, second segment half as long as first. (Figs. 1215, 1216 from type material of *E. rachelae*.) Total length about 3 mm.

First antenna quite similar to that of *E. acuta*, but with different apical armature (Fig. 1220). In *E. acuta* subapical spines close to long seta on anterior margin extending usually up to tip of appendage, in *E. similis* much shorter and more slender. Second antenna as in *E. acuta*. Mandible with six teeth, otherwise similar to that of *E. acuta*. Also similar: first and second maxillae and maxilliped. The structure and armature of four pairs of biramous swimming legs much as in *E. acuta*, except for relatively larger exopod of second leg. Fig. 1223 showing medial corner of base of endopod of first leg with characteristic cuticular flaps. Fifth leg differing in outline (Fig. 1226) from that of *E. acuta* but with similar cuticular adornment on both sides (omitted in drawing). Uropod (Fig. 1227) subtriangular, with straight lateral and slightly curving medial margins, one short apical and one subapical spine and two slender setae, one lateral and one medial; ventral surface with rounded, semicircular or pointed flaps of cuticle.

Male (Fig. 1217): Similar to female in general structure, but narrower, with fourth free segment considerably reduced and three-segmented abdomen. Segments cylindrical, elongated. No cuticular adornment on dorsal shield or terga observed. Total length about 2 mm.

First antenna (Figs. 1218, 1219) with very pronounced geniculate flexion between two-segmented base and distinctly three-segmented shaft; first segment of base (Fig. 1218) with one small seta on distal end; second segment with unciform spine at point of flexure on outer margin; proximal to base of spine four slender setae in one group and three setae on dorsal side of margin; distal to base of spine six more setae; short spine and two setae on dorsal side of segment; first segment with wavy sculpturing on surface near distal end; three setae on ventral side of segment, one short and one longer claw-like spine on anterior margin, three more setae and one long spine on distal margin; second segment with one seta and one shorter spine, both on or near distal margin; third segment

(Fig. 1219) with a bifid spiniform process in midlength of anterior margin. End of segment turned up in anterior direction, rounded with proximal part of upturned portion in form of flange dotted with refractile spots; apical armature essentially similar to that of female, with four apical setae, one short near that group and five on posterior side of apex; in addition one seta on dorsal side and two very prominent, robust setae, extending past tip of appendage (aesthetes?). Second antenna as in *E. similis* but medial side of third segment (Fig. 1221) with more abundant triangular processes (perhaps individual variation?).

First four pairs of swimming legs closely similar to those of *E. acuta*. Some details of third and fourth exopod shown in Fig. 1224. Fifth leg similar to that of *E. acuta*. Distal segment of abdomen with only two cuticular flaps on ventral surface (Fig. 1228). Uropod with very long apical spines, shorter distolateral spines and pair of setae (one medial and one lateral) each.

Comments: The author was able to examine specimens (both whole mounts and free) from the type material of *E. rachelae* and could find no differences between them and those of *E. similis*, other than those easily attributable to individual variation. It appears from the comparison of *E. rachelae* with *E. similis* made by the describer of the former (Green, 1958) that he had to depend on rather inadequate descriptions of the older species for that purpose.

E. similis was first found in British waters and the areas immediately adjacent, parasitic on *Raja radiata* and *R. fullonica*. (Specimens described as *E. rachelae* came from *Torpedo nobiliana*, perhaps another reason for their assignment to another taxon. No subsequent records from that fish have been made.) Boxshall (1974a) found it on *Raja montagui* and *R. naevus* in the North Sea. Delamare Deboutteville and Nunes-Ruivo (1958) discovered it on *R. asterias* in the Mediterranean, while this author was surprised to see it on *R. rhina* and *R. stellulata* in the Pacific, off Vancouver Island.

Eudactylina acanthii A. Scott, 1901

(Figs. 1229–1239)

Female (Figs. 1229, 1230): Dorsal shield of cephalothorax longer than broad, with transversely truncated posterior and rounded anterior margin. Tergum of first free segment somewhat broader than shield, with rounded posterior extremity; terga of second and third free segments subquadrate in dorsal aspect, with indistinct lateral margins. Fourth free segment smaller than two preceding ones, its tergum with rounded posterior extremity. (Author was unable to find any cuticular adornments on terga of this species.) Genital segment much smaller than fourth free segment, broader than long. Abdomen two-segmented, second segment narrower but not much shorter than first. Total length about 2 mm.

First antenna (Fig. 1231) with two-segmented base not clearly delimited from shaft, latter also indistinctly segmented; basal segment with single setule on outer margin; second segment of base with six setae on or near outer margin, one short spine and one slender seta on dorsal surface near junction with following segment and with large unciform spine on outer margin near junction with shaft. In specimens examined by author shaft divisible into two parts, basal of which indistinctly two-segmented, both together with 11 setae of various lengths, mainly on outer margin, but also on dorsal and ventral surfaces and one on distal margin; armature of terminal segment (Fig. 1232) divisible into segmental and apical; anterior margin of segment with two short, sturdy setae and one long (aesthete?) extending beyond tip of appendage; one long seta on ventral surface and one long and slender on posterior margin. Apex with three long setae and one short (truly apical, in contrast with those of *E. acuta*, Fig. 1181, or *E. similis*, Fig. 1220); five slender setae in groups of three and two on posterior side and one short, blunt seta on dorsal side of apex. Second antenna (Fig. 1233) with second segment devoid of processes on posterior margin but with batch of sharp denticles on medial surface; third segment with posterior process much reduced, though accompanied by two slender setae and with semicircular flaps on medial surface; no cuticular armature on fourth segment; unciform terminal claw with auxiliary spine shorter than half length of claw (Fig. 1234). Mouth similar to that of other species. Mandible with six teeth but otherwise as in type species. First maxilla also as in type species. Second maxilla (Fig. 1235) with terminal claw with two rows of denticles and tuft of setae on medial side of base, sparser and coarser than in type species; posterior half of brachium with subtriangular, sometimes fairly long, cuticular flaps, much more substantial and less membranous than in type species; some denticles scattered also on basal part of brachium and on lateral surface of lacertus. Maxilliped essentially similar to that of type species though apparently without transverse flange on base of spoon-shaped immovable limb of chela. First four pairs of swimming legs biramous, all rami three-segmented. Armature formula as shown below.

	Endopod			Exopod		
	1	2	3	1	2	3
Leg 1	0-0	0-0	3	0-1	0-1	3
Leg 2	0-0	0-0	3	0-1	0-1	3
Leg 3	0-0	0-0	3	0-1	0-1	3
Leg 4	0-0	0-0	3	0-1	0-1	3

All segments of rami and ventral surfaces of sympods covered with subtriangular flaps (Figs. 1236–1238), particularly along lateral margins of rami; flaps of stiff cuticle, large and obscuring smaller marginal spines; latter apparently without denticulation or other secondary structures. Fifth leg unarmed on lateral surface except for several scattered very fine setules; on medial surface about five transverse rows of setules. Ventral surface of second abdominal segment (Fig. 1239) with small, scattered denticles. Uropod with similar denticles on ventral surface, terminal spine about $\frac{2}{3}$ its own length, two shorter posterolateral spines and one posteromedial; one slender seta on lateral and one on medial margin.

Male: Unknown.

Comments: While studying the swimming legs of *E. acanthii*, observers should beware of mistaking the cuticular flaps for segmental armature of rami. Some spines are of comparable size and can easily be misinterpreted, and vice versa, as shown in Figs. 1236–1238.

This parasite of *Squalus acanthias*, the common dogfish, was first discovered in the Irish Sea by A. Scott (1901b) and has been recorded since from many localities in the North Atlantic. It occurs throughout the British waters and along the Atlantic seaboard of North America. In the North Pacific it has been found in the Sea of Japan and in the Vancouver Island region, where its presence was confirmed by the present author. (It is of interest to note that the North Pacific dogfish, previously identified as *Squalus suckleyi* (Girard, 1854), is now recognized as synonymous with *S. acanthias*. The presence of the same parasite species on both tends to support that view.) The only record from the southern hemisphere known to the author is that published by Nunes-Ruivo (1956), who found it off the coast of Angola, also on *S. acanthias*.

Eudactylina insolens T. Scott and A. Scott, 1913

(Figs. 1240–1251)

Female (Figs. 1240, 1241): In many specimens dorsal shield of cephalothorax apparently broader anteriorly, with rounded anterior and truncated posterior margins. Terga of free segments poorly developed, apparently without cuticular adornment. First free segment as broad as, or slightly narrower than, shield; second larger than first, broader than cephalothorax; third and fourth larger still, indistinctly delimited from each other. Genital segment small, subspherical. Abdomen two-segmented, with second segment narrower than first. Total length about 2 mm.

First antenna (Figs. 1242, 1243): Two basal segments similar to those of other species; large unciform spine of distal segment of base with several denticles on convex margin; shaft clearly divided into three segments; first segment with three setae on ventral surface (Fig. 1242), four on anterior margin and two on dorsal surface (one very robust, though short); second segment with only one seta on ventral surface; distal segment with moderately long seta on ventral surface, near base of prominent blunt seta (aesthete?) extending beyond tip of segment; three short, strong setae on dorsal surface, in addition to small, scattered denticles (Fig. 1243) (usually six) near apex; fine slender seta on posterior margin; apical armature of usual three long setae and one shorter in centre of apex and five slender, long setae, in groups of three and two, on posterior side of apex. Second antenna (Fig. 1244) similar to that of preceding species; second segment without posterior process, with triangular denticles on medial surface; few crescentic cuticular flaps on medial surface of third segment, posterior process longer than in *E. acanthii* and bifid; auxiliary spine of terminal claw less than $\frac{1}{2}$ length of claw. Mandible with five teeth (total length about 80 μ). First maxilla (Fig. 1246) similar to that of type species. Second maxilla with only two rows of denticles on terminal claw (Fig. 1245), few setae near base of claw and prominent, tooth-like cuticular flaps on posterior margin of brachium and on lacertus. Maxilliped similar to that of type species. First four pairs of swimming legs biramous; rami of first leg two-segmented, those of remaining three-segmented. Armature formula as below.

	Endopod			Exopod		
	1	2	3	1	2	3
Leg 1	0-0	2	—	0-1	4	—
Leg 2	0-0	0-0	1	modified		
Leg 3	0-0	0-0	1	0-1	0-1	3
Leg 4	0-0	0-0	1	0-1	0-1	3

Sympods of four pairs two-segmented, their ventral surfaces with cuticular flaps, as in type species. Lateral and apical setae of first leg with narrow strips of membrane along lateral margins (Fig. 1247), though apparently not so on remaining pairs. Exopod of second leg modified, twice as long as endopod, basal segment constituting most of its length, with digitiform process on apex (Fig. 1248); short second segment with similar, but smaller process; terminal segment with three processes, at least one of them with row of denticles. Legs 3 and 4 similar to each other (endopod, Fig. 1249; exopod, Fig. 1250). Fifth leg with uneven, rough cuticle on lateral surface, but apparently without cuticular flaps; fine setae in transverse rows on medial surface. Ventral surface of second segment of abdomen with subtriangular flaps (Fig. 1251). Uropod short (Fig. 1251), with long apical spine and shorter posteromedial and posterolateral spines; slender setae on medial and lateral margin and short, spiniform cuticular structures on distal half of ventral surface and on lateral margin.

Male: Unknown.

Comments: In their description of the maxilliped of this species, Scott and Scott (1913, Pl. LXXI, Fig. 8) show a curious frilly margin at the base of the movable arm of the chela. This author was unable to find it, though a cuticular artifact, caused by fixation, might occur in that place. In the author's material the endopod of the first maxilla (Fig. 1246) had a short bifid seta at the base of the long apical one. Scott and Scott (1913, Pl. LXXI, Fig. 6) show a more usual set of two short setae at that place. More material is needed to resolve this discrepancy.

E. insolens resembles in many respects *E. acuta* and *E. similis*, but differs from them in the rather ill-defined lineaments of the free segments, lack or scarcity of cuticular adornment of terga and shield, and in particular, in the apical armature of the first antenna. No other species known to this author has scattered, minute denticles on the terminal segment of that appendage (Fig. 1243).

E. insolens is a fairly rare parasite. Known so far only from British waters, where it has been recorded from widely separated localities, it is a parasite of the tope, *Galeorhinus galeus*.

Eudactylina minuta T. Scott, 1904

(Figs. 1252–1265)

Female (Figs. 1252, 1253): Dorsal shield long, slightly broader posteriorly, with truncated posterior margin. Cuticular adornment along lateral and anterolateral margins of shield and terga. Latter well developed, clearly delimited. Tergum of first free segment broader than cephalothorax shield, rounded anteriorly, with straight posterior margin; that of second similar but broader. Tergum of third free segment subcircular, longer than that of second. Fourth free segment smaller than three preceding segments. Genital segment of similar size, not clearly delimited from two-segmented abdomen. Total length about 1.5 mm.

First antenna (Fig. 1254) with two-segmented base and indistinctly segmented shaft (Fig. 1255); first segment of base with short seta on distal outer segment and with tubercular prominence on distal inner margin (Fig. 1256); second segment with three short setae and one long on outer margin, prominent spine and long seta on dorsal surface and powerful unciform spine with minute denticles along convex margin; proximal half of shaft with two setae on ventral surface, one on dorsal and four on anterior margin (including one very short, digitiform, Fig. 1255); powerful spine on anterior margin and one smaller spine at its base on ventral surface; distal half with one prominent seta on ventral surface, three short, digitiform on dorsal (Fig. 1255), one slender on posterior margin and two long ones on anterior margin (one aesthete?); centre of apex with three long setae and one short, posterior side with five slender setae (in groups of two and three). Rostral tine and subrostral plate

(Fig. 1256) broadly similar to those of type species. Second antenna (Fig. 1257) with unarmed second segment; posterior process of third segment prominent, with broad base and tapering tip; a row of spiniform setules on medial surface; unciform terminal claw with slender tip; auxiliary spine longer than half length of claw. Mouth tube and mandible similar to those of type species (number of mandibular teeth not determined). First maxilla similar to that of *E. acuta*. Second maxilla brachiform, its terminal claw narrow and nearly straight, apparently with multiple divisions at tip; brachium with cuticular flaps along posterior margin. (Descriptions of oral appendages after Scott and Scott, 1913.) Maxilliped (Fig. 1258) chelate; claw partly covered by flap of cuticle; two setae on distal half of opposable limb of chela, one on proximal.

First four pairs of legs with two-segmented sympods, biramous. Rami of first pair two-segmented, those of following three three-segmented. Armature formula below.

	Endopod			Exopod		
	1	2	3	1	2	3
Leg 1	0-0	2	—	0-1	3	—
Leg 2	0-0	0-0	2	modified		
Leg 3	0-0	0-0	1	0-1	0-1	3
Leg 4	0-0	0-0	1	0-1	0-1	3

Ventral surfaces of all four pairs of legs, except for exopod of second leg, with many subtriangular cuticular flaps, solid and more easily observable than in other British species. Details of armature sometimes obscured by these flaps. Segmental spines of first and second leg smooth, terminal spines of third and fourth leg (Figs. 1261–1264) with needle-like denticles along lateral margins. Exopod of second leg (Fig. 1259) with elongated first segment, its ventral surface with several rows of fine prickle-like spinules pointing laterally and one distolateral spine; second segment with similar large spine; third segment (Fig. 1260) oval, with one curving, long, medial spine, one lateral spinule and one distal broad-based denticle. Fifth leg with broad and rounded tip; armature similar to that of corresponding leg of type species. Ventral surface of second abdominal segment with two irregular rows of subtriangular flaps (Fig. 1265). Uropod (Fig. 1265) with short, stout terminal and similar posterolateral spine; slender seta on lateral margin and cuticular flaps on ventral surface and margins.

Male: Unknown.

Comments: Although known since nearly the beginning of this century, *E. minuta* has been seen on only a few occasions and only in British waters, where it occurs on the gills of *Dasyatis pastinaca*, itself an uncommon inhabitant of this region. Future investigations might extend both the host and geographic ranges of *E. minuta*, though at least one other species of stingray is now known to harbour a different *Eudactylina* (*Dasyatis kuhlii* in Queensland, Australia, parasitized by *E. papillosa* Kabata, 1970).

Genus Nemesis Risso, 1826

Female: Eudactylinidae. Dorsal shield long, with overhanging lateral and posterior margins, often with embossed ventrolateral surfaces. Four free thoracic segments often with ill-defined terga, separated from one another by transverse constrictions; as broad as, or broader than, cephalothorax. Genital segment subspherical, smaller than preceding segment. Abdomen two- or three-segmented. Cuticular flaps absent. First antenna straight, of 10 to 15 segments. Second antenna four- or five-segmented, with unciform terminal claw. Mouth tube and oral appendages siphonostome. Maxilliped subchelate. First four pairs of legs biramous, rami two-segmented; those of first leg forming prehensile apparatus, others small, with reduced armature. Fifth leg vestigial, one-segmented. Uropod small, with reduced armature.

Male: Similar to female but slender and elongate. Border between fifth leg-bearing segment and genital segment often indistinct. Latter with two ventral genital plates. Abdomen three- or four-segmented. First two pairs of legs similar to those of female, third and fourth sometimes with modified endopod spines. Sixth leg occasionally present. Leg armature and uropod better developed than in female.

Type species: *Nemesis lamna* Risso, 1826.

Comments: Eight currently upheld species of this genus have been described so far. One, originally described as *Pagodina carchariae-glauci* by Hesse (1884) and later transferred to *Nemesis* must be considered a species inquirenda, its description being insufficient for identification. *N. vermi* A. Scott, 1929, has been reduced to the status of a morphological form of *N. lamna*. The distribution of the valid species is given in table 14.

Table 14. Geographic distribution of *Nemesis*.

Species	NA	SA	NP	SP	IND
N. atlantica Wilson, 1922	x				
N. lamna Risso, 1826	x	x	x	x	x
N. macrocephala Shiino, 1957			x		
N. pilosa Pearse, 1951	x				
N. robusta (van Beneden, 1851)	x	x		x	x
N. tiburo Pearse, 1952	x				
N. versicolor Wilson, 1913	x				

NA—North Atlantic; SA—South Atlantic; NP—North Pacific; SP—South Pacific; IND—Indian Ocean.

In addition to species listed in the table, two records of unnamed males of *Nemesis* have been published (Pillai, 1967c; Kabata, 1970c).

Only a small part of the literature dealing with this genus, even of quite recent date, is useful for comparative purposes. Much more work, particularly on the details of appendages, is required to give us a satisfactory knowledge of these parasites. One morphological criterion used for discrimination between various species of *Nemesis* is the relative widths of the cephalothorax, free thoracic segments and genital segment (Shiino, 1957; Hewitt, 1969a). This character, however, discriminates only between two groups of species, one (containing most species of the genus) with the fourth free segment much narrower than the preceding three and the other (*N. lamna*) in which all four free segments are of about the same width. While this can be useful for identification of the British species (one of them being *N. lamna*), it is of little help otherwise. Moreover, suggestions have been made that the width of the free segments might be at least partly influenced by the age of the parasite and by the identity of its host. It is also subject to individual variation and, in any case, can be used only for females. Cressey (1967c) used as a discriminant the spinulation of the second antenna, a procedure justly criticized by Hewitt (1969a) as unreliable, in view of considerable individual variation of these structures.

In British waters the genus is represented by two species: *N. lamna* f. *vermi* A. Scott, 1929; Hewitt, 1969 and *N. robusta* (van Beneden, 1951).

Key to females of British species

Terga of four free segments poorly developed, without posterolateral projections, fourth segment about as broad as third (Fig. 1269) . . . . **N. lamna** f. **vermi**

Terga of four free segments well developed, with protruding posterolateral corners; fourth segment much narrower than third (Fig. 1296) . . . . **N. robusta**

The males can be distinguished by the number of abdominal segments, three in *N. lamna* f. *vermi* and four in *N. robusta*.

Nemesis lamna f. vermi A. Scott, 1929; Hewitt, 1969

(Figs. 1266–1294)

Syn.: *Nemesis robustus*; of Travis Jenkins (1925)
Nemesis vermi A. Scott, 1929
Nemesis lamna vermi A. Scott, 1929; of Hewitt (1969a)

Female (Figs. 1268, 1269): Cephalothorax with poorly developed dorsal shield, long, somewhat narrower posteriorly. Four free segments separated by deep transverse constrictions, about as wide as cephalothorax, each slightly broader posteriorly, except fourth; latter with rounded anterolateral corners. Terga poorly developed, particularly that of fourth free segment. Genital segment partly fused with first abdominal segment, followed by two more, much narrower abdominal segments.

Genital segment much smaller than fourth free segment, with lateral vaginal openings and smaller dorsal openings of oviducts. Total length about 9 mm.

First antenna (Figs. 1270–1272) long, slender, tapering to its tip, not clearly segmented (literature data quote most frequently 13 or 14 segments); segments 2–6 particularly badly delimited (Fig. 1271); basal segment long, with some cuticular sculpturing, armed with one short seta on anterior margin; second segment with three setae on anterior margin and dorsal surface, third with five setae, fourth with one, fifth with five, sixth and seventh with two and remaining segments (except for terminal) with one; terminal segment with prominent aesthete on anterior margin (Fig. 1272), one short seta near its base and another longer on opposite side; one slender seta at midlength of posterior margin; apical armature of one short seta on anterior side of apex, three long and two shorter setae in centre of apex and five slender setae, in groups of two and three, on posterior side of apex. Second antenna (Figs. 1273, 1274) five-segmented; basal segment short, unarmed; second longer, unarmed; third with papilliform process on posterior margin, surmounted by short spinule and oval swelling proximal to process, numerous minute denticles on medial surface; fourth segment long, unarmed, slightly tapering; unciform claw with moderately broad base and gently tapering, curving tip; on inner margin of base two setae and one papilliform swelling (Fig. 1274). Mouth cone short, broad at base, with lateral rows of short, stiff setae. Mandible of two parts, shorter basal and longer terminal, with dentiferous margin carrying nine short, broad teeth (Fig. 1275), diminishing in size towards base. First maxilla biramous (Fig. 1276), with apparently two-segmented sympod; rami arranged dorsoventrally, one-segmented; dorsal (endopod) shorter and more robust, with two long setae about equally long, ventral seta with two rows of fine pinnules (partly omitted in Fig. 1276); ventral (exopod) slightly longer and more slender, surmounted by two short setae and one longer, apparently unarmed. Second maxilla (Figs. 1277, 1278) brachiform; lacertus robust, constituting about half length of appendage, unarmed; brachium with patch of small, sharp denticles on lateral surface near tip (Fig. 1278), tuft of setae at base of terminal claw and prominent swelling on lateral side of tip, its surface covered with slender, distally pointed, spiniform setae; claw rather blunt, tapering and curving, with sharp denticles on surface. Maxilliped (Fig. 1279) subchelate, very prominent; corpus arising from strongly sclerotized, cylindrical base, tapering and flexing in outward arch; inflated myxal area at proximal end of arch with short spinule; subchela long, slender, with shaft and claw indistinctly divided; former with two short setae on inner margin and with equally short barbel at base of claw.

First four pairs of legs biramous, sympods two-segmented, rami two-segmented. Armament formula as follows.

	Endopod		Exopod	
	1	2	1	2
Leg 1	0-0	2(3)	0-1	3(4)
Leg 2	1-1	5	1-1	6
Leg 3	1-1	6	1-1	6
Leg 4	1-1	5	1-1	6

All sympods (Figs. 1281, 1282) with broad band of denticles along distal margin of first segment; other denticles, minute prickles and fine setules scattered over ventral surface; bands of denticles also on parts of distal margin of second segment. First leg (Fig. 1281) modified as prehensile organ; basal segment of endopod extended on ventral side to form fork with distal segment; elongated and strongly curved basal segment of exopod fitting, when closed, into fork, with prominent, spinulated distal protuberance; spines of distal segment with serrated margins; medial margin of basal segment of exopod with long, fringe-forming setae (fringe cut at half length in Fig. 1281); prominent seta at base of endopod with long rows of pinnae (omitted in drawing). Next three pairs similar to one another (Fig. 1282). Basal joints of both rami with posterolateral and posteromedial corners extended into strong spines, usually not clearly delimited from segments themselves (Figs. 1283, 1284); terminal spines short, often on dorsal surface and impossible to see without dissection (Figs. 1285–1287). Fifth leg (Fig. 1288) one-segmented, short, rounded, with three terminal setae and small denticles on lateral surface. Uropod (Fig. 1289) fairly long, cylindrical, with transverse bands of denticles on ventral surface, one strong, short, apical spine, three distolateral and two distomedial, subapical setae.

Male (Figs. 1266, 1267) generally similar to female. Free thoracic segments slightly tapering posteriorly; fifth leg-bearing segment indistinctly delimited from genital segment, latter with two semicircular genital plates on ventral surface (Fig. 1280), usually larger than preceding segment. Abdomen three-segmented, segments progressively diminishing in size towards posterior extremity. Total length 7–8 mm.

Appendages very similar to those of female, except for swimming legs. Armature formula as below.

	Endopod 1	Endopod 2	Exopod 1	Exopod 2
Leg 1	0-0	2(3)	0-0	3(4)
Leg 2	1-1	5	1-1	6
Leg 3	1-1	4-1-1	1-1	7
Leg 4	1-1	3-1-1	1-1	7

First and second legs similar to those of female, though generally more slender. Third endopod (Figs. 1290, 1291) with one spine longer than distal segment, powerful, curving inwards, tapering; medial spine of basal segment short. Fourth leg similar to third, but medial spine of basal endopod segment replaced by long digitiform, blunt process (tactile?) (Fig. 1293); exopod as in third leg (Fig. 1292). Fifth leg similar to that of female (Fig. 1294), relatively smaller and with fewer surface denticles. Sixth leg not observed.

Comments: Although A. Scott (1929) thought that this species might prove to be only a variety of *N. robusta*, its slenderness prompted him to place it in a new taxon, *N. vermi*. The evidence of its affinity with *N. lamna* (the structure of its appendages) escaped the discoverer's notice. Another possible reason for Scott's decision was the identity of the host of his slender *Nemesis*, the basking shark, *Cetorhinus maximus*, from which *Nemesis* was recorded only twice before (Travis Jenkins, 1925; Fage, 1923). Hewitt (1969a) examined specimens of both *N. lamna* and *N. vermi* taken in New Zealand waters and decided that they are conspecific, since they differ only in the relative width of their free thoracic segments. To underline their relationship, he treated them as subspecies *N. lamna lamna* and *N. lamna vermi*. It is this author's view, however, that this difference does not merit a subspecific status and that the "narrow" *N. lamna* is no more than a morphological form occurring on the basking shark.

The greatest difficulty in a morphological study of this species is presented by the segmentation of the first antenna. This is due to the more or less complete fusion of segments 2 to 6 (Fig. 1271), which can be delimited from one another only by viewing the appendage in an optical section. The specimens at the author's disposal had 14-segmented first antennae (both sexes).

A noteworthy morphological feature is the modification of the third and fourth legs of the male, probably an adaptation for maintaining hold of the female during the process of fertilization.

So far, the records of *N. lamna* f. *vermi* are restricted to those of Travis Jenkins (1925) and A. Scott (1929) from British waters, that of Hewitt (1969a) from the New Zealand region and that of Yamaguti (1939b) from Japan. The host was in all instances the same. The "wide" form of the species, *N. lamna* f. *lamna*, is widespread, as shown in Table 14.

Nemesis robusta (van Beneden, 1851)

(Figs. 1295–1308)

Syn.: *Ergasilina robusta* van Beneden, 1851
Pagodina robusta van Beneden, 1853
Nemesis robuta Beneden; of Brian (1906)
nec *Nemesis robustus*; of Travis Jenkins (1925, etc.)
Nemesis pallida Wilson, 1932
Nemesis aggregatus Cressey, 1967

Female (Figs. 1295, 1296): Cephalothorax with suboval dorsal shield, rounded anterior and flattened posterior margin. First three free thoracic segments with well developed terga, broader than cephalothorax, with prominent posterolateral corners, all of about equal width. Fourth free segment about half width of three preceding ones, its tergum with rounded posterior margin, in some specimens overlapping genital segment. Latter small, broader than long, with broad transverse slits of oviduct orifices. Abdomen three-segmented, segments progressively narrower. Total length about 5 mm.

Beginning with first antenna and up to maxilliped appendages very similar to those of *N. lamna*, with few differences in detail. Protuberance on inner margin of terminal claw of second antenna about twice as long as that of *N. lamna* (Fig. 1274); mandible with only eight teeth (nine in *N. lamna*); denticulation of second maxilla (Fig. 1299) much more prominent than that of type species. Armature of first four thoracic legs as below.

	Endopod 1	Endopod 2	Exopod 1	Exopod 2
Leg 1	0-0	2	0-1	3(4)
Leg 2	1-1	5	1-1	7(8)
Leg 3	1-1	4	1-1	7
Leg 4	1-1	4	1-1	8

Spinulation on legs much coarser than in *N. lamna*, spinules and denticles larger and more abundant. First endopod with tip of basal segment covered by spinules, often flexed inwards (as in male, see Fig. 1303). Second leg (Fig. 1300) and remaining legs similar to those of *N. lamna*. (It should be noted that in this species armature tends to vary from one specimen to another so that setation formula cannot be taken as specific diagnosis.) Fifth leg as in preceding species. Uropod (Fig. 1302) long, with pedunculate base; surfaces unarmed, one terminal and one subterminal spine, two distolateral and one distomedial slender setae.

Male (Figs. 1297, 1298) similar to female, except for posterior half of body. Fourth free segment almost completely fused with genital segment, delimited from it by shallow groove on ventral surface of body but overlapping it with posterior margin of tergum. Genital segment with ventral genital plates. Abdomen four-segmented, segments progressively smaller in posterior direction. Total length about 3 mm.

Appendages very similar to those of female, except for swimming legs. Armature formula below.

	Endopod 1	Endopod 2	Exopod 1	Exopod 2
Leg 1	0-0	3	0-1	3(4)
Leg 2	1-1	5(6)	1-I	4+IV
Leg 3	1-1	5	1-I	5+III
Leg 4	1-1	5	1-I	5+III

First leg (Fig. 1303) closely resembling that of female, except for additional spine on second segment of exopod. Second to fourth legs with relatively long, pinnate setae (pinnules omitted in Fig. 1304). Lateral processes of basal segments of rami not clearly delimited from rest of segment. Long medial seta of basal segment of first endopod (Fig. 1304) unarmed. Distal segment of second exopod with four medial pinnate setae and four lateral, short spines. (Individual variation encountered in numbers of both setae and spines.) Third and fourth legs similar to second, with differences in armature. Distal segments of third and fourth endopod (Fig. 1305) with seta second from lateral side unarmed, much thicker than other setae and with rounded apex. Lateral margins of segments with fringes of fine setules. Exopods of legs 3 and 4 with distal segments (Fig. 1306) bearing four setae and three short spines with delicately denticulated margins. Fifth leg (Fig. 1307) one-segmented, with unarmed surface and three short terminal setae. Uropod (Fig. 1308) relatively more prominent than that of female, with three long pinnate setae (some pinnules omitted in drawing), two distomedial and two distolateral setules and fringe of fine setules on medial margin.

Comments: Apart from the early difficulties with the generic name, this species has had a fairly simple taxonomic history. Its large size and characteristic appearance rendered it easy to identify. The apparent distinctness of *N. robusta* became less obvious, however, as more species were described and found to be very much like their longer-known relative. It is very unfortunate that the present state of our knowledge forces us to concentrate on gross morphology in our search for specific discriminants. An attempt to distinguish *N. robusta* from other species of the genus with the aid of biometric appraisal was made by Hewitt (1969a) who concluded that two of them (*N. pallida* and *N. aggregatus*) are synonymous with it. This author accepts Hewitt's judgement, if only because there is no available evidence that can be used to refute it. It is to be hoped that more detailed morphological studies will clarify the problem of specific relationships within the genus *Nemesis* and allow us to gauge the true status of *N. robusta*.

As defined at present, *N. robusta* is a common and ubiquitous parasite of numerous elasmobranch species. Among sharks its recorded hosts are: *Alopias vulpinus*, *Aprionodon isodon*, *Carcharias taurus*, *Carcharinus leucas*, *C. limbatus*, *C. milberti*, *C. obscurus*, *Carcharodon carcharias*, *Galeocerdo cuvieri*, *Hexanchus griseus*, *Lamna nasus*, *Mustelus canis*, *M. equestris*, *M. mustelus*, *M. plebejus*, *Mustelus* sp.,

Negaprion brevirostris, Prionace glauca, Scoliodon terraenovae, Sphyrna lewini, S. tiburo, S. zygaena and *Sphyrna* sp. It is also known to be carried by the following rays: *Dasyatis aspersa, D. pastinaca, Laeviraja oxyrhynchus* and *L. macrorhynchus*.

N. robusta has been recorded from the northern and southern Atlantic, the Mediterranean and Adriatic, the Indian Ocean and the Pacific (New Zealand region). It was found in British waters by Leigh-Sharpe (1936a) on *Alopias vulpinus*.

Family Kroyeriidae fam. nov.

As stated earlier (p. 245), the main characteristics of this family are: (1) the presence of three leg-bearing segments between the cephalothorax and the genital complex; (2) the possession of four pairs of biramous, primitive swimming legs (occasionally with a vestigial fifth in the male); (3) a chelate second antenna. The general habitus of the female is dominated by the length of the cylindrical genital complex, in some species constituting about $\frac{3}{4}$ of the total length of the female. Only two genera have these characters in common. One of them, *Kroeyerina* Wilson, 1932, contains only three species: two known from a single find, the other recorded on only two occasions. All three records come from the Atlantic coast of the USA. *Kroeyerina* differs from the second genus, *Kroyeria* van Beneden, 1853, in the absence of posterior sinuses and dorsal stylets from the dorsal shield of the cephalothorax. *Kroyeria*, the only genus of the family represented in British waters, contains 20 nominal species (not all of them valid).

The history of Kroyeriidae was discussed together with that of Dichelesthiidae (p. 238). The diagnosis of the family is as follows.

Female: First leg-bearing segment fused with cephalothorax. Dorsal shield subtriangular, with or without posterior sinuses and dorsal stylets. Three free thoracic segments between cephalothorax and genital complex. Latter consisting of genital segment and segment preceding it, together constituting more than half of body length, slender, cylindrical. Abdomen short, cylindrical, one- to three-segmented. First antenna six- or seven-segmented, slender, setiferous. Second antenna chelate. Mouth and mouth parts siphonostome. Maxillipeds subchelate. Four pairs of biramous swimming legs, rami three-segmented, with well developed armature. Uropods elongate.

Male: Similar to female but with relatively much shorter genital complex and relatively longer, three-segmented abdomen. Fifth pair of legs sometimes present, vestigial and setiform.

Type genus: *Kroyeria* van Beneden, 1853.

Genus Kroyeria van Beneden, 1853

Female: Kroyeriidae. Dorsal shield of cephalothorax with posterior sinuses; dorsal stylets present. Abdomen one- to three-segmented. Male: Similar to female; abdomen three-segmented. Fifth leg vestigial and setiform, or absent.

Type species: *Kroyeria lineata* van Beneden, 1853.

Comments: The literature dealing with this species contains references to 19 nominal species (with an additional description of an unnamed male by Capart, (1953)). Three of these species (*K. acanthiasvulgaris* Hesse, 1879, *K. galeivulgaris* Hesse, 1884 and *K. scyllicaniculae* Hesse, 1879) are known only from descriptions which are insufficient for comparison with any known species and must be treated as species inquirendae. Three others (*K. aculeata* (Gerstaecker, 1854), *K. gracilis* Wilson, 1932 and *K. sublineata* Yamaguti and Yamasu, 1959) are now relegated to synonymy. *K. trecai* Delamare Deboutteville and Nunes-Ruivo, 1953, has never been fully described and remains a nomen nudum. Many of the remaining species are known from descriptions of doubtful accuracy but must be accepted on their face value in the absence of conclusive evidence regarding their validity.

The systematics of the genus have suffered in the past from our insufficient attention to those morphological details that can be accepted as definite specific discriminants. Delamare Deboutteville and Nunes-Ruivo (1953) suggested that the length of the dorsal stylets, commonly used as a morphologically significant character, is too variable in most (though not all) instances to be validly used in the specific diagnosis. They also rejected for this purpose the cephalic appendages, too uniform within the genus to be useful for species identification. The proportions of the body, especially of the genital complex and free thoracic segments, were held by them to be unreliable, as was the segmentation of the abdomen. The latter tends to become indistinct in the older females, up to the point of complete disappearance. The authors recommended, as the best morphological criterion, the chaetotaxy of the four swimming legs, with the proviso that only the fully developed females should be used for this purpose. Delamare Deboutteville's and Nunes-Ruivo's views, fully shared by this author, mean the rejection of an earlier attempt (Wilson, 1922a) at systematizing identification of the species of *Kroyeria*. Kabata and Gusev (1966) provided a key to eight species of this genus based on the armature of the endopod and uropod. This rather incomplete key can be used as a basis for further development.

Kroyeria occurs in the Atlantic, Pacific and Indian Oceans. The distribution of its currently accepted species is shown in Table 15.

Table 15. Distribution of *Kroyeria*.

Species	NA	SA	NP	SP	IND
K. carchariaeglauci Hesse, 1879	x	x	x	x	x
K. dispar Wilson, 1935	x				x
K. echinata Rangnekar, 1956					x
K. elongatus Fukui, 1965			x		
K. elongata Pillai, 1967					x
K. gemursa Cressey, 1967	x				x
K. lineata van Beneden, 1853	x		x		
K. longicauda Cressey, 1970	x				
K. praelongacicula Lewis, 1966			x		
K. spatulata Pearse, 1948	x				x
K. sphyrnae Rangnekar, 1957	x				x
K. minuta Pillai, 1968					x
K. papillipes Wilson, 1932	x				

NA—North Atlantic; SA—South Atlantic; NP—North Pacific; SP—South Pacific; IND—Indian Ocean.

This list is far from complete. Another species is now being described from the Indian Ocean and intensive search, both there and in other areas, is certain to bring more to light.

Two points of nomenclature must be raised here. The name *K. elongatus* Fukui, 1965, must be amended to make the gender of the specific name agree with that of the generic one. Homonymy thus arises between Fukui's species and *K. elongata* Pillai, 1967. Should both species stand the test of validity on re-examination, Pillai's name will have to be changed. The other point concerns the spelling of the generic name. In the original description the spelling was *Kroyeria*. There is no reason to alter it, although it does not correspond with the spelling of Krøyer, its namesake. The matter does not fall within the scope of Article 32(c) of the International Code of Zoological Nomenclature. (On the other hand, the original spelling of *Krøyerina* must be altered to *Kroeyerina*, in accordance with that Code.)

The genus is represented in British waters by only one species, *K. lineata* van Beneden, 1853.

Kroyeria lineata van Beneden, 1853

(Figs. 1082–1100)

Syn.: *?Lonchidium aculeatum* Gerstaecker, 1854
Lonchidium lineatum; of Bassett-Smith (1899)
Kröyeria lineata Beneden; of Brian (1906)
?Kröyeria aculeata Gerstaecker; of Brian (1906)
Krøyeria lineata Beneden; of Wilson (1932)
Krøyeria sublineata Yamaguti and Yamasu, 1959

Female (Fig. 1082): Dorsal shield of cephalothorax (Fig. 1083) somewhat reminiscent of those of *Caligus* species, but with central zone delimited by prominent sutures converging posteriorly; eyes in centre of that zone; lateral zones with scattered small, rough tubercles. Dorsal stylets (Fig. 1024) extending posteriorly to about anterior margin of second free thoracic segment, their bases broad and tips tapering and curving inwards, apparently capable of dorsiflexion. Three free thoracic segments increasing in length and slightly in width posteriorly, their terga with medial cuticular ribs and chevron-shaped transverse ribs; all segments well delimited. Genital complex cylindrical, constituting almost $\frac{2}{3}$ of total body length, slightly narrower than cephalothoracic shield, in some specimens with slightly protruding posterolateral corners, carrying openings of oviducts. Abdomen less than half width of genital complex, tapering posteriorly, indistinctly divided into two segments (in some specimens no division observable). Total length 3.8–4.5 mm.

First antenna (Fig. 1085) indistinctly segmented (seven or eight segments), with only clear division between basal, shorter part and rest of appendage; armature difficult to determine; anterior margin of basal part with about six setae of various lengths and distally with three fused cuticular ridges; anterior margin of distal part with about 10 setae, additional seta on dorsal surface; apical armature (Fig. 1086) with prominent subapical aesthete, one apical seta equalling aesthete in length and thickness, and shorter seta divided near base into two branches, one much thinner than other; three subapical setae on dorsal side of apex and five long slender setae on ventral side, in two groups of two and three; posterior margin close to apex bearing another long, slender seta. Second antenna (Fig. 1087) chelate and prehensile, apparently four-segmented. Basal part indistinctly two-segmented, its surface with system of sclerites allowing great freedom of movement; third segment forming corpus of chela, distally extending into rigid arm, with tip cupped to receive opposable arm of chela; latter robust, curving, with prominent seta on inner margin near base and flange near tip. Mouth tube with two distolateral fringes of setules on labium (Fig. 1088) and apex slightly incised, with two short flaps of membrane. Mandible of two parts, total length about 50 μ; dentiferous margin (Fig. 1089) with six teeth, small apical tooth followed by largest of series, then two small and two large. (Scott and Scott (1913) showed six teeth of uniform size.) First maxilla (Fig. 1090, disarrayed in drawing) biramous; exopod shorter and more robust with two short, slender setae apically; endopod longer, more slender, its apical setae much more sturdy and long, arising from common base. Second maxilla (Fig. 1091) brachiform; lacertus robust, with ridges of thick cuticle, otherwise unarmed; brachium with inflated distal end; posterior margin and adjacent areas with densely packed, minute prickles, outer margin close to base of claw with large tuft of long, fine setae; claw slightly curved, robust, with rounded tip and margins armed with rows of minute prickles; tip with one to three setules (natural number probably three, setae often broken off). Maxilliped (Fig. 1092) subchelate, with apparently two-segmented corpus, short basal and longer, slightly tapering terminal, both apparently unarmed; subchela very long, not divided into shaft and claw, unarmed, with pronounced curvature of distal half and ill-defined remnants of membranous flange on parts of inner margin.

All four pairs of swimming legs biramous, rami three-segmented (Fig. 1093); legs of each pair connected by interpodal bar, with outgrowths at each end; outgrowths increasing in length from first (Fig. 1094), to second (Fig. 1095) and succeeding. Sympods two-segmented, with surfaces unarmed; second segments with one seta medial to base of endopod and other lateral to base of exopod. Armature of rami as shown below.

	Endopod			Exopod		
	1	2	3	1	2	3
Leg 1	1-0	0-0	6	1-0	1-0	6
Leg 2	1-0	0-0	6	1-1	1-0	7
Leg 3	1-0	0-0	4	1-1	1-0	7
Leg 4	1-0	1-0	3	1-1	1-0	7

Exopods of all four pairs almost identical (Fig. 1096), that of first leg having one seta less; lateral margins with narrow strips of membrane, basal segments with very small distolateral seta; lateral seta on distal segment very small, unarmed, second seta longer, with one strip of marginal membrane, third seta full length, with one strip of membrane and one row of pinnules; other setae pinnate. Endopod of second leg identical with that of first. Endopod of third leg (Fig. 1097) with four apical setae, that of fourth with three. Only fourth leg with seta on medial margin of second endopod segment (Fig. 1098); all endopod setae pinnate. Uropod (Fig. 1099) long, narrow, with fringes of setules on medial margins; two longer apical setae, three shorter posterolateral and one posteromedial.

Male (Fig. 1100) with anterior half closely similar to that of female. Genital complex cylindrical, about as long and wide as three free thoracic segments combined. Abdomen three-segmented, narrow. Total length 2.7 mm (after Scott and Scott, 1913).

Appendages much as in female. Fifth leg, on lateral margin of genital segment, represented by three short setae. Uropod relatively more prominent. (The author had no opportunity to examine specimens of male of this species, description *fide* Scott and Scott (1913).)

Comments: The taxonomic difficulties referred to in the comments on the genus (p. 264) were not without influence on the past record of this species. Gerstaecker's (1854) description of *Lonchidium aculeatum* was not detailed enough to satisfy Brian (1906) that the species was distinct from van

Beneden's *K. lineata*. Delamare Deboutteville and Nunes-Ruivo (1953) shared Brian's views, though they listed *K. aculeata* as a separate taxon. Under the circumstances, a tentative synonymy is indicated until the matter can be clarified beyond doubt. Yamaguti and Yamasu (1959) described *K. sublineata* from Japan, a species they believed very similar to *K. lineata*. The differences they accepted as decisive in establishing it as a new species were the size (4.5–4.6 mm), which they stated to be much smaller, and the setation of the second endopod. However, the female of *K. lineata* falls quite definitely within the size range of the Japanese species. It appears certain that the Japanese workers used Scott and Scott (1913) as a source for their comparison between the two species. The author examined specimens of *K. lineata* from the collection of the British Museum and found no differences between the armature of the swimming legs of that species and that of *K. sublineata*. Yamaguti and Yamasu were misled by an incorrect label in Scott and Scott's (1913) Pl. 70, Fig. 9, which reads "Foot of second pair" but shows that of the fourth pair. The only difference between *K. lineata* and *K. sublineata* appears to be in the number of mandibular teeth. Although the pattern of dentition appears to be similar (even to the presence of the small apical tooth), *K. sublineata*, according to its original description, has two basal teeth more than *K. lineata*. The difference is not in itself sufficient as a specific discriminant and the author proposes that *K. sublineata* be relegated to synonymy with *K. lineata*.

K. lineata has been recorded from several species of sharks (*Galeorhinus galeus*, *Mustelus canis*, *M. equestris*, *M. manazo*, *M. mustelus*, *Prionace glauca* and *Sphyrna zygaena*). Most of its records come from the Atlantic seaboard of Europe, south of the base of the Jutland peninsula. One find was made off the Atlantic coast of the USA (Wilson, 1932). Relegation of *K. sublineata* to synonymy with this species extends its distribution range to Japanese waters. In British waters it was found in the Irish Sea (Scott and Scott, 1913) and off the coast of Norfolk (Hamond, 1969).

Family Pseudocycnidae comb. nov.

The morphology and development of the systematic concept of this family were discussed together with those of Dichelesthiidae (p. 240). One might add here that Wilson (1922a) was first to propose a separate subfamily Pseudocycninae, which he later (1932) upgraded to the familial status. He was followed by Yamaguti (1939b). In a later publication, Yamaguti (1963) again presented Pseudocynidae as a new taxon, diregarding both Wilson's and his own previous work.

Although only nine nominal species can be considered for inclusion in this family (not more than six of them valid), a great deal of confusion exists regarding the generic affinities of these species. Compounded by the absence of adequate descriptions, these difficulties cannot be resolved in our present state of knowledge. The author was compelled, due to lack of material, to give less than full attention to the intrafamilial problems of Pseudocynidae. On the basis of the scanty data at his disposal, he includes in it the genus *Cybicola* Bassett-Smith, 1898 (with three species: *C. armatus* (Bassett-Smith, 1898), *C. buccatus* (Wilson, 1922) and *C. elongatus* Pearse, 1951), *Pseudocycnoides* Yamaguti, 1963 (with *P. scomberomori* (Yamaguti, 1939) and *P. rugosa* Kensley and Grindley, 1973) and *Pseudocycnus* Heller, 1865. (Heegaard's (1962) *Paracycnus lobosus* is probably synonymous with Yamaguti's (1939) *Pseudocycnoides scomberomori.*) These opinions are of necessity tentative.

In their new combination, the diagnosis of Pseudocycnidae is as follows.

Female: Cephalothorax with or without first leg-bearing segment. Region between anterior tagma and genital complex two- or three-segmented, ill-defined and without clear intersegmental borders. Genital complex incorporating fourth leg-bearing segment, subcylindrical, elongate. First antenna uniramous, setiferous; second antenna three-segmented, with unciform terminal claw. Mouth tube and oral appendages siphonostome. Maxilliped subchelate. From three to five pairs of variously modified or vestigial legs.

Male: Similar to female, but with much smaller genital complex. With or without prominent lateral processes of fourth legs.

Type genus: *Pseudocycnus* Heller, 1865.

Pseudocycnidae are ubiquitous parasites of marine teleosts, mainly scombrids. They are represented in British waters by *Pseudocycnus*.

Genus Pseudocycnus Heller, 1865

The diagnosis of this genus is the same as that of its only species, *P. appendiculatus*, and can be found in the description of that species. The author accepts Hewitt's (1969c) and other writers' decision to synonymize *P. spinosus* and *P. thunnus* with *P. appendiculatus*. He does not accept Kirtisinghe's relegation of *Cybicola* to synonymy with *Pseudocycnus*.

Pseudocycnus appendiculatus Heller, 1865

(Figs. 1309–1329)

Syn.: *Pseudocycnus spinosus* Pearse, 1952
Pseudocycnus thunnus Brandes, 1955

Female (Fig. 1309): Cephalothorax fused with first leg-bearing segment, dorsal shield (Fig. 1310) not well developed, its anterior part semicircular in dorsal aspect, posterior expanding, posterolateral corners merging with bases of maxillipeds (obscuring contours of shield, when pointing posteriorly). Posterior end of cephalothorax forming narrow neck, imperceptibly fusing with first free thoracic segment. Latter transversely oval, with anterior end narrow, meeting neck of cephalothorax; width slightly greater than that of dorsal shield. Second thoracic free segment slightly broader than first, transversely oval, with borders marked by transverse constrictions. Fourth leg-bearing segment fused with genital complex, its presence evident from vestigial digitiform legs covered by cuticular sheath of that complex. Latter subcylindrical, more than twice as long as combined lengths of cephalothorax and free segments. Dorsal surface with two parallel rows of depressions lined with heavy cuticle; posterior extremity with two semi-hoops of heavy cuticle with well delimited margins, extending from near mid-dorsal line (Fig. 1311) to ventrolateral sides of complex (Fig. 1312). Oviduct orifices partly covered by dorsal ends of semi-hoops (Fig. 1311 showing base of egg sac emerging from beneath bar). Posterior extremity of genital complex separated by deep constriction from one-segmented abdomen. Latter small, with anal slit in centre of posterior margin; posterolateral corners passing into prominent uropods (Fig. 1309), about half length of genital complex, digitiform, slightly tapering to extremities, apparently unarmed. (Shiino (1959c) found two small apical denticles on tips, not located by present author.) Anterior quarter of freshly preserved specimens bright red on dorsal surface, remainder yellow-brown (Hewitt, 1969b). Total length (with uropod) 7–24 mm.

First antenna (Fig. 1313) indistinctly segmented (Shiino (1959c) and Hewitt (1969c) describe it as four-segmented, but it might be five-segmented); no trace of segmentation in basal part, slightly less than half appendage, with at least two short setae and one prominent process on anterior margin and another, slender seta on posterodorsal wall; middle part (one- or two-segmented) with five setae on anterior margin; terminal part apparently two-segmented, first segment unarmed; apical armature of three groups of setae; three short setae on anterior side, three slightly longer in central position and three much longer on posterior side of apex; one short and two longer setae on posterior margin of terminal segment and one short papilla on dorsal surface near apex. Second antenna (Fig. 1314) apparently three-segmented, subchelate; first and second segment not clearly delimited, forming powerful corpus antennae, slightly tapering distally, with definite inward flexure; third segment, subchela, not divided into shaft and claw, with subtriangular dentiform process on inner side of base and smaller, similar process some distance from base (Hewitt, 1969c, described "two spines near midpoint of inner margin"); tip tapering, curving inwardly. Mouth tube siphonostome with base width equalling about half of tube length; converging margins of labrum with central concavities; tip rounded. Mandible distinctly bipartite, with basal part shorter; dentiferous margin with six subequal teeth (Fig. 1315) and (in all specimens seen by this author) with shallow apical notch; traces of damage on margin proximal to basal tooth often present. First maxilla (Fig. 1316) very small, difficult to dissect and study, cylindrical, apparently unsegmented; proximal part with heavy, pitted cuticle; two terminal conical processes, one much smaller than other, both with attenuated tips; on medial side of tip small dentiform process. Ventral surface of cephalothorax posterior to base of first maxilla with plaque of heavy cuticle, well delimited, larger than appendage and with central subspherical papilla (Fig. 1316, p). Second maxilla (Fig. 1317) brachiform, with very powerful, unarmed, inflated lacertus and subcylindrical brachium becoming thicker towards distal end; no definite border between it and claw (Fig. 1318); distal end with row of strong terminal denticles (about seven) and row of subterminal denticles, as well as much smaller denticles irregularly scattered in the region; claw relatively short, slightly curving, with sharp, tapering tip and several strong denticles on inner margin, forming direct continuation of brachial denticles; small denticles might be present on medial and lateral walls; claw capable of great mobility, in some specimens situated at right angles to long axis of brachium. Maxilliped (Fig. 1319) subchelate, large and powerful; corpus possibly two-segmented (transverse shallow constriction near base), with noticeable lateral inflation and transverse cuticular bar at midlength; distal end tapering subchela (Fig. 1320) much shorter than corpus, with broad base, curving sharp tip, strong subapical denticle and another at midlength.

First leg small, often concealed by maxillipeds; structure aberrant (Fig. 1321); sympod unsegmented, with prominent rounded process medial to base of endopod and similar, smaller process at base of exopod; two transverse rows of sharp, slender denticles on ventral surface; endopod (not shown in full view in Fig. 1321) short, apparently two-segmented, subcylindrical, with two terminal papilliform processes and two short, stiff distomedial setules; exopod unsegmented, larger than sympod, boot-shaped. Second leg (Fig. 1322) much larger than first, biramous; sympod unsegmented, with very prominent projection on lateral margin, one long slender seta on lateral side of exopod and three patches of denticles on ventral surface; rami one-segmented (Fig. 1323); endopod short, with three powerful terminal spines and slender, needle-like denticles on lateral side of ventral surface; exopod longer, with four terminal spines (lateral with serrated margins) and surface denticles along bases of spines and on lateral half. Third leg conical, uniramous; ventral surface of ramus distally with transverse rows of slender denticles (Fig. 1324); three strong apical spines with small slender spinules along bases. Fourth leg (Fig. 1310) vestigial, with single short seta at base. Fifth leg (Fig. 1312) single seta in anterolateral corners of genital complex.

Male (Fig. 1325) differing from female in absence of long genital trunk. Cephalothorax comparatively shorter, free thoracic segments narrower than dorsal shield, very indistinctly delineated; genital complex tapering posteriorly, indistinctly delimited from one-segmented abdomen. Latter short, with broader posterior extremity and long, digitiform uropods. Total length 3–5 mm.

First antenna (Fig. 1326) relatively longer than that of female, more slender and more clearly five-segmented; basal segment with long, pointed and apparently soft papilla, two dentiform processes on anterior margin and one seta on ventral surface; second segment with one dentiform process and one seta; third with one dentiform process and one seta on anterior margin and small papilla on dorsal surface; fourth with dorsal papilla; apical armature of fifth segment similar to that of female appendage. Second antenna (Fig. 1327) with long, slender claw armed with more prominent denticles. Second maxilla comparatively longer, more slender and with longer denticles. Maxilliped similar to that of female, but with patch of denticulation on medial margin near base of subchela; latter (Fig. 1328) with indistinct border between shaft and claw, slender seta near base and subtriangular denticle at base of claw.

First leg relatively large, biramous; sympod unsegmented, with long, slender seta near base of exopod (Fig. 1329), a soft lobe with several denticles upon it on medial side of endopod and with large spine on medial margin (about as large as endopod and spoon-shaped); exopod unsegmented, longer than endopod, with four terminal processes and fifth subterminal, spiniform (ghosted in Fig. 1329); endopod with similar armature. Second leg similar to that of female but with larger rami and terminal spines; lateral seta of sympod almost reaching tips of rami, denticulation more abundant; no process on lateral side of sympod. Third leg similar to that of female. Fourth leg flagelliform (Fig. 1325) at tip of long, digitiform process on lateral side of genital complex. Fifth leg vestigial, setiform, posterior to fourth on flank of genital complex.

Comments: This species, ubiquitous and with a broad range of hosts, is equally broadly variable in many of its morphological characteristics, including size. Specimens described by Pearse (1952) as *P. spinosus* were not more than 7.3 mm long; those described by Brandes (1955) as *P. thunnus* were up to 24 mm long. Those identified by various authors as *P. appendiculatus* vary in length from 13 to 17 mm. Age and host influences are manifest in this, as well as in other variable characteristics (length of genital complex, number of dorsal depressions on it, etc.). Other differences are merely discrepancies in the descriptions. When postmortem contractions introduce cuticular wrinkles and similar artifacts, the absence of clear segmentation in the region between cephalothorax and genital complex leads to many errors of interpretation. This applies also to the segmentation of the first antenna. The author was able to examine type material of *P. thunnus* and found no differences between it and *P. appendiculatus*. It seems that here, as in other instances, the describer (Brandes, 1955) suffered from the necessity of comparing his specimens with poor descriptions of related taxa.

The host list of *P. appendiculatus* includes *Coryphaena* sp., *Euthynnus affinis*, *E. alleteratus*, *Katsuwonus pelamis*, *Kishinoella tonggol*, *Parathunnus sibi*, *Sarda chiliensis*, *S. sarda*, *Thunnus albacares*, *T. alalunga*, *T. obesus* and *T. thynnus*. Its distribution is as wide as that of its hosts, mainly the big scombrids. It visits the North Sea and British waters with the tuna (*Thunnus thynnus*).

Family Hatschekiidae fam. nov.

Morphology

The genera belonging to this family (*Hatschekia*, *Prohatschekia* Nunes-Ruivo, 1954, *Congericola* van Beneden, 1851, *Pseudocongericola* Yu, 1933 and *Bassettithia* (Stebbing, 1900) Wilson, 1922), are characterized by loss of definite segmentation in the region between the cephalothorax and genital complex. This region, forming a neck-like constriction, might be variously subdivided by deeper constrictions into several parts, commonly interpreted as segments. The true segmentation is, however, often uncertain. The fixation of specimens frequently causes contraction (compare Fig. 1101 with Fig. 1102), resulting in the appearance of many spurious constrictions, apt to be misinterpreted as natural features. Only a detailed study of the ontogeny of various members of the family will clarify this point. In the meantime, the safest interpretation of the segmentation in the neck region is afforded by the positions of the segmental appendages, in this instance the swimming legs. Using them as indicators, one quickly realizes that the first leg-bearing segment is always fused, at least partially, with the cephalothorax and that the following segments are progressively incorporated into the genital complex. In *Congericola* (Figs. 1158, 1159) there are two pairs of legs in the neck region, which suggests the two-segmented structure of that part. The same is true of *Pseudocongericola, Bassettithia* and *Prohatschekia*. In *Hatschekia* the neck can be two- or one-segmented, or absent.

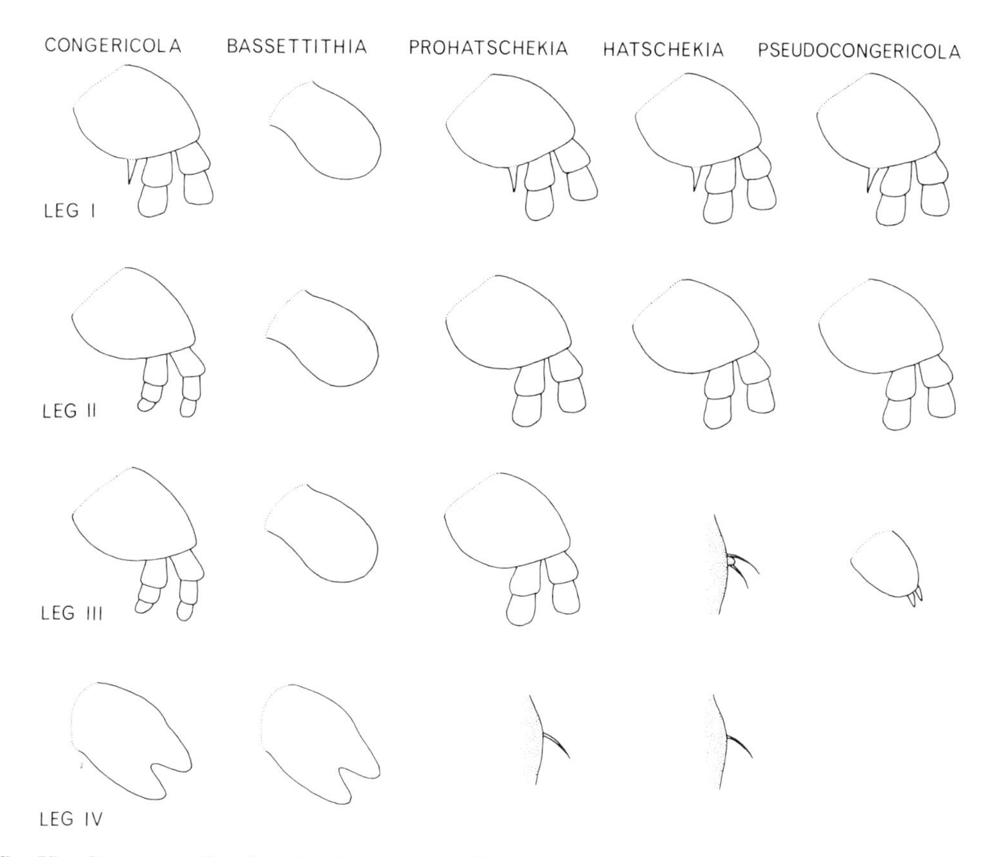

Text fig. 55. Structure of swimming legs among Hatschekiidae, illustrating reduction and regression from posterior end.

The structure of the appendages is not well known for all the species in this family, but the essential features of the antennae and oral appendages appear to be shared by them all. The same is not true of the structure of the swimming legs. The five genera included in the family form a series illustrating a progressive reduction in size, simplification in structure and gradual disappearance of the legs (Text fig. 55). As has been mentioned earlier (p. 38), this process parallels the loss of segmentation and the incorporation of segments into the genital complex. The appendages are best preserved in *Congericola*, which has three pairs of biramous legs, with two-segmented rami in the first and three-segmented in the next two pairs. The fourth pair has lost its original structure, becoming a simple lobe with a bifid distal end. The position is uncertain in the genus *Bassettithia*, which differs from *Congericola* in having no rami on the first three pairs of legs but otherwise appears to be strikingly similar to it. Since both are parasitic on *Conger* and since *Bassettithia* is known from but a single record, one must not discount the possibility that it is no more than a teratological specimen of *Congericola*. In *Prohatschekia* the first three pairs of legs are biramous, all three with one- or two-segmented rami, but the fourth has been reduced to a single small seta (not observed in *P. sebastisci* (Yamaguti, 1939) or *P. laguncula* Shiino, 1957). In *Hatschekia* only the first two pairs of legs remain biramous, their rami not exceeding two segments and varying in armature. The remaining two pairs are vestigial, consisting of one or two, rarely more, small setae. Finally, in *Pseudocongericola* the third leg is reduced to a small rounded lobe, with two apical projections, while the fourth has not been found and is presumably absent.

Their segmentation and general habitus make the members of this family rather similar to Pseudocycnidae. Some of the more elongate species of *Hatschekia* would be difficult to separate clearly from that family. However, the absence of maxillipeds (so far as is known, in both sexes), the type of the second maxillary claw and the small size of the uropods separate Hatschekiidae from Pseudocynidae without much difficulty.

The diagnosis of the new family is as follows.

Female: Dorsal shield present or absent. Cephalothorax usually separated from genital complex by neck-like constriction, indistinctly segmented, consisting of one to three segments, but sometimes absent (in some species of *Hatschekia*). Genital complex subcylindrical to subspherical, sometimes asymmetrical, with or without posterolateral lobes. Abdomen small, one-segmented, partly or completely incorporated into genital complex. First antenna three- to nine-segmented, setiferous; second antenna with unciform terminal segment, with or without papilla at base. Mouth and oral appendages siphonostome. Maxillipeds absent. First leg usually biramous (simple lobe in *Bassettithia*), rami one- or two-segmented; second leg similar to first, rami one- to three-segmented; third leg similar to second, simple lobe or vestigial and setiform; fourth leg bifid lobe, setiform or absent (in *Pseudocongericola*).

Male: Often unknown. Similar to female but with relatively smaller genital complex.

Type genus: *Hatschekia* Poche, 1902.

By far the most common and abundant genus of the family is *Hatschekia*, which will be discussed in more detail below. *Prohatschekia* contains only three species and is an apparently rare parasite of the marine teleosts. The remaining three genera, five species in all, are parasitic on conger eels (*Congericola*, *Bassettithia*) or their relatives, the false conger eels of the genus *Muraenesox* (*Pseudocongericola*). The British fauna includes only *Hatschekia* and *Congericola*. These genera can be distinguished from each other as follows.

Legs 1–3 biramous, leg 4 a bifid lobe . . . . **Congericola**
Legs 1 and 2 biramous, legs 3 and 4 vestigial, setiform, often not detectable . . . . **Hatschekia**

Genus Hatschekia Poche, 1902

Female: Hatschekiidae. Cephalothorax fused with first leg-bearing segment, separated by constriction from following body regions. Neck region indistinctly one- or two-segmented, or absent. Genital complex subcylindrical to subspherical, sometimes asymmetrical. Abdomen one-segmented, partially or completely fused with posterior extremity of genital complex. First antenna indistinctly segmented, consisting of five to seven segments; second antenna three-segmented, with terminal unciform claw and small to prominent parabasal papilla. Mouth and oral appendages siphonostome, mandible with few teeth or toothless. Maxilliped absent. First and second pairs of legs biramous, rami one- or two-segmented. Third and fourth pairs vestigial, consisting of one or more setules. Uropod present.

Male: Similar to female but with shorter genital complex and commonly larger uropods.

Type species: *Hatschekia hippoglossi* (Cuvier, 1830)

Comments: The name *Hatschekia* was proposed by Poche (1902) for copepods previously placed in the genus *Clavella*, which they shared with some lernaeopodid species still known under that name. The composite nature of that genus was changed and the confusion clarified, by removal from it of non-lernaeopodid species. *Hatschekia*, which gathered these species, was included in the family Dichelesthiidae and Yamaguti (1963) left it in it when he revised the family. It was placed together with the genus *Pseudohatschekia* (now *incerte sedis*), *Dichelesthium*, *Pseudocongericola*, *Cybicola* (a pseudocycnid), *Prohatschekia* and *Metahatschekia* (synonym of *Congericola*).

The species of *Hatschekia* display a fair degree of variability in their general appearance, differing from one another in the shape of the cephalothorax (and the structure of its dorsal shield), as well as that of the genital complex. (Their differences in the segmentation of the neck region have already been mentioned.) The trunk-like complex is usually a cylinder, varying in length from one species to another; it may be oval, subcircular, pyriform or irregular. In most instances it is smooth but in some species protuberances occur at either end of the genital complex, or on its dorsal surface. The greatest variety occurs, however, in the structure of the first two swimming legs and of the uropods, aiding a careful observer in identification. Care must be exercised in the study of these very small and delicate appendages, easily damaged by handling. It seems reasonable to suppose that these legs are on the way to becoming reduced to the structure now shown by the fourth and third pairs; the decline in their condition might be expressed as individual variation in structure. Any conclusions based on these appendages must, therefore, involve examination of reasonably abundant material.

Hatschekia, as presently constituted, contains more than 80 species and new ones are being added, as research develops in the tropical and subtropical regions of the world's oceans. Even now, by far the greatest number of species occurs in these waters, dropping in higher latitudes. All species are parasites of marine teleosts. As far as can be judged, most species are specific to one or several species of fish, the hosts belonging usually to one genus, less often to related genera.

British waters are inhabited by six species of *Hatschekia*, which can be identified with the aid of the key below.

Key to British species of *Hatschekia*

1. Cephalothorax with conical posterodorsal process (Fig. 1127) . . . . **H. pagellibogneravei**
   Cephalothorax without posterodorsal process (Figs. 1104, 1142) 2
2. Anterior end of genital complex with transverse pair of rounded dorsal swellings (Fig. 1141) . . **H. mulli**
   Anterior end of genital complex smooth 3
3. First antenna without spine on inner margin of basal segment (Fig. 1107) . . . . **H. hippoglossi**
   First antenna with small or large spine on inner margin of basal segment (Fig. 1119) 4
4. Cephalothorax with protuberances on lateral margins (Figs. 1136, 1137) . . . . **H. labracis**
   Cephalothorax with evenly rounded lateral margins 5
5. First maxilla with endopod of two subequal processes, similar to endopod (Fig. 1155); neck region often with indistinctly delimited single segment (Fig. 1152) . . . . **H. pygmaea**
   First maxilla with sturdy endopod with bifid tip and slender process near base (Fig. 1121); no segment discernible in neck region (Fig. 1116) . . . . **H. cluthae**

Hatschekia hippoglossi (Cuvier, 1830)

(Figs. 1101–1115)

Syn.: *Clavella hippoglossi* Cuvier, 1830

Female (Figs. 1101, 1102): Dorsal shield of cephalothorax rounded in dorsal aspect, with characteristic pattern of ridges (Fig. 1105); lateral margins somewhat curved ventrally (Fig. 1104). Second pedigerous segment separated from cephalothorax and genital segments by shallow constrictions, often ill-defined. Constriction between that segment and genital trunk often neck-like, imperceptibly expanding into genital complex. Latter dorsoventrally flattened, with nearly parallel lateral margins, in adult specimens six to seven times longer than cephalothorax. Posterolateral corners with short, rounded lobes, usually not protruding beyond posterior margin of abdominal region. Abdomen one-segmented, not clearly delimited from trunk, small, with well developed uropods. Entire cuticle covered with sparse denticulation (Fig. 1114). Total length 3.9–5.8 mm.

First antenna (Fig. 1107) indistinctly six-segmented (in some specimens segmentation obscure); basal segment comprising $\frac{1}{3}$ of total length of appendage, with two subequal setae on ventral wall and seven prominent, broad setae on anterior margin; second segment with six similar, though smaller, setae; third with one seta on anterior margin, one short seta on ventral surface and one on posterior margin; fourth segment with two setae on anterior margin; fifth with a prominent seta (aesthete?) on distal margin, extending beyond tip of appendage, two shorter setae on anterior margin and two slender ones near posterior margin, at junction with terminal segment; latter with two long and one short apical setae, one short on anterior margin and three (one robust, two slender) long setae on posterior margin. Second antenna three-segmented (basal segment obscured in Fig. 1106); basal segment short, second long, distally tapering; claw long, sharply curved, distal half covered with delicate longitudinal ridges. Parabasal papilla well developed (Fig. 1106, pb), with broad basal part made up of two tubercles and distal soft, conical part; lateral tubercle of basal part denticulated, proximal bearing extremely fine setules (indicated by dots in drawing); no armature on distal part. Mandible (Fig. 1108) short, flat, apparently unsegmented, armed with three subtriangular teeth. First maxilla (Fig. 1109) biramous; endopod short and squat, bifid, with distal setiform outgrowth between two rounded apices; exopod longer and more slender, with lateral short and medial long processes tapering to fine points. Second maxilla (Fig. 1110) brachiform, with short basal segment proximal to lacertus (not shown in drawing); lacertus broad, sturdy, with short setiform spine on inner margin near base; brachium long, slender, with short seta at distal end of inner margin; terminal claw short, bifid, moderately curving, with short seta at about midlength of inner margin.

First two pairs of legs with bases of sympod inflated into prominent ventrolateral swellings (Fig. 1104), second much larger than first (Fig. 1112). Armature of rami as below.

	Endopod		Exopod	
	1	2	1	2
Leg 1	0-0	4	0-1	4
Leg 2	1-0	4	0-1	5

Sympod of first leg with long process at medial side of endopod, both sympods with setae on lateral side of exopod. Rami of both first and second legs (Figs. 1111, 1113) often draped in loose folds of cuticle; ventral surfaces of basal segments of exopods with patches of denticulation; some denticles also on medial margin of first exopod (both segments.) Third leg, about 0.5 mm from base of second leg, reduced to two subequal setae (Fig. 1114). Fourth leg similar to third, but missing in many specimens. Uropod (Fig. 1115) cylindrical, slightly tapering distally; apex prolonged in setiform outgrowth, with short process on either side; one long, slender seta on dorsal side and two (one short, one longer) on lateral margin.

Male (Fig. 1103, *fide* Scott and Scott, 1913) similar to female but much smaller. Cephalothorax constituting about $\frac{1}{3}$ of total length. Genital complex correspondingly shorter. Uropod comparatively more prominent than in female. Length about 1.5 mm.

Comments: The description of *H. hippoglossi* is usually credited to Krøyer (1837), although it appeared first, under the name *Clavella hippoglossi*, in Cuvier's (1830) third volume. Guérin-Méneville (1829–1843) published its drawing also prior to Krøyer's work. It is difficult to see why the species should be associated with Krøyer's name, when he himself credited it to Cuvier.

H. hippoglossi lives on a single host species, the halibut (*Hippoglossus hippoglossus*). Although it might be found on fish of various sizes and ages, it is most common on the younger halibut. This author's own experience suggests that *H. hippoglossi* very seldom occurs on large and old halibut. Having examined 697 "chicken halibut" (size 29–63 cm; age 2–5 years) in 1959–1960, he found 186 infected specimens (26.9%). The intensity of infection was from one to six copepods per fish, the parasites being attached mainly to the first two gill arches.

The distribution of *H. hippoglossi* is as wide as that of its host, finds having been reported from the European Atlantic seaboard and British waters, as well as from the Barents Sea, Greenland and the Atlantic coast of Canada and the USA.

Hatschekia cluthae (T. Scott, 1902)

(Figs. 1116–1126)

Syn.: *Clavella cluthae* T. Scott, 1902

Female (Figs. 1116, 1117): Cephalothorax rounded; dorsal surface subdivided in mid-dorsal line by thin rib-like suture (Fig. 1118), with one longitudinal bar on each side, both with obliquely transverse bifurcation anteriorly; posterior and lateral margins smoothly rounded, slightly flexed ventrally in some specimens (Fig. 1117). No trace of segmentation in neck region, genital complex apparently beginning directly behind posterior margin of cephalothorax, elongate, dorsoventrally flattened, with slightly convex lateral margins and truncated posterior extremity. Abdomen well delimited from genital complex, very small, subquadrangular. Total length 1.3–1.5 mm.

First antenna (Fig. 1119) indistinctly segmented, only clear segmental boundary between basal segment and next; basal segment heavily sclerotized along inner margin, carrying claw-like, prominent process distally; long, slender seta on dorsal wall, seven setae (three slender, four broad) on anterior margin; remainder of appendage probably of five segments, in some specimens indistinctly separated by shallow constrictions; second segment with five setae (three short and two long) on anterior margin; third with very long and robust seta on posterior margin and three (one very short) on anterior margin; fourth with single seta on anterior margin; fifth with three slender setae on posterior margin and one long seta (aesthete?) and one shorter on anterior; distal segment with three long and one short apical setae, one long on anterior margin and four (one robust, three slender) on posterior margin. Second antenna (Fig. 1120) similar to that of preceding species, but without striation on distal half of claw. Parabasal papilla subspherical, with anterior conical process and smaller, conical lateral process. Mouth and mandible similar to that of type species. First maxilla (Fig. 1121) with exopod bipartite, apparently consisting of two setiform, sturdy and sabequal processes; endopod shorter, much more robust, with bifid tip and long setiform process on lateral side of base. Second maxilla similar to that of type species, with terminal claw (Fig. 1122) armed with fairly long, fine setule on inner margin.

Sympods of first two, biramous pairs of legs without powerful basal swellings, two-segmented. Armature of rami as follows.

	Endopod		Exopod	
	1	2	1	2
Leg 1	0-0	6	0-1	6
Leg 2	1-0	5	0-1	5

Distal segment of first endopod with four terminal, one medial and one lateral setae; that of first exopod with five terminal and one medial (Fig. 1123). Distal segment of second endopod with three terminal, one medial and one lateral setae; that of second exopod with four terminal and one medial (Fig. 1124); all setae unarmed. Third leg (Fig. 1125) on ventrolateral wall of genital complex, about 0.35 mm from base of second leg, small tubercle with subapical conical outgrowth and two short setae on lateral side. Fourth leg reduced to single seta, 0.4 mm from third. Third and fourth legs frequently missing. Uropod (Fig. 1126) with robust apical setiferous process, two shorter distolateral setae, one long, slender distomedial and two (one very short) setae on lateral margin.

Male: Unknown.

Comments: *H. cluthae*, a close relative of *H. labracis*, can be distinguished from it by the setation of its legs, as well as by the shape of its cephalothorax (Fig. 1118). It is possible that Hesse (1879a) saw it on the labrids of the Atlantic coast of France, but his descriptions of *Hatschekia* from these fish cannot be used for comparison and almost all of his species should be treated as species inquirendae.

All records of this species come from British waters and all from *Ctenolabrus rupestris*. This author found it also on *Labrus bergylta* in Scottish waters, but it was very rare and produced infections of low intensity.

Hatchekia labracis (van Beneden, 1871)

(Figs. 1136–1140)

Syn.: *Clavella labracis* van Beneden, 1871
Cycnus Labri mixti Hesse, 1879
nec *Hatschekia labracis* (van Beneden, 1870); of Yamaguti (1939b)
Hatschekia labri-mixti (Hesse, 1878); of Yamaguti (1963)

Female (Fig. 1136): Cephalothorax with characteristic lateral protrusions (Fig. 1137) and prominent mid-dorsal suture flanked by broad, posteriorly converging bars. Genital trunk separated from cephalothorax by broad, shallow constriction, about four times longer than cephalothorax, dorsoventrally flattened, with slightly convex lateral margins and rounded posterior extremity. Abdomen not clearly delimited from trunk, small, one-segmented, with small uropods. Total length 1.0–1.2 mm.

First antenna similar to that of *H. cluthae*, second differing from it mainly in shape of parabasal papilla (Fig. 1139), soft, subdivided distally into three rounded lobes. Mouth and mandible as in *H. cluthae*. First maxilla (Fig. 1138) with both rami bifid, tips of unequal length; second maxilla as in other British species.

Structure and armature of first two pairs of swimming legs as in *H. cluthae*. Third leg (Fig. 1140) 0.25 mm from base of second, consisting of small papilla bearing apical seta, with two setae lateral to base, all three unarmed. Fourth leg about 0.22 mm from base of third, represented by single seta. Uropod as in *H. cluthae*.

Male: Unknown.

Comments: This species is one of several parasitic on fishes of the family Labridae. It is best distinguished from its relatives by the shape of its cephalothorax (Fig. 1137). Hesse (1879a) described *Cycnus Labri mixti* (sic), showing in his Plate 19, Fig. 13 a highly stylized drawing of a cephalothorax resembling that of *H. labracis*. Since the host of Hesse's species was identical with that of *H. labracis*, it can be assumed that Hesse's species was identical with van Beneden's *H. labracis*.

The list of hosts of this species comprises *Labrus bergylta*, *L. mixtus* and *L. trimaculatus*. Having examined many specimens of labrids in Scottish waters (particularly from the west coast areas), this author found it only on *L. mixtus*. Specimens of this fish sometimes carry a heavy burden of copepods, numbers over 100 being not uncommon. The highest count encountered by the author was 267 on a single fish.

The distribution of *H. labracis* appears to be confined to British waters and the adjacent coastal regions of Europe. (Yamaguti (1939, 1963) identified *Hatschekia* found on a labrid, *Choerodon azurio*, from the Inland Sea of Japan, as *H. labracis*. His illustrations show, however, that the parasite of that fish has a completely different shape of cephalothorax. Moreover, its first antenna does not possess a prominent spine on the basal segment, characteristic of *H. labracis*. The Japanese parasite of the labrids is clearly not conspecific with *H. labracis*.)

Hatschekia mulli (van Beneden, 1851)

(Figs. 1141–1151)

Syn.: *Clavella Mulli* van Beneden, 1851
Clavella crassa Valle, 1879 (*fide* Brian, 1906)
nec *Hatschekia mulli*; of Causey (1953a)

Female (Fig. 1141): Cephalothorax oval, width greater than length (Fig. 1143), with slight bulge on anterior margin in antennary region; sclerotized suture pattern poorly developed. Genital complex separated from cephalothorax by unsegmented neck region, subcylindrical, about seven times longer than cephalothorax (when not contracted); directly behind neck region prominent swelling (Fig. 1142), on dorsal surface of genital complex, in some specimens subdivided longitudinally into two oval swellings. Posterior extremity (Fig. 1144) with tubercular outgrowths in lateral corners, sometimes slightly peduncular; oviduct openings close to tubercles. Total length 2.5–2.7 mm.

First antenna (Fig. 1145) resembling that of *H. pagellibogneravei*, with only one distinct subdivision between basal and succeeding segments; lacuna of thin cuticle on tip of inner margin of basal segment, seven setae of varying lengths on anterior margin and two slender setae on dorsal wall; anterior margin of distal unsegmented part with eight setae, posterior with one seta; apical armature with four apical setae, three on anterior side (one aesthete?) and four on posterior side (three slender). Second antenna (Figs. 1146, 1147) with short basal segment, long and tapering second segment and unciform terminal claw-like third, distinguished by prominent, blunt protrusion at base of inner margin (Fig. 1147); parabasal papilla (Fig. 1146, pb) prominent, long and conical, with tip extended into long, slender process. Mouth tube as in other species. Mandible (Fig. 1149) very small and difficult to observe, its dentiferous margin apparently displaced to broad side of blade, armed with three teeth. First maxilla (Fig. 1148) with bifid exopod and apparently single-tipped endopod. Second maxilla as in other species.

First two pairs of legs biramous, with two-segmented, scantily armed rami. Armature formula below.

	Endopod		Exopod	
	1	2	1	2
Leg 1	0-0	1	0-1	3
Leg 2	0-0	2	0-1	3

Medial seta on distal segment of second exopod apparently distinct from others, short, blunt, cylindrical. (The author was unable to trace third and fourth legs in the material available for examination, which does not preclude the possibility that they are present, at least in some specimens.)

Male: Unknown.

Comments: *H. mulli* is readily identifiable by its slenderness (Fig. 1141), the two subspherical swellings on the dorsal wall of the genital complex (sometimes present as one larger swelling) directly behind the neck region and by the short subspherical posterolateral processes of the trunk (Fig. 1144). Originally described from the Belgian coast by van Beneden (1851a), where it had been found on *Mullus surmuletus*, it was later recorded from many localities along the coasts of south-western Europe, as well as from the Mediterranean and Adriatic. All records from these localities are on the original host or *M. barbatus*. Nunes-Ruivo (1954) discovered it on *Pseudupeneus prayensis*, another mullid fish, off the coast of Africa. Her find was later confirmed by another record from that region (Capart, 1959). Causey (1953a) identified (with some uncertainty) as *H. mulli* a *Hatschekia* found on the gills of *Scomberomorus cavalla*, a scombrid fish, caught off the coast of Louisiana. Causey's Fig. 14, illustrating his find, makes it clear that it could not have been *H. mulli*, a conclusion further supported by the identity of the host and the locality of its capture.

In British waters, *H. mulli* was recorded previously only once, having been found on the gills of an unidentified red mullet at Plymouth (Bassett-Smith, 1899). This find was subsequently registered by several authors, the name of the mullet being given as *Mullus barbatus*. This scarcity is due to the fact that the red mullet is far from common in British waters, except for the southern shores. That *Hatschekia mulli* occurs around the British Isles more frequently than that single record suggests, can be seen from the fact that the author, having looked at four specimens of *Mullus surmuletus* taken in the vicinity of Aberdeen in 1961, found three of them infected with *H. mulli* (intensity from two to six copepods per fish).

Hatschekia pagellibogneravei (Hesse, 1879)

(Figs. 1127–1135)

Syn.: *Cycnus Pagelli Bogneravei* Hesse, 1879
Hatschekia (Cycnus) pagelii bogueravei Hesse?; of Goggio (1906)
Hatschekia cornigera T. Scott, 1909

Female (Fig. 1128): Cephalothorax rounded in dorsal aspect (Fig. 1129), with well developed mid-dorsal suture and two parallel bars connected with it by transverse bars; posterior margin with prominent posterodorsal, conical protuberance (Fig. 1127); neck region indistinctly segmented, marked by ill-defined, shallow constriction, in some specimens resembling segmental boundaries. Genital complex long and slender (about eight times length of cephalothorax), subcylindrical, with parallel lateral margins and rounded posterior extremity. Abdomen not distinctly delimited from trunk, small. Total length 2.2–2.6 mm.

First antenna (Fig. 1131) indistinctly segmented, with lacuna of thin cuticle at tip of inner margin of basal segment, with heavily sclerotized rims; basal segment with seven setae on anterior margin and two on dorsal surface; seven setae on anterior margin of distal, unsegmented part and one seta on posterior margin; apical armature of four central setae, three anterior (one aesthete?) and four (three slender) on posterior side; three more slender subapical setae. Second antenna (Fig. 1132) with slender posterior half of second segment, its cuticle bearing large area covered with minute depressions; terminal claw with robust base and very slender, sharply curved distal half; parabasal papilla soft, lobate, small. Mandible and second maxilla similar to those of type species. First maxilla (Fig. 1130) with both rami bifid.

First two pairs of legs small, with segmental divisions of rami not very distinct. Segmental divisions of sympods (Figs. 1133, 1134) marked by heavily sclerotized bars. Armature formula as follows.

	Endopod		Exopod	
	1	2	1	2
Leg 1	0-0	5	0-1	6
Leg 2	1-0	4	0-1	4

Distal segment of first exopod with four terminal and two small medial setae (Fig. 1133); corresponding segment of second exopod (Fig. 1134) with two terminal and two medial setae; proximal segment of second exopod with robust, claw-like spine. Uropod (Fig. 1135) cylindrical, with prominent apical processes, one posterolateral and three lateral setae, as well as one slender dorsal seta.

Male: Unknown.

Comments: The description of this species (Hesse, 1879a) was not accompanied by any illustrations. It failed to mention the species' most salient feature, the pointed conical protuberance on the posterolateral part of its cephalothorax (Fig. 1127). That it could serve as a guide to at least tentative identification is shown, however, by Goggio (1906), who redescribes it with unmistakable drawings. T. Scott's (1909) much more euphonious name, *H. cornigera*, must, therefore, be reduced to synonymy.

Valle (1880, 1882) mentions, without describing, *H. sargi*, taken from *Sargus solviani* in the Adriatic. A description of that species is given by Brian (1906), who found it on the gills of several species of *Sargus*. *H. sargi* bears close similarity with *H. pagellibogneravei*, but neither do Brian's drawings show nor does his description mention the posterodorsal process of the cephalothorax so that its relationship, if any, to the present species cannot be ascertained.

It appears that the most common host of this species is *Pagellus bogaraveo*; its distribution extends over the Mediterranean and Atlantic coasts of Europe, the North Sea and the British waters. Nunes-Ruivo (1954) recorded it, however, from *Dentex maroccanus* and *Sargus vulgaris* taken off the coast of Senegal.

Hatschekia pygmaea T. Scott and A. Scott, 1913

(Figs. 1152–1157)

Syn.: *Clavella labracis* (?) P. J. van Beneden; of A. Scott (1907)

Female (Fig. 1152): Cephalothorax transversely oval, with three more or less parallel, poorly developed bars on dorsal surface; ventral surface inflated in oral region (Fig. 1153); neck region not clearly demarcated, apparently one-segmented. Genital complex subcylindrical, about as wide as cephalothorax and between three and four times as long, with rounded posterior extremity. Abdomen very small, not clearly delimited from genital segment. Total length (after Scott and Scott, 1913) about 1.0 mm; specimen examined by author 1.52 mm.

First antenna resembling that of *H. labracis*, but spine on tip of inner margin of basal segment small or absent (author was unable to locate one in only specimen available; Scott and Scott (1913) showed one present in their Plate XXXVI, Fig. 6). Second antenna with short basal, slender and tapering second and unciform third segment (Fig. 1154). Parabasal papilla (pb) small, conical. Mandible not examined; according to Scott and Scott (1913), armed with only two teeth. First maxilla (Fig. 1155) biramous, both rami bifid, with subequal apices. Second maxilla as in type species.

First two pairs biramous, rami two-segmented. Armature in formula below referring to specimen examined by author and to corresponding illustrations by Scott and Scott (1913).

	Scott and Scott (1913)				Author's specimen			
	Endopod		Exopod		Endopod		Exopod	
	1	2	1	2	1	2	1	2
Leg 1	0-0	5	0-1	6	0-0	3	0-1	3
Leg 2	1-0	4	0-1	4	0-0	2	0-1	3

In both legs distal segment of endopod with loose cuticle draped on ventral surface in imitation of segmental divisions. Third leg reduced to two minute setules; fourth not located by author.

Male: Unknown.

Comments: The author was unable to compare the only specimen he found with the type material, apparently no longer extant. A general resemblance to the original description and illustrations prompted its identification as *H. pygmaea*, in spite of the fact that the setation of the biramous swimming legs was different. Much more material is needed to determine whether these differences result from natural variability or are due to loss of setae caused by handling of the author's specimen.

Originally recorded as a doubtful identification of *C. labracis* by A. Scott (1907), *H. pygmaea* was given its specific name by Scott and Scott (1913). Although, in their book, these authors credited it to T. Scott alone, no reasons were given for their decision and the species should be designated as having been established by them both.

In common with its two congeners parasitic on British Labridae, this species is characterized by the presence of a prominent spine on the basal segment of the first antenna, though in the specimen examined by this author no spine was located. The species is rare, having been so far recorded only on two occasions, once in the Solway Firth (A. Scott, 1907) and once in Plymouth (Leigh-Sharpe, 1933). In both instances the host was *Crenilabrus melops*. The author found a single specimen on *Ctenolabrus rupestris* off the west coast of Scotland.

Yamaguti (1963) listed it as occurring also on *Labrus bergylta* in Roscoff, France. The author was unable to trace the source of this record.

Genus Congericola van Beneden, 1854

Female: Hatschekiidae. Cephalothorax subcircular, sometimes narrower anteriorly, with dorsal shield covering only anterior part; neck region more or less distinctly two-segmented. Genital complex about twice as long as combined lengths of cephalothorax and neck region, subcylindrical, tapering or rounded at posterior extremity; with or without posterior processes. Abdomen one-segmented, small, not delimited clearly from genital complex. First antenna uniramous, indistinctly four- to nine-segmented, setiferous. Second antenna uniramous, two-segmented, subchelate. Mouth siphonostome, mandible toothless. First maxilla biramous, rami bifid; second maxilla brachiform, four-segmented, with bifid terminal claw. Three pairs of biramous legs, first leg with rami two-segmented, second and third with rami three-segmented; fourth leg unsegmented, bifid. Uropods present.

Male: Cephalothorax similar to that of female, more elongated. Two segments distinctly and one indistinctly delimited between cephalothorax and genital complex. Latter small, roughly ovoid. Abdomen well delimited, one-segmented, subquadrate. Uropods well developed. Appendages as in female, but swimming legs relatively better developed.

Type species: *Congericola pallidus* van Beneden, 1854.

Comments: This genus currently contains three species, only one of which (the type species) has been correctly recognized. Yamaguti and Yamasu (1959) described *Metahatschekia congri* which is obviously congeneric with *Congericola pallidus*, though it differs from that species in the segmentation of the first antenna, the armature of the first three pairs of legs, the structure of the fourth leg and the presence of posterior processes on the genital complex. This author suggests that *Metahatschekia congri* be transferred to the genus *Congericola*, as *C. congri* (Yamaguti and Yamasu, 1959). Another species was described and illustrated by Hewitt (1969a) as *C. pallidus*. Comparison of Hewitt's illustrations with Figs. 1158–1174 of this book shows that they represent two distinct species, differing from each other in the segmentation of the first antenna and the armature of the three pairs of swimming legs.* All three species are parasitic on *Conger*, *C. congri* on *Conger conger in* Japanese waters and Hewitt's species on *Conger vereauxi* in New Zealand. Only the type species occurs in the British fauna.

*See notes added in proof.

Congericola pallidus van Beneden, 1854

(Figs. 1158–1174)

Syn.: *Congericola pallida* van Beneden, 1854
Cycnus pallidus v. Bened.; of Richiardi (1880)
Congericola pallidus van Beneden, 1854; of Yamaguti (1963)
nec *Congericola pallidus* van Beneden, 1854; of Hewitt (1969a) and of Kazachenko (1974)

Female (Figs. 1158, 1159): Cephalothorax subcircular in dorsal aspect, with dorsal shield covering anteromedial part; lateral corners protruding ventrally, posterior margin slightly bulging (Fig. 1158), not well delimited from second leg-bearing segment. Two free thoracic segments narrower than cephalothorax, indistinctly divided by broad, shallow constrictions. Genital segment subcylindrical, anteriorly about as broad as cephalothorax, in some specimens with slightly uneven lateral margins, rounded at posterior extremity. Abdomen small, one-segmented, merging with genital complex in dorsal aspect (Fig. 1159) but well set off in lateral (Fig. 1158). Total length 3.1–4.1 mm.

First antenna indistinctly segmented (Fig. 1160); in some specimens five segments can be distinguished, progressively narrower from base to distal end; basal segment usually well delimited, with six setae on anterior margin and one on dorsal surface; second segment with nine setae on anterior and one on posterior margin; third with single seta on anterior margin; fourth with two setae (one aesthete?) on anterior margin near tip; terminal segment (Fig. 1161) with two setae on anterior margin near base, one slender seta on posterior margin and one on dorsal surface; apical armature of three long setae in centre of apex, one shorter on anterior side and two slender ones on posterior side. Second antenna with one-segmented corpus (Fig. 1162), robust, tapering distally, with sclerotized bar on medial margin; subchela strong, very sharply bent, with short seta near base; tip with fine longitudinal striations (omitted in drawing). Mouth tube typical for Hatschekiidae. Mandible unsegmented, flat, toothless, with ventrally curving tip (Fig. 1163). First maxilla (Fig. 1164) biramous; endopod robust, with bluntly rounded tip surmounted by two short processes; exopod slender, with two long apical processes. Second maxilla (Fig. 1165) with two-segmented lacertus; basal segment short, robust, unarmed, second segment long, slender, with small tubercle on inner margin near base; brachium even more slender, subcylindrical, with spiniform seta near apex, at base of terminal claw; latter (Fig. 1166) bifid, tines subequal, small seta between their bases.

Sympods of first three pairs of legs inflated, apparently unsegmented (Fig. 1167), subspherical. First leg with slender, long seta on lateral side of base of exopod and with prominent, robust process on medial side of endopod; second and third legs with lateral setae only. Armature as shown below.

	Endopod			Exopod		
	1	2	3	1	2	3
Leg 1	1-0	5,IV	—	0-I	4,III	—
Leg 2	1-0	0-II	4,II	0-I	0-I	4,II
Leg 3	1-0	0-II	3,I	0-I	0-0	4,I

All setae medial to spines, very long and slender, unarmed; spines of endopod not clearly delimited from their segments, with fairly blunt tips, those of exopod longer, with fine, abruptly tapering tips. First leg (Fig. 1168) with fine serrations on lateral margins of distal segments. Second leg (Fig. 1169) with similar serrations on second and third endopod and second exopod segments; third leg (Fig. 1170) on second and third endopod segments. Second segments of endopods of second and third legs with highly sclerotized lateral parts of ventral surfaces. Fourth leg (Fig. 1171) unsegmented, lobate, with medial lobe smaller than lateral. Uropod (Fig. 1172) cylindrical, with slightly pedunculate base, strong apical spine flanked by two short spiniform setae, with long slender seta dorsally; one long and one short seta on lateral margin.

Male (Fig. 1173) not examined by author, description *fide* van Beneden (1854). Cephalothorax oval in dorsal aspect, with dorsal shield covering its major part; separated by constriction from following segment. Segmentation much better preserved than in female, two leg-bearing segments being clearly delimited and third partly fused with genital complex, separated from it only by shallow groove. Genital complex slightly ovate, with indentation in centre of posterior margin. Abdomen subquadrangular, short, one-segmented, well delimited from genital complex. Uropod as long as genital complex, with well developed armature. Total length between $\frac{1}{4}$ and $\frac{1}{5}$ of female.

Structure of appendages not determined.

Comments: In spite of the long period during which it has been known, this species has never been adequately described. Van Beneden's (1854) illustrations fall short of the modern acceptable standards of accuracy, introducing some confusion into our views on the morphology of *Congericola*. A puzzling feature, for example, is van Beneden's drawing of the second antenna (Fig. 1174) of the male, showing a short basal segment which has not been observed by any subsequent authors. Neither has it been found in the other two species of *Congericola*. On the other hand, it is present in the genus *Hatschekia*. It is also difficult to accept van Beneden's statement that the male appendages are very similar to those of the female. His own drawing (Fig. 1174) shows that at least the first pair of legs differs considerably from that of the female.

A note of uncertainty has been injected into the taxonomy of this species by Edwards' (1840) description of *Cycnus gracilis*, a parasite of an unidentified fish. *C. gracilis* was later recorded by Heller (1865), who gave its description, by Valle (1880) and Brian (1898, 1906), its hosts having been given variously as "a hake" and *Cerna gigas*. *C. gracilis* greatly resembles *Congericola pallidus* in its general features. The name *Cycnus*, having been pre-empted, was replaced by *Congericola* and Yamaguti (1963) lists *Congericola gracilis* as the second species of the genus. It is quite possible that *C. gracilis* is identical with *C. pallidus*. (The host affinity argues against it, but it is not impossible that the host records were in error. No other record from hake has been made of a copepod resembling *C. gracilis*. Indeed, Brian's record is the last made.) At this time there is no possibility of making a detailed comparison of appendages. Should some new material of *C. gracilis* come to hand and should that species prove identical with *C. pallidus*, it would have to be accepted as a senior synonym, predating van Beneden's name by 14 years.

C. pallidus occurs on the gills of *Conger conger*. It is widespread along the coasts of Europe from the western Mediterranean to the southern North Sea. No records have been made so far in Scandinavian waters, but around the British Isles it appears to be quite common, ranging from the English Channel to the Irish Sea and Scottish waters.

Family Lernanthropidae fam. nov.

All genera now included in this family were previously members of Anthosomatidae, a family abolished by the present author. Their most important common character is the segmentation of the body, in the overwhelming majority of species coupled with the presence of a plate on the fourth leg-bearing segment. These genera formed within Anthosomatidae a close-knit group, showing no affinity with other members of that spurious taxon. Hewitt (1968) first drew attention to this fact, though he left *Lernanthropodes* out of the group.

Although no doubt can be cast on the close relationship between various genera of Lernanthropidae, the precise nature of that relationship and, in particular, the dividing lines between them have not been altogether clear. The first to have been described was the type genus, *Lernanthropus*. Based on a scantily described species, never recorded subsequently, this genus was unmistakable enough to have withstood the test of time. With the continuing addition of species, it became, however, a catch-all taxon for all lernanthropids. Krøyer (1837) described *Aethon quadratus*, an obvious lernanthropid, but his description was fairly sketchy and not easily accessible to other workers. It was soon either forgotten or consigned to synonymy with *Lernanthropus*. Von Nordmann's (1864) description of *Norion extensus* was also largely ignored. The first lernanthropid genus which removed some species from *Lernanthropus* was *Sagum* Wilson, 1913, based largely on the nature of the egg sacs, coiled under the dorsal plate rather than straight and long, as in *Lernanthropus*. Yamaguti and Yamasu (1960) attempted to separate from *Sagum* another genus, *Pseudolernanthropus*. The only difference they used in justifying their action was the fact that in some species of *Sagum* the lateral lobe of the third leg (exopod) was fused with the dorsal plate of the fourth leg-bearing segment, while remaining partly separate in others. Pillai and Sebastian (1967), having put up a convincing argument against the Japanese authors' decision, reduced *Pseudolernanthropus* to synonymy with *Sagum*. Hewitt (1968), however, following Yamaguti and Yamasu in his acceptance of the third leg as a generic discriminant, established *Paralernanthropus* for those species with coiled egg sacs and without foliaceous expansion of that leg. He described also several new species of *Aethon*, bringing out the structure of the second leg as an important generic characteristic. Among species with straight egg sacs, Bere (1936) distinguished two genera, isolating from *Lernanthropus* those species with a ventral, rather than dorsal, plate and grouping them in *Lernanthropodes*.

There appears, then, to exist a divergence of views as to what should constitute a valid generic discriminant in this group. Although Hewitt (1968) was right in stating that the final choice remains largely a subjective decision, this author believes that the acceptance of characteristics less directly adaptive or non-adaptive (at least apparently so) would point a way out of the difficulties. He proposes, therefore, the following key to separate the genera of Lernanthropidae.

1.	Egg sacs straight	2
	Egg sacs coiled	3
2.	Fourth segment with dorsal plate	***Lernanthropus***
	Fourth segment with ventral plate	***Lernanthropodes***
3.	Second leg absent	***Norion***
	Second leg present	4
4.	Second leg unsegmented, bifid	***Aethon***
	Second leg biramous, rami one-segmented	***Sagum***

As mentioned earlier (p. 38), the gradual incorporation of segments into the genital complex from the most posterior forwards is accompanied by changes in the structure, or suppression, of the segmental appendages. In Lernanthropidae, where only traces of segmentation remain in some species, all but the first leg-bearing segment have become integrated with the genital complex. The fifth leg was the first to be incorporated, followed by the fourth, third, etc. The fifth leg is either greatly simplified and reduced or absent. The fourth, though much enlarged, is also greatly simplified and without any traces of segmentation. The third, serving as an important aid in attachment, is variously modified; the great variety in its structure renders it less than useful for taxonomic purposes. The exopodal lobe of this leg, being fused with the dorsal plate or free or, indeed, absent, is useless for discrimination between genera. Among the species of *Lernanthropus*, where no question exists as to the validity of their generic affiliation, some are with exopodal lobes fused, others not fused, some have well developed medial (sympodial) lobes, others have none. This leads the author to the rejection of differences in the structure of the third leg as a generic discriminant. (He rejects, in consequence, both Yamaguti and Yamasu's (1960) genus *Pseudolernanthropus* and Hewitt's (1968) genus *Paralernanthropus*.) The fourth and fifth leg are also of very limited value in generic diagnosis.

The matter is different when one comes to consider the first and second pairs of legs, the former a part of the cephalothorax, the latter the most recent acquisition of the genital complex. Both limbs appear uninvolved in the development or maintenance of the host–parasite system, so that any differences in their structure from one species to another are likely to be attributable to deeper genetic causes. This is particularly true of the second leg, which has been made the basis for generic discrimination in the key above. A particularly striking example is the loss of these legs in *Norion*, an event out of sequence, as it were, and a rare occurrence in the parasitic copepods of fishes.

The diagnosis of Lernanthropidae runs as follows.

Female: Cephalothorax fused with first leg-bearing segment, dorsally with well developed shield curved ventrally on each side. Remaining segments fused with genital complex, either completely or with traces of segmentation on dorsal or lateral walls, with or without anterolateral aliform processes. Fourth leg-bearing segment with large or small dorsal or ventral plate (absent in few species). In some species deep constriction between fourth and fifth leg-bearing segments. Abdomen small, indistinctly one- or two-segmented. Egg sacs straight or irregularly coiled under plate of fourth leg-bearing segment, uniseriate. First antenna uniramous, usually indistinctly segmented, with or without parabasal flagellum. Second antenna subchelate. Mouth and mouth parts siphonostome. Maxilliped subchelate. First leg small, biramous, rami one-segmented; second similar to first (unsegmented in *Aethon*, absent in *Norion*); third modified into variously shaped, plate-like structure, with or without foliaceous outgrowths of sympod and exopod; latter free or fused partly or completely with dorsal plate of fourth leg-bearing segment; fourth unsegmented, bilobed, large or small, with or without filiform tips; fifth uniramous, small or absent. Uropods present.

Male: Cephalothorax with dorsal shield flat, no plate on fourth leg-bearing segment. Abdomen one-segmented. Third and fourth legs uni- or bilobed, lobes flat or vermiform, comparatively long. Fifth leg absent.

Type genus *Lernanthropus* Blainville, 1822.

As proposed by this author, the family Lernanthropidae contains five genera: *Lernanthropus*, *Lernanthropodes*, *Norion*, *Aethon* and *Sagum*. *Lernanthropodes* comprises only three species, one occurring in the Bay of Mexico, the other two in the Indian Ocean, all parasitic on teleosts. *Norion* has only two species, one Indian and the other Pacific; *Aethon* has four, one in the West Indies, the remaining three in New Zealand waters; *Sagum* has seven species, scattered in various parts of the world's oceans. All are gill parasites of teleost fishes. The only genus represented in the British fauna is *Lernanthropus*.

Genus Lernanthropus de Blainville, 1822

Female: Lernanthropidae. Beginning with second leg-bearing segment, all segments fused with genital complex, without traces of segmentation on dorsal wall. Fourth leg-bearing segment with large or small dorsal plate (without plate in few species). Abdomen small, one -or two-segmented. Egg sacs straight, uniseriate. First antenna with parabasal flagellum. Second leg larger than, but similar to, first. Fourth leg large or small, bilobed, unsegmented, with or without marginal armature.

Male: As in family diagnosis.

Type species: *Lernanthropus musca* de Blainville, 1822.

Comments: This genus is the most abundant representative of its family and belongs to the most common genera of parasitic copepods. More than 100 species have been described so far and it appears certain that the list is not complete. All species are parasitic on the gills of marine teleosts, most of them inhabiting warmer waters. The number of species decreases with higher latitudes and lower water temperatures. Some species of *Lernanthropus* are strictly specific, but many are parasitic on several species of fish belonging to one genus, or on several genera of one family. *Lernanthropus* is well adapted to attachment to the gill filaments of its hosts. The filaments are gripped by the second antennae with the assistance of the maxillipeds and modified third legs. These are pushed in between the filament to which the parasite becomes attached and the two neighbouring filaments.

The British fauna contains two species of this genus, which can be identified as follows.

Fifth leg lanceolate, extending beyond tips of uropods (Fig. 1064) . . . . ***L. gisleri***
Fifth leg very short lobe, with apical setule (Fig. 1074) . . . . ***L. kroyeri***

Lernanthropus gisleri van Beneden, 1852

(Figs. 1055–1070)

Syn.: *Lernanthropus Gisleri* van Beneden, 1852
Lernanthropus Thompsoni Brian, 1898

Female (Figs. 1055–1057): Cephalothorax with dorsal shield about equally broad and long in dorsal aspect, with sides produced into anteroventral flanges; inside walls of flanges (as well as surfaces of appendages) covered with fine, crowded setules, mainly omitted in drawings. Anterior part of genital complex narrower, posteriorly broadening to bases of third legs. Posterior part of genital complex cylindrical (Fig. 1056) with deep constriction dividing pregenital trunk from genito-abdominal region (Fig. 1064). Dorsal plate of fourth leg-bearing segment broader posteriorly, usually with evenly rounded posterior margin and lateral margins turning slightly ventrad, completely covering abdomen and uropods from dorsal side, but leaving fourth legs largely exposed. Abdomen very small, not clearly delimited from genital area. Total length without fourth legs, up to 15 mm; with fourth legs, up to 32 mm.

First antenna (Fig. 1058) indistinctly segmented, in author's specimens apparently consisting of seven segments, with geniculate flexion between first and second; first segment with one seta on anterior margin, second with two on anterior and one on posterior margin; third apparently unarmed; fourth with two on anterior margin and fifth with one; sixth with one seta distally on posterior margin and two (one long aesthete?) on anterior; terminal segment (Fig. 1059) with group of five slender setae subapically on posterior side, one short subapical and three longer apical setae on anterior side. Parabasal flagellum (Fig. 1058) with short inflated base and long robust whip about $\frac{1}{2}$ length of first antenna. Second antenna (Fig. 1060) with corpus indistinctly subdivided into short basal and longer second segment; latter curving inwards, tapering and with small papilliform process on inner margin; subchela not divided into shaft and claw, with strong base and tapering, curved tip. Mouth tube (Fig. 1061) long, with broad base and tapering tip, with labrum longer, rounded, unarmed and labium shorter, armed with two sharp protruding processes. Mandible (Fig. 1062) siphonostome, two-segmented, with tip curving to fit inner contour of mouth, dentiferous margin with series of seven similar teeth. First maxilla (Fig. 1063) with inflated base covered by fine setules; endopod subcylindrical, slightly tapering distally, with three subequal terminal processes (setule covering of endopod partly omitted in drawing); exopod much smaller, with inflated base and short, sturdy process. Second maxilla (Fig. 1065) brachiform, with lacertus robust, unarmed, brachium long, cylindrical, distally with sharp process on inner margin; terminal claw (Fig. 1066) with one short, dentiform and another longer, blade-like process at base, row of subtriangular, sharp denticles on inner margin and few similar denticles on lateral surface. Maxilliped (Fig. 1067) subchelate; corpus sturdy, suboval, with sharp outgrowth in myxal area; shaft constituting $\frac{2}{3}$ of subchela, with short seta on inner margin; claw with denticle at base, with fine longitudinal ridges.

First leg (Fig. 1068) with inflated sympod, its surface with dense covering of setules; one sharp, sturdy process medial to base of endopod, long seta lateral to base of exopod; rami one-segmented; endopod tapering, with one long apical seta and denticles on distal half of surface on both margins; exopod larger, broader distally, with five short spines and some denticles on distal half of medial margin. Second leg (Fig. 1069) larger, similar, but without sympodial process medial to endopod, with only four spines on exopod and short apical seta on endopod. Third leg (Figs. 1055–1057) long, splayed obliquely outwards, with triangular process on dorsal margin; bases of third legs not meeting at midventral line. Fourth leg (Fig. 1056) bifid, lobes not reaching base of appendage, lanceolate, more than half their length protruding from under dorsal plate. Fifth legs (Fig. 1064) uniramous, lanceolate, with narrow bases, tips protruding beyond level of uropods. Latter long, slender, apparently unarmed.

Male (Fig. 1070) not seen by author, description *fide* Heider (1879). Cephalothorax shield flat, no dorsal plate on fourth leg-bearing segment. Third legs bifid, unsegmented, unarmed; fourth similar but longer than third. Total length up to 8 mm, including fourth legs.

Comments: *L. gisleri* has been recorded from many parts of the Atlantic, the Mediterranean and the coast of Belgium where it was first discovered. Yamaguti (1936) found it in Japanese waters. This author discovered it in the collection made in southern Australian waters. The list of hosts is as follows: *Sciaena* sp., *S. antarctica*, *Argyrosomus regium*, *Umbrina cirrhosa*, *U. steindachneri*, *U. valida*, *Johnius hololepidotus*, *Corvina nigra*, *C. cameronensis*, *Cynoscion nebulosus* (Sciaenidae); *Scatophagus multifasciatus* (Scatophagidae); *Lichia amia* (Trachinotidae); *Centropomus undecimalis* (Centropomidae); *Polydactylus quadrifilis* (Polynemidae) and *Paralichthys* sp. (Bothidae).

It is clear that *L. gisleri* is predominantly parasitic on sciaenid fishes. It is of interest to note that its association with the members of this family appears to be most intimate along the west coast of Africa, where all but one of its known hosts belong to Sciaenidae. On the American side of the Atlantic they are divided about equally between Sciaenidae and fishes of other families. Records such as those of Causey (1953a, 1953b) on a flatfish require confirmation.

The author found *L. gisleri* on *Argyrosomus regium* off the east coast of Scotland, the first British record of this species. It is worth noting that the British specimens of *L. gisleri* were much larger (length 9.6–11.0) than those he examined from the coasts of Africa (length without fourth legs 5.8–6.5). The inference that some species might differ considerably in size, depending on the geographic area, was discussed by Kabata and Gusey (1966).

Lernanthropus kroyeri van Beneden, 1851

(Figs. 1071–1081)

Syn.: *Lernanthropus Kröyeri* van Beneden, 1851
?nec *Lernanthropus Krøyeri* Beneden, 1851; of Nordmann (1864)
Lernanthropus krøyeri Beneden, 1851; of Wilson (1922a)

Female (Figs. 1071–1073): Cephalothorax with dorsal shield narrower anteriorly, posterior margin slightly concave, posterolateral corners rounded, anterolateral extended ventrally as prominent, rounded lobes (Fig. 1071). Deep constriction between cephalothorax and pregenital trunk. Latter with prominent, rounded anterolateral corners (Figs. 1072, 1073) and slightly convex lateral margins. Dorsal plate of fourth leg-bearing segment well delimited from third legs, broader posteriorly, in some specimens with posteromedian notch, often somewhat asymmetrical. Genito-abdominal tagma (Fig. 1074) small, with abdomen not distinctly delimited, subquadrangular. Oviduct orifices surmounting inflated, tubercular processes, anterior to vaginal openings. Total length up to 21 mm, according to Scott and Scott (1913); van Beneden (1851a) quotes 4–5 mm length; author's specimens were 5.2–7.0 mm long with fourth legs.

Appendages similar to those of *L. gisleri* (e.g. first maxilla, Fig. 1075; second maxilla, differing only in absence of short basal denticle, Fig. 1076, or maxilliped, Fig. 1077). Mandible in some specimens with nine teeth. Armature spines of legs with serrations along margins. Third legs protruding posteroventrally, parallel to each other (Fig. 1073). Fourth leg similar to that of *L. gisleri*, but less slender. Fifth leg (Fig. 1074) stumpy, short lobe surmounted by short seta. Uropod fusiform, apparently unarmed. Setal covering of body less dense than in *L. gisleri*.

Male (Fig. 1078): Cephalothorax with flat dorsal shield, about equally long and broad. Genital complex short, narrower posteriorly. Abdomen one-segmented, partially fused with genital complex. Total length (including fourth legs) 3.4 mm.

Cephalothoracic appendages similar to those of female, except for first leg (Fig. 1079). Sympod slightly inflated, with short, sturdy, setiferous process on medial side of endopod and slender, plumose seta on lateral side of exopod; endopod subcylindrical, one-segmented, with denticulation on medial margin and part of ventral surface, distally with long denticulated spine; exopod one-segmented, broad, distally with five blunt spines and some fine denticles. Second leg (Fig. 1080) with modified exopod; latter with peduncular base, broad distally; with broad flange and four spines at apex, flange with serrated margin. Third leg unsegmented, bifid (Fig. 1078), medial lobe much shorter than lateral. Fourth leg resembling that of female, but shorter and stouter. Uropod (Fig. 1081) subcylindrical, distally tapering, with two short apical setae, one short and one long on lateral margin and one long on dorsal surface.

Comments: Wilson (1922a) asserted that von Nordmann's (1864) description of *L. kroyeri* is incompatible with this species. It is, indeed, difficult to find much resemblance between Nordmann's drawings on the one hand and the author's own, on the other. The matter cannot be considered as definitively settled, however. Some of the differences are clearly due to Nordmann's methods of illustration. For example, the truncation of the fourth legs is obviously an intentional abbreviation. Only re-examination of Nordmann's specimens could clear up the problem of their identity.

L. kroyeri has been recorded from many localities along the coast of Europe, from the Adriatic to the southern North Sea. British records extend as far North as the Irish Sea and the coast of Norfolk. The author found it off the east coast of Scotland. The only host in all these waters appears to be *Dicentrarchus labrax*. An unusual record (Bere, 1936) comes from the Gulf of Mexico where the species was identified as being parasitic on *Lutianus griseus*.

Family Pennellidae

Morphology

Pennellidae are virtually unique among copepods parasitic on fishes in that they include intermediate hosts in their life cycles. (Some observations suggest that intermediate hosts occur also in the life cycle of *Lernaea*.) Hatching out as nauplii, they develop in the course of their brief free-swimming life into copepodids (Fig. 1338) which must find an intermediate host to continue their development. The sexes meet and mate on that host, which can be a teleost fish or an invertebrate (Cephalopoda, Pteropoda). Impregnated females enter upon another brief spell of free swimming in search of their definitive host, another teleost or, exceptionally, a marine mammal. For the males the life cycle ends with mating, though they might go through the motions of seeking out and attaching to the same species of definitive host. No further development takes place and death follows within a short time.

Most members of the family are mesoparasitic. Their bodies, up to and including the anterior part of the genital complex, are propelled into the tissues of the host by vigorous growth. The breakdown of these tissues is effected by means not presently understood. The posterior end of the genital complex, however, remains above the surface of the host (though in *Phrixocephalus* Wilson, 1908, its contact with the outside is temporarily interrupted (see Kabata, 1969a)).

All female pennellids lack maxillipeds. They are among the largest parasitic copepods known. It is tempting to speculate how far these two facts influenced the evolution of mesoparasitism. Large size means a considerable increase in the forces threatening to dislodge the copepod from the host. Powerful prehensile organs are required to maintain a constant grip on the slippery substrate. The second antennae and the maxillipeds are the only appendages capable of fulfilling this role. The absence of the latter places a great burden on the former. Ever deeper penetration of the second antennae, ever more intimate contact between the anterior end of the parasite and the tissues of the fish, might provide a convenient starting point for the evolution of the mesoparasitic way of life.

Up to the moment of attachment to the definitive host, the female pennellid retains its generalized structure (Figs. 1342, 1382) which identifies it clearly as a member of its suborder and imparts to it the power of free motion. The demands of the host–parasite relationship result in sweeping morphological changes. In general, the parasite must: (i) establish contact with the site within the host that is suitable as a source of food; (ii) ensure that the contact, once established, will be permanently retained; (iii) be able to shed its offspring into the external environment. For Pennellidae this means: (i) penetration deep into the host in search of a target site; (ii) development of anchoring holdfast organs; (iii) retention of contact between the genital orifices and the external environment.

All these points are achieved by differential growth. A close examination of the pennellid morphology enables one to identify the location of the development centres, the combined activity of which is responsible for the definitive habitus of each species. The centres are: cephalic, main thoracic, one or two subsidiary thoracic, main genital, subsidiary genital (often several subcentres), main abdominal and subsidiary abdominal (usually two, sometimes several). Not all centres are present, or active, in every species. Those present vary greatly in the extent of their activity. In all instances, however, the combined result of that activity is a typical pennellid facies in which most segmental boundaries are lost, and the limits of the individual tagmata can be fixed only by morphological landmarks such as the positions of the thoracic appendages or of the genital orifices.

The location of the development centres is shown diagrammatically in Text fig. 56, representing a generalized pennellid female at the second free-swimming stage. The cephalic centre (ceph.) is responsible for the development of the cephalic papillae, present in some genera and sometimes reaching a relatively large size and great degree of structural complexity. The main thoracic centre (main th.) causes extension of the cephalothorax in length and is never very active. The subsidiary thoracic centre (subs. th.) is located in the segment which in other families (and in pennellid males) carries the maxillipeds; its activity results in the development of the holdfast. There is another subsidiary centre, the oral centre, at the base of the mouth, a small, proboscis-like mouth being produced by its activity. The main genital centre (main gen.) is the seat of the pennellid gigantism, responsible for the phenomenal expansion in the length and girth of the genital complex. The growth induced by this centre in some genera is uniform, resulting in a long, cylindrical shape; in others the *dorsal* side grows more vigorously, causing a sigmoid curvature when counterbalanced by the more vigorous development of the *ventral* surface of the abdomen. The subsidiary genital centre (subs. gen.) is very diffuse, usually localized along the lateral margins of the tagma and active in developing the genital holdfast of some genera. The main abdominal centre (main abd.) causes elongation of the abdomen, while the diffuse subsidiary abdominal (subs. abd.) produces lateral abdominal processes.

Growing within the fish, the parasite encounters the resistance of the host tissues. Pushing through them, the anterior part of the pennellid female is moulded by the pressures exerted upon it, the slender processes of its holdfast being truncated or deflected from their courses by obstacles such as bones of hard connective tissue. The externally protruding genital trunks of species that become attached to the exposed surfaces suffer no such limitations but even that body region cannot fail to be affected by the conditions encountered by the anterior part. The food source tapped by the mouth determines the growth of the trunk and the amount of activity of the various development centres. The inhabitants of enclosed habitats such as the buccal or branchial cavities are subject to space restrictions that determine the definitive shape of their trunks and often impose malformations. In consequence of these environmental difficulties, there is a very broad range of morphological variability in Pennellidae that has plagued taxonomists since the first pennellid was discovered. It is known that Pennellidae reflect in their appearance not only the genetic differences between individuals, and the changes imposed by age and sexual maturity, but also such circumstances as the place of attachment to the host and the identity of that host. The sum total of these differences is the existence of the so-called "biological forms", first described for *Peniculus* by Delamare Deboutteville and Nunes-Ruivo (1951a) but obviously present also in other genera.

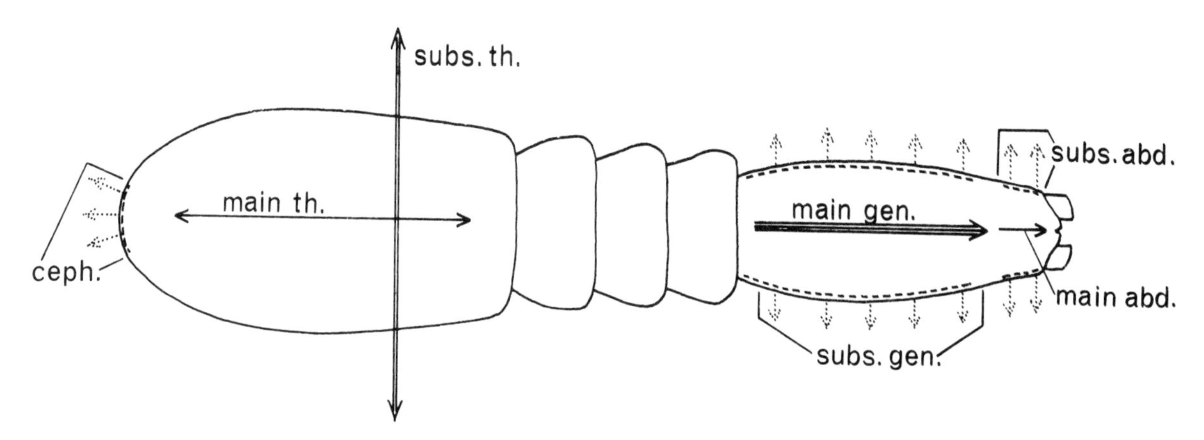

Text fig. 56. Development centres in Pennellidae.
ceph. — cephalic centre; main abd. — main abdominal centre; main gen. — main genital centre; main th. — main thoracic centre; subs. abd. — subsidiary abdominal centre; subs. gen. — subsidiary genital centre; subs. th. — subsidiary thoracic centre.

It is important to stress that the appendages of Pennellidae remain at the stage of development attained by the female prior to attachment to the definitive host. The swimming legs have been reported to undergo partial atrophy of their musculature, but the cuticular exoskeleton of these and of all the other appendages is not affected. Due to the gigantic growth of the genital complex and, in some genera, the development of an equally impressive holdfast, the appendages are dwarfed into insignificance, but they remain otherwise unaltered. Hence the term "rückschreitende Metamorphose" (regressive metamorphosis), sometimes applied to their post-attachment development, is patently incorrect.

As constituted here, the family contains 16 genera. (Leaving out *Collipravus* Wilson, 1917, as genus inquirendum.) *Peniculus* (Text fig. 57) is characterized by an oval, relatively short cephalothorax with prominent proboscis-like mouth cone. Two narrow free thoracic segments (second and third leg-bearing segments) are interposed between the cephalothorax and the genital complex, forming a neck-like structure with almost completely obliterated segmental boundaries. The fourth leg-bearing segment is much broader than the preceding two and is almost completely fused with the genital complex. That tagma, subcylindrical and long, constitutes by far the greatest part of the body. The abdomen, fused with the genital complex, is only a small tubercle bearing a dorsoventral anal slit.

Peniculus can be regarded as the most primitive genus of Pennellidae. Its morphological condition is directly related to its ectoparasitic mode of life, now alien to almost all its relatives. *Peniculus* is usually attached to the fins of its host by the second antennae. These appendages have extremely flexible basal joints and enable the cylindrical body of the parasite to rotate at least 180° around its longitudinal axis in the swirling stream of water washing past the flanks of the fish. (The tissue reactions of the host very often produce a tumour enclosing part of or even the entire cephalothorax of *Peniculus*, a fact that does not make it any the less ectoparasitic.) As there is no need to develop a holdfast, the subsidiary development centre of the thorax remains inactive and there is no appreciable increase in the size of that tagma. The main genital development centre produces substantial elongation in the genital trunk region as a direct response to the reproductive pressure common to all parasites. Since, however, no part of the trunk is buried in the host tissues, it can remain comparatively short; in the absence of any need to conserve space it has remained straight and simple, of about the same diameter throughout.

Similar to *Peniculus* is the genus *Peniculisa* (Text fig. 57), distinguished from it by the presence of two digitiform posterior processes, produced by the subsidiary genital development centre and having no obvious adaptive significance. Its free thoracic segments have retained their boundaries somewhat better than those of *Peniculus*.

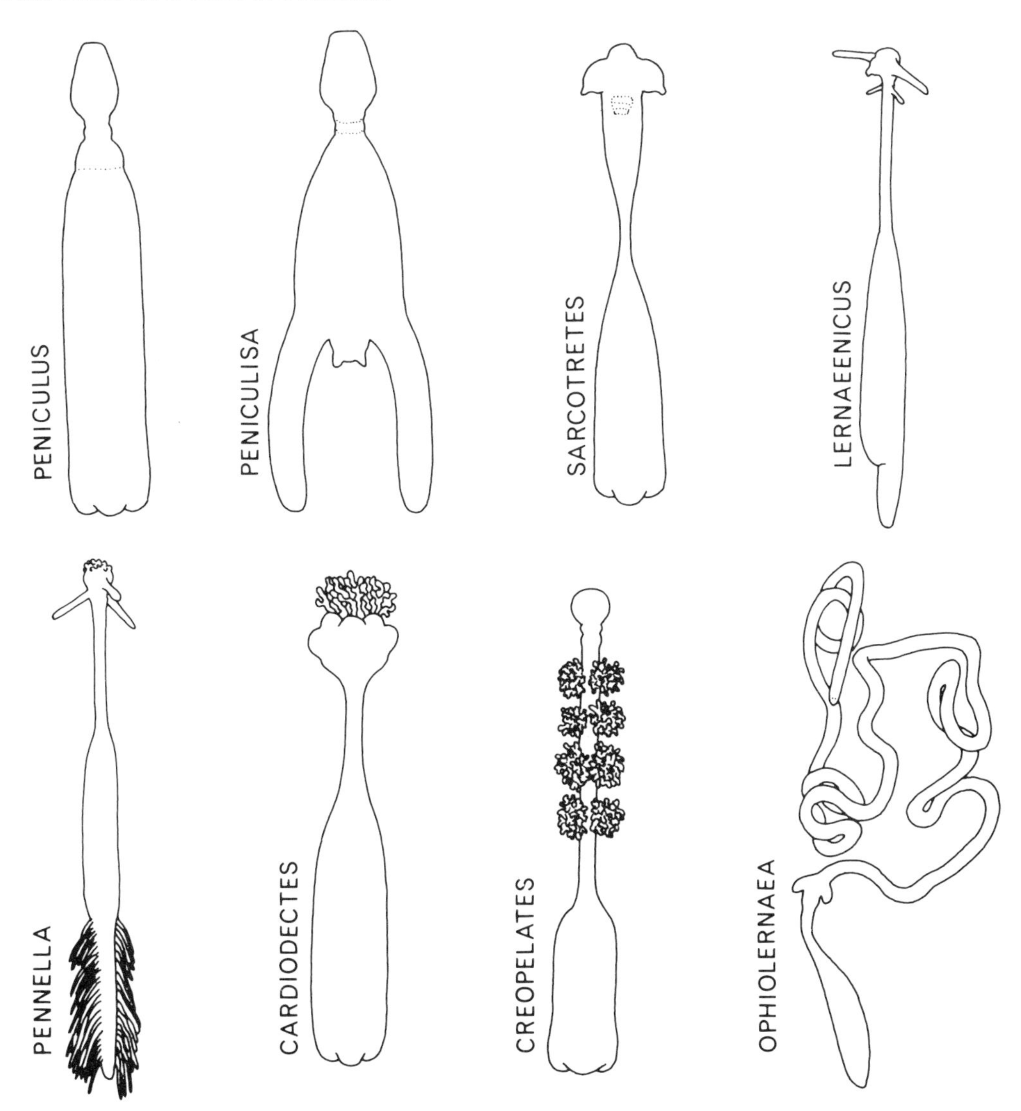

Text fig. 57. Morphological types of adult female Pennellidae. (See also Text fig. 58.)

Among the morphologically simplest pennellid genera with a mesoparasitic mode of life is *Sarcotretes* Jungersen, 1911 (Text fig. 57). A combination of three salient features distinguishes it from other pennellids. The activity of the subsidiary thoracic growth centre produces large lateral lobes, differing in structure from species to species and possessed of a broad range of intraspecific variability. These lobes form the holdfast organ. There are only three pairs of biramous legs, the terga of their segments still recognizable on the dorsal side of the body close behind the lateral lobes. The remainder of the body is a long trunk, rounded in cross-section, gradually narrowing towards its midlength and equally gradually expanding in diameter between that point and the posterior extremity. This trunk constriction is the third salient feature unique to *Sarcotretes*.

Closely resembling *Sarcotretes* is the genus *Lernaeenicus* Le Sueur, 1824 (Text fig. 57). The characteristic features of its gross morphology include the holdfast, produced by the subsidiary thoracic development centre. That anchoring structure varies, depending on the species, from a single dorsal process or a single pair of lateral processes, to a multiplicity of elaborately branching rootlets. In some species the cephalic development centre is also active, producing small tufts of antennary processes or lobes on the dorsal side of the cephalothorax in its antennary region. In other species, the subsidiary genital development centre produces also pairs of lateral processes in the anterior part of the genital trunk. There are four pairs of swimming legs. The genital complex is greatly elongated, its anterior part usually narrow, while the posterior part expands into a solid trunk and imperceptibly joins the unsegmented, sometimes quite long, abdomen, its boundary deducible only by an abrupt narrowing of the trunk and by the position of the oviduct orifices.

Pennella Oken, 1816 (Text fig. 57) bears a general resemblance to *Lernaeenicus*, from which it is distinguished by the luxuriant development of the antennary processes (Fig. 1426), indicating vigorous activity of the cephalic development centre. On the other hand, the subsidiary genital segment is inactive and there are no paired lateral processes in the anterior part of the genital trunk. The only such pair is that produced by the subsidiary thoracic centre and it, together with the antennary processes, acts as the holdfast organ. The most important characteristic of the genus, however, and its main distinction from *Lernaeenicus* is the presence of two rows of lateral abdominal processes, products of the subsidiary abdominal development centre. Sometimes profusely branched, these processes form a brush-like margin on each side of an elongate abdomen and have imparted, by their appearance, the name to the genus.

In *Cardiodectes* Wilson, 1917 (Text fig. 57) the subsidiary thoracic development centre produces a holdfast consisting of irregular lateral lobes. Also very prominent is the action of the cephalic development centre, as evidenced by the extent of development of the antennary processes, varying in shape and mode of branching from one species to another. All the remainder of the body is the genital complex, narrow anteriorly, expanding posteriorly into a long cylindrical trunk, with the abdomen represented only by a small, rounded, posterior swelling.

Creopelates Shiino, 1953 (Text fig. 57) buries in its host more than half its own total length. The cephalothorax is small, rounded, its development centres being clearly inactive or almost so. The holdfast organ is produced by the subsidiary genital centre of development. It consists of several rather irregular groups of branching processes arranged along the anterior part of the genital complex, which itself forms a long, narrow neck anteriorly and expands posteriorly into a cylindrical trunk, with a small anal tubercle representing the abdomen.

Perhaps the most extraordinary pennellid genus is *Ophiolernaea* Shiino, 1958 (Text fig. 57). Its subsidiary thoracic development centre produces a pair of moderately large, simple lateral lobes as the only holdfast present. Its oral centre of development, however, produces an enormous proboscis, separating the oral area and the buccal orifice from the rest of the cephalothorax. The proboscis is several times longer than the remainder of the body. Its vagaries around and through the viscera of the host fish have been described by Shiino (1958) and Ho (1966b). There is little activity of the genital development centres, the genital complex producing a moderately long, subcylindrical trunk with short abdomen.

Peroderma Heller, 1865 (Text fig. 58) is unique in that the longitudinal axis of its cephalothorax and of the short anterior part of the genital complex are at nearly right angles to the remainder of the body, i.e. the genital trunk. Moreover, the trunk and the anterior part are not arranged in series, the position of the anterior part being subterminal. The trunk is cigar-shaped, its anterior end thicker. The holdfast organ is provided by the activity of the cephalic centre of development, the antennary processes; it arises from a common stalk and branches profusely to form a tangled tuft, insinuating itself into a large area of the visceral cavity of the host fish.

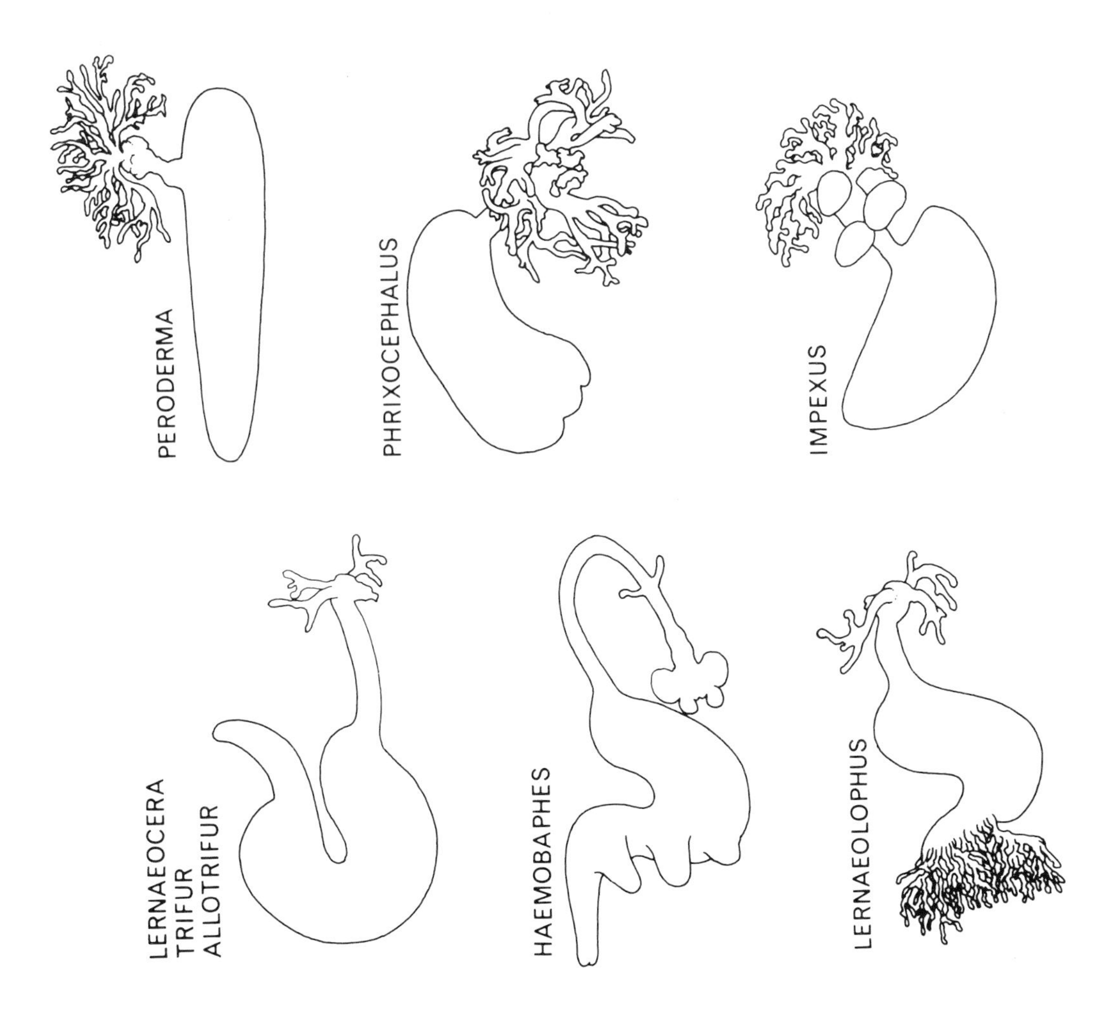

Text fig. 58. Morphological types of adult female Pennellidae. (See also Text fig. 57.)

Phrixocephalus (Text fig. 58) shows some superficial resemblance to *Peroderma*. The angle between the axes of the cephalothorax and the trunk part of the genital complex varies from one specimen to the next, depending largely on the angle at which the parasite emerged from the host (for ontogeny of the genus see Kabata, 1969a). In some individuals the narrow, neck-like part of the genital complex might be slightly subterminal in position vis-a-vis the trunk but this is only an individual characteristic resulting from the particular instance of the host–parasite relationship. An extremely well developed holdfast is produced by the subsidiary thoracic development centre, with secondary lateral pairs of processes developing in the anterior genital area. The cephalic development centre is also usually active, in some species producing very large and complex antennary processes.

To the same morphological type belongs also *Impexus* Kabata, 1972 (Text fig. 58). The holdfast of this genus consists of three components, developed by the cephalic, subsidiary thoracic and subsidiary genital centres respectively. The first produces antennary processes, undoubtedly playing a dominant part in anchoring the parasite within the host. The second produces a pair of prominent lateral oviform lobes, while the third contributes two pairs of similar lobes some distance behind the first pair. The anterior part of the genital complex is thin and round in cross-section, subterminally placed on the trunk, its long axis oblique to that of the trunk. The trunk is plano-convex, elliptical, its flat side ventral. The extent of the range of morphological variability cannot be determined until more material comes to hand. So far, only one specimen of *Impexus* has been found (Kabata, 1972b).

The remaining five genera are distinguished by genital trunks with pronounced sigmoid curves. At least two of them occur exclusively in the branchial cavities of their host fishes where space is at a premium and where this type of space-saving through folding is an obvious advantage. It is as yet not possible to determine whether the other three genera, found more usually on the outer surfaces of their hosts, derive from their gill-dwelling relatives and have secondarily colonized the open outer spaces. This phylogenetic relationship is at least a possibility.

In three genera of this group (*Lernaeocera* Blainville, 1822; *Trifur* Wilson, 1917 and *Allotrifur* Yamaguti, 1963 (Text fig. 58) there is a holdfast produced by the subsidiary thoracic development centre, a neck-like anterior part of the genital complex and a sigmoid trunk with varying degrees of flexion. *Allotrifur* is distinguished apparently by the possession of only three pairs of legs, *Lernaeocera* differs from *Trifur* in the extent of development of the holdfast and, particularly, in the unique looping of its egg sacs around a central solid axle. The egg sacs of *Trifur* are the more usual spiral coils. *Trifur* also differs from *Lernaeocera* in certain anatomical details, especially in the size and position of its genital organs.

Haemobaphes Steenstrup and Lütken, 1861 (Text fig. 58) is distinguised from *Lernaeocera*, *Trifur* and *Allotrifur* by having at least three pairs of lateral trunk processes, developed by the subsidiary genital centre. Its holdfast consists of lateral lobes, differing in shape and number from species to species. It also possesses a well-developed system of processes produced by the subsidiary genital centre on the narrow, anterior part of the genital complex.

Finally, *Lernaeolophus* Heller, 1865 (Text fig. 58) closely resembles *Lernaeocera*, except for the two rows of abdominal processes similar to those occurring in *Pennella* and produced by the subsidiary abdominal development centre. The subsidiary thoracic centre also produces, in addition to the typical holdfast, nine prominent tubercles surrounding the mouth cone. The cone itself is retractable and can be completely withdrawn (see Kabata, 1968c).

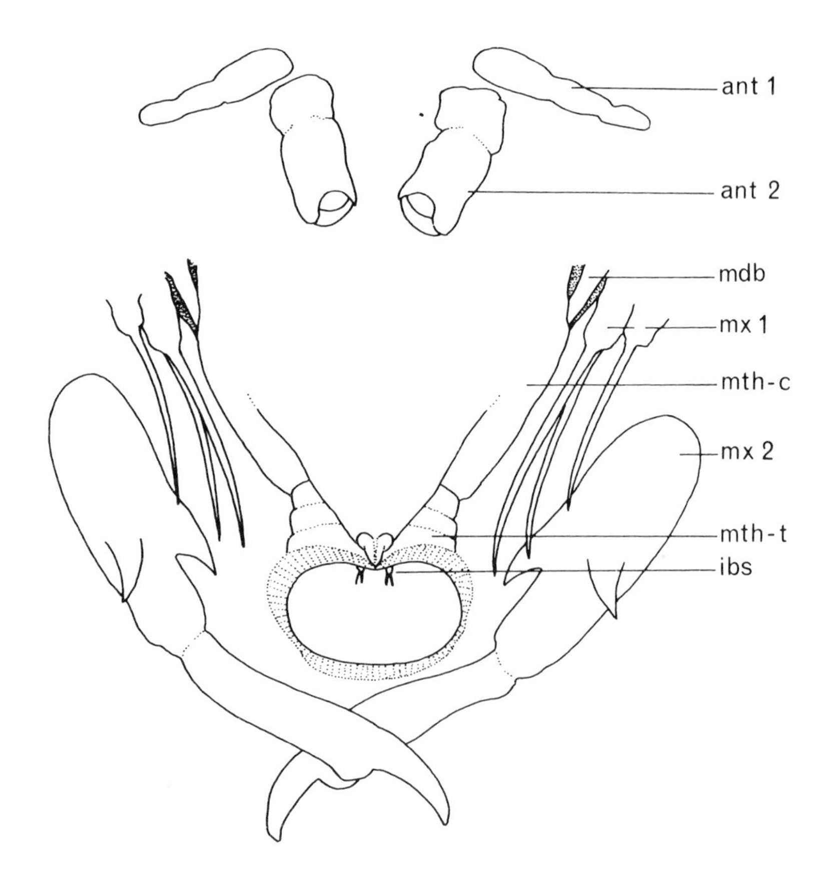

Text fig. 59. Diagram of cephalothoracic appendages of Pennellidae.
ant1 — first antenna; ant2 — second antenna; ibs — intrabuccal stylets; mdb — mandible; mth-c — mouth tube; mx1 — first maxilla; mx2 — second maxilla.

The activity of various development centres, causing a radical metamorphosis of the pennellid female, leaves its appendages dwarfed and apparently crowded together in three groups. The two pairs of antennae are separated by a short space from the buccal region (except in *Ophiolernaea*); the mouth cone (Text fig. 59) is much more prominent than in a pre-metamorphosis female and is flanked by the mandibles and first maxillae. The second maxillae are posterolateral to the base of the cone. The third group of appendages, some distance posterior to the buccal region, are the swimming legs. These are usually arranged so that the first two pairs are very close together (i.e. their positions remain the same as in the pre-metamorphosis female), whereas the third and the fourth, when present, are separated from the first two by small distances. In some genera all four pairs remain in their original positions.

It was mentioned earlier that, prior to the metamorphosis that ensues after the female pennellid becomes attached to the definitive host, its structure can be regarded as generalized. The pre-metamorphosis females are known for only few genera but all those we do know are so similar to one another as to be difficult, often impossible, to tell apart. The same is true of the males, who greatly resemble both one another and the pre-metamorphosis females.

The cephalothorax of the male (Figs. 1341, 1381) is longer than broad, either oval or with a slight indentation near the posterior extremity. It is dorsoventrally flattened and shows a pair of refractile eye spots in the anterior half of the dorsal surface. Posterior to the cephalothorax are three well delimited, leg-bearing, thoracic segments, followed by the genital complex, an unsegmented tagma, oval and seldom larger than the two posterior leg-bearing segments put together. The abdomen is one-segmented, subquadrangular, furnished with well-developed uropods. The appendages of the male are similar to those of the female, but the maxillipeds are always present, their bases posterior to those of the second maxillae.

History and systematics

Perhaps the most obvious fact about Pennellidae is that they are large. Even their smallest members are easily noticeable with the naked eye. Their largest (e.g. some species of *Pennella*) are veritable giants, reaching a phenomenal length of two feet. Moreover, many Pennellidae are attached to the external surfaces of their hosts and are fully exposed to view. It is not surprising, therefore, that they are among the earliest copepods known. Baird (1850) and Wilson (1917) quoted at length references to these parasites in publications of classical antiquity and of medieval centuries. Not many parasites can claim the distinction of having been written about in sonorous Latin verse. The fact that a presumed pennellid was mentioned by Aristotle himself, had to be acknowledged by the golden-tongued scholars of those times.

The beginnings of a systematic approach to Pennellidae was plagued by a nomenclatural misunderstanding and by a morphological similarity between two unrelated genera of convergently evolving mesoparasites. Both points of error involved *Lernaeocera* (Simphonostomatoida) and *Lernaea* (Cyclopoida). The latter genus was established by Linnaeus, who described *Lernaea cyprinacea* in his 10th edition of Systema Naturae (1758). In the 12th edition (1766–1768) he added to it *Lernaea branchialis*. The latter species was made the type of a new genus *Lerneocera* by Blainville (1822). (Following Illg's (1955) proposal to bring it into harmony with the currently widespread spelling, the name was amended to *Lernaeocera* (cf. Anonymous, 1957)). Unfortunately, Blainville transferred also *Lernaea cyprinacea* to his genus. His error was repeated by Desmarest (1825), Nordmann (1832), Burmeister (1835), Krøyer (1837) and Edwards (1840). The last-named author, aware of the incompatibility of those two species, suggested that they should belong to different genera. For reasons we cannot fathom he believed that the earlier authors applied the name *Lernaeocera* only to animals with symmetrical, paired, soft holdfast processes and simple, multiseriate egg sacs. In fact, it is clear from Blainville's (1822) text that the diagnosis of *Lernaeocera* applied to species with three branching holdfast processes and "twisted or coiled" egg sacs. Blainville himself never saw *Lernaea cyprinacea* except in Linnaeus' drawing which, incidentally, did not depict egg sacs. The effect of Edwards' textual inaccuracy was an exchange of names between *Lernaeocera* and *Lernaea*, an exchange that was to remain unobserved until Wilson (1917) restored the appropriate names to their rightful owners. Even Wilson, however, was misled into believing that *Lernaeocera* and *Lernaea* were members of the same family. The superficial resemblance, stemming from the similarity of the host–parasite relationships, still continues to exact its influence on some taxonomists.

The first attempt at grouping pennellid copepods in an assemblage of apparently related species was made by Burmeister (1835). As a starting point to a review of the gradual unfolding of our concept of the family, Burmeister's definition of his Penellina and his key to its genera are given below.

"Penellina. The characteristics of this family are highly striking. All genera are distinguished by soft body, not covered by horny shield; they are irregular in shape, expanded in length more than in width, rounded. One sees no segmentation marked by divisions or constrictions; on the other hand, the body is often bent at an angle in one or several places. Mouth is at the anterior, somewhat thinner end, protrudes forwards in the form of a cone and shows very small, sclerotized jaws and feelers. In other species also a pair of prehensile feet, covered with hooks. Near the mouth there are also fleshy, branching processes, sometimes hardened at their ends; these make attachment possible. The egg sacs, very close to posterior end but at some distance from it, are in most cases string-like, either straight (*Penella*) or coiled (*Lernaea*), sometimes even saccular (*Lernaeocera*)". Note the reversal of generic names.

"The four genera known to me can be differentiated as follows:

A. Trunk with more or less angular bends, of uneven thickness; with branching arms at anterior end.

a. Three long, hard, main arms around the mouth, both anterior arms, or all three arms branching. Egg sacs spirally coiled.

1. *Lernaea* Oken, Cuv. (*Lernaeocera* Blainville, von Nordm.) Species belonging here: *L. branchialis* auct. (*Lernaea gobina* Fabr., Müll.), *L. cyclopterina* Müll., *L. surrirensis* Bl.

b. Four soft, fleshy main arms around the mouth, anterior ones forked. Egg sacs saccular and cylindrical.

2. *Lernaeocera* Blainv., v. Nordm. (*Lernaea* Linn., Cuv.) Here belong two species: *L. cyprinacea* (*Lernaea cyprinacea* Linn.) and *L. esocina* Burm. (*L. cyprinacea* v. Nordm.)

B. Trunk straight, of uniform thickness; four pairs of cuticular flaps at neck-like anterior end.

a. Without arms or feathered tail.

3. *Peniculus* v. Nordm. (*P. fistula* v. Nordm.)

b. With arms and feathered tail.

4. *Penella* Oken, Cuv., v. Nordm. (*Lernaeopenna* Blainv.) Species: *P. filosa* Cuv. (*Pennatula filosa* Gmel., Guérin), *P. sagitta* (*Pennatula sagitta* Linn., Lam.), *P. diodontis* Cham. et Eisenh."

Burmeister's Penellina was clearly a composite taxon combining Pennellidae with the genus *Lernaea*. The present-day genus *Haemobaphes* appeared in it disguised as *Lernaea cyclopterina*. The separation of the genera by the shape of the trunk (straight versus sigmoid) originated a taxonomic trend which is still being followed. It should be kept in mind, however, that, although it is a legitimate morphological discriminant, its value as a sign of phylogenetic relationships is questionable. There are too many indications of convergent evolution to accept mere resemblance in the gross morphology of these parasites as evidence that they are closely related. Another point of interest in Burmeister's key was the "cuticular flaps" present in *Peniculus* and *Pennella*. They were the sympods of the four pairs of swimming legs, from which the fragile rami were broken off in the course of dissection. This not unusual occurrence created some difficulties for early taxonomists, whose optical equipment could not reveal to them the true nature of these structures. The small size of the sympods themselves rendered the legs difficult to see, particularly when remnants of the host tissue still clung to the anterior part of the copepod's genital complex.

Edwards (1840), for example, made the presence or absence of legs (the true nature of which he clearly understood) the main character distinguishing between two groups of his "Lerneoceriens". The genera *Penella* and *Lernaeonema* (= *Lernaeenicus*) were distinguished by their possession of legs, while *Lernaeocera* and *Lernaea* were supposedly devoid of them. *Peniculus* was dropped from the taxon.

Baird (1850) placed Edwards' taxon in his order Lerneadae, as a tribe Anchoraceracea, and divided it into two families. Unfortunately, he viewed taxonomy exclusively from the point of view of British fauna, so that his tribe contained only three genera. The family Penelladae was represented by *Lernaeonema* (= *Lernaeenicus*) (though the name suggests that *Penella* also belonged here), while Lernaeoceradae housed the ill-assorted companions, *Lernaeocera* and *Lernaea*. The presence or supposed absence of legs was used in the Edwardsian manner. Baird's taxonomy agrees closely with that of Dana (1853) though the American carcinologist treated Pennellidae and Lernaeoceridae as subdivisions of the family Penellidae. For the first time, the name of the family was used in its present form.

In the widely acclaimed paper of Steenstrup and Lütken (1861), pennellids contained *Lernaea* (in its incorrect meaning), *Pennella*, *Lernaeonema* and *Lernaeonicus* [sic]. The authors distinguished between the last two genera by the length of the abdomen, as indicated by the position of the genital orifices. *Lernaea* (sensu Linnaeus, 1758) was left out, because Steenstrup and Lütken used egg sacs as one of their most important distinguishing characters and on that basis *Lernaea* was clearly not compatible with other genera of the group.

Krøyer (1863) depended to a great extent on the work of his illustrious compatriots, when he used egg sacs in the following scheme.

A. Eggs sacs string-like, eggs uniseriate. Pennellini. *Pennella*, *Lernaea* (incorrectly named), *Haemobaphes*, *Lernaeonema*, *Lernaeenicus*, and *Staurosoma* (the last-named genus is not a member of this order and is a parasite of invertebrates).

B. Egg sacs saccular, eggs multiseriate. Lernaeocerini. *Lernaeocera* (incorrectly named), *Tanypleurus*, *Herpyllobius* and *Lernaeonema musteli* v. Ben.

There is no taxonomic merit in this scheme. Although the copepods with multiseriate eggs might not be related to those with uniseriate eggs, it does not follow that all species with multiseriate eggs are related to one another. The same applies to species with uniseriate eggs. The consequence of Krøyer's scheme was that copepods such as the cyclopoid *Lernaea* (under its incorrect name *Lernaeocera*) was lumped together with *Tanypleurus*, a siphonostome related to Lernaeopodidae (see Kabata, 1969b) and the highly aberrant *Herpyllobius*, a parasite of the polychaetes.

Heller (1865) retained the division of his Lernaeina into two groups by the type of the egg sacs, those with saccular egg sacs being referred to as Lernaeocerina *sensu stricto*, those with filiform egg sacs as Pennellina *sensu stricto*. His innovations came mainly in subdividing these two groups. Lernaeocerina were divided by the type of holdfast, *Lernaeocera* (= *Lernaea*) with "head provided with two arms, simple or lobate, on both sides", the other group with "head devoid of lateral arms". That group included *Therodamas* Krøyer, 1863, and *Naobranchia* Hesse, 1863, both now type genera of independent families not related to Pennellidae. Pennellina were divided into groups with straight and sigmoid trunks. Within the former the first division was by the presence or absence of the abdominal brush, *Pennella* falling into the first, *Peniculus*, *Lernaeenicus* and *Peroderma* into the second category. The sigmoid trunk group was subdivided by the structure of the anterior part of the body into those with and those without branching holdfasts. The former genera were represented by *Lernaea* (= *Lernaeocera*) and *Lernaeolophus*, the latter by *Haemobaphes*. An interesting implication of Heller's key is that he considered the presence of the abdominal brush less important than the shape of the trunk and did not see fit to link closely *Lernaeolophus* and *Pennella*.

In this he was entirely correct. The discussion of pennellid morphology should have demonstrated sufficiently clearly that the members of this family have a general tendency to develop branching outgrowths, the location and degree of development of those outgrowths depending mainly on the type of host-parasite relationships of the individual genera. The work of Monterosso (1925) has shown that the cephalic processes of *Peroderma* contain two vessels or ducts with different reactions to stains. He suggested that the ducts might represent efferent and afferent vessels involved in some sort of general circulation of the body fluids. This author found a similar internal structure of the processes produced by the subsidiary thoracic centre of *Phrixocephalus*, but not in *Cardiodectes*. In both instances the processes were embedded in the tissues of the host. The abdominal brush processes of *Pennella* remain outside the host. Their function might be expected to differ from those of the other two genera, yet they possess a similar internal structure. Until more is known about the physiology of pennellid copepods and the part played in it by these processes, their possible taxonomic value cannot be assessed and they are best left out of systematic discussions. There is certainly no reason to associate *Pennella* and *Lernaeolophus* in a close relationship because both genera possess the abdominal brush.

A good deal of original thought went into the systematic arrangement of the pennellids by Gerstaecker (1866–1879). Since his work in some ways foreshadowed the present-day concepts of this family, it is worth quoting at length. Gerstaecker defined his family Lernaeodea as follows.

"Body of sexually mature forms of both sexes quite similar to those of preceding families. Anterior feelers short, setiform" (= first antennae), "posterior projecting beyond anterior end of cephalothorax" (= second antennae), "short, hooked. Maxillipeds weak, of similar size, claw-like. Four pairs of well-developed legs, first two pairs biramous. Genital segment of female much enlarged; postabdomen unsegmented. Eye median, with refractile bodies.

— Juvenile stages between cyclops and mature stage with coiled anterior filament. Mature female long, vermiform, almost unsegmented, in part with neck-like constriction anterior to genital segment; some with lappet-shaped processes on anterior end of body (cephalothorax). Egg sacs double, string-like or oviform. Eyes persistent but relatively minute due to absence of continuous growth.
— Sessile parasites."

Gerstaecker divided Lernaeodea into three groups as follows.

Legs distant from one another; egg sacs long, saccular, multiserial (Lernaeocerina).
Legs close together, directly behind head; neck without clear segmentation; egg sacs narrow, string-like (Lernaeina).
Legs at short distances from one another, both posterior pairs separated from anterior; neck distinctly segmented; egg sacs narrow, string-like (Peniculina).

As in Heller's taxonomy, Lernaeocerina contained *Lernaeocera* (= *Lernaea*) and *Therodamas* (Gerstaecker doubted if *Naobranchia* really did belong here.) Peniculina contained only *Peniculus*. All other genera belonged to Lernaeina and were divided into two groups, depending on the relationship of the main axis of the cephalothorax and that of the trunk. *Peroderma* was in a group by itself, all other genera were divided into three groups by the shape of the genital complex, as follows.

"Genital segment completely or nearly completely straight, without sharp curves or twists." (Here were placed *Pennella*, *Lernaeonema* and *Lernaeenicus*. For an author who placed great importance on legs, Gerstaecker had a less than accurate understanding of these appendages. His distinction between *Lernaeonema* and *Lernaeenicus* was largely based on the supposed reduction of the legs of the latter genus to "basal parts". Clearly, he was rather unfortunate in his dissections of this genus and suffered under the misapprehensions of the early authors as regards the true relationships between *Lernaeenicus* and *Lernaeonema*.) "Genital segment short and broad, square, sharply set off from the thin neck." (Here belonged *Echetus* Krøyer, 1863 and *Lophoura* Kölliker, 1853. The first of these genera is a caligid and the second a sphyriid. The reasons for their inclusion in this family are completely obscure.) "Genital segment inflated, strongly sigmoid and twisted around its axis." (This group housed *Lernaeolophus*, *Lernaea* (= *Lernaeocera*) and *Haemobaphes*.)

Gerstaecker added also a "doubtful" member of the family, the genus *Pseudulus* Nordmann, 1864, a copepod never recorded subsequently and of completely obscure affinities.

In the second decade of this century Wilson (1917) revised the pennellids under the name Lernaeidae and divided the family into four subfamilies with the aid of the following key.

Trunk straight; swimming legs widely separated, the posterior pair close to the vulvae; egg cases sacklike; eggs multiseriate . . . . **Lernaeinae**
Trunk straight; first two pairs of swimming legs close together and near the head, the others at short intervals; egg cases filiform and straight or coiled into regular spirals; eggs uniseriate . . . . **Lernaeenicinae**
Trunk with sigmoid curve; all the swimming legs close together and near the head; egg cases filiform and convolute or coiled into spirals; eggs uniseriate and very numerous . . . . **Lernaeocerinae**
Trunk straight; all the swimming legs close together and near the head; egg cases filiform, straight and very long; eggs uniseriate; abdomen with a row of feather-like processes on either side . . . . **Pennellinae**

The genera were distributed among the four subfamilies as follows.

Lernaeinae: *Lernaea*, *Leptotrachelus* Brian, 1903 and *Therodamas*.
Lernaeenicinae: *Peniculus*, *Peroderma*, *Cardiodectes*, *Lernaeenicus*, *Sarcotretes*, *Phrixocephalus* and *Collipravus* n.gen.
Lernaeocerinae: *Lernaeocera*, *Lernaeolophus*, *Haemobaphes*, *Haemobaphoides* Scott and Scott, 1913 and *Trifur* n.gen.
Pennellinae: *Pennella* and *Pegesimallus* Krøyer, 1863.

(The position of *Pegesimallus* in the family Lernaeidae was considered by Wilson to be very questionable. The single specimen described under this name by Krøyer (1863) displayed no characteristic features that could place it with any certainty anywhere among the parasitic copepods. Even its possession of the abdominal brush could have been an artifact, an encrustation of its posterior extremity by coelenterate epizoans.)

It is clear from the key to the subfamilies that Wilson's Lernaeinae are an odd component in the family. The multiserial arrangement of eggs separates them from the other three subfamilies. They are also alone in having their swimming legs spaced out along the entire length of the trunk. This indicates that the trunk consists of the leg-bearing segments of the cephalothorax, each segment contributing about the same amount to the length of the trunk and the genital region being crowded at the posterior extremity. This type of trunk does not occur in the suborder Siphonostomatoida. Curiously, Wilson ignored the structure of the mouth parts in his key (although it was dealt with at some length in his revision). The type of the mandible alone suffices to demonstrate that the place of *Lernaea* is among the cyclopoid copepods. The same is true of *Therodamas*. As for *Leptotrachelus* (renamed *Areotrachelus* by Wilson (1924b)), it is counted among the lernaeids, though its position is as uncertain as that of *Pegesimallus*, no mouth parts or other appendages having been described.

Wilson's revision became a pattern for most subsequent authors. One notable exception was Gurney (1933), whose comments on Lernaeidae were very similar to those in the preceding paragraph. He went so far as to suggest that *Lernaea* should be removed from the family containing *Lernaeocera* and that a family Lernaeoceridae should be established. Unfortunately, he made no formal proposal to this effect, nor did he venture to give a definition of Lernaeoceridae, such definition being outside the scope of his book on freshwater copepods. Gurney's comments had no effect on the development of the concept of Pennellidae. As an example of the modern treatment of this family one might quote Markevich (1956) and Yamaguti (1963).

Markevich (1956) gave the following diagnosis for his Lernaeidae (in which he included *Lernaeocera*, *Haemobaphes*, *Pennella*, *Lernaeenicus* and *Peniculus*).

"Female. Body of mature female very long, more or less cylindrical, frequently vermiform, bent in a figure S. It is unsegmented and can be tentatively divided into three parts: cephalothorax, consisting of head and one or two anterior thoracic segments fused with it; free thorax, often narrow and neck-like; posterior (inflated) part, or trunk proper. Shape of head varies greatly from species to species. It is most frequently spherical or can be dorsoventrally flattened. On the head there are often various (lobose, spherical or shapeless) outgrowths which aid the parasite in maintaining firm hold on its host. Thorax unsegmented but segmental boundaries can still be recognized by positions of legs. Posterior part of trunk large, usually consisting of fused fifth thoracic segment, genital segment and abdomen. Beginning of abdomen is marked by points of attachment of egg sacs. Abdomen reduced in some species, often with various processes or with a true furca. In many other species, however, there are no abdominal processes. Egg sacs of varied shapes: cylindrical, cigar-shaped, string-like, thread-like. The last-named can either be rolled into irregular balls or form regular spirals.

"Antennules very small, carrying smaller or greater number of setae. In some species they are one-segmented. Antennae modified into organs of attachment, pincer-shaped. Buccal apparatus usually suctorial, with rudimentary appendages. Their number varies in different species of Lernaeidae from three to five pairs.

"Male. Body cyclopiform with distinct segmentation. Abdomen ending in well developed caudal rami. Cephalic appendages generally resembling those of female. Short distance behind mouth tube are maxillipeds, serving as attachment organs. From three to five pairs of swimming legs. Sometimes sixth pair of greatly reduced appendages on genital segment. Males, as a rule, are free-living."

It is clear that Markevich based his work on Wilson's revision. Restricting his scope to the species occurring within the boundaries of the USSR, he gave a rather incomplete representation of the family. His most significant innovation was the absence of the subfamilial groupings. Although this author agrees that no sufficient grounds exist currently for dividing pennellids in this way, Markevich's decision had one adverse effect. *Lernaea* was again closely associated with the true pennellid genera, without even the artificial separation afforded by Wilson's Lernaeinae.

The present end-point of the Wilsonian school of thought on this family is represented by the work of Yamaguti (1963), who established a superfamily Lernaeoidea, divided into three families. Perodermatidae (with *Peroderma* its only genus) was separated from the rest by the subterminal position of the anterior genital complex. Pennellidae (*Pennella* and *Lernaeolophus*) was distinguished by the presence of the abdominal brush. All other genera were left in Lernaeidae. As many as six subfamilies were recognized in that family with the aid of the following key.

Trunk horseshoe-shaped, no trace of mouth parts or legs; egg case cylindrical . . . . . . ***Lernaeosoleinae***
Trunk straight; legs widely separated, first two pairs biramose, usually wide apart on anterior portion of neck, posteriormost pair close to vulvae; egg cases sac-like; eggs multiseriate . . . . . . ***Lernaeinae***
Trunk straight, three pairs of legs biramose, on trunk; egg cases elongate; eggs multiseriate . . ***Therodamasinae***
Trunk straight; first two pairs of legs together near head, uni- or biramose, others at short intervals; egg cases filiform, straight or spirally coiled; eggs uniseriate . . . .

A. Head and thorax without lateral processes or horns and in line with axis of trunk; first three or four pairs of legs composed of basal plates only . . . . ***Peniculinae***

B. Cephalothorax with lateral horns or processes, or both, often bent at an angle with neck; at least first two pairs of legs biramose, others uniramose . . . . ***Lernaeenicinae***

Trunk with sigmoid curve; all legs close together near head; egg cases filiform, convoluted or spirally coiled; eggs uniseriate, very numerous . . . . ***Lerneocerinae***

In common with the previous accounts of the family, this concept of the pennellid group fails to take note of the most important morphological criterion: the structure of the buccal appendages. Hence Yamaguti's retention of the lernaeid genera, as well as of *Therodamas* and *Mugilicola* Tripathi, 1960, in the family. The inclusion of *Lernaeosolea* Wilson, 1944s is completely arbitrary also, since no description of the mouth parts of this copepod exists and the best one can do in its case is to treat it as incerte sedis. On the other hand, little understanding has been shown of the extraordinary morphological plasticity of the pennellids, resulting from the nature of their host-parasite relationships, and its impact on the activity of various development centres. *Lernaeolophus* and *Pennella*, the genera with only the abdominal brush in common, are isolated in an artificial family. The presence of the cephalic holdfast of *Peroderma* is ignored and the subterminal position of its anterior genital complex given undue prominence. (Incipient displacement of this part of the body from its terminal position occurs also in some species of *Phrixocephalus* and in *Impexus*.)

With the foregoing in mind, it is entirely clear that the need exists for a redefinition of the family Pennellidae, stripped of its cyclopoid hangers-on. The author's own familial diagnosis is given below.

Pre-metamorphosis female with oval cephalothorax, three free leg-bearing thoracic segments and cylindrical genital complex fused with small abdomen.

Post-metamorphosis female with oval or subspherical cephalothorax in line with long axis of genital complex (*Peniculus*, *Peniculisa*) or at right angles to it, with or without dendriform cephalic processes, with or without dendriform or lobiform posterolateral and/or posterodorsal processes. Genital complex of one of following types: straight and cylindrical, with or without narrow anterior part, with or without tufts of dendriform processes or pairs of lateral, simple or branching processes on that part, with or without constriction at midlength; oval and plano-convex, narrow anterior part with long axis oblique to that of posterior part, with two pairs of oviform swellings; straight and subcylindrical, somewhat narrower posteriorly, with narrow cylindrical part subterminal and at nearly right angles to it; irregularly sigmoid, with or without posterolateral lobiform processes. Abdomen unsegmented, cylindrical or conical, with or without lateral rows of simple or branching processes forming abdominal brush, abdomen sometimes greatly reduced. Egg sacs straight, spirally coiled, or in irregular loops around central axis. Eggs flat, uniserial. First antenna uniramous, indistinctly segmented, setiferous; second antenna chelate. Mouth cone ending in cylindrical tube, sometimes retractable (*Lernaeolophus*), in *Ophiolernaea* at tip of very long proboscis. Mandible siphonostome, sometimes toothless (e.g. *Phrixocephalus*). First maxilla with endopod twice as long as exopod, with two long setae on former and one on latter; second maxilla brachiform. Maxilliped absent. Four or five pairs of swimming legs, numbers of biramous and uniramous legs differing from genus to genus. Uropods small or absent.

Male: Resembling pre-metamorphosis female, but with shorter and broader genital complex, relatively larger uropods. Subchelate maxillipeds present.

Parasitic on marine teleosts.

Type genus: *Pennella* Oken, 1816.

It was with some reluctance that the author decided to adopt the name Pennellidae for this family. During the last decade many workers (present author included) gradually began to use for it the name Lernaeoceridae. When one looks at the history of the family from the point of view of nomenclatural priority, however, one finds that the latter name must yield to Pennellidae. Burmeister (1835), who was the first to treat this family as a separate taxon, used for it the name Pennellina. In doing this, he anticipated by five years Edwards (1840) who applied the name Lernaeoceriens to a virtually identical taxon. Baird (1850) referred to this family as Penelladae and Dana (1852, 1853) gave to it its present name Pennellidae. The term Lernaeoceridae was forgotten until Gurney (1931) suggested its use for a taxon containing *Lernaeocera* but not *Lernaea*. He made no definite proposals and produced no diagnoses. The fact that Yamaguti (1963) decided to restrict the use of Pennellidae to an artificial taxon containing only *Pennella* and *Lernaeolophus* has no bearing on this issue. There appears to be no reason to alter the first name validly applied to this family, a name to be used *sensu* Dana, with the exclusion of *Lernaea*.

The family is represented in British waters by four genera, identifiable with the aid of the following key.

Key to British genera

1. Genital complex with sigmoid curve (Figs. 1337, 1394) 2
   Genital complex straight (Figs. 1398, 1423) 3

2. Conical and irregular processes on abdomen and posterior end of genital complex (Fig. 1394) ***Haemobaphes***
   No processes on abdomen or posterior part of genital complex (Fig. 1337) . . . . ***Lernaeocera***

3. Abdominal brush absent (Fig. 1398) . . . . ***Lernaeenicus***
   Abdominal brush present (Fig. 1423) . . . . ***Pennella***

Genus Lernaeocera de Blainville, 1822

Post-metamorphosis female: Pennellidae. Outlines of cephalothorax obliterated by thoracic holdfast, normally of two lateral and one dorsal process, all profusely branched (one or more outgrowths sometimes suppressed); cephalic processes present or absent. Genital complex with narrow cylindrical anterior part and inflated trunk with sigmoid curve. Abdomen unsegmented, cylindrical, straight or arching, delimited from genital complex only by more or less abrupt narrowing of diameter; tip rounded. Uropods often lost. Egg sacs irregularly looped around central axis, held to it by thin membrane; eggs flat, uniseriate. First antenna not clearly segmented, setiferous; second antenna chelate. Mouth cone with irregular cuticular ridges along posterior side of base. Mandible two-segmented, dentiferous. First maxilla small, exopod half size of endopod, former bearing one, latter two setae. Second maxilla brachiform. Maxilliped absent. Four pairs of swimming legs close together, directly behind cephalothorax; first two pairs biramous, pairs 3 and 4 uniramous.

Male: Cephalothorax oval, with truncated posterior end. Three free leg-bearing thoracic segments. Genital complex oval, shorter than combined length of free segments. Abdomen one-segmented, subquadrate. Uropods shorter than $\frac{1}{2}$ length of abdomen. Appendages similar to those of female. Maxilliped present, subchelate.

Type species: *Lernaeocera branchialis* (L.).

Comments: The members of this genus, like their relatives in the family Pennellidae, are endowed with a very broad range of morphological plasticity. Their anterior parts are, moreover, moulded and modified by the conditions within the host tissues, while their posterior, externally projecting trunks are exposed to various environmental pressures which also result in distortions of the general sigmoid shape. The rich variety of habitus resulting from this situation is reflected by the literature on the genus, containing many descriptions of taxa based on unusually shaped trunks or oddly developed holdfasts. Most of them can be placed in synonymy with the three species which are the only ones with any claims to validity. Some, however, are so distorted as to be clearly teratological. These can be only identified as *Lernaeocera* sp. In this category are *Lernaea lumpi* T. Scott, 1901, later transferred to the genus *Saucissona* Leigh-Sharpe, 1935. (The genus contains also *S. sauciatonis* Leigh-Sharpe, 1935, another misshapen *Lernaeocera*.) Other species (*L. rigida* (Krøyer, 1863), *L. lotellae* (Thompson, 1889) and, probably, *L. godfroyi* (Quidor, 1913)) belong to the genus *Trifur*. *L. centropristi* Pearse, 1947, was identified as *Lernaeenicus* (cf. Kabata, 1965b).

The distribution of *Lernaeocera*, stripped of its invalid species, is restricted to the North Atlantic. The report of Hart (1946) of the occurrence of *Lernaeocera* sp. on *Merluccius hubbsi* in Patagonian waters was not corroborated by later work in that area (Szidat, 1955; Evdokimova, 1973). Hart's copepod was probably *Trifur tortuosus* Wilson, 1917. Houdemer's (1938) report of a *Lernaeocera* sp. from Indochina has also not been confirmed.

The three valid species are: *L. branchialis* (L.), *L. lusci* (Bassett-Smith, 1896) and *L. minuta* (T. Scott, 1900). All three occur in British waters.

Key to the species of *Lernaeocera*

1. Antennary processes present (Figs. 1371, 1384) 2
   Antennary processes absent (Fig. 1349) . . . . ***L. branchialis***

2. Ovigerous female seldom exceeding length of 5 mm (measured from anterior end of cephalothorax along axis of cylindrical neck) (Fig. 1383) . . . . ***L. minuta***
   Ovigerous female usually longer than 5 mm (measured from anterior end of cephalothorax along axis of cylindrical neck) (Figs. 1369–1372) . . . . ***L. lusci***

Lernaeocera branchialis (L.)

(Figs. 1330–1368)

Syn.: *Lernaea branchialis* L., 1767
Lernaea gobina O. F. Müller, 1776
Lernaeocera branchialis Linn. Gmel.; of Blainville (1822)
nec *Lernaeocera branchialis* v. Nordm.; of Burmeister (1835)
Lernaea branchialis var. *sigmoidea* Steenstrup and Lütken, 1861
Lernoela branchialis; of Candiotti (1910)
Lernaeocera megacephala A. Scott, 1929
Lernaeocera wilsoni Schuurmans Stekhoven, 1936
Lernaeocera obtusa Kabata, 1957
?*Lernaeocera caparti* Machado Cruz, 1959

Pre-metamorphosis female (Fig. 1342): Cephalothorax oval, ventrodorsally slightly depressed, with truncated posterior end, rounded anterior margin, slightly convex lateral margins and sometimes small posterolateral lobes; indentations near bases of first antennae, two small setules on each lateral margin; paired eye in centre of anterior half. First free thoracic (second leg-bearing) segment narrower than cephalothorax, short, broader posteriorly than anteriorly; second free thoracic segment as wide as, or somewhat narrower than, first, often slightly longer; third free segment about $\frac{1}{2}$ width of first. Genital complex fused with abdomen, cylindrical, its cuticle profusely wrinkled transversely, its length equal to or much larger than that of rest of body (depending on stage of development). Total length from about 1.5 mm upwards.

First antenna (Fig. 1350) not distinctly segmented, with right-angle bend at base and more or less definite constriction separating terminal quarter; tip rounded; anterior margin between basal flexure and distal constriction with slender, fairly long setae (usually over 20); apical armature (Fig. 1351): anterior side of apex with one long, robust, blunt seta (aesthete?), three similarly long setae with sharp tips (two of them possibly fused at base) and three shorter, slender setae; posterior side with six long, slender setae, one of them split along almost its entire length. Second antenna (Fig. 1353) three-segmented; basal segment short, robust, second segment long, with dentiform process in distal corner; terminal segment unciform claw, with short but sturdy seta at base and sharp tip fitting into cavity in dentiform process of second segment. Mouth cone (Fig. 1355) of two partially fused lips; labium much broader than labrum, distally with mouth tube (Fig. 1357) of three telescoping rings and fringed with delicate marginal membrane; inner surface of labrum with paired buccal stylets (Fig. 1356), broader at base, their blunt tips surmounted by two setiform processes; labium with short distal barbel (Fig. 1355). Mandible (Fig. 1358) two-segmented; proximal segment cylindrical, shorter than distal; latter flat, its dentiferous margin (Fig. 1359) with 8 fairly uniform teeth. First maxilla (Fig. 1354) short; endopod cylindrical, its tip extending into two long, setiform processes; exopod papilliform, small, ending in one setiform process. Second maxilla (Fig. 1360); lacertus robust, with small dentiform process on medial margin near base and two large unciform processes, one on medial margin in distal half, other on ventral surface at that level; brachium broader posteriorly, with rows of spiniform setules; claw short, tapering, with parallel rows of setules (in Fig. 1360 brachium and claw turned away from observer).

Four pairs of swimming legs, pairs 1-2 biramous, 3-4 uniramous. Sympods two-segmented, all four with slender seta lateral to base of exopod, first also with similar seta medial to base of endopod. All rami two-segmented. Armature formula below.

	Endopod		Exopod	
	1	2	1	2
Leg 1	1-0	7	1-I	5,II
Leg 2	1-0	7	1-I	6,I

Leg 3	. . . . . —	—	0-0	5,I
Leg 4	. . . . . —	—	0-0	4,I

Setae are all pinnate, spines (Figs. 1365, 1366) of distal segments with serrated margins. Details of leg structure in Figs. 1363–1366. Uropod (Fig. 1368) very short, with three long posterior setae, one short ventromedial and two short posterolateral setae (one long seta missing in drawing).

Metamorphosing female: Earliest metamorphosing stage ("Pennella" stage) with rudiments of three branches and elongate, though still straight, genital complex. Later stages (Figs. 1345–1348) with progressively developing sigmoid curvature of genital complex.

Post-metamorphosis female: Two forms recognizable, with many intermediate stages.

L. branchialis f. *branchialis* (Figs. 1334–1337). Genital complex usually with three points of flexure: between anterior cylindrical part and trunk, about middle of trunk and between trunk and abdomen; angle between long axes of posterior half of trunk and abdomen usually less than 90°; cylindrical anterior part thin and short.

L. branchialis f. *obtusa* (Figs. 1330–1333). Genital complex usually with only two points of flexure: in middle of trunk and between trunk and abdomen; angle between long axes of posterior half of trunk and abdomen usually more than 90°; cylindrical anterior part of genital complex usually longer and thicker than in preceding form.

Total length of each form (measured along straight line between anterior and posterior ends) reaching up to 60 mm.

Appendages of both forms identical with those of pre-metamorphosis female, but uropods missing.

Male (Fig. 1341): Cephalothorax and three free leg-bearing segments similar to those of pre-metamorphosis female. Genital complex oval, as long as, or shorter than, combined length of free thoracic segments. Abdomen one-segmented, subquadrate, about half length of genital complex. Length 1.0–1.5 mm.

Second antenna (Fig. 1352) with claw more slender than that of female. Second maxilla (Fig. 1361) without unciform processes on lacertus. Maxilliped (Fig. 1362) subchelate; corpus long, slender, with simple spine in myxal area; subchela almost as long as corpus, with very short claw; slender barbel about as long as claw, similar seta some distance from base of claw and small tubercle between two setae. Uropod (Fig. 1367) more prominent than that of female.

Comments: *L. branchialis* is the first member of its family to have its two-host cycle fully elucidated. The discovery of its astonishing transformation from a free-swimming copepodid into a worm-like protrusion from the gill cavity of its final host was made by Metzger (1868). More often than not the intermediate host of the copepodid stage (Fig. 1338) is a flatfish on which the pre-metamorphosis part of the life cycle is completed. This part involves four chalimus stages (two of them shown in Figs. 1339, 1340) and mating (Figs. 1343, 1344). The impregnated female (Fig. 1342) parasitizes predominantly gadid fishes.

L. branchialis is a serious economic pest. Its depredations have been dealt with by Mann (1952–53, 1960) and Kabata (1958a). Although the presence of more than 700 juveniles on the intermediate host appears to produce no ill-effects, more than one adult female (largest number recorded was 11 on a single host) can have devastating, if not directly lethal, effects on the definitive host.

The morphological differences between *L. branchialis* f. *branchialis* and *L. branchialis* f. *obtusa*, in their adult stages, have been described above. The differences are at least in some measure host-dependent. It has been found (Kabata, 1958a) that this parasite attaches itself preferentially to the ventral end of the gill arch, to which it migrates from the gill filament, the point of initial contact with the host. Only a few parasites remain in the central or dorsal parts of the arch, where they reach maturity. The majority migrate to the base of the arch and penetrate from there inwards, to embed their anterior ends in the ventral aorta of the fish, or even in the bulbus arteriosus or ventricle of the heart itself. Living mainly on cod, *Gadus morhua*, or whiting, *Merlangius merlangus*, *L. branchialis* f. *branchialis* has a short distance to cover in its search for the ventral aorta, hence its cylindrical neck tends to be short. In contrast, *L. branchialis* f. *obtusa*, often attacking old and large haddock. *Melanogrammus aeglefinus*, is forced to grow over a longer distance and tends to develop a longer cylindrical neck. This form only seldom penetrates as far as the heart of the fish.

There are few differences between the pre-metamorphosis stages of the two forms. Slight differences in size, proportions and pigmentation have been described by Kabata (1957).

L. branchialis has a very broad range of hosts, both intermediate and definitive. Many of these hosts are, however, only occasionally parasitized and do not play an important part in the life of the species. The host affiliations of *L. branchialis* were discussed in detail by Kabata (1960b). Tables 16 and 17 give enumeration of the intermediate and definitive hosts respectively.

Table 16. Intermediate hosts of *Lernaeocera branchialis*. (Common hosts marked with asterisks.)

Host species	Family	Reference
*Platichthys flesus**	Pleuronectidae	Claus, 1868; Pedashenko, 1898; A. Scott, 1901a; Polyanski, 1955
*P. f. bogdanovi**	Pleuronectidae	Shulman and Shulman-Albova, 1953
*Liopsetta glacialis**	Pleuronectidae	Shulman and Shulman-Albova, 1953
*Microstomus kitt**	Pleuronectidae	Kabata, 1957; 1958a
Limanda limanda	Pleuronectidae	Shulman and Shulman-Albova, 1953
Pleuronectes platessa	Pleuronectidae	Oorde-de Lint and Stekhoven, 1936; Gouillart, 1937
Solea solea	Soleidae	Sproston, 1941
Scophthalmus maximus	Bothidae	Gouillart, 1937
Cyclopterus lumpus	Cyclopteridae	Metzger, 1868; Sproston, 1941; Shulman and Shulman-Albova, 1953; Polyanski, 1955; Reed and Dymond, 1951
Agonus cataphractus	Agonidae	Kabata, 1958a
Zoarces viviparus	Zoarcidae	Polyanski, 1955
Myoxocephalus scorpius	Cottidae	Polyanski, 1955; Kabata, 1958a
Callionymus lyra	Callionymidae	Kabata, 1958a

Table 17. Definitive hosts of *Lernaeocera branchialis*. (Common hosts marked with asterisks.)

Host species	Family	Reference
*Gadus morhua**	Gadidae	Wilson, 1917; Stekhoven, 1936a; Shulman and Shulman-Albova, 1953; Polyanski, 1955; Kabata, 1957, 1958a
*Pollachius pollachius**	Gadidae	Gouillart, 1937; Sproston and Hartley, 1941
Pollachius virens	Gadidae	Boxshall, 1974a
*Merlangius merlangus**	Gadidae	Desbrosses, 1948; Stekhoven, 1936a; Kabata, 1957; Sproston and Hartley, 1941
*Gadus ogak**	Gadidae	Wilson, 1917
Trisopterus minutus	Gadidae	Olsson, 1869
*Boreogadus saida**	Gadidae	Shulman and Shulman-Albova, 1953
*Eleginus navaga**	Gadidae	Shulman and Shulman-Albova, 1953
Melanogrammus aeglefinus	Gadidae	Polyanski, 1955; Kabata, 1957, 1958a
Phycis blennoides	Gadidae	Fox, 1945
Molva molva	Gadidae	Polyanski, 1955; Kabata, 1965f
Merluccius merluccius	Merluccidae	Osorio, 1892; Oorde-de Lint and Stekhoven, 1936; Machado Cruz, 1959
Ammodytes tobianus	Ammodytidae	Stekhoven, 1936a
Dicentrarchus labrax	Serranidae	Oorde-de Lint and Stekhoven, 1936
Serranus cabrilla	Serranidae	Radulescu et al., 1972
Pholis gunnellus	Blennidae	Oorde-de Lint and Stekhoven, 1936; Hansen, 1923
Labrus mixtus	Labridae	Olsson, 1869
Callionymus lyra	Callionymidae	Hansen, 1923; Stekhoven, 1936a
Conger conger	Congridae	Oorde-de Lint and Stekhoven, 1936
Pleuronectes platessa	Pleuronectidae	Markevich, 1956
Solea solea	Soleidae	Stekhoven, 1936a

The plasticity of these host affiliations is best demonstrated by the fact that in different areas different host species might be preferentially infected. For example, in the southern North Sea the common life cycle of *L. branchialis* involves *Platichthys flesus* (intermediate) and *Merlangius merlangus* and *Pollachius pollachius* (definitive). In the northern North Sea, on the other hand, although all these three species are present, they are not commonly infected, the dominant life cycle involving *Microstomus kitt* (intermediate host) and *Melanogrammus aeglefinus* (definitive host). In both regions *Gadus morhua* is infected with equal frequency. North of British waters, in the Faroes and in Iceland waters, *Melanogrammus aeglefinus* carries *L. branchialis* only very rarely. The life cycle there is based on *Microstomus kitt* (intermediate) and *Gadus morhua*. Further north more northern species of flatfishes become common intermediate hosts and northern gadid species common definitive hosts.

The distribution of *L. branchialis* appears to be restricted to the North Atlantic and adjacent seas. Brian (1906) recorded it from *Merluccius* in the Mediterranean. Gusev (1951) found three species of fishes in the North West Pacific infected with the juveniles of this species. The adult female, usually easily detectable, has never been found in the Pacific. It appears probable that Gusev's records deal with a juvenile pennellid other than *Lernaeocera* (possibly *Haemobaphes* sp.).

Lernaeocera lusci (Bassett-Smith, 1896)

(Figs. 1369–1382)

Syn.: *Lernaea lusci* Bassett-Smith, 1896
Lernaeocera phycidis Leigh-Sharpe, 1933
Lernaeocera mulli Leigh-Sharpe, 1935
?*Lernaeocera brevicollis* Schuurmans Stekhoven, 1935

Pre-metamorphosis female (Fig. 1382): Closely similar to corresponding stage of *L. branchialis* in general appearance, as well as in structure of appendages. Total length from 1.1 mm.

Metamorphosing female: Earliest stage ("Pennella") with straight genital complex, increasing in length with age. Antennary processes and thoracic holdfast in rudimentary condition. At stage showing beginnings of sigmoid flexion (Fig. 1375) both holdfast and antennary processes present as three simple tubercular outgrowths (Fig. 1376). Holdfast branches continually elongating (Figs. 1377, 1379), antennary processes increasing in size (Fig. 1378). Sigmoid flexure of genital complex becoming more pronounced and irregular.

Post-metamorphosis female (Figs. 1369–1374): General appearance resembling that of *L. branchialis*. Mouth tube at apex of long proboscis. Holdfast with one dorsal and two lateral branches, often unevenly developed; dorsal branch frequently elongated (following course of branchial vessel) (Fig. 1369), lateral reduced (Fig. 1373) or absent (Fig. 1371); dichotomous branching of holdfast sparse (Fig. 1372) or fairly profuse (Fig. 1374). Antennary processes (Fig. 1371, ap) fused at base, consisting of two branches, simple or subdivided. Anterior cylindrical part of genital complex differing in length depending on distance between point of penetration and site of major blood vessels used as source of food. Sigmoid part of trunk varying in thickness, with slight tubercular inflation near oviduct orifices. Shallow constriction dividing abdomen from genital complex. Abdomen unsegmented, of varying length and thickness, ending bluntly, sometimes with terminal dorsoventral groove; its long axis always at more than 90° angle to that of posterior part of genital complex. Appendages similar to those of *L. branchialis*. Total length (measured along straight line between anterior and posterior end) 6–15 mm.

Male (Fig. 1381): Closely resembling that of *L. branchialis* in general appearance and structure of appendages. Total length about 1 mm.

Comments: The morphological similarity between *L. lusci* and *L. branchialis* is so close as to have caused some doubts whether the former is not simply another form of the latter. With the aid of standard examination methods one cannot detect many appreciable differences between the two. Heegaard (1947a) denied the validity of *L. lusci*. The similarity is particularly close between the pre-metamorphosis stages (compare Figs. 1381 and 1382 with Figs. 1341 and 1342). Though somewhat smaller, the juveniles of *L. lusci* are virtually indistinguishable from those of its relative. The differences become more apparent in the post-metamorphosis females. The most important of them is the presence of the antennary processes in *L. lusci*. No traces of these structures have been found by the author in *L. branchialis*. The length of the proboscis offers another, though less reliable, recognition mark. In *L. lusci* it is relatively longer than in *L. branchialis*.

The appendages of *L. lusci* remain largely unchanged throughout ontogeny. Having largely reached their definitive development at the copepodid stage, they only increase in size during succeeding stages, up to the stage of attainment of sexual maturity by both sexes. The only possible exception is the second antenna, which in the copepodid (Fig. 1380) is comparatively more robust than in the pre-metamorphosis adults and has a papilliform outgrowth at the base of the claw, unknown in adult copepods. (A similar change occurs in *L. branchialis*.)

Unlike its relative, *L. lusci* very commonly becomes attached to the gill arches, its holdfast embedding in the branchial vessels. The morphological consequences of this site of attachment have been described above. Attachment to other sites on the fish allows more balanced development of the holdfast. Scott and Scott (1913) described one specimen of this species attached "behind and a little below the base of the pectoral fin", with its head buried in the musculature of the fish. Boxshall (1974a) found one on the operculum and on "the body surface".

The life history of *L. lusci* has been worked out in some detail by Slinn (1970), who found that the parasite is able to complete its life cycle on *Solea solea*, a fish that might be expected to act rather as an intermediate host. No other intermediate hosts have been recorded.

The most common host of *L. lusci* is *Trisopterus luscus*. Other species have also been found harbouring this copepod, among them most frequently *Phycis blennoides* (cf. Leigh-Sharpe, 1933; O'Riordan, 1961, 1962, 1964). Leigh-Sharpe (1934, 1936a) found it also on *Molva molva* and *Brosme brosme*. Bresciani and Lütken (1962) recorded it on *Merlangius merlangus* and *Trisopterus minutus*, A. Scott (1929) on *Callionymus lyra*, Kabata (1963a) on *Trisopterus esmarkii*, Schuurmans Stekhoven (1936b) on *Gairdropsarus mediterraneus*, *Liparis liparis*, *Agonus cataphractus* and *Melanogrammus aeglefinus*. Boxshall (1974a) added to the list *Trachinus vipera*, *Ciliata mustela* and *Hippoglossoides platessoides*. (Some of these records might be erroneous. The parasite is not always easy to identify and the host list shows an overlap with that of *L. branchialis*.)

The distribution of *L. lusci* covers the European Atlantic seaboard. The most northern record comes from the Barents Sea (Markevich, 1956), the most southern from the coast of Morocco (Leigh-Sharpe, 1936a). *L. lusci* is fairly common in British waters, particularly on the gills of *Trisopterus luscus*, though it is not as abundant as *L. branchialis*.

Schuurmans Stekhoven (1935) described *Lernaeocera brevicollis* from *Myoxocephalus scorpius* taken off the coast of Belgium. The main reason for establishing a new species for the only specimen ever found on this fish was the presence in its holdfast of four branches, all underdeveloped. Antennary processes were present. It is very probable that this specimen was *L. lusci* with teratological development of its holdfast. This type of development is not unknown in Pennellidae and in all other respects the specimen was closer to *L. lusci* than to other species of *Lernaeocera*. Also synonymous with *L. lusci* is *Lernaeocera mulli* Leigh-Sharpe, 1935, found on a single occasion on *Mullus barbatus*, an unusual host for the species.

Lernaeocera minuta (T. Scott, 1900)

(Figs. 1383, 1384)

Syn.: *Lernaea minuta* T. Scott, 1900

Pre-metamorphosis and metamorphosing female unknown.

Post-metamorphosis female (Fig. 1383): General appearance and structure of appendages resembling those of *L. lusci*. Antennary processes present (Fig. 1384), proboscis similar to that of *L. branchialis* rather than that of *L. lusci*. Cylindrical part of genital complex short, its sigmoid part as in *L. lusci*. Abdomen as in *L. lusci*.

Total length (measured in straight line from anterior to posterior end) 4–7.5 mm.

Male: Unknown.

Comments: The main characteristic of this species is its small size, a criterion of doubtful taxonomic significance. In most other characters it resembles quite closely *L. lusci*, from which it is distinguished by the relative shortness of its proboscis. Both species possess antennary processes. Although for the sake of convenience the species is retained here as an independent taxon, its validity still requires confirmation. The small size of the parasite and its relative scarcity make this a problem of some difficulty.

L. minuta parasitizes *Pomatoschistus minutus*, the host-parasite system being similar to those of other species of the genus and their respective hosts. Markevich (1956) recorded it also on *Myoxocephalus quadricornis*, probably a fortuitous occurrence.

Outside British waters it occurs throughout the North Sea, one instance being reported from the Baltic (Markevich, 1956).

Genus Haemobaphes Steenstrup and Lütken, 1861

Pre-metamorphosis female: Cephalothorax oval, dorsoventrally slightly flattened, with truncated posterior end; anterior margin rounded, lateral slightly convex; small lobes in posterolateral corners. Three free leg-bearing thoracic segments. Genital complex fused with abdomen, cylindrical, with cuticle transversely wrinkled. Small uropods present. First antenna indistinctly segmented, setiferous. Second antenna cheliform. Mouth tube similar to that of *Lernaeocera*. Mandible two-segmented, dentiferous. First maxilla small, endopod cylindrical, with two long setiferous processes; exopod shorter than endopod, with one setiform process. Second maxilla brachiform. Four pairs of swimming legs, two biramous and two uniramous; rami two-segmented.

Post-metamorphosis female: Cephalothorax with holdfast consisting of symmetrically arranged lobes, differing in shape and number from species to species. Antennary processes absent. Cylindrical part of genital complex with part of holdfast anteriorly and with one or more regular or irregular pairs of secondary holdfast processes at different levels. Trunk sigmoid, with one or more pairs of lateral lobose outgrowths. Abdomen unsegmented, cylindrical, with one or more pairs of lateral processes of various shapes; apex often bilobed. Uropods absent. Appendages as in pre-metamorphosis female but distances between second and third, as well as between third and fourth pairs of legs much greater. Egg strings coiled in regular spirals.

Male: Cephalothorax and three free thoracic segments similar to those of female. Genital complex about as long as free thoracic segments combined, lateral margins convex, anterior and posterior margins transversely truncated. Abdomen one-segmented, length less than ½ that of genital complex, broader than long. Appendages similar to those of female. Second maxilla without prominent unciform processes on brachium. Maxilliped subchelate, subchela about as long as corpus. Uropods comparatively larger than those of female.

Type species: *Haemobaphes cyclopterina* (Fabricius, 1780).

Comments: In common with other pennellid genera, *Haemobaphes* is endowed with extensive morphological plasticity. Its general appearance is the outcome of the superimposition of the often overwhelming local environmental influences on the "standard" morphology. In particular, the length of the cylindrical part of the genital segment depends greatly on the distance required to reach a suitable blood vessel. Mesoparasitic inhabitants of the branchial cavity, the members of this genus usually become attached to the lower parts of the gill arches of their hosts, the growing anterior part of the body following the course of the branchial vessel and making a sharp turn into the ventral aorta. Hence most *Haemobaphes* specimens have hairpin bends in their cylindrical "necks" (Fig. 1388). Security of attachment is assisted by the subsidiary holdfast processes along the course of the "neck".

The most significant distinguishing features of the species of this genus are the shape, size and arrangement of the holdfast lobes, as well as the processes of the abdomen and the genital complex. In assessing the taxonomic values of these structures, however, one must pay attention to the age of the specimen, since age differences influence these structures quite considerably. It is unfortunate that *Haemobaphes* is very difficult to dissect out without damage. Most collections of its species examined by the author consisted of beheaded specimens, a fact rather hampering taxonomic studies.

The genus *Haemobaphes* currently contains six species, one of them of questionable validity. *H. diceraus* Wilson, 1917, *H. enodis* Wilson, 1917, *H. intermedius* Kabata, 1967 and *H. theragrae* Yamaguti, 1939, occur on teleost fishes in the northern Pacific. Kabata (1967) suggested that *H. theragrae* might be synonymous with *H. diceraus*. Further study is required.

Two remaining species, *H. ambiguus* T. Scott, 1900, and *H. cyclopterina* (Fabricius, 1780) occur in the northern Atlantic, both being present in the British fauna.

The author refrains from including in the genus *H. dilectus* Leigh-Sharpe, 1934, a taxon based on a single decapitated specimen that bears no resemblance to a typical member of the genus. It can be placed in several other pennellid genera (Kabata, 1967; Yamaguti, 1963), but has no place in *Haemobaphes*. Also excluded is *Collipravus parvus* Wilson, 1917, a species transferred to *Haemobaphes* by Delamare Deboutteville and Nunes-Ruivo (1955). Reasons for this decision were given by Kabata (1967), but the author now believes that any decision as to the generic validity of this species must be reinvestigated.

Key to British species

Cylindrical part of genital complex with two pairs of subsidiary holdfast processes directly behind main holdfast (Fig. 1389) and one pair near its bend. Only two pairs of trunk outgrowths and three of abdominal outgrowths near base (Fig. 1388) . . . . ***H. cyclopterina***

Cylindrical part of genital complex with many irregular subsidiary holdfast processes along it course (Fig. 1393); many pairs of outgrowths on trunk and abdomen (Figs. 1395–1397). . . . . ***H. ambiguus***

Haemobaphes cyclopterina (Fabricius, 1780)

(Figs. 1385–1391)

Syn.: nec *Lernaea cyclopterina* Müller, 1776 (nomen nudum)
Lernaea cyclopterina Fabricius, 1780
Schisturus cyclopterinus Oken, 1816
Anops (Lern.) cyclopterina Oken, 1816
Lernaeocera cyclopterina; of Blainville (1822)
Haemobaphes cyclopterinus; of Bassett-Smith (1899)
nec *Haemobaphes cyclopterina*; of Wilson (1912)

Post-metamorphosis female (Figs. 1385, 1388): Cephalothoracic holdfast (Figs. 1389–1391) of two pairs of lobes: irregularly conical pair lateral to oral region (Fig. 1389, a) and much larger subreniform pair (Fig. 1389, b) extending from level of anterior limit of tagma to third leg-bearing segment and forming lateral margin of cephalothorax (latter pair probably formed from segment anterior to first leg-bearing segment). Second leg-bearing segment extending laterally into rounded protuberances (Figs. 1390, 1391, c) dorsal to posterior limits of lobes *b*. Third leg-bearing segment with similar protuberances (Figs. 1390, 1391, d) and with smaller ventral lobes on either side of third pair of legs (Fig. 1389, e). Fourth leg-bearing segment with lateral digitiform processes (Figs. 1389–1391, f), pair of similar ventrolateral processes (Figs. 1389, 1390, g), latter with papilliform outgrowths on or near medial side of base. Cylindrical part of cephalothorax with hairpin bend; near bend two pairs of lateral processes (only one visible on Fig. 1388); some of these suppressed or divided in some specimens. Trunk inflated, curving upon itself, together with abdomen forming sigmoid curvature; longitudinal groove in posterodorsal surface (Fig. 1386); lateral walls with up to three low convexities; near region of oviduct orifices pair of lateral swellings (Figs. 1385–1388, a), always protruding dorsolaterally and (not in all specimens, see Fig. 1385) ventrolaterally. Abdomen unsegmented, subcylindrical, with a pair of lateral swellings near its base (Figs. 1385–1388, b) protruding dorsolaterally and ventrolaterally, latter protrusions sometimes bifid (Fig. 1388), seldom absent (Fig. 1386); tip tapering, bifid. Total length of one artificially straightened specimen (from Gooding and Humes, 1963) 25.2 mm.

First antenna obscurely segmented, resembling that of *Lernaeocera* in its armature. Second antenna chelate, with two-segmented, robust corpus, claw sharp, curving, its basal seta not observed. Mouth tube similar to that of *Lernaeocera*, but with distal ring and mid-anterior plaque much broader. Mandible two-segmented, with seven teeth in dentiferous margin. First and second maxilla much as in *Lernaeocera*.

First two pairs of swimming legs biramous, close together. Third and fourth legs uniramous, at some distance from each other and from first two pairs. All rami two-segmented; armature as in *Lernaeocera*. Uropods absent.

Male: Unknown.

Comments: The best recent description of the species is that of Gooding and Humes (1963), on which the above is largely based. The author suggests, however, that the larger pair of lobes (lobes *b*) of the cephalothoracic holdfast is a product not of the first leg-bearing segment, as supposed by these workers, but of that which in the males of the family bears the maxilliped. This segment appears to be responsible for the formation of the holdfast in other pennellids and very likely plays the same part in *Haemobapes*.

Although *H. cyclopterina* is not a rare species, its life history is not yet known. Heegaard (1947a) described it up to the stage of the free-swimming copepodid. Judging from the fragmentary information on the life cycle of this genus gathered by Kabata (1967), *H. cyclopterina* probably has in its development an intermediate host, or hosts.

The host list of this copepod is shown in table 18.

Table 18. Hosts of adult *Haemobaphes cyclopterina*.

Host	Family	Reference
Boreogadus saida	Gadidae	Sars (1886); Vanthöffen (1897)
Merlangius merlangus	Gadidae	Steenstrup and Lütken (1861)
Mallotus villosus	Osmeridae	Hansen (1923); Fröiland (1974)
Hippoglossoides platessoides	Pleuronectidae	Heegaard (1947a)
Lumpenus lumpretaeformis	Lumpenidae	Gooding and Humes (1963)
Pholis gunnellus	Pholidae	T. Scott (1902)
Pholis fasciatus	Pholidae	Steenstrup and Lütken (1861)
Gymnelis viridis	Zoarcidae	Stimpson (1864)
Lycenchelys sarsi	Zoarcidae	Bresciani and Lützen (1962)
Lycenchelys verrilli	Zoarcidae	Rathbun (1884); Wilson (1917, 1932)
Lycenchelys vahli	Zoarcidae	Prefontaine and Brunel (1962)
Agonus cataphractus	Agonidae	T. Scott (1891, 1892); A. Scott (1929)
Artediellus uncinatus	Cottidae	Prefontaine and Brunel (1962)
Myoxocephalus lilljeborgi	Cottidae	Steenstrup and Lütken (1861)
Myoxocephalus scorpioides	Cottidae	Wilson (1920b)
Myoxocephalus scorpius	Cottidae	Steenstrup and Lütken (1861); A. Scott (1904); Stephenson (1940)
Myoxocephalus scorpius groenlandicus	Cottidae	Steenstrup and Lütken (1861)
Taurulus bubalis	Cottidae	Leigh-Sharpe (1933a)
Eumicropterus spinosus	Cyclopteridae	Fabricius (1780); Baird (1861)
Capreproctus longipinnis	Cyclopteridae	Brunel (1970)
Icelus bicornis	Icelidae	Hansen (1923)
Liparis lineatus	Liparidae	Sars (1886)
Liparis reinhardti	Liparidae	Hansen (1923)
Liparis tunicatus	Liparidae	Hansen (1923)
Sebastes marinus	Scorpaenidae	Steenstrup and Lütken (1861)

In addition, the Zoological Museum of the University of Copenhagen has a small collection of this species from *Gasterosteus aculeatus*, taken in an undetermined locality. Including this collection, the host fishes of this parasite come from 13 families. Some of the records, however, particularly older ones, are not above suspicion. (This is the reason for giving authorities for each one.) In particular, the author never found whiting, *Merlangius merlangus*, infected with this copepod, though he examined many thousand specimens of this species. The pennellid occurring on whiting is usually *Lernaeocera branchialis*.

All records enumerated above come from the North Atlantic, scattered along both its coasts and across the northern islands and decreasing with latitude. Some authors claimed to have found *H. cyclopterina* in the North Pacific, the list of their records being quoted in Gooding and Humes (1963). These records have been discussed and rejected by Kabata (1967). It is also this author's view that *H. diceraus*, a Pacific species, although similar to *H. cyclopterina*, is sufficiently distinct from it to be retained as a separate species. The view is based on comparisons suggested by Gooding and Humes (1963).

Haemobaphes ambiguus T. Scott, 1900

(Figs. 1392–1397)

Syn.: *Haemobaphoides ambiguus* (T. Scott, 1900)

Post-metamorphosis female (Fig. 1394): Cephalothoracic holdfast (Fig. 1393) consisting of two pairs of large and two of small lobes; pair anterolateral to oral region irregularly subdivided and forming anterior margin of cephalothorax (Fig. 1393, b); maxilliped segment with large, irregular lobes (Fig. 1393, c), anteriorly overlapping lobes *b* from dorsal side (Fig. 1394), posteriorly almost meeting each other on ventral surface of cephalothorax. Oral region in centre of concavity formed by lobes *b* and *c*, with two pairs of much smaller, irregular lobes (Fig. 1393, a, a′) in immediate vicinity. Posterior to cephalothoracic holdfast three pairs of lateral lobes produced by second to fourth leg-bearing segments; lobes sometimes extending ventromedially and often partly fused (better seen in dorsal view, Fig. 1394). Cylindrical part of thorax fairly short, thin, with hairpin bend; posterior half with irregular group of processes, simple or branching, large or small, varying in number (Fig. 1393). Trunk curved upon itself, with several pairs of protuberances (Fig. 1397); dorsal side with two prominent (d, e) and varying number of smaller outgrowths, posterolateral surfaces with two pairs of conical outgrowths (f, h); anteroventrally, near base of cylindrical thorax, outgrowths (g) in lateral quarters. Abdomen subcylindrical, unsegmented, with two papilliform outgrowths (i) on ventral surface near base and lateral pair (k) slightly posterior to *i*; small round swelling (m) on ventral surface about same level; near midlength two pairs (n, o) varying in shape from digitiform to aliform and bifid; tip inflated, bifid. Egg sacs in regular spirals. Total length (artificially straightened) 15–30 mm.

Appendages as in *H. cyclopterina*.

Male: Unknown.

Comments: The complicated shape of a fully adult female of this species varies fairly broadly due to individual differences in the development of various trunk and abdominal processes. The description above is based on Fig. 1397, representing a specimen seen also in ventral view in Fig. 1396. Two other specimens are shown in Figs. 1394 and 1395. They differ from each other and from the third specimen in several points which are quite clear from the figures. In particular, Figs. 1396 and 1397 show a smooth abdominal surface between protuberances *k* and *n*, whereas in Figs. 1394 and 1395 additional protuberances are present between these two. Fig. 1394, on the other hand, shows the absence of *d* and *m*. There are many similar differences.

The most valuable recent paper on this species was published by Delamare Deboutteville and Nunes-Ruivo (1955) who suggested suppression of the genus *Haemobaphoides* Scott and Scott, 1913, established for the accommodation of this species. Their action was supported by Kabata (1967). Delamare Deboutteville and Nunes-Ruivo were the first to describe several stages in the metamorphosis of the female (Fig. 1392). Their description shows that the metamorphosis progresses in a certain order. The first signs of development are the growth of the holdfast and the lobes of the three free segments (cephalothoracic subsidiary centre), and elongation of the thorax (main cephalothoracic centre) accompanied by curvature in a ventral direction and the appearance of processes, later to be found on the cylindrical part of the thorax. Trunk formation and elongation of the abdomen come later in the sequence.

H. ambiguus is a rare species, much more limited both in its range of hosts and distribution than *H. cyclopterina*. It lives mainly in the gill cavity of *Callionymus maculatus*. Most of its finds have been made in British waters, though Delamare Deboutteville (1950) and Delamare Deboutteville and Nunes-Ruivo (1955) found it on two occasions off the Mediterranean coast of France (Banyuls) on *C. festivus* and *C. belenus*. It is interesting to note that it has never been found in Britain (or elsewhere) on the much more common species of dragonet, *Callionymus lyra*, closely related to *C. maculatus*.

Genus Lernaeenicus Le Sueur, 1824

Pre-metamorphosis female: Cephalothorax oval, dorsoventrally flattened, with rounded anterior and truncated posterior end, with first leg-bearing segment incorporated. Three free leg-bearing segments, progressively diminishing in size. Genital complex fused with abdomen, cylindrical, shorter than cephalothorax. Uropods present. First antenna uniramous, setiferous, indistinctly segmented. Second antenna chelate, very broad, with slender claw. Mouth tube of pennellid type but very prominent (width about $\frac{1}{2}$ that of cephalothorax, length about $\frac{1}{3}$ that of cephalothorax), with irregular rows of denticles across posterior side of base; ventral surface of cephalothorax posterolateral to mouth tube with or without spinulation. Mandible two-segmented, dentiferous. First maxilla with two setiform processes on longer endopod and one on shorter exopod. Second maxilla brachiform. Maxilliped absent. Four pairs of swimming legs, first two biramous, others uniramous.

Post-metamorphosis female: Holdfast of lateral and/or dorsal processes (from single pair to many), simple or branched, or of inflated lobes, or both. Antennary processes absent, but small lobes sometimes present around oral area. Cylindrical part of cephalothorax long, smooth or nodular, passing gradually into trunk. Latter straight, cylindrical and thicker than anterior part, differing in length from species to species, abruptly narrowing at level of oviduct orifices. Abdomen varying from papilliform to cylindrical, from much shorter to much longer than trunk. Uropods sometimes absent. Egg strings long, straight. Appendages as in pre-metamorphosis female. All four pairs of legs close together, directly behind cephalothorax.

Male: Cephalothorax and free leg-bearing segments as in female. Genital complex shorter than combined length of three free segments, length greater than width, delimited from abdomen by constriction, deep on lateral sides, shallow on dorsal and ventral. Abdomen more than ½ length of genital complex, sometimes with more or less distinct transverse constriction in anterior half (segmental boundary?). Uropods present. Appendages similar to those of female, but second maxilla without unciform spines on brachium. Maxilliped present, subchelate.

Type species: *Lernaeenicus radiatus* Le Sueur, 1824.

Comments: The genus *Lernaeenicus* abounds in nominal species, many of which have been very poorly described. Some are known from single records. A thorough revision of the genus is badly needed, but it is clearly outside the scope of this book. The author tentatively accepts 26 species as valid. Most of them have fairly small known distribution ranges, none inhabits more than one ocean. The Atlantic representatives of the genus comprise *L. affixus* Wilson, 1917, *L. encrasicoli* (Turton, 1807), *L. multilobatus* Lewis, 1959, *L. polyceraus* Wilson, 1917, *L. procerus* (Leidy, 1889), *L. radiatus* Le Sueur, 1824, *L. sprattae* (Sowerby, 1804), *L. triangularis* Heegaard, 1966, *L. gonostomae* Kensley and Grindley, 1973, and *L. longiventris* Wilson, 1917. The last-named has been found along the Atlantic coast of both Americas and *L. gonostomae* in South Africa. All the others occur in the North Atlantic. *L. encrasicoli* has been found also in the Mediterranean. The copepod fauna of the Mediterranean-Adriatic region contains also *L. gracilis* (Heller, 1865) and *L. vorax* Richiardi, 1877. Equalling the Atlantic group in number of species is the Indian Ocean group, comprising *L. alatus* M. P. Rangnekar, 1960–61, *L. anchoviellae* Sebastian and George, 1964, *L. bataviensis* Sebastian, 1966, *L. hemirhamphi* Kirtisinghe, 1956, *L. seeri* Kirtisinghe, 1934 and *L. stromatei* Gnanamuthu, 1953. A single find of *L. cerberus* Leigh-Sharpe, 1927, has established that species in the Suez area of the Red Sea. Two species are known from single finds in Indonesian waters: *L. gnathonicus* Leigh-Sharpe, 1934 and *L. gnavus* Leigh-Sharpe, 1934. The Pacific Ocean is much poorer in *Lernaeenicus*, most records coming from the Asian side of the ocean. Three species have been discovered in Japanese waters: *L. ater* Shiino, 1958, *L. quadrilobatus* Yamaguti and Utinomi, 1953 and *L. sayori* Yamaguti, 1939. (Two descriptions, that of *L. lesueurii* (Edwards, 1840) and *L. tricerastes* Stuardo and Fagetti, 1960, are inadequate to the point where the species they describe have to be treated as species inquirendae.)

The way in which the species of the genus are distributed in the Atlantic and Indian Oceans, with only few occurring in the western Pacific, might indicate that the centre of origin of *Lernaeenicus* was somewhere in the Tethys Sea.

The literature contains several other species of *Lernaeenicus*. *L. surrirai* (Blainville, 1822) is a species inquirenda. *L. abdominalis* (Edwards, 1840) has never been illustrated, but is probably not a member of this genus, since the description refers to its S-shaped body. *L. centropristi* (Pearse, 1947) is known only from its pre-metamorphosis stages and could be merely a juvenile of another species. Finally, there exist two nomina nuda: *L. labracis* Richiardi, 1880 and *L. sargi* Richiardi, 1880.

The main morphological differences between species of this genus are in the structure of the holdfast. Among the simplest holdfasts are those consisting of only one pair of unbranched processes (in *L. sprattae*, Fig. 1398) and they increase in complexity to the bewildering thicket of tangled processes of *L. ramosus*. Irregularly subdivided, inflated lobes represent another type of holdfast. Other morphological differences are chiefly in the proportions of the cylindrical part of the thoracic "neck", the trunk and abdomen. In *L. longiventris*, for example, the proportions of these three parts of the body are about 60 : 15 : 25, in *L. radiatus* 50 : 40 : 10 and in *L. quadrilobatus* 40 : 60 : 0 (its abdomen being a mere papilliform vestige). In *L. seeri*, on the other hand, the length of the abdomen exceeds the combined length of thoracic neck and trunk.

The life history of *Lernaeenicus* has not yet been pieced together, but all the fragmentary information available indicates that its cycle is very similar to that of *Lernaeocera*. At any rate, the stages hitherto described, i.e. a male and a free-swimming pre-metamorphosis female (Kabata, 1963b, 1965b), and a female in the early stages of metamorphosis (Gurney, 1947; Carvalho, 1957) are very much like the corresponding stages of *Lernaeocera*.

There are two species of *Lernaeenicus* in the British fauna.

Key to British species

Holdfast of two lateral processes, with small semispherical dorsal swelling; neck nodose (Figs. 1398, 1399)

. . . . ***L. sprattae***

Holdfast of two lateral (Fig. 1420) or two lateral and one dorsal process (Fig. 1421); neck smooth (Fig. 1418) . . . . ***L. encrasicoli***

Lernaeenicus sprattae (Sowerby, 1806)

(Figs. 1398–1417)

Syn.: *Lernaea spratta* Sowerby, 1806
Lernaea cyclophora Blainville, 1822
Lernaeocera surrirensis Blainville, 1823
Lernaea ocularis Cuvier, 1830
Lerneonema monillaris Edwards, 1840
Lerneonema spratta Baird, 1850
Lerneonema bairdii Salter, 1850
Lerneonema monillaris Steenstrup and Lütken, 1861
Lerneaenicus sprattae (Sow.); of Olsson (1869)
Foroculum spratti Thompson; of Bassett-Smith (1899)
Lernaeenicus sardinae Baudouin, 1904
Lernaeenicus sardinae var. *moniliformis* Baudouin, 1911

Pre-metamorphosis female (Fig. 1415): Cephalothorax oval, dorsoventrally flattened, with truncated posterior extremity. Three free leg-bearing segments progressively smaller in posterior direction. Genital complex fused with abdomen, cylindrical, with rounded tip. Uropods as in *Lernaeocera*. Total length about 1.6 mm.

Appendages as in post-metamorphosis female. Onset of metamorphosis marked by elongation of genital complex (Fig. 1416) and of swellings at posterolateral corners of cephalothorax (Fig. 1417).

Post-metamorphosis female (Fig. 1398): Cephalothorax suborbicular in dorsal view, with digitiform processes projecting backwards from posterolateral corners; dorsal surface sometimes, though not always, with semispherical or posteriorly pointed protuberance (Figs. 1399, 1400); in some specimens two anterolateral shallow swellings; ventral surface at acute angle to dorsal (Fig. 1400), but angle between them variable from specimen to specimen; protuberances present lateral to mouth region or in posterolateral corners, sometimes absent. Thoracic neck (Fig. 1399) much shorter than trunk, anteriorly broader, narrowing rapidly, covered with irregular nodosities part of its length, gradually expanding into trunk. Latter long, cylindrical, abruptly narrowing at level of oviduct orifices. Abdomen (Fig. 1410) short, with truncated or rounded end. Uropods as in *Lernaeocera*, often missing. Total length 18–25 mm.

First antenna (Fig. 1402) obscurely segmented, distal quarter marked off by more or less distinct transverse constriction; anterior margin with 20 naked setae of different lengths, two long, slender pinnate setae on anterior margin, one on dorsal surface; apical armature (Fig. 1401) of 13 setae: apex and anterior side of tip with five long (one aesthete?) and two short setae, ventral side with two long and slender setae, posterior with four (two fused at bases) similar setae. Second antenna (Fig. 1403) three-segmented; basal segment short and squat, unarmed; second with prominent, triangular, concave outgrowth in distal corner; slight inflation beneath base of claw, with heavy cuticular flange close by; claw abruptly curving, slender, with short seta at base. Mouth tube very broad, with several rows of denticles (Fig. 1404) along posterior side of base. Mandible (Figs. 1406, 1407) two-segmented, flat, tip twisted round long axis; dentiform margin with six very small, rounded teeth, recessed from margin. First maxilla (Fig. 1405) with endopod carrying 2 long, slender setiform processes; exopod very small, with one similar process. Second maxilla (Fig. 1408) brachiform; lacertus robust, with one unciform spine on lateral surface and one on ventral margin at same level; brachium cylindrical, with delicate membraneous crista (Fig. 1409); claw (Fig. 1409) curving and tapering, with small notch on convex margin near tip, one strip of striated membrane along that margin and one on each side (only one visible in Fig. 1409). Maxilliped absent.

First and second swimming legs biramous (Fig. 1411), third and fourth uniramous. Sympods and rami two-segmented. First segment of sympod unarmed, second with slender seta lateral to base of exopod (Fig. 1413), first leg also with similar seta medial to base of endopod (Fig. 1411). Armature formula below.

	Endopod		Exopod	
	1	2	1	2
Leg 1	1-0	7	1-I	5,II
Leg 2	1-0	7	1-0	6,I
Leg 3	—	—	0-0	5,I
Leg 4	—	—	0-0	4,I

All endopod setae pinnate. Lateralmost setae of exopod with pinnules along one margin and a strip of membrane along another (Figs. 1412, 1414).

Male: Unknown.

Comments: Because of its characteristic thoracic neck covered with nodosities and because of its unusual habitat, the eye of its host, *L. sprattae* presented few difficulties for those who wished to identify it. The fact that it accumulated an impressive list of synonyms is largely attributable to the inadequate circulation of the early literature and the relatively slow rate of exchange of scientific information. Some workers were misled by its morphological variability. Baudouin (1904) described specimens from the sardine as a species distinct from that parasitic on the sprat. Defending his views for nearly two decades, he kept describing forms intermediate between *L. sprattae* and his *L. sardinae*. The description of *L. sardinae* var. *moniliformis* (cf. Baudouin, 1910), a halfway form between the two, was followed by that of another intermediate form, halfway between a "typical *L. sardinae* and the 1910 'variety' " (Baudouin, 1918). A good example of persistent misunderstanding occurs in the work of Blainville (1822), who took Sowerby's (1804–1806) figure representing the parasite *in situ* in the eye of the sprat as being a species distinct from that shown in the figure of the parasite dissected out of the eye. The former one became *Lernaea cyclophora*, the latter was merely referred to as "Lerneidé articulée" (the lines indicating the anterior part of the trunk became the egg sacs in his interpretation). The mistake was pointed out by Baird (1850), but it continued to haunt the literature. Later authors wrote of *Lernaea articulata*, while Yamaguti (1963) debated the identity of *L. cyclophora*, a species possessing "a discoid head without lateral horns". He was not aware of the fact that this "discoid head" was the eye of the fish.

As might be expected, the morphological plasticity of *L. sprattae* is particularly evident in the cephalothoracic area. For example, the angle between the dorsal and ventral surfaces of that tagma depends largely on the angle of contact between it and the host tissue; the shape of the holdfast processes and other cephalothoracic swellings, as well as of the nodosities on the thoracic neck, are influenced by the position of the parasite within the eye.

The eye of the host is the principal habitat of *L. sprattae*. The copepod pierces the cornea and embeds in the fundus of the eye, either through the pupillary opening or by piercing the iris. How this is accomplished is unknown, but it is possible that this stage in the life cycle resembles the corresponding stage in that of *Phrixocephalus cincinnatus* Wilson, 1908, as described by Kabata (1969a). The copepod, however, has also been found embedded in the dorsal musculature of its host fish, most frequently near the base of the dorsal fin, with its head sometimes penetrating through the musculature and entering the visceral cavity.

The free-swimming male of *Lernaeenicus*, discovered in Oslo Fiord and described by Kabata (1963b), might possibly belong to this species. Since it could equally well belong to *L. encrasicoli*, its description was not included here.

L. sprattae parasitizes *Sprattus sprattus*, but occurs also on *Sardina pilchardus* and, occasionally, on *Clupea harengus*. Its distribution range covers the European Atlantic seaboard. It has been found also in the Danish Straits and in the western Mediterranean. It is quite common in British waters.

Lernaeenicus encrasicoli (Turton, 1807)

(Figs. 1418–1422)

Syn.: *Lernaea encrasicoli* Turton, 1807
Lerneonema encrasicoli, Turton; of Baird (1850)

Pre-metamorphosis female: Unknown.

Post-metamorphosis female (Fig. 1418): Cephalothorax usually at about right angles to thoracic neck, moderately short, with mouth tube usually in line with its long axis (Fig. 1422). In some specimens semispherical swellings lateral to antennary region (Fig. 1421). Holdfast of two lateral horn-like processes, simple and blunt, uneven, occasionally with subdivided tips (Fig. 1420); often with third, similar dorsal process (Figs. 1419, 1421) of variable length. Swimming legs close together, at point of flexure between cephalothorax and neck (Fig. 1422). Neck cylindrical, smooth, often twisted, imperceptibly passing posteriorly into genital trunk. Latter much thicker than neck, cylindrical, long, narrowing at point of junction with abdomen. Junction marked by oviduct orifices. Abdomen short, conical. Total length up to 30 mm.

Appendages as in *L. sprattae*.

Male: Unknown.

Comments: Unlike its sibling species, *L. sprattae*, this copepod had not acquired long synonymy, though its specific name was spelled in several different ways (*encrasicola*, *encrasicholi*, *encrasicolus*). Its morphological plasticity matches that of *L. sprattae* but the absence of nodosities on the thoracic neck helps to distinguish it from that species. *L. encrasicoli* is a parasite of several clupeid fishes (*Engraulis encrasicolus*, *Sprattus sprattus*, *Sardina pilchardus* and *Alosa alosa*). Its distribution is largely coincident with that of *L. sprattae*. In British waters it occurs commonly on the sprat, often together with *L. sprattae*. It lives only occasionally in the eye of its host, its most common habitat being the dorsum of the fish where it deeply burrows in the musculature, causing extensive lesions and sometimes penetrating into the visceral cavity.

Genus Pennella Oken, 1816

Pre-metamorphosis female: Cephalothorax (including first leg-bearing segment) dorsoventrally flattened, with angular anterior, convex lateral and transversely truncated posterior margins; small rounded lobes in posterolateral corners. Three well delimited leg-bearing segments, gradually diminishing in size. Genital complex completely fused with abdomen, about as long as three free segments combined, subcylindrical, with numerous transverse wrinkles. Posterior extremity bluntly rounded, sometimes slightly bifid. Uropods present. First antenna uniramous, obscurely segmented, setiferous. Second antenna chelate. Mouth pennellid, mandible dentiferous. First maxilla small; endopod with two, exopod with one setiform process. Second maxilla brachiform. Maxilliped absent. Four pairs of swimming legs, two biramous, two uniramous; all rami two-segmented.

Post-metamorphosis female: Cephalothorax subspherical; antennary processes well developed, in some species imparting to cephalothorax cup-like appearance. Holdfast of two lateral processes, in some species with third dorsal process, in others varying in number from two to three processes, unbranched, of variable lengths and angles. Thoracic cylindrical neck short or long, imperceptibly passing into genital trunk. Latter cylindrical, often transversely ridged, of varying length, abruptly narrowing at level of oviduct orifices. Abdomen cylindrical, often transversely striated, always shorter than trunk; abdominal brush present, its processes single or variously branched, branching complexity increasing with age. Uropods present. Egg sacs long, straight. Appendages as in pre-metamorphosis female, but third and fourth pairs of legs possibly further apart from each other and from first two pairs than at younger stages.

Male: Cephalothorax similar to that of pre-metamorphosis female; often relatively broader, sometimes with two pairs of posterolateral lobes (one marking anterior level of first leg-bearing segment, now completely fused with cephalothorax). Three leg-bearing segments, progressively smaller, with fairly prominent, rounded posterolateral lobes. Abdomen one-segmented, subquadrangular, shorter than genital complex. Uropods present. Appendages as in female, but subchelate maxilliped present.

Type species: *Pennella sagitta* L., 1758.

Comments: The taxonomy of the genus *Pennella* is probably in a poorer state than that of most pennellid genera. There are two main reasons for this: firstly, the species belonging to this genus are not particularly common, so that no author had an opportunity of studying large collections of *Pennella*; secondly, *Pennella* shares with other Pennellidae a great degree of morphological plasticity, differences being host-induced as well as influenced by the age of the parasite. When one adds to this the fact that the appendages of *Pennella* have never been thoroughly examined for possible fine differences between species, one completes a picture which continues to perplex the students of Pennellidae.

As an introduction to the morphology of the genus, the author proposes to discuss briefly some features of the type species, *P. sagitta*. A young post-metamorphosis female is shown in Fig. 1423. The ventral surface of its cephalothorax is framed in two lateral rows of prominent, irregularly branching papillae which make it somewhat cup-shaped. The entire space between them is filled with similar, though smaller, antennary processes that surround and hide the oral region (Fig. 1424). Further posteriorly, two small peg-like papillae can be seen in most specimens. From the dorsal side (Fig. 1425) one can see both pairs of antennae, marking the dorsal limit of the "cup". All these features are emphasized in older specimens (Fig. 1426). (In young post-metamorphosis females the tergal plates of the free thoracic segments can still be recognized (Fig. 1425).) Similarly, changes take place in the abdominal brush. Its processes in the very young female (Fig. 1427) are short, blunt and simple; in older specimens (Fig. 1428) they are much longer, more slender and occasionally with secondary branches. (It is to be noted that the abdominal brush, in spite of its name, extends to the posterior part of the genital trunk, on which two pairs of processes occur.) The abdomen itself becomes relatively longer in the more mature females.

The appendages are very typical of the family and can be closely compared with those of *Lernaeocera* (q. v.). The first antenna differs from that of *Lernaeocera* only in the relatively greater length of its apical armature. The second antenna (Fig. 1429) is distinguished by the thickness of its basal segment and relatively short, robust claw. The first maxilla (Fig. 1430) is distinguished by the length of its endopod. The brachium of the second maxilla (Fig. 1431) is very short, with two spinulated elevations in the area of the crista, the size of the spinules differing from one to the other elevation; the claw is long, largely covered by spinules.

There is little reliable information on the structure of other appendages and the author could not secure them for a detailed study. The females have no maxilliped. Examination of the pre-metamorphosis stages (Wierzejski, 1877) shows that the structure of the swimming legs is similar to that of other Pennellidae, i.e. the first and second pairs are biramous, while the remaining two pairs are uniramous. (Brian's (1906) rather diagrammatic drawing illustrating a pre-metamorphosis female of *Pennella* with all four pairs biramous is patently incorrect.)

More than 30 nominal species of *Pennella* have been described from various parts of the world, five of them parasitic on cetacean hosts. The taxonomy of the genus is in a deplorable condition. It is not possible to assess currently how many of these species are valid, since the majority of descriptions are poor and no reliable criteria for identification can be established. Some attempts have been made to arrive at these criteria (Quidor, 1913; Leigh-Sharpe, 1928; Delamare Deboutteville and Nunes-Ruivo, 1951b). The shape of the head, the head and neck to trunk and abdomen length ratio and the character of the abdominal processes were proposed as possible discriminants, all with a degree of doubt that makes one treat them with caution. Some differences in the shape of the head might, indeed, be used to discriminate between species or groups of species. For example, *P. sagitta* is distinguished by having a "cup-shaped head" (Figs. 1425, 1426), in contrast to the "spherical head" of *P. filosa* (Figs. 1433, 1434). The relative size of the antennary processes is also different in those two species, being much larger in the former. Some differences in the shape and size of the abdominal processes might also be taxonomically useful. In *P. sagitta* they remain simple, with few exceptions (Fig. 1428), whereas in other species (Figs. 1435, 1436) they might show irregular multiple branching. As for the length ratio of neck to trunk–abdomen, it is probably too much influenced by the site of attachment and the depth of penetration of the copepod to be of any real help in taxonomy.

Only one species of *Pennella* has been reported with any certainty from British waters (though under different names). Some records of unspecified *Pennella* species have also been made (Leigh-Sharpe, 1933a).

Pennella filosa (L. 1758)

(Figs. 1432–1439)

Syn.: *Pennatula filosa* L., 1758
Lernaea cirrhosa La Martiniere, 1798
Pennella orthagorisci Wright, 1870
Pennella Costai Richiardi; of Valle (1882)
Pennella fibrosa of Smiles; of T. Scott (1905)
Lerneopenna Bocconii Blainville; of Brian (1906)
Pennella rubra Brian, 1906

Pennella plumosa Linton, 1925
Pennella germonia Leigh-Sharpe, 1931

Pre-metamorphosis female: Unknown.

Post-metamorphosis female (Fig. 1432): Cephalothorax subspherical, imperceptibly merging with cylindrical neck; antennary processes of more or less even size and irregularly papilliform, covering anterior (Fig. 1433) and most of ventral surfaces (Fig. 1434); holdfast of two lateral horns, often with third dorsal horn present; lateral horns differing in length from very short (Fig. 1437), through moderately long (Figs. 1433, 1434) to very long (Fig. 1439); dorsal horn, when present, shorter than lateral ones (Figs. 1437–1439). Thoracic neck cylindrical, imperceptibly passing into trunk. Latter cylindrical, thicker and either longer or shorter than neck, transversely ridged or striated, abruptly narrowing at point of junction with abdomen, marked by position of oviduct orifices. Abdomen subcylindrical, shorter than trunk. Processes of abdominal brush irregularly branched (Figs. 1435, 1436 show two parts of one such process), with secondary and tertiary branches present. Total length about 200 mm.

Appendages not known in detail, but probably similar to those of type species. (Though distances between second and third legs and between third and fourth legs greater than between first and second pair.)

Male: Unknown.

Comments: The great morphological variability of this species has resulted in many misidentifications. In the absence of sufficient comparative material, some writers tended to describe different specimens as different species. Although their conclusions were usually tentative, their new names became embedded in the literature, further to compound the difficulties. In this account *P. orthagorisci* Wright, 1870 and *P. germonia* Leigh-Sharpe, 1931, are considered synonymous with *P. filosa*, though they have both been treated as independent species by many writers. This author did not find any features in either of them that could not be accommodated within the variability range of Linnaeus' species. Neither were their hosts different.

The hosts of *P. filosa* comprise the large scombrids of the genus *Thunnus* and its relatives, as well as the sunfishes of the genus *Mola*. The copepod has been recorded in both the Atlantic and Pacific Oceans. In British waters its records come mainly from *Mola mola*, but scombrids carrying it have also been taken in this region (*Thunnus thynnus*, *T. alalunga*).

(The difficulty in dealing with *Pennella* is exemplified by the record of *P. fiosa* from *Prognichthys* (= *Hirundichthys*) *rondeleti* and *Hyporhamphus unifasciatus* taken in the Gulf of Mexico. Causey (1955) identified specimens from these hosts as *P. filosa*, although the fishes belong to Exocoetidae, a family whose members are more likely to carry *P. exocoeti* (Holten, 1802). Moreover, the specimens were beheaded and thus devoid of their most important taxonomic clues. The reasons for assigning them to *P. filosa* are difficult to understand).

Family Sphyriidae

Morphology

This small group of siphonostome copepods appears to constitute one of the end-points of the evolutionary path. Like Pennellidae, they have become extensively modified to succeed in that kind of sessile mode of life that has been referred to earlier as mesoparasitism. A substantial part of the body of an adult sphyriid female is embedded in the tissues of the host. Although smaller than the part remaining outside, morphologically it represents a more significant part of the organism, consisting of the entire cephalothorax and at least two free thoracic segments, while the exposed part is composed only of one free thoracic segment and the genito-abdominal complex. The sphyriids are mainly parasitic on the outer surfaces of the fish (though some occur also in the branchial cavities) and their adaptation to life in a constant stream of water has caused them to become elongated and slender. Only the posterior part of the body, required to accommodate the reproductive organs, has retained any substantial bulk.

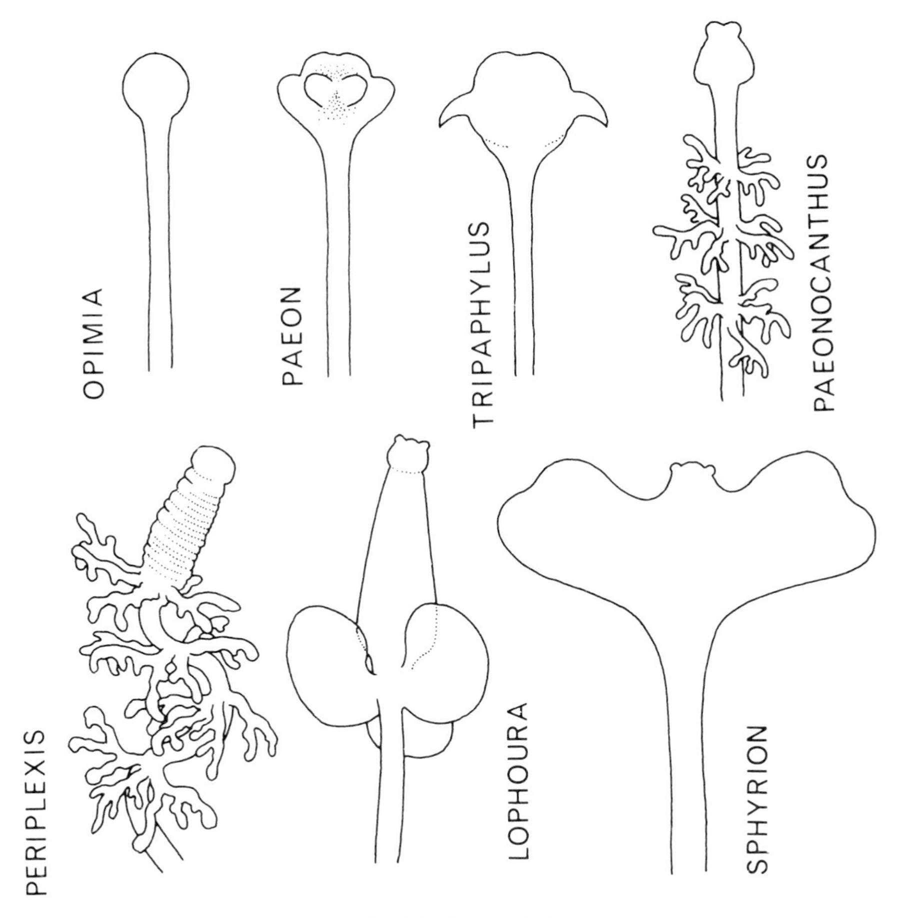

Text fig. 60. Anterior end of body and holdfast of various sphyriid females.

The body of an adult sphyriid female can be divided into three topologically recognizable parts: the broad anterior part of a shape differing from one species to another, the narrow and cylindrical part sometimes armed with complicated holdfast processes, and the posterior part usually broadly expanded and flattened dorsoventrally. A complete loss of external segmentation and of most, or all, swimming legs makes it difficult to determine the morphological content of these three parts. The difficulty is aggravated by the fact that next to nothing is known about the ontogeny of sphyriid species, so that one cannot extrapolate from the pre-adult stages to determine homologies of various parts of the adult female.

The interpretation of the morphology of these copepods is based here on the fragmentary observations of Wilson (1919) who described finding the first and fourth pairs of swimming legs on a young post-metamorphosis female *Lophoura*. Using their positions as landmarks, one can roughly determine the borders between the three major parts. The anterior, broad part is the cephalothorax and it ends at about the point where it narrows into the cylindrical thoracic neck. The neck is formed by the three free segments of the pre-metamorphosis stage, the posterior part of the third segment expanding into the flat genito-abdominal complex, varying in size and outline from one species to the next. The only recognizable trace of abdomen is a small perianal elevation, seen in some species. All sphyriids possess a pair of prominent posterior processes derived from modified uropods and always dorsal to the egg sacs (contrary to observations of Kabata (1965c)).

The cephalothorax is at its simplest in the monotypic genus *Opimia* Wilson, 1908 (Text fig. 60), in which it is a smooth, subspherical expansion of the anterior end of the body. In the genus *Paeon* Wilson, 1919 (Text fig. 60) it expands laterally and is usually broader than long; it is equipped with varying numbers of rounded or pointed protuberances arranged in bilaterally symmetrical order. The genus *Tripaphylus* Richiardi, 1878, has a transversely oval cephalothorax with horn-like lateral processes. *Paeonocanthus* Kabata, 1965, has a relatively small cephalothorax, longer than broad, with two rounded tubercles anteriorly and posterior end broader than anterior. In *Periplexis* Wilson, 1919, the cephalothorax is a transversely wrinkled cylinder, its major part representing the first leg-bearing segment (judging from the position of the maxillipeds). Somewhat similar is the cephalothorax of *Lophoura* Kölliker, 1853; it differs from that of *Periplexis* by having fewer or no transverse wrinkles and sometimes in having a slight increase in diameter of its posterior end. Finally, in *Sphyrion* Cuvier, 1830, the cephalothorax has become greatly extended laterally and aliform in appearance, the extension occurring mainly in the first leg-bearing segment.

The thoracic neck is simply a more or less sclerotized cylindrical tube, either straight or twisted in various ways, depending on its position in the host and the path it travelled through the host's tissues at the time of attachment. In *Paeonacanthus* and *Periplexis* (Text fig. 60) it is abundantly armed with branching holdfast processes, arranged in more or less regular rings at varying intervals along the anterior part, immediately behind the cephalothorax. Occasionally these processes might be simple, but usually some division is present. In all the remaining genera (Text fig. 60) the neck is unarmed.

The genito-abdominal trunk is always dorsoventrally flattened, though the extent of depression in *Opimia* is only slight. In all of them the expansion of the cylindrical neck in the posterior direction is gradual, so that the anterolateral corners of that tagma slope gently away. The posterolateral corners are usually evenly rounded (though they are provided with lobes in *Paeon versicolor* Wilson, 1919). A slight elevation in the midposterior position is bisected by the anal slit and marks the last trace of abdomen.

The posterior processes in *Opimia*, *Paeon*, *Tripaphylus* and *Paeonocanthus* are simple, long cylinders with peduncular bases, easily broken off in fixed specimens. In *Periplexis* these cylinders are more robust and divided into three parts by deep transverse constrictions. In *Lophoura* they are slender and completely covered by numerous secondary cylindrical processes arising from their walls (Fig. 1458). In *Sphyrion*, on the other hand, the processes divide close to the bases, both branches subdividing repeatedly until a profusely branching structure is produced, somewhat resembling a bunch of grapes. The size and complexity of the processes increase with the age of the copepod (comp. Fig. 1440 with 1444).

The appendages of the pre-metamorphosis female are unknown. In the post-metamorphosis female they are greatly reduced and simplified, mainly persisting as semispherical swellings, sometimes with remnants of apical armature.

Sphyriid males bear a striking resemblance to those of the genus *Lernaeopoda* and their allies within the family Lernaeopodidae, such as the males of the *Charopinus*-group species. The bases of their second maxillae are more extensively fused together than those of their lernaeopodid counterparts; the same is true of the bases of the maxillipeds. Otherwise the two sets of males can easily be mistaken for each other, particularly when one looks at their appendages (antennae, maxillae, mandibles) constructed on the same general plan.

History and systematics

The family Sphyriidae contains about 25 nominal species. With our present state of knowledge it is not possible to decide how many of them are valid taxa but it is quite likely that not many more than

15 could stand up to a close scrutiny.

Only four sphyriid species were known in the 19th century. Being relatively scarce, they did not obtrude themselves on the attention of systematists. The older literature, therefore, does not dwell on their systematics to any extent. Burmeister (1835) surmised that *Sphyrion* fitted somewhere between *Lernaea* and *Lernaeocera*, while Edwards (1840) tentatively placed it between his "Penelles" and "Lernees". Steenstrup and Lütken (1861) took the view that *Sphyrion* is closer to Chondracanthidae, considering it to be similar to *Chondracanthus* (= *Lernentoma*) *triglae*, a rather strained comparison. Krøyer (1863) preferred to ally *Sphyrion* with the pennellids. A similar view was held by Quidor (1912) who placed *Sphyrion* and *Lophoura* in Lernaeidae (= Pennellidae) because both exhibited torsion (a rotation around the longitudinal axis of the body), allegedly absent elsewhere. In the same family Wilson (1908) placed his genus *Opimia*.

The family Sphyriidae was established by Wilson (1919) in recognition of the unique features of its members. Wilson believed it to be closely related to Lernaeidae (= Pennellidae), although he pointed out that the life history of the male resembles that of the male lernaeopodid.

It is obvious that the investigators looking at the possible affinities of Sphyriidae were impressed by the general habitus of the post-metamorphosis females. It is equally obvious that this habitus is a direct outcome of the mesoparasitic way of life, to which Sphyriidae have become so splendidly adapted. In this they parallel some members of the family Pennellidae and have come to bear a superficial resemblance to them. It suffices to compare some species of *Lophoura* (Sphyriidae) with certain species of *Lernaeenicus* (Pennellidae) to understand why affinities between their families were postulated. Even the fact that Sphyriidae form the long cylindrical neck from the anterior part of the genito-abdominal complex, while in Pennellidae it involves the free leg-bearing thoracic segments, presented no obstacle in this scheme. After all, Pennellidae were seen as a component of Lernaeidae, in which the trunk developed from precisely these segments. The cyclopoid nature of the Lernaeidae proper was disregarded or went unnoticed.

This author believes that the closest known relatives of Sphyriidae are Lernaeopodidae, in particular the *Charopinus*-group of that family. The gross differences between them reflect their specialized types of host-parasite relationship, the compact browsing Lernaeopodidae being tethered by their long second maxillae, the mesoparasitic Sphyriidae elongating their bodies in the course of their evolution into burrowing parasites.

Clues to the affinities between these two families can be found mainly in the structure of the males, the resemblance being evident in their general morphology, the structure of their appendages and their life history, inasmuch as it is known. Some idea of their resemblance can be gained by comparison of a male *Tripaphylus* (Sphyriidae) (Fig. 1461) with that of *Charopinus* (Lernaeopodidae) (Fig. 1587). The pre-metamorphosis stages of the female sphyriids are as yet unknown (except for some nauplii) and the appendages of the adult are greatly reduced and cannot be used as indicators of relationships but the affinities between Sphyriidae and Lernaeopodidae are suggested by one morphological feature, hitherto disregarded in discussion of the taxonomic position of the sphyriids. All species of Sphyriidae are equipped with posterior processes, structures flanking the perianal elevation and representing modified uropods. Among parasitic copepods only some Lernaeopodidae possess uropods modified in a similar manner. The dorsal position of these structures is typical of the *Charopinus*-branch of that family (see p. 346), hence the suggestion of a possible relationship between it and the sphyriids.

The intrafamilial relationships of Sphyriidae are obscure, but their genera can be arranged in a series, from those most simple structurally to those with a more complicated structure, as described above and illustrated in Text fig. 60. It is worth noting here that a rough sort of parallelism can be drawn between the morphological complexity and host affiliation of Sphyriidae. The genera with simple, cylindrical posterior processes and with necks devoid of holdfast outgrowths can be construed as being the most primitive among Sphyriidae. They are: *Opimia*, *Paeon* and *Tripaphylus*. All their species are parasitic on elasmobranchs. *Paeonocanthus antarcticensis* (Hewitt, 1965), the sole known species of that genus and the only other species with cylindrical posterior processes, is parasitic on *Bathylagus antarcticus*, a teleost. This copepod, however, has a cylindrical neck with a profuse armature of subdivided holdfast outgrowths. In *Periplexis*, another monotypic genus, the posterior processes are subdivided into three and the neck has armature similar to that of *Paeonocanthus*. Its host is also a teleost. Also teleostean are the hosts of *Sphyrion* and *Lophoura*, two genera with posterior processes of great complexity. (It might be mentioned that the advanced nature of these two genera is evident also from the structure of their alimentary canal, the presence of numerous gut diverticula being unique to *Sphyrion* and *Lophoura*.)

The three genera with representatives in British waters will be discussed at greater length below. Of the remaining genera, *Opimia exilipes* Wilson, 1908, has been recorded only twice along the eastern seaboard of the United States. *Paeon* contains five species. *P. elongatus* Wilson, 1932, *P. ferox* Wilson, 1919, *P. vaissieri* Delamare Deboutteville and Nunes-Ruivo, 1954 and *P. versicolor* are known from the Atlantic (though Lewis (1966) recorded *P. vaissieri* from Hawaiian waters). *P. lobatus* Kirtisinghe, 1964, occurs in the Indian Ocean. *Paeonocanthus antarcticensis* has been found only in the Antarctic region, with one record from the North Atlantic (62°N 33°W). *Periplexis lobodes* Wilson, 1919, is another inhabitant of the Atlantic coastal waters of the United States.

The genus *Paeonodes* Wilson, 1944, placed by its discoverer in Sphyriidae was shown by Hewitt (1969b) to be a poecilostome and was transferred to Therodamasidae.

Key to British genera

1. Posterior processes simple, cylindrical (Fig. 1459) . . . . *Tripaphylus*
   Posterior processes profusely subdivided 2
2. Posterior processes with long central stalks (Fig. 1458) . . . . *Lophoura*
   Posterior processes branching close to base (Fig. 1445) . . . . *Sphyrion*

Genus Sphyrion Cuvier, 1830

Pre-metamorphosis female: Unknown.

Post-metamorphosis female: Cephalothorax transversely expanded into pointed or rounded processes, smooth or provided with tubercular warts or outgrowths; oral region and entire cephalic part centrally situated. Free thoracic segments forming cylindrical, straight or twisted neck of variable length, devoid of external segmentation and sometimes exhibiting torsion, posteriorly joining genito-abdominal trunk and slightly expanding in some specimens. Trunk suborbicular or pyriform, dorsoventrally flattened, with uropods modified into profusely branching posterior processes. Egg sacs long and straight; eggs spherical, multiserial. First and second antennae reduced to tubercular swellings. Mouth tube siphonostome. Structure of mandible unknown. First and second maxillae conical or digitiform, very small. Maxillipeds subchelate. In mature specimens swimming legs absent.

Male: Habitus resembling that of lernaeopodid males, body indistinctly divided by transverse constriction into cephalothorax and genito-abdominal trunk. Latter apparently devoid of uropods, with orifices of alimentary canal and genital tract displaced towards centre of ventral surface. First antenna uniramous, apparently three-segmented, with well developed apical armature. Second antenna biramous, with two-segmented sympod; exopod one-segmented, rounded, spiniferous; endopod obscurely segmented, its apical armature of lernaeopodid type. Mouth tube siphonostome. Mandible unknown. First maxilla digitiform; second maxilla subchelate. Maxilliped subchelate, fused with opposite member at base. Swimming legs absent.

Type species: *Sphyrion laevigatum* (Quoy and Gaimard, 1824).

Comments: Of the nominal species of this genus only two can be considered valid, all the others being either species inquirendae or reducible to synonymy with these two. The morphological variability, common among the extensively modified copepods, led some earlier workers to erroneous descriptions of species with untenable validity. The two valid species are: *S. laevigatum* and *S. lumpi* (Krøyer, 1845), the latter occurring in the Atlantic Ocean, the former in other areas (though some records suggest its occurrence also in the southernmost part of the Atlantic). *S. lumpi* is quite common in British waters.

Sphyrion lumpi (Krøyer, 1845)

(Figs. 1440–1450)

Syn.: *Lestes lumpi* Krøyer, 1845
Lesteira lumpi Krøyer; of Krøyer (1863)
?*Sphyrion norvegicum* Thor, 1900

Post-metamorphosis female (Fig. 1440): Cephalothorax trasversely elongated by development of two lateral processes (with expanded and sometimes slightly bifid tips), formed mainly by maxilliped-bearing segment. Segments anterior to maxilliped form subquadrangular cephalic region (Figs. 1441, 1442) with rounded protuberances in anterolateral corners; two subspherical swellings close to each other on ventral surface anterior to maxillipeds. (Cephalothorax often rotated from dorsoventral plane of body due to torsion resulting from attachment.) Free leg-bearing thoracic segments fused together without trace of segmentation into cylindrical neck of variable length. Neck expanding posteriorly into suborbicular (Fig. 1440) or pyriform (Fig. 1444) genito-abdominal complex, flattened dorsoventrally, with midposterior perianal elevation (Fig. 1445) and posterolateral slit-like oviduct orifices (Fig. 1443). Uropods modified into repeatedly branching processes with first division close to base (Fig. 1445). Total length 45–60 mm.

First antenna (Fig. 1446, antl) semispherical, with short apical process, apparently bifid at tip. Second antenna similar, though larger, swelling, with two apical processes (vestiges of rami?). First maxilla (Fig. 1447) digitiform, apically blunt, unarmed. Second maxilla (Fig. 1448) rounded tubercle tipped with two very small, setiform processes. Maxilliped (Fig. 1449) subchelate; corpus suboval, with prominent pointed spine at base and similar, though somewhat curved, spine on medial margin close to base of subchela; latter not divided into claw and shaft, with broad base, triangular spine on medial margin near base, sharp curvature near midlength and tapering slender tip.

Male (*fide* Wilson, 1919; Fig. 1450): Cephalothorax small, separated from trunk by transverse constriction, with somewhat inflated ventral surface. Trunk oviform, with slight swelling in midventral area around genital openings. Total length 2 mm.

First antenna three-segmented; apical armature consisting apparently of two setae. Second antenna with two-segmented sympod; exopod one-segmented with at least three large apical spines; endopod with obscure segmentation, its armature consisting of hook (homologous with hook 1 of Lernaeopodidae), robust spine (spine 2 of Lernaeopodidae) and ventral spine (spine 4 of Lernaeopodidae). First maxilla with exopod greatly reduced or absent; endopod with two terminal papillae. Second maxilla subchelate, corpus with prominent spine near base of subchela, latter claw-like, undivided, very sharply curved, with blunt tip. Maxilliped with obscure segmentation, inflated at base, cylindrical distally, with extension of anteromedial corner into subtriangular process accommodating tip of subchela; latter undivided, curved, with sharp, tapering tip.

Comments: The characteristic morphology of *S. lumpi* makes it easily identifiable, hence the taxonomy at the specific level is not as confused as that of the type species of the genus (mainly because of the smaller range of individual differences in the shape of the cephalothorax). *S. norvegicum* Thor, 1900, was inadequately described and its host and locality were not recorded. Its specific name suggests that it might have been found off the coast of Norway. Should this be the case, then it might be *S. lumpi*, since no other species of *Sphyrion* is known to occur in that area. At the generic level, the names used by Krøyer (the first one having been changed by Krøyer himself as preoccupied) had to yield to Cuvier's chronologically prior name.

The post-metamorphosis female undergoes marked morphological changes with age. These changes might be seen as the final stages of the process of metamorphosis itself, though they do not introduce dramatic overall changes in the habitus of the copepod. They include gradual expansion of the lateral processes, initially no more than small, dorsolateral, rounded swellings; increasing progressively in size, they also acquire bulbous, sometimes bifid tips. The ventral swellings anterior to the maxilliped, at the beginning hardly smaller than the rudimentary lateral processes, are left behind and dwarfed in this process. The cylindrical neck elongates, though the extent of that increase in length varies greatly, so as to lead some writers to the recognition of "short-necked" and "long-necked" varieties. The genito-abdominal complex grows in size and tends to become pyriform (in the younger females it tends to be suborbicular). Finally, the posterior processes, originally simple and subcylindrical, begin to divide, eventually reaching their definitive complexity. (A part of this process can be seen by comparing Figs. 1440, 1441 with Fig. 1444.)

Although the pre-metamorphosis development is unknown, its early stages have been described. The development of the eggs was reported on by Squires (1966), while intraovular stages were described by Krøyer (1863) and, more recently, by Jones and Matthews (1968). The latter work is particularly interesting in that it describes in some detail the appendages of the pre-copepodid stage referred to as metanauplius. Fig. 2E of that paper shows within the naupliar cuticle the second antenna of the copepodid, clearly illustrating its similarity to the corresponding limb of the male and to those of Lernaeopodidae. Fig. 2C shows that between the second antennae and clearly identifiable maxillipeds there are present the second maxillae, as well as another pair of appendages, about to be greatly reduced. Between the latter, on the ventral surface of the body, there are two minute bifid bars of unknown homology. Should the lateral pair prove to be the mandibles (as one might suspect from the fact that in the early nauplius, Fig. 1A, they occupy the normal position of these appendages), then the first maxillae are either missing or represented by the medial pair of minute plates. Be it as it may, one pair of appendages appears to be wanting, or much reduced. The difficulty might perhaps be resolved when a detailed study of the free-swimming copepodid is made. Until then, it must be kept in mind that the homology of the appendages is in some doubt and the terminology here applied to them is only tentative.

S. lumpi has a broad range of hosts, the species harbouring it representing four orders. Two orders contribute one species each (*Anarhichas lupus* of Perciformes and *Reinhardtius hippoglossoides* of Pleuronectiformes). The order Gadiformes is represented by six species: *Antimora australis*, *A. rostrata*, *Gadus morhua*, *Laemonema laureysi*, *Macrurus* sp. and *Nematonurus goodi*, while four species come from the order Scorpaeniformes: *Cottunculoides inermis*, *Cyclopterus lumpus*, *Sebastes marinus* and *S. viviparus*. One characteristic that many of these species tend to have in common is their deep water habitat. Relatively few of them occur commonly in shallow waters. From this point of view, *S. lumpi* is a true member of its family.

The range of distribution of *S. lumpi* covers most of the North Atlantic and adjacent seas, with some records (Barnard, 1948; Capart, 1959) from the southern half of that ocean. Although the parasite has received its name from its association with *Cyclopterus lumpus*, it does not appear to be common on that host. From the author's own experience, by far the most common hosts of *S. lumpi* are species of the genus *Sebastes*. Some hosts, at least, must be considered as accidental and do not appear to play an important part in the life of this copepod. Here belong, for example, *Gadus morhua* and *Reinhardtius hippoglossoides*, both recorded infected with *S. lumpi* off the coast of Norway by Berland (1969).

In British waters, *S. lumpi* has been found on *Cyclopterus lumpus*, *Anarhichas lupus* and *Sebastes marinus*, the last-named often bearing this copepod when examined in the catches in British fish markets.

Genus Lophoura Kölliker, 1853

Post-metamorphosis female: Cephalothorax equipped with various sets of lobes and protuberances on its cephalic part, often set off by transverse constriction; remainder cylindrical, elongate, sometimes transversely wrinkled. Free thoracic segments fused into cylindrical neck, about as thick as, or thinner than, cephalothorax; neck holdfast near posterior end of cephalothorax, consisting of rounded knobs, long processes or branched horns. Posterior end of neck expanding into genito-abdominal complex, subquadrangular to suborbicular, dorsoventrally flattened, with elevated perianal area in mid-posterior point. Uropods modified into posterior processes with cylindrical central stalk and numerous clavate or subcylindrical lateral outgrowths. Egg sacs cylindrical, eggs multiseriate. First and second antennae reduced to small conical or semispherical swellings. Mouth tube siphonostome, very small. Mandible unknown. First and second maxillae small swellings surmounted by simple spinules. Maxilliped vestigial or absent. No swimming legs.

Male: Resembling those of more primitive Lernaeopodidae. Cephalothorax small, its postero-lateral region distended by testes, its posteroventral surface bearing cylindrical protuberance surmounted by maxillipeds. Trunk separated by shallow transverse constriction from cephalothorax and indistinctly subdivided by similar, shallow constrictions; increasing in girth posteriorly, with rounded posterior extremity. Uropods vestigial, spiniform or tubercular. First antenna uniramous, with well developed apical armature. Second antenna biramous, exopod smaller than endopod. Mouth tube siphonostome. Mandible unknown. First maxilla with reduced exopod, two terminal endopod papillae, each tipped with two short processes. Second maxilla and maxilliped subchelate. No swimming legs.

Type species: *Lophoura edwardsi* Kölliker, 1853.

Comments: Kölliker's (1853) description of *L. edwardsi* established a new genus the validity of which has never been in question. Some later authors changed the spelling of the name by omitting from it the letter "u". Bassett-Smith (1899) quoted it in that version. When Poche (1902) subjected Bassett-Smith's work to relentless criticism, he pointed out that the name *Lophura* had been preoccupied and proposed its replacement by *Rebelula*. His proposal was widely accepted, until Yamaguti (1963) reverted to the use of *Lophoura*. In this he was unquestionably correct. Poche's remarks were addressed to the inaccurately spelled name that differed by one letter from the original. According to the International Code for Zoological Nomenclature, article 56(a), a difference of one letter suffices to prevent homonymy. Since *Lophoura* was never preoccupied and since it is the first name given to this taxon, it must remain as its valid name.

Lophoura contains nine species, all quite rare and less than adequately known, though usually fairly easy to distinguish. The most useful distinguishing characteristic is the type of the neck holdfast. *L. bouvieri* (Quidor, 1912), *L. gracilis* (Wilson, 1919) and *L. edwardsi* have lobose holdfasts of two to four lobes; *L. caparti* (Nunes-Ruivo, 1962) appears to have a single bulbous holdfast; *L. laticervix* Hewitt, 1964 and *L. tripartita* (Wilson, 1935) have holdfasts of rather short, trifid outgrowths. In *L. cardusa* (Leigh-Sharpe, 1934), *L. kamoharai* (Yamaguti, 1939) and *L. cornuta* (Wilson, 1919) the holdfasts consist of long, irregularly branched processes, *L. cornuta* differing from the other two in having a particularly thick and profusely wrinkled cephalothorax. (Hewitt (1964b) pointed out, however, that such wrinkling might be nothing more than an artifact.)

Of the nine species, five parasitize fishes belonging to the family Macrouridae (Gadiformes) (*L. bouvieri*, *L. cardusa*, *L. edwardsi*, *L. kamoharai* and *L. laticervix*). Two species (*L. cornuta* and *L. gracilis*) are parasitic on synaphobranchid eels (Anguilliformes). *L. caparti* was found on *Epigonus* (Apogonidae) and *L. tripartita* on *Calamus* (Sparidae), both members of Perciformes.

Five species occur in the Atlantic (*L. bouvieri*, *L. caparti*, *L. edwardsi*, *L. gracilis* and *L. tripartita*), one of them (*L. bouvieri*) having been also recorded in the Indian Ocean, off the coast of the Sudan. *L. cardusa* is known only from the same ocean, Macassar area. Three species have Pacific distribution, two with records from Japan (*L. cornuta* and *L. kamoharai*) and one from New Zealand (*L. laticervix*). The genus is represented in the British fauna by its type species, *L. edwardsi*.

Lophoura edwardsi Kölliker, 1853

(Figs. 1451–1458)

Syn.: *Lophura edwardsi*; of Carus (1885) and auct.
Rebelula edwardsi (Kölliker, 1853); of Poche (1902)

Post-metamorphosis female (Fig. 1451): Cephalic region of cephalothorax sharply set off from remainder, broader posteriorly, with rounded posterolateral corners, small rounded anterolateral protuberances and ventromedial buccal swelling; rest of cephalothorax cylindrical or broader posteriorly, smooth or wrinkled, abruptly narrowing at junction with neck (Figs. 1452–1455). Latter cylindrical, less than $\frac{1}{2}$ thickness of cephalothorax, anteriorly with holdfast organ of two to four rounded or irregular processes of various lengths but always shorter than cephalothorax. Genital complex sharply delimited from neck, subquadrangular (Fig. 1451) or subcircular (Fig. 1456), with two longitudinal rows of depressions (four to seven pits in each) on dorsal surface; centre of posterior margin with perianal swelling on dorsal side (Fig. 1456) and two smaller swellings ventrolateral to it housing slit-like oviduct orifices (Fig. 1457). Lateral to perianal swelling posterior processes of structure typical of genus, with 16 to 18 cylinders on each process (Fig. 1458). Total length (without posterior processes) about 30 mm.

First and second antennae reduced to wart-like protuberances. Mandible unknown. First maxilla diminutive, tipped by short process. Second maxilla, maxilliped and swimming legs not discernible in post-metamorphosis female.

Male: As in generic diagnosis. Total length about 3 mm.

Comments: The scantiness of the description above reflects our insufficient knowledge of the morphology of this species. This insufficiency is due to the fact that the pre-metamorphosis stages of the female have not yet been found. Even mature adults must be considered rare. Nunes-Ruivo (1953) found only four specimens in 108 fish of the host species. Many specimens have been collected decapitated or mutilated, because the parasite is very deeply embedded in the tissues of the host and held fast by a strong connective tissue capsule, which makes it difficult to extricate it without inflicting some damage.

The usual site of attachment of *L. edwardsi* is the dorsum of the fish, commonly near the bases of fins, the neck of the parasite penetrating the musculature, past the vertebral column, between the ribs and often near the visceral cavity.

L. edwardsi appears to have a very limited range of hosts, being known mainly from the macrurid species, *Coelorinchus coelorinchus*. Brian (1908, 1912) reported it from *Macrurus atlanticus* and *Macrurus* sp., but the identity of these hosts cannot be considered incontrovertibly established. The recorded hosts are deep-water fish, all finds coming from depths between 200 and 500 m, with one specimen having been taken at 1,448 m.

The distribution of *L. edwardsi* covers the eastern North Atlantic and stretches as far south as the coast of Angola. The species was discovered in, and subsequently repeatedly recorded from, the western Mediterranean. There are no records from the American side of the Atlantic. In British waters, *L. edwardsi* was reported off the south-west coast of Ireland and west coast of Scotland.

Genus Tripaphylus Richiardi, 1878

Pre-metamorphosis female: Unknown.

Post-metamorphosis female: Cephalothorax subspherical or hexagonal with rounded corners, broader than wide; single short and pointed process on either side; cephalic area either slightly set off or indistinguishably fused with remainder. Neck sharply delimited from cephalothorax, round in cross-section, narrow anteriorly, gradually expanding posteriorly, imperceptibly passing into broad, flat trunk. Posterior processes simple, cylindrical. Egg strings straight, with round multiserial eggs. First antennae uniramous, second biramous. Mouth tube siphonostome. Mandible unknown. First maxilla apparently with three terminal papillae, second unknown. Maxilliped large, subchelate. Swimming legs absent.

Male: Cephalothorax narrow, oblong, somewhat inflated dorsally, separated from trunk by narrow transverse constriction. Trunk broader posteriorly, with rounded posterior extremity and cylindrical uropods. Antennae and buccal appendages as in female. Second maxilla and maxilliped fused at bases, subchelate. Swimming legs absent.

Type species: *Tripapylus musteli* (van Beneden, 1851).

Comments: The generic name first appeared in an anonymous publication (Anon., 1878) reporting on the proceedings of the Natural History Society of Toscania. The reporter used the name *Trypaphylum* and related it to van Beneden's (1851b) description of *Lerneonema musteli*, which Richiardi proposed to transfer to the new genus. When Richiardi (1880) used the name himself for the first time, he spelled it *Tripaphylus* and referred to it as a new genus. Although the 1878 spelling was validly published within the meaning of article 9 of the International Code of Zoological Nomenclature, this author prefers to use Richiardi's 1880 version because the earlier one was published anonymously and was never used by Richiardi. It can be construed as lapsus calami (article 32(a)(ii)), if only because one other name mentioned in the same paragraph is spelt incorrectly.

Until very recently, only the type species of this genus was known. The anonymous publication (Anon., 1878) mentioned that Richiardi had collected several intact female specimens, but no description of these specimens was ever published, as far as this author is aware. No other find of intact females has been made, all available material, like that examined by the author (Fig. 1459), consisting of decapitated specimens. Scott and Scott (1913) never saw the cephalothorax of *Tripaphylus*, but their description includes "head rounded and furnished with cartilaginous horns". This statement, based on an unknown source of information, proved impossible to verify, though it was repeated by several subsequent authors.

Kirtisinghe (1964) described a second species, *Trypaphylum hemigalei*, from the gills of *Hemigaleus balfouri* collected in the fish market of Colombo. He considered it a new species because its cephalothorax was nearly hexagonal, wider than long and because the specimens were smaller than those described in European waters. In the absence of a detailed description of the cephalothorax of *T. musteli*, however, it is not possible to make any valid comparison. The validity of *T. hemigalei* must await the discovery and careful description of the anterior end of *T. musteli*. In the meantime, *T. hemigalei* provides the best information available on the morphology of the genus *Tripaphylus*.

Tripaphylus musteli (van Beneden, 1851)

(Figs. 1459–1461)

Syn.: *Lerneonema musteli* van Beneden, 1851
Trypaphylum musteli; of Anon. (1878)
Lernaeenicus musteli; of Bassett-Smith (1899)

Post-metamorphosis female (Fig. 1459): Cephalothorax (*fide* Scott and Scott, 1913) rounded, with sclerotized horns. Neck narrow anteriorly, gradually widening in posterior direction, imperceptibly passing into genito-abdominal trunk. Latter dorsoventrally flattened, longer than broad, with rather uneven lateral margins slightly diverging posteriorly; posterolateral corners rounded, small tubercle in centre of posterior margin bisected by anal slit (Fig. 1460). Posterior processes cylindrical. Total length (without posterior processes) about 45 mm.

No reliable information available on appendages.

Male (Fig. 1461): As in generic diagnosis.

Comments: As mentioned in the comments on the genus, this species has never been fully described. More material is needed.

T. musteli is parasitic on the gills of *Mustelus*, a genus of sharks. The specific identity of the host is impossible to determine, since sharks of this genus have been commonly confused by many investigators. Hence the literature offers no clues as to which species are involved. The parasite is rare, its scanty records come from the area extending from British waters, along the western European seaboard, to Italy.

Family Lernaeopodidae

Morphology

A decisive influence on the morphology of Lernaeopodidae has been exerted by the adaptation of their second maxillae to fulfil the function of an "anchor chain" connecting the body of the parasite with the bulla, the anchor-like organ of attachment. The development of these appendages, often gigantic in comparative scale, has made them into one of the major parts in the structural plan. The other parts have developed in the wake of changes caused by this modification.

The body of a lernaeopodid female consists of three major parts: the cephalothorax, the second maxillae and the trunk. These parts are concerned with nutrition, prehension and reproduction, respectively. All three parts radiate outwards from a common meeting point at the base of the cephalothorax. Uniformly present throughout the family, the cephalothorax, second maxillae and trunk differ from one genus to another in the extent of their development and in their interrelationship. Taking these differences into account, one can recognize among Lernaeopodidae six variations of the structural plan. According to these variations, currently known lernaeopodid genera are arranged in Table 19 in six groups, from A to F.

Table 19. Lernaeopodid genera arranged in groups according to their structural plan.

Type	Genus
A	*Achtheres, Albionella, Basanistes, Cauloxenus, Coregonicola, Dendrapta, Lernaeopoda, Lernaeopodina, Ommatokoita, Pseudotracheliastes, Salmincola, Schistobrachia, Thysanote*
B	*Acespadia*
C	*Advena, Anaclavella, Brianella, Brachiella, Alella, Charopinopsis, Charopinus, Clavella, Clavellodes, Clavellopsis, Clavellistes, Clavellomimus, Eubrachiella, Neobrachiella, Nudiclavella, Proclavellodes, Pseudocharopinus, Tracheliastes*
D	*Clavellisa, Euclavellisa*
E	*Vanbenedenia*
F	*Nectobrachia*

What might be considered the basic structural plan is shown diagrammatically in Text fig. 61A. It is a lernaeopodid with a short cephalothorax (1), carrying buccal apparatus (2) and the two pairs of antennae at its tip. The cephalothorax is inclined ventrally to the long axis of the trunk (5). Powerful, contractile second maxillae (4), about as long as, or longer than the trunk, are attached at the base of the cephalothorax, near the point of its junction with the trunk. The maxillipeds (3), a pair of appendages which play an important part in prehension, are found on the ventral surface of the cephalothorax, posterior to the buccal apparatus. In small forms, the maxillipeds fill a good deal of space between the buccal cone and the bases of the second maxillae. A good example of this structural plan is shown in Text fig. 62A, representing *Salmincola edwardsii* (Olsson, 1869). (In this drawing the different parts of the body are numbered in the same way as in Text fig. 61.)

The genera built upon plan A are quite a heterogenous assembly. They include a group of freshwater parasites such as *Salmincola* Wilson, 1915, *Achtheres* von Nordmann, 1832 and *Basanistes* von Nordmann, 1832, the first two of them being represented in British fauna. Here also belongs a curious form, *Cauloxenus stygius* Cope, 1871, found on only one occasion on a blind, cave-dwelling fish in the U.S.A. Yet another genus constructed upon plan A is *Coregonicola* Markevich, 1936, a group of Siberian species parasitic on coregonid fishes and characterized by extreme elongation of their trunks. Some other species of type A have developed second maxillae of truly phenomenal length. (*Lernaeopodina longibrachia* (Brian, 1912), for example, has arms ten times longer than its trunk and swings on them like a fob on a chain.) The remaining genera of this group are all marine, parasitizing (with the exception of *Thysanote* Krøyer, 1864) various kinds of elasmobranch fishes.

The only example of structural plan B is *Acespadia pomposa* Leigh-Sharpe, 1925 (Text figs. 61B, 62B), one specimen of which was taken on a whale shark in the East China Sea. The difference between *Acespadia* and species with type A structure depends on the presence of a narrow cephalothoracic "neck", which in *Acespadia* is interposed between the broad base of the cephalothorax and the point of origin of the second maxillae. The anterior part of the cephalothorax is not unlike that occurring in type A; it is inclined ventrally to the long axis of the trunk, though the

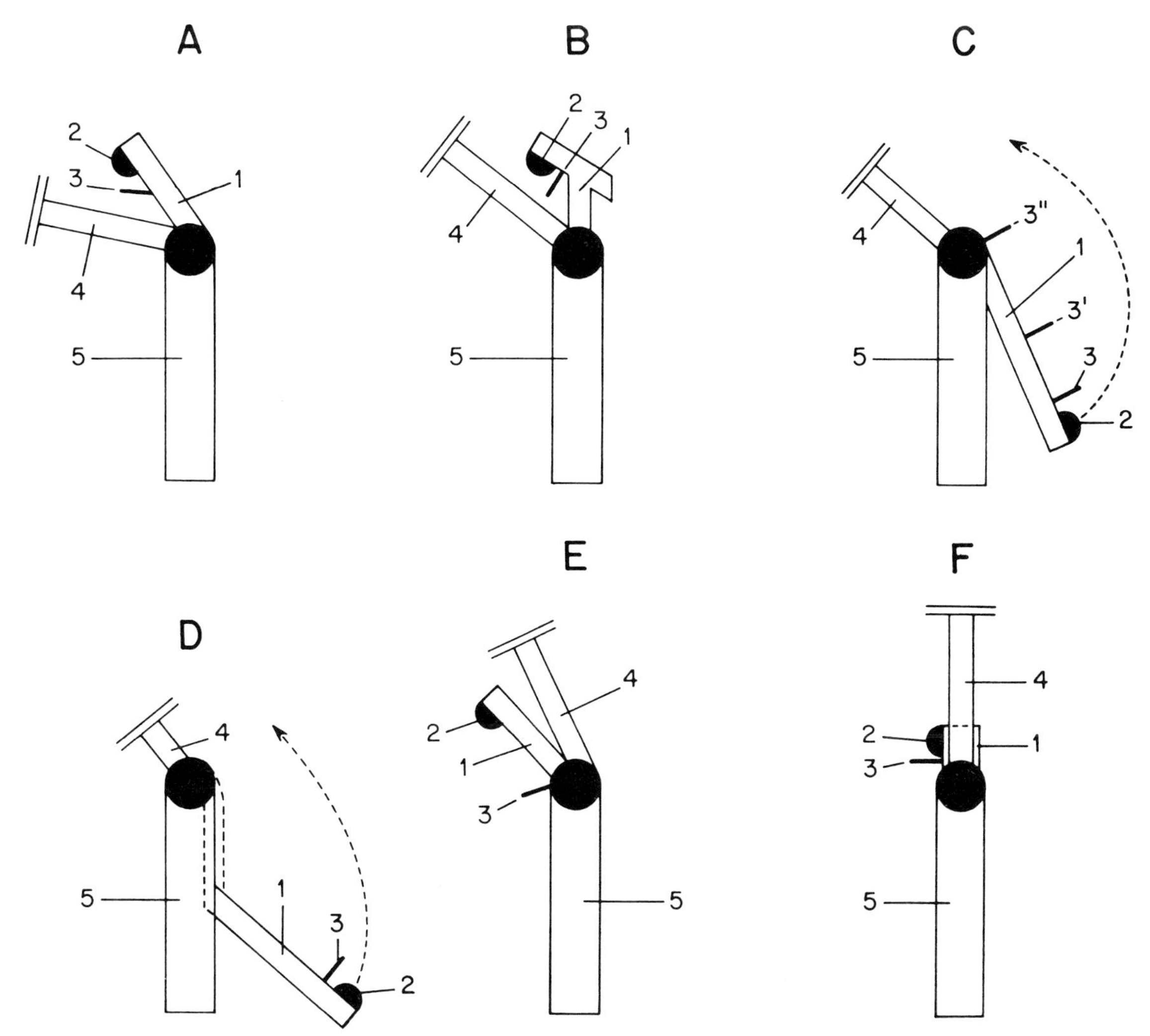

Text fig. 61. Structural plans of lernaeopodid females.
1 — cephalothorax; 2 — buccal apparatus; 3,3′,3″ — maxilliped; 4 — second maxillae; 5 — trunk. (See also Text fig. 52)

"neck" itself might be somewhat deflected towards the dorsal side, as shown in Leigh-Sharpe's drawing (Text fig. 62B).

A large group of genera, all marine except *Tracheliastes* von Nordmann, 1832, is built upon the plan shown in Text fig. 61C. Like all the others, this plan can be derived from plan A, from which it differs mainly in the length of the cephalothorax. The latter has become cylindrical and long, the length differing from one genus to another. The buccal region and its entire apparatus, as well as the antennae, have been removed, by this process, far from the base of the cephalothorax to a point near the apex. It appears as if the short neck of *Acespadia* continued to elongate until it reached or exceeded the length of the trunk. It is quite common for lernaeopodids of type C, when at rest, to carry their cephalothoraces reflected along the dorsal surface of the trunk, as shown by the drawing of *Neobrachiella chevreuxii* (van Beneden, 1891) in Text fig. 62C. The long, cylindrical cephalothorax, however, is equipped with powerful musculature and is capable of executing complicated movements in all directions. In some genera it is regularly found in the anterior position, in line with the long axis of the trunk. In others it curves in a ventral direction over the bases of the second maxillae.

Genera belonging to type C show many variations in the proportions of their three components. This is particularly true for the second maxillae (4). Some genera (e.g. *Brachiella* Cuvier, 1830, *Charopinus* Krøyer, 1863, *Pseudocharopinus* Kabata, 1964) include species with second maxillae of moderate length, about the same as the trunk. Other species have these appendages of greater length, although they do not attain that of some species of *Lernaeopodina* Wilson, 1915. At the other end of the scale there are genera with second maxillae much reduced in length (e.g. *Clavellodes* Wilson, 1915, *Alella* Leigh-Sharpe, 1925) or completely reduced (*Clavellopsis laciniata* (Krøyer, 1864)).

Another interesting variable in the structure of type C genera is the position of the maxillipeds (Text fig. 61C, *3*). The most common position of these appendages is shown by 3. In the genera *Charopinus* and *Brianella* Wilson, 1915, however, the process of elongation of the cephalothorax has left them, as it were, halfway behind, as shown by 3′ in Text fig. 61C. In *Tracheliastes* they remain near the bases of the second maxillae (3″).

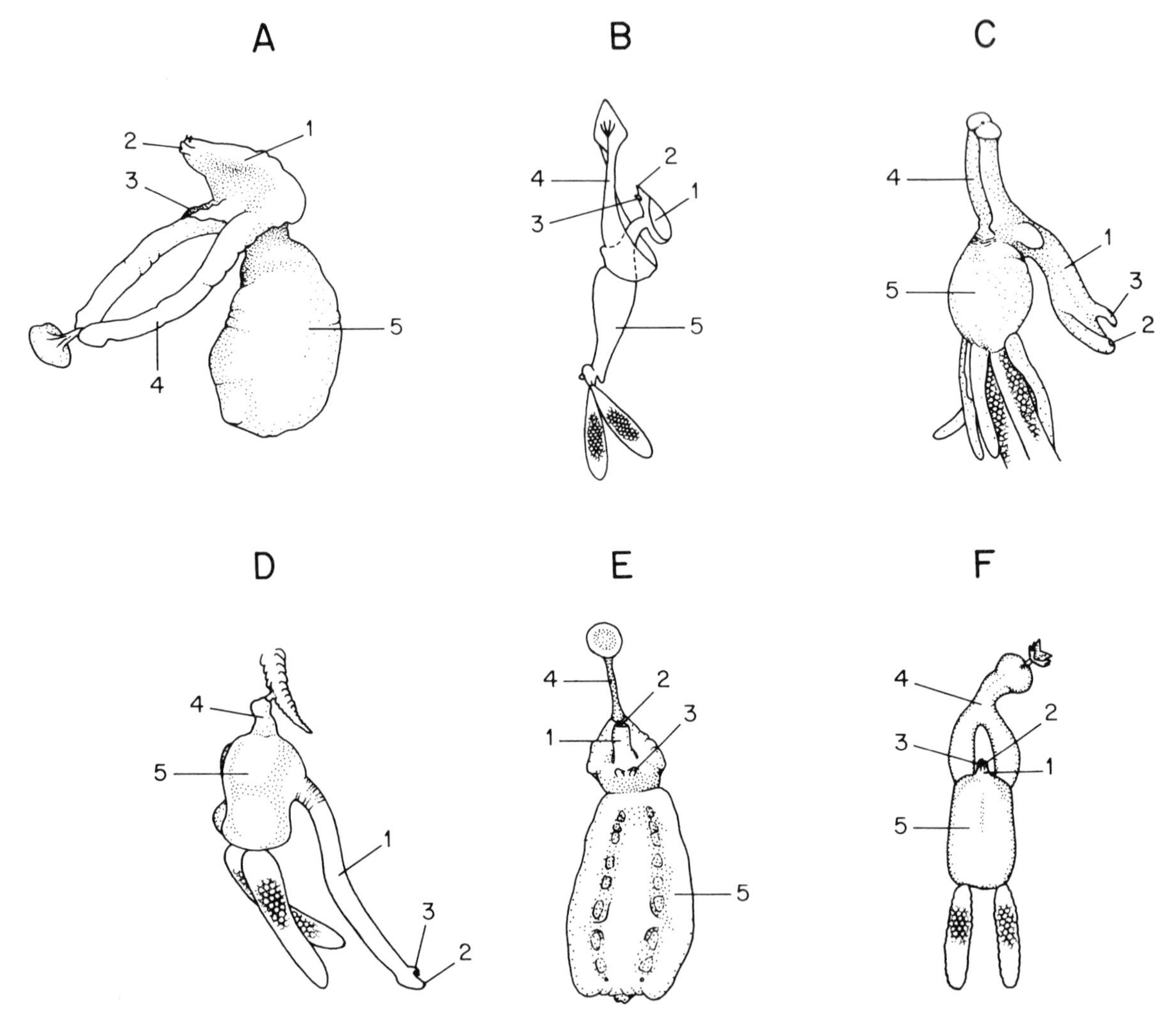

Text fig. 62. Examples of structural plans of lernaeopodid females shown in Text fig. 61. A. *Salmincola edwardsii*; B. *Acespadia pomposa*; C. *Neobrachiella chevreuxii*; D. *Clavellisa emarginata*; E. *Vanbenedenia chimaerae*; F. *Nectobrachia indivisa*. Lettering as in Text fig. 61.

Structural plan D (Text fig. 61D) is represented by only two genera, *Clavellisa* Wilson, 1915 and *Euclavellisa* Heegaard, 1940, the former alone occurring in British fauna. In these parasites, the dorsally deflected cephalothorax appears to have fused with and merged into the dorsal surface of the trunk, or otherwise has become separated from its more normal position near the base of the second maxillae. The cephalothorax of the lernaeopodids of type D is one of the longest in the family, being usually much longer than the trunk, as exemplified by *Clavellisa emarginata* (Krøyer, 1837) (Text fig. 62D), and being capable of almost snake-like movements. The second maxillae of *Clavellisa* and *Euclavellisa* are short, partially fused and often covered by a common cuticular sheath.

Plan E (Text fig. 61E) is represented by three curious and scarce lernaeopodid species, grouped in the genus *Vanbenedenia* Malm, 1860 (Text fig. 62E) and parasitic on chimeroid fishes. Although it appears to bear a close resemblance to type A, the differences between types A and E are, in fact, greater than the diagrams can demonstrate. *Vanbenedenia*, like the genera of type C, has a cylindrical cephalothorax, though it is rather shorter than in that type. Unlike any other genus, it has its cephalothorax permanently positioned on the *ventral* side of the second maxillae, which form a yoke above it and prevent it from passing between them to the dorsal side. The buccal region of *Vanbenedenia* is retractable and often withdraws out of sight. This unique feature on occasions has led to assertions that the adults of *Vanbenedenia* are devoid of mouthparts. Another feature characteristic of type E and shared by it only with *Tracheliastes* (type C), is the general configuration of the base of the cephalothorax. Whereas in most genera this can be compared with the structure seen in *Salmincola* (Text fig. 62A), here the base forms a kind of "torso" anteriorly dividing into three parts (cephalothorax proper and two second maxillae) and posteriorly narrowing into a distinct waist.

Structural type F (Text fig. 61F) is also represented by a single genus, *Nectobrachia* Fraser, 1920. Its morphological uniqueness gained it Markevich's (1946) recognition as a type of a subfamily of Lernaeopodidae, Nectobrachiinae. The hallmark of the type is the cephalothorax of *Nectobrachia* (Text fig. 62F), almost completely absorbed into one structural mass with the trunk, from which it emerges only as a small cone situated between the bases of the powerful, partially fused second maxillae.

Let us now review the three component parts of lernaeopodid structure in some detail, beginning with the cephalothorax. A tagma consisting of five cephalic segments and one thoracic, the cephalothorax carries on its ventral and anterior surfaces the following appendages: the first antennae, the second antennae, the mandibles, the first maxillae, the second maxillae and the maxillipeds. There are no traces left of the original segmentation. In the process of fusion, the appendages of the cephalothorax have remained in their original segmental sequence, with the exception of the maxillipeds and second maxillae. Due to the powerful development and specialization of the second maxillae and the process of elongation of the cephalothorax, these two pairs of appendages are either at the same level (e.g. in *Tracheliastes* and *Vanbenedenia*), or have reversed their positions, so that in many genera the maxillipeds are anterior to the second maxillae.

As stated earlier, the shape of the cephalothorax differs, sometimes extensively, from one structural plan to the next. The differences in the structure of its apex are, however, insignificant. With the exception of *Nectobrachia*, in which this tagma is represented by little more than a terminally located mouth cone, the buccal apparatus is subterminal in position and its component appendages are arranged in much the same way. This is exemplified by the buccal region of *Salmincola carpionis* (Krøyer, 1837), as shown in Text fig. 63. The centre of the region is occupied by the mouth cone (mc). Ventrolateral to it, quite close to the walls of the cone, are the first maxillae (mx1). Immediately dorsal to them are the bases of the mandibles. These appendages, however, are inclined in a medial direction and are lost from sight, having passed inside the mouth cone. In Text fig. 63 we can see their tips (mdb) protruding towards the buccal opening from within the cone. The latter is often situated on a slight eminence. Dorsal to this eminence is another one, the frontal dome, separated by a transverse groove. The frontal dome forms the anterior margin of the cephalothorax. The lateral ends of the groove between the two are occupied by the bases of the second antennae (ant 2), appendages which might point straight forward, as they do in *Achtheres*, or be folded, when at rest, across the anterior margin of the cephalothorax, as they often do in many more advanced genera. Dorsomedial to the second antennae and roughly dorsal to the mouth cone are the first antennae (ant 1). (It must be remembered that the relative positions of the individual appendages are not infrequently affected to some extent by fixation, by distension of the cuticle associated probably with the process of partial moulting and by the mobility of the appendages themselves.) All the appendages, mouth cone included, are moved by extrinsic musculature. The muscles which operate these appendages cross the cephalothorax obliquely in a dorsal direction and are attached to the underside of the dorsal shield, a plaque of thick cuticle which covers the dorsal side of the anterior part of the cephalothorax. All Lernaeopodidae (again with the exception of *Nectobrachia*), have this shield, though it is not always clearly developed and recognizable. In some species the shield is nothing more than an area of thicker cuticle without clearly delimited boundaries. In others one can recognize the lateral limits by the presence of robust, slightly raised ridges (often beginning at the bases of the second antennae), the posterior limit remaining relatively ill-defined. In others still, all the borders are clearly outlined. The outlines of the shield vary from one species to another and can be

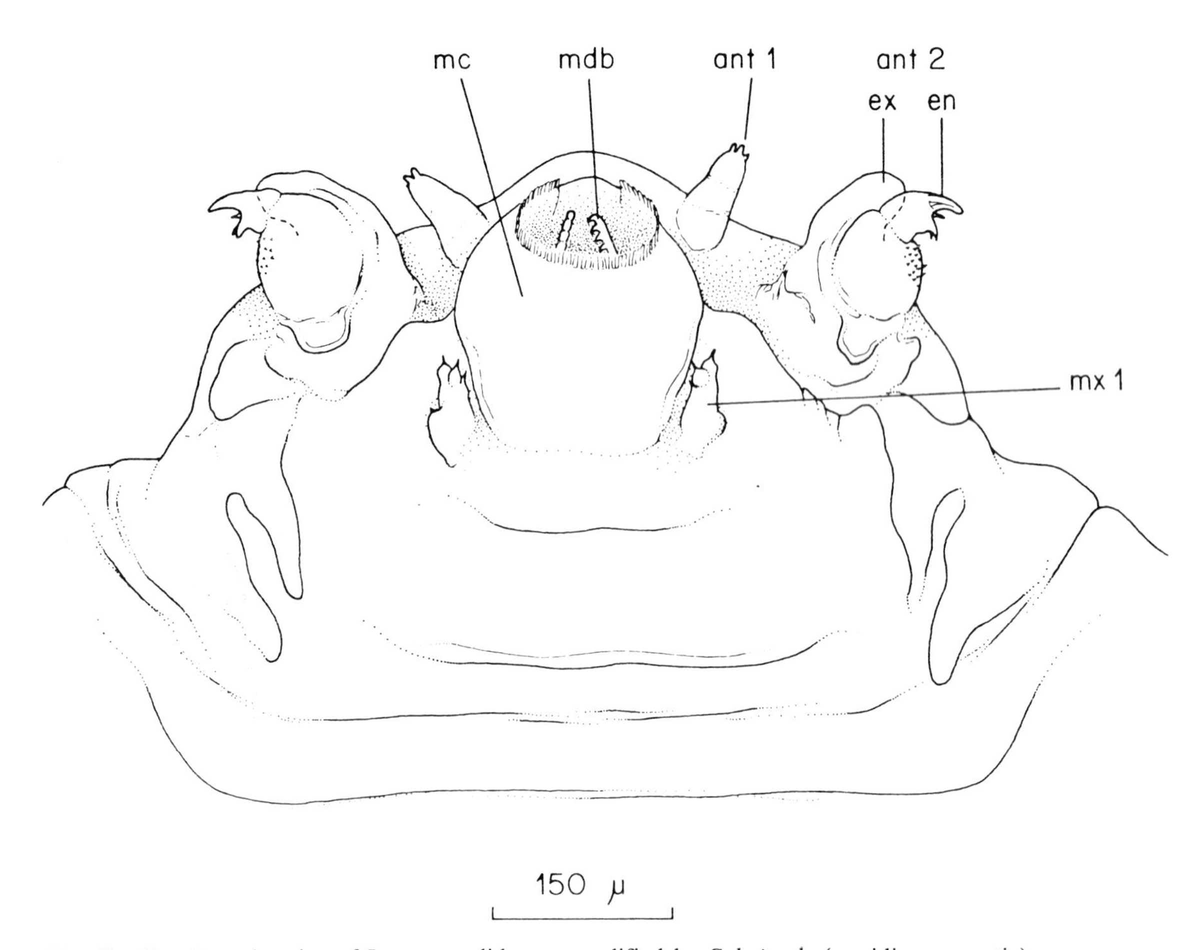

Text fig. 63. Buccal region of Lernaeopodidae, exemplified by *Salmincola* (semidiagrammatic).
ant1 — first antenna; ant2 — second antenna; en — endopod; ex — exopod; mc — mouth cone; mdb — mandible; mxl — first maxilla.

used sometimes as a diagnostic taxonomic characteristic. In some species (e.g. *Clavellodes rugosa* Krøyer, 1837) (Fig. 1916), the posterior margin is transverse and cuts right across the cephalothorax. In others (e.g. *Neobrachiella chevreuxii* (Fig. 1797)), the shield looks like a wedge tapering posteriorly and with a suggestion of subdivision along the mid-dorsal line. In the cephalothoraces of type A the shield extends along much of the dorsal surface and usually has no clearly demarcated margins, whereas in those of type C it covers only the area bearing the appendages ventrally. In these species the shield can sometimes appear wider than the cylindrical cephalothorax and create an impression of a distinct "head" carried at the end of a long "neck". The anterior margin of the shield is usually straight (it might be slightly concave or convex) and transverse. In *Clavellodes rugosa* (Fig. 1916), however, the anterior margin of the shield is tripartite, the central part protruding farther forward than the two shorter lateral parts.

In species belonging to type C, the cephalothorax is always a cylinder of fairly uniform width, ending in an enlarged head or tapering slightly and with appendages quite inconspicuous on the ventral surface. It is usually smooth, though in *Clavellodes rugosa* it is covered with a profusion of transverse wrinkles. In *Eubrachiella* Wilson, 1915 and some species of *Neobrachiella* n. gen. (e.g. *N. impudica* (von Nordmann, 1832) (Fig. 1779)), a transverse welt runs across the dorsal side of the cephalothorax, near its base. In lernaeopodids built upon plan A, the cephalothorax is an oval or triangle, flattened in a dorsoventral direction, with smooth outlines marked anteriorly by the protruding second antennae (e.g. Fig. 1489). Species of the genus *Basanistes* have humps on the dorsal side of the cephalothorax, imparting to them a peculiar appearance, unique among the members of the family.

The configuration of the area in which the three structural components meet (marked by a black circle in Text fig. 61) can be of four different types. The most common is that found in species of type A, in which the second maxillae are laterally attached to the broad base of the cephalothorax and immediately posterior to them the beginning of the trunk narrows into a waist-like groove. The second most common type is exemplified by some species of type C, such as *Clavella alata* Brian, 1906 (Fig. 1890) or *Neobrachiella chevreuxii* (Fig. 1796). In this configuration the second maxillae and cephalothorax are stretched out along one line, with the trunk arranged at an angle to them. Here also there is often a waist-like demarcation between these two tagmata and the trunk. As mentioned earlier, in the third type, exemplified by *Vanbenedenia* (Fig. 1553) and *Tracheliastes* (Fig. 1517), the base of the cephalothorax can be likened to a torso, a broad, flat area, anteriorly giving rise to the narrow cylinders of the "neck" and the second maxillae, and posteriorly pinched in by the usual constriction. The fourth type is found only in *Nectobrachia*, where the mouth cone and the second maxillae appear to arise directly and separately from the trunk.

The second maxillae, constituting the next major structural element in the morphology of Lernaeopodidae (Text figs. 61 and 62, *4*), in most species of the family are modified into a pair of cylindrical appendages and vary from one species to another mainly in their length and extent of fusion. These two variable characteristics suffice to introduce a great variety into the "arms" of Lernaeopodidae. As mentioned earlier, they can be so long as to exceed the length of the trunk by many times (e.g. in *Lernaeopodina longibrachia*) or they can be completely incorporated into the anterior part of the trunk (e.g. in *Clavellopsis laciniata*). In most lernaeopodids the second maxillae are separate from each other, meeting only at the tips where they are permanently joined to the bulla, the organ of attachment. In some species, however, they are fused completely (e.g. in *Brianella corniger* Wilson, 1915) or partly (e.g. in *Nectobrachia indivisa* Fraser, 1920). On the posterior side of the second maxilla in many species a nipple-like opening of an excretory duct can be seen near the base, the nipple sometimes becoming quite prominent (e.g. in *Clavella alata*, Fig. 1891). In the genus *Alella* (Fig. 1978) the bases of the second maxillae form lateral swellings, from which the genus derives its name. Species of the genus *Thysanote* are distinguished by the presence of fimbriate processes at about midlength of their second maxillae. These processes differ from species to species in number, length and grouping; they can also vary with age. Pronounced intraspecific variations occur in similar, though single, processes in *Neobrachiella impudica* (Figs. 1790–1792), to the confusion of taxonomists. Structural modifications occur also at the tips of the second maxillae. In *Neobrachiella insidiosa* (Heller, 1865), for example, the tips are divided into digitform lobes (Fig. 1816), whereas in *Clavella stellata* (Krøyer, 1838) subspherical swellings are carried in this position (Fig. 1904). In several species, in the so-called *Charopinus*-group, the tips have evolved into attachment organs, more or less elaborate, taking over from the bulla, the more usual device of attachment. In British fauna this type is represented by *Schistobrachia ramosa* (Fig. 1598).

The bulla, an organ of attachment unique to Lernaeopodidae, is formed in the frontal region of the head and extruded in a complicated process, followed by implantation in a cavity scooped out in the tissues of the host, transfer from the head to the tips of the second maxillae and inflation within the host tissues. (For details of this process, see Kabata and Cousens, 1973).

The structure of the bulla of *Pseudocharopinus dentatus* (Wilson, 1912) is shown in section in Text fig. 64. The two main parts of this organ are the anchor and the manubrium (c). The latter is perforated by two parallel ducts (k) which receive the terminal plugs of the second maxillae (e) and convey fluids produced by glandular cells associated with these appendages. The fluids pass through the manubrium into the anchor; their further fate is not known with any exactitude. In lernaeopodids parasitic on freshwater fishes the manubrial ducts are in communication with each other via a broad, looped duct running through the anchor. Alternate contractions of the second maxillae propel the secretions to and fro through the bulla. This process seems to facilitate some exchange between the parasite and the fish, but the nature of that exchange is still unknown.

In lernaeopodids parasitic on marine teleosts, the manubrial ducts break up into numerous capillary vessels distributed usually around the periphery of the bulla. In those parasitic on elasmobranchs the ducts end blindly, communicating directly with the core of the bulla (Text fig. 64, h) which can be either fibrous with interstitial spaces, as in *Pseudocharopinus dentatus*, or granular and porous as in *Lernaeopodina longimana* (Olsson, 1869).

The shape of the bulla varies from species to species. The most common one is that shown in Text fig. 64, but some bullae have much elongated manubria (e.g. in *Vanbenedenia chimaerae* (Heegaard, 1962)), and in others clear division between the two parts has been lost. This can be due to the

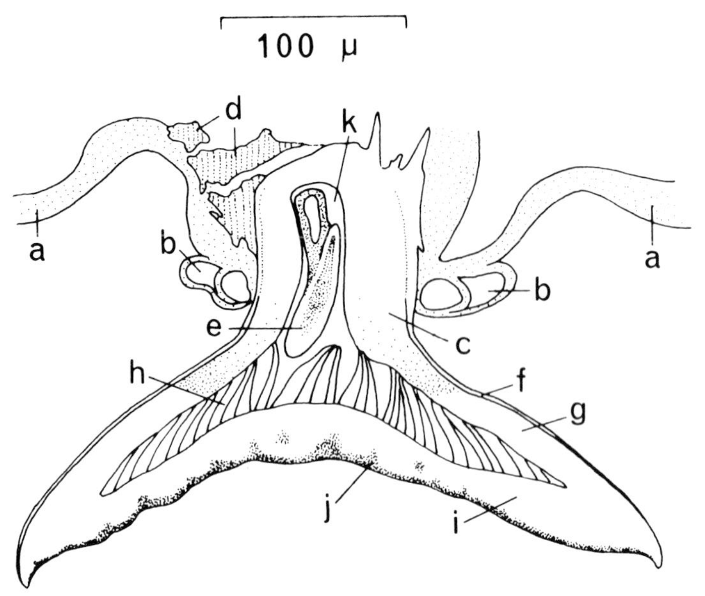

Text fig. 64. Section through the bulla of *Pseudocharopinus dentatus* (diagrammatic). From Kabata and Cousens (1972).

a — cuticle of second maxilla; b — circummanubrial ring; c — manibrium; d — cement substance; e — maxillary plug; f — pellicle; g — supra-anchoral layer; h — central layer; i — subanchoral layer; j — accumulation of secretion; k — manubrial duct.

reduction either of the anchor, as in *Salmincola thymalli* (Kessler, 1868) (Fig. 1487), or of the manubrium (e.g. in *Ommatokoita elongata* (Grant, 1827)) (Fig. 1533). The anchor can also vary in shape quite extensively. *Salmincola stellatus* Markevich, 1936, has a stellate bulla, *Coregonicola baicalensis* Koryakov, 1951, an irregular one, *Nectobrachia indivisa* a palmate one and *Neobrachiella robusta* (Wilson, 1912) a cup-shaped one with circular flanges.

The species of the *Charopinus*-group tend to lose their bullae. Often, however, a trace of it remains, as for example a small link connecting the second maxillae of *Schistobrachia ramosa* (Krøyer, 1864), or a diminutive bulla of the young adult *Dendrapta cameroni* (Heller, 1949) (cf. Kabata and Cousens, 1972). In the latter species, the bulla is discarded with age and replaced by the dendritic branches of the appendages themselves.

The trunk, the third major part of the lernaeopodid body, is associated mainly with reproduction, though it also accommodates the alimentary canal. In most, perhaps all, Lernaeopodidae it consists of four recognizable parts. The first three of them represent three thoracic segments, whereas the posterior one is a complex consisting of a thoracic segment and no longer recognizable vestiges of abdomen. Externally there are no visible signs of the divisions between these four parts. The trunk musculature, however, still retains its segmental arrangement. Contractions of longitudinal trunk muscles throw into folds the cuticle between points of muscle attachments. Grooves in the trunk surface that appear in consequence have been occasionally interpreted as external segmentation.

The trunk is usually flattened in a dorsoventral direction, the degree of flattening varying from one genus, and often from one species, to another. These differences can be due to the methods of preservation and post-mortem changes, to the degree of the parasite's maturity or to its individual characteristics. Some of them, however, are specific and although the shape of the trunk has to be used cautiously in the taxonomy of Lernaeopodidae, it can, none the less, be so used.

In the genus *Coregonicola*, the species of which are parasitic on coregonid fishes of the Palearctic, the trunk has become extremely elongated. The members of this genus are usually found upon the flanks of their fish hosts and their elongation might be viewed as an adaptation to living in a constant and rapid stream of water. It has been found (Kabata, 1965d) that the trunk of *Coregonicola* owes its length to the elongation of one segment only, the first segment of this tagma. Other Lernaeopodidae have long trunks, but none as long as that of *Coregonicola*. In most of them, however, (e.g. *Vanbenedenia kroeyeri* Malm, 1860 (Fig. 1553)), all component parts contribute more or less equally to the length of the tagma. Some species of type C have trunks of orbicular shape or even trunks broader than long. This tendency has reached its extreme in the genus *Clavellisa*. In certain species of this genus the trunk appears somewhat bilobed, being slightly shorter in mid-dorsal line than along its lateral margins.

The trunk cuticle is smooth and seems to be uniformly thick. In *Tracheliastes* (Fig. 1517) and *Venbenedenia* (Fig. 1553) one finds, on the dorsal and ventral surfaces, two rows of depressions containing thicker, scale-like cuticle with a slightly corrugated surface. These "crypts" provide points of attachment for the dorsoventral musculature of the trunk.

In most Lernaeopodidae the lateral margins of the trunk are straight and more or less parallel to each other. On occasions, however, their outlines are sinuous, suggesting segmental subdivisions (e.g. *Tracheliastes* (Fig. 1517)). Like the folds of the ventral and dorsal walls, these wavy outlines are affected, if not caused by, the state of contraction of the longitudinal musculature of the trunk and cannot be considered as a permanent feature. In *Neobrachiella triglae* (Claus, 1860) (Figs. 1846–1848) the lateral margins are made more complicated by the presence of several rounded protuberances, a feature even more prominently developed in the South African species, *Eubrachiella sublobulata* Barnard, 1955. The number and shape of these protuberances vary somewhat from one individual to the next, but can be considered as a specific constant. Perhaps the most peculiar modification of the lateral margins occurs in *Thysanote multifimbriata* Bassett-Smith, 1898. On the posterior corners of its flanks there are two short, cylindrical processes, branching into several long filamentous whips resembling African fly-swatters. Lateral prominences feature also on the lateral margins of species belonging to the genus *Basanistes*, a freshwater genus occurring in the Palearctic.

The most complicated, and structurally the most interesting, part of the trunk is its posterior extremity. It carries the following structures: anus, genital process, genital orifices and posterior processes.

The anal slit, always found in the centre of the posterior extremity and usually slightly displaced dorsalwards, marks the area of the obsolete abdomen. No true and separate abdomen is present in any member of the family. In some species of the genus *Achtheres* the central part of the posterior extremity protrudes backwards in the shape of an irregular cone, sometimes with a suggestion of subdivision. This cone (Fig. 1501) has been considered as abdomen by some authors (Wilson, 1915). In all Lernaeopodidae the openings of the vaginal ducts are normally associated with the perianal area, being placed somewhat ventral to the anal slit. They might or might not be at the tip of a special process, but their normal position (also in related families) is in front of the abdomen, at the posterior limit of the genital segment. In *Achtheres* these openings are at the tip of the so-called abdomen ventral to the anal slit, indicating by analogy that the area is composed of the reduced abdomen as well as part of the genital segment.

The name abdomen has also been used in the past for the genital process. Since, however, it is traversed by ducts which open at its tip and lead from it to the seminal receptacle, it cannot correspond with a true abdomen. The genital process is a cylindrical or clavate outgrowth located in the centre of the posterior margin of the trunk and immediately ventral to the anal slit. It varies in length from one species to the next. In *Salmincola salmoneus* (L.), for example, it is not more than a nipple-like protrusion with two minute openings at its tip (Figs. 1474, 1475). In *Clavella adunca* (Strøm, 1762) it is a fairly short, usually club-shaped, though sometimes subspherical structure (Figs. 1870, 1871). In *Neobrachiella chevreuxii* it is quite long and may equal the length of the posterior processes (Figs. 1795, 1796). In some species the size of the genital process varies with age. For example, it is seldom found in old females of *Salmincola salmoneus* (L.). In *Neobrachiella impudica* it may be absent altogether (Fig. 1790), or short (Fig. 1791), or even moderately long, causing some inconvenience to taxonomists. In many lower genera of the family there is no genital process at any stage. In members of these genera, genital orifices are found directly in the ventral surface of the posterior end of the trunk, a little in front of the anal slit.

The openings of the oviducts are usually found in the posterolateral corners of the trunk or in their immediate proximity. Externally they are slits with heavily reinforced rims, their outlines suggesting the existence of interior structures involved in oviposition. The slits run in a dorsoventral direction, sometimes converging slightly at their ventral ends. Their position can be immediately located in the ovigerous females, because they provide the points of attachment for the egg sacs, which are always conspicuous. The position of the oviduct opening in relation to the posterolateral corner of the trunk varies with the maturity of the female. In young specimens they tend to be nearer the lateral limits of the posterior extremity, but as the trunk becomes distended with eggs its posterolateral corners tend to round out and extend further from the positions of the oviduct openings.

In addition to the genital process, the posterior extremity of the lernaeopodid trunk is often equipped with outgrowths to which is applied the general term of posterior processes. These processes are always paired and vary in length, position and number. Three types are distinguished: modified uropods, posterior processes proper and fimbriate processes. The fact that posterior processes arising on either side of the perianal region are modified uropods has been pointed out by Nunes-Ruivo (1966a). She showed that some of them have even retained short setae of the original armature and resemble the uropods of the male. The origin of these processes is well illustrated by the situation in *Clavellisa emarginata* (Fig. 1949). Kabata and Gusev (1966) demonstrated the existence of vestigal uropods in *Eubrachiella gaini* (Quidor, 1912), a member of a genus previously believed to have no posterior processes of any kind. By following a series of specimens of increasingly older age, they traced the modification of the uropods to the condition of barely noticeable swellings on the posterior margin of the trunk.

In species of the genus *Charopinus* and related genera the uropods are fairly long (Figs. 1567 and 1578), and often remain equipped with several setae even after reaching the definitive stage of their development. In *Dendrapta cameroni longiclavata* Kabata and Gusev, 1966, a parasite of rays in the North Pacific, the length of the uropods equals or even exceeds that of the trunk.

There exist two groups of Lernaeopodidae distinguished by the position of their uropods. In one group the uropods are ventral to the egg sacs (or genital openings, when egg sacs are not present), in the other they are dorsal. A cursory examination shows little difference between, for example, *Lernaeopoda* Blainville, 1822, and *Lernaeopodina* (Wilson, 1915), except for the fact that in the former the uropods are ventral, and in the latter dorsal, to the egg sacs. As will be mentioned later, this fact is of great help in determining at least some of the interrelationships within the family.

In contrast to the uropods, the posterior processes proper are entirely new in origin. They are cylindrical, digitiform or subspherical extensions of the posterior extremity of the trunk. In *Neobrachiella albida* (Rangnekar, 1956), they are flat and foliaceous. It is worth pointing out that these processes, contrary to the illustrations of some authors, are mere outpushings of cuticle confluent with the lumen of the body cavity, and not articulating segments, distinct from the rest of the trunk. An important, if disconcerting, characteristic of the posterior processes is their potential variability. Its extent varies enormously and must be considered as a distinctly specific feature. Some species might be rigidly uniform in this respect. *Neobrachiella robusta*, on the other hand, comprises forms without posterior processes, as well as others with these processes very well developed (Kabata, 1970d). An assessment of the extent of their variability is possible, however, only when abundant material becomes available to the examiner, a condition only rarely fulfilled.

The fimbriate processes are a modification of posterior processes known only in the genus *Thysanote*. Most of them are branched fairly irregularly, either dichotomously or not. Intraspecific variation of branching is common. The fimbriate processes achieve different degrees of complexity in different species of the genus, but their most abundant development has been reached in *Thysanote fimbriata* (Heller, 1865), in which they are fused to form a ragged-hemmed skirt. A similar skirt is present in *T. lobiventris* (Heller, 1865), differing from the one of the previous species in that it is formed of four, not two, groups of coalesced processes.

It is interesting to review the occurrence of the posterior processes throughout the family. Modified uropods, as well as posterior processes proper, are invariably absent from the adults of its freshwater members. Genital processes, when present, are also usually poorly developed in this branch of the family. Lernaeopodids parasitic on elasmobranchs have no posterior processes proper, but their uropods are usually well developed and can be either dorsal or ventral in position.

The parasites of marine teleosts can be divided into two groups of genera, sometimes referred to as the *Clavella* and *Brachiella* groups respectively. The former group comprises genera with usually poorly developed or absent posterior processes. Exceptions are provided by two monotypic genera, *Clavellopsis* Wilson, 1915, and *Advena* n. gen. (Figs. 1996–1998), in which posterior processes proper are well developed. *Clavellisa*, another member of the group, includes species with vestigial uropods on the dorsal side of the anal tubercle (Fig. 1949).

The posterior processes are most highly developed in the *Brachiella*-group. As mentioned earlier, the number and size of these processes vary from species to species. Some have only modified uropods, often poorly developed and in most instances ventral to the egg sacs. (The author knows only one exception, *Neobrachiella robusta*, with uropods dorsal to egg sacs. There might be a few more.) Other species, in addition to uropods have a pair of posterior processes of varying lengths in the posterolateral corners of the trunk. In yet other species there are two, three or even four pairs of such processes, frequently of different lengths. *Neobrachiella suplicans* (Barnard, 1955) is unique in having a pair of foliaceous processes which branch into two quite close to their bases. Spherical processes occur in *Eubrachiella sublobulata* and *Neobrachiella papillosa* (Pearse, 1952). Finally, to this group belongs also the genus *Thysanote* with fimbriate posterior processes.

As regards the egg sacs of Lernaeopodidae, they are of fairly uniform structure throughout the family. They are most commonly cylinders of varying lengths, straight or slightly curving, filled with multiple rows of subspherical eggs. *Clavella gracilis* Hansen, 1923, has only two rows of eggs in its sacs and the recently discovered female of *Clavellistes lampri* (Scott and Scott, 1913) is unique in having uniserial egg sacs (Fig. 1713). Some species of the genus *Clavellisa* have short, almost spherical egg sacs.

Fully developed female Lernaeopodidae have lost almost all traces of swimming appendages (exception being provided by minute setiform vestiges in some species of *Clavellisa*). Other appendages will be reviewed in the section dealing with the phylogeny of this family.

The two most important facts about the lernaeopodid males, their diminutive size (particularly in the *Clavella*-group) and ephemeral life-span, contribute to the difficulties one encounters in the study of their morphology. For almost a quarter of the currently known genera of Lernaeopodidae no males have yet been described.

Even without knowing the morphology of the lernaeopodid males, however, one could advance some fairly accurate suggestions about it. The male does not have to possess structures designed to maintain permanent attachment to the host. Its attachment to the female is aided by its own small size. There is, therefore, no need for special modifications of the prehensile appendages. The brevity of its life makes it unnecessary for the male to possess systems of organs normally employed in nutrition. Hence reduction of the alimentary canal (in the *Clavella*-group the gastro-intestinal tract is a blind sac and there is no functional anus) and some simplification of the buccal apparatus. The reproductive effort of the male is usually concluded with a single act of fertilization (some males fertilize more than one female), for which it uses one batch of spermatophores. There is no sustained effort, similar to that involved in the production of eggs and egg sacs by the female. Consequently, one does not expect hypertrophy of the reproductive part of the body and formation of a massive trunk.

In short, the morphology of the male can be envisaged as being roughly similar to that of the female prior to the attainment of the definitive stage of development. The male simply stops short of the "regressive metamorphosis". In many respects, therefore, the structure of the male can be seen as more primitive than that of the female, except for local specializations required for the performance of exclusively male functions.

Males of the family Lernaeopodidae can be divided into three types, as shown in Table 20.

Table 20. Division of lernaeopodid genera according to the structure of the male.

Type	Genus
A	*Achtheres, Albionella, Brachiella, Charopinopsis, Charopinus, Clavellistes, Dendrapta, Eubrachiella, Lernaeopoda, Lernaeopodina, Neobrachiella, Ommatokoita, Pseudocharopinus, Salmincola, Schistobrachia, Thysanote, Tracheliastes*
B	*Advena, Alella, Clavella, Clavellisa, Clavellodes, Clavellomimus*
C	*Euclavellisa, Nectobrachia*

Type A, of which there are several variants, is exemplified by the male *Salmincola* (Fig. 1485). Its body is divided into anterior cephalothorax and posterior genito-abdominal complex, the counterpart of the female trunk. Both parts are of about the same size and are in line with each other, so that the male is straight, or nearly so. The cephalothorax resembles that of the female and its ventral surface is equipped with the same series of appendages. The trunk is devoid of appendages, except for uropods and genital folds, covering the orifices of the seminal ducts and situated ventrally near the posterior extremity. All known males of freshwater Lernaeopodidae (*Achtheres, Tracheliastes*) resemble that of *Salmincola* very closely and it is possible that those still undiscovered also share this similarity.

Another variant of type A is shown by the male of *Lernaeopoda*, a genus parasitic on elasmobranchs (Figs. 1662, 1685). Essentially similar to males of the freshwater Lernaeopodidae, the male *Lernaeopoda* is distinguished by a more inflated cephalothorax which is usually inclined towards the ventral side at a slight angle (this is often affected by fixative and subject to post-mortem variations). Another feature of this male is grossly enlarged uropods, usually deflected forwards from their point of origin, their tips diverging (Fig. 1663). Also characteristic is the presence of one or two pairs of vestigial thoracic legs, usually present in the form of small setae on the flanks of the trunk near its junction with the cephalothorax.

Males of the so-called *Charopinus*-group of genera also belong to type A. They might be compared to male *Lernaeopoda* with increased emphasis on each individual feature, except for the uropods. The latter are not inflated and point in a posterior direction. The angle of flexion between cephalothorax and trunk is sharper in these males than in *Lernaeopoda*.

The last two examples of type A to be mentioned are the males of *Brachiella* (Figs. 1746, 1747) and of *Neobrachiella* (Figs. 1771, 1788). The obvious similarities of these two males are accompanied by differences that make them distinctive from each other. The differences are marked particularly in the structure of the prehensile appendages (second maxillae and maxillipeds) and in the genital armature of the posterior extremity. Males very similar to those of *Neobrachiella* occur also in some related genera.

Lernaeopodid male type B is shown in Fig. 1881, representing the male of *Clavella adunca*. The striking difference between types A and B is mainly due to the reduction of the trunk in the latter. The cephalothorax remains essentially unaltered, except for some crowding of the appendages on its ventral surface. The same, however, is not true of the posterior part of the body. In *Clavella* Oken, 1816, the trunk is represented only by a small, posteriorly rounded sac. The male of *Clavellisa* (Figs. 1950, 1968) differs from that of *Clavella* mainly in the position of the genital orifice. In *Clavella* this orifice is far forward, between the bases of the last pair of appendages, whereas in *Clavellisa* it is usually rather behind that level, at the tip of a more prominent tubercle.

The differences between the males belonging to this type are fairly small. They depend mainly on the shape of the posterior part of the body. In *Clavellodes* (Figs. 1927, 1928) this part forms an elongated sac, with the appendages crowded together on a very small part of the body surface. In *Alella* (Fig. 1990) the "sac" is less deep, whereas in *Advena* (Fig. 2007) it appears to be missing altogether.

Finally, the only genera with males belonging to type C are *Euclavellisa* and *Nectobrachia*, the males of which appear to occupy a position intermediate between types A and B. Neither of them occurs in British fauna. The male of *Euclavellisa*, described by Heegaard (1940) from Western Australia and never recorded since, resembles that of *Clavellisa* in all respects, except for the presence of a short but apparently distinctly segmented trunk, curving in a posterior direction from the point at which the genital tubercle is present in *Clavellisa*. In *Nectobrachia*, a short, conical process at the same point also appears to correspond to a trunk, especially as it is armed with vestigial appendages of unknown nature.

Wilson (1915) suggested a tendency to "folding" among male Lernaeopodidae. He arranged them in a series, beginning with the completely straight *Achtheres* (Fig. 1516), going through *Neobrachiella* (Fig. 1788), with a widely obtuse angle between its cephalothorax and trunk, and *Lernaeopodina* (Fig. 1647), in which these two tagmata are almost at right angles to each other, to species which had completed the process of "folding". The ultimate was reached by genera such as *Alella* (Fig. 1990) and *Clavellodes* (Fig. 1927), in which the axes of these two tagmata appear to be almost parallel to each other. Wilson supported his theory by the study of sections of the male *Clavella* which showed quite definitely that the reproductive organs run a semicircular course through the "sac". Similarly, the alimentary canal loops along its course, its terminal, blind end being anterior to the arc it ascribes in the vestigial trunk. Gurney (1934) suggested that the abbreviation of the male is due to simple reduction in the size of the trunk. In his study of the development of *Clavella*, he claimed to have observed the remnants of degenerating trunk tissues in the position where the posterior part of it could be expected to develop. The evidence from Wilson's section, however, cannot be easily refuted. It is true that some reduction in the length of the trunk, not accompanied by "folding", seems to occur in males of type A. For example, compared with the male of *Lernaeopoda*, the trunk of the male *Lernaeopodina* (Fig. 1647) is shorter in proportion to the total length. It is shorter still in the male *Ommatokoita* Leigh-Sharpe, 1926 (Fig. 1544), in which the cephalothorax constitutes more than half of the total length. The same, however, does not apply to males of type B.

A recent description of the life history of *Salmincola californiensis* (Dana, 1852) (cf. Kabata and Cousens, 1973) might perhaps be consulted to provide a possible reason for the selective tendency to shortening of the male trunk associated with the forward shift in the position of the genital apparatus. According to these authors, the male *Salmincola* attaches itself to the genital process of the female with its maxillipeds, the last pair of appendages in the series, in such a manner that they enclose a fairly broad space between them and the female. The rather long trunk of the male is then brought forward by "folding" so that its tip and the genital orifices push through the "window" between the maxillipeds. A spermatophore is then extruded and inserted into one of the vaginal orifices. This process appears to present some difficulties and is not always successfully accomplished. It is usual for the male to repeat trunk flexions several times before a single spermatophore is successfully transferred. Type B males do not have the same problem. The position of the genital orifice is such that the spermatophore on emerging finds itself in the immediate vicinity of the maxilliped "window" and can be easily pulled through it by the second maxillae without any effort on the part of the posterior end of the body. These males must, therefore, be considered as a type advanced phylogenetically, in comparison with the more usual males of type A.

With the exception of the second maxillae, the appendages of male Lernaeopodidae are similar structurally to those of the females. The second antennae tend to be less modified and frequently retain greater prehensile powers. The second maxillae are subchelate and often linked together by a strip of cuticle, for which the author proposes the name *tympanum*. Of unknown function, the tympanum is often ornamented in various ways and may carry nipple-like outgrowths (Figs. 1670, 1690). Some males have retained vestiges of their swimming legs.

History and systematics

The first known report on, and description of, a lernaeopodid copepod, is that by Gissler (1751). The species he dealt with was referred to by him as "lax lusen" or the salmon louse. Although this uncomplimentary term has been at various times applied also to a caligid (*Lepeophtheirus salmonis*), in this instance the louse in question was a species now known as *Salmincola salmoneus*. Gissler's lax lusen appeared in the 10th edition of Systema naturae (1758) and in the 2nd edition of Fauna Suecica (1761), both Linnaean classics, under the binomen *Lernaea salmonea*.

This binomial debut of Lernaeopodidae proved to have long-lasting effects on their taxonomy. It ushered in the "lernaean period" of their history, which was to last until almost the middle of the 19th century. This period was characterized by a great deal of chaos resulting largely from the rather scholastic attitudes of contemporary savants. Often the arguments and conclusions of one author centred around an illustration or text of another, rather than around new specimens or observations.

Oken (1816), in his Lehrbuch der Naturgeschichte, arranged Lernaeopodidae in very haphazard fashion, placing them in two different sections, one on p. 182, the other on p. 357. The latter section was further subdivided into two "Sippschaften" (tribes). Tribe I included in the genus *Clavella* two species, *C. uncinata* and *C. clavata*. The latter species is a member of the genus *Acanthochondria*, belonging to a poecilostome family Chondracanthidae. Tribe II is also a heterogenous assembly of species, now distributed between Lernaeidae, Pennellidae and Chondracanthidae. In addition, it includes two lernaeopodids, *Salmincola salmoneus* and *Basanistes huchonis*.

In Cuvier's (1830) Règne Animal, les Lernees are divided into several groups, lernaeopodids occurring in almost all of them mixed indiscriminately with the chondracanthids and dichelesthiids. Les Anchorelles contain *Anchorella adunca* (Strøm, 1762), while les Clavelles have among them *Clavella uncinata* (Müller, 1776). (More than a century later, Dollfus (1953) recognized that these two taxa represent the same species and restored the chronologically prior name *adunca* to *C. uncinata*.) The genus *Clavella* Oken, 1816, hovered between Lernaeopodidae and Dichelesthiidae until Poche (1902) removed the "dichelesthiids" from *Clavella* and placed them in the now well known and abundant genus *Hatschekia*.

The type genus of the present Lernaeopodidae, *Lernaeopoda*, was established by de Blainville (1822) as a part of his "les Lernees". The old lax lusen emerged in his system as *Lernaeopoda salmonea* and was considered a member of this genus until Wilson (1915) established *Salmincola*, making *L. salmonea* its type species.

Burmeister (1835), grouping his Lernaeoda, came much closer than any of his predecessors to the concept of Lernaeopodidae, as we see them today. Lernaeoda contained three divisions. Division A contained species "With simple, sucker-like attachment organ at junction of neck and trunk", exemplified by *Anchorella uncinata* (= *Clavella adunca*). Division B "With elongate, arm-like attachment organ, fused at the tip", contained all other lernaeopodids known to Burmeister (genera *Tracheliastes*, *Brachiella*, *Lernaeopoda*, *Achtheres* and *Basanistes*). Division C was subdivided into (a), containing present day's Chondracanthidae and (b) with two "dichelesthiid" genera (*Lernanthropus* and *Epachthes*, the latter synonymized with the former).

Burmeister's grouping was a step forward in recognizing the lernaeopodids as a distinct group. No longer sharing genera with chondracanthids and "dichelesthiids", they are linked with those unrelated groups at what we might now consider an equivalent of the familial level. Lernaeoda might therefore be considered as a precursor of Lernaeopodidae.

Disregarding division C, it is interesting to note the criteria used by Burmeister for intrafamilial arrangement of the lernaeopodids. The distinction between divisions A and B was based on the structure of the second maxillae (Haftorganen). Within division B two groups were established. One (a) contained species with "cephalothorax neck-like, elongate", the other (b) with "cephalothorax short, ovoid or heart-shaped; hooked claw-feet close in front of arms". The shape of the cephalothorax has remained important to subsequent taxonomists as a diagnostic feature, though it was sometimes disregarded at the generic level, where it really matters, or overstressed at the level of higher taxa, where it is of no significance.

By removing Burmeister's division C from Lernaeoda, Edwards (1840) was the first systematist to recognize Lernaeopodidae as a natural unit, the species of which are related to one another and differ from the members of all other families in the modification of their second maxillae for the purpose of permanent attachment. It is he, therefore, who should be credited with the authorship of this family. Edwards also gave to the family the French version of its familial name, Lernéopodiens. His intrafamilial organization followed that of Burmeister's in using the structure of the second maxillae for the first division into groups. As the second he used the positions of the first and second maxillipeds (= second maxillae and maxillipeds) relative to each other.

Baird's (1850) concept of relationships within this group of copepods differed from those of other systematists as regards ranks of his proposed taxa. He included copepods "attached to their prey by means of two long appendages which arise from the thorax" in Anchoracarpacea, a tribe of the order Lerneadae. (It is interesting to note that he considered "these arm-shaped appendages" to be, probably, thoracic feet.) Anchoracarpacea were divided into two families, as follows:

(1) Lerneopodadae (Arm-shaped appendages long, wide apart from each other at their base, and united only at the tip.)
(2) Anchorelladae (Arm-shaped appendages very short, and united to each other from the base, so as to resemble a single organ.)

In so arranging this group, Baird deliberately split Edwards' (1840) Lernéopodiens, an act that remains without acceptance so far. It might be noted that the definition of Lerneopodadae coincides with Burmeister's (1835) Lernaeoda, division B, whereas Anchorelladae correspond with division A of that taxon.

In its present form, the name of the family appeared first in Dana (1852). This has prompted some authors (e.g. Yamaguti, 1963) to credit Dana with the authorship of the family, to which he had merely applied another version of the familial name. Krøyer (1863) used the name Lernaeopodina, in its plural sense, as the family name. This was similar in usage to his Chondracanthina, Caligina, etc. In the view of the present author, this does not constitute preoccupation of the name for use as a generic name. (Wilson later used *Lernaeopodina* as a genus group name.)

Heller (1865), who also used Lernaeopodina as the familial name, subdivided the family into two categories, according to the presence or absence of "maxillipeds" (his term for the second maxillae). Category I (with second maxillae) comprised true lernaeopodids, whereas category II consisted of Krøyer's genus *Silenium* (considered as identical with *Herpyllobius*) and *Tanypleurus*, both non-lernaeopodid. The latter genus is now a type of the family Tanypleuridae Kabata, 1969, related to, but distinct from, Lernaeopodidae.

The true lernaeopodids were then subdivided by Heller into groups 1 and 2, corresponding with Burmeister's groups B and A respectively, according to the structure of the second maxillae. Further subdivision of group 2 proceeded in three steps, based on the shape of the cephalothorax (short and oval or heart-shaped, as opposed to long and cylindrical), length of trunk and presence or absence of trunk segmentation. The last step was augmented by taking into account the position of the maxillipeds in relation to the bases of the second maxillae. Heller also noted the shape of the bulla (as a distinction between *Lernaeopoda* and *Charopinus*) and played down the importance of the posterior processes. But for the addition of *Silenium* (considered as identical with *Herpyllobius*) and

Tanypleurus and removal of Burmeister's division C from the family, Heller followed him quite closely in arranging genera within Lernaeopodidae.

Gerstaecker (1866–1879) reversed the order of importance of intrafamilial differences, splitting lernaeopodids first by the shape of the cephalothorax, then by the way in which it joined the trunk (imperceptibly or not) and, finally, by the structure of the second maxillae. In this grouping *Thysanote*, *Basanistes*, *Vanbenedenia*, *Charopinus*, *Achtheres* and *Lernaeopoda* belonged to one group, while the other comprised *Tracheliastes*, *Brachiella* and *Anchorella*. Very tentatively, *Herpyllobius* was mentioned as perhaps also a member of the family.

All the above intrafamilial arrangements were clearly meant more as a key to identification of genera than as a guide to phylogenetic relationships. Curiously enough, though morphological criteria used by various authors for the fragmentation of the family were largely similar, the positions of individual genera tended to change from one author to the next. Such lack of consistency reflected insufficient knowledge of lernaeopodid morphology, as well as the biology of the family.

The first author to attempt a revision of Lernaeopodidae in the 20th century was Wilson (1915), who divided it into four subfamilies, as follows.

"Maxillipeds inside the second maxillae and the two close behind the mouth tube; cephalothorax neither narrowed nor flattened, in line with the trunk or inclined forward . . . . **Lernaeopodinae**

Maxillipeds inside of second maxillae and the two removed a considerable distance behind the mouth tube; cephalothorax much narrowed and flattened; inclined backward, or arched dorsally . . **Tracheliastinae**

Maxillipeds removed some distance behind the mouth tube, second maxillae an equal distance behind the maxillipeds; cephalothorax neither narrowed nor flattened, in line with second maxillae and at right angles to the trunk axis . . . . **Brianellinae**

Maxillipeds close to the mouth tube, second maxillae removed a considerable distance behind them; cephalothorax narrowed and wormlike, in line with the trunk or inclined backward . . . **Clavellinae**"

Wilson's grouping of Lernaeopodidae into subfamilies is notable as the first expression of the fact that within this unit many species are related to one another with a different degree of closeness. It might be construed as the first approach to a family tree. The key shows, however, that the grouping is still essentially "pigeon-holing", based on two morphological characters of adult females, with no regard to possible parallelism or convergence in the evolution of various species under similar circumstances of life.

At the generic level, however, Wilson depended heavily on the morphology of the male as a discriminant. He was the first to establish the principle that lernaeopodid species with different males belonged to different genera, regardless of the morphology of the female. It was unfortunate that this undeniably correct view was not coupled with sufficient knowledge of male morphology. Consequently, unnecessary multiplication of genera took place, based on old and incorrect illustrations, post-mortem artifacts, etc. To such spurious genera belong *Epibrachiella*, *Parabrachiella* and *Probrachiella*. On the other hand, distinctness of the males of *Brachiella* and *Neobrachiella* passed unnoticed.

The knowledge of lernaeopodid morphology increased rapidly during the early part of the 20th century, but Wilson's divisions have not been essentially affected. One significant difference was the addition by Markevich (1956) of two more subfamilies, which can be distinguished by the following key.

1 (2) Cephalothorax completely fused with trunk, without traces of subdivision . . . . **Nectobrachiinae**

2 (1) Cephalothorax clearly delimited from trunk

3 (4) Trunk and egg sacs enclosed in a common cuticular sheath . . . . **Naobranchiinae**

4 (3) Trunk and egg sacs not enclosed in a common cuticular sheath

5 (8) Second maxillipeds situated between first maxillipeds

6 (7) First and second maxillipeds close to the mouth cone; cephalothorax oviform . . . . . **Lernaeopodinae**

7 (6) First and second maxillipeds at long distance from the mouth cone; cephalothorax relatively narrow and long, approximately cylindrical in shape . . . . **Tracheliastinae**

8 (5) Maxillipeds in other positions

9 (10) Second maxillipeds at some distance from mouth cone; first maxillipeds about the same distance behind the second; cephalothorax in line with the first maxillipeds and at right angles to the axis of the trunk . . . . **Brianellinae**

10 (9) Second maxillipeds close to the mouth cone; first maxillipeds at a considerable distance behind them; cephalothorax narrow, vermiform, in line with the axis of the trunk or deflected dorsally . . **Clavellinae**

It is clear that Markevich's key is an elaboration on Wilson's theme and introduces few new ideas of relationship between members of the family. His addition of the subfamily Naobranchiinae is adopted from Yamaguti (1939c), with a reduction in status. This subfamily, containing only one genus *Naobranchia*, was removed from the subfamily Clavellinae by Yamaguti and placed in a separate family, Naobranchiidae. Subfamily Nectobrachiinae was first proposed by Markevich (1946) to accommodate a monotypic genus *Nectobrachia* Fraser, 1920. Gurney (1934) suggested that Clavellinae also should be upgraded to the family level, though he did not make an explicit statement on this subject or define the family; he simply used the familial name Clavellidae. His suggestion was not followed.

Markevich's innovations were not approved by Yamaguti (1963), the most recent reviewer of Lernaeopodidae. He retained the four-partite structure of the family. *Nectobrachia* he consigned to the subfamily Lernaeopodinae and maintained his earlier view on the position of *Naobranchia* as a type genus of a separate family.

The main reason for placing *Naobranchia* in Lernaeopodidae is the fact that it employs its modified second maxillae as organs of attachment to the host. Its structural plan is superficially similar to that of the long-necked lernaeopodids. Here, however, ends the resemblance between them. *Naobranchia* possesses no bulla, either functional or vestigial, at any stage of its life cycle. It maintains its hold on a gill filament of its host by enclosing it in an annulus formed by flat second maxillae. Its egg sacs are incorporated into the trunk to form pouches in the posterolateral corners of that tagma. Taking also into account some differences in the structure of the appendages, this writer is inclined to follow Yamaguti's view on *Naobranchia* and to remove it from Lernaeopodidae and from this work.

Phylogeny

The best way to gain some appreciation of intrafamilial relationships within Lernaeopodidae is to reconstruct, as far as possible, the evolutionary paths that brought the extant lernaeopodids to their present-day condition. To achieve any success in this type of exercise, one must break away from the frame of mind of an arranger manoeuvring long rows of glass vials. It is necessary to think of each species as a living functional unit, its individuals performing tasks designed to ensure survival of each of them and of the species as a whole. These tasks must be performed within the framework of selective pressures imposed on each parasite, pressures channelling the evolution of the species into adaptive modifications and changes correlated with them. When considering the possible position of each species within the phylogenetic tree of the family, it is necessary to go beyond its morphology. One must try to deduce reasons for the morphological condition of the species. The outcome of such attempts cannot ever result in the attainment of absolutely certain conclusions. We have, however, at our disposal hitherto neglected evidence that can shed a great deal of light on the problem of lernaeopodid phylogeny.

The first category of evidence suitable for this purpose is presented by comparative morphology, understood in its functional sense. In this category are not only comparisons of various structures occurring among Lernaeopodidae, but also comparisons of morphological features of Lernaeopodidae with the corresponding features found in other families, especially those that can be classified as more primitive and more likely to bear closer resemblance to the condition they had in ancestral forms. The second category is afforded us by the still scanty knowledge of the processes of ontogeny within the family. The earliest developmental stages of Lernaeopodidae, the free-swimming copepodids, resemble one another much more closely than do their adult stages. The comparison of the adult with the copepodid stage (even if not of its own species) can give us some measure of the distance travelled by individual species along the path of evolution.

In short, to arrange Lernaeopodidae in a phylogenetic tree with some hope of probability, one must find means of determining how primitive, or otherwise, is each feature one tries to use. This, indeed, is the procedure that the author has tried to follow in reasoning his way to the concept of lernaeopodid phylogeny.

As stated earlier, the key fact in the morphology, as well as in the biology, of Lernaeopodidae is the modification of their second maxillae as arm-like organs of attachment. Although this modification is unique to the family, it has its source in the function of the second maxillae that its members share with other siphonostomes. Many parasites belonging to that sub-order are known to use their second maxillae to handle their frontal filaments when the latter are being extruded during the process of attachment to the fish. Lernaeopodids simply go a step further. Their second maxillae remain associated with the filament, more or less permanently, throughout the course of ontogeny. In the female this association is then transferred to the bulla, the adult attachment organ. As mentioned earlier, in some genera the second maxillae eventually supersede this organ and become directly embedded in the host's tissues, taking over the process of anchoring the parasite to the fish.

The biological consequences of this mode of attachment are profound. It makes Lernaeopodidae intermediate between those copepods that, like Caligidae, are able to move freely over the surface of the fish and those that, like Pennellidae, bury their anterior extremity in the tissues of the host, leaving only their reproductive trunks exposed to the external environment. It combines the security afforded by attachment with relative freedom of movement. The former depends on the strength of the second maxillae and the solidity of their fusion with the bulla, the latter is determined by the length and contractility of the second maxillae, or the cephalothorax, or both.

Living above the surface of the host, Lernaeopodidae have to feed on its superficial tissues (mucus, epithelial cells). When attached to the gills, or other areas with ample superficial vascularization, they feed also on blood. One can, then, see them as browsers.

Bearing the above in mind, it is possible to visualize selective pressures acting upon the lernaeopodid copepods. To begin with, one can imagine the appearance of a primitive lernaeopodid. This hypothetical ancestral animal, already attached to the host by its second maxillae, probably consisted of two discernible tagmata with a division between them at or near the segment bearing these appendages. The tagma anterior to that segment was shorter than that posterior to it. At the level of demarcation between them some means of flexion or articulation was likely to develop (perhaps in the form of a constriction), due to the fact that the pressure of the ever-present current of water would tend to push both the anterior and the posterior extremity backwards from the point of attachment. The posterior part was likely to be larger than the anterior since it housed the reproductive apparatus, usually generously proportioned in animal parasites. The uropods were likely to be reduced, as they are in most permanently attached parasitic copepods. One would not expect any processes or outgrowths to be present on either the posterior tagma (the trunk) or the anterior (the cephalothorax). The latter had probably already assumed some degree of dorsoventral flattening and developed a hard covering on its dorsal surface, to allow secure attachment of the strong musculature operating the cephalothoracic appendages. In all, this description fits most closely the genera *Salmincola* (Fig. 1462) and *Achtheres* (Fig. 1500) among living Lernaeopodidae.

The success of any browsing animal depends in a large measure on the size of the area it can cover in its search for food. Of course, the larger the animal, the greater are its food requirements and the larger is the area needed to satisfy them. For a lernaeopodid, the size of the area its mouth can reach can be enlarged by increasing the length of its second maxillae. The length of the latter determines the radius of the circle (or, at least, the semicircle) within which the mouth parts of the parasite can be brought in contact with the surface of the host. One could expect, therefore, that among the evolutionary trends determining the progress of Lernaeopodidae there would be a tendency towards the elongation of their second maxillae. We can best assess this length by its relationship to the length of the trunk. The more primitive parasites would tend to have second maxillae shorter than, or about as long as, the trunk. Again, examples of this can be found among *Salmincola* and its relatives, as well as among *Lernaeopoda* (Figs. 1675, 1676) or *Charopinus* (Fig.1577). The advanced species would tend to have much longer second maxillae, those of some species of *Lernaeopodina* and *Neobrachiella* exceeding by many times the length of the trunk. Individuals of species so endowed are capable of swinging along the arc determined by the length of the "tether". (These parasites are more likely to occur on the surface of the host rather than in its branchial or buccal cavity, in which space is at a premium.) The radius of the arc can be altered by the contraction of the maxillary musculature, so that one must be careful, when examining fixed specimens, not to underestimate the length of these appendages, often contracted in fixative.

The ability to extend the size of the browsing surface and to use it in a more flexible manner can be further improved by increasing the length of the cephalothorax. The parasites with a very long attachment "tether" and short cephalothoraces can be fancifully compared to short-necked giraffes. One can postulate the existence of a tendency to elongation of the cephalothorax in lernaeopodid evolution. Species with relatively long cephalothoraces and second maxillae can, then, be considered as more highly evolved. The long cephalothorax must also be considered as less primitive than the short one because it constitutes a very rare departure from the generally short cephalothoraces of the siphonostome copepods. Following this line of reasoning one must conclude that, for example, *Brachiella* (Fig. 1733) is more highly advanced than *Salmincola* (Fig. 1487).

The retention of some freedom of movement, while theoretically advantageous, might also be viewed as a mixed blessing. If nothing else, it certainly would consume some part of the energy at the parasite's disposal. Elongation of the cephalothorax only, without a concomitant increase in the length of the second maxillae, might ensure that a sufficiently large area of the host's surface would fall within the reach of the parasite's mouth. Some parasites, then, might evolve by developing long cephalothoraces without any extension of their "tethers". Once the latter are rendered unnecessary, a tendency might arise towards shortening them. This line of evolution is represented by the genus *Clavella* (Fig. 1861) and, among British fauna, culminates in the genus *Clavellisa*, the species of which have very short second maxillae and inordinately long cephalothoraces (Fig. 1938).

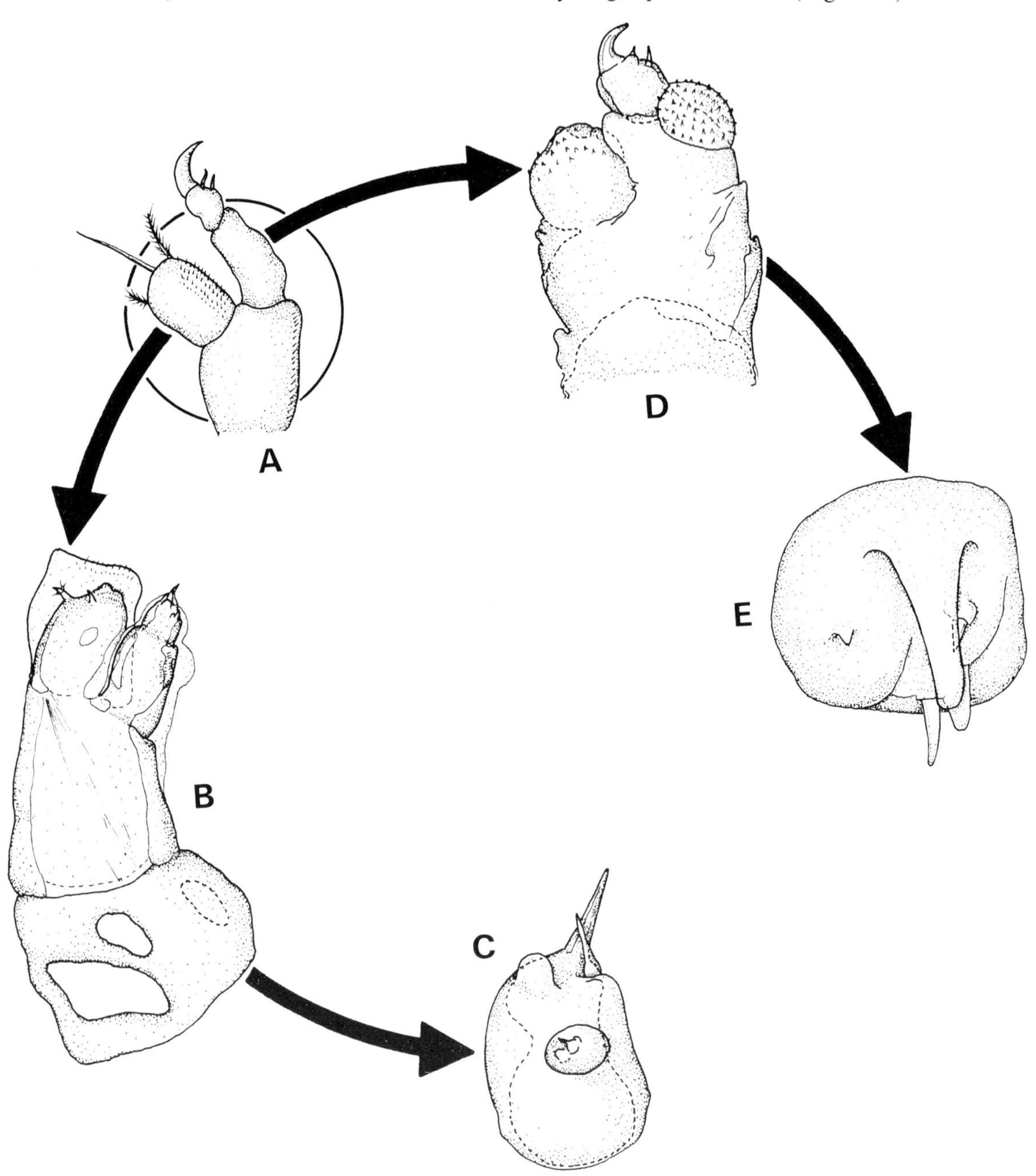

Text fig. 65. Types of second antennae in Lernaeopodidae. A. Copepodid stage; B. *Lernaeopoda*, entire antenna; C. *Lernaeopoda*, terminal segment of endopod; D. Primitive type (e.g. *Tracheliastes* or *Vanbenedenia*), entire antenna; E. As D, apex of endopod.

The specialized nature of the elongated second maxillae and cephalothorax is further suggested by the fact that they have evolved much further away from the condition found in the early stages of ontogeny than those which have retained a relatively modest length.

In trying to determine the level of the evolution of individual appendages, one must also look at them from the point of view of the function they are called upon to perform, as well as by measuring the differences between the adult and juvenile conditions. This is well illustrated by the example of the second antenna, a prehensile appendage of great importance to the parasite during the period of its initial contact with, and attachment to, the host. During the copepodid stage the second antenna is biramous and well suited to serve the purpose of prehension, as can be seen in Text fig. 65A. Its exopod is one-segmented, bulbous and equipped with three long terminal setae. The endopod is two-segmented. Its distal segment is armed with a powerful hook and usually three additional setae, one of them being present at the base of the hook and the other two on the ventral side of the segment. The prehensile capacity of the appendage is due to its endopod and, in particular, to its hook which is able to execute grasping movements. The great mobility and prehensile powers of the endopod are truly impressive (Kabata and Cousens, 1973).

The second antenna in the copepodid stage is separated by a very short space from the second maxillae and maxillipeds, both also capable, grasping appendages. The latter, in particular, continue to develop into powerful limbs during the course of ontogeny, until they become dominant in the set of cephalothoracic limbs (if one disregards the arm-like second maxillae). Situated close behind the mouth cone, they maintain contact with the surface of the host during feeding. In so doing they render the second antennae redundant. Indeed, in those lernaeopodid species in which the elongation of the cephalothorax does not increase the distance between the second antennae and the maxillipeds, the former invariably become modified into sensory appendages and lose their original prehensile function. The modification involves elongation of the sympod, which is now arranged along the same axis as the exopod (Text fig. 65B). The endopod has been pushed aside and left behind in the process of growth. Its apical armature (Text fig. 65C) has become simplified, though its individual components are still recognizable. This type of second antenna prevails in all genera with short cephalothoraces (e.g. *Lernaeopoda*), except for the primitive few (e.g. *Salmincola*) that have retained both these pairs of appendages in functional condition as prehensile limbs. In those genera with elongated cephalothoraces, in which elongation did not affect the spatial relationship between the second antenna and the maxillipeds, the latter also become exclusively concerned with prehension and the former specialize as sensory structures (e.g. *Clavella* or *Neobrachiella*). (The males, however, retain their mobility. Since their mode of progression calls for alternate engagement of both prehensile appendages of the copepodid stage, their second antennae remain functionally unmodified (Text fig. 65D).) In other genera the elongation of the cephalothorax profoundly affects the spatial relationship between these two pairs of appendages. The second antennae are removed a long distance from the bases of the second maxillae and from the maxillipeds (e.g. *Tracheliastes* or *Vanbenedenia*). The latter may also find themselves placed halfway between the base and the apex of the cephalothorax (e.g. *Charopinus* or *Brianella*). In either case they are not capable of rendering any assistance in holding the mouth cone close to the surface of the host, while feeding. This function is left to the second antennae alone. Indeed, in these genera the second antennae remain prehensile and but little modified and their apical armature undergoes no reduction (Text fig. 65E). The maxillipeds, on the other hand, suffer the fate of all appendages no longer useful and become reduced to a lesser or greater extent. In *Vanbenedenia* (Fig. 1563) they still retain much of their original appearance, though their subchelae are relatively reduced. In *Tracheliastes* (Fig. 1528) they are completely degenerate, having become mere outpushings of cuticle, devoid of any intrinsic musculature.

The above example allows us to conclude that the genera with modified second antennae are more likely to be advanced than those that have retained them in a condition resembling the corresponding appendages of the copepodid stage. Unless, of course, this retention was made necessary by functional neutralization of the maxillipeds, in itself a result of profound structural modification.

Another appendage that offers us some insight into the phylogenetic condition of the individual lernaeopodid genera is the mandible. In most copepods belonging to Siphonostomatoida this appendage is armed with a series of teeth borne along one margin of the distal blade. The teeth are short and curve slightly in the direction of the base of the appendage. They tend to diminish in size from the distal to the proximal, but are otherwise undifferentiated. This condition is found in all families of the suborder, including those that can be presumed to be more primitive than Lernaeopodidae (e.g. Caligidae and their allies). It is reasonable, therefore, to view those lernaeopodid mandibles that closely resemble this general siphonostome condition as being more primitive than those that have departed from it. The extent of modification might be indicative of the evolutionary progress of the species.

This primitive siphonostome condition is found in the mandibles of *Salmincola* (Figs. 1476, 1493, 1494) and other freshwater genera not present in British fauna. Among the marine lernaeopodids it occurs in otherwise highly advanced *Advena* (Fig. 2002). The first signs of change are indicated by the mandibles of other British freshwater genera (*Achtheres* (Fig. 1511), and *Tracheliastes* (Fig. 1526)). A similar condition occurs in the marine genus *Vanbenedenia* (Fig. 1562). In all these genera, the uniform series of teeth is broken by the intercalation of a secondary tooth, smaller than the others, usually near the distal end of the series.

Modification of mandibular dentition can proceed in two ways. The distal end of the series of teeth might become further differentiated by the addition of more secondary teeth, though their number never exceeds three (Fig. 1538). The definitive composition of the dentiferous margins of mandibles developing along this line is, therefore, one consisting of three pairs of alternating primary and secondary teeth, followed proximally by a series of progressively smaller basal teeth. The number and positions of the three types of teeth in the dentiferous margin can be recorded in a mandibular formula as follows: P1, S1, P1, S1, P1, S1, B5 (referring to Fig. 1538). This formula is an English version of that proposed by Kurtz (1877). P denotes primary, S secondary and B basal teeth. Most advanced lernaeopodid genera have mandibles with dentition similar to that shown in Fig. 1538, differing only sometimes in the absence of one secondary tooth and in the number of basal teeth.

In the genus *Clavella*, however, modification has been less regular. In addition to the insertion of secondary teeth, the distal end of the blade has become abruptly narrowed (Fig. 1875). Sometimes spaces that should be occupied by secondary teeth remain empty, forming diastemes (Fig. 1876). The dentition shown in Fig. 1875 can be recorded in the following mandibular formula: P1, S1, P1, S1, P1, S1, B2. The underlined part of the formula indicates dentition carried on the narrow part of the blade. Diastemes are indicated by hyphens in appropriate places.

Finally, the first antenna is also an appendage that might be of some help in a study of phylogeny. Although it is not possible to determine what is its primitive condition and what should be considered an advanced one, our understanding of intrafamilial relationships benefits from a closer look at the first antenna and, in particular, at its apical armature.

In view of its very small size, the first antenna has mainly been very difficult to study and has, therefore, been neglected by investigators. Dissection often damages its very fine apical setae. They are also frequently damaged or worn down *in vivo*, with resultant apparent morphological modifications. Consequently, literature on Lernaeopodidae offers but little that would help us to systematize our knowledge of their first antenna.

The first antenna probably consisted originally of four segments. At least this is the largest number now identifiable with any certainty. They are now largely obsolete and their former junctions are sometimes indicated by setae arising from the wall of the appendage. Usually the first two segments are fused into a moderately inflated base, the distal border of which is marked by a long and slender seta, arising on the dorsomedial side and referred to here as the *whip* (Fig. 1571, *w*). The cylindrical distal part consists of two segments, often fused to complete obliteration of the boundary between them. In some species this boundary is still recognizable and might be marked by a short single seta, to which the name *solus* will be applied in the following descriptions (Fig. 1571, *s*). Accounts by various authors are at variance with this writer's interpretation of segmentation of the first antenna. These discrepancies are in most instances caused by the interpretation of artifacts (e.g. cuticular wrinkles or superficial damage) as segmental boundaries.

Our knowledge of the apical armature is far from perfect. Even in large first antennae the length of the apical setae is seldom much greater than 50 μ. Allowing for accidental damage, different angles of observation and other variables impossible to avoid with currently available methods of studies, one is often dependent as much on one's familiarity with the family as on actual observation for correct interpretation of the morphology of the antennary apex. This in itself introduces dangers of misinterpretation, difficult to guard against. The following account of the apical armature should be seen with these reservations in mind.

The apical armature of lernaeopodid first antennae differs from species to species. Most species,

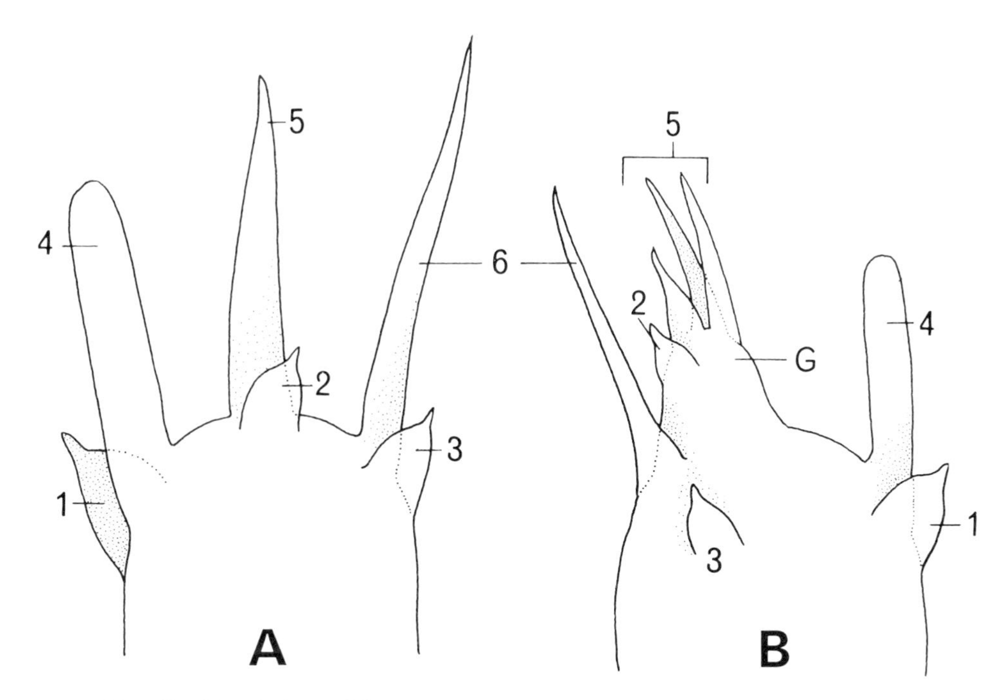

Text fig. 66. Apices of first antenna in Lernaeopodidae. Explanations in text.

however, can be related in this respect to one of the two basic types illustrated in Text fig. 66. The armature present in all species of the *Charopinus*-branch of the family is shown in Text fig. 66A. It consists of three tubercles, sometimes with fine tips, and three setae. (Numerical designations of these structures are based on this author's earlier work (Kabata, 1963c, 1964a, etc.).) The medial side of the tip is occupied by a sturdy, digitiform seta 4 (perhaps an aesthete). At its side is a tubercle 1, sometimes of considerable size. The lateral side of the tip gives rise to a flagelliform seta 6, usually the longest seta of the set. It also has at its base a tubercle, mostly smaller than 1 and labelled tubercle 3. The centre of the tip is occupied by a seta intermediate in size and shape between the two others, seta 5. Tubercle 2 is a short distance away from its base.

The second type is illustrated in Text fig. 66B. It differs from the first type in having a large part of its apex occupied by a swelling, or *gibber* (G). Setae 4 and 6, with their attendant tubercles 1 and 3 respectively, are present in the same positions. The tip of the gibber is surmounted by several setae, their number and size varying from species to species, but not exceeding four. Sometimes tubercle 2 is found on the side of the gibber, or near its tip. All setae arising from the gibber can be considered as corresponding with seta 5 and referred to as complex 5. This type of apical armature is typical of the *Lernaeopoda* and *Brachiella* branches of the family. The size and position of the gibber differs (or appears to differ because of conditions of observation), the size being sometimes much reduced. In most members of these branches, however, the apical armature can be derived from this type.

A great variety of apical armature is displayed by the *Clavella*-branch. In the genus *Clavella* some of the tubercles or setae are missing in certain species, either in the females or males. Seta 5 in this genus is divided for most of its length, forming two branches arising from a common base (Fig. 1873). In genera such as *Clavellisa* (Fig. 1941) and *Clavellodes* (Fig. 1918) it is difficult to find any resemblance of the apical armature to either of the two types. The same is true of the freshwater branch of the family, in which the first antenna is often degenerate and the apical armature of the adult female either reduced or absent (Figs. 1471, 1523).

With the above discussion in mind, one can now propose a phylogenetic tree for Lernaeopodidae (Text fig. 67). Whenever possible, examples to illustrate it will be drawn from genera represented in the British fauna and figured in this account. Otherwise references will be made to appropriate literature sources.

The ancestral lernaeopodid probably was a marine animal. In its appearance it might have been not unlike the present day *Salmincola*, from which it likely differed in having shorter and possibly clearly jointed second maxillae, as well as a relatively longer, more fusiform cephalothorax. It was externally segmented and possessed vestigial swimming legs. Its uropods were also probably retained in some form right through the adult stage.

It is not possible to determine the identity of the hosts of primitive Lernaeopodidae. Neither can we fix in time the period during which these parasites made their first appearance. In view of the fact that many of the currently extant, fairly primitive lernaeopodids are parasitic upon the elasmobranchs, it is permissible to presume that the early stages of lernaeopodid evolution passed in association with this group of fishes.

The first major cleavage in the evolutionary line of the family occurred when a lernaeopodid became adapted to life on a teleost fish, a primitive salmonoid. The present consensus of opinion regarding the evolution of Salmonidae is that they are of freshwater origin and that their migratory habits are a consequence of overcrowding resulting from a population explosion during the interglacial periods of Pleistocene. The immediate ancestor of *Salmincola* came in contact with and became adapted to parasitism on the new arrivals into the marine habitat. With them, this primitive pre-*Salmincola* invaded, survived in and colonized the rivers of the northern hemisphere, giving rise to other freshwater genera.

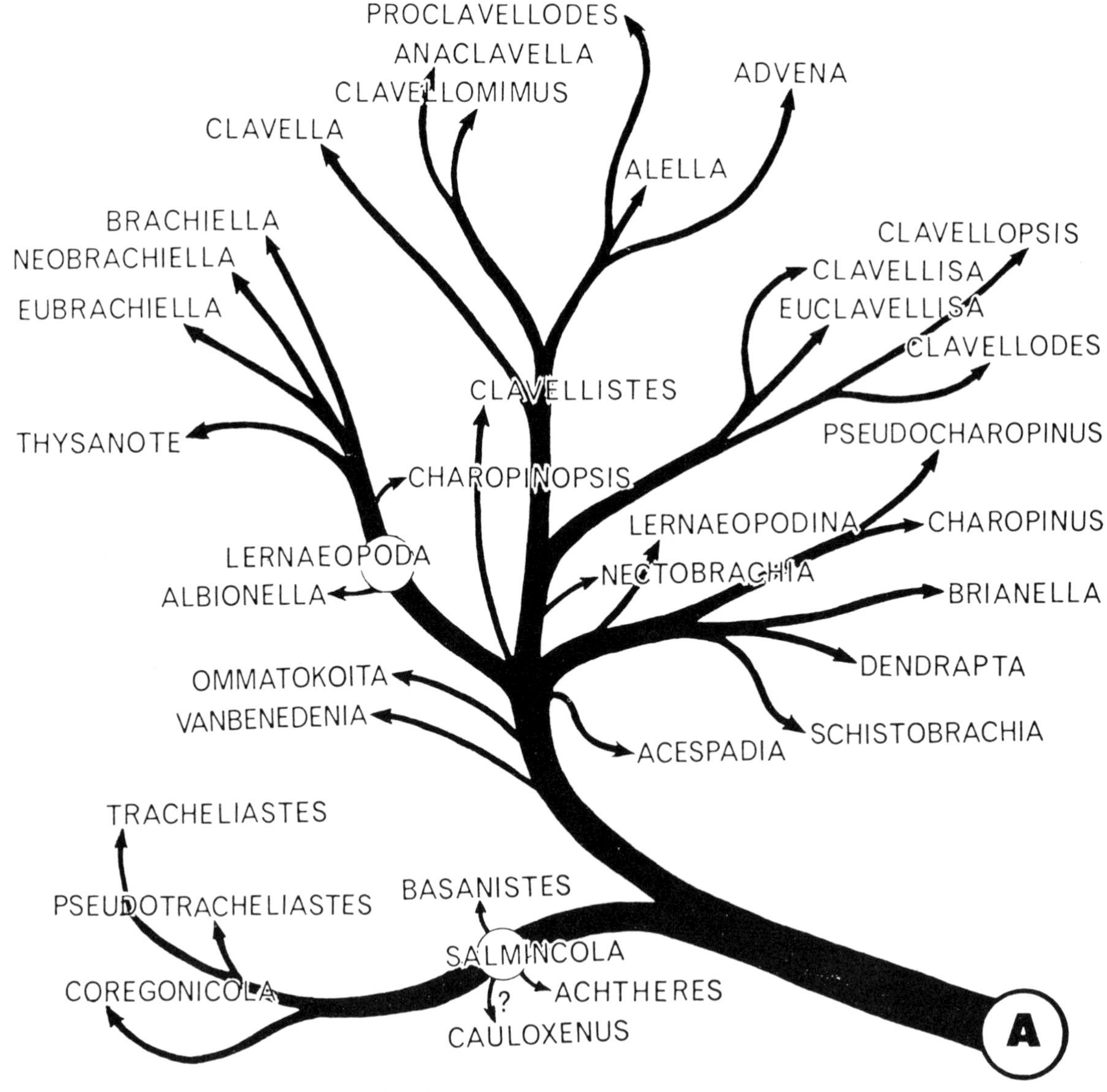

Text fig. 67. The family tree of Lernaeopodidae.

Examined in the light of the foregoing arguments, *Salmincola* is indeed a primitive genus. Its species have moderately long second maxillae (Fig. 1462) and their second antennae retain prehensile function (Fig. 1477) regardless of the fact that the maxillipeds are well developed and situated close behind the mouth cones (Fig. 1464). The dentiferous margins of their mandibles (Fig. 1476) have not altered from the general siphonostome condition. The trunks remain simple sacs of varying shapes (Figs. 1462, 1463, 1486, 1487), devoid of outgrowths, protuberances or posterior processes. This list of primitive characters is, however, at odds with two features which could have only resulted from modification of the ancestral condition. One of them is the condition of the first antenna (Fig. 1471). Its aberrant nature can be seen in adult females, in which it is short and peg-like, devoid of segmentation and almost completely denuded of apical armature. In this it differs not only from all the marine genera of the family but also from its own copepodid and young female stages (Kabata and Cousens, 1973). While young females still retain a set of apical setae (Fig. 1472) that can be related to the type shown in Text fig. 66A, most of them disappear with the progress of maturity. By comparison with the structure of the same appendage in marine lernaeopodids, the other appendage of *Salmincola* that can be judged modified is the first maxilla (Fig. 1484). Although this appendage, like that of other primitive lernaeopodids, has three apical papillae, its exopod has been reduced to a small denticle arising from a swelling on the lateral wall. (Among the marine lernaeopodids this condition is very rare. In *Nectobrachia indivisa* and *Vanbenedenia kroeyeri*, however, even the vestigial denticle has disappeared from the lateral wall.)

In Text fig. 67 the position of *Salmincola* is circled, suggesting that this genus is the central form in the development of the freshwater branch of Lernaeopodidae. The modern *Salmincola* is a typical freshwater animal, capable of temporarily surviving, but not of reproducing, in sea water. From the earliest period of its freshwater history, it speciated, producing 16 species within the genus (Kabata, 1969d), as well as branching off into other genera. One of these genera is *Basanistes*, comprising few species and differing from *Salmincola* mainly by its possession of prominent swellings on its trunk and cephalothorax (Markevich, 1956). It is known only from the Palearctic, where it parasitizes salmonoid fishes. Another genus of this branch with representative species in the British fauna is *Achtheres*. The appearance of *Achtheres* species (Fig. 1500) is much like that of *Salmincola*. That it has not been derived from the modern *Salmincola* directly is suggested by its possession of a more primitive first antenna (Figs. 1506, 1507). On the other hand, the mandible of *Achtheres* (Fig. 1511) possesses one secondary tooth, the first sign of departure from the primitive siphonostome condition and an advance on the condition present in *Salmincola*. As far as one can judge from a sketchy description (Cope, 1872), the aberrant genus *Cauloxenus* Cope, 1872, known from a single North American record, is also a near relative of *Salmincola*.

Continuing evolution in the freshwater environment resulted in the appearance of genera with morphology advanced above the fairly primitive level of *Salmincola* and its allies. While some species evolved, as might be expected, along the lines predictable for the family, one Palearctic genus, *Coregonicola*, took an unusual turn and has become aberrant rather than advanced. Although they retain the primitive features of their ancestors in a fairly unaltered condition (Kabata, 1965d), the members of the genus *Coregonicola* have developed trunks of enormous length, acquired, as mentioned earlier, by elongation of the most anterior segment of that tagma. On the other hand, the genera *Pseudotracheliastes* Markevich, 1956 and *Tracheliastes*, illustrate steps in the gradual increase in the length of the cephalothorax, typical of the evolutionary progress of this family (Markevich, 1956). *Tracheliastes* might be considered the most advanced known genus of the freshwater Lernaeopodidae (Figs. 1517, 1518). Apart from the primitive condition of the second antennae (Fig. 1524), appendages of its species are modified. The mandible (Fig. 1526) has one secondary tooth, the first maxilla (Fig. 1527) has lost its third terminal papilla and its exopod, and the first antenna (Fig. 1523) is reduced in size and has vestigial apical armature. No freshwater lernaeopodid has developed any posterior processes (unless one considers as such the vestigial perianal outgrowths of *Achtheres percarum* von Nordmann, 1832 (Fig. 1504)).

Taken as a whole, the freshwater branch of Lernaeopodidae exhibits evolutionary trends that are common to other branches of the family. The more advanced species have long cephalothoraces and their mandibles have departed from the primitive condition. No freshwater lernaeopodid has modified second antennae. The extent of the advance is small, however.

The close degree of relationship between various freshwater genera of the family becomes obvious when one looks at the known males of the genera. We still lack descriptions of the male *Basanistes*, *Coregonicola* and *Pseudotracheliastes*. The differences between the males of *Salmincola*, *Achtheres* and

Tracheliastes (Figs. 1485, 1516 and 1529 respectively) are small enough to place them in the same genus. Most of the differences between them, claimed by various authors, can be attributed to accidents of fixation or other post-mortem artifacts. All three males are characterized by possession of relatively small cephalothoraces arranged, when fully relaxed, along one axis with the trunk. While we are still unable to make detailed comparisons between the morphology of their appendages, they can all, without hesitation, be consigned to the same type.

The host affiliations of the branch are also interesting. Predominant among them are salmonids, associated with *Salmincola* and *Basanistes*, whereas *Coregonicola* parasitizes mainly coregonids. The genus *Pseudotracheliastes* lives on another group of ancient fishes, Acipenseridae of the genus *Huso* and *Acipenser*, as well as on *Silurus glanis*. The more advanced genus *Tracheliastes* is parasitic mainly on more specialized fishes of the family Cyprinidae. On the other hand, the fairly primitive genus *Achtheres* must have become early adapted to the percomorph teleosts, now constituting the majority of hosts for its members. Even *Achtheres*, however, numbers fishes as primitive as silurids among its hosts.

An exact reconstruction of the early evolution of the marine stem of Lernaeopodidae is impossible. There exist, however, two lernaeopodids that might be considered as survivors of that level in evolution. One of them is *Vanbenedenia kroeyeri* (Fig. 1553). The primitive features of this parasite (and of its two known congeners) are manifest in the absence of posterior processes and in a mandible with only one secondary tooth (Fig. 1562). Its cephalothorax, however, is cylindrical and moderately elongated, with a retractable mouth cone. Maxillipeds remain at the base of the cephalothorax and are functionally retarded (Fig. 1554). The first maxilla recalls the same appendage in the freshwater lernaeopodids in having a completely suppressed exopod. The hosts of *Vanbenedenia* are chimaerid holocephalans.

Another extant offshoot of the primitive marine stock is *Ommatokoita elongata* (Fig. 1530), the only member of its genus. Its primitive character is attested to by the brevity of its cephalothorax (Fig. 1531) accompanied by an unmodified, prehensile second antenna. Its first maxilla is also primitive (Figs. 1539, 1540), equipped with three terminal papillae, and with the exopod in the lateral position. The absence of posterior processes is also a primitive feature. On the other hand, the mandible of *Ommatokoita* (Fig. 1538) has reached the most advanced lernaeopodid condition, its dental formula including three distal pairs of alternating primary and secondary teeth. As might be predicted from the preponderance of its primitive features, *Ommatokoita* is parasitic on sharks.

Probably near the point of the triple split in the lernaeopodid tree (Text fig. 67) is *Acespadia pomposa*, a peculiar, poorly known parasite found on a single occasion on a whale shark in the China Sea. The absence of posterior processes suggests its primitive condition. As far as one can judge from a not altogether satisfactory description, its cephalothorax (Text fig. 61B and 62B) is a cross between that of *Salmincola* and, for example, *Clavella*, the ovate terminal part perched at right angles to the tip of a short cylindrical neck. It has a primitive first maxilla. The condition of the mandible and of the second antenna has not been clearly described. Also puzzling is the condition of the tips of the second maxillae. More information is required before one can find a place for *Acespadia* with greater probability.

The marine stem of the lernaeopodid evolution divided into three branches at a point in time not much later than that characterized by the level of development of *Vanbenedenia* or *Ommatokoita*. This separation occurred at, or soon after, the evolutionary stage at which Lernaeopodidae began to develop their first pair of posterior processes in the form of modified uropods. At the same time their mandibles began with increasing frequency to develop a type of dentition similar to that mentioned above for *Ommatokoita* (Fig. 1538). Of the three branches, one remained faithful to its ties with the elasmobranch hosts, the second passed through an "elasmobranch phase" before giving rise to parasitism on advanced teleosts, and the members of the third accomplished their transfer to teleosts from the outset.

The branch that remained parasitic on elasmobranchs has been sometimes referred to in the past as the *Charopinus*-group (Kabata, 1964a). The two hall-marks of this branch are the possession by its members of a single pair of posterior processes, *dorsal* to the orifices of the oviducts (hence to egg sacs, when the latter are present) and a tendency to evolve complicated and unusual attachment organs.

A primitive condition in this group is exemplified by the members of the genus *Lernaeopodina* (Fig. 1635). Some species of this genus have already developed very long second maxillae (e.g. *L. longibrachia*); in most they are of considerable length, as in *L. longimana*, the British representative of the genus. They still have short, ovate cephalothoraces and prehensile second antennae (Fig. 1642). (Indeed, the second antenna remains at that level of organization in the entire branch, with very few and slight exceptions.) The orifices of the oviducts are clearly ventral to the posterior processes, the latter equally clearly representing modified uropods. In mandibular dentition (Fig. 1643), *Lernaeopodina* is at one with other members of the group. Its relatively primitive character is shown by the structure of the bulla (Figs. 1635, 1636), unaffected by the tendency towards modification. On the other hand, this fairly primitive member of the *Charopinus*-branch is unique in having a reduced third terminal papilla on the first maxilla (Fig. 1644). In contrast to other lernaeopodids with third maxillary papilla reduced, in *Lernaeopodina* this is not accompanied by reduction in size and shift from lateral to ventral in the position of the exopod.

At an early stage, the *Charopinus*-branch became differentiated into two divisions. The first of them comprises two genera, *Charopinus* and *Pseudocharopinus*. The latter is the more primitive of the two and the bulla of its species has retained its button-like shape (Fig. 1621). They have progressed beyond the level of organization exemplified by *Lernaeopodina* and have long, cylindrical cephalothoraces on which the maxillipeds remain close behind the mouth tube. The condition of the cephalothorax is similar in the genus *Charopinus*, though here the maxillipeds have become separated from the oral region in the process of elongation (Fig. 1566). This difference in the relative position of the maxillipeds has also affected the morphology of the second antennae, which in the genus *Pseudocharopinus* have a slightly weaker endopod (Fig. 1616) than is found in *Charopinus* (Fig. 1581). The most striking difference between these two genera, however, is in the structure of their attachment organ. Unlike *Pseudocharopinus*, the genus *Charopinus* is characterized by a long, massive bulla (Figs. 1570, 1577), in *Ch. dalmanni* (Retzius, 1829) devoid of the stalk-like manubrium.

The second division of the *Charopinus*-branch has been less prompt in developing the elongated cephalothorax. Of its three genera, only *Brianella* has a cylindrical cephalothorax. *Schistobrachia* Kabata, 1964, and *Dendrapta* Kabata, 1964, retain primitively short cephalothoraces. All of them, however, clearly display the tendency of the branch to evolve a mode of attachment in which the bulla becomes gradually obsolete and its function is taken over by the modified tips of the second maxillae. In *Schistobrachia* (Fig. 1598) the tips of both second maxillae have developed a pair of two long processes acting as an anchoring device, while the bulla is present only as a minute organ linking the two appendages together. *Dendrapta* has acquired an enormous holdfast consisting of a tangle of dendritic processes sprouting from the tips of the second maxillae (Kabata, 1964a). In *Brianella* the second maxillae have fused and their tips have long expansions that function in place of the bulla (Wilson, 1915). In this long-necked genus the elongation of the cephalothorax has also been accomplished at the expense of displacing the maxillipeds, though the latter are less degenerate than those of *Charopinus*.

The cohesion of the *Charopinus*-branch is evident from the great resemblance among the males of its genera. All of them (Fig. 1587) are characterized by large, inflated and ventrally flexed cephalothoraces and by relatively reduced trunks. The great resemblance of these males is also shown by the fact that in the past all of them were placed in the genus *Charopinus*.

The overwhelming majority of species comprising this branch of the family are parasitic on Hypotremata (exceptions being *Pseudocharopinus bicaudatus* (Krøyer, 1837), living in the spiracles of the dogfish, *Squalus acanthias*, and *Schistobrachia chimaerae* (Yamaguti, 1939), a parasite of a holocephalan fish). In contrast, the branch exemplified by the genus *Lernaeopoda* has members mainly parasitic on sharks (though here also there are a few exceptions; there are records of *Lernaeopoda* living on myliobatid rays).

Though generally similar to the short-necked members of the *Charopinus*-branch, those of the *Lernaeopoda*-branch can be distinguished from them by the position of the posterior processes. They invariably possess only one pair (modified uropods), always *ventral* to the egg sacs (Fig. 1649).

The *Lernaeopoda*-branch hitherto comprised only the genus from which it takes its name. Indeed, there is nothing obvious in the morphology of the females belonging to it which could suggest otherwise. The same, however, is not true of the males. They fall into two categories: one is exemplified by the male of *Lernaeopoda bidiscalis* Kane, 1892 (Figs. 1662, 1663) and the male of *L. galei* Krøyer, 1837 (Fig. 1685). The distinguishing features of the latter are: the shape of the cephalothorax (characterized by a distinct dorsal swelling) and the shape and position of the posterior processes, always inflated and deflected in an anterodorsal direction along the lateral walls of the trunk. The other type is that shown in Figs. 1701, 1702. Its cephalothorax is devoid of a prominent dorsal swelling and is usually less distinctly separated from the trunk by a shallow constriction. Its posterior processes are not inflated, cylindrical, and point in a posteroventral direction. As was the case with the first type, more than a single species has males of the second type (*L. globosa* Leigh-Sharpe, 1918 (Figs. 1701, 1702), *L. centroscyllii* Hansen, 1923 (cf. Kabata, 1964b), *L. longicaudata* Hansen, 1923, and others). Since Wilson's (1915) work it has been recognized that the type of male is of paramount importance in the taxonomy of Lernaeopodidae. In view of these distinct differences in male morphology, it is proposed later in this book that the genus *Lernaeopoda* should be subdivided and that those species with male morphology of the second type should be placed in a new genus *Albionella*.

The *Lernaeopoda*-branch, now consisting of two genera, has attained about the same level of development as the *Charopinus*-branch. The big difference between them is that the former has given rise to another branch of the lernaeopodid family, referred to here as the *Brachiella*-branch. The members of this branch are all parasitic on the teleost fishes, often of phylogenetically advanced groups, and are rich in species and in variability of morphological characteristics.

Before outlining the features of the relationships within the *Brachiella*-branch, mention must be made of a monotypic genus *Charopinopsis* Yamaguti, 1963, which constitutes a stage intermediate between the *Lernaeopoda*- and *Brachiella*-branch. The cephalothorax of *Charopinopsis quaternia* (Wilson, 1935) is cylindrical but only moderately elongated, much shorter than in typical representatives of the *Brachiella*-branch. It has a long trunk with two modified uropods ventral to the egg sacs and with the posterolateral corners prolonged into prominent pointed processes (Kabata, 1964a). Its first maxilla has lost the third terminal papilla and its exopod is no longer lateral but ventral in position. Two secondary teeth are present in the mandible. (The name of the genus, suggesting a relationship with the *Charopinus*-branch, was chosen by Yamaguti (1963) in the mistaken belief that the species has no bulla.)

The taxonomic position within the *Brachiella*-branch has long been confused and variously interpreted. This is mainly due to the fact that, in addition to morphological features they have in common, the members of the branch display a bewildering array of very variable characteristics, particularly in the structure of the trunk and its processes, both posterior and otherwise. This alone suffices to cause many difficulties in the orderly arrangement of the species into genera.

Three facts from the taxonomy of this group are worth recounting as a good illustration of these difficulties. The first one, mentioned earlier, was the creation by Wilson (1915) of spurious genera through an overhasty application of his own rule about the morphology of the males. The second was the establishment by Heegaard (1947b) of the genus *Isobrachia*, rather tentatively proposed and based unfortunately on the examination of a defective male specimen. Yamaguti (1963) followed this by transferring into *Isobrachia* many species, thus further complicating the already chaotic taxonomic situation. (Although Pillai (1968b) proved to this writer's satisfaction that Heegaard's *Isobrachia appendiculata* was a *Brachiella* as then constituted, the fact has not yet been generally recognized.) The third serious reason for confusion is the difficulty in drawing a dividing line between some members of the *Brachiella*- and *Clavella*-branch, unrelated but superficially similar lernaeopodids distinguishable chiefly by the morphology of their males. Since for many species the males have not yet been discovered, the difficulty is seriously compounded. A vexing expression of this difficulty is Yamaguti's (1963) interpretation of the genus *Clavellopsis* which will be discussed at the appropriate place.

The characteristic features of the females belonging to the *Brachiella*-branch are: cylindrical cephalothorax (either long or moderately long) with or without a well-developed dorsal shield, uropods usually, but not always, ventral to the egg sacs (though absent in e.g. mature *Eubrachiella gaini*), second antenna no longer prehensile, mandible with at least two but mainly three secondary teeth (only one in *Neobrachiella regia* (Lewis, 1967)), more or less reduced third terminal papilla on the first maxilla. In addition to the modified uropods, trunks are often equipped with one or more pairs of posterior processes of varying position and structure. Other outgrowths may be present on the trunks as well. These trunk features, however, cannot be taken as taxonomically significant at the generic level and even at the species level they have to be used with caution, since they might be subject (at least in some species) to individual and age variations.

With the foregoing in mind, we can divide the branch into four genera, which fall into two distinct groups, one made up of three genera and the other of one genus. The former comprises *Brachiella*, *Eubrachiella* and *Neobrachiella*. Genera *Parabrachiella*, *Probrachiella*, *Epibrachiella* and *Brachiellina* are rejected and *Brachiella* is divided into two. The division is based on the differences in the type of the male. The male of the type species, *Brachiella thynni* Cuvier, 1830, is illustrated in Figs. 1746–1756.

Its most easily observable characteristic feature is the posterior end of the trunk, ventrally carrying a set of external genitalia (Fig. 1756). The first maxilla (Fig. 1752) has three terminal papillae of equal sizes. The maxilliped (Fig. 1754) is long, slender and has a very characteristic subchela, much better developed than in males of other genera of the branch. Comparing these features with the corresponding features of the male *Neobrachiella* (Fig. 1788), one finds that the posterior end of the latter is devoid of external genitalia (Fig. 1804). The first maxilla is either completely devoid of the third terminal papilla, or has it reduced to a small seta (Fig. 1808). Finally, the structure of the maxilliped (Fig. 1831) is very different from that of the male *Brachiella*. Taking into account the phylogenetic importance of the third papilla on the first maxilla, one cannot but conclude that the two males must not be placed in the same genus. (Differences of a similar order can be found between the females of *Brachiella* and *Neobrachiella*.) The distinctness of the genus *Eubrachiella* is also based primarily on the morphology of the male (Kabata and Gusev, 1966; Kabata et al., 1971), its female resembling those of the other genera, though it shows a tendency towards reduction in size and number of the posterior processes and to the development of lateral trunk outgrowths.

The second group of this branch comprises the genus *Thysanote*. (The genus *Thysanotella* Wilson, 1915, was justifiably rejected by Pillai (1971).) As has been stated earlier, the distinctive features of *Thysanote* are mainly the number and arrangement of the trunk processes. Its males closely resemble those of the genus *Neobrachiella*. *Thysanote* is not represented in British fauna.

(The phylogenetic relationships within the *Brachiella*-branch, as outlined above, differ from those proposed by Kabata and Cousens (1972). When these authors wrote their comments, their view of the phylogeny of *Brachiella* and its allies was not yet completely formed. It is quite probable that the present account will also be subject to modifications in the future. The lernaeopodids of the *Brachiella*-branch belong to the most actively speciating members of their family, as witnessed by their morphological diversity and greatly varied host affiliations. Besides, there are still many undescribed forms, the knowledge of which might influence the views of future investigators.)

It is convenient at this point to mention an aberrant genus *Clavellistes* Shiino, 1963, whose species are parasitic on the moonfishes of the genus *Lampris* (Lampridae: Acanthopterigii). Of the two known species of the genus, one occurs in British fauna. This species shows a mixture of features, some of which relate it to the *Brachiella*-branch of the family, whereas others are reminiscent of the *Clavella*-branch. *Clavellistes lampri* (Fig. 1711) is primitive in possessing no posterior processes and no secondary teeth in the mandibular dentition. On the other hand, its cephalothorax is cylindrical, even if only moderately long (Fig. 1712), and its first maxilla has only two full-sized terminal papillae. The male is very much like that of *Neobrachiella* (Fig. 1722). It is permissible to conclude, therefore, that the genus evolved by separating from the main trunk of the family at about the same time as did the three major branches, before the development of the secondary teeth in the mandible of most lernaeopodids, particularly those that do not possess abbreviated males.

It is the last major branch of the family, the *Clavella*-branch, that is characterized by morphological abbreviation of the males of its species. Another of its typical features is the comparatively poor development of the second maxillae, accompanied by the advanced condition and great length of the cephalothorax. Posterior processes occur in this branch only very exceptionally, as do uropods. When the latter are present, they occur in a position *dorsal* to the egg sacs, as in *Clavellisa emarginata* (Fig. 1949). This prompted Kabata and Cousens (1972) to suggest that the *Clavella*-branch has evolved from *Charopinus*-like ancestors. It is more probable, however, that the origins of this branch are to be sought further back along the evolutionary tree, particularly in view of the unique type of male morphology, as well as of the totally different course of ontogeny. Moreover, the second maxillae of its members are difficult to compare with those belonging to the *Charopinus*-branch. Finally, it is among the females of this group that one finds uniquely vestigial thoracic legs still present along the margins of the trunk (in some species of *Clavellisa*). It is, indeed, possible that further studies will make it imperative to place the *Clavella*-branch in a separate family. Gurney (1934) made that proposal implicitly some time ago, by referring to *Clavella* as belonging to the family Clavellidae, though he did not make an explicit proposal for the establishment of this family. In the view of this writer such action is rather premature, but the differences between the *Clavella*-branch and the rest of the lernaeopodids should not be lost from sight.

Early in the development of the branch it produced a very aberrant offshoot in the form of the genus *Nectobrachia*, consisting of two species. Its affiliation with the branch is shown by the abbreviated morphology of the male, in which the posterior tagma is still present, however, in the form of a ventral tubercle of noticeable size (Markevich, 1956). The mandible of *Nectobrachia* is without secondary teeth, an indication of a primitive condition, but the reduction of the cephalothorax (Text figs. 61F and 62F), a mere cone on the anterior margin of the trunk, is grossly aberrant.

In the course of its development, the branch divided into two, one of the divisions largely retaining primitive mandibular dentition, the other developing three secondary teeth similar in arrangement to those in the mandibles of other branches. The former again divided to separate the genus *Clavella* Oken, 1816, from the remaining genera. The members of the genus *Clavella* are distinct from all other members of the family in the structure of their second antennae, which have the long axis running through sympod and endopod (Kabata, 1963d), rather than exopod. The mandible of this genus is also unusual in its dental formula. All its members have long cephalothoraces, sometimes longer than the trunks, cylindrical and with moderately developed "head" and dorsal shield. Their second maxillae are short and may, or may not, be fused to each other (e.g. *Clavella stellata*). They may possess a genital process, but never posterior processes. The genera *Anaclavella* Heegaard, 1940, and *Clavellomimus* Kabata, 1969, appear to be closely related to each other and are difficult to distinguish, particularly in view of the fact that the male of the former has not yet been discovered. Both are devoid of posterior processes and have primitive mandibles without secondary teeth. There are three more genera in this division of the *Clavella*-branch, *Alella*, *Proclavellodes* Kabata, 1967, and *Advena*. The first two are distinguished by the modification of their second maxillae (Fig. 1978) and the last by the extension of the posterolateral corners of its trunk into fairly long posterior processes (Fig. 1996). The morphology of the male and the type of mandibular dentition further establish their differences. (The newly-proposed genus *Nudiclavella*, if proved valid, should be placed in this division. So far, however, it has been known only from one female specimen. More information on it is needed than was provided in the original description by Ho (1975).)

The third major division of this branch comprises a group of four genera, each with one or only few species. They are: *Clavellodes*, *Clavellopsis*, *Euclavellisa* and *Clavellisa*. All of them differ from their relatives of the *Clavella*-branch in having an advanced type of mandibular dentition, usually with three secondary teeth present. The genus *Euclavellisa* is also distinguished by the morphology of its male, intermediate between the "normal" male of *Lernaeopoda*- or *Brachiella*-branch, on the one hand, and the "abbreviated" male on the other. The male, while resembling closely that of *Clavellisa* (Figs. 1950, 1968), differs from it in having a fairly long, and apparently segmented, vestige of posterior tagma, situated in the position of the genital tubercle in *Clavellisa* (cf. Heegaard, 1940). The second maxillae, generally poorly developed in these genera, are completely obsolete in *Clavellopsis*, in which the bulla emerges directly from the anterior margin of the trunk. *Clavellopsis* is also the only genus of this group to possess posterior processes (Wilson, 1913, 1915) other than very greatly reduced uropods. (The latter, as mentioned above, occur in some species of *Clavellisa*.)

Many species of the *Clavella*-branch are associated with hosts belonging to Gadidae, though *Clavellisa* is mainly parasitic on clupeid fishes, with some species on scombrids. The hosts, however, include also many teleosts of advanced orders.

Summing up the possible evolutionary trends among Lernaeopodidae, we find that progress has moved towards the development of an increasingly longer, cylindrical cephalothorax and/or of long second maxillae, both increasing the area available for browsing. Other developments include gradual departure of the mandibular dentition from the primitive siphonostome pattern, reduction of the third terminal papilla on the first maxilla, accompanied by a shift from lateral to ventral in the position of its exopod, and change in the function of the second antenna from prehensile to sensory tactile. These changes have been accompanied by the development of posterior processes. What the function of the latter is, or whether, indeed, they have a function at all, is as yet impossible to say.

One important point arises from the proposal of a phylogenetic tree, as shown in Text fig. 67. The tendency to evolve advanced features of a similar nature has become manifest in several branches of the family independently. The specific radiation, therefore, was accompanied by an incipient convergence, or at least parallelism. In consequence, the traditional grouping of the family into subfamilies, having been based on gross morphological similarities of no phylogenetic significance, has little meaning. Should one try to set up subfamilial taxa on the basis of the proposed family tree, one would have to consider five subfamilies, corresponding to five major branches (*Salmincola*, *Lernaeopoda*, *Charopinus*, *Brachiella* and *Clavella*). This arrangement, however, would have to leave several genera incerte sedis. This writer believes that such grouping would add little to our understanding of the intrafamilial relationships of Lernaeopodidae and that it is superfluous.

Key to the genera of British Lernaeopodidae (females)

1. Parasites of freshwater fishes 2
   Parasites of marine fishes 4

2. Cephalothorax as long as, or nearly as long as, trunk, cylindrical (Fig. 1517) . . . . . ***Tracheliastes***
   Cephalothorax much shorter than trunk, flat, suboval or subtriangular (Figs. 1464, 1502) 3

3. Posterior extremity pointed, protruding well beyond the openings of oviducts (Fig. 1501); mandible with one secondary tooth (Fig. 1511) . . . . . ***Achtheres***
   Posterior extremity rounded (Figs. 1463, 1486), not protruding far beyond the openings of oviducts; no secondary teeth in mandible (Fig. 1476.) . . . . . ***Salmincola***

4. Cephalothorax short, dorsoventrally flattened, in line with (Fig. 1531) or ventrally inclined to trunk (Fig. 1598) 5
   Cephalothorax cylindrical, less than $\frac{1}{4}$ of trunk length (Figs. 1555, 1712) 7
   Cephalothorax cylindrical, more than $\frac{1}{4}$ of trunk length 8

5. Posterior processes absent (Fig. 1530) . . . . . ***Ommatokoita***
   One pair of posterior processes (modified uropods) present, *ventral* to egg sacs (Fig. 1674) . . . . . ***Lernaeopoda*** and ***Albionella***
   One pair of posterior processes (modified uropods) present, *dorsal* to egg sacs (Fig. 1599) 6

6. Tips of second maxillae not modified (Figs. 1635, 1636) . . . . . ***Lernaeopodina***
   Tips of second maxillae with two long lateral processes (Fig. 1598) . . . . . ***Schistobrachia***

7. Cephalothorax as long as, or slightly shorter than, and *ventral* to second maxillae Fig. 1555) . . . . . ***Vanbenedenia***
   Cephalothorax much shorter than, and *dorsal* to, second maxillae (Fig. 1711) . . . . . ***Clavellistes***

8. Cephalothorax arising from centre of dorsal surface of trunk (Fig. 1937) . . . . . ***Clavellisa***
   Cephalothorax arising from anterior end of trunk 9

9. Posterior processes absent (Fig. 1870) 10
   One pair of posterior processes (modified uropods) present, *dorsal* to egg sacs (Fig. 1567) 11
   Posterior processes other than above 12

10. Second maxillae fused, with lateral swellings at bases (Fig. 1978) . . . . . ***Alella***
    Second maxillae short but distinguishable (Fig. 1861) . . . . . ***Clavella***
    Second maxillae reduced to a collar round base of bulla (Figs. 1914, 1924) . . . . . ***Clavellodes***

11. Maxillipeds close behind the mouth cone (Fig. 1621) . . . . . ***Pseudocharopinus***
    Maxillipeds about halfway along cephalothorax (Fig. 1566) . . . . . ***Charopinus***

12. One pair of posterior processes extending from posterolateral corners of trunk (Fig. 1996) . . . . ***Advena***
    Posterior processes (modified uropods) *ventral* to egg sacs (Fig. 1759), or two pairs (Fig. 1796) or more (Fig. 1779); first maxilla with two terminal papillae (Fig. 1766) . . . . . ***Neobrachiella***
    Posterior processes two pairs, both originating near the centre of posterior margin (Fig. 1733) and both long; first maxilla with three terminal papillae (Fig. 1742) . . . . . ***Brachiella***

Genus Salmincola Wilson, 1915

Female: Cephalothorax much shorter than trunk, dorsoventrally flattened, inclined obliquely from the long axis of trunk towards ventral side, anteriorly tapering, posteriorly rounded or transversely truncated, separated from the trunk by a distinct constriction. Trunk varying from oval to circular, dorsoventrally slightly flattened, with rounded posterior extremity not protruding far beyond the level of oviduct orifices, devoid of posterior processes. First antennae in mature females with greatly reduced or absent apical armature, short and obscurely segmented or unsegmented. Second antennae prehensile. Labrum with subtriangular tip, unarmed. Mandibles without secondary teeth. First maxillae with vestigial exopod and three terminal papillae. Second maxillae separate from each other, about as long as trunk. Bullae of various shapes, moderately large or very large. Maxillipeds with well developed palps and claws either well developed (subgenus *Salmincola*) or much reduced (subgenus *Brevibrachia*). Thoracic legs and uropods absent.

Male: Cephalothorax constituting about half of total length, in line with trunk and separated from it by shallow constriction posterior to bases of maxillipeds. Posterior extremity ventrally with or without genital plates and terminally with modified uropods. First antennae indistinctly segmented, with well retained apical armature. Second antennae and mouth parts similar to those of the female. Second maxillae short, subchelate, fused at their bases. Maxillipeds larger than but generally similar to the second maxillae. Mediative process present. Thoracic legs absent.

Type species: *Salmincola salmoneus* (L.).

Comments: Systematics of the genus *Salmincola* were reviewed and revised recently (Kabata, 1969d). In view of the general resemblance between various species of *Salmincola* and those of *Achtheres*, the following table was prepared to assist in distinguishing between species of these two genera.

Salmincola	*Achtheres*
Posterior extremity of trunk rounded, not protruding far beyond points of attachment of egg sacs (Fig. 1486).	Posterior extremity of trunk conical or pointed, protruding well beyond points of attachment of egg sacs (Fig. 1501).
First antenna usually with obscure segmentation and with apical armature damaged or absent, seldom intact (Fig. 1471).	First antenna usually distinctly three-segmented, apical armature usually intact (Fig. 1506).
Labrum roughly triangular, with unarmed margins (Fig. 1469).	Labrum armed on the tip with many setules, though sometimes very short and fine (Fig. 1503).
Mandible with primary teeth only (Fig. 1476).	Mandibular teeth include one secondary tooth, third from the tip (Fig. 1511).
First maxilla with greatly reduced exopod (often a mere spinule) (Fig. 1484).	First maxilla with well developed exopod (Fig. 1512).
Subchela of maxilliped with poorly developed armature, often with reduced claw, seldom with spinulation. Palp usually well developed (Figs. 1480, 1481).	Subchela of maxilliped with well developed armature and claw. Palp reduced to a single spine accompanied by a slightly inflated, spinulated pad (Figs. 1513–1515).

Yamaguti's (1963) usage of the generic name *Entomoda* Lamarck, 1816, for *Salmincola* is rejected by this author for reasons stated in an earlier publication (Kabata, 1969d).

The genus *Salmincola* presently includes 16 species. All occur in the northern hemisphere. According to their geographical distribution, Kabata (1969d) divided them into five groups: (i) circumpolar (*S. carpionis*, *S. edwardsii*, *S. extensus* (Kessler, 1868), *S. extumescens* (Gadd, 1901) and *S. thymalli*); (ii) continental Palearctic (*S. coregonorum* (Kessler, 1868), *S. cottidarum* Messjatzeff, 1926, *S. jacuticus* Markevich and Bauer, 1950, *S. longimanus* Gundrizer, 1974, *S. lotae* (Olsson, 1877), *S. nordmanni* (Kessler, 1868), *S. stellatus* Markevich, 1936 and *S. strigatus* (Markevich, 1936)); (iii) continental Nearctic (*S. siscowet* (Smith, 1874)); (iv) Atlantic rim (*S. salmoneus*), and (v) Pacific rim (*S. californiensis* (Dana, 1852)).

The species of the genus *Salmincola* are parasitic mainly on salmonids or related fishes. Salmonidae *sensu stricto* are parasitized predominantly by species of the subgenus *Salmincola* (*Salmincola*), whereas coregonid fishes are parasitized by those belonging to subgenus *Salmincola* (*Brevibrachia*) (cf. Kabata, 1969d). (The latter subgenus, distinguished by greatly reduced subchela of their maxillipeds, comprises *S. nordmanni*, *S. extensus*, *S. jacuticus*, *S. extumescens*, *S. strigatus* and *S. longimanus*.) *S. lotae* lives on a gadoid *Lota lota*, in Europe and Asia, and *S. cottidarum* on various cottid fishes in Lake Baykal. *S. longimanus* is a parasite of the Siberian greyling.

Two species of *Salmincola* occur in British waters.

Key to British species of *Salmincola*

Bulla hoof-shaped (Figs. 1467, 1468); maxilliped palp conical, short (Figs. 1482, 1483) . . . . . . ***S. salmoneus***

Bulla clavate (Figs. 1486, 1487); maxilliped palp long and tapering (Figs. 1496, 1498) . . . . ***S. thymalli***

Salmincola salmoneus (L.)

(Figs. 1462–1485)

Syn.: lax-lusen; of Gissler (1751)
Lernaea salmonea L., 1758
Lernea cyprinacea; of Hermann (1783)
Schisturus salmoneus (L.); of Oken (1816)
Entomoda salmonea (L.); of Lamarck (1816)
Lernaeopoda salmonea (L.); of Blainville (1822)
Basanistes salmoneus (L.); of Edwards (1840)
Pediculus salmonis Gissler; of Baird (1850)
Lernaeopoda carpionis (?)Krøyer; of Baird (1850)
Salmincola salmonea (L.); of Wilson (1915)
Salmincola gordoni Gurney, 1933
Lernaeocera cyprinacea Hermann; of Gurney (1933)

Female (Figs. 1462, 1463): Cephalothorax short (Fig. 1464), varying in length: width ratio and outline (Figs. 1465, 1466). Trunk slightly flattened in dorsoventral plane, round to oval; margin near lateral corners of posterior extremity notched with slit-like oviduct orifices (Fig. 1473); small genital process present (Fig. 1474). (In young females this process sometimes carries remnants of spermatophores (Fig. 1475); it regresses with age and is difficult to find or absent in older females.) Total length of body variable (about 7 mm according to Gurney, 1933).

First antenna (Fig. 1471) short and squat, with obscure segmentation (young females occasionally with first antennae still retaining complete apical armature and reduced whip on dorsomedial side of proximal half); apical armature (Fig. 1472) of digitiform seta 4 on medial side of apex and tubercle 1 at its base, long seta 6 on lateral side of apex and tubercle 3 near it; two setae of unequal length and with inflated bases near centre of apex representing complex 5. Second antenna (Fig. 1477) with sturdy sympod, apparently two-segmented; on lateral wall of second segment patch of small denticles (Fig. 1477) which often break off, leaving scars; exopod bulbous, one-segmented, with obsolete border between it and sympod, tip dentiferous (denticles sometimes difficult to see), with two prominent setiform papillae; endopod two-segmented; basal segment cylindrical, short and with patch of small denticles on ventral wall; distal segment prehensile, distally (Figs. 1478, 1479) with powerful hook 1 with spiniform seta 2 on medial side and papilla 3, with or without nipple, on lateral side; slender conical process 4 on ventral side, with smaller process 5 arising from its side. Mouth siphonostome, lips partially fused; tip of labrum (Figs. 1469, 1470) unarmed, varying somewhat in shape from specimen to specimen. Mandible (Fig. 1476) short, unsegmented and flat, its longitudinal margins thin and blade-like; dentiferous margin with series of seven almost uniform teeth. First maxilla (Fig. 1484) with three well developed terminal papillae, ventralmost usually shorter than other two, all surmounted by robust, flexible setae; exopod represented by small lateral swelling with one minute, dentiform setule at apex. Second maxilla (Figs. 1462, 1463) uniform in thickness, cylindrical, separate from opposite member except at tip; latter may or may not form small collar round base of bulla. Bulla (Figs. 1467, 1468) ovate in supra-anchoral plane, plano-convex, anchoral surface flatly adhering to surface of host; manubrium arising from supra-anchoral surface near broader end of bulla at angle of about 45°; broader end of anchor often indented. Maxilliped (Figs. 1480–1483) close behind mouth tube (Fig. 1464), often reaching level of its tip; corpus (Fig. 1480) robust, with prominent conical palp in myxal area (in young females palp surmounted by three outgrowths (Fig. 1482), but often damaged, with reduced apical armature (Fig. 1483) or without any); subchela short, with small seta near base (Fig. 1481), ending in gently hooked claw; at base of claw, on concave margin papilliform barb, sometimes with finely elongated tip.

Male (Fig. 1485): Cephalothorax short and narrow, joining trunk at shallow constriction posterior to level of cephalothoracic appendages. Both tagmata with same longitudinal axis. Trunk cylindrical, posteriorly with pair of genital plates and modified uropods. Thoracic legs absent. Total length not more than 2 mm.

Appendages mainly similar to those of pre-metamorphosis female. First antenna usually clearly three-segmented, with well developed apical armature. Second antenna, mouth tube, mandible and first maxilla as in female. Second maxilla and maxilliped subchelate. Prominent lateral swellings on both sides of second maxillae. Mediative process present between maxillipeds.

Comments: *S. salmoneus* is a parasite specific to fishes of the genus *Salmo*. Its most common host is *Salmo salar*, though it is often carried by various species of trout. A number of unusual hosts have also been recorded (e.g. *Thymallus thymallus*, *Salvelinus alpinus* and *Oncorhynchus keta*). Although it is possible that accidental infections of such hosts with *S. salmoneus* might occur, these reports are probably based on incorrect identifications.

Distribution of *S. salmoneus* is co-extensive with that of range of its host species. It occurs commonly in freshwater habitats of Europe and the Atlantic seaboard of North America. It is widespread both in Great Britain (records ranging from Plymouth to Scotland) and in Ireland.

As the oldest species of its genus, *S. salmoneus* has undergone many changes in its generic status. The history of the species, however, is almost free of complications at the species level. The synonomy of *S. salmoneus* includes only one other species name, *S. gordoni* Gurney, 1933. The specimens on which Gurney (1933) based this species lived on hosts other than *Salmo salar*. This fact in itself could be sufficient to account for the slight difference, i.e. "very much larger" cephalothorax, between them and those living on salmon. The comparison of Gurney's drawings of *S. salmoneus* (Gurney, 1933, fig. 1984, p. 349) and *S. gordoni* (Gurney, 1933, fig. 2010, p. 358) is at variance with this statement. The relative sizes of the cephalothoraces in these specimens certainly do not warrant separation in different taxa. The differences are well within the range of variability of the species. Gurney stated: "The two species are, however, very close in other respects and it is possible that this species should be regarded as a subspecies, or even as the parent form, of *S. salmonea*." It is not clear what Gurney meant by "parent form", but certainly no justification can be found for the continued recognition of *S. gordoni*. It has been proposed recently, therefore (Kabata, 1969d), that it be relegated to synonymy with *S. salmoneus*.

Salmincola thymalli (Kessler, 1868)

(Figs. 1486–1499)

Syn.: *Lernaeopoda thymalli* Kessler, 1868
Lernaeopoda clavigera Olsson, 1872
Salmincola baicalensis Messjatzeff, 1926
Salmincola thymalli thymalli; of Markevich (1937)
Salmincola thymalli baicalensis; of Markevich (1937)
Salmincola thymalli mongolicus; of Gundrizer (1972)

Female (Figs. 1486, 1487): Cephalothorax (Fig. 1488) fairly long, with wide posterior end, tapering anterior tip (Fig. 1489). Trunk varying from elongate oval to almost orbicular shape, barely longer than broad. Length of mature female about 3 mm (Gurney, 1933).

Second antenna (Figs. 1490–1492) similar to that of type species, varying in details of structure from one geographical area to another; distal segment of endopod dominated by hook 1, its median spiniform seta robust, with inflated base; in Palearctic specimens tubercle 3 usually large, with very prominent nipple, in Nearctic much smaller; ventral processes 4 and 5 in Palearctic specimens smooth, in Nearctic with crenated surfaces; armature of exopod also variable, consisting either of very small spiniform processes (Fig. 1491) or of relatively sparse but much larger ones (Fig. 1492); former type apparently predominating in Palearctic region. Mouth tube and tip of labrum resembling those of *S. salmoneus*. Mandible (Figs. 1493, 1494) with series of seven teeth rapidly decreasing in size towards proximal end; in some specimens only six normal teeth present, first normal tooth preceded by small denticle. First maxilla (Fig. 1495) with dorsal and central terminal papillae relatively larger than in most other species of *Salmincola*; ventral papilla very small, its apical seta short; exopod simple dentiform outgrowth arising from lateral wall, without any noticeable swelling. Second maxilla (Figs. 1486, 1487) of uniform diameter and moderate length, often producing marked inflation around base of bulla. Latter clavate (distinguishable from all other species of genus), usually fairly long, often as long as second maxilla itself. (In some specimens, however, it can be only half as long as those depicted in drawings. This largely depends on conditions at the site of attachment.) Maxilliped (Figs. 1496–1498) short and squat, perching on large and inflated base; palp long and slender, carrying on its apex one or two small denticles (some palps are damaged or broken, often only half as long as that shown in Fig. 1498; these can usually be recognized by examination of tip); subchela with strong, though slender, claw (Fig. 1497), sharply curved in its distal half; diameter of claw often irregular, barb long and cylindrical, tipped with fine point.

Male (Fig. 1499 (*fide* Gurney, 1933)) resembles that of *S. salmoneus* and probably also of other species. Apparent trunk segmentation, described and illustrated by Gurney, very likely post-mortem artifact, cannot be taken into account in diagnosis of species. Length about 1 mm (after Gurney, 1933).

Comments: The distribution of *S. thymalli* is circumpolar. In view of the wide distribution range, some differences in morphology between populations from widely separated localities are not surprising. Some morphological differences, however, have no regional significance. For example, the armature of the exopod of the second antenna in the Nearctic region can be either similar to, or quite different from, that found in the Palearctic region. Other differences between specimens from these two regions appear to be more constant. The author was unable to find any Old World specimens with crenation of the ventral process 4 of the endopod of the second antenna, whereas the New World specimens invariably had these processes distinctly crenated. It is tempting to speculate on the possibility of incipient speciation caused by geographical separation of these populations. Markevich (1937, 1956) treated *S. thymalli* as consisting of two subspecies, *S. thymalli thymalli* and *S. thymalli baicalensis*. (Petrushevski and Bauer (1948) referred to the latter as *S. thymalli* v. *baicalensis*.) Gundrizer (1972) recognized yet another subspecies, *S. thymalli mongolicus*. The basic similarities of these taxa to one another are, however, compelling. They certainly outweigh by far any differences which, in any case, appear to be rather inconstant. On the whole, these infraspecific groupings appear to be of little help for our understanding of this polymorphic species.

S. thymalli is a parasite of the genus *Thymallus*. *T. thymallus*, *T. arcticus*, *T. brevirostris* and *T. signifer* have been recorded as its hosts. Other hosts comprise *Salvelinus alpinus*, *Coregonus lavaretus* and *Prosopium* sp. The British host of this parasite is *Thymallus thymallus*.

Genus Achtheres Nordmann, 1832

Female: Cephalothorax much shorter than trunk, dorsoventrally flattened, inclined obliquely from long axis of trunk towards ventral side, anteriorly tapering, posteriorly rounded or transversely truncated, separated from trunk by distinct constriction. Trunk dorsoventrally slightly flattened, its posterior extremity conical and protruding markedly beyond level of oviduct orfices. First antenna three- or four-segmented, with well developed apical armature. Second antenna prehensile. Labrum with transversely truncated tip provided with marginal fringe of setae. Mandible with one secondary tooth, third from tip. First maxilla with vestigial exopod on lateral side and with three well developed terminal papillae. Second maxillae separate from each other, slightly shorter than, or as long as, trunk. Bullae usually plano-convex, circular, with moderately large anchors and short manubria. Medial margin of corpus maxillipedis armed with spinulated pad and papilliform process, its subchela distally armed with barb and additional structures at base of claw. No thoracic legs or uropods.

Male: Cephalothorax constituting about half of total length, in line with trunk and separated from it by shallow constriction posterior to bases of maxillipeds. Trunk with reduced uropods. First antenna long, clearly segmented, with well developed apical armature. Second antenna and mouth parts similar to those of female. Second maxilla and maxilliped short, subchelate. Mediative process present. No thoracic legs.

Type species: *Achtheres percarum* Nordmann, 1832.

Comments: The true nature of the genus *Achtheres* remained unclear until recently, because it had never been precisely diagnosed and delimited from *Salmincola*, its close relative. Recent revision of *Salmincola* (cf. Kabata, 1969d) has resulted in transfer of several species of *Achtheres* to that genus, on purely morphological grounds. In consequence, a much-reduced *Achtheres* emerged as a group of six species with very similar morphology, geographical distribution and host affiliations. In fact, its type species differs from the other five much more than they do among one another. With the exception of that species, the remaining ones are Nearctic and parasitic mainly on silurid and ictalurid fishes. (There exist some old records of coregonid hosts that cannot be verified.) The one remaining difficulty is caused by *A. corpulentus* Kellicott, 1880, a parasite of coregonids in North America. Kabata (1969d) attempted to explain this incongruity by pointing out distinct similarities between *A. corpulentus* and *Salmincola extumescens* (Gadd, 1901). It is also significant that, with Gadd's (1901) description of *S. extumescens*, records of *A. corpulentus* all but disappeared from the literature. It is, however, not possible to resolve this problem beyond question on the basis of presently available evidence.

A table for distinguishing between *Achtheres* and *Salmincola* has been presented earlier (p. 352). In addition, it should be noted that the second antennae of *Achtheres* spp. have exopods armed with few, relatively large denticles (Figs. 1508, 1509), whereas in *Salmincola* the same ramus is usually armed with numerous and minute denticles (Fig. 1477). It is common to find in preserved specimens of *Achtheres* second antennae pointing straight out from their bases in an anterior direction. In contrast, the second antennae of *Salmincola* spp. are folded inwards more often than not and meet each other in the centre of the anterior margin of the cephalothorax.

Only the type species, *A. percarum*, has been found in British waters.

Achtheres percarum Nordmann, 1832

(Figs. 1500–1516)

Syn.: *Achtheres sandrae* Gadd, 1904
Achtheres sibirica Messjatzeff, 1926

Female (Figs. 1500–1502): Cephalothorax short, inclined obliquely towards ventral side from long axis of trunk, in some specimens at right angles to that axis, dorsoventrally flattened; in dorsal view, posterior margin of cephalothorax only slightly broader than anterior, its lateral margins concave. Transverse constriction posterior to bases of second maxillae separating cephalothorax from trunk. Latter always longer than broad, with mildly rounded lateral margins, slightly flattened dorsoventrally, often with several transverse markings roughly coinciding with divisions between four trunk segments; at level of oviduct orifices, narrowing to form conical projection, markedly extending beyond that level; apex of projection varying depending on age and maturity of specimen, pierced by vaginal orifices (Fig. 1505), with attached remnants of spermatophores (Fig. 1504); on each side of apex single subterminal process pointing in posterior direction; varying from short conical protuberances (Fig. 1505) to digitiform tapering outgrowths extending well beyond posterior margin (Fig. 1504). General appearance of preserved specimens depending on condition of trunk, contractile tagma capable of great changes of shape. Posterior cone of trunk also greatly contractile (cf. Figs. 1501, 1502). Length 2.5–5 mm.

First antenna (Fig. 1506) long, slender, with uncertain segmentation; basal half formed by fusion of two segments, in some specimens not discernible, in others with border line better preserved; distal end of this part with short whip on dorsomedial side; terminal half of two clearly delimited cylindrical segments; first segment with solus at distal end, second with rounded apex; apical armature (Fig. 1507) of tubercle 1 at base of digitiform seta 4, flagelliform seta 6 at lateral side, and complex 5 of three robust, unequally long setae; no trace found of tubercles 2 and 3. Second antenna (Figs. 1508–1510) prehensile; exopod (Fig. 1508) bulbous, one-segmented, fused with sympod, about as long as endopod; apical armature of five large denticles arranged in dorsoventral line, with two additional and much smaller denticles, one on each side of the five (Fig. 1509); endopod two-segmented, its basal segment with patch of denticles on ventral wall (Fig. 1508); distal segment (Fig. 1510) capable of executing rotating movements around longitudinal axis of ramus, armed with large hook 1, with spiniform seta 2 on its lateral side; papilla 3 on medial side not well developed; ventral process 4 in form of inflated spinulated pad, with long, smooth, tapering process 5 arising from its medial side. Mouth tube of usual lernaeopodid type; tip of labrum (Fig. 1503) rounded but with rectangular central flap, transversely truncated anteriorly and with fringe of short fine setules, in older specimens sometimes difficult to find; no distinct rostrum present. Mandible (Fig. 1511) with nine teeth, third from tip being secondary; four proximal teeth distinctly smaller than remainder (except for secondary tooth). First maxilla (Fig. 1512) with three terminal papillae: ventralmost papilla smaller than other two and with longest terminal seta; corpus short and squat, with lateral swelling surmounted by dentiform seta marking position of exopod. (In some specimens examined by the writer, two such dentiform setae were seen in this position.) Second maxilla (Figs. 1500–1502) cylindrical, of uniform diameter, separate from its opposite number except at tips; terminal swellings or collar around manubrium; in completely relaxed specimens second maxillae about equal to length of trunk, but can contract to less than half that size (Fig. 1502), becoming covered with profuse transverse wrinkling. Bulla (Fig. 1500) with button-shaped, plano-convex anchor and short central manubrium. Maxilliped (Figs. 1513–1515) with robust corpus and moderately short subchela; myxa with raised pad covered by numerous fine denticles; ventral to and slightly projecting beyond pad is papilliform outgrowth with fine tip and inflated base (Fig. 1515); short seta on ventral surface of subchela, near its base; at base of claw, distal end of subchela armed with barb, about half length of claw and with inner surface covered by rounded denticles; another and shorter outgrowth present on inner side of barb, with inflated base and surface covered by similar denticles; close to base of barb, short conical setiform process points obliquely ventrally; claw curving slightly in distal half, with rather blunt tip and uneven diameter (Fig. 1514).

Male (Fig. 1516) (*fide* Gurney, 1933): Cephalothorax shorter than trunk. Latter fusiform, posteriorly with uropods, anteriorly separated from cephalothorax by narrow, transverse constriction. First antenna long, cylindrical and indistinctly segmented. Second antenna and oral appendages similar to those of female. Second maxilla and maxilliped subchelate. Mediative process present. No thoracic legs. Length under 2 mm.

Comments: The taxonomic history of *A. percarum* has been relatively free of controversy. The main difficulty encountered in its definitive diagnosis was due to some morphological differences in specimens from different areas and parasitic on different hosts. Those differences prompted a description of *A. sandrae* and are noticeable enough to have caused Gurney (1933) to speculate on relationship between it and *A. percarum*. A careful morphometric study of Kozikowska et al. (1956 (1957)), however, demonstrated that these two taxa represent but a single, though morphologically variable, species.

A. percarum is a Palearctic species, parasitic on fishes of the genera *Perca* and *Lucioperca* in Europe and Asia. Its southernmost records in Europe are from northern Italy (Garbini, 1895) and Hungary (Geyer, 1939). In Asia it occurs as far south as Kazakhstan (Agapova, 1966) and Uzbekistan (Sukhenko and Allamuratov, 1967), and as far east as Lake Baykal (Zayka, 1965). Although a freshwater species, *A. percarum* has been recorded from brackish water localities, usually near mouths of rivers (Baltic: Kozikowska (1958), Grabda (1961); Black Sea: Osmanov (1940); Aral Sea: Osmanov (1959); Sea of Azov: Kamenev (1959)).

In Great Britain the occurrence of *A. percarum* on *Perca fluviatilis* was not confirmed until the sixth decade of this century (Harding and Gervers, 1956). Gurney (1933) considered its presence in this country as "very doubtful". The record of Scott and Scott (1913) was certainly in error. It is plain from their illustrations that these authors dealt with *Salmincola salmoneus* (not *S. thymalli*, as suggested by Gurney (1933)). By now it has been reported from several localities in England, its northernmost finds being in Yorkshire (Fryer, 1969b). Kennedy (1974) listed also *Cottus gobio* as its British host.

Genus Tracheliastes Nordmann, 1832

Female: Cephalothorax cylindrical, with most appendages confined to small portion at tip. Dorsal shield absent or ill-defined. Narrow constriction between cephalothorax and trunk. Latter much longer than broad, dorsoventrally flattened. Lateral margins nearly parallel, posterior margin truncated. No posterior processes. First antenna reduced, unsegmented or indistinctly segmented, with vestigial apical armature. Second antenna prehensile. Mandible with third distal tooth secondary. First maxilla reduced in size. Second maxilla forming broad "shoulders" with base of cephalothorax, cylindrical, separate from opposite member. Bulla small. Maxilliped vestigial, at base of cephalothorax.

Male: Cephalothorax constituting about half of total length, in parallel with trunk. Trunk fusiform, posteriorly provided with genital plates. Antenna and oral appendages similar to those of female. Second maxilla and maxilliped close together, short and subchelate. No thoracic legs found.

Type species: *Tracheliastes polycolpus* Nordmann, 1832.

Comments: The genus *Tracheliastes* is exclusively Palearctic. It currently numbers seven species, two of which (*T. gigas* Richiardi, 1880 and *T. mourkii* Hoffmann, 1881) have not been adequately established and must be ranked as species inquirendae. The validity of the most recent addition to the genus, *T. chondrostomi* Hanek, 1969, should also be re-examined, since its description is very superficial and its host, *Chondrostoma nasus*, is known also as a host of *T. polycolpus*. Of the remaining four species, *T. sachalinensis* Markevich, 1936, has so far been recorded only from Sakhalin and *T. longicollis* Markevich, 1940, occurs in the basin of the river Amur and other localities in that region. *T. polycolpus* and *T. maculatus* Kollar, 1835, are common in the western part of the Palearctic.

All species of *Tracheliastes* parasitize predominantly fishes of the family Cyprinidae. The genus is represented in British waters by *T. polycolpus*.

Tracheliastes polycolpus Nordmann, 1832

(Figs. 1517–1529)

Syn.: *T. polycolpus* var. *phoxini* Vejdovsky, 1877
T. polycolpus var. *baicalensis* Messjatzeff, 1926
T. polycolpus var. *kessleri* Messjatzeff, 1926

Female (Figs. 1517, 1518): Cephalothorax cylindrical, about $\frac{2}{3}$ length of trunk, uniform diameter except at tip; latter tapering, usually deflected dorsally from point of origin; dorsal shield not noticeable; base of cephalothorax fused with those of second maxillae into broad structure, separated from trunk by deep transverse constriction. Trunk dorsoventrally flattened, much longer than broad, with uneven lateral (Fig. 1517) and truncated posterior margin (Fig. 1522); in many specimens lateral indentations near posterior extremity; posterolateral lobes formed by these constrictions carrying dorsoventral oviduct orifices; in centre of posterior margin, ventral to anal slit, small genital process present; dorsal surface of trunk with two longitudinal rows of shallow, irregular depressions lined with thick cuticle. Length about 5–6 mm.

First antenna (Fig. 1523) much reduced and unsegmented, slightly pedunculate; apical armature in adult specimens badly damaged or absent; on medial side of apex seta 4 commonly retained; seta 6 on opposite side less commonly present; small tubercles, remnants of other elements of armature, also present in varying numbers. Second antenna (Fig. 1524) large and very well developed, in most specimens projecting straight beyond anterior margin of head; long sympod indistinctly two-segmented; exopod about as long as, or slightly shorter than, endopod, digitiform, one-segmented, with two prominent, tapering papillae, one near apex, other subapical; basal segment of endopod fused with sympod, ventrally with strongly spinulated pad; terminal segment of endopod (Fig. 1525) dominated by particularly strong hook 1, capable of vigorous grasping movements; relatively short spiniform seta 2 and conical papilla 3 on medial and lateral sides of hook respectively; on ventral side inflated bulbous swelling 4 covered with numerous denticles, pointing basalwards; setiform process 5 on medial wall of swelling. Mouth cone of usual structure; tip of labrum (Fig. 1521) rounded, with short rostrum, devoid of fringe of setae. Mandible (Fig. 1526) with nine teeth; third distal tooth secondary, distalmost also slightly reduced, others more or less uniform in size and shape. First maxilla (Fig. 1527) simplified, uniramous, tipped with two setae arising directly from apex, without papillae. Second maxilla (Figs. 1517–1520) cylindrical, slightly tapering towards tip, separate from opposite member; at tips, near base of manubrium, some specimens with two small, papilliform processes (Fig. 1519); no collar round base of bulla. Bulla (Figs. 1519, 1520) with small anchor, and very short manubrium, plano-convex, resembling inverted cone. Maxilliped (Fig. 1528) between bases of second maxillae, minute and functionless, devoid of intrinsic musculature, round in cross-section, tapering to its tip; subchela short, with vestigial claw not clearly demarcated, usually pointing distalwards; two duct orifices in myxal area.

Male (Fig. 1529) resembling those of other freshwater lernaeopodid genera. Cephalothorax about half of body length, in line with trunk. Latter cylindrical, posteriorly with genital plates and short uropods. Antenna and oral appendages similar to those of female. Second maxilla and maxilliped short, subchelate. Length about 0.5 mm.

Comments: Males of the species have been recorded only on very few occasions (Markevich, 1937; Yamaguti, 1940) and have not been examined by this author. Their descriptions are not detailed enough to provide accurate representation of the structure of their appendages.

T. polycolpus is parasitic predominantly on fins of cyprinid fishes, its bulla attached to fin rays (sometimes scales) by a fine layer of adhesive substance covering the subanchoral surface (Kabata and Cousens, 1972). Its hosts are drawn from many cyprinid genera (*Leuciscus*, *Rutilus*, *Phoxinus*, *Idus*, *Squalius*, *Capoeta*, *Chondrostoma*, *Barbus*, *Abramis*, *Pseudaspius*, *Varicorhinus* and *Rhodeus*). It has also been found on two species of the genus *Gobio* (Gobiidae).

Distribution of *T. polycolpus* extends across the entire Paleartic, from western Europe (Belgium:

Capart, 1944) to the river Amur (Akhmerov, 1959), Mongolia (Hanek and Dulmaa, 1970), Manchuria and Sakhalin (Yamaguti, 1940). Its range reaches south as far as Kazakhstan (Sidorov, 1959; Agapova, 1966) and Turkmenia (Sharapova, 1963). In Europe it has been recorded as far south as Hungary (Pónyi and Molnar, 1969).

In Great Britain it was first recorded by Gurney (1933). Specimens identified by him came from the Edinburgh Zoological Park and were found on fish imported from the Netherlands only a few weeks earlier. Only recently (Aubrook and Fryer, 1965) was *T. polycolpus* found in natural habitats in this country. All finds came from Yorkshire rivers, where *Leuciscus leuciscus*, *L. idus* and *L. cephalus* were infected.

Genus Ommatokoita Leigh-Sharpe, 1926

Female: Cephalothorax much shorter than trunk, dorsoventrally flattened. Trunk much longer than broad, with rounded posterior extremity. No posterior processes. First antenna three-segmented, with well developed apical armature. Second antenna prehensile. Mandible with three secondary teeth. First maxilla with lateral exopod and three terminal papillae. Second maxilla long, cylindrical, separate from opposite number. Bulla plano-convex, without manubrium. Maxilliped well developed, subchelate.

Male: Cephalothorax more than half length of body, poorly separated from, and forming obtuse angle with, trunk. Latter fusiform, posteriorly rounded, with genital plates and uropods. Two pairs of vestigial thoracic legs. Parasitic on sharks.

Monotypic. Type species: *Ommatokoita elongata* (Grant, 1827).

Ommatokoita elongata (Grant, 1827)

(Figs. 1530–1552)

Syn.: *Lernaea elongata* Grant, 1827
Lernaeopoda elongata (Grant, 1827); of Krøyer (1838)
Ommatokoita superba Leigh-Sharpe, 1926

Female (Fig. 1530): Cephalothorax much shorter than trunk, dorsoventrally flattened, slightly inclined ventrally to long axis of trunk (Fig. 1531), circular in dorsal view, with strong though ill-defined dorsal shield covering about $\frac{1}{4}$ of dorsal surface; laterally, second maxillae arising from margins of cephalothorax. Trunk demarcated from cephalothorax by very narrow and shallow groove, much longer than broad; anterior third of trunk gradually expanding towards full width, posterior $\frac{2}{3}$ with roughly parallel lateral margins; posterolateral corners rounded; centre of posterior margin (Fig. 1532) with anal tubercle protruding in posterior direction; slit-like oviduct orifices immediately lateral to anal tubercle; no posterior processes. Length of body about 20–25 mm, length of second maxillae similar or greater, length of egg sacs usually greater than that of trunk.

First antenna (Fig. 1535) indistinctly three-segmented (proximal half bearing traces of fusion of two original segments); basal part with long slender whip on dorsomedial side; second segment cylindrical, short, with solus near junction with terminal segment; latter cylindrical, with gently tapering, rounded tip bearing apical armature (Fig. 1536): robust digitiform seta 4 on medial side and tubercle 1 at its base, on lateral side seta 6, stronger and shorter than usual in family; centre of apex occupied by small gibber surmounted by three setae, about equally long though of different thicknesses. Second antenna prehensile, resembling that of *Salmincola* in size of sympod and size relationships of rami; exopod one-segmented, bulbous, with tip covered by fine spinules and equipped with two papillae; basal segment of endopod with batch of spinules on ventral wall; distal segment (Fig. 1537) with strong hook 1, spiniform seta 2 and papilla 3 near dorsal side of apex; ventral wall with bulbous spinulated swelling 4 and tapering unarmed process 5. (In Fig. 1537 full curvature of hook 1 is not shown due to tilt of segment away from observer.) Mouth of usual structure; tip of labrum (Fig. 1534) rounded and equipped with abundant tuft of flat setae; no rostrum discernible. Mandible (Fig. 1538) of advanced type, with formula: P1, S1, P1, S1, P1, S1, B5; basal teeth diminishing slightly in size in proximal direction. First maxilla (Figs. 1539, 1540) with three terminal papillae surmounted by robust, flexible setae; exopod lateral, arising at about half length of appendage, long, cylindrical, with three apical setae of unequal size; at about exopod level, dorsal margin of appendage with batch of fine denticles; another batch on medial wall at base of dorsalmost seta and not visible to observer in situ. Second maxilla (Fig. 1530) cylindrical, often longer than trunk (true length difficult to determine due to great contractility of appendage), diameter uniform throughout, except at tip; latter fused at base of bulla with opposite number; two digitiform processes near tip (Fig. 1533). Bulla (Fig. 1533) plano-convex, circular, with manubrium not visible externally (Kabata and Cousens, 1972). Maxilliped (Fig. 1541) close to mouth cone, sometimes extending as far as its tip; corpus squat and robust, myxa with two pads of fine though blunt denticles; between pads rounded, papilliform swelling surmounted by short tapering process (Fig. 1542); subchela short, with claw half its length and short seta near base on ventral wall; distal end of inner margin of subchela slightly protruding, with pad surrounded by closely set row of denticles; barb at base of claw (Fig. 1543) short and tapering; claw slender, gently curving, with some longitudinal ridges on its walls.

Male (Fig. 1544): Cephalothorax more than half of total length, poorly delimited from trunk, both tagmata forming arch towards ventral surface; part of dorsal surface covered by strong dorsal shield. Mouth cone terminal, surrounded by antennae and oral appendages. Trunk fusiform, posteriorly tapering. Two pairs of thoracic legs on anterolateral margins, posterior extremity with genital plates and uropods. Length about 2.5 mm.

First antenna similar to that of female. Second with exopod slightly smaller than endopod (equal size in female); distal segment of endopod (Fig. 1545) with long spiniform seta 2 and process 5, and with comparatively small swelling 4. Mouth cone and tip of labrum similar to those of female. In specimens examined by this author, mandibular formula differing from that of female by absence of first secondary tooth (P2, S1, P1, S1, B5) (possibly constant feature); flange proximal to dentiferous margin of mandible with several cuticular ridges on its walls (Fig. 1546). First maxilla (Fig. 1547) resembling that of female, but slender, with relatively longer terminal papillae; no denticles observed on medial wall. Second maxilla (Fig. 1549) short and squat, linked with its opposite number by tympanum with reinforced free margin; myxa with papilliform palp covered with prominent denticles; no demarcation between subchela and claw, tip tapering and curving, sharp-pointed; small depression on dorsal side of palp for reception of tip of subchela. Maxilliped (Fig. 1548) short and squat; myxa raised in sharp point towards tip of subchela; at base of myxa small digitiform palp, its surface covered by minute setules; subchela indistinctly separated from claw, border marked only by abrupt change in diameter; anterior margin of proximal half with slender seta; short setiform process (barb?) on inner margin at presumed junction of claw with subchela; claw short, rapidly tapering and sharply curving. Thoracic legs (Fig. 1550) short, papilliform, surmounted by fairly long, unarmed seta; another seta at base, shorter than papilla, possibly arising from its wall. Uropod conical (Fig. 1551), with two apical setae, short and flat, and another on dorsal wall, only slightly shorter than itself. Genital plates (Figs. 1544, 1552) flat and bifid.

Comments: In the course of its approximately 150 year old history, *O. elongata* was moved twice from genus to genus. Leigh-Sharpe (1926b) recognized its unique features (advanced mandibular dentition co-existing with primitive condition of posterior extremity of trunk) by placing it in a genus of its own. It is unfortunate that at the same time he decided to create a new species (*O. superba*) for specimens he had collected from the corneal surface of the Greenland shark, *Somniosus microcephalus*. His decision was influenced by Scott and Scott (1913), whose description of *Lernaeopoda elongata* (= *Ommatokoita*) differed from his own observations. Moreover, Scott and Scott gave *Lamna nasus* as the host of *O. elongata* and quoted Norman who found it on *Mustelus mustelus*.

However, the only serious difference between figures shown in this account and those of Scott and Scott (1913) is in the segmentation and armature of the first antenna, features difficult to study and often given only a cursory treatment by earlier authors. With our present knowledge of frequently erroneous early descriptions and host identifications, we can now safely assume that there is only one lernaeopodid species infecting the cornea of the North Atlantic shark, *Somniosus microcephalus*. It is possible that the same species occurs occasionally on *Lamna nasus* (and perhaps other species of sharks), though the author never saw it on that fish in spite of examining many specimens in various localities in the North Atlantic.

On the cornea of *Somniosus microcephalus* this species is very common. Berland (1961), having examined 1,505 specimens of this shark, found that 84.4% had parasites on both eyes and 14.5% on one eye. Only 1.1% were free of the copepod.

Under different synonyms, *S. microcephalus* was found to carry *O. elongata* in Greenland (Miers, 1880; Stephensen, 1913, 1943; Hansen, 1923), Iceland (Stephensen, 1940), Barents Sea (Polyanski, 1955; Markevich, 1956), Faroe Is. (Stephensen, 1929), Danish Straits (Olsson, 1869) and off the Belgian coast (van Beneden, 1871). T. Scott (1900) found it in Scottish waters on *Lamna nasus*. Canon Norman's find on *Mustelus mustelus* was made off Polperro (Cornwall) and is probably the southernmost record of the species.

Genus Vanbenedenia Malm, 1860

Female: Cephalothorax cylindrical, with antennae and buccal apparatus confined to apical region. Base of cephalothorax fused with bases of second maxillae to form broad "shoulders". Latter delimited from trunk by very deep transverse constriction. Trunk much longer than broad, with roughly parallel lateral margins and transversely truncated posterior margin. No posterior processes or uropods. First antennae short, unsegmented or obscurely segmented, with well developed apical armature. Second antennae prehensile. Mandibles with third distal tooth secondary. First maxillae with three terminal papillae and either with or without exopod. Second maxillae cylindrical, about as long as cephalothorax, joined together by tapering tips on its dorsal side. Bulla subspherical with long or moderately long manubrium. Maxilliped regressed, separated from buccal region, at base of cephalothorax.

Male: Cephalothorax about half of total body length, in line with long axis of trunk, anteriorly with terminal mouth cone, buccal appendages and antennae, posteriorly passing into trunk without clear demarcation. Trunk fusiform, tapering posteriorly and with uropods. Antennae and buccal appendages similar to those of female. Second maxillae and maxillipeds short, subchelate, on posterior half of ventral surface of cephalothorax.

Type species: *Vanbenedenia kroeyeri* Malm, 1860.

Comments: Of the three species currently belonging to this genus, only *V. kroeyeri*, a North Atlantic species, is fairly well known. *V. chimaerae* (Heegaard, 1962), originally described as *Tracheliastes*, was recorded only once from the claspers of *Chimaera ogilbyi*, taken off the coast of southern Australia. Its position in the genus and its development were discussed by Kabata (1964c). The third species, *V. grandis* (Wilson, 1915), was assigned to this genus by Kabata and Bowman (1961) and is known only from specimens belonging to the collection of the Smithsonian Institution, Washington, D.C. Neither host nor place of origin of specimens have been recorded.
The only described male of *Vanbenedenia* is that of *V. chimaerae*.

Vanbenedenia kroeyeri Malm, 1860

(Figs. 1553–1565)

Female (Fig. 1553): Cephalothorax cylindrical, much shorter than trunk, with antennae and buccal apparatus at anterior end; buccal cone retractable. Base of cephalothorax fused with those of second maxillae into broad tagma (Figs. 1554, 1555). Trunk separated from cephalothorax by deep transverse constriction, more than four times longer than combined length of cephalothorax and second maxillae; lateral margins of dorsoventrally flattened trunk roughly parallel to each other, though converging slightly at anterior and posterior extremities (in young adult females trunk much shorter and narrower, being sometimes barely twice as long as anterior tagma (Kabata, 1958b)); two longitudinal rows of depressions along dorsal and ventral surfaces of trunk (Fig. 1553) marking positions of attachment of dorsoventral musculature; posterior margin transversely truncated, with dorsoventral anal slit in the centre and oviduct orifices on either side of slit. Egg sacs cylindrical, usually longer than trunk, straight. Eggs spherical, multiserial. Length of adult female 23–28 mm.

First antenna (Fig. 1557) short, unsegmented or obscurely segmented, with slightly inflated base and tip, narrower at midlength; on dorsomedial side of inflated basal part short, robust whip; apical armature (Fig. 1558) of four setae: seta 4 on medial side of apex, long and more slender seta 6 on lateral side, apex occupied by double seta 5, in some specimens on top of small gibber-like swelling; occasionally (Kabata and Bowman, 1961) a small tubercle 1 at base of seta 4. Second antenna (Fig. 1560) with robust, indistinctly two-segmented sympod (in preserved specimens position of this appendage is such that exopod always appears lateral to endopod); second segment of sympod with spiny pad on lateral wall; exopod much smaller than endopod, roughly pyriform and with several digitiform processes at tip (usually four, sometimes five); occasionally some of them appearing bifid or carrying secondary outgrowths; endopod two-segmented; terminal segment (Fig. 1561) with very powerful hook 1 and well developed spiniform seta 2; tubercle 3 often prominent and with prominent nipple; ventral side of segment provided with fairly short and rounded process 4; process 5, longer and tapering, arising from its medial side. Mouth cone of usual structure; tip of labrum subtriangular (Fig. 1556), with slender, pointed rostrum and abundant fringe of fine setae. Mandible (Fig. 1562) with secondary third distal tooth; teeth proximal to secondary tooth varying from four to six, latter number being more common; teeth tending to diminsh in dimensions towards base of appendage. First maxilla (Fig. 1559) uniramous, short; three terminal papillae with robust flexible setae; ventralmost papilla smaller than other two. Second maxilla (Figs. 1554, 1555) slightly longer than cylindrical part of cephalothorax, cylindrical, tapering towards tip, laterally often with small conical swelling (Fig. 1554) and with duct orifice opening ventrally at base of cylindrical part; members of this pair abut on each other along most of their lengths, so as to prevent cephalothorax, which is ventral to them (Fig. 1555), from passing to dorsal side. Bulla (Figs. 1554, 1555) with fairly short manubrium and subspherical or lenticular anchor, heavily sclerotized, brown in colour. Maxilliped (Fig. 1563) at base of cephalothorax, separated by long distance from other appendages, with reduced claw; corpus well developed, heavy, with distal part sclerotized, myxa with prominent medial process meeting tip of subchela; latter short, not clearly delimited from short and blunt-ended claw; at base of claw short, peg-like barb (Fig. 1564).

Male of the species was not observed. The first report of a male (Malm, 1863) was in error, as can be seen from the illustration depicting it with second maxillae attached to frontal filament. This suggests that the report refers, in fact, to a late chalimus stage, not an adult. The same is true of the alleged male described by Kabata (1958b). When found, the male in all likelihood will not differ greatly from the late chalimus stage, such as is shown in Fig. 1565.

Comments: *Vanbenedenia kroeyeri* is a rare parasite. Kabata (1960a) found only three infected specimens in a haul of 104 *Chimaera monstrosa* at the southern end of the Faroe Islands. Since *Ch. monstrosa*, its only known host species, is also rather rare and since the copepod lives only attached to its dorsal spine, it is not often encountered. There are, as yet, no records of it from British waters. However, *Ch. monstrosa* is a well-known member of British fish fauna and discovery of *V. kroeyeri* here is only a matter of time.

Following its description (Malm, 1860), *V. kroeyeri* was recorded several times in fairly rapid succession (Malm, 1863; Olsson, 1869; van Beneden, 1871; Carus, 1885). There followed a long period during which no more was heard of it, so that its 20th century finds amounted almost to a rediscovery. Kabata (1958b) found it in the Skagerrak region and within a few months Tambs-Lyche (1958) took it near the coast of West Norway. Since then, Bresciani and Lützen (1962) listed it in the fauna of Sweden and Dienske (1968) recorded it off the Dutch coast.

All records of the species come from the Atlantic sea-board of Europe, with the exception of Kabata's (1960a) record in the Faroes.

A unique feature in the biology of *V. kroeyeri* is the fact that the larvae attach their frontal filaments not directly to the host but to the surface of the maternal individual, on which they develop up to the stage when they become capable of movement and attachment to the host. This preferential attachment to the cephalothoracic region of the females was the main reason why they were mistaken for adult males. (A similar phenomenon occurs in *V. chimaerae*.)

Genus Charopinus (Krøyer, 1863)

Female: Cephalothorax cylindrical, about $\frac{2}{3}$ as long as trunk. Head not clearly delimited, dorsal shield indistinct. Trunk slightly flattened dorsoventrally, longer than broad, with greatest width usually near posterior extremity. First antenna indistinctly three- or four-segmented, with well developed apical armature. Second antenna prehensile. Mandible with three secondary teeth. First maxilla with three terminal papillae and well developed exopod in lateral position. Second maxilla as long as, or longer than, trunk, separate along its entire length. Bulla relatively large, transversely elongate, with or without manubrium. Maxilliped reduced, about halfway between tip and base of cephalothorax. Modified uropod fusiform, about as long as cephalothorax, dorsal to egg sacs.

Male: Cephalothorax curving ventrally, its anterior portion at nearly right angles to posterior, large mouth cone subterminal and flexing ventrally to axis of anterior part of cephalothorax. Trunk indistinctly delimited from, and smaller than, cephalothorax, without distinct division between them. Antennae and oral appendages similar to those of female, second maxilla and maxilliped relatively large, of about equal size, subchelate. Two pairs of vestigial thoracic legs present. Posterior extremity with external genitalia and prominent uropods.

Type species: *Charopinus dalmanni* (Retzius, 1829).

Comments: Until its recent revision (Kabata, 1964d), the genus *Charopinus* was a catch-all taxon, with 14 species very heterogeneous in morphology and often united by little more than the similarity of the males. As presently constituted, it contains only two species in addition to its type: *Ch. dubius* T. Scott, 1900 and *Ch. parkeri* (Thomson, 1890). (Two species inquirendae should also be tentatively included in this genus: *Ch. schahriar* Leigh-Sharpe, 1938 and *Ch. dasyatis* Pearse, 1952. Both were based on male specimens that could belong to any genus of the *Charopinus*-branch.)

Of the three species of the genus, *Ch. dalmanni* and *Ch. dubius* occur mainly in the North Atlantic and adjacent waters (with one South African record), whereas *Ch. parkeri* is known, so far, only from the New Zealand region.

In view of the superficial resemblance existing between genera *Pseudocharopinus* and *Charopinus*, the following list of characters has been made to help distinguish between them.

Charopinus	*Pseudocharopinus*
Dorsal shield indistinct or absent, no well-delimited head (Fig. 1567).	Dorsal shield well developed, head distinctly delimited (Fig. 1610).
Base of second antenna not heavily sclerotized.	Base of second antenna heavily sclerotized.
Maxilliped distant from mouth cone, reduced (Figs. 1577, 1586).	Maxilliped close to mouth cone, well developed (Figs, 1621, 1628).
Bulla boat-shaped or transversely elongate, sometimes not sclerotized (Fig. 1577).	Bullar orbicular or plano-convex and round, always sclerotized in adult females (Fig. 1610).

Key to British species

Dorsal surface of cephalothorax with a conical protuberance (Fig. 1567) . . . . ***Ch. dalmanni***

Dorsal surface of cephalothorax smooth, second maxillae linked at tips to produce manubrium of bulla (Fig. 1577) . . . . ***Ch. dubius***

Charopinus dalmanni (Retzius, 1829)

(Figs. 1566–1576)

Syn.: *Lernaea dalmanni* Retzius, 1829
Lernaeopoda dalmanni Retzius; of Nordmann (1832)
Charopinus dalmanni Retzius; of Krøyer (1863)
nec *Lernaeopoda dalmanni* Retzius; of Brian (1898)
Stylophorus hippocephalus Hesse, 1879

Female (Figs. 1566, 1567): Cephalothorax cylindrical, almost as long as trunk, with conical swelling on dorsal side near apex, tapering tip and conical swellings on lateral margins of base. In preserved specimens cephalothorax usually reflected along dorsal side of trunk. Latter as long as, or longer than, broad, its anterolateral and posterolateral corners gently rounded, its anterior part, directly posterior to bases of second maxillae, in form of narrow neck; lateral margins more or less parallel to each other; posterior extremity with transversely truncated margin and prominent anal tubercle slightly dorsal to that margin. Modified uropods on both sides of tubercle. No genital process. Total length about 47 mm (Scott and Scott, 1913).

First antenna (Fig. 1571) indistinctly four-segmented, two proximal segments often indistinctly delimited from each other, basal segment slightly inflated; dorsomedial wall of second segment with long and slender whip (w); division between third and fourth segments also indistinct, border between them marked by short and dentiform solus (s); apex rounded, without gibber (Fig. 1572); apical armature of six elements: tubercles 1, 2 and 3, as well as setae 4, 5 and 6; complex 5 represented by single robust seta, about as long as seta 4. Second antenna (Fig. 1573) with robust, two-segmented sympod, distal segment armed with pad of denticles on medial wall and indistinctly fused with rami; exopod one-segmented, bulbous, its surface armed with numerous denticles and two long papillae; endopod two-segmented, basal segment with inflated pad of denticles on ventral wall; distal segment armed with strong hook 1, spiniform seta 2 and ventrally with inflated, cushion-like pad 4, covered by denticles; medial surface of pad with process 5. In some specimens tubercle 3 also noticeable on medial side of base of hook. Mouth tube of usual structure, tip of labrum with abundant setae and weak rostrum. Mandible (Fig. 1574) long and flat, with dental formula: P1, S1, P1, S1, P1, S1, B4, the basal two teeth much smaller than others (except secondary), no ridges or corrugations on ventral margin of mandible. First maxilla (Fig. 1575) long, with three terminal papillae well developed, of about equal size and tipped by similar setae; exopod lateral, conical, with two distal setae of equal size; dorsal margin of appendage proximal to base of exopod with batch of small denticles. Second maxilla about as long as, or slightly longer than, trunk, gently tapering anteriorly, usually with profuse, transverse wrinkles; tip (Figs. 1568, 1569) expanded, cup-shaped, with irregular margins (damaged in drawings) and with inner central orifices of ducts (Fig. 1568); tips linked with bulla close to, but separately from, each other (Fig. 1570). Bulla transversely elongated, fusiform. (Most specimens examined by the author were damaged, with tips of second maxillae severed and bulla missing.) Maxilliped (Fig. 1576) reduced, separated from mouth cone by distance equal to $\frac{1}{3}$ of cephalothorax length; subchela vestigial, but retaining reduced barb at base of indistinctly delimited, degenerate claw; latter non-functional, at obtuse angle to corpus. Uropod (Figs. 1566, 1567) almost as long as trunk, cylindrical and curving.

Male: The author was unable to examine male specimens of this species. They are very similar to those of the following species. For habitus drawings and general description consult Vogt (1877b).

Comments: The taxonomic history of *Ch. dalmanni*, in spite of its length, is almost free of confusion at the specific level. At the generic level, it shows the rather common transition from the generalized *Lernaea* of the early workers towards its present status. Placed in *Lernaeopoda* by von Nordmann (1832) and Krøyer (1837), it was soon found to be ill-matched with other members of that genus. Turner and Wilson (1862) suggested that it should be transferred to a new genus and Krøyer (1863–64) did it, independently and almost simultaneously. Hesse (1879b) was alone in not recognising the well-established taxon and proposing for it new specific and generic names. His decision was not upheld by later authors.

Ch. dalmanni is a widespread, if not a common, species. Its most usual host is *Raja batis* and its distribution range is largely coincident with the range of that fish. It is found in the North Atlantic, off the coasts of both Europe and North America. It also occurs in adjacent seas (Barents Sea, Norwegian Sea, North Sea). The Mediterranean has been distinguished by records, usually old, of *Ch. dalmanni* occurring on hosts other than *R. batis* and not identifiable with certainty. (For example, Valle (1880) recorded it on *Dasybatis clavata* and *Laeviraja oxyrhynchus* in the Adriatic and Brian (1898) on *Torpedo narce* off Genoa, a clear case of misidentification, as can be seen from his illustrations.) A recent record (Pogoreltseva, 1970) extends the range of *Ch. dalmanni* to the Black Sea, where it was found on an unidentified elasmobranch. Wilson (1915) mentions its occurrence on "barndoor skate" off the coast of the U.S.A. The habitats of *Ch. dalmanni* are the gills, nasal fossae and spiracles of its host. Sometimes adult females are found near the oesophogeal orifice.

In British waters it can be found wherever *Raja batis* is taken, though it is never common. The author never found it on other species of *Raja* in this region.

Charopinus dubius T. Scott, 1900

(Figs. 1577–1597)

Syn.: ?*Brachiella parkeri* (Thomson, 1890); of Bainbridge (1909)

Female (Fig. 1577): Cephalothorax cylindrical, as long as, or slightly shorter than, the trunk, without distinct dorsal shield or conical tubercle on dorsal side near apex; base devoid of lateral conical swellings. Trunk as long as, or longer than, broad, with corners rounded and lateral margins somewhat convex; posterior margin (Fig. 1578) rounded and equipped, in addition to anal tubercle and uropods, with distinct genital process, resembling anal tubercle in shape, though smaller. Length of trunk 8.5 mm, cephalothorax 6 mm, second maxilla 11 mm (Scott and Scott, 1913).

First antenna (Fig. 1579) with basal part somewhat longer than that of *Ch. dalmanni* and with few or no traces of segmental borders; whip long and slender; terminal part cylindrical, not divided into two component segments (border between them marked only by fairly prominent solus); apical armature (Fig. 1580) of three tubercles (1, 2 and 3) and three setae (4, 5 and 6), similar to those of *Ch. dalmanni*; one-seta complex 5 distinguished by well set off fine tip. Second antenna similar to that of *Ch. dalmanni* but longer and more slender; exopod (Fig. 1582) one-segmented, bulbous, irregularly lobate and apparently soft, apex covered with sharp, minute denticles and armed with two prominent, long papillae; distal segment of endopod (Fig. 1581) with hook 1 slender and long (in Fig. 1581 it points slightly away from the observer and its curvature is not fully revealed); setiform seta 2 present, but tubercle 3 not apparent at base of hook 1; ventral, inflated process 4 with digitiform unarmed process 5 also present and covered with minute denticles; long strip of denticulation along ventral wall, proximal to processes 4 and 5. Mouth cone of usual structure, tip of labrum (Fig. 1583) covered with abundant flat setae, usually completely covering rather weak rostrum. Mandible (Fig. 1584) with dental formula P1, S1, P1, S1, P1, S1, B4; second and third primary teeth most prominent, basal teeth progressively smaller from distal to proximal; ventral blade with uneven surface, covered with irregular ridges. (These appear black in phase contrast illumination and were originally described (Kabata, 1964a) as "dark streaks"). First maxilla (Fig. 1585) long, with well-developed three terminal papillae; dorsalmost papilla larger than other two; exopod conical, long, lateral, ending in two setae of equal size and with slightly inflated bases; dorsal margin of appendage with small batch of minute denticles at about level of base of exopod. Second maxilla (Fig. 1577) longer than trunk, capable of considerable contraction, usually tapering at tip to half its basal diameter, separate from its opposite number; tip slightly inflated, attached to short, dark-coloured manubrium of bulla. Latter (Fig. 1577) spindle-shaped, hard and brittle; oval central part surrounded by low ridge. Maxilliped (Fig. 1586) removed from mouth cone about $\frac{1}{3}$ distance of length of cephalothorax, reduced; subchela not clearly distinguishable from corpus, short, armed with basal seta and weak barb; claw degenerate, with slightly inflated base, barely longer than barb; on medial margin of corpus setiform outgrowth, apparently prominent nipple-like orifice of some duct; entire appendage clearly non-functional. Uropod (Fig. 1578) cylindrical, curving, more than half length of trunk.

Male (Fig. 1587): Cephalothorax more than half body length, divisible into anterior and posterior parts, at obtuse angle to each other; anterior part inflated, flexed ventrally to long axis of rest of body, posteriorly with rounded, prominent swelling, anteriorly with very large mouth cone, antennae and buccal appendages; posterior part cylindrical, slightly tapering posteriorly; very prominent second maxillae and maxillipeds on ventral surface, with bifid mediative process (Fig. 1593) between bases of maxillipeds. Trunk cylindrical, not distinctly delimited from, and more slender than, posterior part of cephalothorax. Two pairs of thoracopods present. Uropods and external genitalia on posterior extremity. Length about 2 mm.

First antenna (Fig. 1588) resembling that of female but even less distinctly segmented; tubercles 1, 2 and 3, as well as solus, of about equal size and with rounded tips; whip long, slender, with slightly inflated base. Second antenna (Fig. 1589) similar in structure, but not in proportion of component parts, to that of female; hook and process 5 relatively more prominent than in female; no ventral strip of denticles on ventral wall of distal segment of endopod, though process 4 and ventral wall of basal segment denticulated; endopod less prominent than exopod; latter regularly conical, with very prominent papillae, as well as with marginal denticulation. Tip of labrum (Fig. 1595) with rostrum more prominently displayed than in female; marginal setae often divided into two distinct, though confluent, groups. Mandible (Fig. 1590) with dental formula less regular than in female (in some specimens at least it is P1, S2, P1, S1, B5; it appears as if second primary tooth were missing from its place between first and second secondary); basal teeth five instead of four, their size not much smaller than that of primary teeth. (Males of this species are difficult to find. Only examination of more abundant material will show true range of diversity of their dental formulae.) First maxilla (Fig. 1594) similar to that of female, but with relatively smaller exopod and with different proportions of three terminal papillae. Second maxilla (Fig. 1591) short and squat, linked with its opposite number by cuticular tympanum with reinforced free margin; medial margin of corpus with a prominent tubercle covered by fine denticles; on ventral surface near base of tubercle, duct orifice present; subchela without distinctly delimited claw, slightly curving, with tapering, fine tip; near base on ventral surface a small tubercle (presumably homologous with basal seta of female). Maxilliped (Fig. 1592) fused at base with opposite number, with mediative process at point of fusion; distal part strongly sclerotized, with corpus longer than broad and medial anterior elongation; tip of elongation containing a cavity for accommodation of tip of subchela; rim of cavity reinforced, inner surface covered with transverse cuticular ridges; subchela short, with broad base and rapidly tapering tip, with distal curve, claw not delimited; short seta (homologous with barb of female) on inner margin of subchela. Thoracopods (Fig. 1596) papilliform, with two semispherical swellings at base of papilla; tip of papilla with long, flagelliform seta. External genitalia of one genital plate (Fig. 1597, gpl) and one genital process (Fig. 1597, gp) on each side, genital orifices covered by the genital plates. Uropod (Fig. 1597) cylindrical, with two short setae on lateral margin (in some specimens additional small tubercle might indicate remnant of another, damaged seta) and with finely attenuated, setiform tip; small prickles covering cuticle near tip.

Comments: Like the preceding species, *Ch. dubius* lives within the branchial and buccal cavities, as well as in the nasal fossae of its hosts. The latter comprise various species of the genus *Raja* and related genera of elasmobranchs. *Raja circularis*, *R. fullonica*, *R. naevus*, *R. batis*, *R. montagui*, *R. maculosa* and *Laeviraja macrorhynchus* have been recorded harbouring *Ch. dubius*. Its distribution range covers the eastern North Atlantic, North, Irish and Barents Seas. An isolated, surprising record places it also off Port Elizabeth, South Africa, where it was allegedly found on an unidentified skate.

The reliability of many of these records is open to doubt. The author found this species almost exclusively on *Raja naevus* (a species in the past often confused with *R. circularis*) in Scottish waters, where it is quite common. On other species of *Raja* it seems to be very rare and it has never been found by the author on *R. batis*. A. Scott (1929) took it in the Irish Sea as far south as Caernarvon Bay (on *R. fullonica*). Records farther south than that do not appear to be well authenticated. Bainbridge's (1909) record of tentatively identified *Brachiella parkeri* (= *Ch. parkeri*) from *Raja oxyrinchus* (locality unspecified) is, most probably, also that of *Ch. dubius*.

Genus Pseudocharopinus Kabata, 1964

Female: Cephalothorax cylindrical, about $\frac{2}{3}$ as long as trunk. Head distincly delimited, well developed dorsal shield present. Trunk flattened dorsoventrally, longer than broad. First antenna indistinctly three- or four-segmented, with well-developed apical armature. Second antenna prehensile. Mandible with three secondary teeth. First maxilla with lateral exopod and three equally well developed terminal papillae. Second maxillae separate from each other, of different lengths. Bulla relatively small, lenticular or plano-convex and round, with short manubrium. Maxilliped close behind mouth cone, well developed and functional. Modified uropod cylindrical, of varying lengths, dorsal to egg sacs.

Male: Very similar to that of *Charopinus*.

Type species: *Pseudocharopinus dentatus* (Wilson, 1912).

Comments: The species presently comprising this genus were originally described mainly as members of the genus *Charopinus*, or of *Brachiella* (*P. bicaudatus* was first known as *Lernaeopoda* and *P. dasyaticus* (Rangnekar, 1957) as *Clavellopsis*). The differences between them and the species of *Charopinus* are set out on p. 363; those between them and the members of the *Brachiella* and the *Clavella* branches of Lernaeopodidae are obvious enough to make their enumeration unnecessary.

The genus can be divided into three groups, according to the host affiliation of its species. Two species are parasitic on dogfishes (*Squalus* and *Etmopterus*): *P. bicaudatus* (Krøyer, 1837) has been recorded in the North and South Atlantic, as well as from the Pacific; the only record of *P. squalii* (Wilson, 1944) from *Squalus mitsukurii* from Japan calls for confirmation and the species cannot be considered as fully established. Two other species (the type species and *P. markewitschi* (Gusev, 1951)) live on skates of the genus *Raja*, in the East Pacific and Indo-Pacific regions respectively. Other species parasitize hypotrematous hosts of genera other than *Raja* (*Dasyatis*, *Trygon*, *Myliobatis*, *Torpedo*, *Narcina* and *Pteroplatea*) and are fairly limited in distribution. *P. concavus*

(Wilson, 1913) is known only from Jamaica, *P. pteroplateae* (Yamaguti and Yamasu, 1959) only from Japan and *P. malleus* (Rudolphi in Nordmann, 1832) from the Atlantic and the Mediterranean seaboard of Europe. Three species are known from the Indian Ocean and Australian waters: *P. dasyaticus* (Rangnekar, 1957), *P. dasyaticus* (Pillai, 1962) and *P. narcinae* (Pillai, 1962). Due to the transfer to the genus *Pseudocharopinus* of *Clavellopsis dasyaticus* and *Charopinus dasyaticus*, the latter becomes a junior secondary homonym. Consequently, this author proposes that the name *Pseudocharopinus dasyaticus* (Pillai, 1962) be rejected, in accordance with the Code of Zoological Nomenclature, article 59(b). He further proposes that this name should be replaced by *Pseudocharopinus pillaii* nom. nov.

Key to British species

Dorsal shield roughly in line with long axis of cephalothorax (Figs. 1621, 1631) . . . . ***P. malleus***
Dorsal shield at about right angles to long axis of cephalothorax (Fig. 1610) . . . . ***P. bicaudatus***

Pseudocharopinus bicaudatus (Krøyer, 1837)

(Figs. 1610–1620)

Syn.: *Lernaeopoda bicaudata* Krøyer, 1837
Brachiella pastinacae Baird; of Kurz (1877)
Brachiella pastinaca van Beneden, 1851; of T. Scott (1904)
Charopinus bicaudatus (Krøyer); of Wilson (1915)

Female (Fig. 1610): Cephalothorax of two parts, varying in dimensions; anterior part substantially expanded and covered with well developed dorsal shield, at right angles to posterior, neck-like part, short in Fig. 1610, but often noticeably longer; dorsal shield with clearly demarcated margins, longitudinally divided into two parts in mid-dorsal line (Fig. 1611). Trunk of two distinct parts: anterior narrow, neck-like (Fig. 1611), beginning behind bases of second maxillae, and posterior roughly quadrangular, much broader than anterior and broader than long, with rounded corners and often concave dorsal surface (Figs. 1610, 1612); posterior margin of trunk with anal tubercle, slightly dorsal and central, and with two small swellings accommodating genital orifices, sometimes very inconspicuous (Fig. 1612). Anterior part of body (cephalothorax and narrow part of trunk) often flexing dorsally to long axis of broad part of trunk, as in Fig. 1610. Total length 5.5–6.0 mm; length of dorsal shield 1.7 mm; of cephalothorax 2.0–2.2 mm; of second maxillae 1.5 mm; of trunk 2.3 mm; of uropod 1.6 mm; of egg sacs 3.0 mm.

First antenna (Fig. 1614) indistinctly segmented, of two recognizable parts: basal inflated and terminal cylindrical; long and slender whip on dorsomedial wall of former; dentiform solus on wall of latter indicating border between its two segments; apical armature (Fig. 1615) of seta 4, single 5 and flagelliform 6, as well as tubercles 1, 2 and 3. Second antenna with heavily sclerotized sympod and hinge-like flexion between it and rami; exopod one-segmented, bulbous, denticulated, often overlapping endopod; latter two-segmented, proximal segment with ventral patch of denticles; distal segment (Fig. 1616) with relatively small, though well developed, hook 1 and spiniform seta 2; tubercle 3 usually prominent; inflated ventral wall 4, with fairly large, though sparse, denticles and process 5 unarmed, arising from wall of 4. Tip of labrum (Fig. 1613) rounded, well equipped with flat, short setae; rostrum concealed, if present. Mandible (Fig. 1617) with dental formula P1, S1, P1, S1, P1, S1, B5; series of basal teeth rapidly decreasing in size in proximal direction; largest tooth usually second primary; ventral flat margin of appendage with corrugated cuticle. First maxilla (Fig. 1618) long, surmounted by three relatively short terminal papillae of about equal size and armed with similar setae; exopod lateral, conical and armed with two short setae with inflated bases; patch of fine denticles on dorsal margin of appendage below level of base of exopod. Second maxilla (Figs. 1610, 1611) usually much shorter than trunk, slender, with many transverse wrinkles, separate from opposite number; tips inflated to form jointly collar round manubrium of bulla. Latter small, plano-convex, orbicular, with short manubrium. Maxilliped (Fig. 1619) with corpus sturdy and long, slender subchela not reaching tip of buccal cone; myxa with two slightly raised, denticulated pads, one at base of subchela, other near base of corpus; just distal to latter one short seta arising from small swelling; basal seta of subchela very short, its inner margin near base of claw armed with minute, scattered denticles; claw (Fig. 1620) short, mildly curved, usually with blunt tip and with two secondary denticles at midlength of concave margin; at base short, tapering barb, about half as long as claw. Uropod (Figs. 1610, 1612) more than half as long as broad part of trunk, cylindrical, with rounded tip, somewhat curving, dorsal to egg sacs.

Male: Not examined. Like all males of *Charopinus* branch of family, it is largely similar to that of *Charopinus dubius* (cf. Figs. 1587–1597).

Comments: Almost fifty years had elapsed since *Lernaeopoda bicaudata* was described by Krøyer (1837) and briefly mentioned by Edwards (1840), before Kurz (1877) wrote about it under the name *Brachiella pastinacae* Baird. In so doing, he not only confused two distinctly different species but ascribed the authorship of *B. pastinacae* to Baird (1850), who made no mention of it in his monograph. Kurz's drawings (1877, Pl. XXV, Figs. 2, 3) clearly prove that he studied *P. bicaudatus*.

Some doubts as to the generic affinity of the species were voiced both by the discoverer (name *Lernaeopoda* was followed by a query in the original description) and by Edwards (1840), but Wilson (1915) was the first to recognize its place in the genus *Charopinus*, as then constituted. At the specific level, confusion prevailed until Capart (1946) unravelled it by demonstrating that *Ch. pastinacae* and *Ch. bicaudatus* are two distinct species of the same genus. Even then, residual confusion remained manifest in the way in which the species was quoted. For example, Delamare Deboutteville and Nunes-Ruivo (1952) refer to it, in synonymy, as *Brachiella pastinaca* Krøyer, Bainbridge, though Bainbridge (1909) credited it to van Beneden.

In part at least, this confusion was due to the fact that, in his original description, Krøyer (1837) recorded *Trigla gurnardus* as the only host of the species known to him. There are no subsequent records of *P. bicaudatus* from non-elasmobranch hosts and Krøyer's entry is very probably a lapsus (one of several in host identifications made in his 1837–1838 and 1863–1864 papers) which distracted the attention of workers concerned with copepods parasitic on elasmobranchs.

P. bicaudatus is a parasite commonly occurring in the spiracles of dogfishes of the genus *Squalus*. Its records come most abundantly from the European Atlantic sea-board, but it has been found off the Atlantic coast of North America (Wilson, 1932), in the Gulf of Mexico on *Etmopterus schultzi* (Causey, 1955, tentative) and off the western shores of Africa on *Squalus fernandinus* (cf. Capart, 1959). Pacific records come from the far-eastern sea-board of the U.S.S.R. (Gusev, 1951) and from the coast of the U.S.A. (Wilson, 1935b), both involving species of dogfish (*Squalus acanthias* and a "dog shark" respectively). This author found *P. bicaudatus* in the spiracles of young *Squalus acanthias* taken off the shores of British Columbia, Canada.

More puzzling are records of the occurrence of *P. bicaudatus* on hosts other than dogfishes. Many of them can be dismissed as referring to van Beneden's species *Brachiella pastinacae*, but some remain difficult to explain. Kurz (1877) clearly dealt with *P. bicaudatus*, yet he claimed to have found it in the spiracle of *Myliobatis aquila*, an eagle ray, more usually associated with *Pseudocharopinus malleus*. The illustration of Scott and Scott (1913, Pl. LXIV, Fig. 8) unmistakably shows a female *P. bicaudatus*, but its host is given as *Trygon* sp. taken in Scottish waters.

In the author's own experience, *P. bicaudatus* in British waters, while quite common, is restricted to the spiracles of *Squalus acanthias*. It is possible that it occurs occasionally on other hosts, but they cannot play an important part in the biology of the species.

Pseudocharopinus malleus (Rudolphi in Nordmann, 1832)

(Figs. 1621–1634)

Syn.: *Brachiella malleus* Rudolphi in Nordmann, 1832
Brachiella lophii Edwards, 1840
Brachiella pastinacae van Beneden, 1851
nec *Brachiella pastinacae* Baird; of Kurz (1877)
Lernaeopoda dalmanni Retzius; of Brian (1898)
nec *Brachiella pastinaca* van Beneden; of Bainbridge (1909)
nec *Brachiella pastinaca* van Beneden, 1851; of Scott and Scott (1913)
nec *Clavella lophii* (Edwards, 1840); of Scott and Scott (1913)
Charopinus malleus (Rudolphi); of Wilson (1915)
Charopinus pastinacae (P. J. van Beneden, 1851)
Eubrachiella lophii (Milne-Edwards, 1840); of Yamaguti, 1963

Female (Figs. 1621, 1622, 1631): Cephalothorax cylindrical, about as long as trunk, anteriorly expanded into large head (sometimes comprising $\frac{1}{2}$ of cephalothorax), roughly in line with long axis of cephalothorax; head covered by elongate dorsal shield (Fig. 1623) with distinct margins, mid-dorsal partition and anterior triangular recess. Trunk twice as long as broad, anteriorly narrow and cylindrical, gradually broader towards posterior extremity; posterolateral corners rounded, posterior margin with central tubercle housing genital orifices and, dorsal to it, anal tubercle (Fig. 1630). Total length 5.4 mm; cephalothorax length 3 mm; uropod 0.7 mm; egg sacs 1.7 mm (from Capart, 1949).

First antenna (Fig. 1624) indistinctly segmented, of two recognizable parts: basal part, much larger than terminal, inflated and bifid due to presence of basal papilla projecting dorsally from near base; basal papilla smaller than rest of this part but often appearing larger due to post-mortem distention of cuticle; long, slender whip near apex of basal part; cylindrical terminal part not clearly divided into two segments, with solus on lateral wall; apical armature (Figs. 1625, 1632) similar to that of *P. bicaudatus*, with three setae and three tubercles arranged in standard pattern. Second antenna similar to that of *P. bicaudatus*, though apparently less well equipped with spinulated pads. Mouth tube as in *P. bicaudatus*. Mandible (Fig. 1626) also resembling corresponding appendage of that species, with dental formula P1, S1, P1, S1, P1, S1, B5; distalmost tooth of basal series about as large as primary teeth, proximal much smaller. First maxilla (Figs. 1627, 1633) resembling that of *P. bicaudatus* but often with longer terminal setae and with patch of denticles on dorsal margin either much weaker or absent. Second maxilla (Figs. 1621, 1622, 1631) separate from its opposite number, shorter than trunk, often transversely wrinkled, inflated at apex to form collar round manubrium of bulla. Latter as in *P. bicaudatus*. Maxilliped (Figs. 1628, 1634) with slender, distally tapering corpus often reaching tip of mouth cone; myxa with one short seta at about midlength; denticulated pads weaker than in *P. bicaudatus*, basal sometimes apparently absent; subchela slender, little more than ½ length of corpus, with fairly long basal seta, barb ½ length of claw and distal part of inner margin beset with denticles; claw (Fig. 1629) short, blunt, often with indistinct transverse constriction near base, without secondary teeth, its curvature varying from specimen to specimen. Uropod (Figs. 1621, 1622, 1630, 1631) shorter than ½ length of trunk, fusiform, cylindrical or tapering, with rounded or sharp tips.

Male: Generally resembling that of *Charopinus dubius* (cf. Figs. 1587–1597).

Comments: The early history of this species is linked with the Mediterranean and Adriatic Seas. Its first specimens were discovered in 1817 by Rudolphi in the buccal cavity of *Torpedo marmorata*, in Rimini. The discoverer, however, produced no description himself, but sent his material to von Nordmann, "nebst einigen an Ort und Stelle gemachten Bemerkungen". Nordmann's (1832) description appears to be at least partly based on Rudolphi's notes; to him also is given the credit of the authorship. (Although Rudolphi's proposed manuscript name for this species appears to have been *Dirhynchus malleus*, this name is not mentioned in von Nordmann's text and cannot be considered as a part of the synonymy of *P. malleus*, contrary to views of Valle (1880).)

The hallmark of this species is its first antenna, with the peculiarly modified basal part (Fig. 1624). It has previously been noted by Kabata (1964a, Fig. 50) and, of the older authors, by Vogt (1877b, Pl. IV, Fig. 1). Similar first antennae were described by Pillai (1962) for two other species of *Pseudocharopinus* from Indian waters, both distinct from *P. malleus*.

P. J. van Beneden (1851a) described *Brachiella pastinacae* from the nasal fossa of *Trygon pastinaca* (= *Dasyatis pastinaca*), caught off the Belgian coast. The name of van Beneden's species was subsequently used for a parasite found on other hosts and, in view of the fact that it was often confused with Krøyer's *Pseudocharopinus bicaudatus* (see p. 368), the records are virtually impossible to disentangle.

Capart (1949) established that *P. pastinacae* is distinct from *P. bicaudatus*, having selected and described neotypes of this species. (Original material was destroyed during World War II.) The present author had an opportunity of examining one of Prof. Capart's whole-mounted specimens (Fig. 1631). He also examined a specimen recently collected by Dr. K. MacKenzie (Aberdeen) from the nasal fossa of *Dasyatis pastinaca* in the North Sea (Fig. 1621). A comparison of Fig. 1631 with 1621, of Fig. 1632 with 1627 and of Fig. 1634 with 1628 leaves no doubt that Capart's and Mackenzie's specimens are conspecific. Both, however, have first antennae with basal papillae unmistakably placing them in *Pseudocharopinus malleus*. This significant resemblance, in the absence of significant differences between *P. pastinacae* and *P. malleus*, makes it necessary to relegate the former to synonymy with the latter.

There can be no doubt that Edwards' (1840) *Brachiella lophii* also belongs to this species. The position of the uropods and the structure of the cephalothorax place it firmly in *Pseudocharopinus*, while the figure showing the first antenna proves its identity with *P. malleus*. The identification of

Lophius sp. as its host must have been in error.

Understood in these terms, *P. malleus* has a distribution range which covers the western Mediterranean and Adriatic, as well as the Atlantic coast of Europe, and which extends as far north as the range of *Dasyatis pastinaca*. Its southernmost record appears to be from *Trygon margarita* and *T. marmorata*, captured off the coast of Senegal (Capart, 1953). As far as it is possible to gather from the confused literature, *P. malleus* parasitizes also species of the genera *Myliobatis*, *Rhinoptera* and *Torpedo*. (It was found by the research staff of the Marine Laboratory, Aberdeen, on the tongue of *Torpedo marmorata* in South Minch, in August 1968.) It has not been found outside its Mediterranean-Atlantic range.

Genus Schistobrachia Kabata, 1964

Female: Cephalothorax short, subtriangular in outline, dorsoventrally flattened. Trunk much longer than broad, usually cylindrical anteriorly and dorsoventrally flattened posteriorly, with more or less parallel cylindrical margins. First antenna indistinctly segmented, with well developed apical armature. Second antenna prehensile. Mouth cone of lernaeopodid pattern. Mandible with three primary teeth. First maxilla with lateral exopod and three terminal papillae. Second maxilla as long as, or longer than, trunk, separate from its opposite number, of same diameter throughout, apically inflated, with one or more pairs of long, or short, processes on lateral side of apex. Maxilliped with robust corpus and long subchela, close behind mouth cone. Uropods dorsal to egg sacs.

Male: Cephalothorax longer than trunk, of two parts; anterior inflated and at obtuse angle to long axis of body, posterior in line with trunk. Latter indistinctly segmented or unsegmented, with well developed uropods and external genitalia. Antennae and buccal appendages similar to those of female, second maxillae and maxillipeds subchelate.

Type species: *Schistobrachia ramosa* (Krøyer, 1863).

Comments: Only three species of this genus have been discovered so far, in widely scattered localities. *S. ramosa*, the type species, has been found in the northern Atlantic, parasitizing species of the genus *Raja*, *S. chimaerae* (Yamaguti, 1939) was recorded on a single occasion on a species of *Chimaera* in Japanese waters (one of very few copepods parasitic on holocephalan fishes). The third species, *S. tertia* Kabata, 1970, is known only off the west coast of Vancouver Island, on the Canadian Pacific coast, living on two local species of *Raja* (*R. rhina*, *R. binoculata*).

Schistobrachia ramosa (Krøyer, 1863)

(Figs. 1598–1609)

Syn.: *Charopinus ramosus* Krøyer, 1863

Female (Fig. 1598): Cephalothorax short, subtriangular, dorsoventrally flattened, covered with indistinct dorsal shield, inclined ventrally to long axis of body, imperceptibly merging with trunk. Latter much longer than broad, of two indistinctly delimited parts, anterior narrow and cylindrical, posterior broad and dorsoventrally flattened, with more or less parallel lateral margins; posterior extremity (Fig. 1599) with rounded lateral corners, margin with central prominent anal tubercle and two lateral tubercles carrying openings of oviducts. Length of cephalothorax 1.3 mm, trunk 6.3 mm, second maxillae 6.7 mm, uropods 1.0 mm and egg sacs 5.6 mm (after Markevich, 1956).

First antenna (Fig. 1600) of two parts, not distinctly segmented; basal part with broad base, tapering, with long and slender whip near apex; distal part cylindrical, with poorly marked border between its two segments and solus in position of former border; apex rounded; apical armature (Fig. 1601) of three setae (digitiform 4, single 5 and flagelliform 6) and three tubercles (1, 2 and 3, the first of them more massive than other two). Second antenna resembling that of *Charopinus* more than that of *Pseudocharopinus*; sympod long and not heavily sclerotized; spinulation of rami extensive, spinules fine; endopod relatively long, two-segmented; distal segment (Fig. 1602) with strongly flexed hook 1 and noticeable tubercle 3; ventral wall of that segment very strongly inflated, covered with sparse denticles (corresponding to process 4) and with process 5 on its wall; exopod (Fig. 1603) one-segmented, bulbous, in some specimens incipiently lobate, with margins well covered by spinules and with two long papillae. Mouth cone as in other members of the *Charopinus* branch. Mandible (Fig. 1604) with dental formula P1, S1, P1, S1, P1, S1, B4; flat ventral margin of appendage apparently devoid of cuticular corrugations. First maxilla (Fig. 1605) with three terminal papillae longer than in *Pseudocharopinus*; central papilla often somewhat longer than other two; exopod lateral, long, with two apical setae; dorsal margin above base of exopod with swelling covered by minute denticles. Second maxilla (Fig. 1598) longer than trunk, of more or less even diameter throughout, separate from its opposite number, distally inflated and linked with tip of other second maxilla by vestigial bulla; two long, cylindrical processes arising from lateral side of inflated tip; about ½ as long as second maxilla, possessing some mobility and ability to interlock. Maxilliped (Fig. 1606) with relatively long narrow corpus and long subchela; myxa inflated (Fig. 1607), armed with small conical process pointing distalwards, and with slender, short seta on inflated base; subchela long, with tip of claw reaching nearly to base of appendage, basal seta of subchela short, distal part of its inner margin with sparse, slender denticles and distally protruding papilla at base of barb; barb about ½ length of claw; latter (Fig. 1608) curved, tapering, with rounded tip and one secondary tooth on proximal half of inner margin; shallow constriction not far from base of claw. Uropod (Fig. 1599) arising from lateral wall of anal tubercle, short, cylindrical or fusiform, with rounded or tapering tip.

Male: (Fig. 1609): Cephalothorax of two indistinctly divided parts, anterior inflated, with rounded dorsal swelling, and posterior cylindrical, imperceptibly joining trunk. Latter narrow, cylindrical, tapering to posterior extremity. Antennae and buccal appendages similar to those of female. Second maxilla and maxilliped subchelate, resembling those of male *Charopinus dubius* (cf. Figs. 1587–1597). Posterior extremity with cylindrical, well developed uropods.

(The author was unable to find traces of thoracopods on trunk, but they are probably present and similar to those of other males of the *Charopinus*-branch.)

Comments: Krøyer's (1863) description of this parasite as a second species of the newly founded genus *Charopinus* was in a large measure responsible for the subsequent taxonomic confusion surrounding both this genus and its relatives. *Ch. ramosus*, a parasite with short, flat, ventrally inclined cephalothorax, was linked in one genus with *Ch. dalmanni*, the cephalothorax of which was cylindrical, long and deflected dorsally along the surface of the trunk. To accommodate them both, generic diagnosis had to be so inclusive as to allow the admission of many diverse types of morphology and was not amended until a fairly recent revision (Kabata, 1964a).

The most unique feature of the genus *Schistobrachia* is in the mode of attachment of its members, in which the bulla of the young adult is replaced by the maxillary processes acting as both anchors and cross-beams (Kabata and Cousens, 1972).

S. ramosa is mainly a North Atlantic species. It has also been recorded in adjacent areas such as the Barents Sea (Polyanski, 1955), North Sea (Scott and Scott, 1913, and many other records), or the Irish Sea (A. Scott, 1929). Its most western authenticated record is from Iceland (Hansen, 1923; Stephenson, 1940). Tremblay and Lapointe (1938) recorded its occurrence in the Gulf of St. Lawrence on *Raja scabrata*. There are no subsequent records of this species from the shores of North America, however, and *R. scabrata* is known as the host of another and superficially similar lernaeopodid, *Dendrapta cameroni* (Heller, 1949). The Canadian record must, therefore, be considered doubtful. There are no reports of the occurrence of *S. ramosa* in the Mediterranean but a recent record (Pogoreltsova, 1970) mentions its presence on an unspecified elasmobranch in the Black Sea. More information is needed to confirm it. Kensley and Grindley (1973) recorded this species from *Raja batis* in South Africa, without giving a description of appendages. The habitus drawing shows many differences between the South African specimen and the North Atlantic ones. The host is also unusual. Closer examination of this record is required.

All hosts of *S. ramosa* belong to the genus *Raja*: *R. clavata*, *R. fullonica*, *R. radiata* and *R. maculata* were found to harbour it, usually in the buccal cavity near the bases of the gill arches or on the gills themselves. *S. ramosa* is not uncommon, but cannot be considered abundant. Polyanski (1955) found it on 13.3% of *R. radiata* examined in the Barents Sea. This level of infection is fairly typical of other regions.

Genus Lernaeopodina Wilson, 1915

Female: Cephalothorax much shorter than trunk, flat, ventrally inclined to long axis of trunk. Latter much longer than broad, usually with narrower part behind bases of second maxillae. Posterior extremity usually rounded, centrally with small anal tubercle. First antenna indistinctly segmented, with well developed apical armature. Second antenna prehensile. Labrum with well developed rostrum, mandible with three primary teeth. First maxilla with small lateral exopod and two terminal papillae. Second maxilla as long as, sometimes much longer than, trunk, separate except at tip. Bulla small, suborbicular, with short manubrium. Uropods short but thick, fusiform, dorsal to egg sacs.

Male: Cephalothorax of two indistinctly delimited parts, anterior inclined ventrally to long axis of posterior, latter in line with cylindrical, obscurely segmented trunk. Antennae and buccal appendages similar to those of female. Second maxilla and maxilliped subchelate. Uropods well developed.

Type species: *Lernaeopodina relata* Wilson, 1915.

Comments: Only five species of *Lernaeopodina* are currently known, most of them from either one or few records. The type species was described from the Atlantic coast of the U.S.A., where Wilson (1915) found it once on *Raja laevis*. *L. spinacis* (Brian, 1908) was found, on one occasion also, in the low latitudes of the North Atlantic, attached to *Etmopterus spinax*, a deep-water species of dogfish. *L. longibrachia* (Brian, 1912) is less rare; Brian (1912) found it in the same general area as the preceding species, parasitic on *Etmopterus pusillus* and *Centrophorus squamosus*. It occurs also on chimerid fishes (Holocephala). The present author reported it (Kabata, 1969e) from the cornea of *Hydrolagus affinis* taken in the Newfoundland area. He also has some specimens sent to him by Dr. B. Berland (Bergen, Norway), who took them from the same host in the Bay of Biscay; also some from Dr. R. Overstreet (Ocean Springs, Mississippi, U.S.A.), who collected them in the Gulf of Panama from an unspecified chimerid host, attached behind the pectoral fin.

The only species of the genus occurring outside the Atlantic Ocean is *L. pacifica* Kabata and Gusev, 1966, described from *Raja kenojei* taken off Sakhalin and reported also near the coast of British Columbia on *Raja rhina* (cf. Kabata, 1970d).

British fauna includes only one species of this genus, *L. longimana* (Olsson, 1869).

Lernaeopodina longimana (Olsson, 1869)

(Figs. 1635–1647)

Syn.: *Lernaeopoda longimana* Olsson, 1869
Lernaeopoda cluthae T. Scott, 1900
?*Lernaeopoda similis* Scott and Scott, 1913
Lernaeopodina longimana (Olsson, 1869); of Wilson (1915)
Lernaeopodina cluthae (T. Scott, 1900); of Wilson (1915)
?*Lernaeopodina similis* (Scott and Scott, 1913); of Wilson (1915)

Female (Figs. 1635–1637): Cephalothorax much shorter than trunk, ventrally inclined to long axis of trunk, without distinct dorsal shield, slightly flattened in dorsoventral plane; anterior part of its dorsal surface tending to bulge in prominent swelling (Figs. 1635, 1636), though in some specimens remaining fairly flat (Fig. 1640). Trunk much longer than broad (Fig. 1637), its anterior part narrow and almost cylindrical, its posterior part more than twice as broad as anterior and flattened; lateral expansion beginning at about midlength of trunk, occasionally quite abrupt, though more commonly gradual (Fig. 1637); posterior extremity with gently rounded lateral corners (Fig. 1639); in many specimens oviduct openings on prominent tubercles, genital slits with heavily sclerotized margins; size of tubercles varying with condition of trunk. Uropods directly posterior to them, fusiform, with fine tips (though blunt in some specimens). In midposterior point small anal tubercle. Size of female varying from one population to another. In some, total length 7–8 mm (Olsson, 1869), in others only about 5 mm (T. Scott, 1900).

First antenna with obscure segmentation, short and squat, with fairly short whip and relatively long solus; apex (Fig. 1641) rounded; lateral side of apex with very robust, blunt seta 4 and, close to its base, tubercle 1; central in position single seta of complex 5, only slightly longer than 4 but slender and with sharp tip; tubercle 2 characteristically slender and appearing like short seta; tubercle 3 conical and small, near base of slender seta 6, characteristically short, absent from many specimens. Second antenna (Fig. 1642) long and well developed, prehensile; sympod indistinctly two-segmented, distal segment with two patches of denticles, one ventrolateral and one medial;

exopod bulbous, one-segmented, slightly longer than endopod, abundantly covered with fine denticles; near its apex two soft, tapering, fairly long papillae; endopod two-segmented: basal segment cylindrical, not well delimited from sympod, its ventral surface covered by denticles smaller and much more closely packed than those of exopod; armature of distal segment consisting of slender, mobile hook 1, spiniform seta 2 almost as long as hook and small papilla 3 on opposite side of hook's base; ventral wall 4 only slightly inflated and intensely denticulated, with robust, sharp-pointed process 5 on its side. Labrum (Fig. 1638) broad and bluntly rounded, with fringe of fine, short setae and prominent, broad rostrum, only partly covered by fringe. Mandible (Fig. 1643) with dental formula P1, S1, P1, S1, P1, S1, B5; series of basal teeth slender and narrow, usually protruding beyond ventral margin of appendage; ventral margin itself corrugated and covered with gross cuticular ridges. First maxilla (Fig. 1644) unique in *Charopinus*-branch, with only two terminal papillae; dorsalmost third papilla reduced to small, slender seta, displaced to lateral side of central papilla; two full-sized papillae with short, robust, flexible setae; exopod lateral, short, conical, tipped with two short setae; dorsal margin of appendage with two patches of denticles, one at level of exopod, other near base of terminal papilla. Second maxilla (Figs. 1635, 1636) about as long as, or longer than, trunk, strongly contractile. (In Fig. 1635 it is clearly contracted, as evident from abundant transverse wrinkling.) Two members of pair separate from each other; tip slightly expanded at base of short manubrium. Bulla small (Fig. 1635, 1636), with lenticular orbicular anchor, not much wider in diameter than combined tips of both second maxillae. Maxilliped (Fig. 1645) with long, slender corpus, in some specimens reaching level of tip of mouth cone (Fig. 1640); myxa with short, strong seta at about midlength, transverse patch of very fine denticles at base of subchela; latter long and rather uneven with short basal seta and short claw; claw (Fig. 1646) only slightly curved, blunt, without secondary teeth; at its base robust, blunt barb, about $\frac{3}{4}$ as long as claw; inner margin of subchela with strong denticles for about $\frac{1}{4}$ length.

Male (Fig. 1647): Cephalothorax resembling those of other males in *Charopinus*-branch, but trunk differing from them by presence of distinct segmentation. (Author was unable to examine a male specimen of this species and draws on description of Scott and Scott (1913), not altogether detailed.) Judging by the fact that males within genera of *Charopinus*-branch are usually very similar from one species to another, it can be assumed that segmentation of trunk is, in fact, not as distinct as shown in Fig. 1647. Male *L. pacifica* Kabata and Gusev, 1966, at any rate, has no clear segmentation of trunk. For detail of appendages consult Kabata and Gusev (1966). Total length about 1 mm (Markevich, 1956).

Comments: The taxonomic history of this species is beset with difficulties which it shares with those of several others described by early authors, particularly when their descriptions were sketchy, published in a periodical accessible with difficulty and in a language not widely known. Olsson's (1869) description was scanty and written in Latin, his illustrations generalized and not very helpful for comparative purposes. Only one subsequent report of his species was published (Bere, 1930b). All other authors followed T. Scott (1900), who described *L. cluthae* from the same host species. Not surprisingly, Scott found his specimens sufficiently different from, though similar to, Olsson's drawings to permit erection of a separate species. Following the publication of the well known Scott and Scott (1913) monograph, the presence of *L. cluthae* became generally recognized, whereas *L. longimana* all but disappeared from the literature. (Mentioned only by Wilson (1915) and by Yamaguti (1963).)

The present author, having examined Olsson's type material and many other specimens of *Lernaeopodina* from several species of *Raja*, *R. fullonica* (type host) included, found no morphological differences sufficient to postulate existence of two species of *Lernaeopodina* parasitic on Atlantic *Raja*. In particular, all specimens show a characteristic inflation of the anterior part of the cephalothorax. A similar inflation was illustrated by Bere (1930b) for her *L. longimana*. On the other hand, no other species (e.g. *L. longibrachia* and *L. pacifica*) have this swelling of the anterior cephalothorax. The only differences observed between females from various host individuals were those of length, a feature by itself insufficient to postulate specific distinctions between them. (Similar differences occur e.g. in *Salmincola edwardsii* (cf. Kabata, 1969d).) Consequently, it is proposed that the older name *L. longimana* (Olsson, 1869) be applied to *Lernaeopodina* parasitic on *Raja* in the north Atlantic and that the name *L. cluthae* (T. Scott, 1900) be relegated to synonymy with it.

There remains the problem of a third, inadequately described *Lernaeopodina* from *Raja batis* captured in the Firth of Clyde. *L. similis*, found only once, by the admission of its discoverers (Scott and Scott, 1913) is very similar to *L. cluthae* (= *L. longimana*). Only one damaged specimen of *L. similis* is now extant. Through the kindness of Dr. J. P. Harding (British Museum Nat. Hist.), the author was able to examine its photographs. The similarity between it and the other *Lernaeopodina* from *Raja* is, indeed, close enough to consider it tentatively synonymous with *L. longimana*, but a definitive decision on its validity is not, and might never be, possible.

It seems to have been forgotten that in his original description of *Lernaeopoda longimana*, Olsson (1869) believed it to be a subspecies of *Lernaeopoda galei* Krøyer, 1837. All subsequent authors accorded it the status of an independent species. Indeed, it is no longer in the same genus with *Lernaeopoda galei*.

L. longimana is a North Atlantic species. Most of its records come from the Atlantic sea-board of Europe, though Bere's (1930b) record extends its range to the shores of North America. It was found as far North as Iceland (Hansen, 1923) and the Barents Sea (Polyanski, 1955), but does not appear to extend to the south of British waters. It is not uncommon. (Polyanski (1955) found it on 20% of *R. radiata* examined.) Its known hosts are: *Raja batis*, *R. erinacea*, *R. fullonica*, *R. laevis*, *R. naevus*, *R. montagui* and *R. radiata*. Its usual habitat is on the gill arches and in the buccal cavity near the bases of the gill arches.

Genus Lernaeopoda Blainville, 1822

Female: Cephalothorax short, dorsoventrally flattened, ventrally inclined to long axis of trunk, with strong, well delimited dorsal shield, separated from trunk by transverse constriction. Trunk externally unsegmented, dorsoventrally flattened, either broader than long or much longer than broad, with intermediate shapes. No posterior processes, but modified uropods present, short or long, always ventral to egg sacs. First antenna indistinctly four-segmented, with well developed apical armature, usually with gibber. Second antenna long, with exopod usually longer than endopod. Tip of labrum with copious fringe of setae and weak rostrum. Mandible with three secondary teeth. First maxilla with lateral exopod and three terminal papillae. Second maxillae separate, shorter or longer than trunk. Bulla small, plano-convex or lenticular, with circular anchor and short manubrium. Maxilliped subchelate, strong, close to mouth cone.

Male: Cephalothorax slightly less than $\frac{1}{2}$ of body length, inclined ventrally to long axis of trunk, separated from it by deep transverse constriction, posterodorsally inflated into prominent, rounded swelling. Trunk fusiform, externally unsegmented, posteriorly with external genitalia and large, anterodorsally deflected, modified uropods. Antennae and oral appendages resembling those of female, but endopod of second antenna relatively better developed. Second maxilla and maxilliped subchelate. Two pairs of vestigial thoracic legs present.

Type species: *L. brongniartii* Blainville, 1822.

Comments: In common with some other old-established genera, *Lernaeopoda* suffers from the fact that the abundant literature on its subject is often superficial and confused, sometimes simply erroneous. The difficulty started with the type species, described in a manner sufficient to set vague boundaries to the generic diagnosis, but woefully insufficient to identify the species itself. Blainville (1822) knew neither the host of the parasite, nor the locality whence it came. It was never found again.

Wilson (1915) made a valiant effort to clarify the taxonomy of *Lernaeopoda* and related genera, greatly improving the position by his removal of species now accommodated in *Salmincola*, *Lernaeopodina* and elsewhere. The morphology of the male proved to be of great help in his understanding of relationships within the group. Since his work, however, other nominal species were added to the genus *Lernaeopoda*, sometimes inadequately described, at other times incorrectly diagnosed. The taxonomic position of the genus is still unsatisfactory and calls for a thorough revision. While this task lies outside the scope of this monograph, it is hoped that these comments will be helpful to future workers wishing to undertake it.

This author's own list of species now believed to be members of the genus *Lernaeopoda* contains 16 names. In addition to the unidentifiable type species, only two others can be accepted without hesitation as members of the genus. They are: *L. bidiscalis* Kane, 1892, and *L. galei* Krøyer, 1837, both present in British waters. Three are definitely not *Lernaeopoda*. *L. sebastis* Krøyer, 1863, is probably a damaged specimen of *Dendrapta cameroni* (Heller, 1949) with a confused host record. *L. upenaei* Pillai, 1962, is a *Neobrachiella*, as is clear from the morphology of its male (and from its non-elasmobranch host). *L. stromatei* Gnanamuthu, 1950, also a parasite of teleost fish, bears a general resemblance to *Lernaeopoda*, but the structure of its posterior extremity shows that its affinities lie elsewhere. Six species must be considered as species inquirendae, either because of inadequate descriptions (*L. bivia* Leigh-Sharpe, 1930, *L. musteli* Thomson, 1889, *L. mustelicola* Leigh-Sharpe, 1919, *L. obesa* Brian, 1924 (nec Krøyer, 1837) and *L. scoliodontis* Kirtisinghe, 1964) or because their males are unknown (*L. oviformis* Shiino, 1956).

The remaining four species (*L. centroscyllii* Hansen, 1923, *L. etmopteri* Yamaguti, 1939, *L. longicaudata* Hansen, 1923 and *L. globosa* Leigh-Sharpe, 1918) are herewith transferred to a new genus *Albionella*, on the basis of the morphology of their males.

The male of *L. brongniartii* was not described. Since the morphology of the male plays an important part in the taxonomy of Lernaeopodidae, Wilson (1915) designated the male of *L. bidiscalis* as the type for males of the genus. The presently known males fall into two groups, easily distinguishable by their general appearance. The type male is shown in Figs. 1662, 1663. It is clearly different from the male shown in Figs. 1701, 1702. The differences between them can be tabulated as follows:

Lernaeopoda	*Albionella*
Cephalothorax shorter than trunk, with prominent posterodorsal swelling.	Cephalothorax about as long as trunk, without posterdorsal swelling.
Deep constriction between cephalothorax and trunk.	Separation between cephalothorax and trunk indistinct.
Uropods inflated, pointing anterodorsally.	Uropods not inflated, pointing posteriorly.

It might also be worth mentioning that a species described as *Lernaeopodella major* by Heegaard (1962) might prove to be either a *Lernaeopoda* or an *Albionella* (depending on the still unknown morphology of its male). Its generic distinctness is doubtful.

The two definitely valid species of *Lernaeopoda* occurring in British waters can be distinguished with the aid of the following key:

L. galei	*L. bidiscalis*
Trunk longer than broad (Fig. 1674).	Trunk length equal to, or less than, its width (Fig. 1648).
Dorsal shield of cephalothorax without distinctive colour	Dorsal shield of cephalothorax brick-red in its anterior part.
Tips of second maxillae not greatly expanded (Fig. 1676).	Tips of second maxillae expanded into prominent semicircles (Fig. 1651).

Lernaeopoda galei Krøyer, 1837

(Figs. 1674–1690)

Syn.: *Achtheres selachiorum* Kurz, 1877
Lernaeopoda selachorum Kurz (?); of Vejdovsky (1822)
Achtheres galei Krøyer; of Carus (1855), Brian (1906)
Lernaeopoda scyllii Richiardi, 1880; of Brian (1899, 1906)
Lernaeopoda musteli Thomson, 1890; of Brian (1899, 1906), Bassett-Smith (1899)
Lernaeopoda scyllicola Leigh-Sharpe, 1916
Lernaeopoda mustelicola Leigh-Sharpe; of Brian (1944)

Female (Figs, 1674, 1675): Cephalothorax short, dorsventrally flattened, inclined ventrally to long axis of trunk; strong, clearly delimited dorsal shield with margins better defined laterally than posteriorly, bisected by groove in mid-dorsal line; posterior margin of cephalothorax rounded, slightly inflated, anterior narrower, transversely truncated. Constriction between cephalothorax and trunk deep and well marked laterally (Fig. 1674), but shallow dorsally and ventrally (Fig. 1676), differing from one specimen to another. Trunk about three times longer than cephalothorax, dorsoventrally flattened, anteriorly narrower, often showing signs of division by tranverse groove short distance behind cephalothorax (this feature, often absent or poorly expressed, is sometimes considered as segmental boundary between first and second trunk segments). Lateral margins of trunk posteriorly diverging, but often converging towards posterior extremity (Fig. 1674); posterolateral corners rounded; posterior margin transversely truncated, even, with anal slit seldom elevated on tubercle and oviduct orifices near posterolateral corners. No genital process (Fig. 1677). Total length (uropods included) 13–14 mm, uropods 2 mm (Scott and Scott, 1913). Length 15–16 mm (Markevich, 1956).

First antenna (Fig. 1678) indistinctly four-segmented, division between first and second segments not always discernible; ventromedial wall of short second segment with moderately long whip; basal segment inflated, unarmed; penultimate segment cylindrical (apparently divided in two in Fig. 1678), armed with dentiform solus; apical armature (Fig. 1679) comprising digitiform seta 4 on medial side of apex, with subterminal tubercle 1 near its base, flagelliform seta 6 with accompanying tubercle 3 and three slender setae in complex 5; all three appearing shorter and less well developed than either 4 or 6; tubercle 2 present, gibber possibly present. Second antenna with long, strongly sclerotized sympod; long axis of appendage running through sympod and exopod; latter one-segmented, larger than endopod, covered with fine denticles, armed with two terminal and one subterminal papillae; endopod (Fig. 1680) reduced, two-segmented, slightly to one side of main axis; proximal segment with small ventral patch of delicate denticles, distal with hook 1 (in this species either straight or only slightly hooked and reduced), spiniform seta 2, and ventrally small protuberance and seta equal to 2 in length but more slender; ventral wall with denticles similar to those of proximal segment. (Homology of ventral structure is doubtful, especially since the small protuberance is missing in some specimens. Probably longer seta represents process 5, the remaining structures belonging to process 4.) Tip of labrum similar to that described for *L. bidiscalis*, but marginal fringe not distributed into two groups. Mandible (Fig. 1681) with dental formula: P1, S1, P1, S1, P1, S1, B4; teeth of basal series long, slender. First maxilla (Fig. 1682) with lateral, well developed exopod armed with two longer setae inflated at base, and sometimes with third, small seta, difficult to detect; distal end of endopod wider than base, armed with three terminal papillae, short but robust, each surmounted by single, flexible seta of its own length. No denticulation observed. Second maxilla in fully relaxed specimens (Fig. 1676) almost as long as trunk, cylindrical, slightly tapering distally, with tip slightly expanded, separate from its opposite number. When contracted (Fig. 1675), it can be only ½ length of trunk, profusely wrinkled. (This is important to note, because earlier authors used the length of the second maxilla as a taxonomic feature to distinguish between species.) Bulla with short manubrium and small plano-convex anchor, circular in outline. Maxilliped (Fig. 1683) long, fairly slender; myxa with short seta and denticulated pad; another pad at base of subchela; latter with basal seta at about ½ length and denticulation at tip of inner margin with fine, crowded denticles; barb short, slender at tip, not reaching half length of claw; claw (Fig. 1684) with strong curvature close to base, with uneven diameter and often slightly sinuous outline; on lateral margin, near base, two secondary teeth, conical and blunt; at same level, nearer concave margin of claw, two more present. (In specimens from *Galeus* there are usually two pairs, as shown in Fig. 1684, but in those from *Scyllium* only one of the second pair might be present, or only the first pair.)

Male (Fig. 1685): Cephalothorax less than ½ length of body, ventrally inclined to long axis of trunk, posteriorly rounded and well delimited from trunk by deep constriction; dorsal shield indistinct; anterior part with mouth cone, flexed obliquely ventrad from long axis of cephalothorax; posterior part ventrally inflated at bases of second maxillae and maxillipeds. Trunk fusiform, posteriorly tapering. For description of posterior extremity see male of *L. bidiscalis* (p. 378). Length about 2 mm (Leigh-Sharpe, 1918).

First antenna (Fig. 1689) similar to that of female, but with more clearly developed gibber and tubercle 2 closer to bases of setae of complex 5. Mandible (Figs. 1686, 1687), in some specimens lacking first primary tooth (this might be due to damage in vivo or might be accounted for by developmental aberrations). Second maxillae and maxilliped similar to those of male *L. bidiscalis*. Tympanum of second maxilla (Fig. 1690) with more prominent papillae displaying subapical, nipple-like swellings, usually two each. Mediative process (Fig. 1688) distinctly bifid and with smaller branch bearing small, blunt corrugations. Two pairs of vertigial thoracopods represented by single setae.

Comments: Although Kurz (1877) confused the generic identity of this species, most authors had no difficulty in recognizing it as *Lernaeopoda*. Where the specific identity is concerned the matter is far less simple. Synonymy of *L. galei* presents a confused story. The main source of confusion was created by the inadequacy of the original description, by equally inadequate subsequent descriptions of new species of *Lernaeopoda* parasitic on mustelid and galeid sharks, as well as on scyllid dogfishes. Difficulties were further augmented by uncertainties about the identity of the hosts. Brian (1899, 1906) believed he recognised Richiardi's (1880) undescribed *L. scyllii* in his specimens from *Scyllium*, but changed his view when dealing with specimens from *Mustelus* collected in Argentina. Although he thought them identical with the former, he changed their name to *L. mustelicola* Leigh-Sharpe, 1919. An impression is created that sometimes the identity of the host influenced the examiners' views on the identity of the specimens before them.

Leigh-Sharpe's (1916) description of *L. scyllicola* from *Scyllium canicula* distracted attention from the taxonomic problems of *L. galei*. Due to the abundance in British waters both of the host and of its lernaeopodid parasite, Leigh-Sharpe's name *L. scyllicola* became well known and tended to replace *L. galei* in the literature. (A notable exception to this trend can be found in the work of A. Scott (1929), who identified as *L. galei* specimens from both *Mustelus* and *Scyllium* in the Irish Sea, long after *L. scyllicola* was described.) It should be noted that Leigh-Sharpe's decision to treat the parasites of *Scyllium* as a new species was taken on the basis of their supposed differences from *L. galei* (as represented by Wilson (1915) and Scott and Scott (1913)) in the structure of mandibles and maxillipeds. Unfortunately, (Leigh-Sharpe, 1918a) was unable to arrive at correct decisions on the basis of material at his disposal.

Having examined *Lernaeopoda* from both *Galeus* (=*Galeorhinus*) and *Scyllium* (=*Scylliorhinus*), this author found no differences between them other than those that can be accounted for by individual variation, easily accommodated within intraspecific boundaries. (In fact, other than slight differences in the proportions of the body, the numbers of secondary teeth on the claw of the maxilliped was the only worthwhile difference.) He proposes, therefore, that *L. scyllicola* be relegated to synonymy with *L. galei*. It is also his view that *L. scyllii* is synonymous with this species.

In this combination, *L. galei* is a species abundantly occurring on small sharks along the Atlantic sea-board of Europe, as well as in the Mediterranean. A recent record (Pogoreltseva, 1970) extends its range to the Black Sea. Barnard (1955a) mentions its occurrence in South African waters on *Mustelus laevis* and Brian (1944) off the Atlantic coast of South America. Capart (1953, 1959) found it on *Paragaleus gruveli* off the coast of Senegal (Africa). To the North, Iceland was mentioned as its locality by Hansen (1923) and the Barents Sea by Markevich (1956).

The common hosts of *L. galei* comprise sharks of the genera *Mustelus* and *Galeus*. Species of the genus *Scyllium* are also commonly parasitized. There are some records of unusual hosts. Kurz (1877) reported *Myliobatis aquila* and van Beneden (1871) *Dasyatis pastinaca* as harbouring *L. galei*.

L. galei is very common in British waters, where it can be found in the cloacal region, less commonly on the claspers of the male, exceptionally in other habitats, on *Scylliorhinus caniculus*, *Mustelus mustelus* and *Galeorhinus galeus*, wherever they occur.

One comment on the male of this species applies equally well to that of *L. bidiscalis*. Although on occasions the male is found attached near the posterior extremity of the female, it is much more common to find it attached as shown in Figs. 1674, 1675. This position "on the left shoulder" of the female might be considered as a generic characteristic of *Lernaeopoda*. Males of the closely related *Albionella* have never been observed so attached. The reason for this type of association between the sexes is not known, though Gray (1926) suggests that it might be the result of the male's feeding on the secretion of the so-called "bromatophore glands" of the female, the ducts of which are situated in the groove between the cephalothorax and trunk.

Lernaeopoda bidiscalis Kane, 1892

(Figs. 1648–1673)

Female (Figs. 1648–1650): Cephalothorax short, dorsoventrally flattened, obliquely ventral, or at right angles, to long axis of trunk with heavily sclerotized, sharply delimited dorsal shield, anteriorly brick-red (distinctive coloration, retained in most fixed and preserved specimens, extending also to anterior margin of cephalothorax and basal parts of second antennae); lateral margins of cephalothorax almost parallel to each other, posterior margin rounded, usually with prominent dorsal swelling. Trunk separated from cephalothorax by deep constriction running transversely behind bases of second maxillae. Trunk subquadrate, often rather broader than long, dorsoventrally flattened, with rounded anterolateral and posterolateral corners and gently convex lateral margins; dorsal and ventral surfaces of trunk with two parallel rows of depressions, but without definite external segmentation; no anterior, neck-like narrowing; in some specimens, posterolateral corners in shape of small rounded lobes (Fig. 1652); anal slit in centre of posterior margin, flanked by uropods, fusiform, usually tapering posteriorly, longer than $\frac{1}{4}$ of trunk length. Length without uropods, 5.5 mm, with uropods 7 mm (Scott and Scott, 1913).

First antenna (Fig. 1654) apparently four-segmented; basal segment inflated and unarmed; second segment cylindrical, distally with moderately long, robust whip: penultimate segment cylindrical, narrower than preceding one, distally with dentiform solus; terminal segment armed with apical armature. (The author was unable to determine with accuracy the details of apical armature, due to poor condition of first antennae in all specimens he examined. Tubercles 1, 2 and 3 were present in all, as were setae 4 and 6. Of complex 5 either two or three setae were present. In some specimens no details of complex 5 were possible to ascertain. By analogy with other species of *Lernaeopoda*, as well as with male of *L. bidiscalis*, three setae should be present in this complex.) Second antenna (Fig. 1655) long, slender, with heavily sclerotized sympod clearly divisible into two segments (basal often coloured brick-red); exopod longer and more robust than endopod, covered with marginal denticulation and armed with two papillae, bulbous and one-segmented; endopod two-segmented, with small patch of denticles on ventral wall of proximal segment; distal segment with reduced apical armature, comprising small hook 1, spiniform seta 2, ventral patch of denticles and, in some specimens, barely noticeable process probably homologous with process 5; in other specimens two small processes corresponding to processes 4 and 5. Tip of labrum (Fig. 1653) bluntly rounded; rostrum present, though usually largely covered by marginal setae; latter abundant, broad and flat, longest at anterolateral corners, shorter at mid-dorsal line. Mandible (Fig. 1656) with dental formula P1, S1, P1, S1, P1, S1, B5; teeth of basal series diminishing towards base of appendage. First maxilla (Figs. 1657, 1658) with exopod lateral, cylindrical, with rounded apex, armed with two short setae, one terminal and one subterminal; endopod distally broader than at base, with three terminal papillae of about equal size, all armed with short, soft, flexible setae; no marginal denticulation observed. Second maxilla (Figs. 1649, 1651) short, cylindrical, even in relaxed specimens never as long as trunk; members of pair free of each other, except at tips; latter greatly expanded to form flat, semicircular structures, together creating collar round base of bulla. (In many specimens, unlike one shown in Fig. 1651, no gap found between two halves of collar.) Bulla with short manubrium and plano-convex anchor, often with slightly concave subanchoral surface; diameter of anchor barely $\frac{1}{2}$ that of collar. Maxilliped (Fig. 1659) long, with robust, tapering corpus; myxa with short, proximally inflated seta and patch of denticles (Fig. 1661); other patch transversely crescentic at base of subchela; marginal denticles often fused at bases, in some specimens with gaps in "fence", individual scattered denticles in centre; subchela $\frac{1}{2}$ as long as corpus, with basal seta and with denticulation on distal half of inner margin; barb (Fig. 1660) short, tapering, less than $\frac{1}{2}$ length of claw; latter gently curved, tapering, with blunt tip; at level of barb two lateral secondary teeth.

Male (Figs. 1662, 1663): Cephalothorax less than $\frac{1}{2}$ length of body, fusiform, without clearly delimited dorsal shield but with prominent posterodorsal swelling, inclined obliquely to long axis of trunk, slightly inflated in posterior part of ventral surface, at bases of second maxillae and maxillipeds. Border between cephalothorax and trunk formed by transverse groove behind bases of maxillipeds. Trunk fusiform, posteriorly tapering, with inflated uropods pointing anterodorsally and with full set of external genitalia (Fig. 1673) consisting of pair of genital plates, prominent and lateral to genital processes. (Leigh-Sharpe (1916), in his description of male *L. scyllicola*, described also presence of a median genital process with 2 pairs of spines. These were not observed by present author.) Length about 3.5 mm.

First antenna (Fig. 1664) similar to that of female. (Male specimens, examined by author, had apical armature (Fig. 1665) intact.) Tubercles 1, 2 and 3 and setae 4 and 6 present; complex 5, consisting of three setae, at tip of distinct gibber. Second antenna (Fig. 1666) relatively shorter and less slender than that of female; exopod with prominent papillae but apparently without denticulations; endopod two-segmented, much better developed than in female, retaining its prehensile function; hook 1 robust, curved, dominating apical armature, spiniform seta 2 present, in some specimens suggestion of tubercle 3 at base of hook 1; ventral wall without denticulations, but with long process 4 and much shorter, not always observable, process 5. Tip of labrum similar to that of female. Mandible (Fig. 1667) with dental formula: P1, S1, P1, S1, B5; two secondary teeth barely smaller than primary, dentiferous part of mandible narrower than remainder, but without abrupt transition in width; entire appendage long and slender. First maxilla (Fig. 1668) similar to that of female, but relatively much longer and more slender, particularly three terminal papillae and their setae. Second maxilla (Fig. 1670) short, squat, with subchela broad at base, short and tapering at tip, unarmed, without well-delimited claw; corpus subcircular, with round protuberance in anteromedial corner; latter (Fig. 1671) covered with prominent denticles, supporting tip of subchela; second maxillae arising from common base and linked by strong tympanum with reinforced free margin; tympanal papillae smaller than corresponding structures in *L. galei*. Maxillipeds (Fig. 1669) also arising from common base; corpus nearly oval, articulating distally with subchela; myxa prolonged into two processes, distal larger, with concave surface, ridged and accommodating tip of subchela; proximal process smaller, smooth, subconical; subchela without clearly delimited claw, though base of latter marked by small denticle on concave margin; another outgrowth (seta?) on convex margin of subchela; tip marked by delicate longitudinal ridges. Mediative process (Fig. 1672) with two unequal arms, larger slightly bifid. Two pairs of thoracopods represented by single setae on ventrolateral surface of trunk, close to its junction with cephalothorax.

Comments: In contrast with the preceding species, the identity of *L. bidiscalis* has never been in doubt. It was always readily recognized due to its characteristic morphology, particularly the brick-red colour of the dorsal shield, the brevity and unusual width of trunk and the characteristically prominent collar around the manubrium of the bulla, a feature that gave the species its name.

L. bidiscalis is a parasite of sharks of the genus *Galeorhinus*. There are also some records of its occurrence on *Mustelus*. O'Riordan (1966) reported on specimens in a museum in Dublin, apparently collected from *Scylliorhinus*. Distribution range of *L. bidiscalis* covers the Atlantic coast of Europe, though it has been found also in the Mediterannean and in Icelandic waters. One suspect record comes from the coast of California, where Wilson (1935b) claimed to have found it on *Galeus californicus*.

L. bidiscalis is not rare in British waters, though, its hosts not belonging to abundant species, it cannot be considered common. It has been found in many localities around the coast of the British Isles, in some instances accompanied on the same individual hosts by *L. galei*. In contrast with the latter species, however, it is usually parasitic only on the tips of the claspers of male sharks, often inflicting extensive damage.

Lernaeopoda mustelicola Leigh-Sharpe, 1919 (species inquirenda)

The original description of this species is based on two female specimens found on *Mustelus vulgaris*, captured in Plymouth in August 1918. Unfortunately, neither the description nor the accompanying figures are helpful for comparative purposes. It is the view of this author that *L. mustelicola* is very probably another synonym of *L. galei*. This view is indirectly supported by the fact that the species has not been seen again since its discovery. (Brian's (1944) record of *L. mustelicola* from *Mustelus asterias* taken in Argentinian waters is accompanied by illustrations sufficiently detailed to show that it was *L. galei*.) The author was unable to trace a Swedish record of *L. mustelicola* mentioned by Markevich (1956).

In view of the fact that Leigh-Sharpe's specimens were not preserved, it is unlikely that the identity of this species can be definitively ascertained. The author proposes, therefore, to accord to it the status of species inquirenda.

Genus Albionella gen. nov.

Female: Diagnosis as for *Lernaeopoda*.

Male: Cephalothorax about $\frac{1}{2}$ of total body length, inclined ventrally to long axis of trunk, with distinct dorsal shield but without posterodorsal inflation. Mouth cone roughly in line with long axis of cephalothorax. Trunk separated from cephalothorax by shallow, often indistinct groove, fusiform, posteriorly with uropods and external genitalia. Antennae and oral appendages similar to those of female. Second maxilla and maxilliped subchelate. Two pairs of vestigial thoracic legs. Uropods not inflated, pointing posteriorly.

Type species: *Albionella globosa* (Leigh-Sharpe, 1918).

Comments: The very close similarity between the females of the genus *Albionella* and those of

Lernaeopoda caused them to be included in the same genus until now. Lack of adequate information about the morphology of the males compounded the difficulties. What is now known about males currently placed in *Lernaeopoda* suffices, however, to transfer to the genus *Albionella*, in addition to the proposed type species, also *Lernaeopoda etmopteri* Yamaguti, 1939, *L. longicaudata* Hansen, 1923 and *L. centroscyllii* Hansen, 1923.

Albionella etmopteri is known so far only from Japanese waters, where it parasitizes *Etmopterus lucifer*, a small shark species, and from South Africa, where it was found on *Etmopterus* sp. In Japan was recorded also *A. longicaudata* (on *Centrophorus acus* and *C. atromarginatus*), originally discovered in Icelandic waters on *C. squamosus*. *A. centroscyllii* was found so far only on *Centroscyllium fabricii* in Iceland and the Newfoundland region.

It might be noted that *Lernaeopoda musteli*, described by Thomson (1890) in New Zealand, seems to resemble *Albionella etmopteri* more than it does any other species of either *Albionella* or *Lernaeopoda*, at least as shown in Thomson's Pl. XXVIII, Fig. 9. Further studies in New Zealand waters might cause relegation of this species to synonymy with *A. etmopteri*.

Albionella globosa (Leigh-Sharpe, 1918)

(Figs. 1619–1710)

Syn.: *Lernaeopoda globosa* Leigh-Sharpe, 1918

Female (Figs. 1691, 1692): Cephalothorax short, flat, oblique or at right angles to long axis of trunk, with definite though not heavy dorsal shield and moderate posterodorsal swelling. Border between cephalothorax and trunk formed by shallow groove. Trunk of variable shape, from short and squat to long and fairly slender, always of greater length than width and with varying degree of dorsoventral flattening; in some specimens indistinct transverse division near anterior end of trunk might indicate obsolete segmental boundary (feature unreliable for taxonomic purposes and dependent on maturity and condition of specimen at fixation); posterior extremity of trunk (Fig. 1691) broader than its anterior part, posterolateral corners rounded, in some specimens lobate; centre of posterior margin inflated into anal tubercle flanked by uropods, fusiform, tapering, about $\frac{1}{4}$ as long as trunk, always ventral to egg sacs. Length of trunk about 2 mm, of cephalothorax about 1 mm (Leigh-Sharpe, 1918).

First antenna (Fig. 1693) indistinctly four-segmented; proximal half of two segments, base not greatly inflated, near junction with distal half very long whip, reaching to tip of appendage (in some specimens shorter due to damage in vivo or caused by handling); third segment cylindrical, distally with dentiform solus; apical armature (Fig. 1694) dominated by robust digitiform seta 4 and very long, flagelliform seta 6; tubercle 1 at base of seta 4; four short, setiform structures, presumably representing complex 5; no distinct tubercle 2 and no tubercle 3 present. Second antenna with long axis running through sympod and between rami, rather than through sympod and exopod, as in *Lernaeopoda*, rami less unequal in size than those of latter genus; exopod one-segmented, bulbous, denticulated and with two prominent papillae; endopod two-segmented, distal segment (Fig. 1695) with better developed armature than in *Lernaeopoda*; hook 1 strong, spiniform seta 2 with inflated base, tubercle 3 very prominent and with apical nipple; ventral wall without denticulation, but with prominent process 4, apically tapering, inflated at base. Tip of labrum as in *Lernaeopoda*. Mandible (Fig. 1696) with dental formula: P1, S1, P1, S1, P1, S1, B4; appendage much broader than in British species of *Lernaeopoda*, with dentiferous margin relatively shorter and size difference between primary and secondary teeth more marked; no ridges or corrugations on ventral margin of appendage. First maxilla (Fig. 1697) with cylindrical, prominent, lateral exopod, armed with three long, slender apical setae; distal end of endopod less extended in width than in *Lernaeopoda*, with three terminal papillae longer, more slender and armed with relatively longer setae; dorsal margin of appendage, at level of exopod, inflated and covered with minute denticles; similar denticulation on dorsal margin of exopod. Second maxilla (Figs. 1691, 1692) usually longer than rest of body, robust at base, tapering distally, often found with abundant transverse wrinkles, separate from opposite number except at tip; tips expanded into flat semicircular pads forming moderately prominent collar round manubrium of bulla, short and usually completely obscured. Anchor of bulla circular, of smaller diameter than collar, plano-convex. Maxilliped (Fig. 1698) slender and long, with corpus slightly narrowing anteriorly. Longitudinal myxa (Fig. 1699) with distal seta; patch of denticles obliquely transverse near base of subchela; latter $\frac{1}{2}$ length of corpus, with prominent basal seta and denticles on medial margin; claw (Fig. 1700) short, gently curving and sharp, with one very prominent, sharp secondary tooth on inner margin; barb straight, sharp, ½ length of claw.

Male (Figs. 1701, 1702): Cephalothorax inclined ventrally to long axis of trunk, without prominent posterodorsal inflation or definite dorsal shield, not sharply delimited from trunk. Latter fusiform, often with varying number of indistinct transverse grooves on dorsal surface as pronounced as that dividing it from cephalothorax, length of trunk about equal to that of cephalothorax; posterior extremity with well developed external genitalia (Fig. 1710) and uropods. Length less than 1 mm.

First antenna (Fig. 1703) resembling that of female in segmentation and length of whip; apical armature prominent, of composition similar to that of female, but one of four setae of complex 5 shorter and apparently representing tubercle 2. (Author was unable to observe solus on penultimate segment, but it is probably present, though difficult to see.) Second antenna (Fig. 1704) long; exopod shorter than endopod, bulbous, one-segmented, with two prominent setiform papillae but apparently with no trace of denticulation; endopod two-segmented, without denticulation, distal segment armed as in female. Tip of labrum as in female. Mandible (Fig. 1705) resembling that of female in shape, but with dental formula P2, S1, P1, S1, B6; no secondary tooth between first two primary ones, number of basal teeth greater than in female. (The author had limited material at his disposal. It is possible that some individual variation exists in mandibular dentition of males of this species.) First maxilla (Fig. 1706) very much like that of female, but apparently without denticulation on either endopod or exopod. Second maxilla (Figs. 1708, 1709) similar to those of *Lernaeopoda*, distinguishable from them by smaller and fewer denticles in myxal area and by absence of papillae on tympanum; short oblique ridges, however, present in distolateral corners of tympanum; subchela short, curving, with delicate longitudinal striation on tip. Maxilliped much like that described for *Lernaeopoda* (cf. Fig. 1669), with mediative process present between two maxillipeds. Process bifid (Fig. 1707), its branches short, inflated and bulbous. Uropod (Fig. 1710) with slightly penduncculate base, tapering at tip and armed with three setae, one long on dorsal surface and two short ones on lateral margins. Two pairs of vestigial thoracic legs represented by single setae (Fig. 1702).

Comments: *Albionella globosa* has not been found outside British waters, though its distribution probably covers the entire, or major part of, range of its host, *Scylliorhinus caniculus*. Its secluded habitat in the nasal fossae of the host prevented its earlier discovery and makes it difficult for casual observers to find. This author found it on very rare occasions attached outside the nasal fossa, on the side of the host's head.

The specific name was given to *A. globosa* because of the short and squat trunk of the type specimens. Figs. 1691, 1692 show that the name was not altogether fortunate.

It is interesting to note that *A. globosa*, living in the confined space of the nasal fossae, developed differently from another species of lernaeopodid, living under similar conditions. *Pseudocharopinus bicaudatus*, inhabiting the spiracles of *Squalus acanthias*, has characteristically short second maxillae (cf. Fig. 1610), whereas *Albionella globosa* has second maxillae longer than its entire body. Possibly differences in the sizes of both species, in relation to the spaces they inhabit, can account for these morphological dissimilarities.

Genus Clavellistes Shiino, 1963

Female: Cephalothorax moderately long, cylindrical, with antennae and oral appendages confined to anterior part. Latter with well delimited dorsal shield. Trunk longer than wide, dorsoventrally flattened, with very large genital process but without posterior processes. First antenna three-segmented, with well-developed apical armature. Second antenna with reduced endopod. Mandible without secondary teeth. First maxilla with ventral exopod and two terminal papillae. Second maxilla short or long, separate and slender or fused and much inflated. Maxilliped subchelate, close to mouth cone. Uropods absent. Egg sacs with uniserial or multiserial eggs.

Male: Cephalothorax less than ½ total length, flat, with indistinct dorsal shield, separated from trunk by shallow transverse groove. Trunk fusiform, posteriorly tapering, with uropods but without external genitalia. First antenna as in female. Second antenna prehensile. Mandible and first maxilla as in female, second maxilla and maxilliped subchelate. No thoracic legs.

Type species: *Clavellistes shoyoae* Shiino, 1963.

Comments: Shiino (1963b) established this genus for a species found on a single occasion "around the eye" of the moonfish, *Lampris regius,* in the Pacific. In his view, it represented a mosaic of features of *Clavella*-type and *Brachiella*-type lernaeopodids, so that it could not be placed in any existing genus of this family. Shiino was correct, *Clavellistes* is a valid taxon.

British fauna contains the only other known species of this genus, *C. lampri* (Scott and Scott, 1913). The female of this species is described below for the first time.

Clavellistes lampri (Scott and Scott, 1913)

(Figs. 1711–1732)

Syn.: *Lernaeopoda* sp. T. Scott, 1901
?*Lernaeopoda lampri* Scott and Scott, 1913
Thysanote lampri; of Leigh-Sharpe (1933)
Andropoda lampri (Scott and Scott, 1913); of Kabata (1965e)

Female (Fig. 1711): Cephalothorax cylindrical, moderately long (Fig. 1711 shows it foreshortened, it is half as long again, seen in full length); anterior part covered by well developed, heavy dorsal shield (Fig. 1712), with nearly parallel lateral margins and ill-defined posterior margin. Trunk not clearly delimited from cephalothorax, much longer than broad, narrower anteriorly, broader posteriorly, dorsoventrally flattened, with rounded posterolateral corners, without posterior processes but with very large genital process; latter (Fig. 1713) cylindrical, with bilobed tip, each lobe with one vaginal orifice. Measurements of two specimens: cephalothorax length 1.24 mm (1.20–1.28); cephalothorax width 0.60 mm; trunk length 4.30 mm (4.00–4.60); trunk width 0.92 mm; genital process length 1.28 mm (1.20–1.36); genital process width 0.36 mm; egg sac length 6.60 mm (6.00–7.20), egg sac diameter 0.36 mm. Numbers of eggs in three intact sacs examined: 35, 36 and 38.

First antenna (Fig. 1714) three-segmented; first segment comprising almost $\frac{1}{2}$ total length of appendage, slightly inflated, with moderately long whip; second segment short, cylindrical, with solus; third segment also cylindrical, with gibber and well developed apical armature (Fig. 1715) comprising seta 4 with robust base and slender, rounded tip; at its base conical tubercle 1; flagelliform seta 6 on opposite side of gibber; complex 5 with either two (Fig. 1715) or three (Fig. 1714) setae; tubercle 2 on side of gibber. (Lack of material prevented author from establishing whether Fig. 1714 represents an abnormal first antenna.) Second antenna (Fig. 1716) short and robust; sympod two-segmented, heavily sclerotized, with distinct hinge between segments; boundaries between second segment of sympod and rami, as well as between segments of endopod, obsolete; exopod one-segmented, larger than endopod, conical and apparently unarmed; endopod two-segmented, first segment unarmed, second with much reduced hook 1, spiniform seta 2 and at least one process on ventral wall (details impossible to ascertain). Tip of labrum (Fig. 1717) tapering, with broad, prominent rostrum, partly covered by short fringe of fine setae. Mandible (Fig. 1718) of primitive type, with dental formula P10; teeth sharp and curving, only slightly diminishing towards base; ventral margin smooth. First maxilla (Fig. 1719) with exopod slightly displaced from ventral margin to lateral surface, cylindrical, as long as terminal papillae, with two short setae on apex; two terminal papillae with soft, flexible, long setae; lateral to base of dorsal papilla slender seta, as long as papilla; no denticulation observed. Second maxilla (Fig. 1711) cylindrical, slender, as long as, or longer than, trunk, tapering at tip; members of this pair appearing partly fused. (Bulla missing in all four specimens examined.) Maxilliped (Fig. 1720) with broad corpus and short subchela; myxa with one slender seta and patch of fine denticles, proximal to it; other patch at base of subchela; latter with small basal seta and peg-like denticles covering about $\frac{1}{2}$ of inner margin; distal process overlapping base of claw; barb short, slender; claw (Fig. 1721) $\frac{1}{2}$ length of subchela, with curvature in proximal half, slender, with sharp tip and no secondary teeth. Egg sacs (Fig. 1813) with single rows of slightly flattened eggs.

Male (Fig. 1722): Cephalothorax about $\frac{1}{3}$ length of body, flat, with indistinct dorsal shield, inclined ventrally to long axis of trunk, separated from latter by shallow groove. Trunk fusiform, unsegmented, posteriorly tapering (Fig. 1723), with uropods but no external genitalia. Length 4.27 mm (3.78–4.83) (Kabata, 1965e).

First antenna (Fig. 1724) similar to that of female; no solus on second segment; third segment with poorly developed gibber (Fig. 1725), three setae in complex 5 and third tubercle (Fig. 1725) possibly homologous with tubercle 3. Second antenna (Fig. 1726) prominent, prehensile, with indistinctly segmented sympod and endopod much longer than exopod; latter one-segmented, bulbous, with distal half of its surface denticulated; at least one (possibly two) papillae near tip; ventral surface of endopod's basal segment with inflated, denticulated pad; armature of distal segment with sharply curved hook 1, slender, spiniform seta 2 and prominent tubercle 3; process 4 on ventral wall slender, spiniform; no denticulation on ventral wall. Tip of labrum (Fig. 1728) with well developed rostrum; setae of fringe much shorter than those of female. Mandible (Fig. 1729) slender, with dental formula P10. First maxilla (Fig. 1727) very similar to that of female; in some specimens (seen under phase contrast illumination) exopod with numerous refractile spots of unknown nature. Second maxilla (Fig. 1732) with squat, subcircular corpus; distomedial margin pulled out into triangular process with cavity in anterior surface for reception of tip of subchela; latter short, with curving distal $\frac{1}{2}$ and tapering, sharp tip; strong, sharp secondary teeth at about midlength of inner margin. Maxilliped (Figs. 1730, 1731) with corpus obliquely slanted to base of appendage, medial margin forwards (Fig. 1730); corpus squat, with triangular extension of medial margin; latter with distal, conical cavity and single, small tooth; subchela long, slender, with sharp tip and one secondary tooth on distal $\frac{1}{2}$ of inner margin. No thoracic feet found. Uropod (Fig. 1723) flat, lanceolate.

Comments: T. Scott (1901) found "a few male specimens" of what appeared to be a species of *Lernaeopoda* on the gills of *Lampris guttatus*, taken in Shetland waters. These specimens, at first described as *Lernaeopoda* sp., were later designated as ? *L. lampri* by Scott and Scott (1913). The authors were rather doubtful as to the generic identity of the parasite and considered that in "the form of the various appendages [it] shows a close relationship with the Lernaeopodidae, closer perhaps with *Charopinus* than with *Lernaeopoda*". There were no later records of this parasite, but Leigh-Sharpe (1933a) listed it in his list of British parasitic Copepoda under the name *Thysanote lampri*, giving no reasons for this transfer.

The present author examined 17 male specimens of the same parasite from *Lampris guttatus*, taken in the Faroe Islands. Detailed study suggested that *L. lampri* cannot be placed in any known lernaeopodid genus. Unaware of Shiino's (1963) recent work, the author (Kabata, 1965a) suggested a new genus for these males and named it *Andropoda*. It is now clear that *Andropoda* is synonymous with *Clavellistes*.

The existence of a species known for seven decades from its males alone is unique among Lernaeopodidae. It was only on 22 February 1971 that Dr. K. MacKenzie from the Marine Laboratory, Aberdeen, found four female specimens on the gills of *L. guttatus* taken in the North Sea. Although the specimens were in poor condition when the author had the chance to examine them, and although they were not accompanied by males, they were clearly conspecific with *Clavellistes lampri*, as shown by their mandibles, first antennae and first maxillae.

C. lampri, alone among Lernaeopodidae, has uniserial egg sacs. (Previous lowest number for rows of eggs was established by *Clavella gracilis* Hansen, 1923, with two rows per sac.) The most unusual feature of *C. lampri*, however, is the apparently common existence of solitary males of a size, by lernaeopodid standards, really gigantic. The problem of solitary males in Lernaeopodidae was discussed by Quidor (1910), who found an oversized juvenile male of *Brachiella* sp. attached to the gills of the host; he also recorded a giant male *Neobrachiella insidiosa* (Heller, 1865) attached to the cephalothorax of the female. According to Quidor, the reproductive organs of the giant males were atrophied or missing. The author's own study of sectioned specimens of giant males of *C. lampri* proved inconclusive, their condition being insufficiently fresh to make out details of the anatomy of the reproductive system. It might be suggested that the unattached males of this species, and perhaps also of others, grow to larger sizes than those which find their mates. The males of *C. shoyoae*, normally attached to females, were certainly no giants. More material is needed to solve this interesting problem.

Genus Brachiella Cuvier, 1830

Female: Cephalothorax cylindrical, either shorter or longer than trunk, with well developed dorsal shield on its anterior quarter. Trunk as long as, or longer than, wide, dorsoventrally flattened, posteriorly with long uropods and one pair of long ventrolateral processes. First antenna four-segmented, with poorly developed apical armature. Second antenna with greatly reduced endopod nearly at right angles to long axis of appendage. Mandible with three secondary teeth. First maxilla with lateral exopod and three terminal papillae. Second maxilla slender, cylindrical, separate from its opposite number, about as long as, or somewhat shorter than, trunk, only slightly expanded at apex. Bulla with short manubrium and small, circular, plano-convex anchor. Maxilliped subchelate, with robust corpus and weak claw.

Male: Cephalothorax less than $\frac{1}{2}$ of body length, with indistinct dorsal shield and pronounced posterodorsal inflation, slightly inclined ventrally to long axis of trunk, separated from latter by deep transverse constriction behind level of maxillipeds. Trunk fusiform, tapering posteriorly, with well developed uropods and external genitalia. First antenna as in female. Second antenna prehensile, with endopod well developed and about as long as exopod. Buccal apparatus as in female. Second maxilla and maxilliped subchelate. No thoracic legs.

Type species: *Brachiella thynni* Cuvier, 1830.

Comments: The taxonomic history of the genus *Brachiella* is probably the most involved of all the lernaeopodid genera. The reasons for it are three-fold: firstly, until recently not enough attention has been paid to the structure of the female appendages, only the second maxilla and perhaps maxilliped receiving any attention; secondly, undue importance has been given to the gross morphology of the female second maxilla and, thirdly, not enough has been known about the morphology of the male. As a consequence, our concept of the genus *Brachiella* remained vague. Since the *Brachiella*-group of the family is one of the most highly advanced and apparently very actively speciating, it abounds in species, the large number and great morphological variety of which add to the difficulties of a sensible taxonomic arrangement.

As mentioned earlier, Wilson (1915) attempted to divide the assemblage of species placed in the genus *Brachiella* into several genera, using male morphology as the basic discriminant. This sound principle was not, however, correctly applied, due to the inadequate state of current knowledge of the lernaeopodid male. All but one of Wilson's genera are rejected by this author.

It was also the morphology of an abnormal (probably damaged) male that prompted Heegaard (1947b) to propose the genus *Isobrachia*. Heegaard's initiative was followed up and developed by Yamaguti (1963), who enlarged *Isobrachia* in a most perplexing manner, using the morphology of the female second maxilla as the main basis for consigning to that genus species previously placed in *Brachiella* and other genera. The invalid nature of *Isobrachia* was suspected by Pillai (1968b), whose doubts were well founded. *Isobrachia* is invalid, its species must be distributed anew.

As sometimes happens, the type species of the genus, *B. thynni*, when examined more closely, bears only superficial resemblance to species later assigned to the genus. In fact, among the many species now included in it, there is only one that can be considered as its congener. It is *B. magna* Kabata, 1968, a parasite of *Scomberomorus commersoni* discovered off the Queensland coast of Australia. In spite of the fact that its male is still unknown, the details of the female morphology suffice to prove that it is closely related to *B. thynni*. (The hosts of these two species are also systematically close.)

All other species must be removed to different genera. Most will find their place in the genus *Neobrachiella*, to be proposed below. *Brachiella lithognatha* from South Africa (Kensley and Grindley, 1973) is probably a type of a new genus, certainly not belonging to the *Brachiella*-group, judging from the morphology of its abbreviated male.

The morphological differences between *Brachiella* and newly-proposed *Neobrachiella* are many and clear enough to make one wonder why they were not observed earlier. They are set out in table form below:

Female

Brachiella	*Neobrachiella*
First antenna with poorly developed apical armature (Fig. 1738).	First antenna with well-developed armature (Fig. 1761).
Second antenna with endopod at nearly right angles to long axis of appendage (Fig. 1739).	Second antenna with endopod not much more deflected from long axis of appendage than exopod (Fig. 1762).
First maxilla with exopod lateral and three terminal papillae (Fig. 1742).	First maxilla with exopod ventral and two (sometimes reduced third) terminal papillae (Figs. 1766, 1839).

Maxilliped with short, poorly developed claw without secondary teeth (Fig. 1745).

Maxilliped with long claw, usually with secondary teeth on it (Fig. 1768).

Male

First antenna with poorly developed apical armature (Fig. 1749).

First antenna with well-developed apical armature (Fig. 1772).

Endopod of second antenna as long as, or shorter than exopod (Fig. 1750).

Endopod of second antenna longer than exopod (Fig. 1773).

Subchela of maxilliped heavily armed, with well-delimited claw (Fig. 1755).

Subchela of maxilliped not armed, claw not delimited (Fig. 1775).

External genitalia present (Fig. 1756).

External genitalia absent (Fig. 1771).

The type species, *B. thynni*, is the sole representative of this genus in British waters.

Brachiella thynni Cuvier, 1830

(Figs. 1733–1756)

Syn.: *Thynnicola ziegleri* Miculicich, 1905

Female (Fig. 1733): Cephalothorax cylindrical, varying in length from shorter to longer than trunk, with slightly expanded anterior end covered by dorsal shield; latter with anterior margin broader than posterior (Fig. 1736) and lateral margins converging in posterior direction. Trunk varying in shape, but usually somewhat longer than broad, expanding posteriorly, its greatest width at level of posterior extremity; dorsal surface often with two parallel rows of depressions; ventral surface with three pairs of swellings, transverse depressions between them giving appearance of external segmentation; posterolateral corners rounded; posterior extremity with anal slit in the centre (Fig. 1734) and four very long processes. Dorsal pair, modified uropods, with spherical swellings on medial side of base, long, cylindrical, distally tapering (Fig. 1734); ventral pair (Fig. 1735) smaller and more slender than dorsal, arising from the medial side of last pair of trunk swellings; between latter (Fig. 1735) genital process, incorporated into trunk but distinguishable from it; in some specimens close above it small conical outgrowth, to which male is usually found attached. Length (without posterior processes) 9–23 mm; posterior processes (dorsal) 6–10 mm.

First antenna (Fig. 1737) indistinctly four-segmented, division between first and second segment sometimes obsolete; first segment comprising about ½ length of appendage, inflated, apparently without whip; second segment (when discernible) cylindrical, distally tapering, unarmed; third segment narrower than second, cylindrical, unarmed; terminal segment cylindrical, longer than penultimate, with rounded tip; latter (Fig. 1738) with strong gibber, but with much-reduced apical armature (it is possible that specimens examined by the author had damaged first antennae; it is also possible that this condition, similar to that commonly occurring in the genus *Salmincola*, might be accepted as another generic characteristic of the genus *Brachiella*); of latter clearly recognizable only seta 4 and tubercle 1 at base of gibber; on tip of gibber three structures, two short setae and one tubercle; setae probably vestiges of complex 5, tubercle homologous with 2. (It is not possible to determine true composition of apical armature on evidence available.) Second antennae (Fig. 1739) with long axis passing through sympod and exopod, though latter somewhat inclined to long axis of former; segmentation of sympod obscure and joint between its base and cephalothorax indistinct, as is also division between it and rami, particularly exopod; latter one-segmented, bulbous, its apex armed with numerous minute denticles, as well as with usual two papillae; endopod displaced from long axis of appendage, at about right angles to it, two-segmented; basal segment with patch of denticles on ventral wall; armature of distal segment much reduced, particularly hook 1, no longer curved, slender, about as large as spiniform seta 2; on ventral wall small, pointed process 4; details of armature difficult to ascertain. Tip of labrum (Fig. 1740) triangular, with small but distinct rostrum at apex; marginal setae short, rather sparse, leaving rostrum uncovered. Mandible (Fig. 1741) with three secondary teeth, its dental formula P1, S1, P1, S1, P1, S1, B5; blade of appendage broad and robust. First maxilla (Fig. 1742) with slender sympod, lateral, cylindrical endopod, long and armed with two subequal setae on apex; endopod broad, with three terminal papillae of about equal size, tipped with short setae; dorsal margin inflated above level of exopod; region of inflation beset with sparse but fairly strong denticles. Second maxilla (Fig. 1733) cylindrical, as long as, or longer than, trunk, separate from opposite member, with slightly inflated tip forming small collar round base of bulla. Bulla with short manubrium and small, circular, plano-convex anchor. Maxilliped (Fig. 1743) large (in some specimens visible from dorsal side, as in Fig. 1736), with corpus broad at base, strongly converging apically; myxa often with prominent inflation at about midlength (in some specimens less well developed); tip of inflation covered with sparse, blunt denticles; another patch of denticles at base of subchela; between two patches (Fig. 1744) short seta, mounted on papilliform base; subchela cylindrical, half as long as corpus, curving, basally with short, slender seta, distally with long batch of blunt denticles on inner margin; barb at base of claw (Fig. 1745) short, flexible, with slightly inflated base; claw short, barely curving, with blunt tip and rather uneven diameter.

Male (Figs. 1746, 1747): Cephalothorax flat, short, ovoid in dorsal outline, without clearly defined dorsal shield, slightly inclined ventrally to long axis of trunk (in some specimens long axes of cephalothorax and trunk are nearly in line with each other), separated from latter by deep transverse constriction behind level of maxillipeds. Trunk fusiform, posteriorly tapering, with external genitalia and uropods on posterior extremity. In some males (Fig. 1747) trunks curving more or less in ventral direction. Length 1–2 mm.

First antenna (Fig. 1748) similar to that of female, but division between segments 1 and 2 very indistinct, sometimes impossible to see; basal two segments comprising about half of appendage, inflated, with short, robust whip at distal end; third segment cylindrical (author was unable to observe solus in specimens he examined, but Lewis (1967) reports its presence on third segment); apical armature (Fig. 1749) retained better than in female appendage, but still reduced; digitiform seta 4 and tubercle 1 at base of gibber, tubercle 3 on wall of gibber; apex of gibber with four setae, longest of them probably seta 6, remainder homologous with complex 5. Second antenna (Fig. 1750) primitive, prehensile, with rami of about equal size; segmentation of sympod obsolete, border between it and exopod not visible; exopod cylindrical, with conical, rounded apex, provided with two long papillae with inflated bases; no spinulation observed; endopod two-segmented: basal segment cylindrical, with transverse patch of fine denticles on ventral wall; terminal (Fig. 1751) with prominent though slender hook 1 and robust spiniform seta 2 at its base; on ventral wall slender process 4, resembling in size and shape seta 2; no process 5 and no spinulation observed. Buccal apparatus resembling that of female. First maxilla (Fig. 1752) also similar to that of female, but endopod relatively broader and terminal papillae longer and more slender; no dorsal denticulation. Second maxilla (Fig. 1753) short, with nearly circular corpus, connected with opposite member by posteriorly bifurcate, heavily sclerotized tympanum, with maxillary duct opening near medial margin; subchela short, its distal half at right angles to proximal, with short, sharp tip and inflated base; no secondary armature, no delimited claw. Maxilliped (Fig. 1754) long and slender, usually protruding laterally from its base and visible in dorsal aspect (Fig. 1766); corpus with well armed medial margin, near base of corpus transversely oval myxa with narrow band of denticulation across it; near midlength another patch of denticles, just proximal to short seta; very large patch of similar denticles at base of subchela; latter bent at nearly right angles near midlength; inner margin with patch of denticulation (Fig. 1755) consisting of flat, plaque-like denticles. At proximal end of denticles short, fine-tipped seta. Claw short, well delimited, resembling feline claws. Uropod (Fig. 1756, ur) cylindrical, tapering, unarmed. External genitalia consisting of flat, oval, but sharp-tipped genital processes (Fig. 1756, gp) and pair of flat, squamous genital plates (Fig. 1756, gpl), situated on ventral surface proximal to anal slit.

Comments: *Brachiella thynni* is a very characteristic, easily recognizable parasite, with large females that seldom fail to attract the attention of even casual observers. With the exception of Miculicich (1905), nobody has ever been unable to identify it correctly. It is a common parasite of various species of tuna and related scombrids, usually found attached on the outer surfaces (often near bases of fins) of their hosts. Of the non-scombrid hosts, it was recorded on *Argyrosomus regium* by Bassett-Smith (1899) and on *Pomatomus saltatrix* by Bere (1936).

B. thynni is known to occur in all localities where one finds its large scombrid hosts. It is, therefore, an inhabitant of every ocean in low and moderate latitudes. Though infrequently, it visits British waters with tuna.

Neobrachiella gen. nov.

Female: Cephalothorax cylindrical, anteriorly with more or less expanded head and well developed dorsal shield, posteriorly with or without transverse dorsal welt. Border between cephalothorax and trunk marked by bases of second maxillae, followed or not by very shallow constriction or cylindrical, short, neck-like anterior part of trunk. Latter greatly varied in shape, subcircular to elongated, with or without lateral, dorsal and ventral, usually tubercular, outgrowths; posterior extremity with one or several pairs of processes, varying in length and arrangement, one of them always modified uropods. First antennae indistinctly four-segmented, with well developed apical armature. Second antenna with sympod-exopod long axis; endopod much smaller than exopod, its long axis usually in line with that of exopod. Mandible with three (rarely fewer) secondary teeth. First maxilla with ventral exopod and either two, or two and reduced third, terminal papillae. Second maxilla either short or long, fused or separate, with or without collar round manubrium of bulla. Latter usually with short manubrium and small, circular, flat anchor (stelliform in *N. stellifera* (Heegard, 1962)). Maxilliped subchelate, its claw well developed, with secondary teeth.

Male: Cephalothorax about as long as, or somewhat shorter than, trunk, flat, oval in dorsal aspect, its anterior part covered with indistinct dorsal shield. Trunk separated from cephalothorax by transverse constriction behind maxillipeds, its anterior part waist-like, expanding into fusiform, posteriorly tapering main part; posterior extremity with uropods but no external genitalia. First antenna as in female. Second antenna prehensile, its endopod as long as, or longer than, exopod. Buccal apparatus similar to that of female, but mandible sometimes with fewer secondary teeth (at least one always present). Second maxilla and maxilliped subchelate. No thoracic legs.

Type species: *Neobrachiella merluccii* (Bassett-Smith, 1896).

Comments: The genus *Neobrachiella* is established to accommodate species previously placed in *Brachiella*, but differing from its type species, *B. thynni*, as indicated in p. 384. *Neobrachiella* receives also species of *Parabrachiella*, *Probrachiella*, *Epibrachiella* and *Brachiellina*, all genera founded on insufficient grounds and on various occasions recommended by this author for suppression.

The precise composition of *Neobrachiella* is difficult to establish, mainly because of the similarity borne by some of its species to species of the *Clavella*-group. Often only the morphology of the male can determine whether a species should belong to the *Clavella-* or *Brachiella*-group. Since males are not always known, doubts must be entertained as to the generic affiliation of some species. The matter can be clarified only by extensive and painstaking work, largely outside the scope of this monograph. For our purpose it is sufficient to point out that *Neobrachiella* is a genus of many species, widespread in all seas and oceans, parasitic on teleost fishes mainly of phylogenetically advanced groups.

In British waters, *Neobrachiella* is represented by seven species.

Key to females of British species of *Neobrachiella*

1. One pair of posterior processes 2
   More than one pair of posterior processes 3
2. Trunk subcircular or broader than long (Fig. 1758); transverse welt on dorsal side of cephalothorax (Fig. 1757); length of dorsal shield more than ½ that of trunk (Fig. 1757) . . . . . . **N. bispinosa**
   Length of trunk more than twice its width (Fig. 1832); no transverse welt on dorsal side of cephalothorax (Fig. 1832); dorsal shield less than ¼ length of cephalothorax . . . . . . **N. rostrata**
3. Two pairs of posterior processes 4
   Three pairs of posterior processes; trunk with many tubercular outgrowths (Figs. 1846–1848) . . **N. triglae**
   Four pairs of posterior processes; dorsal pair usually partly fused (Figs. 1777–1779) . . . . . **N. impudica**
4. Tips of second maxillae ending in five short digitiform processes each (Fig. 1816) . . . . . . **N. insidiosa**
   Tips of second maxillae undivided 5
5. Process 4 of second antenna prominent, covered with tubercles (Fig. 1799); second maxillae very short, apparently fused (Fig. 1795); dorsal shield roughly triangular (Fig. 1797) . . . . . . **N. chevreuxii**
   No tubercles on process 4 of second antenna; second maxillae clearly separate; dorsal shield with concave lateral margins (Fig. 1823) . . . . . . **N. merluccii**

Males of the genus *Neobrachiella* closely resemble one another and can be distinguished only by small details of morphology of appendages. In the account below, full description will be given for the male of *N. bispinosa*. Other males will be described by comparing them with it.

Neobrachiella merluccii (Bassett-Smith, 1896)

(Figs. 1821–1831)

Syn.: *Brachiella merlucci* Bassett-Smith, 1896
?*Brachiella impudica*; of Belloc (1929)

Female (Figs. 1821, 1822): Cephalothorax cylindrical, its anterior part with well developed dorsal shield; latter (Fig. 1823) broad anteriorly, with concave lateral margin and indistinct posterior end; no dorsal welt at base of cephalothorax; in most specimens cephalothorax flexed ventrally over bases of second maxillae, its anterior end facing ventral surface of trunk. Border between cephalothorax and trunk marked by bases of second maxillae, short part of trunk behind them cylindrical and narrow, expanding laterally. Trunk roughly trapezoid, with narrower anterior and broader posterior end, with four rounded corners, dorsoventrally flattened, with truncated posterior margin; proportions of length to width varying; on each side of anal tubercle cylindrical posterior process, varying in length but usually at least as long as $\frac{1}{2}$ of trunk and posteriorly somewhat tapering; ventral to anal slit genital process (Fig. 1821) of moderate length, less than $\frac{1}{2}$ length of posterior processes; lateral to it modified uropods, as long as posterior processes, though often of narrower diameter. Length about 8 mm (Scott and Scott, 1913).

First antenna four-segmented, with slightly inflated basal part and prominent whip at distal end of second segment, border between first and second indistinct; third segment with small solus; distal segment with well developed apical armature (Fig. 1824), comprising digitiform seta 4 with accompanying tubercle 1; no gibber found in specimens examined by author; on opposite side of apex slender seta 6; complex 5 apparently of four structures, two setae and two resembling elongated tubercles. (Homology of any of these with tubercles 2 or 3 could not be determined.) Second antenna resembling those of *N. insidiosa* and *N. rostrata*. Tip of labrum blunt, with rostrum and fringe of marginal setae. Mandible (Fig. 1825) with dental formula P1, S1, P1, S1, P1, S1, B5; secondary teeth tending to form units with preceding primary teeth; fifth basal tooth often overlapped by thin ventral margin of slender appendage. First maxilla (Fig. 1826) long, slender, with exopod slightly displaced from ventral margin to lateral surface, long, cylindrical, distally with two subequal setae; endopod short, narrow, with two long, terminal papillae tipped by long, flexible setae; at base of dorsal papilla another and much reduced papilla present, its apical seta reaching no further than distal end of dorsal papilla; no denticulation observed. Second maxilla (Fig. 1822) cylindrical, shorter than trunk, with rounded tip, separate from its opposite number. Bulla with short manubrium and small, circular anchor. Maxilliped (Fig. 1828) with robust corpus, with traces of subdivision near basal end. Medial margin with slightly inflated myxa beset with fine, small denticles; just distal to them short seta with inflated base; another patch of fine denticles near base of subchela; latter fairly long, straight, with basal seta and denticulated distal end of inner margin; barb (Fig. 1827) stout, $\frac{1}{2}$ as long as claw; some indistinct transverse grooves at base of claw; latter gently curving, with tapering, fine, slightly rounded tip; near midlength of inner margin a prominent secondary tooth, with one more on both sides of claw some distance above inner margin.

Male (Fig. 1829): Cephalothorax about half of total body length, flat, with fairly well developed dorsal shield, forming obtuse angle with long axis of trunk (angle varying according to position at time of fixation). Waist-like constriction separating cephalothorax and trunk. Latter fusiform, with somewhat inflated dorsal surface and rounded posterior extremity equipped with uropods, but no external genitalia. Length about 1 mm.

First antenna as in female. Second antenna prehensile, with prominent, mobile hook 1 at end of second segment of endopod; latter larger than one-segmented exopod. Buccal apparatus similar to that of female. Mandibular formula P1, S1, P1, S1, P1, —, B4/5; basal teeth varying in number; third secondary tooth occasionally replaced by diasteme. First maxilla (Fig. 1830) similar to that of female, but third terminal papilla more reduced, its seta barely reaching $\frac{1}{2}$ length of dorsal papilla. Second maxilla (Fig. 1831) with short, squat corpus, linked with opposite member of pair by narrow, thin tympanum, apparently without reinforced margin or surface ornaments; subchela short, with stout base and slender, curving posterior half, without delimited claw and devoid of secondary armature; corpus with conical distomedial process, covered by fine prickles (with duct orifice proximal to its base); subcircular sclerite of unknown nature on ventral surface of corpus close to base of subchela. Maxilliped similar to those found in other species. No thoracic legs observed. Uropods short, cylindrical, tapering, apparently unarmed.

Comments: The choice of *N. merluccii* as the type species for *Neobrachiella* was largely dictated by two facts. Firstly, it is instantly recognizable due to its characteristically flexed position. Rigidly selective in the choice of their habitat on the host, the females of *N. merluccii* are invariably attached to, or near, tips of gill rakers (Fig. 1822) and give the impression of being impaled upon them, their anterior and posterior extremities pointing towards the gill arch. Secondly, the taxonomic history of the species is unclouded by any misunderstandings. Apart from being transferred to *Neobrachiella* by the present author, this species was always recognized by others as *Brachiella merluccii*.

Neobrachiella merluccii is a parasite of the genus *Merluccius*, the hake. It is common on the European hake, *Merluccius merluccius* (recorded under diverse names by different authors), occurring in the eastern North Atlantic and the Mediterranean. It also extends south along the western coastal areas of Africa. Capart (1959) found it there on *Merluccius poli*. There are no records of *N. merluccii* from the Atlantic coast of America.

In view of this definite distribution range, it is strange to find Thompson and Scott (1903) recording it from the Indian Ocean (Ceylon), from a very atypical host, *Sciaena diacanthus*. Since, however, no subsequent workers rediscovered it from that area and host (though Kirtisinghe (1964) and Pillai (1967a) mention the old record), one must view the validity of this record as very doubtful.

Belloc's (1929) record of *Brachiella impudica* from hake refers most probably to this species.

Neobrachiella bispinosa (von Nordmann, 1832)

(Figs. 1757–1776)

Syn.: *Brachiella bispinosa* von Nordmann, 1832
Lernaeopoda obesa Krøyer, 1837
Brachiella obesa (Kr. sub *Lernaeopoda*); of Olsson (1869)
Anchorella ovalis Krøyer; of van Beneden (1871)
nec *Anchorella ovalis* Krøyer, 1837
Brachiella ovalis (Krøyer); of T. Scott (1901)
nec *Lernaeopoda obesa* Brian, 1924

Female (Figs. 1757–1759): Cephalothorax cylindrical, anteriorly expanded to form head, covered by dorsal shield; latter well delimited, anteriorly broader than posteriorly, with slightly converging lateral margins (Fig. 1770); head constituting more than ½ length of cephalothorax and often flexing ventrally to long axis of body, degree of flexure differing from one specimen to another (comp. Fig. 1757 with Fig. 1759); near base of cephalothorax transverse dorsal welt of varying degree of prominence. Border between cephalothorax and trunk marked by bases of second maxillae. Beyond that level trunk (Fig. 1757) continuing cylindrical, neck-like part, soon expanding laterally into broad, oval, dorsoventrally flattened main part; shape of trunk varying with age and stage of maturity, from fairly narrow, longitudinally oval (Kabata, 1973b) to subcircular or pyriform (Fig. 1758), often broader than long; posterior extremity rounded, with anal slit in centre, often raised on fairly prominent tubercle; ventral to anus genital process (Fig. 1769), cylindrical, varying in length, in younger specimens sometimes absent (Kabata, 1973b). Uropods short, cylindrical, distally tapering, ventral to egg sacs. Length about 3.5 mm.

First antenna (Fig. 1760) indistinctly four-segmented, first two segments more or less fused; basal segment inflated, unarmed, second segment cylindrical, distally with whip its own length; third segment narrower than second, with small solus; distal segment slightly tapering; tip (Fig. 1761) with prominent gibber; apical armature comprising very strong seta 4, tubercle 1 at its base with slender apical part, on opposite side of gibber long seta 6, with base broader than in other species, complex 5 of three subequal setae on tip of gibber, near tip of gibber also tubercle 2, smaller than but resembling tubercle 1 in shape. Second antenna (Fig. 1762) with almost rectangular, apparently unsegmented sympod, heavily sclerotized and with hinge-like articulation between it and cephalothorax; long axis of appendage passing through sympod and exopod; latter one-segmented, bulbous, with tip covered by diminutive denticles and armed with at least one (probably two) papillae; in most specimens, exopod inclined obliquely to long axis of appendage; endopod two-segmented: basal segment apparently unarmed; distal (Fig. 1763) with hook 1 relatively reduced and no longer curved, but dominating apical armature; spiniform seta 2 strong, ¾ length of hook 1; ventral wall of segment denticulated, with two spiniform processes, 4 and 5, arising from lateral and ventral sides of segmental wall. Tip of labrum (Fig. 1764) truncated, with very prominent rostrum and well developed fringe of setae, barely reaching ½ length of rostrum. Mandible (Fig. 1765) relatively broad, with dental formula P1, S1, P1, S1, P1, S1, B3/4; secondary teeth much smaller than primary, basal teeth either three or four, progressively smaller towards base. First maxilla (Fig. 1766) with exopod ventral, cylindrical, prominent, with rounded tip and two setae, one apical, other subapical; two terminal papillae long, surmounted with setae not exceeding their own length, tapering, flexible; at base of dorsal papilla slender seta, reaching ½ length of papilla; no denticulation observed. Second maxilla (Fig. 1759) cylindrical, shorter than trunk, separate from its opposite number, with rounded tip. Bulla with short manubrium and subcircular, plano-convex anchor of narrow diameter. Maxilliped (Fig. 1767) with robust, gently tapering corpus. Myxa with patch of denticulation and short, tapering seta on mild eminence, another patch at base of subchela; latter strong, with slight flexure in proximal half; basal seta prominent; distal half of inner margin with finely denticulated patch, extending anteriorly on small outgrowth (Fig. 1768); barb strong, with slender tip, reaching slightly beyond ½ length of claw; latter with flexure in proximal half, near base, and with slender, sharp tip; two secondary teeth prominent at about ½ length, near inner margin.

Male (Fig. 1771): Cephalothorax nearly half length of total body, inclined ventrally to long axis of trunk, oval in dorsal aspect, with anterior part covered by ill-defined dorsal shield, posterodorsally with prominent swelling. Waist-like constriction of anterior part of trunk marking borderline between it and cephalothorax. Posterior to constriction, trunk fusiform with rounded posterior extremity, tapering and provided with uropods, but no external genitalia. Length about 1 mm.

First antenna similar to that of female, but differing from it in details of apical armature (Fig. 1772); tubercles relatively more prominent and with blunt tips; tubercle 3 present at base of seta 6; gibber relatively smaller; complex 5 apparently of only two setae. (Possibly third seta was broken in specimen at author's disposal.) Second antenna (Fig. 1773) with slender, indistinctly segmented sympod, poorly delimited from rami; exopod shorter than endopod, digitiform, robust, with rounded apex, apparently without denticles, but with one apical and one subapical papilla; endopod two-segmented; basal segment cylindrical, with patch of small denticles on ventral wall; distal segment with well developed apical armature; dorsal hook 1 strong, mobile, gently curved, with slender spiniform seta 2 on one side of base and very prominent tubercle 3 at other, tipped with nipple; ventrally, wall of segment inflated and denticulated, with digitiform process 4; process 5 not observed. Buccal apparatus as in female, but mandible (Fig. 1774) relatively more slender; dental formula P1, S1, P1, S1, B4; no third pair of primary and secondary teeth. (In specimens examined by the author ventral margin below basal teeth had a notch in it, the nature of which could not be determined. Possibly an artifact, such a notch has been observed by the author also in other Lernaeopodidae.) Second maxilla (Fig. 1776) with squat, short corpus; medial margin with prominent, tubercular myxa; subchela short, with robust base and slender tip, sharply flexed at about midlength; claw not delimited; no secondary armature. Maxilliped (Fig. 1775) with strong, quadrangular corpus, with prominent subtriangular outgrowth in distal half of medial margin. Distal surface of outgrowth concave, partly covered with diminutive denticles; one larger denticle on margin, near tip of claw, when closed; two transverse rows of denticles near base of subchela; latter with broad base and slender tip, with curvature of about 90° at midlength; no delimited claw, but small secondary tooth on inner margin near position where such delimitation usually occurs. No thoracic legs.

Comments: The history of this species involves three specific names, as well as some confusion as to the host affiliation. In his original account of *Brachiella bispinosa*, von Nordmann (1832) stated that he was not sure of its origin but he "supposed with some probability" that it came from a species of *Gadus*. The supposition was erroneous. No true *Gadus* has been known to harbour any *Neobrachiella*.

Krøyer (1837), describing *Lernaeopoda obesa*, also could not remember the host of its single specimen, but thought it came from *Squalus acanthias*. Olsson (1869) recognized Krøyer's species in a parasite he collected from *Trigla gurnardus* and transferred it to the genus *Brachiella*.

Edwards (1840) was the first to associate von Nordmann's *Brachiella bispinosa* with the genus *Trigla* as hosts. From that date onwards, many authors recorded the occurrence of *B. bispinosa*, or *B. obesa*, or both, on the gills of various species of Triglidae. The present author (Kabata, 1973b) re-examined the problem of the specific identity of *Brachiella* possessed of only one pair of posterior processes and parasitic on Triglidae. He assigned them to *B. bispinosa*, reducing Krøyer's species to the status of a junior synonym.

Van Beneden (1871), illustrating (Pl. II, Fig. 8) the parasite referred to by him as *Anchorella ovalis*

Krøyer, unmistakably had before him *N. bispinosa*. Unfortunately, that figure induced other authors, notably T. Scott (1901), A. Scott (1904) and Scott and Scott (1913) to use the name *Brachiella ovalis* in their descriptions and records of *N. bispinosa*. Richiardi (1880) contributed to the confusion surrounding this species by naming, without describing, *Brachiella obesa* n. sp. from *Trigla corax* in the Mediterranean. Brian (1906, 1924) referred to *Brachiella* found on the gills of *Trigla* sp. as *Brachiella obesa* Richiardi, though he, too, failed to produce descriptions or figures. There can be little doubt that *B. ovalis* Richiardi is yet another synonym of *N. bispinosa*, but in view of the fact that it has never been described it has been left out of the synonymy.

In the choice of its hosts, *N. bispinosa* is restricted to species of the genus *Trigla* and its relatives. Numerous records of this parasite failed to locate it on any other host. This makes it all the more probable that the early reports from *Gadus* and dogfish are instances of mislabelled records. *N. bispinosa* occurs in British waters, off the North Sea and Atlantic coast of Europe and in the western Mediterranean. It has not been found off the coast of Northern Europe or along the Atlantic shores of North America.

Neobrachiella chevreuxii (van Beneden, 1891)

(Figs. 1795–1809)

Syn.: *Brachiella chevreuxii* van Beneden, 1891
Brachiella (*neglecta* Richiardi)?; of Brian (1898)
Brachiella sp. (*Anchorella*); of Brian (1902)
Clavella scianae Brian, 1906
?*Anchorella sciaenophila* Heller, 1865; of Brian (1906)
Brachiella sciaenae (Brian, 1924)
Brachiella scianophila Heller, 1865; of Wilson (1915)
Brachiella macrura Wilson, 1920

Female (Figs. 1795, 1796): Cephalothorax cylindrical, as long as, or longer than, trunk, its anterior quarter slightly expanded to form head, covered by subtriangular, well delimited dorsal shield (Fig. 1797) with transversely truncated anterior and posteriorly converging lateral margins, slightly subdivided by mid-dorsal groove; at base of cephalothorax noticeable semispherical, lateral swellings. Shallow constriction beyond level of bases of second maxillae separating cephalothorax from trunk. Latter variable in shape, but usually about as long as broad, slightly narrower anteriorly, with rounded anterolateral corners; posterior extremity (Fig. 1795) transversely truncated, with slightly raised perianal region in centre; ventral to anal slit cylindrical genital process, varying in length from ¼ length of trunk to as long as trunk, flanked on either side by uropod modified into cylindrical process, in some specimens reaching length of trunk, slightly tapering distally, with rounded tip; posterolateral corners with one posterior process each, processes similar to, and as long as, or longer than, uropods, sometimes exceeding length of trunk. (For variability in shape of trunk, see Kabata, 1966.) Length of cephalothorax 1.8 mm, trunk 1.8 mm, posterior process 2.4 mm.

First antenna indistinctly four-segmented, basal segment inflated, indistinctly fused with second; latter with whip; third segment armed with solus; no gibber at apex of terminal segment; apical armature (Fig. 1798) comparatively short and squat; digitiform seta 4 with rounded tip, slightly tapering, with small, conical tubercle 1 at base; slender and long seta 6 on opposite side of apex; centre with three setae, presumably comprising complex 5; tubercles 2 and 3 not observed. Second antenna similar to that of preceding species, with long axis passing through sympod and exopod, with reduced endopod pushed aside; basal segment of two-segmented endopod with small patch of blunt, rounded denticles on ventral wall; terminal segment (Fig. 1799) with hook 1, robust at base, sharp-tipped and straight; tubercle 3 not observed; spiniform seta 2 shorter than hook 1, slender; process 4 on ventral wall very prominent (size varies from specimen to specimen) and armed with blunt denticles; on its side conical unarmed process 5. Buccal apparatus similar to that of preceding species. Tip of labrum evenly rounded, with fairly short rostrum. Mandible (Fig. 1800) with dental formula P1, S1, P1, S1, P1, S1, B4; first primary tooth small, primary and secondary teeth forming pairs, basal teeth diminishing towards base, last overlapped by thin ventral margin of appendage; margin with some irregular cuticular ridges. First maxilla (Fig. 1801) sturdy; exopod displaced slightly from ventral margin to lateral surface, apically with two subequal setae; two full-sized terminal papillae, long, with relatively short setae; third short papilla, less than ½ length of dorsal papilla, together with its seta barely reaching base of seta of dorsal papillae; group of small denticles on medial surface close to base of dorsal and reduced papillae (difficult to see from lateral side). Second maxilla (Figs. 1795, 1796) short, slender, close to opposite number, in some specimens covered apparently by same cuticular sheath (common artifact of preservation), but not fused, tapering towards tip, without apical expansion; prominent nipple-like duct on posterior surface, near base. Bulla as in preceding species. Maxilliped (Fig. 1802): corpus broad, strong, with indistinct surface transverse markings in proximal half; myxa with two prominent swellings covered by small denticles, distal swelling some distance from base of subchela; between swellings short, pointed seta on inflated base; subchela fairly short, slightly tapering distally, with basal seta and distal part of inner margin with pad-like inflation, covered by crowded, blunt denticles (Fig. 1803); barb short, not reaching $\frac{1}{2}$ length of claw; latter gently curved, tapering, with rounded tip; near base two secondary teeth, close to inner margin.

Male: Cephalothorax about $\frac{1}{2}$ total body length, oval in dorsal aspect, covered with ill-defined dorsal shield, separated from trunk by waist-like constriction and inclined to it obliquely towards ventral side. Trunk fusiform, with rounded, tapering posterior extremity, without external genitalia, but with uropods (Fig. 1804) armed with small outgrowths on distal half. Length about 1 mm.

First antenna similar to that of female. Second antenna prehensile, with long sympod indistinctly two-segmented and exopod shorter than endopod; exopod (Fig. 1805) long, cylindrical, with rounded apex, armed with two papillae, one apical and one subapical, no denticulation observed; endopod two-segmented, proximal segment cylindrical, unarmed, longer than distal; distal segment (Fig. 1806) with very characteristic apical armature: hook 1 with stout base, tapering, gently curved, with spiniform seta 2 at base; tubercle 3 not seen; ventrally, powerful process 4 present, as long as hook 1, slightly tapering, covered by sparse spinulation; process 5 setiform, as long as hook 1, with fine tip, unarmed. Buccal apparatus similar to that of female, but dental formula of mandible (Fig. 1807) differing from that of female by absence of first primary and one basal tooth (S1, P1, S1, P1, S1, B4); basal teeth about as large as primary; notch in ventral margin near proximal basal tooth. (It is possible that specimens examined by author had partly worn mandibles. First primary tooth sometimes wears out in females, but this is less likely in short-lived males.) First maxilla (Fig. 1808) differing from that of female in having completely ventral exopod, no third papilla (only seta in its position) and no denticulation on medial wall. Second maxilla (Fig. 1809) distinguished by double tubercle covered by fine denticles and located in distomedial corner of corpus; pointed tip of subchela fitting into deep concavity in distal surface of corpus, next to double papilla. Maxilliped similar to that of preceding species. No thoracic legs.

Comments: The history of this species poses a taxonomic problem, not fully resolved and—in view of the destruction of the type material—not resolvable. It involves the nominal species *B. sciaenophila*, *B. chevreuxii*, *B. sciaenae* and *B. macrura*. It includes also incorrect host records. The oldest of these species, *B. sciaenophila*, is known to us from a single habitus drawing (Heller, 1865), showing four posterior processes reminiscent of our species, but devoid of a genital process. The specimen thus illustrated came from *Sciaena* sp. from the Indian Ocean. It was never redescribed, though some authors used its name, either with hesitation (Brian, 1906), or assertively (Causey, 1960; Wilson, 1915).

Brachiella (originally *Clavella*) *sciaenae* was described by Brian (1906) on the basis of specimens resembling *B. sciaenophila* (as far as could be judged from Heller's drawing), but with a genital process. Wilson (1920a) proposed the name *B. macrura* for a similar parasite. None of these workers took note of van Beneden's (1891) *B. chevreuxii*, probably because of the inordinately large maxillipeds it was alleged to have and because its host was recorded as a *Squalus*, an obvious impossibility. Some doubts were, however, entertained as to the true relationship between *B. chevreuxii* and *B. macrura*, the possibility that the latter was synonymous with the former being suspected (Brian, 1939). Capart (1959) thought also that *B. macrura* might be identical with *B. sciaenae*.

The present author (Kabata, 1966) proved to his own satisfaction that *B. macrura* is a junior synonym of *B. chevreuxii*. He was not inclined to reduce *B. sciaenae* also to synonymy with it, because specimens of this species at his disposal had the second maxillae encased in a common cuticular sheath. It is now known that swelling of the cuticle resulting from fixation and preservation can obliterate the gap between the second maxillae. Even if pressed closely together, separate second maxillae within a single cuticular sheath are no evidence of fusion. Specimens of one species might, or might not, show this condition of the second maxillae, depending on how they were preserved. There being no difference, other than in the condition of the second maxillae, between *B. chevreuxii* and *B. sciaenae*, the author now suggests that the latter be reduced to synonymy with the former.

There remains the question of Heller's *B. sciaenophila*. It is entirely possible that it is identical with *B. chevreuxii*. Absence of the genital process might be attributed to causes other than specific distinction. The author believes, however, that in view of the inadequacy of its description, *B. sciaenophila* should be treated, at best, as a species inquirenda. Nothing is to be gained by using its name rather than that of the chronologically younger but better documented *B. chevreuxii*.

Neobrachiella chevreuxii is a parasite of sciaenid and lutianid teleosts, widespread in the Atlantic, the Mediterranean and (if Heller's record is to be included) the Indian Ocean. Most of its records come from the Mediterranean and the Atlantic sea-board of Europe. Off the Atlantic coast of North America it was recorded by Bere (1936) and Causey (1953b, 1960). The western shores of Africa have also been recorded as part of its range (Capart, 1953, 1959; Kabata and Gusev, 1966). Its hosts include *Sciaena* sp., *Argyrosomus regium*, *Neomaenis fulgens*, *Otolithus* sp., *Otolithus senegalensis*, *O. brachygnathus*, *Pogonias cromis*, *Roncador stearnsi*, *Atractoscion equidens*, *Umbrina valida*, *Corvina cameronensis*, *Larimus fasciatus*, *Lutianus griseus* and *Johnius hololepidotus*.

This author found it on the gills of *Argyrosomus regium* taken in Scottish waters.

Neobrachiella impudica (von Nordmann, 1832)

(Figs. 1777–1794)

Syn.: *Brachiella impudica* von Nordmann, 1832
Brachiella impudica var. *parva* Bassett-Smith, 1896
Thysanote impudica (Nordmann); of Bassett-Smith (1899)
Epibrachiella impudica (Nordmann); of Wilson (1915)
nec *Brachiella impudica* (Nordmann); of Belloc (1929)
Brachiella parva Nunes-Ruivo, 1957

Female (Figs. 1777–1779, 1789–1794): Cephalothorax cylindrical, as long as, or shorter than, trunk, its anterior half expanded slightly in girth to form head, inclined either obliquely or at right angles to long axis of posterior part of cephalothorax, dorsally covered with well defined dorsal shield and often with posterodorsal swelling marking its end; dorsal surface at base with prominent transverse welt (Fig. 1778). Border between trunk and cephalothorax marked by bases of second antennae. Trunk anteriorly with short cylindrical part, of about same diameter as cephalothorax, expanding into broad main part (in some specimens, particularly young females (Figs. 1789, 1790), transition from cylindrical part to broad part of trunk less abrupt, trunk gradually flaring out to its full width); trunk dorsoventrally flattened, in young females longer than broad, in fully adult about as long as broad, with rounded anterolateral corners (Fig. 1779) (in some specimens these corners with small, conical or slightly pedunculate processes on dorsal surface); posterior extremity transversely truncated, with four pairs of posterior processes; dorsal side (Fig. 1779) in centre with one pair of processes, length varying with age of female, either fused partially or separate to base (Fig. 1794); posterolateral corners each with one pair of processes, one dorsal and one ventral, also differing in length with age (Figs. 1777, 1793); in centre of ventral surface fourth pair, modified uropods, flanking genital process, when present. Uropods usually shorter than other pairs in old females, of same size in younger (Fig. 1780), sometimes (Fig. 1790) small and slender. Genital process absent in some females (Fig. 1790), or short (Fig. 1777) or of moderate length. Length up to 6.5 mm (Scott and Scott, 1913).

First antenna (Fig. 1781) four-segmented, basal segment indistinctly fused with second, inflated; second segment with long, slender whip reaching distal end of third segment; latter cylindrical, with dentiform solus at distal end; terminal segment rounded, with not well delimited gibber; apical armature (Fig. 1782) comprising powerful digitiform seta 4 on medial side, with tubercle 1 at base; on opposite side slender flagelliform seta 6; tubercles 2 and 3 not observed; complex 5 on apex of gibber, of four setae, grouped in one long and one shorter pair. Second antenna with heavy sympod and long axis passing through sympod and exopod; latter bulbous, with two large papillae and tip armed with denticles; endopod (Fig. 1783) two-segmented: basal segment cylindrical, apparently unarmed; terminal with hook 1 (unevenly twisted), spiniform seta 2 and ventral wall denticulated and inflated; one or two setiform outgrowths on ventral wall, probably homologous with processes 4 and 5. (Due to small size, definitive details could not be ascertained; one or both of processes 4 and 5 easily damaged by handling.) Buccal apparatus as in preceding species. Mandible broad, with dental formula P1, S1, P1, S1, P1, S1, B4 (Fig. 1784); considerable discrepancy in size between primary and secondary teeth, basal teeth long, slender. First maxilla (Fig. 1785) with ventral exopod, apically with two subequal setae; two terminal papillae long, with their apical setae shorter than themselves; at base of dorsal papilla one seta, shorter than papilla, sometimes with slightly inflated base suggesting vestige of third papilla; dorsal margin of appendage with pronounced curvature in distal half. Second maxilla (Figs. 1777, 1778, 1790, 1791) cylindrical, either shorter or longer than trunk, robust or slender (partly depending on state of contraction), with or without very small apical expansion; in some, especially older females, second maxillae with either one (Fig. 1790) or two (Fig. 1791) processes, cylindrical and of varying lengths; two members of this pair separate from each other. Bulla (Fig. 1791) with short manubrium and narrow, subcircular anchor. Maxilliped (Fig. 1787) with long, robust corpus; myxa with two patches of denticles, one in distal, other in proximal half; latter often difficult to find, perhaps absent; between them short, though robust, tapering seta; subchela cylindrical, about ½ length of corpus, with slight flexure in or near midlength, with basal seta and denticulated distal half of inner margin; barb strong, with slightly inflated base (Fig.1786); claw moderately short, strongly tapering, with sharp tip; pair of secondary teeth in distal half of claw, near inner margin.

Male (Fig. 1788): Cephalothorax about half length of body, dorsoventrally flattened, covered with ill-defined shield, inclined ventrally to long axis of trunk, with posterolateral swelling, separated from trunk by waist-like constriction. Trunk fusiform, expanding in posterior direction from narrow anterior part, with tapering, rounded posterior extremity, equipped with uropods. No external genitalia. Length less than 1 mm (about 0.6; Nordmann, 1832).

Appendages comparable to those of other males of the genus.

Comments: *Neobrachiella impudica* belongs to those species of Lernaeopodidae, of which the most characteristic feature is their morphological variability. Difficulties encountered by earlier authors in its identification and—even more—in the assessment of its generic status, are in the main due to that factor. The original description (von Nordmann, 1832), by presenting rather stylized drawings, augmented these difficulties. In spite of this, *N. impudica*, as evidenced by its synonymy, had a fairly straightforward taxonomic history. Its posterior extremity, regardless of its variability, could not be easily confused with that of other species. It is the second maxilla that provoked the misunderstandings and doubts. The presence of the lateral processes on this appendage caused *N. impudica* to be classed as *Thysanote*. The absence of these processes in some specimens resulted in their separation from those with the processes in another taxon, first designated as *Brachiella impudica* var. *parva* by Bassett-Smith (1896b) and later upgraded by Nunes-Ruivo (1957a) to a full species, *Brachiella parva*. The male of *N. impudica*, as illustrated by von Nordmann (1832), gave reason for the creation of a patently spurious genus *Epibrachiella*. Its similarity to other males then classed as *Brachiella* was quite obvious to Matthews (1923). It is also evident from Fig. 1788.

The type material of the species, its morphological variability and its taxonomic history have been described and discussed in detail by the present author (Kabata, 1973b). They are sufficiently well summarized in the synonymy above. This author has suggested also that von Nordmann's original record of *N. impudica* from *Gadus aeglefinus* was in error. Von Nordmann had the identity of the host on the authority of Dr. Mehlis, who obviously recorded it incorrectly. Although some later authors repeated this record in their account of the species, *N. impudica* was never found on a gadid host. Indeed, the fishes of that group are parasitized by the *Clavella*-group of the family and do not harbour *Brachiella* or its relatives. (The only fish known to the author to be infected by both *Brachiella* and *Clavella* is the hake, *Merluccius merluccius*.)

N. impudica parasitizes exclusively teleosts belonging to Triglidae and has been found on the gills of several triglid species. Its distribution range is similar to that of another *Neobrachiella* parasitic on Triglidae, *N. bispinosa*. It covers the North Sea, Atlantic coast of Europe and western part of the Mediterranean, its southernmost record being from Rio de Oro in West Africa (Capart, 1941). It is absent from the American side of the Atlantic and does not extend north of the North Sea.

N. impudica is common in British waters, where it occurs on most of the local species of Triglidae, the gurnards.

Neobrachiella insidiosa (Heller, 1865)

(Figs. 1810–1820)

Syn.: *Brachiella insidiosa* Heller, 1865
Parabrachiella insidiosa (Heller, 1865)
?*Anchorella appendiculata*; of A. Scott (1904)
Parabrachiella australis Wilson, 1923
Brachiella australis (Wilson, 1923)
Brachiella lageniformis Szidat, 1955

Female (Figs. 1810, 1811): Cephalothorax long, cylindrical, of about same diameter throughout, with head only slightly expanded, involving no more than anterior $\frac{1}{4}$ of cephalothorax, covered by well developed dorsal shield; in preserved specimens often flexed ventrally over bases of second maxillae. Very shallow transverse constriction below level of second maxillae demarcating between cephalothorax and trunk. Latter varying with age and maturity of female, increasing in length and width of anterior part; in young females (Fig. 1811) anterior part sloping away from cephalothorax, in older (Fig. 1810), trunk more commonly with shoulder-like anterolateral corners; trunk longer than wide, with lateral margins either nearly parallel, or slightly diverging posteriorly, dorsoventrally flattened (in live specimens very flat); posterolateral corners of trunk with posterior processes, differing in length with age of female, in young specimens shorter than uropods, in older about twice as long. Uropods ventral to egg sacs, slightly tapering, with rounded tips. Noticeable anal tubercle in centre of posterior margin. Length (including posterior processes) up to 14 mm (Scott and Scott, 1913).

First antenna indistinctly four-segmented, with whip and solus resembling those of other species in *Neobrachiella*; apex (Fig. 1812) with gibber, apical armature comprising digitiform seta 4, with tubercle 1 at base, conical and small; on opposite side slender seta 6; complex 5 of three structures: two setae of equal length, shorter than 4, and large tubercle arising from, and extending, tip of gibber; tubercles 2 and 3 not observed. Second antenna closely resembling that of *N. impudica*. Buccal apparatus similar to that of preceding species, but labrum (Fig. 1813) with slender, pointed tip and prominent rostrum; marginal fringe abundant, covering more than half of rostrum. Mandible broad, with dental formula: P1, S1, P1, S1, P1, S1, B5 (in some specimens, as shown in Fig. 1814, first secondary tooth sometimes absent); basal teeth decreasing in size towards base, proximal tooth overlapped by ventral margin of appendage. First maxilla (Fig. 1815) uniformly slender, with exopod displaced slightly from ventral margin to lateral surface, armed with two subequal apical setae; terminal papillae long, tipped with long setae; third seta at base of dorsal papilla, not reaching its distal end; no denticulation observed. Second maxilla usually about as long as cephalothorax, separate from opposite number, cylindrical, near base with prominent nipple-like duct orifice; tip expanded into semicircular swelling, subdivided into five digitiform lobes; tips of both second maxillae (Fig. 1816) forming ten-fingered collar round manubrium of bulla. Latter with short manubrium and narrow, plano-convex anchor. Maxilliped (Fig. 1817) long and comparatively slender; myxa with short, slender seta on conspicuous base near midlength and with two denticulated fields, proximal below seta, covering subspherical swelling, distal close to base of subchela, extensive; subchela short, with hump-like curvature in proximal half, with short basal seta and major part of inner margin covered by strip of crowded, blunt denticles (Fig. 1818); claw well delimited, fairly long, with slender tip and pair of teeth near midlength; barb about half as long as claw.

Male (Fig. 1819) typical for *Neobrachiella*. (Fig. 1819 showing effects of violent contraction produced by fixative.) When relaxed, male comparable to that shown in Fig. 1788. Appendages, with exception of second maxilla and maxilliped, resembling those of female. Apical armature of first antenna (Fig. 1820) differing from that of female by presence of tubercle 3 at base of flagelliform seta 6; gibber extended by tubercular swelling beyond bases of setae of complex 5. Length under 1 mm.

Comments: *N. insidiosa* is a highly polymorphic species, exclusively specific to the genus *Merluccius*. It is distributed with various species of *Merluccius*, in the Atlantic and Pacific. Its morphological variability prevented Wilson (1923b) and Szidat (1955) from recognizing the identity of their specimens with the old-established species from the European hake, *Merluccius merluccius*, and caused them to propose new taxa for the parasite of the South African and Argentinian hake species (*M. capensis* and *M. gayi* respectively). The matters were further complicated by the existence of *Parabrachiella*, a spurious genus established by Wilson (1915) on an artifact in male morphology. *N. insidiosa* was hidden in that genus from the attention of those workers who recognized their material as *Brachiella*. Kabata (1970e) proposed that *Parabrachiella* be suppressed, but he did not recognize at that time that, beside *N. merluccii*, there exists only one species of *Neobrachiella* parasitic exclusively on hake.

In spite of their morphological differences, specimens of *N. insidiosa* have a unique feature in common, which sets them apart from other species of *Neobrachiella*. The tips of their second maxillae are divided into five digitiform processes each (Fig. 1816). The females, as well as the males, are also nearly identical in the structure of their appendages. The differences between them are those of size and of the morphology of the posterior extremity; in particular, their posterior processes can be long, short, or absent.

The author is currently engaged, in collaboration with Dr. Ho (Long Beach, California), in assessing the importance of parasitic copepods as indicators in the zoogeography of *Merluccius*. In a joint paper a proposal will be made to recognize several forms of *N. insidiosa*, differing in the morphology of the posterior extremity, as well as in host affiliation and geographical distribution. Trinomial nomenclature will be used to designate these forms as follows: *N. insidiosa* f. *insidiosa*, parasitic on *Merluccius merluccius* and other species of this genus along the European and African Atlantic sea-board and distinguished by well developed posterior processes; *N. insidiosa* f. *lageniformis*, living on the gills of *M. gayi* along the Atlantic coast of South America and completely devoid of posterior processes; and *N. insidiosa* f. *pacifica*, on the gills of *M. hubbsi* along the Pacific coast of South America and on those of *M. productus* along the Pacific coast of North America. The posterior processes of the last-named form of *N. insidiosa* are variable in length and may be absent in younger specimens, but are never as long as those of the European-African form.

This type of intraspecific variability in the development of the posterior processes has not been frequently observed among Lernaeopodidae. One example was described by Kabata (1970d) in *Brachiella robusta* (Wilson, 1912) (= *Neobrachiella robusta*). Like *N. insidiosa*, that parasite of the teleost genus *Sebastes* in the North Pacific was found to exist in three morphological forms, the differences being confined to the posterior extremity of the trunk. Intraspecific variability might be more common than is suspected and should be remembered when systematics of the *Brachiella*-group are being considered.

N. insidiosa f. *insidiosa* occurs in British waters on *Merluccius merluccius*. It is always attached to the gill filaments (unlike *N. merluccii*, parasitizing gill rakers of the same host). It is rather scarce in comparison with the latter species.

Neobrachiella rostrata (Krøyer, 1837)

(Figs. 1832–1845)

Syn.: *Brachiella rostrata* Krøyer, 1837
Parabrachiella rostrata (Krøyer), 1837; Wilson, 1915

Female (Fig. 1832): Cephalothorax cylindrical, slender, slightly shorter than trunk, only small anterior part slightly expanded to form head, dorsally covered by well developed dorsal shield; latter with truncated anterior margin and slightly converging lateral margins, indistinctly divided in mid-dorsal line (Fig. 1833). Trunk beginning at bases of second maxillae, dorsoventrally flattened, much longer than broad; posterior extremity varying with age; in young females (Fig. 1835) often with rounded and slightly inflated posterolateral corners, coinciding with the greatest width of trunk; ventrally, conical genital process usually present, flanked by prominent uropods, always ventral to egg sacs; in older females (Fig. 1834) posterolateral corners not as prominent, greatest width of trunk anterior to their level. Uropods fusiform, tapering, with rounded tips. Length about 15 mm (uropods excluded) (Scott and Scott, 1913).

First antenna indistinctly four-segmented, with first two segments forming basal part, distally with long slender whip; third segment cylindrical, with solus; distal segment with poorly developed gibber; apical armature (Fig. 1837) with digitiform seta 4 and tubercle 1 at its base; slender seta 6 on opposite side, apparently without associated tubercle 3; gibber occupied by complex 5 of two setae about as long as seta 4, though less robust, and of one slender, conical structure; tubercle 2 on side of gibber. Second antenna similar to that of *N. impudica*. Labrum with moderately pointed tip, rounded and with rostrum partly covered by marginal fringe. Mandible (Fig. 1838) very broad, with dental formula P1, S1, P1, S1, P1, S1, B5; characteristic feature large size of secondary teeth, only slightly smaller than primary ones; distalmost basal tooth largest in entire appendage, remaining teeth of series reducing to size of secondary teeth; two proximal teeth of basal series overlapped by thin ventral margin of mandible. First maxilla (Fig. 1839) with sympod comprising more than half length of appendage, exopod slightly displaced to lateral surface and with two subequal apical setae; two full-sized terminal papillae at apex of short endopod, with setae their own length; at base of dorsal papilla a small third papilla, together with apical seta about equalling length of dorsal papilla. Second maxilla (Fig. 1832) cylindrical, slender, slightly tapering, as long as, or shorter than, trunk, separate from opposite number, very slightly expanded at tip (Fig. 1836), enclosing base of manubrium. Bulla with short manubrium and very small plano-convex anchor (not more than twice width of manubrium). Maxilliped (Fig. 1840) with robust, fairly long corpus, indistinctly divided by transverse groove in proximal half. Myxa with two denticulated pads: proximal just below transverse division, somewhat inflated; distal large, transversely elongated, close to base of subchela; between them short, tapering seta on inflated base; subchela cylindrical, fairly slender, reaching level of myxal seta, armed with short basal seta and with protruding distal part of inner margin beset by crowded, blunt denticles; barb tapering, less than half length of claw; latter (Fig. 1841) with several incomplete, transverse constrictions near base, gently curving, with slender tip; three secondary teeth, one in proximal half near base on inner margin, two other on either side, at same level.

Male (Fig. 1842): Cephalothorax about half of total body length, with fairly well defined dorsal shield, with posterodorsal inflation, separated from trunk by waist-like constriction. Trunk fusiform, posteriorly tapering, with uropods but no external genitalia. Length about 2 mm (Scott and Scott, 1913).

First antenna similar to that of female. Second antenna (Fig. 1843) prehensile, with endopod longer than exopod; latter one-segmented, bulbous, apparently without denticulation but with two subterminal papillae; endopod two-segmented; basal segment cylindrical, with patch of denticles on ventral wall; apical armature of distal segment with robust hook 1, slender spiniform seta 2, but no tubercle 3; central wall of distal segment denticulated, inflated into conical process 4, with spiniform process 5 on its side. Buccal apparatus as in female. First maxilla (Fig. 1844) with shorter sympod, longer exopod and reduced third terminal papilla; only simple seta at base of dorsal papilla. Second maxilla as in males of other *Neobrachiella* species. Maxilliped (Fig. 1845) suboval, with semicircular flap in distomedial quarter, concave on dorsal side, with concave surface covered by small denticles (ghosted in drawing); subchela without distinctive claw, with robust base and tapering, slender, sharp tip; small denticle on inner margin near base of subchela.

Comments: The taxonomic history of this species involves only one change of the generic status besides the present transfer to *Neobrachiella*.

Resembling in many respects *N. insidiosa* parasitic on the hake, *N. rostrata* differs from it in having no digitiform lobes at the tips of its second maxillae. Like the parasite of the hake, it is strictly limited to parasitism on one host species, the halibut, *Hippoglossus hippoglossus*. (Early records of its occurrence on *Pleuronectes pinguis* (= *Reinhardtius hippoglossoides*) and *Scophthalmus maximus* have not been substantiated by later finds.) It has been found in most areas inhabited by its host, though it cannot be considered a common species. Known from the North Sea, it extends south to the coast of Belgium, visits with the halibut the Danish Straits, lives in the northern waters (Barents Sea, Iceland, Greenland, Faroes) and off the Atlantic coast of North America. It has not been found in the North Pacific, however, on the closely related halibut species *H. stenolepis*.

Little is known about its choice of site on the host, the gills being usually given as its habitat. The present author found it most commonly (though not exclusively) attached to the pseudobranch of its host.

Neobrachiella triglae (Claus, 1860)

(Figs. 1846–1860)

Syn.: *Brachiella triglae* Claus, 1860
Anchorella triglae Claus; Kurz, 1877
Clavella triglae Claus, 1860; of Brian (1906)

Female (Figs. 1846–1848): Cephalothorax cylindrical, about as long as trunk, slightly tapering towards apex but with head slightly enlarged, covered by dorsal shield; latter (Fig. 1849) with anterior margin slightly broader than indistinct posterior and with posteriorly converging lateral margins; in most fixed specimens cephalothorax arching anterodorsally, its apex in close contact with area of tips of second maxillae (Fig. 1848); transition from cephalothorax to trunk indistinct.

Trunk as broad as, or slightly broader than, long, armed with array of processes and protuberances in fairly constant pattern, varying only slightly from specimen to specimen; lateral to base of cephalothorax large swelling subdivided into four subspherical tubercles (Fig. 1848, a); on dorsolateral margin, posterior to *a*, two similar tubercles, usually one larger than other, sometimes fused (Fig. 1847, b); ventral to group *a*, on lateral margin of trunk, four-square group of four subspherical tubercles, slightly larger than those in group *a* (Fig. 1848, c) (this group sometimes with only three tubercles, one being suppressed or two adjoining ones fused); anteroventral surface, at level of second maxillae with very large swelling, extending posteriorly to about ½ length of trunk, anteriorly projecting in front of margin (Fig. 1848, d); anterior end of swelling simple (right swelling *d* in Fig. 1846) or divided into two subspherical tubercles (Fig. 1848); lateral margins tapering posteriorly (Fig. 1846); posterior extremity with two pairs of processes, short, robust, with rounded tips, one pair posteroventral, other posterodorsal; anal slit slightly displaced dorsally (Fig. 1847); ventral to anus short conical genital process (Fig. 1850), shorter than posterior processes, flanked by uropods. Latter longer than posterior processes, cylindrical, with fine, tapering tips. Egg sacs dorsal to uropods. Length about 4.5 mm (Scott and Scott, 1913).

First antenna (Fig. 1851) four-segmented, basal two segments indistinctly fused, base of appendage inflated, about four times wider than terminal segment; whip slender, reaching distal end of third segment; latter cylindrical, with dentiform solus; distal segment with broad, low gibber; apical armature (Fig. 1852) with digitiform seta 4 and conical tubercle 1 at its base, prominent and fine-tipped; flagelliform seta 6 on opposite side, much longer than others; apex with three, nearly equal, strong setae of complex 5 and shorter, unusually slender tubercle 2; tubercle 3 not observed. Second antenna with heavily sclerotized sympod, indistinctly segmented and with two rami, long axis passing through sympod and exopod; latter much larger than endopod (Fig. 1853), one-segmented, with rounded tip armed with two papillae and scattered, small denticles; endopod two-segmented: terminal segment with straight, somewhat reduced, spiniform seta 2 at its base and inflated ventral wall beset with spinules and armed with two spiniform processes probably homologous with processes 4 and 5. Tip of labrum (Fig. 1855) with narrow medial protrusion, extending into well developed rostrum; marginal fringe restricted to tip of central protrusion. Mandible (Fig. 1854) with dental formula P1, S1, P1, S1, P1, S1, B5; primary teeth prominent, sharp, secondary teeth much smaller, slender; basal teeth all smaller than primary, decreasing in size towards base, first of them overlapped by fine ventral margin of appendage; in some specimens notch in that margin, near basal teeth. First maxilla (Fig. 1856) very slender and long, with long, cylindrical exopod slightly displaced from ventral margin to lateral surface, armed with two apical, almost equal setae; two full-sized, long terminal papillae, with apical setae shorter than themselves; much reduced third papilla at base of dorsal papilla, together with its slender seta barely reaching tip of latter; no denticulation found. Second maxilla (Figs. 1846, 1848, 1857) almost completely obsolete, in most specimens recognizable only as irregular, flat apical expansion, completely enclosing part of host tissues with inserted bulla. Latter (Fig. 1858) with short manubrium and small anchor, usually oval in outline, plano-convex, with long diameter in line with, and flat surface attached to, supporting rod of gill filament; manubrium usually displaced towards posterior half of bulla. Maxilliped (Fig. 1859) with long, relatively narrow corpus; myxa with two patches of denticles, proximal on subcircular swelling in lower half of corpus, distal more extensive near base of subchela; between them short, slender seta on inflated base; subchela cylindrical, with basal seta on ventral side and densely denticulated distal part of inner margin; latter (Fig. 1860) extending distally in denticulated process overlapping barb; barb with stout base, reaching well into distal half of claw; two secondary teeth at level of tip of barb, near inner margin of claw.

Male: This author was unable to examine male specimens of *N. triglae*. Apparently the male resembles closely other males of the genus and measures less than 1 mm in length (Kurz, 1877).

Comments: *N. triglae* was originally described as *Brachiella* by Claus (1860), who based his taxon on the male only. When Kurz (1877) found later both the males and females of this species, he felt unable to place it in *Brachiella* because of the unusual morphology of the female. Referring to it as "höchst abenteuerlich geformt" (of highly bizarre shape), he placed it in *Anchorella* and added comments on the "cauliflower-like outgrowths" being one of the hall-marks of parasitic adaptations. In the light of contemporary knowledge, however, his drawings of the male (Figs. 22, 23) clearly indicate that Claus had made a correct taxonomic decision. All subsequent authors, until now, followed Claus' usage.

In common with *N. bispinosa* and *N. impudica*, this species is parasitic exclusively on teleost fishes belonging to Triglidae. It has been recorded on the gills of five species of gurnards but probably occurs on others. Its distribution range extends over the North Sea and British waters, and as far south as the Atlantic coast of France. There are, so far, no records of *N. triglae* from the coast of the Iberian Peninsula, but it has been found in the western Mediterranean and the Adriatic. It is in the latter sea that *N. triglae* was found by Kurz.

Genus Clavella Oken, 1816

Female: Cephalothorax cylindrical, either slightly shorter or longer than trunk, with distinguishable dorsal shield and either slightly expanded or not expanded head, usually deflected along dorsal surface of trunk. Latter dorsoventrally flattened, suborbicular, subquadrate or elongate, with or without genital process, without uropods or posterior processes. First antenna obscurely segmented or unsegmented, with well developed but variable apical armature. Second antenna with sympod-endopod long axis, exopod either greatly reduced or obsolete. Mandible with variable dental formula, sometimes without clear distinction into primary and secondary teeth or with diastemes present. First maxilla with ventral exopod and two terminal papillae. Second maxilla very short, either apparently fused with opposite number or clearly separate (*C. stellata*), with prominent duct orifice on posterior surface, sometimes with terminal swellings (*C. stellata*) or squamae (*C. squamigera*). Maxillipeds subchelate.

Male: No division into cephalothorax and trunk, posterior extremity folded ventrally forwards and completely fused with cephalothorax into oval, unsegmented body. First antenna similar to that of female. Second antenna with small, one-segmented exopod and longer, two-segmented endopod with reduced apical armature. Buccal apparatus similar to that of female. Second maxilla and maxillipeds subchelate. No thoracic legs, uropods or external genitalia.

Type species: *Clavella adunca* (Strøm, 1762).

Comments: The taxonomic history of *Clavella* was described in detail by Wilson (1915). It involves two synonyms: *Lerneomyzon* Blainville, 1822, and *Anchorella* Cuvier, 1830. A good deal of confusion arose because of the uncertainty of a dividing line between *Clavella* and the latter of these two. Krøyer (1837), by making *C. hippoglossi* the type species of *Clavella*, introduced into this genus a number of species now belonging to the genus *Hatschekia* (Hatschekiidae) and added another layer of confusion that lasted until the beginnings of this century, when Poche (1902) took all of them out again. Later developments consisted mainly of the pruning down of *Clavella* by the establishment of separate genera for some of its species (*Clavellisa*, *Clavellodes*, *Clavellopsis* and *Alella*). One must also disregard nine nomina nuda at one stage or another associated with the generic name *Clavella*, one of them proposed by van Beneden (1871) and the remainder by Richiardi (1880).

In addition to two species inquirendae (*C. characis* (Richiardi, 1880); Brian, 1906 and *C. pagri* Krøyer, 1863) with doubtful generic affiliation, *Clavella* comprises currently 19 species with reasonable claims on its membership. They parasitize teleost fishes belonging to four orders and are widespread in both the Atlantic and Pacific Oceans. The parasites of this genus are traditionally considered as living exclusively, or predominantly, on the gadoid teleosts. This view, however, is not entirely justified. Of the 19 species included in *Clavella* by this author, only 12 are parasites of fishes belonging to the order Gadiformes. Of these, only eight (*C. adunca*, *C. alata*, *C. canaliculata*, *C. irina*, *C. perfida*, *C. squamigera*, *C. stellata* and *C. tumida*) parasitize fishes of the suborder Gadoidei. Two (*C. levis and C. porogadi*) have hosts from the suborder Ophioidei and one from Zoarcoidei and Macrouroidei (*C. pinguis*). Six species parasitize members of the order Perciformes. Two of them (*C. insolita* and *C. stichaei*) have hosts from the suborder Blennioidei, one (*C. bowmani*) from Notothenioidei and two (*C. ovata* and *C. scombropis*) from Percoidei. One species is known exclusively and, very unusually, from Notacanthiformes, the spiny eels (*C. gracilis*). At least two species straddle the ordinal barriers in their choice of hosts. *C. parva* parasitizes fishes from the orders Perciformes and Scorpaeniformes, whereas the hosts of *C. stichaei* come from Gadiformes and Perciformes.

Although *C. adunca* is quite a ubiquitous species, the genus can be regarded as largely limited to the northern hemisphere. This is illustrated in table 21.

Table 21. Geographic distribution of *Clavella*.

	North Atlantic	North Pacific	South Atlantic	South Pacific	Indian Ocean
C adunca (Strøm, 1762)	+	+		+	+
C. alata Brian, 1906	+				
C. bowmani Kabata, 1963			+		
C. canaliculata Wilson, 1915		+			
C. gracilis Hansen, 1923	+		+		
C. insolita Wilson, 1915	+				
C. irina Wilson, 1915		+			
C. levis Wilson, 1915	+				
C. ovata Yamaguti, 1939		+			
C. parva Wilson, 1912		+			
C. perfida Wilson, 1915		+			
C. pinguis Wilson, 1915	+				
C. porogadi Nunes-Ruivo, 1964					+
C. recta Wilson, 1915		+			
C. scombropis Yamaguti, 1939		+			
C. squamigera Wilson, 1915	+				
C. stellata (Krøyer, 1838)	+				
C. stichaei (Krøyer, 1863)	+				
C. tumida Wilson, 1915	+				

It is interesting to note that, whereas in the North Atlantic seven out of ten species are parasitic exclusively on Gadiformes and one more choses its hosts from both this order and Perciformes, of the eight species occurring in the North Pacific only four have gadiform hosts (three, if one discounts the ubiquitous *C. adunca*).

Three species of *Clavella* are known from British waters.

Key to British species of *Clavella*

1. Tips of second maxillae with two spherical swellings each (Fig. 1904); bulla flat, with radially arranged marginal lacunae (Fig. 1911) . . . . **C. stellata**
   Tips of second maxillae without spherical swellings 2
2. Bases of second maxillae with prominent papillae (Fig. 1891); dorsal shield tapering anteriorly (Fig. 1892) . . . . **C. alata**
   No prominent papillae at bases of second maxillae (Figs. 1862, 1864); dorsal shield slightly broader anteriorly (Fig. 1869) . . . . **C. adunca**

Clavella adunca (Strøm, 1762)

(Figs. 1861–1888)

Syn.: *Lernea adunca* Strøm, 1762
Lernaea cyclopterina O. F. Müller, 1776
Lernaea uncinata O. F. Müller, 1776
Schisturus uncinatus L.u.; of Oken (1816)
Clavella uncinata Oken, 1816
Lernaeomyzon uncinata Müller; of Blainville (1822)
Anchorella lagenula Cuvier, 1830
Ancorella uncinata von Nordmann, 1832
Anchorella brevicollis Edwards, 1840
Anchorella agilis Krøyer, 1863
Anchorella uncinata (Müll.); of Olsson (1869)
Anchorella uncinata, Müll.; of van Beneden (1871)
Anchorella quadrata Bassett-Smith, 1896
Anchorella rugosa, var. T. Scott, 1900

Clavella uncinata (O. F. Müller); of M. Rathbun (1905)
Anchorella uncinata Nordmann; of Brian (1905)
Clavella uncinata Müller; of Brian (1906)
Clavella (Anchorella) uncinata, Müll.; of Brian (1908)
Clavella brevicollis M. Edwards; of Scott and Scott (1913)
Clavella dubia Scott and Scott, 1913
Clavella lophii (M. Edwards); of Scott and Scott (1913)
Clavella quadrata Bassett-Smith; of Scott and Scott (1913)
Clavella sciatherica Leigh-Sharpe, 1918
Clavella iadda Leigh-Sharpe, 1920
Clavella devastatrix Leigh-Sharpe, 1925
Clavella invicta Leigh-Sharpe, 1925
Clavella typica Leigh-Sharpe, 1925
Clavella deliciosa Leigh-Sharpe, 1933
Clavellina uncinata Dogiel, 1936
Clavellina brevicollis Dogiel, 1936

Female (Figs. 1861–1866): Cephalothorax cylindrical, as long as, or longer than, trunk, with anterior slightly expanded end covered by a hard dorsal shield (Fig. 1869); latter with truncated anterior margins and lateral margins converging posteriorly; posterior margin indistinct. Trunk separated from cephalothorax by more or less developed constriction, dorsoventrally flattened, of greatly variable shape, from narrow and moderately long (Fig. 1865) to very wide (Figs. 1862, 1863), with all intermediate stages (Fig. 1864); posterior extremity with rounded or slightly lobate lateral corners, dorsally displaced anal slit and ventral genital process, either short and subspherical (Fig. 1870) or long and clavate (Fig. 1871); slits of oviduct orifices on either side of genital process. No uropods or posterior processes. Length of trunk up to 7 mm.

First antenna (Fig. 1872) indistinctly bipartite, with shorter and slightly inflated basal part and longer, cylindrical terminal part; short, slender whip on dorsomedial wall near junction of basal and terminal parts; apical armature (Fig. 1873) with powerful digitiform seta 4 and relatively large, fine-tipped tubercle 1 at its base; complex 5 of one seta, subdivided near base into two subequal branches, tubercle 2 at base of 5, tubercle 3, longer and more slender than others, on opposite side of apex, no seta 6. Second antenna (Fig. 1874) uniramous, with sympod-endopod long axis and completely obsolete exopod; sympod heavily sclerotized, apparently unsegmented, articulating with also apparently unsegmented endopod; latter with rounded apex armed with four papilliform processes (three in one group, fourth at some distance from them) and covered by small denticles. Tip of labrum (Fig. 1880) rounded, with prominent rostrum and only poorly developed marginal fringe. Mandible (Figs. 1875, 1876) with variable dental formula, either P1, S1, P1, S1, P1, S1, B2 (Fig. 1875), or P1, S1, P1, –, P1, –, B2 (Fig. 1876) (differences due to frequent absence of secondary teeth resulting in appearance of diastemes); blade narrowing abruptly at level of basalmost primary tooth. First maxilla (Fig. 1877) with ventral exopod and two terminal papillae; exopod conical or cylindrical, with two subequal apical setae, short, slender seta at base of dorsal terminal papilla; dorsal margin inflated at about exopod level, with sparse denticulation. Second maxilla (Figs. 1861, 1866) very short, fused with, though distinguishable from, opposite number; tip expanded to form moderately broad collar round base of bulla. Latter clavate (Fig. 1861), with short manubrium imperceptibly merging with subspherical to obovate anchor. Maxilliped (Fig. 1878) with robust corpus; myxa with denticulated area and stout, short seta at about midlength; subchela cylindrical, about $\frac{1}{2}$ length of corpus, with slender basal seta and sparse denticulation on distal part of inner margin, prolonged into rounded extension beyond base of barb; latter reaching $\frac{1}{2}$ length of claw (Fig. 1879); claw slightly curved, tapering, in some specimens with one secondary tooth at about midlength of inner margin.

Male (Fig. 1881): Body not divisible into cephalothorax and trunk, ovate, with tapering anterior and broadly rounded posterior end; ventral surface carrying all appendages, dorsal surface curving around posterior extremity and coming close to bases of maxillipeds; no dorsal shield. (Earlier report (Kabata, 1964d) suggested that reduced shield is present near anterior end. This and similar reports are based on artifacts, mainly cuticular folds caused by fixation and subsequent handling.) Length about 1 mm.

First antenna (Fig. 1882) similar to that of female but with relatively more inflated base and longer whip; apical armature (Fig. 1883) distinguished by large size of tubercles, slenderness of setae and presence of large seta 6. Second antenna (Fig. 1884) apparently unsegmented, biramous, with obsolete borders between sympod and rami; exopod short, with rounded tip armed by sparse, minute denticles and with one long slender seta on lateral wall; endopod twice as long as exopod, cylindrical, with obsolete segmentation; two patches of denticulation on ventral wall corresponding with two segments; apical armature much reduced: hook 1 spiniform, sometimes with tubercle 3 at base, never with seta 2; two slender processes on ventral side (homologous with processes 4 and 5) as long as hook 1. Mouth cone similar to that of female. Mandible (Fig. 1885) with aberrant, short dentiform margin, armed with about seven blunt teeth of unequal sizes, all pointing distalwards. First maxilla (Fig. 1886) similar to that of female, but with relatively longer exopod and terminal papillae; seta at base of dorsal papilla much smaller than in female; no dorsal denticulation. Second maxilla (Fig. 1888) with short, squat, subcircular corpus; distomedial corner with concave extension accommodating tip of subchela; irregular ridges and grooves on surface, near lateral margin, subchela without delimited claw, unarmed, with broad base and curving, tapering tip. Maxilliped (Fig. 1887) cylindrical, with blunt end; lip-like shelf with reinforced, serrated margin, wide at distomedial corner, narrowing posteriorly along medial surface of corpus; subchela moving at right angles to long axis of corpus, its tip pointing inwards; no delimited claw. No thoracic legs, anus, uropods or posterior processes.

Comments: Having been known for over two centuries, *C. adunca* has accumulated a long list of synonyms, reflecting its interesting and confused taxonomic history. The early history of Lernaeopodidae being what it is, one is not surprised to see that the generic designation of this species changed on several occasions. More unusual is the widespread confusion that followed it at the specific level. The first, still not totally resolved, difficulty was the relationship between *C. adunca* (Strøm, 1762) and *C. uncinata* (O. F. Müller, 1776). That the two are identical was suggested by Wilson (1915), who eventually equated them, relegating *adunca* to synonymy with *uncinata*, though the former was the senior synonym. Only recently (Dollfus, 1953) was this situation reversed. Since then, some authors (Nunes-Ruivo, 1957b; Kabata, 1960c, 1963c) have adopted Dollfus' usage as conforming with the rule of chronological priority, but many others still use the name *C. uncinata*. The second taxonomic difficulty, stemming from the extraordinary morphological variability of the species, arose mainly due to the splitting of *C. adunca* into six species by Leigh-Sharpe (1918b, 1920, 1925, 1933). Other spurious species based on specimens of *C. adunca* were described by Bassett-Smith (1896a) and Scott and Scott (1913). The process of gathering unnecessarily scattered fragments was initiated by Poulsen (1939), continued by Nunes-Ruivo (1957b) and concluded by Kabata (1960c, 1963c). It is now known that *C. adunca* is a very polymorphous species, with great range of morphological variability and appropriately wide range of host fishes. This plasticity has been given recognition in the earlier work of this author by the use of trinomial nomenclature for various morphological types of the species, the third name being the original name of the earlier spurious species. Thus, *C. iadda* became *C. adunca* f. *iadda*, etc.

The most common hosts of *C. adunca* are fishes belonging to the family Gadidae, at least ten species of this family having been recorded as harbouring it on the gills, fins, skin or in the buccal cavity. There are also some records of unusual host affiliations. Heller (1866) and Wilson (1932) reported finding it on *Merluccius* sp. in the Adriatic and Woods Hole region respectively. Their records were mentioned later (Shiino, 1956) but were never corroborated by other finds in those localities. Brian (1898) claimed to have found *C. adunca* on *Sargus rondeleti*, the specimen having been identified by a British scientist, I. C. Thompson. Hansen (1923) mentioned its occurrence on *Sebastes marinus* and *Macrurus fabricii*, the former probably a wrongly recorded host name. Gusev (1951) listed among its hosts *Hexagrammos octogrammus* and *Pleurogrammus* sp.. Bassett-Smith (1896a) described his *C. quadrata* (= *C. adunca*) from *Callionymus lyra*. *Lycodes frigidus* was recorded by Wilson (1922b), who also listed (Wilson, 1923a) *Doydixodon fasciatus*.

Although this list does not claim to be comprehensive, it gives an idea of the adaptability of *C. adunca*, making its morphological variability more acceptable and easily understood.

C. adunca has been most frequently recorded in the North Atlantic region (the Mediterranean included). Most of its hosts live in that region. In the North Pacific it occurs mainly on *Gadus macrocephalus*, off both the American and Asian coasts (Gusev, 1951; Shiino, 1956; Kabata, 1970d). It was recorded off the Pacific coast of South America (Wilson, 1923a) but its southernmost record comes from the Antarctic, where it was found on *Macrurus whitsoni* and *Trematomus loennbergi*, both unusual hosts, by Kabata and Gusev (1966). The sole record from the Indian Ocean was by

Thompson and Scott (1903), who found it on *Gazza minuta* off Ceylon. Later workers in that area failed to locate it again.

In British waters, *C. adunca* is a common parasite of gadoid hosts, the principal commercial species *Gadus morhua*, *Merlangius merlangus* and *Melanogrammus aeglefinus*, being extensively infected. Among other hosts can be mentioned *Pollachius pollachius*, *P. virens* and *Trisopterus luscus*. Since some observers find it difficult to identify its fairly common developmental stages, it was deemed worth while to include Figs. 1867 and 1868, showing the pupal and early post-pupal stages of *C. adunca*, as well as Fig. 1866, showing a young adult female with disproportionately large head and still undeveloped trunk, and with male attached.

In its host-parasite relationships *C. adunca* also shows a range of differences. For example, on haddock (*Melanogrammus aeglefinus*) a prominent tumour of attachment is provoked by this parasite (Nunes-Ruivo, 1957b), though it is barely evident on cod (*Gadus morhua*). On both cod and haddock *C. adunca* attaches itself to gill bars, as well as to the filaments, but it occurs only very exceptionally on the gill filaments of whiting (*Merlangius merlangus*). On that fish it infects almost exclusively the bars and rakers of the gills. The possibility of antagonistic relationships existing between *C. adunca* and the monogenean *Diclidophora merlangi*, living on the filaments of the gills of whiting, was suggested by this author (Kabata, 1960c) and denied by Smith (1969).

Clavella alata Brian, 1906

(Figs. 1889–1901)

Syn.: *Alella alata* (Brian, 1906); of Leigh-Sharpe (1925)

Female (Figs. 1889, 1890): Cephalothorax cylindrical, longer than trunk, of about same diameter throughout, head tapering, covered by dorsal shield with narrow anterior margin and convex lateral margins; posterior margin indistinct (Fig. 1892); prominent papillae on lateral sides of base of cephalothorax (Fig. 1891). Trunk divided from cephalothorax by transverse constriction, in some specimens (Fig. 1890) quite deep. Trunk ovate or pyriform, only slightly flattened dorsoventrally, with rounded posterolateral corners; dorsoventral slits on these corners marking orifices of oviducts; posterior margin transversely flattened, sometimes with slight central swelling of perianal area. Genital process (Fig. 1893) ventral to anus, subspherical or somewhat clavate. Position of oviduct orifices varying, depending on degree of lateral expansion of trunk, determined by age and stage of maturity (cf. Figs. 1889 and 1893). Length rarely exceeding 3 mm (length of egg sacs also about 3 mm).

First antenna indistinctly bipartite, basal part inflated, with dorsomedial whip, terminal part cylindrical; apical armature (Fig. 1894) dominated by digitiform seta 4; tubercle 1 at its base relatively long, cylindrical, its rounded apex with small, fine point; centre of apex occupied by robust complex 5, comprising one seta split to base into two noticeably subequal branches; on opposite side of apex very long, slender, flagelliform seta 6; tubercles 2 and 3 absent. Second antenna (Fig. 1895) with sympod-endopod long axis; sympod apparently unsegmented, heavily sclerotized; exopod displaced from main axis of appendage, relatively large but much smaller than endopod, one-segmented, bulbous, armed with at least one papilliform process and covered by sparse, small denticles; endopod indistinctly two-segmented, basal segment longer than exopod, unarmed, terminal segment bifid, one of its branches with rounded tip, unarmed, other of similar shape but covered with fine denticles; no other apical armature. Labrum similar to that of *C. adunca*. Mandible (Fig. 1896) with dental formula $\underline{\text{P1, S1, P1, —}}$, P1, —, B2; diastemes due to absence of second and third secondary teeth; third primary largest tooth of entire series; first three teeth on narrower distal part of blade, but narrowing less noticeable than in *C. adunca*. First maxilla (Fig. 1897) with cylindrical, slightly tapering exopod and endopod with two terminal papillae; exopod ventral, with two apical, subequal setae; dorsal terminal pailla longer than ventral, both with long, sturdy but flexible setae; no seta observed at base of dorsal papilla; dorsal denticulation absent. Second maxilla (Figs. 1899–1891) short (seldom reaching length of ½ that of trunk), slender, cylindrical and tapering, closely apposed to opposite member so as to appear fused (though Fig. 1891 shows that no fusion exists); posterolateral side of base with very prominent papilla (Fig. 1891), barrel-shaped and often extending beyond level of lateral margins of cephalothorax; tip only slightly expanded. Bulla with short manubrium and subspherical, clavate anchor, resembling that of *C. adunca*. Maxilliped (Fig. 1898) with robust corpus; myxa with denticulated patch and one short seta at about midlength; subchela cylindrical, somewhat tapering, straight, with basal seta and distal swelling on inner margin; latter with few small denticles and long, sturdy barb reaching near tip of claw (Fig. 1899); claw sturdy, with robust base and tapering, curving tip; in specimens examined by this author claws devoid of secondary teeth.

Male (Fig. 1900) similar to that of *C. adunca*, its short, ovate, posteriorly rounded body not divisible into cephalothorax and trunk. Whip of first antenna (Fig. 1901, w) very long and slender, its apical armature apparently with complex 5 comprising only one seta (two in *C. adunca*). Second antenna with shorter and more squat endopod than in preceding species, its apical armature much reduced and consisting of three processes, with obscure homology. Other appendages similar to those of male *C. adunca*. No thoracic legs, anus, uropods or posterior processes. Males of this species usually less than 0.5 mm long.

Comments: *C. alata* was transferred by Leigh-Sharpe (1925), together with *Clavellodes canthari* (Heller, 1865), as the only two species of the new genus *Alella*. Although the structure of its second antenna places it unmistakably in the genus *Clavella* and although this has been pointed out by Nunes-Ruivo (1966b), some modern authors prefer to follow Leigh-Sharpe's usage (Yamaguti, 1963).

There are only two known host fishes for *C. alata*, both belonging to the gadiform genus *Phycis*. With the exception of one record of *Phycis mediterranean* (cf. Nunes-Ruivo, 1953), all finds of this parasite are from *P. blennoides*, clearly indicating that this fish is the main host of *C. alata*.

The distribution range of *C. alata* coincides with that of its main host and extends over the North Sea and British waters, and south to the coasts of Spain and the Mediterranean, at least to the longitude of Italy. Brian (1906) described it first from the vicinity of Naples.

Clavella stellata (Krøyer, 1838)

(Figs. 1902–1913)

Syn.: *Anchorella stellata* Krøyer, 1838

Female (Figs. 1902, 1903): Cephalothorax cylindrical, somewhat shorter than trunk, tapering anteriorly, head region covered with anteriorly narrower, well-delimited dorsal shield. Transverse groove below level of second maxillae separating trunk from cephalothorax. Trunk dorsoventrally flattened, at least twice (sometimes three times) longer than wide, with shoulder-like anterolateral corners and convex lateral margins (Fig. 1902); posterior extremity transversely truncated, with central eminence around anus, with two slit-like orifices of oviducts in lateral quarters, and with rounded posterolateral corners. No genital process, uropods or posterior processes. (Careless fixation grossly distorts shape of trunk, making it fusiform, narrow, and with artificially prominent areas around oviduct orifices. Cf. Kabata, 1962b, Fig 1.) Length of trunk (10 specimens) 3.1–4.6 (mean 3.7) mm (Kabata, 1962b).

First antenna indistinctly segmented, with base broader than tip; whip apparently absent; apical armature (Fig. 1905) comprising digitiform seta 4, very prominent cylindrical tubercle 1 with fine tip at its base and bipartite central seta 5 with subequal branches. Second antenna (Fig. 1906) with characteristic right-angle bend near obsolete border between sympod and rami; exopod short, unarmed, in some specimens appearing bilobed; endopod very long, indistinctly segmented, cylindrical, with rounded tip; apical armature of spinulated ventral pad and two setiform processes; tip of appendage facing inwards and meeting its opposite number near centre of anterior margin of cephalothorax. Mandible (Fig. 1907) with dental formula usually P1, S2, P1, S1, B2; in many specimens no secondary tooth between basals and nearest primary, but usually no diasteme; first two secondary teeth not much smaller than first primary, but third secondary, when present, very small; very abrupt change in width of blade at level of nearest primary tooth, part distal to it about $\frac{1}{3}$ width of that proximal to it. First maxilla (Fig. 1908) long, slender, with cylindrical ventral exopod and two terminal papillae; exopod with rounded tip and two subequal apical setae, longer reaching bases of terminal papillae; latter long, slender, with apical setae their own length; small seta at base of dorsal papilla; above level of exopod base extensive spinulation of dorsal margin and lateral surface. Second maxilla (Fig. 1903) cylindrical, short, separate from opposite number, with characteristically modified tip (Fig. 1904); pair of subterminal spherical swellings extending posteriorly and posterolaterally, followed by abrupt tapering; this terminal peduncle expanding again into semicircular half-collar with inflated rim. Bulla with short manubrium and flat, subcircular anchor (Fig. 1911) with seven slit-like lacunae in marginal ring, additional lacuna cutting into margin at midposterior point; manubrium displaced from centre to posterior half of anchor, ducts spreading radially from it through anchor. Maxilliped (Fig. 1910) with long, comparatively slender corpus; two setae with inflated bases in myxal area; no spinulation observed; subchela cylindrical, straight, with prominent basal seta and small patch of comparatively large denticles at distal end of inner margin (Fig. 1909); barb with greatly inflated proximal and slender distal half, reaching farther than $\frac{1}{2}$ length of claw; latter long, slender, gently curved and with finely tapering tip.

Male (Fig. 1912): Body ovate, short, not divisible into cephalothorax and trunk. Length about 0.5 mm (Kabata, 1962b).

First antenna with long sturdy whip; apical armature comprising setae 4 and 6, complex 5 of double seta, as well as tubercles 1, 2 and 3. Second antenna (Fig. 1913) biramous, short, with obsolete intersegmental boundaries; exopod shorter than endopod, with rounded tip, armed with apical conical protuberance and long lateral seta; tip of endopod with sharply conical apical process and two longer, slender, subapical processes, all of uncertain homology; no denticulation observed on either ramus. Mandible greatly reduced, with only two apical teeth, both pointing obliquely forwards. First maxilla similar to that of female; in some specimens shorter seta of exopod, basal seta of dorsal terminal papilla and dorsal marginal spinulation not detectable. Second maxilla resembling that of male *C. adunca* but with longer, more slender corpus and without surface sculpturing near dorsal margin. Maxilliped also resembling that of *C. adunca*. No thoracic legs, anus, uropods or posterior processes.

Comments: Originally described as *Anchorella*, *C. stellata* is distinctive enough to be easily identifiable. Consequently, it has never been confused with other species and its taxonomic history, except for the original generic appellation, contains no other synonyms.

C. stellata parasitizes only one host species, the European hake, *Merluccius merluccius*. With this host, it is widespread along the Atlantic sea-board of Europe, being quite common in British waters, particularly on hake of younger age groups.

C. stellata is unique among its congeners in the manner of its association with the host fish. Its site of predilection is a narrow strip of the external surface of the host, close to and in parallel with the rims of the opercula. More than 80% of *C. stellata* found by this author were attached to that site, the remainder being found usually between the bases of the pectoral fins, on the ventral surface of the fish. Only very occasionally is *C. stellata* found elsewhere (surface of the operculum, dorsal surface of head, skin posterior to bases of pectoral fins, etc.).

The site of predilection of *C. stellata* imposes on it the necessity of attachment to scales; hence the broad and flat anchor of its bulla. The subanchoral surface of the bulla adheres to the scale of the host with the aid of a thin layer of adhesive substance. It appears that its solvent action causes it to dissolve, and mix with, the substance of the scale, the surplus of the combined material being drawn into the marginal lacunae of the anchor and riveting it to the scale. The supra-anchoral surface is then covered by the proliferating epithelium of the fish, completing its attachment (Kabata, 1970a; Kabata and Cousens, 1972).

Genus Clavellodes Wilson, 1915

Female: Cephalothorax cylindrical, longer than trunk, with considerably expanded head. Trunk dorsoventrally flattened, its width equal to, or greater than, length. Genital process, posterior processes and uropods absent. First antenna indistinctly segmented, with well developed apical armature. Second antenna with sympod-exopod long axis and reduced endopod. Labrum without rostrum. Mandible with four secondary teeth. First maxilla with lateral exopod and two full-sized and one reduced terminal papillae. Second maxilla vestigial, in form of collar round manubrium of bulla. Latter with short, plano-convex anchor and short manubrium. Maxilliped subchelate.

Male: Body oval, with ventral surface transversely anterior and dorsal surface elongate, arched, its anterior half nearly parallel with posterior half. First antenna similar to that of female. Second antenna with well developed endopod. Mandible with one secondary tooth. First maxilla similar to that of female. Second maxilla and maxilliped subchelate. Genital area on posteriorly pointing tubercle, close to base of maxillipeds. No thoracic legs, anus, uropods or posterior processes.

Type and only species: *Clavellodes rugosa* (Krøyer, 1837).

Comments: In addition to the type species, Wilson (1915) placed in it *Clavella macrotrachelus* Brian, 1906, and *Anchorella intermedia* Quidor, 1906. In so doing, he established a genus with the type species clearly different from its other members. In particular, *Clavella macrotrachelus*, a species of uncertain validity, resembled some similar (some, perhaps, even identical) taxa and caused them to be eventually included in *Clavellodes*. Later (Leigh-Sharpe, 1925), one of these species, *C. canthari* (Heller, 1865) was made the type species of the genus *Alella*. It is now known that five of the species, placed in *Clavellodes* at different times, are members of, or synonymous with, members of the genus *Alella*. They are: *A. canthari* (Heller, 1865), *A. ditrematis* (Yamaguti, 1939), *A. macrotrachelus* (Brian, 1906), *A. pagelli* (Krøyer, 1863) and *A. pterobrachiata* (Kabata, 1968). *Anchorella intermedia* Quidor, 1906, was placed by this author in a recently proposed genus *Clavellomimus* (cf. Kabata, 1969e).

Clavellodes rugosa (Krøyer, 1837)

(Figs. 1914–1936)

Syn.: *Anchorella rugosa* Krøyer, 1837
Anchorella (?) *rugosa* Krøyer; of T. Scott (1900)
nec *Anchorella rugosa* var.; of T. Scott (1900)
Clavella rugosa (Krøyer); of Scott and Scott (1913)

Female (Figs. 1914, 1915): Cephalothorax cylindrical, of about same diameter throughout, as long as, or longer than, trunk, usually with profuse transverse wrinkles; head slightly expanded, sharply delimited from rest of cephalothorax, with heavy dorsal shield, anterior margin (Fig. 1916) tripartite, lateral quarters recessed and marked off by longitudinal grooves, lateral margins usually converging posteriorly, posterior margin truncated and sharply delimited. Trunk dorsoventrally flattened, subquadrangular, with rounded corners and re-entrant margins, often with longitudinal dorsal depression and corresponding ventral flexure. Anal tubercle prominent in centre of posterior margin (Fig. 1917), flanked by oviduct orifices. No genital process, uropods or posterior processes. Cephalothorax length 4.5–5.0 mm, trunk length up to 4 mm (Scott and Scott, 1913; Markevich, 1956).

First antenna (Fig. 1919) apparently tripartite; basal part inflated, with short dorsomedial whip, middle part cylindrical, with slender solus about as long as whip in some specimens; apical armature (Fig. 1918) comparatively short and squat, comprising digitiform seta 4, large and fine-tipped tubercle 1 at its base, long seta 6 with accompanying tubercle 3 and complex 5 of four setae, two fused at base and two separate (one of latter with inflated base, perhaps homologous with tubercle 2). Second antenna (Fig. 1920) with apparently unsegmented, heavy sympod, sympod-exopod long axis, large, inflated, strongly denticulated exopod and very small (not more than 50 μ long) endopod; latter two-segmented, basal segment cylindrical, apparently unarmed, distal segment with hook 1, spiniform seta 2 and, ventrally, long, slender processes 4 and 5; denticulation not observed. Labrum (Fig. 1921) either without rostrum or with one covered by fringe of leaf-like cuticular flaps extending along entire anterior margin. Mandible (Fig. 1922) with comparatively short dentiferous margin; dental formula $\underline{\text{P1, S1,}}$ P1, S1, P1, S1, P1, S1; in distal and penultimate pair secondary teeth much smaller than primary, two proximal secondary teeth long and slender, particularly most proximal tooth. First maxilla (Fig. 1923) with lateral exopod, two large and one reduced terminal papillae; exopod prominent, with two subequal, subterminal setae, its surface partly covered by comparatively large, sharp denticles pointing distalwards; extensive denticulation on dorsal margin at, and distal to, level of exopod base; vestigial terminal papilla with differentiated seta not reaching tip of dorsal papilla. Second maxilla (Fig. 1924) almost obsolete, leaf-like lobe arising directly from surface of cephalothorax-trunk junction, with opposite number forming collar round bulla. Latter (Figs. 1914, 1915) with very short manubrium and small, plano-convex anchor. Maxilliped (Fig. 1925) with short, stocky corpus; two short conical processes in myxal area, apparently without denticulation; subchela short, with stubby basal seta and inner margin densely denticulated for more than half its length; barb strong, extending to second half of claw; latter (Fig. 1926) short, with blunt, curving tip.

Male (Figs. 1927, 1928) bearing general resemblance to those of most other genera of *Clavella*-group, but differing from them in body symmetry. (Whereas in males of e.g. *Clavella*, line drawn through bases of appendages forms very acute angle with line linking it with posteriormost point of body, in *Clavellodes* these lines are at, or nearly at, right angles to each other.) Shape of body as in generic diagnosis. Length about 1.15 mm (Markevich, 1956).

First antenna (Fig. 1929) similar to that of female; basal part less inflated, whip reduced to small conical process; no solus found on middle part; apex with gibber (Fig. 1930) occupied by two short, rounded processes and two setiform longer ones, all representing complex 5; also present digitiform seta 4 and long seta 6, with accompanying tubercles 1 and 3; tubercles with rounded tips. Second armature (Fig. 1932) comprising hook 1, spiniform seta 2 and, ventrally, inflated denticulated between distal segment of sympod and rami; exopod slightly larger than endopod, with finely denticulated apex and at least one papilliform process; endopod two-segmented, basal segment cylindrical, with ventral patch of denticulation, distal with well developed, functional set of apical armature (Fig. 1932) comprising hook 1, spiniform seta 2 and, ventrally, inflated denticulated process 4 and slender, long process 5; tubercle 3 not observed. Labrum with very small, barely detectable rostrum, covered by fringe of marginal setae. Mandible (Fig. 1933) with dental formula P2, S1, P1, B3 (in Fig. 1933 two distal primary teeth appear to point obliquely anteriorly due to displacement of appendage from horizontal plane); two basal teeth very slender (perhaps in progress of reduction). (The author had at his disposal limited material on which to base his observation on this appendage. It is possible that some variations exist in dental formula. More studies are needed.) First maxilla (Fig. 1934) with lateral exopod and two terminal papillae, differing from female in absence of denticulation on surface of exopod and in complete suppression of vestigial third papilla, represented at base of dorsal papilla by small, slender seta. Second maxilla (Fig. 1936) with short, robust corpus; small pointed process on medial margin; subchela not divided into claw and shaft, with tapering curving tip and devoid of secondary armature. Maxilliped (Fig. 1935) with slender base armed with prominent, pedunculate, lateral papilla; corpus subcircular, with small hollow at distomedial quarter for reception of tip of subchela; latter short, truncated, with stubby, blunt end slightly curved, armed with minute prickles; no distinct claw, no secondary armature observed. No anus, external genitalia, uropod or posterior processes. Small mediative process between bases of second maxillae and maxillipeds (Fig. 1928).

Comments: Although in its general morphological characters *C. rugosa* resembles closely a typical structure of any member of the *Clavella*-group, many of its small distinctive features make it readily identifiable. Its taxonomic history, as a result, is fairly straightforward, though it includes the more or less usual changes of generic names, associated with the gradually improving state of our knowledge of this group. The only confusion in which *C. rugosa* was involved was caused by a clerical error and is described in the comments to the description of *Clavellisa emarginata*. This purely unintentional slip, however, caused *C. rugosa* to be relegated to synonymy with *Clavellisa emarginata* by Bassett-Smith (1899). His action remained completely isolated and had no taxonomic repercussions.

C. rugosa parasitizes marine catfishes of the genus *Anarhichas*. Attached to the gills of its host, it has been recorded on *A. lupus*, *A. minor* and *A. latifrons* (= *Lycichthys denticulatus*). There are few data as to its incidence on these fishes. Bazikalova (1932) found it on 33% of *A. lupus* examined in the Barents Sea, whereas Polyanski (1955) found it on 20% of this fish in the same area. The distribution range of *C. rugosa* extends over the North Atlantic, reaching as far South as the coast of Belgium and stretching into the Danish Straits. It includes the North, Norwegian and Barents Seas, covers Faroes, Iceland, Greenland and reaches the Atlantic coast of North America. *C. rugosa* is quite common in British waters, particularly on the gills of *A. lupus*.

Baird (1850) recorded *Anchorella rugosa* (= *C. rugosa*) from *Gadus cellarius* (presumably a corruption of *G. callarias*) from Larne and T. Scott (1900) mentions "what appears to be the same species of *Anchorella*" found on cod. Later findings do not confirm those records and Scott and Scott (1913) omit cod from their list of hosts for this copepod.

Genus Clavellisa Wilson, 1915

Female: Cephalothorax cylindrical, longer than trunk, with or without expanded head, with indistinct dorsal shield. Base of cephalothorax at or near centre of dorsal surface of trunk, separated by large gap from bases of second maxillae on anterior margin of trunk. Latter dorsoventrally flattened, subquadrangular, subcircular, often broader than long, Posterior processes absent, genital process and uropods present or absent. First antenna obscurely segmented, with well developed apical armature. Second antenna with reduced endopod, usually with sharp flexure between sympod and rami. Mandible with at least two secondary teeth. First maxilla with lateral exopod and either two or three terminal papillae. Second maxilla shorter than trunk, either separate from, or partially fused with, opposite number. Maxilliped subchelate. Vestigial thoracopods present or absent. Egg sacs subcylindrical to subspherical.

Male: Body oval, not divisible into cephalothorax and trunk, with prominent genital tubercle (including anus?) posterior to bases of maxillipeds. Uropods present or absent. No external genitalia or posterior processes. First antenna similar to that of female. Second antenna with endopod longer than exopod but with reduced apical armature. Buccal apparatus similar to that of female. Second maxilla and maxilliped subchelate.

Type species: *Clavellisa spinosa* Wilson, 1915.

Comments: Although *Clavellisa* is a genus with all the typical features of the *Clavella*-group of the family, it is distinguished by many unusual structural details that set it apart from the other members of that branch. The most striking morphological characteristic of *Clavellisa* is the presence of a gap between the base of the cephalothorax and the bases of the second antennae (Text fig. 61D). The presence of uropods and of vestigial thoracic feet in some species is very unusual for lernaeopodid females. Details of structure of appendages are also distinctive (e.g. armature of the first antennae, difficult to homologize with any other lernaeopodid genus).

With a single exception, members of this genus are parasitic on teleost fishes of the order Clupeiformes, mainly herrings, shads and anchovies. *C. scombri* (Kurz, 1877) parasitizes mackerels (Scombridae). They are predominantly warm-water species and some of them, at least, are capable of withstanding extensive river migrations with their hosts.

There exist 11 nominal species of *Clavellisa*. To their number must be added also *Epiclavella chinensis* Yu, 1933, a lernaeopodid clearly belonging to *Clavellisa* and, like its congeners, parasitic on a clupeiform fish (*Coilia gravi*). Some species were, however, very poorly described and further studies might result in several taxonomic adjustments. For example, *C. ilishae* Pillai, 1963, might have to be reduced to synonymy with *C. hilsa* Tripathi, 1962, known from a very sketchy description; *C. cordata* Wilson, of Pillai (1962) is probably synonymous with very poorly described *C. pellone* Tripathi, 1962.

Two species occur in the eastern North Atlantic and the Mediterranean: *C. emarginata* (Krøyer, 1837), parasitic on various species of shad, and *C. scombri*, a parasite of mackerels. On the other side of the Atlantic, off the shores of North America, *C. spinosa* occurs on the menhaden (*Brevoortia tyrannus*) and *C. cordata* on shads of the genus *Pomolobus*.

In the Asian North Pacific *C. dorosomatis* Yamaguti, 1939, lives on *Dorosoma thrissa* in Japanese waters, *C. scombri* parasitizes *Scomber japonicus* in the same region, whereas *C. chinensis* (Yu, 1933) is known from a single record off the shores of China.

All the remaining species live in Indian waters. Even allowing, therefore, for relegation to synonymy of Pillai's *C. ilishae*, there are at least five species of *Clavellisa* in the region: *C. dussumieriae* Gnanamuthu, 1947 (hosts *Dussumieria hasselti* and *D. acuta*), *C. hilsa* (hosts *Hilsa ilisha*, perhaps also *Ilisha filigera* and *Euplatygaster indica*), *C. obcordata* Rangnekar, 1957 (host *Chatoessus nasus*), *C. pellone* (hosts *Euplatygaster indica*, *Ilisha filigera*) and *C. phasa* (hosts *Coilia dussumieri* and *Setipinna phasa*).

The zoogeography and the host affiliations of the members of the genus *Clavellisa* indicate that the evolution of this genus, apparently one of the older genera of the *Clavella*-group, was associated with the clupeiform fishes, probably in the part of the world ocean now coincident with Indian waters.

Two species of *Clavellisa* occur in British waters: *C. emarginata* and *C. scombri*.

Key to females of British species

Small uropods present on flanks of anal tubercle (Fig. 1949); anterior margin of maxilliped subchela smooth (Fig. 1947) . . . . **C. emarginata**

No uropods; several small denticles on anterior margin of subchela of maxilliped (Fig. 1967) . . . **C. scombri**

Clavellisa emarginata (Krøyer, 1837)

(Figs. 1937–1957)

Syn.: *Anchorella emarginata* Krøyer, 1837
Anchorella ovalis Krøyer, 1837
Anchorella rugosa; of Bassett-Smith (1899) (partim)
Clavella emarginata Krøyer; of Brian (1906)
nec *Brachiella ovalis* (Krøyer); of Scott and Scott (1913)
Clavellisa ovalis (Krøyer), 1837; of Wilson (1915)
Clavellisa emarginata (Krøyer), 1837; of Wilson (1915)
?*Clavellisa emarginata* (Krøyer); of Pillai (1963)

Female (Figs. 1937, 1938): Cephalothorax cylindrical, longer than trunk, of same diameter, except with slightly expanded head, well delimited from rest of cephalothorax; base of cephalothorax near centre of dorsal surface of trunk, with large gap between it and bases of second maxillae; head with quadrangular, not heavily sclerotized dorsal shield (Fig. 1939). Trunk subquadrangular, usually with rounded anterolateral corners and diverging lateral margins, dorsoventrally flattened; posterior margin apparently bilobed, due to presence of two posterolateral swellings; in centre on dorsal side of that margin low anal tubercle (Fig. 1949), flanked by small uropods, cylindrical, with fine tips. Ventral to anal tubercle short conical genital process. Length of trunk 2.5 mm, of cephalothorax 3.5 mm (Scott and Scott, 1913).

First antenna (Fig. 1940) indistinctly three-segmented; basal segment inflated, with long, slender whip distal on dorsomedial side; second segment cylindrical, with solus at distal end; terminal segment cylindrical, with rounded tip; apical armature (Fig. 1941) of seven setae with completely obscure homology; only one long slender seta tentatively homologous with seta 6 and one with seta 4; in addition one strong, long seta present near possible seta 6, three other shorter, more slender setae in centre and one shorter seta on opposite site to possible seta 6. Second antenna (Fig. 1942) with heavily sclerotized, apparently one-segmented sympod. In most specimens right-angled flexure between sympod and rami; exopod one-segmented, bulbous, its surface with numerous small denticles and two short conical papilliform processes; endopod two-segmented, much reduced; basal segment cylindrical, apparently unarmed; distal segment (Fig. 1943) with hook 1 reduced in size; spiniform seta 2 present; on ventral side of apex sturdy process, as long as seta 2, present in some specimens, probably homologous with process 4; sometimes small, nipple-like protuberance present near base of that process. Labrum as in *Clavella adunca*. Mandible (Fig. 1944) short and broad; dental formula P1, S1, P1, S1, P1, S1, P1, S1; primary teeth diminishing in size towards base, but proximal secondary larger than others, slightly shorter than corresponding primary. First maxilla (Fig. 1945) with lateral exopod and two terminal papillae; exopod long, cylindrical, with two subequal, apical setae; two terminal papillae short, robust, with short apical setae; short, but stout seta at base of dorsal papilla; dorsal margin convex; no denticulation observed. Second maxilla (Figs. 1937, 1946) short, cylindrical; base on anterior margin of trunk; free of opposite number, not expanded into collar round base of bulla. Latter with short manubrium and round, plano-convex anchor of small diameter. Maxilliped (Fig. 1947) with short, robust corpus; myxa with two short, cylindrical processes with rounded tips about midlength (in some specimens only one process observable); subchela short and sturdy, cylindrical, with long and slender basal seta; inner margin with broad strip of denticles for more than half its length, denticles sharp, covering particularly ventral side of subchela; barb (Fig. 1948) long, reaching second half of claw, very slender; claw with irregular thickness, curving, with sharp tip; no secondary armature.

Male (Fig. 1950): Body oval, not divisible into cephalothorax and trunk, its dorsal surface extending round posterior extremity and reaching bases of maxillipeds. Generally resembling males of *Clavella*. Prominent genital tubercle posterior to these bases. No uropods observed. No posterior processes or external genitalia. Length up to 0.5 mm (Markevich, 1956).

First antenna (Fig. 1951) three-segmented; basal segment cylindrical, inflated with two lobes, one above other and not completely overlapping (Fig. 1952); upper lobe with long, slender whip on distal surface; middle segment cylindrical, apparently unarmed; distal segment slightly expanded, with rounded apex. Apical armature with uncertain homology, comprising two short, sturdy structures (tubercles?), two robust digitiform setae and two long, slender setae. Second antenna (Fig. 1953) with indistinctly two-segmented sympod; exopod much shorter than endopod, one-segmented, with rounded apex apparently unarmed and with one sturdy, blunt seta on lateral wall; endopod two-segmented: basal segment cylindrical, with transverse ridge on ventral wall; terminal segment with small hook 1, short spiniform seta 2 and inflated, denticulated ventral wall, often with one short, sharp process. Mouth cone similar to that of female. Mandible slender, with dental formula P2, S1, P2 (Fig. 1954). (It is possible that some differences from specimen to specimen can be found in this formula.) First maxilla (Fig. 1955) similar to that of female, but more slender and with base of exopod nearer ventral margin; two terminal papillae comparatively longer. Second maxilla (Fig. 1957) with short, squat corpus; medial margin with two sharp processes, one at about midlength, other nearer base of subchela; distal end of corpus narrowing towards base of subchela; latter without delimited claw or secondary armature; base broad, abruptly tapering to tip, short and gently curving. Maxilliped (Fig. 1956) similar to that of *Clavella*, cylindrical, long, with medial cavity near tip, housing point of subchela. Latter closing in plane at right angles to that of long axis of appendage, curving, with tapering apex and without secondary armature.

Comments: Originally described as *Anchorella* by Krøyer (1837), this species was transferred to *Clavella* by Brian (1906) and to *Clavellisa* by Wilson (1915). In view of its morphological distinctness, *C. emarginata* has seldom been confused with other species. Its taxonomic history contains, however, a peculiar case of confused identity resulting from an editorial mistake. Edwards (1840) mentioned that *Anchorella emarginata* lives "sur les branchies de l'*Anarrhicha* [sic] *lupus*". That his statement was a mere lapsus is evident from the way in which he misspelt the name of the host. In Krøyer's (1837) description of *Anchorella rugosa* the name of the host is also spelt with its last letter left out. Although van Beneden (1851a) realized that Edwards' statement was a mistake, it succeeded in confusing other authors. Olsson (1869) believed that he found *Anchorella emarginata* on the gills of *Anarrhichas lupus* in Oeresund and could not see any difference between this parasite and *A. rugosa*. Kurz (1877) suggested that *A. emarginata* and *A. rugosa* might be two different stages of development of one species. Finally, Bassett-Smith (1899) considered *A. rugosa* a mere synonym of *A. emarginata* and the latter species a parasite of both shads and marine catfishes.

Krøyer's (1837) *Anchorella ovalis* is known from a single specimen. Having examined that specimen, the present author found it identical with *Clavellisa emarginata*. As in several other instances, Krøyer misidentified the host of his specimen, giving its name as *Trigla*. This caused some subsequent confusion, summarized by Nunes-Ruivo (1957a). *C. emarginata* does not live on *Trigla*.

The hosts of *C. emarginata* all belong to the genus *Alosa* and its relatives (*Caspialosa*, *Clupeonella*). With these hosts, it occurs in the North Sea and the adjacent waters. There appear to be no records of more northerly distribution. It is found in the Mediterranean and the Black and Azov Seas. It migrates with its hosts into the lower reaches of the rivers such as the Danube, Southern Bug, Dnieper and Don. Chernova and Sukhenko (1969) found it in a lacustrine environment in Lake Paleostomi (Georgian Republic, USSR), a body of water connected with the Black Sea by a short stream.

There exist some unusual records of *C. emarginata*. Pillai (1962) found it in Indian waters on *Thrissocles malabaricus*. His description, however, discloses some differences between the Indian specimens and those occurring in the West. Thorough comparison is required. Brian (1898) quotes *Pagellus centrodontus* as a host of *C. emarginata*, while Carus (1885) lists also *Atherina hepsetus* and *Scorpaena porcus*. In all likelihood these are incorrect host designations.

In his description of the male *Clavellisa*, Wilson (1915) suggested that it differed from the male *Clavella* mainly in having a dorsal shield. This author (Kabata, 1964d) described a similar structure for the male of *C. emarginata*. It is his present opinion, however, that both he and Wilson erred and that no clearly delimited dorsal shield exists in males of this genus. What was taken for the margin of the shield were cuticular wrinkles caused by treatment and handling of specimens. The only observable difference between the males of *Clavella* and *Clavellisa* is the genital tubercle of the latter, comparatively prominent and often situated nearer the posterior extremity.

Clavellisa scombri (Kurz, 1877)

(Figs. 1958–1974)

Syn.: *Anchorella scombri* Kurz, 1877
Clavella scombri Kurz, 1877; of Brian (1906)
Clavellisa scombri (Kurz), 1877; of Wilson (1915)

Female (Figs. 1958, 1959): Cephalothorax cylindrical, much longer than trunk, its base at or near centre of dorsal surface of trunk; diameter slightly decreasing towards anterior end, head slightly expanded, covered with clearly delimited but not heavily sclerotized dorsal shield (Fig. 1961), with convex anterior and parallel lateral margins (in most fixed specimens cephalothorax flexed in sigmoid curve). Trunk subcircular (Fig. 1960), moderately flattened in dorsoventral plane; posterior extremity with noticeable anal tubercle in centre, flanked with swellings housing orifices of oviducts. No genital process or uropods observed. Length of trunk 2.5 mm, of cephalothorax 6 mm (Scott and Scott, 1913).

First antenna resembling that of *C. emarginata*; homology of apical armature uncertain; apex (Fig. 1962) occupied by six setae, three digitiform and three slender and sharp-pointed; two more long, flagelliform setae in subapical positions. Second antenna as in *C. emarginata*. Labrum with abundant marginal setae on tip, covering rostrum, if present; anterolateral corners of lip with aliform processes unique in Lernaeopodidae. Mandible (Fig. 1963) short, broad, with dental formula P2, S1, P1, S1, P1, S1; no secondary tooth present between first two primaries (this might be a variable feature and requires checking on more abundant material); basalmost secondary much longer than other two. First maxilla (Figs. 1964, 1965) with lateral exopod and three terminal papillae; exopod in form of small swelling covered by sharp denticles, with two subequal setae arising from distal surface; terminal papillae with poorly marked differentiation into setae and papillae proper, long, tapering; at their bases entire appendage somewhat inflated (Fig. 1964). Second maxilla (Figs. 1958–1960) very short, cylindrical, tapering, closely apposed to opposite member. Bulla similar to that of *C. emarginata*. Maxilliped (Fig. 1967) short, broad, medial margin of corpus with one slender seta near midlength; no denticulation observed; subchela cylindrical, tapering distally, with slender, long basal seta; anterior margin with about seven minute denticles, set at wide intervals; almost entire inner margin with strip of closely crowded denticles; barb (Fig. 1966) long, slender, reaching near tip of claw; latter long, slender, gently curving, with relatively blunt tip.

Male (Fig. 1968): Body not divisible into cephalothorax and trunk, oval, with rounded posterior extremity formed by dorsal surface. Generally resembling males of *C. emarginata*, it differs from them in larger size and more posterior position of genital tubercle; tip of tubercle (Fig. 1974) with dorsoventral slit and two flanking lobes; at base of tubercle short uropods with uneven diameter. Length 0.3–0.4 mm.

First antenna resembling that of female, but comparatively squatter and more robust. (The author was unable to determine details of apical armature.) Second antenna (Fig. 1969) with sturdy, obscurely segmented sympod, short exopod and considerably longer endopod; sympod with patch of denticles on lateral wall; exopod one-segmented, with rounded tip and at least one papilliform process, no denticulation observed; endopod two-segmented, basal segment cylindrical, with patch of denticles on ventral wall; terminal segment with hook 1 and spiniform seta 2, with ventral wall inflated and covered by denticulation; no further details ascertained. Buccal apparatus similar to that of female. Labrum with weak rostrum but devoid of lateral aliform processes; marginal fringe of setae present. Mandible of specimen examined by present author with dental formula S1, P1, S3, P1, S1 (Fig. 1970). (This very unusual dental formula might not be constant for the species and requires confirmation by further study.) First maxilla (Fig. 1971) short, with lateral exopod and two terminal papillae; exopod cylindrical, long, with one short apical seta; terminal papillae also short, their setae much longer than their own length; dorsal margin of appendage with noticeable inflation; no denticulation observed. Second maxilla (Fig. 1972) with oval corpus, armed only by small blunt tubercle near midlength of medial margin; subchela with undifferentiated claw, robust at base, tapering towards tip, gently curving and without secondary armature. Maxilliped (Fig. 1973) with subcircular corpus arising out of common base; ventral surface of corpus with deep transverse fold in medial half; distomedial corner shelf-like, with slightly hollow centre; subchela without differentiated claw, truncated, with rounded tip and sharp point at distomedial surface; one short denticle on inner margin; no denticulation observed. Mediative process not observed, but probably present.

Comments: This species is at odds with its congeners in being parasitic on mackerels. It is not common, but has been recorded from a very wide area. On *Scomber scombrus* it can be found in the North Sea and neighbouring areas, as well as in the Mediterranean and Adriatic. In the Mediterranean it has also been found on *Scomber colias*. Yamaguti (1939c) and Shiino (1959c) found it in Japanese waters on the same species of mackerel. It might be supposed that *C. scombri* offers one of the rare instances among parasitic copepods of discontinuous distribution. However, Solonchenko (1968) recently recorded the occurrence of *C. scombri* off the shores of south-western

Africa. The species is relatively rare and the areas between the North Sea and Japanese waters have not been studied very thoroughly. It is possible that further research will result in closing the gap.

It is interesting to observe that the maxilliped of the male *C. scombri* is quite different from that of *C. emarginata.* While that of the latter species closely resembles that of the male *Clavella*, being cylindrical and having its claw twisted at right angles to the long axis of the appendage (Fig. 1956), in *C. scombri* it is more like that of *Clavellodes rugosa*, particularly in having an abbreviated subchela (Fig. 1973). This suggests that a lot of changes must have occurred during the evolution of *Clavellisa* and that they can now be used as an aid in arranging a scheme of intrageneric relationships. For example, males of *C. dussumieriae* have maxillipeds much like those of *C. emarginata* (possibility of close relationship?), whereas those of *C. cordata* and *C. spinosa* appear to occupy intermediate positions, their maxillipeds being cylindrical and long, as in *C. emarginata*, but, judging from Wilson's (1915) drawings, with claws not twisted at right angles.

Genus Alella Leigh-Sharpe, 1925

Female: Cephalothorax cylindrical, as long as, or longer than, trunk, with or without expanded head, with base laterally inflated into aliform expansions of varying shapes. Trunk subcircular or oval, dorsoventrally flattened, with or without genital process. No uropods, or posterior processes. First antenna obscurely segmented, with well developed apical armature. Second antenna with rami of about equal length but more robust exopod. Mandible without secondary teeth. First maxilla fused with base of cephalothorax and folded upon itself in sigmoid fashion. Maxilliped subchelate.

Male: Body not divisible into cephalothorax and trunk, with ventral surface at, or nearly at, right angles to long axis of body, with rounded posterior extremity, resembling males of *Clavellodes* but sometimes comparatively shorter. First antenna similar to that of female. Second antenna with endopod larger than exopod and with reduced apical armature. Buccal apparatus similar to that of female. Second maxilla and maxilliped subchelate. No uropods, posterior processes or external genitalia.

Type species: *Alella pagelli* (Krøyer, 1863).

Comments: Established relatively recently, the genus *Alella* has been surrounded with confusion from its very beginning. Leigh-Sharpe (1925) placed in it two species, without designating or indicating the type. His intention was to establish a taxon for those *Clavella*-like copepods that (i) have aliform expansions at the bases of their cephalothoraces and (ii) are not parasitic on gadoid fishes. One of these species was *Clavellodes canthari* (Heller, 1865), originally described as *Anchorella*, the other was *Clavella alata* Brian, 1906. It became clear later (Kabata, 1963d; Nunes-Ruivo, 1966b) that the latter species is a true *Clavella.* Its second antenna was unmistakably of *Clavella* type and its aliform expansions were nothing more than rather large nipples of duct orifices. In the meantime, Leigh-Sharpe (1936b) added to the genus another species, *Alella berecynthia*, a copepod without aliform expansions and, by definition, not aceeptable in *Alella.*

The matter was further complicated by Yamaguti (1963), who revised *Alella* and made *A. berecynthia* its type species. Not being one of the originally proposed species of *Alella*, *A. berecynthia* was not eligible for subsequent designation as the type species of that genus, in accordance with the International Code for Zoological Nomenclature, article 67 (a) (i) (Kabata, 1969e). Of the other four species placed by Yamaguti in *Alella*, only one (*A. canthari*) fits the generic diagnosis. *Clavella alata* belongs to a different genus, *Clavella* (Cycnus) *canthari grisei* (Hesse, 1879) is a *Hatschekia* and does not belong to this family, whereas *Clavellopsis longimana* Bere is, in fact, a *Neobrachiella.*

Nunes-Ruivo (1966b) proposed a revision of *Alella*, designating *A. canthari* as its type. Her action constituted a valid subsequent designation, but the name of the proposed type species is, unfortunately, a junior synonym of *A. pagelli*, in the combination used by that author and many others. (See comments to species description below.) She included in *Alella A. macrotrachelus* (Brian, 1906) and *A. centrodonti* (van Beneden, 1871). The latter, judging from van Beneden's illustration, is a synonym of *A. pagelli.* As to the former, no definite decision can now be reached on its validity, but it is also a possible synonym of *A. pagelli.* Nunes-Ruivo's (1966b) *Alella elongata* Richiardi and *A. richiardii* nom. nov. are nomina nuda.

To make the genus *Alella* conform with its original generic diagnosis, only one species of all those mentioned above can be retained within it. It is its type species, *Alella pagelli*. Two more species must be transferred to *Alella* from the genus *Clavellodes*. They are: *Alella ditrematis* (Yamaguti, 1939), a parasite of *Ditrema temmincki* and *Hexagrammos otakii* in Japan, and *A. pterobrachiata* (Kabata, 1968), parasitic on *Epinephelus merra* in Queensland, Australia.

British fauna contains only the type species of this genus.

Alella pagelli (Krøyer, 1863)

(Figs. 1975–1995)

Syn.: *Anchorella pagelli* Krøyer, 1863
?*Anchorella canthari* Heller, 1865
Anchorella centrodonti van Beneden, 1871
Anchorella pagelli Krøyer?; of Brian (1898)
Clavella pagelli Krøyer; of Brian (1906)
?*Clavella macrotrachelus* Brian, 1906
Lerneomyzon pagelli Stebbing, 1910
Clavella canthari (Heller, 1865); of Scott and Scott (1913)
Clavella macrotrachelus Brian, 1906; of Scott and Scott (1913)
Clavellodes canthari (Heller); of Monod (1923)
Clavellodes macrotrachelus (Brian); of Monod (1923)
Alella canthari (Heller, 1865); of Leigh-Sharpe (1925)
Clavellodes pagelli (Krøyer, 1863); of Kabata (1964a)

Female (Figs. 1975, 1976): Cephalothorax cylindrical, much longer than trunk, anteriorly slightly tapering and with narrow, long dorsal shield (Fig. 1979) with truncated anterior and converging lateral margins; posterior margin indistinct. Base of cephalothorax fused with second maxillae, laterally with subspherical swelling on each side (Fig. 1978), swellings differing somewhat in size and shape from one specimen to another (usually bilobed). Trunk short oval, dorsoventrally flattened, longer than wide, with rounded posterior corners; posterior extremity (Fig. 1977) with slight inflation around anus in centre. Genital process short, conical. No uropods or posterior processes. Trunk length 1.75 mm, cephalothorax length 2.94 mm (Kabata, 1964a).

First antenna (Fig. 1980) obscurely segmented, apparently bipartite; basal part slightly inflated, tapering distally, with short, slender whip; distal part cylindrical, with rounded tip, unarmed except for apical armature; latter (Fig. 1981) comprising recognizable digitiform seta 4 on medial side of apex and very prominent conical tubercle 1 at its base; of three remaining setae, one opposite 4 and longest probably homologous with seta 6, two others representing elements of complex 5; tubercles 2 and 3 not observed. Second antenna (Figs. 1983–1985) with long sympod and short rami, axis of appendage passing through sympod and exopod; latter one-segmented, with rounded apex, armed with prominent, scattered spinules and one whip-like lateral seta; endopod slender, two-segmented, about as long as exopod (Fig. 1984); basal segment cylindrical, unarmed; terminal segment (Fig. 1985) with apical armature consisting of hook 1, with spiniform seta 2, slender and of equal length, at its base; prominent tubercle 3 also present; ventral wall covered with scattered, sharp, prominent denticles. Labrum (Fig. 1982) with rounded tip and strips of hyaline membrane on lateral sides; rostrum well developed, its proximal half covered by slender setae of short apical fringe. Mandible (Fig. 1986) with eight teeth in poorly differentiated series; largest tooth fourth from tip, three teeth anterior to it of about equal size, four posterior to it gradually decreasing towards base of appendage; no true secondary teeth found. First maxilla (Fig. 1987) long, slender, with ventrolateral exopod; latter slender, tapering, with sturdy apical seta reaching base of ventral terminal papilla; two terminal papillae with apical setae longer than themselves; no trace of third papilla; at, and anterior to, level of exopod, dorsal margin inflated, with several prominent denticles. Second maxilla (Figs. 1975, 1976, 1978) fused with opposite number and with basal inflations of cephalothorax, short, folded upon itself in sigmoid curve (Fig. 1975), forming narrow collar round base of bulla. Latter (Fig. 1975) slender, long clavate, not divisible into manubrium and anchor (with damaged tip in Fig. 1978). Maxilliped (Fig. 1988) sturdy, fairly short; myxa with prominent seta inflated at base and with fine tip; no denticulation observed; subchela straight, cylindrical, with basal seta; distally, its inner and ventral wall beset with scattered, prominent denticles (Fig. 1988), one very large at distomedial corner at base of barb; latter slender, reaching second half of claw (Fig. 1989); claw often with uneven diameter, gently curved, with sharp tip, but without secondary teeth.

Male (Fig. 1990) intermediate in shape between *Clavella* and *Clavellodes*, resembling former in length of body and latter in angle between ventral surface and long axis of body. Not divisible into cephalothorax and trunk, with genital tubercle posterior to bases of maxillipeds. Length (in anteroposterior plane) about 0.45 mm (Kabata, 1964a).

First antenna similar to that of female (Fig. 1991), but with apical armature comprising identifiable setae 4 and 6, as well as two medial setae of complex 5; no tubercles observed. Second antenna (Fig. 1992) with indistinctly segmented sympod, comparatively shorter than in female, and with rami not clearly delimited from sympod; exopod shorter than endopod, one-segmented, bulbous, its surface distally covered by scattered minute denticles and with one fairly long lateral seta; endopod two-segmented, basal segment cylindrical, apparently unarmed (in some specimens single small denticle on ventral wall near border with second segment); latter with rounded tip, armed with three spiniform processes of equal size and uncertain homology (dorsalmost probably homologous with hook 1); ventral wall with widely scattered small denticles. Buccal apparatus similar to that of female. (This author was unable to study in detail dental formula of mandible. It appeared to have two distal primary teeth, followed by at least three secondary and an undetermined number of primaries.) First maxilla (Fig. 1993) similar to that of female but comparatively more slender, with shorter exopod. Second maxilla (Fig. 1995) with oval corpus; subchela short, curving, with sharp tip and apparently without secondary teeth; in distomedial corner of corpus subspherical tubercle near point of purchase of tip of subchela. Maxilliped (Fig. 1994) cylindrical, with distal part twisted medially from long axis and subchela moving in plane at right angles to that axis, closing into cavity with serrated, reinforced margin. No uropods or posterior processes.

Comments: The taxonomy of this species has been uncertain until recently, due to confusion between *Anchorella pagelli* Krøyer, 1863, and *A. canthari* Heller, 1865. Both species were poorly described and illustrated. The type specimen of the former was examined by the author, but that of the latter is no longer extant. The two species are, none the less, distinguishable. They differ from each other most definitely in the type of the bulla, which is long and clavate in *A. pagelli*, while it has a short manubrium and a flat anchor in *A. canthari*. A difference of this kind does not occur within a single species. It is, therefore, most likely that Krøyer's and Heller's descriptions refer to two distinct species, though one cannot completely rule out their identity with each other.

Scott and Scott (1913) described a parasite they named *Clavella canthari* (Heller), quoting Krøyer's name *pagelli* as a junior synonym of this species. Their action was without foundation. Should these species be synonymous, a fact most doubtful, Krøyer's name, antedating that of Heller by two years, should be used. Unfortunately, subsequent to Scott and Scott's monograph, many authors followed their usage, giving wide currency to the name *canthari*, where *pagelli* would have been more appropriate. Barnard (1955a) was alone in suggesting that *canthari* is, in fact, a junior synonym of *pagelli*. When Nunes-Ruivo (1966b) designated *Alella canthari* as the type species of *Alella*, it was clear that she followed Leigh-Sharpe (1925) who, in turn, used it *sensu* Scott and Scott (1913).

As a result of this confusion, the genus *Alella* was given a type species with an incorrect name. Application was made to the International Committee on Zoological Nomenclature (Kabata, 1972c) for redesignation of the type species as *Alella pagelli* (Krøyer, 1863). The application was accepted (Anonymous, 1976). This name is now used for the species at various times identified as both *pagelli* and *canthari*, with different generic names.

Of other taxa that might be confused with *A. pagelli*, note should be made of *Anchorella canthari* van Beneden, 1871, and *Anchorella canthari* Richiardi, 1880, both nomina nuda. Species originally described as *Clavella macrotrachelus* Brian, 1906, is most probably synonymous with this species and has been considered as such by Monod (1923) and Scott and Scott (1913). More work is required, however, before making a definitive decision on this problem.

Alella pagelli lives on the gills of teleost fishes belonging to several genera of the family Sparidae. Its distribution range covers the North Sea and extends south to the Atlantic coast of Europe and the Mediterranean. It was recorded by Barnard (1955a) on the gills of *Pachymetopon bleekeri* (= *Cantharus b.*) and *Pagellus lithognathus* off the Cape of Good Hope, South Africa. (If *Clavella macrotrachelus* is accepted as synonymous with this species, the records extend also to Angola (Nunes-Ruivo, 1956), where it was found on *Smaris melanurus*, and to Japan (Yamaguti, 1939c), where it parasitizes *Sparus longispinis*.) In British waters, *Alella pagelli* occurs on *Spondyliosoma cantharus* and on two species of *Pagellus*, *P. bogaraveo* and *P. erythrinus*.

As might be expected, it has been found that some differences in size and morphological details exist between specimens taken from different parts of the large distribution area. For example, specimens found in British waters on *Spondyliosoma cantharus* were found to be larger than those from the South African hosts (Kabata, 1964e). They also differed in spinulation of the second antenna and of the first maxilla. Such regional differences are known also in other Lernaeopodidae and have been referred to earlier in this work, when discussing *Salmincola thymalli*.

Genus Advena gen. nov.

Female: Cephalothorax cylindrical, longer than trunk, with tapering anterior end covered by dorsal shield. Trunk dorsoventrally flattened, with or without rounded anterolateral swellings, its posterior extremity with genital process and one pair of short or long lateral processes. No uropods. First antenna obscurely segmented, apical armature reduced in number of elements but well developed in size. Second antenna with rami of about equal length. Mandible without secondary teeth. First maxilla with ventral exopod and two terminal papillae. Second maxilla short, fused with its opposite number. Maxilliped subchelate.

Male: Not divisible into cephalothorax and trunk, subcircular or crescentic, with convex dorsal and either concave or flat ventral surfaces. Pair of conical tubercles of unknown nature posterior to bases of maxillipeds. First antenna similar to that of female. Second antenna with endopod much longer than exopod. Buccal apparatus similar to that of female. Mandible without secondary teeth. First maxilla similar to that of female. Second maxilla and maxilliped subchelate.

Type species: *Advena paradoxa* (van Beneden, 1851).

Comments: The identity of this genus is based mainly on the general morphology of its males, unlike any males within the *Clavella*-group of the family to which they clearly belong. The morphology of the female is also characterized by a combination of features which, though not individually unique, do not occur together in any other genus of Lernaeopodidae. They comprise the apical armature of the first antenna, the absence of secondary mandibular teeth and a unique shape of the first maxilla, as well as the presence of a single pair of posterior trunk processes.

In addition to the type species, one other species must be placed in *Advena*. This author proposes that *Clavellopsis saba* Yamaguti, 1939, be transferred to this genus as *Advena saba* (Yamaguti, 1939). *A. saba* is closely related to the type species, with which it shares most of the important features, particularly the morphology of the male. It differs from *A. paradoxa* in having very short posterior processes and no anterolateral trunk swellings. *A. saba* is a parasite of *Scomber colias* (= *S. japonicus*) and has been discovered, so far, only in Japanese waters. The hosts of both species are also closely related.

Only the type species occurs in British fauna.

Advena paradoxa (van Beneden, 1851)

(Figs. 1996–2016)

Syn.: *Anchorella paradoxa* van Beneden, 1851
Clavella paradoxa Beneden van; of Brian (1906)
Clavellopsis paradoxa (van Beneden), 1851; of Wilson (1915)

Female (Figs. 1996–1998): Cephalothorax cylindrical, longer than trunk, with slightly tapering head covered by dorsal shield; latter (Fig. 1999) longer than broad, with transverse anterior, converging lateral and indistinct posterior margins; prominent subspherical, anterolateral swellings at base of cephalothorax. Trunk dorsoventrally flattened, its anterolateral corners formed by subspherical swellings (Figs. 1996, 1997, a); another, similar swelling in midventral line, immediately behind bases of second maxillae (Fig. 1996, b); lateral margins convex, posterolateral corners extending into tapering processes, only slightly shorter than rest of trunk and taking up most of posterior margin; between their bases cylindrical genital process of varying length, with rounded tip. Length of trunk (including posterior processes) about 4 mm.

First antenna obscurely segmented, apparently tripartite, with basal part slightly inflated; neither whip nor solus observed by author; apical armature (Fig. 2000) comprising digitiform seta 4 with stout, blunt-tipped tubercle 1 at its base and complex 5 of one double-branched seta, divided nearly to base; one branch digitiform, as long as 4, other slender and longer (in some specimens slender branch of 5 missing); seta 6, tubercles 2 and 3 not observed. Second antenna with obscurely segmented sympod, sympod-exopod long axis and reduced endopod; exopod (Fig. 2001) longer than endopod, one-segmented, with rounded tip covered by fine denticulation; no papilliform processes observed; endopod two-segmented: basal segment cylindrical, unarmed; terminal segment with hook 1, spiniform seta 2 and patch of large denticles on ventral wall; no other processes found. Labrum with very long, slender rostrum and scanty fringe of marginal setae. Mandible (Fig. 2002) broad, its dentiferous margin with ten undifferentiated teeth more slender near base; one or two basal teeth overlapped by thin anterior margin of appendage. First maxilla (Fig. 2003) with sympod comprising about $\frac{3}{4}$ of appendage; exopod with short apical seta, ventral, cylindrical, its base close to that of ventral terminal papilla; latter posterior to that of dorsal papilla; both papillae cylindrical, long, with short apical setae; no trace of third papilla or denticulation. Second maxilla (Figs. 1996, 1997, 2004) short, thick, fused with other member of pair, subapically swollen into prominent transverse welt on posterior surface; apex (Fig. 2004) attenuated into stalked collar surrounding base of bulla. Latter with short manubrium and very small plano-convex anchor. Maxilliped (Fig. 2005) with long, slender corpus; myxa with short, sturdy seta and patch of denticles; few denticles on dorsal surface nearby; subchela short, tapering, with basal seta, most of inner margin denticulated; claw (Fig. 2006) long, slender, with sharp, curving, tapering tip.

Male (Figs. 2007, 2008): Body not divisible into cephalothorax and trunk, crescentic, with circular dorsal and concave, contracted ventral surface; vestigial dorsal shield marked by two weak longitudinal ridges at anterior end. Close behind bases of maxillipeds pair of short, cylindrical processes (Fig. 2007, ?), possibly housing genital orifices. Length about 1 mm.

First antenna (Fig. 2009) similar to that of female; apical armature comprising digitiform seta 4 with tubercle 1 at its base and double seta of complex 5, with unequal branches and tubercle 2 at base; no seta 6 or tubercle 3. Second antenna (Fig. 2010) with apparently unsegmented sympod constituting more than half of length of appendage, sclerotized, with patch of denticles on ventral wall near border with endopod; exopod half as long as endopod, one-segmented, with rounded tip, entire surface covered with slender, sharp, distally-pointing denticles and with long, sturdy seta at base; endopod two-segmented: basal segment cylindrical, with patch of denticles on ventral wall; terminal segment (Fig. 2011) long, cylindrical, with rounded tip; apical armature comprising hook 1, hard and robust, with prominent tubercle 3 at base; homology of two other processes uncertain (they might represent either seta 2 and process 4, or processes 4 and 5; former is more probable in view of presence of seta 2 in female); ventral wall partly covered by slender, sharp denticles. Buccal apparatus similar to that of female. Mandible (Fig. 2012) slender, its dentiferous margin with a series of eight undifferentiated teeth, gradually narrowing towards base (in Fig. 2012 third tooth from tip broken). First maxilla (Fig. 2013) with ventral exopod and two terminal papillae; sympod comparatively shorter than in female; exopod short, cylindrical, with single apical seta; terminal papillae long, slender, with short terminal setae; no trace of third seta or denticulation. Second maxilla (Fig. 2014) with slender, oval corpus, its basal half partly fused with opposite number; distomedial corner with triangular process, containing hollow to accommodate tip of subchela; lateral to hollow, small lobate papilla; duct orifice proximal to base of triangular process; subchela short, sharply bent in proximal half, with tapering tip; no delimited claw or secondary armature. Maxilliped (Fig. 2015) very long and slender, cylindrical, with slightly inflated base and tip; subchela (Fig. 2016) at right angles to long axis of appendage, undifferentiated, without secondary armature, closing into cavity with serrated, reinforced rim, at tip of maxilliped.

Comments: *Advena paradoxa* is a rare species. The records of its occurrence on its only known host, *Scomber scombrus*, are infrequent and do not extend to the northern part of the hosts's distribution range. The original description is based on specimens taken off the coast of Belgium. It has been found in the English Channel and as far North as the shores of Norfolk (Hamond, 1969), but so far no records have come from higher latitudes. *A. paradoxa* is also known in the Mediterranean, where it appears to be equally rare.

It is interesting to note that *A. paradoxa* has not been found on *Scomber colias* in the part of its range where it overlaps with that of *S. scombrus*. It would appear that these two closely related species of fishes have their own closely related species of *Advena*, a case, possibly, of evolutionary parallelism. Both species of mackerel, however, are infected by *Clavellisa scombri*.

Host-parasite check list

To facilitate checking, both hosts and parasites in this list have been arranged in alphabetical order. Where possible, the names of fishes are used in accordance with Wheeler (1969). The names of parasites considered as doubtful records, either by the literature source or by this author, are preceded by question marks. Unless otherwise indicated, the records refer to adult copepods.

Abramis brama
  Ergasilus sieboldi
Acipenser sturio
  Dichelesthium oblongum
  Lepeophtheirus sturionis
Agonus cataphractus
  Haemobaphes cyclopterina
  Lernaeocera branchialis (larvae)
  ?*L. lusci*
Alopias vulpinus
  Dinemoura producta
  Nemesis robusta
Alosa alosa
  Clavellisa emarginata
  Lernaeenicus encrasicoli
Alosa fallax
  Clavellisa emarginata
Ammodytes tobianus
  Lernaeocera branchialis
Anarchichas lupus
  Clavellodes rugosa
  ?*Lepeophtheirus pectoralis*
  Sphyrion lumpi
Anarchichas minor
  Clavellodes rugosa
Anguilla anguilla
  Ergasilus gibbus
Argyrosomus regium
  Brachiella thynni
  Lernanthropus gisleri
  Neobrachiella chevreuxii
  Sciaenophilus tenuis
Aspitrigla cuculus
  Caligus diaphanus
  C. elongatus
  C. gurnardi
  Lernentoma asellina
  Neobrachiella bispinosa
  N. impudica
  N. triglae
Belone bellone
  Bomolochus bellones
Brama brama
  Caligus elongatus
Brosme brosme
  Caligus curtus
  Lernaeocera lusci
Callionymus lyra
  Clavella adunca
  ?*Lepeophtheirus pectoralis*
  Lernaeocera branchialis
  L. branchialis (larvae)
  L. lusci
Callionymus maculatus
  Chondracanthus ornatus
  Haemobaphes ambiguus
Carassius auratus
  Lernaea cyprinacea
Centrophorus squamosus
  ?*Lepeophtheirus hippoglossi*
Cetorhinus maximus
  Anthosoma crassum
  Dinemoura producta
  Nemesis lamna f. *vermi*
Chimaera monstrosa
  Caligus curtus
  C. elongatus
  Vanbenedenia kroeyeri
Ciliata mustella
  Anchistrotos onosi
  Lernaeocera lusci
  Pseudocaligus brevipedis
Clupea harengus
  Caligus elongatus
  ?*Lepeophtheirus pollachius*
  Lernaeenicus sprattae
Coelorinchus coelorinchus
  Lophoura edwardsi
Conger conger
  Congericola pallidus
  ?*Lepeophtheirus pectoralis*
  Lernaeocera branchialis
Cottus gobio
  Achtheres percarum
Crenilabrus melops
  Hatschekia pygmaea
Crenimugil labrosus
  Caligus elongatus
  Ergasilus nanus
Ctenolabrus rupestris
  Hatschekia cluthae
  H. pygmaea
Cyclopterus lumpus
  Caligus elongatus
  Holobomolochus confusus
  Lernaeocera branchialis
  L. branchialis (larvae)
  Sphyrion lumpi
Cyprinus carpio
  Lernaea cyprinacea
Dasyatis pastinaca
  Eudactylina minuta
  ?*Lernaeopoda galei*
  Nemesis robusta
  Pseudocharopinus malleus
Dicentrarchus labrax
  Caligus elongatus
  C. minimus
  Lernaeocera branchialis
  Lernanthropus kroyeri
Engraulis encrasicolus
  Lernaeenicus encrasicoli
Eutrigla gurnardus
  Caligus brevicaudatus
  C. curtus
  C. diaphanus
  C. elongatus
  C. gurnardi
  Lernentoma asellina
  Neobrachiella bispinosa
  N. impudica
  N. triglae
Gadus morhua
  Caligus curtus
  C. diaphanus
  C. elongatus
  Clavella adunca
  Holobomolochus confusus
  Lepeophtheirus pectoralis
  Lernaeocera branchialis
  Sphyrion lumpi
Gaidropsaurus mediterraneus
  Lernaeocera lusci
  Pseudocaligus brevipedis
Galeorhinus galeus
  Anthosoma crassum
  Eudactylina insolens
  Kroyeria lineata
  Lernaeopoda bidiscalis
  L. galei
  Pandarus bicolor
  Trebius caudatus
Gasterosteus aculeatus
  Haemobaphes cyclopterina
  Lernaea cyprinacea
  Thersitina gasterostei
Glyptocephalus cynoglossus
  Acanthochondria cornuta
Gobius(*Macrogobius*) *paganellus*
  Pseudocaligus brevipedis
Hexanchus griseus
  Demoleus heptapus
  Nemesis robusta
Hippoglossoides platessoides
  Acanthochondria cornuta
  Haemobaphes cyclopterina
  Lernaeocera lusci
Hippoglossus hippoglossus
  Caligus curtus
  ?*C. minimus*
  Hatschekia hippoglossi
  Lepeophtheirus hippoglossi
  ?*L. pectoralis*
  Neobrachiella rostrata

Isurus oxyrinchus
  Anthosoma crassum
  Dinemoura producta
  Echthrogaleus coleoptratus
  Nogagus borealis
  Pandarus bicolor
  Phyllothyreus cornutus
Labrus bergylta
  Caligus centrodonti
  C. labracis
  Hatschekia cluthae
  H. labracis
  Holobomolochus confusus
Labrus mixtus
  Caligus labracis
  Hatschekia labracis
  ?*Lernaeocera branchialis*
Lamna nasus
  Anthosoma crassum
  Dinemoura producta
  Echthrogaleus coleoptratus
  Nemesis robusta
  Ommatokoita elongata
  Phyllothyreus cornutus
Lampris guttatus
  Clavellistes lampri
Lepidorhombus whiffiagonis
  Acanthochondria cornuta
  Caligus diaphanus
Leuciscus cephalus
  Tracheliastes polycolpus
Leuciscus idus
  Lernaea cyprinacea
  Tracheliastes polycolpus
Leuciscus leuciscus
  Tracheliastes polycolpus
Limanda limanda
  Acanthochondria limandae
  Caligus elongatus
  Lepeophtheirus pectoralis
  Lernaeocera branchialis (larvae)
Liparis liparis
  Lernaeocera lusci
Liza ramada
  Caligus elongatus
Lophius piscatorius
  Caligus curtus
  C. elongatus
  Chondracanthus lophii
Luvarus imperialis
  Luetkenia asterodermi
Malacocephalus laevis
  Chondracanthus neali
Melanogrammus aeglefinus
  Caligus curtus
  C. elongatus
  Clavella adunca
  Holobomolochus confusus
  ?*Lepeophtheirus pectoralis*
  Lernaeocera branchialis
  ?*L. lusci*
Merlangius merlangus
  Caligus curtus
  C. elongatus
  Clavella adunca
  Haemobaphes cyclopterina
  Holobomolochus confusus
  Lernaeocera branchialis
  ?*L. lusci*
Merluccius merluccius
  Caligus curtus
  C. diaphanus
  Chondracanthus merluccii
  Clavella stellata
  Lernaeocera branchialis
  Neobrachiella insidiosa
  N. merluccii
Microstomus kitt
  Acanthochondria clavata
  Caligus curtus
  C. elongatus
  Lernaeocera branchialis (larvae)
Mola mola
  Anthosoma crassum
  Caligus elongatus
  Cecrops latreillii
  ?*Echthrogaleus coleoptratus*
  Lepeophtheirus nordmanni
  Orthagoriscicola muricatus
  ?*Pandarus bicolor*
  Pennella filosa
  Philorthagoriscus serratus
Molva dypterygia
  Caligus curtus
Molva molva
  Caligus curtus
  C. diaphanus
  C. elongatus
  Holobomolochus confusus
  Lepeophtheirus pollachius
  Lernaeocera branchialis
  L. lusci
Mullus barbatus
  Hatschekia mulli
  ?*Lernaeocera lusci*
Mullus surmulletus
  Hatschekia mulli
Mustelus mustelus
  Kroyeria lineata
  Lernaeopoda bidiscalis
  L. galei
  Nemesis robusta
  Ommatokoita elongata
  Pandarus bicolor
Mustelus sp.
  Tripaphylus musteli
Myliobatis aquila
  ?*Lernaeopoda galei*
  ?*Pseudocharopinus bicaudatus*
  ?*P. malleus*
Myoxocephalus scorpius
  Haemobaphes cyclopterina
  Lernaeocera branchialis (larvae)
  ?*L. lusci*
Pagellus bogaraveo
  Alella pagelli
  Caligus centrodonti
  C. diaphanus
  C. minimus
  Hatschekia pagellibogneravei
Pagellus erythrinus
  Alella pagelli
  Caligus diaphanus
Perca fluviatilis
  Achtheres percarum
Pholis gunnelus
  Haemobaphes cyclopterina
  Lernaeocera branchialis
Phycis blennoides
  Caligus curtus
  Clavella alata
  Lernaeocera branchialis
  L. lusci
Platichthys flesus
  Acanthochondria cornuta
  Caligus elongatus
  Lepeophtheirus pectoralis
  Lernaeocera branchialis (larvae)
Pleuronectes platessa
  Acanthochondria cornuta
  Caligus curtus
  C. elongatus
  Lepeophtheirus pectoralis
  ?*Lernaeocera branchialis*
  L. branchialis (larvae)
Pollachius pollachius
  Caligus curtus
  C. elongatus
  Clavella adunca
  Holobomolochus confusus
  Lepeophtheirus pollachius
  Lernaeocera branchialis
Pollachius virens
  Caligus curtus
  C. diaphanus
  C. elongatus
  Clavella adunca
  Holobomolochus confusus
  Lepeophtheirus pollachius
  Lernaeocera branchialis
Pomatoschistus minutus
  Caligus elongatus
  Lernaeocera minuta
Prionace glauca
  Anthosoma crassum
  Dinemoura producta
  Echthrogaleus coleoptratus
  Kroyeria lineata
  Nemesis robusta
  Pandarus bicolor
  Phyllothyreus cornutus
Pungitius pungitius
  Thersitina gasterostei
Raja alba
  Trebius caudatus
Raja batis
  Acanthochondrites annulatus
  Caligus curtus
  C. elongatus

Charopinus dalmanni
?*Ch. dubius*
Lernaeopodina longimana
?*Schistobrachia ramosa*
Trebius caudatus
Raja circularis
?*Charopinus dubius*
Raja clavata
Caligus elongatus
Lepeophtheirus pectoralis
Schistobrachia ramosa
Trebius caudatus
Raja fullonica
Caligus curtus
?*Charopinus dubius*
Eudactylina similis
Lernaeopodina longimana
Schistobrachia ramosa
Taeniacanthus wilsoni
Trebius caudatus
Raja microocellata
Trebius caudatus
Raja montagui
Caligus curtus
C. elongatus
?*Charopinus dubius*
Eudactylina similis
Lernaeopodina longimana
Trebius caudatus
Raja naevus
Caligus curtus
C. elongatus
Charopinus dubius
Eudactylina similis
Lernaeopodina longimana
Taeniacanthus wilsoni
Raja oxyrinchus
Caligus curtus
?*Charopinus dubius*
Raja radiata
Acanthochondrites annulatus
Caligus curtus
C. elongatus
Eudactylina similis
?*Lepeophtheirus pectoralis*
Lernaeopodina longimana
Schistobrachia ramosa
Rhinonemus cimbrius
Anchistrotos onosi
Rutilus rutilus
Lernaea cyprinacea
Salmo salar
Caligus elongatus
Lepeophtheirus salmonis
Salmincola salmoneus
Salmo trutta
Caligus elongatus
Ergasilus sieboldi
Lepeophtheirus salmonis
Salmincola salmoneus
Salvelinus alpinus
Salmincola salmoneus
Sarda sarda
Caligus bonito
Pseudocycnus appendiculatus
Sardina pilchardus
Lernaeenicus encrasicoli
L. sprattae
Scomber scombrus
Advena paradoxa
Caligus diaphanus
C. elongatus
C. pelamydis
Clavellisa scombri
?*Lepeophtheirus pectoralis*
Scophthalmus maximus
Caligus curtus
C. diaphanus
C. elongatus
Lepeophtheirus hippoglossi
?*L. pectoralis*
L. thompsoni
Lernaeocera branchialis (larvae)
Scophthalmus rhombus
Acanthochondria solae
Caligus curtus
?*Lepeophtheirus hippoglossi*
L. thompsoni
Scyliorhinus caniculus
Albionella globosa
Lernaeopoda galei
Scyliorhinus stellaris
Pandarus bicolor
Scyliorhinus sp.
Lernaeopoda bidiscalis
Sebastes marinus
Caligus centrodonti
C. curtus
Chondracanthus nodosus
?*Haemobaphes cyclopterina*
Sphyrion lumpi
Sebastes viviparus
Sphyrion lumpi
Serranus cabrilla
Anchistrotos laqueus
Solea solea
Acanthochondria soleae
Bomolochus soleae
Caligus diaphanus
Lepeophtheirus thompsoni
Lernaeocera branchialis
L. branchialis (larvae)
L. lusci
Somniosus microcephalus
?*Dinemoura producta*
?*Lepeophtheirus hippoglossi*
Ommatokoita elongata
Sphyrna zygaena
Kroyeria lineata
Nemesis robusta
Phyllothyreus cornutus
Spondyliosoma cantharus
Alella pagelli
Sprattus sprattus
Lernaeenicus encrasicoli
L. sprattae
Squalus acanthias
Caligus curtus
C. elongatus
Demoleus heptapus
Dinemoura producta
Echthrogaleus coleoptratus
Eudactylina acanthii
?*E. acuta*
Nogagus ambiguus
Pandarus bicolor
Pseudocharopinus bicaudatus
Trebius caudatus
Squatina squatina
Eudactylina acuta
Taurulus bubalis
Haemobaphes cyclopterina
Thunnus alalunga
Brachiella thynni
Pennella filosa
Thunnus thynnus
Branchiella thynni
Caligus coryphaenae
Euryphorus brachypterus
Pennella filosa
Pseudocycnus appendiculatus
Thymallus thymallus
Salmincola thymalli
Tinca tinca
Ergasilus sieboldi
Torpedo marmorata
Pseudocharopinus malleus
Torpedo nobiliana
Eudactylina similis
Trachinus vipera
Lernaeocera lusci
Trachurus trachurus
Caligus diaphanus
C. elongatus.
C. pelamydis
Trigla lucerna
Caligus brevicaudatus
C. diaphanus
C. elongatus
C. gurnardi
Lernentoma asellina
Neobrachiella bispinosa
N. impudica
N. triglae
Trigla lyra
Caligus diaphanus
Lernentoma asellina
Neobrachiella bispinosa
N. impudica
Trigla sp.
Neobrachiella triglae
Trigloporus lastoviza
Caligus diaphanus
Lernentoma asellina
Neobrachiella bispinosa
N. impudica
N. triglae

Trisopterus esmarkii
Lernaeocera lusci
Trisopterus luscus
Caligus elongatus
Clavella adunca
Holobomolochus confusus
Lernaeocera lusci
Trisopterus minutus
Caligus curtus
C. elongatus
?*Lernaeocera branchialis*
?*L. lusci*
Trygon sp.
?*Pseudocharopinus bicaudatus*
Umbrina cirrosa
Caligus minimus
Lernanthropus gisleri
Sciaenophilus tenuis
Xiphias gladius
Philichthys xiphiae
Zeugopterus punctatus
Anchistrotos zeugopteri
Lernaeocera sp. (larvae)

Zeus faber
Caligus elongatus
C. zei
Chondracanthus zei

Zoarces viviparus
Lernaeocera branchialis (larvae)

Bibliography

Abildgaard, P. C. 1794. Beskrivelse af tvende nye insekter henherende under den Linneiske slaegt *Monoculus*, og den Müllerske slaegt *Caligus*. Skr. naturh. Selsk. Kjøbenhavn, 3: 46–54.

Agapova, A. I. 1966. Parazity ryb vodoemov Kazakhstana. [Parasites of fishes of Kazakhstan.] Izd. Nauka Kaz. SSR: Alma Ata, 342 pp.

Akhmerov, A. Kh. 1959. Parazitofauna i zabolevaniya ryb pri akklimatizatsii. Parazity i bolezni amurskikh ryb v period akklimatizatsii v prudovykh khozyaystvakh RSFSR. [Parasite fauna and diseases of fishes during acclimation. Parasites and diseases of the Amur fishes during their acclimation in the pond farms of RSFSR.] Trudy Soveshch. ikhtiol. Kom. (Trudy Sov. Bol. Ryb), 9: 104–109.

Anderson, D. T. and G. T. Rossiter. 1969. Hatching and larval development of *Dissonus nudiventris* Kabata (Copepoda, fam. Dissonidae), a gill parasite of the Port Jackson shark. Proc. Linn. Soc. N.S.W., 93: 476–481.

Anonymous. 1878. Adunanza del di 5 maggio 1878. Processi verbali. Proc. verb. Soc. tosc. Sci. nat., 1878: xvii–xxiv.

Anonymous. 1957. International Commission on Zoological Nomenclature. Opinion 481. Emendation under the plenary powers to *Lernaeocera* of the generic name *Lerneocera* and designation under the same power of a type species in harmony with accustomed usage for the genus so named (Class Crustacea: Order Copepoda). Bull. zool. Nomencl., 17: 1–4.

Anonymous. 1965. International Commission on Zoological Nomenclature. Opinion 732. *Bomolochus* von Nordmann, 1832 (Crustacea, Copepoda): designation of a type-species under the plenary powers. Bull. zool. Nomencl., 22: 88–89.

Anonymous. 1976. International Commission on Zoological Nomenclature. Opinion 1047. *Alella* Leigh-Sharpe, 1925 (Crustacea: Copepoda): designation of type-species under the plenary powers. Bull. zool. Nomencl., 32 : 230–232.

Aubrook, E. W. and G. Fryer. 1965. The parasitic copepod *Tracheliastes polycolpus* Nordmann in some Yorkshire rivers: the first British record. Naturalist, Hull, No. 893: 51–56.

Bainbridge, M. E. 1909. Notes on some parasitic Copepoda with a description of a new species of *Chondracanthus* (*C. inflatus*). Trans. Linn. Soc. Lond., Zool., ser. 2, 11: 45–60.

Baird, W. 1850. The natural history of the British Entomostraca. Ray Society: London, viii + 364 pp.

Baird, W. 1861. Note on *Lernaea cyclopterina* occurring in the gills of the *Cyclopterus spinosus*, a fish from Greenland. Ann. Mag. nat. Hist., ser. 3, 8: 496.

Barnard, K. H. 1948. New records and descriptions of new species of parasitic Copepoda from South Africa. Ann. Mag. nat. Hist., ser. 12, 1: 242–254.

Barnard, K. H. 1955a. South African parasitic Copepoda. Ann. S. Afr. Mus., 41: 223–312.

Barnard, K. H. 1955b. Additions to the fauna-list of South African Crustacea. Ann. Mag. nat. Hist., ser. 12, 10: 1–12.

Bassett-Smith, P. W. 1896a. Notes on the parasitic Copepoda of fish obtained at Plymouth, with descriptions of new species. Ann. Mag. nat. Hist., ser. 6, 18: 8–16.

Bassett-Smith, P. W. 1896b. A list of the parasitic Copepoda of fish obtained at Plymouth. J. mar. biol. Ass. U.K., 4: 155–163.

Bassett-Smith, P. W. 1898. Some new parasitic copepods found on fish at Bombay. Ann. Mag. nat. Hist., ser. 7, 1: 1–17.

Bassett-Smith, P. W. 1899. A systematic description of parasitic Copepoda found on fishes, with an enumeration of the known species. Proc. zool. Soc. Lond., 1899 (2): 438–507.

Baudouin, M. 1904. Le *Lernaeenicus sprattae*, parasite de la sardine en Vendée. C. r. hebd. Séanc. Acad. Sci., Paris, 139: 998–1000.

Badouin, M. 1910. Découverte d'un type de transition entre *Lernaeenicus sardinae* M.B. et *Lernaeenicus sprattae* Sowerby, sur le meme sardine (*Clupea pilchardus* Wal.): *L. sardinae*, variété *moniliformis*. C. r. Ass. fr. Avanc. Sci., 2: 163–167.

Baudouin, M. 1918. Découverte d'une variété de *Lernaeenicus sardinae* M.B., intermédiaire entre le type et la variété *moniliformis*. Bull. Mus. natn. Hist. nat., Paris, 24: 394–396.

Bazikalova, A. Ya. 1932. Materialy po parazitologii murmanskikh ryb. [Data on the parasitology of Murmansk fishes.] pp. 136–153. In: S. Ya. Mittelman (ed.). Sb. nauchno-prom. Rab. Murman. Snabtekhizdat: Moscow and Leningrad.

Belloc, G. 1929. Etude monographique du merlu *Merluccius merluccius* L. Revue Trav. Off. Pêch. marit., 2: 153–199, 238–288.

*Beneden, G. -J. van, 1851a. Recherches sur quelques crustacés inférieurs. Annls Sci. nat., ser. 3, Zool., 16: 71–131.

*Beneden, G. -J. van, 1851b. Note sur un crustacé parasite nouveau, avec l'énumération des espèces de cette classe qu'on observe sur les poissons du littoral de Belgique. Bull. Acad. r. Belg., Cl. Sci., 18: 286–290.

Beneden, P. -J. van, 1853a. Notice sur un nouveau genre de la tribe des Caligiens (genre *Kroyeria* van Ben.)

*Lapsus for P. -J.

Bull. Acad. r. Belg., Cl. Sci., 20: 23–30.
Beneden, P. -J. van, 1853b. Note sur un nouveau genre de crustacé parasite, *Eudactylina*. Bull. Acad. r. Belg., Cl. Sci., 20: 235–239.
Beneden, P. -J. van, 1854. Notice sur un nouveau genre de siphonostome (genre *Congericola*) habitant les branchies du congre. Bull. Acad. r. Belg., Cl. Sci., 21: 583–589.
Beneden, P. -J. van, 1857. Notice sur un nouveau *Dinemoura* provenant de *Scimnus glacialis*. Bull. Acad. r. Belg., Cl. Sci., ser. 1, 24: 226–235.
Beneden, P. -J. van, 1871. Les poissons des côtes de Belgique, leurs parasites et leurs commensaux. Mém. Acad. r. Sci. Lett. Belg., 38: i–xx, 1–100.
Beneden, P. -J. van, 1891. Deux lernéopodiens nouveaux recueillis, l'un aux Açores, l'autre su les côtes du Senegal. Bull. Acad. r. Belg., Cl. Sci., ser. 3, 22: 23–35.
Bere, R. 1930a. Parasitic copepods from the Vancouver Island region. Fish. Res. Bd Can. MS Rep. Ser., Biol. St. No. 259. 3 pp.
Bere, R. 1930b. The parasitic copepods of the fish of the Passamaquoddy region. Contr. Can. Biol. Fish., new ser., 5: 423–430.
Bere, R. 1936. Parasitic copepods from Gulf of Mexico fish. Am. Midl. Nat., 17: 577–625.
Bergsoe, V. 1864. *Philichthys xiphiae* Stp., monographisk fremstillet. Naturh. Tiddskr., ser. 3, 3: 87–103.
Berland, B. 1961. Copepod *Ommatokoita elongata* (Grant) in the eyes of the Greenland shark - a possible case of mutual dependance. Nature, Lond., 191: 829–830.
Berland, B. 1969. En parasitisk copepod, *Sphyrion lumpi* (Krøyer, 1845), funnet pa torsk og blakveite. Fauna, Oslo, 22: 147–152.
Birkett, L. and A. C. Burd, 1952. A new host for the copepod *Anthosoma crassum* (Abildgaard), 1794. Ann. Mag. nat. Hist., ser. 12, 5: 391–392.
Blainville, M. H. D. de, 1822. Mémoire sur les lernées (*Lernaea*, Linn.). J. Phys., 95: 372–380, 437–447.
Bocquet, C. and J. H. Stock, 1963. Some recent trends in work on parasitic copepods. Oceanogr. mar. Biol. ann. Rev., 1: 289–300.
Bowman, T. E. 1971. The case of the nonubiquitous telson and the fraudulent furca, Crustaceana, 21: 165–175.
Boxshall G. A. 1974a. Infections with parasitic copepods in North Sea marine fishes. J. mar. biol. Ass. U.K., 54: 355–372.
Boxshall, G. A. 1974b. The validity of *Acanthochondrites inflatus* (Bainbridge, 1909), a parasitic copepod occurring on *Raja radiata* Donovan, 1806 in the North Sea. J. nat. Hist., 8: 11–17.
Brady, G. S. 1883. Report on the Copepoda collected by H.M.S. Challenger. Rep. Challenger Soc., 8 (23): 1–142.
Brady, G. S. 1910. Die marine Copepoden der deutschen Südpolar-Expedition 1901–1903. I. Uber die Copepoden der Stämmer Harpactoidea, Cyclopoidea, Notodelphyoidea und Caligoidea. Dt. Südpol.-Exped., 11 (5): 497–593.
Brandes, C. H. 1955. Über eine neue Art der parasitischen Copepoden: *Pseudocycnus thynnus* n. sp. Veröff. Inst. Meeresforsch. Bremerh., 3: 190–198.
Bresciani, J. and J. Lützen. 1962. Parasitic copepods from the west coast of Sweden, including some new or little known species. Vidensk. Meddr dansk naturh. Foren., 124: 367–408.
Brian, A. 1898. Catalogo di copepodi parassiti dei pesci della Liguria. Boll. Musei Lab. Zool. Anat. comp. R. Univ. Genova, No. 61: 1–27.
Brian, A. 1899a. Di alcuni crostacei parassiti dei pesci dell'Isola d'Elba. Atti Soc. ligust. Sci. nat. geogr., 10: 3–10.
Brian, A. 1899b. Crostacei parassiti dei pesci dell'Isola d'Elba (II contribuzione). Boll. Musei Lab. Zool. Anat. comp. R. Univ. Genova, No. 85: 1–11.
Brian, A. 1902. Note su alcuni crostacei parassiti dei pesci del Mediterraneo. Atti Soc. ligust. Sci. nat. geogr., 13: 30–45.
Brian, A. 1905. Sui copepodi raccolti nel Golfo di Napoli da Oronzio G. ed Achille Costa. Annuar. R. Mus. zool. R. Univ. Napoli, new ser., 1 (24): 1–11.
Brian, A. 1906. Copepodi parassiti dei pesci d'Italia. Genova. 190 p.
Brian, A. 1908. Note preliminaire sur les copépodes parasites des poissons provenant des campagnes scientifiques de S.A.S. le prince Albert Ier de Monaco ou déposés dans les collections du Musée Océanographique. Bull. Inst. océanogr. Monaco, No. 110: 1–19.
Brian, A. 1912. Copépodes parasites des poissons et des echinides provenant des campagnes scientifiques de S.A.S. le prince Albert Ier de Monaco. Résult. Camp. scient. Prince Albert I, No. 38: 1–58.
Brian, A. 1924. Parasitologia Mauritaniça. Matérieux pour la faune parasitologique en Mauritanie. Arthropoda (1re partie). Copepoda. Bull. Com. Étud. hist. scient. Afr. occid. fr., Année 1924: 365–427.
Brian, A. 1935. I *Caligus* parassiti dei pesci del Mediterraneo (Copepodi). Annali Mus. civ. Stor. nat. Giacomo Doria, 57: 152–211.
Brian, A. 1939. Copépodes parasites recueillis par M. E. Dartevelle à l'embouchure du fleuve Congo. Rev. Zool. Bot. afr., 32: 176–198.
Brian, A. 1944. Copepodes parasitas de peces y cetaceos del Museo Argentino de Ciencias Naturales. An. Mus.

argent. Cienc. nat., 41: 193–220.
Brian, A. 1946. Sulla inesistenza del gen. *Laminifera* "Franz Poche" (fide Ch. Br. Wilson, 1907) e sulla sinonimia della specie *Laminifera doello-juradoi* Brian (1944) colla species "*Phyllothyreus cornutus*" (M.-Edw., 1840). Monitore zool. ital., 55: 142–143.
Brunel, P. 1970. Catalogue d'invertébrés benthiques du golfe Saint-Laurent recueillis de 1951 à 1966 par la Station de Biologie Marine de Grande-Rivière. Trav. Pêcher. Québec, No. 32: 1–55.
Burmeister, H. 1835. Beschreibung einiger neuen oder weniger bekannten Schmarotzerkrebse, nebst allgemeinen Betrachtungen über die Gruppe, welcher sie angehören. Nova Acta Acad. Caesar. Leop. Carol., 17: 269–336.
Calman, W. T. 1926. The Rhynie crustacean. Nature, Lond., 118: 89.
Candiotti, Ch. 1910. Crustacé parasite de la morue. In: Mission arctique commandée par M. Charles Bénard. Soc. Océanogr. Golfe Gascogne, Fasc. V: 1–3.
Canning, E. U., F. E. G. Cox, N. A. Croll and K. M. Lyons. 1973. The natural history of Slapton Ley nature reserve: VI. Studies on the parasites. Field Studies, 3: 681–718.
Capart, A. 1941. Résultats scientifiques des croisières du navaire-école Belge "Mercator". Vol. III. 5. Copepoda parasitica. Mém. Mus. r. Hist. nat. Belg., sér. 2, 21: 171–197.
Capart, A. 1944. Notes sur les copépodes parasites. II. A propos de quatre copépodes parasites des poissons d'eau douce rares ou nouveaux pour la Belgique. Bull. Mus. r. Hist. nat. Belg., 20(22): 1–4.
Capart, A. 1946. Notes sur les copépodes parasites. IV. A propos de *Charopinus pastinacae* (P. J. van Beneden, 1851) parasite de *Dasybatis pastinaca* (L.). Bull. Mus. r. Hist. nat. Belg., 22(10): 1–6.
Capart, A. 1953. Quelques copépodes parasites de poissons marins de la region de Dakar. Bull. Inst. fr. Afr. noire, 15: 647–671.
Capart, A. 1959. Copépodes parasites. Result. scient. Expéd. océanogr. belg. Eaux côt. afr. Atlant. sud (1948–1949), 3(3): 55–126.
Carus, J. V. 1885. Prodromus faunae mediterraneae. Vol. 1. Coelenterata, Echinodermata, Vermes, Arthropoda. E. Schweizerbart: Stuttgart. XI + 524 pp.
Carvalho, J. de Paiva. 1951. Notas sôbre alguns copépodes parasitos de peixes marítimos da costa do Estado de Sao Paulo. Bolm. Inst. Oceanogr., S Paulo, 2: 135–144.
Carvalho, J. de Paiva. 1957. Sôbre estadios larvares de *Lerneenicus longiventris* (Crustacea-Copepoda). Bolm Inst. Oceanogr., S Paulo, 8: 241–253.
Causey, D. 1953a. Parasitic Copepoda from Grand Isle, La. Occ. Pap. mar. Lab. La St. Univ., No. 7: 1–18.
Causey, D. 1953b. Parasitic Copepoda of Texas coastal fishes. Publs Inst. mar. Sci. Univ. Texas, 3: 6–16.
Causey, D. 1955. Parasitic Copepoda from Gulf of Mexico fishes. Occ. Pap. mar. Lab. La St. Univ., No. 9: 1–19.
Causey, D. 1960. Parasitic Copepoda from Mexican coastal fishes. Bull. mar. Sci. Gulf Caribb., 10: 323–337.
Chernova, T. and G. Sukhenko. 1969. Paraziticheskie rakoobraznye ryb ozera Paleostomi. [Parasitic copepods of fishes of Lake Paleostomi.] Problemy Parazitologii (Trudy VI nauch. Konf. Parazit. Ukr. SSR), 6, pt. 2: 276–277.
Claus, C. 1860. Zur Morphologie der Copepoden. I. Eine Hemmungsbildung von *Cyclops*. II. Über den Bau von *Nicothoë*. III. Über die Leibesgliederung und die Mundwerkzeuge der Schmarotzerkrebse. Würtzb. naturw. Z., 1: 20–36.
Claus, C. 1862. Untersuchungen über die Organisation und Verwandschaft der Copepoden. Würtzb. naturw. Z., 3: 51–103.
Claus, C. 1863. Die frei lebenden Copepoden mit besonderer Berücksichtigung der Fauna Deutchlands, der Nordsee und des Mittelmeeres. W. Engelmann: Leipzig, 230 pp., XXXVII pl.
Claus, C. 1864. Beiträge zur Kenntniss der Schmarotzerkrebse. Z. wiss. Zool., 14: 365–382.
Claus, C. 1868. Über die Metamorphose und systematische Stellung der Lernaeen. Sitzungsb. Gesellsch. Beförd. ges. naturw. Marburg, 2: 5–13.
Claus, C. 1875. Neue Beiträge zur Kenntniss parasitischer Copepoden, nebst Bemerkungen über das System derselben. Z. wiss. Zool., 25: 327–361.
Claus, C. 1886. Über *Lerneascus nematoxys*, eine seither unbekannt gebliebene Lernaee. Anz. Akad. Wiss. Wien, No. 25: 231–233.
Claus, C. 1887. Über *Lerneascus nematoxys* Claus und die Familie der Philichthyiden. Arb. zool. Inst. Univ. Wien, 7: 281–315.
Claus, C. 1895. Über die Maxillarfüse der Copepoden. Arb. zool. Inst. Univ. Wien, 11: 49–63.
Cope, E. D. 1872. On the Wyandotte cave and its fauna. Am. Nat., 6: 406–422.
Cornalia, E. 1865. Sulla *Lophoura edwardsii* di Kölliker osservazioni zoologiche e anatomiche. Atti Soc. ital. Sci. nat., 9: 259–268.
Cressey, R. 1967a. Revision of the family Pandaridae (Copepoda: Caligoida). Proc. U.S. natn. Mus., 121 (No. 3570): 1–133.
Cressey, R. 1967b. Genus *Gloiopotes* and a new species with notes on host specificity and intraspecific variation (Copepoda: Caligoida). Proc. U.S. natn. Mus., 122 (No. 3600): 1–22.
Cressey, R. 1967c. Caligoid copepods parasitic on sharks of the Indian Ocean. Proc. U.S. natn. Mus., 121 (No.

3572): 1–21.
Cressey, R. 1968. Caligoid copepods parasitic on *Isurus oxyrinchus*, with an example of habitat shift. Proc. U.S. natn. Mus., 125 (No. 3653): 1–26.
Cressey, R. 1972. Revision of the genus *Alebion* (Copepoda: Caligoida). Smithson. Contr. Zool., No. 123: 1–29.
Cressey, R. 1975. A new family of parasitic copepods (Cyclopoida, Shiinoidae). Crustaceana, 28: 211–219.
Cressey, R. and B. B. Collette. 1970. Copepods and needlefishes: a study in host–parasite relationships. Fishery Bull. Fish Wildl. Serv. U.S., 68: 341–432.
Cressey, R. and E. A. Lachner. 1970. The parasitic copepod diet and life history of diskfishes (Echeneidae). Copeia, 1970 (2): 310–318.
Cressey, R. and C. Patterson. 1973. Fossil parasitic copepods from a Lower Cretaceous fish. Science, 180 (No. 4092): 1283–1285.
Cuvier, G. 1830. La règne animal distribué d'après son organisation, pour servir de base à l'histoire naturelle des animaux et d'introduction à l'anatomie comparée. Nouvelle Edition, vol. 3: XVI + 504 pp. Déterville: Paris.
Dana, J. D. 1852. Conspectus crustaceorum que in orbis terrarum circumnavigatione Carolo Wilkes e classae reipublicae foederate duce. Pars II. Proc. Am. Acad. Arts Sci., 2: 9–61.
Dana, J. D. 1853. United States Exploring Expedition during the years 1838–42 under command of Charles Wilkes, U.S.N., Crustacea, 13: 686–1618.
Dana, J. D. 1855. Atlas. Crustacea. United States Exploring Expedition during the years 1838–42 under the command of Charles Wilkes, U.S.N. [Plates for Dana, 1853].
Delamare Deboutteville, C. 1948. Sur quelques copépodes parasites du squale pélerin [*Cetorhinus maximus* (Gunner)]. Bull. mus. natl. Hist. nat., Paris, sér. 2, 20: 446–447.
Delamare Deboutteville, C. 1950. Copépodes parasites des poissons de Banyuls (1re série). VieMilieu, 1: 305–309.
Delamare Deboutteville, C. 1962. Prodrome d'une faune d'Europe des copépodes parasites des poissons. Les copépodes Phylichthyidae (Confrontation des données actuelles). Bull. Inst. Océanogr. Monaco, No. 1249: 1–44.
Delamare Deboutteville, C. and L. Nunes-Ruivo. 1951a. Existence de "formes biologiques" chez *Peniculus fistula* (Rudolphi) (Copepoda). VieMilieu, 2: 448–458.
Delamare Deboutteville, C. and L. Nunes-Ruivo, 1951b. Étude de *Pennella remorae* Murray et remarques sur la biologie et la systematique des *Pennella* Oken. Revta port. Fac. Cienc. Lisboa, ser. 2, C, 1: 341–352.
Delamare Deboutteville, C. and L. Nunes-Ruivo. 1952. Copépodes parasites des poissons de Banyuls (2me série). VieMilieu, 3: 292–300.
Delamare Deboutteville, C. and L. Nunes-Ruivo. 1953. Copépodes parasites des poissons Méditerranéens (3me série). VieMilieu, 4: 201–218.
Delamare Deboutteville, C. and L. Nunes-Ruivo. 1955. Remarques sur le développement de la femelle d'*Haemobaphoides* (T. Scott) et analyse critique des genres *Haemobaphes* Steenstrup et Lütken, *Haemobaphoides* T. et A. Scott et *Collipravus* Wilson (Crustacea Copepoda). Bull. Soc. zool. Fr., 80: 27–37.
Delamare Deboutteville, C. and L. Nunes-Ruivo. 1958. Copépodes parasites des poissons Méditerranéens (4me série). VieMilieu, 9: 215–235.
Desbrosses, P. 1945. Le merlan (*Gadus merlangus* L.) de la côte française de l'Atlantique. Rev. Trav. Off. Pêches marit., 13: 177–195.
Desmarest, A. G. 1825. Considerations générales sur la classe des crustacés, et description des espèces de ces animaux, qui vivent dans la mer, sur les côtes, et dans les eaux douces de la France. F. G. Levrault: Paris, 446 pp.
Dienske, H. 1968. A survey of the metazoan parasites of the rabbit fish, *Chimaera monstrosa* L. (Holocephali). Neth. J. Sea Res., 4: 32–58.
Dollfus, R. Ph. 1946. Essai de catalogue des parasites poisson-lune *Mola mola* (L. 1758) et autres Molidae. Annls Soc. Sci. nat. Charente-Marit., new ser., 3: 70–76.
Dollfus, R. Ph. 1953. Aperçu général sur l'histoire naturelle des parasites animaux de la morue Atlanto–Arctique *Gadus callarias* L. (= *morrhua* L.). Encycl. biol., 43: 1–426.
Dollfus, R. Ph. 1956. Liste des parasites animaux du hareng le l'Atlantique Nord et de la Baltique. J. Cons. perm. int. Explor. Mer, 22: 58–65.
Dollfus, R. Ph. 1960. Mission Maurice Blanc, Francois d'Auberton (1954). VII. Copépodes parasites de teleostéens du Niger. Bull. Inst. fr. Afr. noire, 22A (1): 170–192.
Dudley, P. L. 1966. Development and systematics of some Pacific marine symbiotic copepods. A study of the biology of the Notodelphyidae, associates of ascidians. Univ. Washington Press: Seattle and London. 282 pp.
Edwards, H. M. 1840. Histoire naturelle des crustacés, comprenant l'anatomie, la physiologie et la classificacion de ces animaux. Vol. 3. Roret: Paris. 605 pp.
Evdokimova, E. B. 1973. Parazitofauna promyslovykh kostistykh ryb patagonskogo shelfa. [Parasite fauna of commercial teleosts of the Patagonian shelf.] (Thesis abstract.) Sci. Council, Zool. Inst. Acad. Sci. U.S.S.R.: Kaliningrad. 18 pp.
Fabricius, O. 1780. Fauna Groenlandica. Hafniae et Lipsiae. XVI + 452 pp.

Fage, L. 1923. Sur deux copépodes [*Dinemoura producta* (Müller) et *Nemesis lamna* Riss.] parasites du pélerin (*Cetorhinus maximus* Gunner). Bull. Soc. zool. Fr., 48: 280–287.

Faure-Fremiet, E. and Y. Guilcher. 1948. *Trochiloides filans* n. sp. infusoire holotriche de la famille de Dysteridae. Bull. Soc. zool. Fr., 72: 106–112.

Ferris. G. F. and L. M. Henry. 1949. The nervous system and the problem of homology in certain Crustacea (Crustacea: Copepoda: Caligidae). Microentomology, 14: 114–120.

Fox. M. H. 1945. Haemoglobin in blood-sucking parasites. Nature, Lond., 156: 475–476.

Fraser, C. M. 1920. Copepods parasitic on fish from the Vancouver Island region. Trans. R. Soc. Can., sect. 5, ser. 3, 13: 45–67.

Friend, G. F. 1941. The life-history and ecology of the salmon gill-maggot, *Salmincola salmonea* (L.) (Copepod crustacean). Trans. R. Soc. Edinb., 60: 503–541.

Fröiland, O 1974. The gill parasite *Haemobaphes cyclopterina* (Copepoda: Lernaeoceridae) in the Barents Sea. Sarsia, 56: 123–130.

Fryer, G. 1956. A report on the parasitic Copepoda and Branchiura of the fishes of Lake Nyasa. Proc. zool. Soc. Lond., 127: 293–344.

Fryer, G. 1957. The feeding mechanism of some freshwater cyclopoid copepods. Proc. zool. Soc. Lond., 129: 1–25.

Fryer, G. 1958. A note on *Lernaea bistricornis* Harding, a parasitic copepod from Lake Tanganyika. Revue Zool. Bot. afr., 58: 3–4.

Fryer, G. 1959. A report on the parasitic Copepoda and Branchiura of the fishes of Lake Bangweulu (N. Rhodesia). Proc. zool. Soc. Lond., 132: 517–550.

Fryer, G. 1961a. Variation and systematic problems in a group of lernaeid copepods. Crustaceana, 2: 275–285.

Fryer, G. 1961b. The parasitic Copepoda and Branchiura of the fishes of Lake Victoria and the Victoria Nile. Proc. zool. Soc. Lond., 137: 41–60.

Fryer, G. 1964. Further studies on the parasitic Crustacea of African freshwater fishes. Proc. zool. Soc. Lond., 143: 79–102.

Fryer, G. 1965a. Crustacean parasites of African freshwater fishes, mostly collected during the expeditions to Lake Tanganyika and to Lakes Kivu, Edwards and Albert by the Institut Royal des Sciences Naturelles de Belgique. Bull. Inst. Sci. nat. Belg., 41 (7): 1–22.

Fryer, G. 1965b. Parasitic crustaceans of African freshwater fishes from the Nile and Niger systems. Proc. zool. Soc. Lond., 145: 285–303.

Fryer, G. 1968a. The parasitic Crustacea of African freshwater fishes; their biology and distribution. J. Zool., Lond., 156: 45–95.

Fryer, G. 1968b. The parasitic copepod *Lernaea cyprinacea* L. in Britain. J. nat. Hist., 2: 531–533.

Fryer, G. 1969a. The parasitic copepod *Ergasilus sieboldi* Nordmann, new to Britain. Naturalist, Hull, No. 909: 49–51.

Fryer, G. 1969b. The parasitic copepods *Achtheres percarum* and *Salmincola gordoni* in Yorkshire. Naturalist, Hull, No. 910: 77–81.

Gadd, P. 1901. Nagra förut aberskrifna, parasitisk lefvandte copepoder. Meddr Soc. Fauna Flora fenn., 27: 98–100, 181–182.

Gadd, P. 1904. Parasit-copepoder i Finland. Acta Soc. Fauna Flora fenn., 26 (8): 1–60.

Garbini, A. 1895. Appunti di carcinologia Veronese. Memorie Accad. Verona (ser. 3), 71 (fasc. 1): 1–94.

Geptner, M. V. 1968. Opisanie i funktsionalnaya morfologiya *Megapontius pleurospinosus* sp. n. iz Tikhogo Okeana i polozhenie roda *Megapontius* v sisteme semeystv gruppy Siphonostoma (Copepoda, Cyclopoida). [Description and functional morphology of *Megapontius pleurospinosus* sp. n. from the Pacific and the position of the genus *Megapontius* in the system of families of the group Siphonostoma (Copepoda, Cyclopoida).] Zool. Zh., 47: 1628–1638.

Gerstaecker, A. 1854. Beschreibung zweier neuer Siphonostomen-Gattungen. Arch. Naturgesch., 20: 185–195.

Gerstaecker, A. 1866–1879. Die Klassen und Ordnungen der Arthropoden. In: H. G. Bronn. Die Klassen und Ordnungen des Thierreichs. Vol. 5, part 1, Crustacea (1st half). C. W. Winter: Leipzig und Heidelberg. 1320 pp.

Geyer, F. 1939. Über parasitische Lernaeopodiden von Balaton-Fischen (Ungarn). Zool. Anz., 127: 145–159.

Giesbrecht, W. 1882. Die freilebenden Copepoden der Kieler Föhrde. Jber. Commn Unters. dt. Meere in Kiel, 4: 85–168.

Giesbrecht, W. 1892. Systematik und Faunistik der pelagischen Copepoden des Golfes von Neapel und der angrenzenden Meeres-Abschnitte. Fauna Flora Golf. Neapel, 19: IX + 831.

Gissler, N. 1751. Ron om laxens natur och fiskande i de norrlandska alfvarna. K. svenska Vetensk-Akad. Handl., 12: 171–190.

Goggio, E. 1906. Intorno al gen. *Clavella* (*Hatschekia* Poche). Archo. zool. ital., 2: 215–225.

Gooding, R. U. and Humes, A. G. 1963. External anatomy of the female *Haemobaphes cyclopterina*, a copepod parasite of marine fishes. J. Parasit., 49: 663–677.

Gouillart, M. 1937. Recherches sur les copépodes parasites (biologie, spermatogénèse, ovogénèse). Trav. Stn

zool. Wimereaux, 12: 308–457.
Grabda, J. 1961. Pasożytnicze widłonogi—Copepoda parasitica ryb Zalewu Wiślanego. Wiad. parazyt., 7 (2 suppl.): 179–181.
Grabda, J. 1963. Life cycle and morphogenesis of *Lernaea cyprinacea* L. Acta parasit. pol., 11: 169–198.
Grabda, J. 1973. Contribution to knowledge of biology of *Cecrops latreillii* Leach, 1816 (Caligoida: Cecropidae), the parasite of the ocean sunfish *Mola mola* (L.). Acta ichthyol. piscat., 3: 61–74.
Gray, P. 1926. On the nutrition of the male *Lernaeopoda scyllicola*. Parasitology, 18: 399–401.
Green, J. 1958. *Eudactylina rachelae* n. sp., a copepod parasitic on the electric ray, *Torpedo nobiliana* Bonaparte. J. mar. biol. Ass. U.K., 37: 113–116.
Guérin-Méneville, F. E. 1829–1843. Iconographie du règne animal de G. Cuvier. Crustacés. J. B. Baillière: Paris and London.
Guiart, J. 1913. Crustacés commensaux et parasites de la baie de Concarneau. Bull. Inst. océanogr. Monaco, No. 264: 1–11.
Gundrizer, A. N. 1972. Parazitofauna mongolskogo khariusa. [Parasite fauna of the Mongolian grayling.] Trudy nauch.-issled. Inst. Biol. Biofiz. gosud. Univ. Tomsk, 2: 99–100.
Gurney, R. 1913. Some notes on the parasitic copepod *Thersitina gasterostei* Pagenstecher. Ann. Mag. nat. Hist., ser. 8, 12: 415–424.
Gurney, R. 1931. British fresh-water Copepoda. Vol. 1, Ray Society: London, LII + 238 pp.
Gurney, R. 1933. British fresh-water Copepoda. Vol. 3, Ray Society: London, XXIX + 384 pp.
Gurney, R. 1934. Development of certain parasitic copepods of the families Caligidae and Clavellidae. Proc. zool. Soc. Lond., 1934: 177–217.
Gurney, R. 1947. Some notes on parasitic copepods. J. mar. biol. Ass. U.K., 27: 133–137.
Gusev. A. V. 1951. Paraziticheskie Copepoda s nekotorykh morskikh ryb. [Parasitic Copepoda of some marine fishes.] Parazit. Sb., 1B: 394–463.
Halisch, W. 1940. Anatomie und Biologie von *Ergasilus minor* Halisch. Z. ParasitKde, 11: 284–330.
Hamond, R. 1969. The copepods parasitic on Norfolk marine fishes. Trans. Norfolk Norwich Nat. Soc., 21: 229–234.
Hanek, G. and W. Threlfall. 1969. *Thersitina gasterostei* (Pagenstecher, 1861) (Copepoda: Ergasilidae) from *Gasterosteus wheatlandi* Putnam, 1867. Can. J. Zool., 47: 627–629.
Hanek, G. and W. Threlfall. 1970a. Parasites of the threespine stickleback (*Gasterosteus aculeatus*) in Newfoundland and Labrador. J. Fish. Res. Bd Can., 27: 901–907.
Hanek, G. and W. Threlfall. 1970b. Parasites of the ninespine stickleback *Pungitius pungitius* (L.) in Newfoundland and Labrador. Can. J. Zool., 48: 600–602.
Hanek, J. and A. Dulmaa. 1970. Parasitic copepods of some fish species from Mongolia. Folia parasit., 17: 77–80.
Hansen, H. J. 1893. Zur Morphologie der Gliedermassen und Mundteile bei Crustaceen und Insekten. Zool. Anz., 16: 193–198, 201–212.
Hansen, H. J. 1923. Crustacea Copepoda II. Copepoda parasita and hemiparasita. Dan. Ingolf-Exped., 3 (7): 1–92.
Harding, J. P. 1950. On some species of *Lernaea* (Crustacea, Copepoda: parasites of freshwater fish.) Bull. Br. Mus. nat. Hist., Zool., 1 (1): 1–27.
Harding, J. P. and F. W. K. Gervers. 1956. Occurrence of *Achtheres percarum* Nordmann in English waters. Nature, Lond., 177: 664.
Hart, T. J. 1946. Report on trawling surveys on the Patagonian continental shelf. 'Discovery' Rep., 23: 223–408.
Hartmann, C. E. R. 1870. Beiträge zur anatomischen Kenntniss der Schmarotzerkrebse. 1. *Bomolochus belones* Burmeister. Arch. Anat. Physiol., 1870: 116–158.
Heckman, R. and D. G. Farley. 1973. Ectoparasites of the western roach from two foothill streams. J. Wildl. Dis., 9: 221–224.
Heegaard, P. 1940. Some new parasitic copepods (Chondracanthidae and Lernaeopodidae) from Western Australia. Vidensk. Meddr dansk. naturh. Foren., 104: 87–101.
Heegaard, P. 1943. Parasitic copepods mainly from tropical and Antarctic seas. Ark. Zool., 34A (18): 1–37.
Heegaard, P. 1945. Discussion of the mouth appendages of the copepods. Ark. Zool. 40A (3): 1–8.
Heegaard, P. 1947a. Contributions to the phylogeny of the arthropods. Copepoda. Spolia zool. Mus. haun., 8: 1–236.
Heegaard, P. 1947b. A new lernaeopodid (*Isobranchia appendiculata* nov. gen., nov. sp.) from the Iranian Gulf. Dan. scient. Invest. Iran, Pt. 4: 239–245.
Heegaard, P. 1962. Parasitic Copepoda from Australian waters. Rec. Aust. Mus., 25: 149–233.
Heegaard, P. 1966. Parasitic copepods from Texas. Vidensk. Meddr dansk. naturh. Foren., 129: 187–197.
Heegaard, P. 1972. Caliginae and Euryphorinae of the Dana Expedition (Crustacea, Copepoda, Caligidae). Steenstrupia, 2: 295–317.
Heider, C. 1879. Die Gattung *Lernanthropus*. Arb. zool. Inst. Univ. Wien, 2: 269–368.
Heinemann, E. 1934. Gross-Schädlinge. Der Kiemenkrebs (*Ergasilus sieboldi* Nordm.) Biologe, 3: 284–287.

Heller, C. 1865.* Crustaceen. Reise der Oesterreichischen Fregatte Novara um die Erde in den Jahren 1857, 1858, 1859. Zool. Theil, 2 (3): 1–280.

Heller, C. 1866. Carcinologische Beiträge zur Fauna der Adriatischen Meeres. Verh. zool.-bot. Ges. Wien, 16: 723–760.

Hermann, J. 1783. Helmintologische Bemerkungen. Zweites Stück. Naturforscher, Halle, 19: 31–59.

Hermann, J. F. 1804. Mémoire Aptérologique. Strasbourg. 144 pp.

Hesse, C. E. 1866. Observations sur des crustacés rares ou nouveaux des côtes de France. 7me article. Annls Sci. nat., ser. 5, Zool., 5: 265–279.

Hesse, C. E. 1871. Observations sur les crustacés rares ou nouveaux des côtes de France. 19me article. Annls Sci. nat., ser. 5, 15: 1–50.

Hesse, C. E. 1873. Mémoire sur des crustacés rares ou nouveaux des côtes de France. 22me article. Annls Sci. nat., ser. 5, Zool., 17: 1–18.

Hesse, C. E. 1879a. Description des crustacés rares ou nouveaux des côtes de France. 29me article. Annls Sci. nat., ser. 6, Zool., 8 (11): 1–34.

Hesse, C. E. 1879b. Description des crustacés rares ou nouveaux des côtes de France décrits sur des individus vivants. Annls Sci. nat., ser. 6, Zool., 8 (16): 1–16.

Hesse, C. E. 1883. Crustacés rares ou nouveaux des côtes de France. 33me article. Annls Sci. nat., ser. 6, Zool., 15 (3): 1–48.

Hesse, C. E. 1884. Description de quatre nouveaux crustacés suceurs de l'ordre des Rostrostomes (nobis), de la famille des Pachycephales, de la tribu des Dichelestiens, appartenant au genre *Kroyeria*, *Eudactylina* et *Pagodina*. Annls Sci. nat., ser. 6, Zool., 16 (3): 1–18.

Hewitt, G. C. 1964a. A redescription of *Gloiopotes huttoni* (Thomson, 1889) with a key to the species of the genus. Trans. R. Soc. N.Z., Zool., 5: 85–96.

Hewitt, G. C. 1964b. A new species of *Lophoura* (Sphyriidae, Copepoda) from New Zealand waters. Trans. R. Soc. N.Z., Zool., 5: 55–58.

Hewitt, G. C. 1967. Some New Zealand parasitic Copepoda of the family Pandaridae. N.Z. Jl mar. freshwat. Res., 1: 180–264.

Hewitt, G. C. 1968. Some New Zealand parasitic Copepoda of the family Anthosomidae. Zool. Publ. Victoria Univ. Wellington, No. 47: 1–31.

Hewitt, G. C. 1969a. Some New Zealand parasitic Copepoda of the family Eudactylinidae. Zool. Publ. Victoria Univ. Wellington. No. 49: 1–31.

Hewitt, G. C. 1969b. A new species of *Paeonodes* (Therodamasidae, Cyclopoida, Copepoda) parasitic on New Zealand freshwater fish, with a re-examination of *Paeonodes exiguus* Wilson. Zool. Publ. Victoria Univ. Wellington, No. 50: 32–39.

Hewitt, G. C. 1969c. *Pseudocycnus appendiculatus* Heller, 1865 (Copepoda, Dichelesthiidae) in New Zealand waters. N.Z. Jl mar. freshwat. Res., 3: 169–176.

Hewitt, G. C. 1971. Species of *Lepeophtheirus* (Copepoda, Caligidae) recorded from the ocean sunfish (*Mola mola*) and their implication for the caligid genus *Dentigryps*. J. Fis. Res. Bd Can., 28: 323–334.

Hewitt, G. C. and P. M. Hine. 1972. Checklist of parasites of New Zealand fishes and of their hosts. N.Z. Jl mar. freshwat. Res., 6: 69–114.

Ho, J.-S. 1961. Parasitic Copepoda, genus *Lernaea*, on Formosan freshwater fishes, with a special reference to *Lernaea parasiluri* Yü. Q. Jl Taiwan Mus., 14: 143–158.

Ho, J.-S. 1966a. Redescription of *Echetus typicus* Krøyer, a caligid copepod parasitic on the red drum, *Sciaenops ocellatus* (Linnaeus). J. Parasit., 52: 752–761.

Ho, J.-S. 1966b. Three species of Formosan copepods parasitic on fishes. Crustaceana, 11: 163–177.

Ho, J.-S. 1967. Cyclopoid copepods of the genus *Telson*, parasitic on uranoscopic fishes in the Gulf of Mexico. J. Parasit., 53: 852–858.

Ho, J.-S. 1969. Copepods of the family Taeniacanthidae (Cyclopoida) parasitic on fishes in the Gulf of Mexico. Bull. mar. Sci. Gulf Caribb., 19: 111–130.

Ho, J.-S. 1970. Revision of the genera of the Chondracanthidae, a copepod family parasitic on marine fishes. Beaufortia, 17: 105–218.

Ho, J.-S. 1971a. *Pharodes* Wilson, 1935, a genus of cyclopoid copepods (Pharodidae) parasitic on marine fishes. J. nat. Hist., 5: 349–359.

Ho, J.-S. 1971b. Parasitic copepods of the family Chondracanthidae from fishes of eastern North America. Smithson. Contr. Zool., No. 87: 1–39.

Ho, J.-S. 1972a. Copepod parasites of California halibut, *Paralichthys californicus* (Ayres), in Anaheim Bay, California. J. Parasit., 58: 993–998.

Ho, J.-S. 1972b. Four new parasitic copepods of the family Chondracanthidae from California inshore fishes. Proc. biol. Soc. Wash., 85: 523–540.

Ho, J.-S. 1972c. Copepods of the family Chondracanthidae (Cyclopoida) parasitic on South African marine

*Some authors mistakenly date this reference 1868. First published in 1865, it was reissued at that later date as a bound volume with other fascicles.

fishes. Parasitology, 65: 147–158.
Ho, J.-S. 1975. Copepod parasites of deep-sea fish off the Galapagos Islands. Parasitology, 70: 359–375.
Horst, R. 1897. *Philorthagoriscus serratus* Kr. (*Dinematura serrata* Kr.). Notes Leyden Mus., 19: 137–144.
Houdemer, F. E. 1938. Recherches de parasitologie comparée Indochinoise. Paris. 235 pp.
Humes, A. G. 1967. *Vahinius petax* n. gen., n. sp., a cyclopoid copepod parasitic in an antipatharian coelenterate in Madagascar. Crustaceana, 12: 233–242.
Humes, A. G. 1970. *Clavissodalis heterocentroti* gen. et sp. nov., a cyclopoid copepod parasitic on an echinoid at Eniwetok Atoll. J. Parasit., 56: 575–583.
Illg, P. L. 1955. Proposed designation under the plenary powers, for the generic name *Lernaeocera* (emend. of *Lerneocera*) Blainville, 1822 (Class Crustacea, Order Copepoda) of a type species in harmony with the current nomenclatorial usage. Bull. zool. Nomencl., 11: 252–255.
Izawa, K. 1969. Life history of *Caligus spinosus* Yamaguti, 1939, obtained from cultured yellow tail, *Seriola quinqueradiata* T. & S. (Crustacea: Caligoida). Rep. Fac. Fish. prefect. Univ. Mie, 6: 127–157.
Izawa, K. 1973. On the development of parasitic Copepoda; I. *Sarcotaces pacificus* Komai (Cyclopoida: Philichthyidae). Publ. Seto mar. biol. Lab., 21: 77–86.
Johnson, S. K. and W. A. Rogers. 1953. Distribution of the genus *Ergasilus* in Gulf of Mexico drainage basins. Bull. Auburn Univ. Agric. Res. Stn, No. 445: 1–74.
Johnston, G. M. D. 1835. Illustrations in British zoology. 33. *Pandarus alatus*. 34. *Pandarus lamnae*. Mag. nat. Hist., 8: 202–204.
Jones, D. H. and B. L. Matthew. 1968. On the development of *Sphyrion lumpi* (Krøyer). Crustaceana, Suppl. 1: 177–185.
Kabata, Z. 1957. *Lernaeocera obtusa* n. sp., a hitherto undescribed parasite of the haddock (*Gadus aeglefinus* L.). J. mar. biol. Ass. U.K., 36: 569–592.
Kabata, Z. 1958a. *Lernaeocera obtusa* n. sp.; its biology and its effects on the haddock. Mar. Res. Scot., 1958 (No. 3): 1–26.
Kabata, Z. 1958b. *Vanbenedenia krøyeri* Malm, 1860: a rare parasitic copepod. Ann. Mag. nat. Hist., ser. 13, 1: 331–336.
Kabata, Z. 1959. Ecology of the genus *Acanthochondria* Oakley (Copepoda parasitica). J. mar. biol. Ass. U.K., 38: 249–261.
Kabata, Z. 1960a. *Vanbenedenia krøyeri* (Copepoda parasitica): taxonomic review and other notes. Ann. Mag. nat. Hist., ser. 13, 2: 731–735.
Kabata, Z. 1960b. On the specificity of *Lernaeocera* (Copepoda parasitica). Ann. Mag. nat. Hist., ser. 13, 3: 133–139.
Kabata, Z. 1960c. Observations on *Clavella* (Copepoda) parasitic on some British gadoids. Crustaceana, 1: 342–352.
Kabata, Z. 1962a. The mouth and the mouth-parts of *Lernaeocera branchialis* (L.), a parasitic copepod. Crustaceana, 3: 311–317.
Kabata, Z. 1962b. The parasitic copepod *Clavella stellata* (Krøyer, 1838), with the description of the male. Crustaceana, 4: 107–117.
Kabata, Z. 1963a. New host for *Lernaeocera lusci* (Bassett-Smith, 1896). Crustaceana, 6: 159–160.
Kabata, Z. 1963b. The free-swimming stage of *Lernaeenicus* (Copepoda parasitica). Crustaceana, 5: 181–187.
Kabata, Z. 1963c. *Clavella* (Copepoda) parasitic on British Gadidae: one species or several? Crustaceana, 5: 64–74.
Kabata, Z. 1963d. The second antenna in the taxonomy of Clavellinae (Copepoda, Lernaeopodidae). Crustaceana, 6: 7–14.
Kabata, Z. 1964a. Revision of the genus *Charopinus* Krøyer, 1863 (Copepoda: Lernaeopodidae). Vidensk. Meddr dansk naturh. Foren., 127: 85–112.
Kabata, Z. 1964b. Redescription of *Lernaeopoda centroscyllii* Hansen, 1923 (Copepoda: Lernaeopodidae). J. Fish. Res. Bd Can., 21: 681–689.
Kabata, Z. 1964c. On the adult and juvenile stages of *Vanbenedenia chimaerae* (Heegaard, 1962) (Copepoda: Lernaeopodidae) from Australian waters. Proc. Linn. Soc. N.S.W., 89: 254–267.
Kabata, Z. 1964d. *Clavellisa emarginata* (Krøyer, 1837): morphological study of a parasitic copepod. Crustaceana, 7: 1–10.
Kabata, Z. 1964e. The morphology and taxonomy of *Clavellodes pagelli* (Krøyer, 1863) (Copepoda, Lernaeopodidae). Crustaceana, 7: 103–112.
Kabata, Z. 1965a. Copepoda parasitic on Australian fishes. IV. Genus *Caligus* (Caligidae). Ann. Mag. nat. Hist., ser. 13, 8: 109–126.
Kabata, Z. 1965b. Systematic position of the copepod *Lernaeocera centropristi*. Proc. zool. Soc. Lond., 144: 351–360.
Kabata, Z. 1965c. Parasitic Copepoda of fishes. Rep. B.A.N.Z. antarct. Exped., 8 (6): 1–16.
Kabata, Z. 1965d. *Coregonicola orientalis* Markevich and Bauer, 1950, a Siberian parasitic copepod. Crustaceana, 8: 225–232.
Kabata, Z. 1965e. *Andropoda*, a new genus of Lernaeopodidae (Copepoda) from the gills of *Lampris luna*

(Gmelin). Crustaceana, 8: 213–221.

Kabata, Z. 1965f. *Lernaeocera* (Copepoda) parasitic on ling (*Molva elongata* Otto). Crustaceana, 9: 104–105.

Kabata, Z. 1966. *Brachiella chevreuxii* van Beneden, 1891 (Copepoda, Lernaeopodidae); a taxonomic problem. Crustaceana, 10: 98-108.

Kabata, Z. 1967. The genus *Haemobaphes* (Copepoda: Lernaeoceridae) in the waters of British Columbia. Can. J. Zool., 45: 853-875.

Kabata, Z. 1968a. Copepoda parasitic on Australian fishes. VIII. Families Lernaeopodidae and Naobranchiidae. J. nat. Hist., 2: 505-523.

Kabata, Z. 1968b. Some Chondracanthidae (Copepoda) from fishes of British Columbia. J. Fish. Res. Bd Can., 25: 321-345.

Kabata, Z. 1968c. The appendages of *Lernaeolophus sultanus* (H. Milne Edwards, 1840) (Lernaeoceridae). Crustaceana, Suppl. 1: 103–111.

Kabata, Z. 1969a. *Phrixocephalus cincinnatus* Wilson, 1908 (Copepoda: Lernaeoceridae): morphology, metamorphosis and host-parasite relationship. J. Fish. Res. Bd Can., 26: 921–934.

Kabata, Z. 1969b. Tanypleuridae fam. nov. (Copepoda: Caligoida), parasitic on fishes in the Canadian Atlantic. J. Fish. Res. Bd Can., 26: 1407–1414.

Kabata, Z. 1969c. Copepoda parasitic on Australian fishes. IX. Family Chondracanthidae. J. nat. Hist., 3: 497–507.

Kabata, Z. 1969d. Revision of the genus *Salmincola* Wilson, 1915 (Copepoda: Lernaeopodidae). J. Fish. Res. Bd Can., 26: 2987–3041.

Kabata, Z. 1969e. Four Lernaeopodidae (Copepoda) parasitic on fishes from Newfoundland and West Greenland. J. Fish. Res. Bd Can., 26: 311–324.

Kabata, Z. 1970a. Crustacea as enemies of fishes. In: S. F. Snieszko and H. R. Axelrod [ed.]. Diseases of fishes. Book 1. T.F.H. Publications: Jersey City. 171 pp.

Kabata, Z. 1970b. Three copepods (Crustacea) parasitic on fishes of the genus *Lepidion* Swainson, 1838 (Pisces: Teleostei). J. Parasit., 56: 175–184.

Kabata, Z. 1970c. Copepoda parasitic on Australian fishes. X. Families Eudactylinidae and Pseudocycnidae. J. nat. Hist., 4: 159–173.

Kabata, Z. 1970d. Some Lernaeopodidae (Copepoda) from fishes of British Columbia. J. Fish. Res. Bd Can., 27: 865–885.

Kabata, Z. 1970e. Discovery of *Brachiella lageniformis* (Copepoda: Lernaeopodidae) in the Canadian Pacific and its significance to zoogeography of the genus *Merluccius* (Pisces: Teleostei). J. Fish. Res. Bd Can., 27: 2159–2165.

Kabata, Z. 1972a. Developmental stages of *Caligus clemensi* (Copepoda: Caligidae). J. Fish. Res. Bd Can., 29: 1571–1593.

Kabata, Z. 1972b. Copepoda parasitic on Australian fishes. XI. *Impexus hamondi* new genus, new species, with a key to the genera of Lernaeoceridae. Proc. biol. Soc. Wash., 85: 317–322.

Kabata, Z. 1972c. *Alella* Leigh-Sharpe, 1925 (Crustacea: Copepoda): a request for designation of type-species. Z.N. (S.) 2006. Bull. zool. Nomencl., 29: 216–217.

Kabata, Z. 1973a. Taxonomic status of *Homoiotes palliata* Wilson, 1905 (Copepoda: Caligidae). J. Fish. Res Bd. Can., 30: 1892–1893.

Kabata, Z. 1973b. *Brachiella impudica* Nordmann, 1832 and *B. bispinosa* Nordmann, 1832 (Copepoda: Lernaeopodidae) redescribed from type specimens. Mitt. zool. Mus. Berl., 49: 1–12.

Kabata, Z. 1973c. The species of *Lepeophtheirus* (Copepoda: Caligidae) from fishes of British Columbia. J. Fish. Res. Bd Can., 30: 729–759.

Kabata, Z. 1974. Mouth and mode of feeding of Caligidae (Copepoda), parasites of fishes, as determined by light and scanning electron microscopy. J. Fish. Res. Bd Can., 31: 1583–1588.

Kabata, Z. and T. E. Bowman. 1961. Revision of *Tracheliastes grandis* Wilson, 1915 (Copepoda, Lernaeopodidae). Crustaceana, 3: 120–126.

Kabata, Z. and B. Cousens. 1972. The structure of the attachment organ of Lernaeopodidae (Crustacea: Copepoda). J. Fish. Res. Bd Can., 29: 1015–1023.

Kabata, Z. and B. Cousens. 1973. Life cycle of *Salmincola californiensis* (Dana, 1852) (Copepoda: Lernaeopodidae). J. Fish. Res. Bd Can., 30: 881–903.

Kabata, Z. and A. V. Gusev. 1966. Parasitic Copepoda of fishes from the collection of the Zoological Institute in Leningrad. J. Linn. Soc. (Zool.), 46: 155–207.

Kabata, Z. and G. C. Hewitt. 1971. Locomotory mechanisms in Caligidae (Crustacea: Copepoda). J. Fish. Res. Bd Can., 28: 1143–1151.

Kabata, Z., A. Raibaut and Oum Kalthoum Ben Hassine. 1971. *Eubrachiella mugilis* n. sp. un copépode parasite de muges de Tunisie. Bull. Inst. Océanogr. Pêche Salammbô, 2: 87–93.

Kamenev, V. P. 1959. Parazitofauna glavneishikh promyslovykh ryb priazovskikh limanov. [Parasite fauna of the principal commercial fishes of the Azov limans.] Trudy Soveshch. ikhtiol. Kom., 9: 158–162.

Kane, W. F. de Wismes. 1892. On a new species of *Lernaeopoda* (*bidiscalis*) from the west coast of Ireland and

Polperro, Cornwall. Proc. R. Ir. Acad., ser. 3, 2: 203–211.

Kašták, V. 1956. Erster Befund des parasiten *Ergasilus nanus* (Copepoda) in der Tschechoslowakei (German summary). Čslka Parasit., 3: 75–76.

Kazachenko, V. N. 1972. Nekotorye parazitícheskie kopepody ryb Tikhego i Indiyskogo Okeanov. [Some parasitic copepods of fishes of the Pacific and Indian Ocean.] Problemy Parazitologii (Trudy 7 nauch. Konf. parazit. Obshch. Ukr. S.S.R.), 1: 323–325.

Kazachenko, V. N. 1974. *Congericola pallidus* Beneden (Copepoda: Eudactylinidae) iz rayona Novoy Zelandii. [*Congericola pallidus* Beneden (Copepoda: Eudactylinidae) from the region of New Zealand.] Izv. tikhookeansk. nauchno-issled. Inst. ryb. Khoz. Okeanogr., 88: 36–41.

Kazachenko, V. N., V. D. Korotaeva and Yu. V. Kurochkin. 1972. Parazitícheskie rakoobraznye nekotorykh ryb Tikhogo Okeana. [Parasitic copepods of some fishes from the Pacific Ocean.] Izv. tikhookean. nauchno-issled. Inst. ryb. Khoz. Oceanogr., 81: 224–238.

Kennedy, C. R. 1974. A check list of British and Irish freshwater fish parasites with notes on their distribution. J. Fish. Biol., 6: 613–644.

Kensley, B. and J. R. Grindley. 1973. South African parasitic Copepoda. Ann. S. Afr. Mus., 62: 69–130.

Kirtisinghe, P. 1964. A review of the parasitic copepods of fish recorded from Ceylon, with description of additional forms. Bull. Fish. Res. Stn Ceylon, 17: 45–132.

Kölliker, A. 1853. Bericht über einige im Herbste 1852 in Messina angestellte vergleichend anatomische Untersuchungen. Z. wiss. Zool., 4: 299–370.

Kozikowska, Z. 1958. Skorupiaki pasożytnicze (Crustacea parasitica) Polski. I. Pasożyty ryb wód ujściowych Odry. Zoologica Pol., 8: 217–270 (for 1957).

Kozikowska, Z., Z. Jara and E. Grabda. 1957. *Achtheres percarum* Nordm. u okonia i sandacza. (Próba wyjaśnienia wzajemnego stosunku form: *percarum* i *sandrae*). Zoologica Pol., 7: 219–267 (for 1956).

Krøyer, H. 1837. Om snyltekrebsene, isaer med hensyn til den danske fauna. Naturh. Tidsskr., 1: 172–208, 253–304, 476–505, 605–628.

Krøyer, H. 1838. Om snyltekrebsene, isaer med hensyn til den danske fauna. Naturh. Tidsskr., 2: 8–52, 131–157.

Krøyer, H. 1863. Bidrag til kundskab om snyltekrebsene. Naturh. Tidsskr., ser. 3, 2: 75–320.

Krøyer, H. 1864. Bidrag til kundskab om snyltekrebsene (slutning). Naturh. Tidsskr., ser. 3, 2: 321–426.

Kurtz, W. 1877. Studien über die Familie der Lernaeopodiden. Z. wiss. Zool., 29: 380–423.

Lamarck, J. B. P. de, 1816. Histoire naturelle des animaux sans vertèbres. Vol. 3. Verdière: Paris. 586 pp.

Lamarck, J. B. P. de, 1818. Histoire naturelle des animaux sans vertèbres. Vol. 5. Déterville et Verdière: Paris. 612 pp.

Lang, K. 1946. A contribution to the question of the mouthparts of the Copepoda. Ark. Zool., 38A (5): 1–24.

Lang, K. 1948. Copepoda "Notodelphyoidea" from the Swedish west-coast with an outline of the systematics of the copepods. Ark. Zool., 40A (14): 1–36.

Latreille, P. A. 1829. Crustacés, arachnides et partie des insectes. In: G. Cuvier. La règne animal. Nouvelle (2nd) édition. Vol. IV. Déterville et Crochard: Paris. XXVII + 584 pp.

Laubier, L. 1968. Sur deux copépodes parasites de la raie *Mobula mobular* en Mediterranée occidentale. Crustaceana, Suppl. 1: 159–171.

Leach, W. E. 1816. Annulosa. pp. 401–453. In: Supplement to the fourth, fifth and sixth editions of the Encyclopaedia Britannica, vol. 1.

Leach, W. E. 1819. Entomostracés, Entomostraca, (Crust.), pp. 524–543. In: F. G. Levrault (ed.). Dictionnaire des sciences naturelles. Vol. 14. Le Normant: Paris.

Leigh-Sharpe, W. H. 1916. *Lernaeopoda scyllicola* n. sp., a parasitic copepod of *Scyllium canicula*. Parasitology, 8: 262–273.

Leigh-Sharpe, W. H. 1918a. The genus *Lernaeopoda* Blainville. Including description of *L. mustelicola* n. sp., remarks on *L. galei* and further information on *L. scyllicola*. Parasitology, 11: 256–266.

Leigh-Sharpe, W. H. 1918b. *Clavella sciatherica* n. sp., a parasitic copepod of *Gadus morrhua*. Parasitology, 11: 118–126.

Leigh-Sharpe, W. H. 1920. *Clavella iadda* n. sp., a parasitic copepod of *Gadus morrhua*. J. mar. biol. Ass. U.K., 12: 332–338.

Leigh-Sharpe, W. H. 1925. A revision of the British species of *Clavella* (Crustacea, Copepoda), with a diagnosis of new species: *C. devastatrix* and *C. invicta*. Parasitology, 17: 194–200.

Leigh-Sharpe, W. H. 1926a. A list of parasitic Copepoda found at Plymouth, with a note on the bulla of *Clavella devastatrix*. Parasitology, 18: 384–386.

Leigh-Sharpe, W. H. 1926b. *Ommatokoita superba* n. gen. et sp., a parasitic copepod of *Somniosus microcephalus*. Parasitology, 18: 224–229.

Leigh-Sharpe, W. H. 1928. The genus *Pennella* (Copepoda) as represented by the collection in the British Museum. Parasitology, 20: 79–89.

Leigh-Sharpe, W. H. 1930. *Chondracanthus neali* n. sp., a parasitic copepod of *Malacocephalus laevis*. Parasitology, 22: 468–470.

Leigh-Sharpe, W. H. 1933a. A list of British fishes with their characteristic parasitic Copepoda. Parasitology,

25: 109–112.
Leigh-Sharpe, W. H. 1933b. A second list of parasitic Copepoda of Plymouth, with a description of three new species. Parasitology, 25: 113–118.
Leigh-Sharpe, W. H. 1934. A third list of parasitic Copepoda of Plymouth with notes. Parasitology, 26: 112–113.
Leigh-Sharpe, W. H. 1935a. *Anchistrotos laqueus* n. sp., a parasitic copepod of *Serranus cabrilla*. Parasitology, 27: 266–269.
Leigh-Sharpe, W. H. 1935b. *Epibrachiella impudica* (Nordmann) (Copepoda). Parasitology, 27: 101–106.
Leigh-Sharpe, W. H. 1936a. Some rare and new parasitic Copepoda, etc. from Brighton and elsewhere. Parasitology, 28: 410–413.
Leigh-Sharpe, W. H. 1936b. New parasitic Copepoda from Naples. Parasitology, 28: 123–127.
Leigh-Sharpe, W. H. 1939. *Anchistrotos zeugopteri* (T. Scott), a parasitic copepod of *Zeugopterus punctatus*. Parasitology, 31: 166–170.
Leigh-Sharpe, W. H. and C. L. Oakley. 1927. Lernentominae, a new subfamily of Chondracanthidae, with a description of *Oralien triglae* (Blainville, 1822). Parasitology, 19: 455–467.
Leigh-Sharpe, W. H. and M. G. L. Perkins. 1924. Some parasitic Copepoda from Iceland. With an account of *Peniculus clavatus*, the conjunctive tubes of *Chondracanthus nodosus* and the males of *Clavella dubia*. Parasitology, 16: 289–295.
Lester, R. J. G. 1974. Parasites of *Gasterosteus aculeatus* near Vancouver, British Columbia. Syesis, 7: 195–200.
Lewis, A. G. 1964a. Caligoid copepods (Crustacea) of the Hawaiian Islands: parasitic on fishes of the family Acanthuridae. Proc. U.S. natn. Mus., 115: 137–244.
Lewis, A. G. 1964b. The caligid genus *Dentigryps* (Crustacea: Caligoida). Proc. U.S. natn. Mus., 115: 347–380.
Lewis, A. G. 1966. Copepod crustaceans parasitic on elasmobranch fishes of the Hawaiian Islands. Proc. U.S. natn. Mus., 118: 57–154.
Lewis, A. G. 1967. Copepod crustaceans parasitic on teleost fishes of the Hawaiian Islands. Proc. U.S. natn. Mus., 121 (No. 3574): 1–204.
Lewis, A. G. 1969. A discussion of the maxillae of the "Caligoida" (Copepoda). Crustaceana, 16: 65–77.
Leydig, F. 1851. Ueber ein neues parasitisches Krustenthier. Arch. Naturgesch., 17: 259–262.
Linnaeus, C. 1758. Systema naturae, per regna tria naturae, secundum classes, ordines, genera, species, cum characteribus, differentiis, synonymis, locis. Editio decima, reformata. Laurentius Salvius: Holmiae. 832 pp.
Linnaeus, C. 1761. Fauna suecica sistens animalia Sueciae regni: mammalia, aves, amphibia, pisces, insecta, vermes. Distributa per classes et ordines, genera et species. Editio altera, auctior. Stockholm. 578 pp.
Linnaeus, C. 1766–1768. Systema naturae, per regna tria naturae, secundum classes, ordines, genera, species, cum characteribus, differentiis, synonymis, locis. Editio duodecima, reformata. Laurentius Salvius: Holmiae. 1327 pp.
Lucas, H. 1887. Sur le *Cecrops latreilli*. Annls Soc. ent. Fr., ser. 6, 7: 31–32.
Machado Cruz, J. A. 1959. *Lernaeocera caparti* sp. nov., copépode parasite de *Merluccius merluccius* (Linné). Publçoes inst. Zool. Dr. Auguste Nobre, 64: 7–12.
Malm, A. W. 1860. Om flere for den skandinaviske fauna nye fiske, krebsdyr og bløddyr. Forh. skand. naturf. Møte, 1860: 616–623.
Malm, A. W. 1863. Nya fiskar, kräft- och blüt-djur för skandinaviens fauna. Göteborgs Vetensk-Samh. Handl., 8: 114–118.
Mann, H. 1952/53. *Lernaeocera branchialis* (Copepoda parasitica) und seine Schadwirkung bei einigen Gadiden. Arch FischWiss., 4: 133–143.
Mann, H. 1960. Schadwirkung des parasitischen Copepoden *Lernaeocera branchialis* auf das Wachstum von Wittlingen. Inf. FischWirt., 7: 153–155.
Margolis, L. 1958. The identity of the species of *Lepeophtheirus* (Copepoda) parasitic on Pacific salmon (genus *Oncorhynchus*) and Atlantic salmon (*Salmo salar*). Can. J. Zool., 36: 889–892.
Margolis, L., Z. Kabata and R. R. Parker. 1975. Catalogue and synopsis of *Caligus*, a genus of Copepoda (Crustacea) parasitic on fishes. Bull. Fish. Res. Bd Can., No. 192: 1–117.
Markevich, A. P. 1931. Parasitische Copepoden und Branchiuren des Aralsees, nebst systematischen Bemerkungen über die Gattung *Ergasilus* Nordmann. Zool. Anz., 96: 121–143.
Markevich, A. P. 1933. Les crustacés parasites des poissons de la Mer Caspienne. Bull. Inst. océanogr. Monaco, No. 638: 1–27.
Markevich, A. P. 1937. Copepoda parasitica prisnikh vod SRSR. [Copepoda parasitica of the fresh waters of the USSR.] Akad. Nauk Ukr. SSR: Kiev. Ukrainian text pp. 3–158; Russian text 159–189; German text 190–222.
Markevich, A. P. 1946. *Nectobrachia indivisa* (Copepoda paras.), osoblivosti ii budovi ta sistematichne polozhenye. [*Nectobrachia indivisa* (Copepoda paras.), its structural features and systematic position.] Nauk. Zap. kyyiv. derzh. Univ., 5: 215–220.
Markevich, A. P. 1951. Materiali do fauni parazitichnikh veslonogikh (Copepoda parasitica) rib Chornogo morya. [Contributions to the study of the fauna of parasitic copepods of fishes in the Black Sea.] Pratsi Inst. Zool. Akad. Nauk Ukr. SSR, 8: 91–99.

Markevich, A. P. 1956. Paraziticheskie veslonogie ryb SSSR. [Parasitic copepods of fishes of the USSR.] Izd. Akad. Nauk Ukr. SSR: Kiev. 259 pp.
Matthews, L. H. 1923. The parasitic copepod *Epibrachiella impudica* (Nordmann) (Copepoda). Parasitology, 27: 101–106.
Metzger, A. 1868. Ueber das Männchen und Weibchen der Gattung *Lernaea* vor den Eintritt der sogenannten rückschreitenden Metamorphose. Translation in Ann. Mag. nat. Hist., ser. 4, 3: 154–157 (1869).
Miculicich, M. 1905. Ein neuer Lernaeopodide. Zool. Anz., 28: 47–51.
Miers, E. J. 1880. On a small collection of Crustacea made by Edward Whimper, chiefly in the North-Greenland seas. J. Linn. Soc., Zool., 15: 59–73.
Möller, H. 1974. Untersuchungen über die Parasiten der Flunder (*Platichthys flesus* L.) in der Kieler Förde. Ber. dt. wiss. Kommn Meeresforsch., 23: 136–149.
Monod, T. 1923. Notes carcinologiques (parasites et commensaux). Bull. Inst. océanogr. Monaco, No. 427: 1–23.
Monod, T. and R. Ph. Dollfus. 1938. Pandarinés peu connus (genus *Phyllothyreus* Norman 1903 et *Gangliopus* Gerstaecker 1854). Annls Parasit. hum. comp., 16: 196–209.
Monterosso, B. 1925. Sur la struttura e la funzione delle appendici rizoidi cefaliche de *Peroderma cylindricum* Heller. Boll. Acad. Sci. nat. Catania (5a), 54: 3–8.
Müller, O. F. 1785. Entomostraca, seu Insecta testacea que in aques Daniae et Norvegiae reperit. F. W. Thiele: Leipzig and Copenhagen. 136 pp.
Nordmann, A. von, 1832. Mikrographische Beiträge zur Naturgeschichte der wirbellosen Thiere., Heft 2, I-XVIII, 1–150. G. Reimer: Berlin.
Nordmann, A. von, 1864. Neue Beiträge zur Kenntniss parasitischer Copepoden. Bull. Soc. imp. Nat. Moscou, 37: 461–520.
Norman, A. M. 1903. New generic names for some Entomostraca and Cirripedia. Ann. Mag. nat. Hist., ser. 7, 11: 367–379.
Norman, A. M. 1905. Museum Normanianum, or a catalogue of the Invertebrata of the Arctic and North Atlantic temperate ocean and Palearctic region, which are contained in the collection of the Rev. Canon A. M. Norman. III. Crustacea. Second edition. 47 pp.
Norman, A. M. and G. S. Brady. 1909. The Crustacea of Northumberland and Durham. Trans. nat. Hist. Soc. Northumb., new ser., 3: 1–168.
Norman, A. M. and T. Scott. 1906. The Crustacea of Devon and Cornwall. W. Wesley and Son: London. 232 pp.
Nunes-Ruivo, L. 1953. Copépodes parasites de poissons. Resultats des campaignes du "Pr. Lacaze Duthiers". VieMilieu, Suppl. 3: 115–138.
Nunes-Ruivo, L. 1954. Parasites de poissons de mer ouest-africains récoltés par M. J. Cadenat. III. Copépodes (2me note). Genres *Prohatschekia* n. gen. et *Hatschekia* Poche. Bull. Inst. fr. Afr. noire, 16: 479–505.
Nunes-Ruivo, L. 1956. Copépodes parasitas de peixes dos mares de Angola. Anais Jta Invest. Ultramar, 9: 8–44 (for 1954).
Nunes-Ruivo, L. 1957a. Lernaeopodidae (Copepoda) parasites des trigles. Revta port. Zool. Biol. gen., 1: 89–107.
Nunes-Ruivo, L. 1957b. Contribution à l'étude des variations morphologiques de *Clavella adunca* (H. Ström), copépode parasite de *Gadus callarias*. Considérations sur quelques *Clavella* parasites des Gadidae. Revta Fac. Cienc. Univ. Lisboa, (2)C, 5: 229–256.
Nunes-Ruivo, L. 1962. Copépodes parasites de poissons des côtes d'Angola (2me série). Mems Jta Invest. Ultramar, ser. 2, 33: 65–86.
Nunes-Ruivo, L. 1966a. Sur la valeur morphologique des "appendices postérieurs" de la femelle de divers genres de Lernaeopodidae (Crustacea, Copepoda). Proc. 1st int. Congress Parasitol., 2: 1080–1081.
Nunes-Ruivo, L. 1966b. Le genre *Alella* Leigh-Sharpe, 1925 (Copepoda, Fam. Lernaeopodidae). Proc. 1st int. Congress Parasitol., 2: 1081–1082.
Oakley, C. L. 1927. see Leigh-Sharpe, W. H. and C. L. Oakley, 1927.
Oakley, C. L. 1930. The Chondracanthidae (Crustacea: Copepoda); with a description of five new genera and one new species. Parasitology, 22: 182–201.
Oken, L. 1816. Lehrbuch der Naturgeschichte. 3 Theil: Zoologie, Abt. 1, Fleischlose Thiere. Leipzig and Jena. 842 + XVIII + XXVIII pp.
Olsson, P. 1869. Prodromus faunae copepodorum parasitantium Scandinaviae. Acta Univ. lund., 5: 1–49.
Oorde-de Lint, G. M. van and Schuurmans Stekhoven, J. H. 1936. Copepoda parasitica. Tierwelt N. -u. Ostsee, 31: 73–197.
O'Riordan, C. E. 1961. Occurrence of *Urophycis blennoides* (Brünn.) off the south coast of Ireland. Ir. Nat. J., 13: 211–213.
O'Riordan, C. E. 1962. Further notes on *Urophycis blennoides* (Brünn.), the greater forkbeard, off the south and southwest coast of Ireland. Ir. Nat. J., 14: 40–42.
O'Riordan, C. E. 1964. Notes on *Urophycis blennoides* (Brünn.), the greater forkbeard, in the National Museum collection. Ir. Nat. J., 14: 207–208.

O'Riordan, C. E. 1966. Parasitic copepods in the collections of the National Museum of Ireland. Proc. R. Ir. Acad., 64, sect. B: 371–378.
Osmanov, S. O. (V.). 1940. Materialy k parazitofaune ryb Chernogo Morya. [Dana on the parasite fauna of fishes of the Black Sea.] Uchen. Zap. leningr. gos. pedag. Inst. Gertsena, 30: 187–265.
Osmanov, S. O. 1959. Parazitofauna i parazitarnye bolezni ryb Aralskogo Morya. [Parasite fauna and parasitic diseases of fishes of the Aral Sea.] Trudy Soveshch. ikhtiol. Kom., 9: 192–197.
Osorio, B. 1892. Appendice ao catalogo dos crustaceos de Portugal existentes no Museu Nacional de Lisboa. J. Sci. math. phy. nat. Lisboa, ser. 2, 8: 233–241.
Pagenstecher, A. H. D. 1861. *Thersites gasterostei* und *Leptodora nicothoe*, eine neue Gattung parasitischer Crustaceen und eine Nematoden. Arch. Naturgesch., 27: 118–126.
Parker, R. R. 1965. A review and redescription of *Caligus gurnardi* Krøyer, 1863 (Copepoda, Caligidae). Crustaceana, 9: 93–103.
Parker, R. R. 1969. Validity of the binomen *Caligus elongatus* for a common parasitic copepod formerly misidentified with *Caligus rapax*. J. Fish. Res. Bd Can., 26: 1013–1035.
Parker, R. R., Z. Kabata, L. Margolis and M. D. Dean. 1968. A review and description of *Caligus curtus* Müller, 1785 (Caligidae: Copepoda), type species of its genus. J. Fish. Res. Bd Can., 25: 1923–1969.
Pearse, A. S. 1947. Parasitic copepods from Beaufort, North Carolina. J. Elisha Mitchell scient. Soc., 63: 1–16.
Pearse, A. S. 1952. Parasitic Crustacea from the Texas coast. Publs Inst. mar. Sci. Univ. Tex., 2: 5–42.
Pedashenko, D. D. 1898. Die Embryonalentwickelung und Metamorphose von *Lernaea branchialis*. Trav. Soc. Nat. St. Pétersb., 26: 247–307.
Pesta, O. 1908. Copepodentypen (Betrachtungen über Körperbau und Lebensweise). Sber. Akad. Wiss. Wien, mathl-naturw. Kl., 118: 561–572.
Pesta, O. 1934. Krebstiere oder Crustacea. I. Ruderfüssler oder Copepoda (4. Monstrilloida, 5. Notodelphyoida, 6. Caligoida, 7. Lernaeoida). Tierwelt Dtl., 29: 1–68.
Petrushevski, G. K. 1957. Parazitofauna seldevykh ryb Chernogo Morya. [Parasite fauna of the clupeid fishes of the Black Sea.] Izv. Vses. nauchno-issled. Inst. ozern. rechn. ryb. Khoz., 42: 303–314.
Petrushevski, G. K. and O. N. Bauer. 1948. Zoogeograficheskaya kharakteristika parazitov ryb Sibiri. [Zoogeographical characteristics of parasites of Siberian fishes.] Izv. Vses. nauchno-issled. Inst. ozern. rechn. ryb. Khoz., 27: 217–231.
Pillai, N. K. 1961. Copepods parasitic on South Indian fishes. Part 1, Caligidae. Bull. cent. res. Inst. Univ. Kerala, ser. C (nat. Hist.), 8: 87–130.
Pillai, N. K. 1962. Copepods parasitic on South Indian fishes: families Lernaeopodidae and Naobranchiidae. J. mar. biol. Ass. India, 4: 58–94.
Pillai, N. K. 1963a. Copepods of the family Taeniacanthidae parasitic on South Indian fishes. Crustaceana, 6: 110–128.
Pillai, N. K. 1963b. Copepods parasitic on South Indian fishes—family Caligidae. J. mar. biol. Ass. India, 5: 68–96.
Pillai, N. K. 1963c. Observations on the synonymy of *Caligus coryphaenae* Stp. & Ltk., Ann. Mag. nat. Hist., ser. 13, 5: 513–522.
Pillai, N. K. 1966. Redescription of *Bomolochus confusus* Stock. Bull. Dept. mar. Biol. Oceanogr. Univ. Kerala, 2: 33–38.
Pillai, N. K. 1967a. Copepods parasitic on Indian marine fishes—a review. Proc. Symp. Crustacea, Mandapam Camp, India, part 5: 1556–1680.
Pillai, N. K. 1967b. Description of a new species of *Anuretes* (Copepoda: Caligidae) and comments on the validity of a few caligid genera. Zool. Anz., 178: 358–367.
Pillai, N. K. 1967c. Three species of dichelesthiid copepods parasitic on South Indian sharks. Zool. Anz., 179: 286–297.
Pillai, N. K. 1968a. Additions to the copepod parasites of South Indian fishes. Parasitology, 58: 9–36.
Pillai, N. K. 1968b. Description of some species of *Brachiella* and *Clavellopsis*, with comments on *Isobrachia* Heegaard. Crustaceana, Suppl. 1: 119–135.
Pillai, N. K. 1971. Notes on some copepods parasites in the collection of the British Museum (N. H.), London. J. mar. biol. Ass. India, 11: 149–174.
Pillai, N. K. and M. J. Sebastian. 1967. Redescription of *Sagum epinepheli* (Yamaguti and Yamasu) with comments of the validity of *Pseudolernanthropus* (Copepoda, Anthosomatidae). Crustaceana, 13: 73–80.
Poche, F. 1902. Bemerkungen zu der Arbeit Herrn Basset-Smith, "A systematic description of parasitic Copepoda found on fishes, with the enumeration of the known species." Zool. Anz., 26: 8–20.
Poddubnaya, A. V. 1973. Izmenchivost i specifichnost lerney prudovykh ryb. [Variability and specificity of *Lernaea* from pond fishes.] Trudy Vses. nauchno-issled. Inst. prud. ryb. Khoz., 22: 159–173.
Poddubnaya, A. V. 1974. Morfologicheskaya izmenchivost roda *Lernaea* prudovykh ryb. [Morphological variability of the genus *Lernaea* from pond fishes.] Tezusy Dokl. VI Vses. Soveshch. Bolezn. Parazit. Ryb (3–5 Apr. 1974): 180–184.
Pogoreltseva, T. P. 1970. Parazitofauna khryashchevykh ryb Chernogo Morya. [Parasite fauna of elasmobranch fishes of the Black Sea.] pp. 106–107. In: A. M. Parukhin (ed.). Voprosy morskoy parazitologii. Naukova

Dumka: Kiev.
Polyanski, Yu. I. 1955. Materialy po parazitologii ryb severnykh morey SSSR. Parazity ryb Barentsova Morya. [Data on parasitology of fishes of the northern seas of the USSR. Parasites of fishes of the Barents Sea.] Trudy zool. Inst. Leningr., 19: 5–170.
Pónyi, J. and K. Molnar. 1969. Studies on the parasite fauna of fish in Hungary. V. Parasitic copepods. Parasit. Hung., 2: 137–148.
Poulsen, E. 1939. Investigations upon the parasitic copepod *Clavella uncinata* (O. F. Müller) in Danish waters. Vidensk. Meddr dansk. naturh. Foren., 10: 223–244.
Prefontaine, G. and P. Brunel. 1962. Liste d'invertebres marins recueillis dans l'estuaire du Saint-Laurent de 1929 à 1934. Naturaliste Can., 89: 237–263.
Quidor, A. 1910. Sur la protandrie chez les Lernaeopodidae. C. r. hebd. Séanc. Acad. Sci., Paris, 150: 1464–1465.
Quidor, A. 1912. Affinités des genres *Sphyrion* (Cuvier) et *Hepatophylus* (g.n.). Archs Zool. exp. gén., ser. 5., 10: 39–50.
Quidor, A. 1913. Copépodes parasites. Sciences naturelles. pp. 197–214. Deuxième Expedition Antarctique Française (1908–1910).
Quignard, J. P. 1968. Rapport entre la présence d'une "gibbosite frontale" chez Labridae (Poissons; Téléostéens) et le parasite *Leposphilus labrei* Hesse, 1866 (Copépode Philichthyidae). Ann. Parasit. hum. comp., 43: 51–57.
Radulescu, I. I., T. T. Nalbant and N. Angelescu. 1972. Nei contribuţii la cunoaşterea parasitofaunei peştilar din Oceane Atlantic. Bul. Cerc. Piscicole, 31: 71–78.
Rathbun, M. J. 1905. Fauna of New England. 5. List of the Crustacea. Occ. Pap. Boston Soc. nat. Hist., No. 7: 1–117.
Rathbun, R. 1884. Annotated list of the described species of parasitic Copepoda (Siphonostoma) from American waters, contained in the U.S. National Museum. Proc. U.S. natn. Mus., 7: 483–492.
Reed, G. B. and J. R. Dymond. 1951. Report Newfoundland Fisheries Research Station. Rep. Fish. Res. Bd Can., 1951: 43–65.
Reichenbach-Klinke, H. H., 1956. Vorläufige Mitteilung über die Parasiten der Fisch des Golfes von Neapel. Pubbl. Staz. zool. Napoli, 30: 115–126.
Reshetnikova, A. V. 1955. K izucheniyu parazitofauny ryb Chernogo Morya. [Contribution to the study of the parasite fauna of fishes of the Black Sea.] Trudy karadag. biol. Sta., 13: 105–121.
Retzius, A. 1829. Beskrifning öfvar en ny skandinavisk *Lernaea* från Nordsjön, kallad *Lernaea dalmanni*. K. svenska Vetensk-Akad. Handl., 1829: 109–119.
Richiardi, A. 1876a. Sopra lo *Spherifer cornutus* Rich. (*Sphaerosoma corvinae* Leydig) e una nuova specie del gen. *Philichthys* Stenstr., *Ph. Sciaenae*. Atti. Soc. tosc. Sci. nat., 2: 99–111.
Richiardi, S. 1876b. Intorno al *Peroderma cylindricum* dell'Heller e sopra due specie nuove dell'genere *Philichthys*. Atti Soc. tosc. Sci. nat., 2: 189–201.
Richiardi, S. 1877a. Descrizione di cinque specie nuove del gen. *Philichthys* et una dei *Sphaerifer*. Atti Soc. tosc. Sci. nat., 3: 166–179.
Richiardi, S. 1877b. Dei filictidi: osservazioni critiche e descrizione di sei specie nuove. Atti Soc. tosc. Sci. nat., 3: 180–194.
Richiardi, S. 1880. Contribuzione alla fauna d'Italia. I. Catalogo systematico di crostacei che vivono sul corpo di animali acquatici. Catalogo degli Espozitioni e della cosa Esposte, Espozitione internationale di Pesca in Berlino. pp. 147–152.
Richiardi, S. 1883. Descrizione di una specie nuova di crostacei parassiti, *Philichthys doderleini*. Atti Soc. tosc. Sci. nat., 3: 279–280.
Risso, A. 1816. Histoire naturelle des crustacés des environs de Nice. Librairie Grecque-Latine-Allemande: Paris. 175 pp.
Risso, A. 1826. Histoire naturelle des principales production de l'Europe méridionale et particulièrement de celles des environs de Nice et des Alpes Maritimes. F. -G. Levrault: Paris, Vol. 5. 402 pp.
Roberts, L. S. 1970. *Ergasilus* (Copepoda: Cyclopoidea): revision and key to species in North America. Trans. Am. microsc. Soc., 89: 134–161.
Ronald, K. 1959. A check list of the metazoan parasites of the Heterosomata. Contr. Dep. Fish. Queb., No. 67: 1–152.
Rose, M. 1933. Copépodes pélagiques. Faune de France, 26: 374 pp.
Sars, G. O. 1886. Crustacea, II. Copepoda, pp. 76–80. List of species observed on the expedition, with remarks on occurrence and distribution. N. -Atl. Exped., Zool., Vol. 15.
Sars, G. O. 1903. Copepoda Calanoida. In: An account of The Crustacea of Norway, with short descriptions and figures of all the species. 4: I–VI, 1–171.
Sars, G. O. 1909. Zoological results of the third Tanganyika Expedition conducted by Dr. W. A. Cunnington, F. Z. S., 1904–1905. Report on the Copepoda. Proc. zool. Soc. Lond., 1909: 31–77.
Sars, G. O. 1915. Copepoda Cyclopoida. Parts VII and VIII. Cyclopidae (concluded), Ascomyzontidae. In: An account of the Crustacea of Norway, with short descriptions and figures of all the species. 6: 81–104.

Sars, G. O. 1918. Copepoda Cyclopoida. Parts XIII and XIV. Lichomolgidae (concluded), Oncaeidae, Corycaeidae, Ergasilidae, Clausiidae, Eunicicolidae, Supplement. In: An account of the Crustacea of Norway, with short descriptions and figures of all the species. 6: 176–225.

Schimkewitsch, W. M. 1896. Studien über parasitische Copepoden. Z. wiss. Zool., 61: 339–362.

Schuurmans Stekhoven, J. H. 1935. Copepoda parasitica from the Belgian coast. Bull. Mus. r. Hist. nat. Belg., 11 (7): 1–13.

Schuurmans Stekhoven, J. H. 1936a. Copepoda parasitica from the Belgian coast, II. (Included some habitats in the North-Sea.) Mém. Mus. r. Hist. nat. Belg., No. 74: 1–20.

Schuurmans Stekhoven, J. H. 1936b. Beobachtungen zur Morphologie und Physiologie der *Lernaeocera branchialis* und *L. lusci*. Z. Parasitenk., 8: 659–696.

Schuurmans Stekhoven, J. H. and E. W. Schuurmans Stekhoven, 1956. Copépodos parasitarios de *Orthagoriscus mola* L. de la costa Argentina. Revta iber. Parasit., 16: 267–275.

Scott, A. 1901a. *Lepeophtheirus* and *Lernaea*. L. M. B. C. Mem. typ. Br. mar. Pl. Anim., No. 6: 54 pp.

Scott, A. 1901b. Some additions to the fauna of Liverpool Bay, collected May 1st, 1900, to April 30th, 1901. Trans. Lpool biol. Soc., 15: 342–353.

Scott, A. 1904. Some parasites found on fishes in the Irish Sea. Trans. Lpool biol. Soc., 18: 113–125.

Scott, A. 1907. Faunistic notes. Trans. Lpool biol. Soc., 21: 191–198.

Scott, A. 1929. The copepod parasites of Irish Sea fishes. Proc. Trans. Lpool biol. Soc., 43: 81–119.

Scott, T. 1891. Additions to the fauna of the Firth of Forth. Rep. Fishery Bd Scotl., 9, pt. 3: 300–310.

Scott, T. 1892. *Haemobaphes cyclopterina*, Fabr., in the Firth of Forth. Ann. Scot. nat. Hist., 1892: 142.

Scott, T. 1893. Additions to the fauna of the Firth of Forth, part V. Rep. Fishery Bd Scotl., 11, pt. 3: 197–219.

Scott, T. 1900. Notes on some crustacean parasites of fishes. Rep. Fishery Bd Scotl., 18, pt. 3: 144–188.

Scott, T. 1901. Notes on some parasites of fishes. Rep. Fishery Bd Scotl., 19, pt. 3: 120–153.

Scott, T. 1902. Notes on some parasites of fishes. Rep. Fishery Bd Scotl., 20, pt. 3: 288–303.

Scott, T. 1904. On some parasites of fishes new to the Scottish marine fauna. Rep. Fishery Bd Scotl., 22, pt. 3: 275–278.

Scott, T. 1905. Observations on some parasites of fishes new or rare in Scottish waters. Rep. Fishery Bd Scotl., 23, pt. 3: 108–119.

Scott, T. 1907. Some additional notes on Copepoda from the Scottish seas. Rep. Fishery Bd Scotl., 25, pt. 3: 209–220.

Scott, T. 1909. Some notes on fish parasites. Rep. Fishery Bd Scotl., 26, pt. 3: 73–77.

Scott, T. and A. Scott. 1913. The British parasitic Copepoda. Vol. I and II. Ray Society: London. 257 pp, 72 plates.

Sharapova, D. 1963. Contribution to the knowledge of the parasite fauna of fishes in Tashkeprinsk Hydroelectric Reservoir. (In Russian.) Izv. Akad. Nauk turkmen. SSR, ser. biol. Nauk, 4: 64–67.

Shen, C. -J. 1957. Parasitic copepods from fishes of China. Part II. Caligoida, Caligidae (1). Acta zool. sin., 9: 351–377.

Shiino, S. M. 1954a. Copepods parasitic on Japanese fishes. 5. Five species of the family Pandaridae. Rep. Fac. Fish. prefect. Univ. Mie, 1: 291–332.

Shiino, S. M. 1954b. Copepods parasitic on Japanese fishes. 6. Two species of the family Cecropidae. Rep. Fac. Fish. prefect. Univ. Mie, 1: 333–345.

Shiino, S. M. 1955. Copepods parasitic on Japanese fishes. 9. Family Chondracanthidae. Sub-family Chondracanthinae. Rep. Fac. Fish. prefect. Univ. Mie, 2: 70–111.

Shiino, S. M. 1957. Copepods parasitic on Japanese fishes. 15. Eudactylinidae and Dichelesthiidae. Rep. Fac. Fish. prefect. Univ. Mie, 2: 392–410.

Shiino, S. M. 1958. Copepods parasitic on Japanese fishes. 17. Lernaeidae. Rep. Fac. Fish. prefect. Univ. Mie, 3: 75–100.

Shiino, S. M. 1959a. Neuer Artname für japanische Exemplare von *Caligus bonito*. Bull. biogeogr. Soc. Japan, 20: 51–57.

Shiino, S. M. 1959b. Revision der auf Goldmakrele, *Coryphaena hippurus* L., schmarotzenden Caligidenarten. A. Rep. prefect. Univ. Mie, sect. 2, nat. Sci., 3: 1–34.

Shiino, S. M. 1959c. Sammlung der parasitischen Copepoden in der Präfekturuniversität von Mie. Rep. Fac. Fish. prefect. Univ. Mie, 3: 334–374.

Shiino, S. M. 1963a. Parasitic copepods of Eastern Pacific fishes. 2. *Lütkenia elongata* sp. nov. Rep. Fac. Fish. prefect. Univ. Mie, 4: 349–356.

Shiino, S. M. 1963b. Parasitic copepods of Eastern Pacific fishes. 4. On *Clavellistes shoyoae* gen. et. sp. nov., Rep. Fac. Fish. prefect. Univ. Mie, 4: 369–375.

Shulman, S. S. and R. E. Shulman-Albova. 1953. Parazity ryb Belogo Morya. [Parasites of fishes of the White Sea.] Izd. Akad. Nauk SSSR: Moscow and Leningrad. 199 pp.

Sidorov, E. G. 1959. Parazitofauna ryb vodokhranilishch tsentralnogo Kazakhstana. [Parasite fauna of fishes in hydroelectric reservoirs of Central Kazakhstan.] Trudy Sov. ikhtiol. Kom. (Trudy Sov. Bol. Ryb), 9: 134–137.

Silas, E. G. and A. N. P. Ummerkutty. 1967. Parasites of scombroid fishes. Part II. Parasitic Copepoda. Proc. Symp. Scombroid Fishes, Mandapam Camp (India), pt. III, pp. 876–993.

Sinderman, C. J. 1961. Parasitological tags for redfish of the western North Atlantic. Rapp. P. -v. Réun. Cons. perm. int. Explor. Mer, 150: 111–117.
Slinn, J. D. 1970. An infestation of adult *Lernaeocera* (Copepoda) on wild sole, *Solea solea*, kept under hatchery conditions. J. mar. biol. Ass. U.K., 50: 787–800.
Smith, J. W. 1969. The distribution of one monogenean and two copepod parasites of whiting, *Merlangius merlangus* (L), caught in British waters. Nytt Mag. Zool., 17: 57–63.
Solonchenko, A. I. 1968. Parazitofauna skumbrii *Scomber colias* Gmelin, obitayushchey v Atlanticheskom Okeane u poberezha yugo-zapadnoy Afriki. [Parasite fauna of *Scomber colias* Gmelin from the Atlantic Ocean off the south-west coast of Africa.] Biol. Morya, Kiev, 14: 90–95.
Sowerby, J. 1804–1806. The British Miscellany: or coloured figures of new, rare, of little known animal subjects; many not before ascertained to be inhabitants of the British Isles; and chiefly in the possession of the author. 77 plates with descriptive letterpress.
Sproston, N. G. 1941. The developmental stages of *Lernaeocera branchialis*. J. mar. biol. Ass. U.K., 25: 441–466.
Sproston, N. G. and P. T. H. Hartley. 1941. The ecology of some parasitic copepods of gadoid and other fishes. J. mar. biol. Ass. U.K., 25: 361–392.
Sproston, N. G., W.-Y. Yin and Y.-T. Hu. 1950. The genus *Lamproglena* (Copepoda parasitica): the discovery of the life history and males of two Chinese species from food fishes revealing their relationship with *Lernaea* and of both to Cyclopoidea. Sinensia, new ser., 1: 51–84.
Squires, H. J. 1966. Reproduction in *Sphyrion lumpi*, a copepod parasitic on redfish (*Sebastes* spp.). J. Fish. Res. Bd Can., 23: 521–526.
Steenstrup, J. D. 1862. *Philichthys xiphiae*, en ny snylter hos svaerd-fisken. Overs. K. danske Vidensk. Selsk. Forh., 1861: 295–305.
Steenstrup, J. J. S. and C. F. Lütken. 1861. Bidrag til kundskab om det aabne havs snyltekrebs og lernaeer samt om nogle andre nye eller hidtil kun ufulstaendigt kjendte parasitiske copepoder. K. danske Vidensk. Selsk. Skr., ser. 5, 5: 341–432.
Stephensen, K. 1913. Grønlands krebsdyr og pycnogonider. (Conspectus crustacearum et pycnogonidorum Groenlandiae). Meddr Grønland, 22: 1–479.
Stephensen, K. 1929. Marine parasitic, semiparasitic, and non-pelagic Crustacea Copepoda. Zoology Faroes, 2, pt. I, sect. 30: 1–18.
Stephensen, K. 1940. Parasitic and semiparasitic Copepoda. Zoology Iceland, 3 (34): 1–24.
Stephensen, K. 1943. The zoology of East Greenland. Marine Ostracoda, parasitic and semi-parasitic Copepoda and Cirripedia. Meddr Grønland, 121 (9): 1–24.
Stimpson, W. 1864. Synopsis of the marine invertebrates collected by the late Antarctic expedition under Dr. I. I. Hayes. Proc. Acad. nat. Sci. Philad., 1863: 138–142.
Stock, J. H. 1953. *Bomolochus soleae* Claus, 1864 and *B. confusus* n. sp.: two hitherto confounded parasitic copepods, with remarks on some other *Bomolochus* species. Beaufortia, No. 24: 1–13.
Stock, J. H. 1959. New host and distribution records of parasitic Copepoda. Bull. aquat. Biol., 1: 43.
Stossich, M. 1880. Prospetto della fauna del Mare Adriatico. Parte 3, Classe 5, Crustacea. Boll. Soc. adriat. Sci. nat., 3: 178–271.
Sukhenko, G. E. and B. Allamuratov. 1967. O paraziticheskikh rakoobraznykh ryb Uzbekistana. [On parasitic copepods of fishes of Uzbekistan.] pp. 505–506. In: Problemy Parazitologii. (Abstracts of papers given at the 5th conference of the Ukrainian Parasitological Society.)
Sumner, F. B., R. C. Osborn and L. J. Cole. 1913. A biological survey of the waters of Woods Hole and vicinity. Section 3. A catalogue of the marine fauna of Woods Hole and vicinity. Bull. Bur. Fish., Wash., 31: 545–794.
Sumpf, K. 1871. Ueber eine neue Bomolochiden-Gattung nebst Bemerkungen über die Mundwerkzeuge der sogenannten Poecilostomen. Inaugural Ph.D. dissertation, Univ. Göttingen. 32 pp.
Szidat, L. 1955. La fauna de parasitos de *Merluccius hubbsi* como caracter auxiliar para la solucion de problemas sistematicos y zoogeograficos del genero *Merluccius* L. Commun. Inst. nac. B. Rivadavia, Zool., 3: 1–54.
Szidat, L. 1956. Ueber die Parasitenfauna von *Percichthys trucha* (Cuv. & Val.) Girard der patagonischen Gewässer und die Beziehungen des Wirtsfisches und seiner Parasiten zur palaerktischen Region. Arch. Hydrobiol., 51: 542–577.
Tambs-Lyche, H. 1958. Zoogeographical and faunistic studies on West Norwegian marine animals. Årbok Univ. Bergen, Naturv. rekke, 1958 (7): 1–24.
Thomas, M. M. 1967. A new parasitic copepod, *Caligus krishnai*, from the mackerel tuna *Euthynnus affinis affinis* (Cantor). J. mar. biol. Ass. India, 9: 137–140.
Thompson, I. C. 1893. Revised report on the Copepoda of Liverpool Bay. Trans. Lpool biol. Soc., 7: 1–56.
Thompson, I. C. and A. Scott. 1903. Report on the Copepoda collected by Professor Herdman, at Ceylon, in 1902. Rep. Govt Ceylon Pearl Oyster Fish. Gulf Manaar, pt. 1, suppl. rep. 7: 1–307.
Thompson, W. 1847. Addition to the fauna of Ireland. Ann. Mag. nat. Hist., ser. 1, 20: 237–250.
Thompson, W. 1856. The natural history of Ireland. Vol. IV. H. G. Bohn: London, 516 pp.
Thomson, G. M. 1890. Parasitic Copepoda of New Zealand, with descriptions of new species. Trans. Proc. N.Z. Inst., 22 (for 1889): 353–376.

Thorell, T. T. T. 1859. Bidrag till kannedomen om krustaceer som levfa i arter af slägtet *Ascidia* L. K. svenska Vetensk-Akad. Handl., 3: 1–84.
Threlfall, W. 1968. A mass die-off of three-spined stickleback (*Gasterosteus aculeatus* L.) caused by parasites. Can. J. Zool., 41: 105–106.
Tiews, K. 1957. Biologische Untersuchungen am roten Thun (*Thunnus thynnus* (Linnaeus)) in der Nordsee. Ber. dt. wiss. Kommn Meeresforsch., 14: 192–220.
Tilesius, W. D. 1815. De cancris camtschaticis, oniscis, entomostracis et cancellis marinis microscopicis noctilucentibus. Mém. Acad. imp. Sci. St. Pétersb. [Zap. imp. Akad. Nauk], 5: 331–405.
Timm, R. 1894. Copepoden und Cladoceren. Wiss. Meeresunters., new ser., 1: 363–404.
Timon-David, P. and J. J. Musso. 1971. Les trematodes digenes du poisson-lune (*Mola mola*) dans le golfe de Marseille (Accacoeliidae, Didymozooidae). Annls Parasit. hum. comp., 46: 233–256.
Travis Jenkins, J. 1925. The fishes of the British Isles, both fresh water and salt. F. Warne & Co. Ltd.: London and New York. 408 pp.
Tremblay, J. L. and C. Lapointe. 1938. Quelques copépodes parasites de l'estuaire de St. Laurent. Annls ACFAS, 4: 100.
Turner, W. and H. S. Wilson, 1862. On the structure of *Lernaeopoda dalmanni* with observations on its larval forms. Trans. R. Soc. Edinb., 23: 77–87.
Valdez, R. A. 1974. Two parasites of threespine stickleback from Amchitka, Aleutian Islands, Alaska. Trans. am. Fish. Soc., 10:632–635.
Valle, A. 1880. Crostacei parassiti dei pesci del mare Adriatico. Boll. Soc. adriat. Sci. nat., 6: 55–90.
Valle, A. 1882. Aggiunte ai crostacei parassiti dei pesci del mare Adriatico. Boll. Soc. adriat. Sci. nat.,7: 1–3.
Vanthöffen, E. 1897. Die Fauna und Flora Grønlands. In: von Drygalski (ed.). Grönland-Exped. Gesells. Erdk. Berlin, 2: 1–383.
Vervoort, W. 1962. A review of the genera and species of Bomolochidae (Crustacea, Copepoda), including the description of some old and new species. Zool. Verh., Leiden, No. 56: 1–111.
Vervoort, W. 1969. Caribbean Bomolochidae (Copepoda: Cyclopoida). Stud. Fauna Curaçao, 27 (105): 1–125.
Veydovsky, F. 1882. On the male of *Lernaeopoda selachiorum* (In Czech). Anz. Ver. Böhm. Ärzte Naturf., 2: 58.
Vogt, C. 1877a. Recherches cotières. Premier mémoire. De la famille des philichthydes et en particulier du léposphile des labres. Mém. Inst. natn. génev., 13: 1–40.
Vogt, C. 1877b. Recherches cotières. Second mémoire. Sur quelques copépodes parasites a males pygmées habitant les poissons. Mém. Inst. natn. génev., 13: 44–104.
Wheeler, A. 1969. The fishes of the British Isles and North-West Europe. Macmillan: London, Melbourne, Toronto. 613 pp.
White, A. 1850. List of the specimens of British animals in the collection of the British Museum. Pt. 4. Crustacea. Siphonostoma. pp. 117–132.
Wierzejski, A. 1877. Ueber Schmarotzerkrebse von Cephalopoden. I. Lerneenlarven (*Pennella varians* Stp. & Ltk.?). Z. wiss. Zool., 29: 562–582.
Williams, I. C. 1963. The infestation of the redfish *Sebastes marinus* (L.) and *S. mentella* Travin (Scleroparei: Scorpaenidae) by the copepods *Peniculus clavatus* (Müller), *Sphyrion lumpi* (Krøyer) and *Chondracanthopsis nodosus* (Müller) in the eastern North Atlantic. Parasitology, 53: 501–525.
Wilson, C. B. 1905. North American parasitic copepods belonging to the family Caligidae. Part 1. The Caliginae. Proc. U.S. natn. Mus., 28: 479–672.
Wilson, C. B. 1907a. North American parasitic copepods belonging to the family Caligidae. Part 2. The Trebinae and Euryphorinae. Proc. U.S. natn. Mus., 31: 669–720.
Wilson, C. B. 1907b. North American parasitic copepods belonging to the family Caligidae. Parts 3 and 4. A revision of the Pandarinae and Cecropinae. Proc. U.S. natn. Mus., 33: 323–490.
Wilson, C. B. 1908. North American parasitic copepods. A list of those found upon the fishes of the Pacific coast, with descriptions of new genera and species. Proc. U.S. natn. Mus., 35: 431–481.
Wilson, C. B. 1910. The classification of the copepods. Zool. Anz., 35: 609–620.
Wilson, C. B. 1911. North American parasitic copepods belonging to the family Ergasilidae. Proc. U.S. nat. Mus., 39: 263–400.
Wilson, C. B. 1912. Parisitic Copepoda from Namaimo, British Columbia, including eight species new to science. Contrib. Can. Biol. Fish., 1906–1910: 85–101.
Wilson, C. B. 1915. North American parasitic copepods belonging to the Lernaeopodidae, with a revision of the entire family. Proc. U.S. natn. Mus., 47: 565–729.
Wilson, C. B. 1917. North American parasitic copepods belonging to Lernaeidae, with a revision of the entire family. Proc. U.S. natn. Mus., 53: 1–150.
Wilson, C. B. 1919. North American parasitic copepods belonging to the new family Sphyriidae. Proc. U.S. natn. Mus., 55: 549–604.
Wilson, C. B. 1920a. Parasitic copepods from the Congo basin. Bull. Am. Mus. nat. Hist., 43 (1): 1–8.
Wilson, C. B. 1920b. Report on the parasitic Copepoda collected during the Canadian Arctic Expedition, 1913–1918. Rep. Can. arct. Exped., 7(L): 1–16.
Wilson, C.B. 1921a. Report on the parasitic Copepoda collected during the survey of the Juan Fernandez

Islands, 1916–1917. The Natural History of Juan Fernandez and Easter Island, 3, Zool., pp. 69–74.
Wilson, C. B. 1921b. New species and a new genus of parasitic copepods. Proc. U.S. natn. Mus., 59: 1–17.
Wilson, C. B. 1922a. North American parasitic copepods belonging to the family Dichelesthiidae. Proc. U.S. natn. Mus., 60 (2400): 1–100.
Wilson, C. B. 1922b. Parasitic copepods in the collection of the Zoological Museum, Kristiania. Meddr zool. Mus., Kristiania, 1922 (4): 1–7.
Wilson, C. B. 1923a. Parasitic copepods in the collection of the Riksmuseum at Stockholm. Ark. Zool., 15 (No. 3): 1–15.
Wilson, C. B. 1923b. New species of parasitic copepods from southern Africa. Meddn Göteborgs Mus. Zool. Avd., 19: 1–11.
Wilson, C. B. 1923c. A new species of parasitic copepod from Florida. Am. Mus. Novit., 80: 1–4.
Wilson, C. B. 1924a. Parasitic copepods from the Williams Galapagos Expedition. Zoologica, N.Y., 5: 211–217.
Wilson, C. B. 1924b. New North American parasitic copepods, new hosts and notes on copepod nomenclature. Proc. U.S. natn. Mus., 64 (17): 1–22.
Wilson, C. B. 1932. The copepods of the Woods Hole region, Massachusetts. Bull. U.S. natn. Mus., 158: 1–635.
Wilson, C. B. 1935a. Parasitic copepods from the Dry Tortugas. Pap. Tortugas Lab., 29: 327–347.
Wilson, C. B. 1935b. Parasitic copepods from the Pacific coast. Am. Midl. Nat., 16: 776–797.
Wilson, C. B. 1936. Plankton of the Bermuda Oceanographic Expedition. IV. Notes on Copepoda. Zoologica, N.Y., 21: 89–93.
Wilson, C. B. 1937. Some parasitic copepods from Panama Bay. J. Wash. Acad. Sci., 27: 423–431.
Wright, E. P. 1877. On a new genus and species belonging to the family Pandarina. Proc. R. Ir. Acad., ser. 2, 2: 583–585.
Yamaguti, S. 1936. Parasitic copepods from fishes of Japan. Part 2. Caligoida, I. Publ. by author: Japan. 22 pp.
Yamaguti, S. 1939a. Parasitic copepods from fishes of Japan, 4. Cyclopoida, II. In: Vol. Jub. Prof. Sadao, 2: 391–415.
Yamaguti, S. 1939b. Parasitic copepods from fishes of Japan. Part 5. Caligoida, III. In: Vol. Jub. Prof. Sadao, 2: 443–487.
Yamaguti, S. 1939c. Parasitic copepods from fishes of Japan. Part 6. Lernaeopodoida, I. In: Vol. Jub. Prof. Sadao, 2: 529–578.
Yamaguti, S. 1940. *Tracheliastes polycolpus* von Nordmann, 1832, parasitic on *Leuciscus waleckii* (Dybowski) from Mandchoukuo and Sakhalin. Annotnes zool. jap., 19: 39–42.
Yamaguti, S. 1963. Parasitic Copepoda and Branchiura of fishes. Interscience Publ.: New York, London and Sydney. 1104 pp.
Yamaguti, S. and T. Yamasu. 1959. Parasitic copepods from fishes of Japan with description of 26 new species and remarks on two known species. Biol. J. Okayama Univ., 5: 89–165.
Yamaguti, S. and T. Yamasu. 1960. New parasitic copepods from Japanese fishes. Publs Seto mar. biol. Lab., 8: 141–152.
Yanulov, K. P. 1963. Parasites as indicators of local redfish stocks. In: Soviet Fisheries Investigations in the Northwest Atlantic [ed. Yu. Yu. Marti] (English translation by B. Hershkowitz, Israel Program for Scientific Translation: Jerusalem). pp. 266–276.
Yin, W.-Y., M.-E. Ling, G.-A. Hsu, I.-S. Chen, P. -R. Kuang and S.-L. Chu. 1963. Studies on the lernaeosis (*Lernaea*, Copepoda parasitica) of freshwater fishes of China. (Chinese, with extensive English summary.) Acta hydrobiol. sin., 1963 (2): 48–117.
Zayka, V. E. 1965. Parazitofauna ryb ozera Baykal. [Parasite fauna of fishes of Lake Baykal.] Nauka: Moscow, 107 pp.
Zmerzlaya, E. I. 1972. *Ergasilus sieboldi* Nordmann, 1832, ego razvitie, biologiya i epizootologicheskoe znachenie. [*Ergasilus sieboldi* Nordmann, 1832, its development, biology and epizootological significance.] Izv. gosud. nauchno-issled. Inst. ozern. rechn. ryn. Khoz., 80: 132–177.

Index of Latin Names of Copepods

Abasia 157, 158, 159, 163, 169
abbreviatus, Caligus 175, 176
abdominalis, Chondracanthus 118
Lernaeenicus 309
acanthii, Eudactylina 251, 252
255–6, 419, Figs. 1229–39
acanthias vulgaris, Kroyeria 264
Acanthocanthopsis 114
acanthocepolae, Taeniacanthus 78
Acanthochondria 12, 96, 98, 99, 100, 104, 105, 106, 108, 111, 112, 113, 123, 124, 126, 129, 130, 335
argutula 96
barnardi 126
bicornis 123
brevicorpa 124
chilomycteri 124, 126
clavata 124, *126–7*, 418, Figs. 282–96
compacta 126
cornuta 123, *124–6*, 129, 417, 418, Figs. 253–81
cornuta f.*cornuta* 126
cornuta f.*flurae* 126
deltoidea 123, 126
depressa 124, 126, 129
depressa var.*oblongata* 124
flurae 124, 126
gemina 126
laemonemae 124
lepidionis 112, 123
limandae 124, *127–8*, 418, Figs. 297–316
platycephala 124
psetti 124
soleae 124, 126, *128–9*, 419, Figs. 317–334
tchangi 124
Acanthochondrites 96, 99, 105, 106, 108, 111, 112, 113, 129, 131
annulatus 112, *130–2*, 418, 419, Figs. 335–356
inflatus 130, 131
japonicus 100, 130, 131
Acanthocolax 55, 56, 57, 58
Acespadia 324, 325, 333, 344, 346
pomposa 324, 326, 346
achantii-vulgaris, Cecrops 230
Achtheinus 212, 213
Achtheres 324, 327, 331, 333, 334, 335, 337, 339, 342, 344, 345, 346, 351, 352, 355, 356
corpulentus 355
galei 375
percarum 345, 355, *356–7*, 417, 418, Figs. 1500–16
sandrae 356, 357
selachiorum 375
sibirica 356
aculeata, Kroyeria 264, 265, 267
aculeatum, Lonchidium 265, 266
Acusicola 89
acuta, Eudactylina 251, *252–4*, 255, 257, 258, 419, Figs. 1175–1212
acutus, Holobomolochus 65
adunca, Anchorella 335
Clavella 6, 126, 331, 334, 335, 399, *400–3*, 404, 405, 409, 417, 418, 420, Figs. 1861–88
adunca f.*iadda, Clavella* 402
Advena 324, 332, 333, 334, 342, 344, 350, 351, 415, 416
paradoxa *415–16*, 419, Figs. 1996–2016
saba 415
aeglefini, Caligus 171
Aethon 109, 110, 240, 242, 244, 282, 283
quadratus 282
affinis, Dinematura 218
Dinemoura 220
Echthrogaleus 220
affixus, Lernaeenicus 309
Afrolernaea 144, 145, 146, 147, 148, 149
longicollis 146, 147
nigeriensis 146, 147
Afrolernaeinae 148
aggregatus, Nemesis 261, 262
agilis, Anchorella 400
akajeii, Trebius 205
alata, Alella 403
Clavella 329, 399, 400, *403–4*, 412, 418, Figs. 1889–1901
Dinemoura 220
Eudactylina 251
alatalongicollis, Pterochondria 97
alatus, Lernaeenicus 309
Pandarus 220
albida, Neobrachiella 332
albidus, Holobomolochus 65
Taeniacanthus 78
Albionella 324, 333, 344, 348, 351, 375, 377, 379, 380
centroscyllii 380
etmopteri 380
globosa 379, *380–1*, 419, Figs. 1691–1710
longicaudata 380
alcicornis, Tanypleurus 19
Alebion 167, 200, 201, 202
Alella 324, 326, 329, 333, 334, 344, 350, 351, 399, 404, 406 412, 413, 414
alata 403
berecynthia 412
canthari 406, 412, 413, 414
centrodonti 412
ditrematis 406, 413
elongata 412
macrotrachelus 406, 412
pagelli 406, 412, *413–15*, 417, 419, Figs. 1975–95
pterobrachiata 406, 413
richiardii 412
Alicaligus 157, 158, 159, 162, 163
tripartitus 162
aliuncus, Caligus 176
Diphyllogaster 202
Allotrifur 291, 292
alter, Mappates 159
ambiguus, Haemobaphes 305, 306, *307–8*, 417, Figs. 1392–7
Haemobaphoides 307
Nogagus *227–8*, 419, Fig. 768
americanus, Caligus 171
amplectens, Ergasilus 88, 89
Anaclavella 324, 344, 350
Anchicaligus 168
Anchistrotos 68, 69, 71, 72, 73, 74, 76, 142
balistae 74
callionymi 74
gobii 74
gracilis 74
hamatus 74
laqueus 74, *76–7*, 419, Figs. 73–5
moa 74
occidentalis 74
onosi 18, 74, *75–6*, 417, 419, Figs. 53–72
ostracionis 74
pleuronichthydis 74
sauridi 74
zeugopteri 74, *77*, 420, Figs. 76–8
Anchoracarpacea 336

Anchoraceracea 294
Anchorastomacea 109
Anchorella 337, 391, 398, 399, 405, 407, 410, 412
adunca 335
agilis 400
appendiculata 395
brevicollis 400
canthari 413, 414
centrodonti 413
emarginata 409, 410
intermedia 406
lagenula 400
ovalis 389, 390, 409, 410
pagelli 413, 414
paradoxa 415
quadrata 400
rugosa 406, 407, 409, 410
rugosa var. 400, 406
sciaenophila 391
scombri 410
stellata 404
triglae 397
uncinata 335, 400, 401
(Anchorella) uncinata, Clavella 401
Anchorelladae 336
anchoviellae, Lernaeenicus 309
Ancorellidae 109
Andreina 96, 99, 104, 105, 106, 107, 108, 111
synapturae 111
Andreinidea 111
Andropoda 383
lampri 382
angularis, Neopontius 29, 30
angustulus, Nogagus 214
annulatus, Acanthochondrites 112, *130-2*, 418, 419, Figs. 335–56
Chondracanthus 130
anonymous, Bomolochus 61
Anops cornuta 124
cyclopterina 306
antarcticensis, Paeonocanthus xii, 317, 318
Periplexis xii
Anteacheres 110, 111
Anteacheridae 111
Anthosoma 23, 84, 164, 238, 239, 240, 241, 242, 243, 244, 245, 246, 248
crassum 18, 19, 40, 42, *248–9*, 417, 418, Figs. 1026–54
smithii 248, 249
Anthosomadae 239
Anthosomatidae 241, 282
Anthosomatinae 239
Anthosomidae 241
Anthosominae 240
Anuretes 159, 160, 161, 163, 167, 168, 169
Anuretinae 168
apodus, Pseudocaligus 187
appendiculata, Anchorella 395
Isobrachia 348
appendiculatus, Pseudocycnus *268–70*, 419, Figs. 1309–29
Apus 164
arcticus, Caligus 179
arcuata, Lernaea 150
ardeolae, Holobomolochus 65
Areotrachelus 144, 145, 146, 148, 149, 297
truchae 145
argentiniensis, Lernaea 150
Argulidae 166
Argulus 164, 166, 238
argutula, Acanthochondria 96
armatus, Cybicola 268
Arnaeus thynni 203
Artacolax 57, 60
articulata, Lernaea 31
arthrosis, Ergasilus 91
Artrotrogidae 31
Artrotrogus 84
Ascomyzon asterocheres 29, 30
simulans 29, 30
Ascomyzontidae 166
asellina, Lernaea 132, 133
Lernentoma *132–4*, 417, 419, Figs. 357–79
asellinum, Medesicaste 132
asellinus, Oralien 132, 133
aspera, Eudactylina 251
Assecula 73
Asterocheres 84
asterocheres, Ascomyzon 29, 30
asterodermi, Luetkenia *236–7*, 418, Figs. 986–1002
Lütkenia 236
ater, Lernaeenicus 309
atlantica, Elytrophora 203
Nemesis 259
attenuatus, Holobomolochus 65
auritus, Ergasilus 86, 87, 88
australis, Brachiella 395
Parabrachiella 395

Baculus 240
bagri, Lernaea 150, 152
baicalensis, Coregonicola 330
Ergasilus 90
Salmincola 354
bairdii, Lerneonema 310
balistae, Anchistrotos 74
Eucanthus 74
Bariaka xii, 242, 243, 244, 250
barbicola, Lernaea 150
barilii, Lernaea 150, 152
barnardi, Acanthochondria 126
Barnardia 114
barnimiana, Lernaea 150, 151, 152
Basanistes 324, 328, 331, 335, 337, 344, 345, 346
huchonis 335
salmoneus 353
Bassettithia 242, 244, 271, 272
basteri, Lernaea 150
bataviensis, Lernaeenicus 309
bella, Caligeria 202, 203, 204
bellones, Bomolochus 61, *64–5*, 66, 417, Figs. 25–37
Parabomolochus 64
benedeni, Caligus 188
Sciaenophilus 188, 189
bengalensis, Lernaea 150
bengoensis, Caligus 176
bennetti, Sciaenophilus 188
Berea 97, 98, 99, 102, 103, 106, 108
berecynthia, Alella 412
bergyltae, Colobomatus 139
bicaudata, Lernaeopoda 367, 368
bicaudatus, Charopinus 367, 368
Pseudocharopinus 347, 366, *367–8*, 369, 381, 418, 420, Figs. 1610–20
bicolor, Caligus 214
Pandarus 213, *214–16*, 225, 228, 417, 418, 419, Figs. 769–800
bicornis, Acanthochondria 123
bicuspidatus, Caligus 171
bidiscalis, Lernaeopoda 347, 374, 375, 376, *377–9*, 417, 418, 419, Figs. 1648–73
bifurcus, Caligus 172

bilobatus, Trebius 205
Binoculus 164
piscinus 196
productus 217
sexsetaceus 223
bispinosa, Brachiella 389, 390
Neobrachiella 387, *389–91*, 394, 399, 417, 419, Figs. 1757–77
bistricornis, Lernaea 150, 152
biuncinata, Thersitina 94
biuncinatus, Ergasilus 94, 95
bivia, Lernaeopoda 375
Blias 97, 98, 99, 102, 103, 105, 106, 107, 108, 110, 111
Bocconi, Lerneopenna 313
Bomolochidae 9, 11, 16, 18, 22, 28, 30, 31, 32, 33, 34, 35, 36 38, 39, 51, 52, 53, 55, 56, 58, 59, 60, 61, 68, 70, 71, 72, 73, 81
Bomolochinae 85
Bomolochoides 60
Bomolochus 10, 11, 28, 29, 39, 55, 56, 57, 58, 59, 60, 61, 65, 66, 74, 76, 83, 84, 85
anonymous 61
bellones 61, *64–5*, 66, 417, Figs. 25–37
concinnus 61
confusus 66
constrictus 61
cuneatus 61
decapteri 61
ensiculus 61
exilipes 61
globiceps 61
hemirhamphi 61
hyporhamphi 61
jonesi 61
megaceros 61
muraenae 73
mycterobius 61
onosi 75
ostracionis 74
parvulus 61, 83
psettobius 61
selaroides 61
sinensis 61
soleae *22, 61, 62–4*, 66, 419, Figs. 1–24
tumidus 61
unicirrus 61
zeugopteri 77
bonito, Caligus 171, *173–4*, 419, Figs. 469–80
borealis, Nessipus 227
Nogagus *227*, 418, Fig. 767
boscii, Pandarus 214
bouvieri, Lophoura 321
bowmani, Clavella 399, 400
braccata, Dinematura 220
braccatus, Echthrogaleus 220
Brachiella 324, 325, 332, 333, 334, 335, 337, 339, 343, 344, 348, 349, 350, 351, 366, 382, 383, 384, 385, 387, 390, 391, 394, 396, 398
australis 395
bispinosa 389, 390
chevreuxii 391, 392, 393
impudica 388, 389, 393
impudica var.*parva* 393, 394
insidiosa 395
lageniformis 395
lithognatha 384
lophii 368, 369
macrura 391, 392
magna 384
malleus 368
merluccii 388, 389
neglecta 391
obesa 389, 390, 391
ovalis 389, 390, 409
parkeri 364, 366
parva 393
pastinaca 367, 368
pastinacae 367, 368, 369
robusta 396
rostrata 396
sciaenae 391, 392
sciaenophila 391, 392, 393
thynni 348, 384, *385–6*, 387, 417, 419, Figs. 1733–56
triglae 397
Brachiellina 348, 387
Brachiochondria 97, 98, 99, 106, 108
Brachiochondrites 99, 105, 106, 108
brachycera, Lernaea 150
brachyptera, Elytrophora 203
brachyptera brachyptera, Elytrophora 203
brachypterus, Euryphorus 202, *203–4*, 419, Figs. 741–66
Bradypontius magniceps 30
branchialis, Caligus 198
Lepeophtheirus 198
Lernaea 293, 294, 300
Lernaeocera 4, 233, 299, *300–3*, 304, 307, 417, 418, 419, 420, Figs. 1338–68
Lernoela 300
branchialis var.*sigmoida, Lernaea* 300
branchialis f.*branchialis, Lernaeocera* 301, Figs. 1334–7
branchialis f.*obtusa, Lernaeocera* 301, Figs. 1330–3
Brevibrachia 36, 351
(Brevibrachia), Salmincola 352
brevicaudatus, Caligus 171, *174–5*, 417, 419, Figs. 481–7
brevicollis, Anchorella 400
Clavella 401
Clavellina 401
Lernaeocera 303, 304
brevicorpa, Acanthochondria 124
brevipedes, Pseudocaligus 187
brevipedis, Caligus 187
Pseudocaligus 186, *187–8*, 417, Figs. 616–27
brevipes, Pseudocaligus 187
Brianella 324, 326, 341, 344, 347
corniger 329
Brianellinae 337
brongniarti, Lernaeopoda 374, 375
buccatus, Cybicola 268

caeruleus, Ergasilus 88
Calanoida 48, 49
Calanus 20, 46
finmarchicus 20
californiensis, Salmincola 17, 334, 352
Caligacea 166
Caligeria 166, 167, 201, 202
bella 202, 203, 204
Caligidae 6, 8, 9, 10, 11, 15, 16, 17, 18, 21, 22, 29, 31, 33, 34, 39, 41, 42, 52, 53, 59, 111, 156, 157, 159, 160, 162, 164, 165, 166, 167, 168, 169, 170, 186, 188, 189, 200, 201, 202, 203, 204, 205, 206, 208, 210, 211, 212, 214, 226, 229, 239, 240, 242, 339, 342
Caligina 165, 167, 211, 212, 336
Caliginae 166, 168, 169, 201
Caligini 167
Caligodes 159, 160, 163, 167, 168
megacephalus 159
Caligoida 29, 52, 111
Caligoidea 31
Caligopsis 41, 158, 161, 163, 169, 187

Caligulina 156, 158, 162, 163
ocularis 164
Caligus 7, 8, 10, 16, 18, 24, 27, 39, 40, 41, 42, 46, 83, 156, 157, 158, 159, 162, 163, 164, 165, 166, 167, 168, 169, 170, 171, 176, 177, 183, 185, 186, 188, 189, 190, 199, 202, 206, 238, 246, 249, 265
abbreviatus 175, 176
aeglefini 171
aliuncus 176
americanus 171
arcticus 179
benedeni 188
bengoensis 176
bicolor 214
bicuspidatus 171
bifurcus 172
bonito 171, *173–4,* 419, Figs. 469–80
branchialis 198
brevicaudatus 171, *174–5,* 417, 419, Figs. 481–7
brevicaudus 175
brevipedis 187
centrodonti 35, 171, *175-6,* 418, 419, Figs. 505–18
chelifer 36, 163
clemensi 17, 181
constrictus 202
coryphaenae 41, 156, 162, 170, 171, *176–7,* 178, 185, 419, Figs. 519–30
crassus 238, 248, 249
curtus 25, 26, 164, 170, *171–3,* 178, 182, 191, 417, 418, 419, 420, Figs. 433–68
diaphanus 171, *178–9,* 417, 418, 419, Figs. 538–48
elegans 171
elongatus 171, 176, *179–80,* 181, 417, 418, 419, 420, Figs. 549–58
epidemicus 171
euthynnus 177
fallax 171
gracilis 198, 199
gurnardi 171, 179, *180–1,* 182, 183, 417, 419, Figs. 559–71
heptapus 223
hippoglossi 191
hippoglossis 191
imbricatus 248, 249
katuwo 184
krishnai 174
kroeyeri 179
kuroshio 173, 174
labracis 171, *181–2,* 183, 418, Figs. 572–81
lacustris 170
latifrons 179
(Lepeophtheirus) pectoralis 190
lobatus 184
lobodes 170
lumpi 179
minimus 162, 171, 181, *182–3,* 417, 418, 420, Figs. 582–91
minimus var. *mugilis* 182, 183
minutus 182, 183
monacanthi 184
mülleri 171
multispinosus 178, 179
nanus 180, 181
nordmanni 193
oblongus 238, 246, 247
ornatus 193
pacificus 196
paradoxus 223
pelamydis 171, *183–4,* 185, 419, Figs. 489–504
piscinus 199
productus 171, *184–6,* 217 Figs. 592–603
quadratus 177, 185
quinqueabdominalis 158
rabidus 179
rapax 179, 180, 181
rissoanus 179
salmonis 196
(Sciaenophilus) benedeni 188
scomberi 183, 184, 185
scombri 183, 184, 185
scutatus 176
spinosus 29
strömii 196
sturionis 197
tesserifer 176
thymni 176
torpedinis 178, 179
trachypteri 179
vespa 196
vesper 196
willungae 171
zei 171, *186,* 420, Figs. 604–15
Calina 168
Calistes 166, 167, 168
callionymi, Anchistrotos 74
cameroni, Dendrapta 330, 371, 375
cameroni longiclavata, Dendrapta 332
canaliculata, Clavella 339, 400
Cancerillidae 24, 46
canthari, Alella 406, 412, 413, 414,
Anchorella 413, 414
Clavella 413, 414
Clavellodes 404, 406, 412, 413
canthigasteri, Taeniacanthus 78
Canthocamptus 37
Canuella 50
caparti, Lernaeocera 300
Lophoura 321
carangis, Midias 202
carassii, Lernaea 153
carchariae, Taeniacanthus 68, 69, 78
carchariae-glauci, Eudactylina 251
Kroyeria 265
Pagodina 259
carcharii-glauci, Pandarus 214
carcharini, Pandarus 214
Cardiodectes 43, 289, 290, 295, 296
cardusa, Lophoura 321
Caritus 158
carpionis, Lernaeopoda 353
Salmincola 327, 352
Catlaphila 23
elongata 24
Catlaphilidae 241
catostomi, Lernaea 150
caudatus, Trebius 205, *206–7,* 417, 418, 419, Figs. 722–40
Cauloxenus 324, 344, 345
stygius 324
Cecropidae 8, 9, 17, 18, 21, 31, 33, 52, 53, 166, 167, 168, 211, 212, 213, 229, 230, 236
Cecropinae 166, 167, 168, 211
Cecrops 8, 164, 165, 166, 211, 212, 229, 230, 233, 238
achantii-vulgaris [sic] 230
desmaresti 230
exiguus 230, 232
latreilli 40, *230–2,* 235, 418, Figs. 905–35
Cecropsina 212
glabra 236, 237
centrodonti, Alella 412
Anchorella 413
Caligus 35, 171, *175–6,* 418, 419, Figs. 505–18
centropristi, Lernaeenicus 309

Lernaeocera 299
centroscyllii, Albionella 380
Lernaeopoda 348, 375, 380
Ceratochondria 96, 99, 100, 101, 103, 106, 108
Ceratocolax 55, 56, 57, 60
euthynni 57
cerberus, Lernaeenicus 309
ceylonensis, Ergasilus 89
chackoensis, Lernaea 153
chaetodontis, Pseudanuretes 170
Chalimus 165, 166
characis, Clavella 399
Charopinopsis 324, 333, 344, 348
quaternia 348
Charopinus 316, 317, 324, 325, 326, 329, 330, 332, 333, 334, 336, 337, 339, 341, 343, 344, 346, 347, 348, 349, 350, 351, 362, 363, 366, 368, 370, 371, 373, 383
bicaudatus 367, 368
dalmanni 347, *363–4*, 365, 371, 419, Figs. 1566–76
dasyaticus 367
dasyatis 363
dubius 363, *364–6*, 368, 369, 371, 419, Figs. 1577–97
malleus 368
parkeri 363, 366
pastinacae 368
ramosus 370, 371
schahriar 363
chelifer, Caligus 36, 163
Chelodiniformis 111
chevreuxii, Brachiella 391, 392, 393
Neobrachiella 325, 326, 328, 329, 331, 387, *391–3*, 417, Figs. 1795–1809
chilomycteri, Acanthochondria 124, 126
chimaerae, Schistobrachia 347, 370
Vanbenedenia 326, 329, 361, 362
chinensis, Clavellisa 408
Epiclavella 408
Chondracanthidae 12, 19, 21, 22, 30, 31, 32, 33, 34, 35, 46, 51, 52, 54, 96, 98, 99, 100, 101, 102, 103, 105, 106, 107, 108, 109, 110, 111, 112, 113, 136, 144, 317, 335, 336
Chondracanthina 110, 336
Chondracanthinae 54, 109, 111, 113, 133
Chondracanthodes 96, 97, 101, 105, 106, 108
Chondracanthopsis 111, 114, 123
nodosus 120
Chondracanthus 5, 40, 96, 97, 98, 100, 101, 102, 104, 105, 106, 108, 109, 110, 111, 112, 113, 114, 119, 122, 123, 124, 125, 130
abdominalis 118
annulatus 130
clavatus 126
cornutus 109, 124
Delarochiana 118
delarochiana 114
depressus 124, 125, 126
depressus var.*oblongatus* 124, 125
distortus 97, 98, 114, 117
flurae 124, 125
gibbosus 118
gibbus 118
gurnardi 132
horridus 97, 114
inflatus 130, 131
laevirajae 131
lepidionis 112
limandae 127
lophii 6, 97, 112, 114, *118–19*, 122, 418, Figs. 175–95
Lophii 118
Lophius 118
merlangi 119
merluccii 114, *119–20*, 418, Figs. 196–212
narium 112
neali 114, *117–18*, 418, Figs. 159–74
nodosus 114, *120–2*, 419, Figs. 213–31
ornatus 97, 98, 114, *122–3*, 417, Figs. 232–52
pallidus 124, 131, 132
palpifer 120
pinguis 97, 98
soleae 128
stramineus 119, 120
Triglae 132
triglae 109, 132, 317
trilobatus 114
triventricosus 112
tuberculatus 109, 116
williamsoni 120
wilsoni 114, 121
xyphiae 119
zei 109, 112, 113, *114–17*, 119, 420, Figs. 129–58
zeus 114
chondrostomi, Tracheliastes 357
cincinnatus, Phrixocephalus 13, 14, 43, 311
cirrosa, Lernaea 313
Clausidiidae 30, 51
clavata, Acanthochondria 124, *126–7*, 418, Figs. 282–96
Clavella 335
clavatus, Chondracanthus 126
Clavella 17, 28, 109, 110, 166, 239, 240, 273, 324, 332, 333, 334, 335, 340, 341, 342, 343, 344, 346, 348, 349, 350, 351, 366, 382, 387, 394, 399, 400, 404, 406, 407, 408, 409, 410, 412, 414, 415
adunca 6, 126, 331, 334, 335, 399, *400–3*, 404, 405, 409, 417, 418, 420, Figs. 1861–88
adunca f.*iadda* 402
alata 329, 399, 400, *403–4*, 412, 418, Figs. 1889–1901
(Anchorella) alata 401
bowmani 399, 400
brevicollis 401
canaliculata 339, 400
canthari 413, 414
characis 399
clavata 335
cluthae 275
crassa 276
(Cycnus) canthari grisei 412
deliciosa 401
devastatrix 401
dubia 401
emarginata 409
gracilis 333, 383, 399, 400
hippoglossi 273, 274, 399
iadda 401
insolita 399, 400
invicta 401
irina 399, 400
labracis 275, 278, 279
levis 399, 400
lophii 368, 401
macrotrachelus 406, 413, 414
Mulli 276
ovata 399, 400
pagelli 413, 414
pagri 399
paradoxa 415
parva 399, 400
perfida 399, 400
pinguis 399, 400
porogadi 399, 400
quadrata 401, 402
recta 400
rugosa 406
sciaenae 391, 392

sciatherica 401
scombri 410
scombropis 399, 400
squamigera 399, 400
stellata 329, 350, 400, *404–5,* 418, Figs. 1902–13
stichaei 399, 400
triglae 397
tumida 399, 400
typica 401
uncinata 335, 400, 401, 402
Clavellidae 338, 349
Clavellina brevicollis 401
uncinata 401
Clavellinae 110, 337, 338
Clavellisa 324, 326, 330, 332, 333, 334, 340, 343, 344, 349, 350, 351, 399, 407, 408, 410, 412
chinensis 408
cordata 408, 412
dorosomatis 408
dussumieriae 408, 412
emarginata 326, 332, 349, 407, 408, *409–10,* 411, 412, 417, Figs. 1937–57
hilsa 408
ilishae 408
obcordata 408
ovalis 409
pellone 408
phasa 408
scombri 408, *410–12,* 416, 419, Figs. 1958–74
spinosa 408, 412
Clavellistes 324, 333, 344, 349, 351, 381, 383
lampri 333, 349, *382–3,* 418, Figs. 1711–32
shoyoae 382, 383
Clavellodes 324, 326, 333, 334, 343, 344, 350, 351, 399, 405, 406, 412, 413, 414
canthari 404, 406, 412, 413
macrotrachelus 413
pagelli 413
rugosa 328, 405, *406–7,* 412, 417, Figs. 1914–36
Clavellomimus 324, 333, 344, 350, 406
Clavellopsis 324, 332, 344, 348, 350, 366, 399
dasyaticus 367
laciniata 326, 329
longimana 412
paradoxa 415
saba 415
clavigera, Lernaeopoda 354
Clavissodalis 69, 70, 71, 72, 73
clemensi, Caligus 17, 181
cleopatra, Lamproglena 143
cluthae, Clavella 275
Hatschekia 273, *275,* 276, 417, 418, Figs. 1116–30
Lernaeopoda 372
coleoptrata, Dinematura 220
Dinemoura 220
coleoptratus, Echthrogaleus 42, *220–2,* 418, 419, Figs. 822–49
coleus, Ergasilus 87, 88
Taeniacanthus 78
Collipravus 288, 296
parvus 305
Collocheres gracilicauda 29, 30
Colobomatus 135, 136, 137, 138, 139
bergyltae 139
compacta, Acanthochondria 126
complexa, Eudactylina 252, 254
concavus, Pseudocharopinus 366
concinnus, Bomolochus 61
confusus, Bomolochus 66
Holobomolochus 63, 65, *66–7,* 417, 418, 420, Figs. 38–52
Congericola xii, 9, 11, 12, 239, 240, 241, 242, 244, 250, 271, 272, 273, 279, 281
congri 279
gracilis 281
kabatai xii
pallida 280
pallidus 131, 241, 279, *280–1,* 417, Figs. 1158–74
congri, Congericola 279
Metahatschekia 242, 279
constrictus, Bomolochus 61
Caligus 202
cordata, Clavellisa 408, 412
Coregonicola 324, 330, 344, 345, 346
baicalensis 330
coregonorum, Salmincola 352
corniger, Brianella 329
cornigera, Hatschekia 277, 278
cornuta, Acanthochondria 123, *124–6,* 129, 417, 418, Figs. 253–81
Anops 124
Entomoda 124
Laminifera 225
Lamproglena 147, 148
Lernaea 124
Lernentoma 124
Lophoura 321
Phyllophora 225
cornuta f.*cornuta, Acanthochondria* 126
cornuta f.*flurae, Acanthochondria* 126
cornutus, Chondracanthus 109, 124
Phyllophorus 225
Phyllothyreus 22, 224, *225–6,* 418, 419, Figs. 878–904
corpulentus, Achtheres 355
corvinae, Sphaerosoma 138
Corycaeidae 84
Corycaeus 20
coryphaenae, Caligus 41, 156, 162, 170, 171, *176–7,* 178, 185, 419, Figs. 519–30
Costai, Pennella 313
cotti, Ergasilus 87
cottidarum, Salmincola 352
cranchii, Pandarus 214
crassa, Clavella 276
crassum, Anthosoma 18, 19, 40, 42, *248–9,* 417, 418, Figs. 1026–54
crassus, Caligus 238, 248, 249
Creopelates 289, 290
cruciata, Lernaea 150
Cryptochondria 97, 99, 100, 104, 105, 106, 108
crypturus, Nessipus 227
ctenopharyngodonis, Lernaea 154
cuneifer, Lepeophtheirus 162
cuneatus, Bomolochus 61
cunula, Ergasilus 88, 89
curticaudis, Paralebion 202
curtus, Caligus 25, 26, 164, 170, *171–3,* 178, 182, 191, 417, 418, 419, Figs. 433–68
Cybicola 242, 243, 244, 268, 273
armatus 268
buccatus 268
elongatus 268
cyclophora, Lernaea 310, 311
Cyclopidae 20
Cyclopoida 19, 23, 24, 29, 46, 48, 49, 50, 51, 52, 85, 111, 139, 142, 239, 293
Cyclops 46, 51, 83
cyclopterina, Anops 306
Haemobaphes 29, 305, *306–7,* 308, 417, 418, 419, Figs. 1385–91
Lernaea 294, 306, 400
Lernaeocera 306
cyclopterinus, Haemobaphes 306
Schisturus 306

Cycnus 109, 240, 281
gracilis 281
Labri mixti 275
Pagelli Bogneravei 633
pallidus 280
(Cycnus) canthari grisei, Clavella 412
(Cycnus) pagelli bogneravei, Hatschekia 277
cylindricum, Peroderma 42
cyprinacea, Lernaea 17, 142, 148, 149, 150, 152, *153–5*, 293, 294, 352, 417, 418, 419, Figs. 380–405
Lernaeocera 153, 294, 353
cyprinaceus, Ergasilus 88
dalmanni, Charopinus 347, *363–4*, 365, 371, 419, Figs. 1566–76
Lernaea 363
Lernaeopoda 363, 368

Dartevellia 159, 160, 161, 163
dasyaticus, Charopinus 367
Clavellopsis 367
Pseudocharopinus 366, 367
dasyatis, Charopinus 363
decapteri, Bomolochus 61
Delamarina nigeriensis 146
Delarochiana, Chondracanthus 118
delarochiana, Chondracanthus 114
Lernacantha 114
Lernacanthus 114
deliciosa, Clavella 401
deltoidea, Acanthochondria 123, 126
Demoleus 208, 209, 210, 211, 212, 214, 222
heptapus 211, 222, *223–4*, 417, 419, Figs. 850–77
latus 211, 222
paradoxus 223
Dendrapta 324, 333, 344, 347
cameroni 330, 371, 375
cameroni longiclavata 332
dentatus, Pseudocharopinus 17, 329, 330, 366
Taeniacanthus 78
denticulatus, Echthrogaleus 220
Dentigryps 164, 170
depressa, Acanthochondria 124, 126, 129
depressa var.*oblongata, Acanthochondria* 124
depressus, Chondracanthus 124, 125, 126
depressus, var.*oblongatus, Chondracanthus* 124, 125
desmaresti, Cecrops 230
devastatrix, Clavella 401
diaphanus, Caligus 171, *178–9*, 417, 418, 419, Figs. 538–48
diceracephala, Lernaea 150
diceraus, Haemobaphes 29, 305, 306
Pseudochondracanthus 97, 98
Dichelesthiidae 11, 18, 19, 22, 31, 32, 34, 43, 52, 53, 84, 148, 166, 167, 238, 239, 240, 241, 242, 243, 244, 245, 246, 249, 250, 264, 268, 335
Dichelesthiina 240
Dichelesthiinae 240
Dichelesthioidea 241, 245
Dichelesthium 23, 39, 84, 164, 166, 238, 239, 240, 241, 242, 243, 244, 245, 246, 273
oblongum 17, 18, 31, *246–8*, 417, Figs. 1003–25
sturionis 246, 247
Dichelestidae 239
Dichelestina 544
Dichelestinae 239
Dichelestium 239
Dicrobomolochus 55, 56, 57, 58
dilectus, Haemobaphes 305
Dinematura 165, 212, 217
affinis 218
braccata 220
coleoptrata 220
elongata 218, 219
gracilis 217
indistincta 220
lamnae 218
producta 217
serrata 234
sexsetacea 223
Dinemoura 165, 166, 185, 208, 209, 210, 211, 212, 214, 217, 221, 235
affinis 220
alata 220
coleoptrata 220
discrepans 210, 211, 217, 219
ferox 210, 217, 219
hamiltoni 217
lamnae 218
latifolia 210, 217
producta 185, 210, *217–19*, 221, 417, 418, 419, Figs. 801-21
Dinemura 165, 238
Diocus 97, 98, 100, 101, 102, 103, 104, 105, 106, 108, 110, 111
diodontis, Pennella 294
Diphyllogaster 161, 163, 168
aliuncus 202
Dirhynchus malleus 369
discrepans, Dinemoura 210, 211, 217, 219
dispar, Kroyeria 265
Disphaerocephalus 97, 111, 114, 123
ornatus 122
Dissonidae 6, 8, 9, 17, 21, 31, 32, 33, 202
Dissonus 6, 7, 9, 11, 33, 40, 201, 202
nudiventris 9
distortus, Chondracanthus 97, 98, 114, 117
ditrematis, Alella 406, 413
duello-juradoi, Laminifera 225
dolabroides, Lernaea 150
Donusa 239, 240
Doridicola 84
dorosomatis, Clavellisa 408
dubia, Clavella 401
dubius, Charopinus 363, *364–6*, 368, 371, 419, Figs. 1577–97
Dufresnii, Lernentoma 118
dussumieriae, Clavellisa 408, 412
Dysgamus 167, 168, 201, 202, 203
longifurcatus 203
sagamiensis 203, 204
Dysphorus 37, 144, 145, 146, 147, 148, 149

Echetinae 168
Echetus 157, 159, 163, 168, 169, 296
echinata, Kroyeria 265
Echinirus 36, 69, 70, 72, 73
Echinosocius 69, 70, 71, 72
Echthrogaleus 209, 210, 211, 212, 213, 214, 217, 219, 220, 222
affinis 220
braccatus 220
coleoptratus 42, *220–2*, 418, 419, Figs. 822–49
denticulatus 220
indistinctus 220
lutkenii 220
pectinatus 220
pellucidus 220
perspicax 220
torpedinis 209, 220
Ectinosoma 50
edwardsi, Lophoura 320, *321–2, 417, Figs. 1451–8*
Rebelula 321
edwardsii, Lepeophtheirus 16
Salmincola 324, 326, 352, 373
elegans, Caligus 171
Lernaea 150, 154

elongata, Alella 412
Catlaphila 24
Dinematura 218, 219
Kroyeria 265
Lernaea 359
Lernaeopoda 359, 360
Luetkenia 236, 237
Ommatokoita 330, 346, *359–61,* 418, 419, Figs. 1530–52
elongatus, Caligus 171, 176, *179–80,* 181, 417, 418, 419, 420, Figs. 549–58
Cybicola 268
Kroyeria 265
Paeon 315, 316, 317, 318
Trebius 205
Elytrophora 167, 201, 202, 203
atlantica 203
brachyptera 203
brachyptera brachyptera 203
hemiptera 203
indica *203*
emarginata, Anchorella 409, 410
Clavella 409
Clavellisa 326, 332, 349, 407, 408, *409–10,* 411, 412, 417, Figs. 1937–57
encrasicoli, Lernaea 311
Lernaeenicus 309, 310, *311–12,* 417, 419, Figs. 1418–22
Lerneonema 311
enodis, Haemobaphes 305
ensiculus, Bomolochus 61
Enterocola 240
Enterocolidae 50
Entepherus 213, 229
laminipes 212
Entomoda 352
cornuta *124*
salmonea 353
Entomostraca 164
Eolicola 84
Epachthes 240, 336
Epibrachiella 337, 348, 387, 394
impudica 393
Epiclavella chinensis 408
epidemicus, Caligus 171
Ergasilidae 4, 15, 16, 18, 23, 28, 29, 30, 31, 34, 36, 51, 52, 53, 56, 59, 72, 81, 82, 84, 85, 86, 87, 111, 139, 166
Ergasilina 83, 84, 165, 238, 239
Ergasilina 59, 239, 240
robusta 261
Ergasilinae 85
Ergasiloidea 84
Ergasiloides 85, 87, 89
Ergasilus 10, 11, 20, 21, 23, 28, 30, 35, 51, 59, 81, 82, 83, 84, 85, 86, 87, 88, 89, 91, 92, 94, 95
amplectens 88, 89
arthrosis 91
auritus 86, 87, 88
baicalensis 90
biuncinatus 94, 95
caeruleus 88
ceylonensis 89
coleus 87, 88
cotti 87
cunula 88, 89
cyprinaceus 88
esocis 90
felichthys 89
funduli 88, 89
gasterostei 94
gibbus 83, 89, *93–4,* 417, Figs. 2017–31
hoferi 90
hypomesi 89
iheringi 89
lizae 92, 93
mendisi 89
mugilis 92
nana 93
nanus 89, *92–3,* 417, Figs. 101–7
nerkae 87
semicoleus 87, 88
sieboldi 18, 82, 83, 87, 89, *90–1, 417, 419, Figs. 88–100*
sieboldi var.*patagonicus* 91
surbecki 90
tenax 88, 89
trisetaceus 90
tumidus 87
wilsoni 87
esocina, Lernaea 153, 154
Lernaeocera 153, 294
esocis, Ergasilus 90
etmopteri, Albionella 380
Lernaeopoda 375, 380
Eubrachiella 324, 328, 333, 344, 348
gaini 332, 348
lophii 368
sublobulata 331, 333
Eucanthus 59
balistae 74
marchesetti 74
Euclavellisa 324, 326, 333, 334, 344, 350
Eudactylina 18, 40, 42, 59, 236, 240, 241, 242, 243, 244, 250, 251, 258
acanthii 251, 252, *255–6,* 419, Figs. 1229–39
acuta 251, *252–4,* 255, 257, 258, 419, Figs. 1175–1212
alata 251
aspera 251
carchariae-glauci 251
complexa 252, 254
insolens 251, 252, *256–7,* 417, Figs. 1240–51
lancifera 251
minuta 251, 252, *257–8,* 417, Figs. 1252–65
musteli-laevis 251
papillosa 251, 258
rachelae 254, 255, Figs. 1215, 1216
similis 251, 252, *254–5,* 257, 419, Figs. 1213–28
spinula 254
squatinae-angeli 251
Eudactylinae 240
Eudactylinella 241, 242, 243, 244
Eudactylinidae xii, 18, 31, 34, 38, 39, 52, 53, 241, 245, 250, 251, 258
Eudactylinodes 241, 242, 243, 244, 250
Eudactylinopsis 242, 243, 244, 250
Euryphora 211
Euryphoridae 8, 9, 17, 21, 31, 33, 52, 53, 168, 170, 200, 201, 202, 205
Euryphorinae 168, 201, 202
Euryphorus 8, 41, 166, 167, 200, 201, 202, 203
brachypterus 202, *203–4,* 419, Figs. 741–66
nordmanni 203
euthynni, Ceratocolax 57
euthynnus, Caligus 177
exiguus, Cecrops 230, 232
exilipes, Bomolochus 61
Opimia 318
exilis, Trebius 205
exocoeti, Pennella 314
extensus, Norion 282
Salmincola 352
extumescens, Salmincola 352, 355

fallax, Caligus 171
felichthys, Ergasilus 89

Macrobrachinus 85
ferox, Dinemoura 210, 217, 219
Paeon 318
fibrosa, Pennella 313
filosa, Pennatula 294, 313
Pennella 294, *313–14*, 418, 419, Figs. 1432–9
fimbriata, Monima 235
Thysanote 332
finmarchicus, Calanus 20
fissifrons, Pandarus 214
fistula, Peniculus 294
fistulariae, Pseudocaligus 187
Fistulata 29
flagellans, Taeniacanthus 78
floridanus, Pandarus 214
flurae, Acanthochondria 124, 126
Chondracanthus 124, 125
Foroculum spratti 310
fortipedis, Pseudanuretes 169
fugu, Pseudocaligus 187
Taeniacanthus 78
funduli, Ergasilus 88, 89

gaini, Eubrachiella 332, 348
galei, Achtheres 375
Lernaeopoda 347, 374, *375–7*, 379, 417, 418, 419, Figs. 1674–90
galeivulgaris, Kroyeria 264
Gangliopus 208, 209, 210, 212
japonicus 208
pyriformis 208, 213
tetrapturi 213
gasterostei, Ergasilus 94
Lernaeocera 153
Thersites 84, 94
Thersitina 81, *94–5*, 417, 418, Figs. 108–28
gasterosteus, Thersites 94
gemina, Acanthochondria 126
gemursa, Kroyeria 265
germonia, Pennella 314
gibbosus, Chondracanthus 118
gibbus, Chondracanthus 118
Ergasilus 83, 89, *93–4*, 417, Figs. 2017–31
Lepeophtheirus 198
gigas, Tracheliastes 357
Gisleri, Lernanthropus 284
gisleri, Lernanthropus *284–5*, 417, 420, Figs. 1055–70
glabra, Cecropsina 236, 237
Lütkenia 236
globiceps, Bomolochus 61
globosa, Albionella 379, *380–1*, 419, Figs. 1691–1710
Lernaeopoda 348, 375, 380
Gloiopotes 167, 200, 201, 202
gnathonicus, Lernaeenicus 309
Gnathostoma 46, 50, 51
gnavus, Lernaeenicus 309
gobii, Anchistrotos 74
gobina, Lernaea 294, 300
godfroyi, Lernaeocera 299
gonostomae, Lernaeenicus 309
gordoni, Salmincola 353, 354
gracilescens, Lepeophtheirus 198
gracilicauda, Collocheres 29, 30
gracilis, Anchistrotos 74
Caligus 198, 199
Clavella 333, 383, 399, 400
Congericola 281
Cycnus 281
Dinematura 217
Kroyeria 264
Lepeophtheirus 198
Lernaeenicus 309
Lophoura 321
Nogagus 218
Taeniacanthodes 68, 69
grandis, Nogagus 224, 225
Vanbenedenia 361
gurnardi, Caligus 171, 179, *180–1*, 182, 183, 417, 419, Figs. 559–71
Chondracanthus 132
Gymnoplea 47, 49

Haemobaphes 291, 292, 294, 295, 269, 297, 299, 303, 305, 306
ambiguus 305, 306, *307–8*, 417, Figs. 1392–7
cyclopterina 29, 305, *306–7*, 308, 417, 418, 419, Figs. 1385–91
cyclopterinus 306
diceraus 29, 305, 306
dilectus 305
enodis 305
intermedius 305
theragrae 305
Haemobaphoides 296, 308
ambiguus 307
Haeniochophilus 158, 161, 162, 163, 169
hamatus, Anchistrotos 74
hamiltoni, Dinemoura 217
haplocephala, Lernaea 150, 152
hardingi, Lernaea 150
Harpacticoida 48, 50
Harpacticus 46
Hatschekia 11, 12, 19, 31, 32, 43, 109, 166, 240, 241, 242, 243, 244, 245, 271, 272, 273, 275, 276, 277, 281, 335, 399, 412
cluthae 273, *275*, 276, 417, 418, Figs. 1116–30
cornigera 277, 278
(*Cycnus*) *pagelli bogneravei* 277
hippoglossi *273–4*, 417, Figs. 1101–19
labracis 273, *275–6*, 278, 418, Figs. 1136–40
labri-mixti 275
mulli 273, *276–7*, 418, Figs. 1141–51
pagellibogneravei 273, 276, *277–8*, 418, Figs. 1127–35
pygmaea 273, *278–9*, 417, Figs. 1152–7
sargi 278
Hatschekiidae 18, 19, 22, 29, 31, 32, 33, 36, 52, 54, 245, 250, 271, 272, 279, 399
Hemicyclops 46, 51
hemigalei, Trypaphylum 322
hemiptera, Elytrophora 203
hemirhamphi, Bomolochus 61
Lernaeenicus 309
heptapus, Caligus 223
Demoleus 211, 222, *223–4*, 417, 419, Figs. 850–77
Hermilius 157, 158, 159, 161, 162, 163, 167, 168, 169
youngi 159
Herpyllobius 148, 295, 336, 337
hesarangattensis, Lernaea 150
Heterochondria 98, 99, 105, 106, 108
longicephala 96
heterotidicola, Lernaeogiraffa 146
hilsa, Clavellisa 408
hippocephalus, Stylophorus 363
hippoglossi, Caligus 191
Clavella 273, 274, 399
Hatschekia *273–4*, 417, Figs. 1101–19
Lepeophtheirus 190, *191–3*, 417, 419, Figs. 648–61
hippoglossis, Caligus 191
hoferi, Ergasilus 90
Holobomolochus 55, 57, 58, 60, 65
acutus 65
albidus 65
ardeolae 65
attenuatus 65

confusus 63, 65, *66–67*, 417, 418, 420, Figs. 38–52
longicaudus 65
nothrus 65
occultus 65
palleucus 65
prolixus 65
scutigerulus 65
spinulus 65
venustus 65
Homoiotes 168
horridus, Chondracanthus 97, 114
huchonis, Basanistes 335
humesi, Scolecicara 69
Humphreysia 97, 98, 99, 103, 104, 105, 106, 108
Hymenopoda 238
hypomesi, Ergasilus 89
hyporhamphi, Bomolochus 61

iadda, Clavella 401
Ichthyotaces 135, 136, 137, 138
Ichthyotacinae 139
iheringi, Ergasilus 89
ilishae, Clavellisa 408
imbricata, Otrophesia 248
imbricatus, Caligus 248, 249
Immanthe 97, 99, 100, 103, 104, 105, 106, 107, 108
Impexus 291, 298
impudica, Brachiella 388, 389, 393
Epibrachiella 393
Neobrachiella 328, 329, 331, 387, *393–4*, 395, 396, 398, 417, 419, Figs. 1777–94
Thysanote 393
impudica var.*parva, Brachiella* 393, 394
incisus, Paranessipus 211
indica, Elytrophora 203
indicus, Taeniacanthus 78
indistincta, Dinematura 220
indistinctus, Echthrogaleus 220
indivisa, Nectobrachia 326, 329, 330, 345
inflata, Lernaea 150
inflatus, Acanthochondrites 130, 131
Chondracanthus 130, 131
Teredophilus 81
innominatus, Lepeophtheirus 194, 195
insidiosa, Brachiella 395
Neobrachiella 329, 383, 387, 388, *395–6*, 397, 418, Figs. 1810–20
Parabrachiella 395
insidiosa f.*insidiosa, Neobrachiella* 396
insidiosa f.*lageniformis, Neobrachiella* 396
insidiosa f.*pacifica, Neobrachiella* 396
insignis, Lepeophtheirus 193
insolens, Eudactylina 251, 252, *256–7*, Figs. 1240–1251
Lernaea 150
insolita, Clavella 399, 400
integra, Luetkenia 213, 236
intercedens, Lamproglena 147, 148
intermedia, Anchorella 406
intermedius, Haemobaphes 305
invicta, Clavella 401
irina, Clavella 399, 400
Irodes 73
Ismaila 111
Isobranchia 348, 384
appendiculata 348

jacuticus, Salmincola 352
japonicus, Acanthochondrites 100, 130, 131
Gangliopus 208
javanicus, Trebius 205
jonesi, Bomolochus 61
jordani, Lamproglena 147
Juanettia 97, 98, 100, 101, 102, 103, 104, 105, 106, 107, 108

kabatai, Congericola xii
Kabataia 8, 9, 205
ostorhinchi 205
kamorahai, Lophoura 321
katoi, Pandarus 214
katuwo, Caligus 184
kirtii, Trebius 205
kitamakura, Taeniacanthus 78
komai, Sarcotaces 135
krishnai, Caligus 174
kroeyeri, Caligus 179
Vanbenedenia 330, 344, 346, *361–2*, 417, Figs. 1553–65
Kroeyerina 241, 242, 243, 244, 250, 264, 265
Kroyeri, Lernanthropus 285
kroyeri, Lernanthropus 284, *285–6*, 417, Figs. 1071–81
Kroyeria 9, 11, 23, 36, 41, 42, 239, 240, 241, 242, 243, 244, 245, 250, 264, 265
acanthias vulgaris 264
aculeata 264, 265, 267
carchariaeglauci 265
dispar 265
echinata 265
elongata 265
elongatus 265
galeivulgaris 264
gemursa 265
gracilis 264
lineata 264, *265–7*, 417, 418, 419, Figs. 1082–1100
longicauda 265
minuta 265
papillipes 265
praelongacicula 265
scyllicaniculae 264
spatulata 265
sphyrnae 265
sublineata 264, 265, 267
trecai 264
Kroyeriidae 18, 31, 34, 52, 53, 245, 264
kuroshio, Caligus 173, 174

labracis, Caligus 171, *181–2*, 183, 418, Figs. 572–81
Clavella 275, 278, 279
Hatschekia 273, *275–6*, 278, 418, Figs. 1136–40
Lernaeenicus 309
labrei, Leposphilus 135, 141
Labri mixti, Cycnus 275
labri-mixti, Hatschekia 275
laciniata, Clavellopsis 326, 329
lacustris, Caligus 170
Laemargus 166, 167, 211, 229, 233
muricatus 232
laemonemae, Acanthochondria 124
laevigatum, Sphyrion 318
laevirajae, Chondracanthus 131
lageniformis, Brachiella 395
lagenula, Anchorella 400
Lernaea 150
lagocephali, Taeniacanthus 78
laguncula, Prohatschekia 272
laminatus, Pseudocaligus 187
Laminifera 224
cornuta 225
duello-juradoi 225
Lamippe 110, 111
lamna, Nemesis 259, 261, 262
lamna lamna, Nemesis 261
lamna f.*lamna, Nemesis* 261
lamna vermi, Nemesis 259, 261

lamna f.*vermi, Nemesis* *259–61*, 417, Figs. 1266–94
lamnae, Dinematura 218
Dinemoura 218
Pandarus 217
lampri, Andropoda 382
Clavellistes 333, 349, *382–3*, 418, Figs. 1711–32
Lernaeopoda 382, 383
Thysanote 382, 383
Lamproglena 84, 142, 143, 144, 146, 147, 148, 149, 238, 239, 240, 250, 341
cleopatra 143
cornuta 147, 148
intercedens 147, 148
jordani 147
markewitschi 143
Lamprogleninae 148
Lamproglenoides 142, 143, 144, 147, 148, 149
vermiformis 144
lancifera, Eudactylina 251
laqueus, Anchistrotos 74, *76–7*, 419, Figs. 73–5
Lateracanthus 97, 100, 101, 104, 105, 106, 108
laticervix, Lophoura 321
latifolia, Dinemoura 210, 217
latifrons, Caligus 179
latifurcatus, Trebius 205
latreilli, Cecrops 40, *230–2*, 235, 418, Figs. 905–35
latus, Demoleus 211, 222
Nogagus 214, 228
Lepeophtheirinae 168
Lepeophtheirus 15, 24, 39, 42, 156, 158, 159, 161, 162, 163, 164, 165, 166, 167, 168, 169, 170, 189, 191, 192, 195
branchialis 198
cuneifer 162
edwardsii 16
gibbus 198
gracilenscens 198
gracilis 198
hippoglossi 190, *191–3*, 417, 419, Figs. 648–61
innominatus 194, 195
insignis 193
nordmanni 41, 190, *193–4*, 418, Figs. 662–74
obscurus 191, 198
ornatus 193
pacificus 196
pectoralis 22, 189, *190–1*, 417, 418, 419, Figs. 628–47
pollachii 194
pollachii forma *harengi* 195
pollachius 190, *194–5*, 196, 198, 417, 418, Figs. 676–88
rhombi 198
salmonis 189, 190, 195, *196–7*, 335, 419, Figs. 689–700
strömii 196
sturionis 162, 189, 195, *197–8*, 417, Figs. 531–7
thompsoni 189, 190, 195, *198–9*, 419, Figs. 701–12
trygonis 199
uenoi 196, 197
vesper 196
(*Lepeophtheirus*) *pectoralis, Caligus* 190
lepidionis, Acanthochondria 112, 123
Chondracanthus 112
Lepidopus 166, 211
Lepimacrus 212
Leposphilus 135, 136, 137, 138, 139
labrei 135, 141
Leptotrachelus 296, 297
Lernacantha delarochiana 114
Lernacanthus delarochiana 114
Lernaea 12, 22, 28, 30, 34, 37, 46, 109, 142, 143, 144, 145, 146, 147, 148, 149, 150, 151, 152, 154, 238, 287, 293, 294, 295, 296, 297, 298, 299, 317, 364
adunca 400
arcuata 150
argentinensis 150
articulata 311
asellina 132, 133
bagri 150, 152
barbicola 150
barilii 150, 152
barnimiana 150, 151, 152
basteri 150
bengalensis 150
bistricornis 150, 152
brachycera 150
branchialis 293, 294, 300
branchialis var.*sigmoida* 300
carassii 153
catostomi 150
chackoensis 153
cirrhosa 313
cornuta 124
cruciata 150
ctenopharyngodonis 154
cyclophora 310, 311
cyclopterina 294, 306, 400
cyprinacea 17, 142, 148, 149, 150, 152, *153–5*, 293, 294, 352, 417, 418, 419, Figs. 380–405
dalmanni 363
diceracephala 150
dolabroides 150
elegans 150, 154
elongata 359
encrasicoli 311
esocina 153, 154
gobina 294, 300
haplocephala 150, 152
hardingi 150
hesarangattensis 150
inflata 150
insolens 150
lagenula 150
lophiara 150, 152
lotae 150
lumpi 299
lusci 303
merluccii 119
minima 150
minuta 304
muilticornis 150
nodosa 120
octocornua 145, 150
ocularis 310
orizophila 150
palati 150
parasiluri 150
pectoralis 190
phoxinacea 150
piscinae 150
polymorpha 150
pomatoides 150
quadrinucifera 154
ranae 153
rhodei 150
salmonea 335, 353
senegali 150
spratta 310
squamicola 150
surrirensis 294
tentaculis quatuor 148, 153
tenuis 150
tilapia 150, 152
tuberosa 145, 149, 150, 152
uncinata 400
variabilis 150

werneri 150
Lernaeenicinae 296, 298
Lernaeenicus 289, 290, 294, 295, 296, 297, 299, 308, 309, 311, 317
abdominalis 309
affixus 309
alatus 309
anchoviellae 309
ater 309
bataviensis 309
centropristi 309
cerberus 309
encrasicoli 309, 310, *311–12*, 417, 419, Figs. 1418–22
gnathonicus 309
gnavus 309
gonostomae 309
gracilis 309
hemirhamphi 309
labracis 309
lesueurii 309
longiventris 309
multilobatus 309
musteli 323
polyceraus 309
quadrilobatus 309
radiatus 309
ramosus 309
sardinae 310, 311
sardinae var.*moniliformis* 310, 311
sargi 309
sayori 309
seeri 309
sprattae 309, *310–11*, 312, 417, 419, Figs. 1398–1417
stromatei 309
surrirai 309
triangularis 309
tricerastes 309
vorax 309
Lernaeidae 12, 19, 20, 35, 36, 37, 38, 51, 52, 53, 54, 111, 141, 143, 147, 148, 166, 167, 296, 297, 317, 335
Lernaeidea 110
Lernaeina 295, 296
Lernaeinae 148, 296, 297
Lernaeocera 12, 13, 15, 37, 46, 146, 148, 291, 292, 293, 294, 295, 296, 297, 298, 299, 303, 304, 305, 306, 309, 310, 313, 317, 420
branchialis 4, 233, 299, *300–3*, 304, 307, 417, 418, 419, 420, Figs. 1338–68
branchialis f.*branchialis* 301, Figs. 1334–7
branchialis f.*obtusa* 301, Figs. 1330–3
brevicollis 303, 304
caparti 300
centropristi 299
cyclopterina 306
cyprinacea 153, 294, 353
esocina 153, 294
gasterostei 153
godfroyi 299
lotellae 299
lusci 299, 300, *303–4*, 417, 418, 419, 420, Figs. 1369–82
megacephala 300
minuta 299, 300, *304*, 418, Figs. 1383, 1384
mulli 303, 304
obtusa 300
phycidis 303
rigida 299
surrirensis 310
wilsoni 300
Lernaeoceradae 294
Lernaeoceridae 294, 297, 298
Lernaeocerina 110, 295, 296
Lernaeocerinae 296, 298
Lernaeocerini 148, 295
Lernaeoda 109, 336, 355
Lernaeodea 295, 296
Lernaeogiraffa 37, 144, 145, 146, 147, 148, 149
heterotidicola 146
Lernaeoida 111
Lernaeoidea 109, 111, 297
Lernaeolophus 291, 292, 295, 296, 297, 298, 299
Lernaeomyzon 133, 399
pagelli 413
Triglae 132
uncinata 400
Lernaeonicus 295
Lernaeopenna 294
Lernaeopoda 316, 324, 332, 333, 334, 335, 336, 337, 339, 340, 341, 343, 344, 347, 348, 350, 351, 364, 366, 368, 374, 375, 376, 377, 378, 379, 380, 381, 382, 383, 389
bicaudata 367, 368
bidiscalis 347, 374, 375, 376, *377–9*, 417, 418, 419, Figs. 1648–73
bivia 375
brongniartii 374, 375
carpionis 353
centroscyllii 348, 375, 380
clavigera 354
cluthae 372
dalmanni 363, 368
elongata 359, 360
etmopteri 375, 380
galei 347, 374, *375–7*, 379, 417, 418, 419, Figs. 1674–90
globosa 348, 375, 380
lampri 382, 383
longicaudata 348, 375, 380
longimana 372, 374
musteli 375, 380
mustelicola 375, 377, *379*
obesa 375, 389, 390
oviformis 375
salmonea 335, 353
scoliodontis 375
scyllii 375, 377
scyllicola 375, 377, 378
sebastis 374
selachorum 375
similis 372
stromatei 375
thymalli 354
upeneai 375
Lernaeopodella major 375
Lernaeopodidae 9, 12, 16, 19, 28, 31, 32, 34, 36, 42, 43, 46, 52, 54, 107, 110, 111, 166, 295, 316, 317, 319, 320, 324, 327, 328, 329, 330, 331, 332, 333, 334, 335, 336, 337, 338, 339, 340, 342, 343, 344, 345, 346, 348, 350, 351, 366, 375, 383, 390, 394, 402, 411, 415
Lernaeopodina 336
Lernaeopodina 324, 326, 332, 333, 334, 336, 339, 344, 346, 347, 351, 371, 372, 373, 374
cluthae 372, 373, 374
longibrachia 324, 329, 346, 372, 373
longimana 329, 346, *372–4*, 419, Figs. 1635–47
pacifica 372, 373
relata 372
similis 372, 374
spinacis 372
Lernaeopodinae 337
Lernaeopodoida 46, 52, 110, 111
Lernaeosolea 298
Lernaeosoleinae 297

Lernanthropidae 22, 31, 32, 39, 52, 54, 245, 282, 283
Lernanthropodes 242, 244, 282, 283
Lernanthropus 9, 11, 12, 19, 32, 38, 39, 43, 109, 239, 240, 241, 242, 243, 244, 245, 282, 283, 336
Gisleri 284
gisleri *284–5,* 417, 420, Figs. 1055–70
Kroyeri 285
kroyeri 284, *285–6,* 417, Figs. 1071–81
musca 109, 283
paradoxus 109
Thompsoni 284
Lernaedae 109, 294, 336
Lerneascidae 139
Lerneascus 135, 136, 137, 138
Lernentoma 99, 106, 108, 109, 111, 132, 133
asellina *132–4,* 417, 419, Figs. 357–79
cornuta 124
Dufresnii 118
Lophii 118
lophii 118
nodosa 120
trigla 132, 317
Lernentominae 54, 98, 99, 111, 113
Lerneonema 294, 295, 296
bairdii 310
encrasicoli 311
monillaris 310
musteli 295, 322, 323
spratta 310
Lerneopenna Bocconi 313
Lerneopodadae 336
Lerneopodidea 111
Lernoela branchialis 300
Lesteira 110, 111
lumpi 318
Lestes lumpi 318
lesueurii, Lernaeenicus 309
levis, Clavella 399, 400
Lichomolgidae 51, 59
Lichomolgus 84
limandae, Acanthochondria 124, *127–8,* 418, Figs. 297–316
Chondracanthus 127
lineata, Kroyeria 264, *265–7,* 417, 418, 419, Figs. 1082–1100
lineatum, Lonchidium 265
litognatha, Brachiella 384
lividus, Pandarus 214
lizae, Ergasilus 92, 93
lobatus, Caligus 184
Paeon 318
lobiventris, Thysanote 332
lobodes, Caligus 170
Midias 170, 201, 202
Periplexis 318
lobosus, Paracycnus 268
Lonchidium aculeatum 265, 266
lineatum 265
longibrachia, Lernaeopodina 324, 329, 346, 372, 373
longicauda, Kroyeria 265
longicaudata, Albionella 380
Lernaeopoda 384, 375, 380
longicaudatus, Trebius 205, 207
longicaudus, Holobomolochus 65
Taeniacanthus 78
longicephala, Heterochondria 96
longichela, Taeniacanthus 78
longicollis, Afrolernaea 146, 147
Tracheliastes 357
longifurcatus, Dysgamus 203
longimana, Clavellopsis 412
Lernaeopoda 372, 374
Lernaeopodina 329, 346, *372–4,* 419, Figs. 1635–47
longimanus, Salmincola 352
longispinosus, Neoergasilus 83
longiventris, Lernaeenicus 309
lophiara, Lernaea 150, 152
Lophii, Chondracanthus 118
Lernentoma 118
lophii, Brachiella 368, 369
Chondracanthus 6, 97, 112, 114, *118–19,* 122, 418, Figs. 175–95
Clavella 368, 401
Eubrachiella 368
Lernentoma 118
Lophius, Chondracanthus 118
Lophoura 20, 110, 296, 315, 316, 317, 318, 320, 321
bouvieri 321
caparti 321
cardusa 321
cornuta 321
edwardsi 320, *321–2,* 417, Figs. 1451–8
gracilis 321
kamorahai 321
laticervix 321
tripartita 321
Lophura 321
edwardsi 321
Lophyropoda 83
lotae, Lernaea 150
Salmincola 352
lotellae, Lernaeocera 299
Luetkenia 212, 213, 229, 230, 236
asterodermi *236–7,* 418, Figs. 986–1002
elongata 236, 237
integra 213, 236
lumpi, Caligus 1769
Lernaea 299
Lesteira 318
Lestes 318
Sphyrion 19, *318–20,* 417, 419, Figs. 1440–50
lusci, Lernaea 303
Lernaeocera 299, 300, *303–4,* 417, 418, 419, 420, Figs. 1369–82
Lütkenia asterodermi 236
glabra 236
lutkenii, Echthrogaleus 220
Nogagus 220

Macrobrachinus 84, 85
felichthys 85
macrocephala, Nemesis 259
macrotrachelus, Alella 406, 412
Clavella 403, 413, 414
Clavellodes 413
macrura, Brachiella 391, 392
macrurus, Sciaenophilus 188
maculatus, Tracheliastes 358
magna, Brachiella 384
magniceps, Bradypontius 30
major, Lernaeopodella 375
malleus, Brachiella 368
Charopinus 368
Dirhynchus 369
Pseudocharopinus xii, 367, *368–70,* 417, 418, 419, Figs. 1621–34
Mappates 157, 158, 161, 162, 163
alter 159
plataxus 162
Mappatinae 168
marchesetti, Eucanthus 74
Markewitschia 86

markewitschi, Lamproglena 143
Pseudocharopinus 366
Medesicaste 96, 99, 100, 106, 107, 108, 110, 111, 133, 134
asellinum 132
triglarum 132
Megabrachinus 84, 85
megacephala, Lernaeocera 300
megacephalus, Caligodes 159
megaceros, Bomolochus 61
Megapontiidae 52
Megapontius 42
mendisi, Ergasilus 89
merlangi, Chondracanthus 119
merluccii, Brachiella 388, 389
Chondracanthus 114, *119–20,* 418, Figs. 282–96
Lernaea 119
Neobrachiella 387, *388–9,* 396, Figs. 1821–31
Metahatschekia 241, 273
congri 242, 279
Metataeniacanthus 68, 69, 71, 72, 73
synodi 68
Micropontiidae 24
Microsetella 50
Midias 201
carangis 202
lobodes 170, 201, 202
minima, Lernaea 150
minimus, Caligus 162, 171, 181, *182–3,* 417, 418, 420, Figs. 582–91
minimus var.*mugilis, Caligus* 182, 183
minor, Pupulina 159
minuta, Eudactylina 251, 252, *257–8,* 417, Figs. 1252–65
Kroyeria 265
Lernaea 304
Lernaeocera 299, 300, *304,* 418, Figs. 1383, 1384
minutus, Caligus 182, 183
Trebius 205
Misophrioida 48, 50
moa, Anchistrotos 74
monacanthi, Caligus 184
Taeniacanthus 78
monillaris, Lerneonema 310
Monima fimbriata 235
Monoculus 196
piscinus 172
Monstrilla 46
Monstrillidae 84
Monstrilloida 48, 49, 50, 51
mourkii, Tracheliastes 357
Mugilicola 298
mugilis, Ergasilus 92
mülleri, Caligus 171
Mulli, Clavella 276
mulli, Hatschekia 273, *276–7,* 418, Figs. 1141–51
Lernaeocera 303, 304
multicornis, Lernaea 150
multifimbriata, Thysanote 321
multilobatus, Lernaeenicus 309
multispinosus, Caligus 178, 179
muraenae, Bomolochus 73
muricata, Orthagoriscicola 232
muricatus, Laemargus 232
Orthagoriscicola *232–4,* 235, 418, Figs. 936–60
musca, Lernanthropus 109, 283
musteli, Lernaeenicus 323
Lernaeopoda 375, 380
Lerneonema 295, 322, 323
Tripaphylus 322, *323,* 418 Figs. 1459–61
Trypaphylum 323
mustelicola, Lernaeopoda 375, 377, *379*
musteli-laevis, Eudactylina 251
Pandarus 214
mycterobius, Bomolochus 61

nana, Ergasilus 93
Nanaspidae 24, 46, 52
nanus, Caligus 180, 181
Ergasilus 89, *92–3,* 417, Figs. 101–7
Naobranchia 295, 296, 338
Naobranchiidae 19, 23, 31, 34, 111, 338
Naobranchiinae 337, 338
narcini, Taeniacanthus 78
narium, Chondracanthus 112
neali, Chondracanthus 114, *117–18,* 418, Figs. 159–74
Nectobrachia 324, 327, 329, 333, 334, 338, 344, 349, 350
indivisa 326, 329, 330, 345
Nectobrachiinae 337, 338
neglecta, Brachiella 391
Nemesis 9, 11, 22, 38, 39, 83, 84, 238, 239, 240, 241, 242, 243, 244, 250, 251, 258, 259, 261, 262
aggregatus 261, 262
atlantica 259
lamna 259, 261, 262
lamna lamna 261
lamna f.*lamna* 261 ·
lamna vermi 259, 261
lamna f.*vermi* *259–61,* 417, Figs. 1266–94
macrocephala 259
pallida 261, 262
pilosa 259
robusta 259, *261–3,* 417, 418, 419, Figs. 1295–1308
robustus 259, 261
robuta 261
tiburo 259
vermi 259, 261
versicolor 259
Neobrachiella 324, 328, 333, 334, 337, 339, 341, 344, 348, 349, 351, 375, 384, 386, 387, 389, 390, 394, 395, 396, 397, 412
albida 332
bispinosa 387, *389–91,* 394, 399, 417, 419, Figs. 1757–77
chevreuxii 325, 326, 328, 329, 331, 387, *391–3,* 417, Figs. 1795–1809
impudica 328, 329, 331, 387, *393–4,* 395, 396, 398, 417, 419, Figs. 1777–94
insidiosa 329, 383, 387, 388, *395–6,* 397, 418, Figs. 1810–20
insidiosa f.*insidiosa* 396
insidiosa f.*lageniformis* 396
insidiosa f.*pacifica* 396
merluccii 387, *388–9,* 396, 418, Figs. 1821–31
papillosa 333
regia 348
robusta 330, 332, 396
rostrata 387, 388, *396–7,* 417, Figs. 1832–45
stellifera 387
suplicans 332
triglae 331, 387, *397–9,* 417, 419, Figs. 1846–60
Neobrachiochondria 97, 98, 99, 103, 104, 105, 106, 108
Neoergasilus 81, 83
longispinosus 83
neopercis, Taeniacanthus 78
Neopontius angularis 29, 30
Nereicola 111
nerkae, Ergasilus 87
Nessipus 209, 210, 212, 227
borealis 227
crypturus 227
orientalis 236
tigris 227
nibae, Sciaenophilus 188
Nicothoe 84, 238

Nicothoidae 84, 166
niger, Pandarus 214
nigeriensis, Afrolernaea 146, 147
Delamarina 146
Nipergasilus 81, 82, 83, 86
nodosa, Lernaea 120
Lernentoma 120
nodosus, Chondracanthus 114, *120–2*, 419, Figs. 213–31
Parapandarus 224, 225
Phyllothyreus 225
Nogagina 212
Nogagus 165, 166, 211, 212, 226, 228
ambiguus *227–8*, 419, Fig. 768
angustulus 214
borealis *227*, 418, Fig. 767
gracilis 218
grandis 224, 225
latus 214, 228
lutkenii 220
productus 218
Nogaus 213, 226, 227, 238
nordmanni, Caligus 193
Euryphorus 203
Lepeophtheirus 41, 190, *193–4*, 418, Figs. 662–74
Salmincola 352
Norion 240, 242, 244, 282, 283
extensus 282
norvegicum, Sphyrion 318, 319
Nothobomolochus 55, 57, 58
nothrus, Holobomolochus 65
Notodelphyidiformes 51
Notodelphyoida 50, 51
Notodelphys 46
Nudiclavella 324, 350
nudiventris, Dissonus 9
nunesi, Trebius 205

obcordata, Clavellisa 408
obesa, Brachiella 309, 389, 390
Lernaeopoda 375, 389, 390
oblongatus, Perissopus 213
oblongum, Dichelesthium 17, 18, 31, *246–8*, 417, Figs. 1003–25
oblongus, Caligus 238, 246, 247
Pandarus 214
obscurus, Lepeophtheirus 191, 198
obtusa, Lernaeocera 300
occidentalis, Anchistrotos 74
occultus, Holobomolochus 65
octocornua, Lernaea 145, 150
ocularis, Caligulina 164
Lernaea 310
Ommatokoita 324, 333, 334, 344, 346, 351, 359, 360
elongata 330, 346, *359–61*, 418, 419, Figs. 1530–52
superba 359, 360
oniscoides, Orbitocolax 59
onosi, Anchistrotos 18, 74, *75–6*, 417, 419, Figs. 53–72
Bomolochus 75
Ophiolernaea 24, 289, 290, 292, 298
Opimia 315, 316, 317
exilipes 318
Opistholernaea 144, 145, 146, 147, 148, 149, 150
Oralien 133
asellinus 132, 133
triglae 132
Orbitocolax 55, 56, 57, 58, 60
oniscoides 59
orientalis, Nessipus 236
orizophila, Lernaea 150
ornatus, Caligus 193
Chondracanthus 97, 98, 114, *122–3*, 417 , Figs. 232–52
Disphaerocephalus 122
Lepeophtheirus 193
orthagorisci, Pennella 313, 314
Orthagoriscicola 8, 211, 212, 213, 229, 230, 232
muricata 232
muricatus *232–4*, 235, 418, Figs. 936–60
szidati 232, 233
wilsoni 232, 233
ostorhinchi, Kabataia 205
ostracionis, Anchistrotos 74
Bomolochus 74
Ostrincola *81, 82, 83*
Otrophesia 249
imbricata 248
ovalis, Anchorella 389, 390, 409, 410
Brachiella 389, 390, 409
Clavellisa 409
ovata, Clavella 399, 400
oviformis, Lernaeopoda 375

Pachycephala 84
pacifica, Lernaeopodina 372, 373
pacificus, Caligus 196
Lepeophtheirus 196
Paeon 315, 316, 317, 318
elongatus 318
ferox 318
lobatus 318
vaissieri 318
versicolor 316, 318
Paeonocanthus xii, 315, 316, 317
antarcticensis xii, 317, 318
tricornutus xii
Paeonodes 36, 318
pagelli, Allela 406, 412, *413–15*, 417, 419, Figs. 1975–95
Clavella 413, 414
Clavellodes 413
Lernaeomyzon 413
Pagelli Bogneravei, Cycnus 633
pagellibogneravei, Hatschekia 273, 276, 277–8, 418, Figs. 1127–35
Pagina 208, 209, 210, 212
Pagodina 239, 240
carchariae-glauci 259
robusta 261
pagri, Clavella 399
palati, Lernaea 150
palleucus, Holobomolochus 65
pallida, Congericola 280
Nemesis 261, 262
pallidus, Chondracanthus 124, 131, 132
Congericola 131, 241, 279, *280–1*, 417, Figs. 1158–74
Cycnus 280
palpifer, Chondracanthus 120
Pandaridae 8, 9, 10, 17, 18, 21, 31, 32, 33, 41, 42, 52, 53, 166, 168, 208, 209, 210, 211, 212, 213, 217, 219, 222, 225, 226, 229
Pandarina 167, 212
Pandarinae 166, 168, 201, 211, 226
Pandarini 167, 212
Pandarus 8, 9, 22, 42, 164, 165, 166, 208, 209, 210, 211, 212, 213, 214, 216, 217, 238
alatus 220
bicolor 213, *214–16*, 225, 228, 417, 418, 419, Figs. 769–800
boscii 214
carcharii-glauci 214
carcharini 214
cranchii 214

fissifrons 214
floridanus 214
katoi 214
lamnae 217
lividus 214
musteli-laevis 214
niger 214
oblongus 214
satyrus 214
sinuatus 214
smithii 214
spinacii-acanthii 214
unicolor 214
zygaenae 214
Pannosus 208, 209, 210, 211, 212
papillipes, Kroyeria 265
papillosa, Eudactylina 251, 258
Neobrachiella 333
Parabomolochus 60
bellones 64
Parabrachiella 337, 348, 387, 396
australis 395
insidiosa 395
rostrata 396
Paracycnus lobosus 268
paradoxa, Advena *415–6*, 419, Figs. 1996–2016
Anchorella 415
Clavella 415
Clavellopsis 415
paradoxus, Caligus 223
Demoleus 223
Lernanthropus 109
Paraergasilus 81, 83, 85, 86
Paralebion 200, 201, 202
curticaudis 202
Paralernanthropus 282, 283
Paranessipus 209, 210, 211, 212
incisus 211
Parapandarus 212
nodosus 224, 225
Parapetalus 159, 163, 167, 168
parasiluri, Lernaea 150
Parataeniacanthus 68, 69, 73
Parechetus 159, 160, 161, 163
parkeri, Brachiella 364, 366
Charopinus 363, 366
parva, Brachiella 393
Clavella 399, 400
parvus, Collipravus 305
Pseudocaligus 187
parvulus, Bomolochus 61, 83
pastinaca, Brachiella 367, 368
pastinacae, Brachiella 367, 368, 369
Charopinus 368
Pseudocharopinus xii, 369
Pectinata 29
pectinatus, Echthrogaleus 220
Taeniacanthus 78
pectoralis, Lepeophtheirus 22, 189, *190–1*, 417, 418, 419, Figs. 628–47
Lernaea 190
Pediculus salmonis 353
Pegessimalus 296, 297
pelamydis, Caligus 171, *183–4*, 185, 419, Figs. 489–504
pellone, Clavellisa 408
pellucidus, Echthrogaleus 220
Peltocephala 166, 167, 211
Penelladae 294, 298
Penellina 294
Peniculina 296
Peniculinae 298
Peniculisa 289, 298
Peniculus 109, 110, 288, 289, 294, 295, 296, 297, 298
fistula 294
Pennatula filosa 294, 313
sagitta 294
Pennella 4, 43, 166, 240, 289, 290, 292, 293, 294, 295, 296, 297, 298, 299, 301, 312, 313
Costai 313
diodontis 294
exocoeti 314
fibrosa 313
filosa 294, *313–14*, 418, 419, Figs. 1432–39
germonia 314
orthagorisci 313, 314
plumosa 314
rubra 313
sagitta 294, 312, 313, Figs. 1423–31
Pennellidae 19, 20, 21, 23, 24, 29, 32, 34, 36, 38, 42, 43, 52, 53, 54, 109, 148, 149, 287, 288, 289, 291, 292, 293, 294, 295, 297, 298, 299, 304, 312, 313, 315, 317, 335, 339
Pennellina 109, 295, 298
Pennellinae 296
Pennellini 295
percarum, Achtheres 345, 355, *356–7*, 417, 418, Figs. 1500–16
perfida, Clavella 399, 400
Periplexis xii, 315, 316, 317
antarcticensis xii
lobodes 318
Perissopus 208, 209, 210, 211, 212
oblongatus 213
Peroderma 290, 291, 295, 296, 297, 298
cylindricum 42
Perodermatidae 297
perspicax, Echthrogaleus 220
petax, Vahinius 24
pharaonis, Sciaenophilus 188
Phagus 73
Pharodidae 19, 31, 34, 100
Pharodinae 111, 112
phasa, Clavellisa 408
Philichthyidae 34, 36, 52, 53, 135, 137, 138, 139, 240
Philichthyidea 139
Philichthys 135, 136, 137, 138, 139, 240
sciaenae 138
xiphiae 22, 42, 110, 138, *139–41*, 420, Figs. 406–32
Philorthagoriscus 213, 229, 230, 234, 237
serratus *234–5*, 418, Figs. 961–85
Pholidopus 212
phoxinacea, Lernaea 150
Phrixocephalus 287, 291, 295, 296, 298
cincinnatus 13, 14, 43, 311
phycidis, Lernaeocera 303
Phyllophora 166, 211, 224
cornuta 225
Phyllophorus 167
cornutus 225
Phyllothyreus 208, 209, 210, 211, 212, 214, 224
cornutus 22, 224, *225–6*, 418, 419 Figs. 878–904
nodosus 225
pillaii, Pseudocharopinus 367
pilosa, Nemesis 259
pinguis, Chondracanthus 97, 98
Clavella 399, 400
Pinnodactyla 238
piscinae, Lernaea 150
piscinus, Binoculus 196
Caligus 199
Monoculus 172

plataxus, Mappates 162
platycephala, Acanthochondria 124
pleuronichthydis, Anchistrotos 74
plumosa, Pennella 314
Podoplea 47, 50
Poecilopoda 83, 166
Poecilostoma 46, 51, 84
Poecilostomatoida 15, 20, 22, 23, 24, 28, 29, 48, 51, 52, 55, 139, 146
pollachii, Lepeophtheirus 194
pollachii forma *harengi, Lepeophtheirus* 195
pollachius, Lepeophtheirus 190, *194–5*, 196, 198, 417, 418, Figs. 676–88
polyceraus, Lernaeenicus 309
polycolpus, Tracheliastes 357, *358–9*, 418, Figs. 1517–29
polycolpus var. *baicalensis, Tracheliastes* 358
polycolpus var. *kessleri, Tracheliastes* 358
polycolpus var. *phoxini, Tracheliastes* 358
polymorpha, Lernaea 150
Polyrhynchus 138, 139
pomatoides, Lernaea 150
pomposa, Acespadia 324, 326, 346
porogadi, Clavella 399, 400
Praecidochondria 97, 100, 101, 105, 106, 108
praelongacicula, Kroyeria 265
Probrachiella 337, 348, 387
Prochondracanthopsis 97, 99, 106, 108
Prochondracanthus 97, 99, 103, 104, 105, 106, 108
Proclavellodes 324, 344, 350
producta, Dinematura 217
  Dinemoura 185, 210, *217–9*, 221, 417, 418, 419, Figs. 801–21
productus, Binoculus 217
  Caligus 171, *184–6*, 217, Figs. 592–603
  Nogagus 218
Prohatschekia 242, 244, 271, 272, 273
  laguncula 272
  sebastisci 272
prolixus, Holobomolochus 65
Prosaetes 212
Protochondracanthoides 114
Protochondracanthus 97, 100, 101, 104, 105, 106, 108
Protochondria 96, 99, 100, 103, 104, 105, 106, 107, 108
Protodactylina xii, 242, 243, 244, 250, 251
psetti, Acanthochondria 124
psettobius, Bomolochus 61
Pseudacanthocanthopsis 97, 98, 99, 104, 105, 106, 108
Pseudanuretes 33, 158, 159, 162, 163, 169
  chaetodonti 170
  fortipedis 169
Pseudergasilus 81, 82, 86, 89
Pseudoblias 96, 99, 105, 106, 108
Pseudobomolochus 60
Pseudocaligus 41, 158, 163, 168, 169, 170, 186, 187
  apodus 187
  brevipedes 187
  brevipedis 186, *187–8*, 417, Figs. 616–27
  brevipes 187
  fistulariae 187
  fugu 187
  laminatus 187
  parvus 187
  similis 186, 187
Pseudocharopinus 19, 28, 324, 325, 333, 344, 347, 351, 363, 366, 367, 369, 370, 371
  bicaudatus 347, 366, *367–8*, 369, 381, 418, 419, 420, Figs. 1610–20
  concavus 366
  dasyaticus 366, 367
  dentatus 17, 329, 330, 366
  malleus xii, 367, *368–70*, 417, 418, 419, Figs. 1621–34
  markewitschi 366
  narcinae 367
  pastinacae xii, 369
  pillaii 367
  pteroplateae 367
  squalii 366
Pseudochondracanthus 99, 106, 108
  diceraus 97, 98
Pseudocongericola 241, 242, 244, 245, 271, 272, 273
Pseudocycnidae 18, 22, 31, 34, 52, 53, 241, 245, 268, 272
Pseudocycninae 240, 268
Pseudocycnoides 242, 243, 244, 268
  rugosa 268
  scomberomori 268
Pseudocycnus 240, 241, 242, 243, 244, 268
  appendiculatus *268–70*, 419, Figs. 1309–29
  spinosus 268, 270
  thunnus 268, 270
Pseudodiocus 97, 100, 101, 104, 105, 106, 108
Pseudoeucanthus 55, 56, 58
Pseudohatschekia 23, 241, 244, 245, 273
Pseudolepeophtheirus 41, 158, 163, 169, 170
Pseudolernanthropus 282, 283
Pseudomyicolidae 30
Pseudonura 164
Pseudopandarus 208, 209, 210, 212
Pseudopetalus 157, 159, 161, 163
Pseudorbitocolax 55, 56, 57, 58
Pseudotaeniacanthus 68, 69, 70, 71, 72, 73
Pseudotracheliastes 324, 344, 345, 346
Pseudulus 110, 296
pterobrachiata, Alella 406, 413
Pterochondria 97, 98, 99, 100, 106, 108
  alatalongicollis 97
pteroisi, Taeniacanthus 78
pteroplateae, Pseudocharopinus 367
Pterygopoda 238
Pumiliopes 55, 56, 57, 58
Pumiliopsis 55, 57, 58
Pupulina 159, 160, 162, 163
  minor 150
purpurocinctus, Rhynchomyzon 29, 30
pygmaea, Hatschekia 273, *278–9*, 417, Figs. 1152–7
pyriformis, Gangliopus 208, 213

quadrata, Anchorella 400
  Clavella 401, 402
quadratus, Aethon 282
  Caligus 177, 185
quadrilobatus, Lernaeenicus 309
quadrinucifera, Lernaea 154
quaternia, Charopinopsis 348
quinqueabdominalis, Caligus 158

rabidus, Caligus 179
rachelae, Eudactylina 254, 255, Figs. 1215, 1216
radiatus, Lernaeenicus 309
ramosa, Schistobrachia 329, 330, *370—371*, 419, Figs. 1598–1609
ramosus, Charopinus 370, 371
  Lernaeenicus 309
ranae, Lernea 153
rapax, Caligus 179, 180, 181
Rebelula 321
  edwarsdi 321
recta, Clavella 400
regia, Neobrachiella 348
relata, Lernaeopodina 372
rhodei, Lernaea 150

rhombi, Lepeophtheirus 198
Rhynchochondria 97, 98, 99, 100, 104, 105, 106, 107, 108
Rhynchomyzon purpurocinctus 29, 30
richiardii, Alella 412
rigida, Lernaeocera 299
rissoanus, Caligus 179
robusta, Brachiella 396
Ergasilina 261
Nemesis 259, *261–3*, 417, 418, 419, Figs. 1295–1308
Neobrachiella 330, 332, 296
Pagodina 261
robustus, Nemesis 259, 261
robuta, Nemesis 261
rostrata, Brachiella 396
Neobrachiella 387, 388, *396–7*, 417, Figs. 1832–45
Parabrachiella 396
rubra, Pennella 313
rugosa, Anchorella 406, 407, 409, 410
Clavella 406
Clavellodes 328, 405, *406–7*, 412, 417, Figs. 1914–36
Pseudocycnoides 268
rugosa var. *Anchorella* 400, 406

saba, Advena 415
Clavellopsis 415
Sabelliphilus 59, 84
sachalinensis, Tracheliastes 357
sagamiensis, Dysgamus 203, 204
sagitta, Pennatula 294
Pennella 294, 312, 313, Figs. 1423–31
Sagum 242, 244, 282, 283
Salmincola 19, 36, 324, 327, 328, 333, 334, 335, 339, 341, 342, 343, 344, 345, 346, 350, 351, 352, 354, 355, 356, 359, 374, 385
baicalensis 354
(Brevibrachia) 352
californiensis 17, 334, 352
carpionis 327, 352
coregonorum 352
cottidarum 352
edwardsii 324, 326, 352, 373
extensus 352
extumescens 352, 355
gordoni 353, 354
jacuticus 352
longimanus 352
lotae 352
nordmanni 352
(Salmincola) 352
salmonea 353, 354
salmoneus 331, 335, 352, *353–4*, 357, 419, Figs. 1462–85
siscowet 352
stellatus 330, 352
strigatus 352
thymalli 330, 352, *354–5*, 356, 415, 419, Figs. 1486–99
thymalli baicalensis 354, 355
thymalli v.*baicalensis* 355
thymalli mongolicus 354, 355
thymalli thymalli 354, 355
(Salmincola), Salmincola 352
salmonea, Entomoda 353
Lernaea 335, 353
Lernaeopoda 335, 353
Salmincola 353, 354
salmoneus, Basanistes 353
Salmincola 331, 335, 352, *353–4*, 357, 419, Figs. 1462–85
Schisturus 353
salmonis, Caligus 196
Lepeophtheirus 189, 190, 195, *196–7*, 335, 419, Figs. 689–700
Pediculus 353
sandrae, Achtheres 356, 357
Sapphirina 20
Sarcotaces 20, 135, 136, 137, 138
komai 135
Sarcotacidae 139
Sarcotacidea 139
Sarcotacinae 139
Sarcotretes 289, 290, 296
sardinae, Lernaeenicus 310, 311
sardinae var. *moniliformis, Lernaeenicus* 310, 311
sargi, Hatschekia 278
Lernaeenicus 309
satyrus, Pandarus 214
sauciatonis, Saucissona 299
Saucissona 299
sauciatonis 299
sauridae, Taeniacanthus 78
sauridi, Anchistrotos 74
sayori, Lernaeenicus 309
schahriar, Charopinus 363
Scheherezade 97, 98, 99, 100, 103, 104, 105, 106, 108
Schistobrachia 324, 333, 344, 347, 351, 370, 371
chimaerae 347, 370
ramosa 329, 330, *370–71*, 419, Figs. 1598–1609
tertia 370
Schisturus cyclopterinus 306
salmoneus 353
uncinatus 400
sciaenae, Brachiella 391, 392
Clavella 391, 392
Philichthys 138
sciaenophila, Anchorella 391
Brachiella 391, 392, 393
Sciaenophilidae 188
Sciaenophilus 159, 160, 163, 167, 168, 169, 170, 188, 189
benedeni 188, 189
bennetti 188
macrurus 188
nibae 188
pharaonis 188
tenuis *188–9*, 417, 420, Figs. 713–21
(Sciaenophilus) benedeni, Caligus 188
sciatherica, Clavella 401
Scolecicara 69, 70, 71, 72, 73
humesi 69
scoliodontis, Lernaeopoda 375
scomberi, Caligus 183, 184, 185
scomberomori, Pseudocycnoides 268
scombri, Anchorella 410
Caligus 183, 184, 185
Clavella 410
Clavellisa 408, *410–12*, 416, 419, Figs. 1958–74
scombropsis, Clavella 399, 400
scutatus, Caligus 176
scutigerulus, Holobomolochus 65
scyllicaniculae, Kroyeria 264
scyllicola, Lernaeopoda 375, 377, 378
scyllii, Lernaeopoda 375, 377
sebastichthydis, Taeniacanthus 78
sebastis, Lernaeopoda 374
sebastisci, Taeniacanthus 78
Prohatschekia 272
seeri, Lernaeenicus 309
selachiorum, Achtheres 375
selachorum, Lernaeopoda 375
selaroides, Bomolochus 61
Selinae 109
Selius 109, 110, 111
semicoleus, Ergasilus 87, 88
senegali, Lernaea 150

sepheni, Trebius 205
Sepicola 84
serrata, Dinematura 234
serratus, Philorthagoriscus *234–35*, 418, Figs. 961–85
sexsetacea, Dinematura 223
Shiinoa 21, 23, 28, 30
Shiinoidae 141
shoyoae, Clavellistes 382, 383
sibirica, Achtheres 356
sieboldi, Ergasilus 18, 82, 83, 87, 89, *90–1*, 417, 419, Figs. 88–100
sieboldi var.*patagonicus, Ergasilus* 91
Silenium 336
similis, Eudactylina 251, 252, *254–5*, 257, 419, Figs. 1213–28
  Lernaeopoda 372
  Lernaeopodina 372, 374
  Pseudocaligus 186, 187
simulans, Ascomyzon 29, 30
sinensis, Bomolochus 61
Sinergasilus 81, 83
sinuatus, Pandarus 214
Siphonostoma 46, 52, 165, 167
Siphonostomata 166, 238
Siphonostomatoida 19, 21, 23, 28, 29, 32, 38, 48, 52, 156, 205, 245, 293, 297, 341
siscowet, Salmincola 352
smithii, Anthosoma 248, 249
  Pandarus 214
soleae, Acanthochondria 124, 126, *128–9*, 419, Figs. 317–34
  Bomolochus 22, 61, *62–4*, 66, 419, Figs. 1–24
  Chondracanthus 128
spatulata, Kroyeria 265
Specilligus 166, 211
Sphaerifer 136, 137, 138, 139
Sphaerosoma 110
  corvinae 138
Sphyriidae xii, 19, 31, 38, 43, 52, 54, 111, 315, 316, 317, 318
Sphyrion 23, 110, 315, 316, 317, 318, 319
  laevigatum 318
  lumpi 19, *318–20*, 417, 419, Figs. 1440–50
  norvegicum 318, 319
sphyrnae, Kroyeria 265
spinacii-acanthii, Pandarus 214
spinacis, Lernaeopodina 372
spinifrons, Trebius 206, 207
spinosa, Clavellisa 408, 412
spinosus, Caligus 29
  Pseudocycnus 268, 270
spinula, Eudactylina 254
spinulus, Holobomolochus 65
Splanchnotrophus 111
spratta, Lernaea 310
  Lerneonema 310
sprattae, Lernaeenicus 309, *310–11*, 312, 417, 419, Figs. 1398–1417
spratti, Foroculum 310
squalii, Pseudocharopinus 366
squamigera, Clavella 399, 400
squamicola, Lernaea 150
squatinae-angeli, Eudactylina 251
Stalagmus 239, 240
Staurosoma 110, 111, 295
stellata, Anchorella 404
  Clavella 329, 350, 400, *404–5*, 418, Figs. 1902–13
stellatus, Salmincola 330, 352
Stellicomitidae 24, 46
stellifera, Neobrachiella 387
stichaei, Clavella 399, 400
Strabax 97, 98, 100, 101, 106, 108, 110, 111
stramineus, Chondracanthus 119, 120
strigatus, Salmincola 352
stromatei, Lernaeenicus 309
  Lernaeopoda 375
strömii, Caligus 196
  Lepeophtheirus 196
sturionis, Caligus 197
  Dichelesthium 246, 247
  Lepeophtheirus 162, 189, 195, *197–8*, 417, Figs. 531–7
stygius, Cauloxenus 324
Stylophorus hippocephalus 363
sublineata, Kroyeria 264, 265, 267
sublobulata, Eubrachiella 331, 333
surbecki, Ergasilus 90
surrirai, Lernaeenicus 309
surrirensis, Lernaea 294
  Lernaeocera 310
superba, Ommatokoita 359, 360
suplicans, Neobrachiella 332
synapturae, Andreina 111
Synestius 159, 160, 162, 163, 167, 168
synodi, Metataeniacanthus 68
szidati, Orthagoriscicola 232, 233
Taeniacanthidae 16, 18, 22, 30, 31, 34, 36, 39, 41, 52, 53, 56, 59, 68, 69, 70, 72, 73, 139
Taeniacanthinae 59, 73, 85
Taeniacanthodes 68, 70, 71, 72, 73
  gracilis 68, 69
  haakeri 68
Taeniacanthus 39, 59, 68, 69, 71, 72, 73, 74, 77, 78, 79, 142
  acanthocepolae 78
  albidus 78
  canthigasteri 78
  carchariae 68, 69, 78
  coleus 78
  dentatus 78
  flagellans 78
  fugu 78
  indicus 78
  kitamakura 78
  lagocephali 78
  longicaudus 78
  longichela 78
  monacanthi 78
  narcini 78
  neopercis 78
  pectinatus 78
  pteroisi 78
  sauridae 78
  sebastichthydis 78
  sebastisci 78
  tetradonis 78
  upenei 78
  wilsoni 78, *79–80*, 419, Figs. 79–87
Taeniastrotos 68, 70, 71, 72, 73
Tanypleuridae 19, 31, 336
Tanypleurus 23, 34, 36, 111, 148, 295, 336, 337
  alcicornis 19
Taurocheros 114, 145, 146, 147, 148, 149
tchangi, Acanthochondria 124
Telsidae 18, 22, 30, 31, 32, 34, 36, 59
Telson 59
tenax, Ergasilus 88, 89
tentaculis quatuor, Lernaea 148, 153
tenuifurcatus, Trebius 205
tenuis, Lernaea 150
  Sciaenophilus *188–9*, 417, 420, Figs. 713–21
Terebellicola 84
Teredophilus 81, 82, 83, 86
  inflatus 81
tertia, Schistobrachia 370

tesserifer, Caligus 176
tetradonis, Taeniacanthus 78
tetrapturi, Gangliopus 213
Thecata 164
Thepesiopsyllus 49
theragrae, Haemobaphes 305
Therodamas 36, 295, 296, 297, 298
Therodamasidae 19, 22, 30, 318
Therodamasinae 297
Thersites 84, 85, 94
gasterostei 84, 94
gasterosteus 94
Thersitina 30, 81, 82, 83, 84, 85, 86, 94, 95
biuncinata 94
gasterostei 81, *94–95*, 417, 418, Figs. 108–28
thompsoni, Lepeophtheirus 189, 190, 195, *198–9*, Figs. 701–12
Thompsoni, Lernanthropus 284
thunnus, Pseudocycnus 268, 270
thymalli, Lernaeopoda 354
Salmincola 330, 352, *354–5*, 356, 415, 419, Figs. 1486–99
thymalli baicalensis, Salmincola 354, 355
thymalli v. *baicalensis, Salmincola* 355
thymalli mongolicus, Salmincola 354, 355
thymalli thymalli, Salmincola 354, 355
thymni, Caligus 176
thynni, Arnaeus 203
Brachiella 348, 384, *385–6*, 387, 417, 419, Figs. 1733–56
Thynnicola ziegleri 385
Thysanote 324, 329, 332, 333, 337, 344, 349, 394
fimbriata 332
impudica 393
lampri 382, 383
lobiventris 332
multifimbriata 331
Thysanotella 349
tiburo, Nemesis 259
tigris, Nessipus 227
tilapia, Lernaea 150, 152
torpedinis, Caligus 178, 179
Echthrogaleus 209, 220
tortuosus, Trifur 299
Tracheliastes 324, 325, 326, 327, 329, 331, 333, 335, 337, 339, 341, 342, 344, 345, 346, 351, 357, 358
chondrostomi 357
gigas 357
longicollis 357
maculatus 358
mourkii 357
polycolpus 357, *358–9*, 418, Figs. 1517–29
polycolpus var. *baicalensis* 358
polycolpus var. *kessleri* 358
polycolpus var. *phoxini* 358
sachalinensis 357
Tracheliastinae 337
trachypteri, Caligus 179
Trebiidae 8, 9, 17, 21, 31, 33, 52, 53, 168, 205
Trebiinae 201, 205
Trebinae 168
Trebius 7, 8, 40, 41, 165, 166, 167, 205, 211
akajeii 205
bilobatus 205
caudatus 205, *206–7*, 417, 418, 419, Figs. 722–40
elongatus 205
exilis 205
javanicus 205
kirtii 205
latifurcatus 205
longicaudatus 205, 207
minutus 205
nunesi 205
sepheni 205
spinifrons 206, 207
tenuifurcatus 205
trecai, Kroyeria 264
trangularis, Lernaeenicus 309
tricerastes, Lernaeenicus 309
Trichthacerus 110, 111
tricornutus, Paeonocanthus xxi
Trifur 291, 292, 296, 299
tortuosus 299
trigla, Lernentoma 132, 317
triglae, Anchorella 397
Brachiella 397
Chondracanthus 109, 132, 317
Clavella 397
Neobrachiella 331, 387, *397–9*, 417, 419, Figs. 1846–60
Oralien 132
Triglae, Chondracanthus 132
Lernaeomyzon 132
triglarum, Medesicaste 132
trilobatus, Chondacanthus 114
Tripaphylus 20, 315, 316, 317, 318, 322
musteli 322, *323*, 418, Figs. 1459–61
tripartita, Lophoura 321
tripartitus, Alicaligus 162
trisetaceus, Ergasilus 90
triventricosus, Chondracanthus 112
truchae, Areotrachelus 145
trygonis, Lepeophtheirus 199
Trypaphylum 322
hemigalei 322
musteli 323
tuberculatus, Chondracanthus 109, 116
tuberosa, Lernaea 145, 149, 150, 152
Tucca 10, 11, 36, 59, 109, 110, 240
Tuccidae 18, 22, 30, 31, 34, 36, 38, 59
Tuccinae 59
tumida, Clavella 399, 400
tumidus, Bomolochus 61
Ergasilus 87
Tuxophorinae 201
Tuxophorus 200, 201, 202
typica, Clavella 401

uenoi, Lepeophtheirus 196, 197
uncinata, Anchorella 335, 400, 401
Clavella 355, 400, 401, 402
Clavellina 401
Lernaea 400
Lernaeomyzon 400
uncinatus, Schisturus 400
unicirrus, Bomolochus 61
unicolor, Pandarus 214
upeneai, Lernaeopoda 375
upenei, Taeniacanthus 78

Vahinius petax 24
vaissieri, Paeon 318
Vanbenedenia 324, 327, 329, 331, 337, 339, 341, 342, 344, 346, 351, 361
chimaerae 326, 329, 361, 362
grandis 361
kroeyeri, 330, 344, 346, *361–2*, 417, Figs. 1553–65
variabilis, Lernaea 150
Ventriculininae 148
venustus, Holobomolochus 65
vermi, Nemesis 259, 261
vermiformis, Lamproglenoides 144
versicolor, Nemesis 259
Paeon 316, 318

vespa, Caligus 196
vesper, Caligus 196
  Lepeophtheirus 196
vorax, Lernaeenicus 309

werneri, Lernaea 150
williamsoni, Chondracanthus 120
willungae, Caligus 171
wilsoni, Chondracanthus 114, 121
  Ergasilus 87
  Lernaeocera 300
  Orthagoriscicola 232, 233
  Taeniacanthus 78, *79–80*, 419, Figs. 79–87

xiphiae, Philichthys 22, 42, 110, 138, *139–41*, 420, Figs. 406–32
xyphiae, Chondracanthus 119

youngi, Hermilius 159

ziegleri, Thynnicola 385
zei, Caligus 171, *186*, 420, Figs. 604–15
  Chondracanthus 109, 112, 113, *114–17*, 119, 420, Figs. 129–58
zeugopteri, Anchistrotos 74, *77*, 420, Figs. 76–8
  Bomolochus 77
zeus, Chondracanthus 114
zygaenae, Pandarus 214

Index of Latin Names of Fishes

Ablennes hians 65
Abramis 358
brama 91, 417
acanthias, Squalus 172, 178, 180, 207, 216, 219, 224, 228, 230, 235, 254, 256, 347, 368, 381, 390, 419
Acanthocybium solanderi 177
Acanthopterygii 349
acarnae, Pagellus 179
Acipenser 346
gueldenstadti 247
nacarii 247
oxyrhynchus 247
ruthenus 247
stellatus 247
sturio 147, 197, 417
transmontanus 197
Acipenseridae 154, 346
Acipenseriformes 154
aculeatus, Gasterosteus 95, 155, 307, 417
acus, Centrophorus 380
Tylosurus 65
acuta, Dussumieria 408
aeglefinus, Gadus 394
Melanogrammus 67, 173, 180, 191, 301, 302, 303, 304, 403, 418
Aetobates narinari 207
affinis, Euthynnus 174, 177, 184, 185, 270
Hydrolagus 372
agassizi, Cratinus 174
Agonidae 302, 307
Agonus cataphractus 302, 304, 307, 417
alalunga, Thunnus 204, 270, 312, 419
alba, Raja 207, 418
albacares, Thunnus 185, 204, 270
albacora, Germo 177
alleteratus, Euthynnus 174, 177, 270
Alopias 220
vulpinus 219, 262, 263, 417
Alosa 92, 410
alosa 312, 417
fallax 183, 417
alosa, Alosa 312, 417
alpinus, Salvelinus 354, 355, 419
americanus, Lophius 199
Pseudopleuronectes 126
amia, Lichia 285
Ammodytes hexapterus 197
tobianus 302, 417
Ammodytidae 302
Anabantidae 154
Anarhichas 407
latifrons 407
lupus 191, 320, 407, 410, 417
minor 407, 417
anastomella, Strongylura 65
angolensis, Sciaena 189
Anguilla anguilla 94, 417
anguilla, Anguilla 94, 417
anguillaris, Clarias 92
Anguillidae 154
Anguilliformes 154, 321
antarctica, Sciaena 285
antarcticus, Bathylagus 317
Antimora australis 320
rostrata 320
Apogonidae 154, 321
Aprionodon isodon 262
aquila, Myliobatis 368, 377, 418
arcticus, Thymallus 355
argalus, Platybelone 65
argentea, Sphyraena 185
Argyrosomus regium 189, 285, 386, 393, 417
Arius venosus 199
Artediellus uncinatus 307
aspera, Dasyatis 263
Aspitrigla cuculus 133, 179, 180, 181, 198, 417
asterias, Mustelus 379
Raja 255
Asterodermus coryphaenoides 237
elegans 237
Atheresthes stomias 13
Atherina hepsetus 410
pontica 95
atlanticus, Macrurus 322
Atractoscion equidens 393
atromarginatus, Centrophorus 380
atun, Thyrsites 184, 186
aurata, Sparus 185
auratus, Carassius 155, 417
australis, Antimora 320
Galeorhinus 216, 249
Autisthes puta 179
Auxis thazard 185
azurio, Choerodon 276

Bagridae 154
balfouri, Hemigaleus 322
Balistes 177, 185
polylepis 185
Balistoidei 185
barbatus, Mullus 277, 304, 418
Barbus 358
Bathylagus antarcticus 317
batis, Raja 132, 172, 180, 207, 364, 366, 371, 373, 374, 418
belenus, Callionymus 308
bellone, Belone 65, 179, 417
Belone bellone 65, 179, 417
svetovidovi 65
bergylta, Labrus 67, 139, 176, 182, 275, 276, 279, 418
bicornis, Icelus 307
bideni, Isurus 249
bilinearis, Merluccius 120, 173
bilineatus, Gramatorcynus 177
binoculata, Raja 370
bipinnulata, Elegatus 177
blanda, Raja 180, 207
bleekeri, Cantharus 414
Pachymetopon 414
Blennidae 302
Blennioidei 399
blennoides, Phycis 173, 302, 304, 404, 418, Fig. 1372
bogaraveo, Pagellus 176, 179, 183, 278, 414, 418
Boreogadus saida 302, 307
Bothidae 285, 302
brachygnathus, Otolithus 393
brachysomas, Calamus 185
Brama brama 180, 184, 417
brama, Abramis 91, 417
Brama 180, 184, 417
Bramidae 184
brandti, Leuciscus 197
brevirostris, Negaprion 263
Thymallus 355
Brevoortia tyrannus 408
Brosme brosme 173, 304, 417
brosme, Brosme 173, 304, 417

bubalis, Taurulus 307, 419
budegasse, Lophius 119

cabrilla, Serranus 77, 302, 419
Calamus 321
brachysomas 185
calcarifer, Lates 179
californicus, Galeus 379
Hexanchus 224
callarias, Gadus 407
Callionymidae 302
Callionymus belenus 308
festivus 123, 308
lyra 123, 191, 302, 304, 308, 402, 417, Fig. 1369
maculatus 123, 308, 417
reticulatus 123
cameronensis, Corvina 285, 393
canadum, Rachycentrodon 177
canicula, Scyllium 377
caniculus, Scyliorhinus 377, 381, 419
canis, Mustelus 216, 262, 267
Cantharus bleekeri 414
cantharus, Spondyliosoma 414, 415, 419
capensis, Merluccius 120, 395
Trigla 133, 175, 184
Capoeta 358
Carangidae 179, 184, 185
Caranx hippos 177, 179
melampygus 177
uraspis 204
Carassius auratus 155, 417
carassius 154
carassius, Carassius 154
Carcharias 249
ferox 249
littoralis 249
taurus 249, 262
carcharias, Carcharodon 219, 249, 262
Carcharinus 216
falciformis 216
floridanus 222, 226
leucas 262
limbatus 262
milberti 222, 226, 262
obscurus 262
Carcharodon carcharias 219, 249, 262
rondelletti 222
Careproctus longipinnis 307
carpio, Cyprinus 155, 417
Caspialosa 91, 92, 410
cataphractum, Peristedion 175
cataphractus, Agonus 302, 304, 307, 417
Catostomidae 154
cavalla, Scomberomorus 174, 184, 185, 277
cavillone, Lepidotrigla 179
cellarius [sic], *Gadus* 407
Centrarchidae 154
centrodontus, Pagellus 176, 410
Centrolabrus exoletus 135
Centrophorus acus 380
atromarginatus 380
granulosus 222
squamosus 193, 372, 380
Centropomidae 179, 185, 285
Centropomus 185
undecimalis 285
Centroscyllium fabricii 380
centrura, Dasyatis 199
cepedianus, Notorhynchus 216
cephalus, Leuciscus 359, 418
Mugil 173, 174, 183
Cerna gigas 281
Cetorhinus maximus 219, 249, 261, 417
Chalcaburnus chalcoides 95
chalcoides, Chalcaburnus 95
Channidae 154
Channiformes 154
Chatoessus nasus 408
chelo, Mugil 92
chilensis, Sarda 174, 270
Chimaera 370
monstrosa 172, 362, 417
ogilbyi 361
Choerodon azurio 276
Chondrostoma 358
nasus 357
Cichlidae 154
Ciliata mustela 76, 187, 304, 417
cimbrius, Rhinonemus 76, 419
circularis, Raja 366, 419
cirrosa, Umbrina 183, 189, 285, 420
Clarias anguillaris 92
clathratus, Paralabrax 185
clavata, Dasybatis 364
Raja 180, 191, 207, 371, 419
Clinus superciliosus 182
Clupea harengus 180, 195, 311, 417
Clupeidae 197
Clupeiformes 408
Clupeonella 410
Cobitidae 154
Coelorhinchus coelorhinchus 322, 417
coelorhinchus, Coelorhinchus 322, 417
Coilia dussumieri 408
gravi 408
colias, Scomber 411, 415, 416
colliei, Hydrolagus 127
commersoni, Scomberomorus 384
conchifer, Zenopsis 116
Conger 272, 279
conger 67, 191, 279, 281, 302, 417
vereauxi 279
conger, Conger 67, 191, 279, 281, 302, 417
Congridae 302
conwaii, Ostorhinchus 205
corax, Trigla 391
Coregonidae 91
Coregonus lavaretus 355
Corvina cameronensis 285, 393
nigra 285
Coryphaena 185, 270
hippurus 177, 185
Coryphaenidae 185
coryphaenoides, Asterodermus 237
Cossyphus diplotaenia 135
Cottidae 302, 307
Cottunculoides inermis 320
Cottus gobio 357, 417
Cratinus agassizi 174
Crenilabrus melops 279, 417
ocellatus 176
pavo 176
Crenimugil labrosus 93, 173, 180, 417
crocodilus, Tylosurus 65
cromis, Pogonias 184, 185, 189, 393
Ctenolabrus rupestris 275, 277, 417
cuculus, Aspitrigla 133, 179, 180, 181, 198, 417
cuvieri, Galeocerdo 262
Cybium 174
Cyclopteridae 302, 307

Cyclopterus lumpus 67, 180, 302, 320, 417
cynoglossus, Glyptocephalus 126, 417
Cynoglossus oligolepis 126
Cynoscion nebulosus 285
nobilis 199
Cyprinidae 91, 154, 346, 358
Cypriniformes 154
Cyprinodontidae 154
Cyprinodontiformes 154
Cyprinus carpio 155, 417

Dasyatis 366
aspera 263
centrura 199
kuhlii 258
pastinaca 258, 263, 369, 370, 377, 417
Dasybatis clavata 364
Dentex maroccanus 278
denticulatus, Lycichthys 407
diacanthus, Sciaena 189, 389
Dicentrarchus labrax 180, 183, 285, 302, 417
diego, Scomber 184
Diodon 232
diplotaenia, Cossyphus 135
Ditrema temmincki 413
Dorosoma thrissa 408
dorsalis, Seriola 185
Doydixodon fasciatus 402
dumeril, Squatina 254
dussumieri, Coilia 408
Dussumieria acuta 408
hasselti 408
dipterygia, Molva 173, 418

Echeneis 177
elegans, Asterodermus 237
Elegatus bipinnulata 177
Eleginus navaga 302
Elopiformes 185
Elops saurus 185
encrasicolus, Engraulis 312, 417
Engraulis 92
encrasicolus 312, 417
Epigonus 321
Epinephelus merra 413
equestris, Mustelus 262, 267
equidens, Atractoscion 393
erinacea, Raja 374
erythrinus, Pagellus 179, 414, 418
esmarkii, Trisopterus 304, 420, Fig. 1373
Esocidae 91, 154
Esox lucius 154
Etmopterus 366, 380
lucifer 380
pusillus 372
schultzi 368
spinax 372
Eucitharus linguatula 126
Eulamia 216
Eumicropterus spinosus 307
Euplatygaster indica 408
Euthynnus affinis 174, 177, 184, 185, 270
alleteratus 174, 177, 270
lineatus 174, 177
pelamis 174
Eutrigla gurnardus 133, 173, 175, 179, 180, 181, 417
Exocoetidae 314
exoletus, Centrolabrus 135

faber, Zeus 116, 117, 180, 186, 420
fabricii, Centroscyllium 380
Macrurus 402
falciformis, Carcharinus 216
fallax, Alosa 183, 417
fasciatus, Doydixodon 402
Larimus 189, 393
Pholis 307
fernandinus, Squalus 207, 368
ferox, Carcharias 249
festivus, Callionymus 123, 308
filigera, Ilisha 408
Fistularidae 187
flesus, Platichthys 91, 126, 128, 129, 180, 191, 302, 418
flesus bogdanovi, Platichthys 302
floridanus, Carcharinus 222, 226
fluviatilis, Gobius 95
Perca 95, 357, 418
frigidus, Lycodes 402
fulgens, Neomaenis 393
fullonica, Raja 79, 172, 207, 255, 366, 371, 373, 374, 419, Fig. 1636

Gadidae 154, 302, 307, 350, 402
Gadiformes 154, 320, 321, 399, 400
Gadoidei 399
Gadus 134, 391
aeglefinus 394
callarias 407
cellarius [sic] 407
macrocephalus 402
morhua 67, 173, 179, 180, 191, 301, 302, 303, 320, 403, 417
ogak 302
Gairdrops mediterraneus 126, 129
Gairdropsarus mediterraneus 187, 304, 417
Galeocerdo cuvieri 262
Galeoidea 249
Galeorhinus 216, 377, 379
australis 216, 249
galeus 207, 216, 249, 257, 267, 377, 417
Galeus 377
californicus 379
vulgaris 132
galeus, Galeorhinus 207, 216, 249, 257, 267, 377, 417
Gasterosteidae 91, 154
Gasterosteiformes 154
Gasterosteus aculeatus 95, 155, 307, 417
wheatlandi 95
gavialoides, Lhotskia 65
gayi, Merluccius 395, 396
Gazza minuta 403
Gempylidae 184
Germo albacora 177
gigas, Cerna 281
glacialis, Liopsetta 191, 302
gladius, Xiphias 126, 138, 141, 420
glanis, Silurus 346
glauca, Prionace 216, 219, 222, 226, 249, 263, 267, 418
glaucus, Isurus 222, 249
glesne, Regalecus 61
Glyptocephalus 126
cynoglossus 126, 417
Gobiesox sanguineus 127
Gobiidae 91, 154, 358
Gobio 358
gobio 92
gobio, Gobio 92
Cottus 357, 417
Gobius 92
fluviatilis 95

(*Macrogobius*) *paganellus* 188, 417
goodi, Nematonurus 320
Gramatorcynus bilineatus 177
granulosus, Centrophorus 222
gravi, Coilia 408
griseus, Hexanchus 224, 251, 262, 417
Lutjanus 174, 286, 393
gruveli, Paragaleus 377
gueldenstadti, Acipenser 247
gunnellus, Pholis 302, 307, 418
gurnardus, Eutrigla 133, 173, 175, 179, 180, 181, 417
Trigla 368, 390
guttatus, Lampris 368, 418
Gymnelis viridis 307
Gymnura micrura 207

halm. Platessa 126
harengus, Clupea 180, 195, 311, 417
hasselti, Dussumieria 408
Hemigaleus balfouri 322
hepsettus, Atherina 410
Heptranchias perlo 249
Hexagrammos octogrammus 402
otakii 413
stelleri 126
Hexanchus 224
californicus 224
griseus 224, 251, 262, 417
hexapterus, Ammodytes 197
hians, Ablennes 65
Hilsa ilisha 408
Hippoglossoides platessoides 126, 304, 307, 417
hippoglossoides, Reinhardtius 193, 320, 397
Hippoglossus hippoglossus 126, 173, 183, 191, 192, 193, 199, 274, 397, 417
stenolepis 126, 397
hippoglossus, Hippoglossus 126, 173, 183, 191, 192, 193, 199, 274, 379, 417
hippos, Caranx 177, 179
hippurus, Coryphaena 177, 185
Hirundichthys 314
Holocephala 372
hololepidotus, Johnius 189, 285, 393
hubbsi, Merluccius 120, 299, 396
Huso 346
huso 247
huso, Huso 247
Hydrolagus affinis 372
colliei 127
Hypoprion signatus 219, 222
Hyporhamphus melanochir 65
regularis 65
unifasciatus 314
Hypotremata 347

Icelidae 307
Icelus bicornis 307
Ictaluridae 154
Idus 358
idus, Leuciscus 155, 359, 418
Ilisha filigera 408
ilisha, Hilsa 408
imperialis, Luvarus 236, 237, 418
incisa, Strongylura 65
indica, Euplatygaster 408
inermis, Cottunculoides 320
Sphaeroides 177, 204
isodon, Aprionodon 262
Isurus 249
bideni 249
glaucus 222, 249
mako 216, 249
nasus 226
oxyrinchus 177, 216, 219, 222, 226, 227, 249, 418

japonicus, Scomber 408, 415
Johnius 189
hololepidotus 189, 285, 393

Katsuwonus 177, 185
pelamis 174, 177, 184, 185, 270
vagans 177
kenojei, Raja 132, 327
keta, Oncorhynchus 354
Kishinoella tonggol 270
kitt, Microstomus 126, 127, 173, 180, 302, 303, 418
kuhlii, Dasyatis 258

labrax, Dicentrarchus 180, 183, 285, 302, 417
Labridae 276, 279, 302
labrosus, Crenimugil 93, 173, 180, 417
Labrus 176
bergylta 67, 139, 176, 182, 275, 276, 279, 418
mixtus 182, 276, 302, 418
trimaculatus 276
Laemonema laureysi 320
Laeviraja macrorhynchus 263, 364
oxyrhynchus 263, 364
laevis, Malacocephalus 118, 418
Mustelus 377
Raja 132, 172, 372, 374
Lamna 222, 249
ditropis 219, 222, 249
nasus 219, 222, 226, 249, 262, 360, 361, 418
Lampridae 349
Lampris 349
guttatus 368, 418
regius 382
lanceolata, Mola 232
Larimus fasciatus 189, 393
lastoviza, Trigloporus 133, 179, 419
Lates calcarifer 179
latifrons, Anarhichas 407
laureysi, Laemonema 320
lavaretus, Coregonus 355
leiura, Strongylura 65
Lepidion 112
Lepidorhombus whiffiagonis 126, 179, 418
Lepidotrigla cavillone 179
leucas, Carcharinus 262
Leuciscus 358
brandti 197
cephalus 359, 418
idus 155, 359, 418
leuciscus 359, 418
rutilus 94
leuciscus, Leuciscus 359, 418
lewini, Sphyrna 263
Lhotskia gavialoides 65
Lichia amia 285
lilljeborgi, Myoxocephalus 307
Limanda limanda 126, 128, 129, 180, 191, 302, 418
limanda, Limanda 126, 128, 129, 180, 191, 302, 418
limbatus, Carcharinus 262
lineatus, Euthynnus 174, 177
Liparis 307
linguatula, Eucitharus 126
Liopsetta glacialis 191, 302
Liparidae 307
Liparis lineatus 307

liparis 304, 418
reinhardti 307
tunicatus 307
liparis, Liparis 304, 418
lithognathus, Pagellus 414
littoralis, Carcharias 249
Liza ramada 180, 418
loennbergi, Trematomus 402
longipinnis, Careproctus 307
longispinis, Sparus 414
Lophius 370
americanus 199
budegasse 199
piscatorius 173, 180, 199, 418
Lota lota 352
lota, Lota 352
lucerna, Trigla 133, 175, 179, 180, 181, 198, 419
lucifer, Etmopterus 380
Lucioperca 357
lucius, Esox 154
Lumpenidae 307
Lumpenus lupretaeformis 307
lumpretaeformis, Lumpenus 307
lumpus, Cyclopterus 67, 180, 302, 320, 417
lupus, Anarhichas 191, 320, 407, 410, 417
luscus, Trisopterus 67, 120, 180, 304, 403, 420, Figs. 1370, 1371
Lutjanidae 185
Lutjanus 174, 185
griseus 174, 286, 393
novemfasciatus 174
Luvarus imperialis 236, 237, 418
Lycenchelys sarsi 307
vahli 307
verrilli 307
Lycichthys denticulatus 407
Lycodes frigidus 402
lyra, Callionymus 123, 191, 302, 304, 308, 402, 417, Fig. 1369
Trigla 133, 179, 419

macrocephalus, Gadus 402
(Macrogobius) paganellus, Gobius 188, 417
macropterus, Neothunnus 177
macrorhynchus, Laeviraja 263, 366
Raja 207
Macrouridae 321
Macrouroidei 399
Macrurus 320, 322
atlanticus 322
fabricii 402
whitsoni 402
maculata, Raja 371
Seriolella 184
maculofasciatus, Paralabrax 185
maculatus, Callionymus 123, 308, 417
Scomberomorus 174, 184, 185
maculosa, Raja 366
Makaira mitsukuuri 204
mako, Isurus 216, 249
malabaricus, Thrissocles 410
Malacocephalus laevis
malleus, Zygaena 74
Mallotus villosus 307
manazo, Mustelus 267
margarita, Trygon 370
marina, Strongylura 65
marinus, Sebastes 173, 176, 307, 320, 402, 419
marmorata, Torpedo 369, 370, 419
Trygon 370
maroccanus, Dentex 278
Masogobius 95
Mastacembalidae 154
maximus, Cetorhinus 219, 249, 261, 417
Pleuronectes 232
Scophthalmus 126, 129, 173, 179, 180, 191, 193, 199, 302, 397, 419
mediterranean, Phycis 404
mediterraneus, Gairdrops 126, 129
Gairdropsaurus 187, 304, 417
melampygus, Caranx 177
melanochir, Hyporhamphus 65
Melanogrammus aeglefinus 67, 173, 180, 191, 301, 302, 303, 304, 403, 418
melanurus, Smaris 414
melops, Crenilabrus 279, 417
Merlangius merlangus 67, 173, 180, 301, 302, 304, 307, 403, 418
merlangus, Merlangius 67, 173, 180, 301, 302, 304, 307, 403, 418
Merlucciidae 302
Merluccius 120, 303, 389, 395, 396, 402
bilinearis 120, 173
capensis 120, 395
gayi 395, 396
hubbsi 120, 299, 396
merluccius 120, 173, 179, 302, 389, 394, 395, 396, 405, 418
poli 389
productus 120, 396
merluccius, Merluccius 120, 173, 179, 302, 389, 394, 395, 396, 405, 418
merra, Epinephalus 413
microcephalus, Somniosus 193, 219, 360, 361, 419
Microgadus tomcod 173
microocellata, Raja 207, 419
Microstomus kitt 126, 127, 173, 180, 302, 303, 418
micrura, Gymnura 207
milberti, Carcharinus 222, 226, 262
milvus, Trigla 179
minor, Anarhichas 407, 417
minuta, Gazza 403
minutus, Pomatoschistus 180, 304, 418
Trisopterus 173, 180, 302, 304, 420
mitsukuuri, Makaira 204
Squalus 366
Tetrapturus 213
mixtus, Labrus 182, 276, 302, 418
Mola 213, 232, 235, 312
lanceolata 232
mola 180, 192, 194, 213, 216, 222, 229, 232, 234, 249, 312, 418
mola, Mola 180, 192, 194, 213, 216, 222, 229, 232, 234, 249, 312, 418
Molva dipterygia 173, 418
molva 67, 173, 179, 180, 195, 302, 304, 418,
molva, Molva 67, 173, 179, 180, 195, 302, 304, 418
monstrosa, Chimaera 172, 362, 417
montagui, Raja 172, 180, 207, 255, 366, 374, 419, Fig. 1635
morhua, Gadus 67, 173, 179, 180, 191, 301, 302, 303, 320, 403, 417
mormyrus, Pagellus 179
Mugil 93
cephalus 173, 174, 183
chelo 92
Mugilidae 92, 93
Mugiloidei 174
Mullidae 197
Mullus barbatus 277, 304, 418
surmulatus 277, 418
multifasciatus, Scatophagus 285
Muraenosox 272

mustela, Ciliata 76, 187, 304, 417
Mustelus 262, 323, 377, 379, 418
laevis 377
manazo 267
mustelus 216, 262, 267, 360, 361, 377, 418
plebejus 262
vulgaris 379
mustelus, Mustelus 216, 262, 267, 360, 361, 377, 418
Myliobatis 366, 370
aquila 368, 377, 418
Myoxocephalus lilljeborgi 307
quadricornis 304
scorpioides 307
scorpius 302, 304, 307, 418
scorpius groenlandicus 307

nacarii, Acipenser 247
naevus, Raja 80, 172, 180, 255, 366, 374, 419
Narcina 366
narce, Torpedo 364
narinari, Aetobates 207
nasus, Chatoessus 408
Chondrostoma 357
Isurus 226
Lamna 219, 222, 226, 249, 262, 360, 361, 418
navaga, Eleginus 302
nebulosus, Cynoscion 285
Negaprion brevirostris 263
negra, Corvina 285
Nematonurus goodi 320
Neogobius 92
Neomaenis fulgens 393
Neothunnus macropterus 177
nigrolineatus, Syngnathus 95
nobiliana, Torpedo 255, 419
nobilis, Cynoscion 199
Notacanthiformes 199
notata, Strongylura 65
Notorhynchus cepedianus 216
Notothenioidei 399
novemfasciatus, Lutjanus 174

obesus, Parathunnus 204
Thunnus 177, 270
obscurus, Carcharinus 262
ocellatus, Crenilabrus 176
occidentalis, Torpedo 220
octogrammus, Hexagrammos 402
ogak, Gadus 302
ogilbyi, Chimaera 361
oligolepis, Cynoglossus 126
Terhops 126
Oligoplites saurus 174
Oncorhynchus 197
keta 354
Onos 76
opercularis, Polydactylus
Ophioidei 399
orientalis, Sarda 174, 177, 185
Thunnus 204
Orthagoriscus 229
Osmeridae 307
Osphronemidae 154
Ostorhinchus conwaii 205
otakii, Hexagrammos 413
Otolithus 393
brachygnathus 393
senegalensis 393
oxyrhynchus, Acipenser 247
Laeviraja 263, 364

oxyrinchus, Isurus 177, 216, 219, 222, 226, 227, 249, 418
Raja 132, 172, 366, 419

Pachymetopon bleekeri 414
Pagellus 414
acarnae 179
bogaraveo 176, 179, 183, 278, 414, 418
centrodontus 176, 410
erythrinus 179, 414, 418
lithognathus 414
mormyrus 179
pallasi, Pleuronectes 126
Paragaleus gruveli 377
Paralabrax clathratus 185
maculofasciatus 185
Paralichthys 205, 285
Parathunnus obesus 204
sibi 177, 204, 270
passer, Platessa 179
pastinaca, Dasyatis 258, 263, 369, 370, 377, 417
Trygon 369
pavo, Crenilabrus 176
pelamis, Euthynnus 174
Katsuwonus 174, 177, 184, 185, 270
Perca 357
fluviatilis 95, 357, 418
Percidae 91, 154
Perciformes 154, 174, 176, 179, 185, 320, 321, 399, 400
Percoidei 174, 185, 399
Peristedion cataphractum 175
perlo, Heptranchias 259
phasa, Setipinna 408
Pholidae 307
Pholis fasciatus 307
gunnellus 302, 307, 418
Phoxinus 358
phoxinus 92
phoxinus, Phoxinus 92
Phycis 404
blennoides 173, 302, 304, 404, 418, Fig. 1372
mediterranean 404
pilchardus, Sardina 311, 312, 419
pinguis, Pleuronectes 397
piscatorius, Lophius 119, 173, 180, 418
Platessa halm 126
passer 179
platessa, Pleuronectes 126, 129, 133, 173, 180, 191, 302, 418
platessoides, Hippoglossoides 126, 304, 307, 417
Platichthys flesus 91, 126, 128, 129, 180, 191, 302, 418
flesus bogdanovi 302
Platybelone argalus 65
platygaster, Pygasteus 95
plebejus, Mustelus 262
Plecoglossidae 154
Pleurogrammus 402
Pleuronectes 126
maximus 232
pallasi 126
pinguis 397
platessa 126, 129, 133, 173, 180, 191, 302, 418
rhombus 232
Pleuronectidae 302, 307
Pleuronectiformes 320
Poecillidae 154
Pogonias cromis 184, 185, 189, 393
poli, Merluccius 389
Pollachius pollachius 67, 173, 180, 195, 302, 403, 418
virens 67, 173, 179, 180, 195, 302, 403, 418
pollachius, Pollachius 67, 173, 180, 195, 302, 403, 418

Polydactylus opercularis 177, 185
quadrifilis 285
polylepis, Balistes 185
Polynemidae 185, 285
Pomatomidae 184
Pomatomus saltarix 174, 184, 386
Pomatoschistus 92
minutus 180, 304, 418
Pomolobus 408
pontica, Atherina 95
porcus, Scorpaena 410
prayensis, Pseudopeneus 277
Prionace glauca 216, 219, 222, 226, 249, 263, 267, 418
Pristiurus 232
productus, Merluccius 120, 396
Prognichthys rondeleti 314
Prosopium 355
Pseudaspius 358
Pseudopeneus prayensis 277
Pseudopleuronectes americanus 126
Pteroplatea 366
punctatus, Zeugopterus 77, 420
Pungitius pungitius 95, 418
pungitius, Pungitius 95, 418
pusillus, Etmopterus 372
puta, Autisthes 179
Pygasteus platygaster 95

quadricornis, Myoxocephalus 304
quadrifilis, Polydactylus 285

Rachycentron canadum 177
radiata, Raja 132, 172, 180, 191, 255, 371, 374, 419
Raja 80, 131, 172, 207, 364, 366, 370, 371, 373, 374
alba 207, 418
asterias 255
batis 132, 172, 180, 207, 364, 366, 371, 373, 374, 418
binoculata 370
blanda 180, 207
circularis 366, 419
clavata 180, 191, 207, 371, 419
erinacea 374
fullonica 79, 172, 207, 255, 366, 371, 373, 374, 419, Fig. 1636
kenojei 132, 372
laevis 132, 172, 372, 374
macrorhynchus 207
maculata 371
maculosa 366
montagui 172, 180, 207, 255, 366, 374, 419, Fig. 1635
naevus 80, 172, 180, 255, 366, 374, 419
oxyrinchus 132, 172, 366, 419
radiata 132, 172, 180, 191, 255, 371, 374, 419
rhina 132, 255, 370, 372
scabrata 371
stellulata 255
ramada, Liza 180, 418
Regalecus glesne 61
regium, Argyrosomus 189, 285, 386, 393, 417
regius, Lampris 382
regularis, Hyporhamphus 65
reinhardti, Liparis 307
Reinhardtius hippoglossoides 193, 320, 397
reticulatus, Callionymus 123
rhina, Raja 132, 255, 370, 372
Rhineodon typus 220
Rhinonemus cimbrius 76, 419
Rhinoptera 370
Rhodeus 358
rhombus, Pleuronectes 232
Scophthalmus 129, 173, 193, 199, 419
Roncador stearnsi 393
rondeleti, Prognichthys 314
Sargus 402
rondelletti, Carcharodon 222
rostrata, Antimora 320
rubrivinctus, Sebastes 197
rupestris, Ctenolabrus 275, 277, 417
ruthenus, Acipenser 247
Rutilus 358
rutilus 94, 155, 419
rutilus, Leuciscus 94
Rutilus 94, 155, 419

saida, Boreogadus 302, 307
salar, Salmo 180, 354, 419
Salmo 197, 353
salar 180, 354, 419
trutta 91, 180, 419
Salmonidae 91, 154, 344, 352
Salmoniformes 154
saltatrix, Pomatomus 174, 184, 386
Salvelinus alpinus 354, 355, 419
sanguineus, Gobiesox 127
Sarda 174, 177
chilensis 174, 270
orientalis 174, 177, 185
sarda 174, 184, 185, 270, 419
velox 174
sarda, Sarda 174, 184, 185, 270, 419
Sardina pilchardus 311, 312, 419
Sargus 278
rondeleti 402
solviani 278
vulgaris 278
sarsi, Lycenchelys 307
saurus, Elops 185
Oligoplites 174
scabrata, Raja 371
Scatophagidae 285
Scatophagus multifasciatus 285
schultzi, Etmopterus 368
Sciaena 285, 392, 393
angolensis 189
antarctica 285
diacanthus 189, 389
Sciaenidae 184, 185, 189, 285
Scoliodon terraenovae 263
Scomber colias 411, 415, 416
diego 184
japonicus 408, 415
scombrus 179, 180, 184, 191, 411, 416, 419
Scomberomorus cavalla 174, 184, 185, 277
commersoni 384
maculatus 174, 184, 185
sierra 185
Scombridae 179, 185, 408
scombrus, Scomber 179, 180, 184, 191, 411, 416, 419
Scophthalmus maximus 126, 129, 173, 179, 180, 191, 193, 199, 302, 397, 419
rhombus 129, 173, 193, 199, 419
Scorpaena porcus 410
Scorpaenidae 121, 307
Scorpaeniformes 320, 399
scorpioides, Myoxocephalus 307
scorpius, Myoxocephalus 302, 304, 307, 418
scorpius groenlandicus, Myoxocephalus 307
Scyliorhinus 377, 379, 419
caniculus 377, 381, 419
stellaris 216, 419
Scyllium 377
canicula 377

Sebastes 121, 122, 320, 396
marinus 173, 176, 307, 320, 402, 419
rubrivinctus 197
viviparus 320, 419
Selene vomer 234
senegalensis, Otolithus 393
Strongylura 65
Seriola dorsalis 185
Seriolella maculata 184
Serranidae 185, 302
Serranus cabrilla 77, 302, 419
Setipinna phasa 408
sibi, Parathunnus 177, 204, 270
sierra, Scomberomorus 185
signatus, Hypoprion 219, 222
signifer, Thymallus 355
Siluridae 91, 154
Siluriformes 154
Silurus glanis 346
Smaris melanurus 414
solanderi, Acanthocybium 177
Solea 129
solea 63, 126, 129, 179, 199, 302, 304, 419
solea, Solea 63, 126, 129, 179, 199, 302, 304, 419
Soleidae 302
solviani, Sargus 278
Somniosus microcephalus 193, 219, 360, 361, 419
Sparidae 179, 185, 321, 414
Sparisoma viridis 121
Sparus aurata 185
longispinis 414
Sphaeroides inermis 177, 204
Sphyraena argentea 185
Sphyraenidae 185
Sphyrna 263
lewini 263
tiburo 263
zygaena 226, 263, 267, 419
spinax, Etmopterus 372
spinosus, Eumicropterus 307
Spondyliosoma cantharus 414, 415, 419
Sprattus sprattus 311, 312, 419
sprattus, Sprattus 311, 312, 419
Squalius 358
Squalus 116, 216, 366, 368, 392
acanthias 172, 177, 180, 207, 216, 219, 224, 228, 230, 235, 254, 256, 347, 368, 381, 390, 419
fernandinus 207, 368
mitsukuuri 366
suckleyi 256
squamosus, Centrophorus 193, 372, 380
Squatina dumeril 254
squatina 254, 419
squatina, Squatina 254, 419
stearnsi, Roncador 393
steindachneri, Umbrina 285
stellaris, Scyliorhinus 216, 419
stellatus, Acipenser 247
stelleri, Hexagrammos 126
stellulata, Raja 255
stenolepis, Hippoglossus 126, 397
stomias, Atheresthes 13
Stromateidae 184
Strongylura anastomella 65
incisa 65
leiura 65
marina 65
notata 65
senegalensis 65
strongylura 65
timucu 65
urvillii 65
strongylura, Strongylura 65
sturio, Acipenser 197, 247, 417
suckleyi, Squalus 256
superciliosus, Clinus 182
surmuletus, Mullus 277, 418
svetovidovi, Belone 65
symmetricus, Trachurus 184
Syngnathus nigrolineatus 95

Taurulus bubalis 307, 419
taurus, Carcharias 249, 262
temmincki, Ditrema 413
Terhops oligolepis 126
terraenovae, Scoliodon 263
Tetraodontidae 187
Tetraodontiformes 177
Tetrapturus mitsukurii 213
thazard, Auxis 185
Theraponidae 179
thrissa, Dorosoma 408
Thrissocles malabaricus 410
Thunnus 204, 314
alalunga 204, 270, 312, 419
albacares 185, 204, 270
obesus 177, 270
orientalis 204
thynnus 174, 177, 185, 204, 232, 270, 312, 419
Thymallidae 91
Thymallus 355
arcticus 355
brevirostris 355
signifer 355
thymallus 354, 355, 419
thymallus, Thymallus 354, 355, 419
Thynnus 119
thynnus 232
thynnus, Thunnus 174, 177, 185, 204, 232, 270, 312, 419
Thynnus 232
Thyrsites atun 184, 186
tiburo, Sphyrna 263
timucu, Strongylura 65
Tinca tinca 91, 419
tinca, Tinca 91, 419
tobianus, Ammodytes 302, 417
tomcod, Microgadus 173
tonggol, Kishinoella 270
Torpedo 366, 370
marmorata 369, 370, 419
narce 364
nobiliana 255, 419
occidentalis 220
Trachinotidae 285
Trachinus vipera 304, 419
Trachurus symmetricus 184
trachurus 177, 180, 184, 419
trachurus, Trachurus 177, 180, 184, 419
transmontanus, Acipenser 197
Trematomus loennbergi 402
Trigla 179, 390, 391, 410, 419
capensis 133, 175, 184
corax 391
gurnardus 368, 390
lucerna 133, 175, 179, 180, 181, 198, 419
lyra 133, 179, 419
milvus 179
Triglidae 133, 134, 175, 179, 181, 184, 390, 394, 399
Trigloporus lastoviza 133, 179, 419
trimaculatus, Labrus 276
Trisopterus esmarkii 304, 420, Fig. 1373
luscus 67, 120, 180, 304, 403, 420, Figs. 1370, 1371

minutus 173, 180, 302, 304, 420
trutta, Salmo 91, 180, 419
Trygon 366, 368, 420
margarita 370
marmorata 370
pastinaca 369
tunicatus, Liparis 307
Tylosurus acus 65
crocodilus 65
typus, Rhineodon 220
tyrannus, Brevoortia 408

Umbridae 154
Umbrina cirrosa 183, 189, 285, 420
steindachneri 285
valida 189, 285, 393
uncinatus, Artediellus 307
undecimalis, Centropomus 285
unifasciatus, Hyporhamphus 314
uraspis, Caranx 204
urvilli, Strongylura 65

vagans, Katsuwonus 177
vahli, Lycenchelys 307
valida, Umbrina 189, 285, 393
Varicorhinus 358
velox, Sarda 174
venosus, Arius 199
vereauxi, Conger 279
verrilli, Lycenchelys 307
villosus, Mallotus 307
vipera, Trachinus 304, 419
virens, Pollachius 67, 173, 179, 180, 195, 302, 403, 418
viridis, Gymnelis 307
Sparisoma 121
viviparus, Sebastes 320, 419
Zoarces 302, 420
vomer, Selene 234
vulgaris, Galeus 132
Mustelus 379
Sargus 278
vulpinus, Alopias 219, 262, 263, 417

wheatlandi, Gasterosteus
whiffiagonis, Lepidorhombus 126, 179, 418
whitsoni, Macrurus 402

Xiphias gladius 126, 138, 141, 320

Zenopsis conchifer 116
Zeugopterus punctatus 77, 420
Zeus faber 116, 117, 180, 186, 420
Zoarces viviparus 302, 420
Zoarcidae 302, 307
Zoarcoidei 399
Zygaena malleus 74
zygaena, Sphyrna 226, 263, 267, 419

Plates

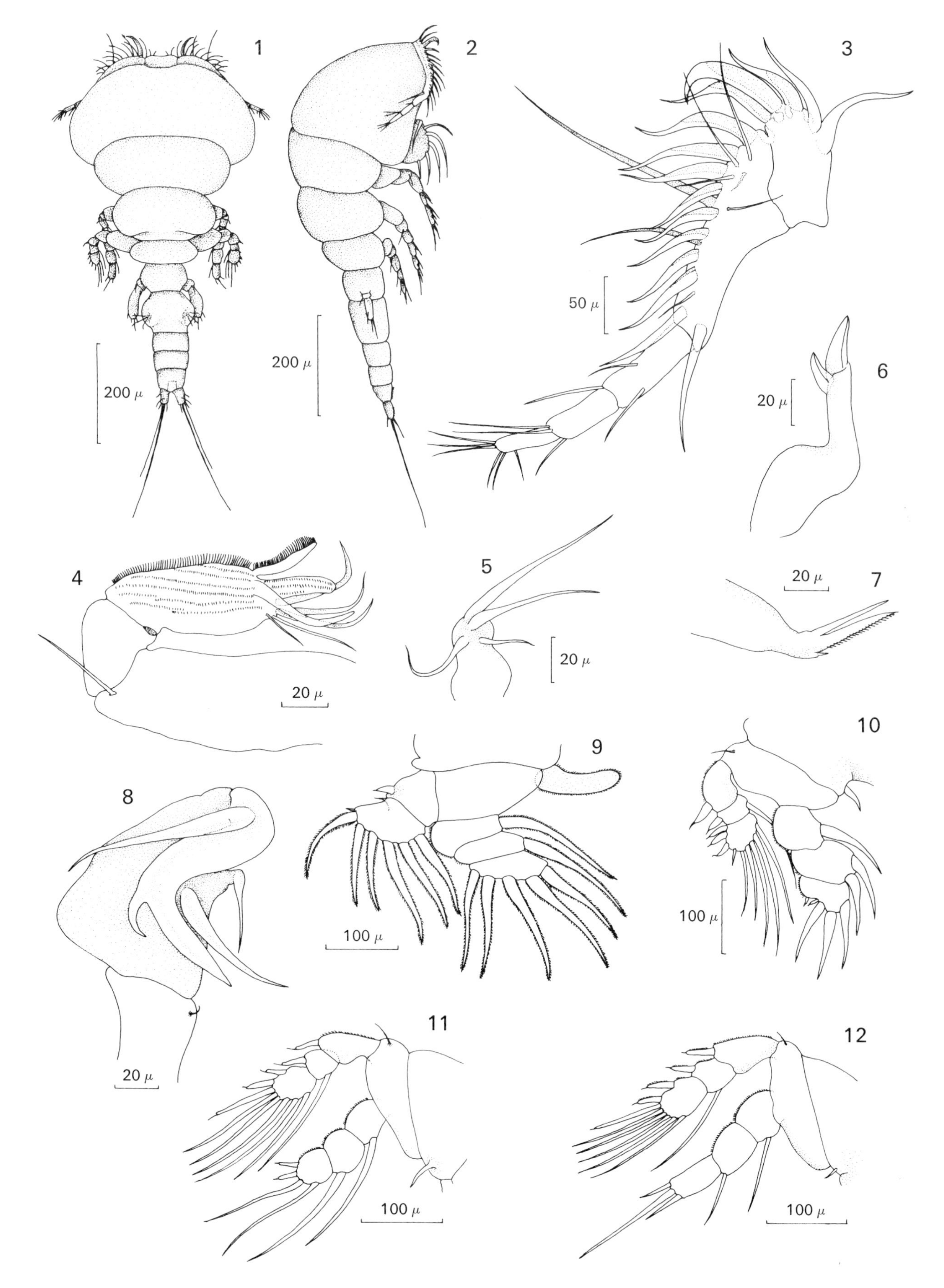

Figs. 1–12.
Bomolochus soleae Claus, 1864. **Fig. 1.** Female, dorsal. **Fig. 2.** Same, lateral. **Fig. 3.** First antenna, ventral. **Fig. 4.** Second antenna, ventral. **Fig. 5.** First maxilla, ventral. **Fig. 6.** Mandible, dorsal. **Fig. 7.** Second maxilla, ventral. **Fig. 8.** Maxilliped, ventral. **Fig. 9.** First leg, ventral. **Fig. 10.** Second leg, ventral. **Fig. 11.** Third leg, ventral. **Fig. 12.** Fourth leg, ventral.
Redrawn and slightly modified from Vervoort (1969).

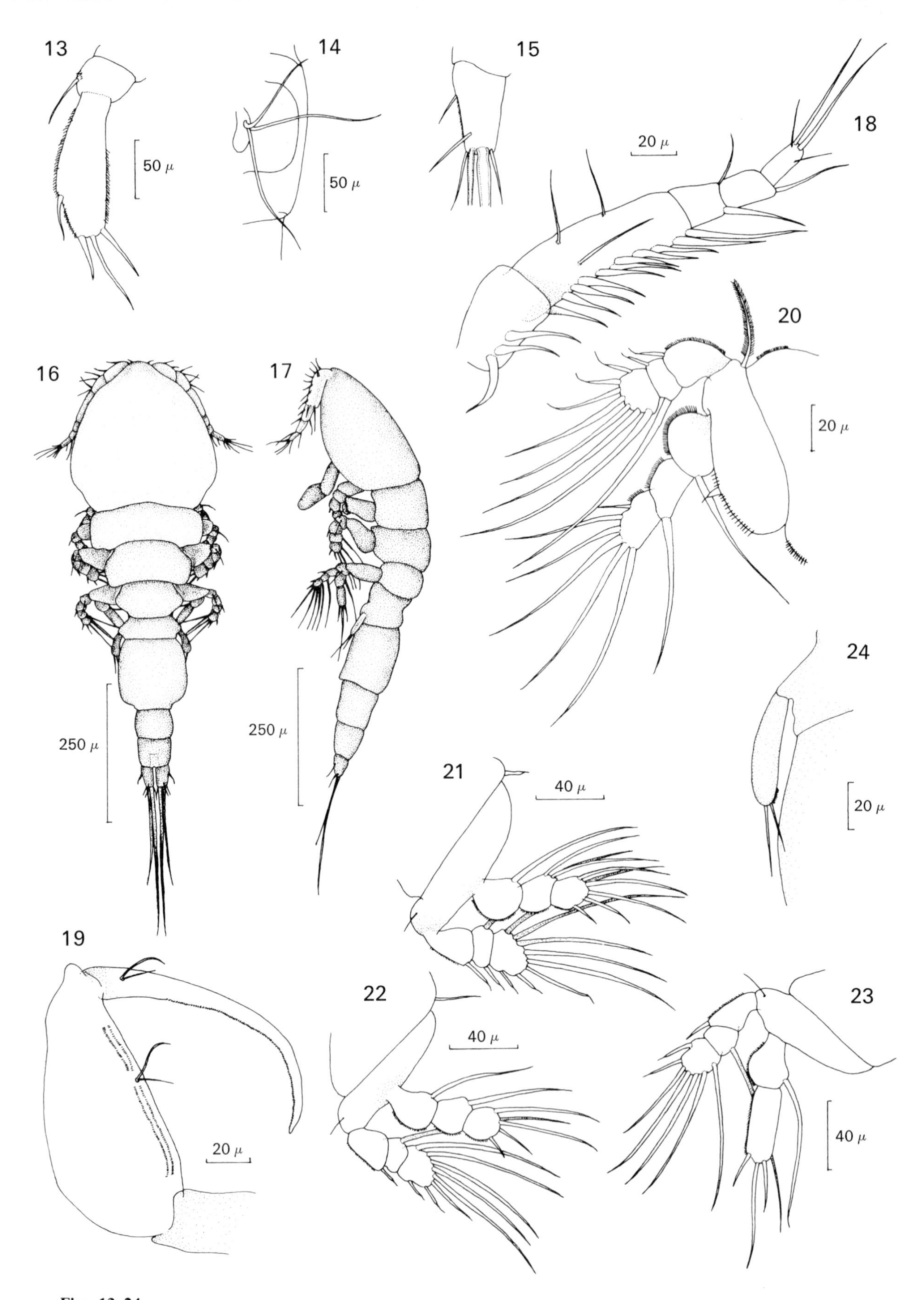

Figs. 13–24.
Bomolochus soleae Claus, 1864. **Fig. 13.** Female, fifth leg, ventral. **Fig. 14.** Sixth leg and region of genital orifice, dorsal. **Fig. 15.** Uropod, dorsal. **Fig. 16.** Male, dorsal. **Fig. 17.** Same, lateral. **Fig. 18.** First antenna, ventral. **Fig. 19.** Maxilliped, ventral. **Fig. 20.** First leg, ventral. **Fig. 21.** Second leg, ventral. **Fig. 22.** Third leg, ventral. **Fig. 23.** Fourth leg, ventral. **Fig. 24.** Fifth leg, ventral.
Redrawn and slightly modified from Vervoort (1969).

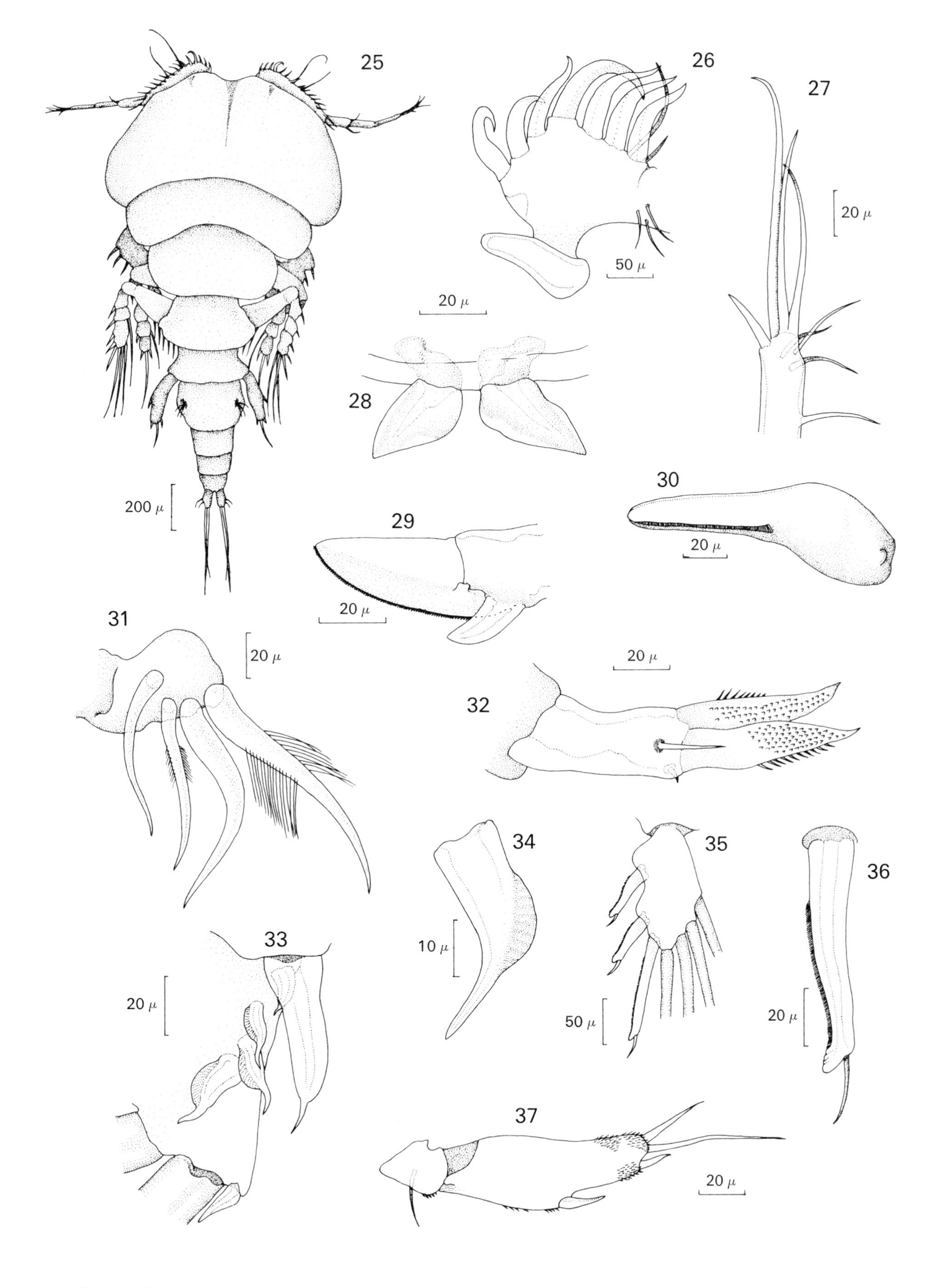

Figs. 25–37.
Bomolochus bellones Burmeister, 1835. **Fig. 25.** Female, dorsal. **Fig. 26.** First antenna, base, ventral. **Fig. 27.** Same, tip, ventral. **Fig. 28.** Rostral tines, ventral. **Fig. 29.** Mandible, tip, dorsal. **Fig. 30.** Paragnath, ventral. **Fig. 31.** First maxilla, ventral. **Fig. 32.** Second maxilla, distal part, ventral. **Fig. 33.** First leg, exopod, details of lateral margin, dorsal. **Fig. 34.** Same, one of spines enlarged. **Fig. 35.** Fourth leg, exopod, distal segment, ventral. **Fig. 36.** Same, spine enlarged. **Fig. 37.** Fifth leg, ventral.

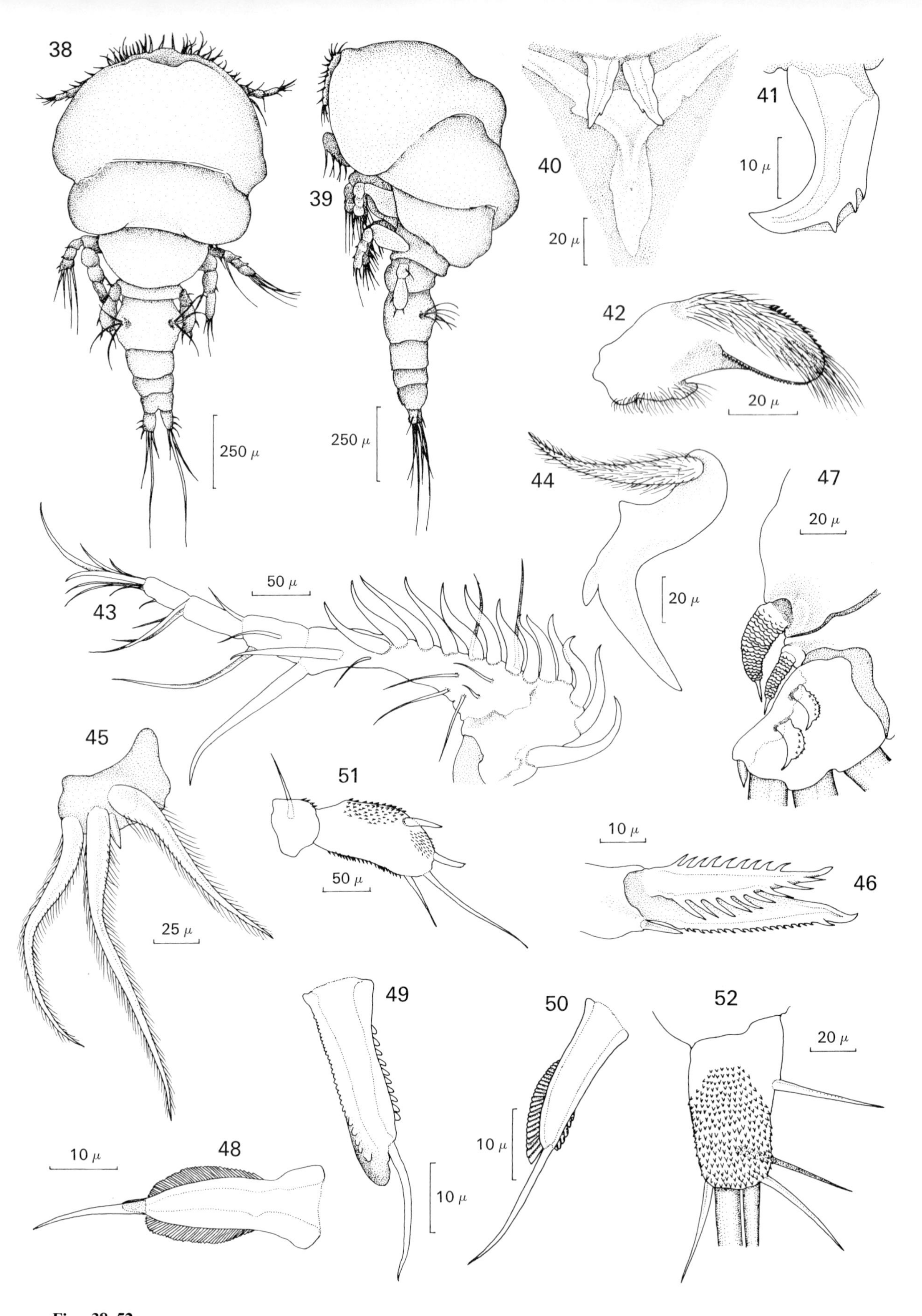

Figs. 38–52
Holobomolochus confusus (Stock, 1953). **Fig. 38.** Female, dorsal. **Fig. 39.** Same, lateral. **Fig. 40**. Rostral tines and subrostral plate, ventral. **Fig. 41.** One rostral tine, lateral. **Fig. 42.** Paragnath, ventral. **Fig. 43.** First antenna, ventral. **Fig. 44.** Maxilliped, claw, ventral. **Fig. 45.** First maxilla, ventral. **Fig. 46.** Second maxilla, distal part, ventral. **Fig. 47.** First leg, detail of lateral margin of exopod, dorsal. **Fig. 48.** Second leg, exopod spine. **Fig. 49.** Third leg, exopod spine. **Fig. 50.** Fourth leg, exopod spine. **Fig. 51.** Fifth leg, ventral. **Fig. 52.** Uropod, ventral.

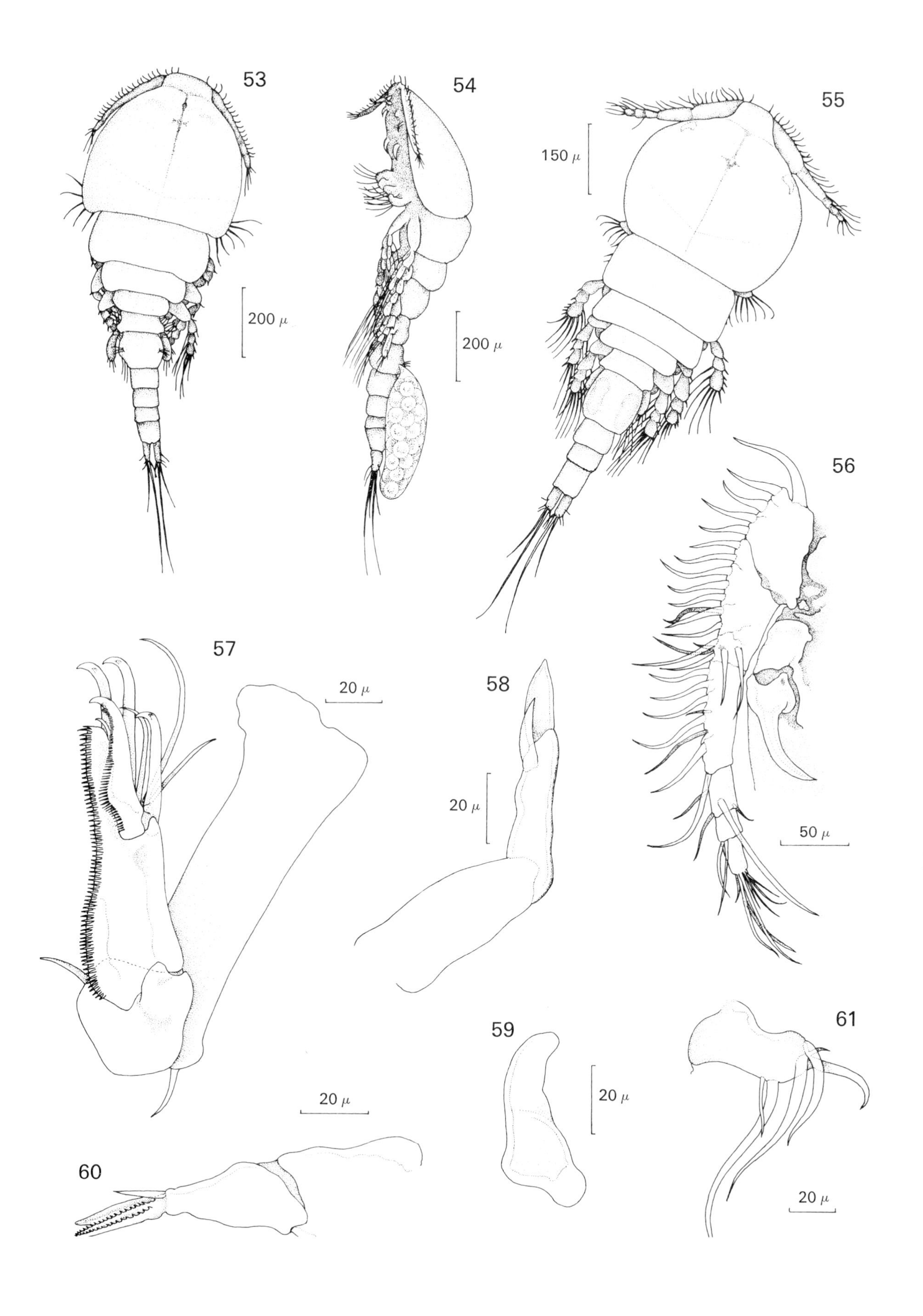

Figs. 53–61
Anchistrotos onosi (T. Scott, 1902). **Fig. 53.** Female, dorsal. **Fig. 54.** Same, lateral. **Fig. 55.** Male, dorsal. **Fig. 56.** Male and female, first antenna, ventral. **Fig. 57.** Second antenna, ventral. **Fig. 58.** Mandible, dorsal. **Fig. 59.** Paragnath, ventral. **Fig. 60.** Second maxilla, ventral. **Fig. 61.** First maxilla, lateral.

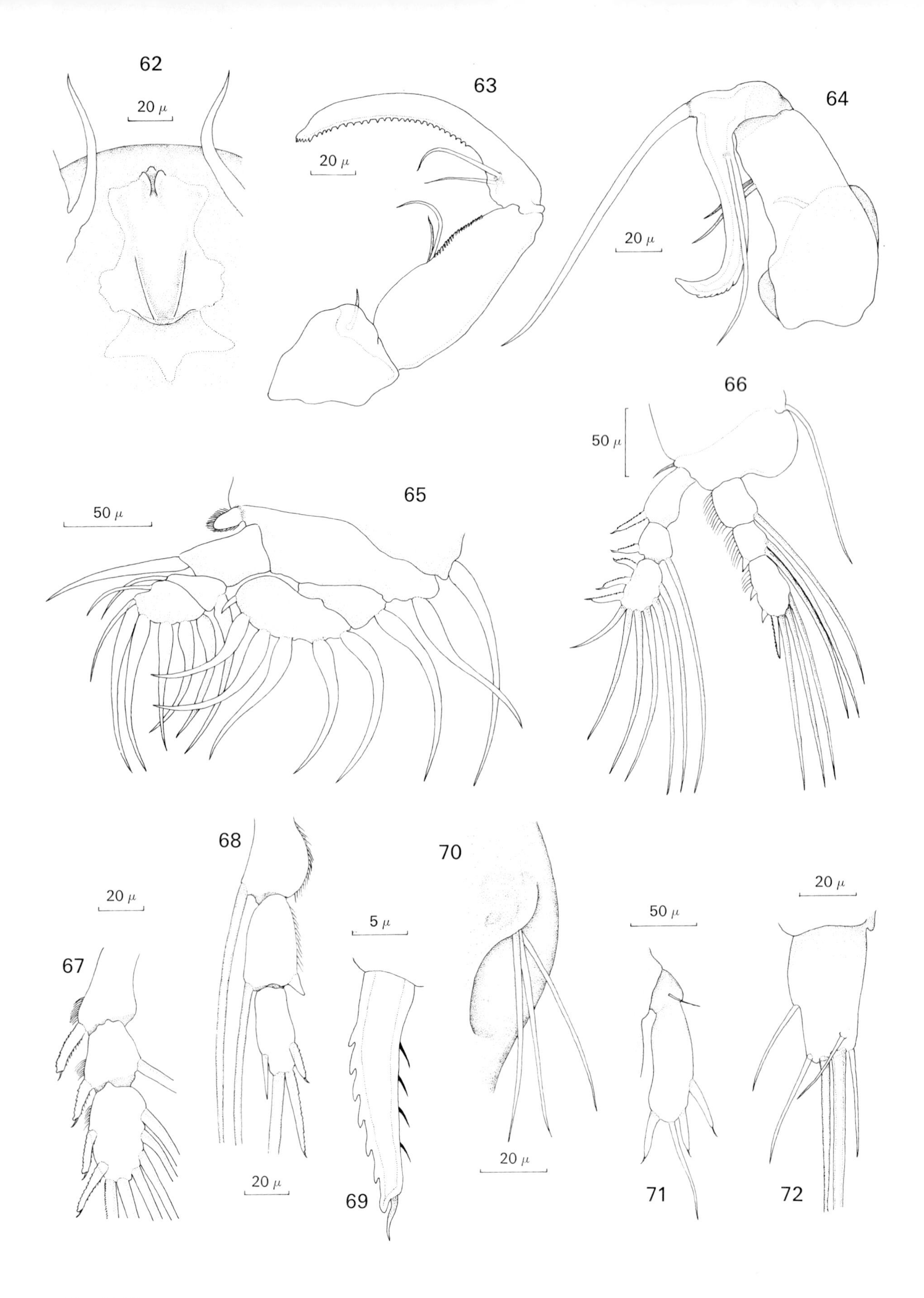

Figs. 62–72
Anchistrotos onosi (T. Scott, 1902). **Fig. 62.** Rostral area, ventral. **Fig. 63.** Male, maxilliped, ventral. **Fig. 64.** Female, maxilliped, ventral. **Fig. 65.** First leg, ventral. **Fig. 66.** Second leg, ventral. **Fig. 67.** Fourth leg, exopod, ventral. **Fig. 68.** Same, endopod, ventral. **Fig. 69**. Legs two-four, exopod spine. **Fig. 70.** Sixth leg, dorsal. **Fig. 71.** Fifth leg, dorsal. **Fig. 72.** Uropod, dorsal.

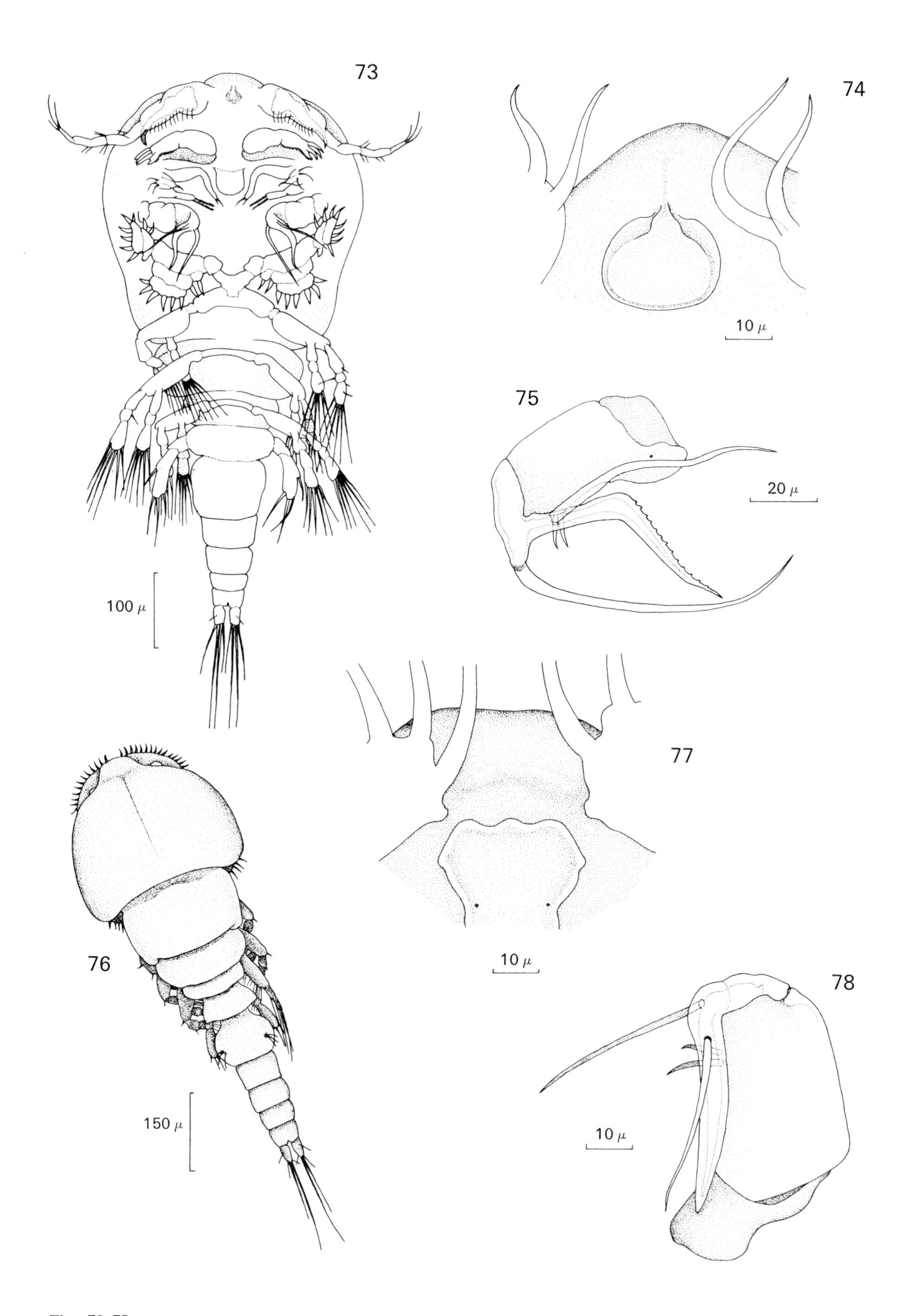

Figs. 73–78
Anchistrotos laqueus Leigh-Sharpe, 1935. **Fig. 73.** Female, ventral. **Fig. 74.** Rostral area, ventral. **Fig. 75.** Maxilliped, ventral.
Anchistrotos zeugopteri (T. Scott, 1902). **Fig. 76.** Female, dorsal. **Fig. 77.** Rostral area, ventral. **Fig. 78.** Maxilliped, ventral.
Fig. 73 redrawn from Leigh-Sharpe (1935a)

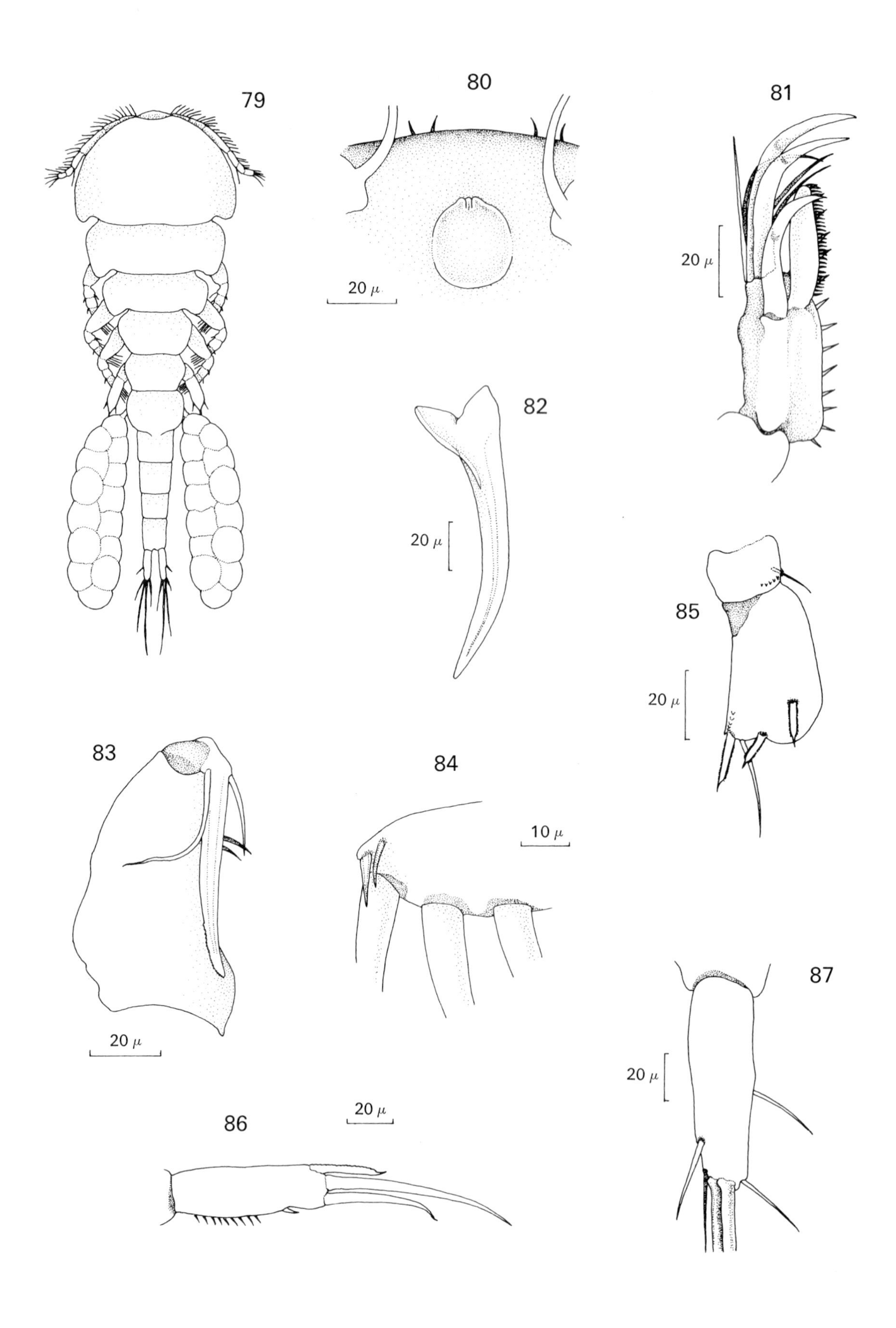

Figs. 79–87
Taeniacanthus wilsoni A. Scott, 1929. **Fig. 79.** Female, dorsal. **Fig. 80.** Rostral area, ventral. **Fig. 81.** Second antenna, tip, dorsal. **Fig. 82.** Lateral spine, ventral. **Fig. 83.** Maxilliped, ventral. **Fig. 84.** First leg, endopod tip, ventral. **Fig. 85.** Fifth leg, ventral. **Fig. 86.** Distal segment of fourth endopod, ventral. **Fig. 87.** Uropod, dorsal.
Fig. 79 redrawn from A. Scott (1929).

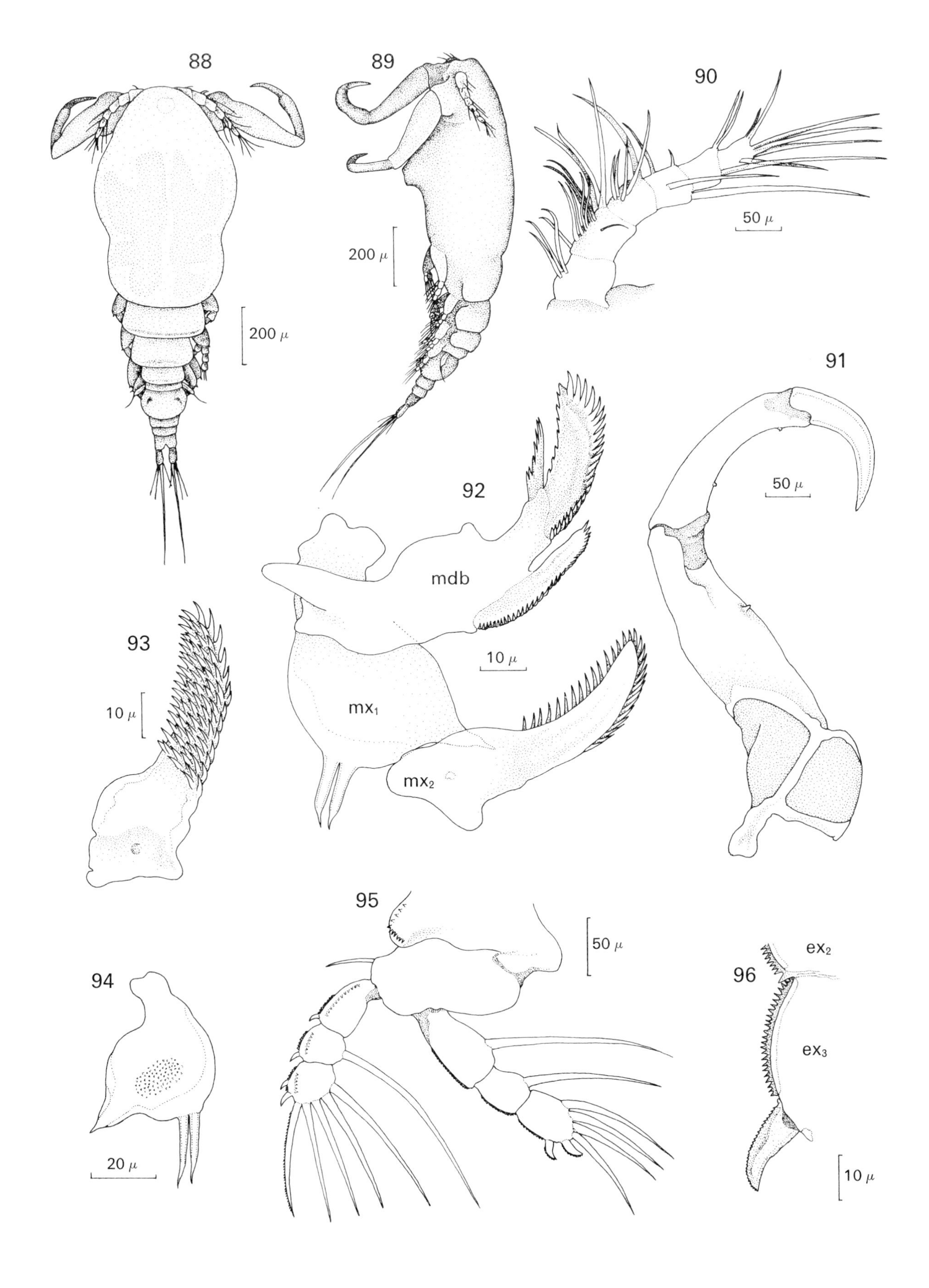

Figs. 88–96
Ergasilus sieboldi von Nordmann, 1832. **Fig. 88.** Female, dorsal. **Fig. 89.** Same, lateral. **Fig. 90.** First antenna, ventral. **Fig. 91.** Second antenna, ventral. **Fig. 92.** Mouth-parts, ventral. **Fig. 93.** Second maxilla, dorsal. **Fig. 94.** First maxilla, dorsal. **Fig. 95.** First leg, entire, ventral. **Fig. 96.** First endopod, detail of lateral margin.
Ex2—exopod segment 2; Ex3—exopod segment 3; mdb—mandible; mxl—first maxilla; mx2—second maxilla.

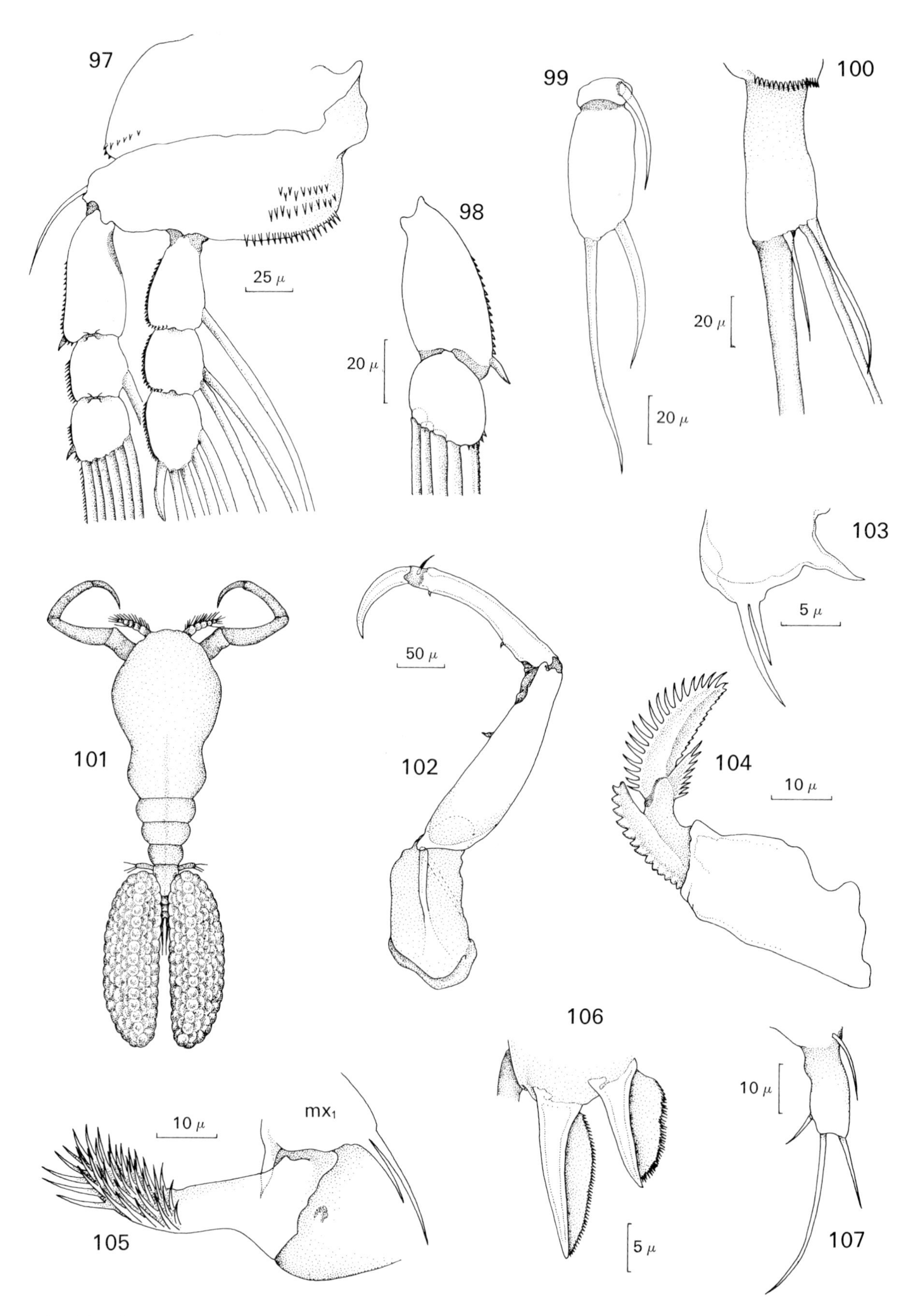

Figs. 97–107.
Ergasilus sieboldi von Nordmann, 1832. **Fig. 97.** Female, second leg, ventral. **Fig. 98.** Fourth leg, exopod, ventral. **Fig. 99.** Fifth leg, dorsal. **Fig. 100.** Uropod, ventral.
Ergasilus nanus van Beneden, 1871. **Fig. 101.** Female, dorsal. **Fig. 102.** Second antenna, ventral. **Fig. 103.** First maxilla, ventral. **Fig. 104.** Mandible, ventral. **Fig. 105.** First and second maxilla, dorsal. **Fig. 106.** First leg, endopod 3, details of lateral margin. **Fig. 107.** Fifth leg, dorsal. mx1—first maxilla.
Fig. 101. redrawn from Scott and Scott (1913).

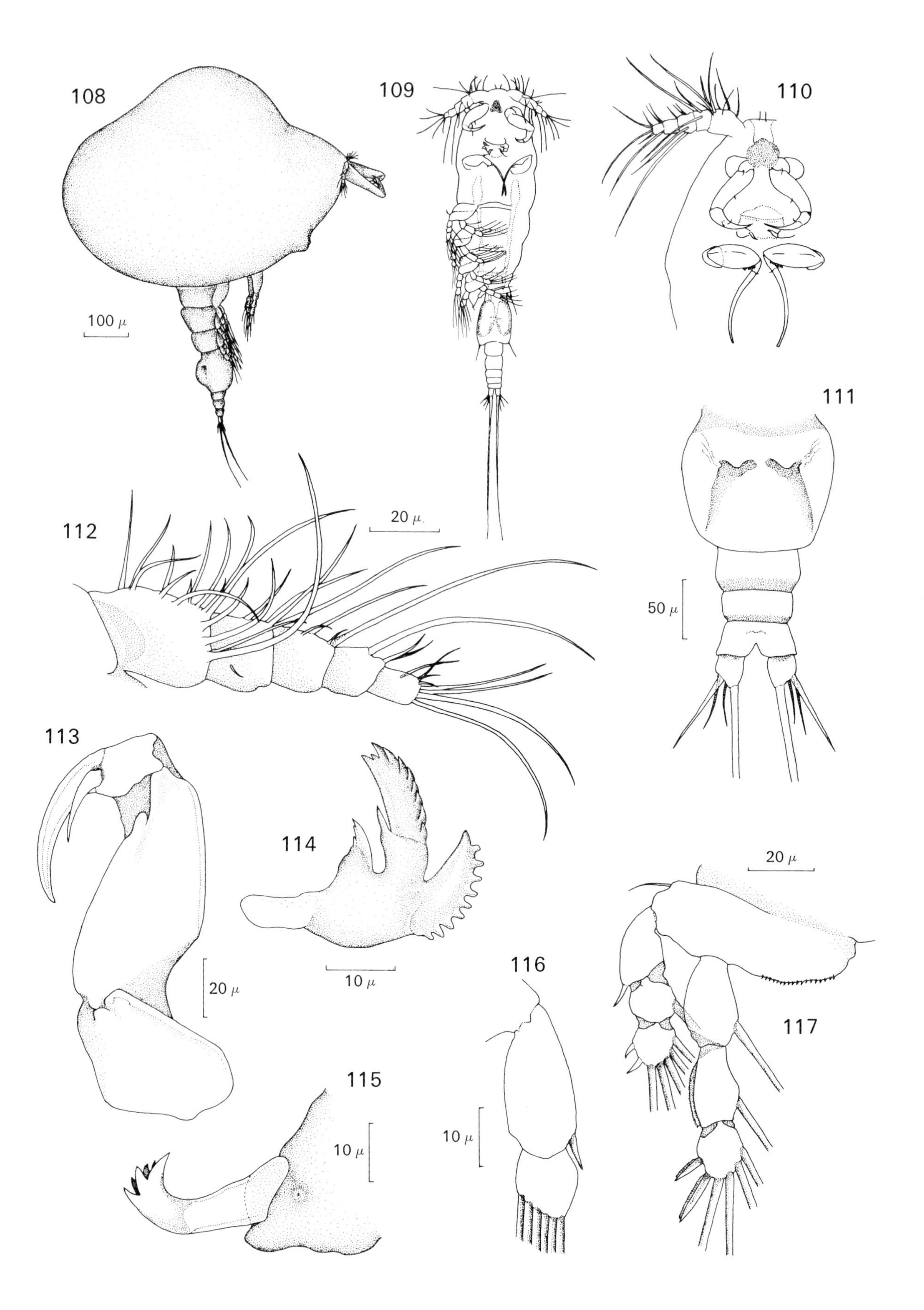

Figs. 108–117

Thersitina gasterostei (Pagenstecher, 1861). **Fig. 108.** Female, lateral. **Fig. 109.** Male, ventral. **Fig. 110.** Same, part of cephalothorax, ventral. **Fig. 111.** Female, posterior part, dorsal. **Fig. 112.** First antenna, dorsal. **Fig. 113.** Second antenna, dorsal. **Fig. 114.** Mandible, ventral. **Fig. 115.** Second maxilla, dorsal. **Fig. 116.** Fourth leg, exopod, ventral. **Fig. 117.** First leg, ventral.

Figs. 109 and 110 redrawn from Gurney (1913).

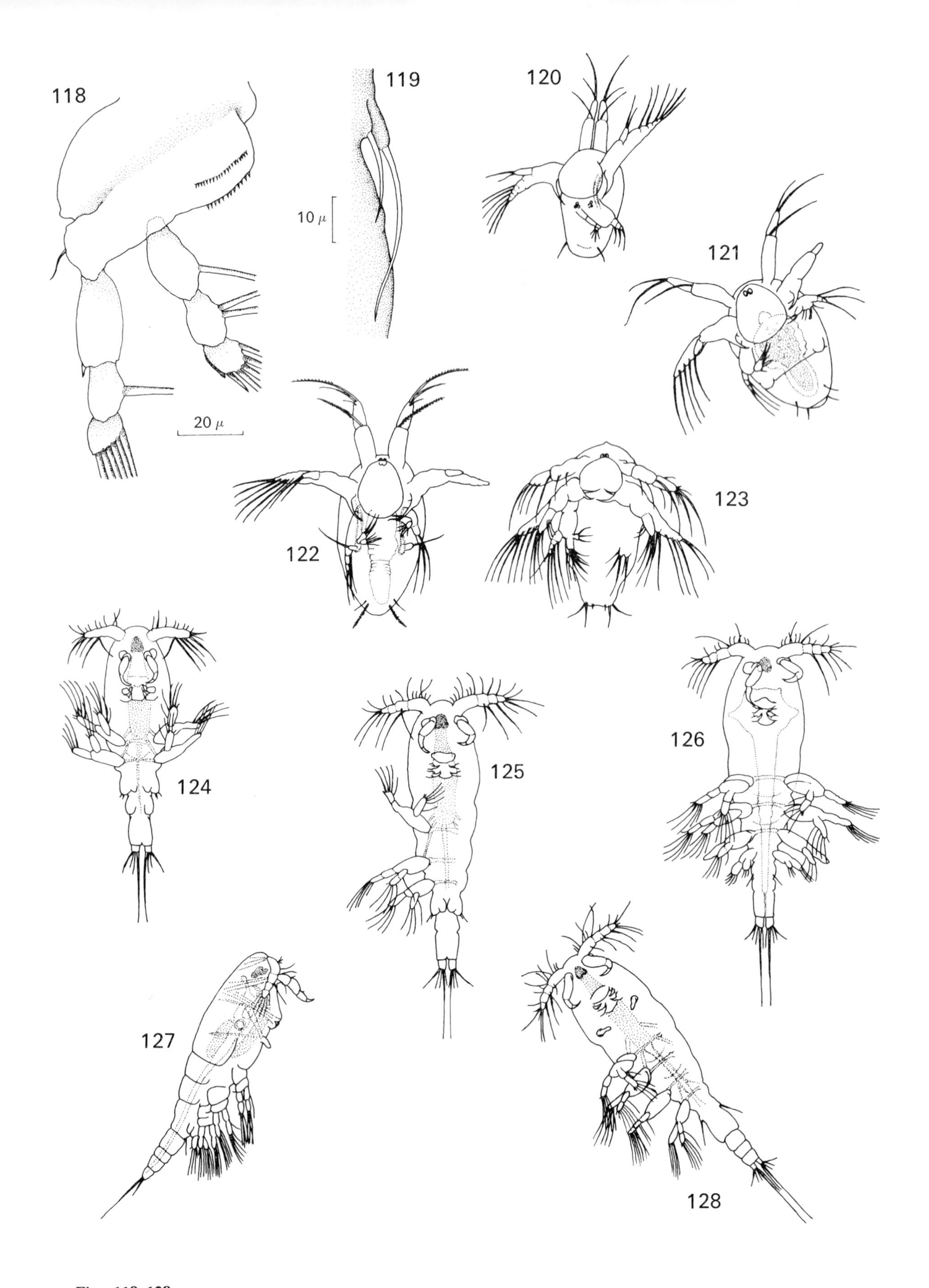

Figs. 118–128.
Thersitina gasterostei (Pagenstecher, 1861). **Fig. 118.** Second leg, ventral. **Fig. 119.** Fifth leg, dorsal. **Fig. 120.** First nauplius, after hatching. **Fig. 121.** Second nauplius, six days old. **Fig. 122.** Third nauplius, eleven days old. **Fig. 123.** Fourth nauplius. **Fig. 124.** First cyclopid stage. **Fig. 125.** Second cyclopid stage. **Fig. 126.** Third cyclopid stage. **Fig. 127.** Fourth cyclopid stage. **Fig. 128.** Fifth cyclopid stage.
Figs. 120—128 redrawn from Gurney (1913).

Figs. 129–133.
Chondracanthus zei Delaroche, 1811. **Fig. 129.** Young adult female, dorsal. **Fig. 130.** Same, ventral. **Fig. 131.** Same, lateral. **Fig. 132.** Older adult female, lateral. **Fig. 133.** Same, contracted specimen, ventral.
Fig. 129 redrawn from Kabata (1970a).

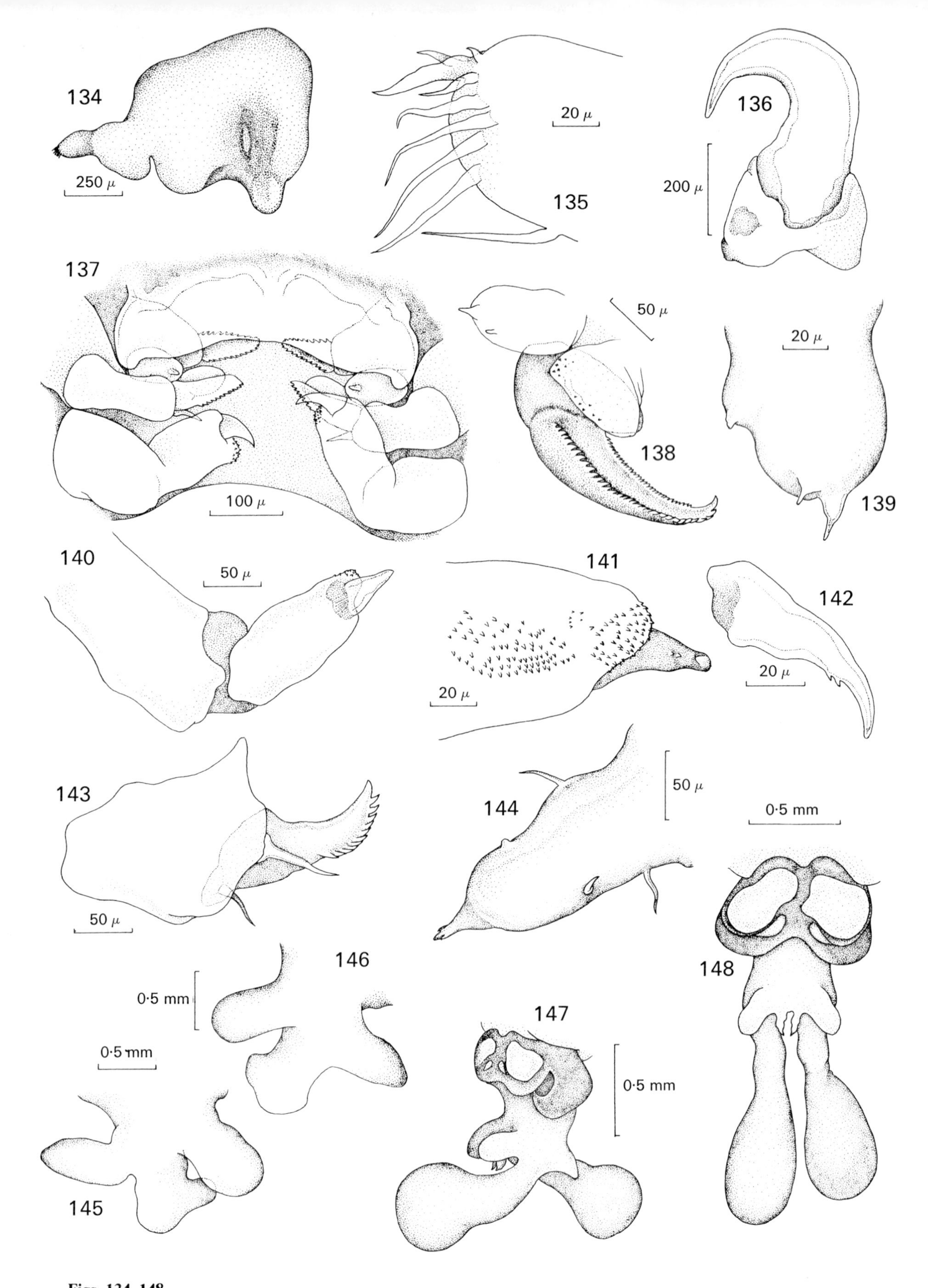

Figs. 134–148.
Chondracanthus zei Delaroche, 1811. **Fig. 134.** Female, first antenna, ventral. **Fig. 135.** Same, tip, ventral. **Fig. 136.** Second antenna, ventral. **Fig. 137.** Buccal region, ventral. **Fig. 138.** Mandible, first maxilla and paragnath *in situ*, ventral. **Fig. 139.** First maxilla, full view. **Fig. 140.** Maxilliped, ventral. **Fig. 141.** Same, tip, dorsal. **Fig. 142.** Same, claw, full view. **Fig. 143.** Second maxilla, ventral. **Fig. 144.** Uropod, lateral. **Fig. 145.** First leg, ventral. **Fig. 146.** Second leg, ventral. **Fig. 147.** Genito-abdominal tagma, lateral. **Fig. 148.** Same, ventral.

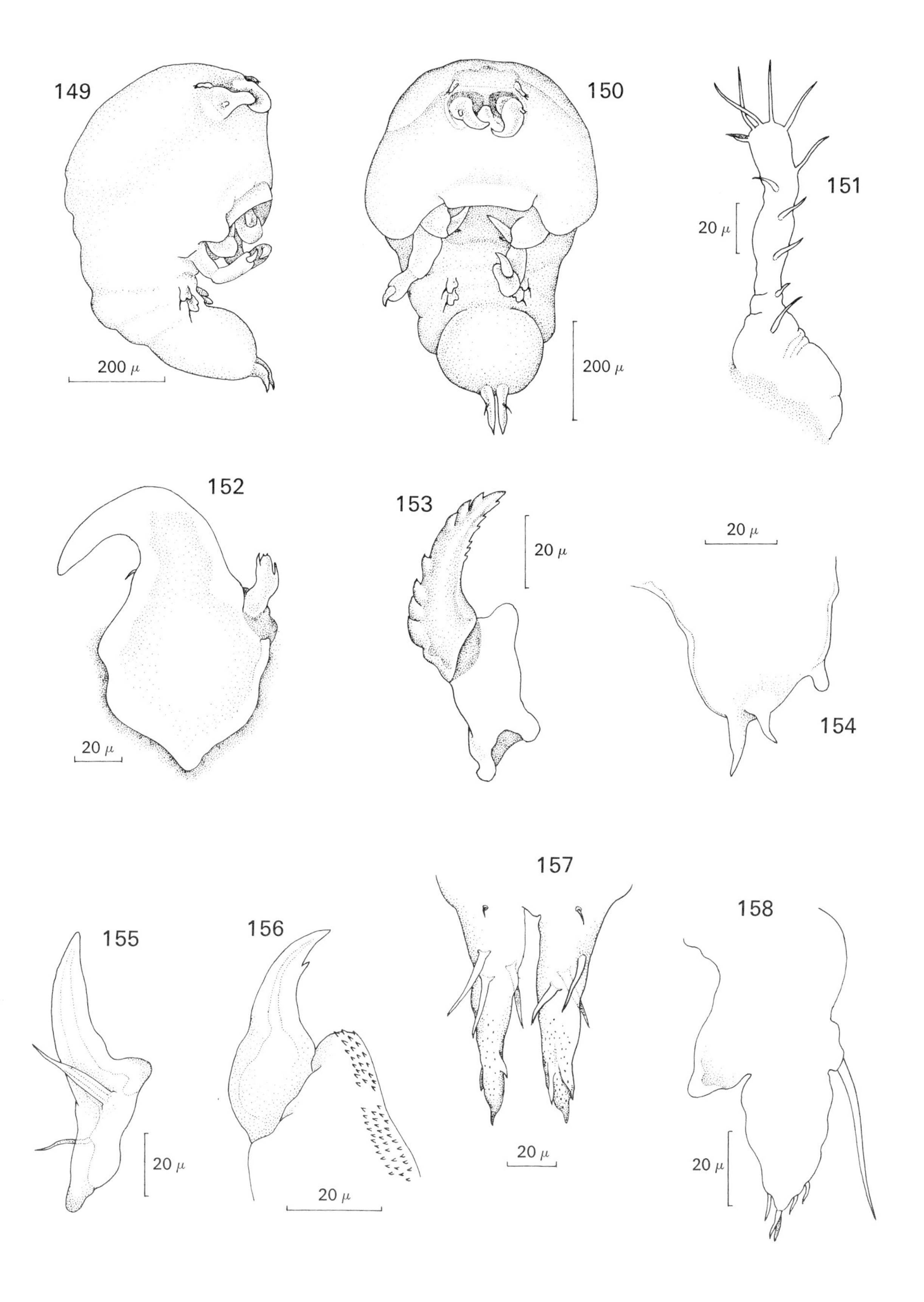

Figs. 149–158.
Chondracanthus zei Delaroche, 1811. **Fig. 149.** Male, lateral. **Fig. 150.** Same, ventral. **Fig. 151.** First antenna, dorsal. **Fig. 152.** Second antenna, dorsal. **Fig. 153.** Mandible, ventral. **Fig. 154.** First maxilla, ventral. **Fig. 155.** Second maxilla, claw, dorsal. **Fig. 156.** Maxilliped, tip, full view. **Fig. 157.** Uropods, ventral. **Fig. 158.** Second leg, ventral.

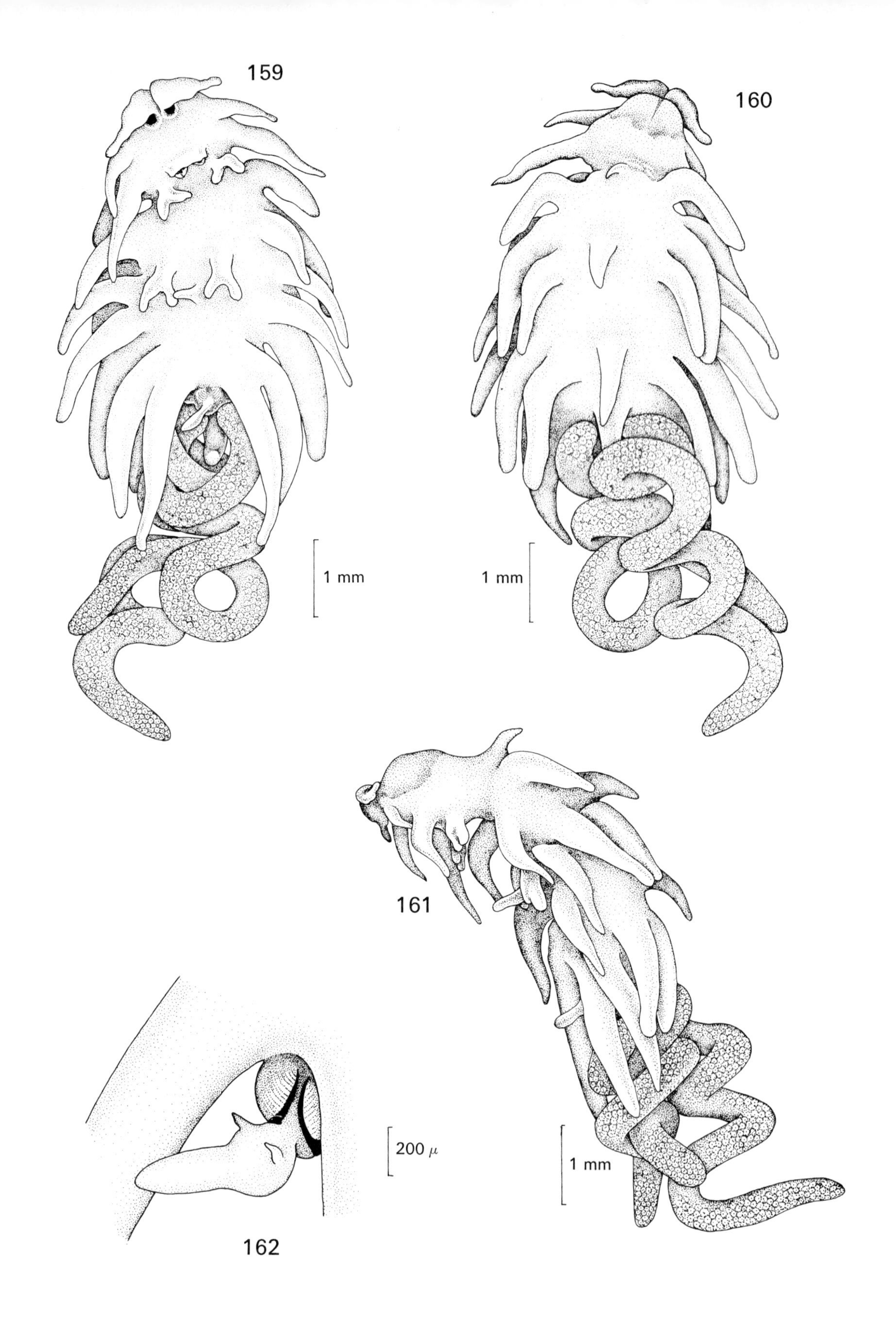

Figs. 159–162.
Chondracanthus neali Leigh-Sharpe, 1930. **Fig. 159.** Female, ventral. **Fig. 160.** Same, dorsal. **Fig. 161.** Same, lateral. **Fig. 162.** Genito-abdominal tagma, ventrolateral.

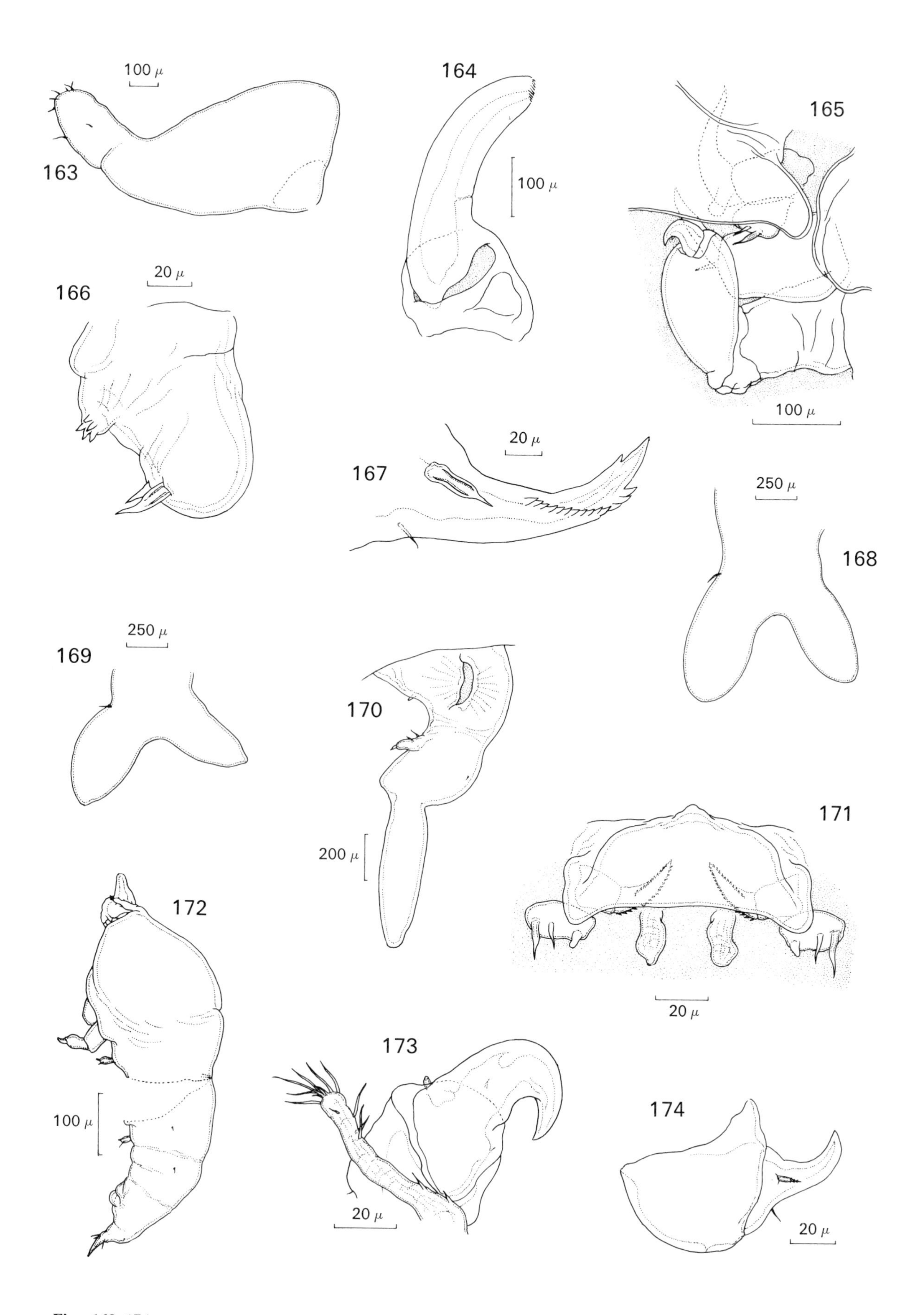

Figs. 163–174.
Chondracanthus neali Leigh-Sharpe, 1930. **Fig. 163.** Female, first antenna. **Fig. 164.** Second antenna (with broken tip). **Fig. 165.** Lateral margin of buccal cavity. **Fig. 166.** First maxilla. **Fig. 167.** Second maxilla, claw. **Fig. 168.** First leg. **Fig. 169.** Second leg. **Fig. 170.** Genito-abdominal tagma, lateral. **Fig. 171.** Male, mouth-parts. **Fig. 172.** Male, lateral (broken near middle). **Fig. 173.** First and second antenna. **Fig. 174.** Second maxilla.
Redrawn from Ho (1972c).

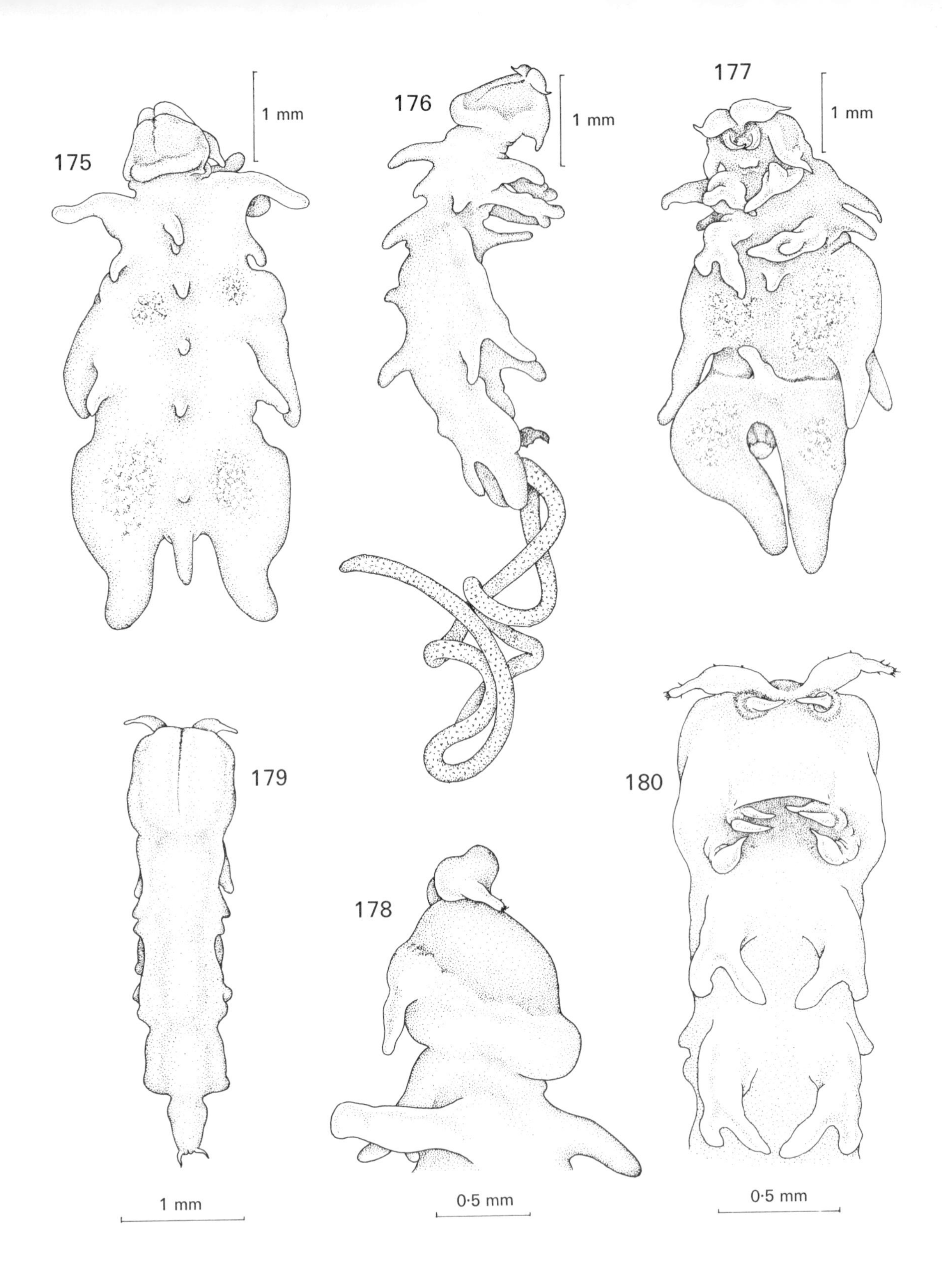

Figs. 175–180.
Chondracanthus lophii Johnston, 1836. **Fig. 175.** Female, dorsal. **Fig. 176.** Same, lateral. **Fig. 177.** Same, ventral. **Fig. 178.** Cephalothorax, lateral. **Fig. 179.** Juvenile female, dorsal. **Fig. 180.** Same, anterior half, ventral.

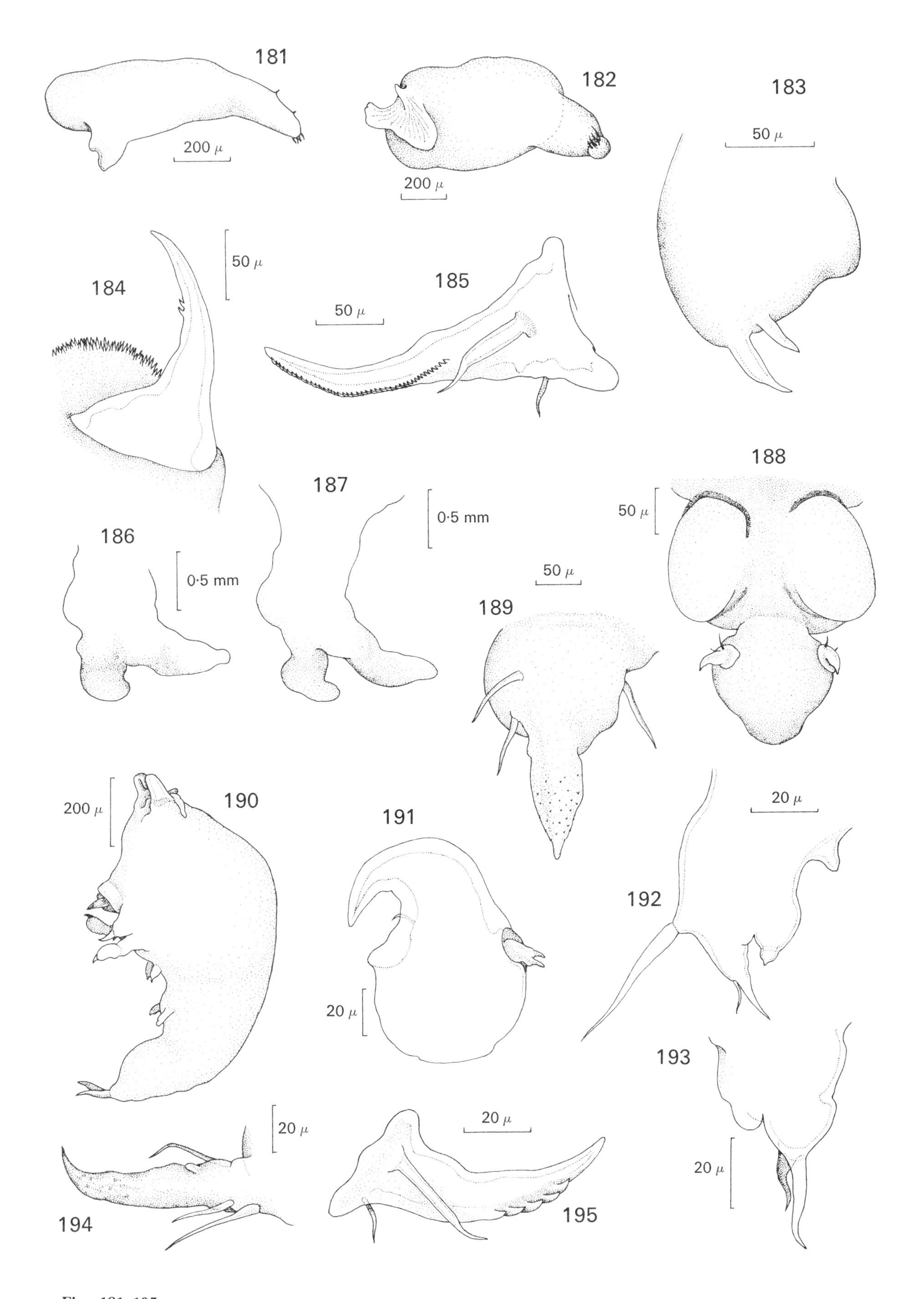

Figs. 181–195.
Chondracanthus lophii Johnston, 1836. **Fig. 181.** Female, first antenna, ventral. **Fig. 182.** Same, proximal surface. **Fig. 183.** First maxilla, ventral. **Fig. 184.** Maxilliped, claw, full view. **Fig. 185.** Second maxilla, claw dorsal. **Fig. 186.** First leg, ventral. **Fig. 187.** Second leg, ventral. **Fig. 188.** Genito-abdominal tagma, ventral. **Fig. 189.** Uropod, ventral. **Fig. 190.** Male, lateral. **Fig. 191.** Second antenna, ventral. **Fig. 192.** First leg, ventral. **Fig. 193.** First maxilla, ventrolateral. **Fig. 194.** Uropod, ventral. **Fig. 195.** Second maxilla, claw, dorsal.

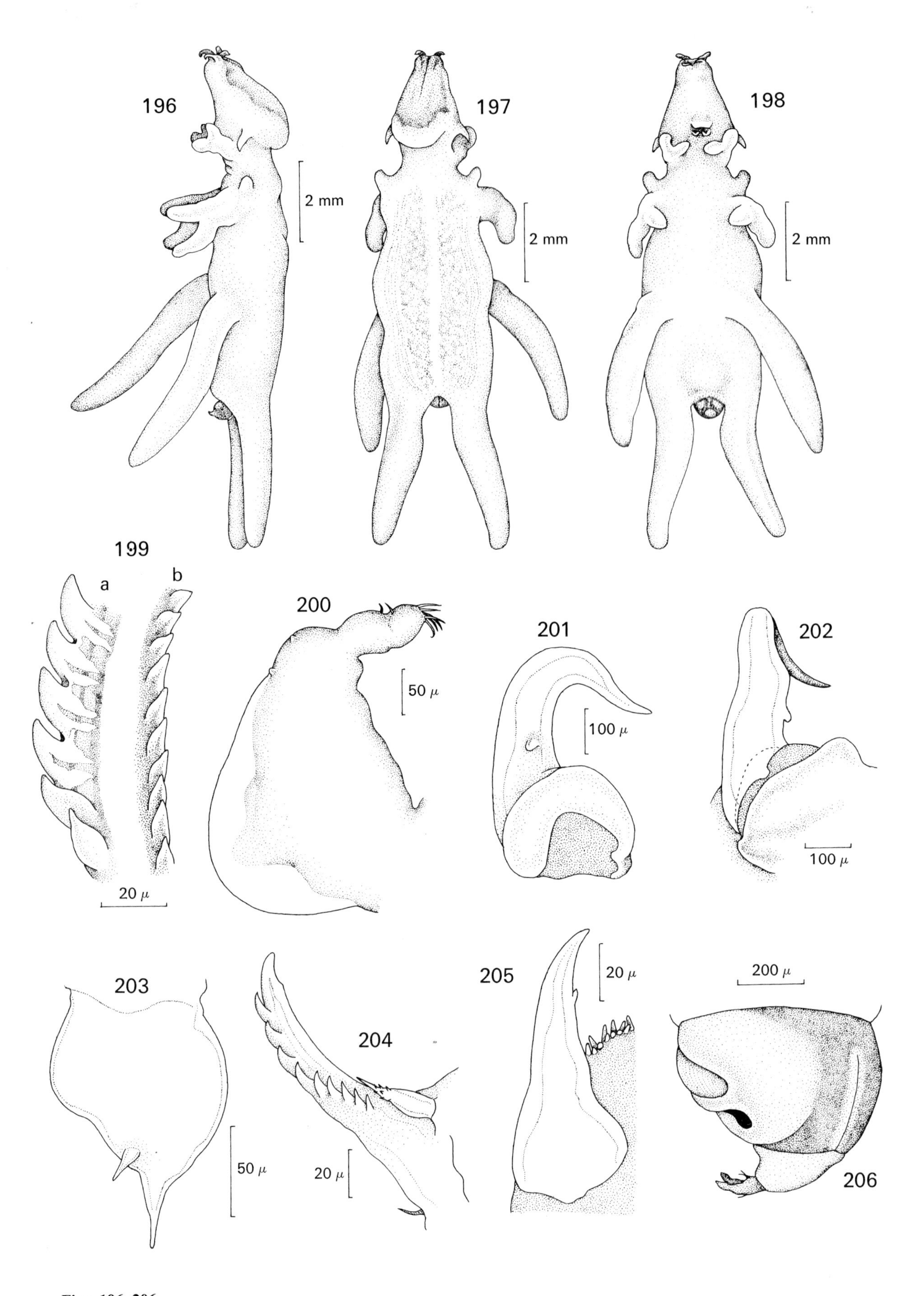

Figs. 196–206.
Chondracanthus merluccii (Holten, 1802). **Fig. 196.** Female, lateral. **Fig. 197.** Same, dorsal. **Fig. 198.** Same, ventral. **Fig. 199.** Details of mandibular margins: (a) outer; (b) inner. **Fig. 200.** First antenna, anterior. **Fig. 201.** Second antenna, ventral. **Fig. 202.** Same, lateral. **Fig. 203.** First maxilla, ventral. **Fig. 204.** Second maxilla, claw, dorsal. **Fig. 205.** Maxilliped, claw, ventral. **Fig. 206.** Genito-abdominal tagma, dorsolateral.

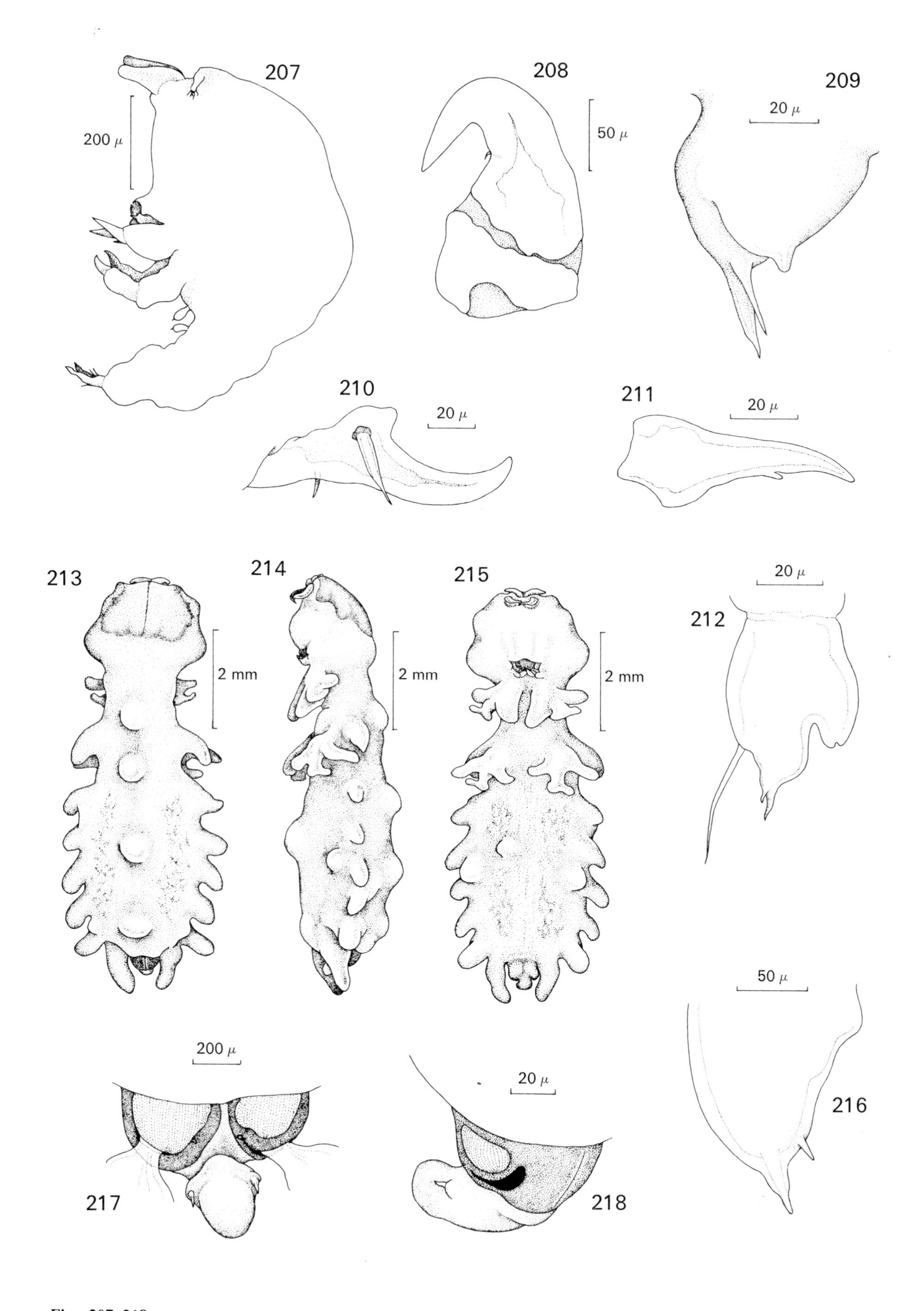

Figs. 207–218

Chondracanthus merluccii (Holten, 1802). **Fig. 207.** Male, lateral. **Fig. 208.** Second antenna, ventral. **Fig. 209.** First maxilla, ventromedial. **Fig. 210.** Second maxilla, claw, dorsal. **Fig. 211.** Maxilliped claw, full view. **Fig. 212.** Second leg, ventral.

Chondracanthus nodosus (Müller, 1776). **Fig. 213.** Female, dorsal. **Fig. 214.** Same, lateral. **Fig. 215.** Same, ventral. **Fig. 216.** First maxilla, ventral. **Fig. 217.** Genito-abdominal tagma, ventral. **Fig. 218.** Same, dorsolateral.

Figs. 219–231.
Chondracanthus nodosus (Müller, 1776). **Fig. 219.** Female, dorsal (contracted specimen). **Fig. 220.** First antenna, anterior. **Fig. 221.** Second maxilla, claw, dorsal. **Fig. 222.** Maxilliped, claw, full view. **Fig. 223.** First leg, ventral. **Fig. 224.** Second leg, ventral. **Fig. 225.** Male, lateral. **Fig. 226.** Second antenna, ventral. **Fig. 227.** Maxilliped claw, full view. **Fig. 228.** First maxilla, ventral. **Fig. 229.** Second leg, ventral. **Fig. 230.** Second maxilla, claw (toothed), dorsal. **Fig. 231.** Same (toothless).

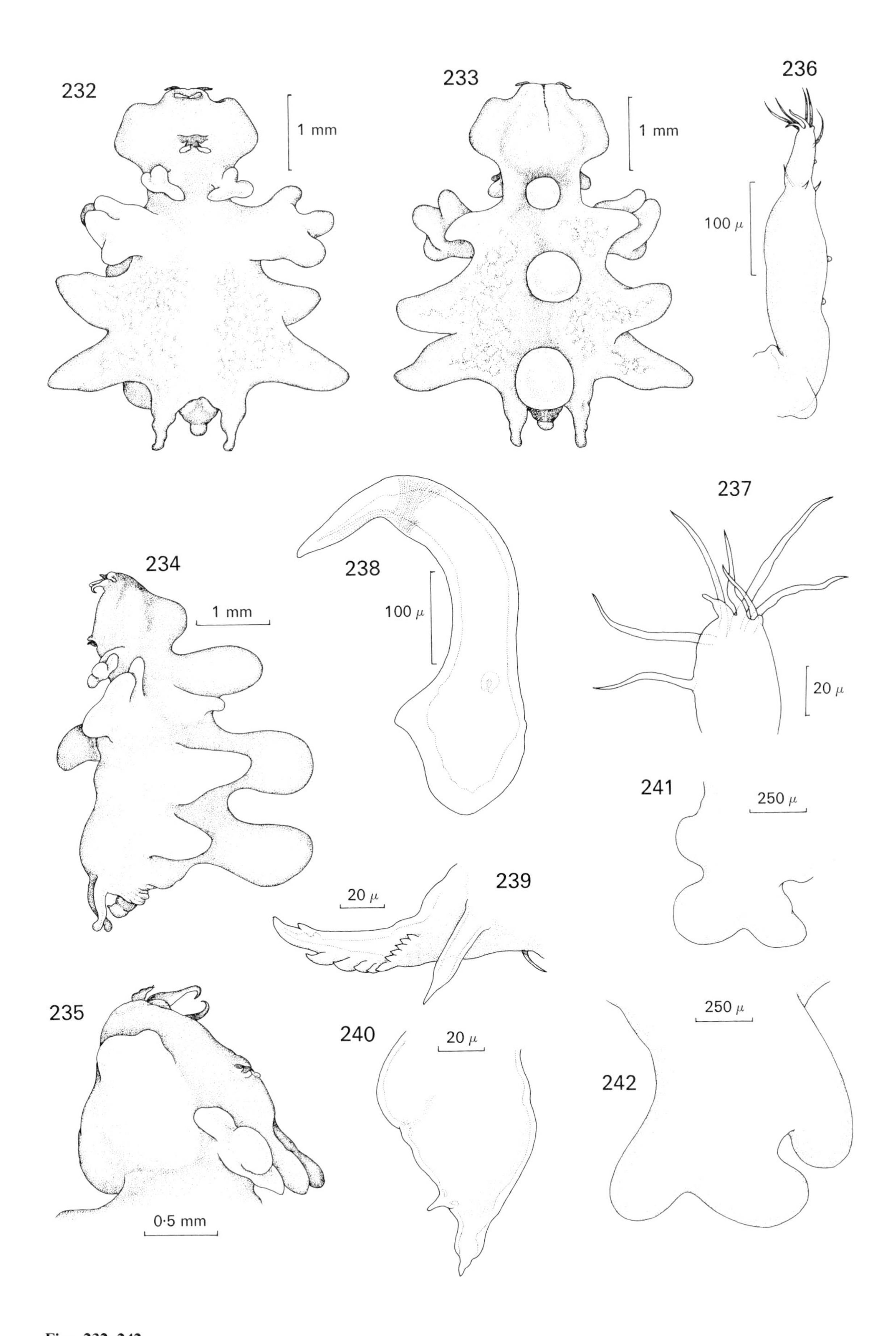

Figs. 232–242.
Chondracanthus ornatus T. Scott, 1900. **Fig. 232.** Female, ventral. **Fig. 233.** Same, dorsal. **Fig. 234.** Same, lateral. **Fig. 235.** Cephalothorax, lateral. **Fig. 236.** First antenna, dorsal. **Fig. 237.** Same, tip, dorsal. **Fig. 238.** Second antenna, dorsal. **Fig. 239.** Second maxilla, claw, dorsal. **Fig. 240.** First maxilla, ventral. **Fig. 241.** First leg, ventral. **Fig. 242.** Second leg, ventral.

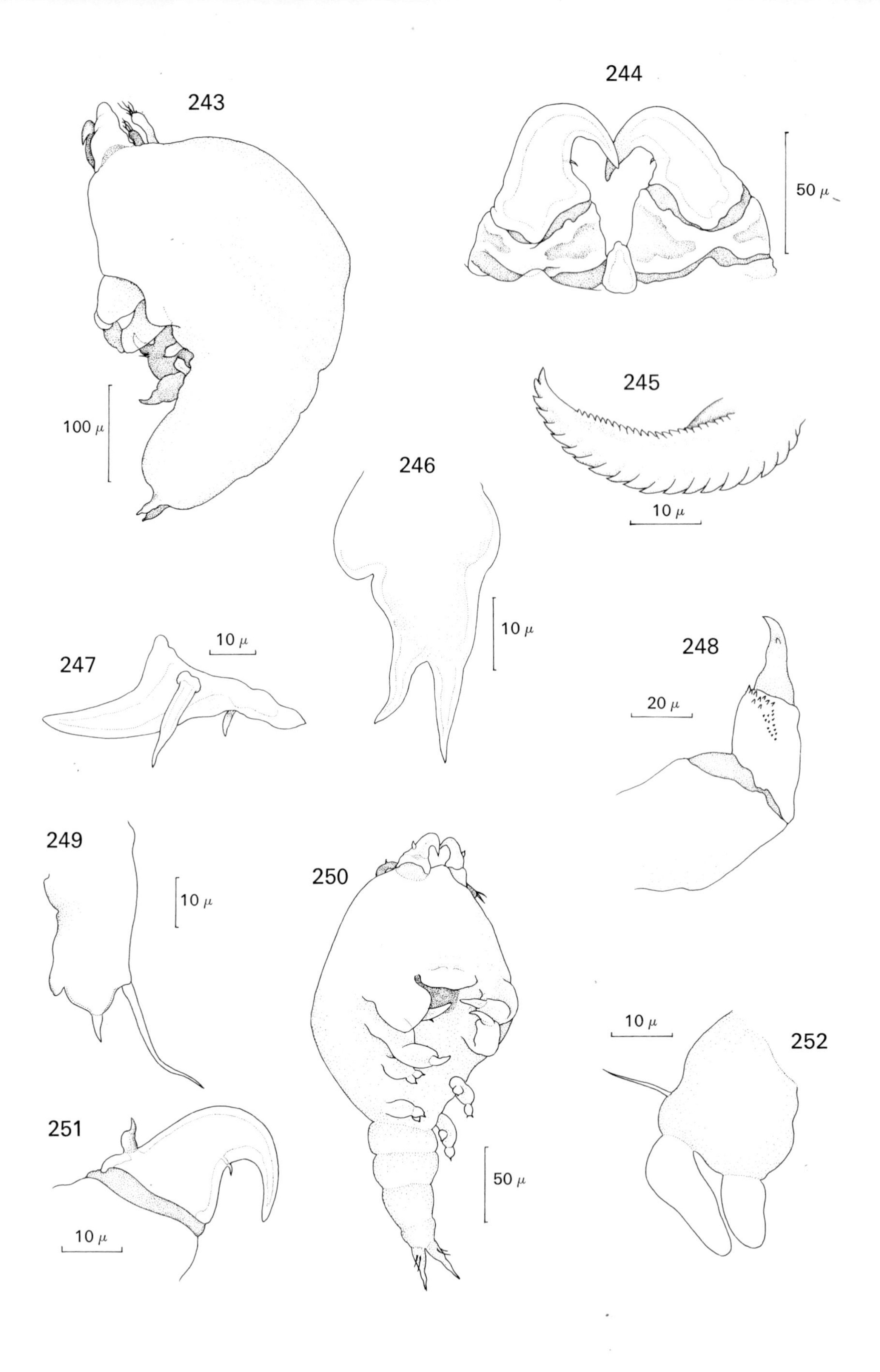

Figs. 243–252.
Chondracanthus ornatus T. Scott, 1900. **Fig. 243.** Male, lateral. **Fig. 244.** Second antennae, ventral. **Fig. 245.** Mandibular blade, dorsal. **Fig. 246.** First maxilla, ventral. **Fig. 247.** Second maxilla, claw, dorsal. **Fig. 248.** Maxilliped, dorsal. **Fig. 249.** First leg, ventral. **Fig. 250.** Juvenile specimen (probably male), ventrolateral. **Fig. 251.** Second antenna of same, ventral. **Fig. 252.** First leg of same, ventral.

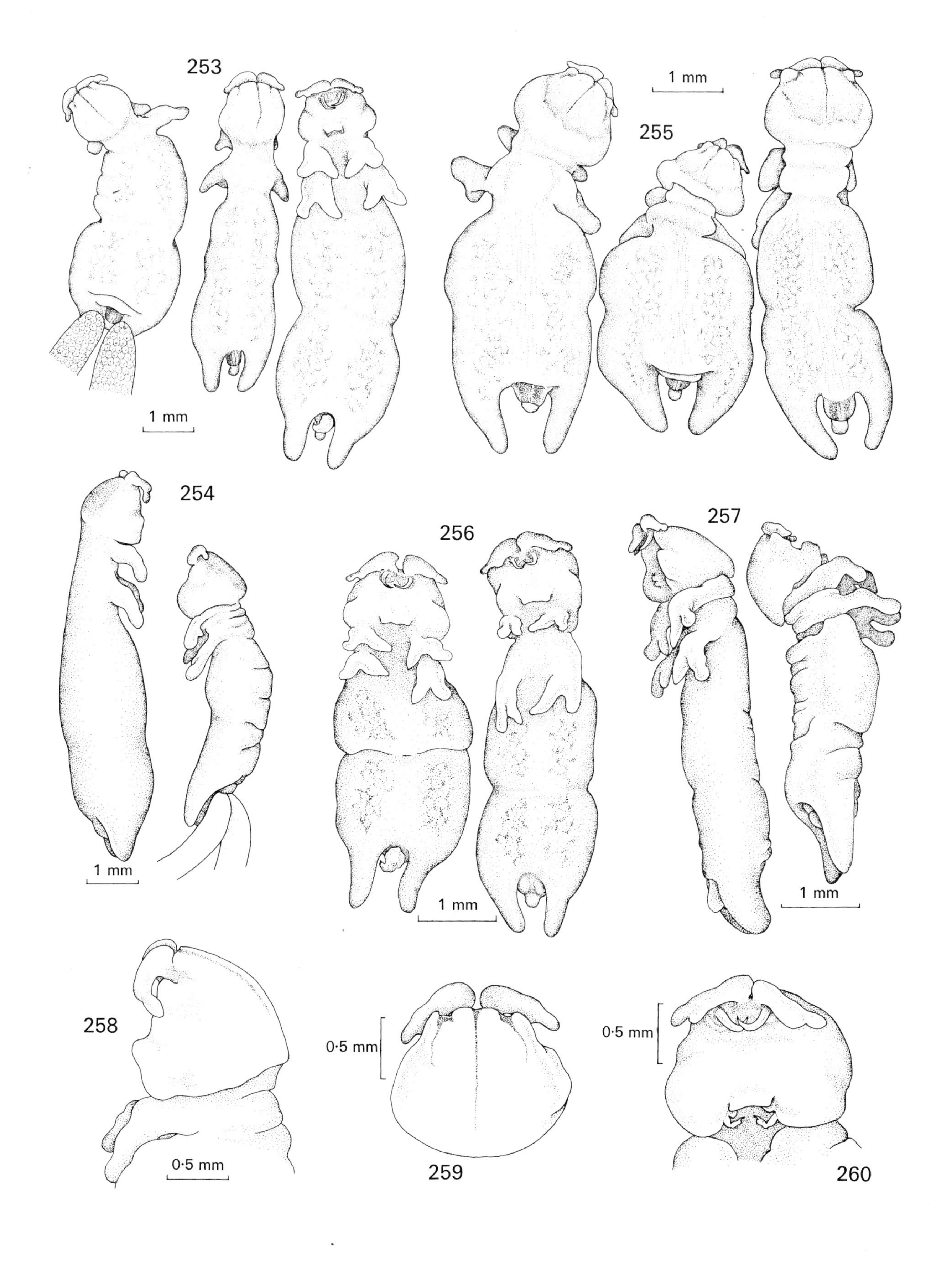

Figs. 253–260.
Acanthochondria cornuta (Müller, 1776). **Fig. 253.** *A. cornuta f. cornuta*, three female specimens (left and centre dorsal, right ventral). **Fig. 254.** Same, two specimens, lateral. **Fig. 255.** *A. cornuta f. flurae*, three specimens, dorsal. **Fig. 256.** *A. cornuta* (*depressa*), two females, ventral. **Fig. 257.** Same, two specimens, lateral. **Fig. 258.** Cephalothorax, lateral. **Fig. 259.** Same, dorsal. **Fig. 260.** Same, ventral.

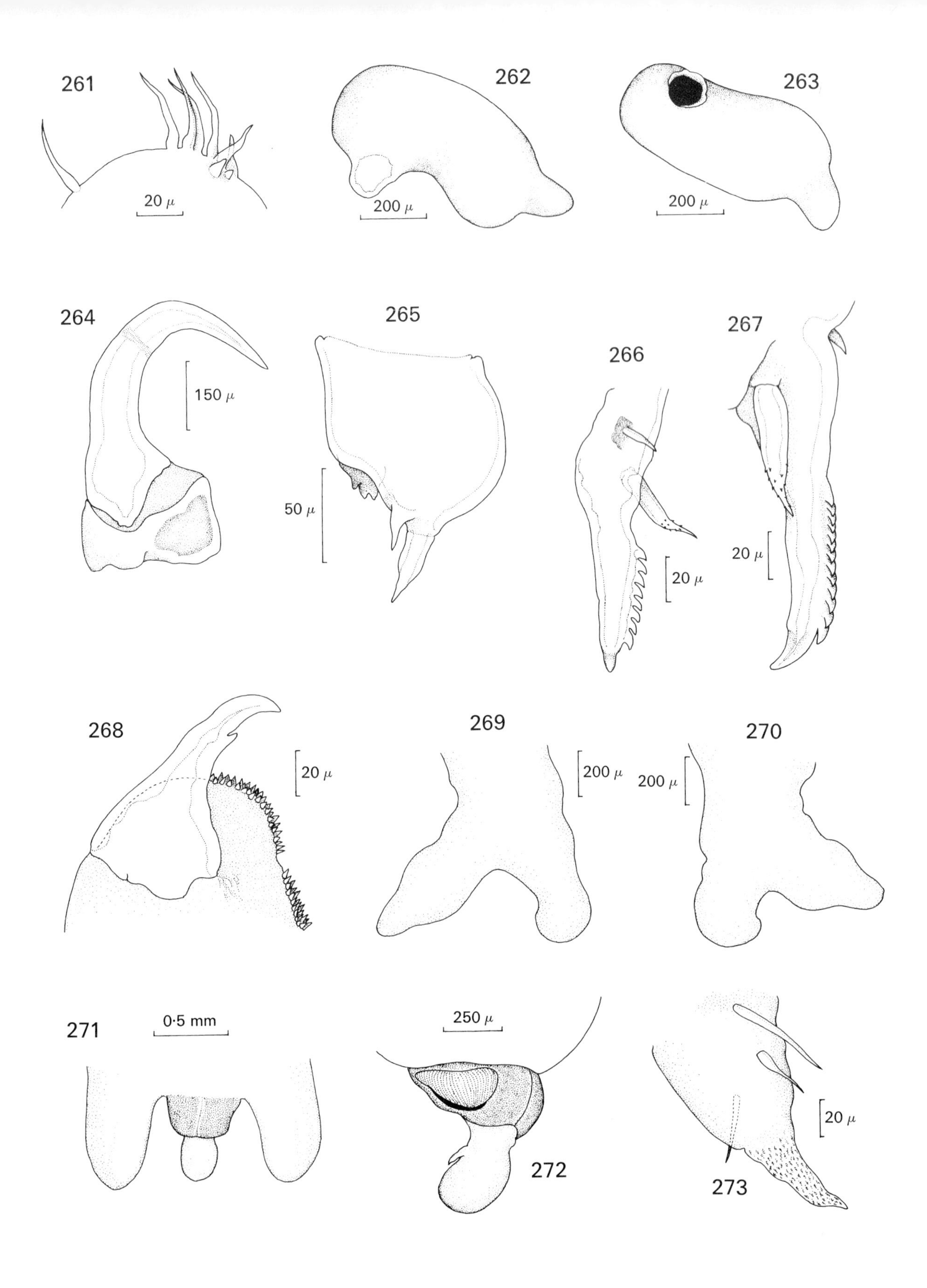

Figs. 261–273.
Acanthochondria cornuta (Müller, 1776). **Fig. 261.** First antenna, tip. **Fig. 262.** Same, entire, ventral. **Fig. 263.** Same, proximal. **Fig. 264.** Second antenna, dorsal. **Fig. 265.** First maxilla, ventral. **Fig. 266.** Second maxilla, claw, posterior. **Fig. 267.** Same, dorsal. **Fig. 268.** Maxilliped claw, full view. **Fig. 269.** First leg, ventral. **Fig. 270.** Second leg, ventral. **Fig. 271.** Posterior extremity of trunk, dorsal. **Fig. 272.** Genito-abdominal tagma, lateral. **Fig. 273.** Uropod, ventral.

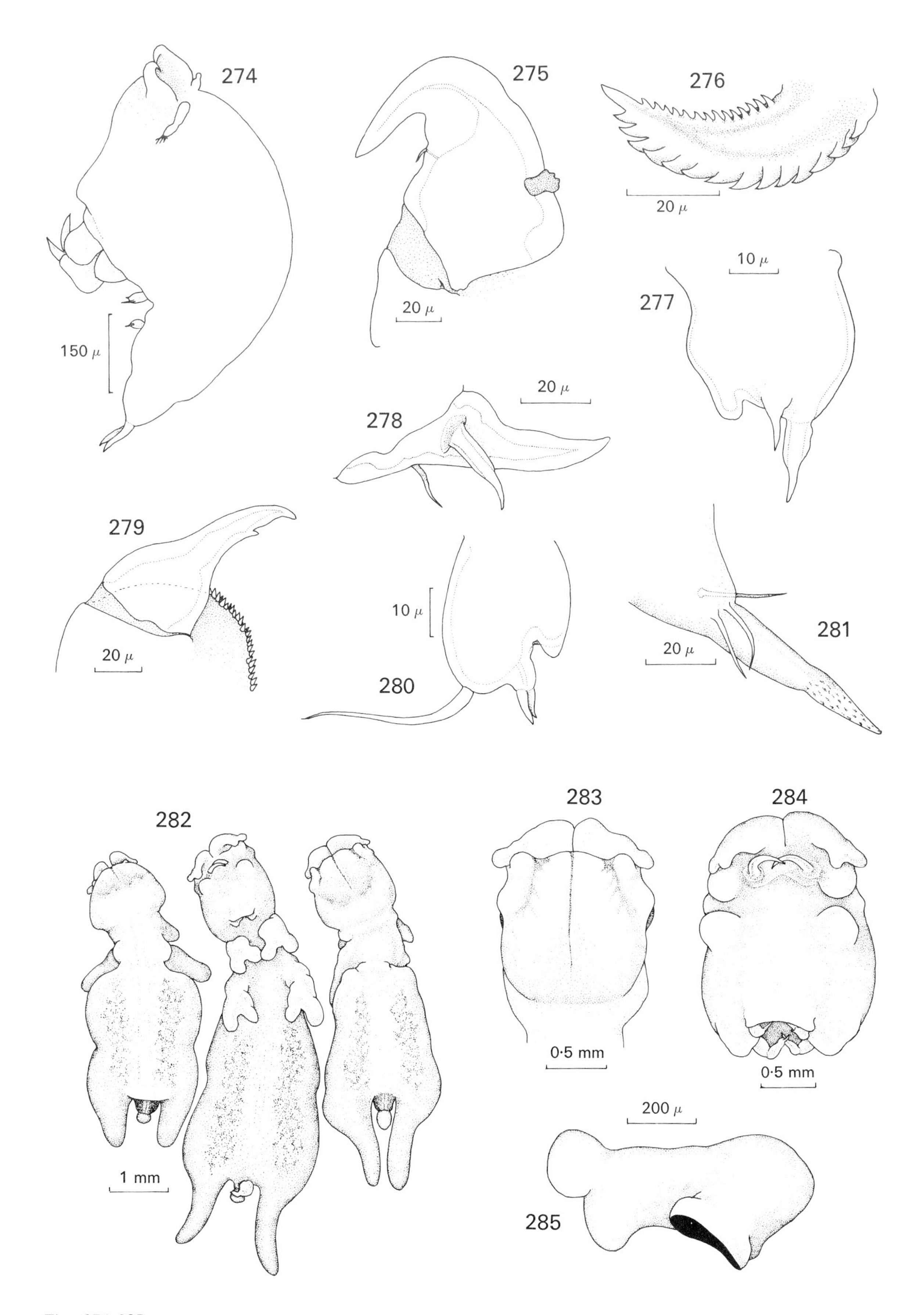

Figs. 274–285.
Acanthochondria cornuta (Müller, 1776). **Fig. 274.** Male, lateral. **Fig. 275.** Second antenna, dorsal. **Fig. 276.** Mandibular blade, dorsal. **Fig. 277.** First maxilla, ventral. **Fig. 278.** Second maxilla, claw, dorsal. **Fig. 279.** Maxilliped claw, full view. **Fig. 280.** First leg, ventral. **Fig. 281.** Uropod, ventral.
Acanthochondria clavata (Bassett-Smith, 1896). **Fig. 282.** Three females, left and right dorsal, centre ventral. **Fig. 283.** Cephalothorax, dorsal. **Fig. 284.** Same, ventral. **Fig. 285.** First antenna, proximal.
Fig. 284 redrawn from Kabata (1968b).

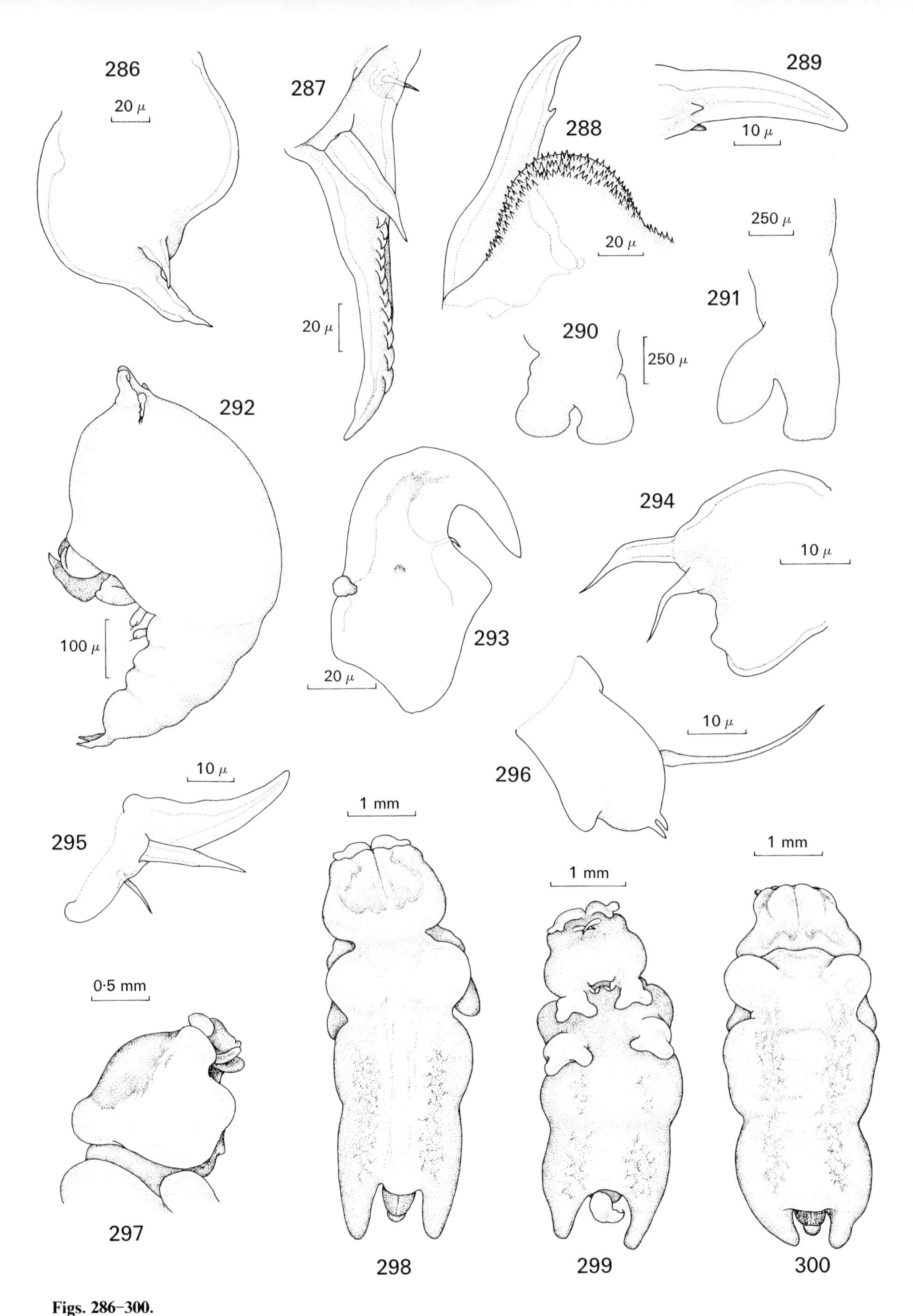

Figs. 286–300.
Acanthochondria clavata (Bassett-Smith, 1896). **Fig. 286.** First maxilla, ventral. **Fig. 287.** Second maxilla, claw, dorsal. **Fig. 288.** Maxilliped claw, dorsal. **Fig. 289.** Same, seen from concave side. **Fig. 290.** First leg, ventral. **Fig. 291.** Second leg, ventral **Fig. 292.** Male, lateral. **Fig. 293.** Second antenna, ventral. **Fig. 294.** First maxilla, ventral. **Fig. 295.** Second maxilla, claw, dorsal. **Fig. 296.** Second leg, ventromedial.
Acanthochondria limandae (Krøyer, 1863). **Fig. 297.** Female, cephalothorax, lateral. **Fig. 298.** Female, entire, dorsal. **Fig. 299.** Same, ventral. **Fig. 300.** Same, dorsal (partly contracted).

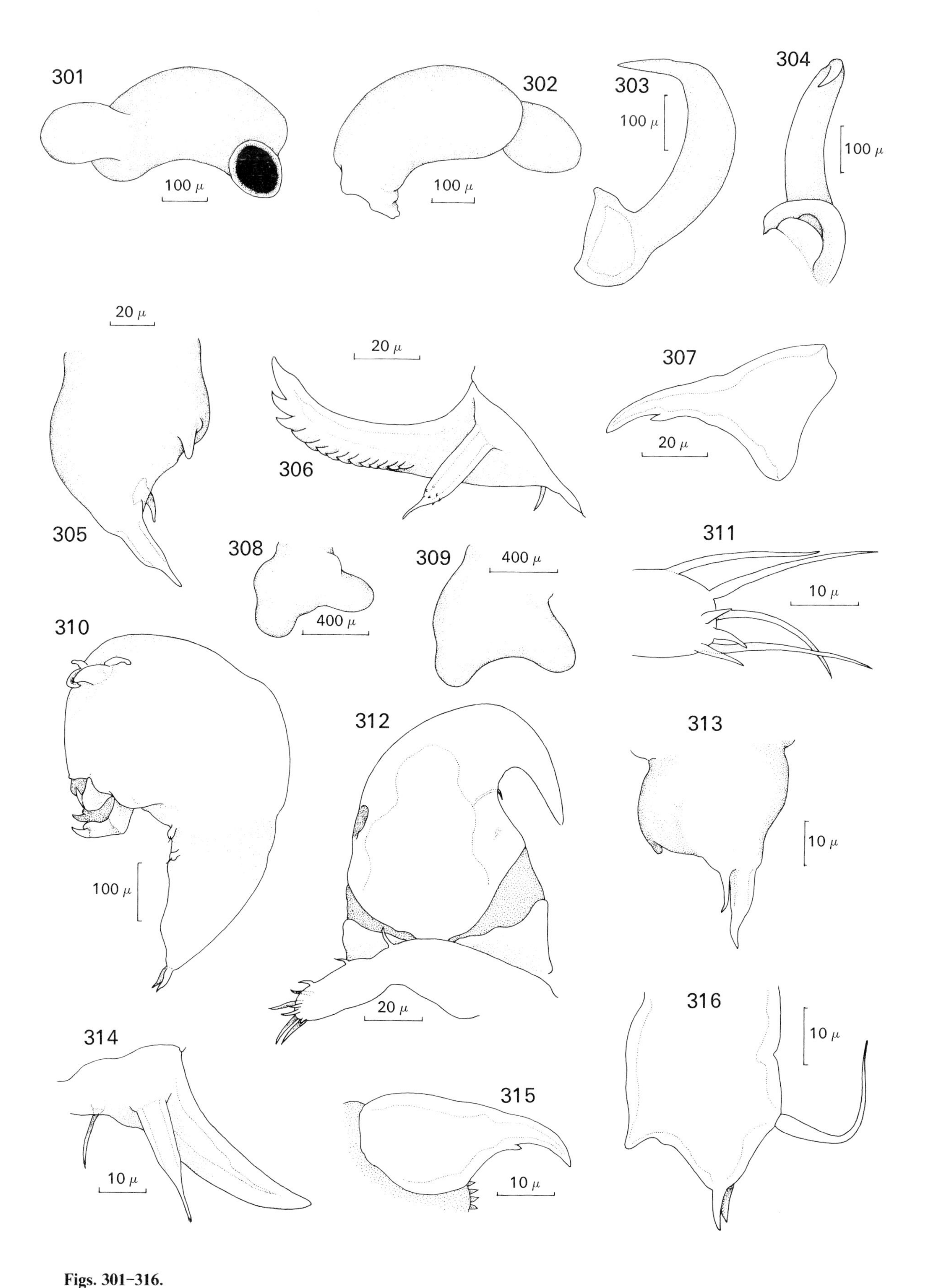

Figs. 301–316.
Acanthochondria limandae (Krøyer, 1863). **Fig. 301.** Female, first antenna, proximal. **Fig. 302.** Same, ventral. **Fig. 303.** Second antenna, hook, ventral. **Fig. 304.** Same, medial. **Fig. 305.** First maxilla, dorsal. **Fig. 306.** Second maxilla, claw, dorsal. **Fig. 307.** Maxilliped claw, full view. **Fig. 308.** First leg, ventral. **Fig. 309.** Second leg, ventral. **Fig. 310.** Male, lateral. **Fig. 311.** First antenna, tip, dorsal. **Fig. 312.** First and second antenna, dorsal. **Fig. 313.** First maxilla, ventral. **Fig. 314.** Second maxilla, claw, dorsal. **Fig. 315.** Maxilliped claw, full view. **Fig. 316.** Second leg, ventral.

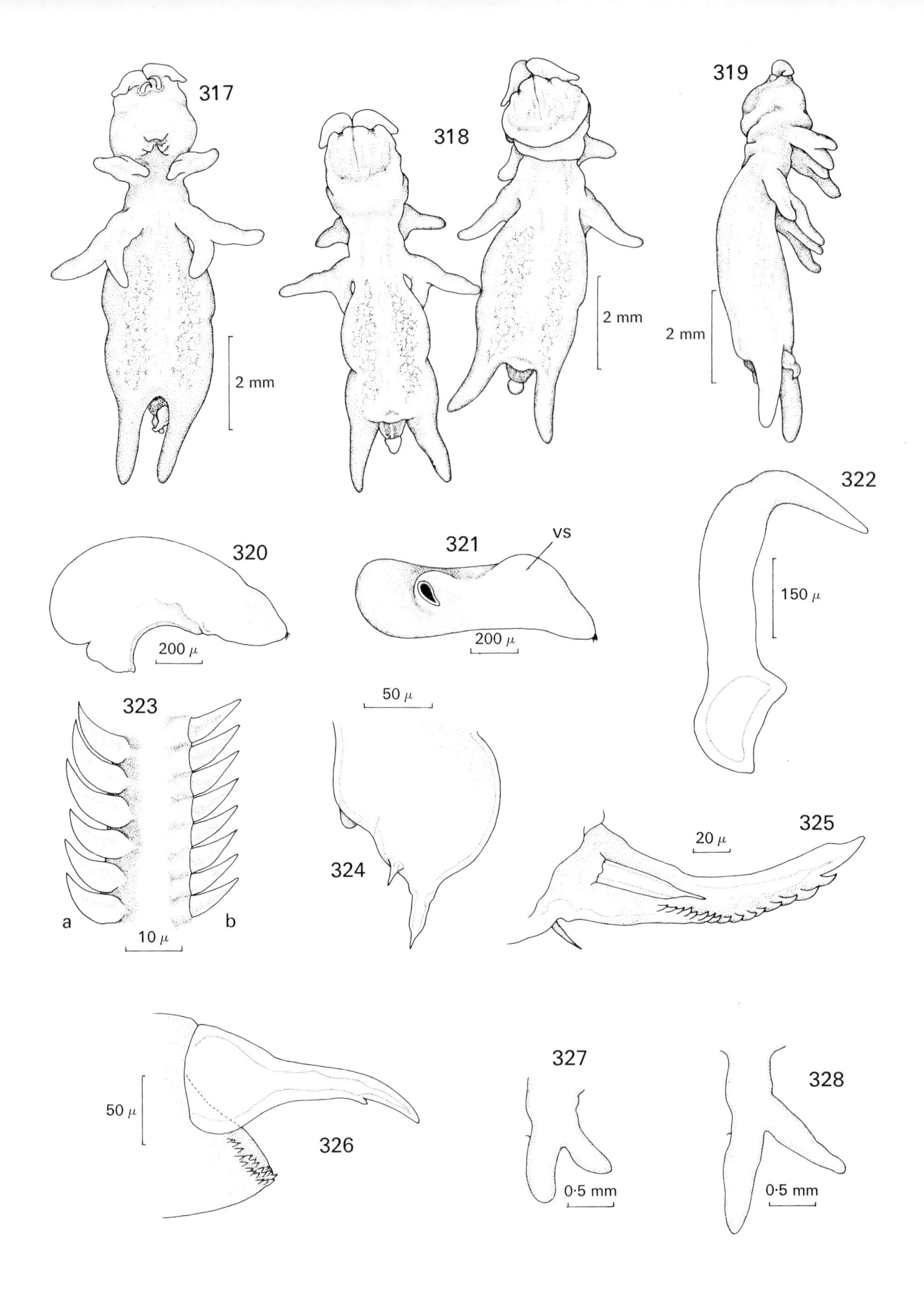

Figs. 317–328.
Acanthochondria soleae (Krøyer, 1838). **Fig. 317.** Female, ventral. **Fig. 318.** Same, two specimens, dorsal. **Fig. 319.** Same, lateral. **Fig. 320.** First antenna, ventral. **Fig. 321.** Same, proximal. **Fig. 322.** Second antenna, ventral. **Fig. 323.** Details of mandibular teeth : (a) outer; (b) inner. **Fig. 324.** First maxilla, ventral. **Fig. 325.** Second maxilla, claw, dorsal. **Fig. 326.** Maxilliped claw, full view. **Fig. 327.** First leg, ventral. **Fig. 328.** Second leg, ventral.
vs—ventral swelling.

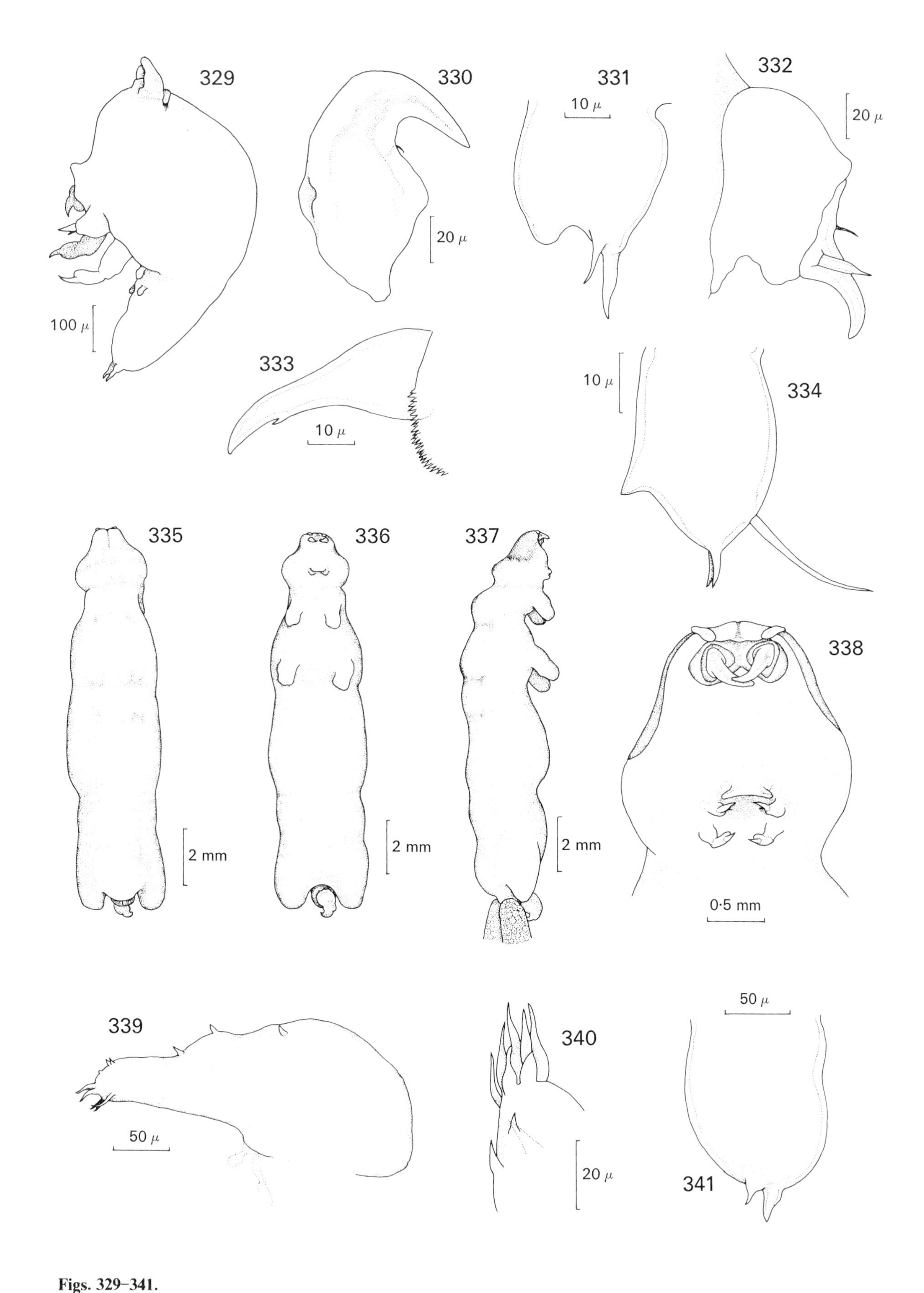

Figs. 329–341.

Acanthochondria soleae (Krøyer, 1838). **Fig. 329.** Male, lateral. **Fig. 330.** Second antenna, dorsal. **Fig. 331.** First maxilla, ventral. **Fig. 332.** Second maxilla, dorsal. **Fig. 333.** Maxilliped claw, full view. **Fig. 334.** First leg, ventral.

Acanthochondrites annulatus (Olsson, 1869). **Fig. 335.** Female, dorsal. **Fig. 336.** Same, ventral. **Fig. 337.** Same, lateral. **Fig. 338.** Cephalothorax, ventral. **Fig. 339.** First antenna, ventral. **Fig. 340.** Same, tip, dorsal. **Fig. 341.** First maxilla, ventral.

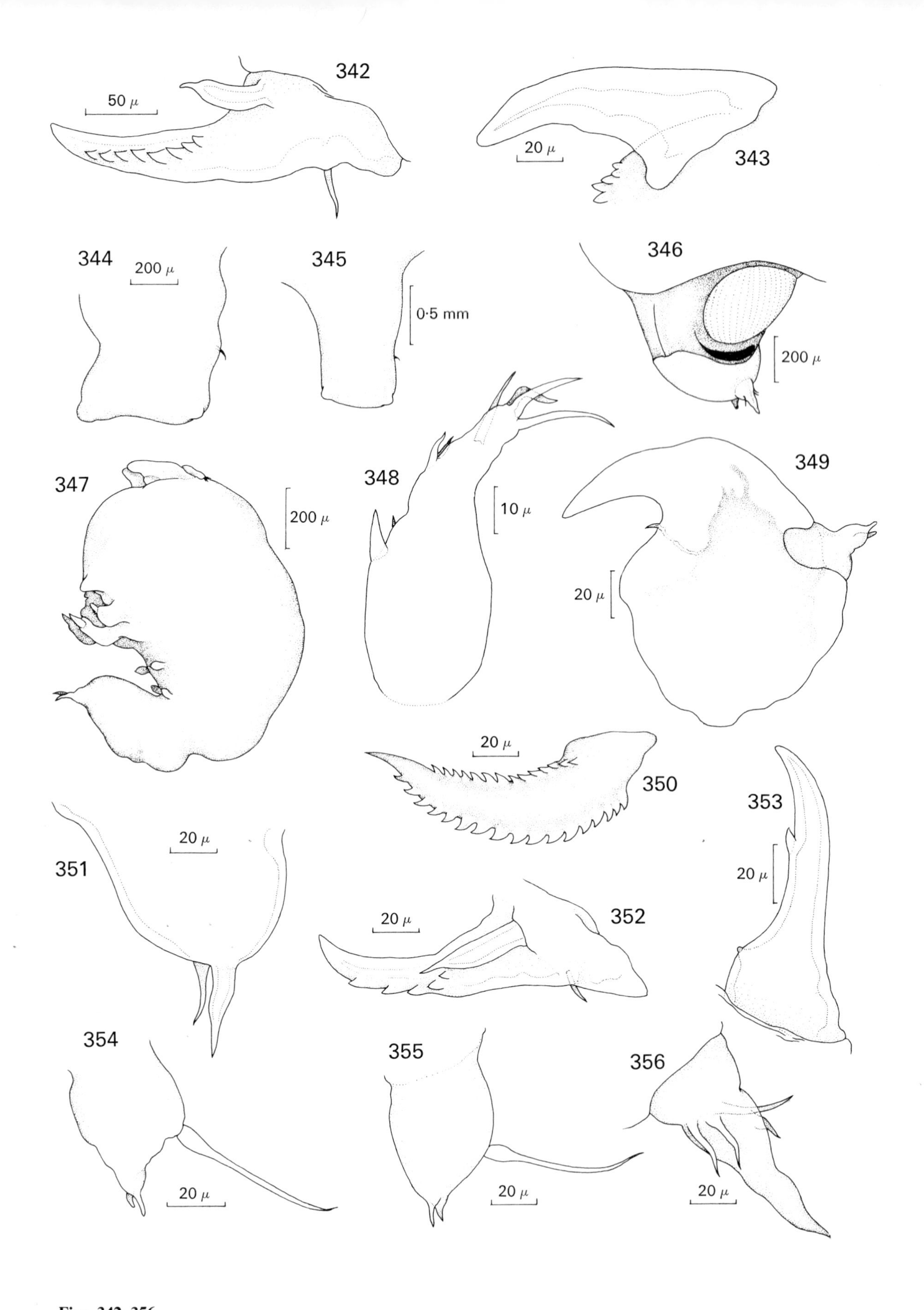

Figs. 342–356.
Acanthochondrites annulatus (Olsson, 1869). **Fig. 342.** Female, second maxilla, claw, dorsal. **Fig. 343.** Maxilliped claw, full view. **Fig. 344.** First leg, ventral. **Fig. 345.** Second leg, ventral. **Fig. 346.** Genito-abdominal tagma, lateral. **Fig. 347.** Male, lateral. **Fig. 348.** First antenna, dorsal. **Fig. 349.** Second antenna, ventral. **Fig. 350.** Mandibular blade, dorsal. **Fig. 351.** First maxilla, ventral. **Fig. 352.** Second maxilla, claw, dorsal. **Fig. 353.** Maxilliped claw, full view. **Fig. 354.** First leg, ventral. **Fig. 355.** Second leg, ventral. **Fig. 356.** Uropod, ventral.

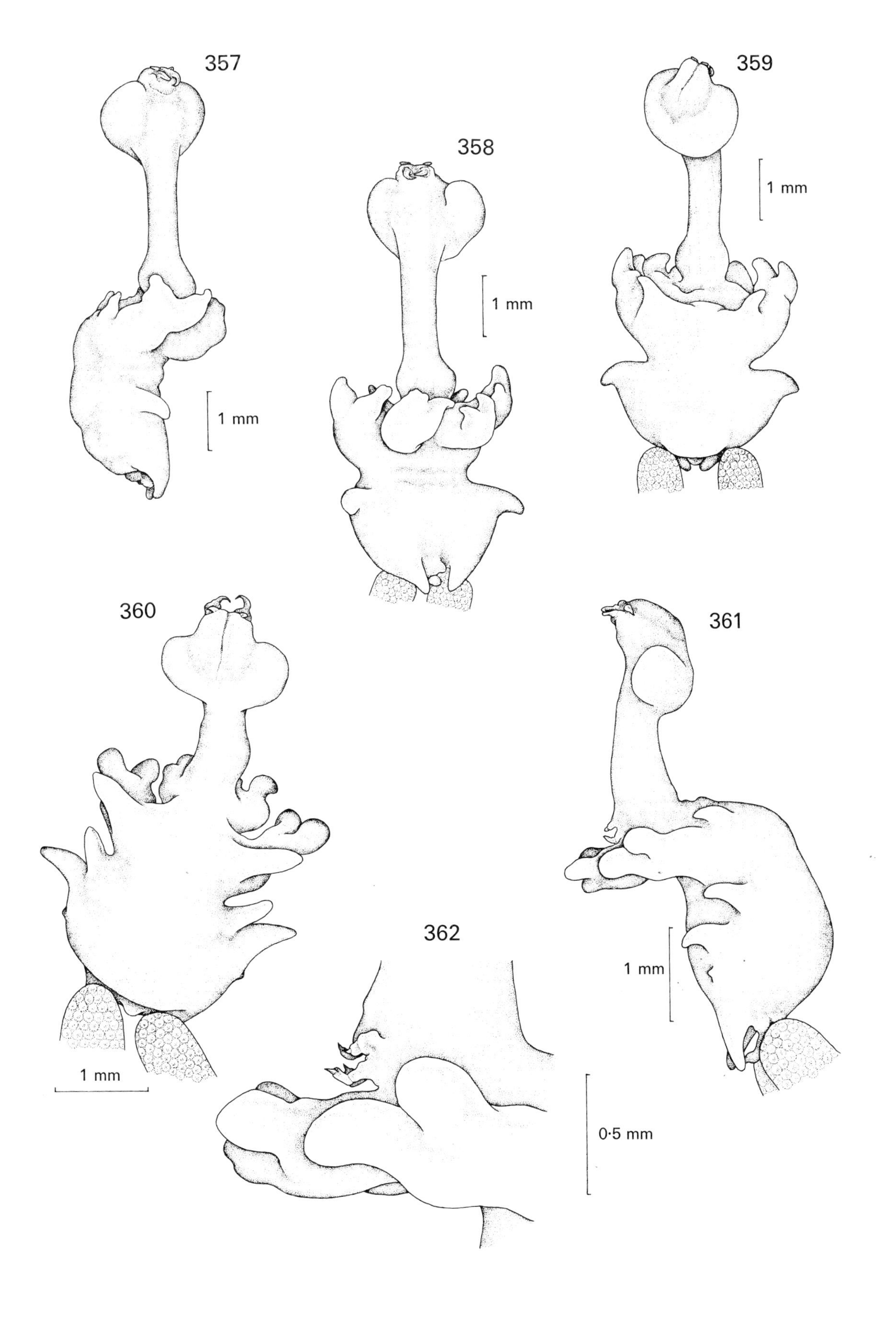

Figs. 357–362.
Lernentoma asellina (L.). **Fig. 357.** Female, lateral. **Fig. 358.** Same, ventral. **Fig. 359.** Same, dorsal. **Fig. 360.** Female with more than one lateral processes, dorsal. **Fig. 361.** Same, lateral. **Fig. 362.** Oral region of **Fig. 361,** enlarged.

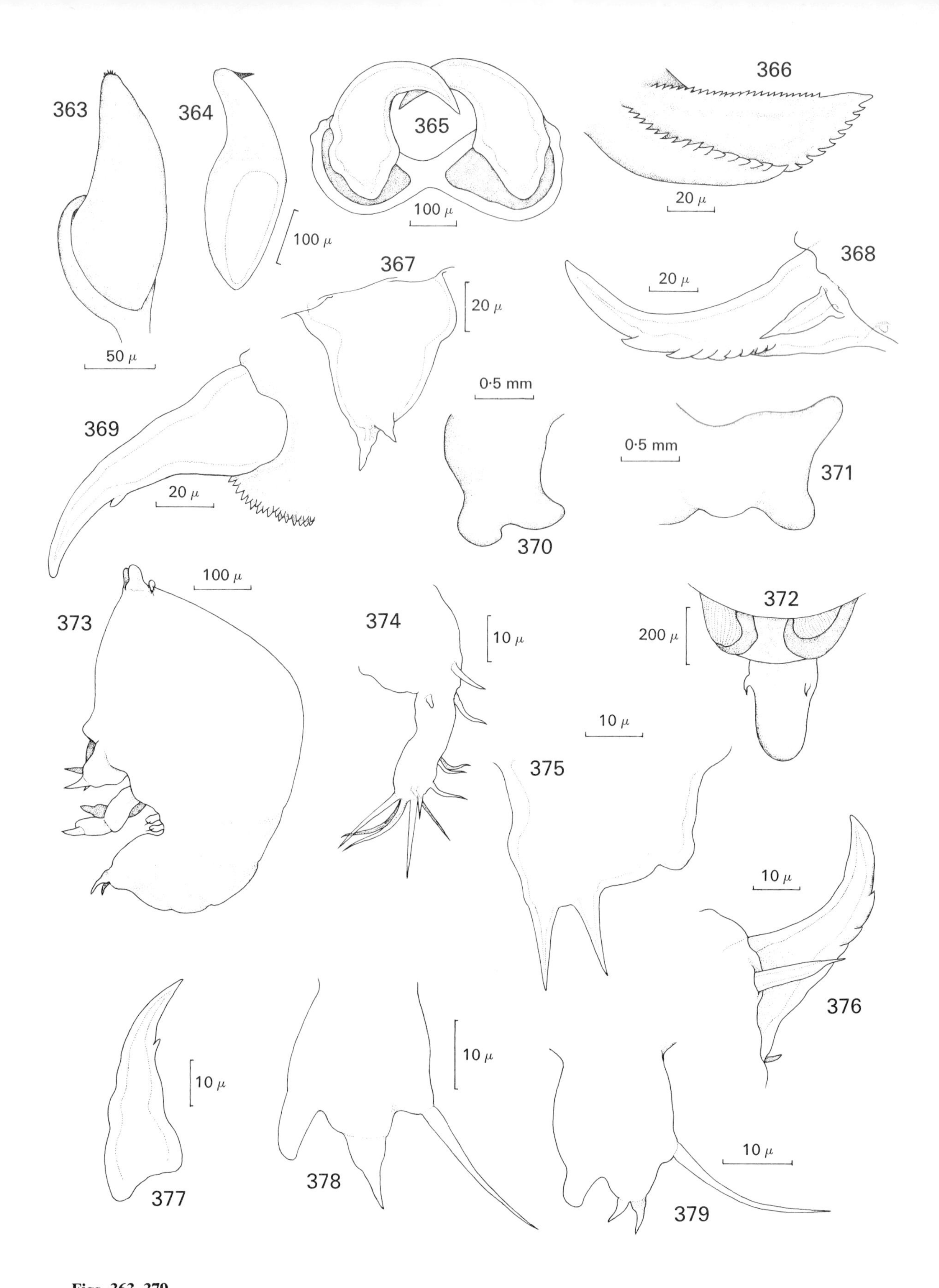

Figs. 363–379.
Lernentoma asellina (L.). **Fig. 363.** Female, first antenna, dorsal. **Fig. 364.** Second antenna, hook, lateral. **Fig. 365.** Both second antennae, ventral. **Fig. 366.** Mandibular blade, dorsal. **Fig. 367.** First maxilla, ventral. **Fig. 368.** Second maxilla, claw, dorsal. **Fig. 369.** Maxilliped claw, full view. **Fig. 370.** First (inner) leg, anterior. **Fig. 371.** Second (outer) leg, anterior. **Fig. 372.** Genito-abdominal tagma, ventral. **Fig. 373.** Male, lateral. **Fig. 374.** First antenna, dorsal. **Fig. 375.** First maxilla, ventral. **Fig. 376.** Second maxilla, claw, dorsal. **Fig. 377.** Maxilliped claw, full view. **Fig. 378.** First leg, ventral. **Fig. 379.** Second leg, ventral.

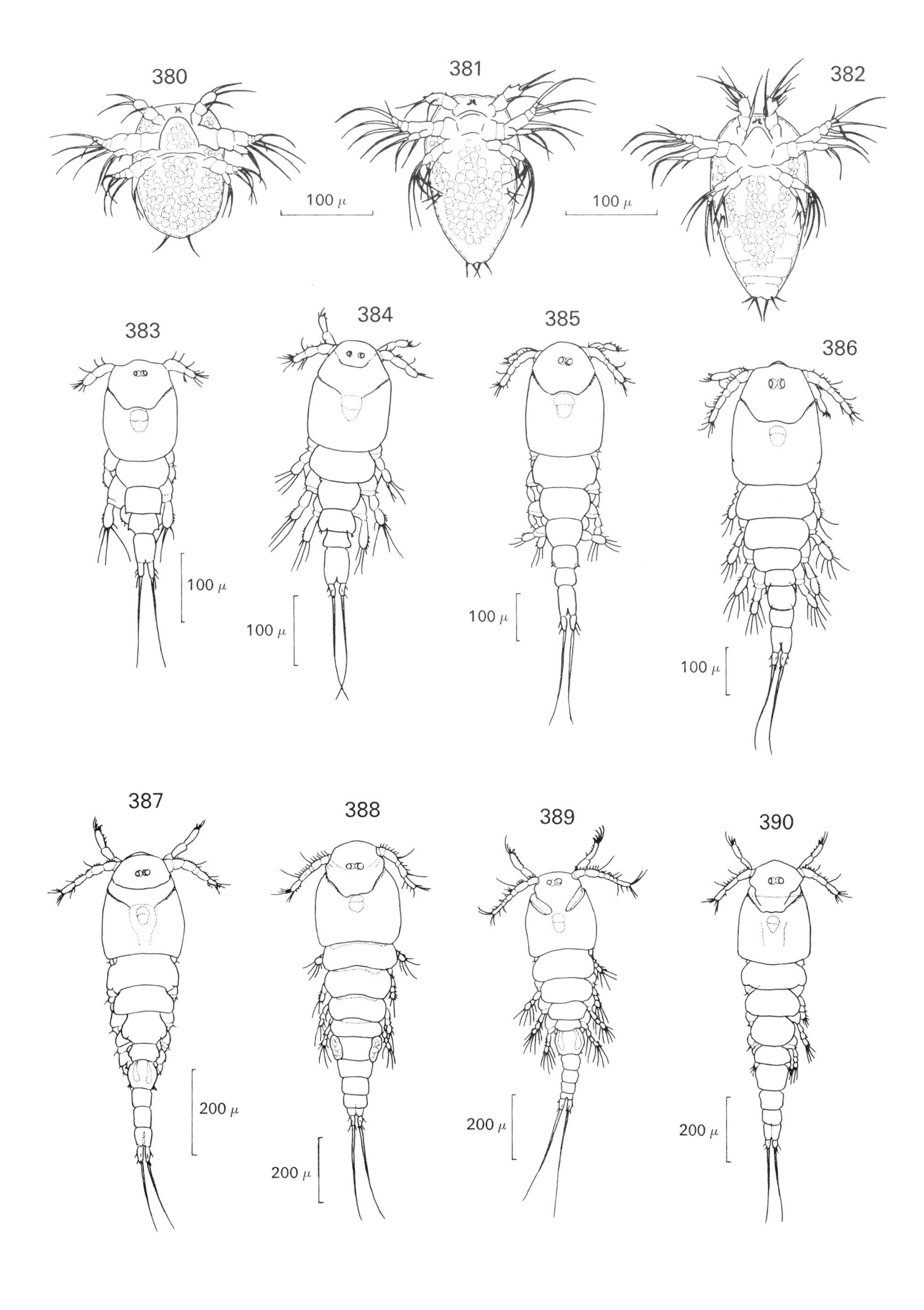

Figs. 380–390.
Lernaea cyprinacea L. **Fig. 380.** Nauplius I. **Fig. 381.** Nauplius II. **Fig. 382.** Nauplius III. **Fig. 383.** Copepodid I. **Fig. 384.** Copepodid II. **Fig. 385.** Copepodid III. **Fig. 386.** Copepodid IV. **Fig. 387.** Copepodid V, male. **Fig. 388.** Cyclopoid, female. **Fig. 389.** Cyclopoid, male. **Fig. 390.** Copepodid V, female.
Redrawn from Grabda (1963).

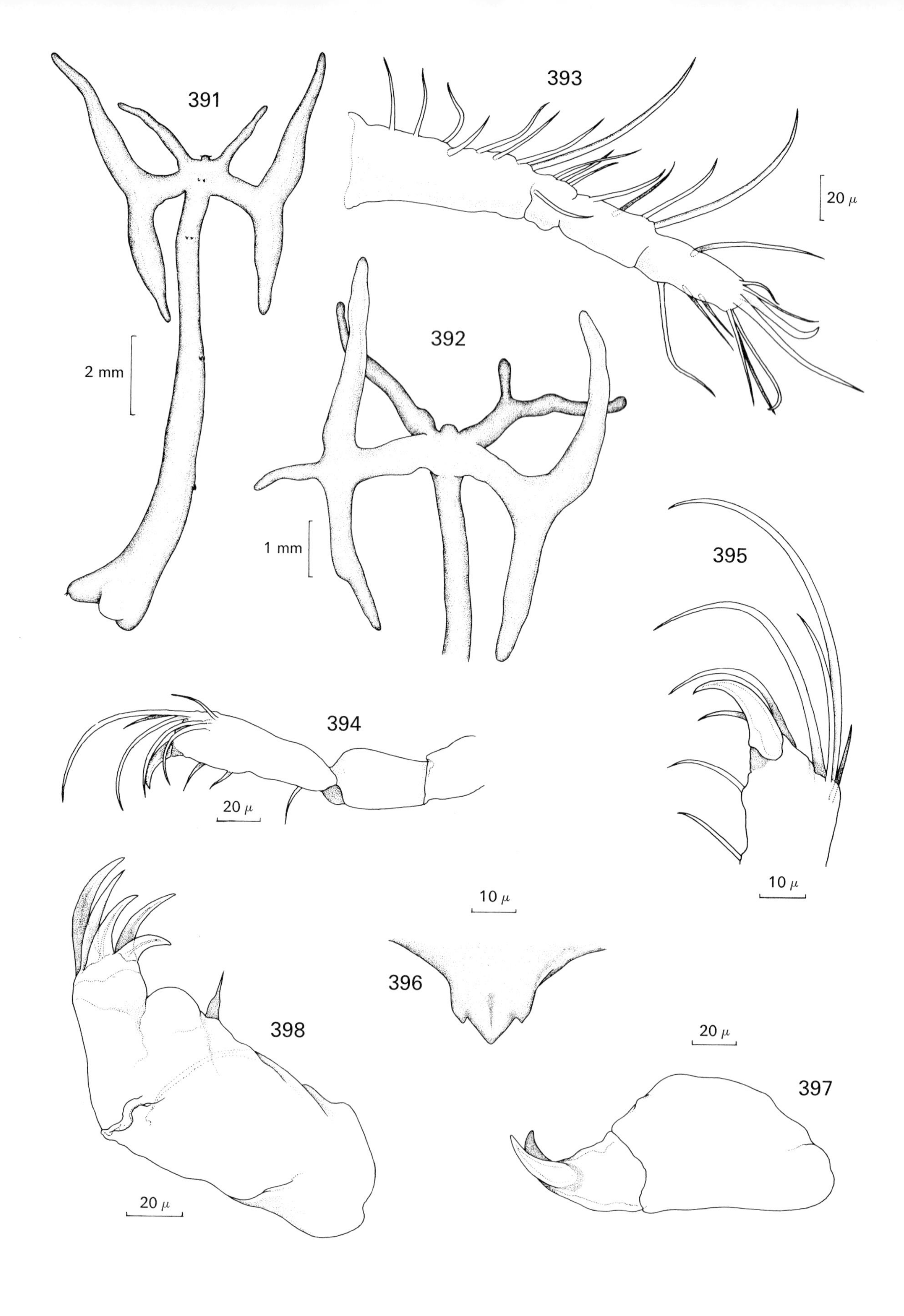

Figs. 391–398.
Lernaea cyprinacea L. **Fig. 391.** Female, cephalothorax ventral, posterior extremity lateral. **Fig. 392.** Cephalothorax, dorsal. **Fig. 393.** First antenna, dorsal. **Fig. 394.** Second antenna, dorsal. **Fig. 395.** Same, tip, ventral. **Fig. 396.** Labrum, outer surface, **Fig. 397.** Second maxilla, ventral. **Fig. 398.** Maxilliped, ventral.

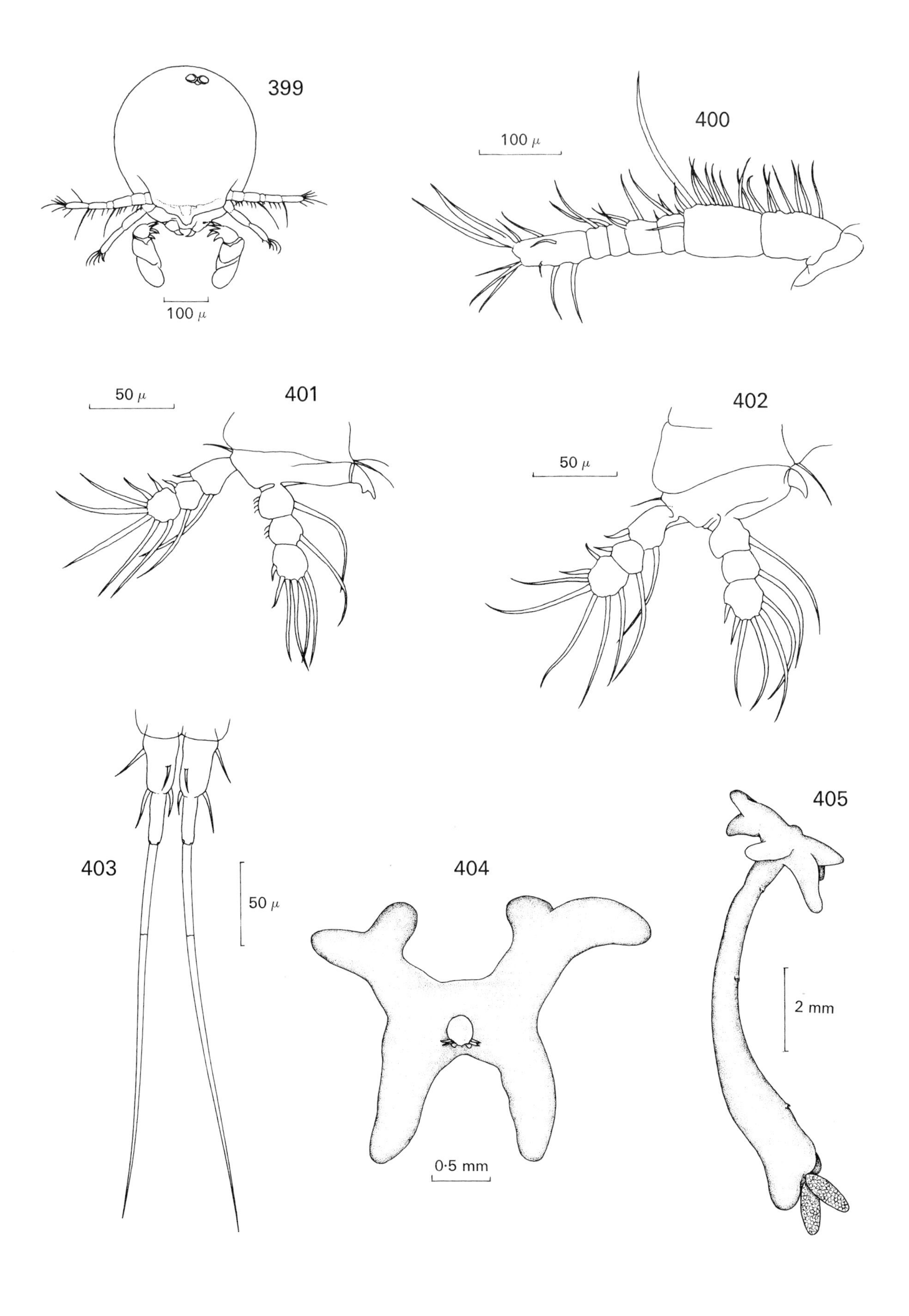

Figs. 399–405.
Lernaea cyprinacea L. **Fig. 399.** Female, cephalothorax, anterior. **Fig. 400.** Male, first antenna. **Fig. 401.** First leg. **Fig. 402.** Female, first leg. **Fig. 403.** Copepodid V, uropod. **Fig. 404.** Young adult female, cephalothorax, anterior. **Fig. 405.** Same, entire, lateral.
Figs. 399–403 redrawn from Grabda (1963).

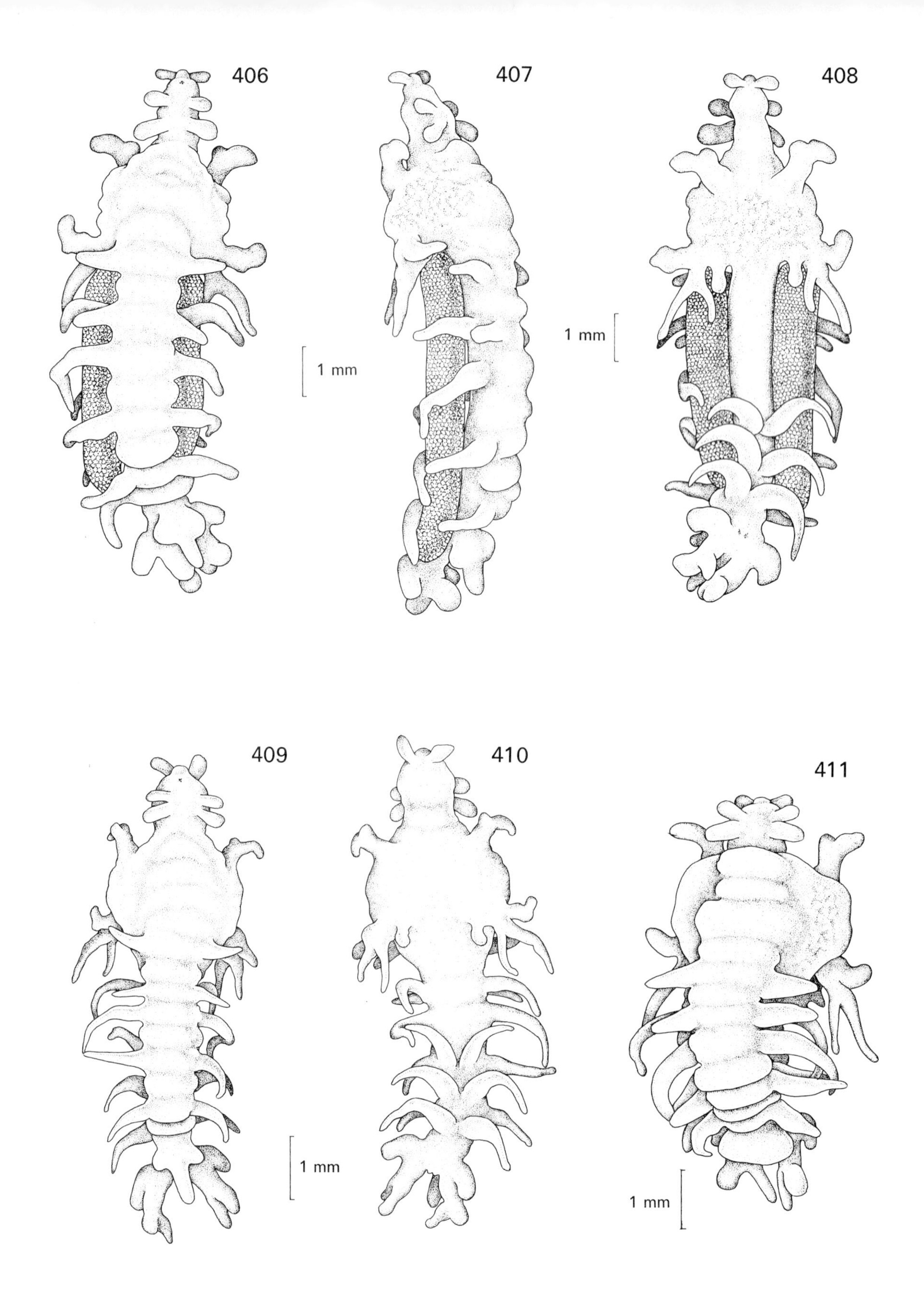

Figs. 406–411.
Philichthys xiphiae Steenstrup, 1862. **Fig. 406.** Ovigerous female, ventral. **Fig. 407.** Same, lateral. **Fig. 408.** Same, dorsal. **Fig. 409.** Non-ovigerous female, ventral. **Fig. 410.** Same, dorsal. **Fig. 411.** Contracted female, ventral.

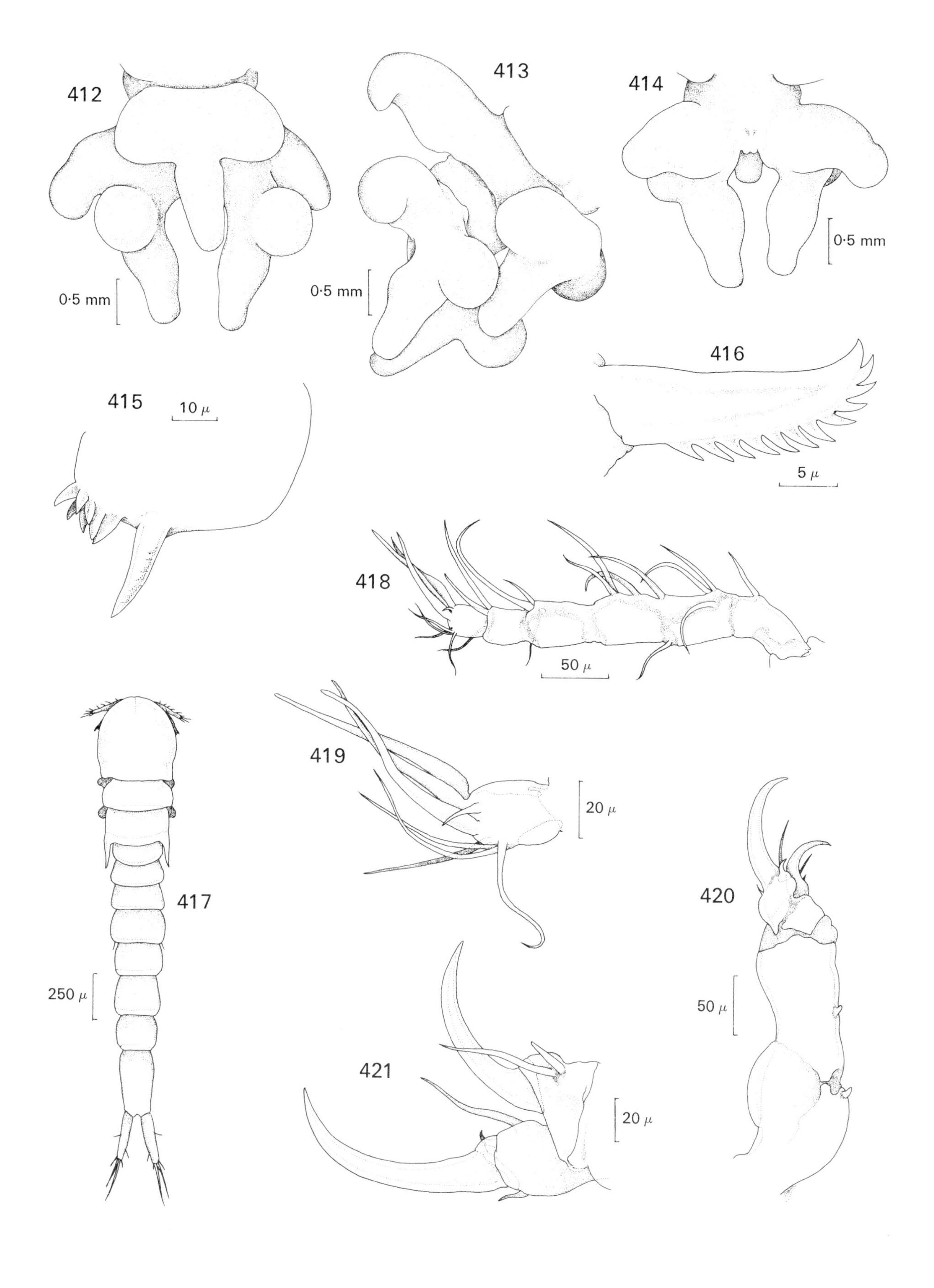

Figs. 412–421.
Philichthys xiphiae Steenstrup, 1862. **Fig. 412.** Posterior extremity of female, ventral. **Fig. 413.** Same, ventrolateral. **Fig. 414.** Same, dorsal. **Fig. 415.** First maxilla, dorsal. **Fig. 416.** Same, claw, posterior. **Fig. 417.** Male, dorsal. **Fig. 418.** First antenna, ventral. **Fig. 419.** Same, tip, ventral. **Fig. 420.** Second antenna. **Fig. 421.** Same, tip, opposite side.

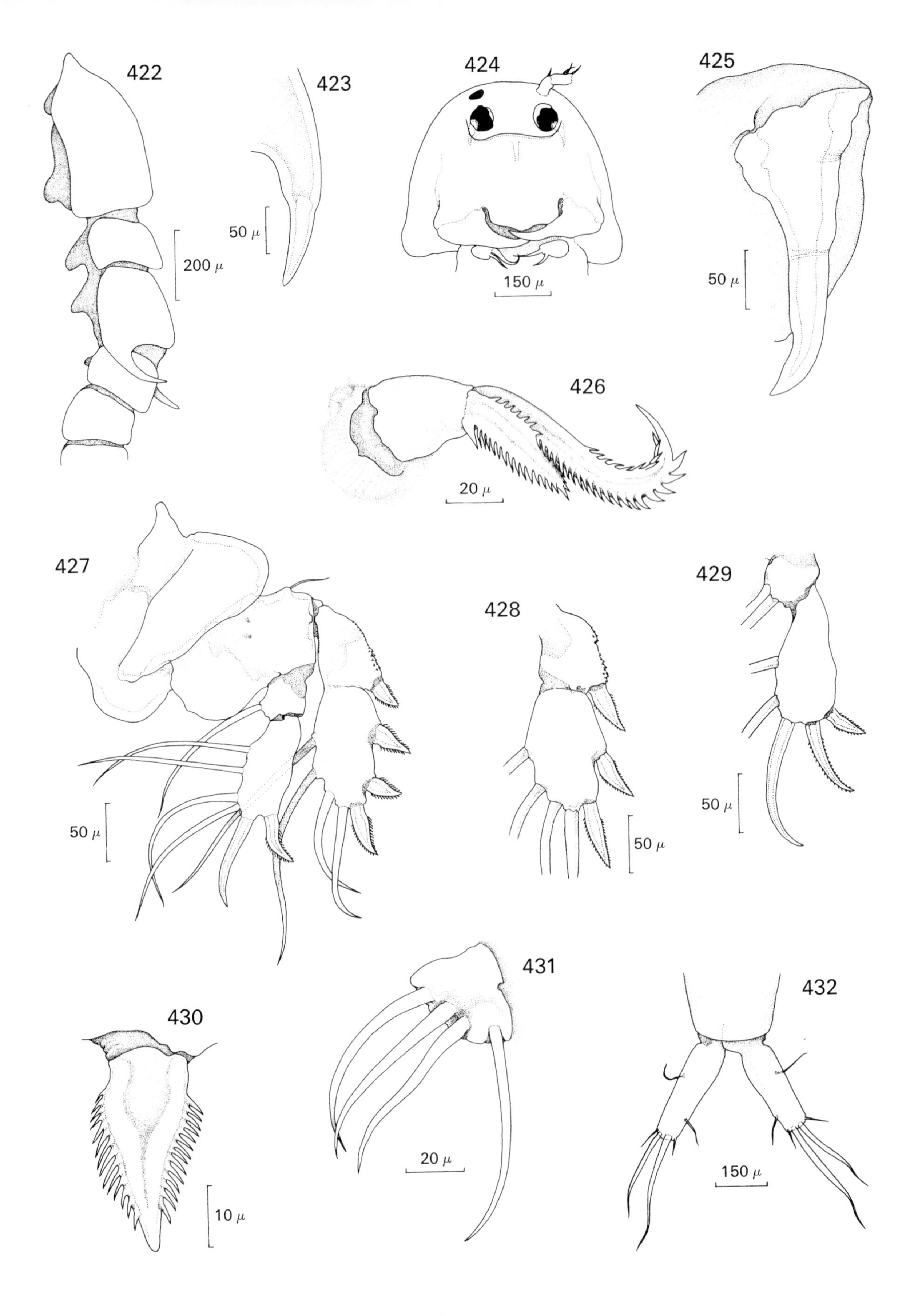

Figs. 422–432.
Philichthys xiphiae Steenstrup, 1862. **Fig. 422.** Anterior half of male, lateral (diagramatic). **Fig. 423.** Male, posterolateral process of second free thoracic segment, dorsal. **Fig. 424.** Cephalothorax, ventral. **Fig. 425.** Second maxilla, ventral. **Fig. 426.** Mandible, ventral. **Fig. 427.** First leg, ventral. **Fig. 428.** Second exopod, ventral. **Fig. 429.** Second endopod, ventral. **Fig. 430.** Exopod spine of first two legs, ventral. **Fig. 431.** Third leg, ventral. **Fig. 432.** Uropods, dorsal.

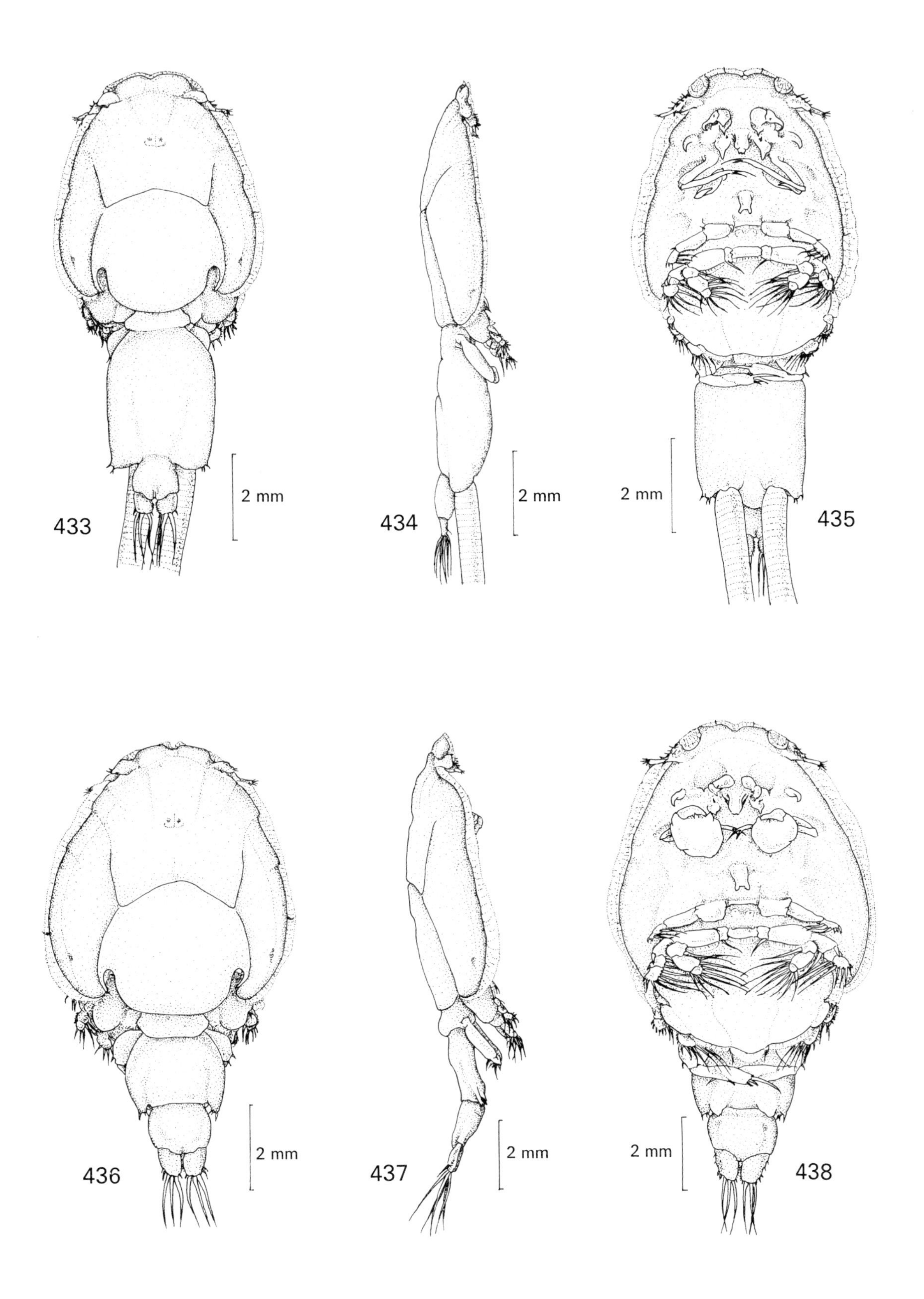

Figs. 433–438.
Caligus curtus Müller, 1785. **Fig. 433.** Female, dorsal. **Fig. 434.** Same, lateral. **Fig. 435.** Same, ventral. **Fig. 436.** Male, dorsal. **Fig. 437.** Same, lateral. **Fig. 438.** Same, ventral.
Redrawn from Parker et al. (1968).

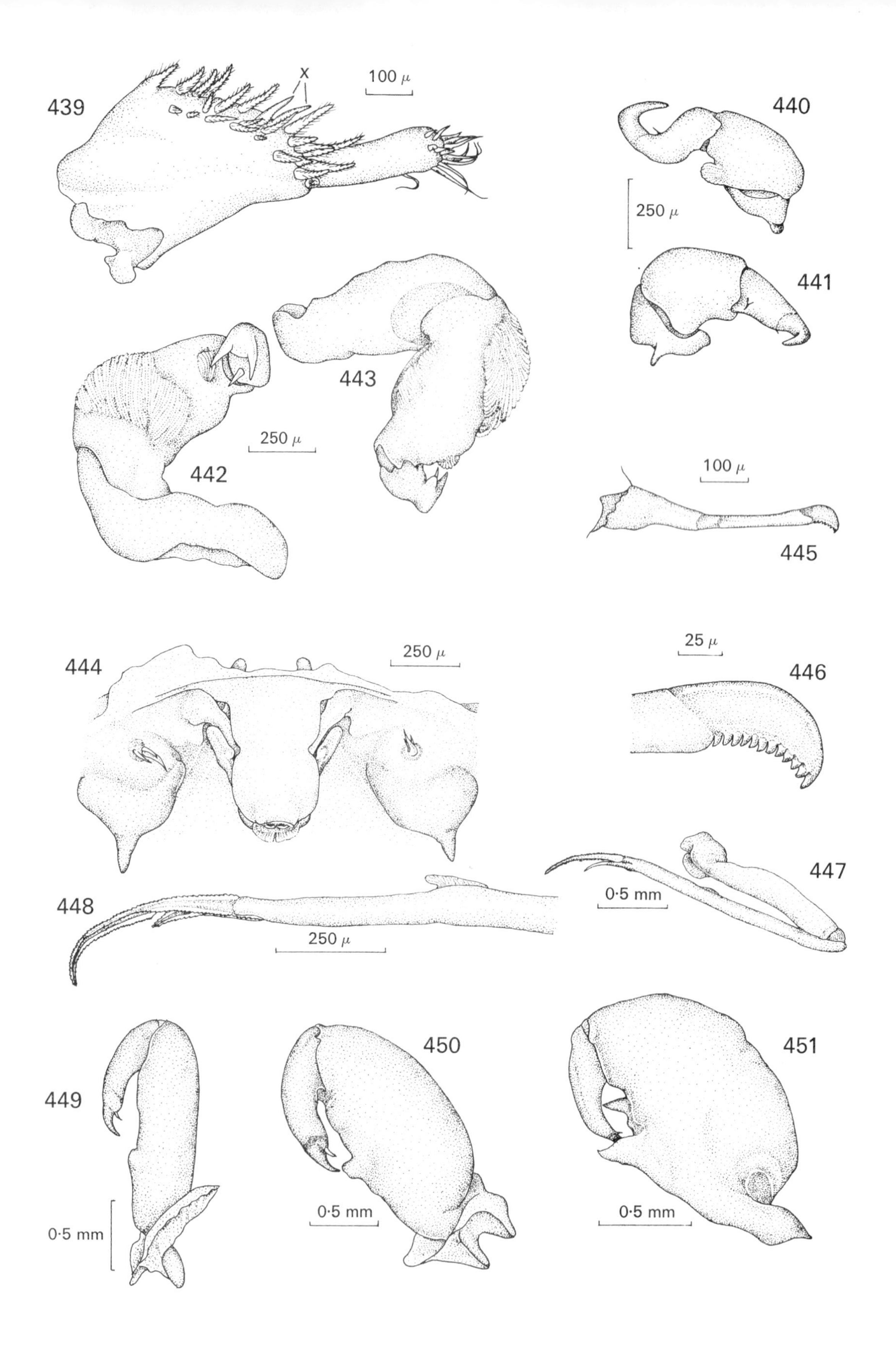

Figs. 439–451.
Caligus curtus Müller, 1785. **Fig. 439.** Male and female, first antenna, ventral. **Fig. 440.** Female, second antenna, dorsomedial. **Fig. 441.** Same, posterior. **Fig. 442.** Male, second antenna, posterior. **Fig. 443.** Same, anterior. **Fig. 444.** Male and female, oral region, ventral. **Fig. 445.** Mandible, lateral. **Fig. 446.** Same, tip. **Fig. 447.** Second maxilla, ventral. **Fig. 448.** Same, brachium enlarged. **Fig. 449.** Female, maxilliped, dorsal. **Fig. 450.** Young male, maxilliped, ventral. **Fig. 451.** Adult male, maxilliped, dorsal.
Redrawn from Parker et al. (1968).

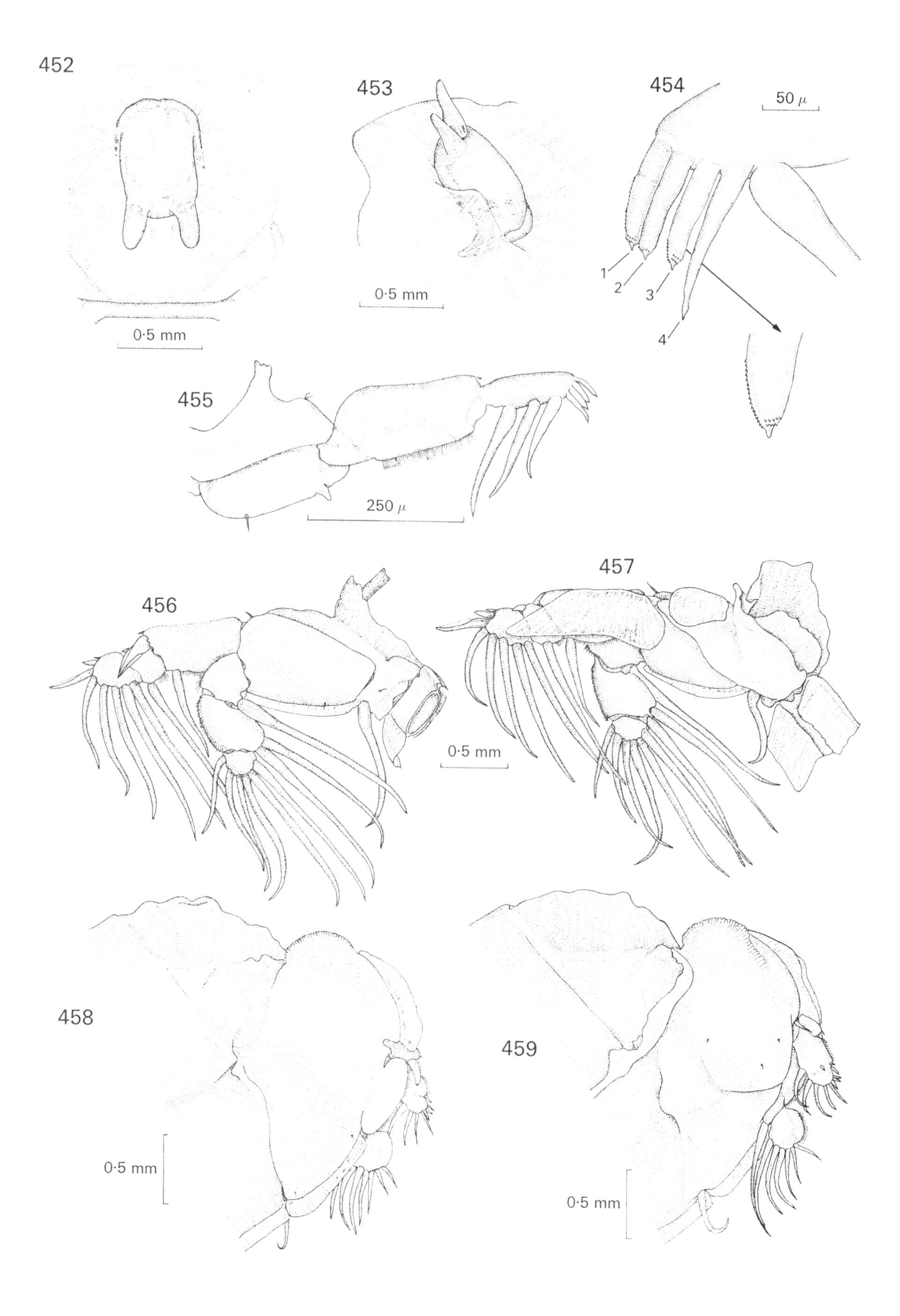

Figs. 452–459.
Caligus curtus Müller, 1785. **Fig. 452.** Male and female, sternal furca, ventral. **Fig. 453.** Same, lateral. **Fig. 454.** Tip of first exopod, ventral (plumes of large seta omitted). **Fig. 455.** First leg, dorsal. **Fig. 456.** Second leg, ventral. **Fig. 457.** Same, dorsal. **Fig. 458.** Third leg, ventral. **Fig. 459.** Same, dorsal.
Redrawn from Parker et al. (1968).

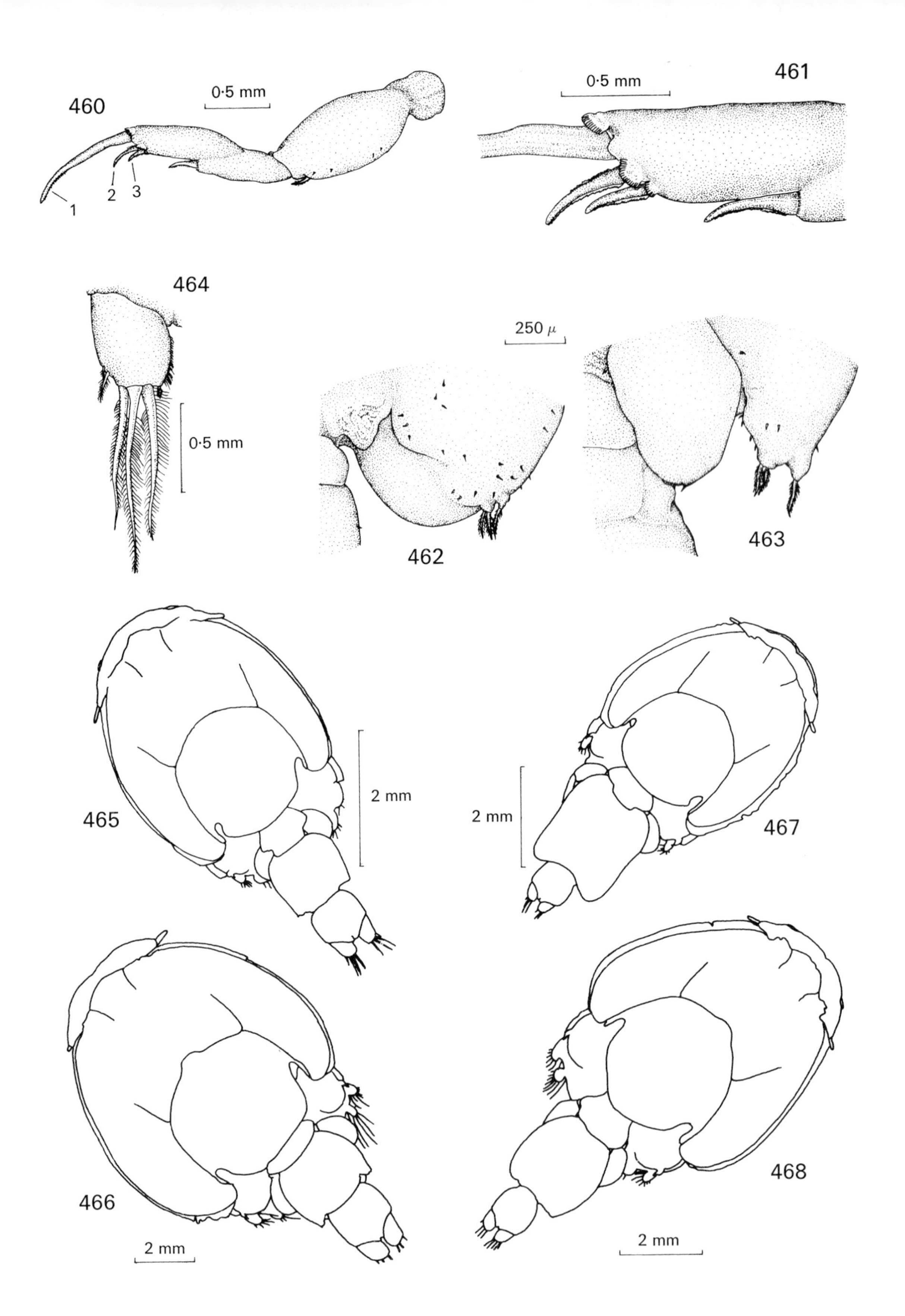

Figs. 460–468.
Caligus curtus Müller, 1785. **Fig. 460.** Male and female, fourth leg, ventral. **Fig. 461.** Same, tip, ventral. **Fig. 462.** Posterolateral corner of female genital complex, ventral. **Fig. 463.** Posterolateral corner of male genital complex, ventral. **Fig. 464.** Male and female, uropod, ventral. **Fig. 465.** Outline of young male, dorsal. **Fig. 466.** Outline of mature male, dorsal. **Fig. 467.** Outline of gravid female, dorsal. **Fig. 468.** Outline of postgravid female, dorsal.
Redrawn from Parker et al. (1968).

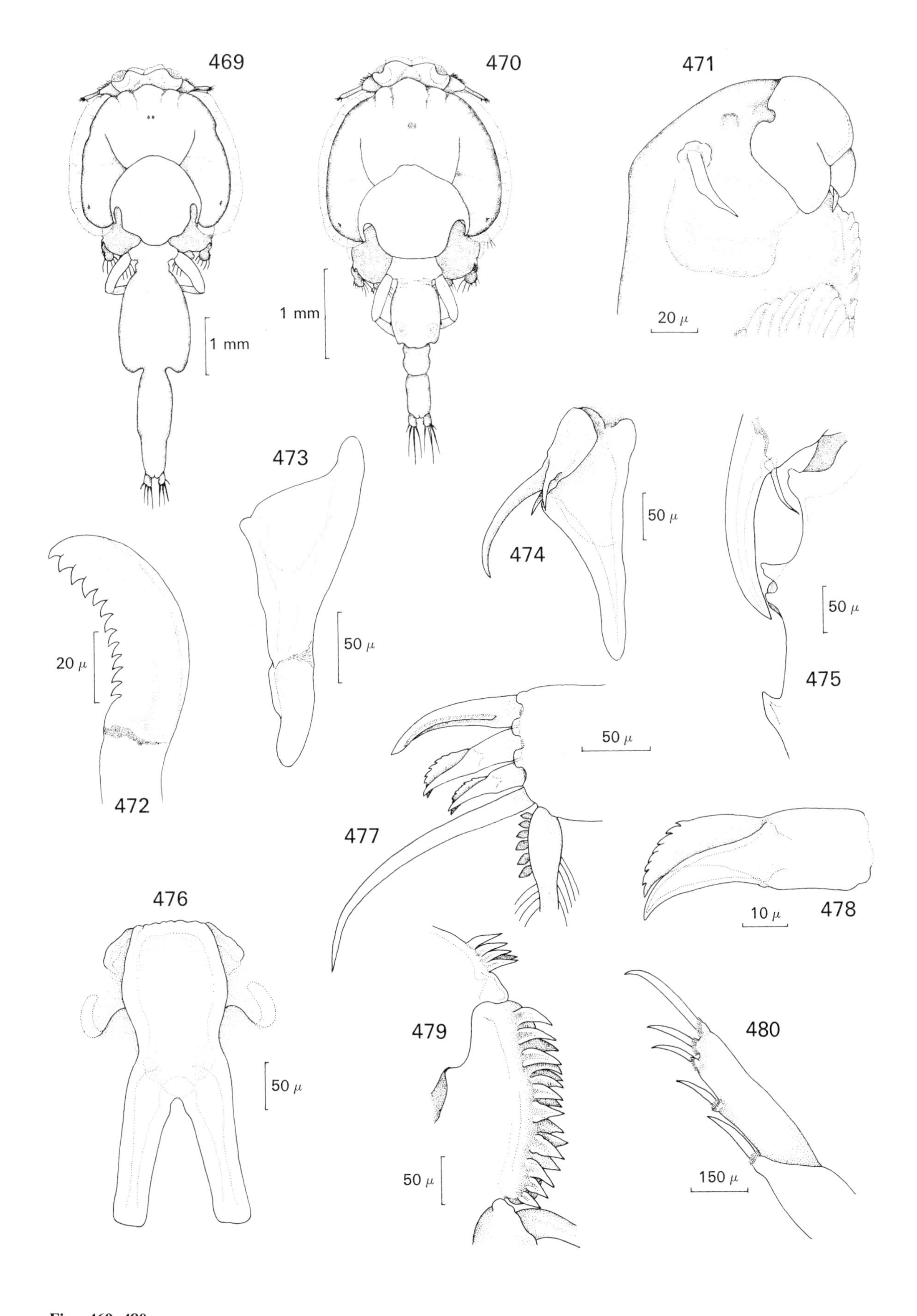

Figs. 469–480.
Caligus bonito Wilson, 1905. **Fig. 469.** Female, dorsal. **Fig. 470.** Male, dorsal. **Fig. 471.** Second antenna, tip. **Fig. 472.** Male and female, tip of mandible, lateral. **Fig. 473.** Male, process of first maxilla, lateral. **Fig. 474.** Female, first maxilla, ventral. **Fig. 475.** Male maxilliped, detail of medial margin and tip of claw, ventral. **Fig. 476.** Sternal furca, ventral. **Fig. 477.** Tip of first exopod, ventral. **Fig. 478.** Same, third spine, ventral. **Fig. 479.** Second endopod, details of lateral margin. **Fig. 480.** Tip of fourth leg, ventral.

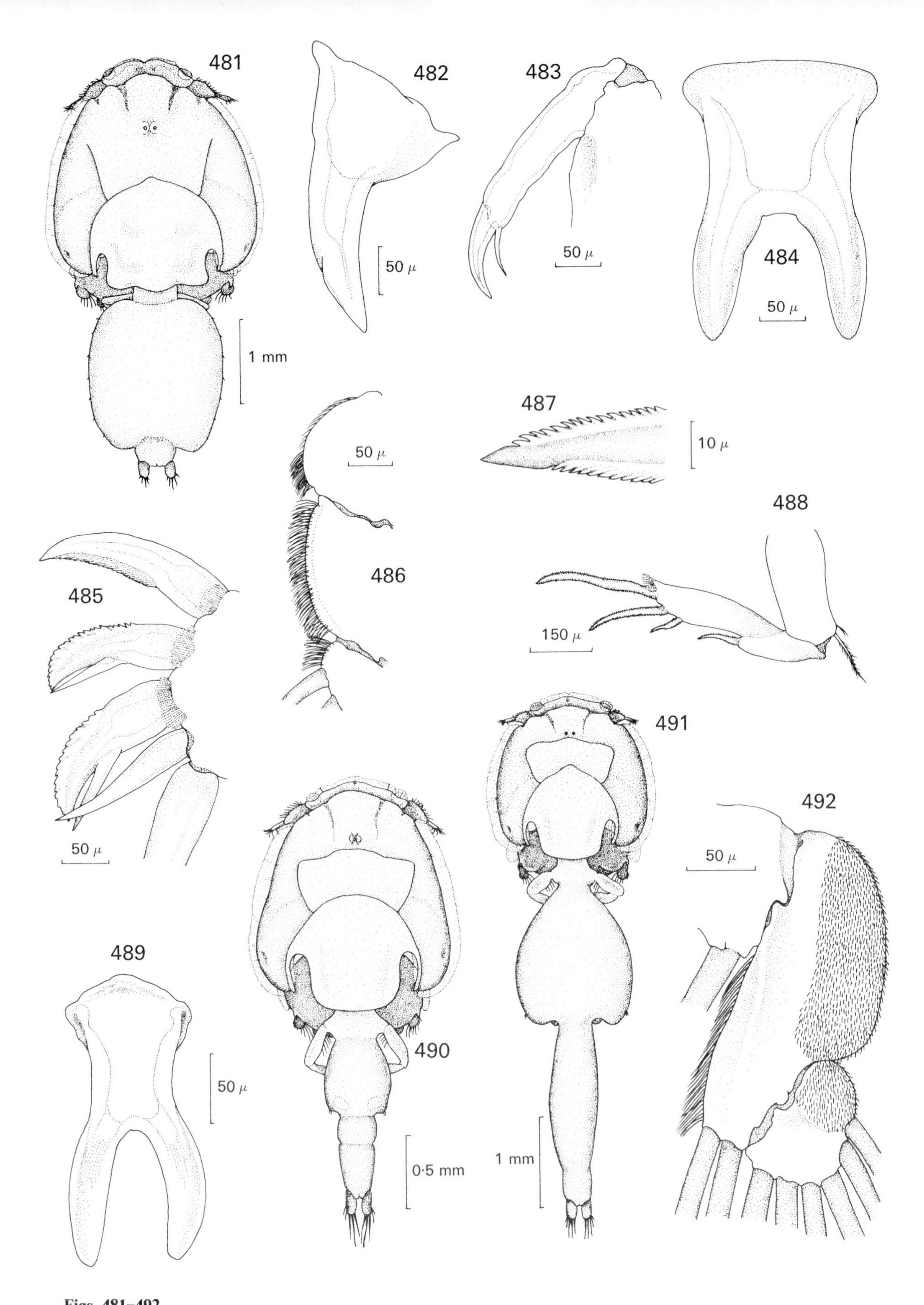

Figs. 481–492.
Caligus brevicaudatus A. Scott, 1901. **Fig. 481.** Female, dorsal. **Fig. 482.** First maxilla, ventral. **Fig. 483.** Subchela of maxilliped, ventral. **Fig. 484.** Sternal furca, ventral. **Fig. 485.** Tip of first exopod, ventral. **Fig. 486.** Second endopod, details of lateral margin. **Fig. 487.** Tip of seta, fourth leg. **Fig. 488.** Fourth leg, ventral.
Caligus pelamydis Krøyer, 1863. **Fig. 489.** Male and female, sternal furca, ventral. **Fig. 490.** Male, dorsal. **Fig. 491.** Female, dorsal. **Fig. 492.** Male and female, second endopod, ventral.

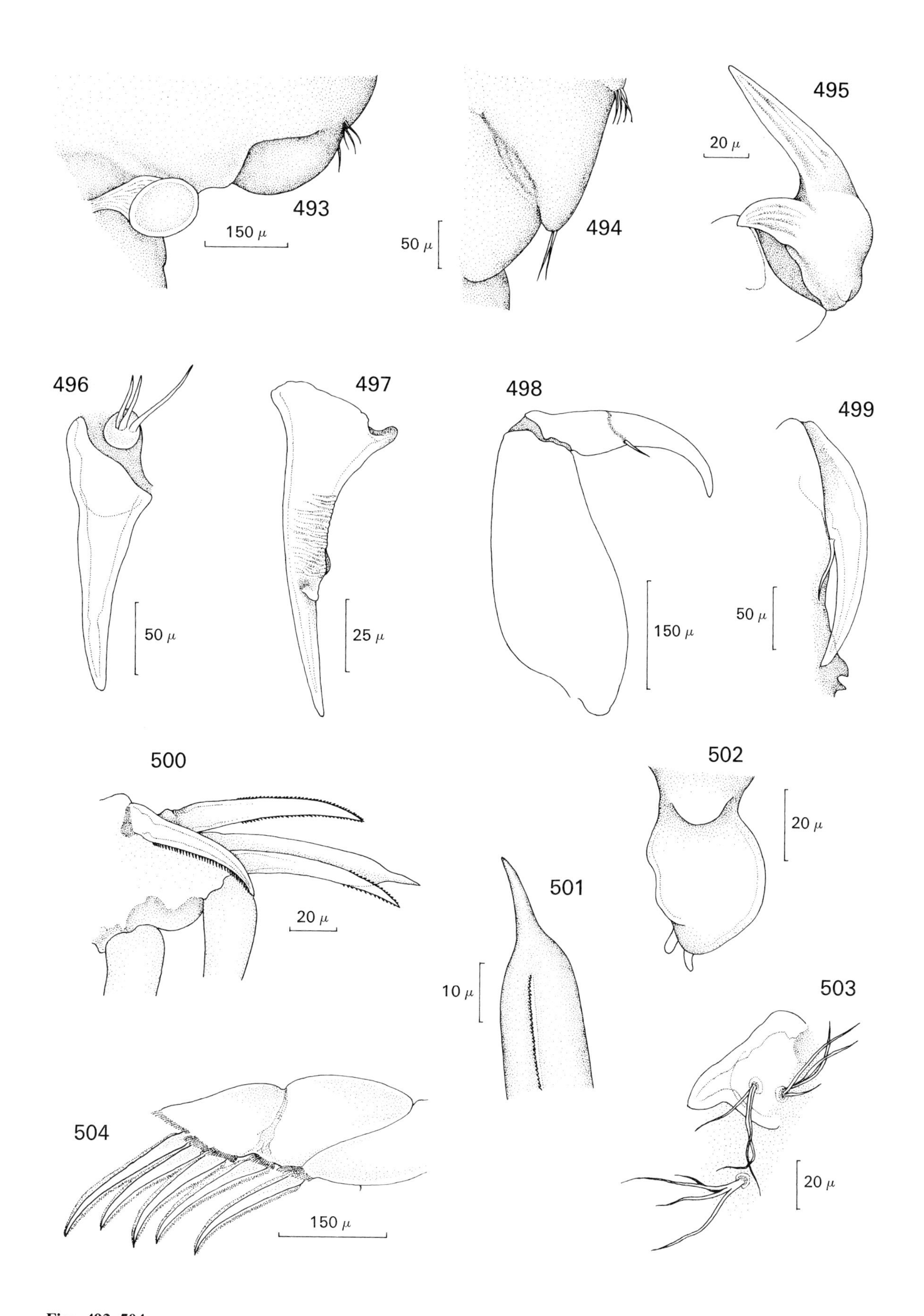

Figs. 493–504.
Caligus pelamydis Krøyer, 1863. **Fig. 493.** Posterolateral corner of female genital complex, ventral. **Fig. 494.** Posterolateral corner of male genital complex, ventral. **Fig. 495.** Male, tip of second antenna. **Fig. 496.** Female, first maxilla, ventral. **Fig. 497.** Male, first maxilla, ventral. **Fig. 498.** Female, maxilliped, ventral. **Fig. 499.** Male, subchela of maxilliped, dorsal. **Fig. 500.** Male and female, tip of first exopod, ventral. **Fig. 501.** First exopod, tip of first spine. **Fig. 502.** First endopod. **Fig. 503.** Postantennary process, ventral. **Fig. 504.** Distal half of fourth leg, ventral.

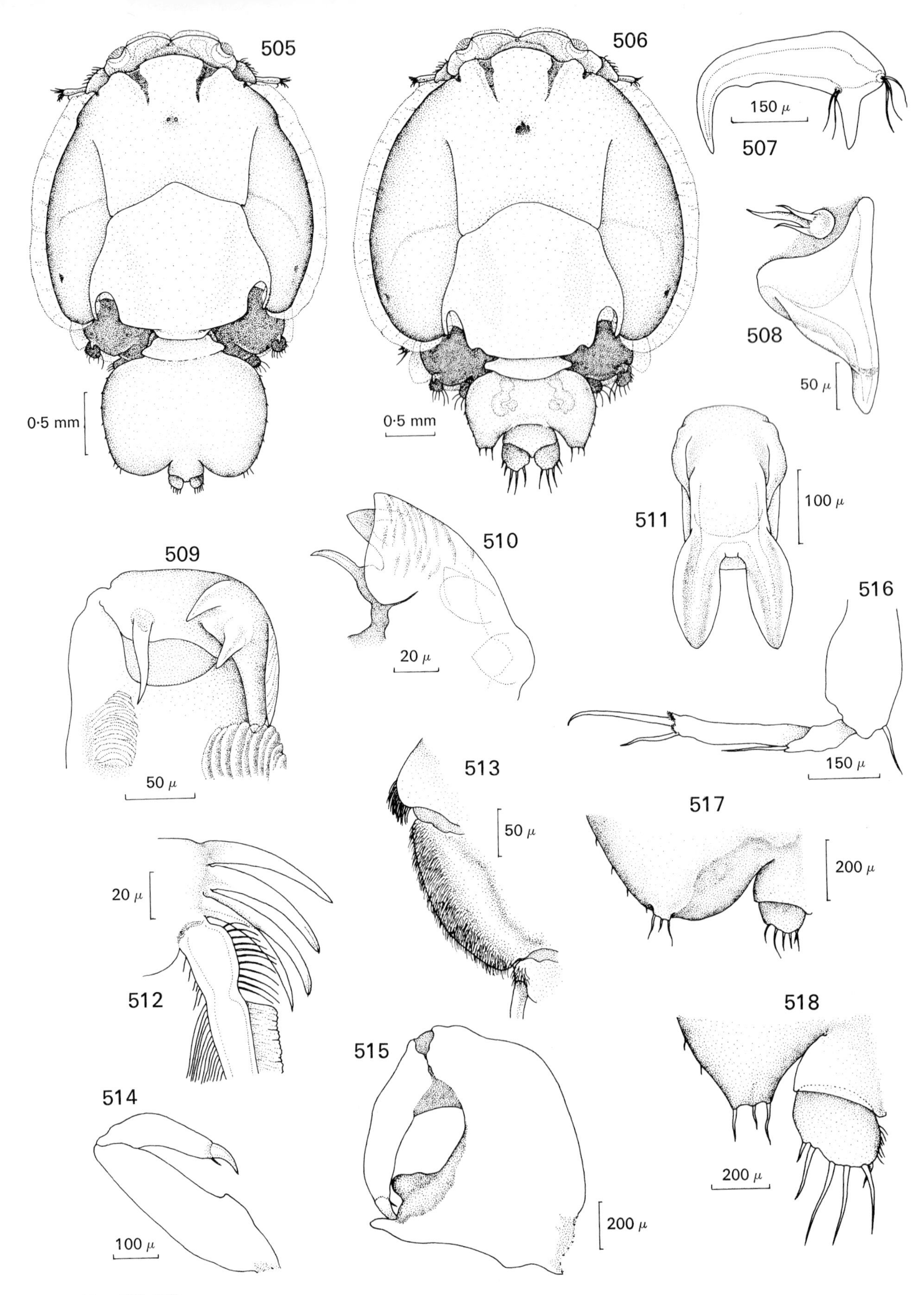

Figs. 505–518.
Caligus centrodonti Baird, 1850. **Fig. 505.** Female, dorsal. **Fig. 506.** Male, dorsal. **Fig. 507.** Male and female, postantennary process, ventral. **Fig. 508.** First maxilla, ventral. **Fig. 509.** Male, tip of second antenna. **Fig. 510.** Same, claw, anterior. **Fig. 511.** Male and female, sternal furca, ventral. **Fig. 512.** Tip of first exopod, ventral. **Fig. 513.** Second endopod, lateral margin, ventral. **Fig. 514.** Female maxilliped, ventral. **Fig. 515.** Male maxilliped, ventral. **Fig. 516.** Fourth leg, ventral. **Fig. 517.** Posterolateral corner of female genital complex, ventral. **Fig. 518.** Posterolateral corner of male genital complex, ventral.

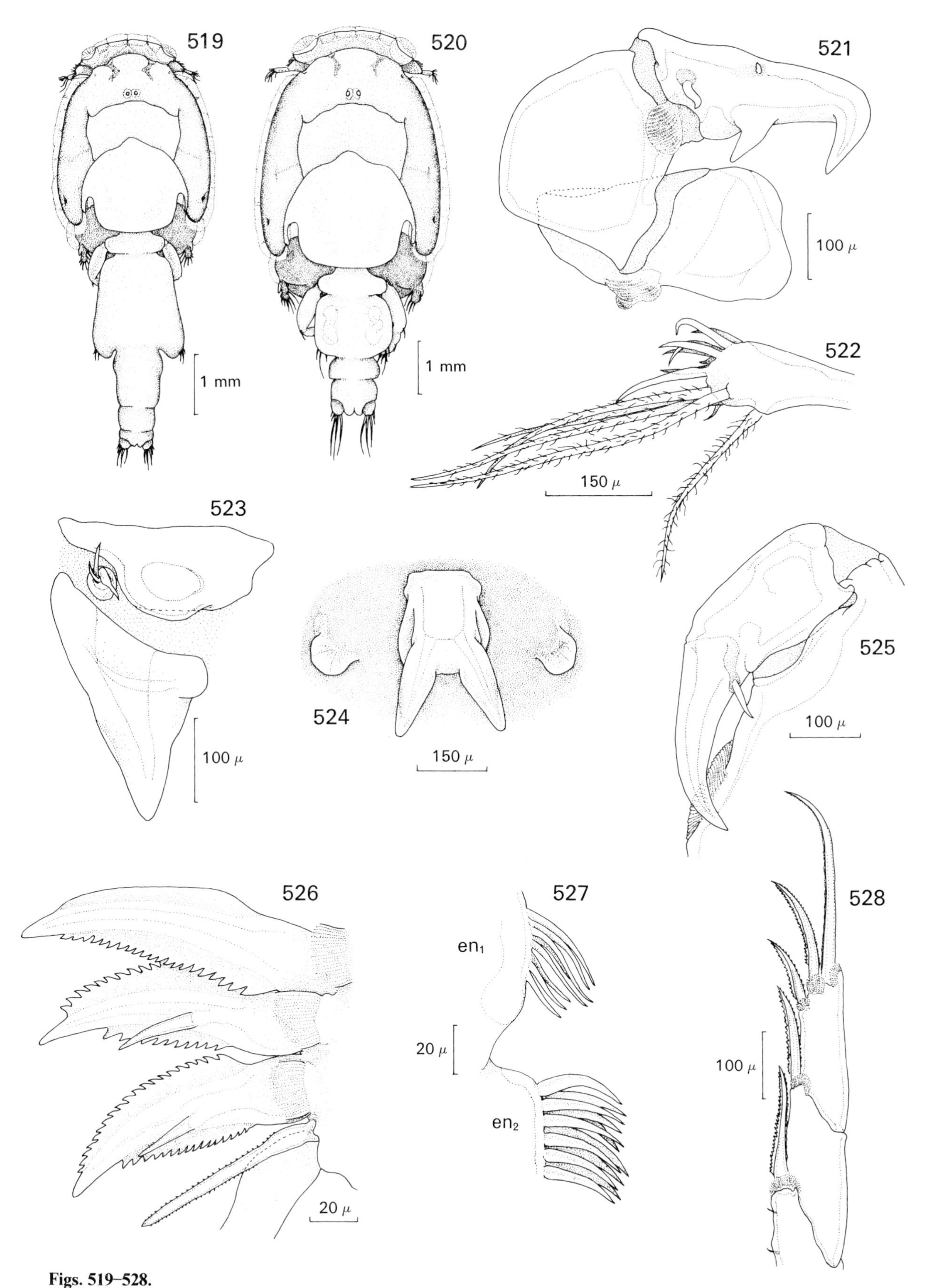

Figs. 519–528.
Caligus coryphaenae Steenstrup and Lütken, 1861. **Fig. 519.** Female, dorsal. **Fig. 520.** Male, dorsal. **Fig. 521.** Male second antenna, ventral. **Fig. 522.** Male and female, tip of first antenna, ventral. **Fig. 523.** First maxilla, ventral. **Fig. 524.** Sternal furca, ventral. **Fig. 525.** Subchela of male maxilliped, ventral. **Fig. 526.** Male and female, tip of first exopod, ventral. **Fig. 527.** Second endopod, detail of lateral margin, ventral. **Fig. 528.** Tip of fourth leg, ventral.
en1—first endopod segment; en2—second endopod segment.
Figs. 521, 523, 525 and 528 modified from unpublished drawings of Dr. A. V. Gusev, Zoological Institute, Leningrad, U.S.S.R.

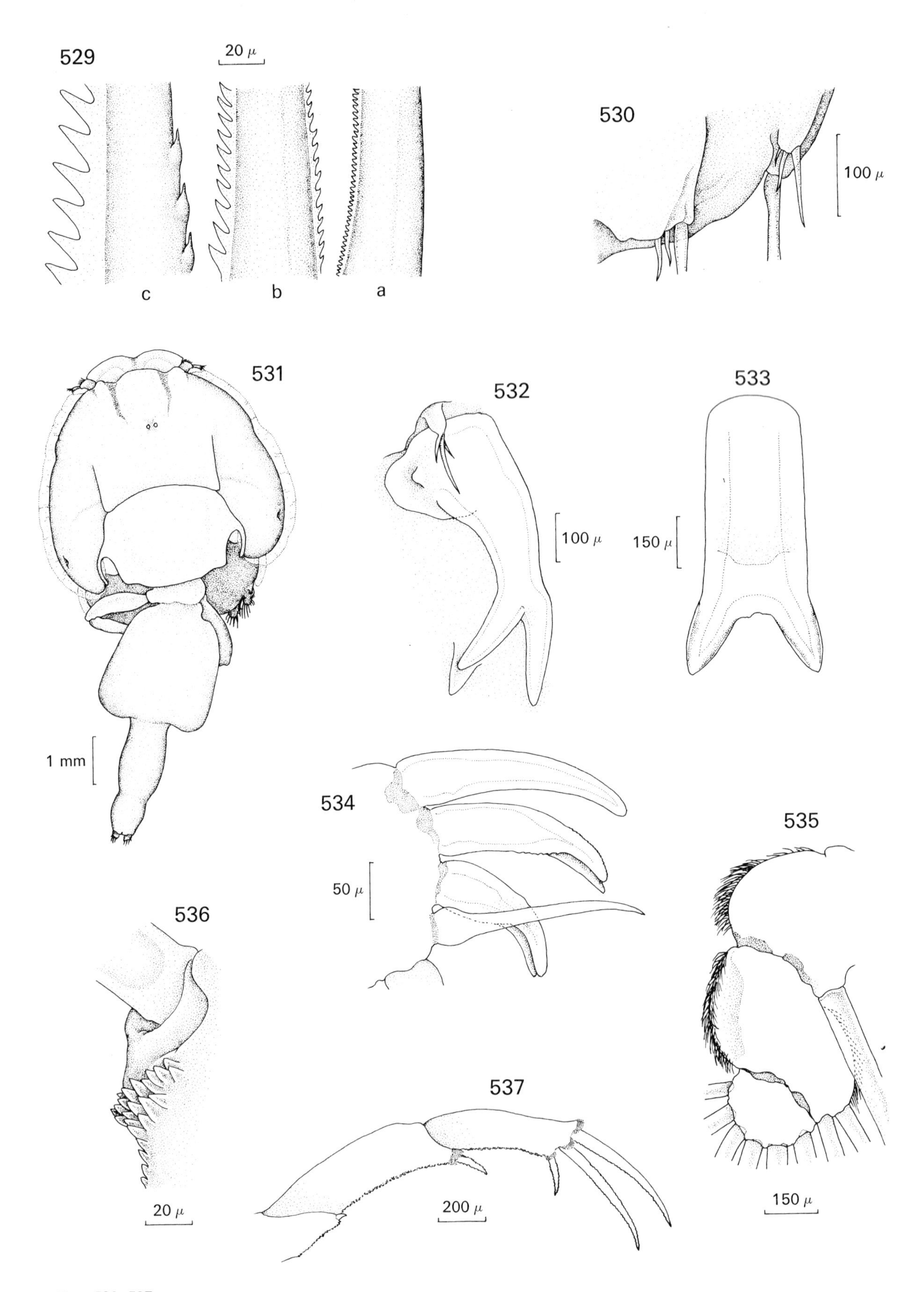

Figs. 529–537.
Caligus coryphaenae Steenstrup and Lütken, 1861. **Fig. 529.** Details of terminal spines of fourth leg (a—first spine; b—second spine; c—third spine). **Fig. 530.** Posterolateral corner of male genital complex, ventral.
Lepeophtheirus sturionis (Krøyer, 1838). **Fig. 531.** Female, dorsal. **Fig. 532.** First maxilla, ventral. **Fig. 533.** Sternal furca, ventral. **Fig. 534.** Tip of first exopod, dorsal. **Fig. 535.** Second endopod, ventral. **Fig. 536.** Fourth leg, base of third spine, dorsal. **Fig. 537.** Distal half of fourth leg, ventral.
Fig. 530 modified from unpublished drawing of Dr. A. V. Gusev, Zoological Institute, Leningrad.

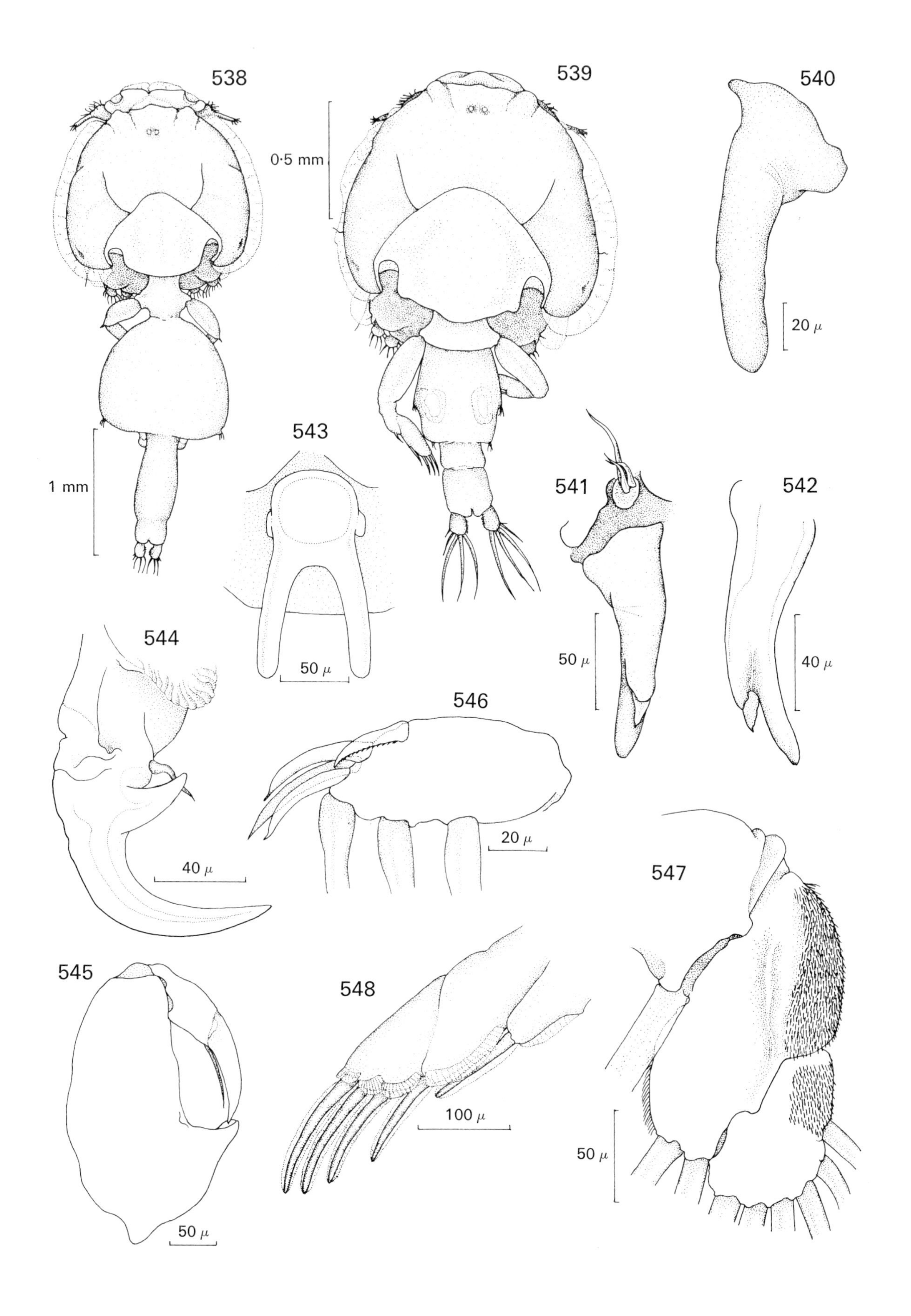

Figs. 538–548.

Caligus diaphanus Nordmann, 1832. **Fig. 538.** Female, dorsal. **Fig. 539.** Male, dorsal. **Fig. 540.** Female, first maxilla, lateral. **Fig. 541.** Male, first maxilla, ventral. **Fig. 542.** Same, lateral. **Fig. 543.** Male and female, sternal furca, ventral. **Fig. 544.** Male, tip of second antenna, ventral. **Fig. 545.** Male, maxilliped, ventral. **Fig. 546.** Male and female, tip of first exopod, ventral. **Fig. 547.** Second endopod, ventral. **Fig. 548.** Tip of fourth leg, ventral.

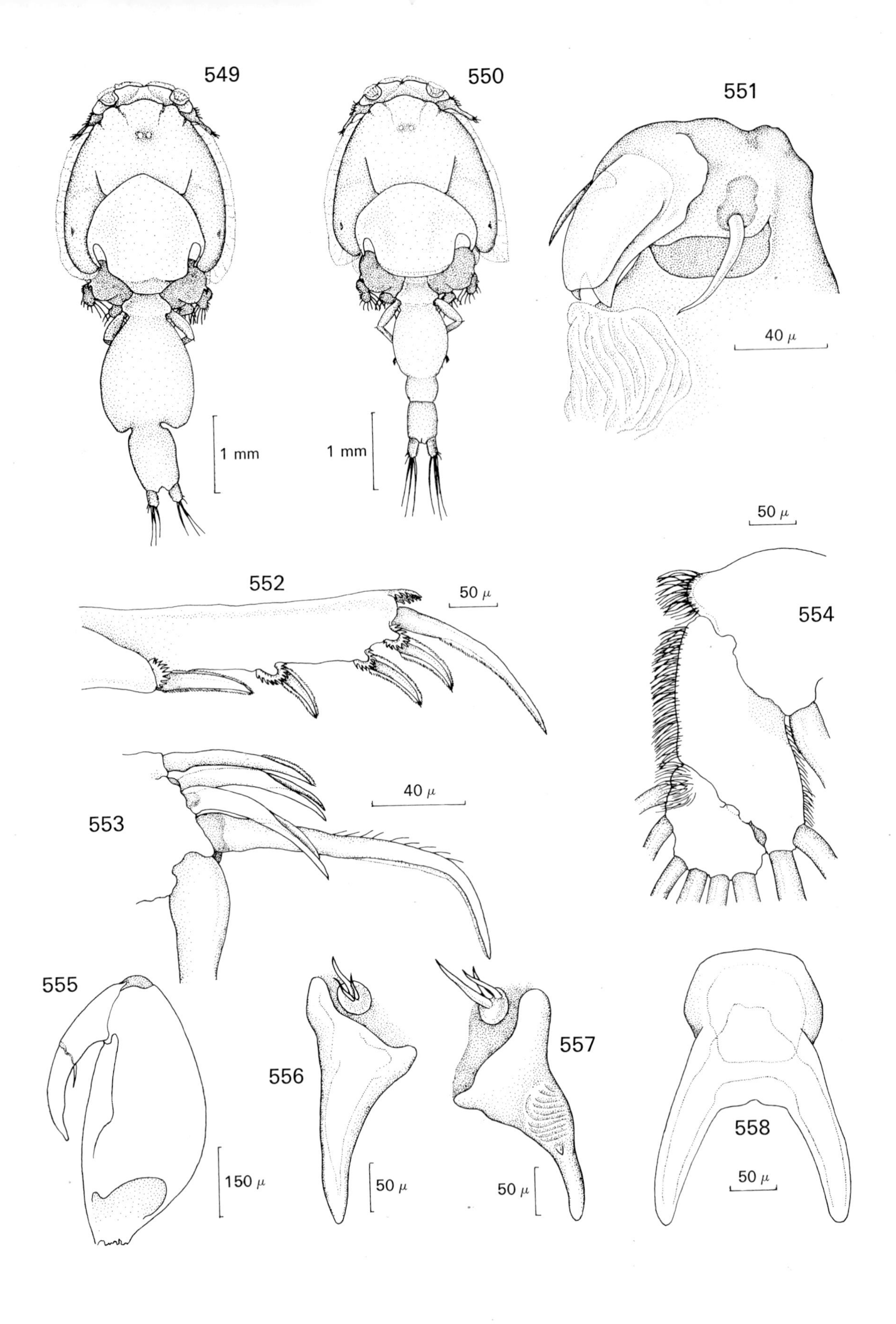

Figs. 549–558.
Caligus elongatus Nordmann, 1832. **Fig. 549.** Female, dorsal. **Fig. 550.** Male, dorsal. **Fig. 551.** Male, tip of second antenna, ventral. **Fig. 552.** Male and female, tip of fourth leg, ventral. **Fig. 553.** Tip of first exopod, ventral. **Fig. 554.** Second endopod, ventral. **Fig. 555.** Maxilliped, dorsal. **Fig. 556.** Female, first maxilla, ventral. **Fig. 557.** Male, first maxilla, ventral. **Fig. 558.** Female, sternal furca, ventral.

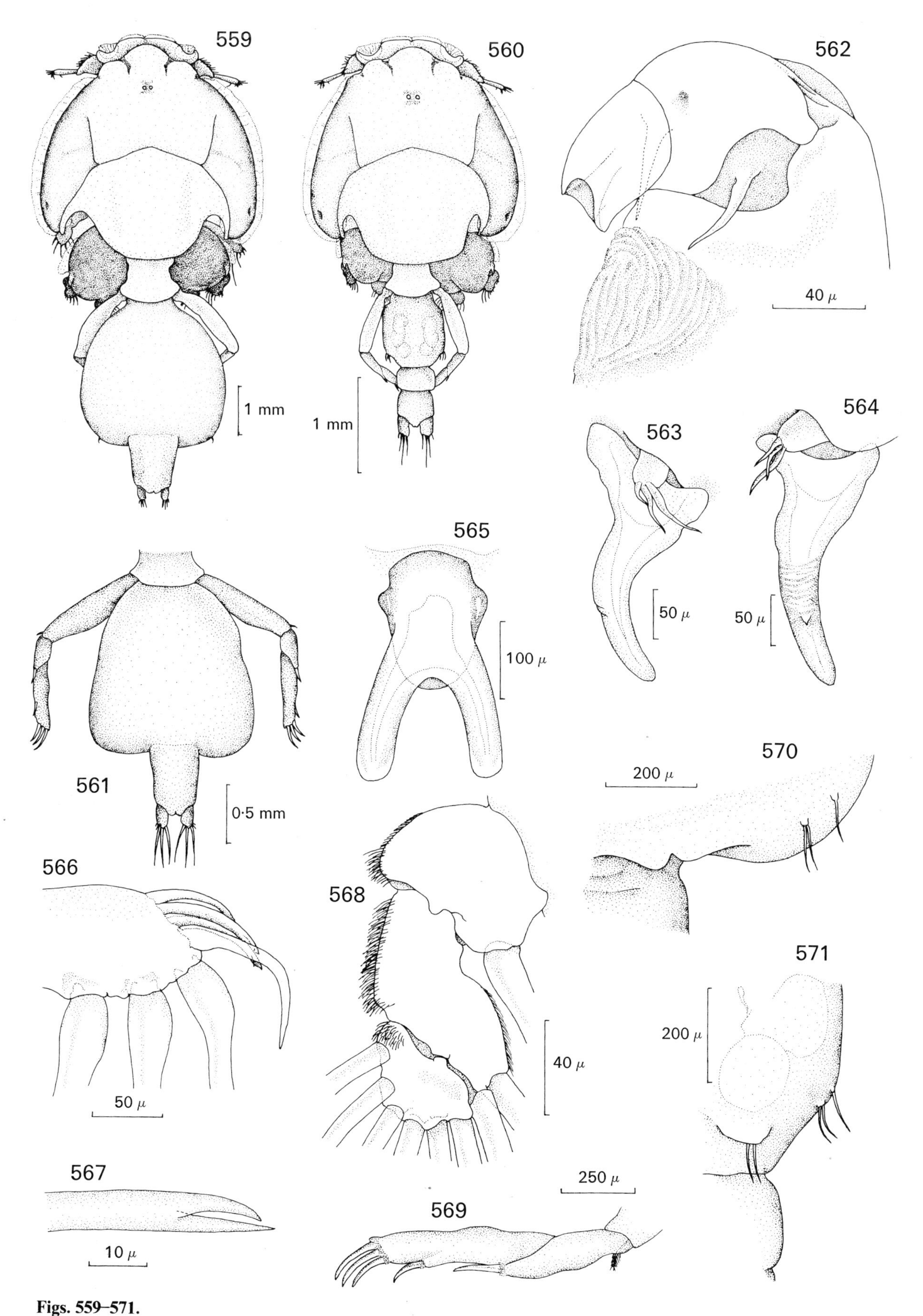

Figs. 559–571.
Caligus gurnardi Krøyer, 1863. **Fig. 559.** Female, dorsal. **Fig. 560.** Male, dorsal. **Fig. 561.** Female, posterior half, dorsal. **Fig. 562.** Male, tip of second antenna, lateral. **Fig. 563.** Female, first maxilla, ventral. **Fig. 564.** Male, first maxilla, ventral. **Fig. 565.** Male and female, sternal furca, ventral. **Fig. 566.** Tip of first exopod, ventral. **Fig. 567.** Same, tip of second spine. **Fig. 568.** Second endopod, ventral. **Fig. 569.** Distal half of fourth leg, ventral. **Fig. 570.** Posterolateral corner of female genital complex, ventral. **Fig. 571.** Posterolateral corner of male genital complex, ventral.

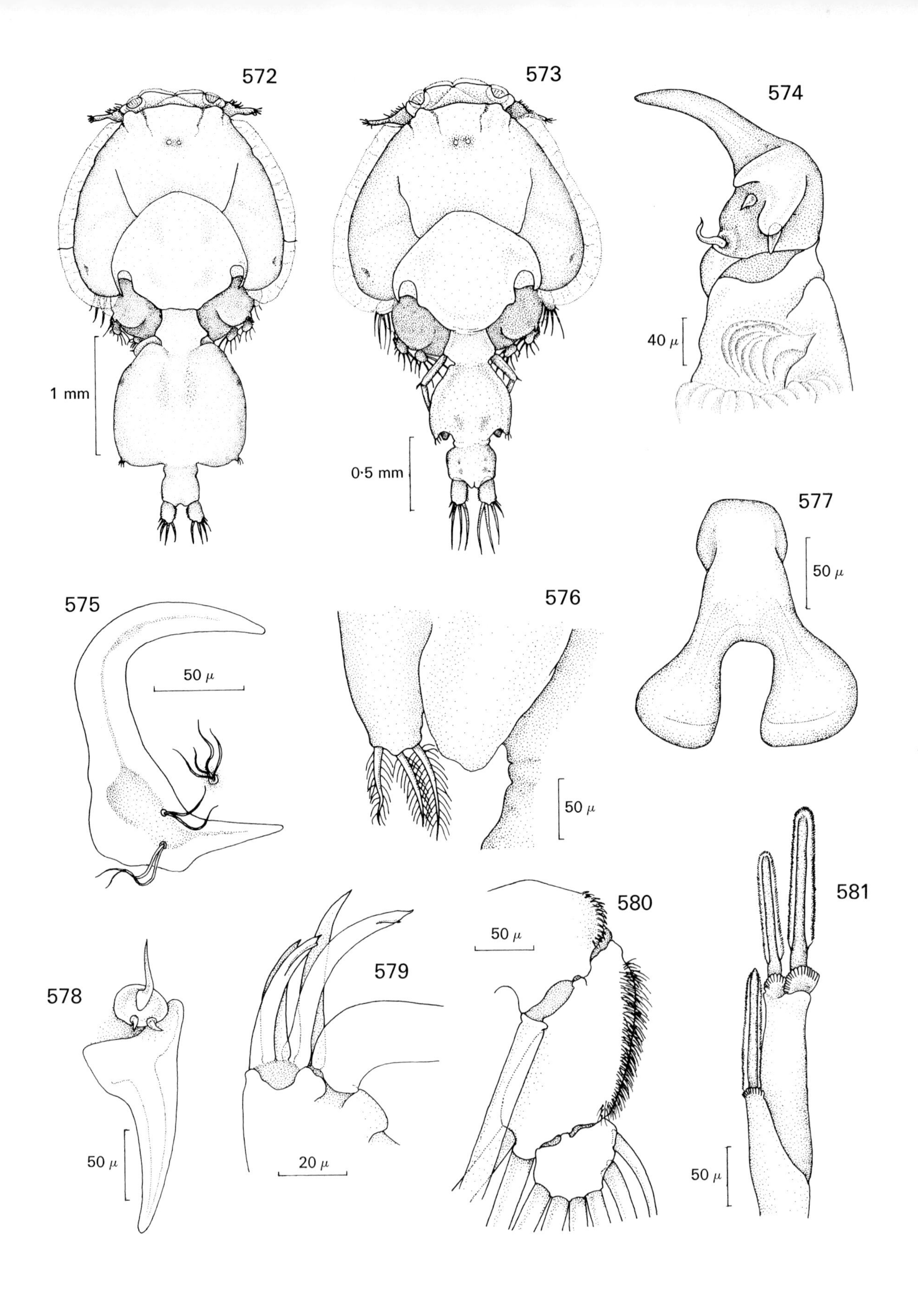

Figs. 572–581.
Caligus labracis T. Scott, 1902. **Figs. 572.** Female, dorsal. **Fig. 573.** Male, dorsal. **Fig. 574.** Male, tip of second antenna, ventral. **Fig. 575.** Male and female, postantennary process, ventral. **Fig. 576.** Posterolateral corner of male genital complex, ventral. **Fig. 577.** Male and female, sternal furca, ventral. **Fig. 578.** First maxilla, ventral. **Fig. 579.** Tip of first exopod, ventral. **Fig. 580.** Second endopod, ventral. **Fig. 581.** Tip of fourth leg, ventral.

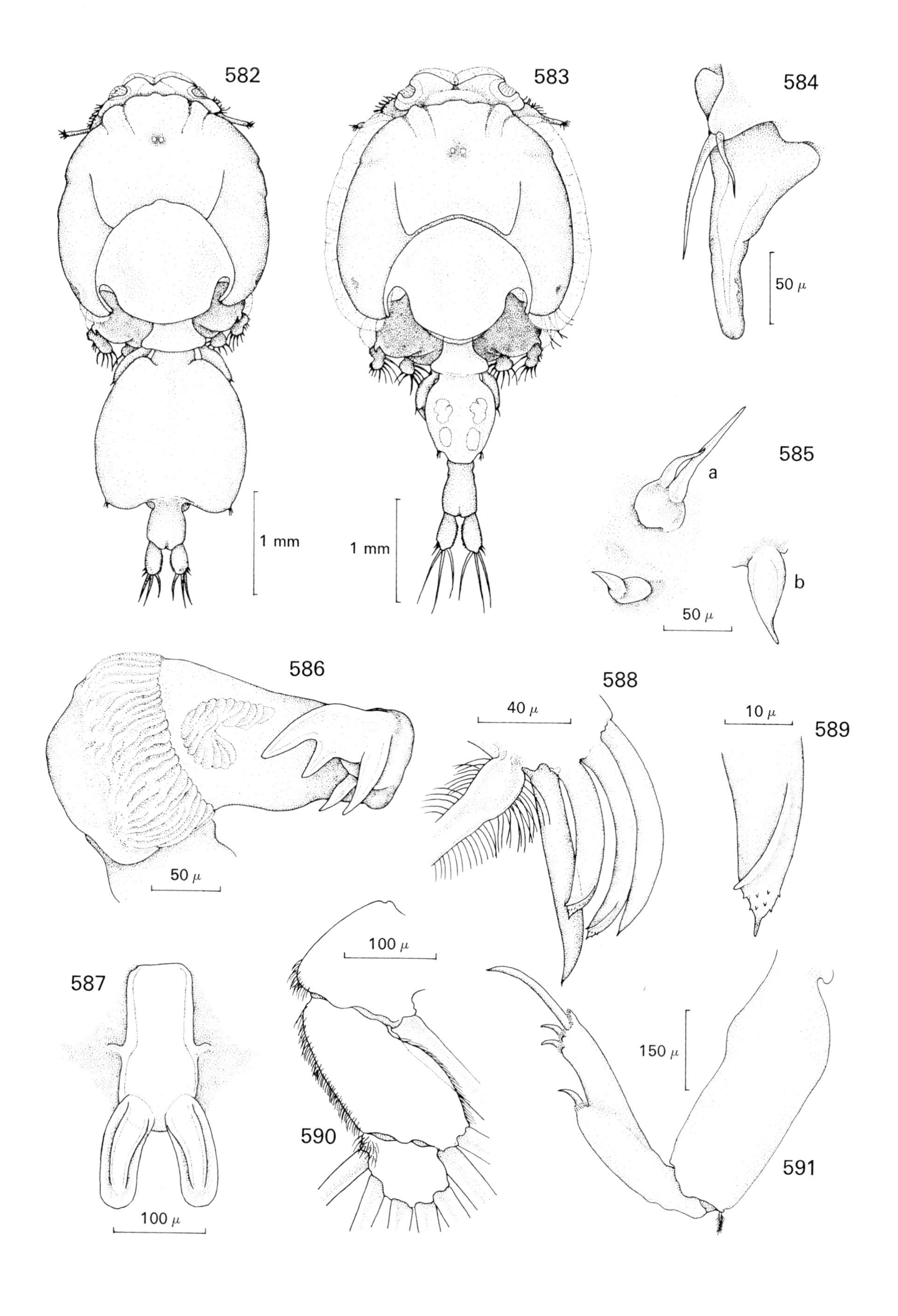

Figs. 582–591.
Caligus minimus Otto, 1821. **Fig. 582.** Female, dorsal. **Fig. 583.** Male, dorsal. **Fig. 584.** Female, first maxilla, ventral. **Fig. 585.** Male, first maxilla: (a) *in situ*, (b) lateral (without papilla). **Fig. 586.** Male, second antenna, distal part, ventral. **Fig. 587.** Male and female, sternal furca, ventral. **Fig. 588.** Tip of first exopod, ventral. **Fig. 589.** Same, tip of third spine, enlarged. **Fig. 590.** Second endopod, ventral. **Fig. 591.** Fourth leg, ventral.

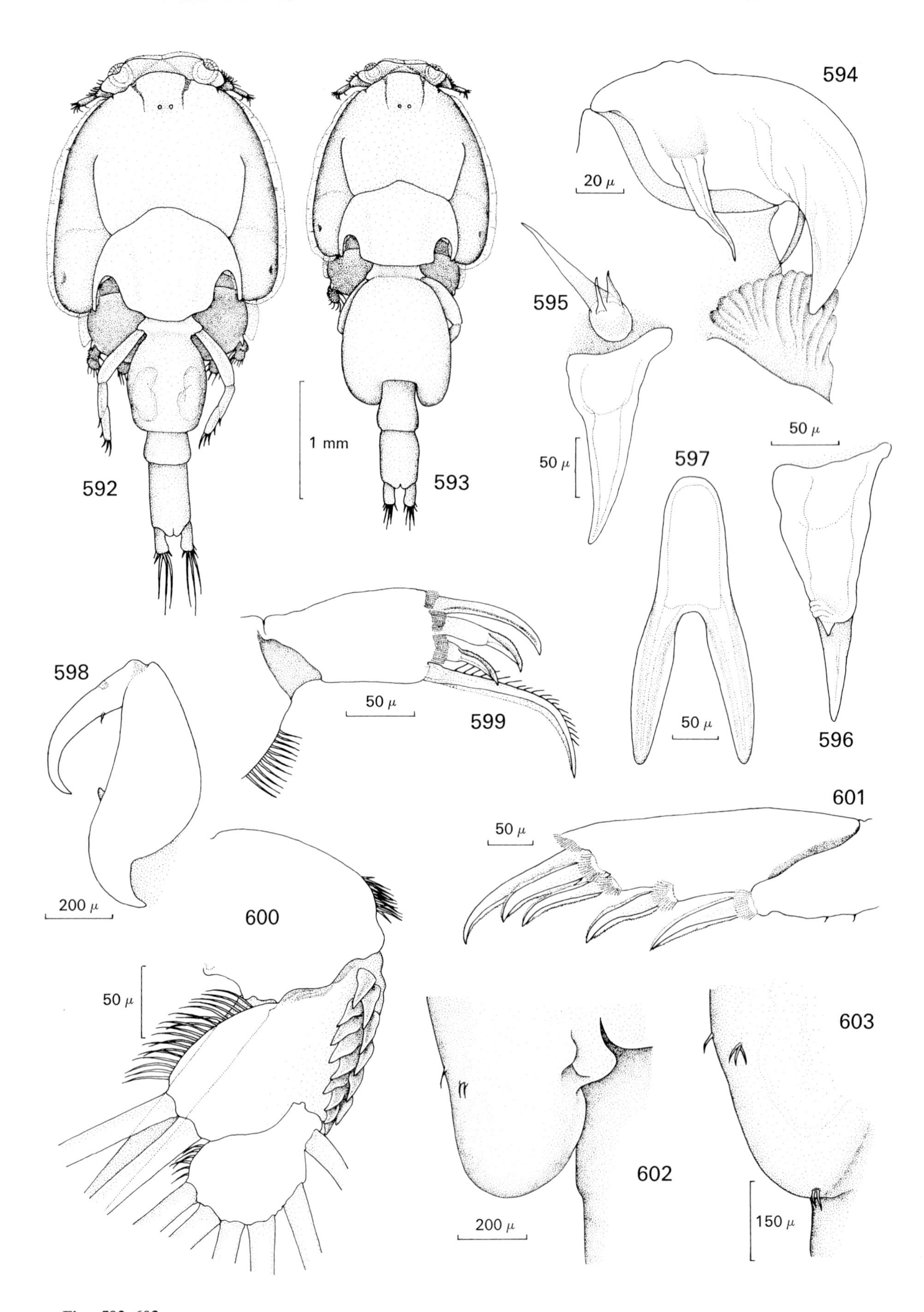

Figs. 592–603.
Caligus productus Dana, 1852. **Fig. 592.** Male, dorsal. **Fig. 593.** Female, dorsal. **Fig. 594.** Male, tip of second antenna, ventral. **Fig. 595.** Female, first maxilla, ventral. **Fig. 596.** Male, first antenna, ventral. **Fig. 597.** Male and female, sternal furca, ventral. **Fig. 598.** Maxilliped, ventral. **Fig. 599.** Tip of first exopod, ventral. **Fig. 600.** Second endopod, dorsal. **Fig. 601.** Tip of fourth leg, ventral. **Fig. 602.** Posterolateral corner of female genital complex, ventral. **Fig. 603.** Posterolateral corner of male genital complex, ventral.

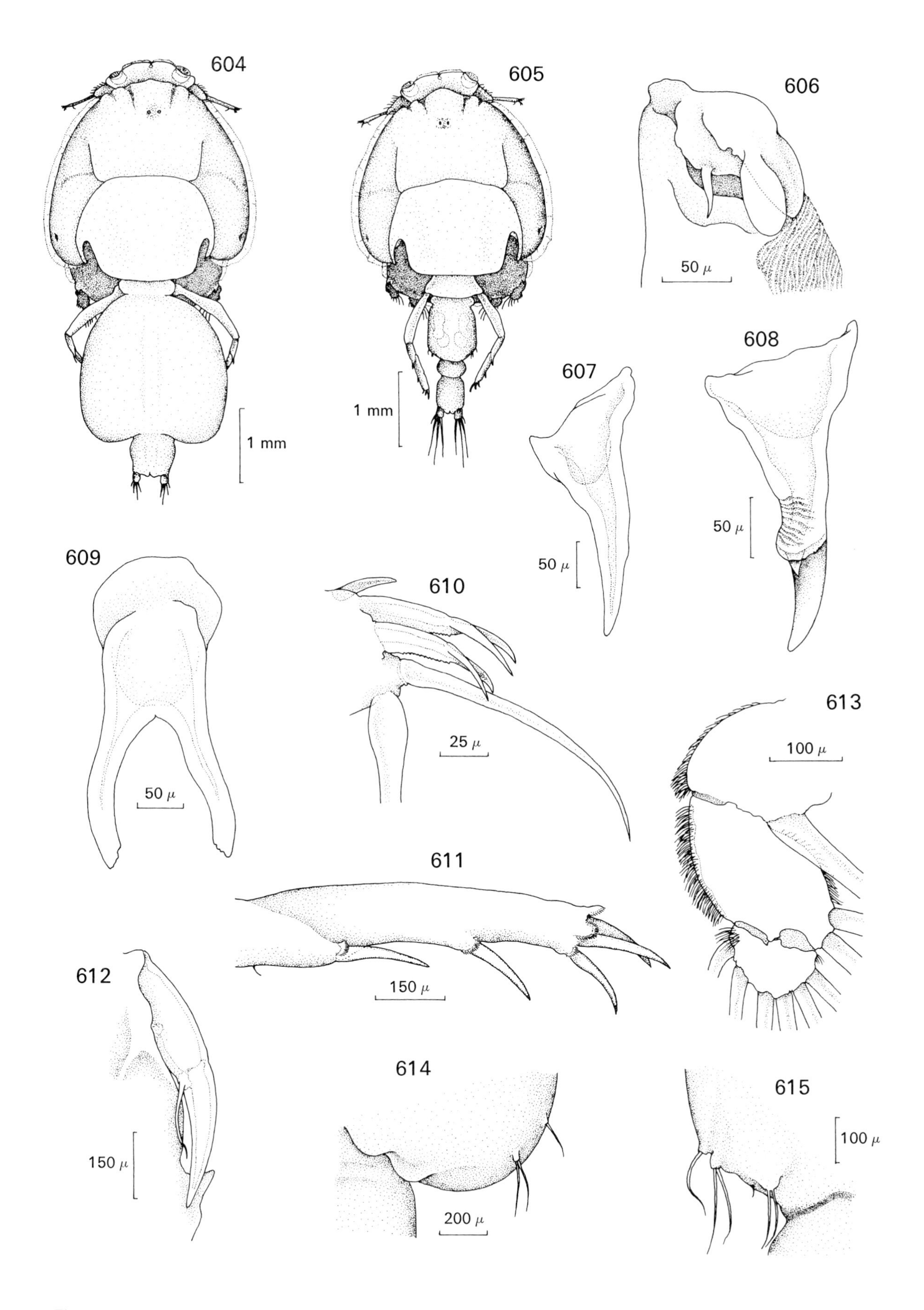

Figs. 604–615.
Caligus zei Norman and T. Scott, 1906. **Fig. 604.** Female, dorsal. **Fig. 605.** Male, dorsal. **Fig. 606.** Male, tip of second antenna, ventral. **Fig. 607.** Female, first maxilla, ventral. **Fig. 608.** Male, first maxilla, ventral. **Fig. 609.** Male and female, sternal furca, ventral. **Fig. 610.** Tip of first exopod, ventral. **Fig. 611.** Distal half of fourth leg, ventral. **Fig. 612.** Subchela of maxilliped, dorsal. **Fig. 613.** Second endopod, ventral. **Fig. 614.** Posterolateral corner of female genital complex, ventral. **Fig. 615.** Posterolateral corner of male genital complex, ventral.

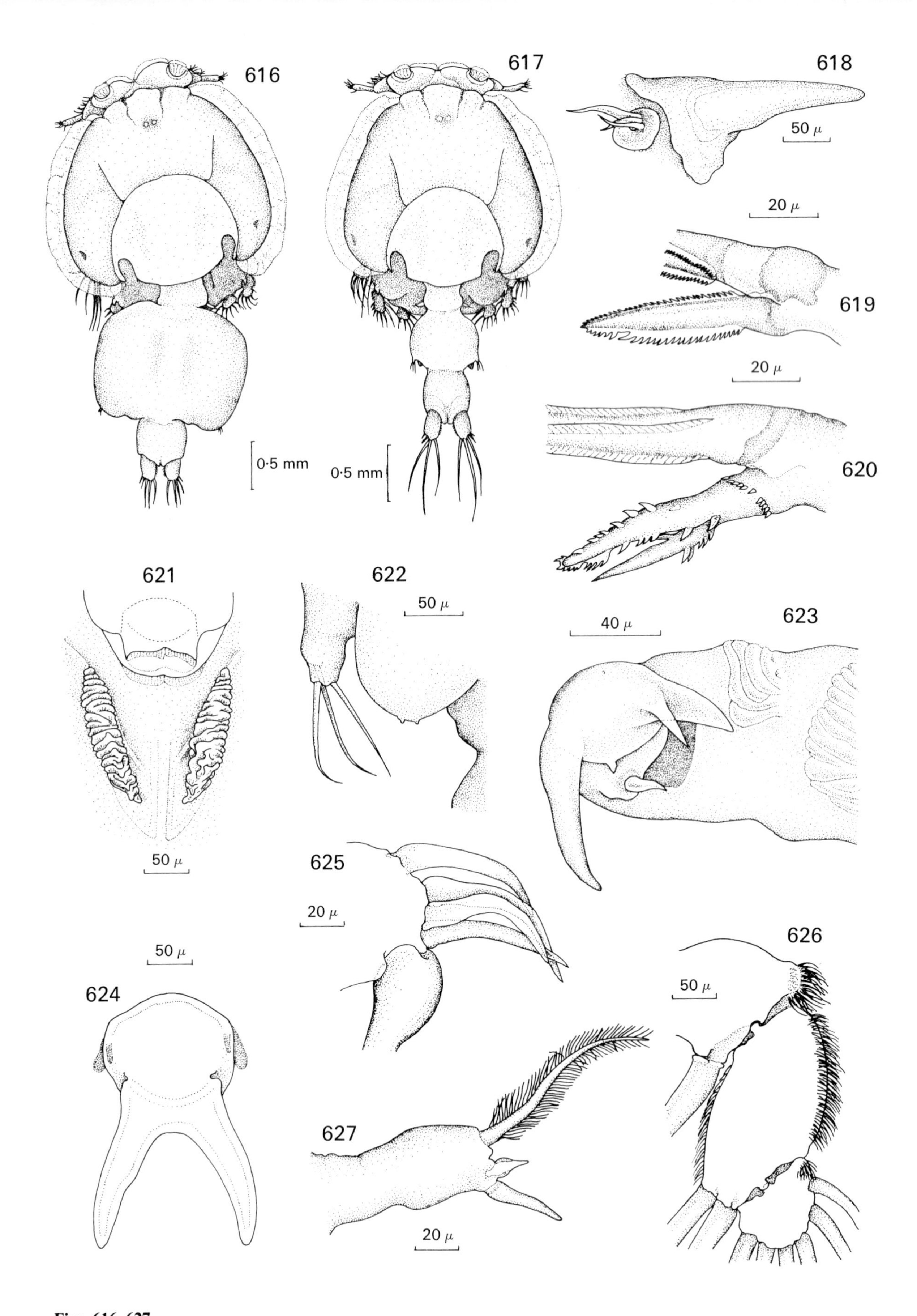

Figs. 616–627.
Pseudocaligus brevipedis (Bassett-Smith, 1896). **Fig. 616.** Female, dorsal. **Fig. 617.** Male, dorsal. **Fig. 618.** Male and female, first maxilla, ventral. **Fig. 619.** Male, second maxilla, tip of brachium. **Fig. 620.** Female, same. **Fig. 621.** Male and female, postoral adhesive pads. **Fig. 622.** Posterolateral corner of male genital complex, ventral. **Fig. 623.** Tip of second antenna, ventral. **Fig. 624.** Male and female, sternal furca, ventral. **Fig. 625.** Tip of first exopod, ventral. **Fig. 626.** Second endopod, ventral. **Fig. 627.** Fourth leg.

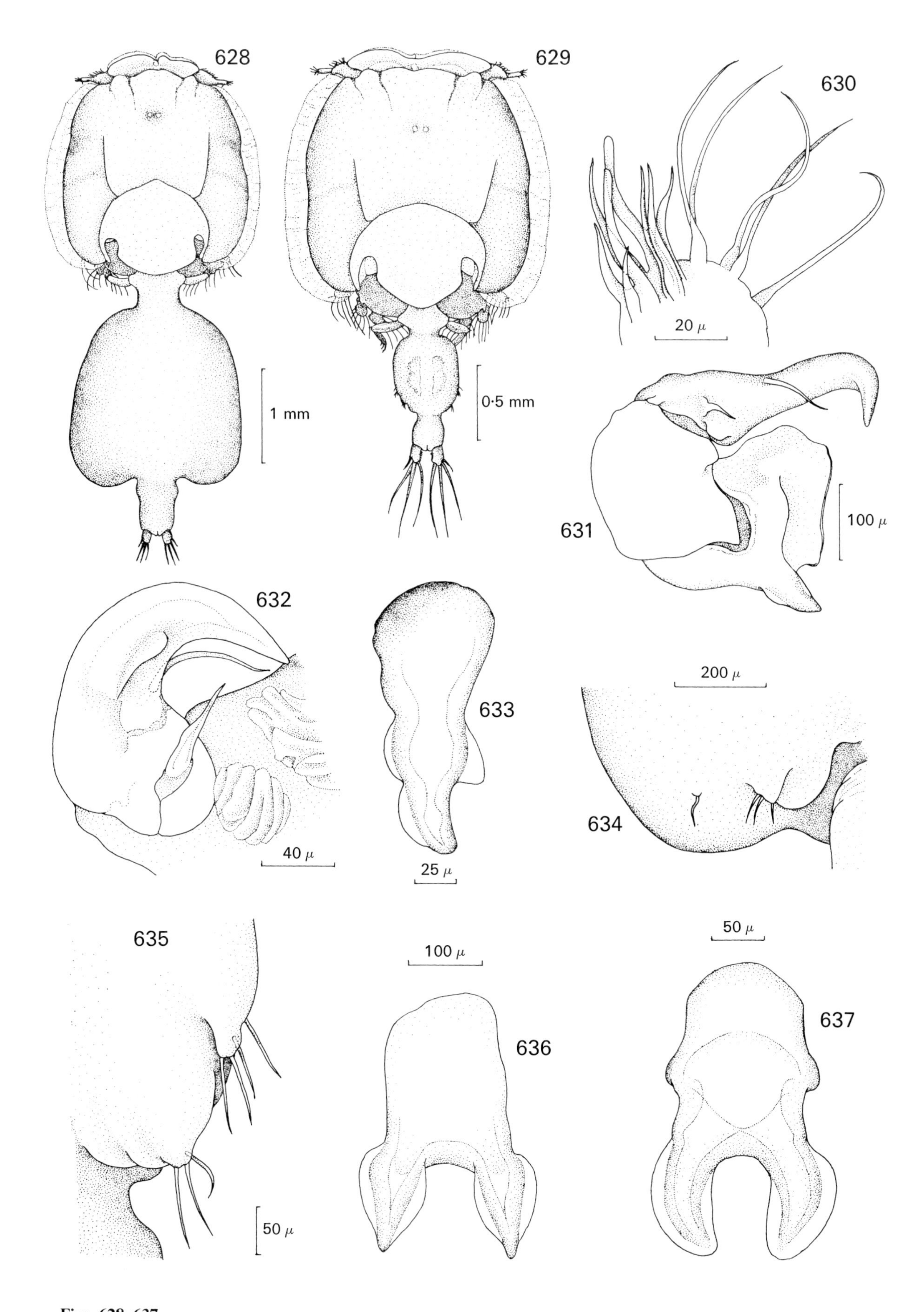

Figs. 628–637.
Lepeophtheirus pectoralis (Müller, 1776). **Fig. 628.** Female, dorsal. **Fig. 629.** Male, dorsal. **Fig. 630.** Male and female, tip of first antenna, ventral. **Fig. 631.** Female, second antenna, ventral. **Fig. 632.** Male, tip of second antenna, ventral. **Fig. 633.** Same, distal surface of claw. **Fig. 634.** Posterolateral corner of female genital complex, ventral. **Fig. 635.** Posterolateral corner of male genital complex, ventral. **Fig. 636.** Female sternal furca, ventral. **Fig. 637.** Male sternal furca, ventral.

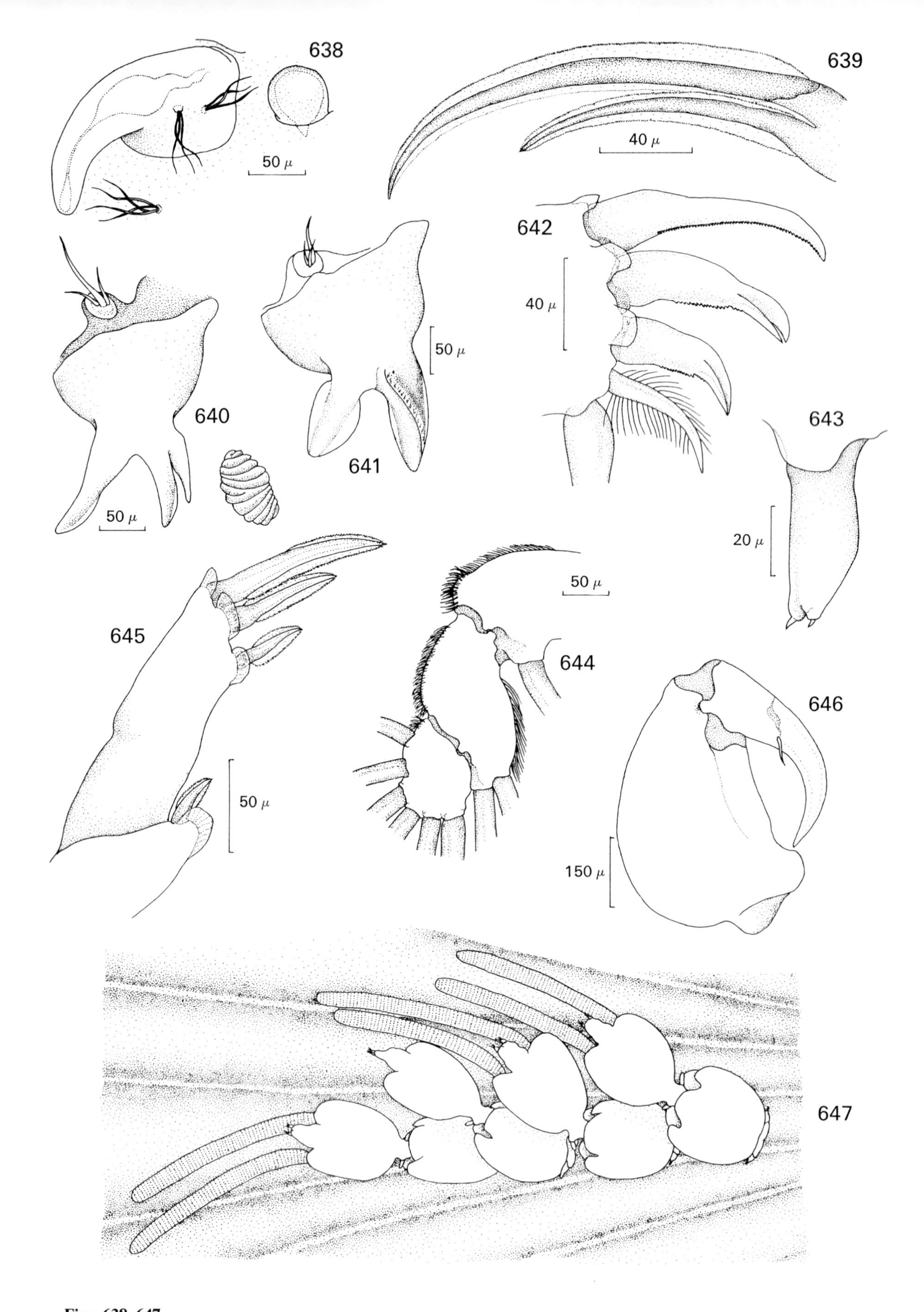

Figs. 638–647.
Lepeophtheirus pectoralis (Müller, 1776). **Fig. 638.** Male and female, postantennary process, ventral. **Fig. 639.** Tip of second maxilla, ventral. **Fig. 640.** Male first maxilla and adhesion pad, ventral. **Fig. 641.** Female first maxilla, ventral. **Fig. 642.** Male and female, tip of first exopod, ventral. **Fig. 643.** First endopod, dorsal. **Fig. 644.** Second endopod, ventral. **Fig. 645.** Tip of fourth leg, ventral. **Fig. 646.** Female maxilliped, ventral. **Fig. 647.** Four females attached to host's fin (semidiagramatic).

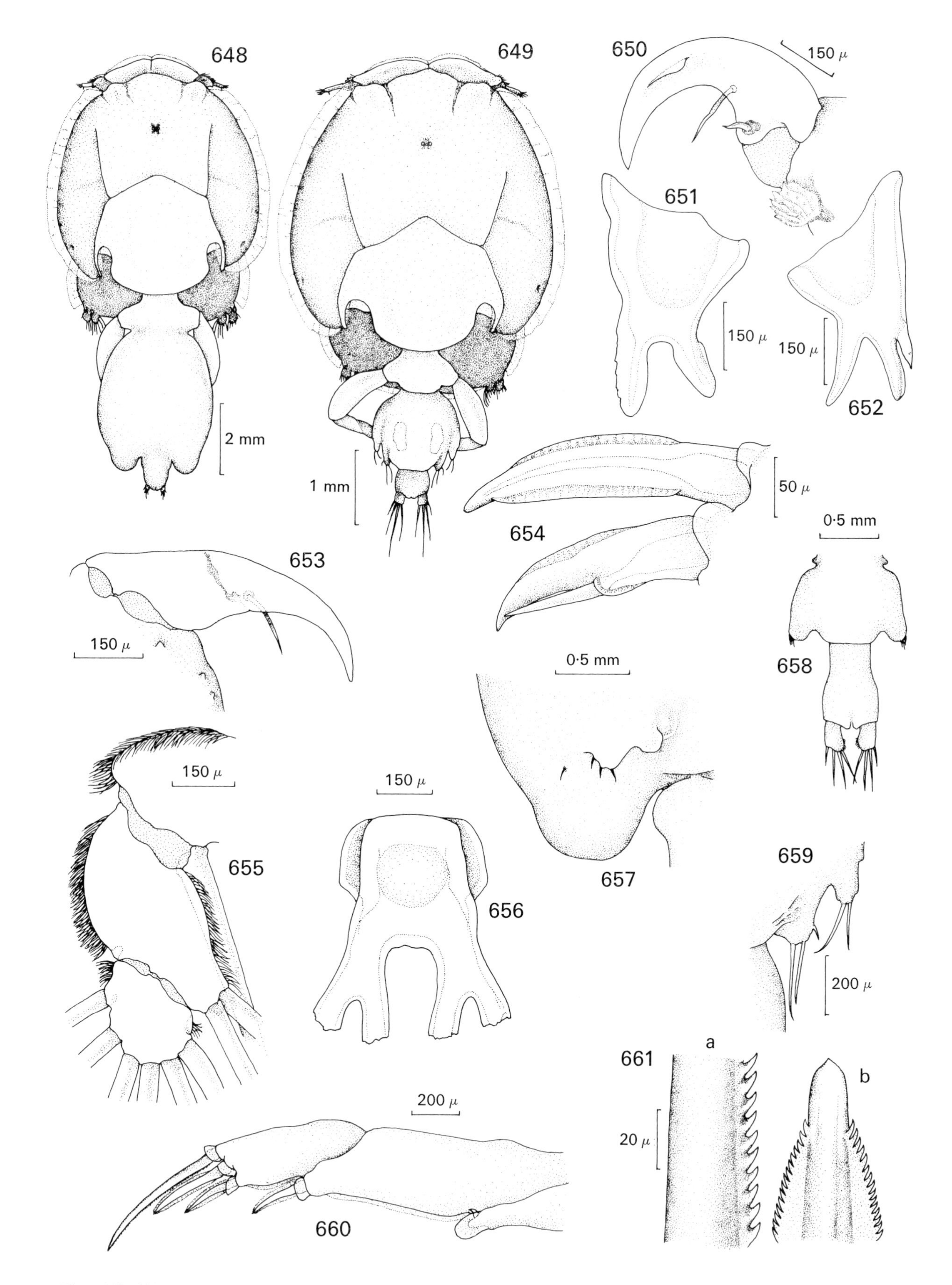

Figs. 648–661.

Lepeophtheirus hippoglossi (Krøyer, 1837). **Fig. 648.** Female, dorsal. **Fig. 649.** Male dorsal. **Fig. 650.** Tip of male second antenna. **Fig. 651.** Female first maxilla, ventral. **Fig. 652.** Male first maxilla, ventral. **Fig. 653.** Subchela of male maxilliped. **Fig. 654.** Male and female, first and second spines of first exopod. **Fig. 655.** Second endopod, dorsal. **Fig. 656.** Sternal furca, ventral. **Fig. 657.** Posterolateral corner of female genital complex, ventral. **Fig. 658.** Genital complex and abdomen of juvenile female. **Fig. 659.** Posterolateral corner of male genital complex, ventral. **Fig. 660.** Male and female, tip of fourth leg, ventral. **Fig. 661.** Details of terminal spines, fourth leg, magnified.

a—first spine of leg 4; b—second spine of leg 4.

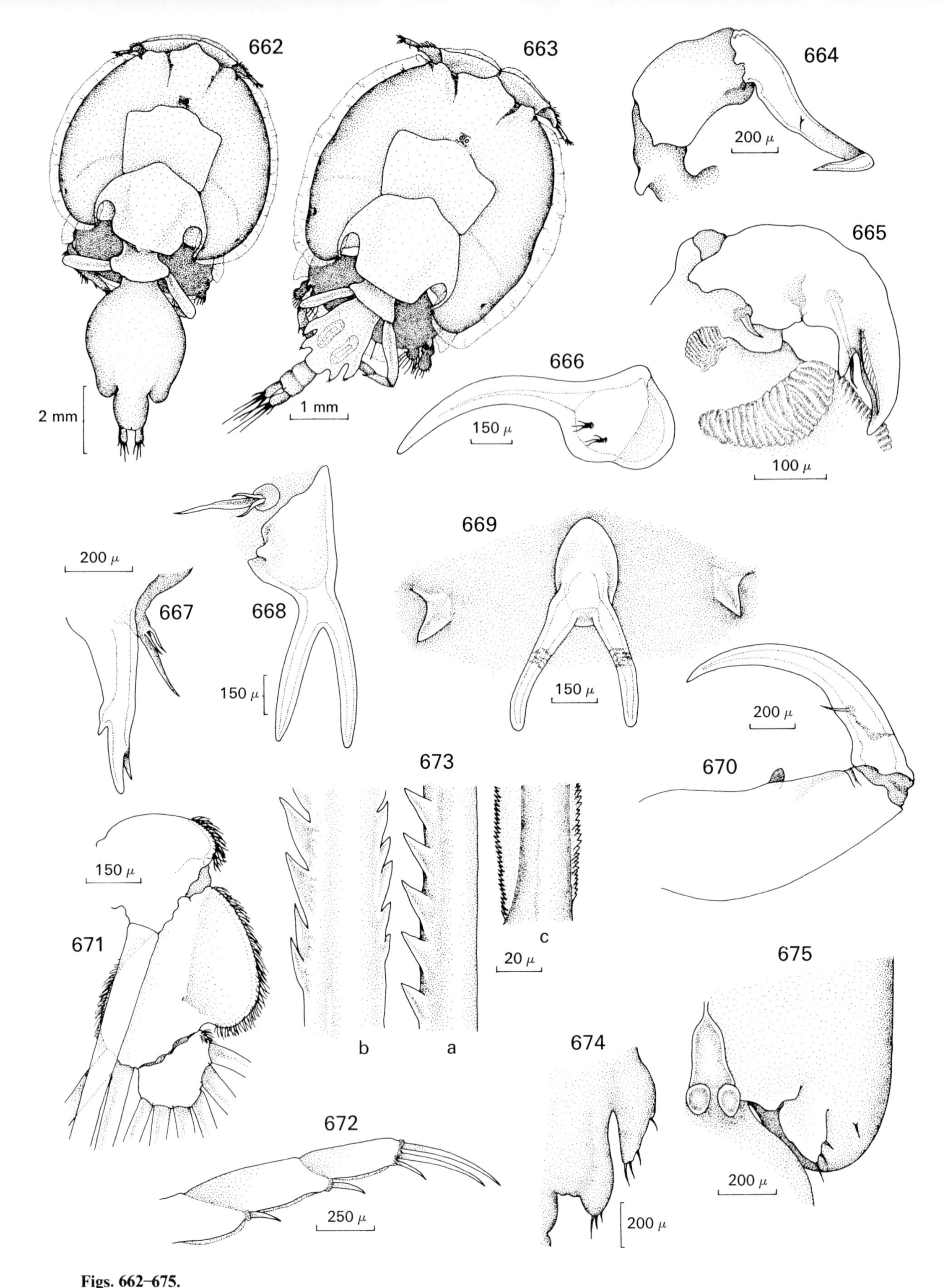

Figs. 662–675.
Lepeophtheirus nordmanni (Edwards, 1840). **Fig. 662.** Female, dorsal. **Fig. 663.** Male, dorsal. **Fig. 664.** Female second antenna, ventral. **Fig. 665.** Tip of male second antenna. **Fig. 666.** Male and female, postantennary process, ventral. **Fig. 667.** Male first maxilla, medial. **Fig. 668.** Female first maxilla, ventral. **Fig. 669.** Male and female, sternal furca, ventral. **Fig. 670.** Male maxilliped, ventral. **Fig. 671.** Male and female, second endopod, ventral. **Fig. 672.** Distal half of fourth leg, ventral. **Fig. 673.** Same, details of terminal armature: a—first spine; b—second spine; c—third spine. **Fig. 674.** Posterolateral corner of male genital complex, ventral. **Fig. 675.** Posterolateral corner of female genital complex, ventral.

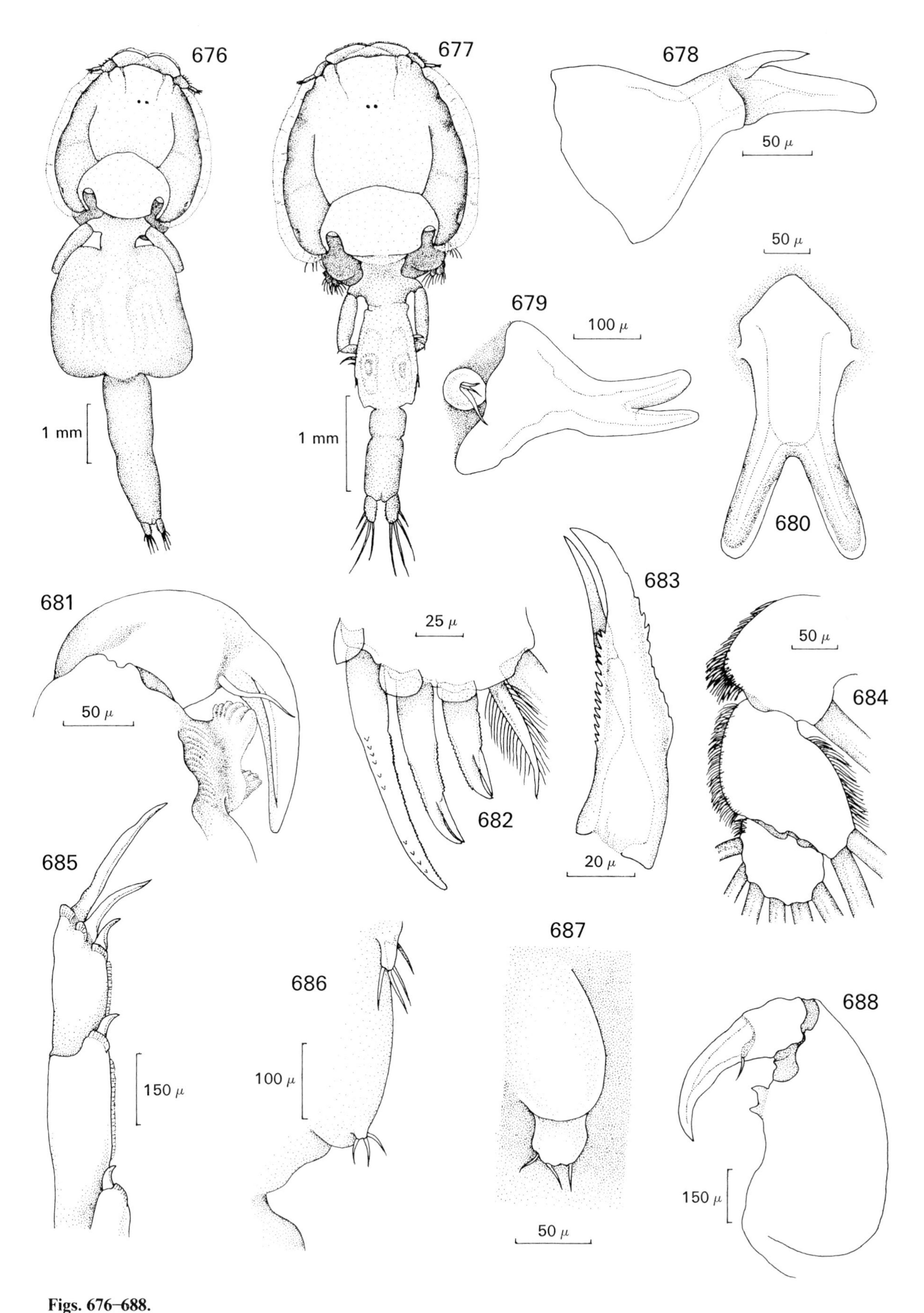

Figs. 676–688.
Lepeophtheirus pollachius Bassett-Smith, 1896. **Fig. 676.** Female, dorsal. **Fig. 677.** Male, dorsal. **Fig. 678.** Male first maxilla, ventral. **Fig. 679.** Female first maxilla, ventral. **Fig. 680.** Male and female, sternal furca, ventral. **Fig. 681.** Tip of male second antenna, posterodorsal. **Fig. 682.** Male and female, tip of first exopod, ventral. **Fig. 683.** Same, second spine, enlarged. **Fig. 684.** Second endopod, ventral. **Fig. 685.** Distal half of fourth leg, ventral. **Fig. 686.** Posterolateral corner of male genital complex, ventral. **Fig. 687.** Female fifth leg, ventral. **Fig. 688.** Male maxilliped, ventral.

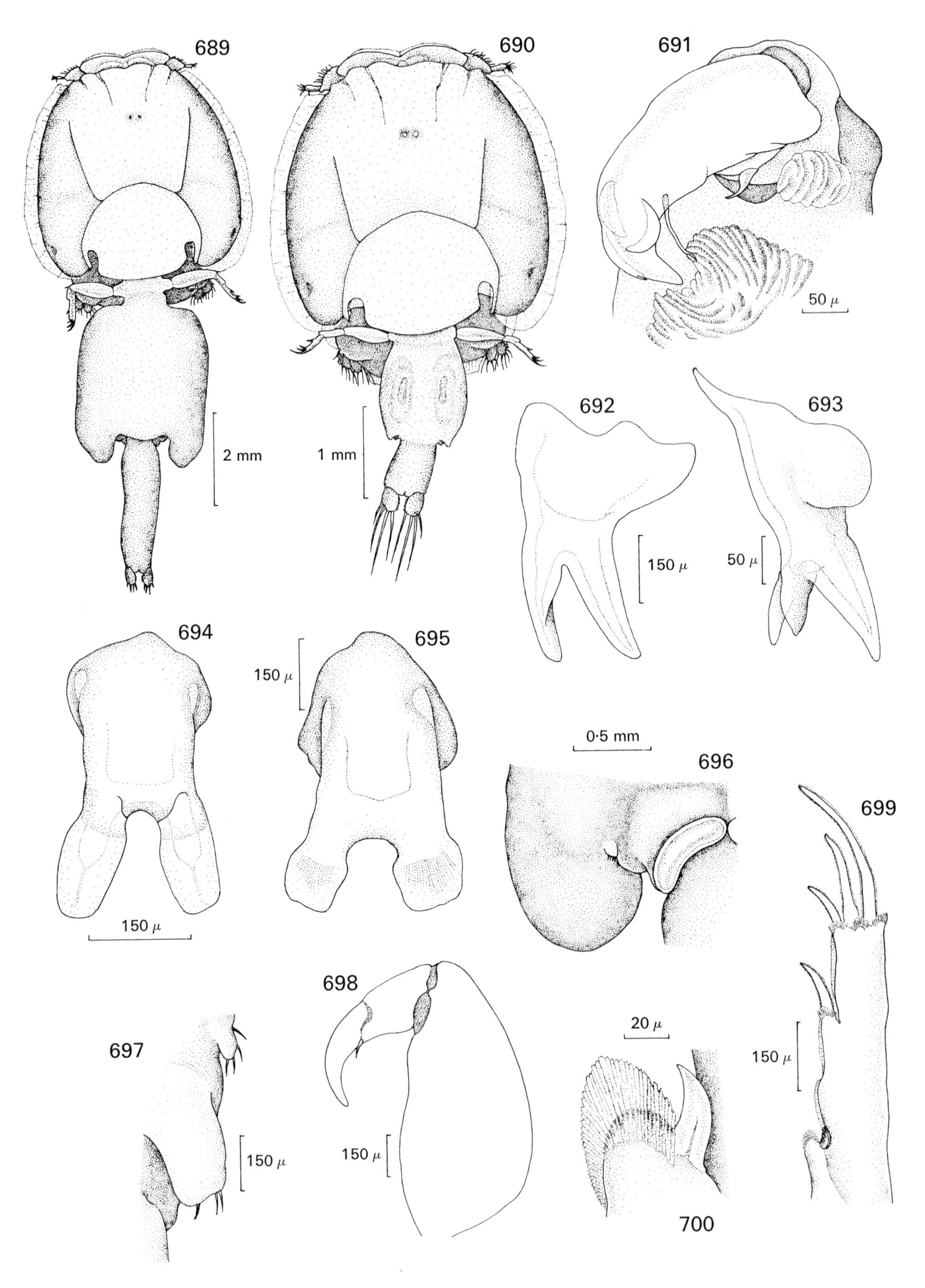

Figs. 689–700.
Lepeophtheirus salmonis (Krøyer, 1838). **Fig. 689.** Female, dorsal. **Fig. 690.** Male, dorsal. **Fig. 691.** Tip of male first antenna. **Fig. 692.** Female first maxilla, ventral. **Fig. 693.** Male first maxilla, lateral. **Fig. 694.** Male sternal furca, ventral. **Fig. 695.** Female sternal furca, ventral. **Fig. 696.** Posterolateral corner of female genital complex, ventral. **Fig. 697.** Posterolateral corner of male genital complex, ventral. **Fig. 698.** Male and female, maxilliped, ventral. **Fig. 699.** Distal half of fourth leg, ventral. **Fig. 700.** Fourth leg, claw of second segment, ventral.

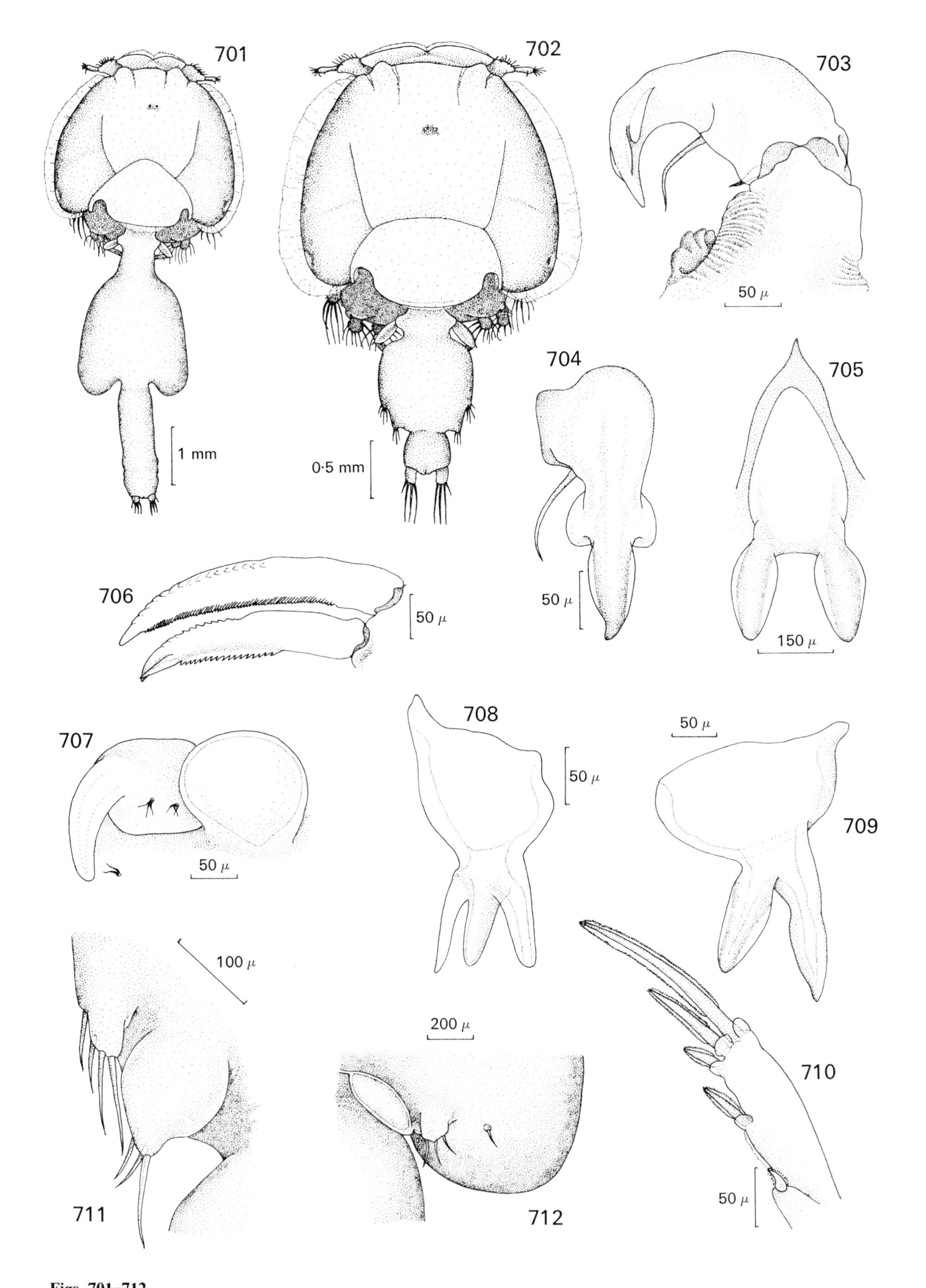

Figs. 701–712.

Lepeophtheirus thompsoni Baird, 1850. **Fig. 701.** Female, dorsal. **Fig. 702.** Male, dorsal. **Fig. 703.** Tip of male second antenna, dorsal. **Fig. 704.** Same, distal surface of claw. **Fig. 705.** Male and female, sternal furca, ventral. **Fig. 706.** First two spines of first exopod, ventral. **Fig. 707.** Postantennary process, ventral. **Fig. 708.** Male first maxilla, ventral. **Fig. 709.** Female first maxilla ventral. **Fig. 710.** Tip of fourth leg, ventral. **Fig. 711.** Posterolateral corner of male genital complex, ventral. **Fig. 712.** Posterolateral corner of female genital complex, ventral.

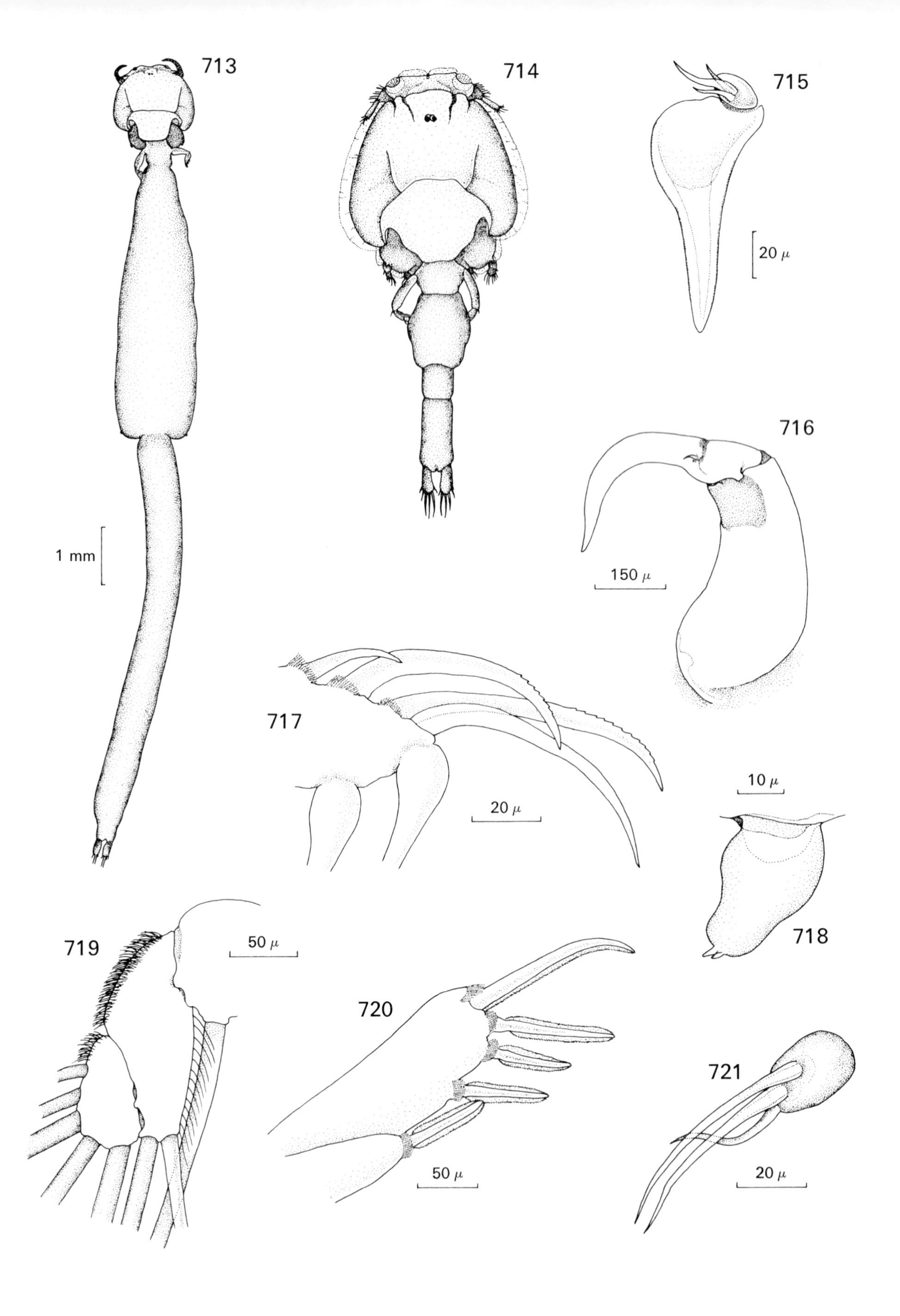

Figs. 713–721.
Sciaenophilus tenuis van Beneden, 1852. **Fig. 713.** Female, dorsal. **Fig. 714.** Male, dorsal. **Fig. 715.** Female first maxilla, ventral. **Fig. 716.** Maxilliped, ventral. **Fig. 717.** Tip of first exopod, ventral. **Fig. 718.** First endopod, ventral. **Fig. 719.** Second endopod, ventral. **Fig. 720.** Tip of fourth leg, ventral. **Fig. 721.** Fifth leg, ventral.
Fig. 714 redrawn from Pillai (1961).

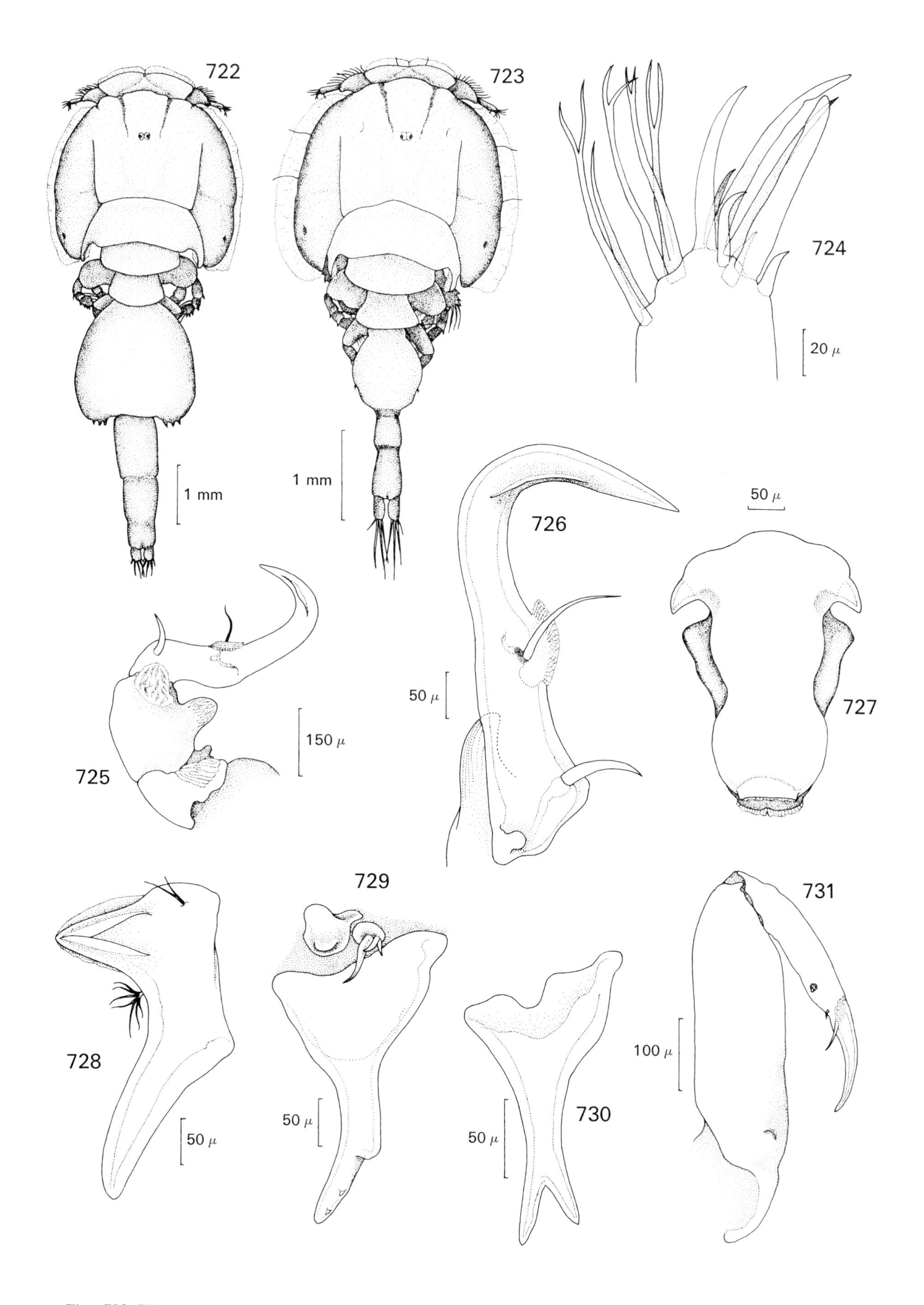

Figs. 722–731.
Trebius caudatus Krøyer, 1838. **Fig. 722.** Female, dorsal. **Fig. 723.** Male, dorsal. **Fig. 724.** Male and female, tip of antenna. **Fig. 725.** Male second antenna. **Fig. 726.** Same, claw, ventral. **Fig. 727.** Male and female, mouth, anterior. **Fig. 728.** Male postantennary process, lateral. **Fig. 729.** Male first maxilla, ventral. **Fig. 730.** Female first maxilla, ventral. **Fig. 731.** Male and female, maxilliped, ventral.

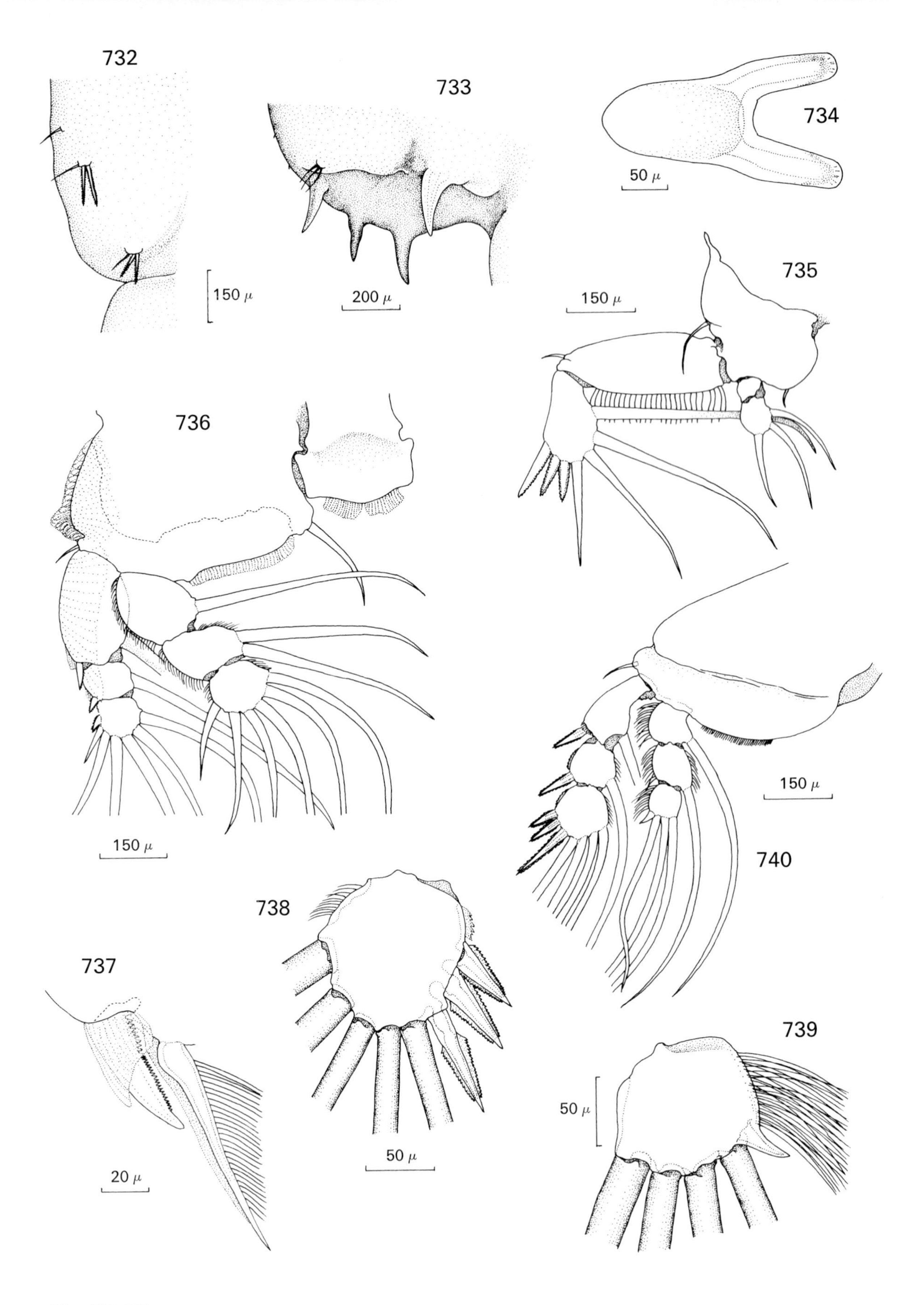

Figs. 732–740.
Trebius caudatus Krøyer, 1838. **Fig. 732.** Posterolateral corner of male genital complex, ventral. **Fig. 733.** Posterolateral corner of female genital complex, ventral. **Fig. 734.** Male sternal furca, ventral. **Fig. 735.** Male and female, first leg, ventral. **Fig. 736.** Second leg, ventral. **Fig. 737.** Same, third segment of exopod, detail of lateral margin. **Fig. 738.** Third leg, distal segment of exopod, ventral. **Fig. 739.** Same, distal segment of endopod, ventral. **Fig. 740.** Fourth leg, ventral.

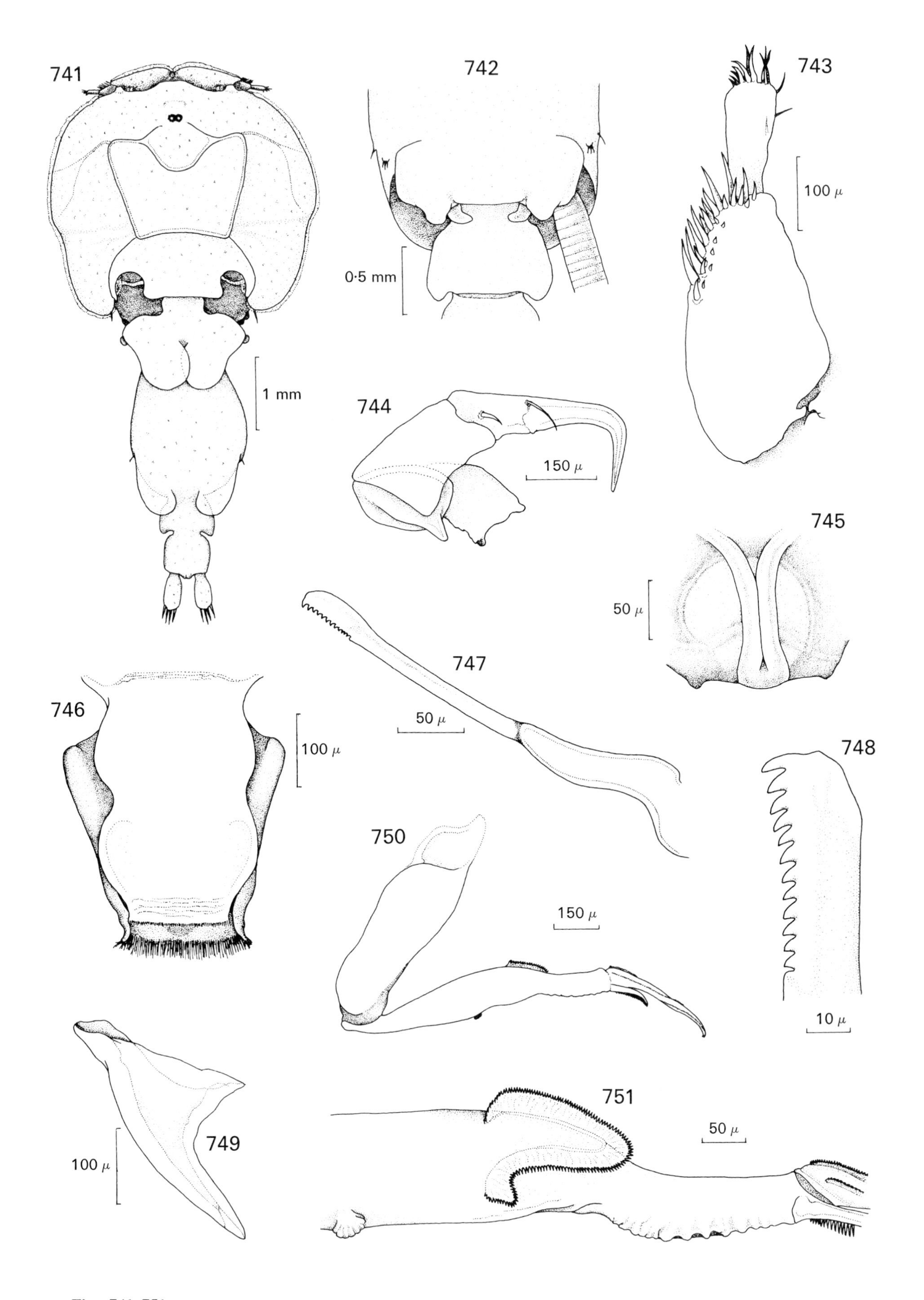

Figs. 741–751.
Euryphorus brachypterus (Gerstaecker, 1853). **Fig. 741.** Female, dorsal. **Fig. 742.** Genital region, ventral. **Fig. 743.** First antenna, ventral. **Fig. 744.** Second antenna, ventral. **Fig. 745.** Detail of frontal plate, dorsal. **Fig. 746.** Mouth, anterior. **Fig. 747.** Mandible, lateral. **Fig. 748.** Same, tip. **Fig. 749.** First maxilla, ventral. **Fig. 750.** Second maxilla, ventral. **Fig. 751.** Same, part of brachium, dorsal.

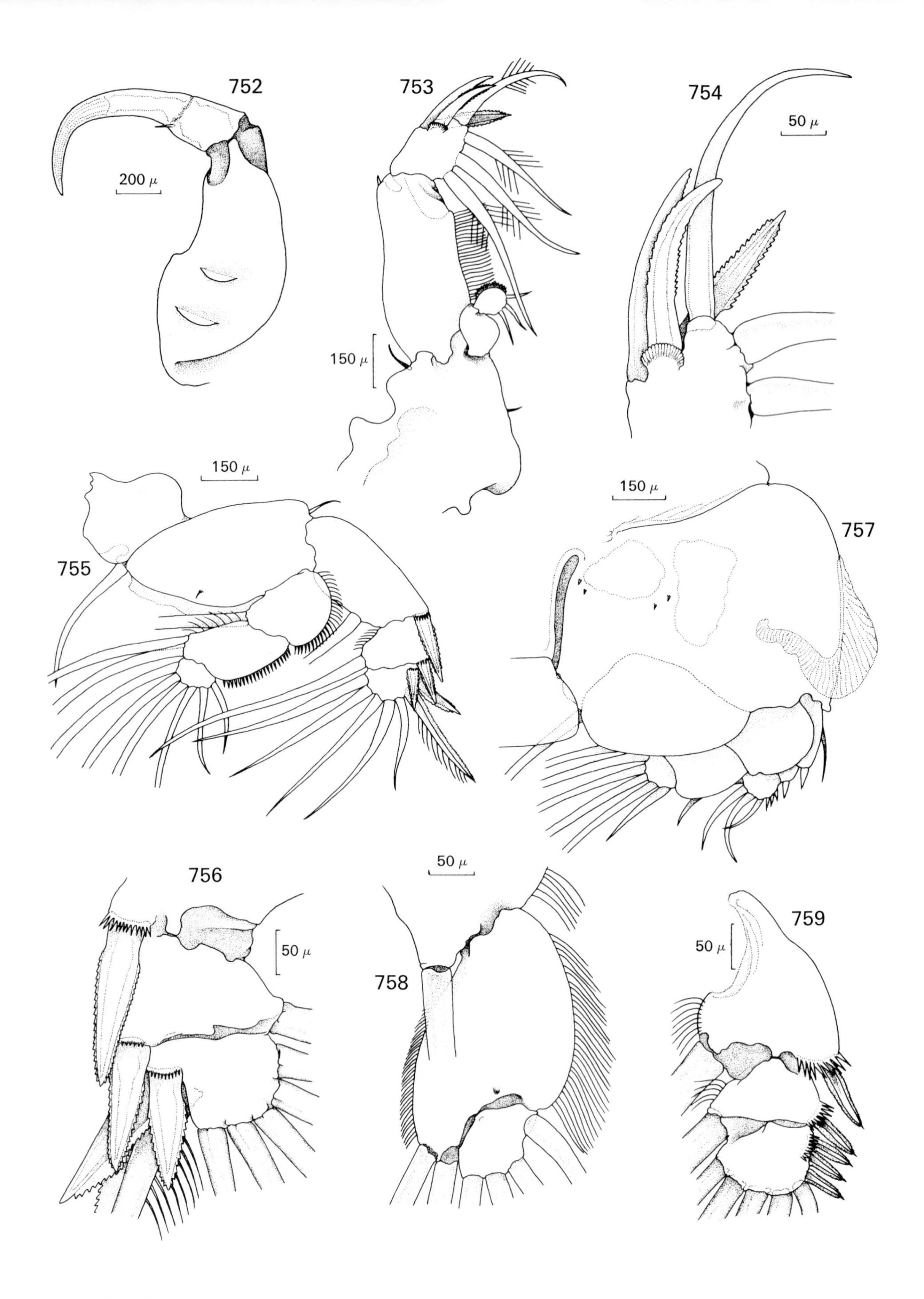

Figs. 752–759.
Euryphorus brachypterus (Gerstaecker, 1853). **Fig. 752.** Female maxilliped, ventral. **Fig. 753.** First leg, ventral. **Fig. 754.** Tip of first exopod, ventral. **Fig. 755.** Second leg, ventral. **Fig. 756.** Detail of second exopod, ventral. **Fig. 757.** Third leg, ventral. **Fig. 758.** Tip of third endopod, ventral. **Fig. 759.** Third exopod, ventral.

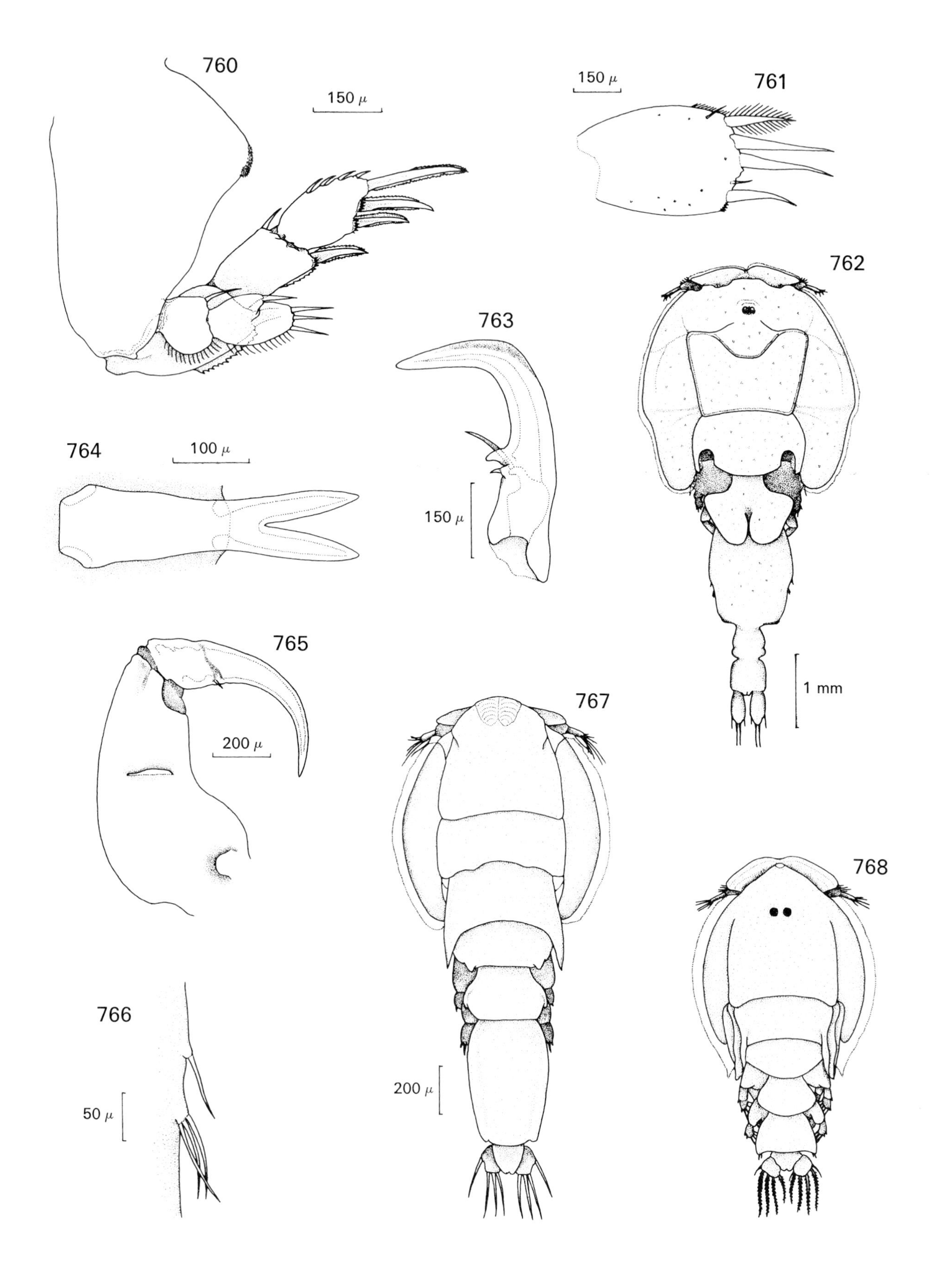

Figs. 760–768.
Euryphorus brachypterus (Gerstaecker, 1853). **Fig. 760.** Female fourth leg, ventral. **Fig. 761.** Uropod, dorsal. **Fig. 762.** Male, dorsal. **Fig. 763.** Claw of male second antenna, ventral. **Fig. 764.** Sternal furca, ventral. **Fig. 765.** Maxilliped, ventral. **Fig. 766.** Fifth leg, ventral.
Nogagus borealis Steenstrup and Lütken, 1861. **Fig. 767.** Male, dorsal.
Nogagus ambiguus T. Scott, 1907. **Fig. 768.** Male, dorsal.
Fig. 768 redrawn from Scott and Scott (1913).

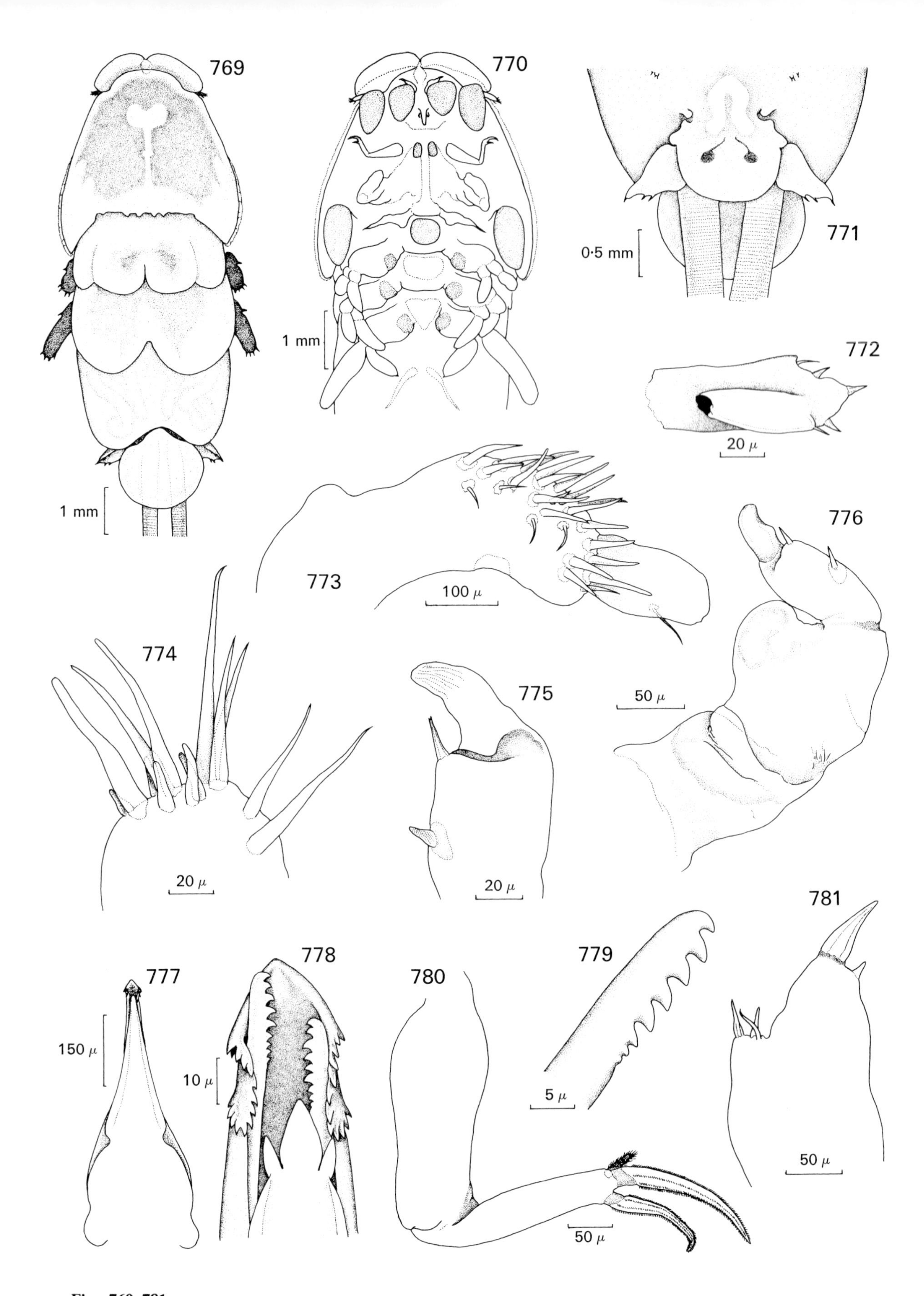

Figs. 769–781.
Pandarus bicolor Leach, 1816. **Fig. 769.** Female, dorsal. **Fig. 770.** Cephalothorax, ventral (semidiagramatic). **Fig. 771.** Posterior extremity, ventral. **Fig. 772.** Uropod, posterior. **Fig. 773.** First antenna ventral (apical armature omitted). **Fig. 774.** Same, apical armature, ventral. **Fig. 775.** Claw of second antenna, lateral. **Fig. 776.** Second antenna, ventral. **Fig. 777.** Mouth, anterior. **Fig. 778.** Same, tip. **Fig. 779.** Tip of mandible, lateral. **Fig. 780.** Second maxilla, ventral. **Fig. 781.** First maxilla, medial.

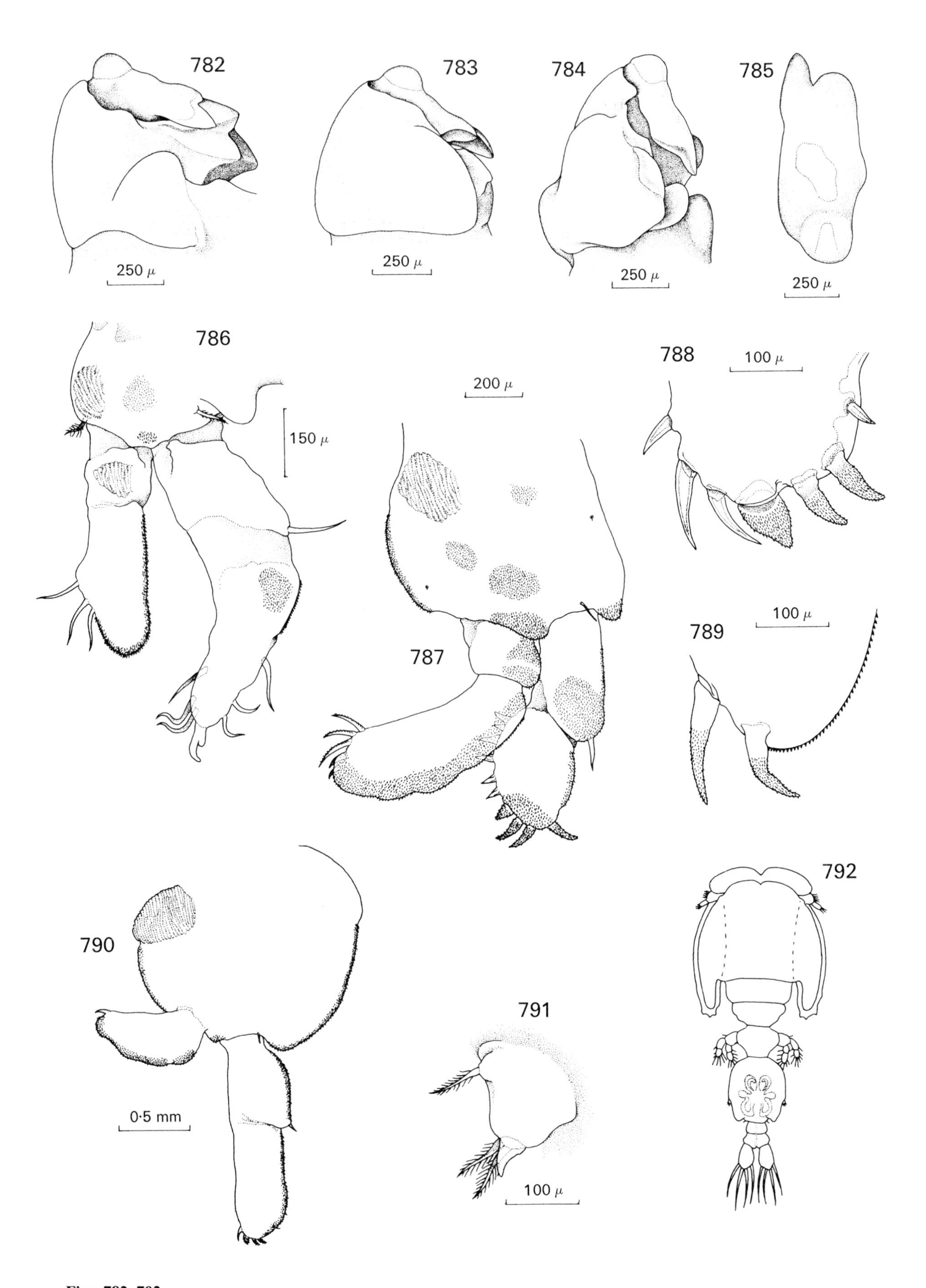

Figs. 782–792.
Pandarus bicolor Leach, 1816. **Fig. 782.** Female maxilliped, posterior. **Fig. 783.** Same, anterior. **Fig. 784.** Same, medial. **Fig. 785.** Same, claw, distal surface. **Fig. 786.** First leg, ventral. **Fig. 787.** Second leg, ventral. **Fig. 788.** Tip of third exopod, dorsal. **Fig. 789.** Tip of third endopod, dorsal. **Fig. 790.** Fourth leg, ventral. **Fig. 791.** Fifth leg, ventral. **Fig. 792.** Male, dorsal.
Fig. 792 redrawn from Scott and Scott (1913).

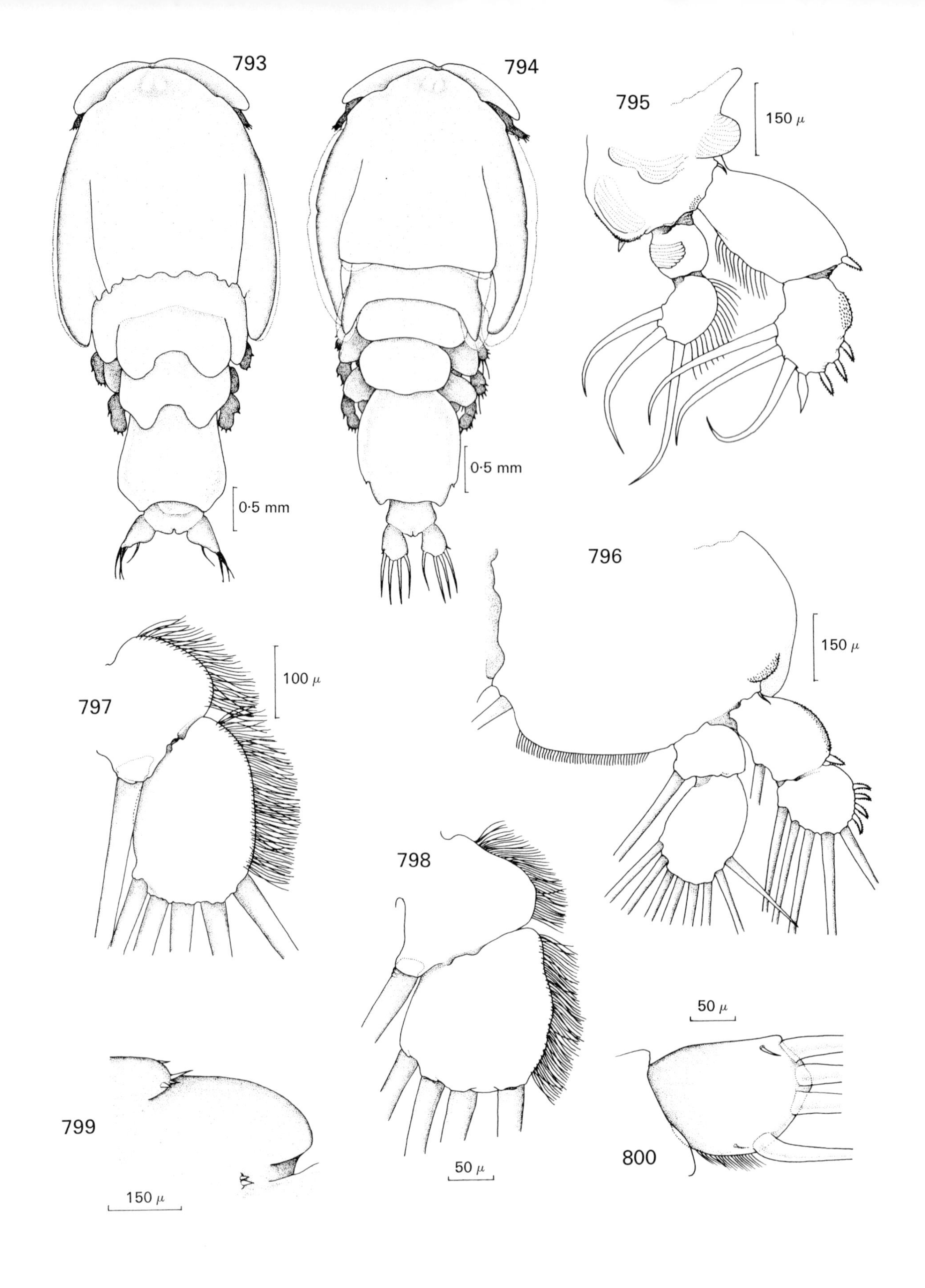

Figs. 793–800.
Pandarus bicolor Leach, 1816. **Fig. 793.** Immature female, dorsal. **Fig. 794.** Male, dorsal. **Fig. 795.** Male first leg, ventral. **Fig. 796.** Second leg, ventral. **Fig. 797.** Third endopod, ventral. **Fig. 798.** Fourth endopod, ventral. **Fig. 799.** Posterolateral corner of genital complex, ventral. **Fig. 800.** Uropod, ventral.

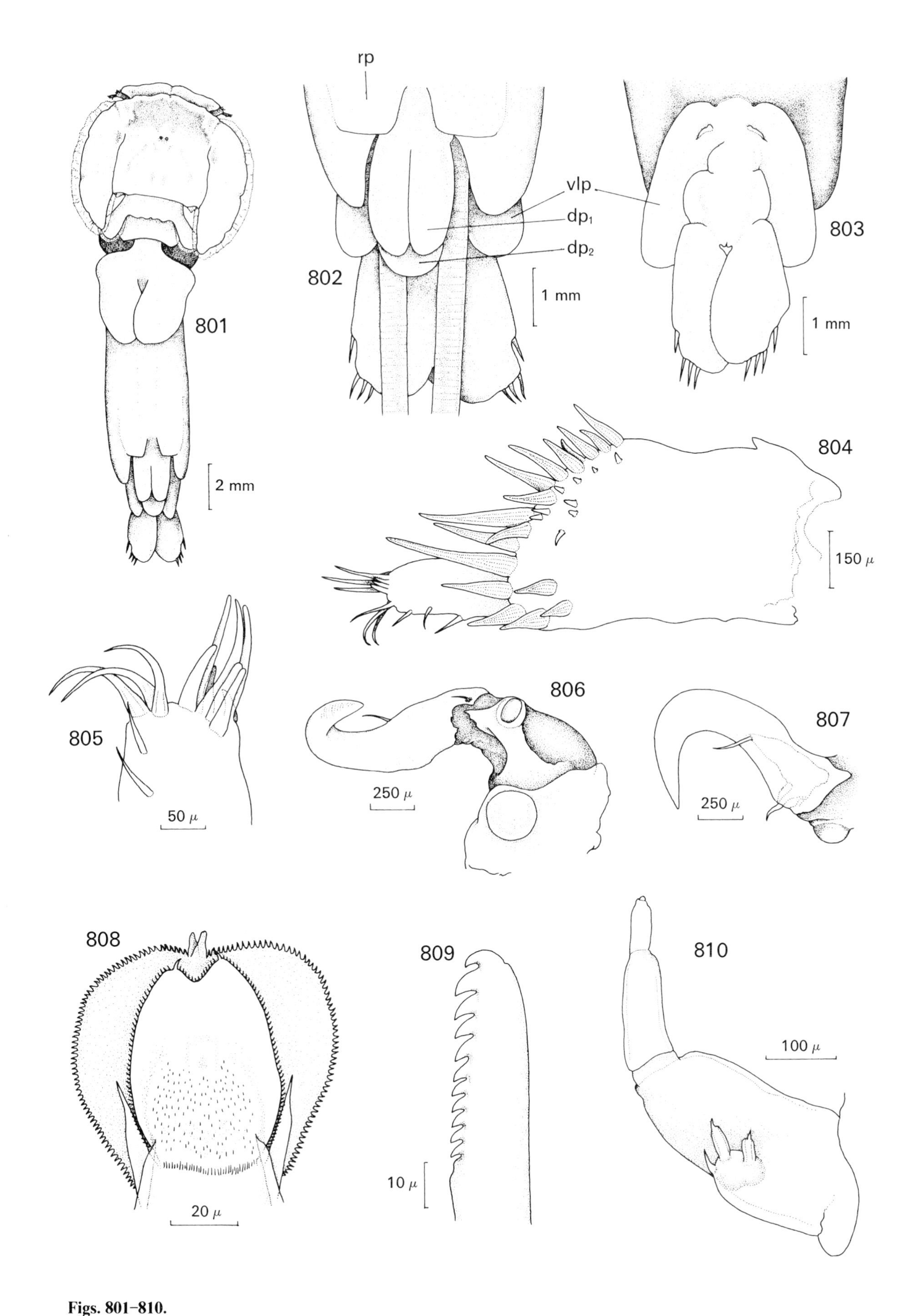

Figs. 801–810.
Dinemoura producta (Müller, 1785). **Fig. 801.** Female, dorsal, **Fig. 802.** Posterior extremity, dorsal. **Fig. 803.** Same, ventral. **Fig. 804.** First antenna, ventral. **Fig. 805.** Same, tip, ventral. **Fig. 806.** Second antenna, ventral. **Fig. 807.** Same, claw, anterior. **Fig. 808.** Tip of mouth, anterior. **Fig. 809.** Tip of mandible, lateral. **Fig. 810.** First maxilla, lateral.

dpl—dorsal plate, paired; dp2—dorsal plate, median; rp—rudimentary plate; vlp—ventrolateral plate.

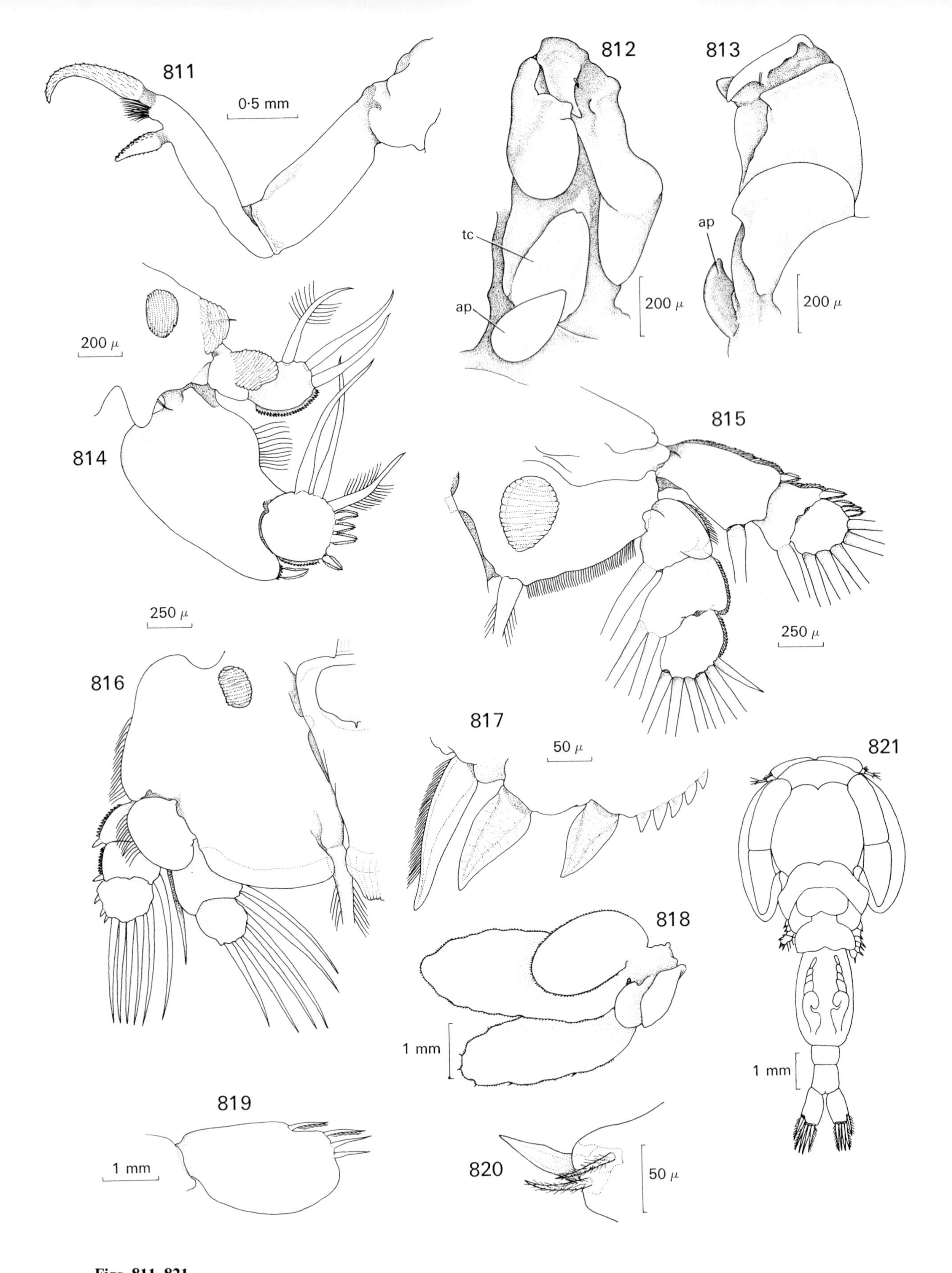

Figs. 811–821.

Dinemoura producta (Müller, 1785). **Fig. 811.** Female second maxilla, ventral. **Fig. 812.** Maxilliped, ventral. **Fig. 813.** Same, medial. **Fig. 814.** First leg, ventral. **Fig. 815.** Second leg, ventral. **Fig. 816.** Third leg, ventral. **Fig. 817.** Third leg, detail of lateral margin of third exopod segment. **Fig. 818.** Fourth leg, ventral. **Fig. 819.** Uropod, dorsal. **Fig. 820.** Fifth leg, medial. **Fig. 821.** Male, dorsal.

Fig. 821 redrawn from Wilson (1923a).

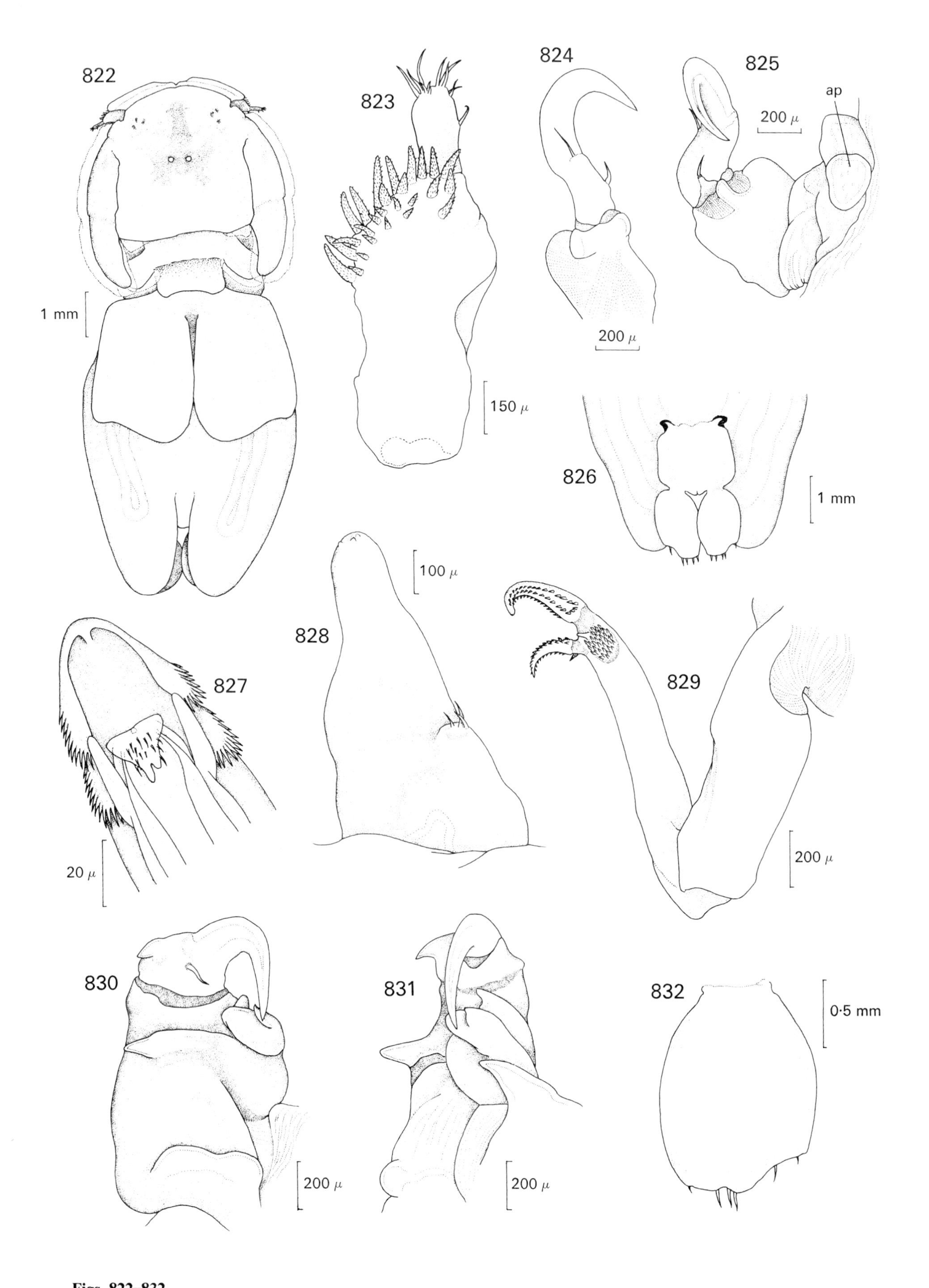

Figs. 822–832.
Echthrogaleus coleoptratus (Guerin-Meneville, 1837). **Fig. 822.** Female, dorsal. **Fig. 823.** Female first antenna, ventral. **Fig. 824.** Claw of second antenna, anterior. **Fig. 825.** Second antenna, ventral. **Fig. 826.** Posterior extremity, ventral. **Fig. 827.** Tip of mouth, anterior. **Fig. 828.** First maxilla, lateral. **Fig. 829.** Second maxilla, ventral. **Fig. 830.** Maxilliped, ventromedial. **Fig. 831.** Same, medial. **Fig. 832.** Uropod, ventral.

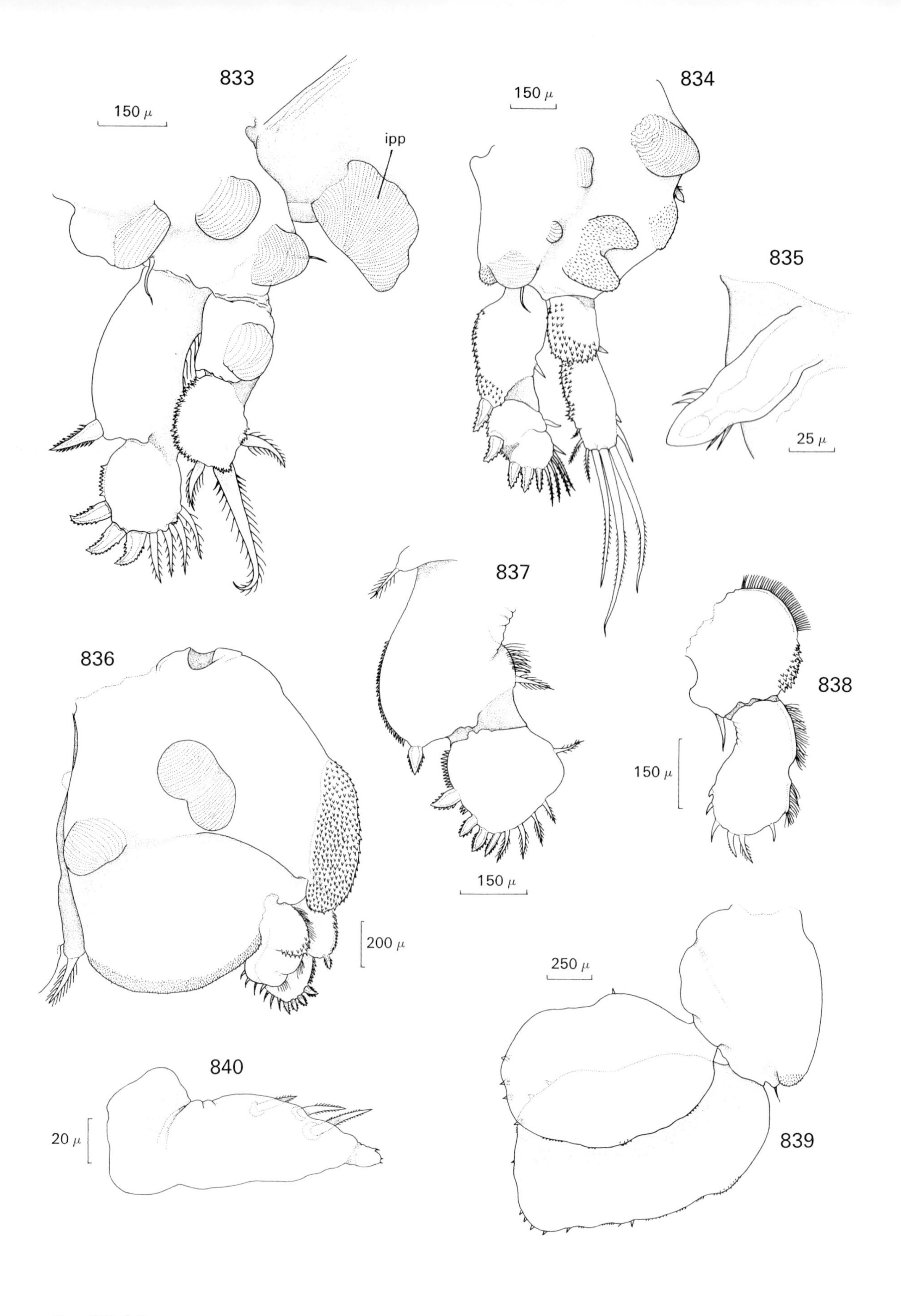

Figs. 833–840.
Echthrogaleus coleoptratus (Guerin-Meneville, 1837). **Fig. 833.** Female first leg, ventral. **Fig. 834.** Second leg, ventral. **Fig. 835.** Same, medial spine of sympod, dorsal. **Fig. 836.** Third leg, ventral. **Fig. 837.** Same, exopod, dorsal. **Fig. 838.** Same, endopod, ventral. **Fig. 839.** Fourth leg, ventral. **Fig. 840.** Fifth leg, ventral.

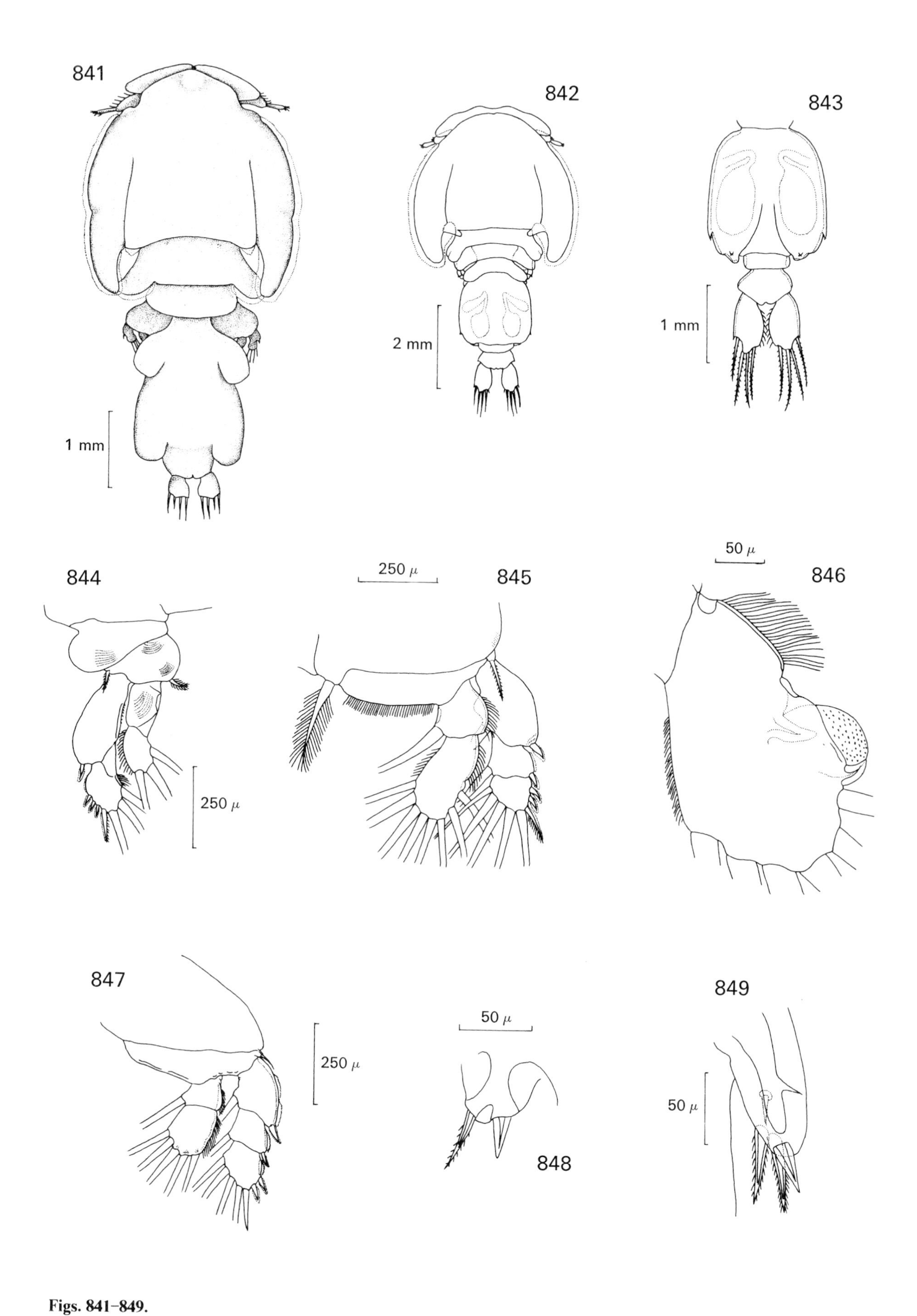

Figs. 841–849.
Echthrogaleus coleoptratus (Guerin-Meneville, 1837). **Fig. 841.** Immature female, dorsal. **Fig. 842.** Male, dorsal. **Fig. 843.** Male posterior extremity, ventral. **Fig. 844.** First leg, ventral. **Fig. 845.** Second leg, ventral. **Fig. 846.** Third leg, detail of endopod. **Fig. 847.** Fourth leg, ventral. **Fig. 848.** Sixth leg, ventral. **Fig. 849.** Fifth leg, ventral.

Figs. 842–849 redrawn from Cressey (1967a).

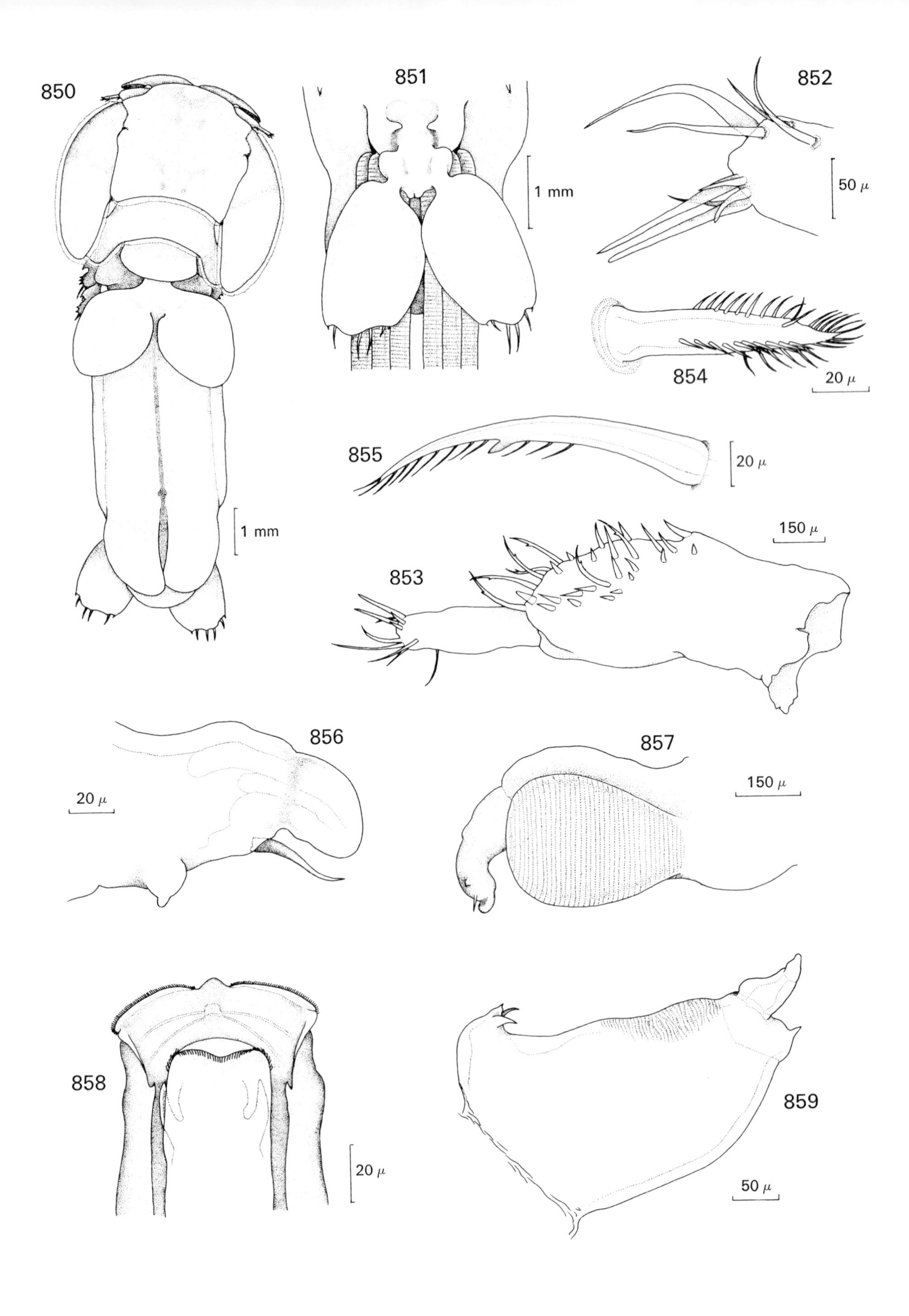

Figs. 850–859.
Demoleus heptapus (Otto, 1821). **Fig. 850.** Female, dorsal. **Fig. 851.** Female posterior extremity, ventral. **Fig. 852.** Tip of first antenna. **Fig. 853.** First antenna, ventral. **Fig. 854.** Seta of first segment of first antenna, enlarged. **Fig. 855.** Same, seta of different type. **Fig. 856.** Tip of second antenna. **Fig. 857.** Second antenna, ventral. **Fig. 858.** Mouth tip, anterior. **Fig. 859.** First maxilla, lateral.

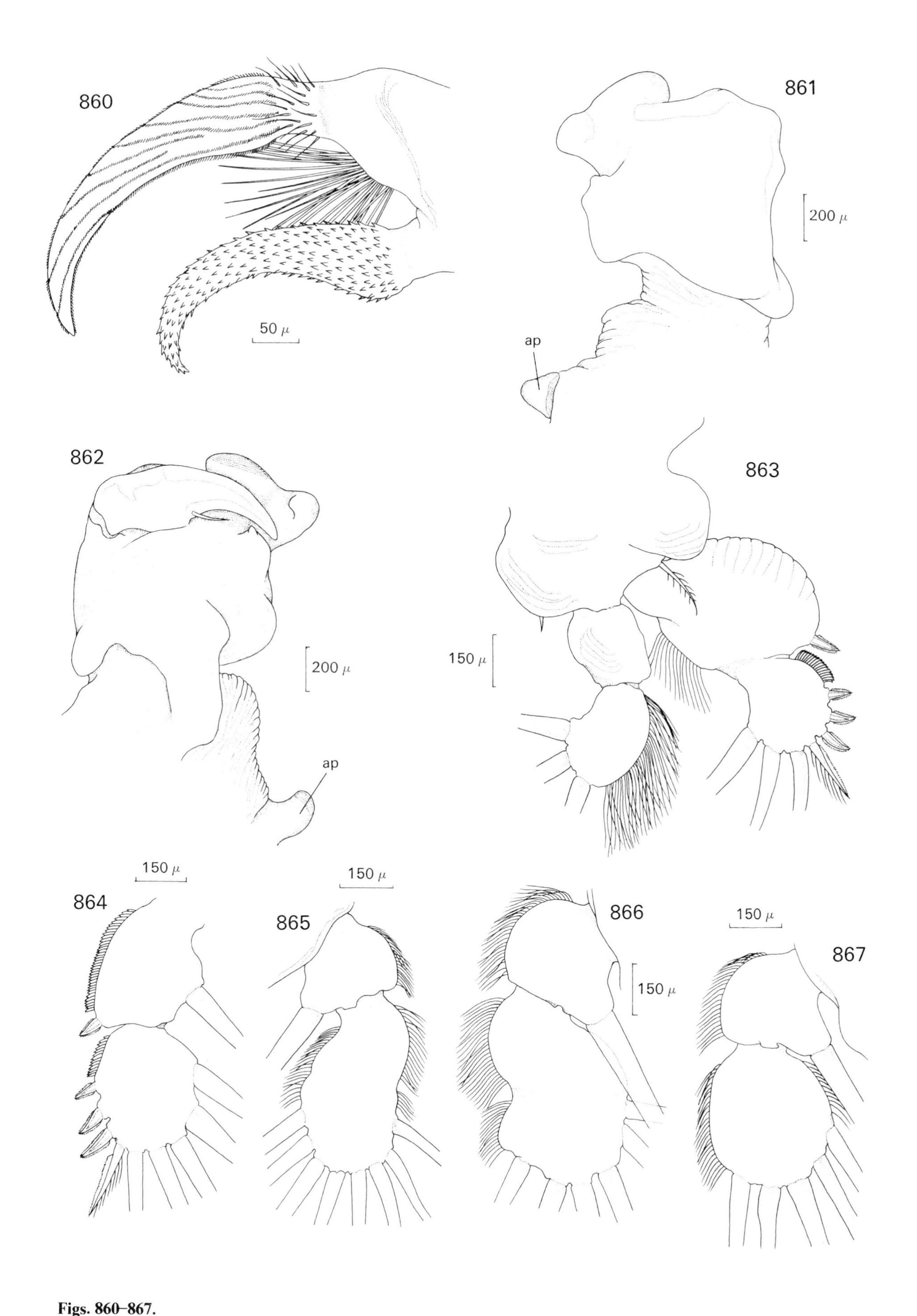

Figs. 860–867.
Demoleus heptapus (Otto, 1821). **Fig. 860.** Female, tip of second maxilla, medial. **Fig. 861.** Maxilliped, lateral. **Fig. 862.** Same, medial. **Fig. 863.** First leg, ventral. **Fig. 864.** Second exopod, ventral. **Fig. 865.** Second endopod, ventral. **Fig. 866.** Third endopod, ventral. **Fig. 867.** Fourth endopod, ventral.

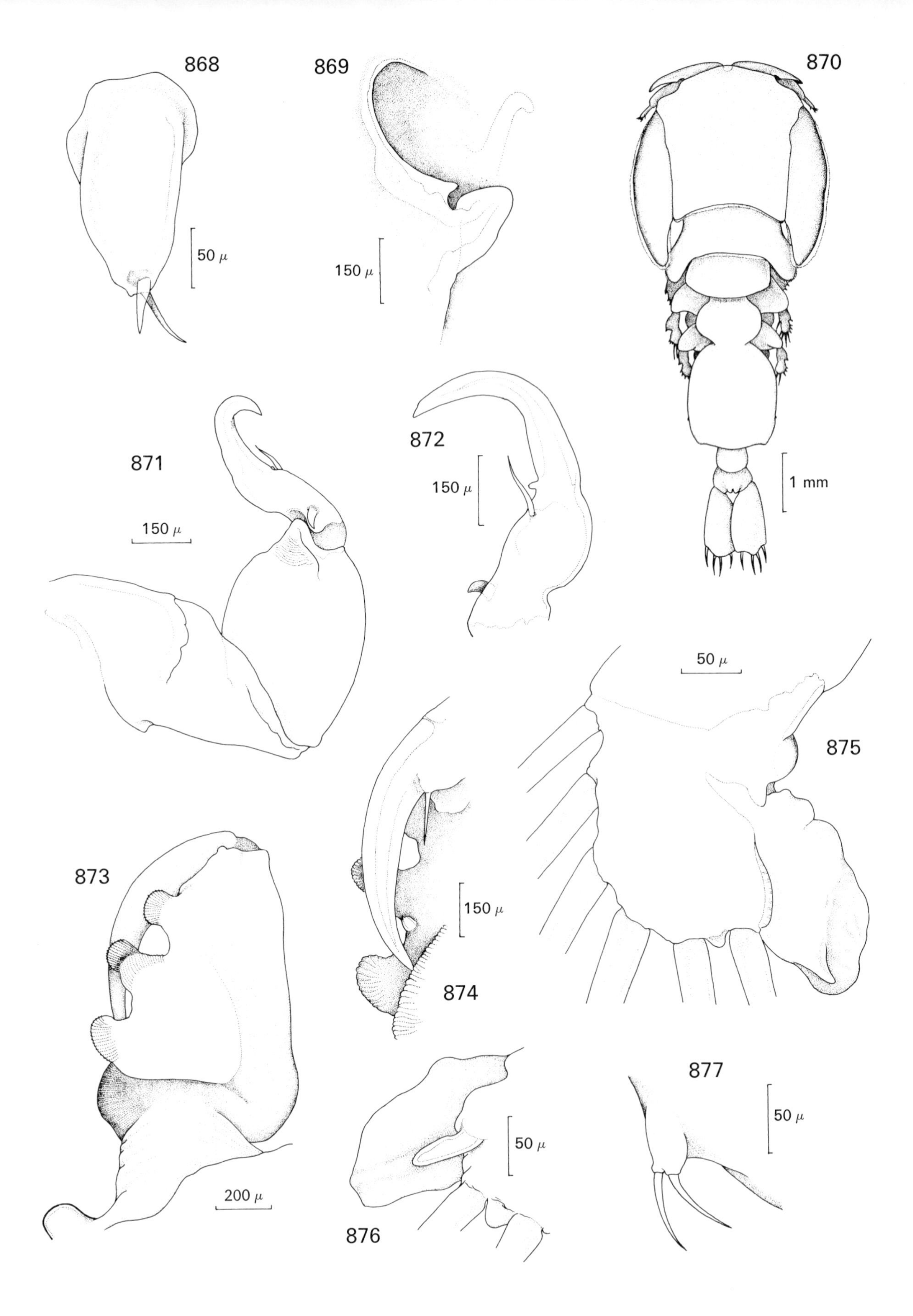

Figs. 868–877.
Demoleus heptapus (Otto, 1821). **Fig. 868.** Female fifth leg, ventral. **Fig. 869.** Orifice of oviduct, ventral. **Fig. 870.** Male, dorsal. **Fig. 871.** Male second antenna, ventral. **Fig. 872.** Same, anterolateral. **Fig. 873.** Maxilliped, lateral. **Fig. 874.** Same, detail of medial margin. **Fig. 875.** Tip of third endopod, dorsal. **Fig. 876.** Same, ventral. **Fig. 877.** Fifth leg, lateral.

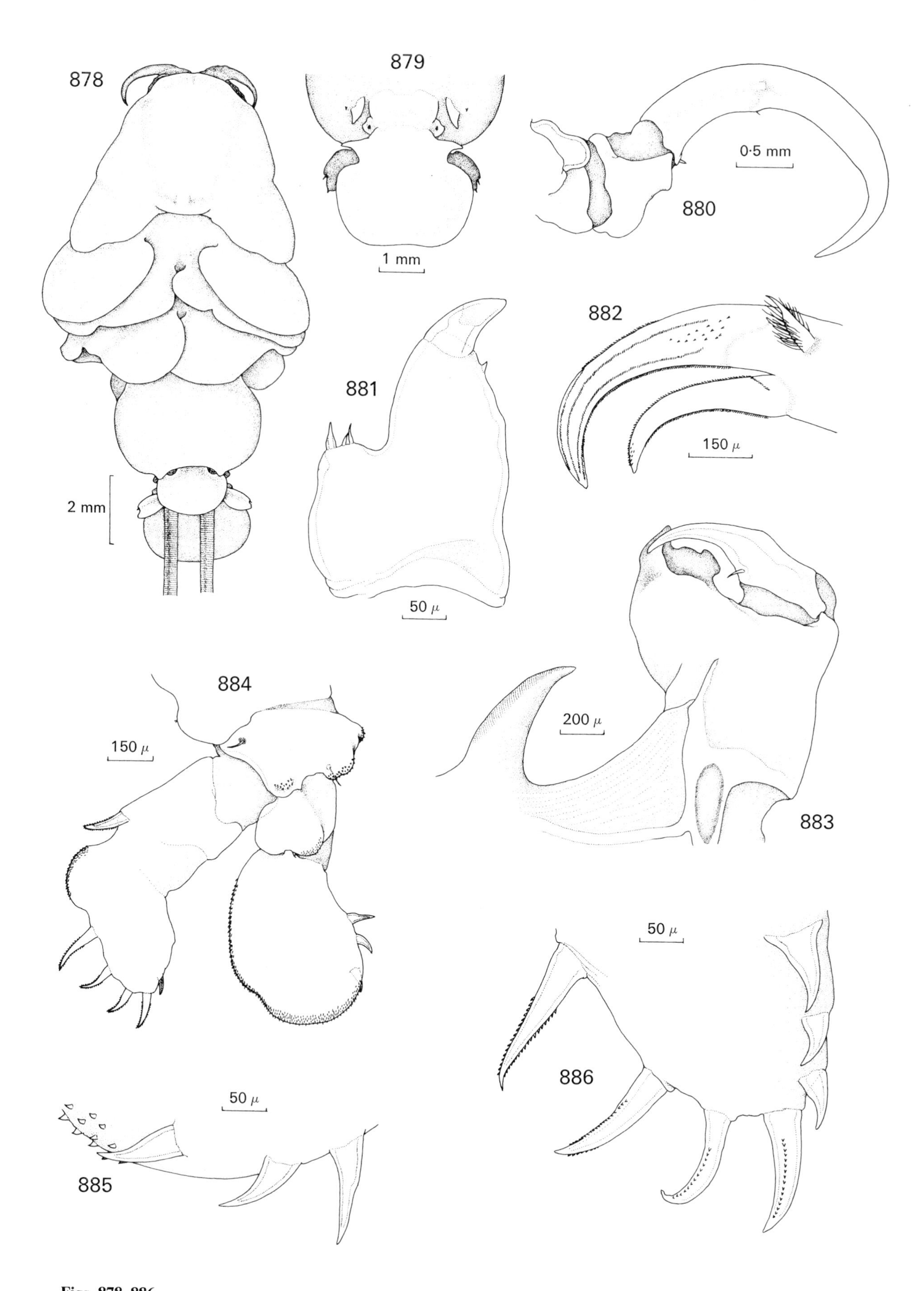

Figs. 878–886.
Phyllothyreus cornutus (Edwards, 1840). **Fig. 878.** Female, dorsal. **Fig. 879.** Female posterior extremity, ventral. **Fig. 880.** Second antenna, ventral. **Fig. 881.** First maxilla, lateral. **Fig. 882.** Tip of second maxilla, dorsal. **Fig. 883.** Maxilliped, medial. **Fig. 884.** First leg, ventral. **Fig. 885.** Tip of first endopod, dorsal. **Fig. 886.** Tip of first exopod, dorsal.

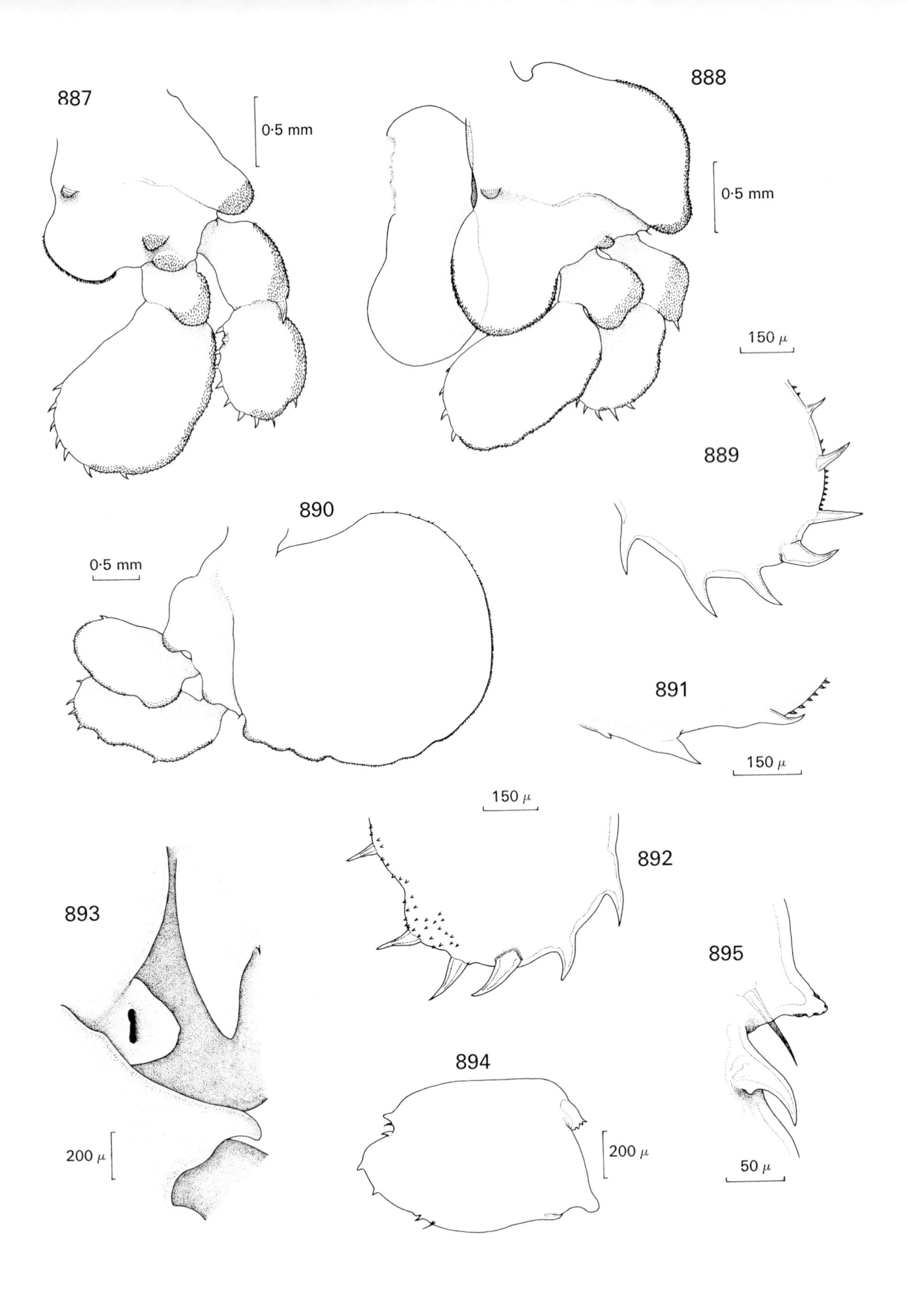

Figs. 887–895.
Phyllothyreus cornutus (Edwards, 1840). **Fig. 887.** Female second leg, ventral. **Fig. 888.** Third leg, ventral. **Fig. 889.** Tip of third exopod, dorsal. **Fig. 890.** Fourth leg, ventral. **Fig. 891.** Margin of fourth endopod, dorsal. **Fig. 892.** Tip of fourth exopod, ventral. **Fig. 893.** Genital orifice, ventral. **Fig. 894.** Uropod, ventral. **Fig. 895.** Same, detail of tip, dorsal.

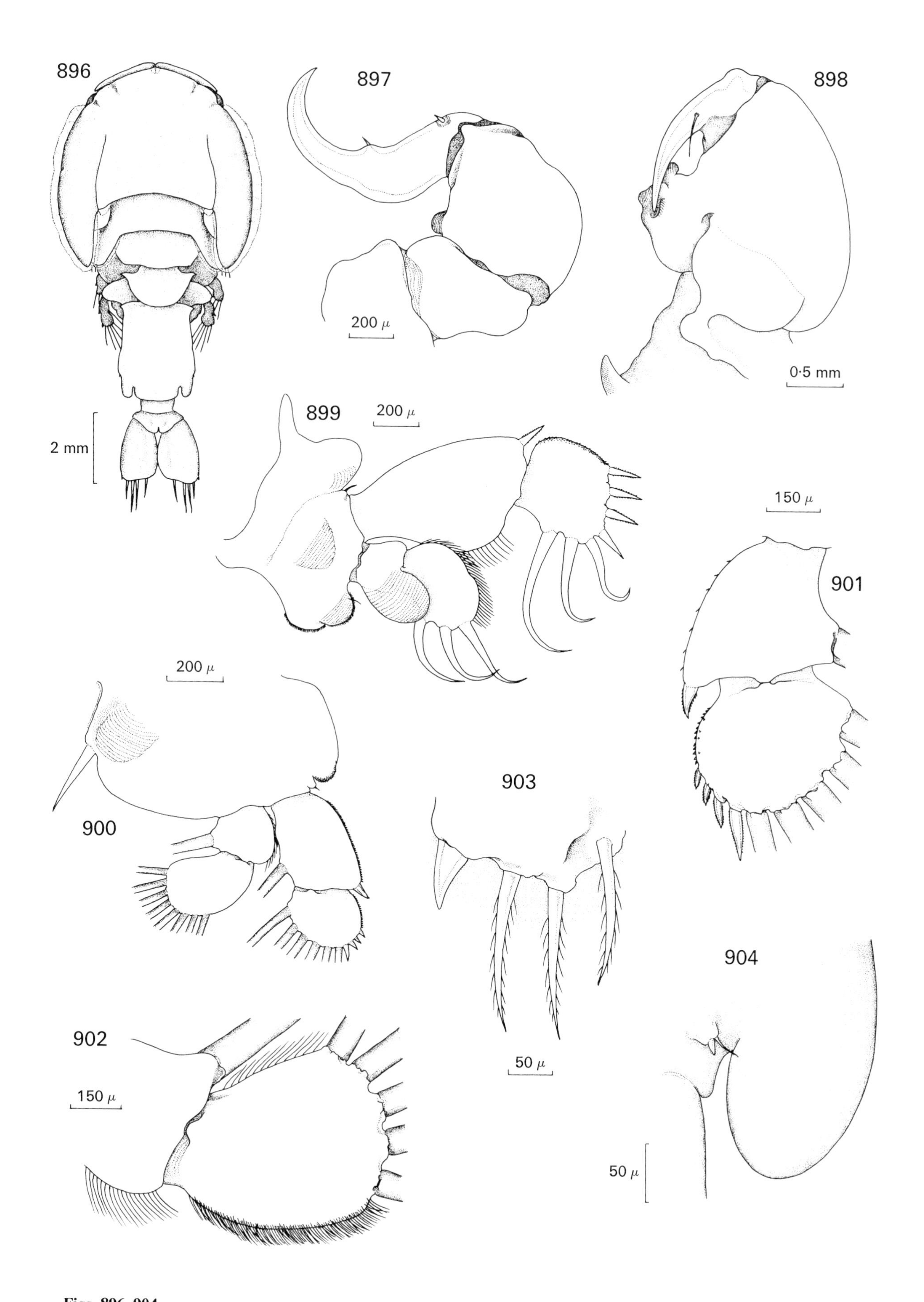

Figs. 896–904.

Phyllothyreus cornutus (Edwards, 1840). **Fig. 896.** Male, dorsal. **Fig. 897.** Male second antenna, ventral. **Fig. 898.** Maxilliped, medial. **Fig. 899.** First leg, ventral. **Fig. 900.** Second leg, ventral. **Fig. 901.** Third exopod, ventral. **Fig. 902.** Fourth endopod, ventral. **Fig. 903.** Fifth leg, ventral. **Fig. 904.** Posterolateral corner of genital complex, ventral.

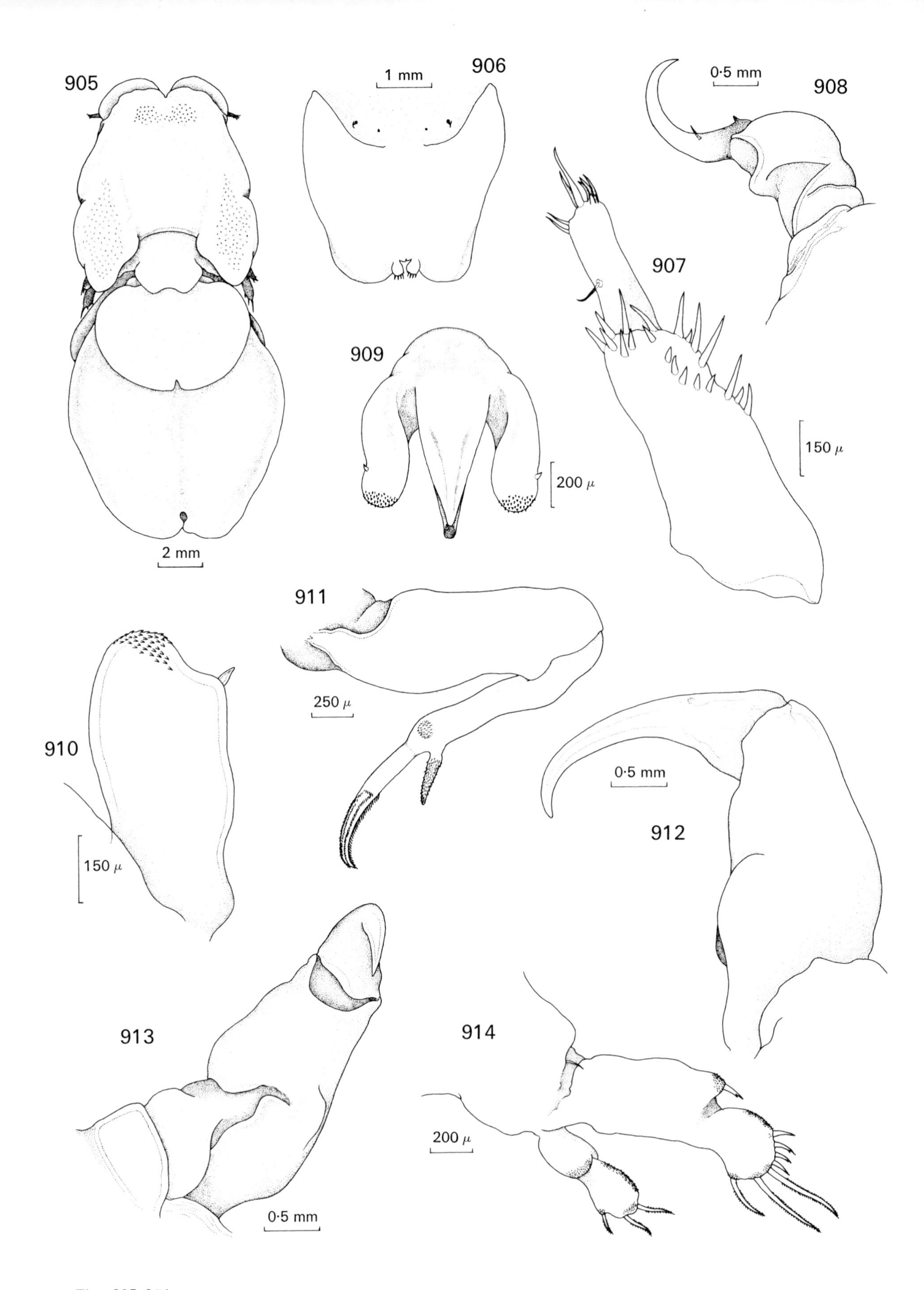

Figs. 905–914.
Cecrops latreillii Leach, 1816. **Fig. 905.** Female, dorsal. **Fig. 906.** Female abdomen, ventral. **Fig. 907.** First antenna, ventral. **Fig. 908.** Second antenna, lateral. **Fig. 909.** Mouth tube and first maxillae, ventral. **Fig. 910.** First maxilla, lateral. **Fig. 911.** Second maxilla, ventral. **Fig. 912.** Maxilliped, medial. **Fig. 913.** Same, ventral. **Fig. 914.** First leg, ventral.

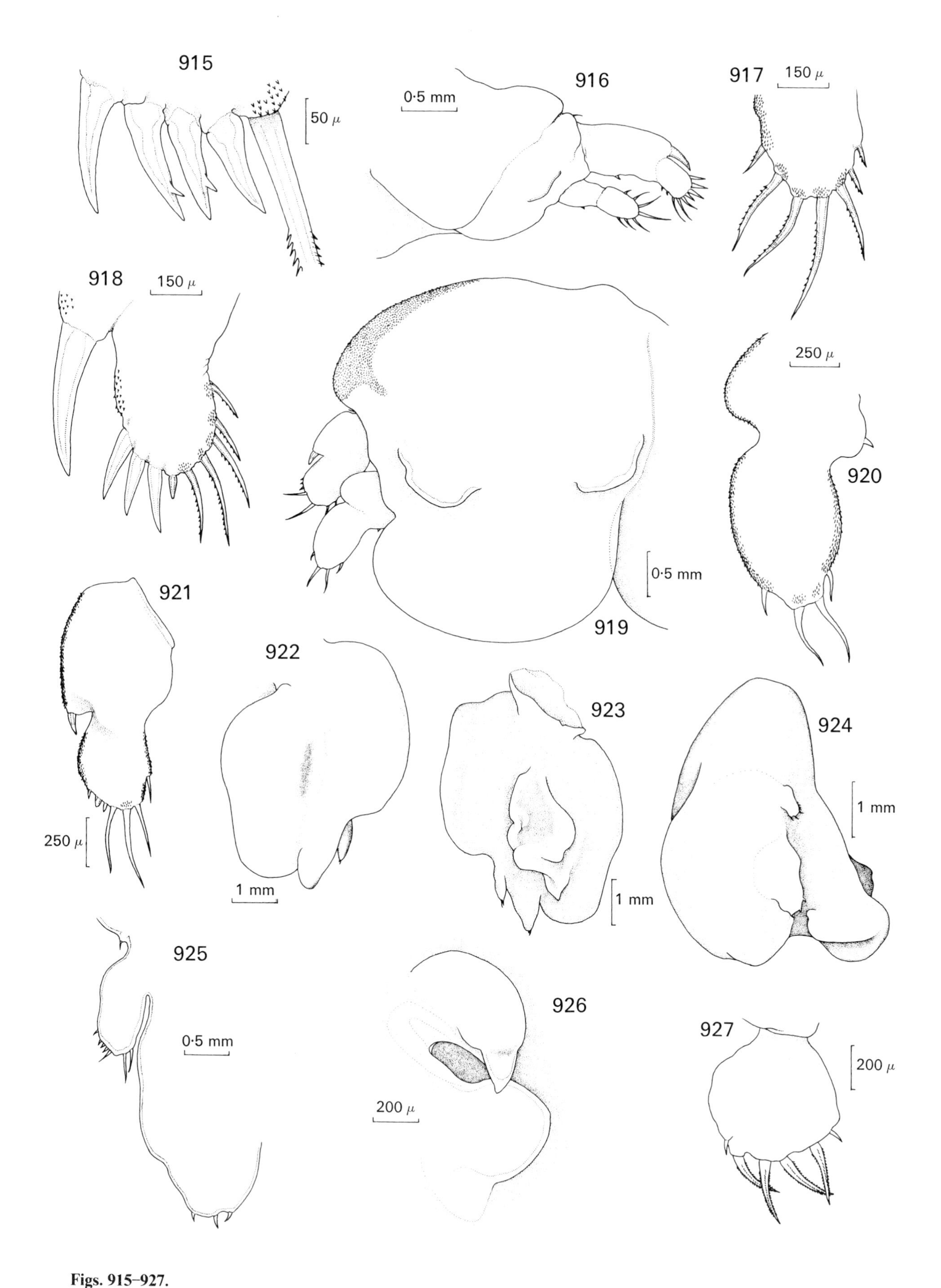

Figs. 915–927.
Cecrops latreillii Leach, 1816. **Fig. 915.** Female, tip of first exopod, ventral. **Fig. 916.** Second leg, ventral. **Fig. 917.** Tip of second endopod, ventral. **Fig. 918.** Tip of second exopod, ventral. **Fig. 919.** Third leg, ventral. **Fig. 920.** Third endopod, ventral. **Fig. 921.** Third exopod, ventral. **Fig. 922.** Fourth leg, ventral. **Fig. 923.** Same, dorsal. **Fig. 924.** Same, lateral (distended). **Fig. 925.** Same, rami. **Fig. 926.** Genital orifice, ventral. **Fig. 927.** Uropod, ventral.

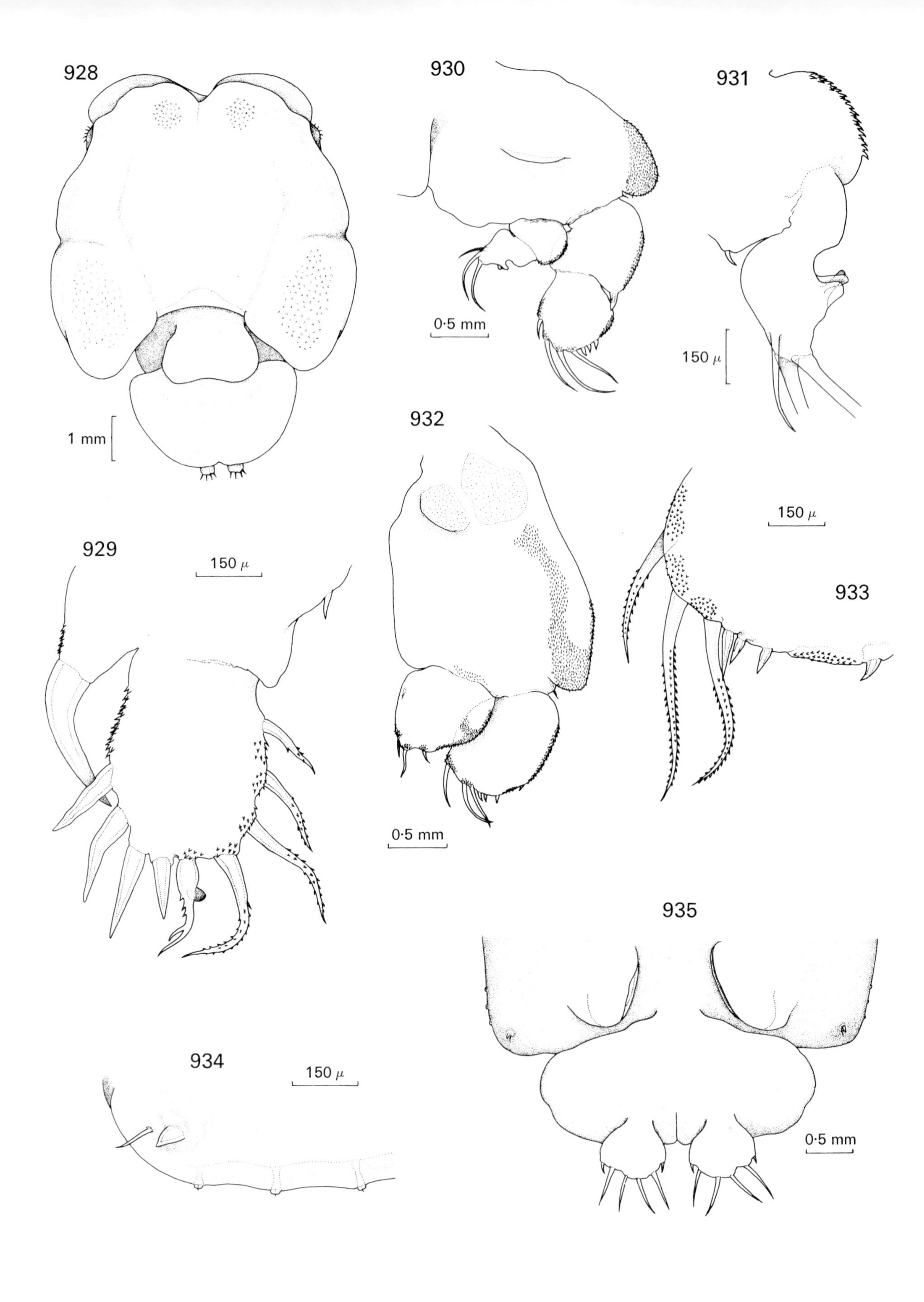

Figs. 928–935.
Cecrops latreillii Leach, 1816. **Fig. 928.** Male, dorsal. **Fig. 929.** Male second leg, tip of exopod, ventral. **Fig. 930.** Third leg, ventral. **Fig. 931.** Third endopod, dorsal. **Fig. 932.** Fourth leg, ventral. **Fig. 933.** Tip of fourth exopod, ventral. **Fig. 934.** Posterolateral corner of genital complex, ventral. **Fig. 935.** Posterior extremity, ventral.

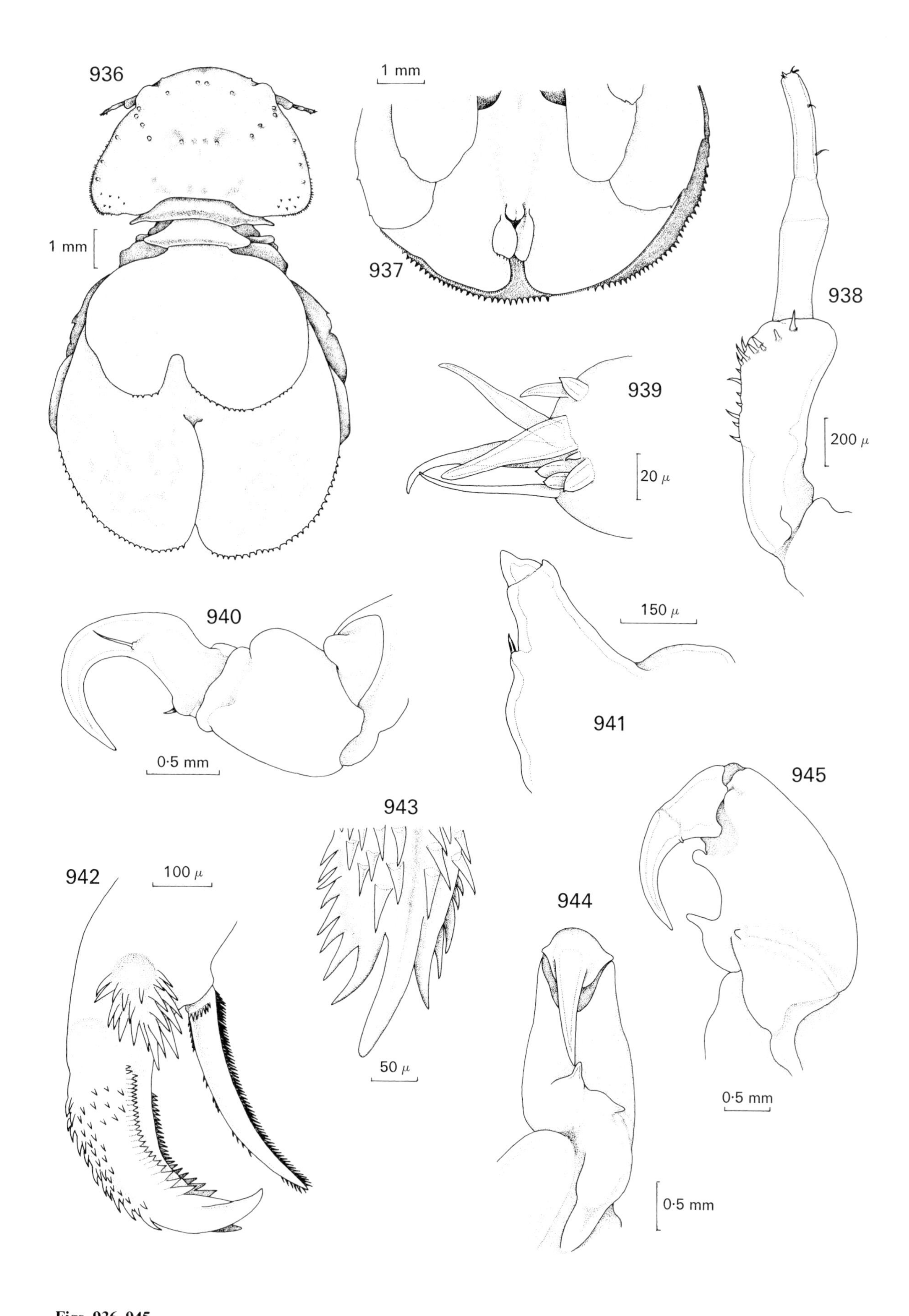

Figs. 936–945.

Orthagoriscicola muricatus (Krøyer, 1837). **Fig. 936.** Female, dorsal. **Fig. 937.** Posterior extremity, ventral. **Fig. 938.** First antenna, ventral. **Fig. 939.** Same, tip, anterior. **Fig. 940.** Second antenna, ventral. **Fig. 941.** First maxilla, medial. **Fig. 942.** Tip of second maxilla, medial. **Fig. 943.** Same, tip of calamus, enlarged. **Fig. 944.** Maxilliped, anterior. **Fig. 945.** Same, medial.

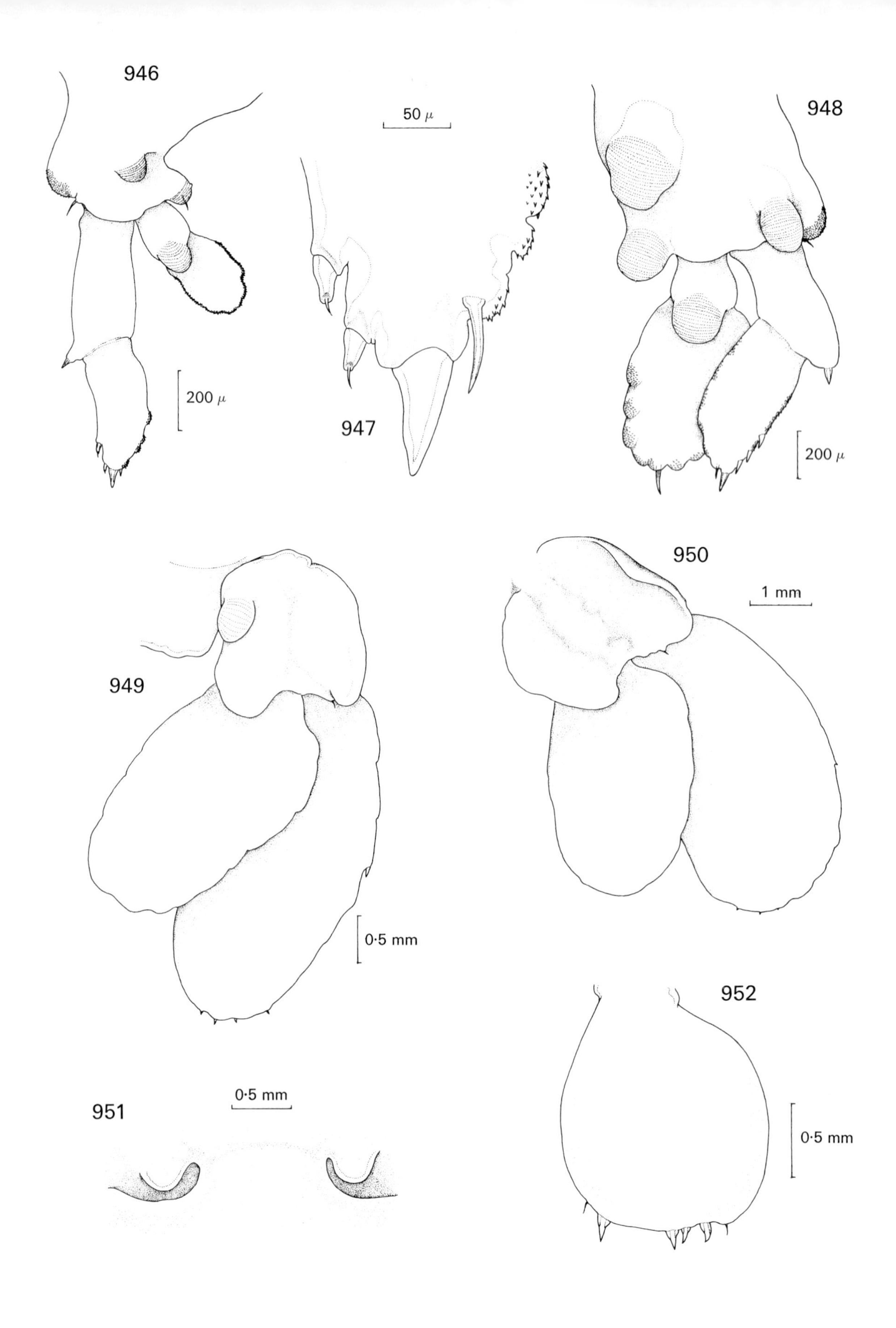

Figs. 946–952.
Orthagoriscicola muricatus (Krøyer, 1837). **Fig. 946.** Female first leg, ventral. **Fig. 947.** Tip of first exopod, dorsal. **Fig. 948.** Second leg, ventral. **Fig. 949.** Third leg, ventral. **Fig. 950.** Fourth leg, ventral. **Fig. 951.** Junction of genital complex and abdomen, ventral. **Fig. 952.** Uropod, ventral.

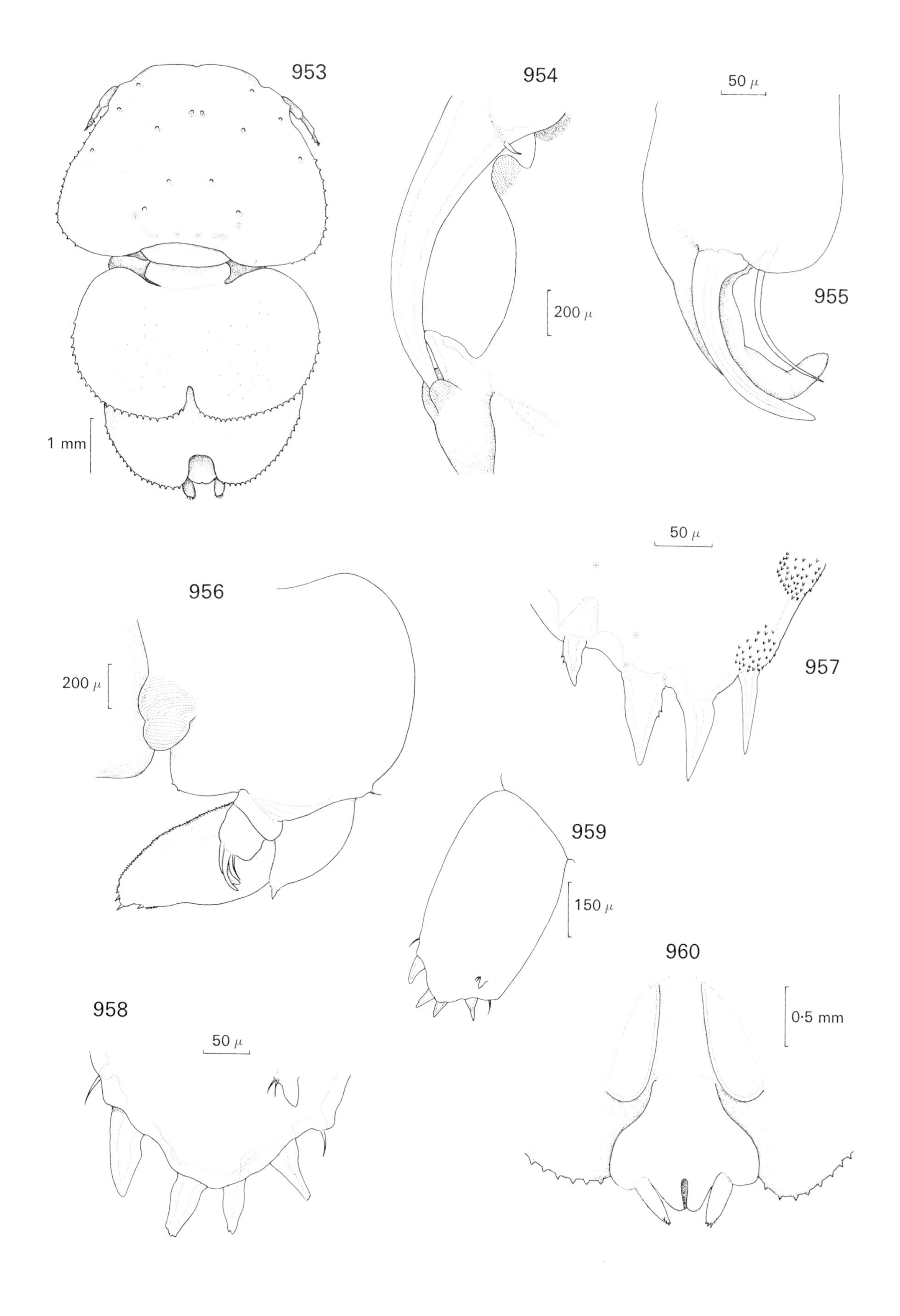

Figs. 953–960.
Orthagoriscicola muricatus (Krøyer, 1837). **Fig. 953.** Male, dorsal. **Fig. 954.** Maxilliped, tip of subchela and part of medial margin. **Fig. 955.** Tip of third endopod, ventral. **Fig. 956.** Third leg, ventral. **Fig. 957.** Tip of third exopod, ventral. **Fig. 958.** Uropod, tip, ventral. **Fig. 959.** Uropod, entire, ventral. **Fig. 960.** Posterior extremity, ventral.

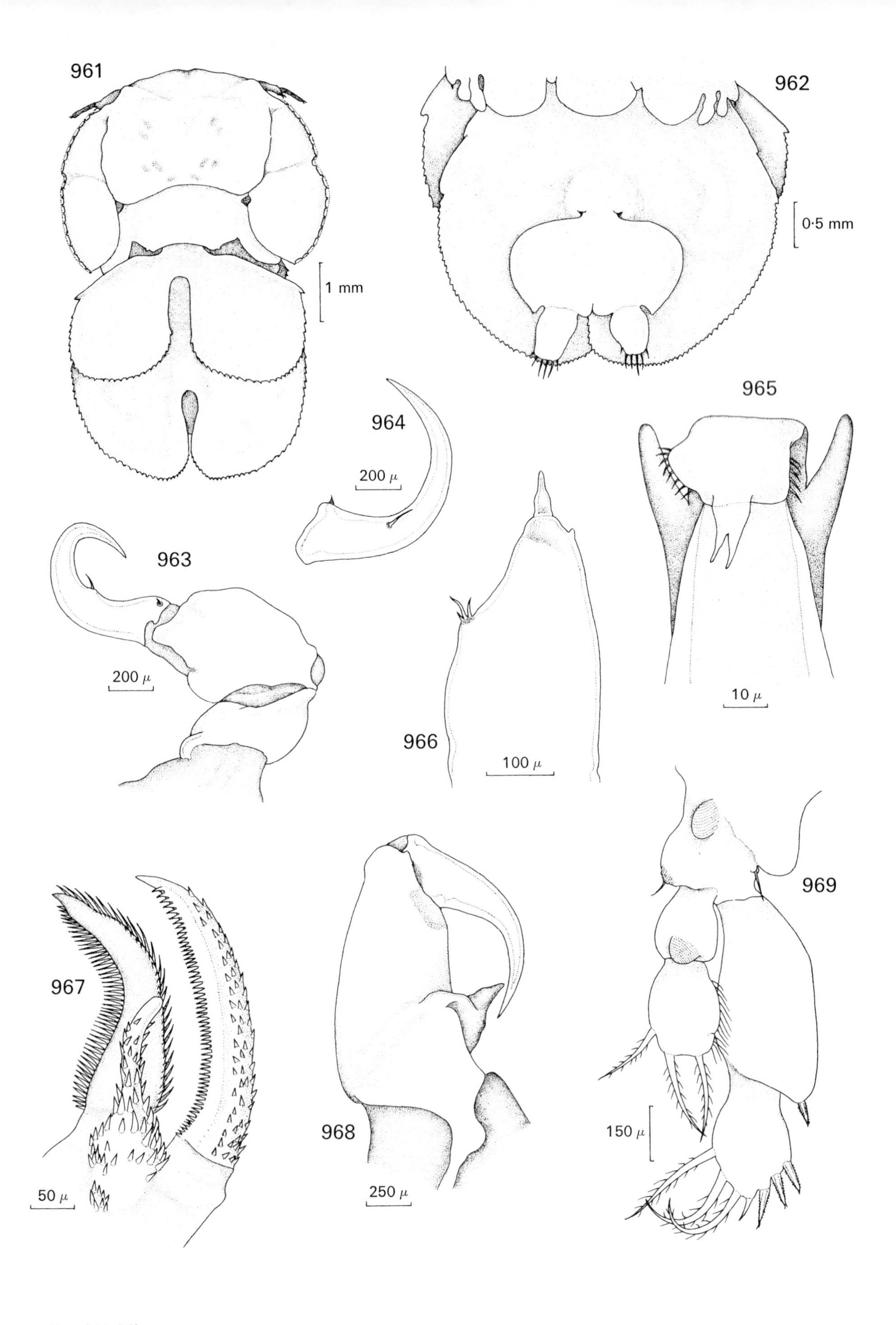

Figs. 961–969.
Philorthagoriscus serratus (Krøyer, 1863). **Fig. 961.** Female, dorsal. **Fig. 962.** Posterior extremity, ventral. **Fig. 963.** Second antenna, ventral. **Fig. 964.** Same, claw, medial. **Fig. 965.** Tip of mouth, anterior. **Fig. 966.** First maxilla, lateral. **Fig. 967.** Tip of second maxilla. **Fig. 968.** Maxilliped, medial. **Fig. 969.** First leg, ventral.

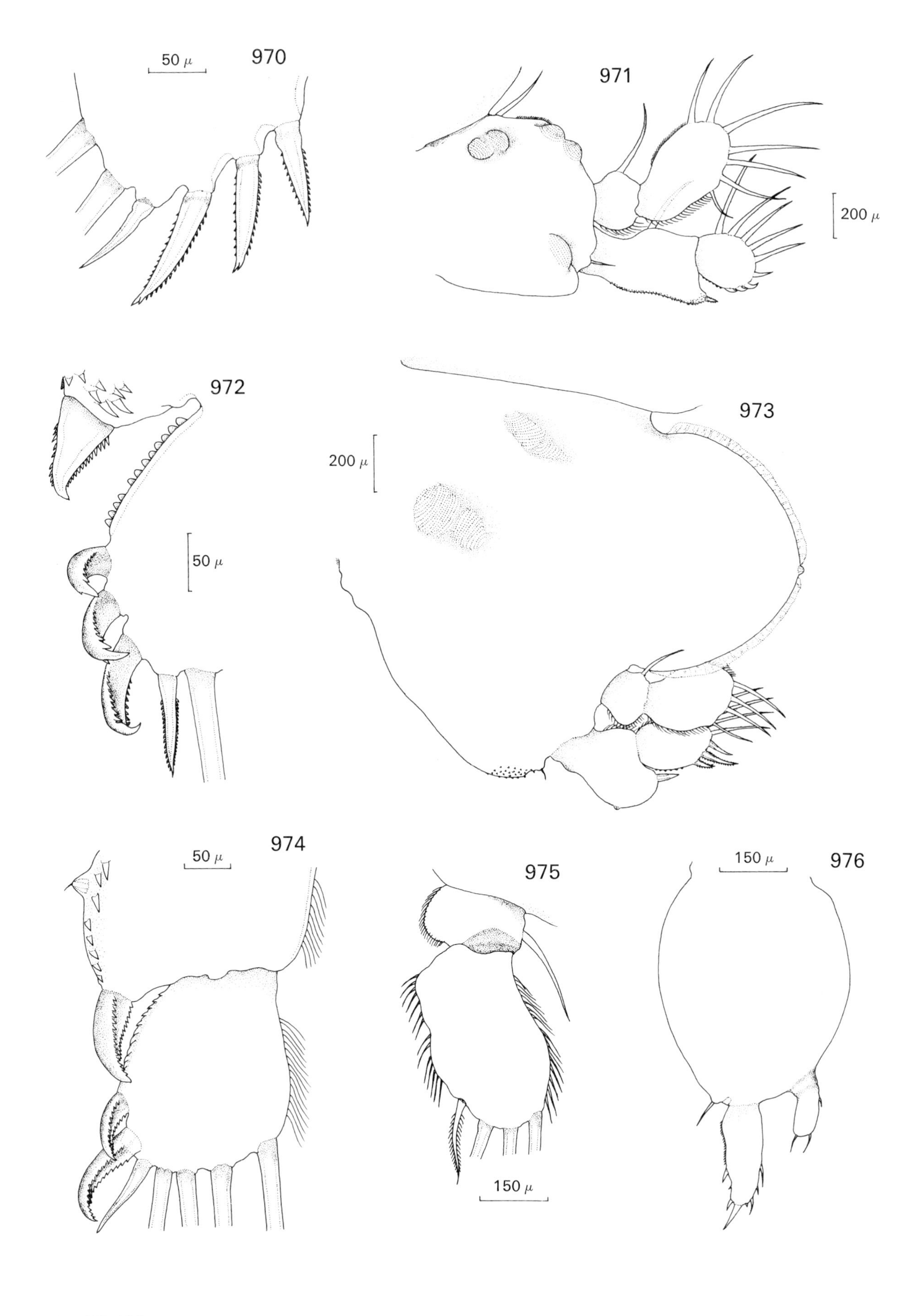

Figs. 970–976.
Philorthagoriscus serratus (Krøyer, 1863). **Fig. 970.** Female first leg, exopod tip, ventral. **Fig. 971.** Second leg, ventral. **Fig. 972.** Same, detail of lateral margin of exopod, dorsal. **Fig. 973.** Third leg, ventral. **Fig. 974.** Tip of third exopod, dorsal. **Fig. 975.** Third endopod, ventral. **Fig. 976.** Fourth leg, ventral.

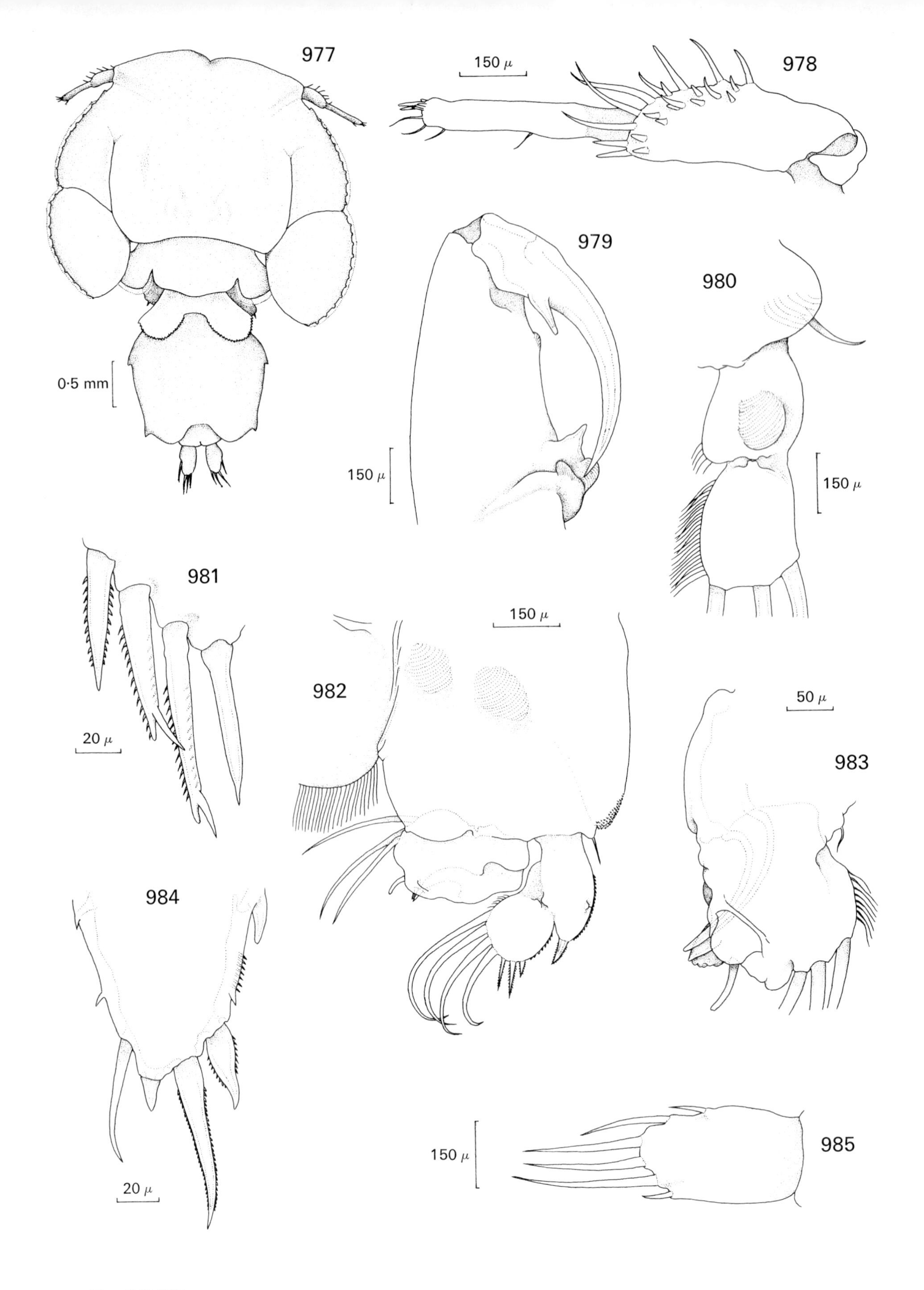

Figs. 977–985.
Philorthagoriscus serratus (Krøyer, 1863). **Fig. 977.** Male, dorsal. **Fig. 978.** First antenna, ventral. **Fig. 979.** Maxilliped, medial. **Fig. 980.** First endopod, ventral. **Fig. 981.** Tip of first exopod. **Fig. 982.** Third leg, ventral. **Fig. 983.** Third endopod, ventral. **Fig. 984.** Tip of fourth exopod, ventral. **Fig. 985.** Uropod, ventral.

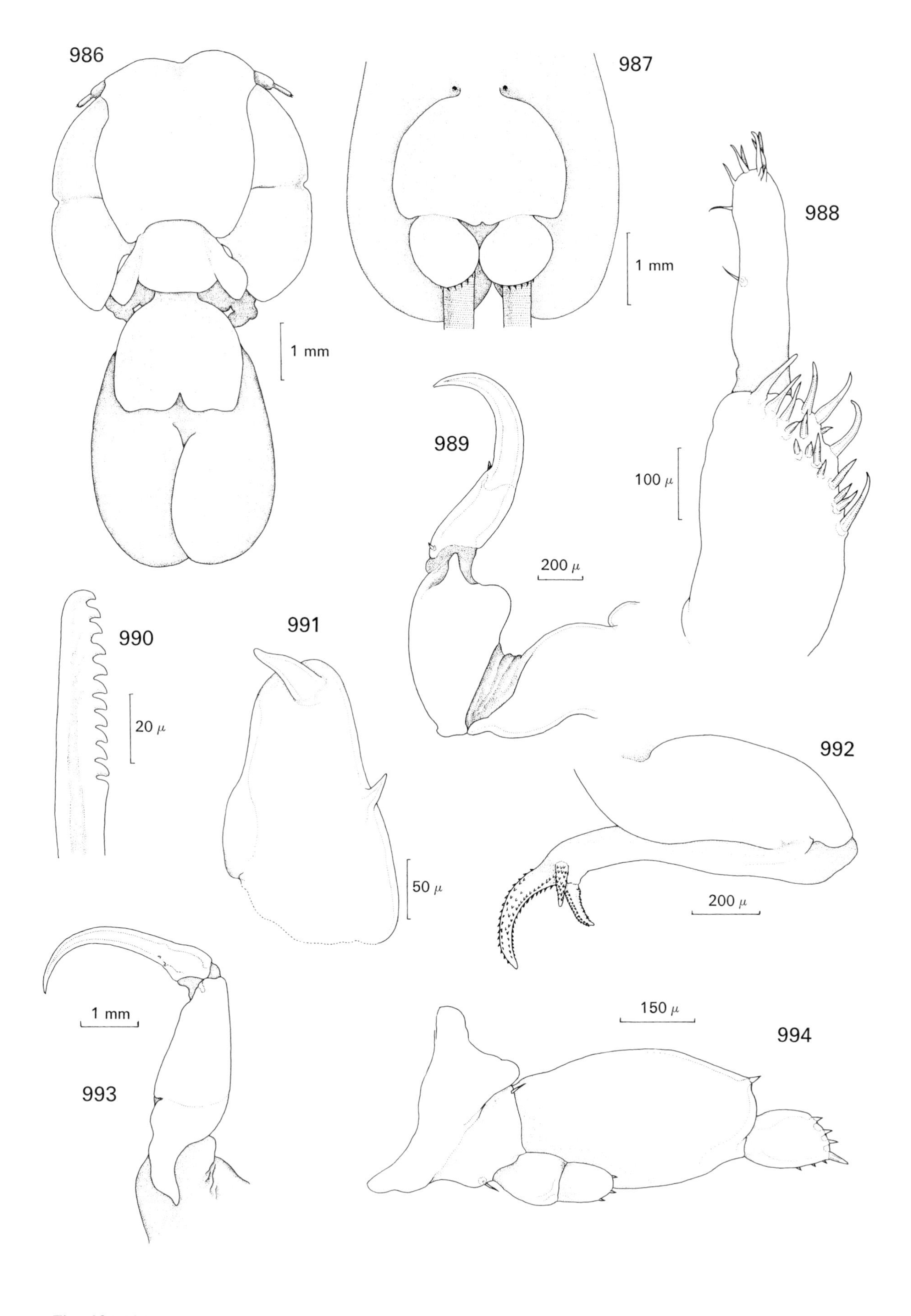

Figs. 986–994.

Luetkenia asterodermi Claus, 1864. **Fig. 986.** Female, dorsal. **Fig. 987.** Posterior extremity, ventral. **Fig. 988.** First antenna, ventral. **Fig. 989.** Second antenna, ventral. **Fig. 990.** Tip of mandible, lateral. **Fig. 991.** First maxilla, lateral. **Fig. 992.** Second maxilla, ventral. **Fig. 993.** Maxilliped, medial. **Fig. 994.** First leg, ventral.

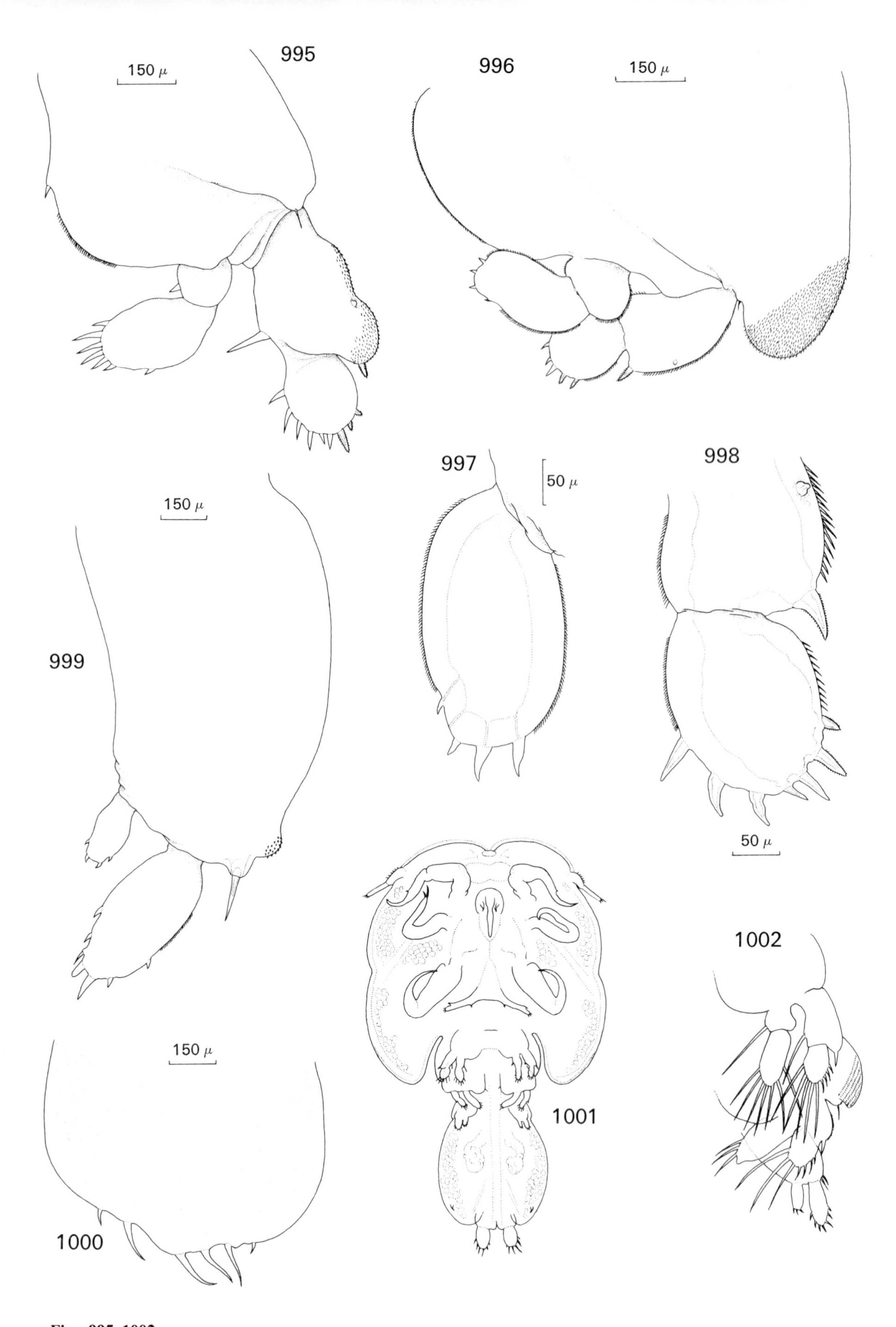

Figs. 995–1002.
Luetkenia asterodermi Claus, 1864. **Fig. 995.** Female second leg, ventral. **Fig. 996.** Third leg, ventral. **Fig. 997.** Tip of third endopod, ventral. **Fig. 998.** Tip of third exopod, ventral. **Fig. 999.** Fourth leg, ventral. **Fig. 1000.** Tip of uropod, ventral. **Fig. 1001.** Male, ventral. **Fig. 1002.** Male legs 2–4, ventral.
Figs. 1001, 1002 redrawn from Claus (1864).

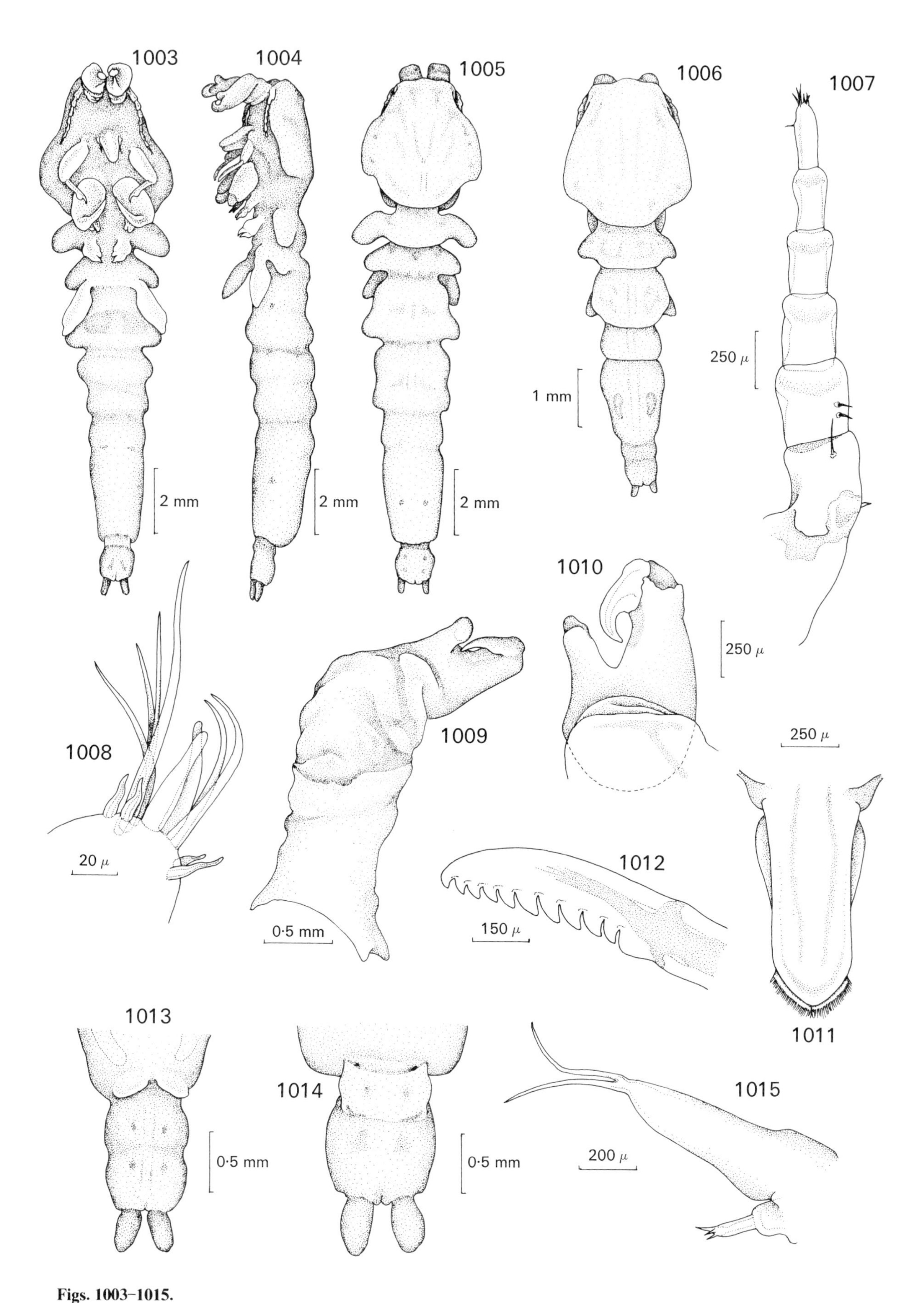

Figs. 1003–1015.
Dichelesthium oblongum (Abildgaard, 1794). **Fig. 1003.** Female, ventral. **Fig. 1004.** Same, lateral. **Fig. 1005.** Same, dorsal. **Fig. 1006.** Male, dorsal. **Fig. 1007.** Male and female, first antenna, ventral. **Fig. 1008.** Same, tip. **Fig. 1009.** Second antenna, extended, medial. **Fig. 1010.** Same, tip, medial. **Fig. 1011.** Mouth, anterior. **Fig. 1012.** Tip of mandible, lateral. **Fig. 1013.** Male posterior extremity, ventral. **Fig. 1014.** Female posterior extremity, ventral. **Fig. 1015.** Male and female, first maxilla, lateral.

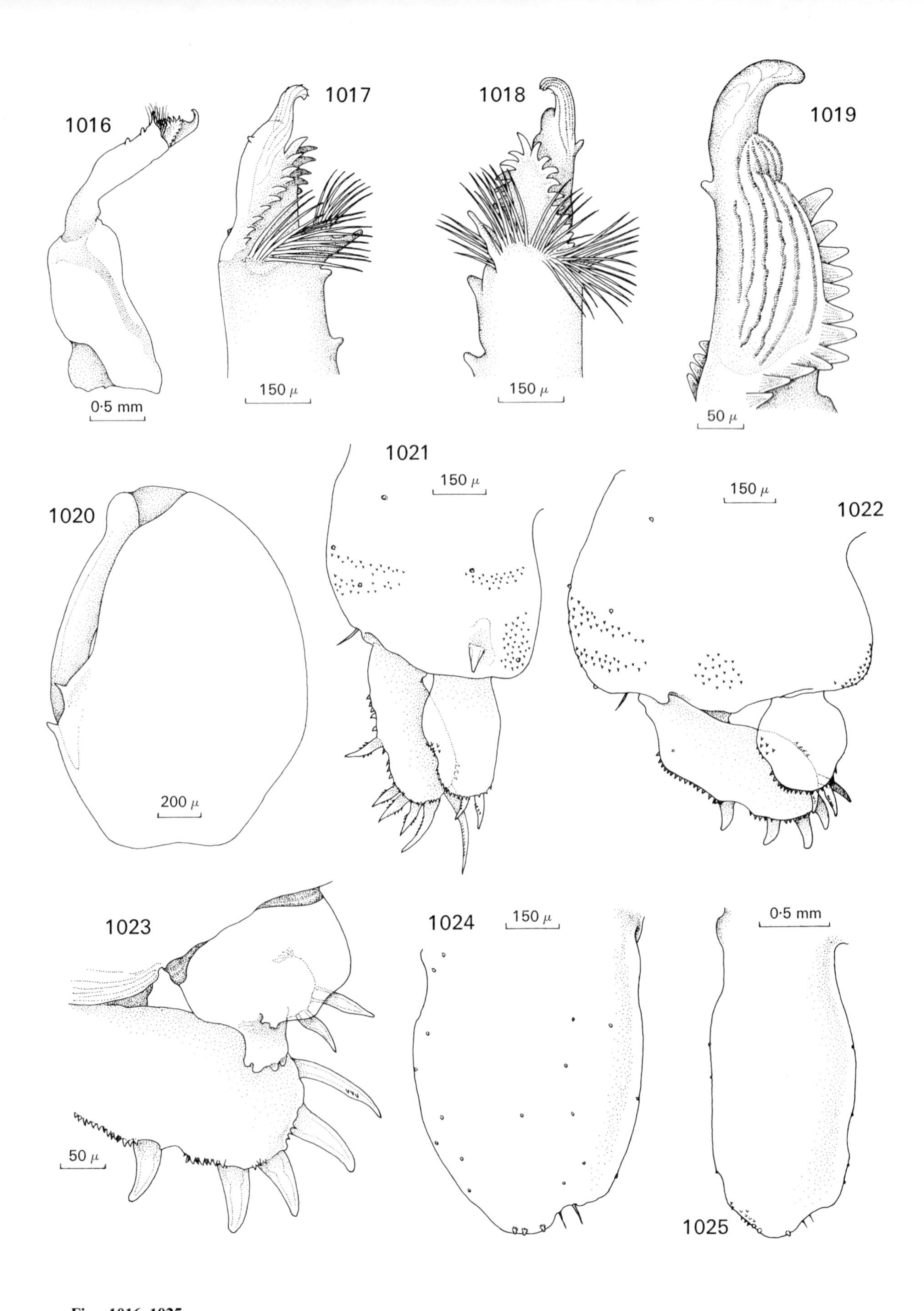

Figs. 1016–1025.
Dichelesthium oblongum (Abildgaard, 1794). **Fig. 1016.** Second maxilla, dorsal. **Fig. 1017.** Same, tip, dorsal. **Fig. 1018.** Same, tip, posteroventral. **Fig. 1019.** Same, tip, anterodorsal. **Fig. 1020.** Maxilliped, posterior. **Fig. 1021.** Female first leg, ventral. **Fig. 1022.** Second leg, ventral. **Fig. 1023.** Male second leg, rami, ventral. **Fig. 1024.** Female third leg, ventral. **Fig. 1025.** Male third leg, ventral.

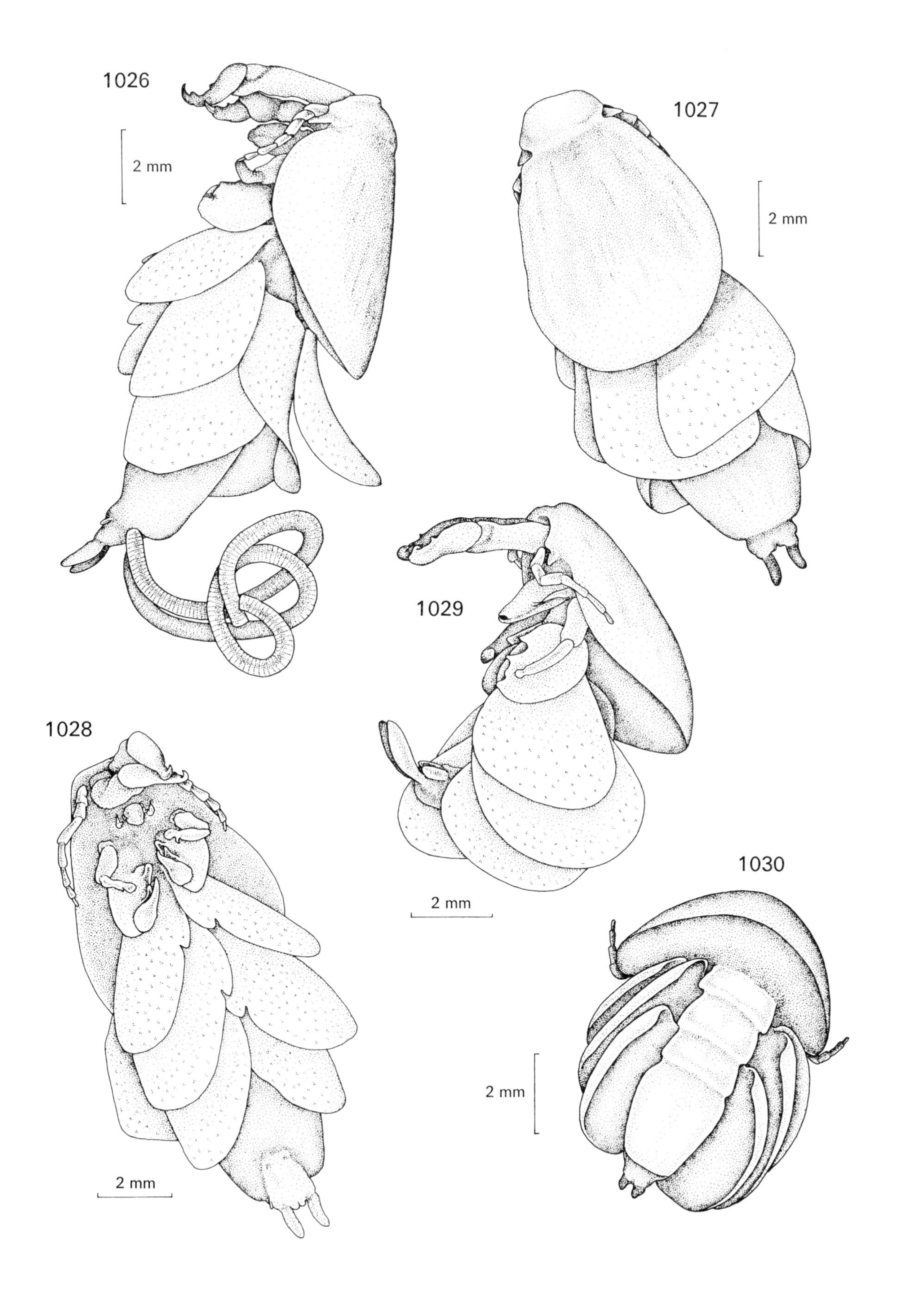

Figs. 1026–1030.
Anthosoma crassum (Abildgaard, 1794). **Fig. 1026.** Female, lateral. **Fig. 1027.** Same, dorsal. **Fig. 1028.** Same, ventral. **Fig. 1029.** Male, lateral. **Fig. 1030.** Same, dorsal.

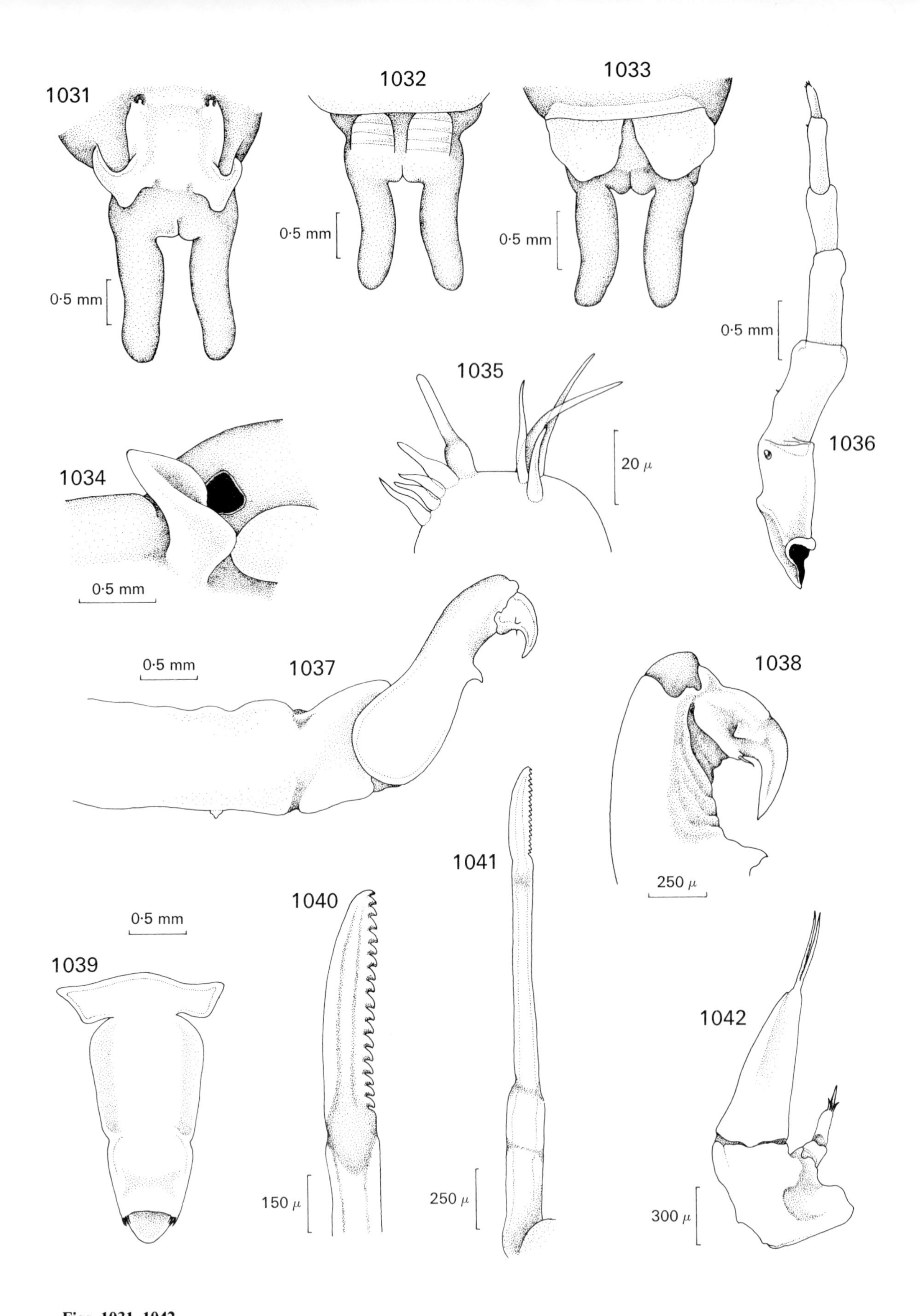

Figs. 1031–1042.
Anthosoma crassum (Abildgaard, 1794). **Fig. 1031.** Female posterior extremity, ventral. **Fig. 1032.** Same, dorsal. **Fig. 1033.** Male posterior extremity, ventral. **Fig. 1034.** Male and female, anterolateral corner of dorsal shield, seen from ventrolateral side. **Fig. 1035.** Tip of first antenna, ventral. **Fig. 1036.** First antenna, ventral. **Fig. 1037.** Second antenna, extended. **Fig. 1038.** Tip of second antenna, medial. **Fig. 1039.** Mouth, anterior. **Fig. 1040.** Tip of mandible, lateral. **Fig. 1041.** Mandible, lateral. **Fig. 1042.** First maxilla, lateral.

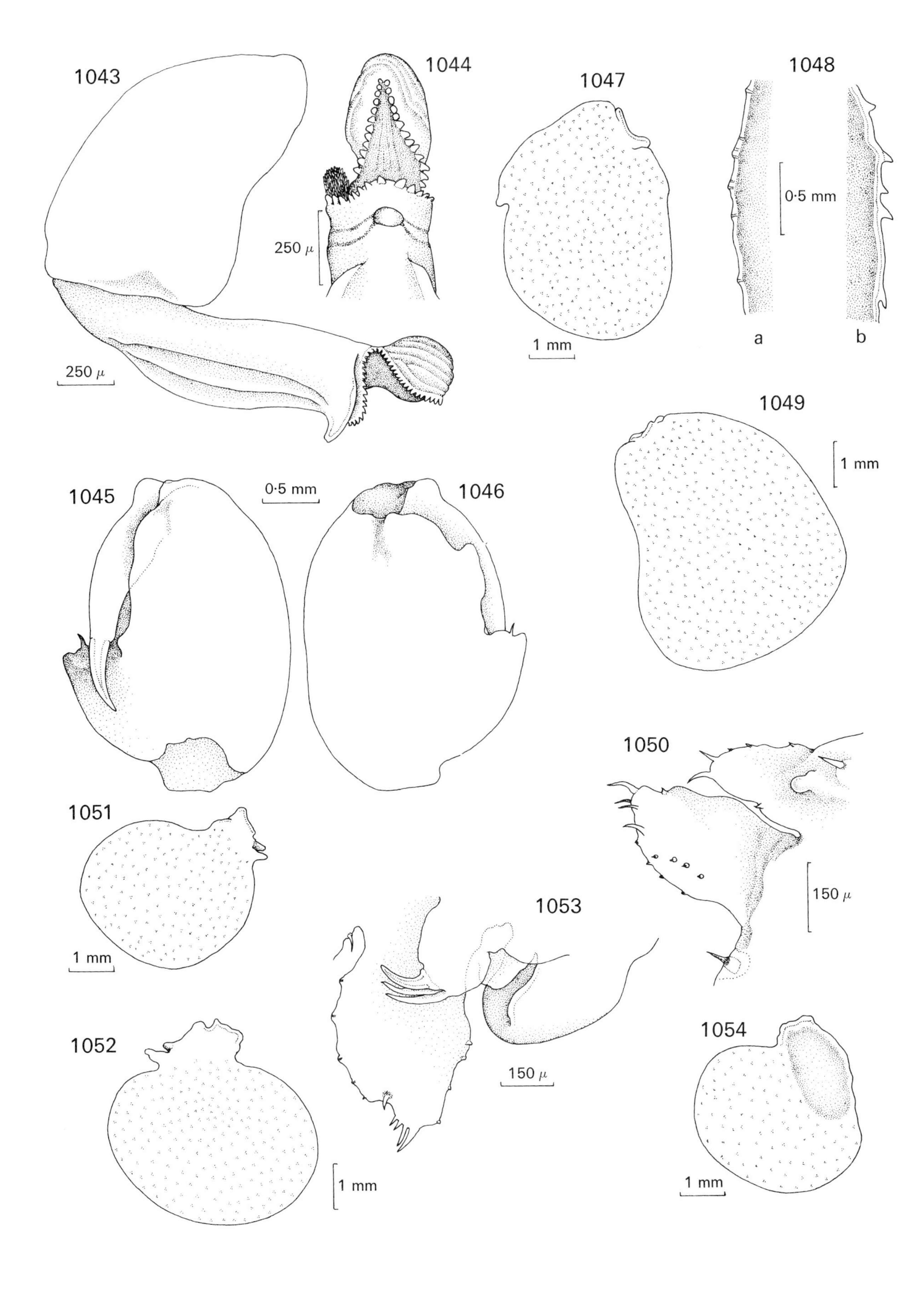

Figs. 1043–1054.
Anthosoma crassum (Abildgaard, 1794). **Fig. 1043.** Male and female, second maxilla, lateral. **Fig. 1044.** Same, tip, posterior. **Fig. 1045.** Maxilliped, dorsal. **Fig. 1046.** Same, ventral. **Fig. 1047.** Female first leg. **Fig. 1048.** Same, details of margins: (a) medial margin; (b) lateral margin. **Fig. 1049.** Third leg. **Fig. 1050.** Male first leg, rami. **Fig. 1051.** Same, entire. **Fig. 1052.** Male second leg. **Fig. 1053.** Same, rami. **Fig. 1054.** Third leg.

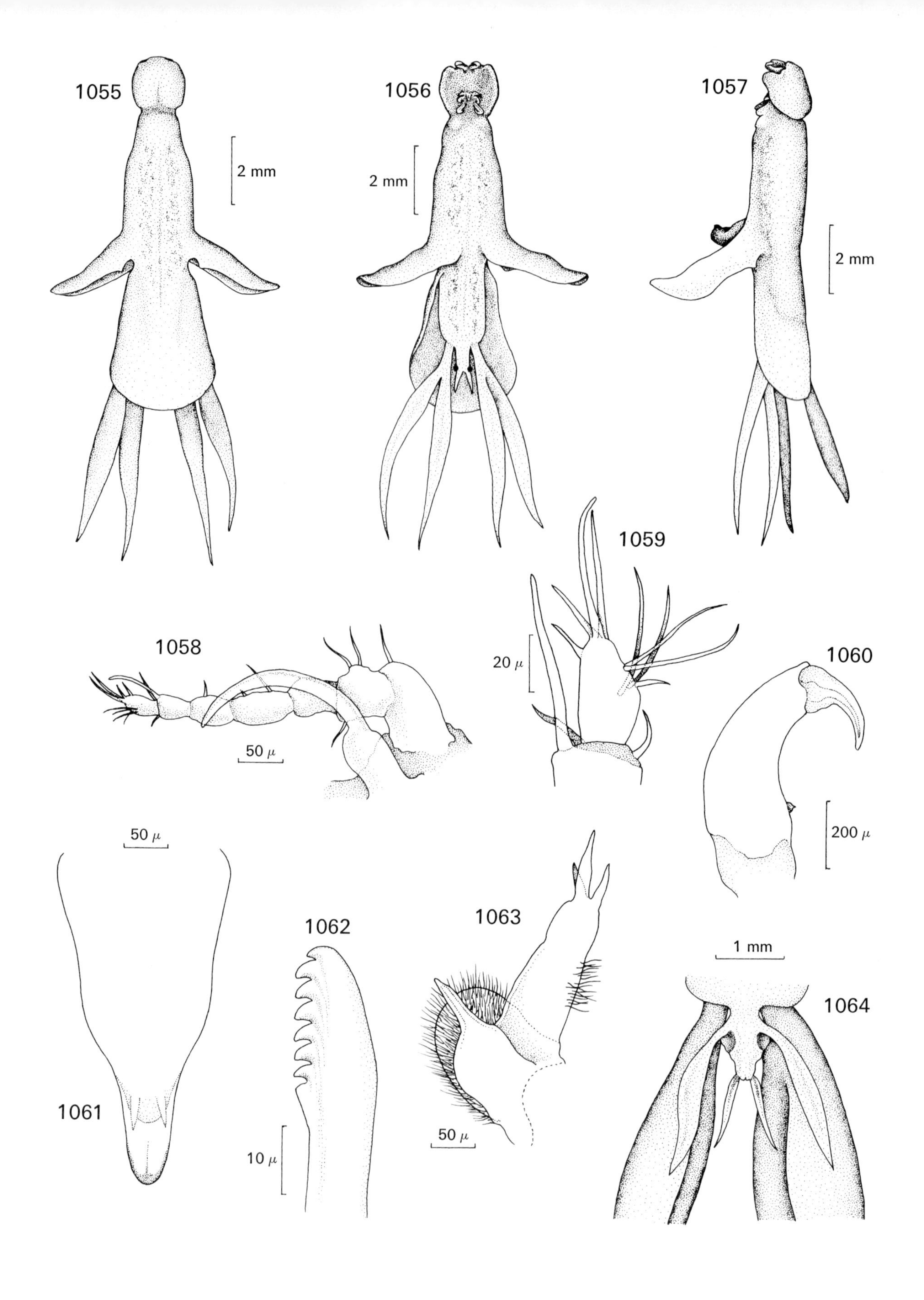

Figs. 1055–1064.
Lernanthropus gisleri van Beneden, 1852. **Fig. 1055.** Female, dorsal. **Fig. 1056.** Same, ventral. **Fig. 1057.** Same, lateral. **Fig. 1058.** First antenna, ventral. **Fig. 1059.** Same, tip, lateral. **Fig. 1060.** Second antenna, ventral. **Fig. 1061.** Mouth, anterior. **Fig. 1062.** Tip of mandible, lateral. **Fig. 1063.** First maxilla, lateral. **Fig. 1064.** Posterior extremity, dorsal (dorsal shield removed).

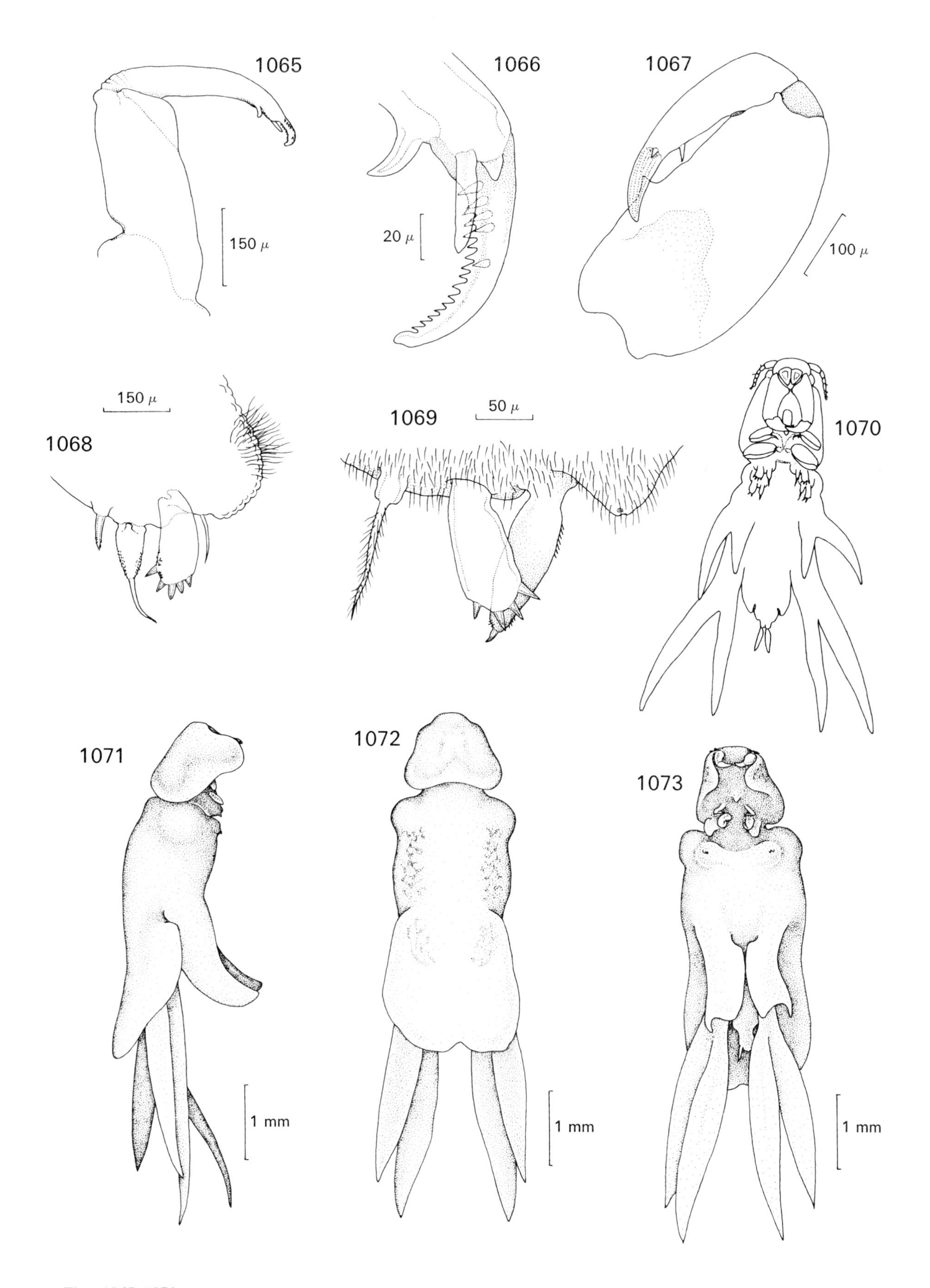

Figs. 1065–1073.
Lernanthropus gisleri van Beneden, 1852. **Fig. 1065.** Female second maxilla, ventral. **Fig. 1066.** Same, tip, dorsal. **Fig. 1067.** Maxilliped, ventral. **Fig. 1068.** First leg, ventral. **Fig. 1069.** Rami of second leg, ventral. **Fig. 1070.** Male, ventral.
Lernanthropus kroyeri van Beneden, 1851. **Fig. 1071.** Female, lateral. **Fig. 1072.** Same, dorsal. **Fig. 1073.** Same, ventral.
Fig. 1070 redrawn from Heider (1879).

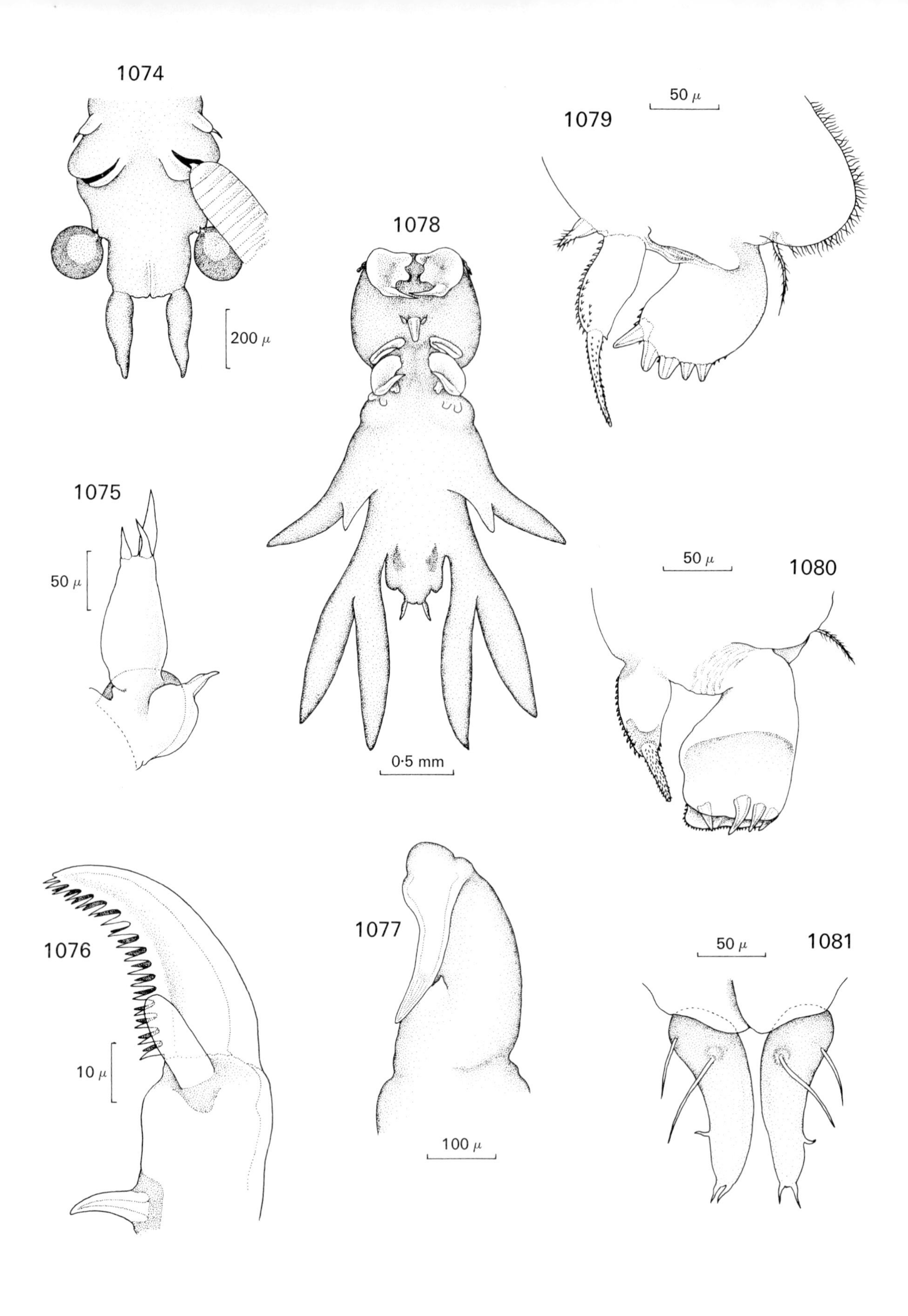

Figs. 1074–1081.

Lernanthropus kroyeri van Beneden, 1851. **Fig. 1074.** Female posterior extremity, dorsal (dorsal shield removed). **Fig. 1075.** First maxilla, lateral. **Fig. 1076.** Tip of second maxilla, dorsal. **Fig. 1077.** Maxilliped, medial. **Fig. 1078.** Male, ventral. **Fig. 1079.** First leg, dorsal. **Fig. 1080.** Rami of second leg, dorsal. **Fig. 1081.** Uropods, dorsal.

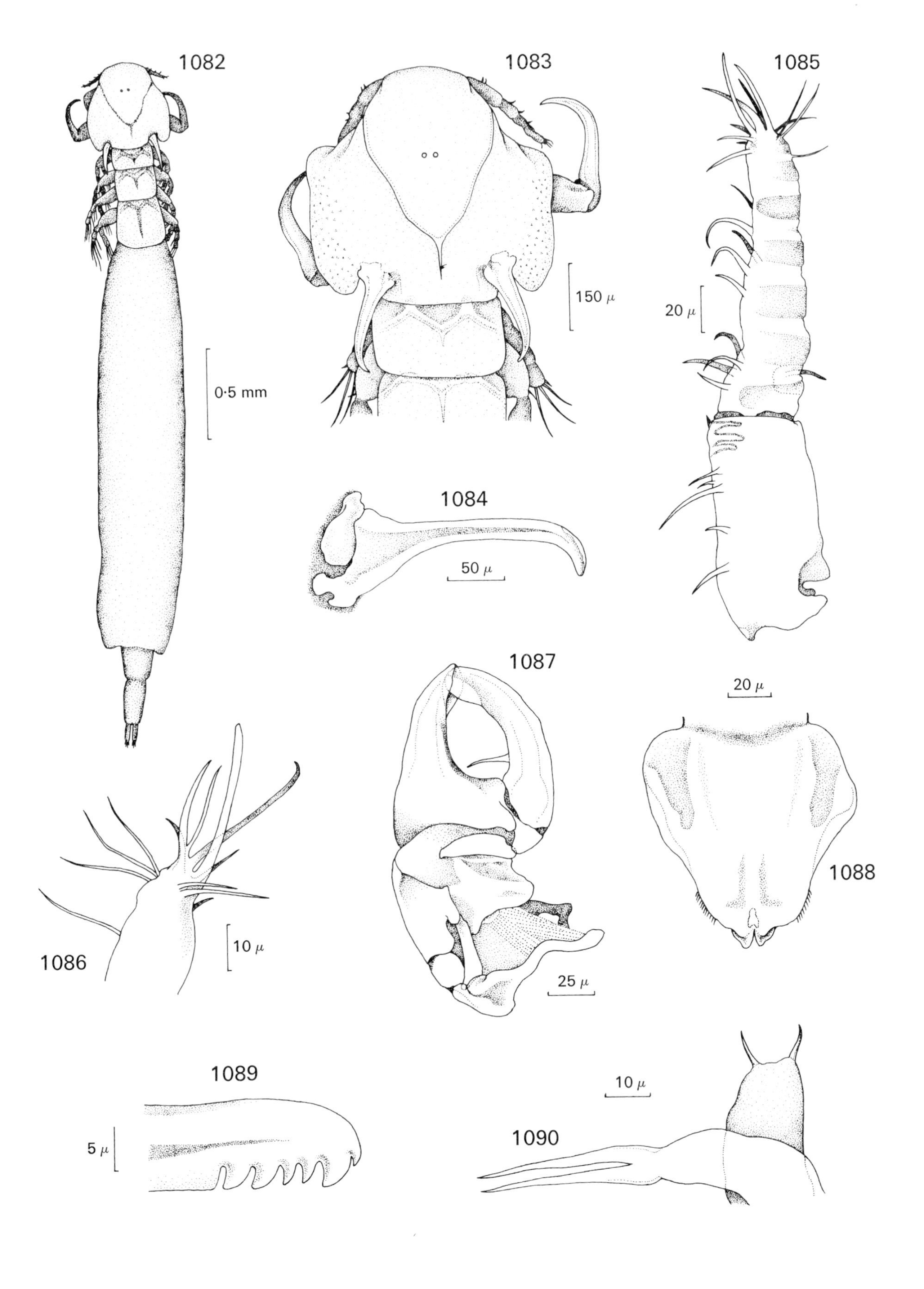

Figs. 1082–1090.

Kroyeria lineata van Beneden, 1853. **Fig. 1082.** Female, dorsal. **Fig. 1083.** Cephalothorax, dorsal. **Fig. 1084.** Dorsal stylet. **Fig. 1085.** First antenna, ventral. **Fig. 1086.** Same, tip, ventral. **Fig. 1087.** Second antenna. **Fig. 1088.** Mouth, posterior. **Fig. 1089.** Tip of mandible, lateral. **Fig. 1090.** First maxilla, ventral.

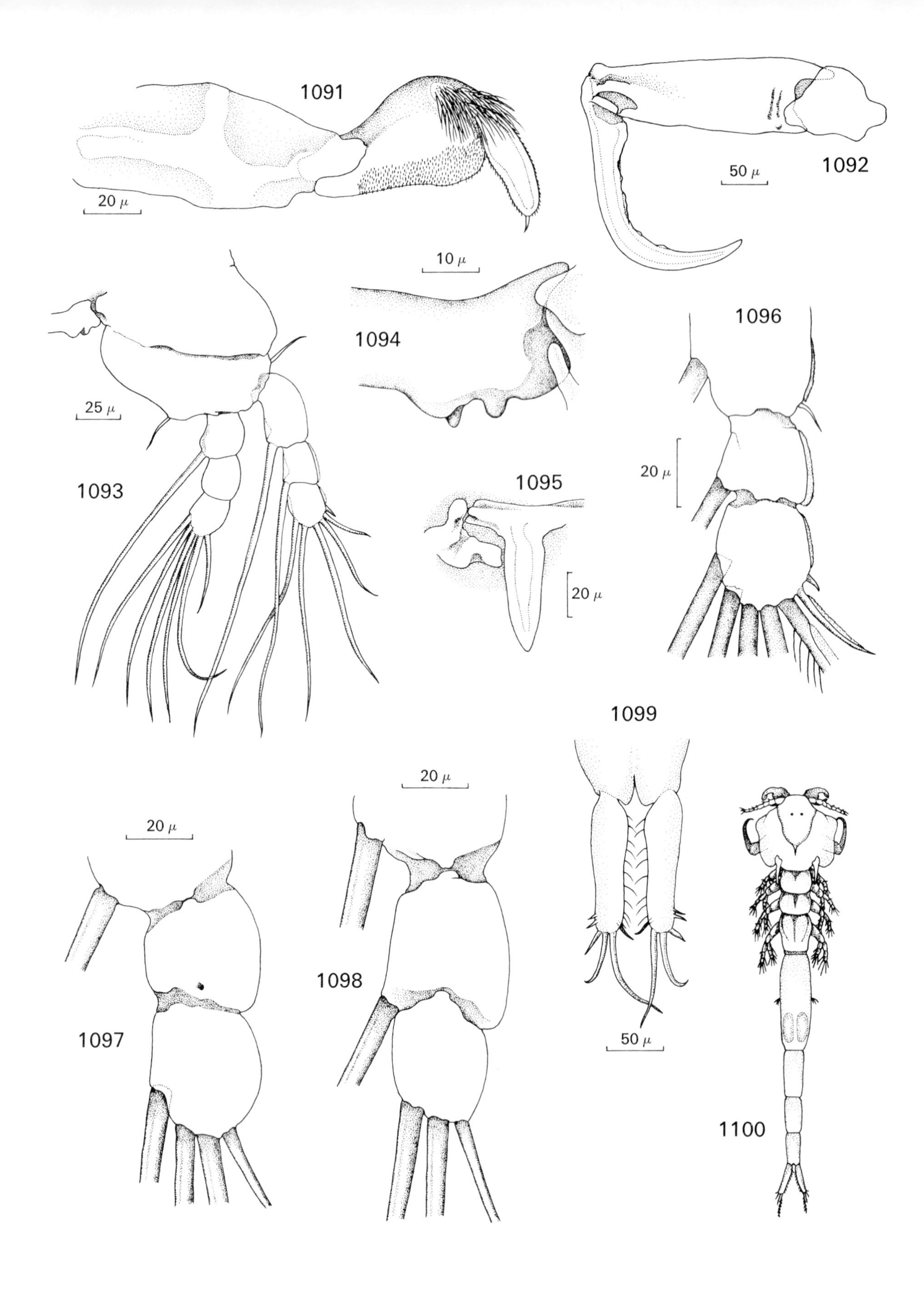

Figs. 1091–1100.
Kroyeria lineata van Beneden, 1853. **Fig. 1091.** Female second maxilla, medial. **Fig. 1092.** Maxilliped, ventral. **Fig. 1093.** First leg, ventral. **Fig. 1094.** End of first interpodal bar, ventral. **Fig. 1095.** End of second interpodal bar, ventral. **Fig. 1096.** Second—fourth exopod, ventral. **Fig. 1097.** Third endopod, ventral. **Fig. 1098.** Fourth endopod, ventral. **Fig. 1099.** Uropods, ventral. **Fig. 1100.** Male, dorsal.
Fig. 1100 redrawn from Scott and Scott (1913).

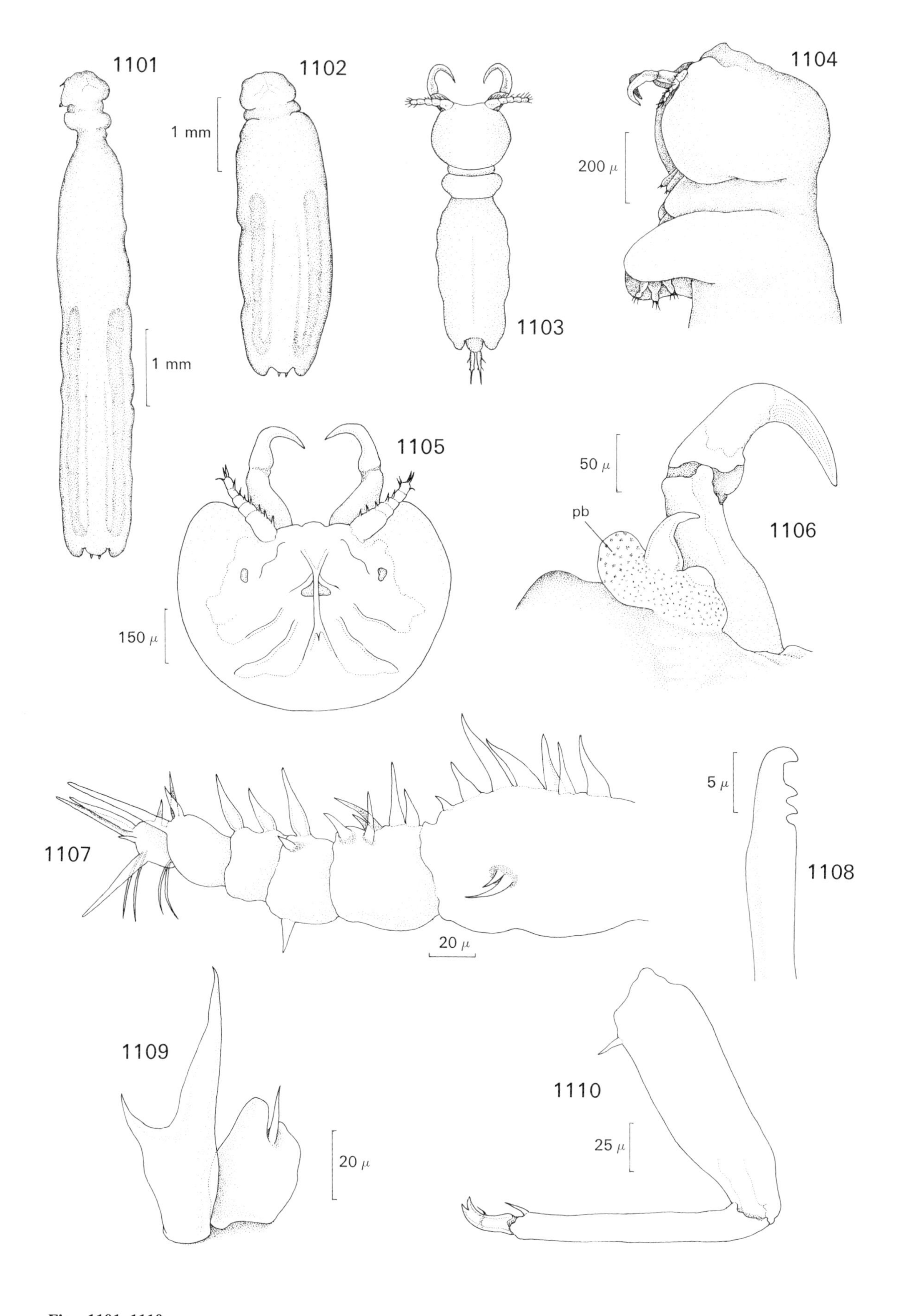

Figs. 1101–1110.
Hatschekia hippoglossi (Krøyer, 1837). **Fig. 1101.** Female, relaxed, dorsal. **Fig. 1102.** Same, contracted. **Fig. 1103.** Male, dorsal. **Fig. 1104.** Female cephalothorax, lateral. **Fig. 1105.** Same, anterodorsal. **Fig. 1106.** Second antenna, ventral. **Fig. 1107.** First antenna, dorsal. **Fig. 1108.** Tip of mandible, lateral. **Fig. 1109.** First maxilla, ventral. **Fig. 1110.** Second maxilla, ventral.
Fig. 1103 redrawn from Scott and Scott (1913).

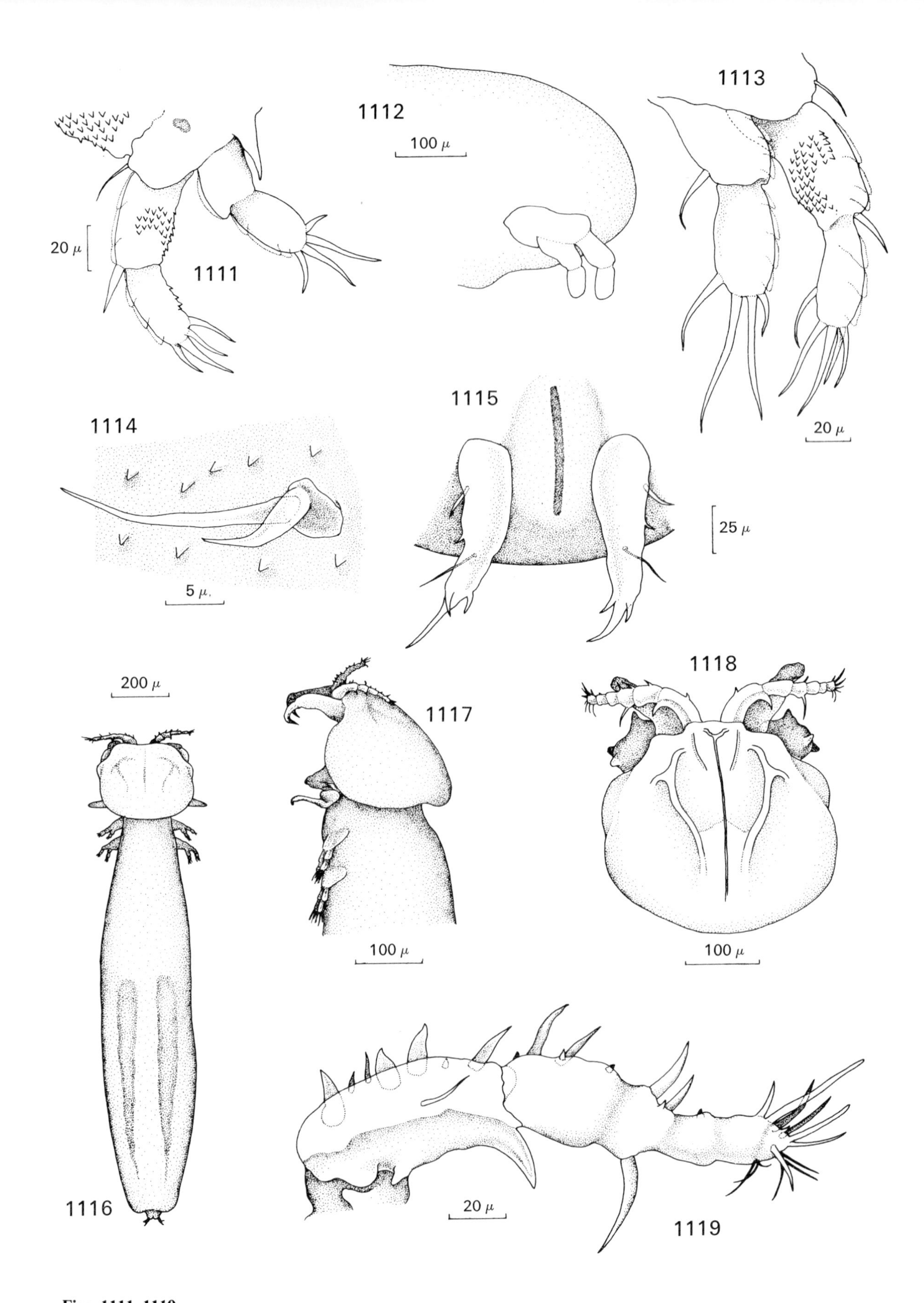

Figs. 1111–1119.
Hatschekia hippoglossi (Krøyer, 1837). **Fig. 1111.** Female, rami of first leg, ventral. **Fig. 1112.** Second leg in outline, ventral. **Fig. 1113.** Rami of second leg, ventral. **Fig. 1114.** Third leg, ventral. **Fig. 1115.** Posterior extremity, ventral.
Hatschekia cluthae (T. Scott, 1902). **Fig. 1116.** Female, dorsal. **Fig. 1117.** Cephalothorax, lateral. **Fig. 1118.** Same, dorsal. **Fig. 1119.** First antenna, posterodorsal.

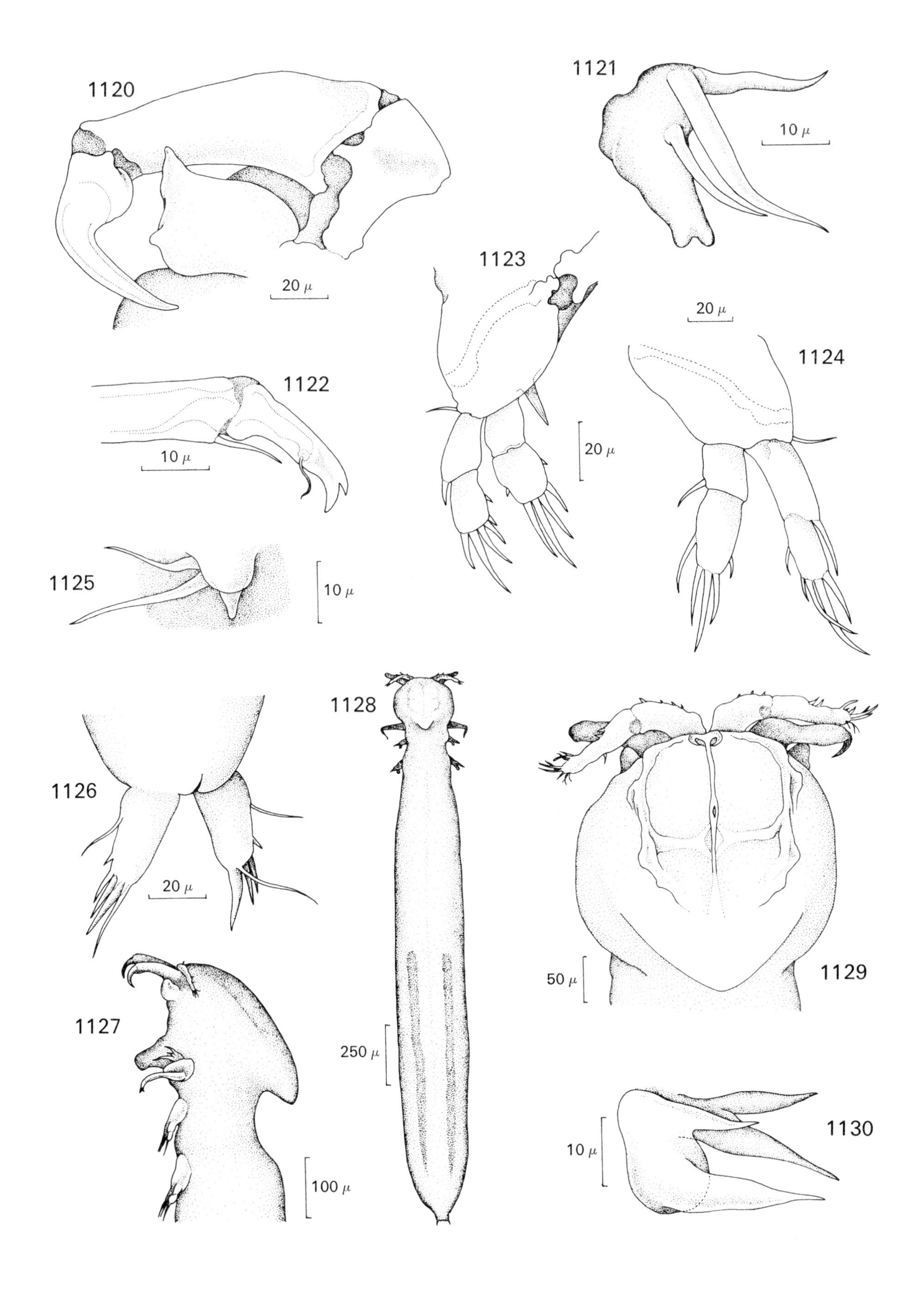

Figs. 1120–1130.
Hatschekia cluthae (T. Scott, 1902). **Fig. 1120.** Second antenna, ventral. **Fig. 1121.** First maxilla, ventral. **Fig. 1122.** Tip of second maxilla, ventral. **Fig. 1123.** First leg, ventral. **Fig. 1124.** Second leg, ventral. **Fig. 1125.** Third leg, ventral. **Fig. 1126.** Posterior extremity, dorsolateral.
Hatschekia pagellibogneravei (Hesse, 1879). **Fig. 1127.** Female cephalothorax, lateral. **Fig. 1128.** Female, dorsal. **Fig. 1129.** Cephalothorax, dorsal. **Fig. 1130.** First maxilla, ventral.

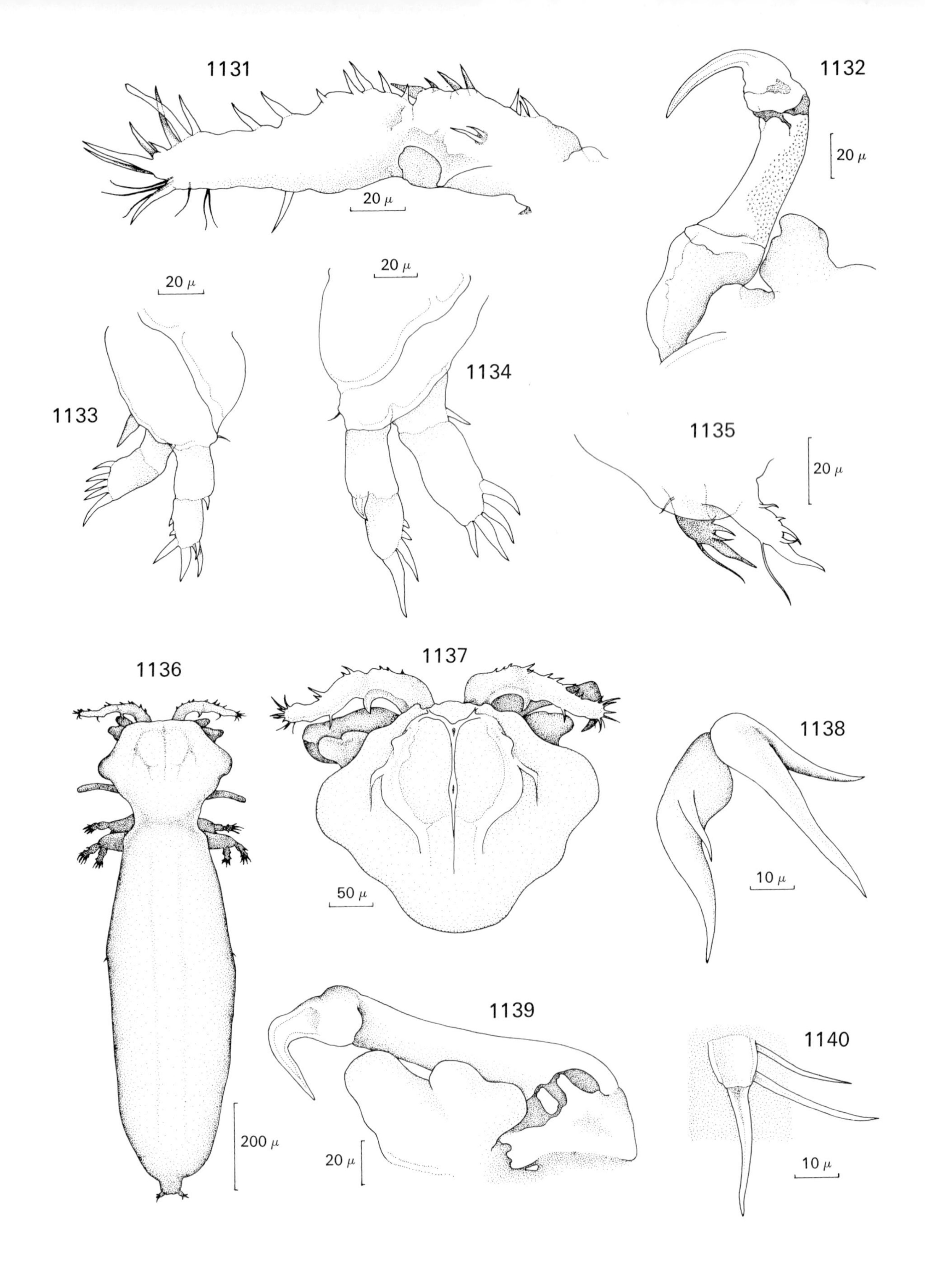

Figs. 1131–1140.
Hatschekia pagellibogneravei (Hesse, 1879). **Fig. 1131.** First antenna, dorsal. **Fig. 1132.** Second antenna, ventral. **Fig. 1133.** First leg, ventral. **Fig. 1134.** Second leg, ventral. **Fig. 1135.** Posterior extremity, dorsolateral.
Hatschekia labracis (van Beneden, 1871). **Fig. 1136.** Female, dorsal. **Fig. 1137.** Cephalothorax, dorsal. **Fig. 1138.** First maxilla, ventral. **Fig. 1139.** Second antenna, ventral. **Fig. 1140.** Third leg, ventral.

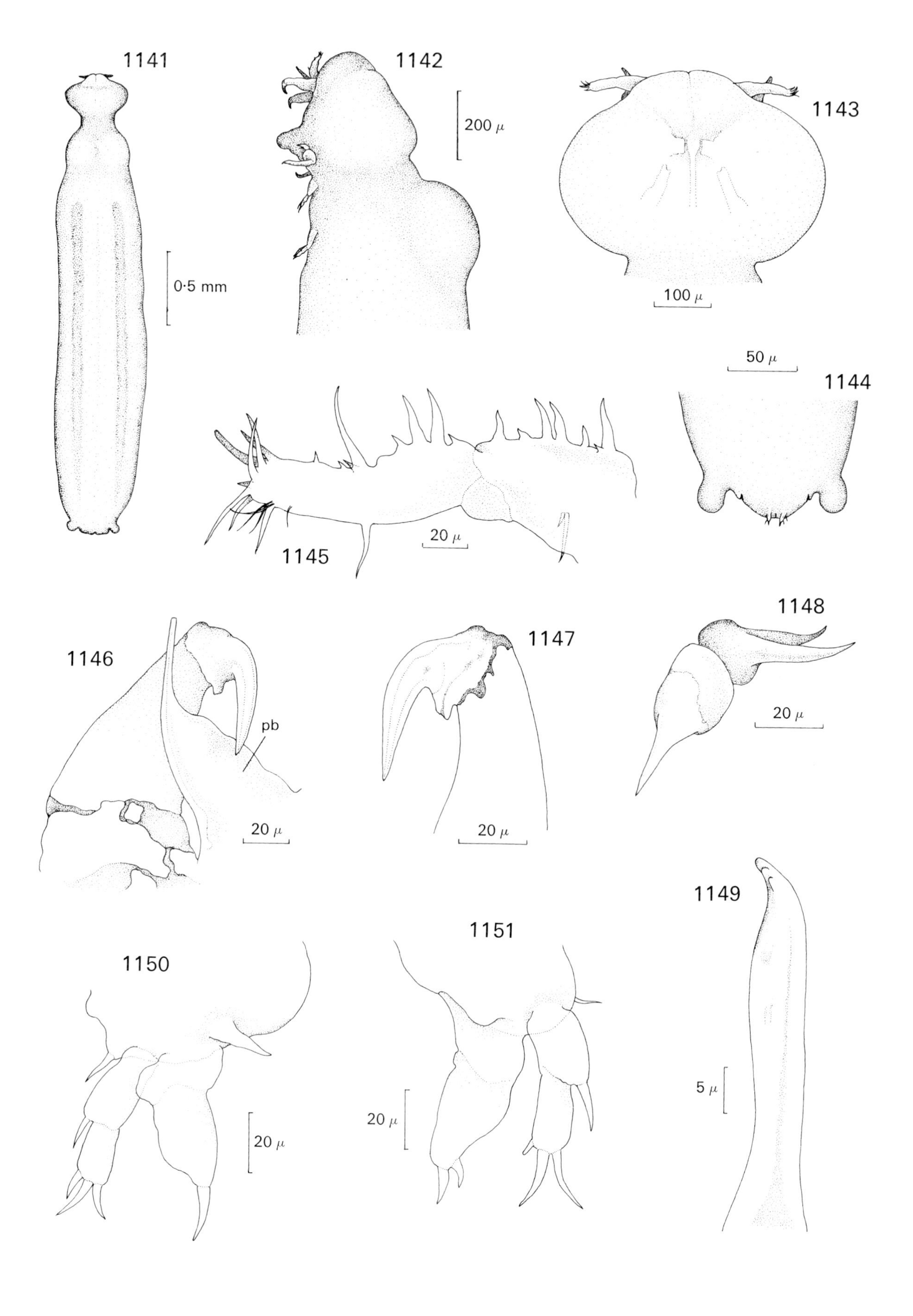

Figs. 1141–1151.
Hatschekia mulli (van Beneden, 1851). **Fig. 1141.** Female, dorsal. **Fig. 1142.** Cephalothorax, lateral. **Fig. 1143.** Same, dorsal. **Fig. 1144.** Posterior extremity, ventral. **Fig. 1145.** First antenna, ventral. **Fig. 1146.** Second antenna, ventral. **Fig. 1147.** Same, claw in full view. **Fig. 1148.** First maxilla, ventral. **Fig. 1149.** Mandible, anterior. **Fig. 1150.** First leg, ventral. **Fig. 1151.** Second leg, ventral.

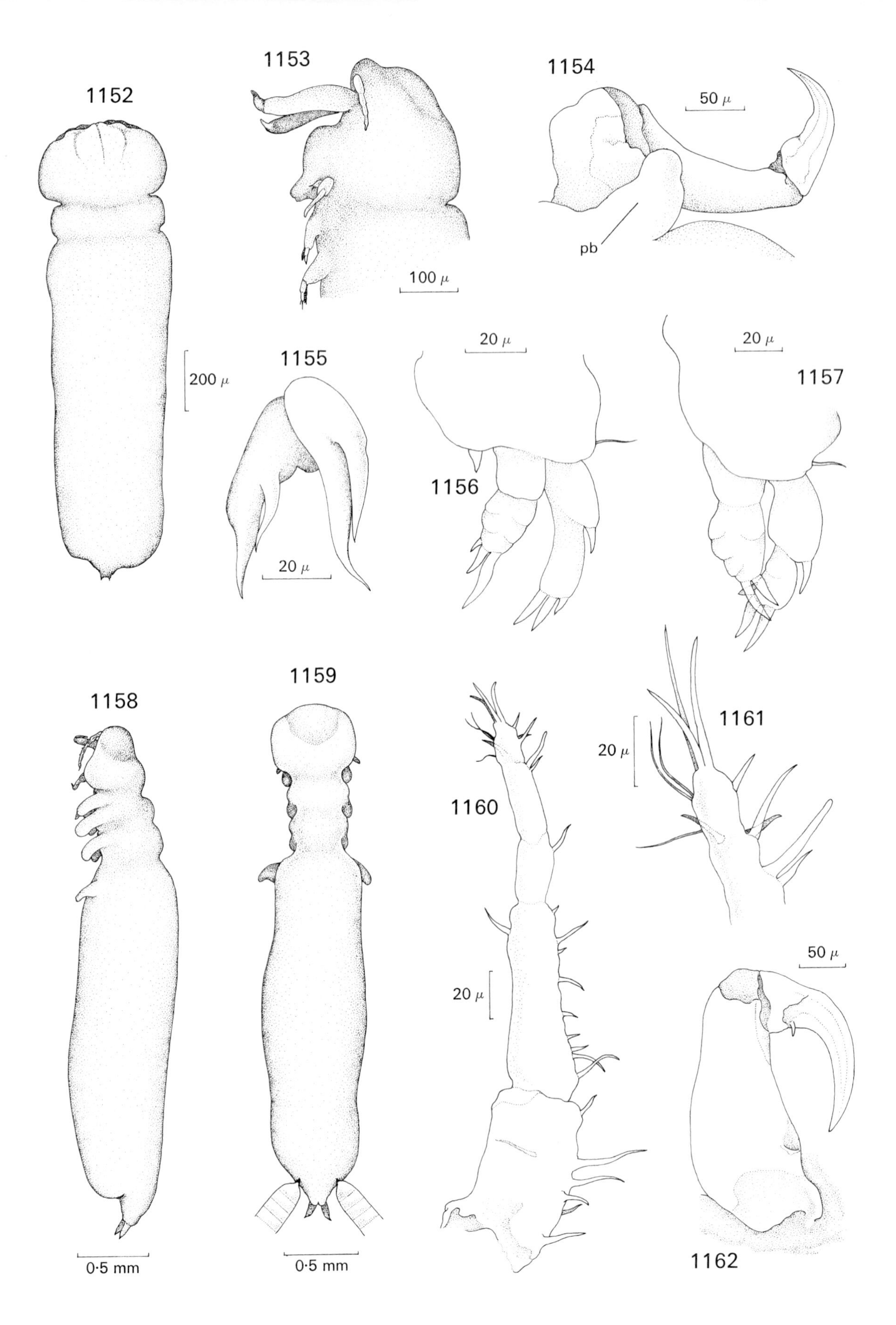

Figs. 1152–1162.
Hatschekia pygmaea T. Scott and A. Scott, 1913. **Fig. 1152.** Female, dorsal. **Fig. 1153.** Cephalothorax, dorsal. **Fig. 1154.** Second antenna, ventral. **Fig. 1155.** First maxilla, ventral. **Fig. 1156.** First leg, ventral. **Fig. 1157.** Second leg, ventral.
Congericola pallidus van Beneden, 1854. **Fig. 1158.** Female, lateral. **Fig. 1159.** Same, dorsal. **Fig. 1160.** First antenna, ventral. **Fig. 1161.** Same, tip. **Fig. 1162.** Second antenna, ventral.

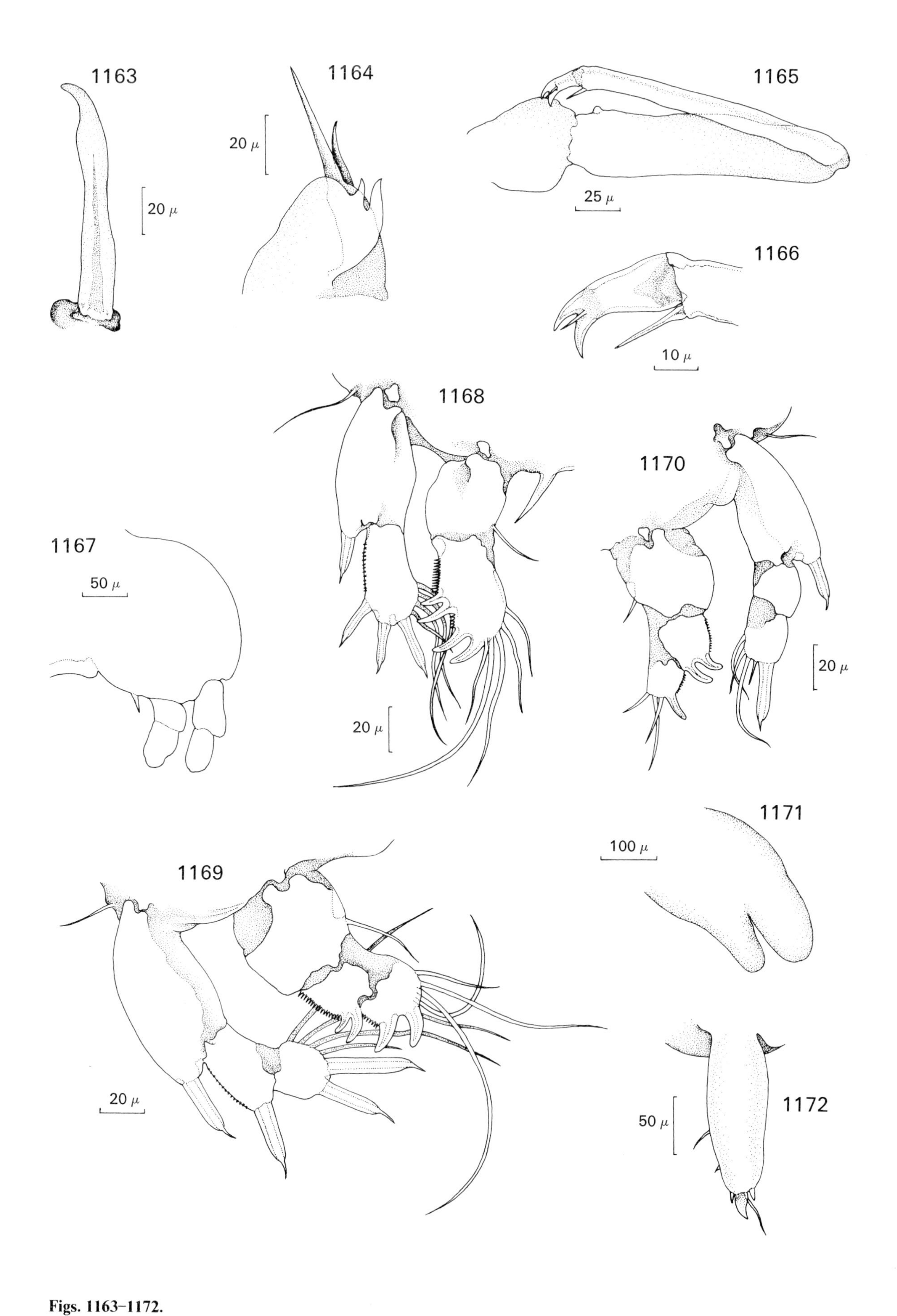

Figs. 1163–1172.
Congericola pallidus van Beneden, 1854. **Fig. 1163.** Female mandible, lateral. **Fig. 1164.** First maxilla, medial, **Fig. 1165.** Second maxilla, ventral. **Fig. 1166.** Same, tip. **Fig. 1167.** First leg in outline, ventral. **Fig. 1168.** Rami of first leg, ventral. **Fig. 1169.** Rami of second leg, ventral. **Fig. 1170.** Rami of third leg, ventral. **Fig. 1171.** Fourth leg, ventral. **Fig. 1172.** Uropod, ventral.

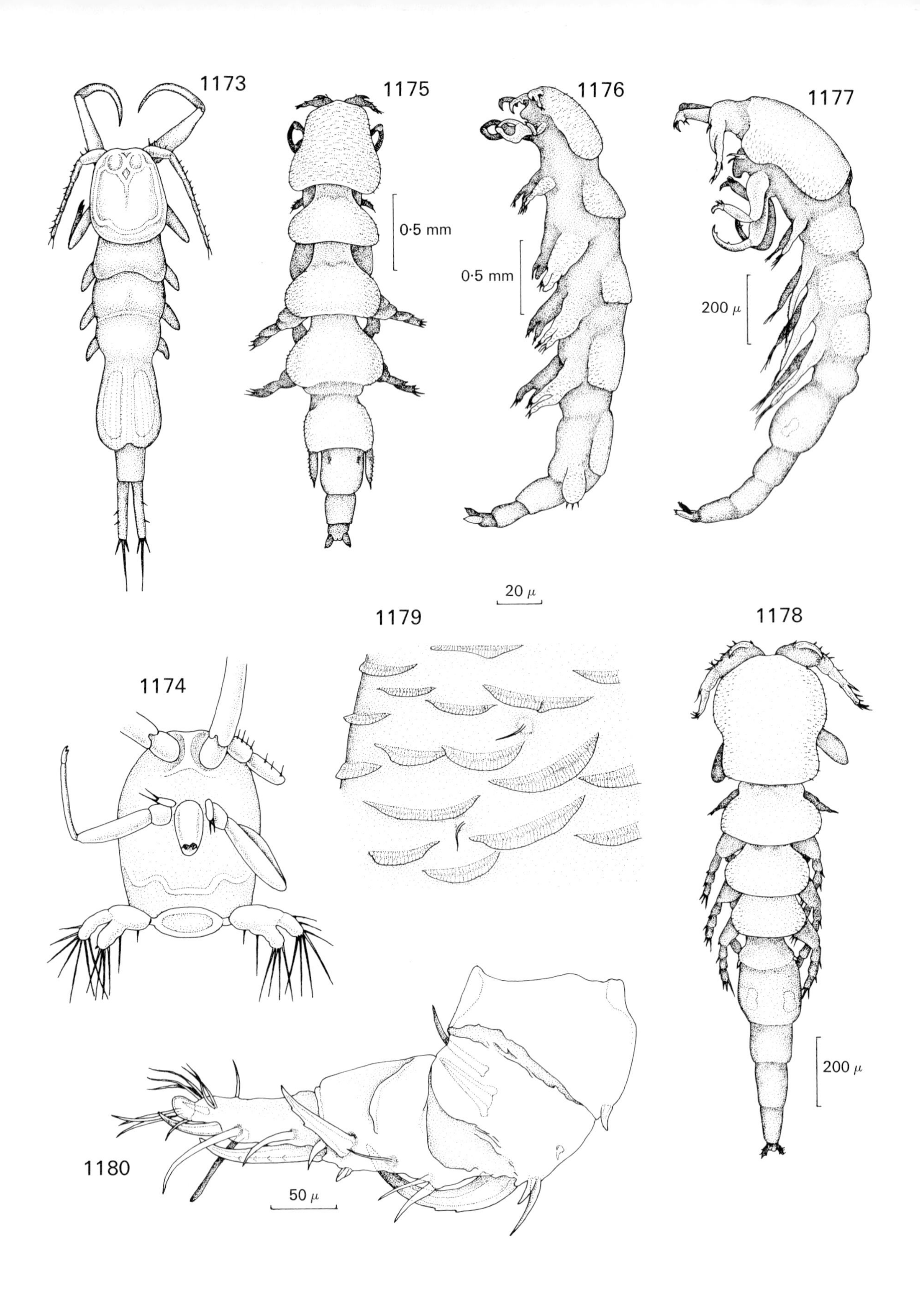

Figs. 1173–1180.
Congericola pallidus van Beneden, 1854. **Fig. 1173.** Male, dorsal. **Fig. 1174.** Same, cephalothorax, ventral. *Eudactylina acuta* van Beneden, 1853. **Fig. 1175.** Female, dorsal. **Fig. 1176.** Same lateral. **Fig. 1177.** Male, lateral. **Fig. 1178.** Same, dorsal. **Fig. 1179.** Male and female, detail of cephalothoracic cuticle. **Fig. 1180.** Female first antenna, ventral.
Figs. 1173, 1174 redrawn from van Beneden (1854).

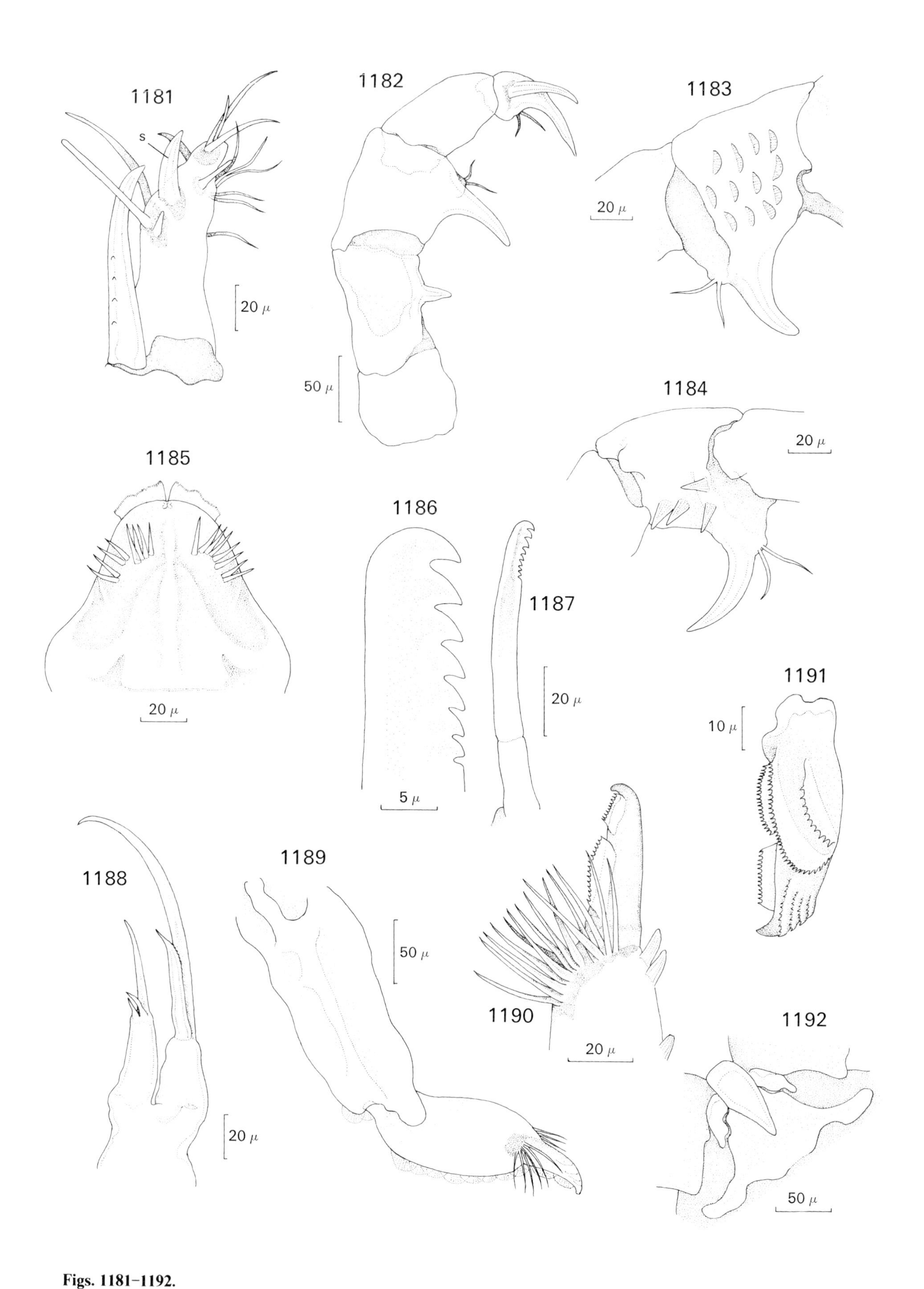

Figs. 1181–1192.
Eudactylina acuta van Beneden, 1853. **Fig. 1181.** Female first antenna, tip, dorsal. **Fig. 1182.** Second antenna, lateral. **Fig. 1183.** Same, second segment, medial. **Fig. 1184.** Same, male. **Fig. 1185.** Mouth, posterior. **Fig. 1186.** Tip of mandible, lateral. **Fig. 1187.** Mandible, lateral. **Fig. 1188.** First maxilla, ventral. **Fig. 1189.** Second maxilla, medial. **Fig. 1190.** Same, tip, anterior. **Fig. 1191.** Claw of second maxilla, medial. **Fig. 1192.** Female, rostrum, ventral.

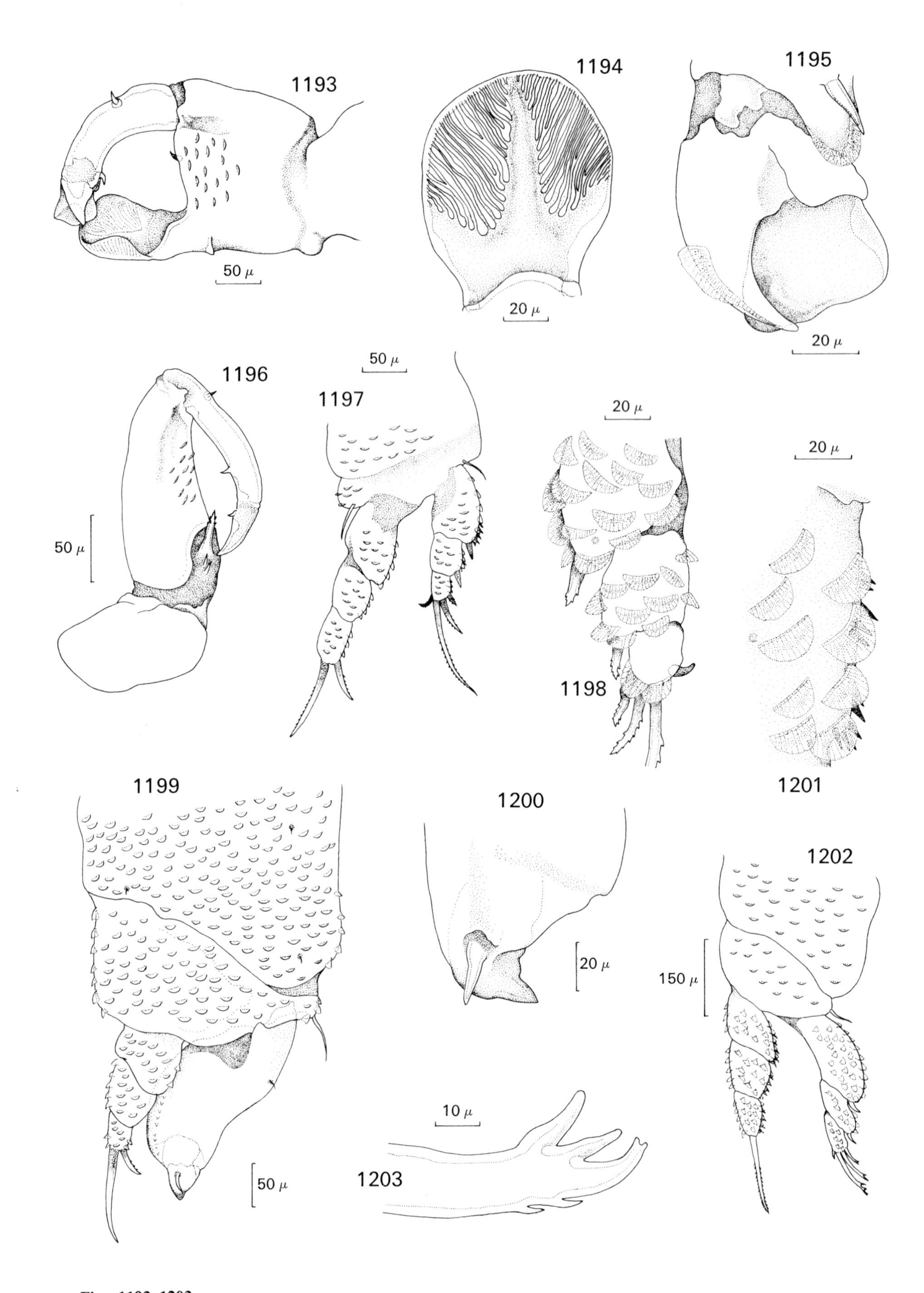

Figs. 1193–1203.
Eudactylina acuta van Beneden, 1853. **Fig. 1193.** Female maxilliped, ventral. **Fig. 1194.** Tip of immovable arm of chela, outer surface. **Fig. 1195.** Tip of opposable arm of chela, ventral. **Fig. 1196.** Male maxilliped. **Fig. 1197.** Female first leg, ventral. **Fig. 1198.** Tip of first exopod, ventral. **Fig. 1199.** Second leg, ventral. **Fig. 1200.** Tip of second exopod, ventral. **Fig. 1201.** Detail of medial margins of second endopod segments, legs three and four, ventral. **Fig. 1202.** Third and fourth leg, ventral. **Fig. 1203.** Same, tip of first exopodal spine.

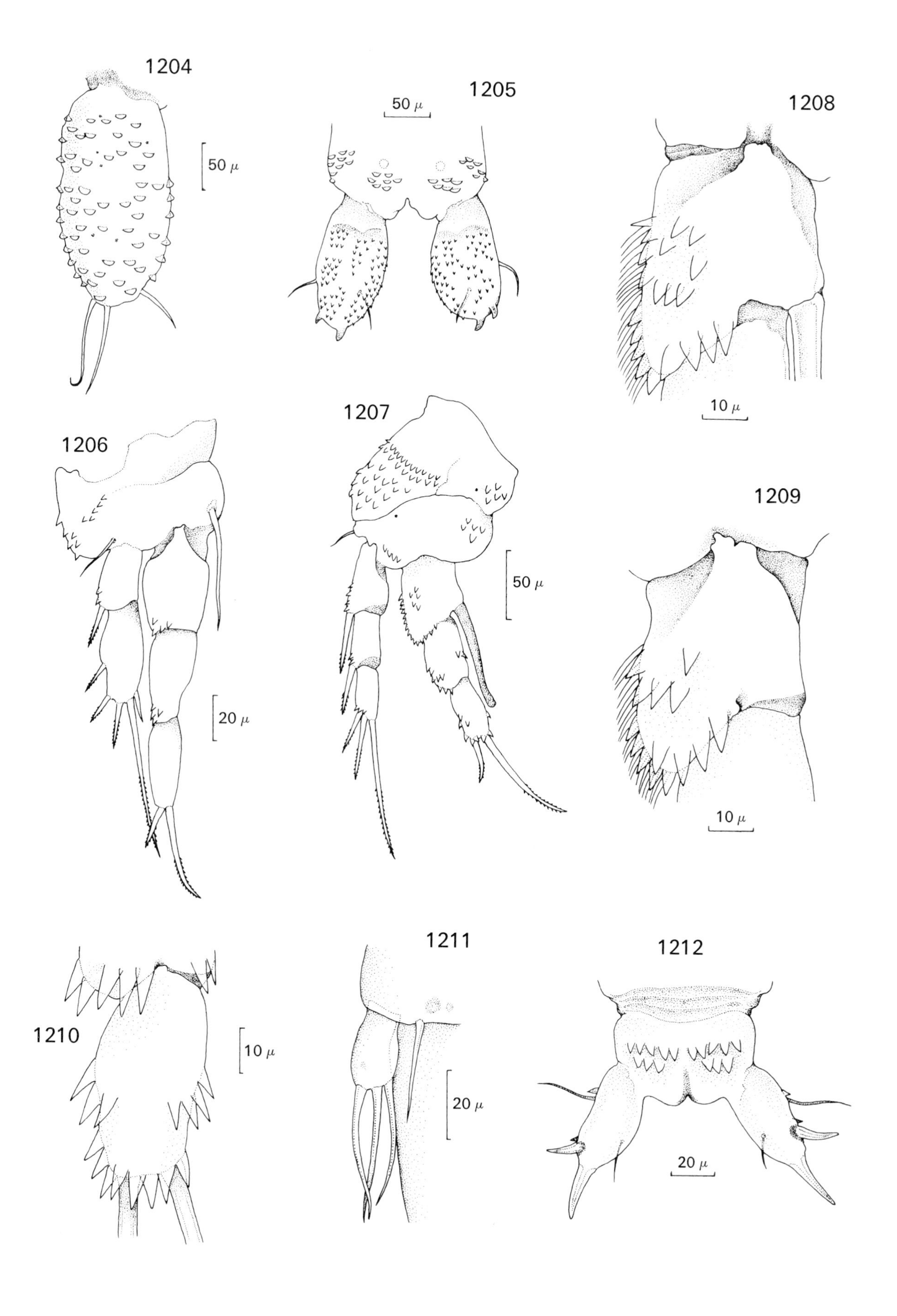

Figs. 1204–1212.
Eudactylina acuta van Beneden, 1853. **Fig. 1204.** Female fifth leg, lateral. **Fig. 1205.** Posterior extremity, ventral. **Fig. 1206.** Male first leg, ventral. **Fig. 1207.** Second leg, ventral. **Fig. 1208.** Base of second endopod, ventral. **Fig. 1209.** Base of third and fourth endopod, ventral. **Fig. 1210.** Tip of endopod 2–4, ventral. **Fig. 1211.** Fifth leg, dorsal. **Fig. 1212.** Posterior extremity, ventral.

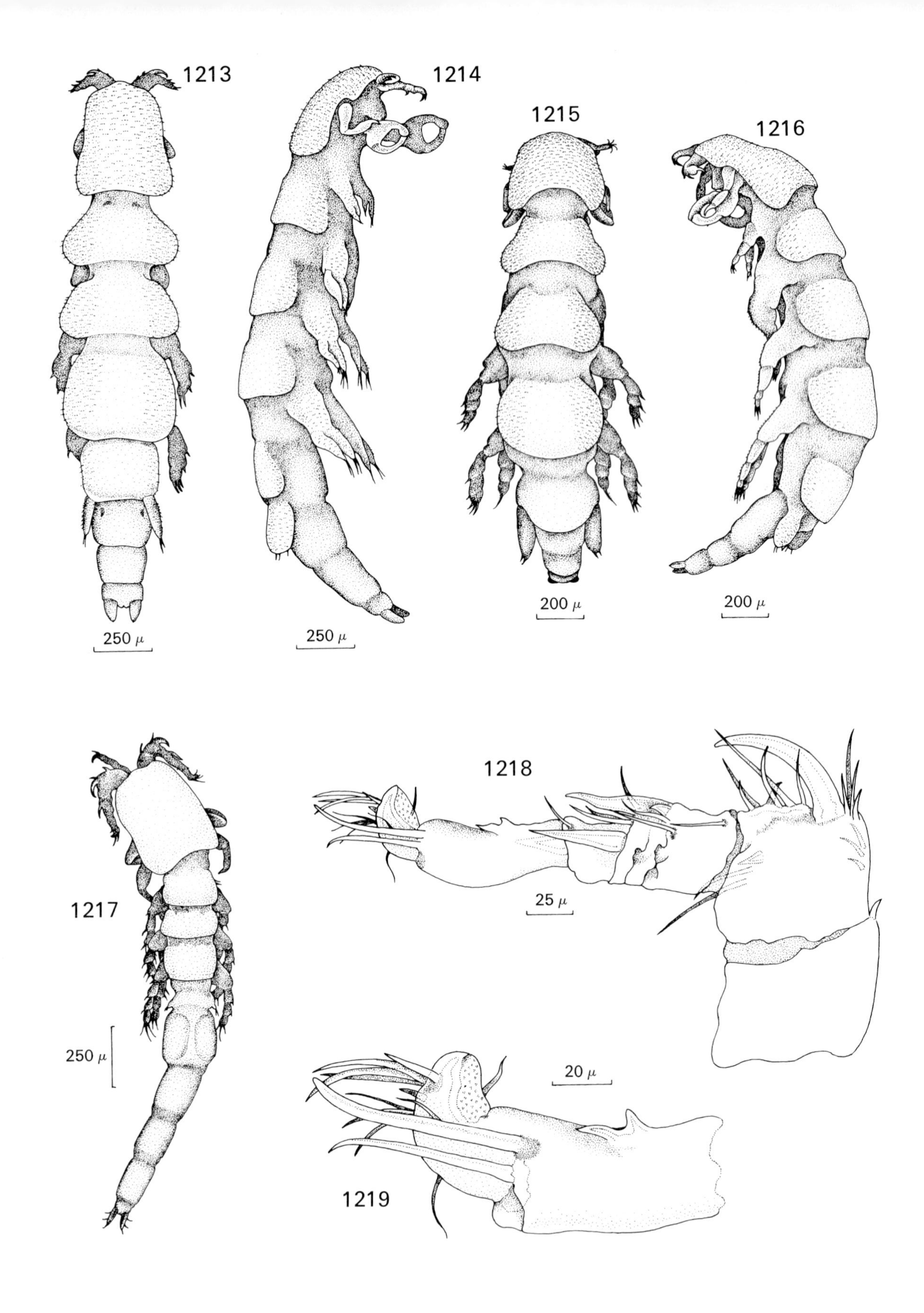

Figs. 1213–1219.
Eudactylina similis T. Scott, 1902. **Fig. 1213.** Female, dorsal. **Fig. 1214.** Same, lateral. **Fig. 1215.** Same, dorsal (specimen described as *E. rachelae*). **Fig. 1216.** Same specimen, lateral. **Fig. 1217.** Male, dorsal. **Fig. 1218.** Male first antenna, ventral. **Fig. 1219.** Same, tip, ventral.

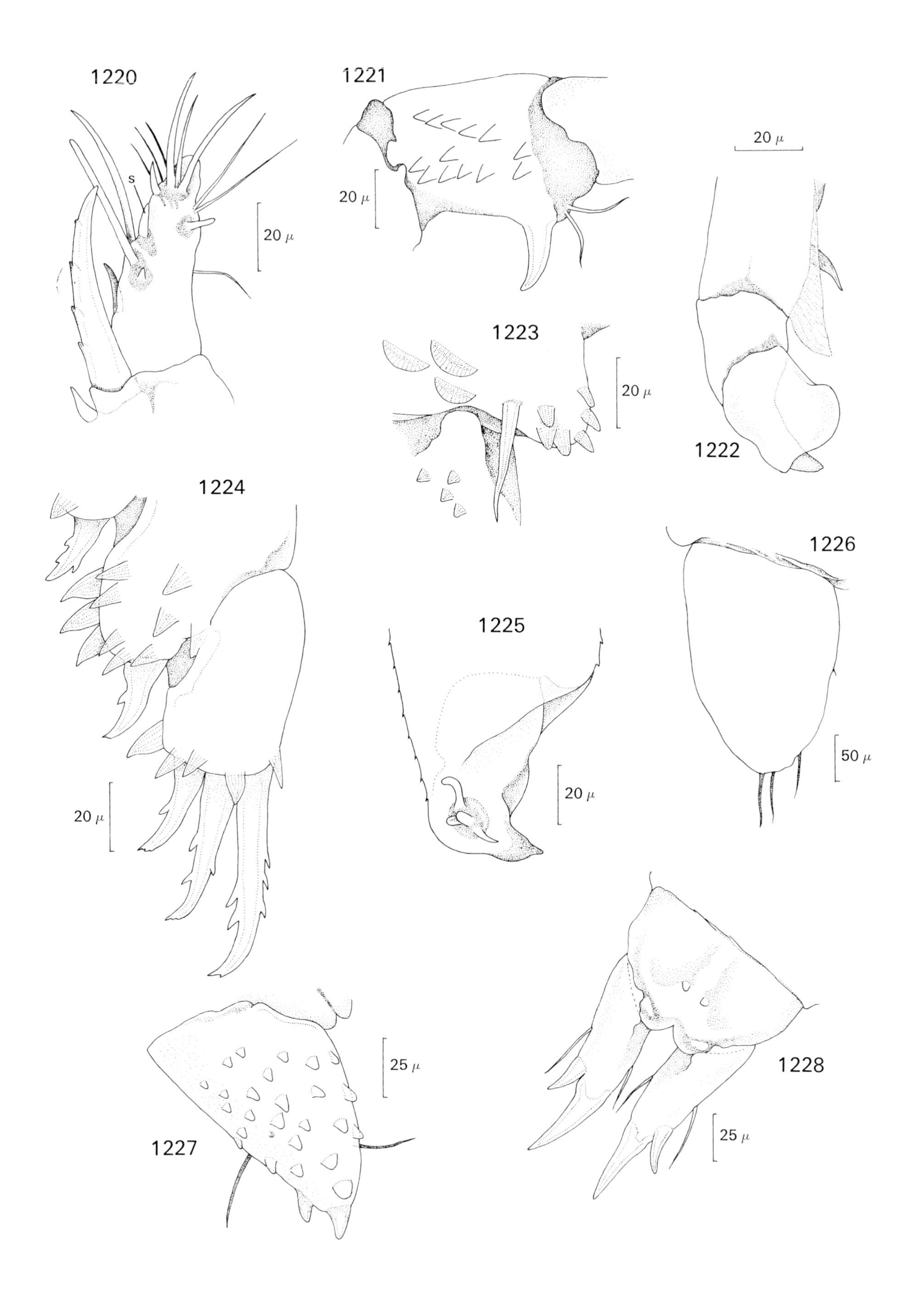

Figs. 1220–1228.
Eudactylina similis T. Scott, 1902. **Fig. 1220.** Tip of female first antenna, dorsal. **Fig. 1221.** Male second antenna, second segment, medial. **Fig. 1222.** Female maxilliped, claw, medial. **Fig. 1223.** Base of first endopod, inner corner. **Fig. 1224.** Male, tip of third and fourth exopod, ventral. **Fig. 1225.** Female, tip of second exopod dorsal. **Fig. 1226.** Outline of fifth leg. **Fig. 1227.** Uropod, ventral. **Fig. 1228.** Male posterior extremity, ventral.

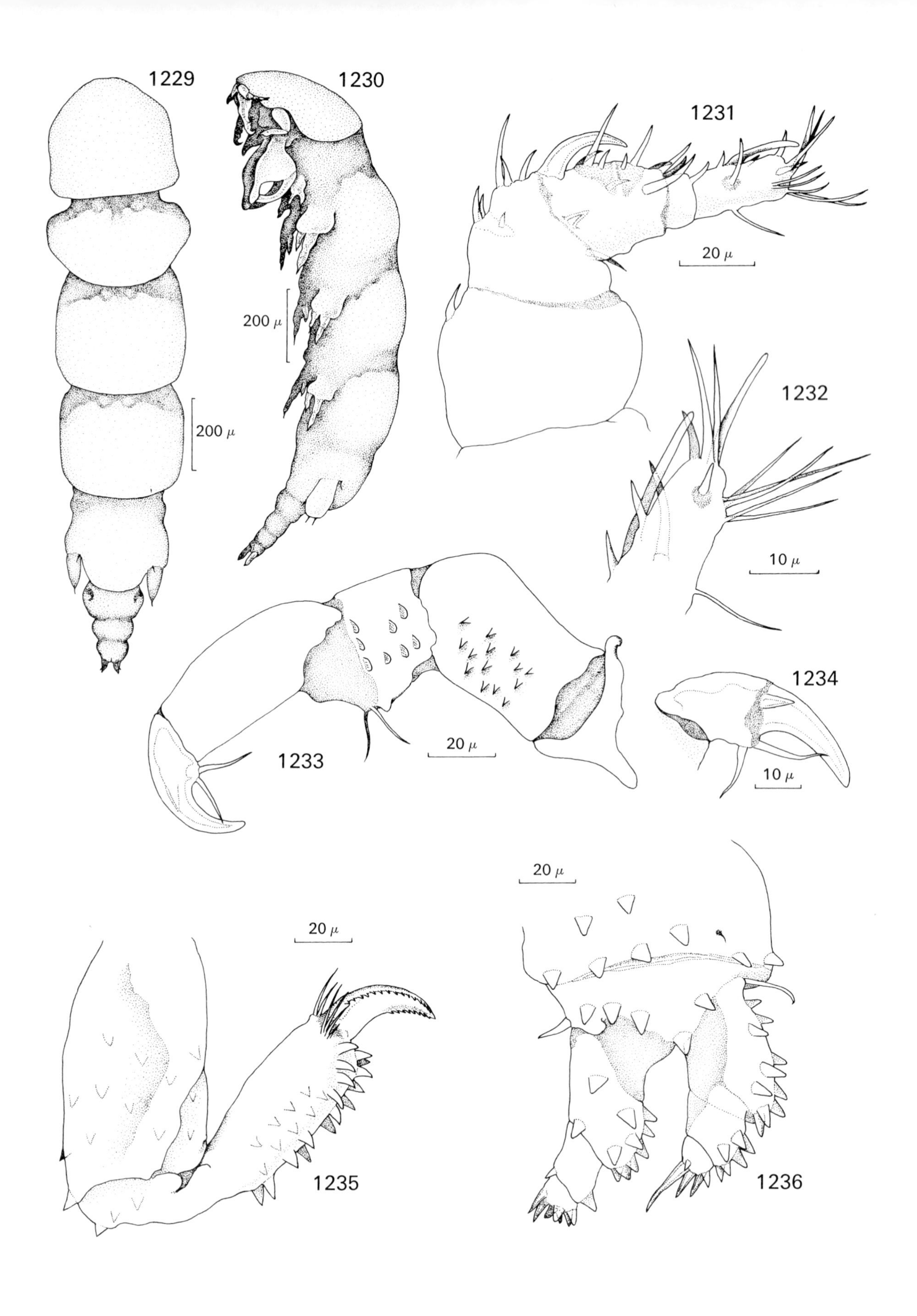

Figs. 1229–1236.
Eudactylina acanthii A. Scott, 1901. **Fig. 1229.** Female, dorsal. **Fig. 1230.** Same, lateral. **Fig. 1231.** First antenna, ventral. **Fig. 1232.** Same, tip, dorsal. **Fig. 1233.** Second antenna, medial. **Fig. 1234.** Same, tip, lateral. **Fig. 1235.** Second maxilla, medial. **Fig. 1236.** First leg, ventral.

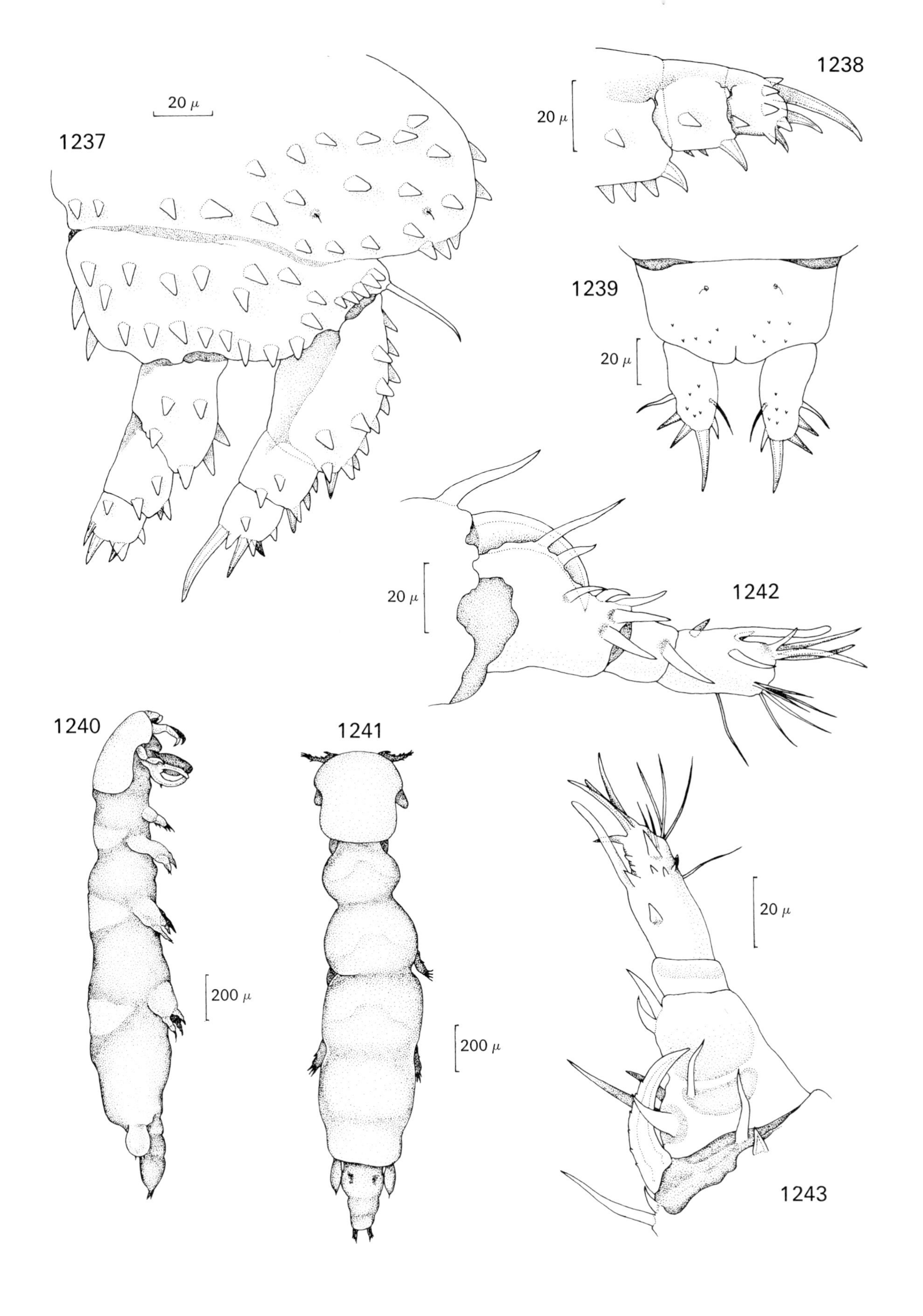

Figs. 1237–1243.
Eudactylina acanthii A. Scott, 1901. **Fig. 1237.** Female, second leg, ventral. **Fig. 1238.** Tip of third and fourth exopod, ventral. **Fig. 1239.** Posterior extremity, ventral.
Eudactylina insolens T. Scott and A. Scott, 1913. **Fig. 1240.** Female, lateral. **Fig. 1241.** Same, dorsal. **Fig. 1242.** Distal part of first antenna, ventral. **Fig. 1243.** Same, dorsal.

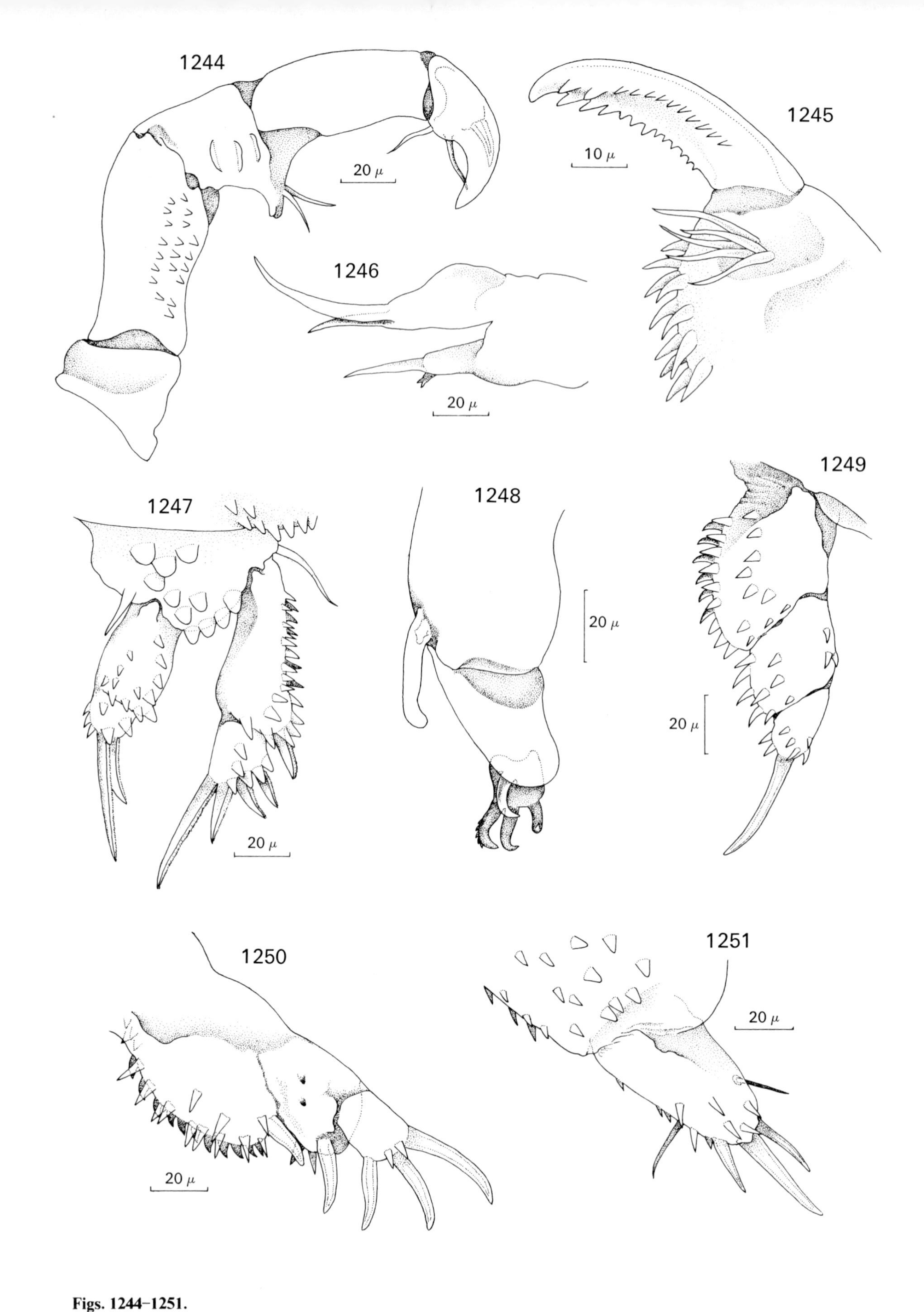

Figs. 1244–1251.
Eudactylina insolens T. Scott and A. Scott, 1913. **Fig. 1244.** Female, second antenna, medial. **Fig. 1245.** Tip of second maxilla, medial. **Fig. 1246.** First maxilla, lateral. **Fig. 1247.** Rami of first leg, ventral. **Fig. 1248.** Tip of second exopod, ventral. **Fig. 1249.** Third and fourth endopod, ventral. **Fig. 1250.** Same, exopod, ventral. **Fig. 1251.** Uropod, ventral.

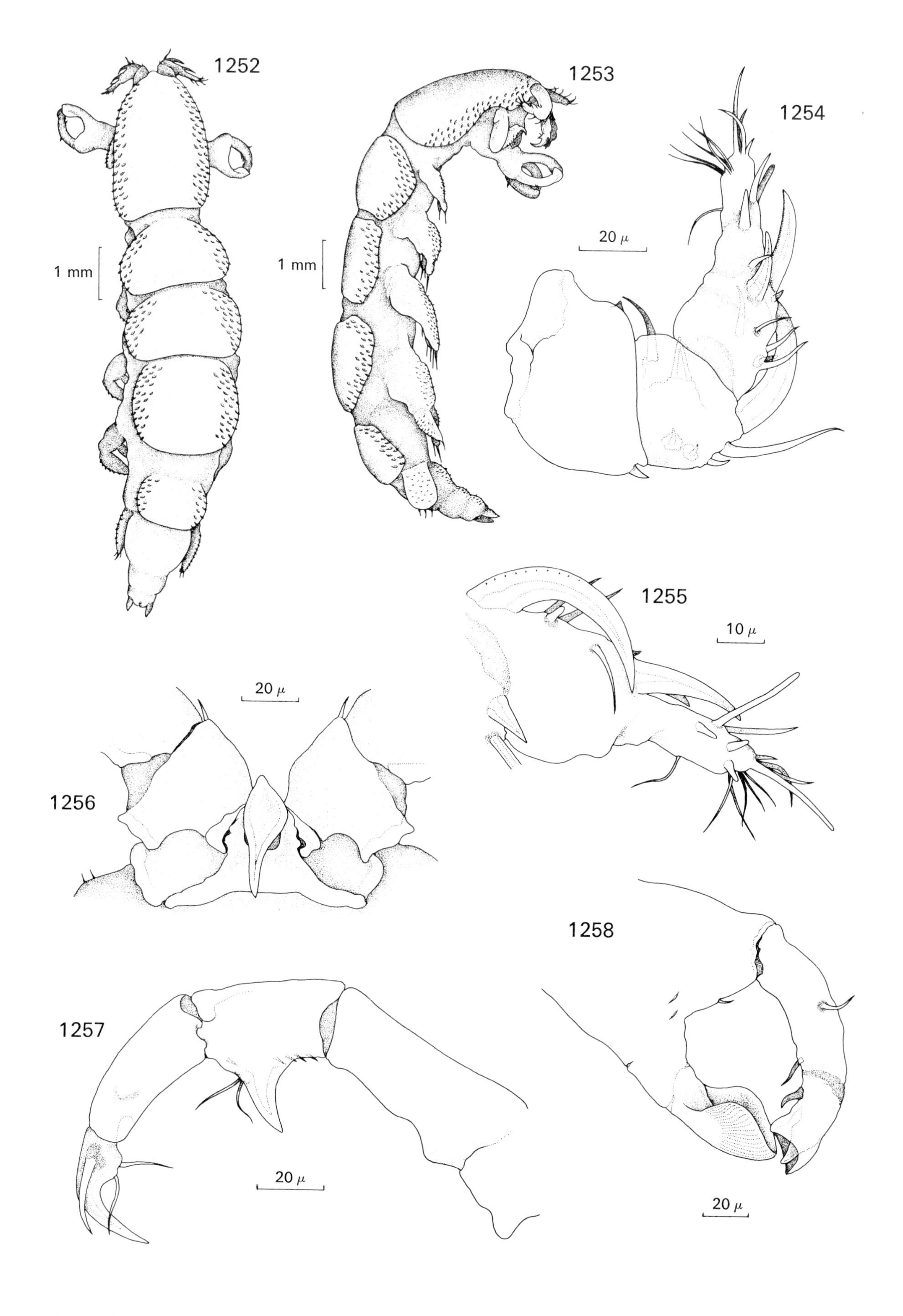

Figs. 1252–1258.
Eudactylina minuta T. Scott, 1904. **Fig. 1252.** Female, dorsal. **Fig. 1253.** Same, lateral. **Fig. 1254.** First antenna, ventral. **Fig. 1255.** Same, tip, dorsal. **Fig. 1256.** Rostrum, ventral. **Fig. 1257.** Second antenna, ventral. **Fig. 1258.** Tip of maxilliped, ventral.

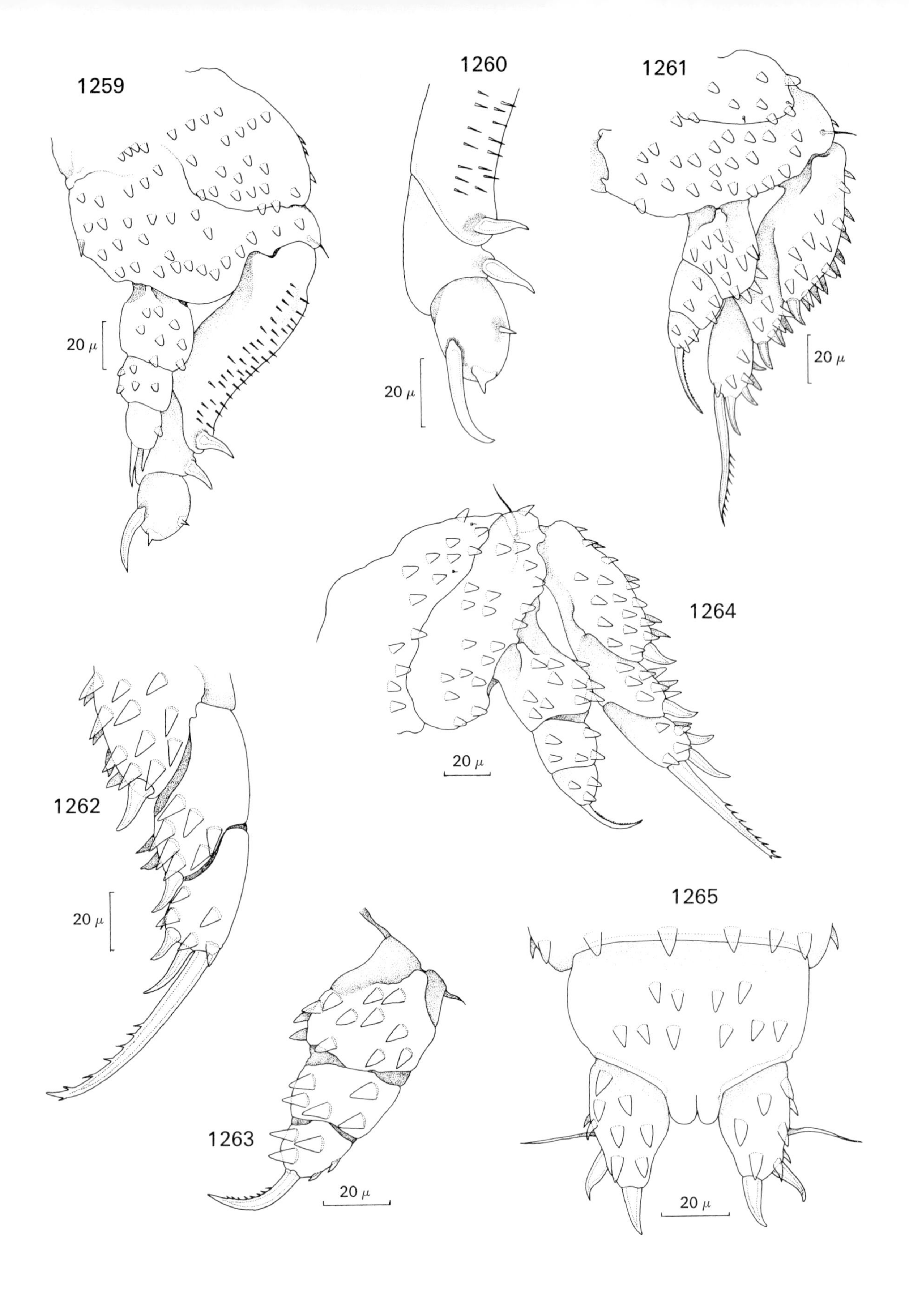

Figs. 1259–1265.
Eudactylina minuta T. Scott, 1904. **Fig. 1259.** Second leg, ventral. **Fig. 1260.** Same, tip of exopod, enlarged. **Fig. 1261.** Third leg, ventral. **Fig. 1262.** Same, tip of exopod. **Fig. 1263.** Same, endopod, ventral. **Fig. 1264.** Fourth leg, ventral. **Fig. 1265.** Posterior extremity, ventral.

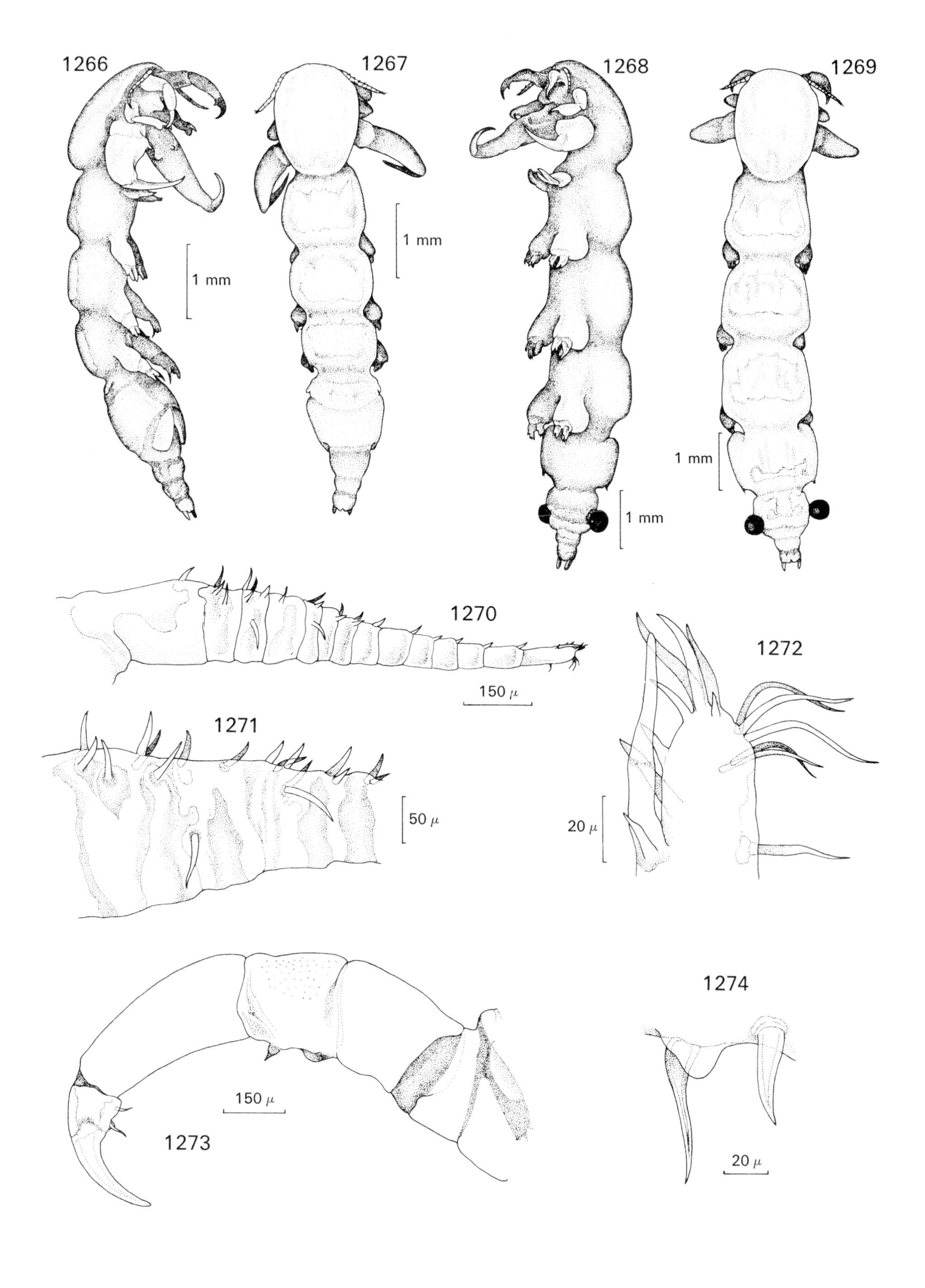

Figs. 1266–1274.
Nemesis lamna vermi A. Scott, 1929; of Hewitt (1969). **Fig. 1266.** Male, lateral. **Fig. 1267.** Same, dorsal. **Fig. 1268.** Female, ventrolateral. **Fig. 1269.** Same, dorsal. **Fig. 1270.** Male and female, first antenna. **Fig. 1271.** Same, fused part distal to basal segment. **Fig. 1272.** Same, tip. **Fig. 1273.** Second antenna, lateral. **Fig. 1274.** Same, detail of inner margin of claw, lateral.

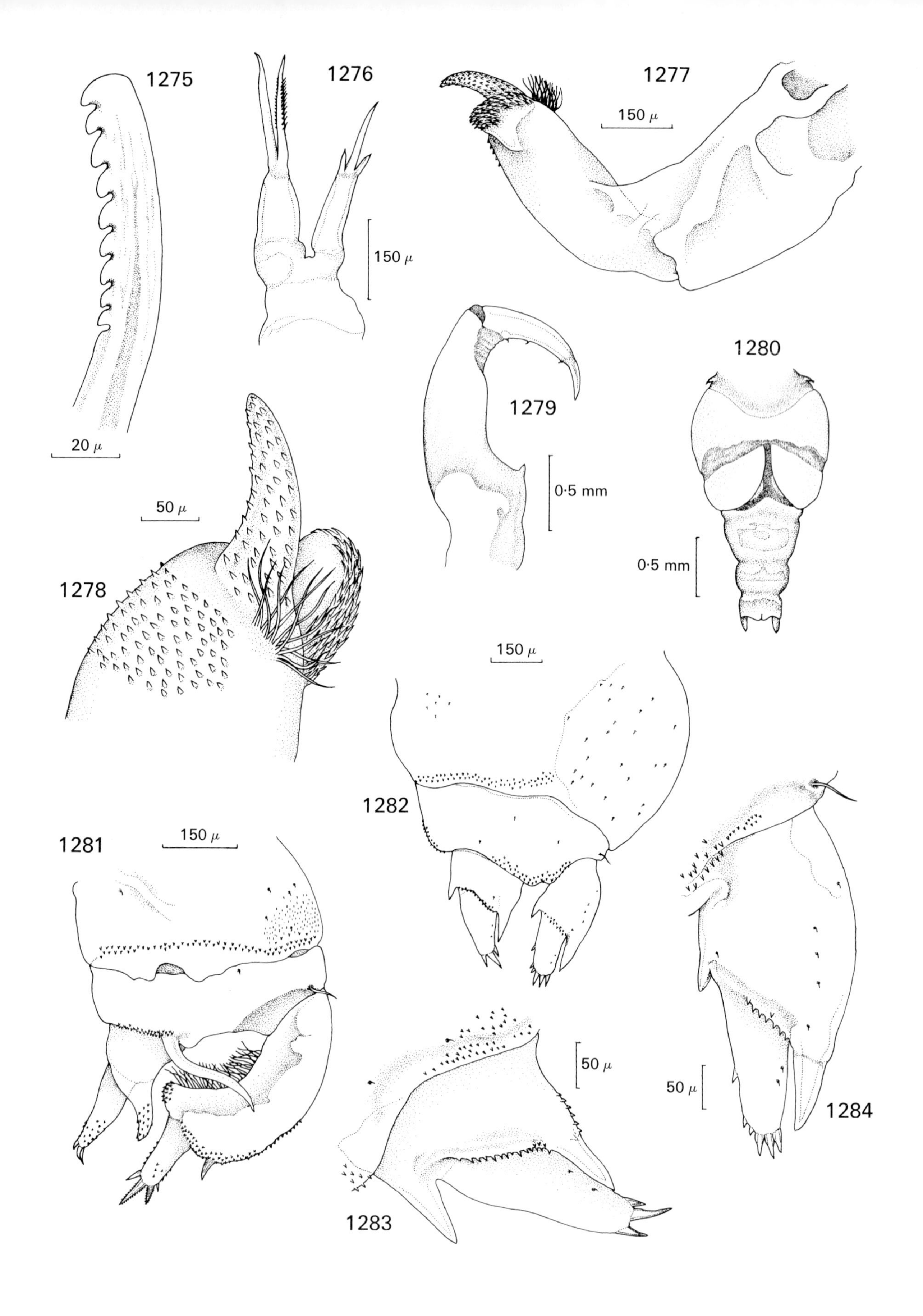

Figs. 1275–1284.
Nemesis lamna vermi A. Scott, 1929; of Hewitt (1969). **Fig. 1275.** Tip of mandible, lateral. **Fig. 1276.** First maxilla, lateral. **Fig. 1277.** Second maxilla, medial. **Fig. 1278.** Same, tip, anterior. **Fig. 1279.** Maxilliped, lateral. **Fig. 1280.** Male, posterior extremity, ventral. **Fig. 1281.** Male and female, first leg, ventral. **Fig. 1282.** Second leg, ventral. **Fig. 1283.** Second endopod, enlarged. **Fig. 1284.** Second exopod, enlarged.

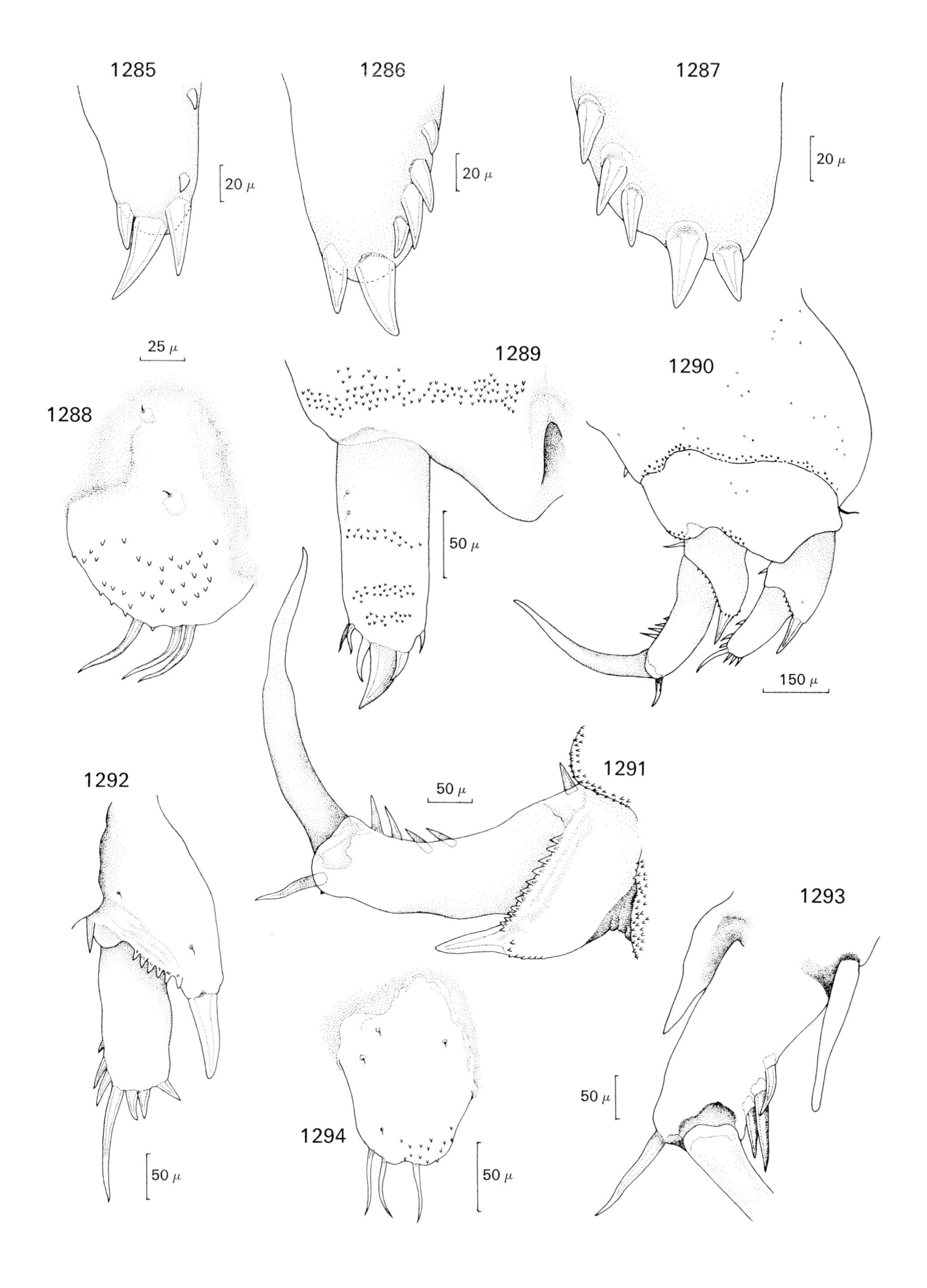

Figs. 1285–1294.
Nemesis lamna vermi A. Scott, 1929; of Hewitt (1969). **Fig. 1285.** Tip of second endopod, dorsal. **Fig. 1286.** Tip of female third endopod, dorsal. **Fig. 1287.** Tip of fourth endopod, dorsal. **Fig. 1288.** Fifth leg, lateral. **Fig. 1289.** Uropod, ventral. **Fig. 1290.** Male, third leg, ventral. **Fig. 1291.** First endopod, ventral. **Fig. 1292.** First exopod, ventral. **Fig. 1293.** Fourth endopod, middle segment, dorsal. **Fig. 1294.** Fifth leg, lateral.

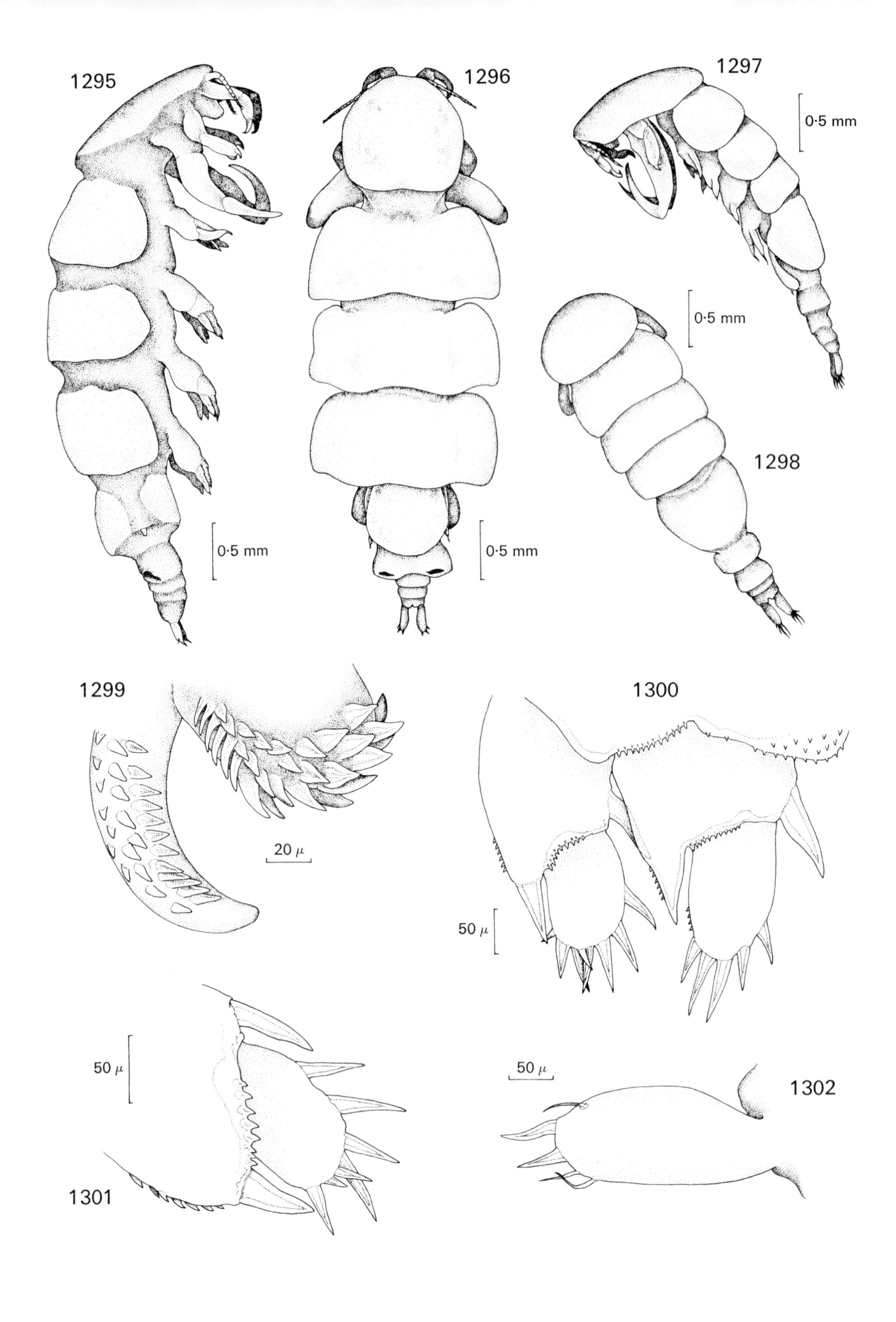

Figs. 1295–1302.
Nemesis robusta (van Beneden, 1851). **Fig. 1295.** Female, lateral. **Fig. 1296. Same, dorsal. Fig. 1297.** Male, lateral. **Fig. 1298.** Same, dorsal. **Fig. 1299.** Male and female, tip of second maxilla, medial. **Fig. 1300.** Rami of female second leg, ventral. **Fig. 1301.** Third exopod, ventral. **Fig. 1302.** Uropod, ventral.

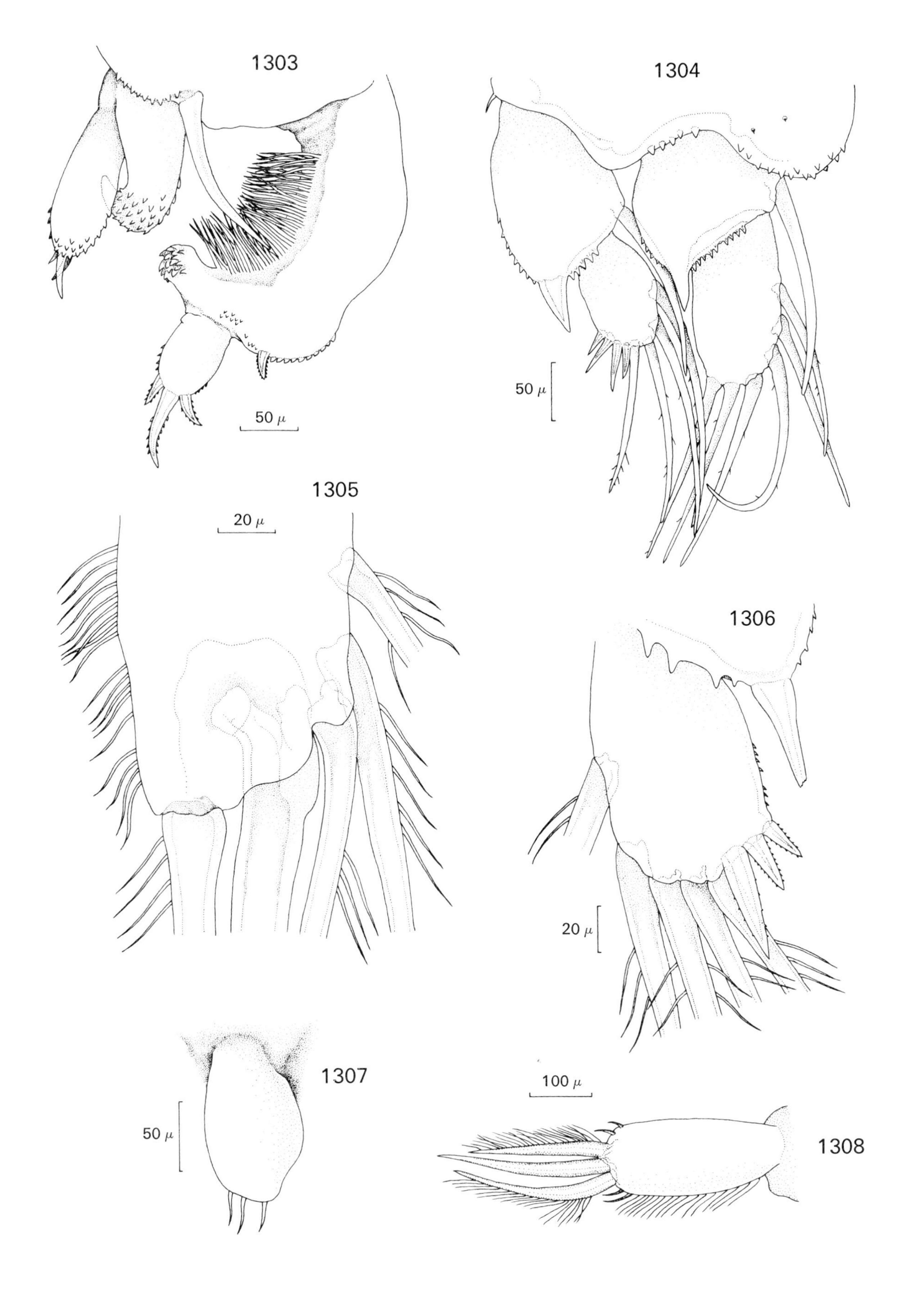

Figs. 1303–1308.
Nemesis robusta (van Beneden, 1851). **Fig. 1303.** Rami of male first leg, ventral. **Fig. 1304.** Rami of second leg, ventral. **Fig. 1305.** Tip of third and fourth endopod, ventral. **Fig. 1306.** Tip of third and fourth exopod, ventral. **Fig. 1307.** Fifth leg, lateral. **Fig. 1308.** Uropod, ventral.

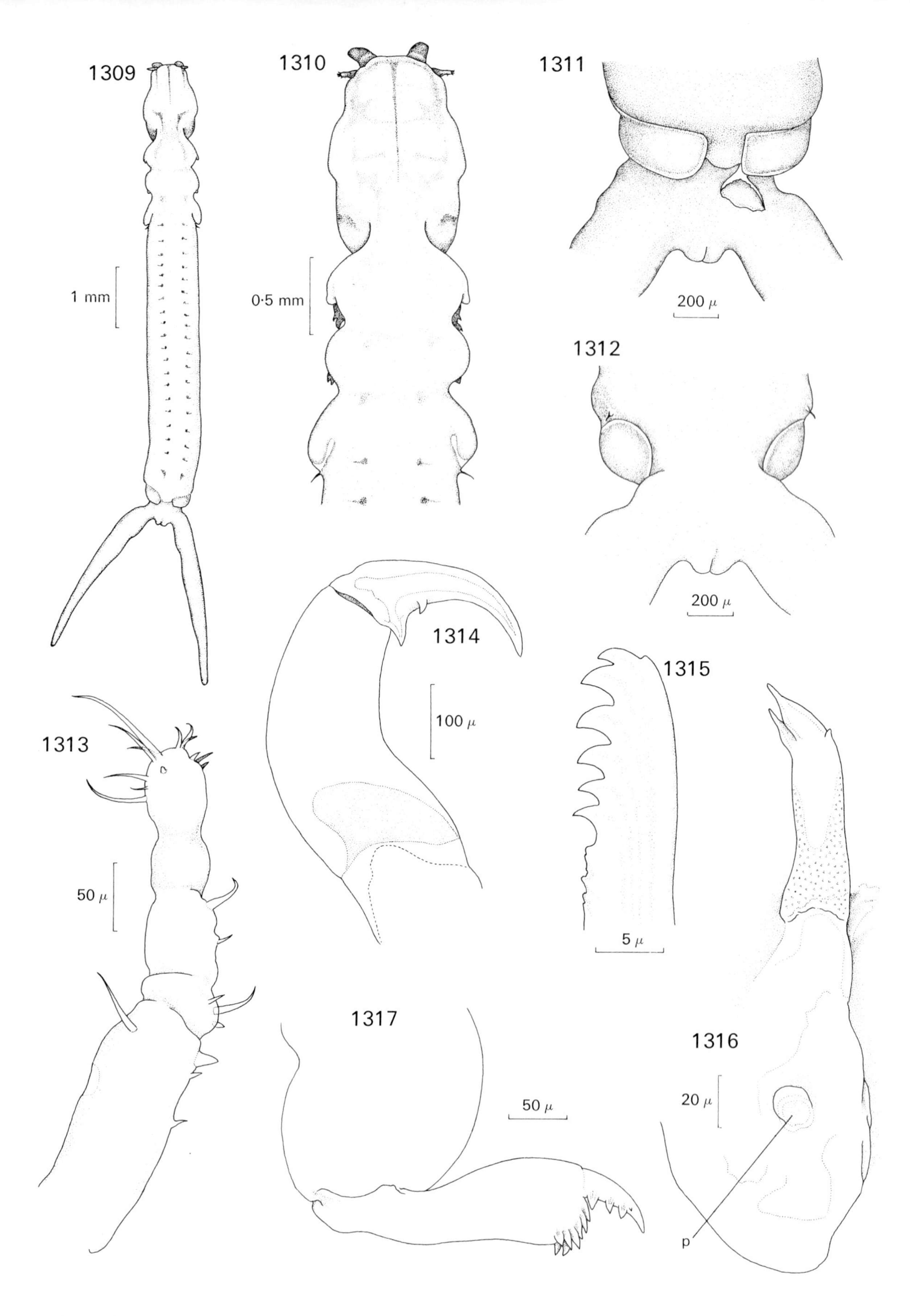

Figs. 1309–1317.
Pseudocycnus appendiculatus Heller, 1865. **Fig. 1309.** Female, dorsal. **Fig. 1310.** Same, anterior end, dorsal. **Fig. 1311.** Genito-abdominal region, dorsal. **Fig. 1312.** Same, ventral. **Fig. 1313.** First antenna, dorsal. **Fig. 1314.** Second antenna, medial. **Fig. 1315.** Tip of mandible, lateral. **Fig. 1316.** First maxilla, ventral. **Fig. 1317.** Second maxilla, lateral.

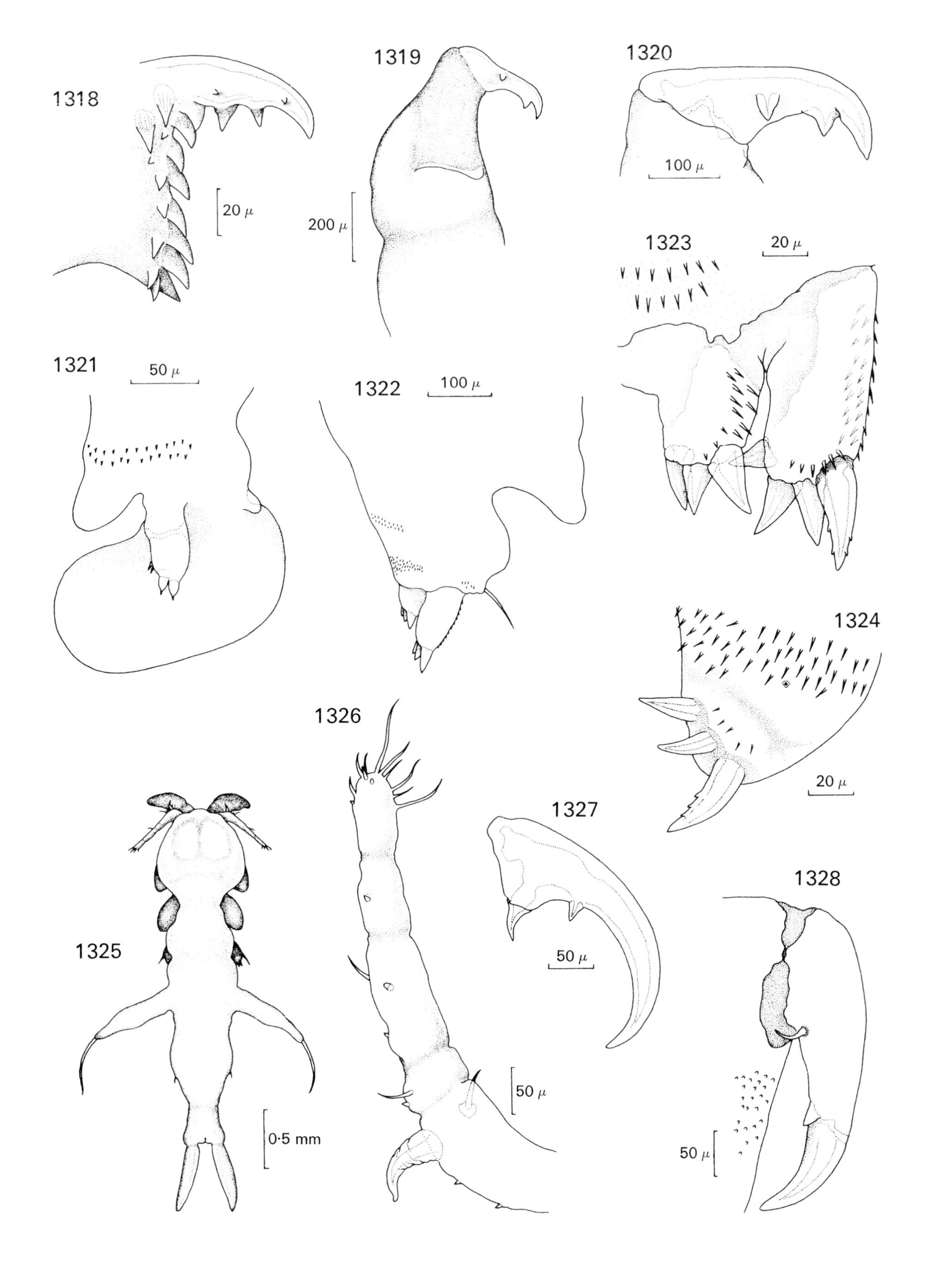

Figs. 1318–1328.
Pseudocycnus appendiculatus Heller, 1865. **Fig. 1318.** Female second maxilla, claw, medial. **Fig. 1319.** Maxilliped, posterior. **Fig. 1320.** Same, claw, enlarged. **Fig. 1321.** First leg, ventral. **Fig. 1322.** Second leg, ventral. **Fig. 1323.** Rami of second leg, dorsal. **Fig. 1324.** Tip of third leg, ventral. **Fig. 1325.** Male, dorsal. **Fig. 1326.** Male first antenna, dorsal. **Fig. 1327.** Claw of second antenna. **Fig. 1328.** Claw of maxilliped.

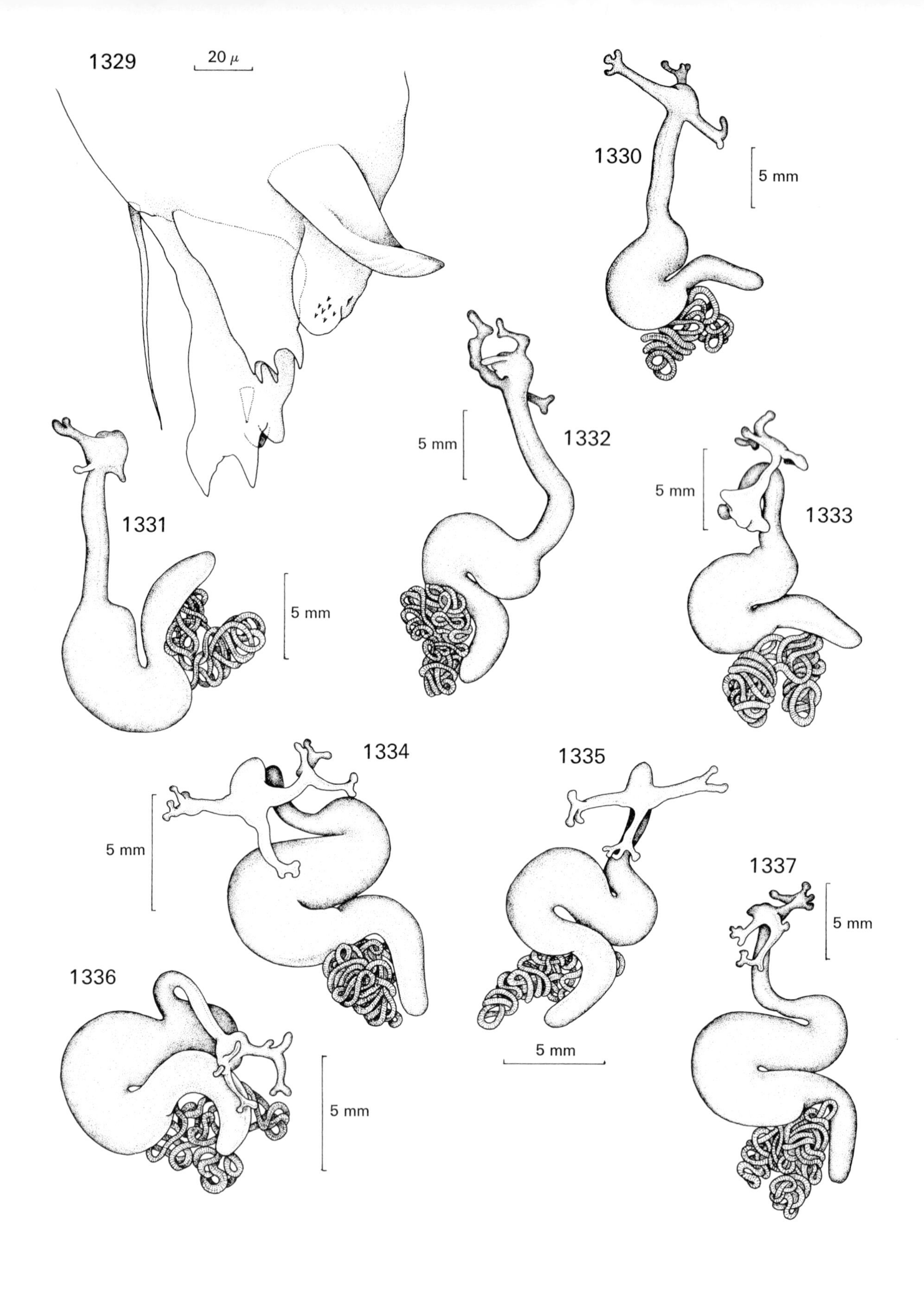

Figs. 1329–1337.
Pseudocycnus appendiculatus Heller, 1865. **Fig. 1329.** Rami of male first leg, dorsal.
Lernaeocera branchialis f. *obtusa* new comb. **Fig. 1330–1333.** Adult female.
Lernaeocera branchialis f. *branchialis* new comb. **Fig. 1334–1337.** Adult female.

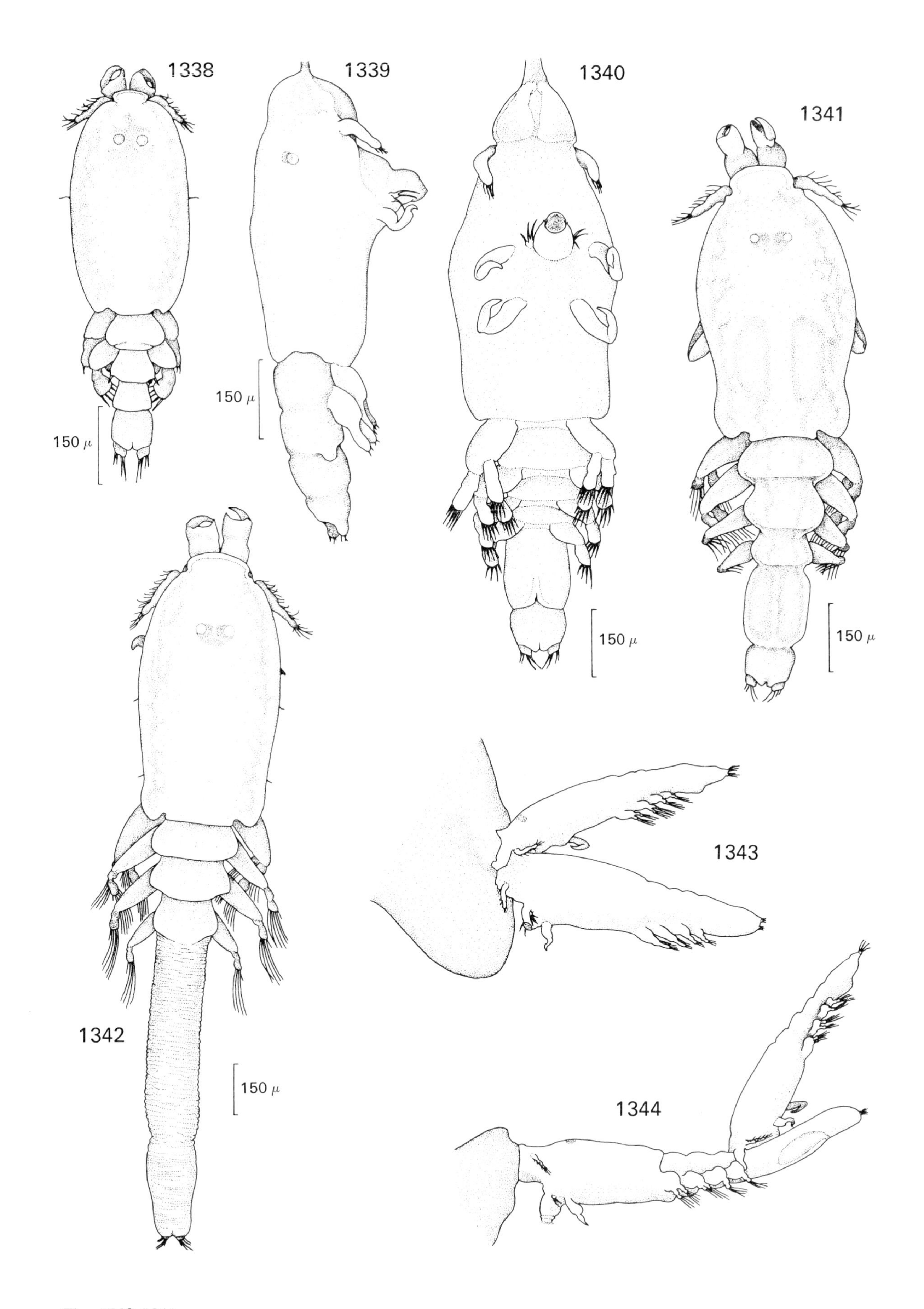

Figs. 1338–1344.

Lernaeocera branchialis (L.). **Fig. 1338.** Copepodid, dorsal. **Fig. 1339.** Chalimus I, lateral. **Fig. 1340.** Male chalimus IV, ventral. **Fig. 1341.** Adult male dorsal. **Fig. 1342.** Mature premetamorphosis female, dorsal. **Fig. 1343.** Male and chalimus IV female, precopulation posture. **Fig. 1344.** Male and female, copulation.
Figs. 1343, 1344 redrawn from Kabata (1958a).

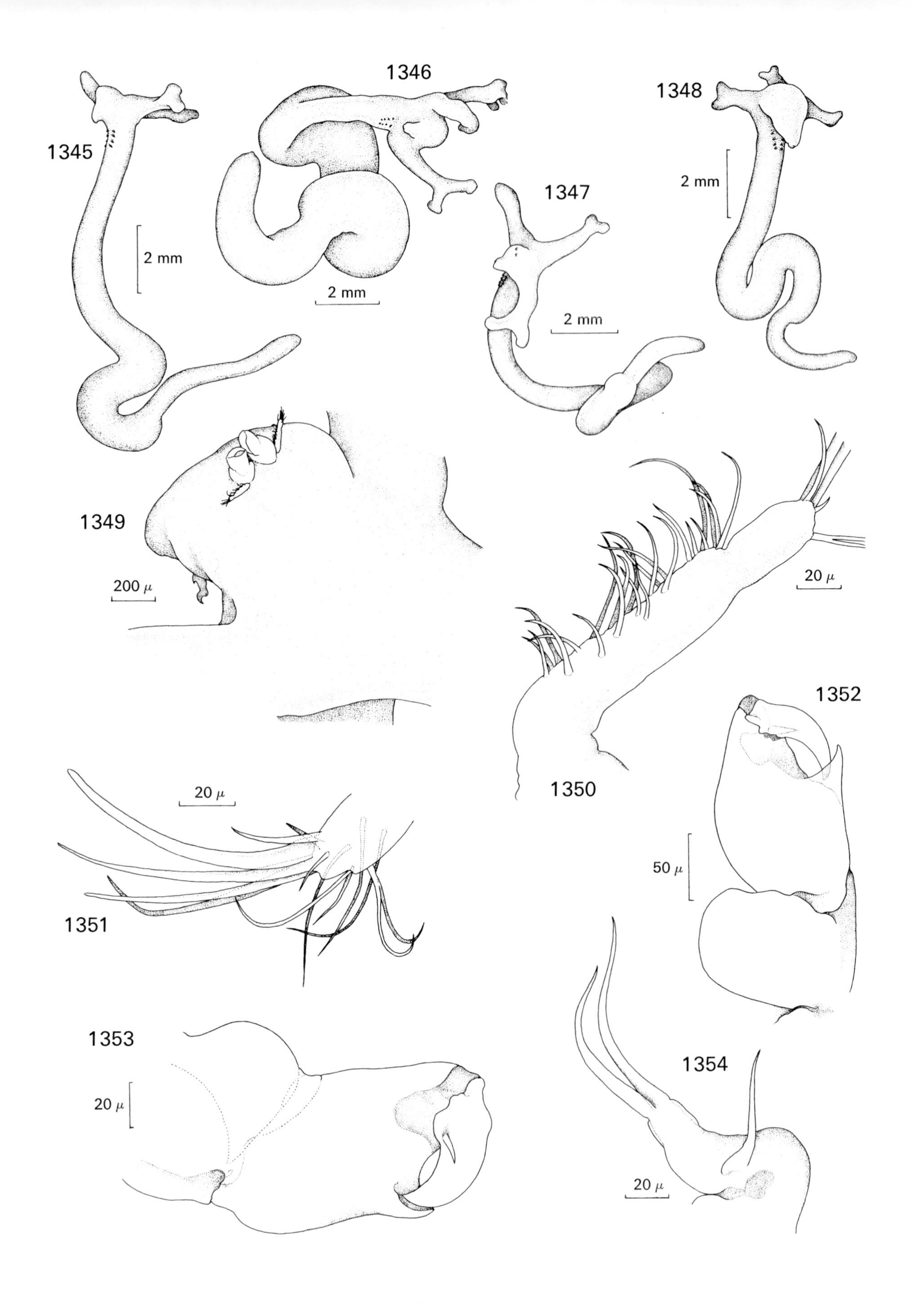

Figs. 1345–1354.

Lernaeocera branchialis (L.). **Figs. 1345–1348.** Young adult females. **Fig. 1349.** Same head, anterodorsal. **Fig. 1350.** First antenna (without apical armature). **Fig. 1351.** Same, apical armature. **Fig. 1352.** Female second antenna, medial. **Fig. 1353.** Female second antenna, medial. **Fig. 1354.** Male and female, first maxilla, lateral.

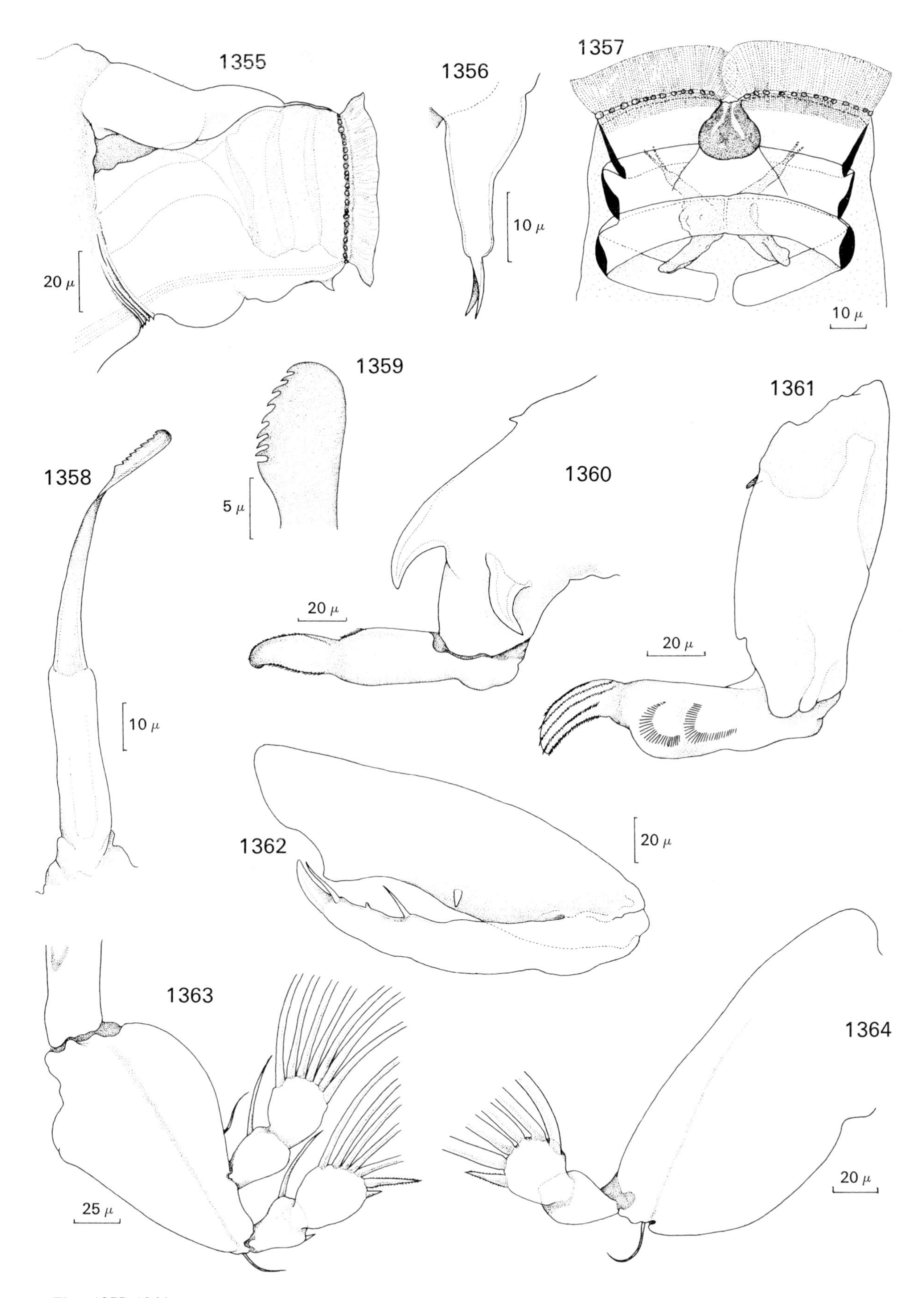

Figs. 1355–1364.
Lernaeocera branchialis (L.). **Fig. 1355.** Male and female, mouth, lateral. **Fig. 1356.** Buccal stylet. **Fig. 1357.** Buccal tube, anterior. **Fig. 1358.** Mandible, lateral. **Fig. 1359.** Tip of mandible, lateral. **Fig. 1360.** Female second maxilla. **Fig. 1361.** Male second maxilla. **Fig. 1362.** Maxilliped. **Fig. 1363.** Male and female, first leg, ventral. **Fig. 1364.** Third leg, ventral.
Fig. 1357 modified, Figs. 1358, 1359 redrawn from Kabata (1962a).

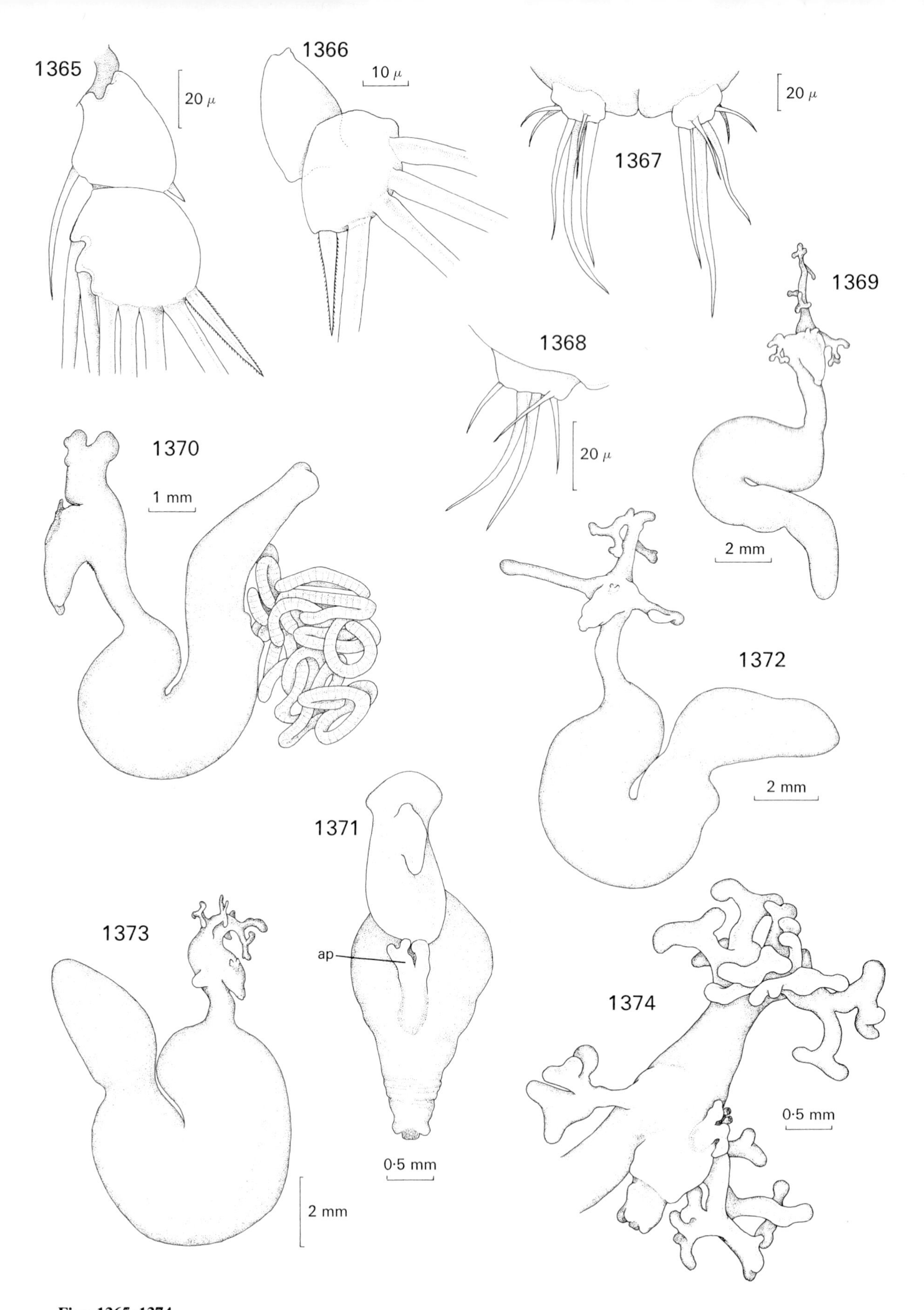

Figs. 1365–1374.
Lernaeocera branchialis (L.). **Fig. 1365.** Male and female, second exopod, ventral. **Fig. 1366.** Fourth exopod, ventral. **Fig. 1367.** Male uropods, dorsal. **Fig, 1368.** Female uropod, ventral.
Lernaeocera lusci (Bassett-Smith, 1896). **Fig. 1369.** Female (specimen from *Callionymus lyra*). **Fig. 1370.** Female (specimen from *Trisopterus luscus*.) **Fig. 1371.** Same head, anterior. **Fig. 1372.** Female (specimen from *Phycis blennioides*). **Fig. 1373.** Female (specimen from *Trisopterus esmarkii*). **Fig. 1374.** Head and holdfast.

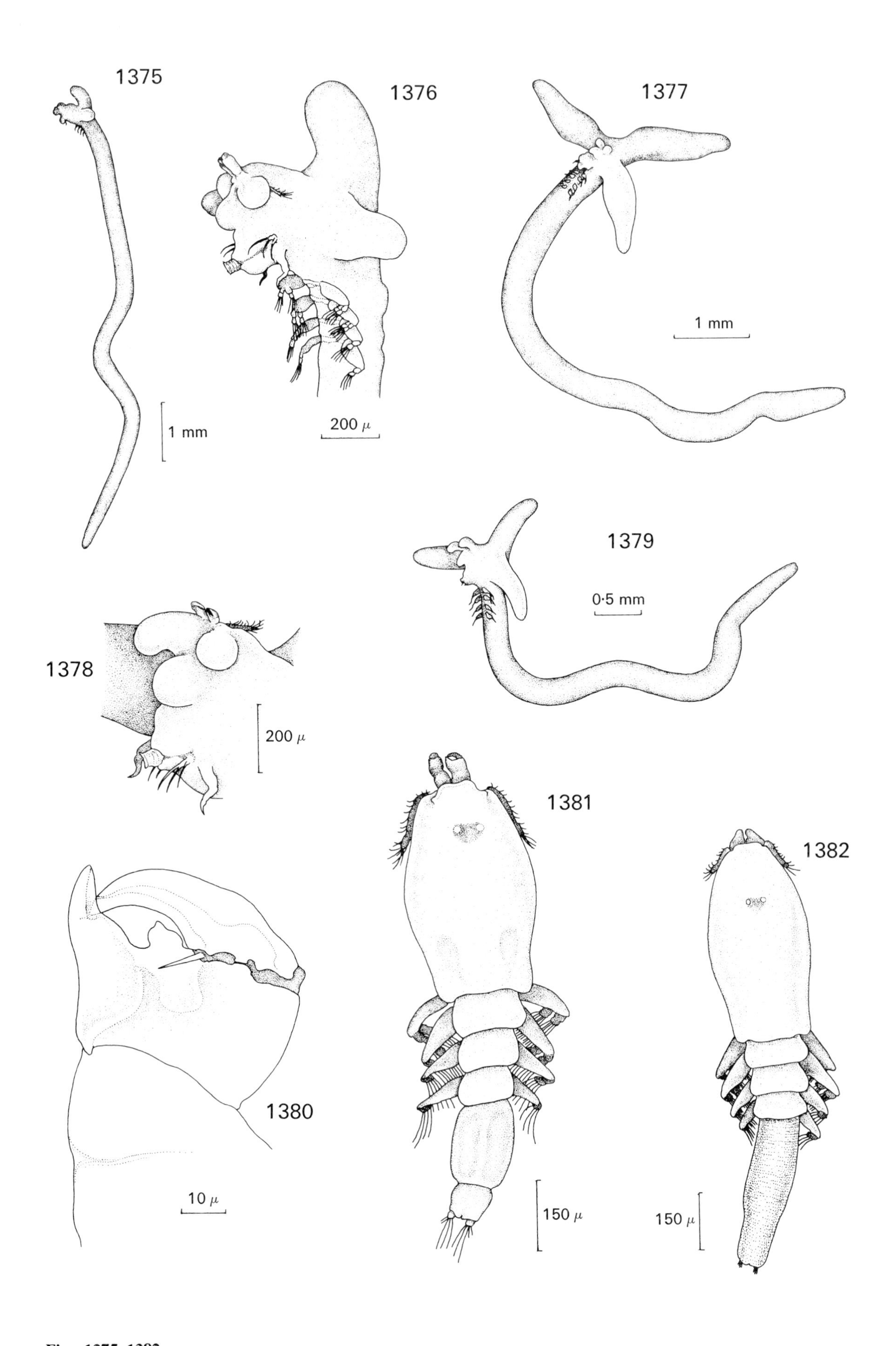

Figs. 1375–1382.
Lernaeocera lusci (Bassett-Smith, 1896). **Fig. 1375.** Very young female, lateral. **Fig. 1376.** Same, cephalothorax, lateral. **Fig. 1377.** Somewhat older specimen, lateral. **Fig. 1378.** Cephalothorax of specimen shown in Fig. 1379, lateral. **Fig. 1379.** Female about as old as that shown in Fig. 1377. **Fig. 1380.** Copepodid second antenna. **Fig. 1381.** Adult male, dorsal. **Fig. 1382.** Free swimming female, dorsal.

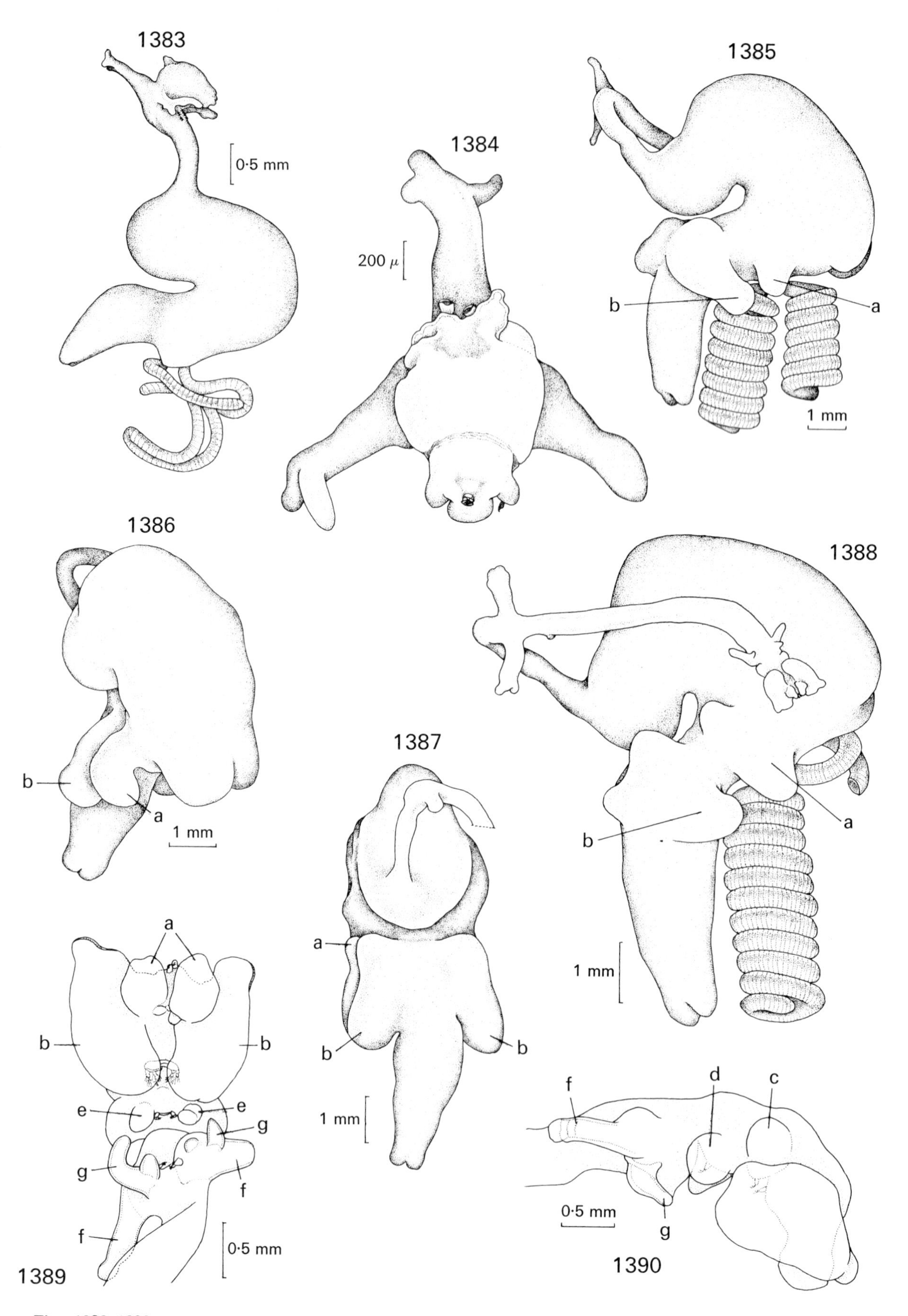

Figs. 1383–1390.
Lernaeocera minuta (T. Scott, 1900). **Fig. 1383.** Female, lateral. **Fig. 1384.** Same, head, anterior.
Haemobaphes cyclopterina (Fabricius, 1780). **Fig. 1385.** Female, lateral. **Fig. 1386.** Same, dorsolateral. **Fig. 1387.** Same, ventral. **Fig. 1388.** Another specimen, lateral. **Fig. 1389.** Cephalothorax, ventral. **Fig. 1390.** Same, lateral.
Figs. 1389, 1390 modified from Goodings and Humes (1963).
Explanations in text.

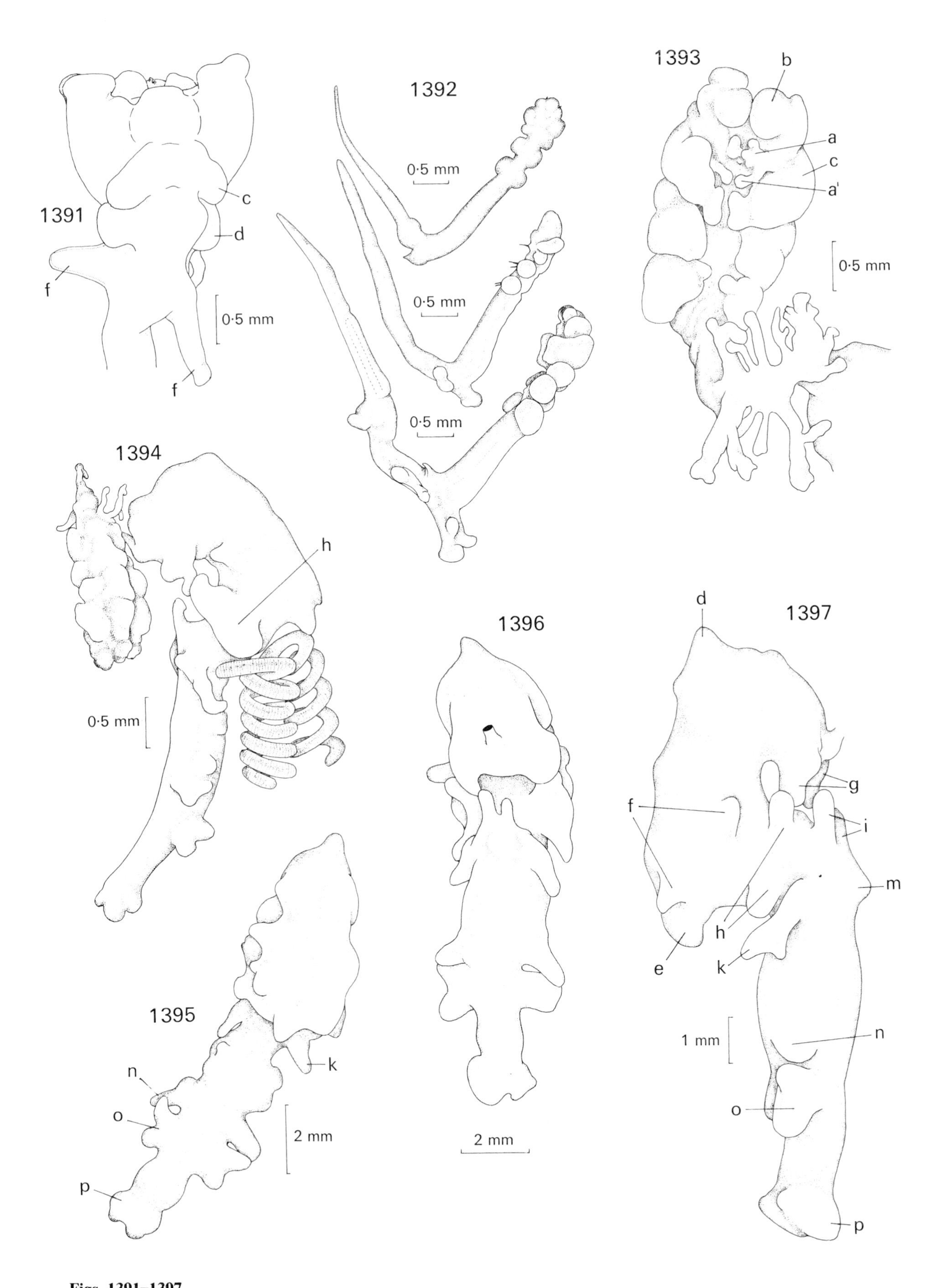

Figs. 1391–1397.
Haemobaphes cyclopterina (Fabricius, 1780). **Fig. 1391.** Female cephalothorax, dorsal.
Haemobaphes ambiguus T. Scott, 1900. **Fig. 1392.** Three stages in metamorphosis of adult female, lateral. **Fig. 1393.** Female cephalothorax, ventral. **Fig. 1394.** Female, lateral. **Fig. 1395.** Trunk, dorsal. **Fig. 1396.** Same, ventral. **Fig. 1397.** Same, lateral.

Fig. 1391 modified from Gooding and Humes (1963); Figs. 1392–1394 redrawn from Delamare Deboutteville and Nunes-Ruivo (1955).

Explanations in text.

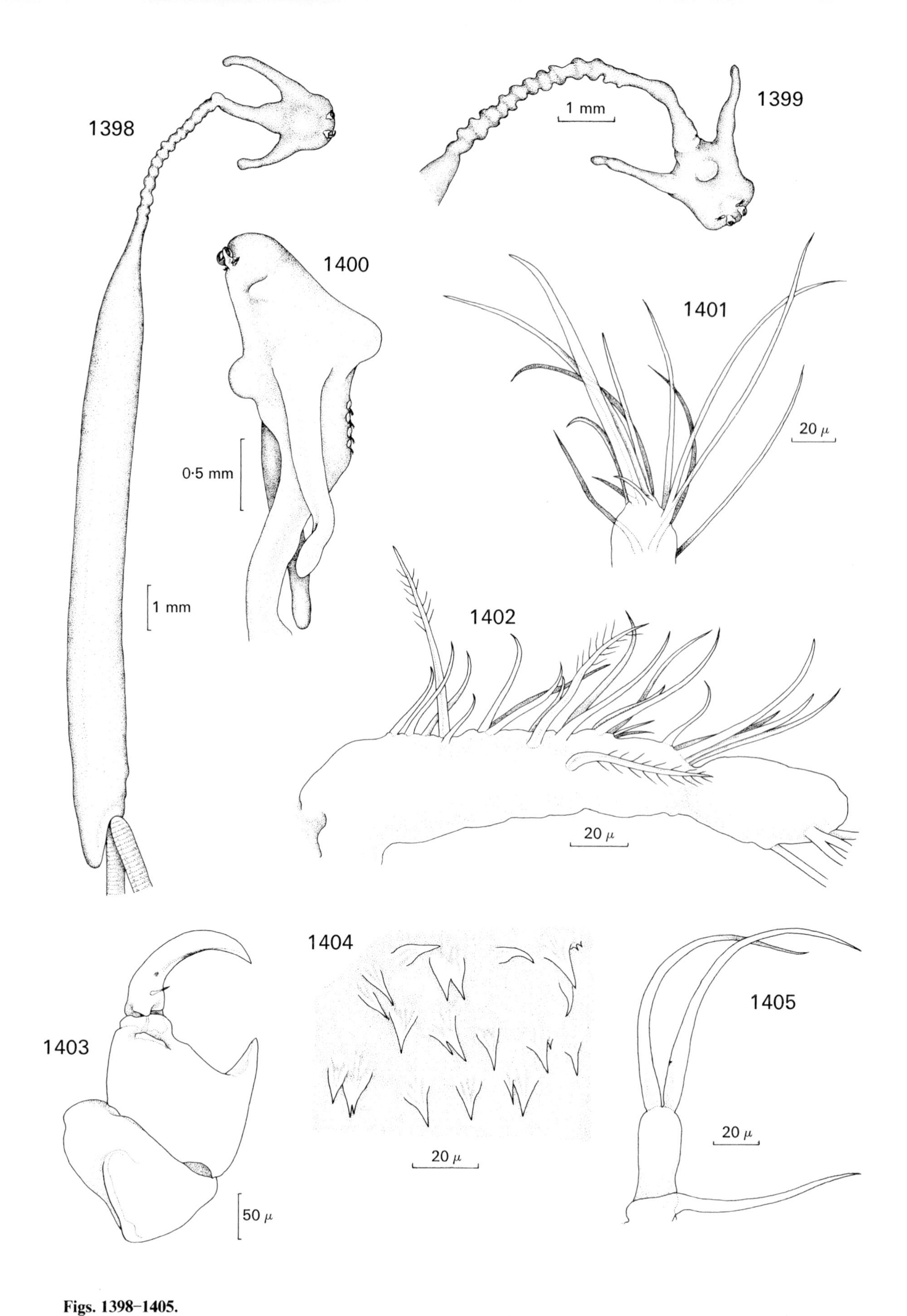

Figs. 1398–1405.
Lernaeenicus sprattae (Sowerby, 1806). **Fig. 1398.** Female, trunk lateral, cephalothorax dorsal. **Fig. 1399.** Same, cephalothorax, dorsal. **Fig. 1400.** Same, lateral. **Fig. 1401.** Tip of first antenna, dorsal. **Fig. 1402.** First antenna (without apical armature), dorsal. **Fig. 1403.** Second antenna. **Fig. 1404.** Denticulation along posterior side of base of mouth. **Fig. 1405.** First maxilla, lateral.

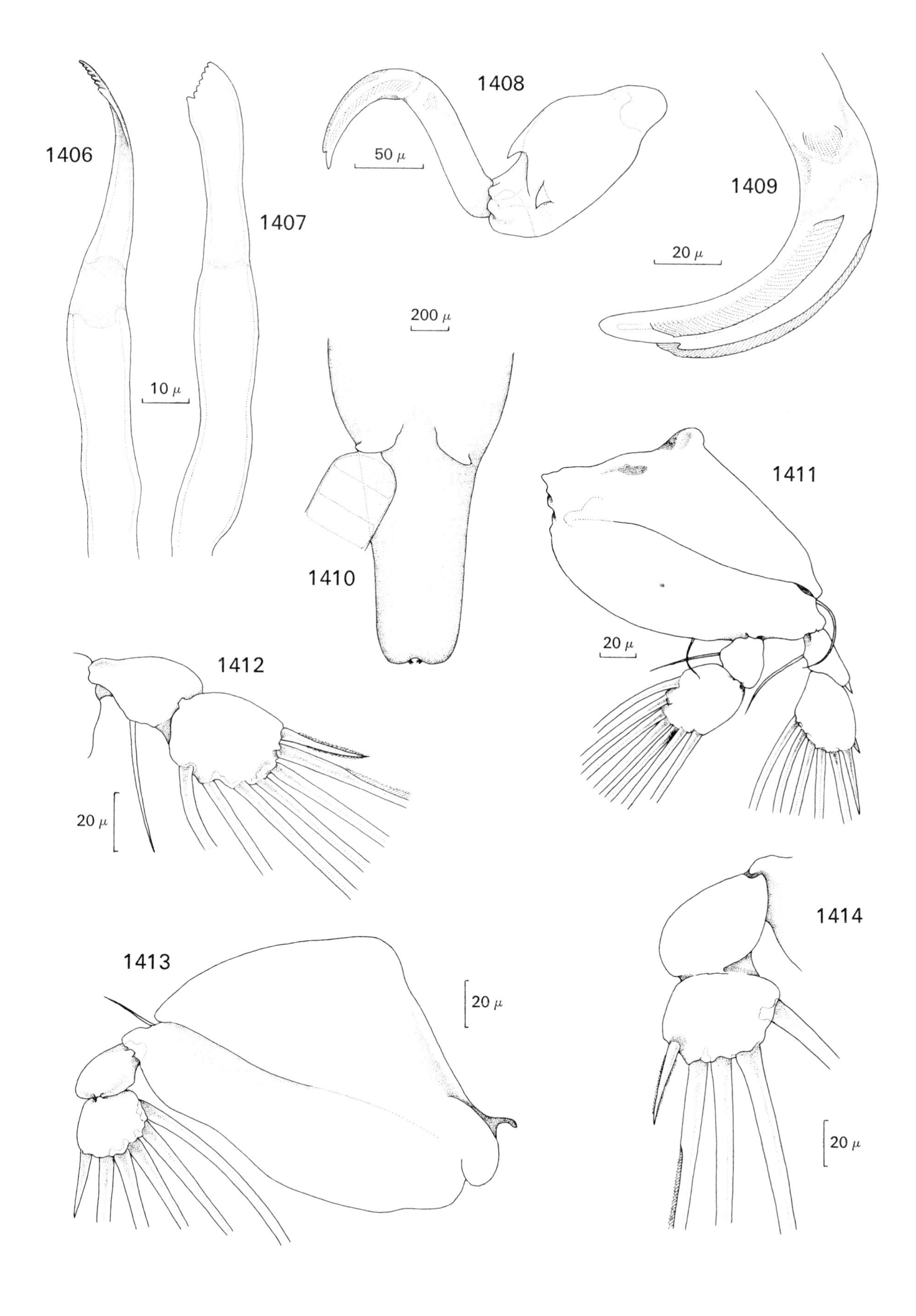

Figs. 1406–1414.
Lernaeenicus sprattae (Sowerby, 1806). **Fig. 1406.** Mandible, lateral (tip ventral). **Fig. 1407.** Same, lateral. **Fig. 1408.** Second maxilla, lateral. **Fig. 1409.** Same, tip, medial. **Fig. 1410.** Posterior extremity of trunk, ventral. **Fig. 1411.** First leg, ventral. **Fig. 1412.** Second exopod, ventral. **Fig. 1413.** Third leg, ventral. **Fig. 1414.** Fourth exopod, dorsal.

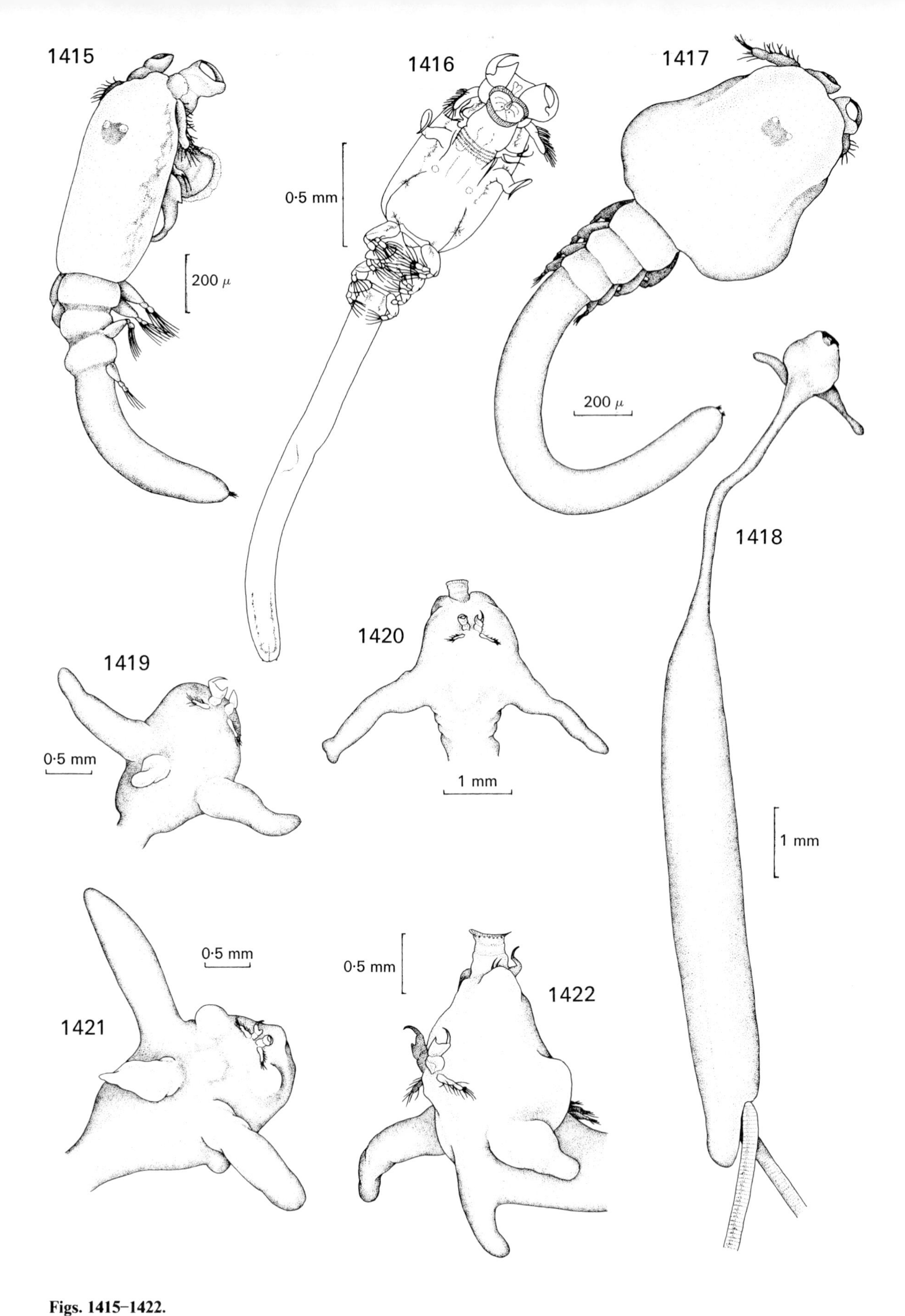

Figs. 1415–1422.

Lernaeenicus sprattae (Sowerby, 1806). **Fig. 1415.** Premetamorphosis female, dorsolateral. **Fig. 1416.** Same, slightly older. **Fig. 1417.** Same, beginning to metamorphose, dorsal.

Lernaeenicus encrasicoli (Turton, 1807). **Fig. 1418.** Female, trunk lateral, cephalothorax ventral. **Fig. 1419.** Cephalothorax, dorsal. **Fig. 1420.** Same, posterodorsal. **Fig. 1421.** Same, dorsolateral. **Fig. 1422.** Same, lateral.

Fig. 1416 redrawn from Gurney (1947).

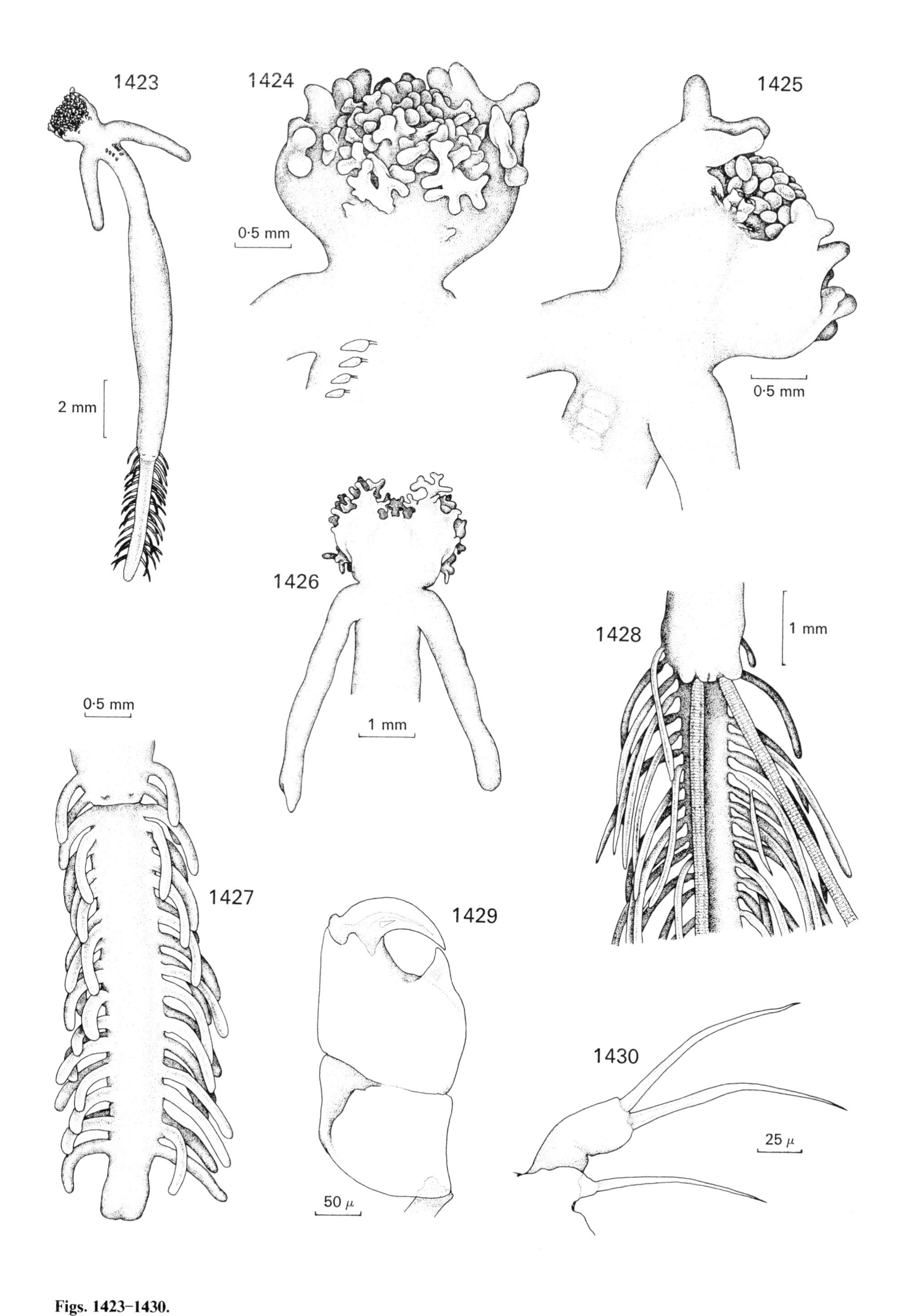

Figs. 1423–1430.
Pennella sagitta (L.). **Fig. 1423.** Adult female, ventral. **Fig. 1424.** Same, cephalothorax, ventral. **Fig. 1425.** Same, dorsal. **Fig. 1426.** Cephalothorax of older specimen, dorsal. **Fig. 1427.** Posterior extremity of specimen shown in Fig. 1423, ventral. **Fig. 1428.** Posterior extremity of older specimen, ventral. **Fig. 1429.** Second antenna, lateral. **Fig. 1430.** First maxilla, lateral.

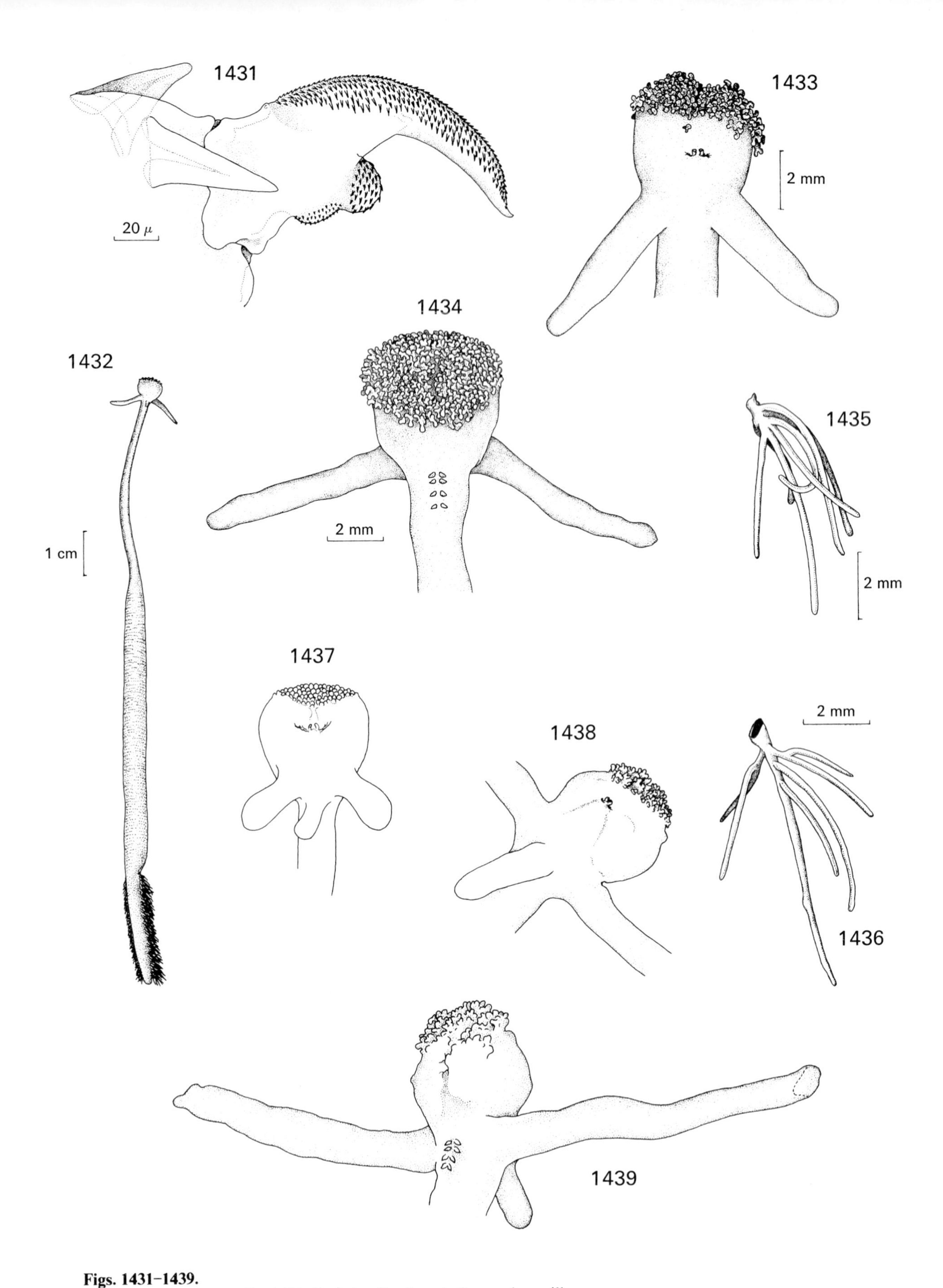

Figs. 1431–1439.
Pennella sagitta (L.). **Fig. 1431.** Female, distal part of second maxilla.
Pennella filosa (L.). **Fig. 1432.** Female, trunk lateral, cephalothorax ventrolateral. **Fig. 1433.** Same, cephalothorax, dorsal. **Fig. 1434.** Same, ventral. **Figs. 1435, 1436.** Two parts of one process of abdominal brush. **Fig. 1437.** Cephalothorax of "three-horned" female, dorsal. **Fig. 1438.** Cephalothorax, dorsal. **Fig. 1439.** Same, ventral.
Fig. 1437 redrawn from Leigh-Sharpe (1928); 1438, 1439 from Nunes-Ruivo (1962).

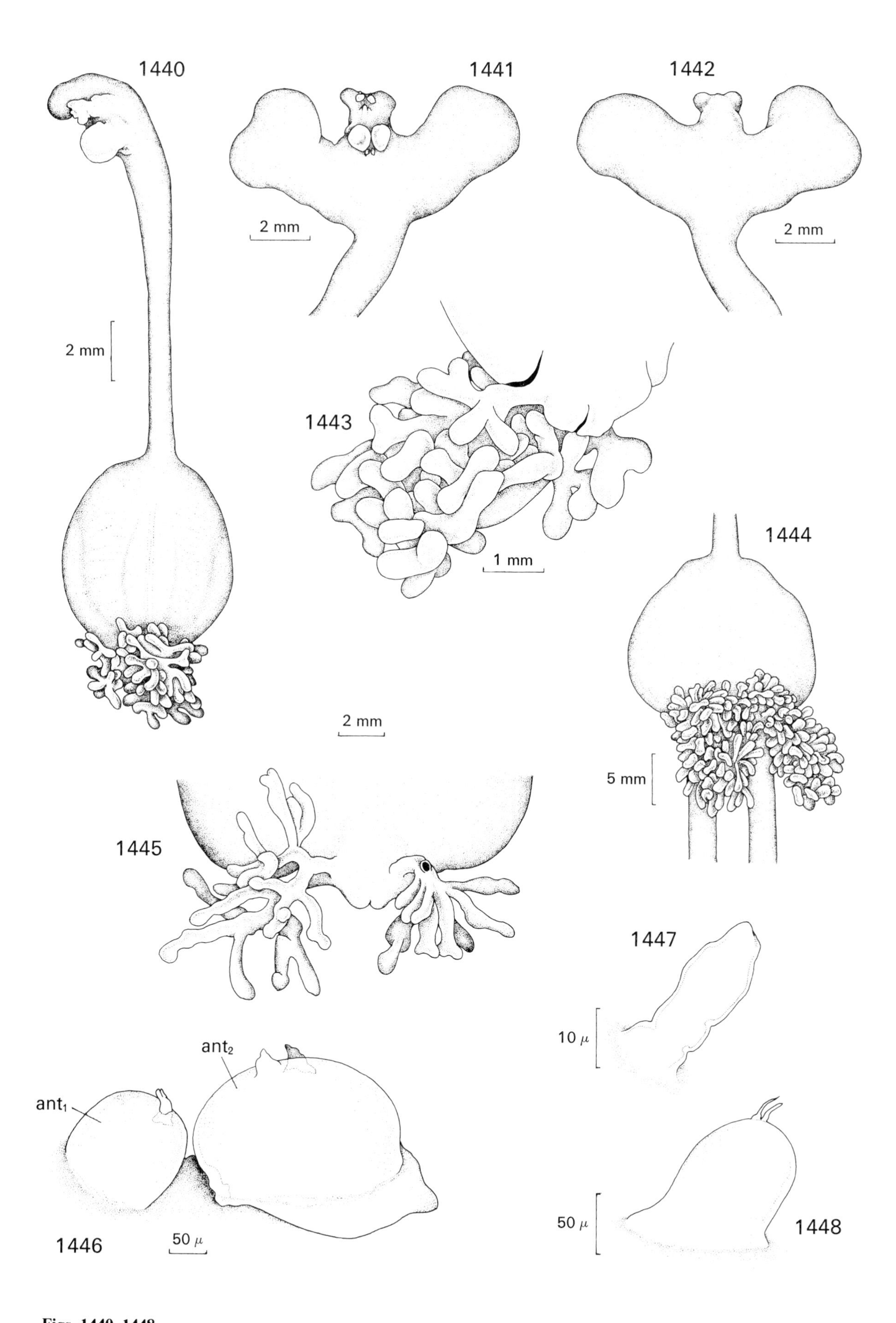

Figs. 1440–1448.
Sphyrion lumpi (Krøyer, 1845). **Fig. 1440.** Young adult female, cephalothorax lateral, trunk dorsal. **Fig. 1441.** Cephalothorax of older female, ventral. **Fig. 1442.** Same, dorsal. **Fig. 1443.** Posterior extremity of specimen from Fig. 1440, ventral. **Fig. 1444.** Trunk of older female, dorsal. **Fig. 1445.** Same, magnified, uropods partly removed. **Fig. 1446.** First and second antenna, lateral. **Fig. 1447.** First maxilla, lateral. **Fig. 1448.** Second maxilla, lateral.

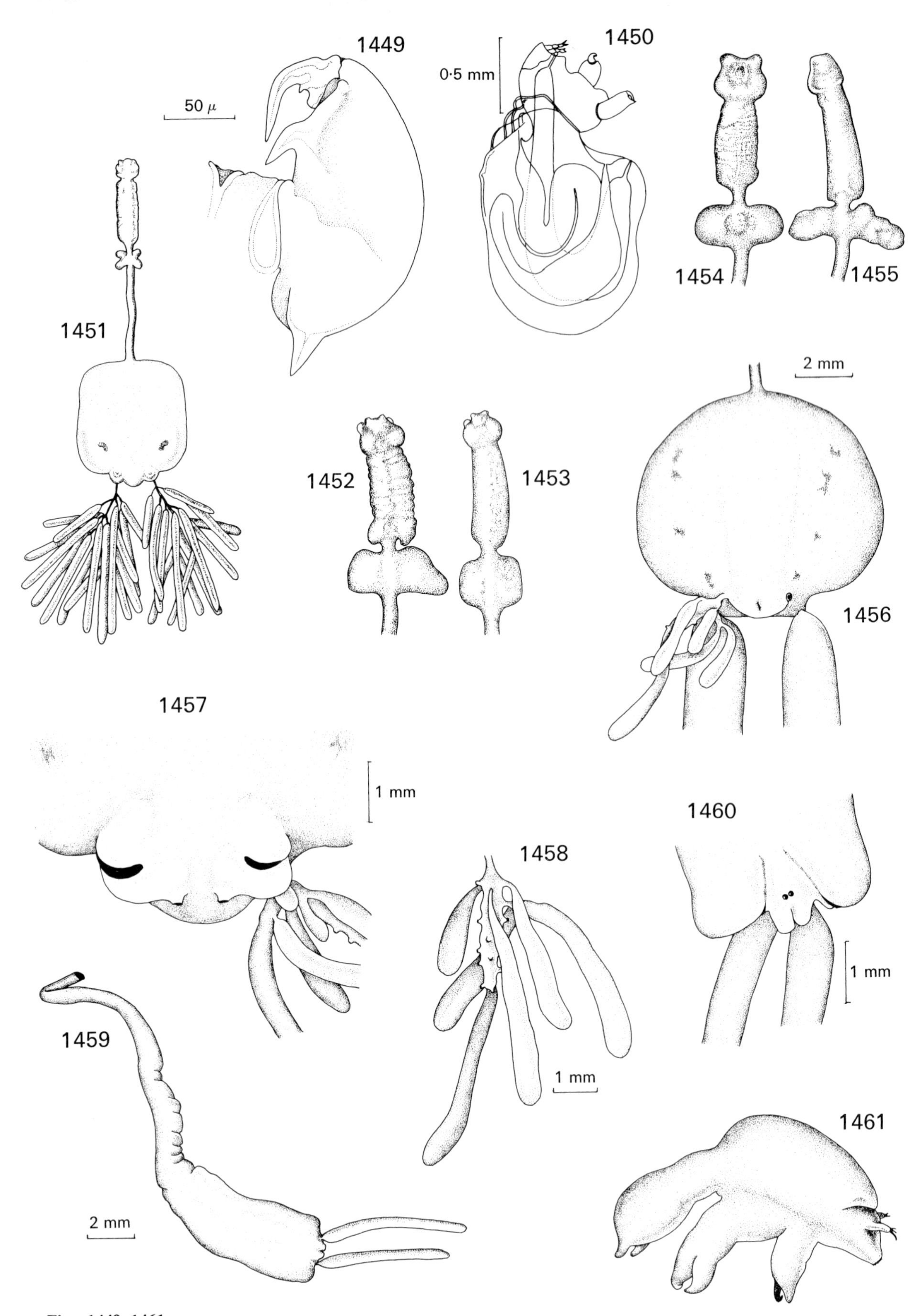

Figs. 1449–1461.
Sphyrion lumpi (Krøyer, 1845). **Fig. 1449.** Female maxilliped, ventral. **Fig. 1450.** Male, lateral.
Lophoura edwardsi Kölliker, 1853. **Fig. 1451.** Female, ventral. **Figs. 1452–1455.** Same, cephalothorax, different views. **Fig. 1456.** Same, trunk, dorsal. **Fig. 1457.** Same, posterior extremity, posteroventral. **Fig. 1458.** Same, part of uropod.
Tripaphylus musteli (van Beneden, 1851). **Fig. 1459.** Decapitated female, trunk dorsal. **Fig. 1460.** Posterior extremity, ventral. **Fig. 1461.** Male, lateral.
Fig. 1450 redrawn from Wilson (1919); Figs. 1452–1455 from Nunes-Ruivo (1953); Fig. 1451 from Cornalia (1865); Fig. 1461 from van Beneden (1851b).

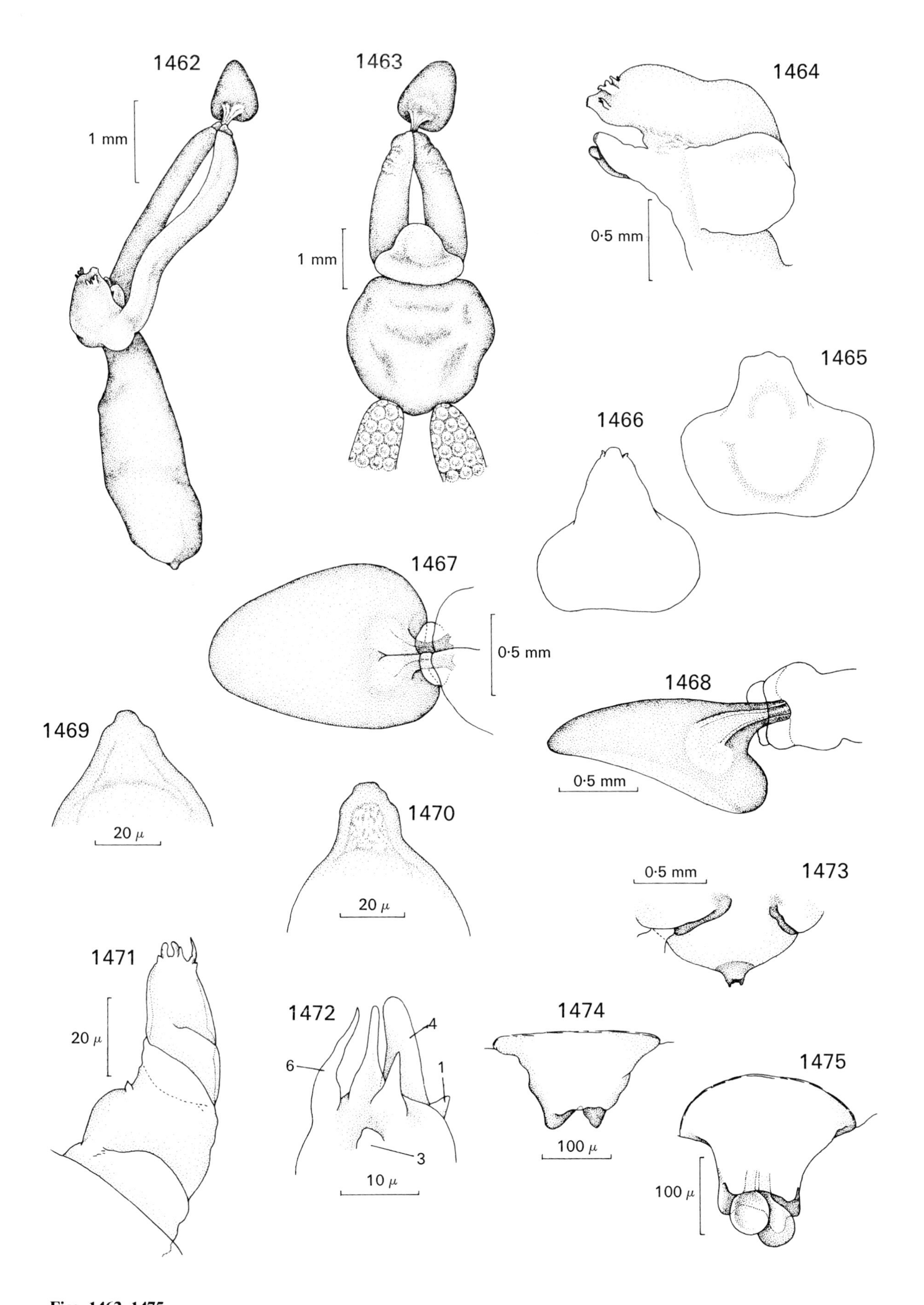

Figs. 1462–1475.
Salmincola salmoneus (L.). **Fig. 1462.** Female, lateral. **Fig. 1463.** Female, dorsal. **Fig. 1464.** Cephalothorax, lateral. **Figs. 1465, 1466.** Same, dorsal outline (diagrammatic). **Fig. 1467.** Bulla, supra-anchoral surface. **Fig. 1468.** Same, lateral. **Figs. 1469, 1470.** Tip of labrum, outer surface. **Fig. 1471.** First antenna, dorsal. **Fig. 1472.** Same, tip, dorsal. **Fig. 1473.** Posterior extremity, ventral **Fig. 1474.** Genital process, ventral. **Fig. 1475.** As above. Fig. 1463 original, others redrawn from Kabata (1969d).

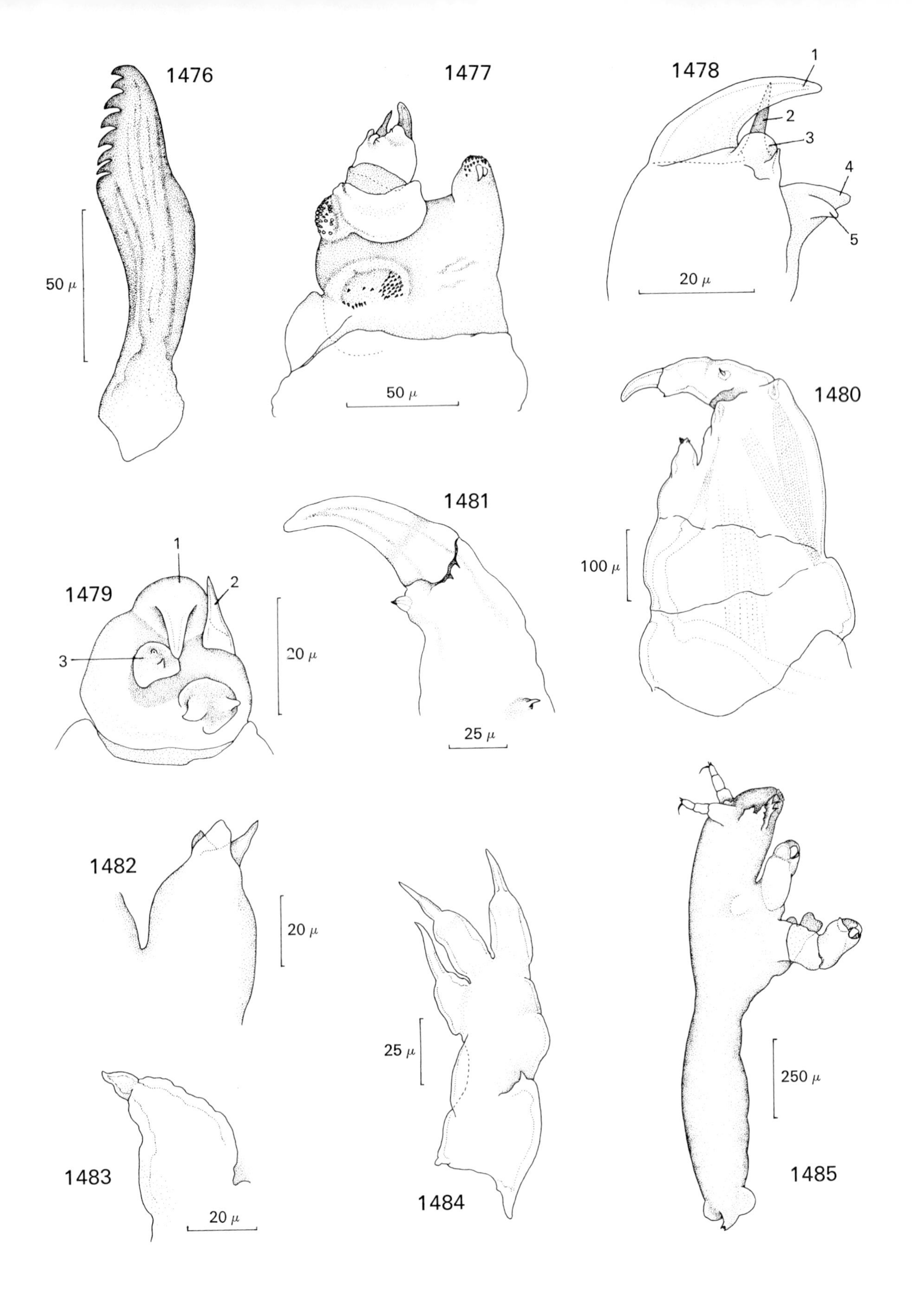

Figs. 1476–1485.
Salmincola salmoneus (L.). **Fig. 1476.** Mandible, lateral. **Fig. 1477.** Second antenna, lateral. **Fig. 1478.** Same, tip of endopod, lateral. **Fig. 1479.** Same, ventral. **Fig. 1480.** Maxilliped, ventral. **Fig. 1481.** Same, claw, full view. **Figs. 1482, 1483.** Same, palp. **Fig. 1484.** First maxilla, lateral. **Fig. 1485.** Male, lateral.
Fig. 1476 original; Fig. 1485 modified from Friend (1941); other figures redrawn from Kabata (1969d).

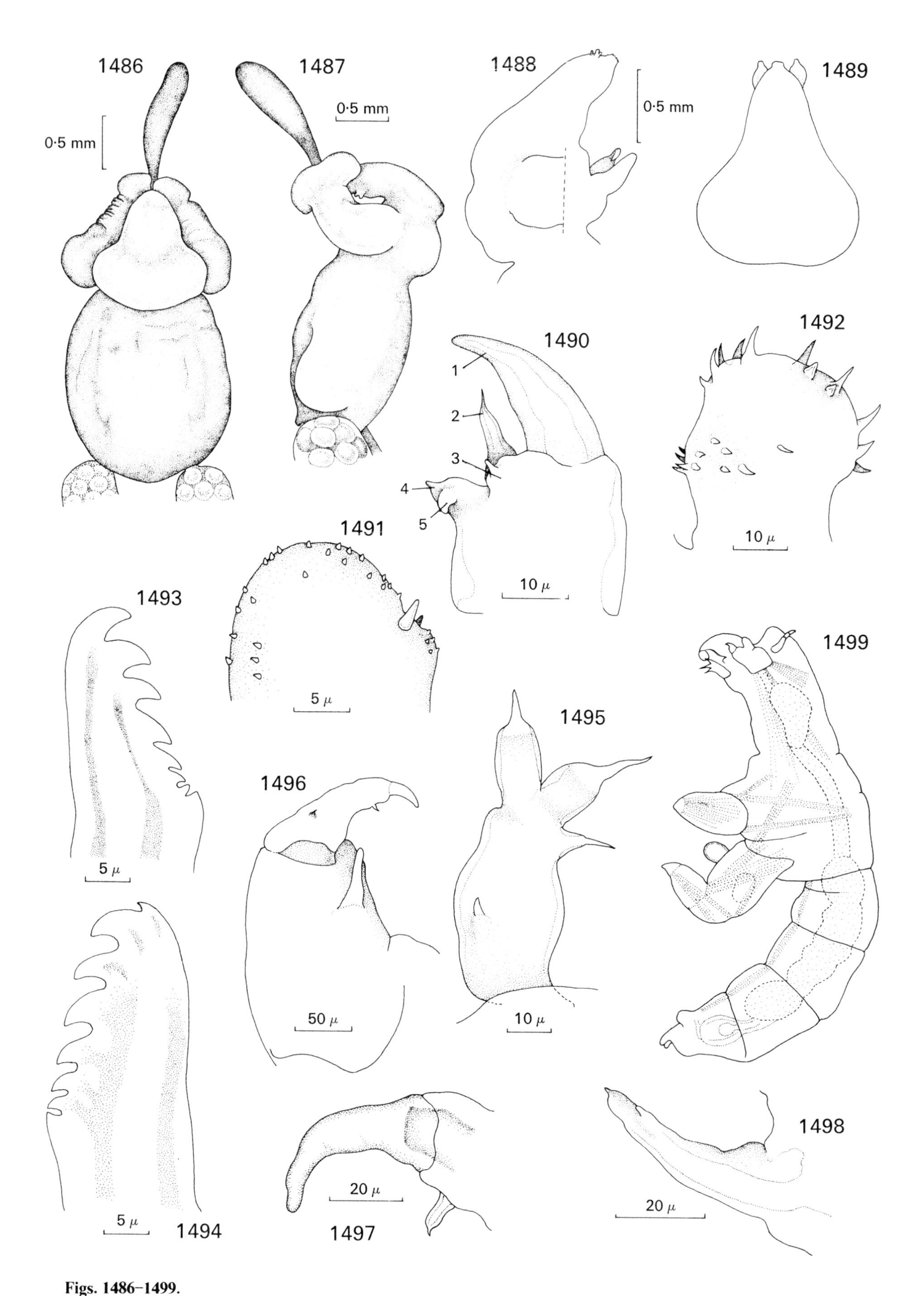

Figs. 1486–1499.
Salmincola thymalli (Kessler, 1868). **Fig. 1486.** Female, dorsal. **Fig. 1487.** Same, lateral. **Fig. 1488.** Cephalothorax, lateral (diagrammatic). **Fig. 1489.** Same, dorsal outline (diagrammatic). **Fig. 1490.** Second antenna, tip of endopod, lateral. **Figs. 1491, 1492.** Same, exopod. **Fig. 1493.** Tip of mandible, lateral. **Fig. 1494.** Same with terminal tooth broken. **Fig. 1495.** First maxilla, lateral. **Fig. 1496.** Maxilliped, ventral. **Fig. 1497.** Same, tip of claw. **Fig. 1498.** Same, palp. **Fig. 1499.** Male, lateral.
Fig. 1499 redrawn from Gurney (1933), others from Kabata (1969d).

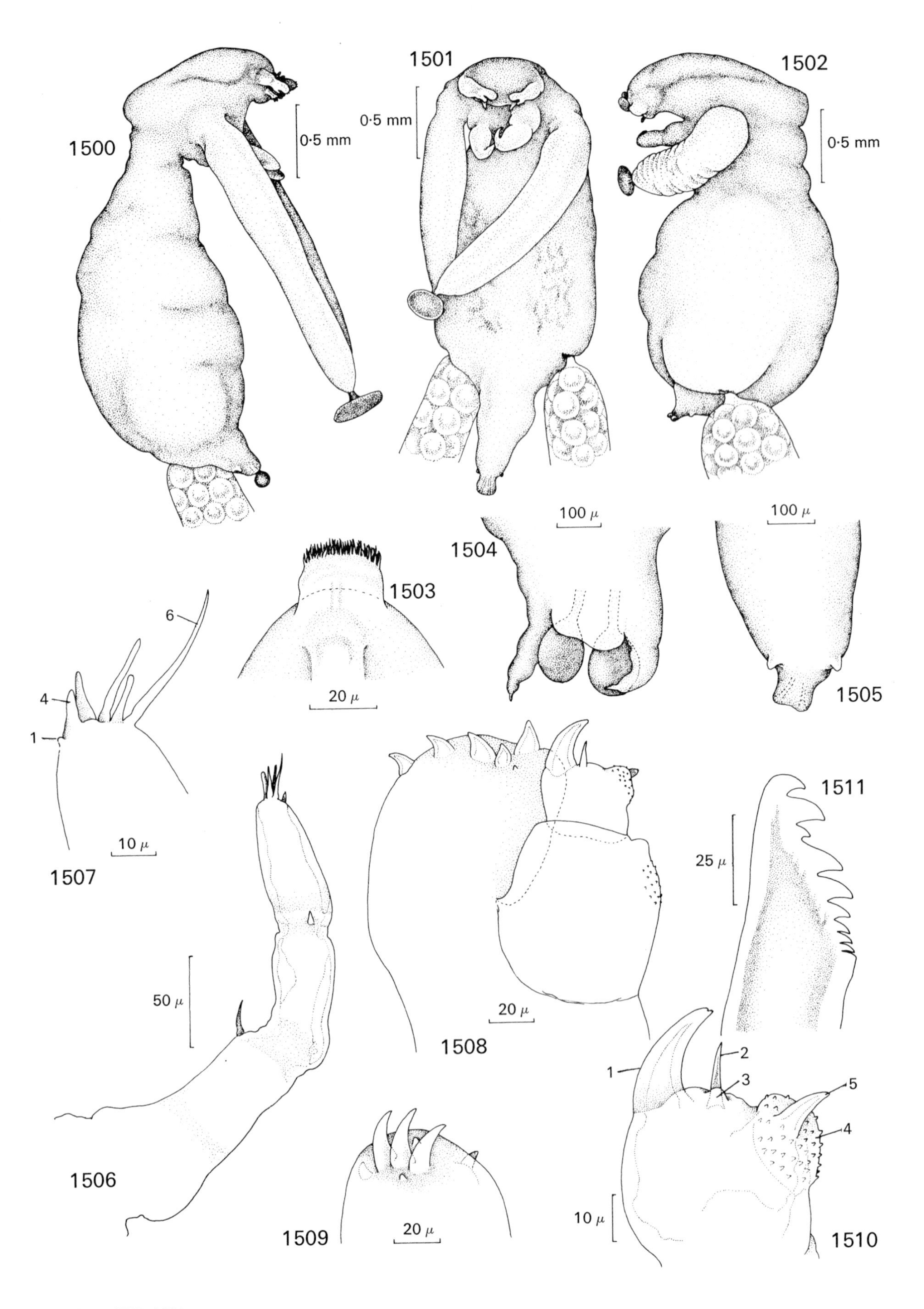

Figs. 1500–1511.
Achtheres percarum Nordmann, 1832. **Fig. 1500.** Female, relaxed, lateral. **Fig. 1501.** Same, ventral. **Fig. 1502.** Female, slightly contracted, lateral. **Fig. 1503.** Tip of labrum. **Figs. 1504, 1505.** Posterior extremity, ventral. **Fig. 1506.** First antenna. **Fig. 1507.** Same, tip. **Fig. 1508.** Distal half of second antenna, medial. **Fig. 1509.** Same, tip of exopod. **Fig. 1510.** Same, tip of endopod. **Fig. 1511.** Tip of mandible, lateral.

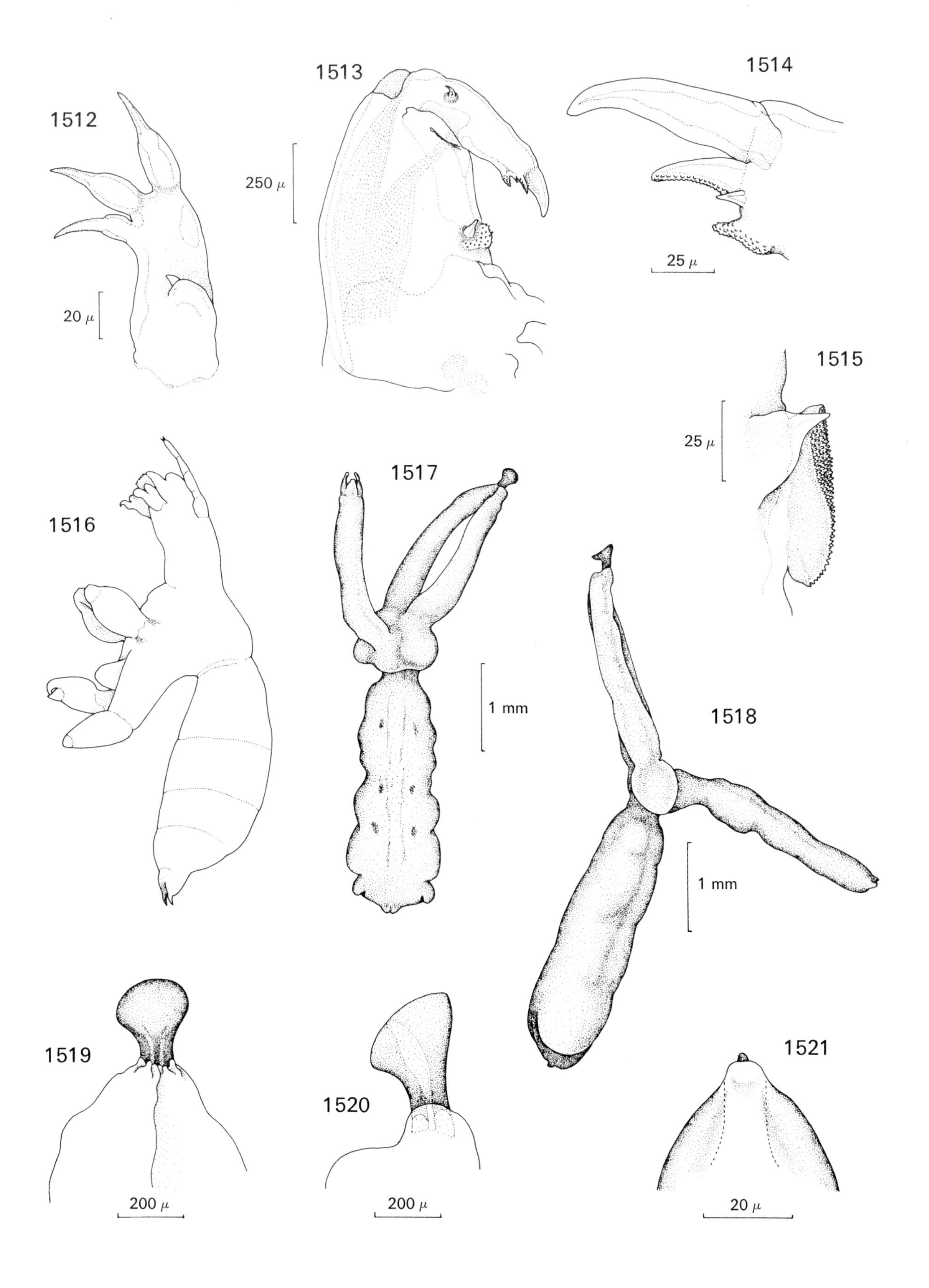

Figs. 1512–1521.
Achtheres percarum Nordmann, 1832. **Fig. 1512.** First maxilla, lateral. **Fig. 1513.** Maxilliped, ventral. **Fig. 1514.** Same, claw, ventral. **Fig. 1515.** Same, myxal area, ventral. **Fig. 1516.** Male, lateral.
Tracheliastes polycolpus Nordmann, 1832. **Fig. 1517.** Female, dorsal. **Fig. 1518.** Same, lateral. **Figs. 1519, 1520.** Bulla, dorsal and lateral. **Fig. 1521.** Tip of labrum, outer surface.
Fig. 1516 redrawn from Gurney (1933).

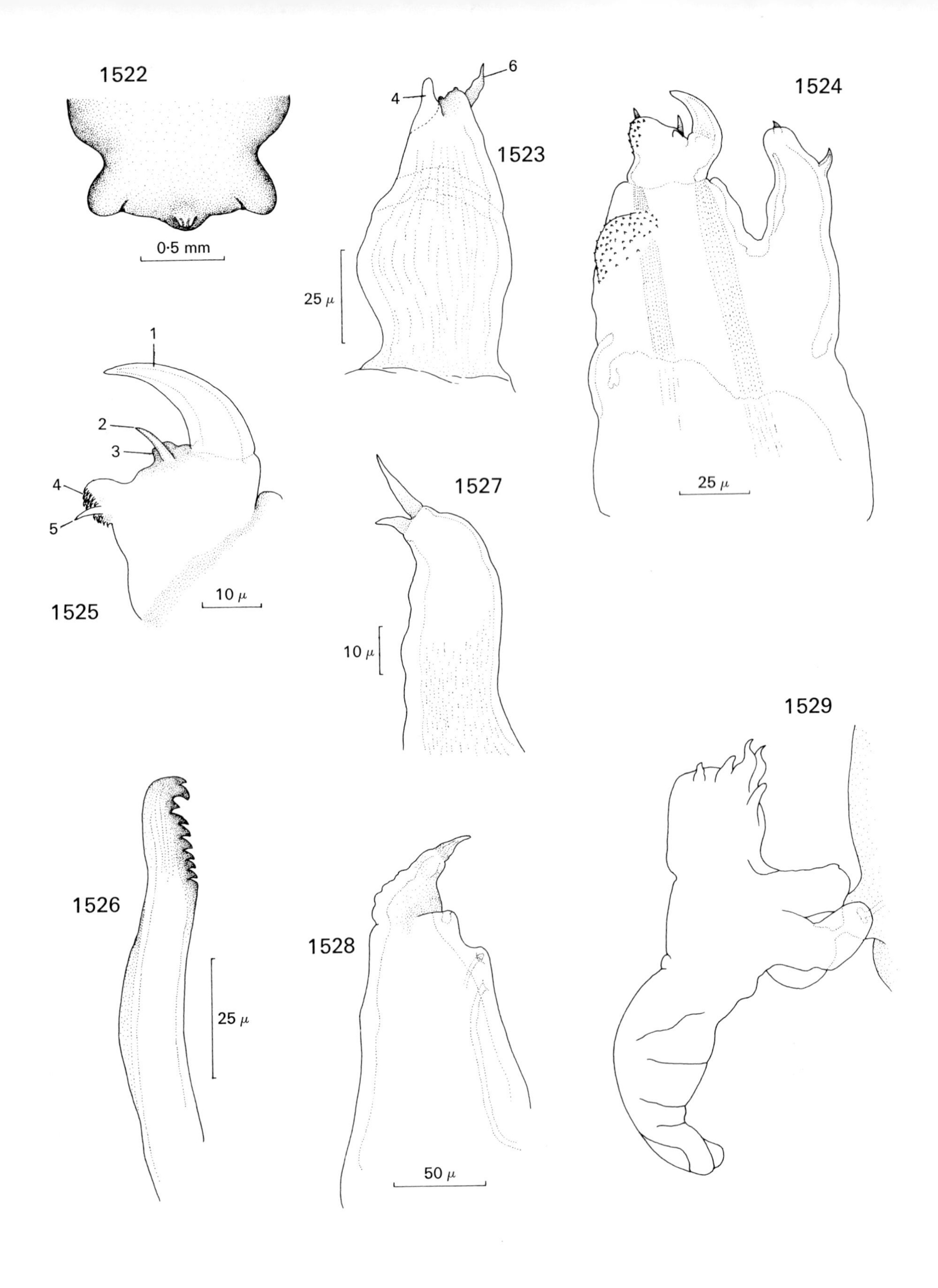

Figs. 1522–1529.
Tracheliastes polycolpus Nordmann, 1832. **Fig. 1522.** Female posterior extremity, ventral. **Fig. 1523.** First antenna, medial. **Fig. 1524.** Second antenna, lateral. **Fig. 1525.** Same, tip of endopod, medial. **Fig. 1526.** Mandible, lateral. **Fig. 1527.** First maxilla, lateral. **Fig. 1528.** Maxilliped. **Fig. 1529.** Male, lateral.
Fig. 1529 redrawn from Markevich (1937).

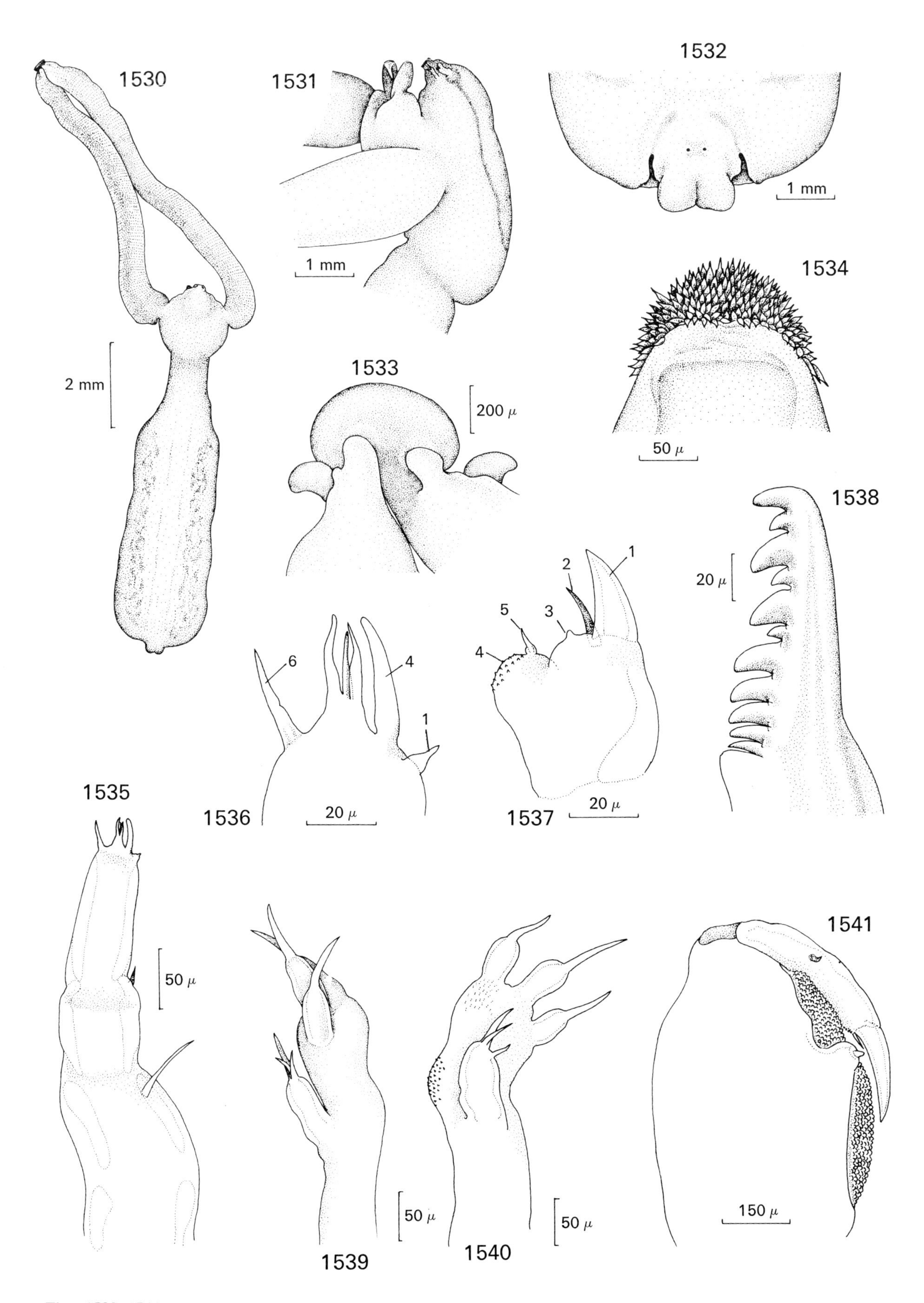

Figs. 1530–1541.
Ommatokoita elongata (Grant, 1827). **Fig. 1530.** Female, dorsal. **Fig. 1531.** Cephalothorax, lateral. **Fig. 1532.** Posterior extremity, ventral. **Fig. 1533.** Tips of second maxillae and bulla. **Fig. 1534.** Tip of labrum, outer surface. **Fig. 1535.** First antenna, dorsal. **Fig. 1536.** Same, tip, dorsal. **Fig. 1537.** Second antenna, tip of endopod. **Fig. 1538.** Tip of mandible, lateral. **Fig. 1539.** First maxilla, ventral. **Fig. 1540.** Same, lateral. **Fig. 1541.** Maxilliped, ventral.

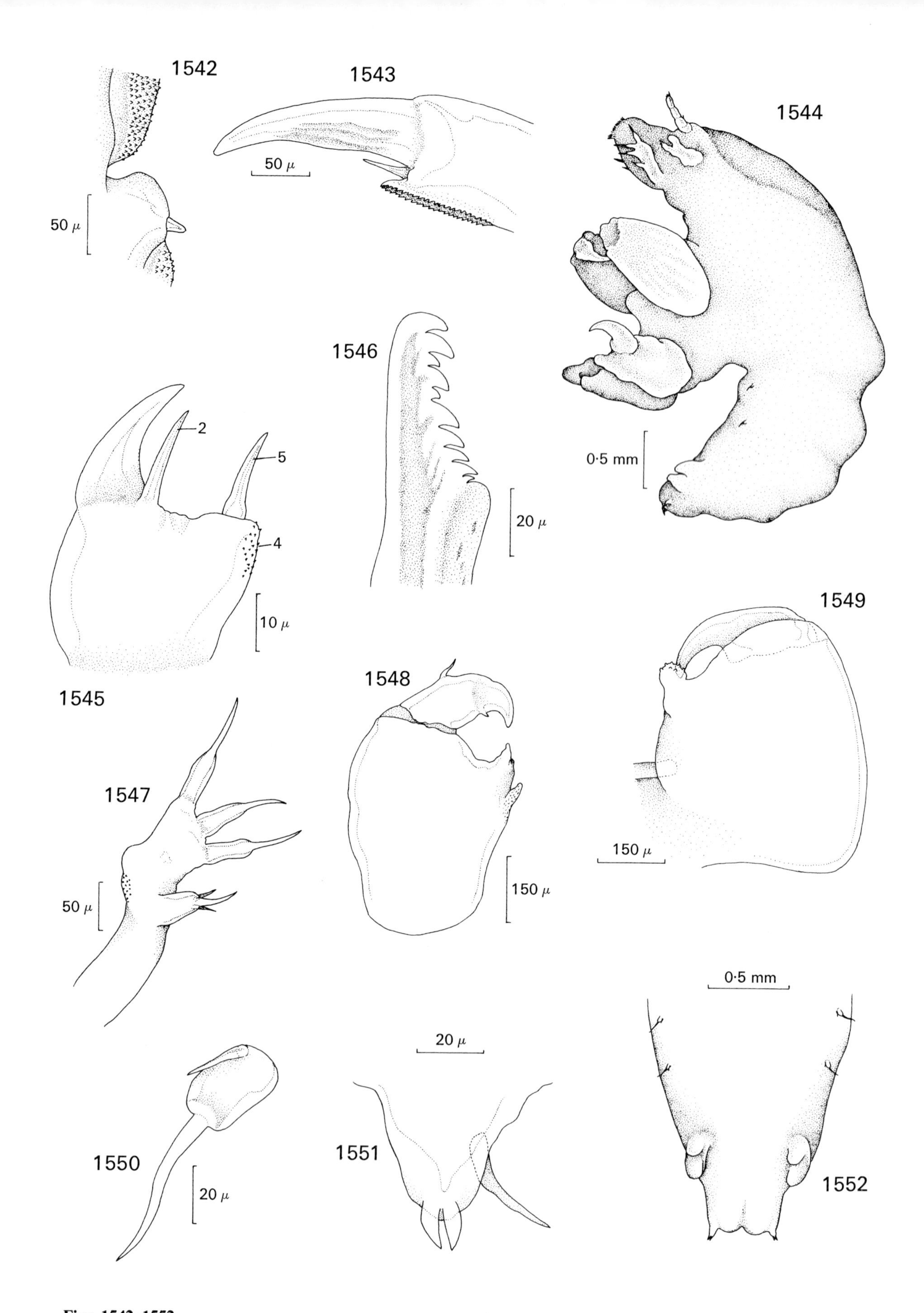

Figs. 1542–1552.
Ommatokoita elongata (Grant, 1827). **Fig. 1542.** Female maxilliped, myxal area. **Fig. 1543.** Same, claw. **Fig. 1544.** Male, lateral. **Fig. 1545.** Male second antenna, tip of endopod. **Fig. 1546.** Tip of mandible, lateral. **Fig. 1547.** First maxilla, lateral. **Fig. 1548.** Maxilliped, ventral. **Fig. 1549.** Second maxilla, ventral. **Fig. 1550.** Vestigial swimming leg, ventral. **Fig. 1551.** Uropod, ventral. **Fig. 1552.** Posterior extremity, ventral.

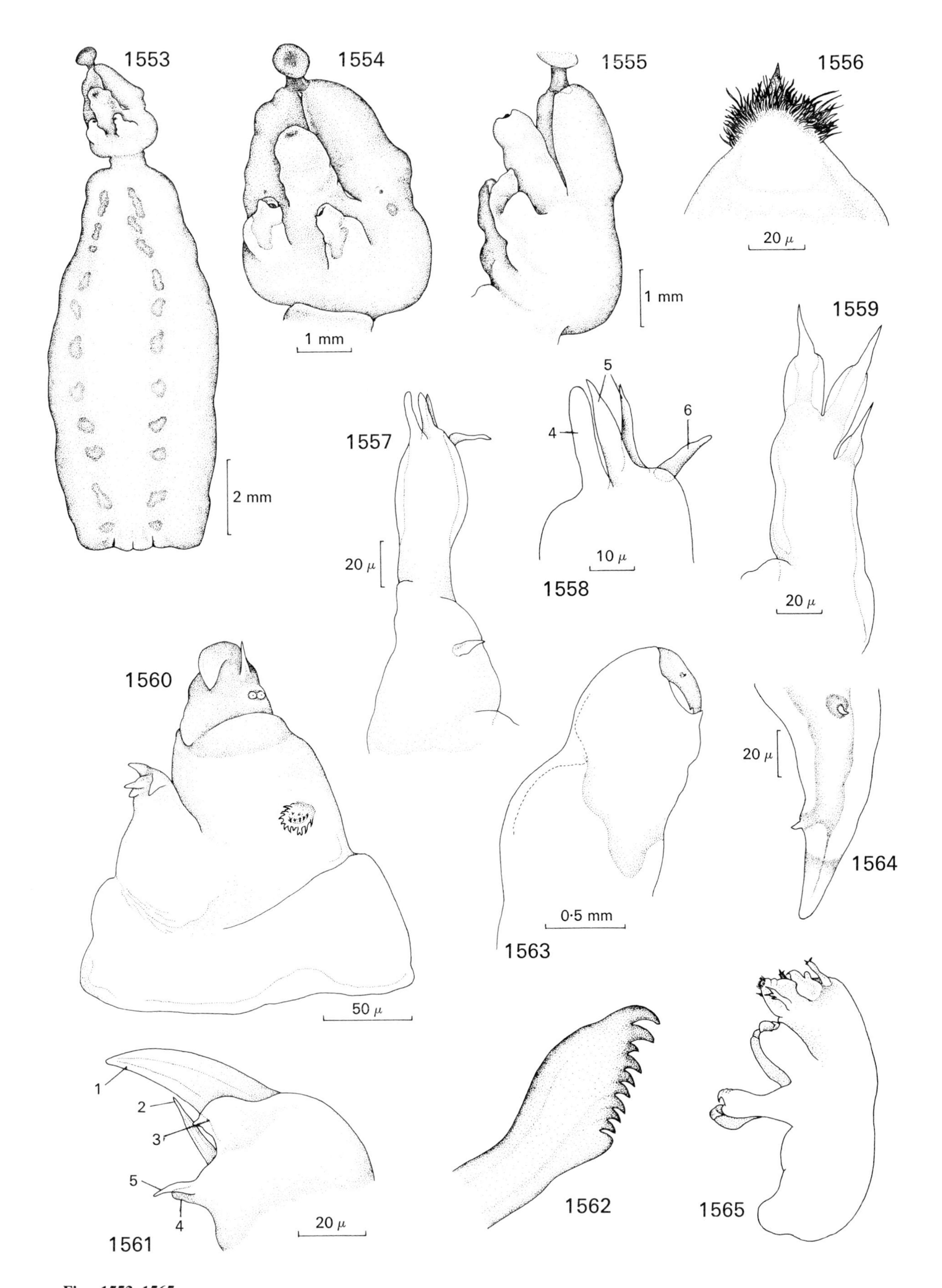

Figs. 1553–1565.
Vanbenedenia kroeyeri Malm, 1860. **Fig. 1553.** Female, ventral. **Fig. 1554.** Cephalothorax, ventral. **Fig. 1555.** Same, lateral. **Fig. 1556.** Labrum, outer surface. **Fig. 1557.** First antenna, dorsal. **Fig. 1558.** Same, tip. **Fig. 1559.** First maxilla, lateral. **Fig. 1560.** Second antenna, lateral. **Fig. 1561.** Same, tip of endopod, lateral. **Fig. 1562.** Mandible, lateral. **Fig. 1563.** Maxilliped, ventral. **Fig. 1564.** Same, claw, ventral. **Fig. 1565.** Juvenile specimen.
Fig. 1562 redrawn from Kabata and Bowman (1961); Fig. 1565 from Kabata (1958b).

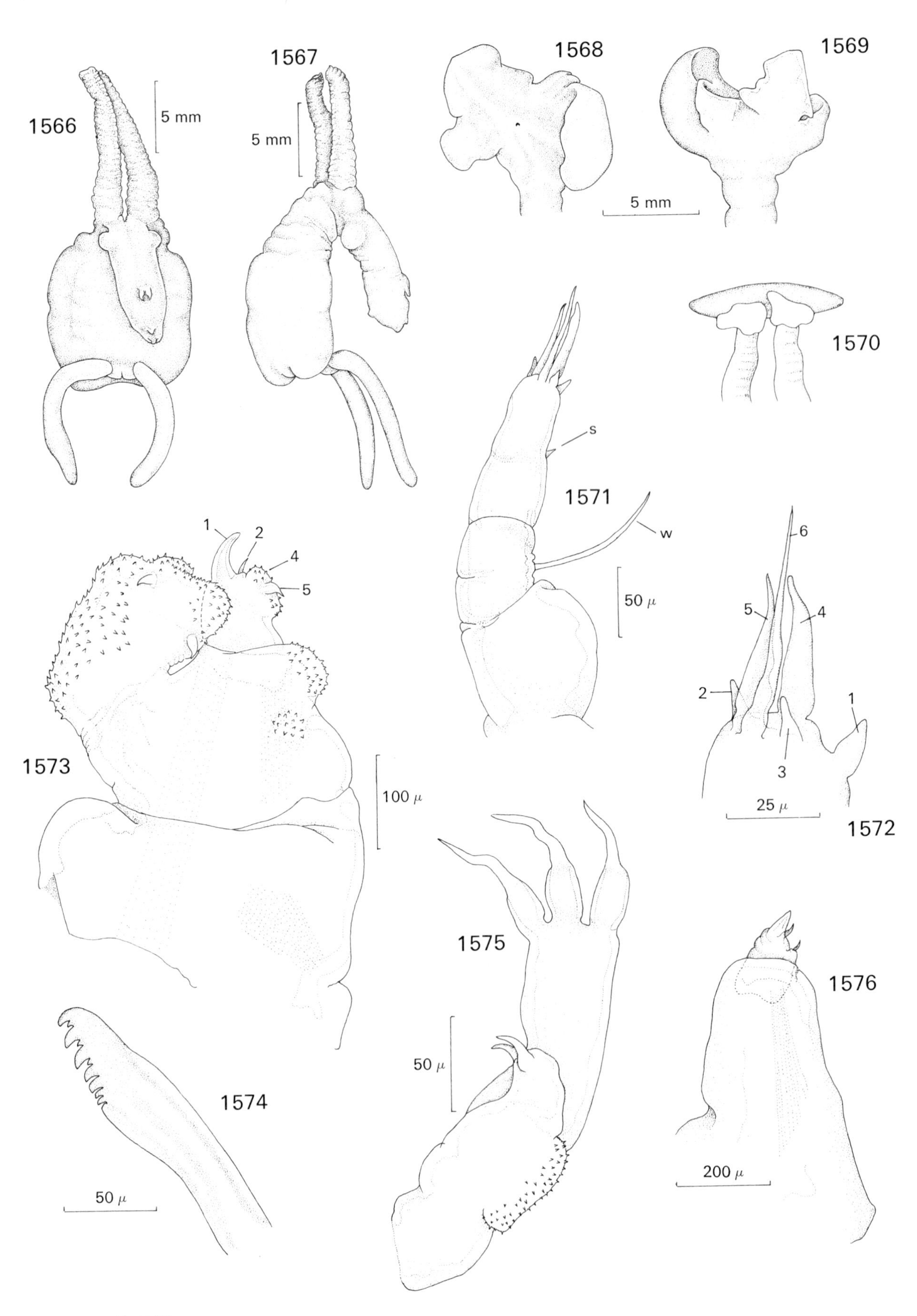

Figs. 1566–1576.
Charopinus dalmanni (Retzius, 1829). **Fig. 1566.** Female, trunk dorsal, cephalothorax ventral. **Fig. 1567.** Same, lateral. **Fig. 1568.** Tip of second maxilla with bulla removed, inner surface. **Fig. 1569.** Same, outer surface. **Fig. 1570.** Tips of second maxillae, with bulla. **Fig. 1571.** First antenna, dorsolateral. **Fig. 1572.** Same, tip. **Fig. 1573.** Second antenna, lateral. **Fig. 1574.** Mandible, lateral. **Fig. 1575.** First maxilla, lateral. **Fig. 1576.** Maxilliped, medial.
All figures redrawn from Kabata (1964a).

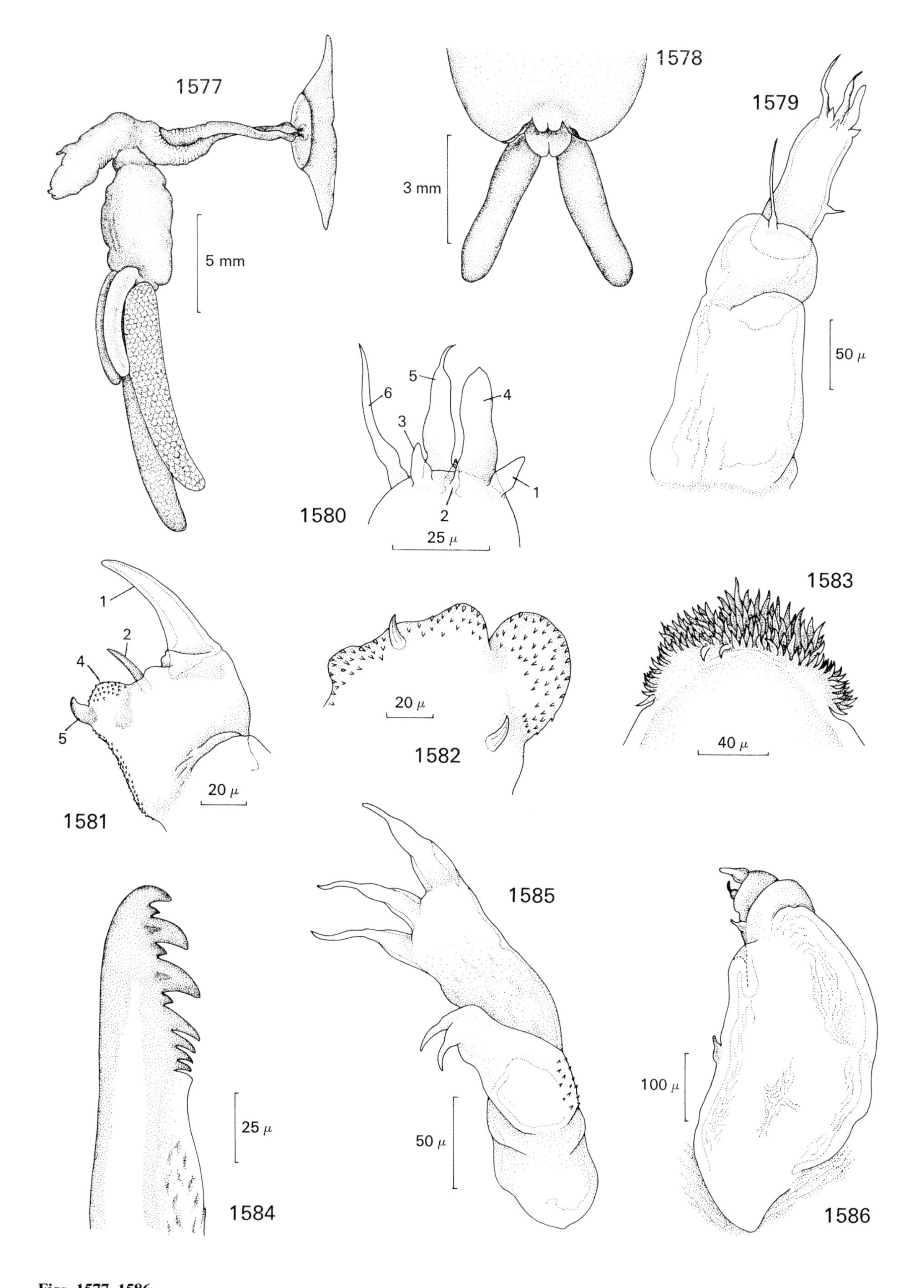

Figs. 1577–1586.
Charopinus dubius T. Scott, 1900. **Fig. 1577.** Female, lateral. **Fig. 1578.** Posterior extremity, ventral. **Fig. 1579.** First antenna, dorsal. **Fig. 1580.** Same, tip (armature slightly foreshortened). **Fig. 1581.** Second antenna, tip of endopod, lateral. **Fig. 1582.** Same, exopod, lateral. **Fig. 1583.** Tip of labrum, outer surface. **Fig. 1584.** Distal half of mandible, lateral. **Fig. 1585.** First maxilla, lateral. **Fig. 1586.** Maxilliped, lateral.
Figs. 1577–1580, 1584–1586 redrawn from Kabata (1964a).

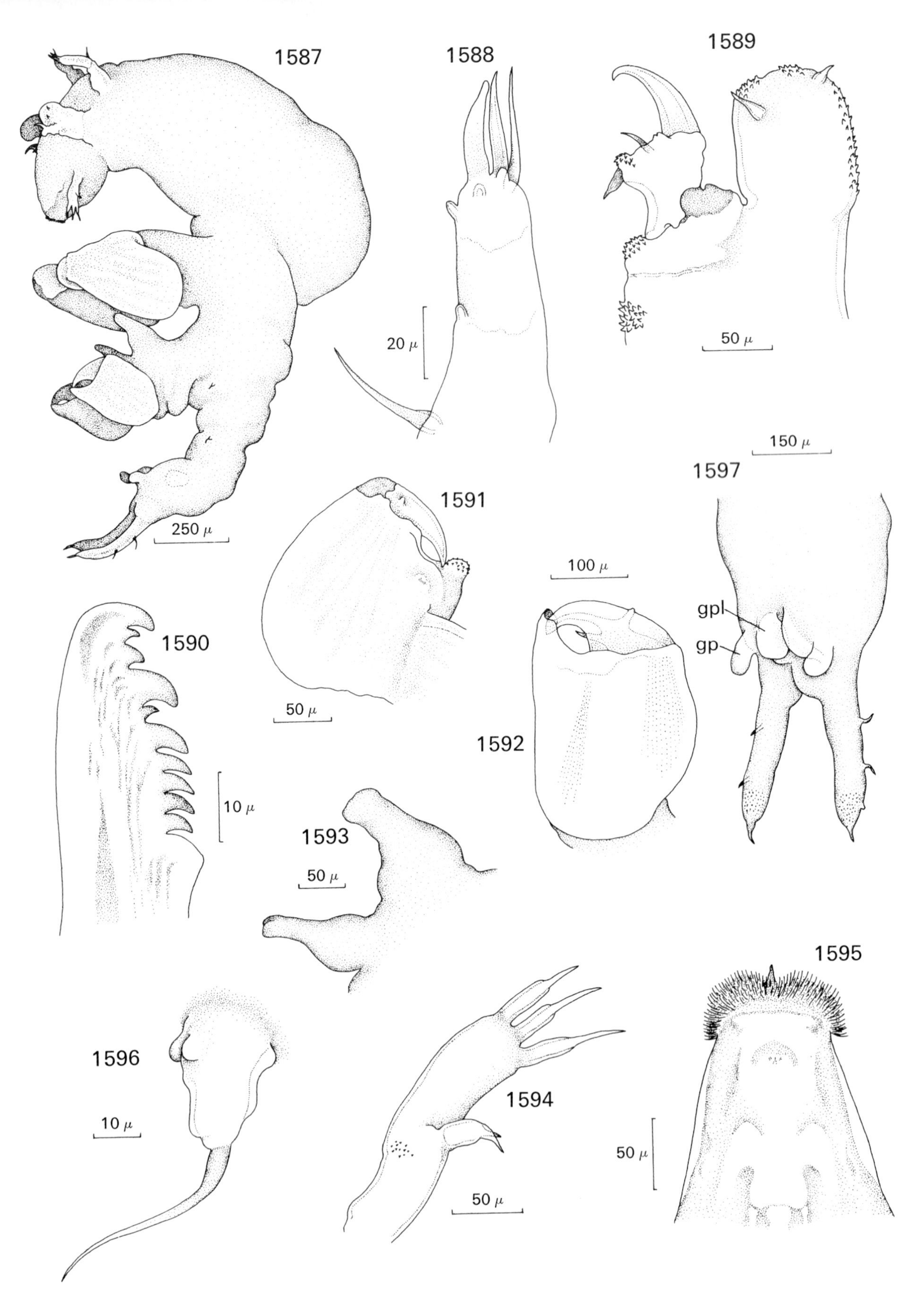

Figs. 1587–1597.
Charopinus dubius T. Scott, 1900. **Fig. 1587.** Male, lateral. **Fig. 1588.** First antenna (without base), ventral. **Fig. 1589.** Second antenna, distal half, lateral. **Fig. 1590.** Tip of mandible, lateral. **Fig. 1591.** Second maxilla, ventral. **Fig. 1592.** Maxilliped, ventral. **Fig. 1593.** Mediative process, lateral. **Fig. 1594.** First maxilla, lateral. **Fig. 1595.** Labrum, outer surface. **Fig. 1596.** Vestigial swimming leg, lateral. **Fig. 1597.** Posterior extremity, ventrolateral.
gpl—genital plate; gp—genital process.

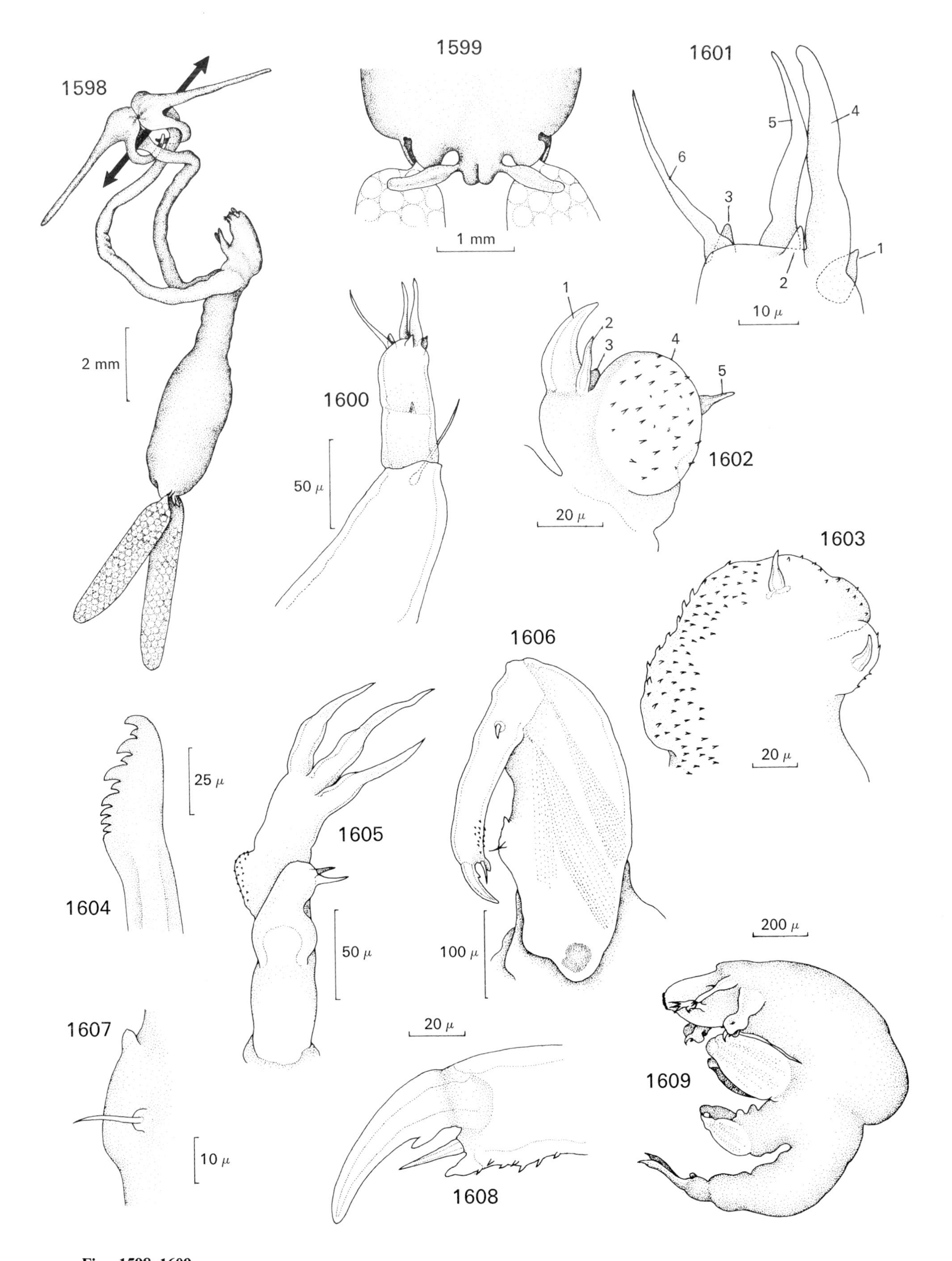

Figs. 1598–1609.
Schistobrachia ramosa (Krøyer, 1863). **Fig. 1598.** Female, lateral. **Fig. 1599.** Posterior extremity, dorsal. **Fig. 1600.** First antenna, ventrolateral. **Fig. 1601.** Same, tip, ventrolateral. **Fig. 1602.** Second antenna, tip of endopod, medial. **Fig. 1603.** Same, tip of exopod, lateral. **Fig. 1604.** Tip of mandible, lateral. **Fig. 1605.** First maxilla, lateral. **Fig. 1606.** Maxilliped, ventral. **Fig. 1607.** Same, myxal area, ventral. **Fig. 1608.** Same, claw, ventral. **Fig. 1609.** Male, lateral.
Figs. 1598–1601, 1604–1606 redrawn from Kabata (1964a).

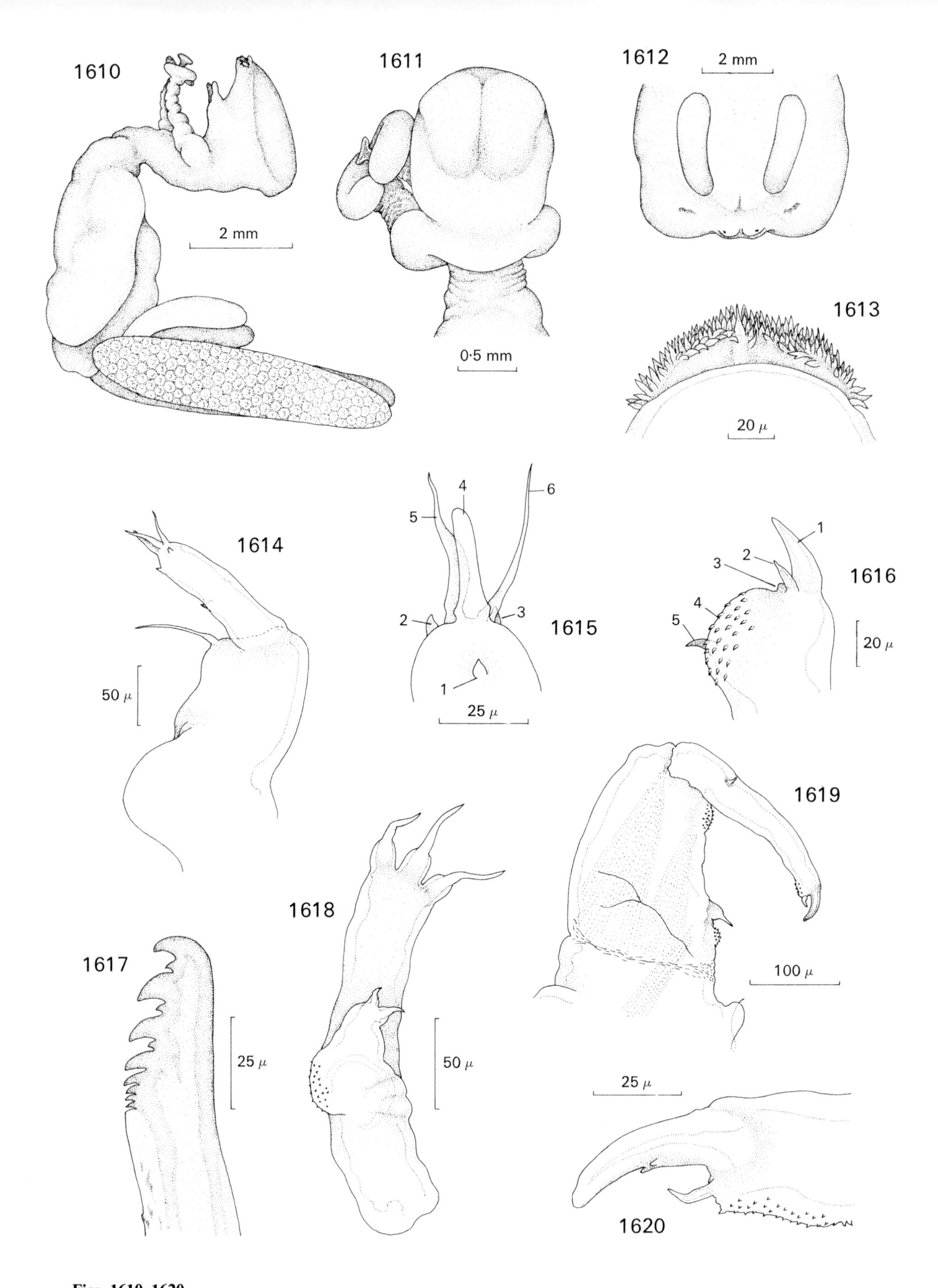

Figs. 1610–1620.

Pseudocharopinus bicaudatus (Krøyer, 1837). **Fig. 1610.** Female, lateral. **Fig. 1611.** Cephalothorax, dorsal. **Fig. 1612.** Posterior extremity, posterodorsal. **Fig. 1613.** Tip of labrum, outer surface. **Fig. 1614.** First antenna, lateral. **Fig. 1615.** Same, tip, medial. **Fig. 1616.** Second antenna, tip of endopod, medial. **Fig. 1617.** Tip of mandible, lateral. **Fig. 1618.** First maxilla, lateral. **Fig. 1619.** Maxilliped, ventral. **Fig. 1620.** Same, claw, ventral.

Figs. 1610, 1612, 1615, 1617–1620 redrawn from Kabata (1964a).

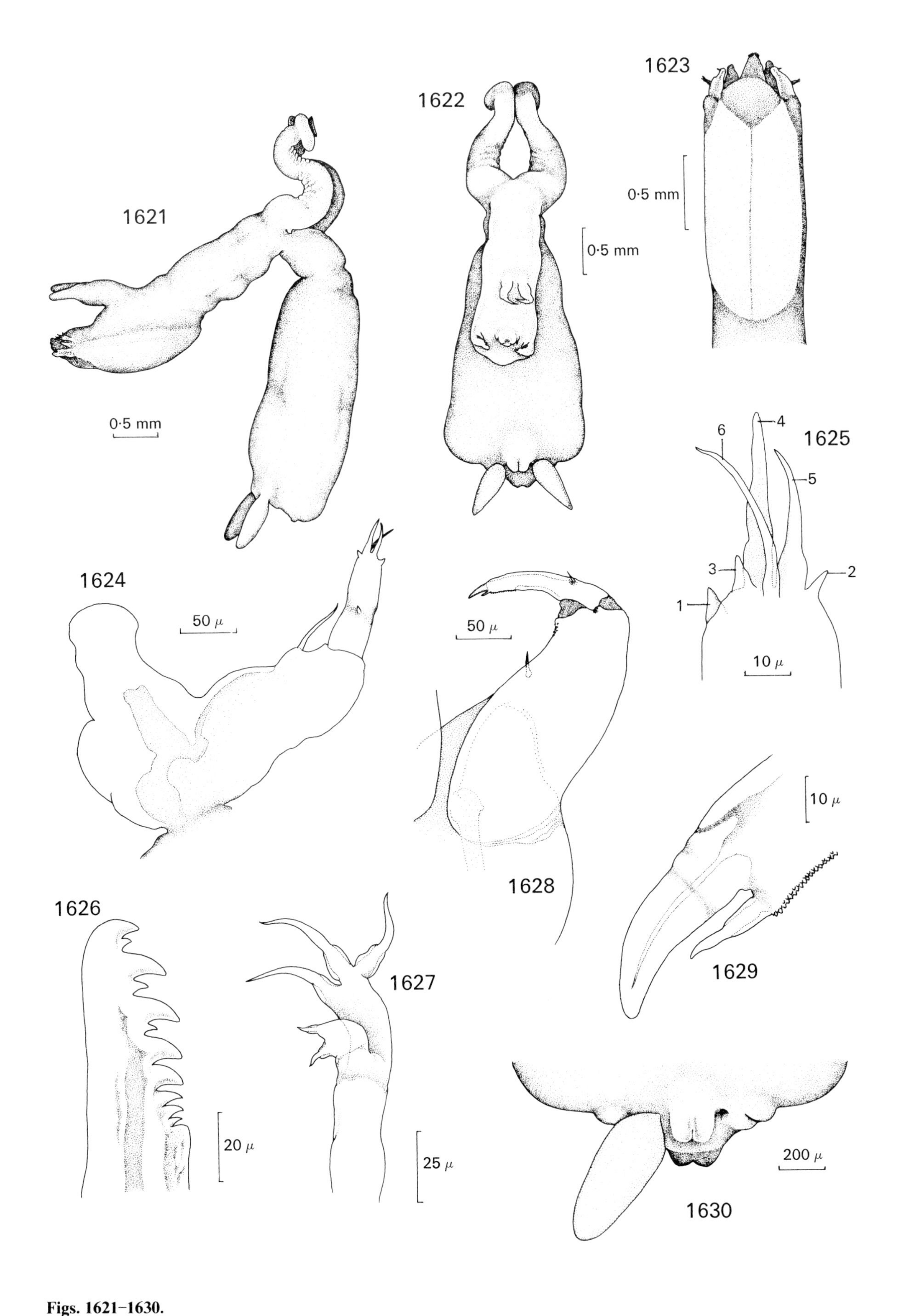

Figs. 1621–1630.
Pseudocharopinus malleus (Rudolphi, in Nordmann, 1832). **Fig. 1621.** Female, lateral. **Fig. 1622.** Same, trunk dorsal, cephalothorax ventral. **Fig. 1623.** Tip of cephalothorax, dorsal. **Fig. 1624.** First antenna, ventral. **Fig. 1625.** Same, tip, lateral. **Fig. 1626.** Tip of mandible, lateral. **Fig. 1627.** First maxilla, lateral. **Fig. 1628.** Maxilliped, ventral. **Fig. 1629.** Same, claw, ventral. **Fig. 1630.** Posterior extremity, dorsal, with one posterior process removed.

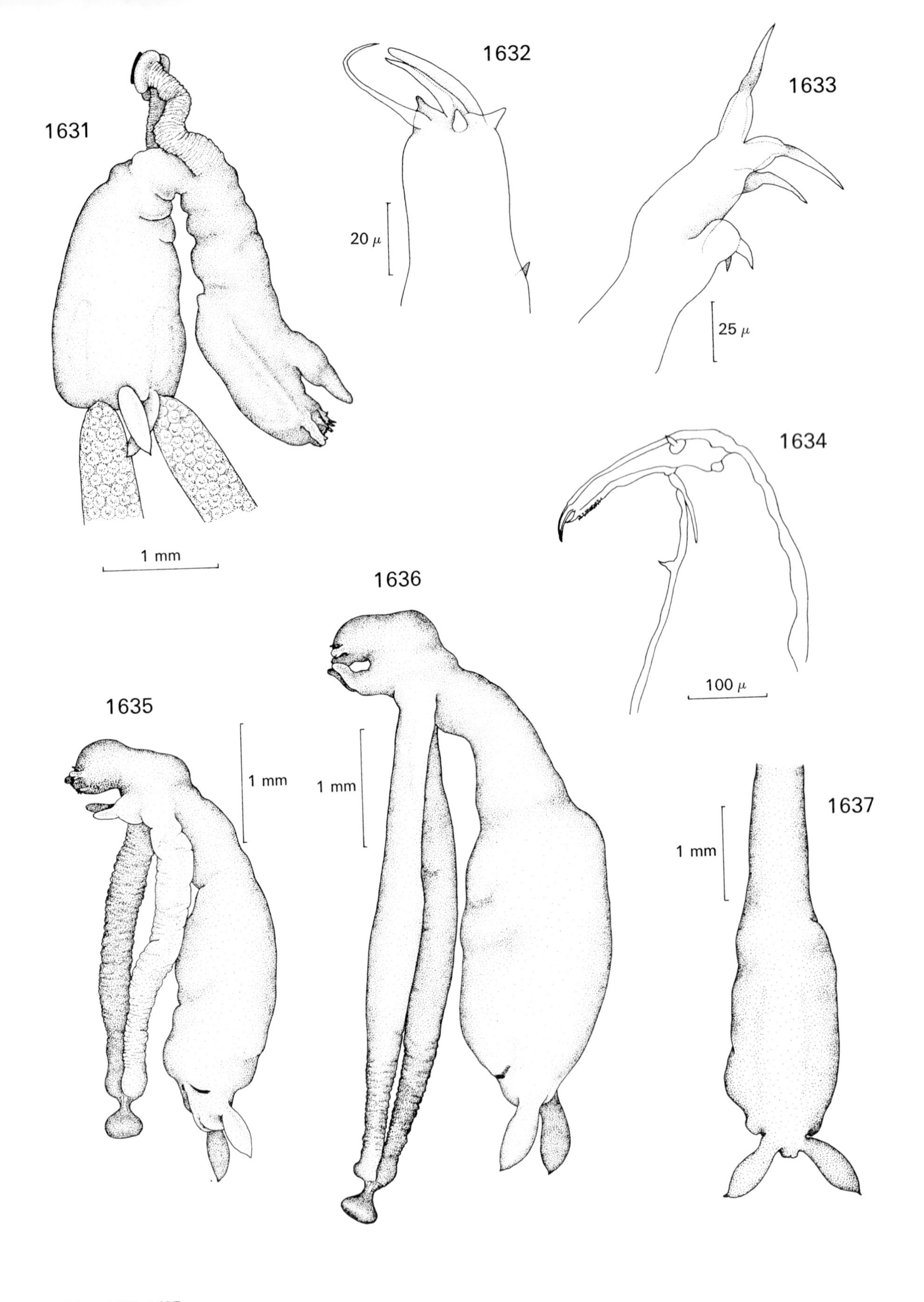

Figs. 1631–1637.
Pseudocharopinus malleus (Rudolphi, in Nordmann, 1832). **Fig. 1631.** Female under coverslip, trunk dorsal, cephalothorax lateral. **Fig. 1632.** Tip of first antenna, lateral. **Fig. 1633.** First maxilla, lateral. **Fig. 1634.** Maxilliped ventral.
Lernaeopodina longimana (Olsson, 1896). **Fig. 1635.** Female, lateral (specimen from *Raja montagui*, Scotland). **Fig. 1636.** Same (specimen from *Raja fullonica*, Skagerrack). **Fig. 1637.** Same, trunk, dorsal.
Fig. 1634 redrawn from Capart (1946).

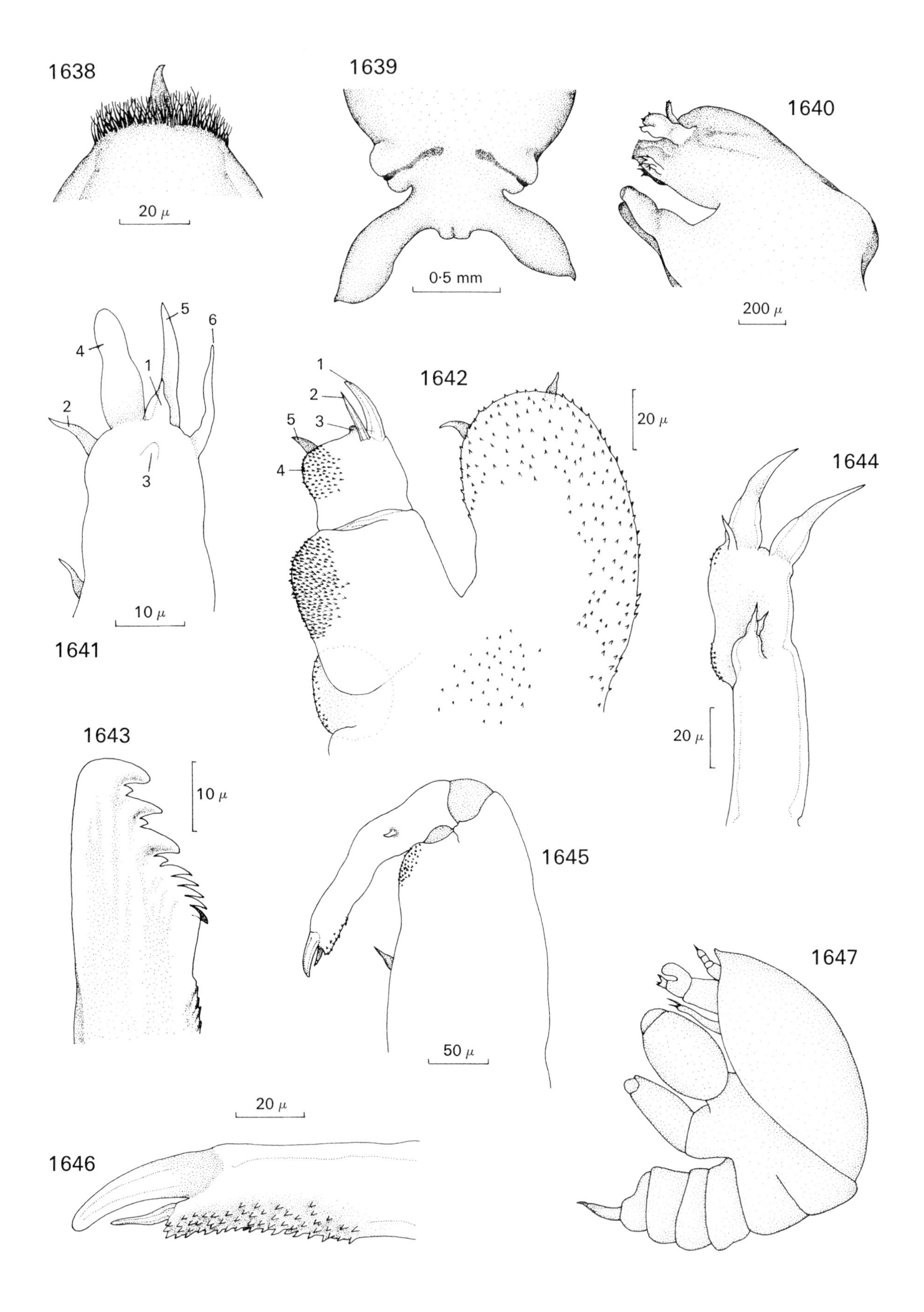

Figs. 1638–1647.
Lernaeopodina longimana (Olsson, 1869). **Fig. 1638.** Female, tip of labrum, outer surface. **Fig. 1639.** Posterior extremity, ventral. **Fig. 1640.** Cephalothorax, lateral. **Fig. 1641.** Tip of first antenna, medial. **Fig. 1642.** Distal half of second antenna, medial. **Fig. 1643.** Tip of mandible, lateral. **Fig. 1644.** First maxilla, lateral. **Fig. 1645.** Maxilliped, ventral. **Fig. 1646.** Same, claw, dorsal. **Fig. 1647.** Male, lateral.
Fig. 1647 modified from Scott and Scott (1913).

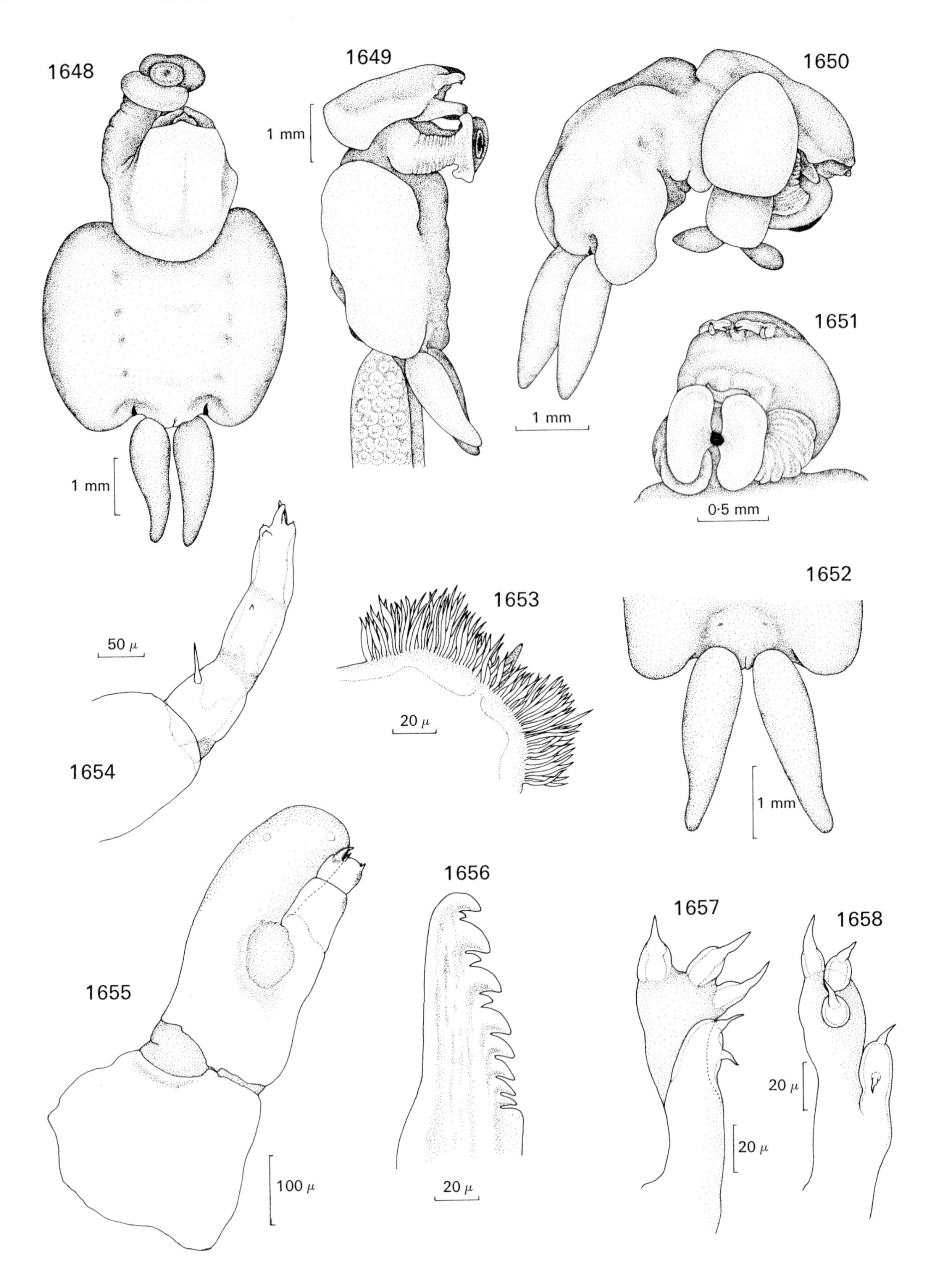

Figs. 1648–1658.
Lernaeopoda bidiscalis Kane, 1892. **Fig. 1648.** Female, dorsal. **Fig. 1649.** Same, lateral. **Fig. 1650.** Female with male attached, lateral. **Fig. 1651.** Cephalothorax, ventral. **Fig. 1652.** Posterior extremity, ventral. **Fig. 1653.** Tip of labrum, outer surface. **Fig. 1654.** First antenna, medial. **Fig. 1655.** Second antenna, ventral. **Fig. 1656.** Tip of mandible, lateral. **Fig. 1657.** First maxilla, lateral. **Fig. 1658.** Same, ventral.

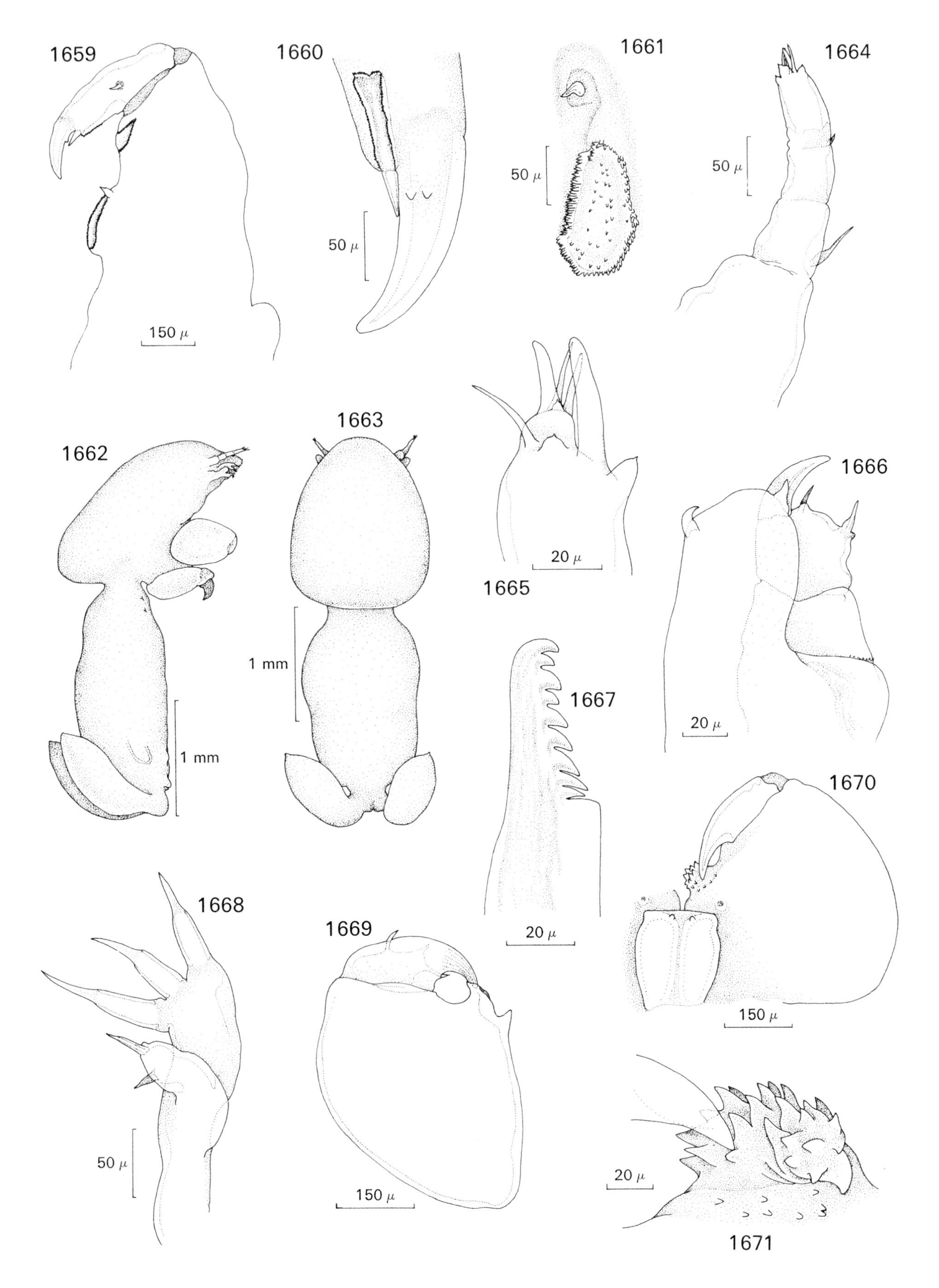

Figs. 1659–1671.
Lernaeopoda bidiscalis Kane, 1892. **Fig. 1659.** Female maxilliped, ventral. **Fig. 1660.** Same, claw, postero-ventral. **Fig. 1661.** Same, myxal area, medial. **Fig. 1662.** Male, lateral. **Fig. 1663.** Same, dorsal. **Fig. 1664.** First antenna, lateral. **Fig. 1665.** Same, tip, medial. **Fig. 1666.** Second antenna, distal half, lateral. **Fig. 1667.** Tip of mandible, lateral. **Fig. 1668.** First maxilla, lateral. **Fig. 1669.** Tip of maxilliped, ventral. **Fig. 1670.** Second maxilla, ventral. **Fig. 1671.** Same, tip of claw and basal tubercle.

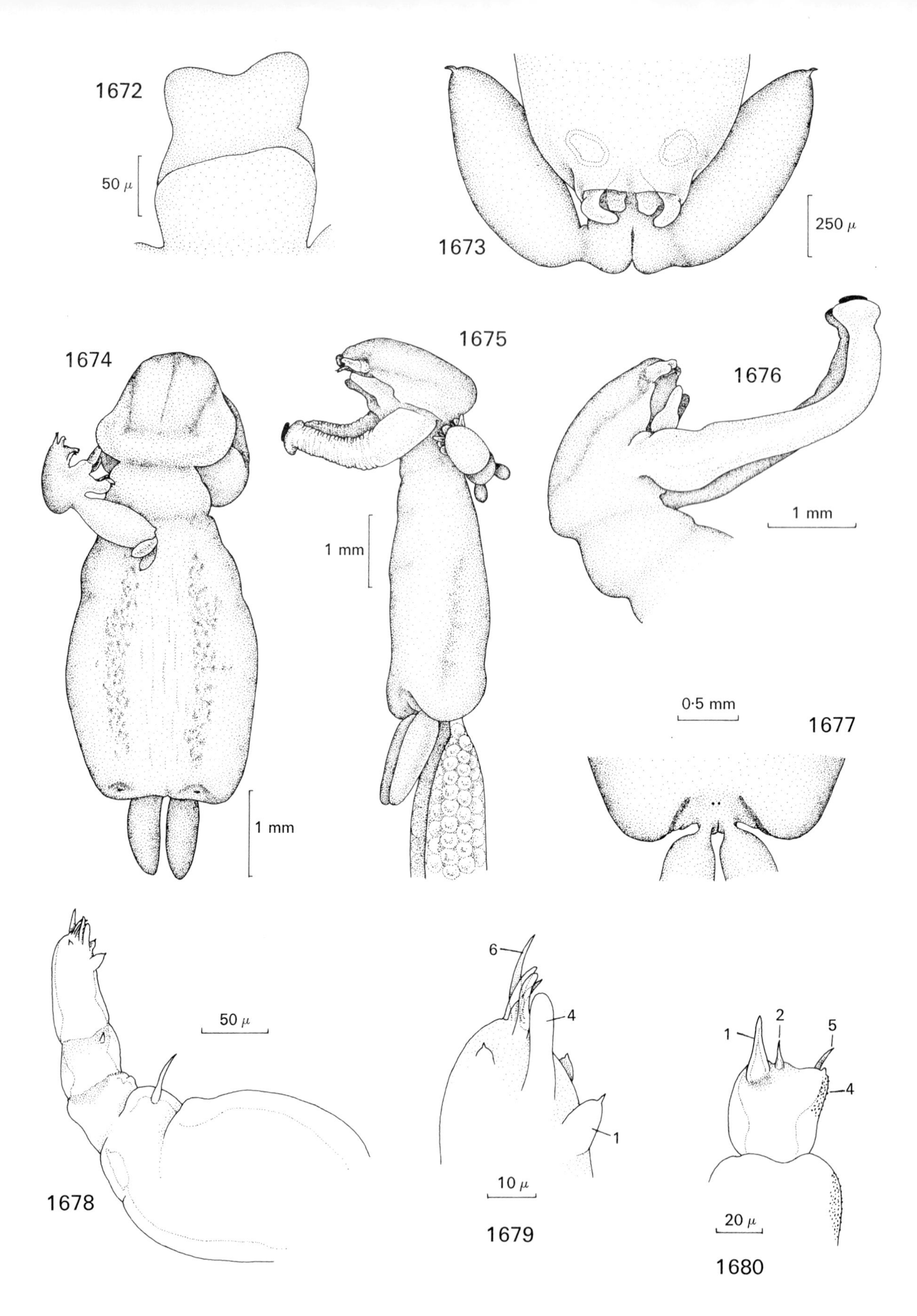

Figs. 1672–1680.

Lernaeopoda bidiscalis Kane, 1892. **Fig. 1672.** Male, mediative process, ventral. **Fig. 1673.** Posterior extremity, ventral.

Lernaeopoda galei Krøyer, 1837. **Fig. 1674.** Female with male attached, dorsal. **Fig. 1675.** Same, lateral. **Fig. 1676.** Cephalothorax, lateral **Fig. 1677.** Posterior extremity, ventral. **Fig. 1678.** First antenna, ventral. **Fig. 1679.** Same, tip, ventral. **Fig. 1680.** Second antenna, tip of endopod.

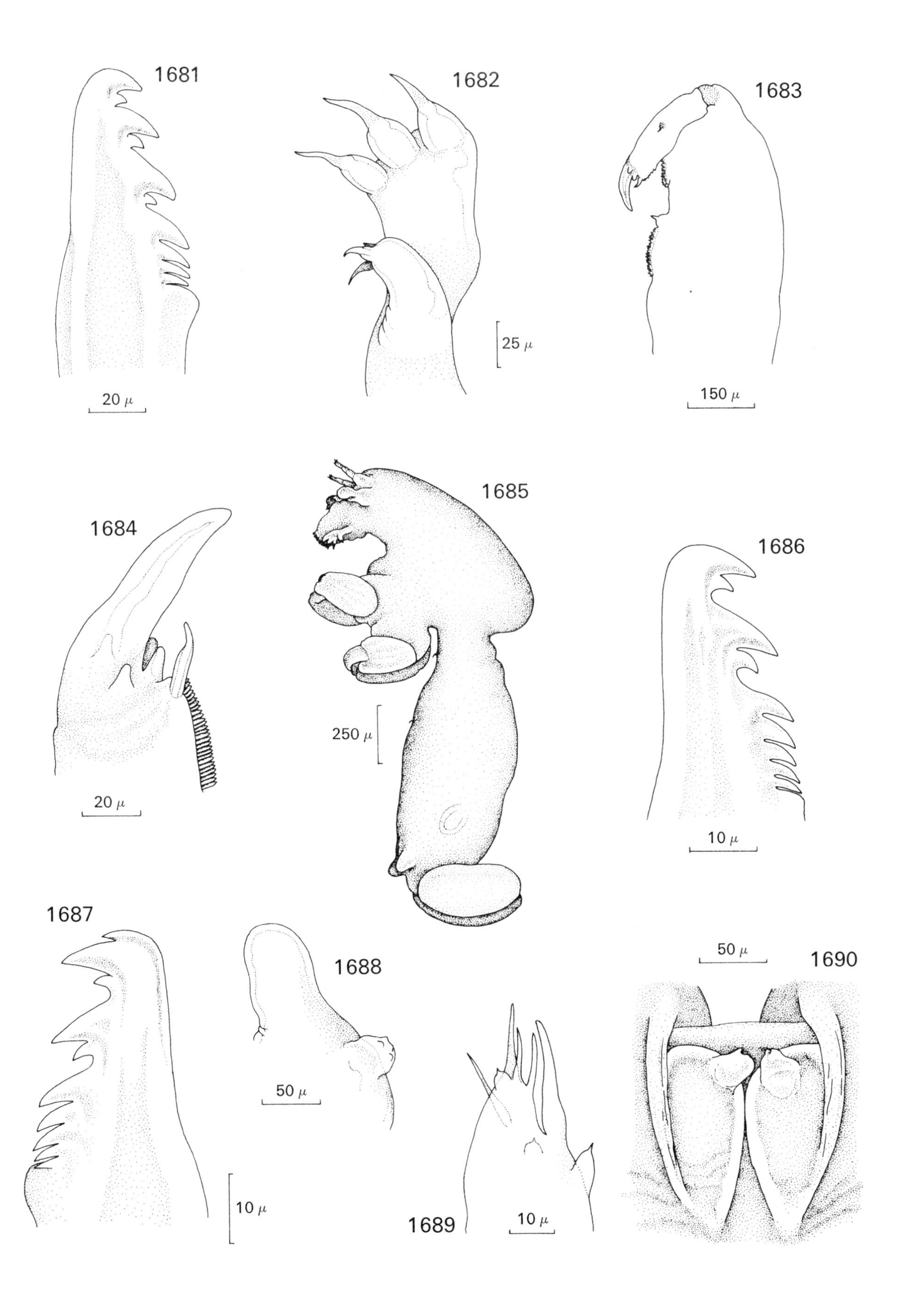

Figs. 1681–1690.
Lernaeopoda galei Krøyer, 1837. **Fig. 1681.** Tip of mandible, lateral. **Fig. 1682.** First maxilla, lateral. **Fig. 1683.** Maxilliped, ventral. **Fig. 1684.** Same, claw, ventral. **Fig. 1685.** Male, lateral. **Fig. 1686.** Male mandible, tip. **Fig. 1687.** Same, with abnormal dentition. **Fig. 1688.** Mediative process, lateral. **Fig. 1689.** Tip of first antenna, medial. **Fig. 1690.** Tympanal membrane of second maxilla, ventral.

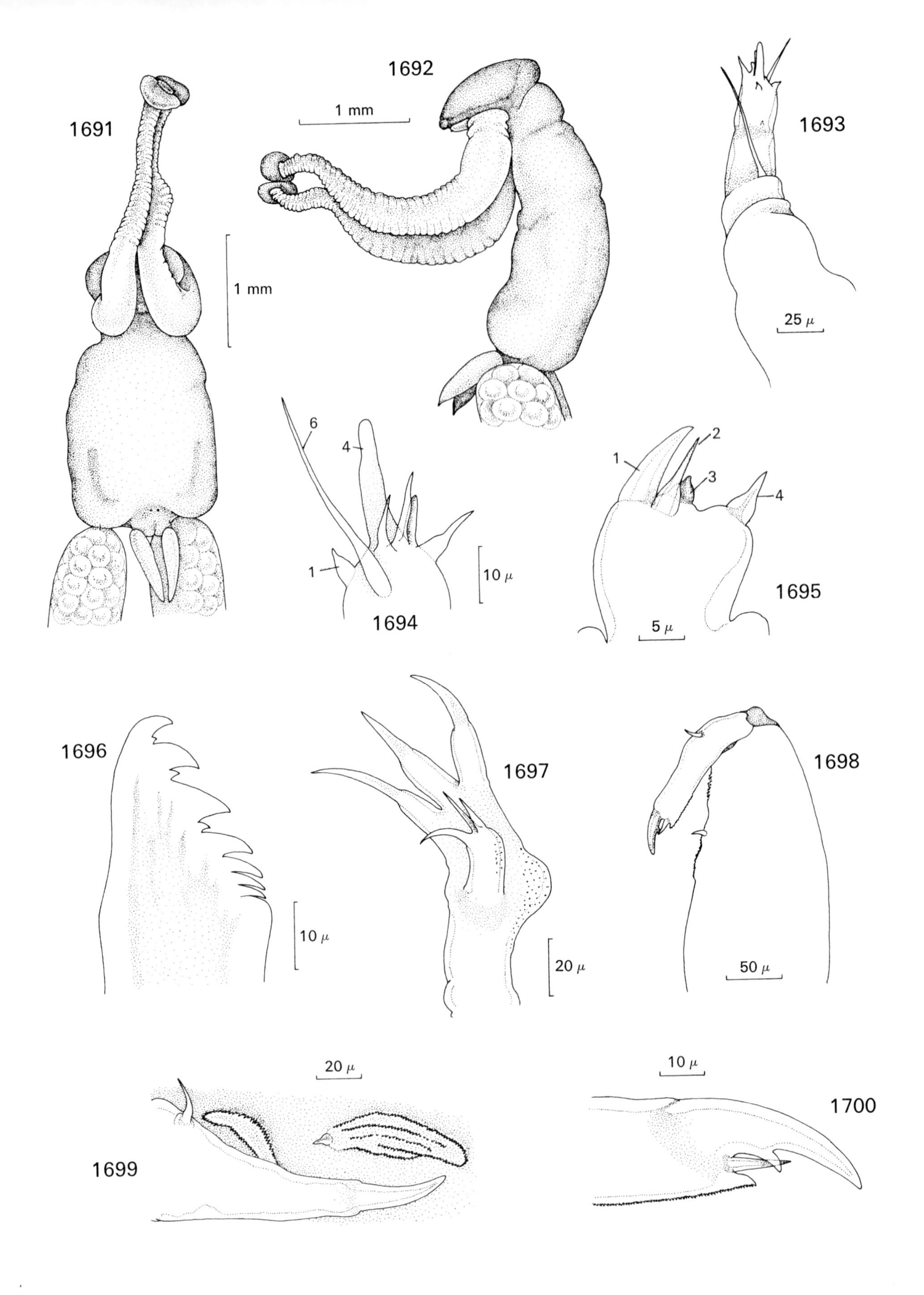

Figs. 1691–1700.
Albionella globosa (Leigh-Sharpe, 1918). **Fig. 1691.** Female, ventral. **Fig. 1692.** Same, lateral. **Fig. 1693.** First antenna, dorsomedial. **Fig. 1694.** Same, tip, lateral. **Fig. 1695.** Second antenna, tip of endopod, medial. **Fig. 1696.** Tip of mandible, lateral. **Fig. 1697.** First maxilla, lateral. **Fig. 1698.** Maxilliped, ventral. **Fig. 1699.** Same, myxal area and tip of subchela, medial. **Fig. 1700.** Same, claw, dorsal.

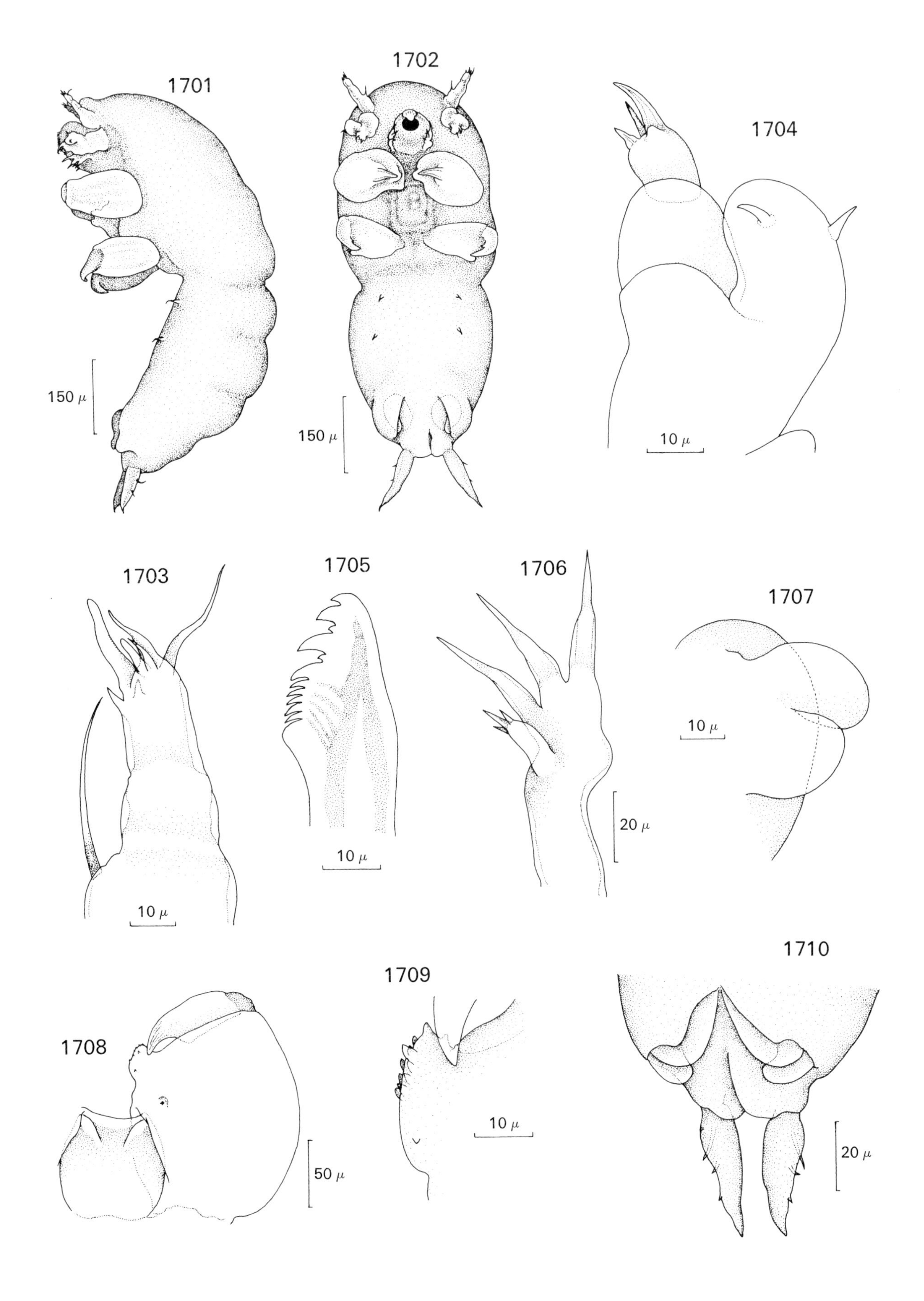

Figs. 1701–1710.
Albionella globosa (Leigh-Sharpe, 1918). **Fig. 1701.** Male, lateral. **Fig. 1702.** Same, ventral. **Fig. 1703.** First antenna, distal half, ventral. **Fig. 1704.** Second antenna, distal half, lateral. **Fig. 1705.** Tip of mandible, lateral. **Fig. 1706.** First maxilla, lateral. **Fig. 1707.** Mediative process, ventrolateral. **Fig. 1708.** Second maxilla, ventral. **Fig. 1709.** Same, tip of claw and basal tubercle, ventral. **Fig. 1710.** Posterior extremity, ventral.

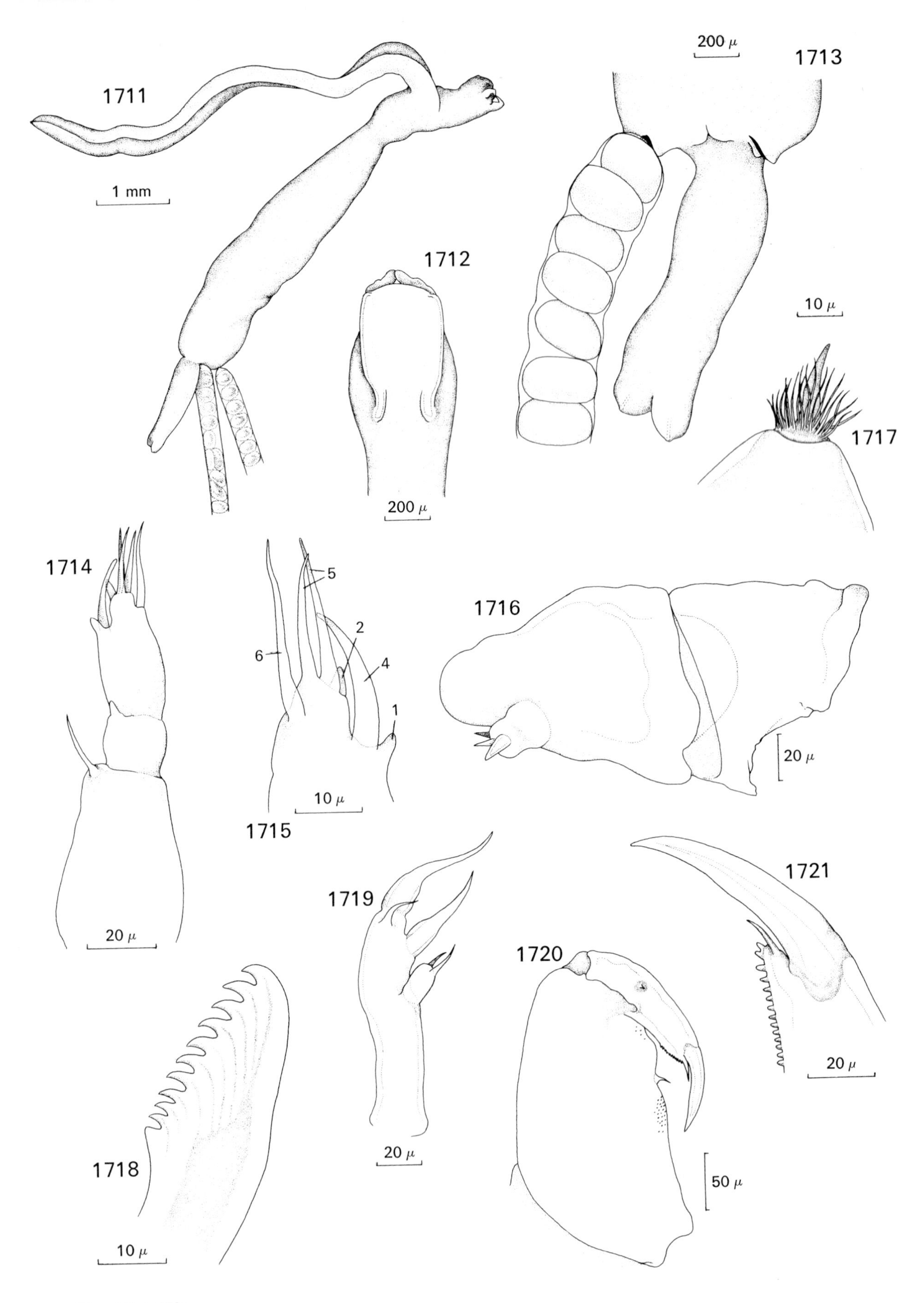

Figs. 1711–1721.
Clavellistes lampri (Scott and Scott, 1913). **Fig. 1711.** Female, lateral. **Fig. 1712.** Cephalothorax, dorsal. **Fig. 1713.** Posterior extremity, dorsal. **Fig. 1714.** First antenna, dorsal. **Fig. 1715.** Same, tip, lateral. **Fig. 1716.** Second antenna, ventral. **Fig. 1717.** Tip of labrum, outer surface. **Fig. 1718.** Tip of mandible, lateral. **Fig. 1719.** First maxilla, lateral. **Fig. 1720.** Maxilliped, ventral. **Fig. 1721.** Same, claw, ventral.

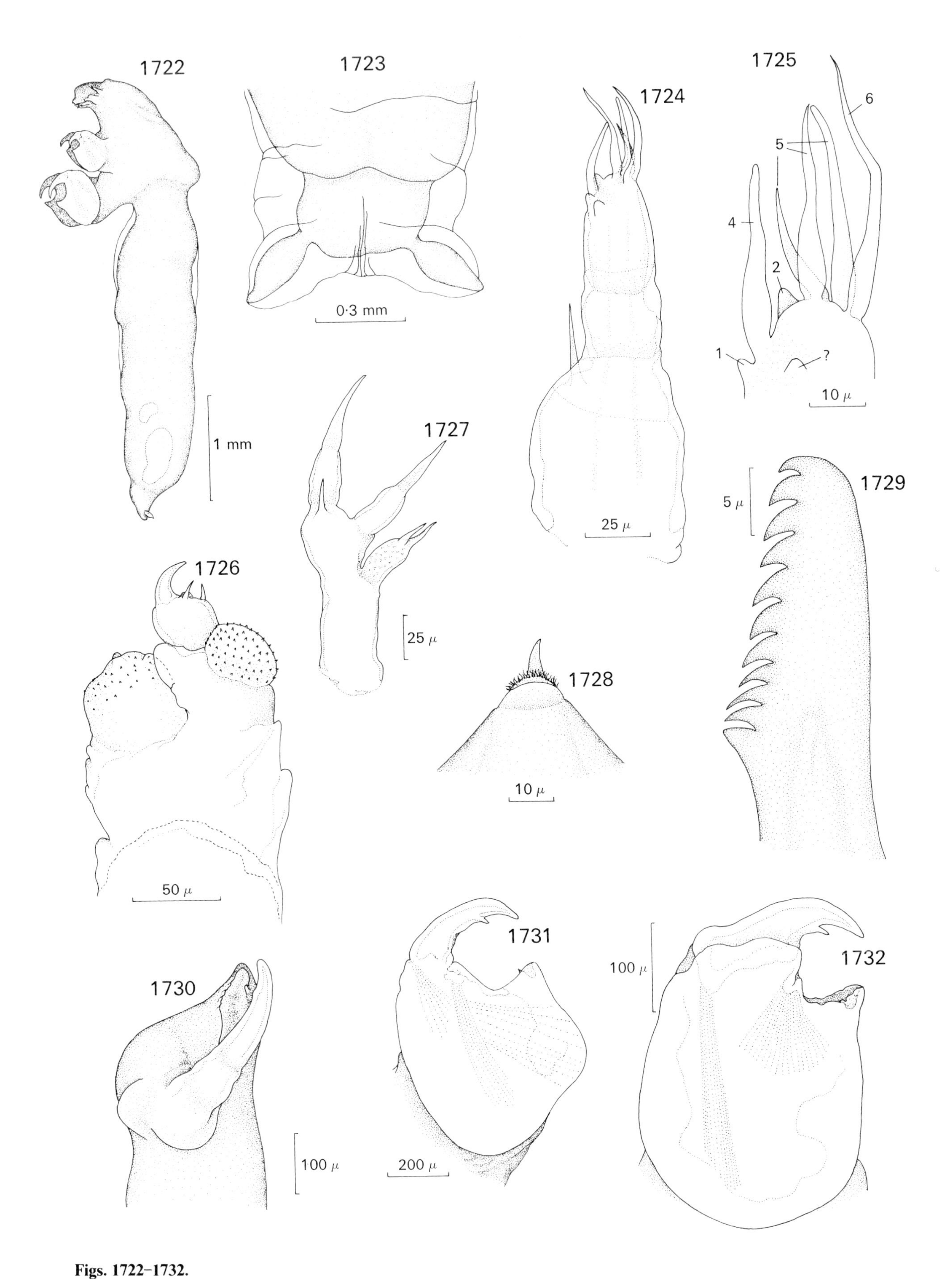

Figs. 1722–1732.
Clavellistes lampri (Scott and Scott, 1913). **Fig. 1722.** Male, lateral. **Fig. 1723.** Posterior extremity, ventral. **Fig. 1724.** First antenna, ventrolateral. **Fig. 1725.** Same, tip, ventrolateral. **Fig. 1726.** Second antenna, medial. **Fig. 1727.** First maxilla, lateral. **Fig. 1728.** Tip of labrum, outer surface. **Fig. 1729.** Tip of mandible, lateral. **Fig. 1730.** Maxilliped, tip, anterior. **Fig. 1731.** Same, ventral. **Fig. 1732.** Second maxilla, ventral.
Fig. 1728 original, all others redrawn from Kabata (1965e).

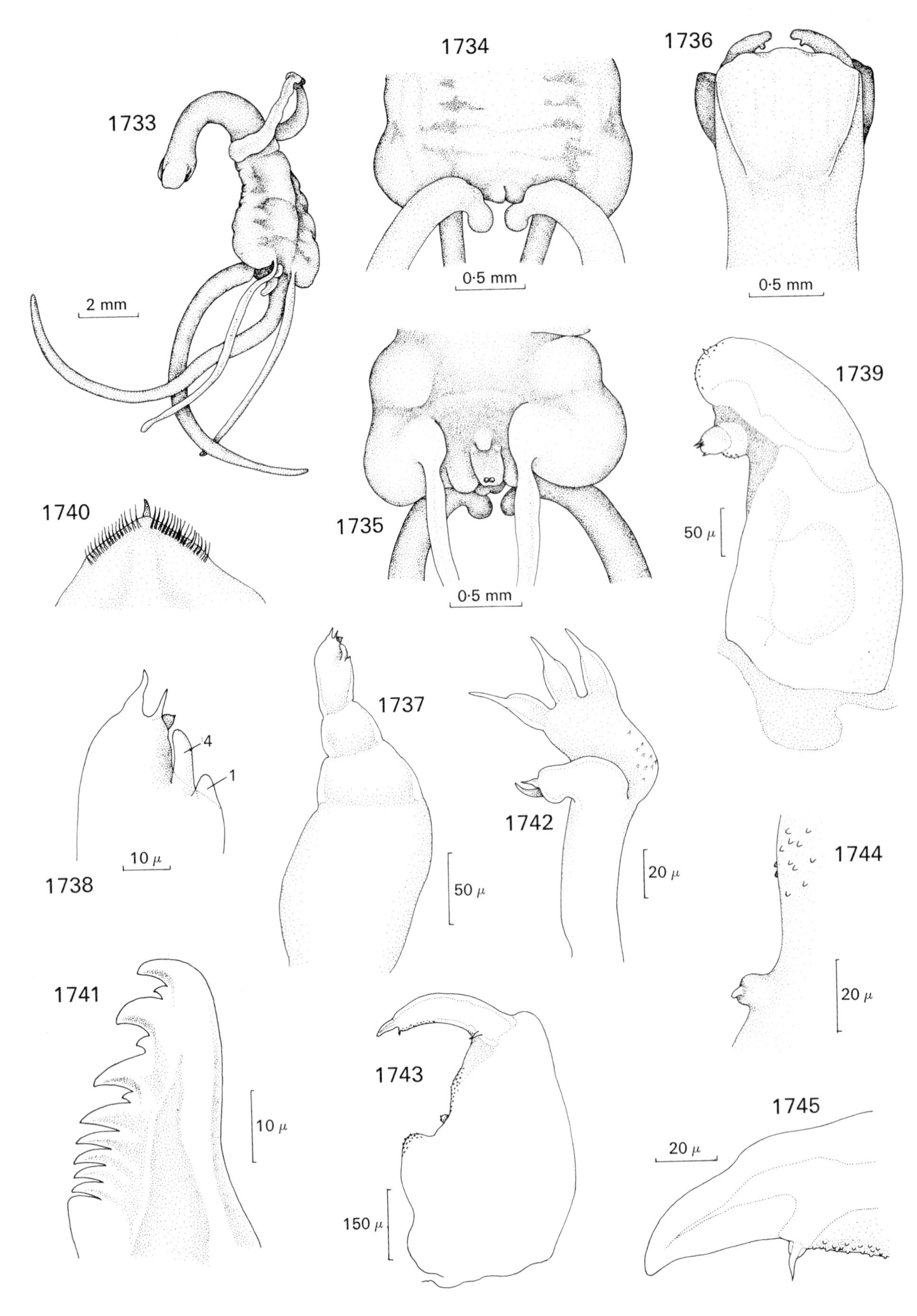

Figs. 1733–1745.
Brachiella thynni Cuvier, 1830. **Fig. 1733.** Female, ventrolateral. **Fig. 1734.** Posterior half of trunk, dorsal. **Fig. 1735.** Same, ventral. **Fig. 1736.** Cephalothorax tip, dorsal. **Fig. 1737.** First antenna, dorsomedial. **Fig. 1738.** Same, tip. **Fig. 1739.** Second antenna, ventral. **Fig. 1740.** Tip of labrum, outer surface. **Fig. 1741.** Tip of mandible, lateral. **Fig. 1742.** First maxilla, lateral. **Fig. 1743.** Maxilliped, ventral. **Fig. 1744.** Same, myxal area. **Fig. 1745.** Same, claw, ventral.

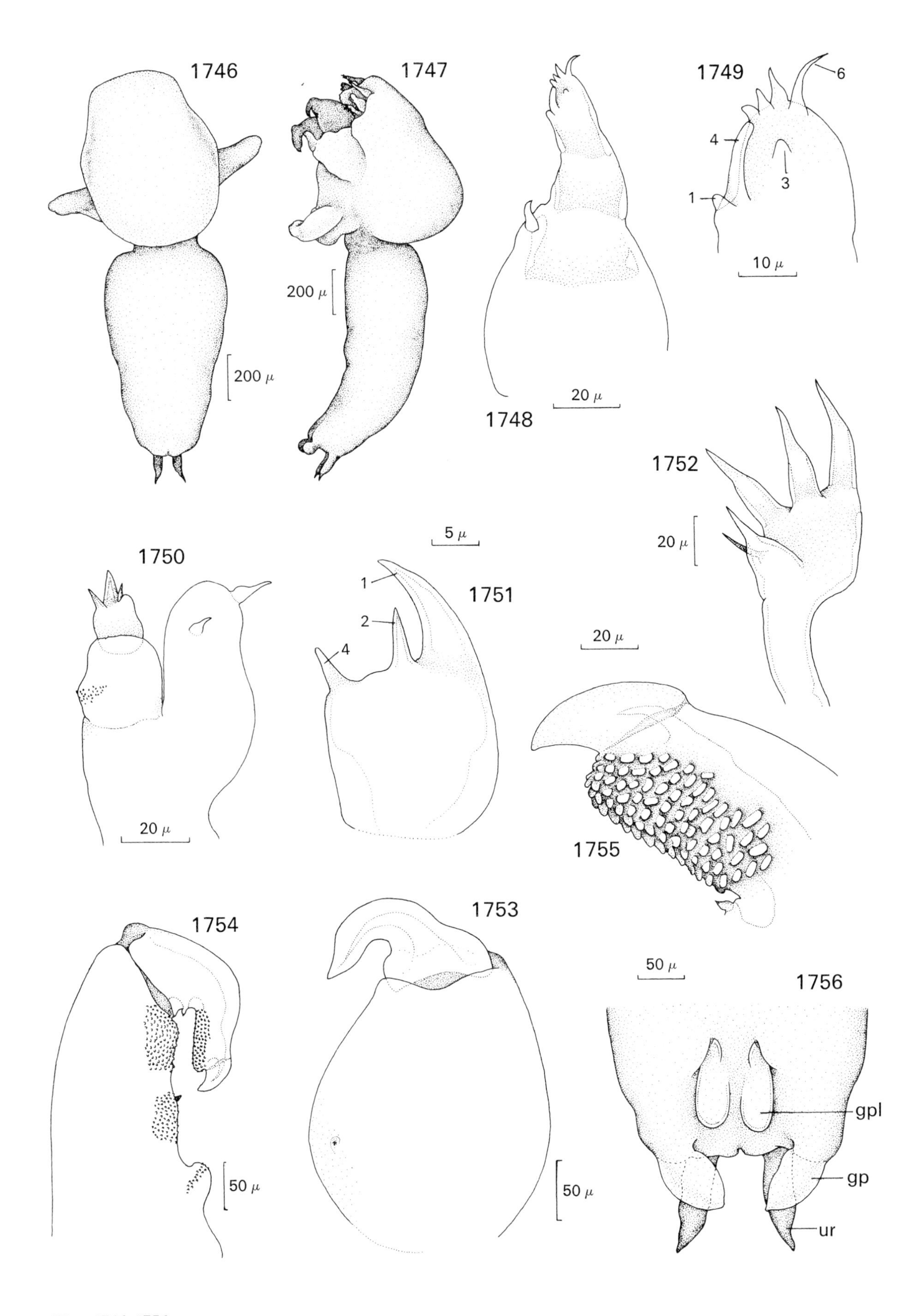

Figs. 1746–1756.
Brachiella thynni Cuvier, 1830. **Fig. 1746.** Male, dorsal. **Fig. 1747.** Same, lateral. **Fig. 1748.** First antenna, dorsomedial. **Fig. 1749.** Same, tip. **Fig. 1750.** Second antenna, distal half, lateral. **Fig. 1751.** Same, tip of endopod. **Fig. 1752.** First maxilla, lateral. **Fig. 1753.** Second maxilla, ventral. **Fig. 1754.** Maxilliped, ventral. **Fig. 1755,** Same, claw, ventral. **Fig. 1756.** Posterior extremity, ventral.
gpl – genital plate; gp – genital process; ur – uropod.

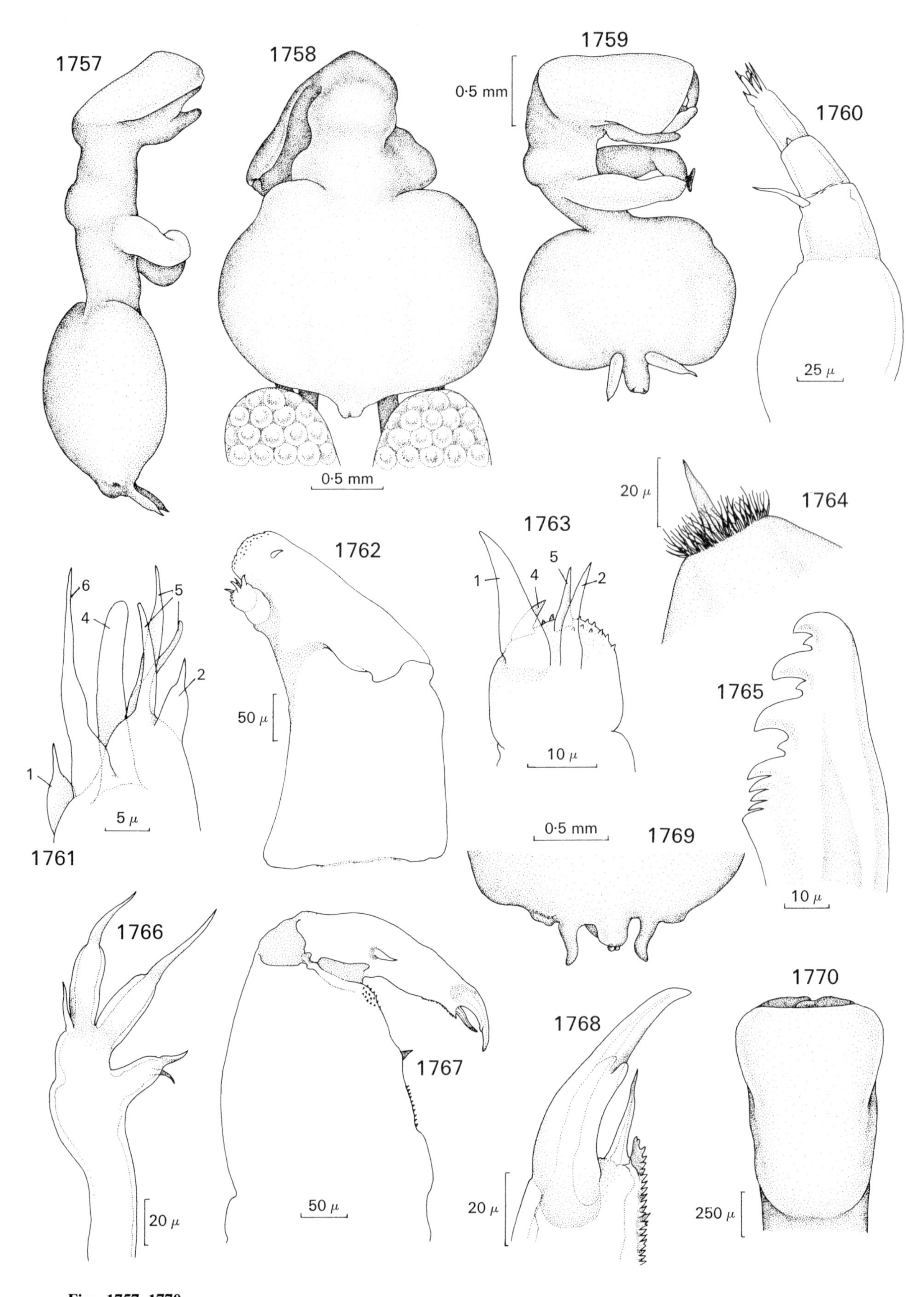

Figs. 1757–1770.
Neobrachiella bispinosa (von Nordmann, 1832). **Fig. 1757.** Female, lateral. **Fig. 1758.** Same, dorsal. **Fig. 1759.** Same, ventral. **Fig. 1760.** First antenna, dorsomedial. **Fig. 1761.** Same, tip, lateral. **Fig. 1762.** Second antenna, ventral. **Fig. 1763.** Same, tip of endopod. **Fig. 1764.** Tip of labrum, outer surface. **Fig. 1765.** Tip of mandible, lateral. **Fig. 1766.** First maxilla, lateral. **Fig. 1767.** Maxilliped, ventral. **Fig. 1768.** Same, claw, ventral. **Fig. 1769.** Posterior extremity, ventral. **Fig. 1770.** Cephalothorax tip, dorsal.

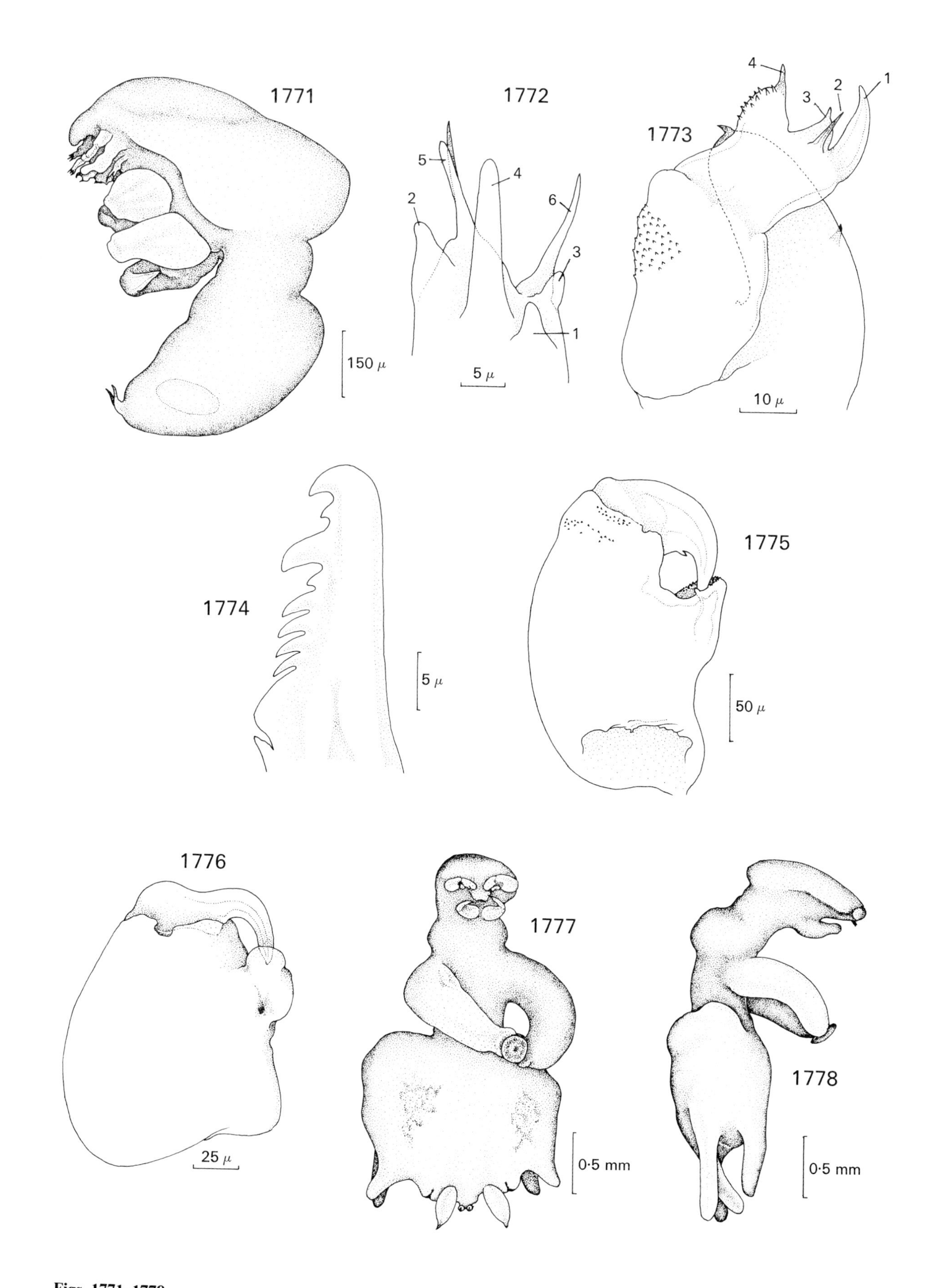

Figs. 1771–1778.
Neobrachiella bispinosa (von Nordmann, 1832). **Fig.1771.** Male, lateral. **Fig. 1772.** Male first antenna, tip, medial. **Fig. 1773**. Distal half of second antenna, medial. **Fig. 1774.** Tip of mandible, lateral. **Fig. 1775.** Maxilliped, dorsal. **Fig. 1776.** Second maxilla, ventral.
Neobrachiella impudica (von Nordmann, 1832). **Fig.1777.** Female, ventral. **Fig. 1778.** Same, lateral.

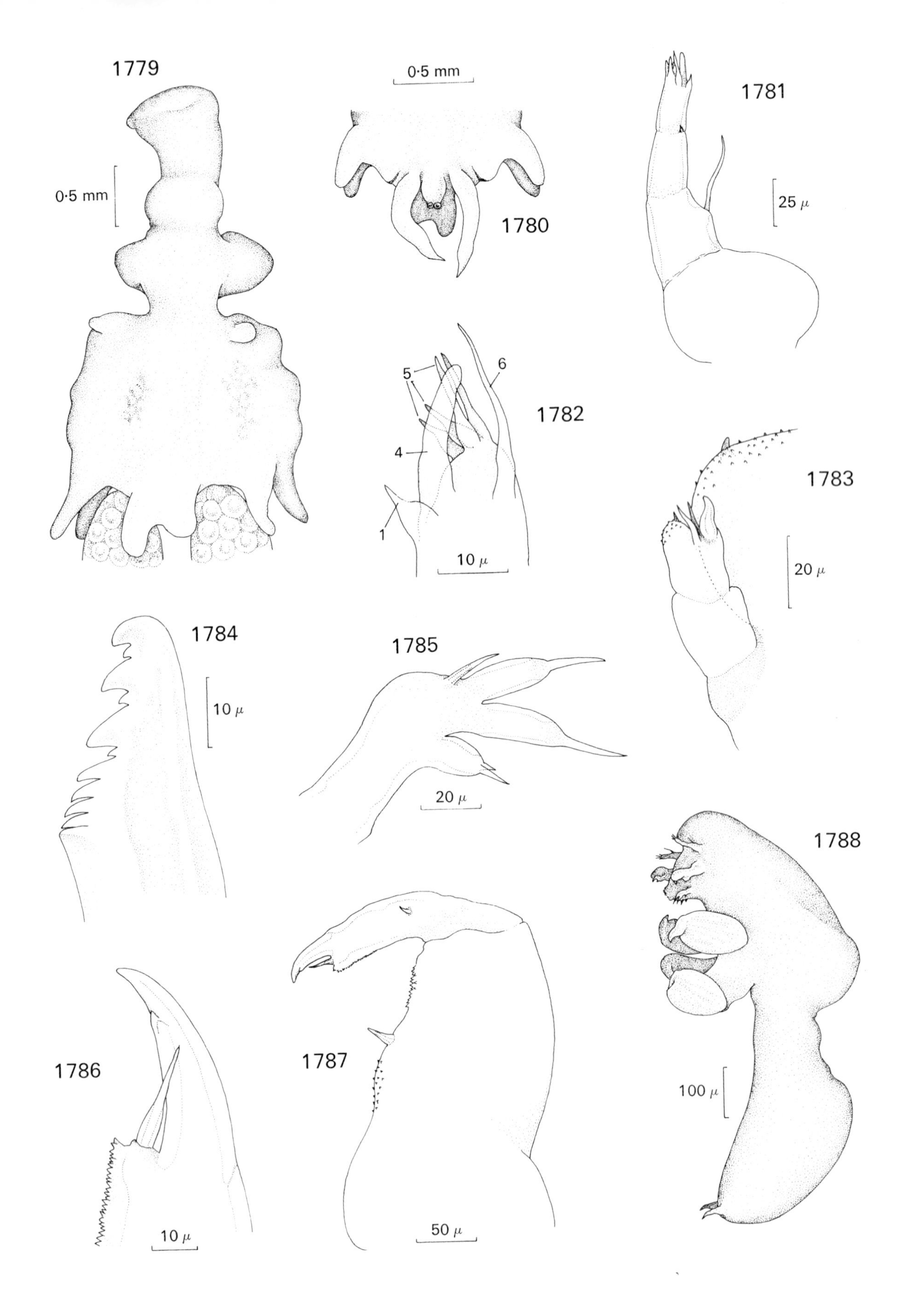

Figs. 1779–1788.
Neobrachiella impudica (von Nordmann, 1832). **Fig.1779.** Female, dorsal. **Fig. 1780.** Posterior extremity, ventral. **Fig. 1781.** First antenna, lateral. **Fig. 1782.** Same, tip, ventromedial. **Fig. 1783.** Endopod of second antenna, ventral. **Fig. 1784.** Tip of mandible, lateral. **Fig. 1785.** First maxilla, lateral. **Fig. 1786.** Claw of maxilliped, ventral. **Fig. 1787.** Maxilliped, ventral. **Fig. 1788.** Male, lateral.

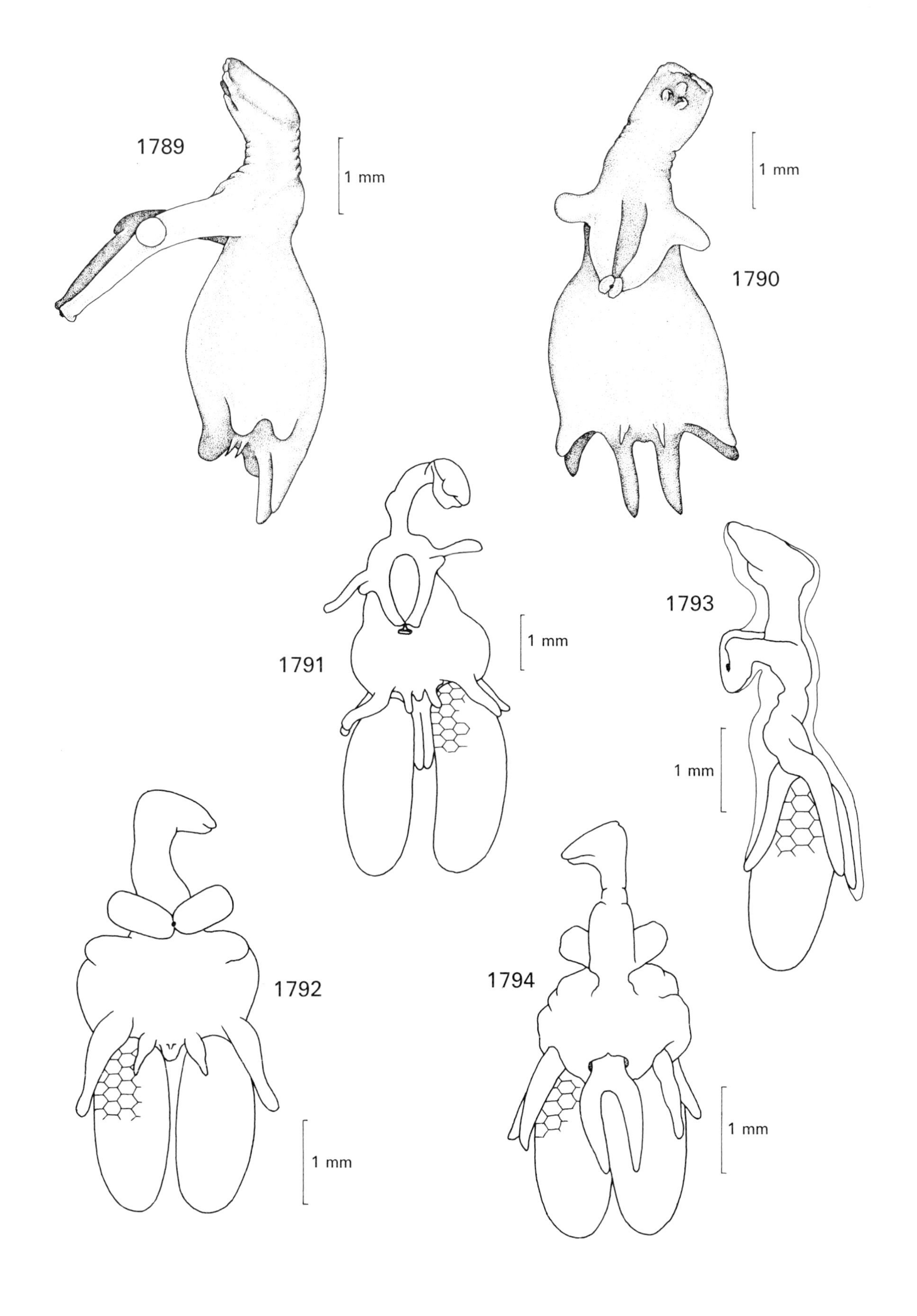

Figs. 1789–1794.
Neobrachiella impudica (von Nordmann, 1832.) **Fig.1789.** Female, lateral (type material). **Fig. 1790.** Same, ventral. **Figs. 1791, 1792.** Female, ventral. **Fig. 1793.** Same, lateral. **Fig. 1794.** Same, dorsal.
Figs. 1789, 1790 redrawn from Kabata (1973b), others from Leigh-Sharpe (1935b).

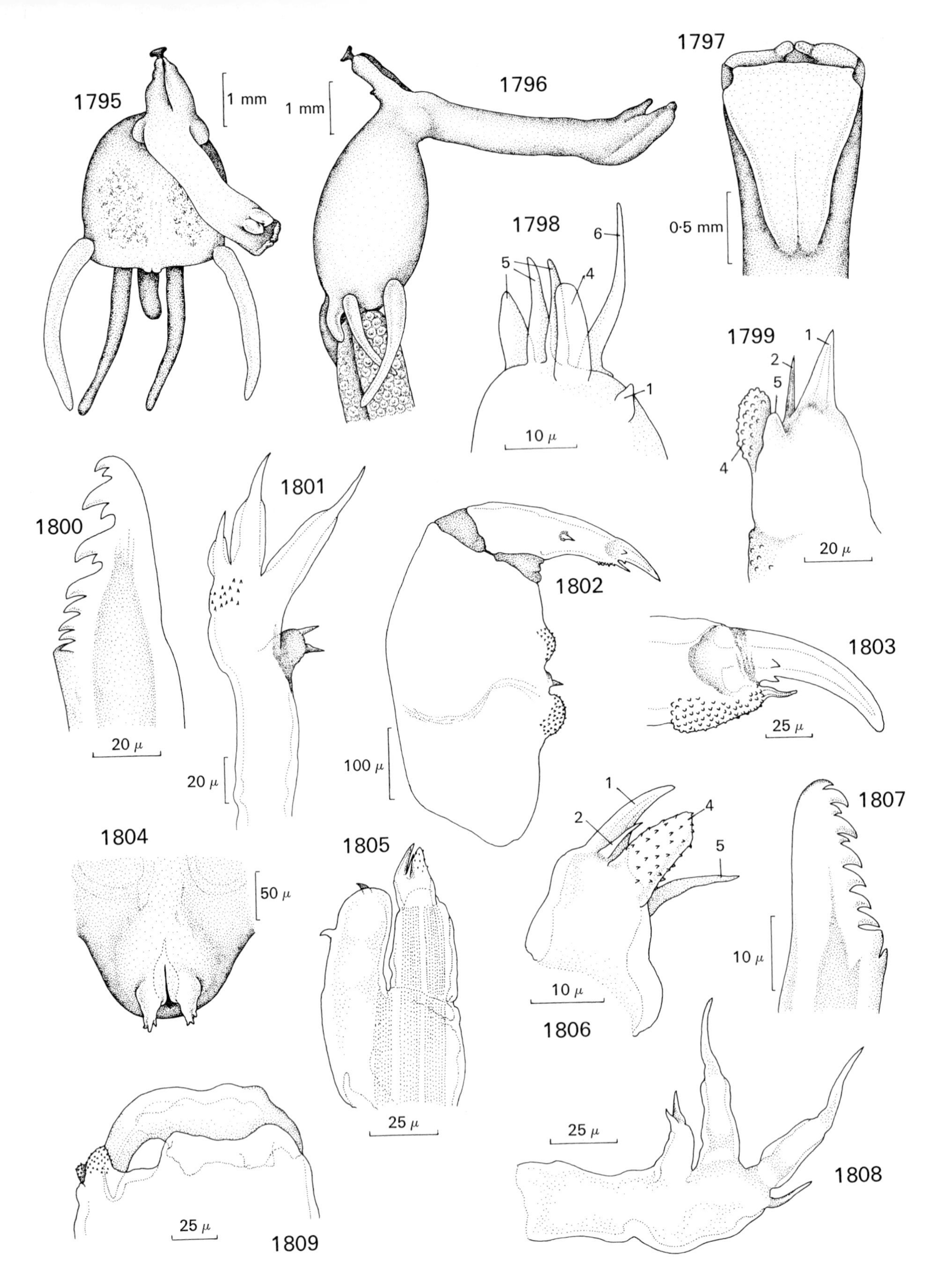

Figs. 1795–1809.
Neobrachiella chevreuxii (van Beneden, 1891). **Fig. 1795.** Female, trunk dorsal, cephalothorax ventral. **Fig. 1796.** Same, lateral. **Fig. 1797.** Tip of cephalothorax, dorsal. **Fig. 1798.** Tip of first antenna, medial. **Fig. 1799.** Second antenna, tip of endopod, ventromedial. **Fig. 1800.** Tip of mandible, lateral. **Fig. 1801.** First maxilla, medial. **Fig. 1802.** Maxilliped, ventral. **Fig. 1803.** Same, claw, dorsal. **Fig. 1804.** Male, posterior extremity, ventral. **Fig. 1805.** Male second antenna (without base), medial. **Fig. 1806.** Same, tip of endopod, ventral. **Fig. 1807.** Tip of mandible, lateral. **Fig. 1808.** First maxilla, lateral. **Fig. 1809.** Claw of second maxilla, ventral.

Figs. 1797, 1798, 1804–1809 redrawn from Kabata (1966).

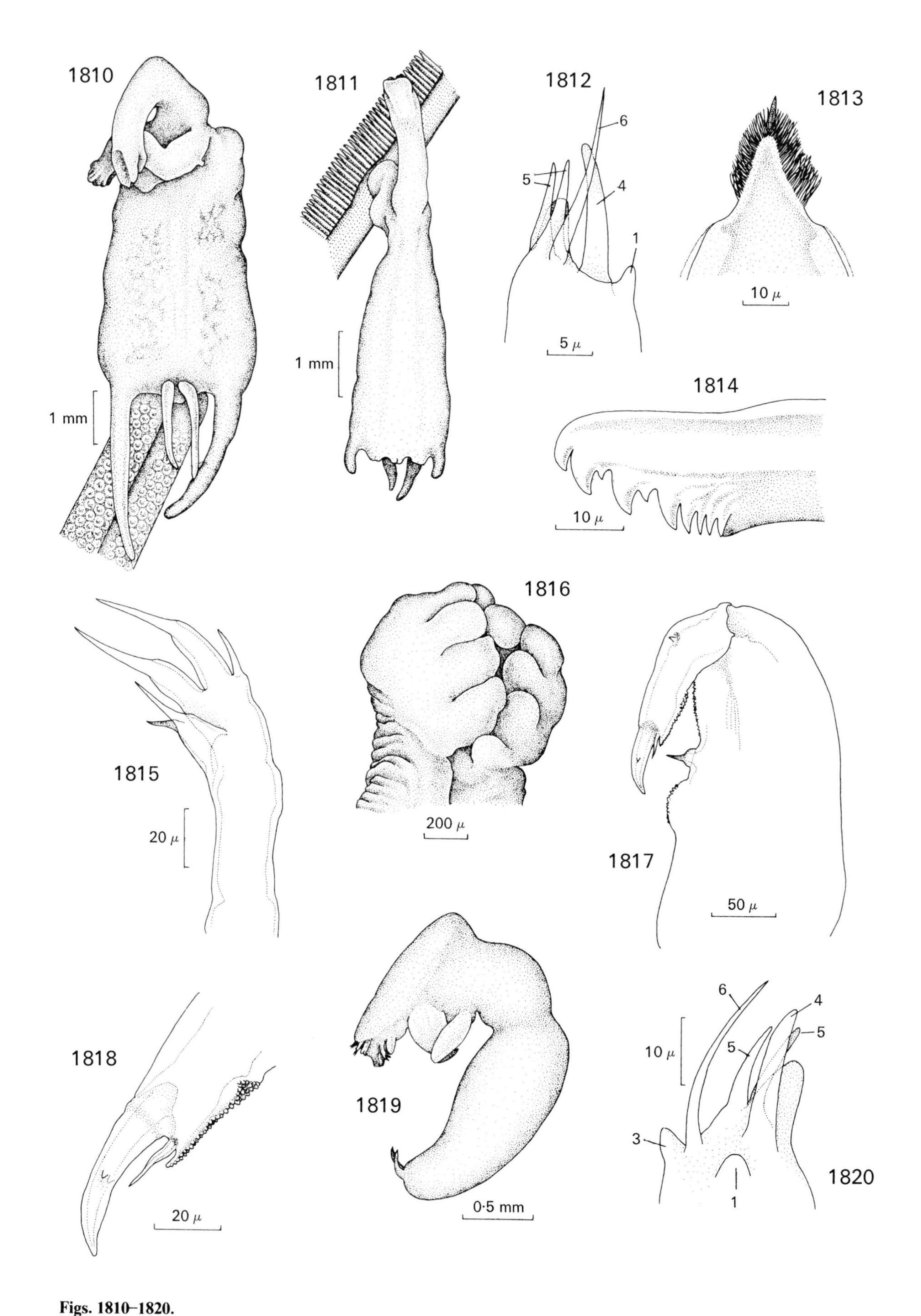

Figs. 1810–1820.
Neobrachiella insidiosa (Heller, 1865). **Fig.1810.** Female, ventral. **Fig. 1811.** Younger female, dorsal. **Fig. 1812.** Tip of first antenna, lateral. **Fig. 1813.** Tip of labrum, outer surface. **Fig. 1814.** Tip of mandible, lateral. **Fig. 1815.** First maxilla, lateral. **Fig. 1816.** Tips of second maxillae. **Fig. 1817.** Maxilliped, ventral. **Fig. 1818.** Same, claw, dorsal. **Fig. 1819.** Male, lateral. **Fig. 1820.** Male first antenna, tip, lateral.

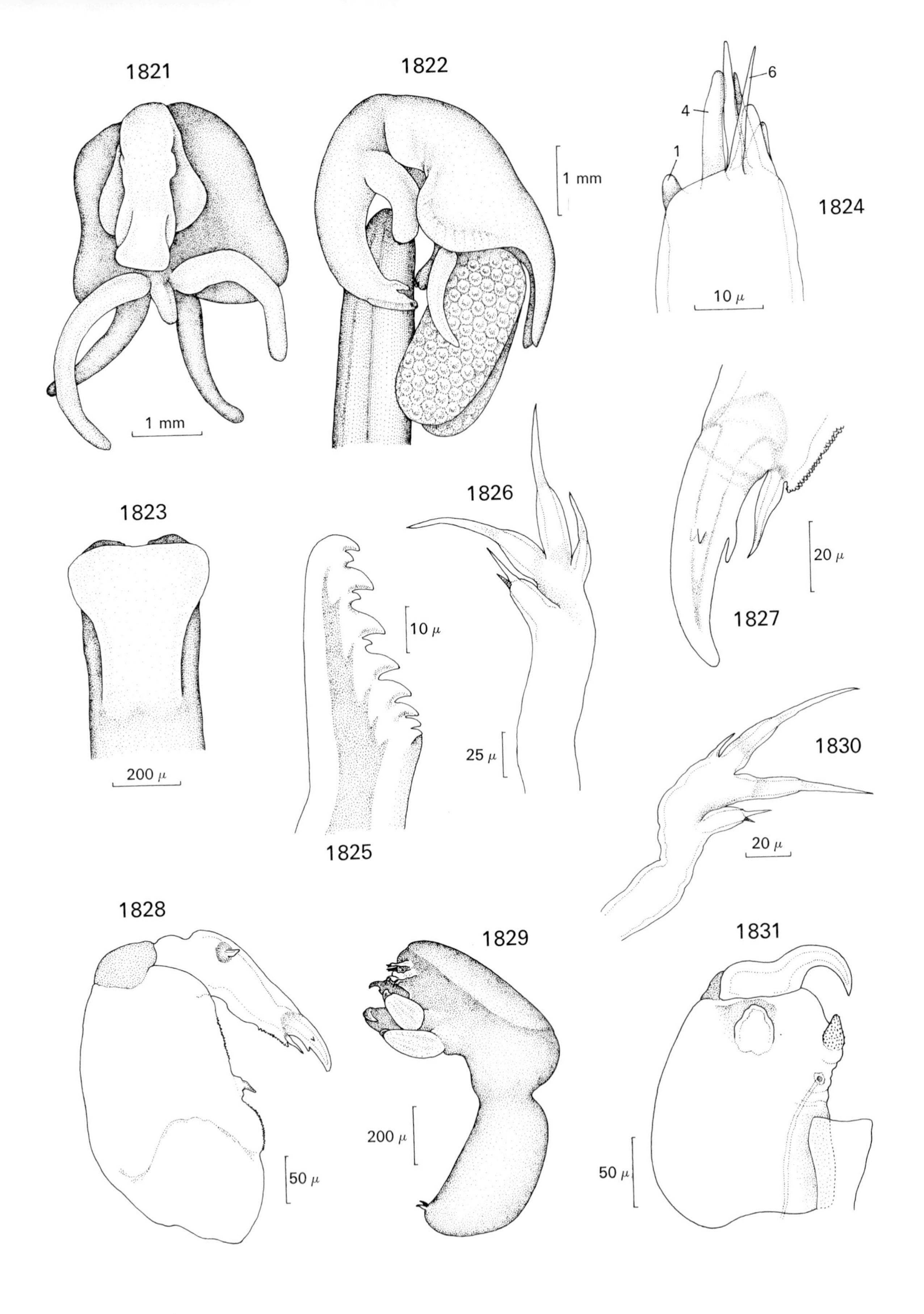

Figs. 1821–1831.
Neobrachiella merluccii (Bassett-Smith, 1896). **Fig. 1821.** Female, trunk ventral, cephalothorax dorsal. **Fig. 1822.** Same, lateral. **Fig. 1823.** Tip of cephalothorax, dorsal. **Fig. 1824.** Tip of first antenna, lateral. **Fig. 1825.** Tip of mandible, lateral. **Fig. 1826.** First maxilla, lateral. **Fig. 1827.** Claw of maxilliped, dorsal. **Fig. 1828.** Maxilliped, ventral. **Fig. 1829.** Male, lateral. **Fig. 1830.** Male first maxilla, lateral. **Fig. 1831.** Second maxilla, ventral.

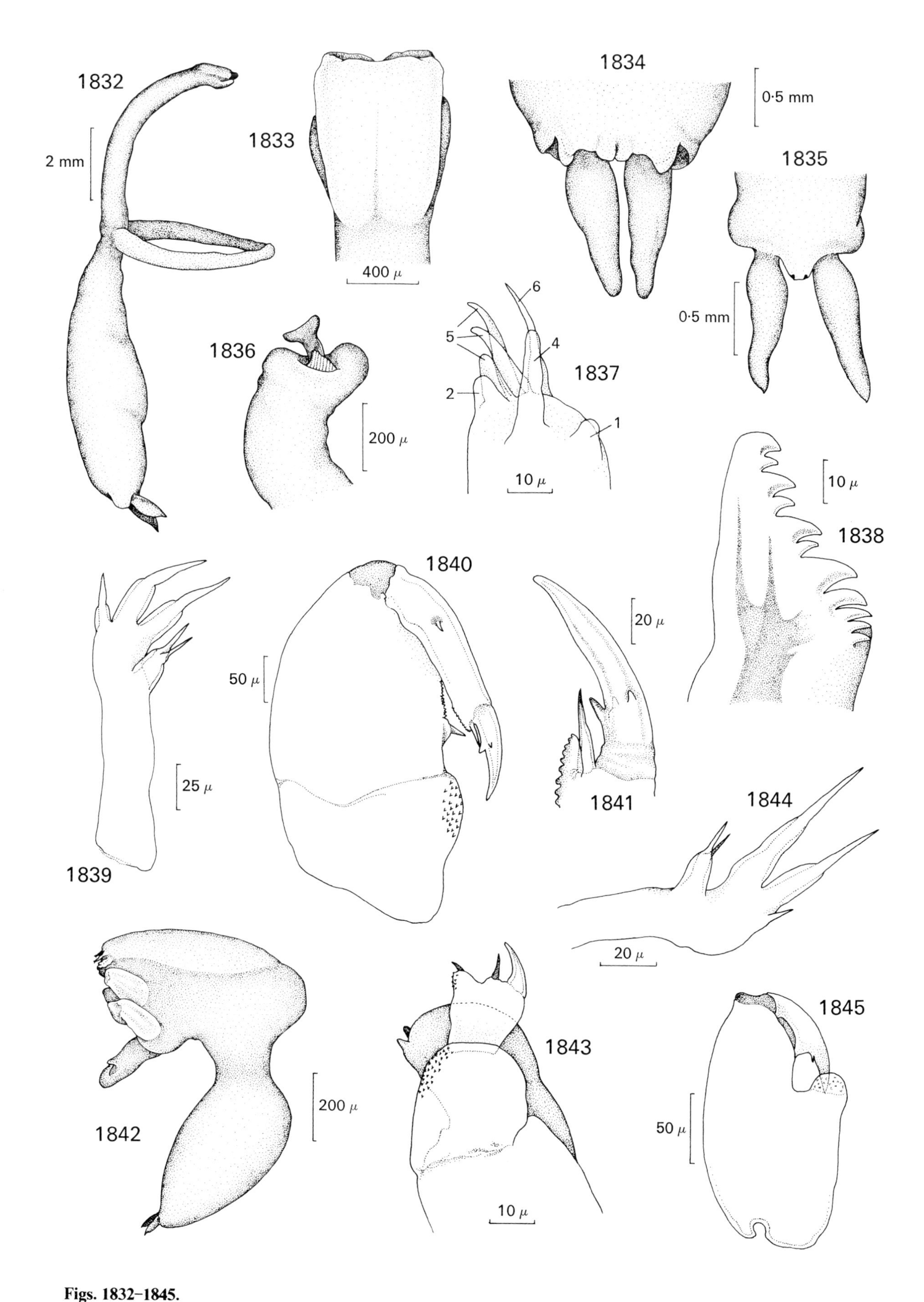

Figs. 1832–1845.
Neobrachiella rostrata (Krøyer, 1837.) **Fig. 1832.** Female, lateral. **Fig. 1833.** Tip of cephalothorax, dorsal. **Fig. 1834.** Mature adult female, posterior extremity, dorsal. **Fig. 1835.** Young adult female, posterior extremity, ventral. **Fig. 1836.** Second maxilla, tip with bulla, medial. **Fig. 1837.** Tip of first antenna, lateral. **Fig. 1838.** Tip of mandible, lateral. **Fig. 1839.** First maxilla, lateral. **Fig. 1840.** Maxilliped, ventral. **Fig. 1841.** Claw of maxilliped, dorsal. **Fig. 1842.** Male, lateral. **Fig. 1843.** Male second antenna, distal half, ventral. **Fig. 1844.** First maxilla, lateral. **Fig. 1845.** Maxilliped, ventral.

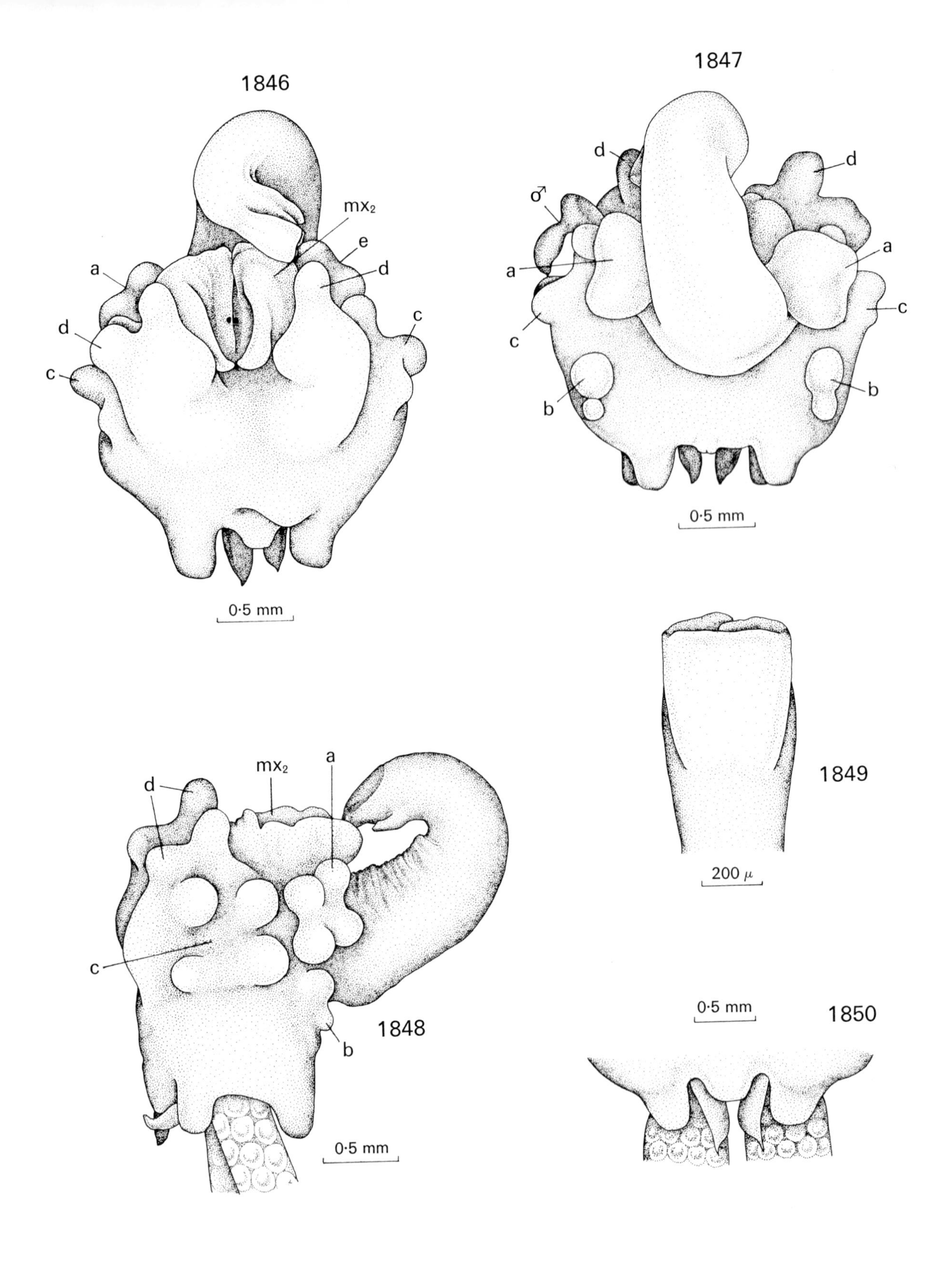

Figs. 1846–1850.
Neobrachiella triglae (Claus, 1860). **Fig. 1846.** Female, ventral. **Fig. 1847.** Same, dorsal. **Fig. 1848.** Same, lateral. **Fig. 1849.** Tip of cephalothorax, dorsal. **Fig. 1850.** Posterior extremity, ventral.
mx2 – second maxilla; a, b, c, d – explanations in text.

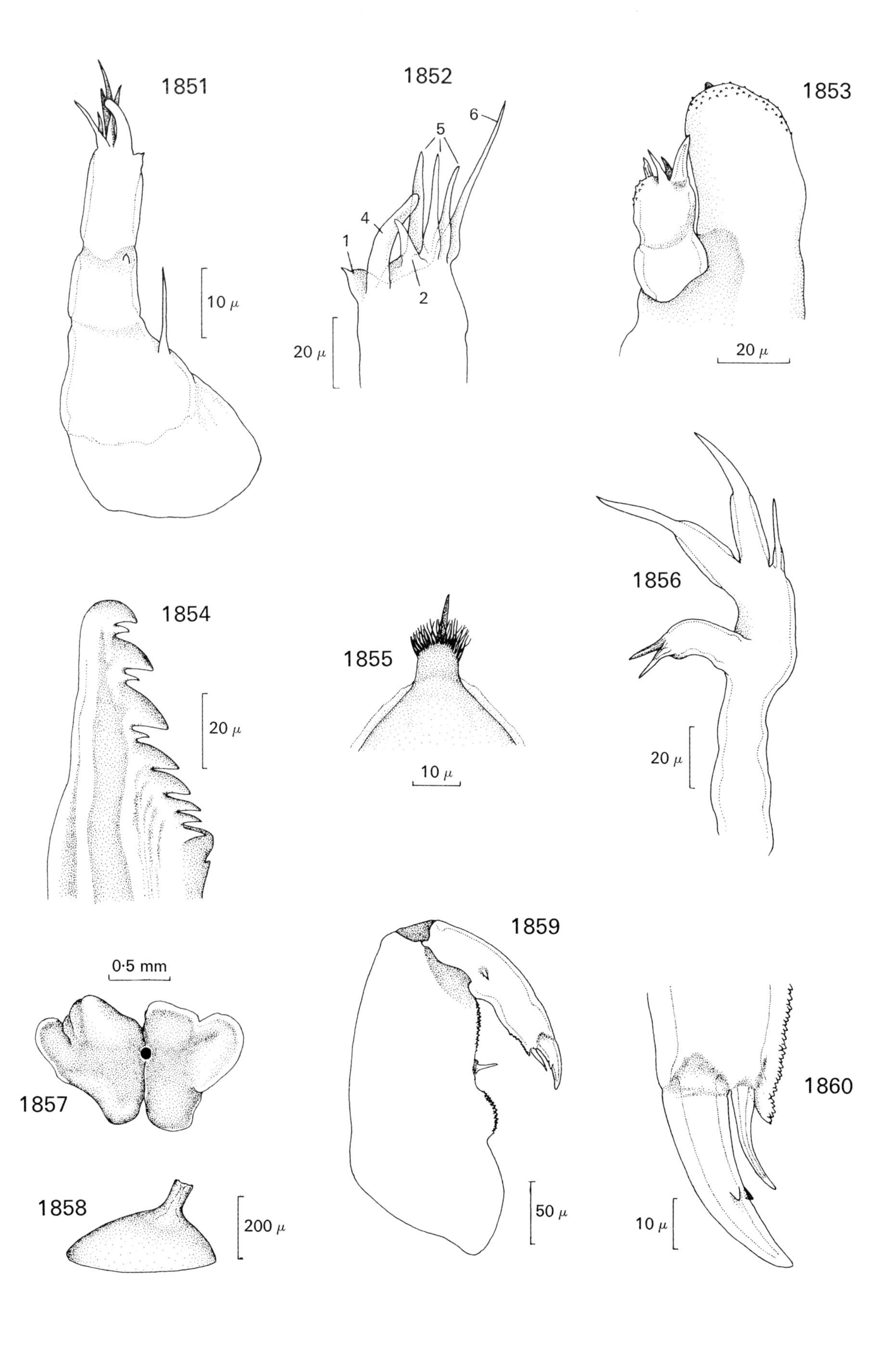

Figs. 1851–1860.
Neobrachiella triglae (Claus, 1860). **Fig. 1851.** Female first antenna, dorsomedial. **Fig. 1852.** Same, tip. **Fig. 1853.** Tip of second antenna, anterodorsal. **Fig. 1854.** Tip of mandible, lateral. **Fig. 1855.** Tip of labrum, outer surface. **Fig. 1856.** First maxilla, lateral. **Fig. 1857.** Tips of second maxillae, anterior. **Fig. 1858.** Bulla, lateral. **Fig. 1859.** Maxilliped, ventral. **Fig. 1860.** Same, claw, ventral.

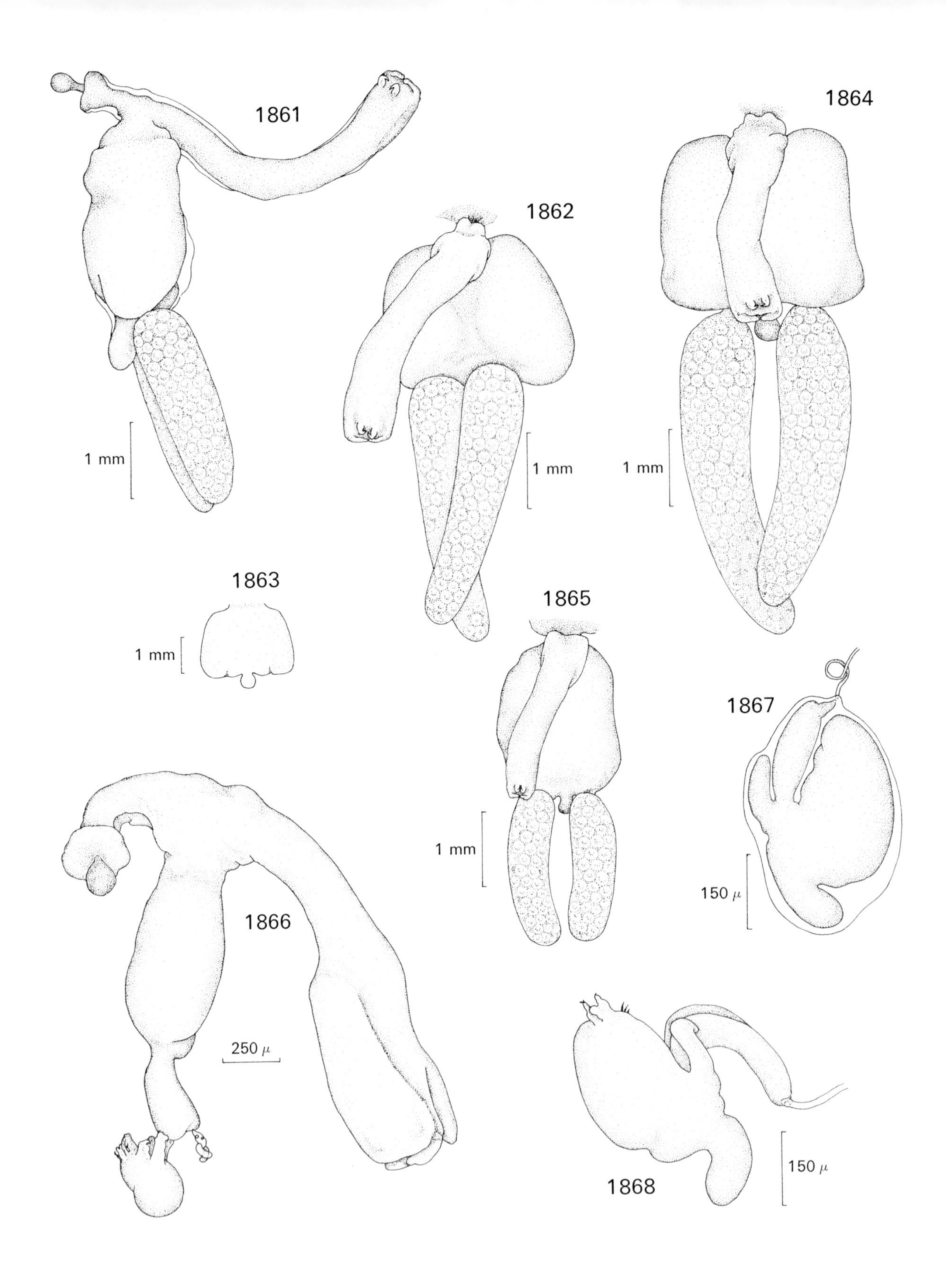

Figs. 1861–1868.
Clavella adunca (Strøm, 1762). **Fig. 1861.** Female *C. adunca f. deliciosa*, lateral. **Fig. 1862.** *C. adunca f. invicta*, trunk dorsal, cephalothorax ventral. **Fig. 1863.** Same, trunk, diagrammatic. **Fig. 1864.** *C. adunca* f. *dubia*, trunk dorsal, cephalothorax ventral. **Fig. 1865.** *C. adunca* f. *brevicollis*, trunk dorsal, cephalothorax ventral. **Fig. 1866.** Young adult female with male and spermatophores attached, lateral. **Fig. 1867.** Pupal stage. **Fig. 1868.** Postpupal stage, after ecclosion.

Figs. 1861–1865 redrawn from Kabata (1963c).

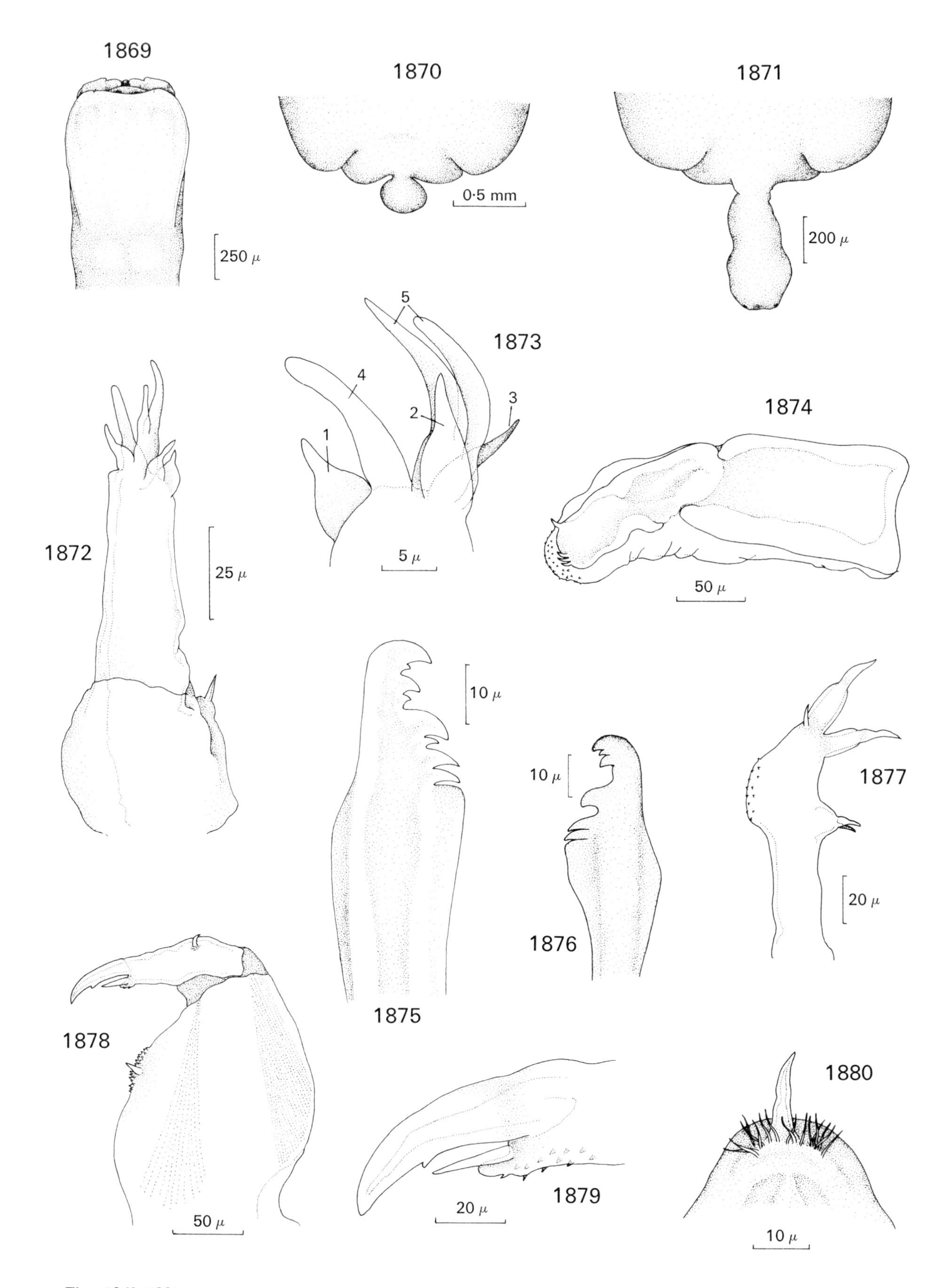

Figs. 1869–1880.
Clavella adunca (Strøm, 1762). **Fig. 1869.** Female, tip of cephalothorax, dorsal. **Fig. 1870.** Posterior extremity, ventral. **Fig. 1871.** Same, with longer genital process. **Fig. 1872.** First antenna, ventrolateral. **Fig. 1873.** Same, tip, lateral. **Fig. 1874.** Second antenna, ventral. **Fig. 1875.** Mandible, lateral. **Fig. 1876.** Same, with different dentition. **Fig. 1877.** First maxilla, lateral. **Fig. 1878.** Maxilliped, ventral. **Fig. 1879.** Same, claw, ventral. **Fig. 1880.** Tip of labrum, inner surface.

Fig. 1869 redrawn from Kabata (1970d); Fig. 1872 from Kabata (1963c); Fig. 1876 from Kabata (1960c); Fig. 1880 from Kabata (1969e).

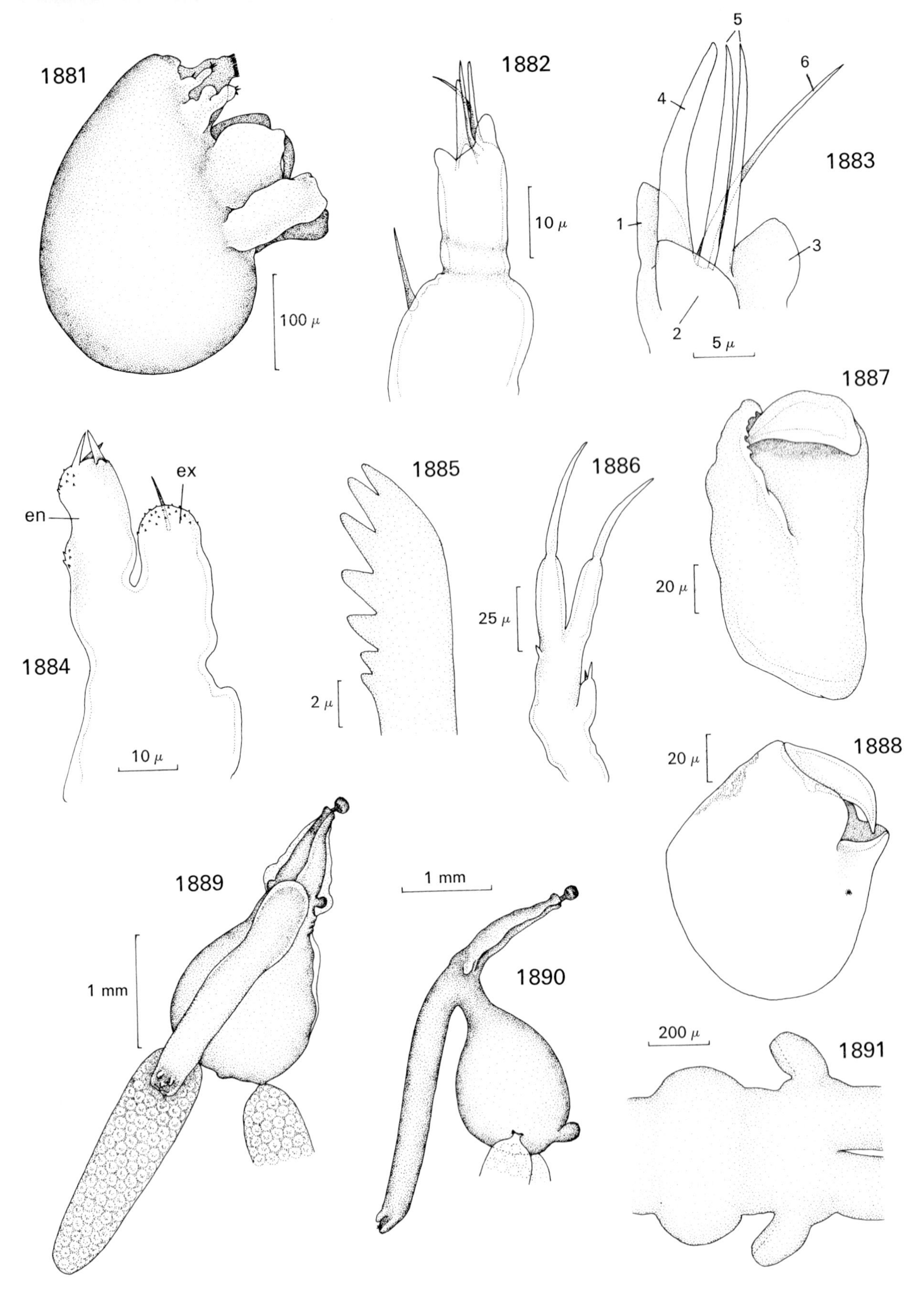

Figs. 1881–1891.
Clavella adunca (Strøm, 1762). **Fig. 1881.** Male, lateral. **Fig. 1882.** Male, first antenna, medial. **Fig. 1883.** Same, tip, medial. **Fig. 1884.** Second antenna, medial. **Fig. 1885.** Tip of mandible, lateral. **Fig. 1886.** First maxilla, lateral. **Fig. 1887.** Maxilliped, medial. **Fig. 1888.** Second maxilla, lateral.
Clavella alata Brian, 1906. **Fig. 1889.** Female, trunk dorsal, cephalothorax ventral. **Fig. 1890.** Same, lateral. **Fig. 1891.** Base of cephalothorax and second maxillae, anterior.
en – endopod; ex – exopod.
Figs. 1885, 1886 redrawn from Kabata (1963c)

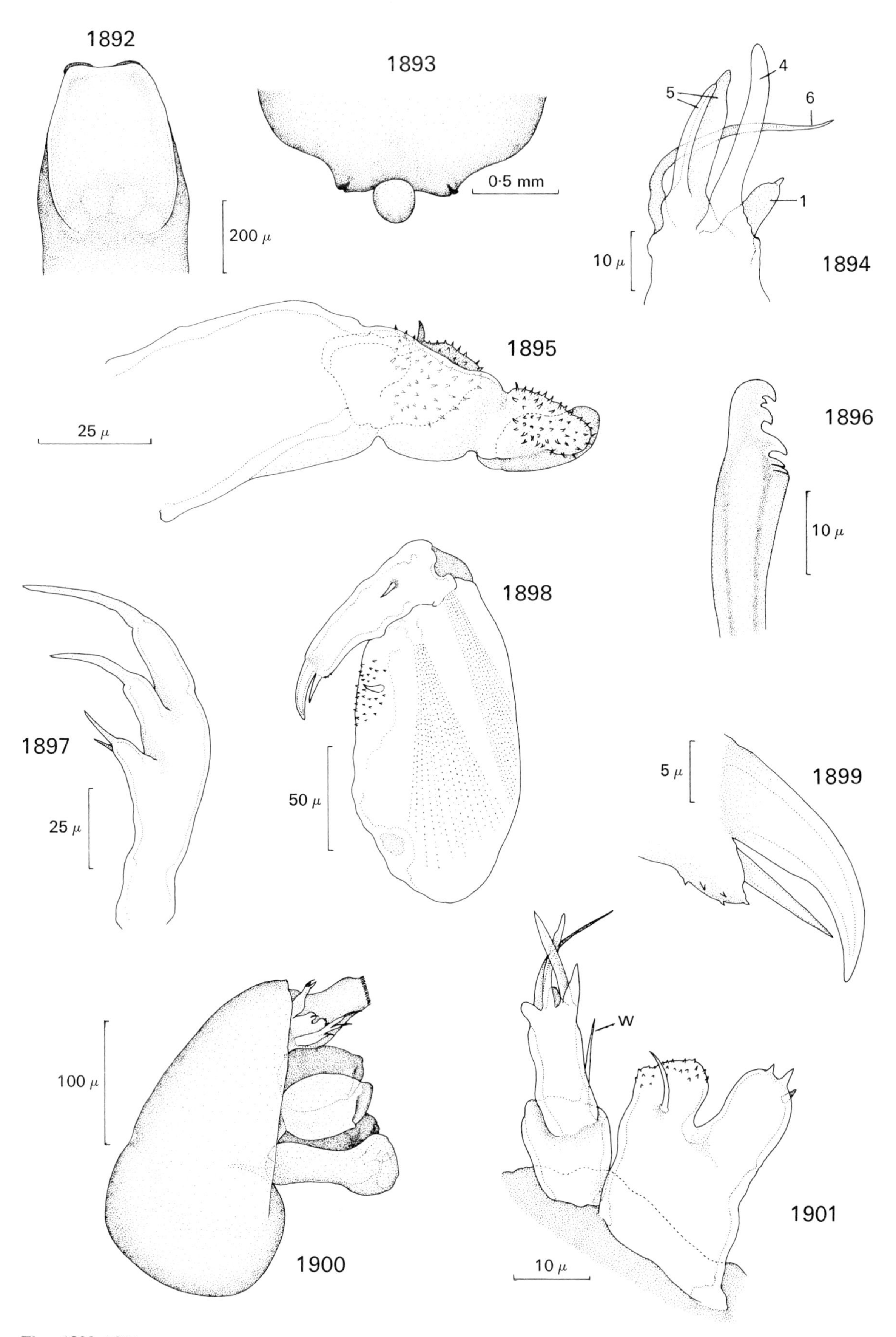

Figs. 1892–1901.
Clavella alata Brian, 1906. **Fig. 1892.** Female, tip of cephalothorax, dorsal. **Fig. 1893.** Posterior extremity, ventral. **Fig. 1894.** Tip of first antenna, ventral. **Fig. 1895.** Second antenna, ventral. **Fig. 1896.** Mandible, lateral. **Fig. 1897.** First maxilla, lateral. **Fig. 1898.** Maxilliped, ventral. **Fig. 1899.** Same, claw dorsal. **Fig. 1900.** Male, lateral. **Fig. 1901.** First and second antenna, lateral.
w – whip of first antenna.
Figs. 1895, 1896 redrawn from Kabata (1963c)

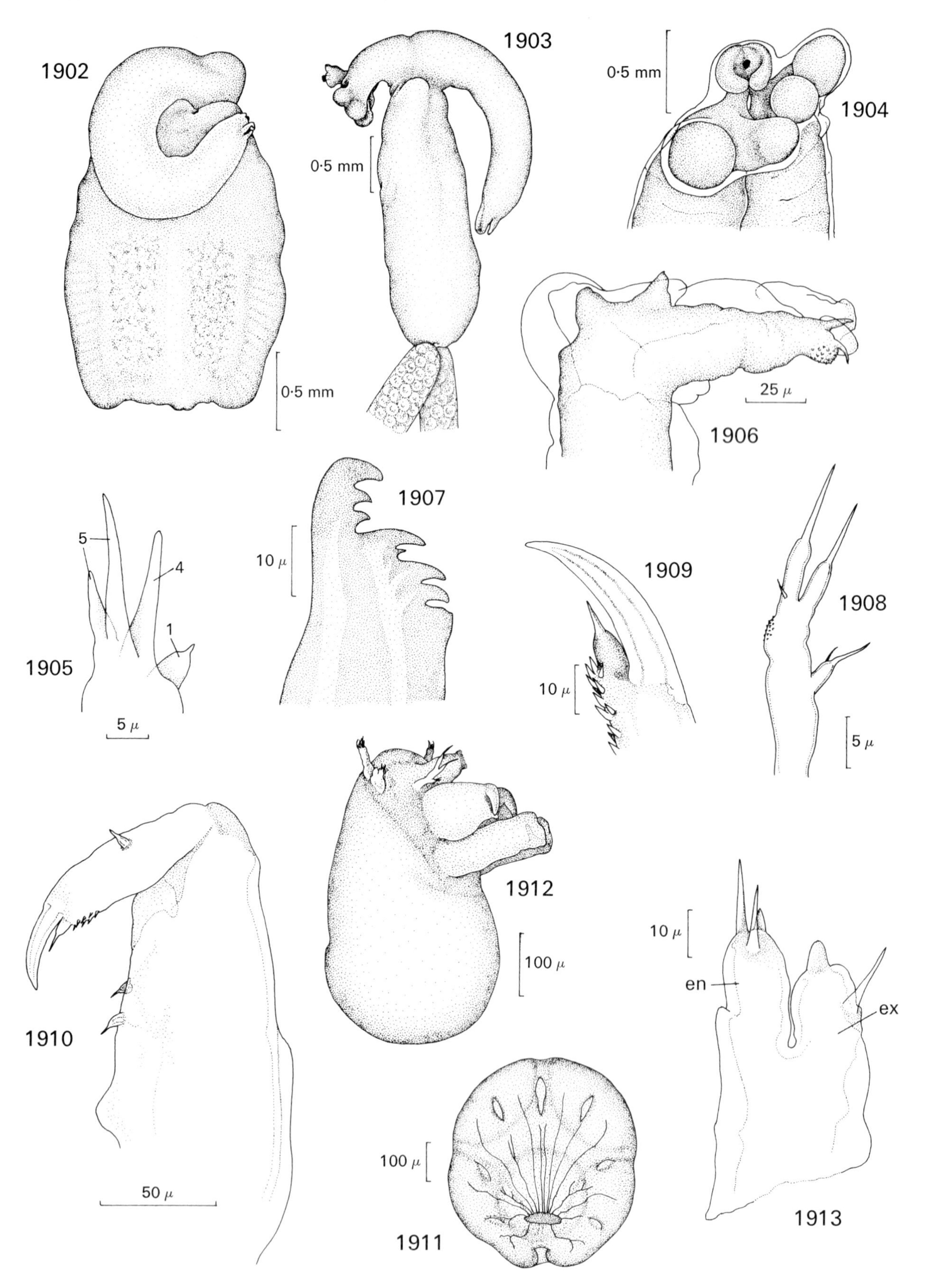

Figs. 1902–1913.
Clavella stellata (Krøyer, 1838). **Fig. 1902.** Female, dorsal. **Fig. 1903.** Same, lateral. **Fig. 1904.** Second maxillae, posterior. **Fig. 1905.** Tip of first antenna, lateral. **Fig. 1906.** Second antenna, ventral. **Fig. 1907.** Tip of mandible, lateral. **Fig. 1908.** First maxilla, lateral. **Fig. 1909.** Claw of maxilliped, ventral. **Fig. 1910.** Maxilliped, ventral. **Fig. 1911.** Bulla, supra-anchoral surface (manubrium removed). **Fig. 1912.** Male, lateral. **Fig. 1913.** Male second antenna, ventrolateral.

en – endopod; ex – exopod.

Figs. 1904, 1906, 1911–1913 redrawn from Kabata (1962b.)

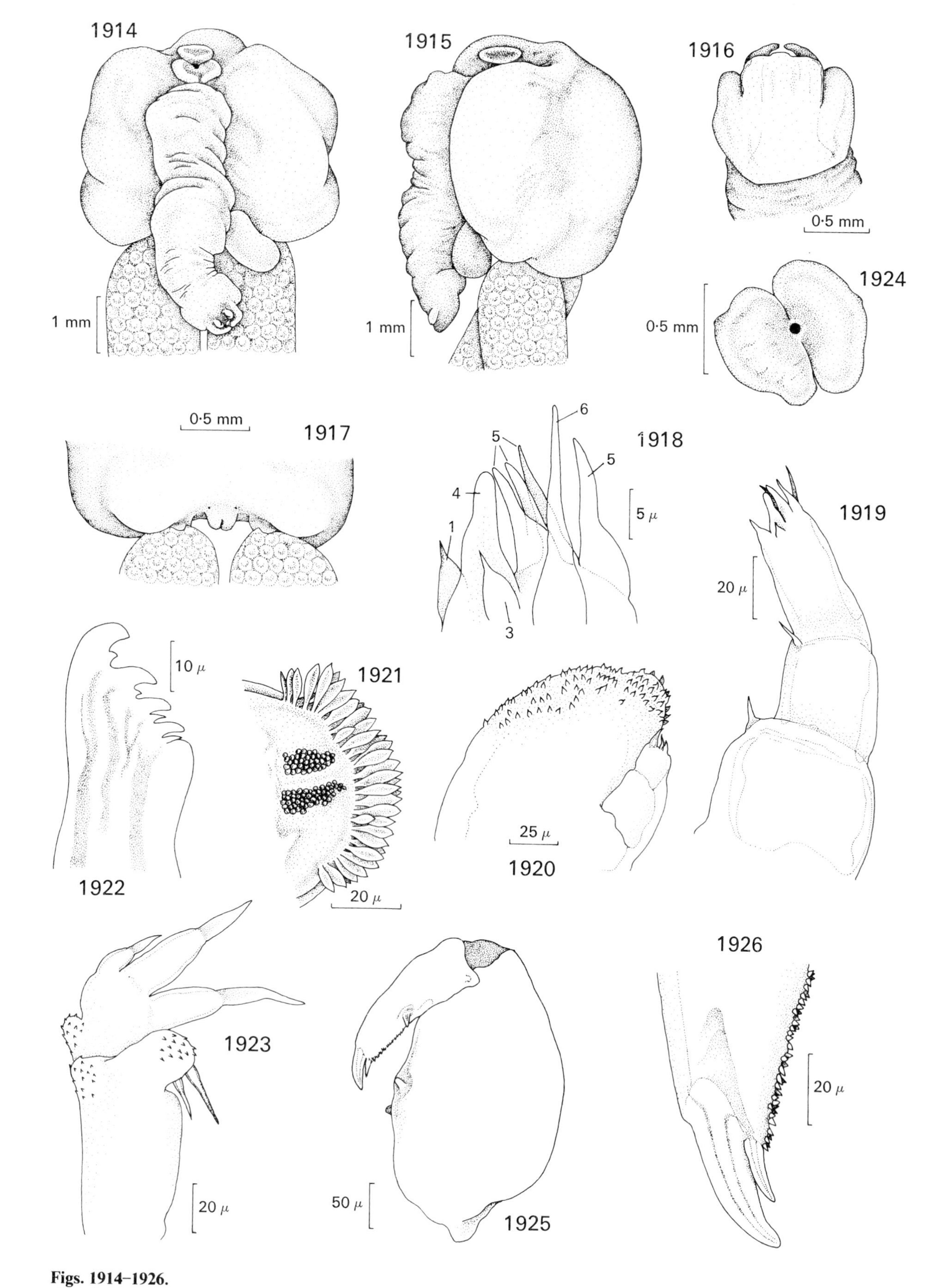

Figs. 1914–1926.
Clavellodes rugosa (Krøyer, 1837). **Fig. 1914.** Female, trunk dorsal, cephalothorax ventral (male attached). **Fig. 1915.** Same, lateral. **Fig. 1916.** Tip of cephalothorax, dorsal. **Fig. 1917.** Posterior extremity, ventral. **Fig. 1918.** Tip of first antenna, lateral. **Fig. 1919.** First antenna, ventromedial. **Fig. 1920.** Tip of second antenna, median. **Fig. 1921.** Tip of labrum, inner surface. **Fig. 1922.** Tip of mandible, lateral. **Fig. 1923.** First maxilla, lateral. **Fig. 1924.** Tips of second maxillae, anterior (bulla removed). **Fig. 1925.** Maxilliped, ventral. **Fig. 1926.** Same, claw, dorsal.

Figs. 1918, 1919 original, others redrawn from Kabata (1969e).

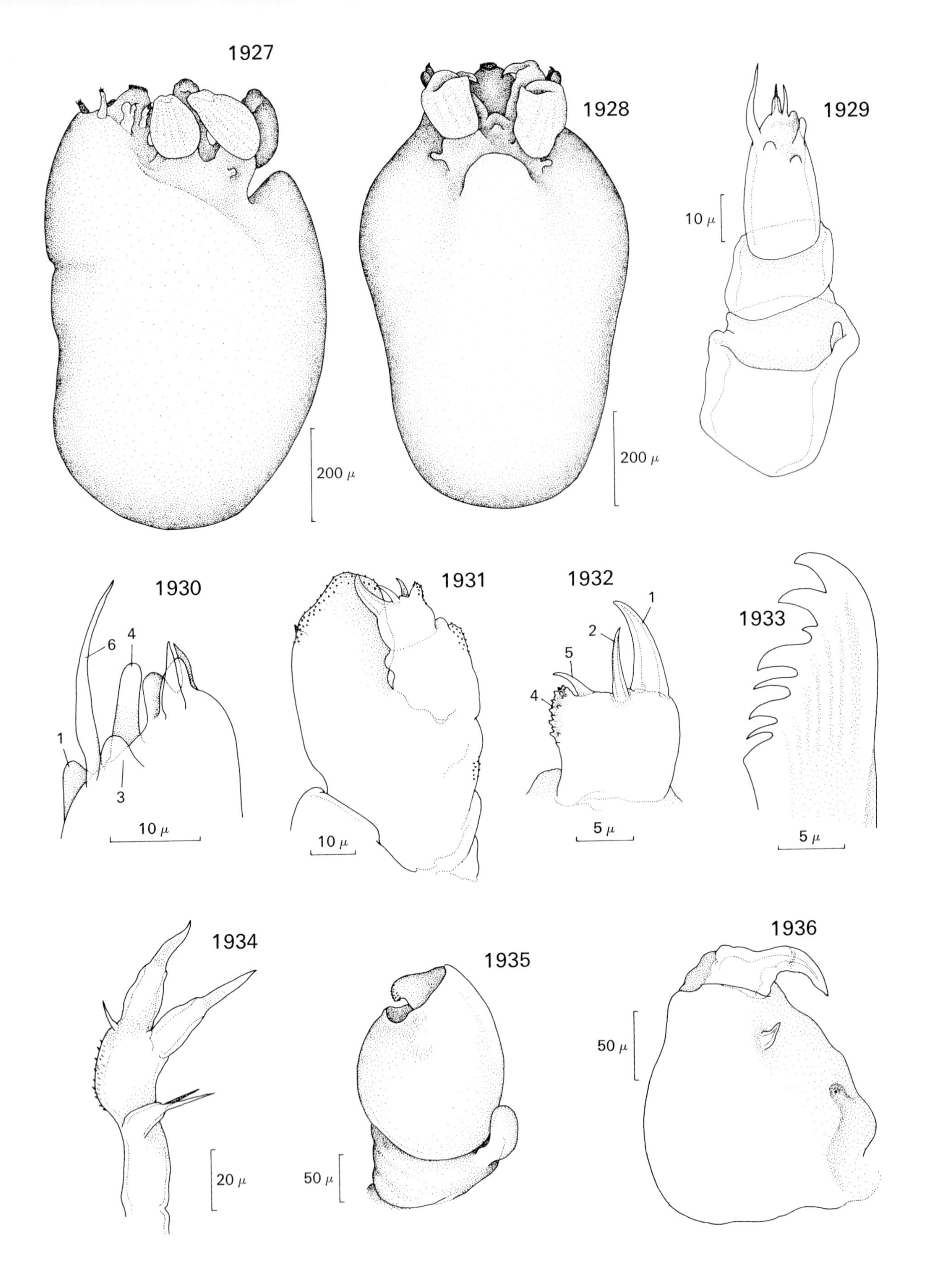

Figs. 1927–1936.
Clavellodes rugosa (Krøyer, 1837). **Fig. 1927.** Male, lateral. **Fig. 1928.** Same, ventral. **Fig. 1929.** First antenna, medial. **Fig. 1930.** Same, tip, lateral. **Fig. 1931.** Second antenna (without base), medial. **Fig. 1932.** Same tip of endopod, medial. **Fig. 1933.** Tip of mandible, lateral. **Fig. 1934.** First maxilla, lateral. **Fig. 1935.** Maxilliped, ventral. **Fig. 1936.** Second maxilla, ventral.

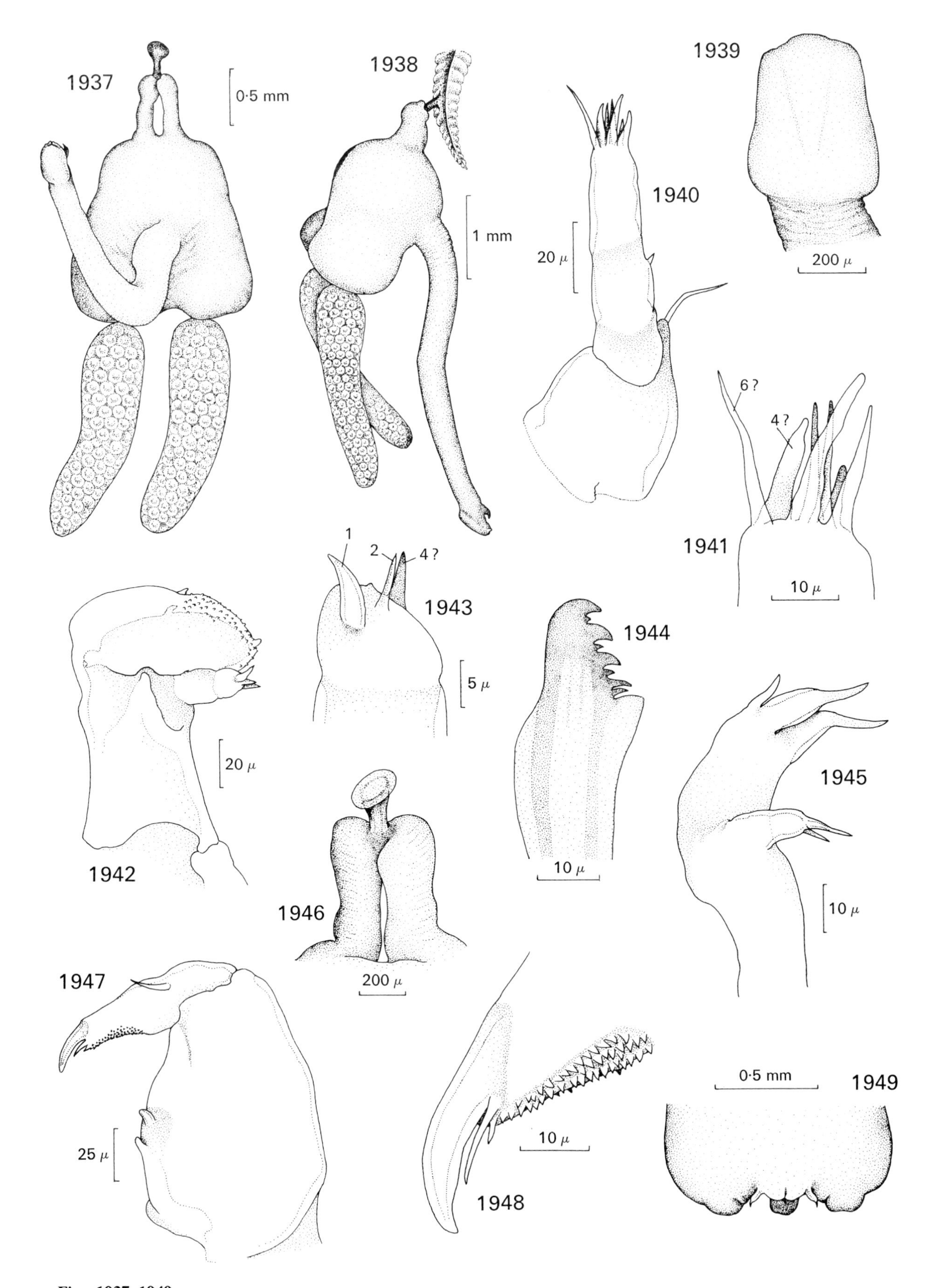

Figs. 1937–1949.
Clavellisa emarginata (Krøyer, 1837.) **Fig. 1937.** Female, dorsal. **Fig. 1938.** Same, lateral. **Fig. 1939.** Tip of cephalothorax, dorsal. **Fig. 1940.** First antenna, ventrolateral. **Fig. 1941.** Same, tip, ventrolateral. **Fig. 1942.** Second antenna, ventral. **Fig. 1943.** Same, tip of endopod. **Fig. 1944.** Tip of mandible, lateral. **Fig. 1945.** First maxilla, lateral. **Fig. 1946.** Second maxillae and bulla, dorsal. **Fig. 1947.** Maxilliped, ventral. **Fig. 1948.** Same, claw, ventral. **Fig. 1949.** Posterior extremity, dorsal.
Figs. 1937, 1938, 1944 and 1949 redrawn from Kabata (1964d).

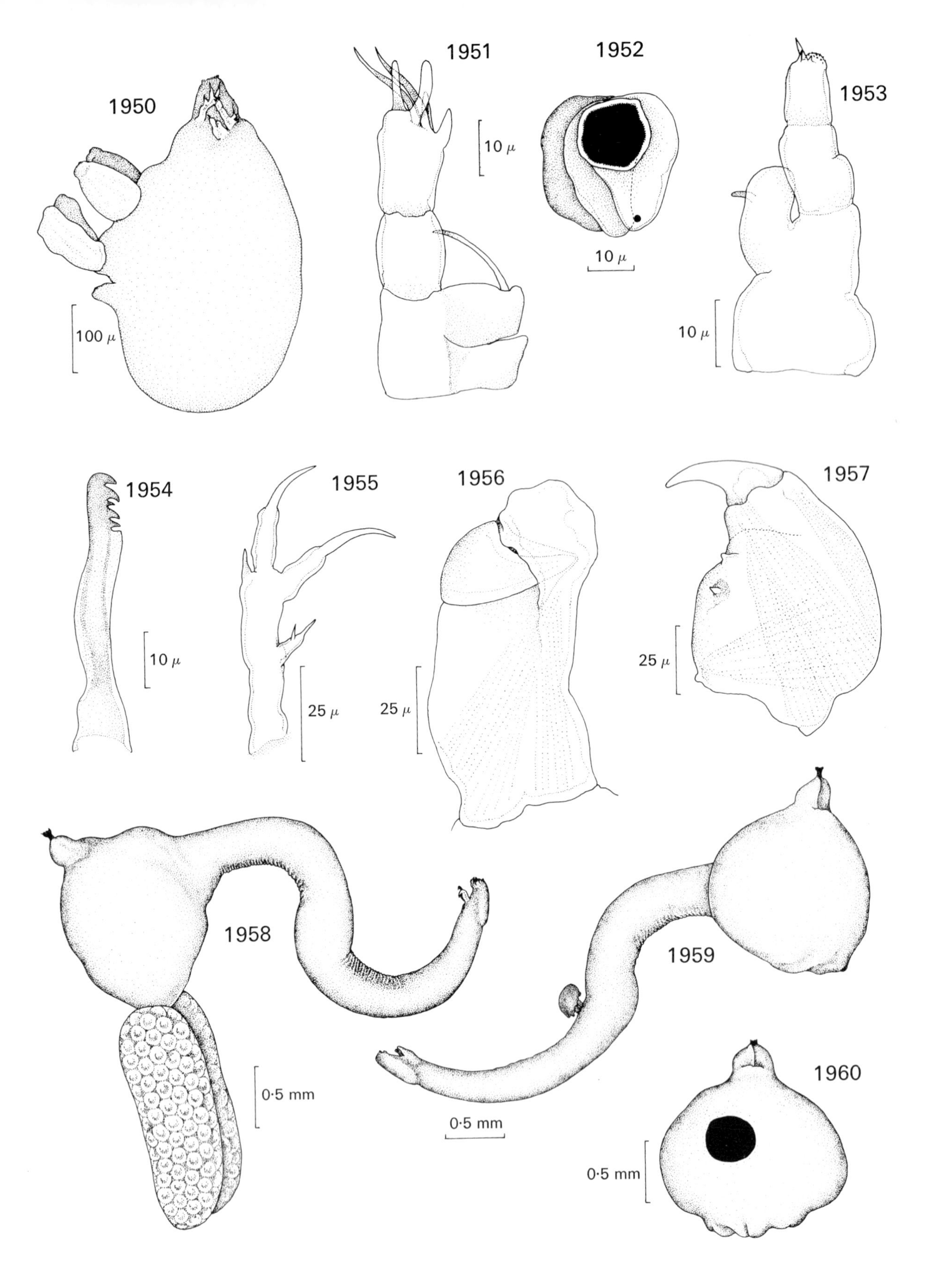

Figs. 1950–1960.
Clavellisa emarginata (Krøyer, 1837). **Fig. 1950.** Male, lateral. **Fig. 1951.** First antenna, ventrolateral. **Fig. 1952.** Same, basal segment, anterior. **Fig. 1953.** Second antenna, lateral. **Fig. 1954.** Mandible, lateral. **Fig. 1955.** First maxilla, lateral. **Fig. 1956.** Maxilliped, medial. **Fig. 1957.** Second maxilla ventrolateral.
Clavellisa scombri (Kurz, 1877). **Fig. 1958.** Female, lateral. **Fig. 1959.** Same, trunk ventrolateral, male attached. **Fig. 1960.** Trunk, dorsal (cephalothorax amputated).
Figs. 1950–1957 modified from Kabata (1964d).

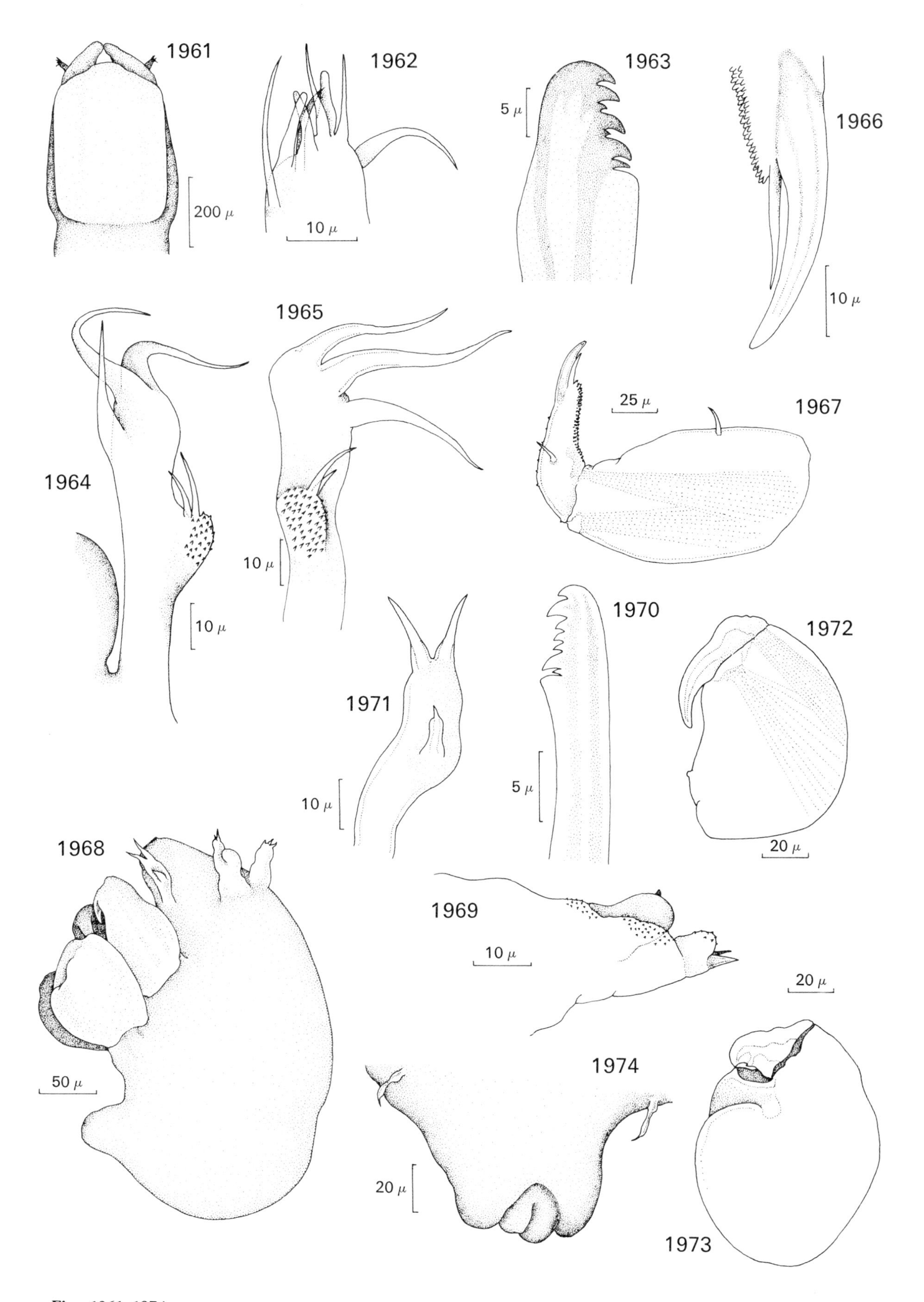

Figs. 1961–1974.
Clavellisa scombri (Kurz, 1877). **Fig. 1961.** Tip of female cephalothorax, dorsal. **Fig. 1962.** Tip of first antenna, medial. **Fig. 1963.** Tip of mandible, lateral. **Fig. 1964.** First maxilla, ventral. **Fig. 1965.** Same, lateral **Fig. 1966.** Claw of maxilliped, ventral. **Fig. 1967.** Maxilliped, ventral. **Fig. 1968.** Male, lateral. **Fig. 1969.** Male second antenna, ventral. **Fig. 1970.** Mandible, lateral. **Fig. 1971.** First maxilla, lateral. **Fig. 1972.** Second maxilla, ventral. **Fig. 1973.** Maxilliped, ventral. **Fig. 1974.** Posterior extremity, anterior.

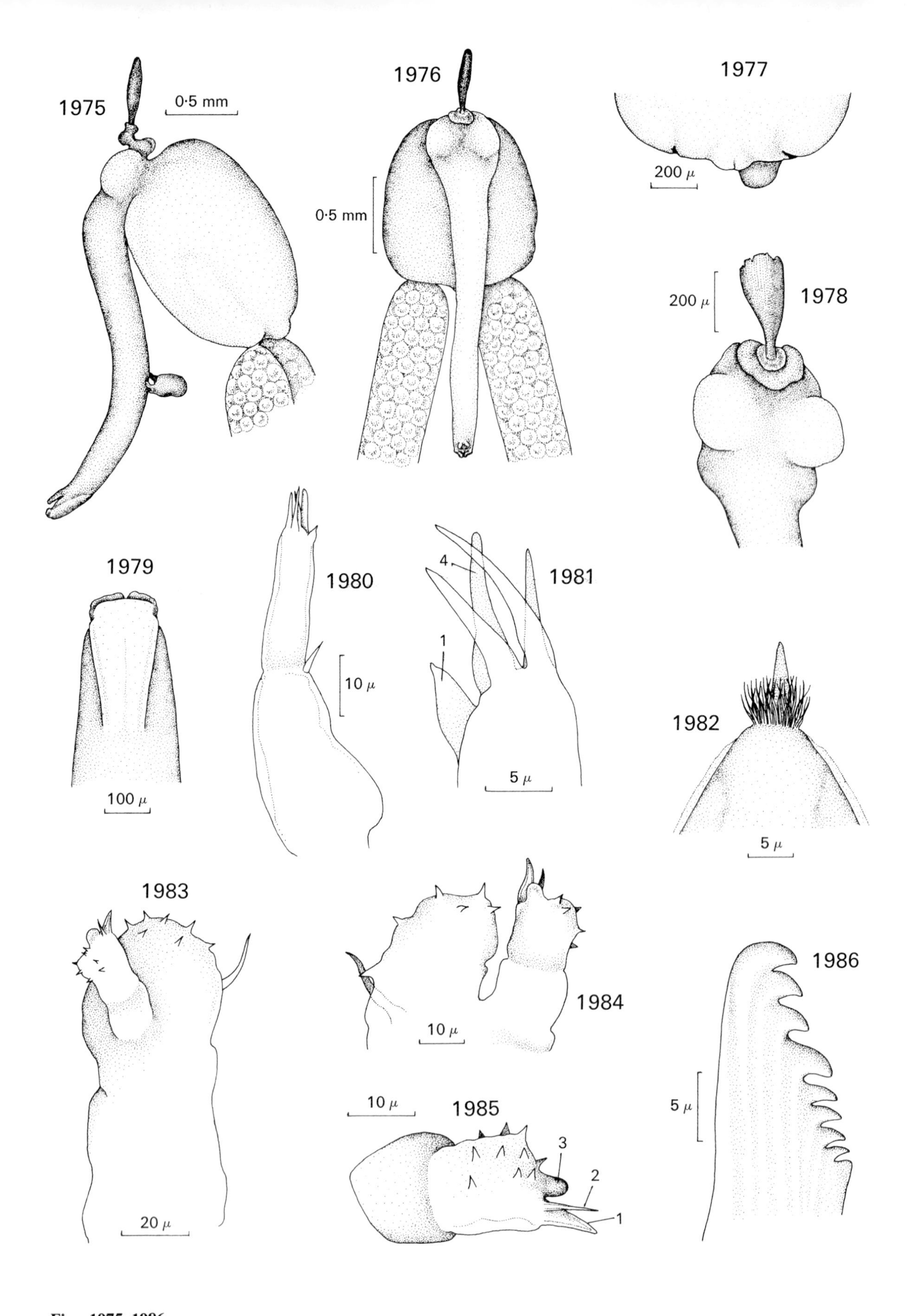

Figs. 1975–1986.
Alella pagelli (Krøyer, 1863). **Fig. 1975.** Female, with male attached, lateral. **Fig. 1976.** Same, trunk dorsal, cephalothorax ventral. **Fig. 1977.** Posterior extremity, dorsal. **Fig. 1978.** Second maxillae, bulla and base of cephalothorax, anterior. **Fig. 1979.** Tip of cephalothorax, dorsal. **Fig. 1980.** First antenna, lateral. **Fig. 1981.** Same, tip, lateral. **Fig. 1982.** Tip of labrum, outer surface. **Fig. 1983.** Second antenna, ventral. **Fig. 1984.** Same, tip, lateral. **Fig. 1985.** Same, endopod, medial. **Fig. 1986.** Tip of mandible, lateral.

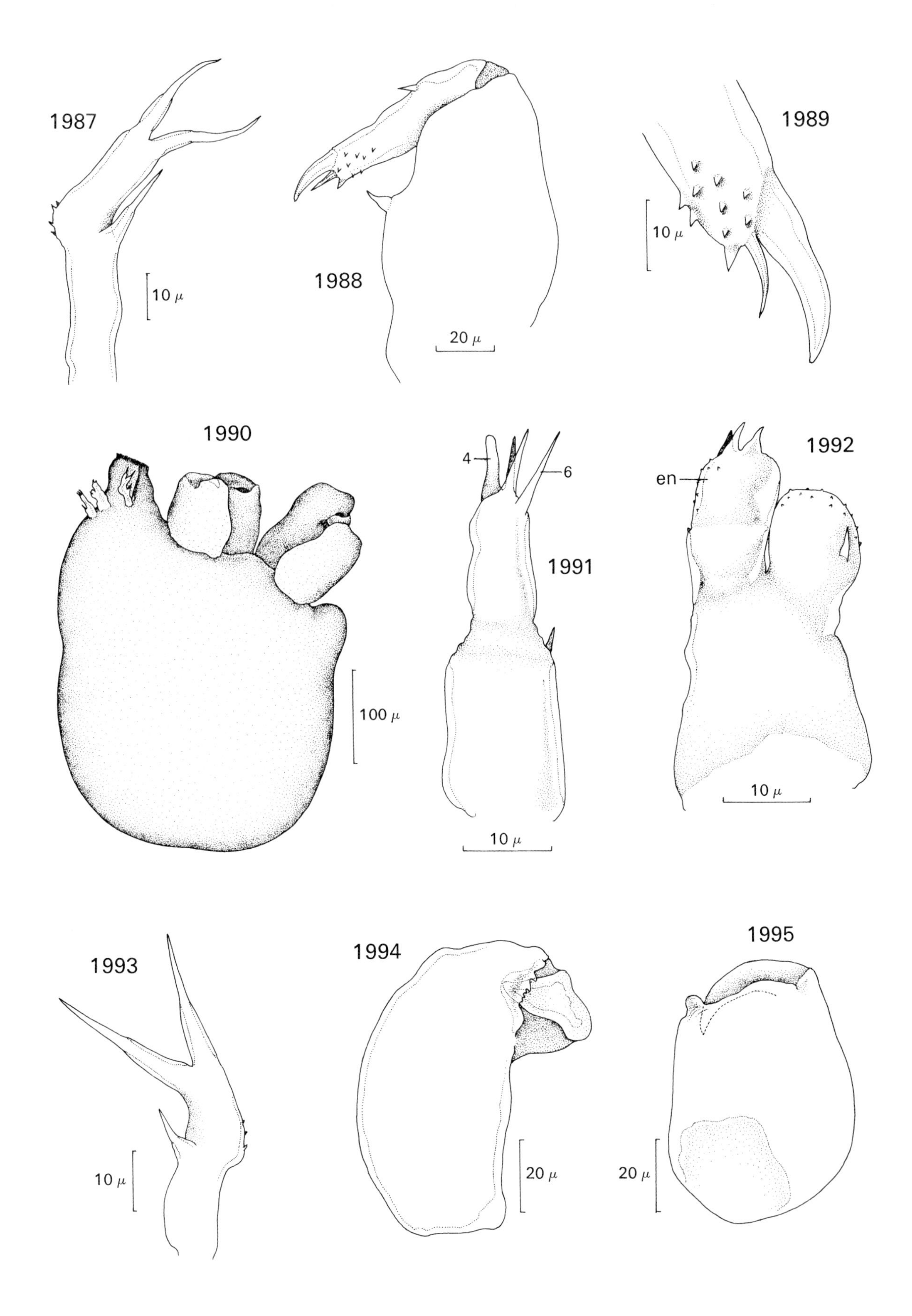

Figs. 1987–1995.
Alella pagelli (Krøyer, 1863). **Fig. 1987.** Female first maxilla, lateral. **Fig. 1988.** Maxilliped, ventral. **Fig. 1989.** Same, claw, ventral. **Fig. 1990.** Male, lateral. **Fig. 1991.** Male first antenna lateral. **Fig. 1992.** Second antenna, lateral. **Fig. 1993.** First maxilla, lateral. **Fig. 1994.** Maxilliped, medial. **Fig. 1995.** Second maxilla, lateral.
en – endopod.
Fig. 1990 redrawn from Kabata (1964e).

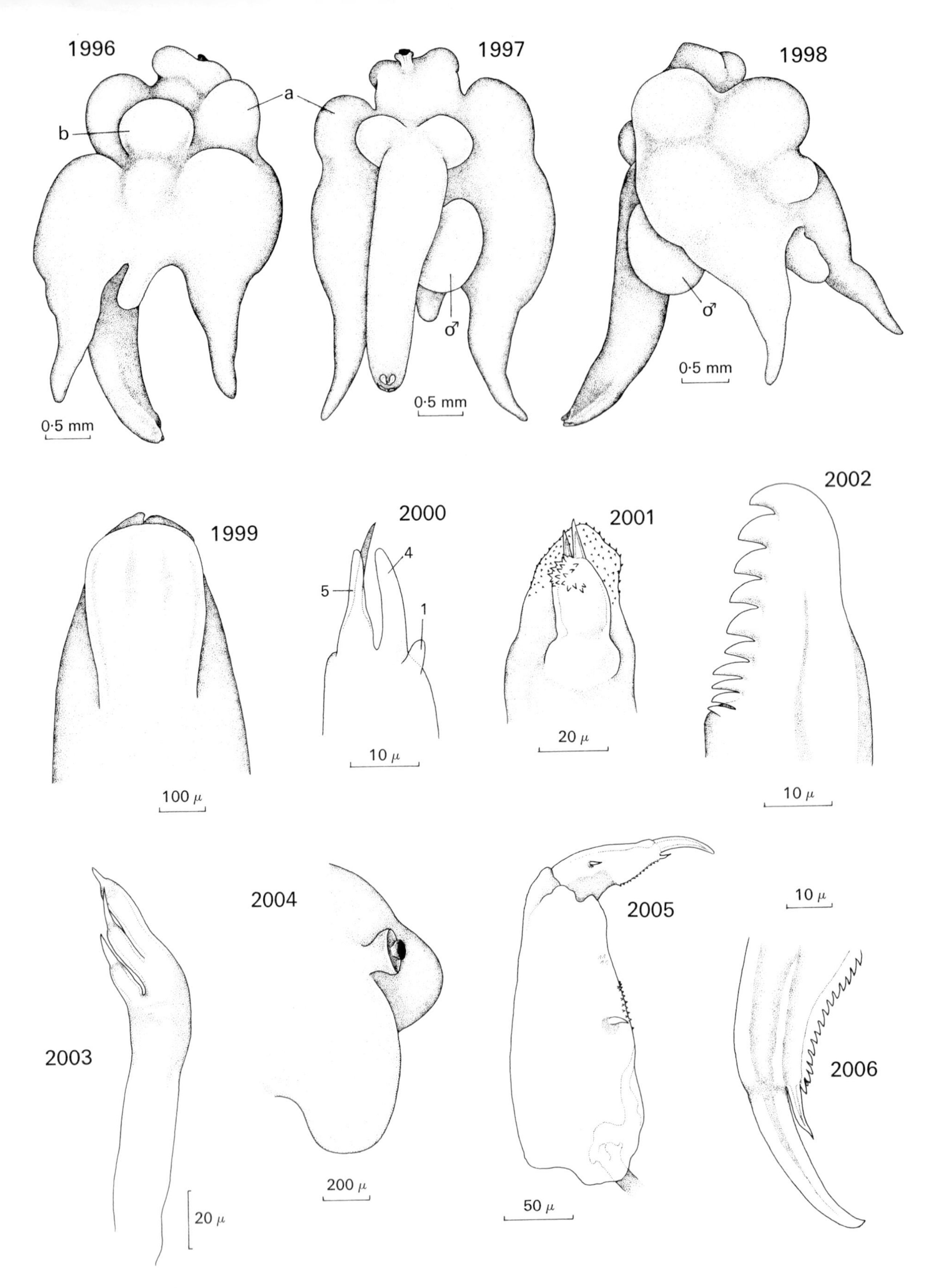

Figs. 1996–2006.
Advena paradoxa (van Beneden, 1851.) **Fig. 1996.** Female, ventral. **Fig. 1997.** Same, trunk dorsal, cephalothorax ventral. **Fig. 1998.** Same, ventrolateral. **Fig. 1999.** Tip of cephalothorax, dorsal. **Fig. 2000.** Tip of first antenna, medial. **Fig. 2001.** Tip of second antenna, ventral. **Fig. 2002.** Tip of mandible, lateral. **Fig. 2003.** First maxilla, lateral. **Fig. 2004.** Second maxillae and bulla, anterolateral. **Fig. 2005.** Maxilliped, ventral. **Fig. 2006.** Same, claw, ventral.
a, b – explanation in text.

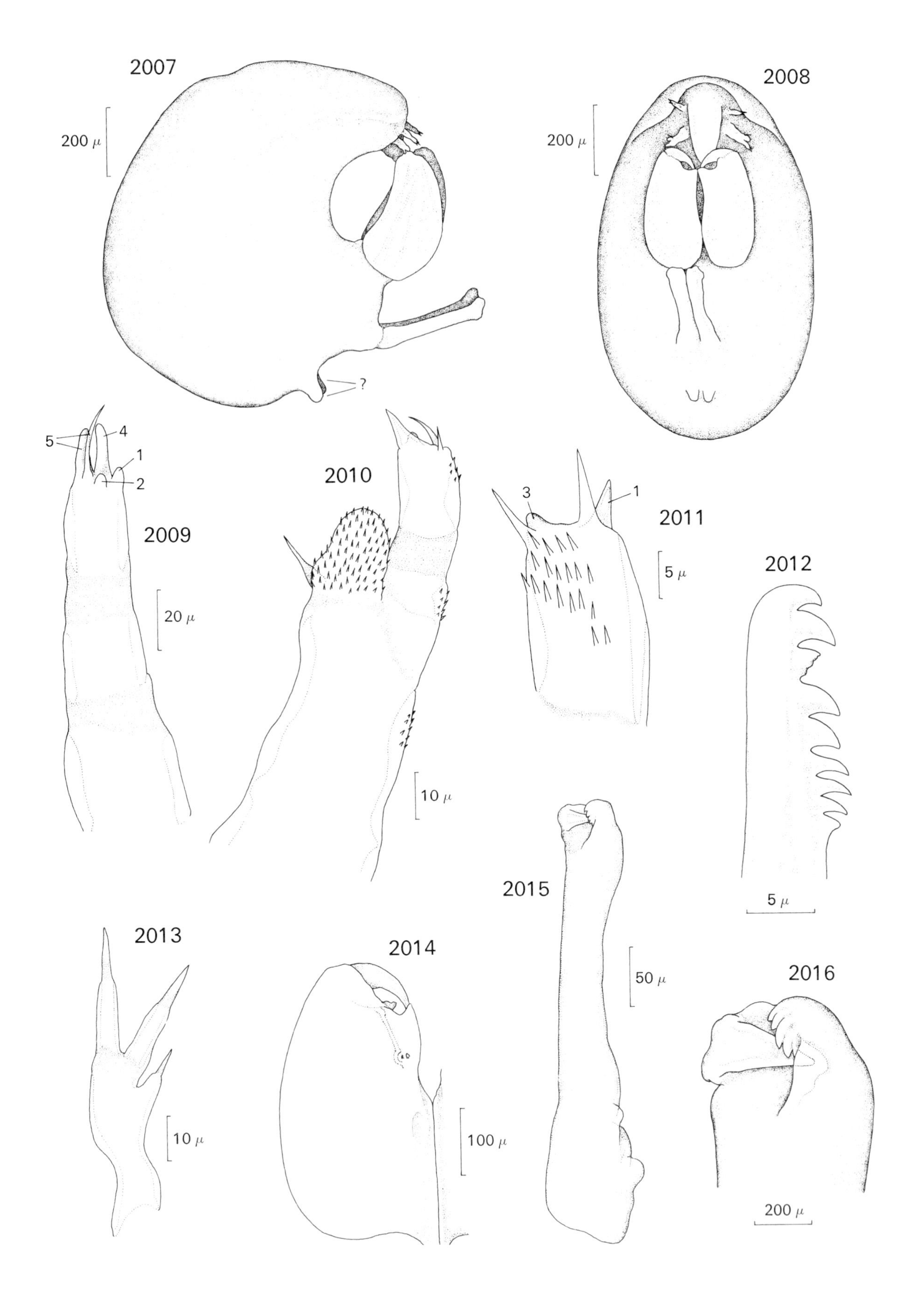

Figs. 2007–2016.
Advena paradoxa (van Beneden, 1851). **Fig. 2007.** Male, lateral. **Fig. 2008.** Same, ventral. **Fig. 2009.** First antenna, dorsolateral. **Fig. 2010.** Second antenna, lateral. **Fig. 2011.** Same, tip of endopod, ventral. **Fig. 2012.** Tip of mandible, lateral (one tooth broken). **Fig. 2013.** First maxilla, lateral. **Fig. 2014.** Second maxilla, ventral. **Fig. 2015.** Maxilliped, medial. **Fig. 2016.** Same, tip.

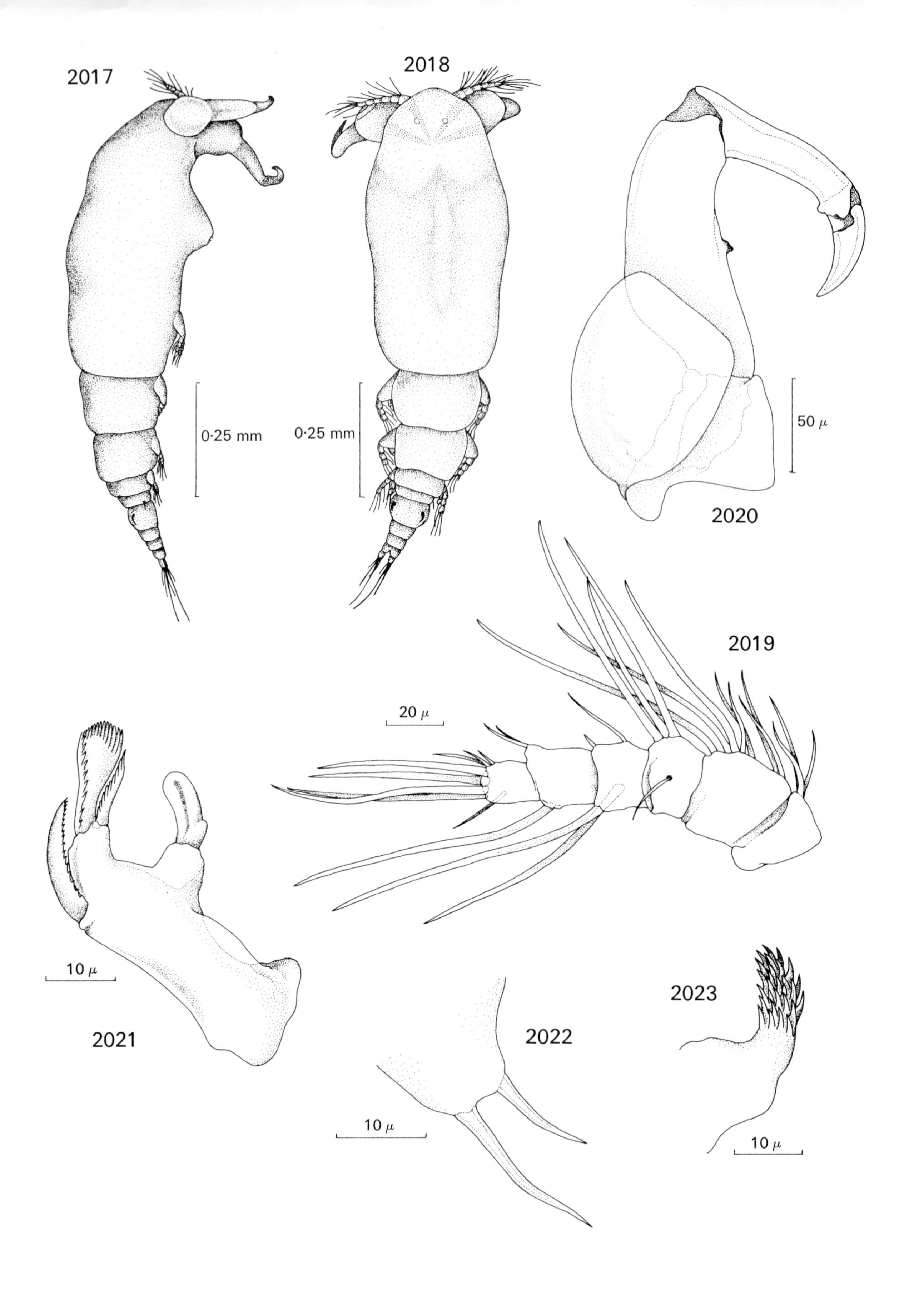

Figs. 2017–2023.
Ergasilus gibbus von Nordmann, 1832. **Fig. 2017.** Female, lateral. **Fig. 2018.** Same, dorsal. **Fig. 2019.** First antenna, ventral. **Fig. 2020.** Second antenna, ventral. **Fig. 2021.** Mandible, ventral. **Fig. 2022.** First maxilla, ventral. **Fig. 2023.** Tip of second maxilla, dorsal.

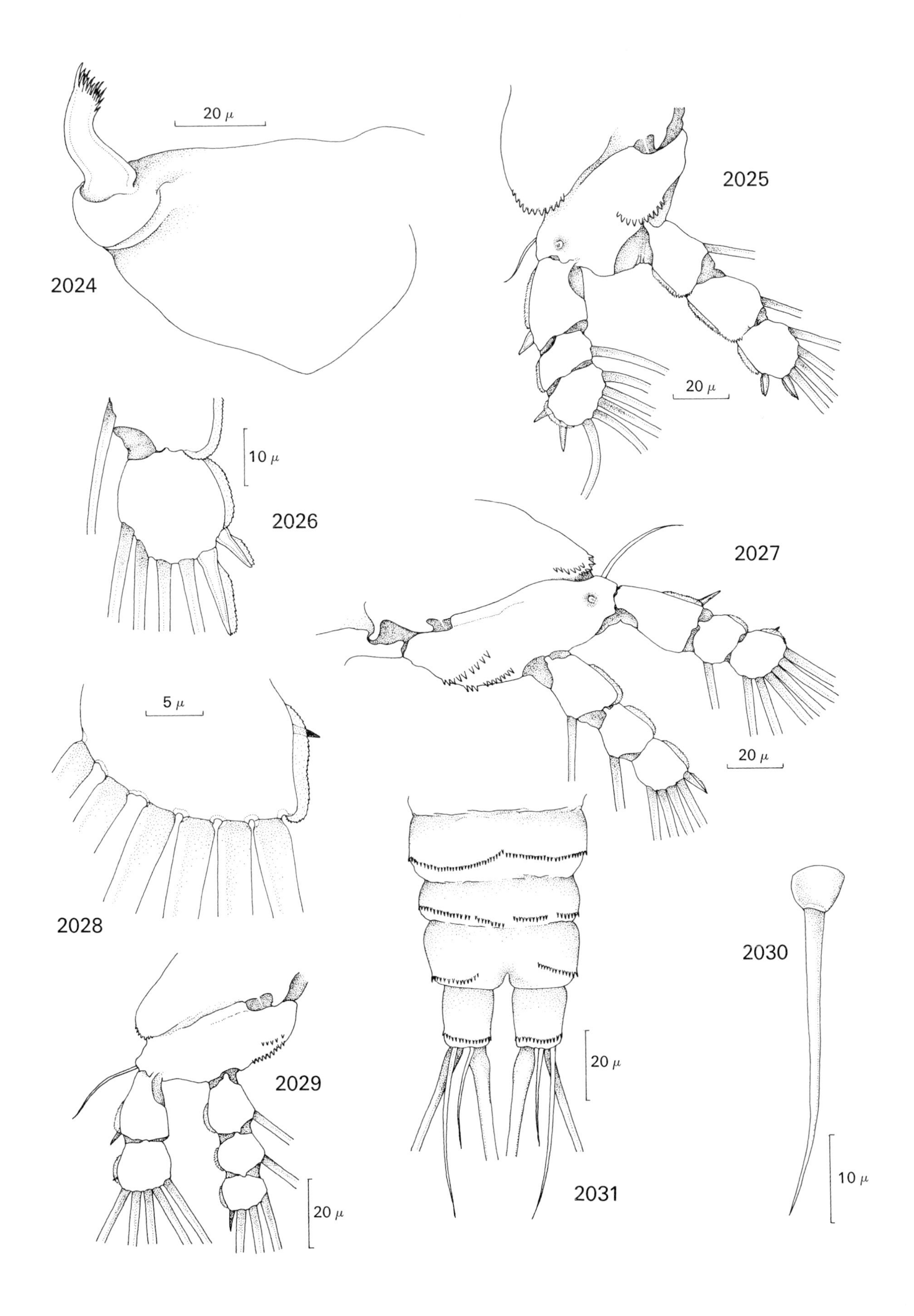

Figs. 2024–2031.
Ergasilus gibbus von Nordmann, 1832. **Fig. 2024.** Second maxilla, ventral. **Fig. 2025.** First leg, ventral. **Fig. 2026.** Tip of first endopod, ventral. **Fig. 2027.** Second leg, ventral. **Fig. 2028.** Tip of second exopod, ventral. **Fig. 2029.** Fourth leg, ventral. **Fig. 2030.** Fifth leg, lateral. **Fig. 2031.** Abdomen and uropods, ventral.